Methods of Air Sampling and Analysis

THIRD EDITION

Methods of Air Sampling and Analysis

THIRD EDITION

JAMES P. LODGE, JR., EDITOR

INTERSOCIETY COMMITTEE
AWMA   ACS   AIChE   APWA
ASME   AOAC   HPS   ISA

Lewis Publishers

Boca Raton   Boston   London   New York   Washington, D.C.

Library of Congress Cataloging-in-Publication Data

Methods of air sampling and analysis.–3rd ed. / James P. Lodge, Jr., editor
     p. cm.
    Bibliography: p.
    Includes index.
    ISBN 0-87371-141-6
    1. Air–Pollution–Measurement–Handbooks, manuals, etc. 2. Air–
Analysis–Handbooks, manuals, etc. I. Lodge, James P. II. Intersociety Committee.
TD890.M488 1988
628.5'3—dc19   88-11462
                                                 CIP

© 1998 by CRC Press LLC
Lewis Publishers is an imprint of CRC Press LLC

No claim to original U.S. Government works
International Standard Book Number 0-87371-141-6
Library of Congress Card Number 88-11462
Printed in the United States of America   13 14 15 16 17 18 19 20
Printed on acid-free paper

PREFACE

This is the third edition of the manual of methods adopted by the Intersociety Committee for a Manual on Methods of Air Sampling and Analysis according to its established procedures. The Intersociety Committee, which was founded in 1960, with its subcommittees consists of several scores of experts currently representing eight major national professional societies concerned with environmental measurements. These experts have volunteered their time and services to evaluate, codify, and publish hundreds of test methods for measurement of air contaminants affecting the health and welfare of industrial workers, as well as of the general public.

While many methods have been published by various agencies, for hundreds of materials agreed-upon methods of adequate accuracy remain to be established. The Committee and subcommittees have served as a workshop and forum for interaction and coordination of professionals from different governmental agencies, academia, and industry, each with knowledge of the methods in use by his organization. In the decade that has passed since the second edition, many changes have occurred in the status of the Committee, its procedures, and the field of atmospheric chemistry.

Financial support from the Environmental Protection Agency and the National Institute for Occupational Safety and Health has ceased, and with it the ability of the Committee to meet frequently, and to convene the subcommittees. Consequently, most of the procedures previously carried out face-to-face have been consigned to mail and, on occasion, telephone. There have been changes in personnel, as noted below. The grand plan for a major program of interlaboratory testing of methods has been replaced by occasional round-robin testing where it was specifically needed,

driven by economic concern for EPA acceptance of commercial instruments, or conducted to resolve dispute. New, more powerful methods have emerged and become widely accepted, and previous concerns have disappeared. Plans to prepublish all methods in "Health Laboratory Science" for comment have not been implemented since the second edition.

While the American Public Health Association had previously given valuable support to the Committee by serving as the contracting and administrating society, we are now an independent organization. Nevertheless, there still appears to be a great need for a manual such as the present one. While instruments for routine air monitoring are today uniquely qualified or disqualified by the Environmental Protection Agency, many local agencies are justifiably reluctant to buy expensive monitoring equipment prior to preliminary tests to assess the nature of their problems. Such tests are generally made with available equipment and manual methods of measurement. For a large number of chemical species there are as yet no approved methods; nevertheless, it is of considerable importance to characterize their concentration in the environment. In the workplace, monitoring instruments are still less common than individual chemical tests on integrated samples spanning a significant portion of the workday. The likelihood that all these methods will be subjected to rigorous interlaboratory comparison is probably declining rather than increasing.

Accordingly, a number of changes have been made in the present manual. The status of methods as "tentative" or "recommended" has been removed throughout. Every method carries statements of precision, accuracy, and interferences from which the reader can quickly learn how much is known about the dependability of that method. In the absence of sponsorship, the requirement to list method numbers according to the several agencies has also ceased. The given methods are advanced by the subcommittees and, where they are knowledgeable, by members of the parent Committee. They are among the best available, whatever their formal status, and those needing EPA and/or NIOSH method numbers are undoubtedly in position to ascertain them.

For further simplification, certain ubiquitous references have been omitted from the bibliographies. For example, virtually every method requires the use of reagent water of some type or other, as that is defined by the American Society for Testing and Materials, and the use of reagent chemicals, as these are defined by the American Chemical Society. Few analysts are unaware of these requirements, and it seems redundant to enter the appropriate references in every bibliography.

We have kept the previously assigned method numbers within the book, and assigned new numbers to added methods. Thus, some numbers are missing due to the judgment of subcommittees that the old methods were no longer in use, or should not be in use. The previous text, where it is retained, has been carefully edited to remove grammatical and typographical errors. Even in cases where old methods are largely retained, they have been carefully checked to be certain that they were clear and easy to follow; the overall result is that few pages of the second edition have survived without at least minor change. A number of new sections have been added to Part I, on general techniques, with the emergence of new laboratory tools.

In a number of situations it has been necessary to go outside the primary membership to reach truly knowledgeable scientists in the use of particular techniques. To give appropriate credit in these cases, a new convention has been adopted. The actual authors of the technique are listed first, followed by the principal reviewers, if any, and then the balance of the subcommittee is listed in alphabetical order. Where changes are minor, the listing of the previous subcommittee, presumably responsible for the original writing, has been preserved.

The original 12 subcommittees have been reduced in number to 9, in effect, by the merger of Subcommittees 4 and 5, and the termination of Committees 7 and 12. The founding chairman of the Intersociety Committee was Dr. Leonard Greenburg. He was succeeded by Dr. E. R. Hendrickson, Professor A. C. Stern, Dr. B. E. Saltzman, Mr. R. F. Toro, Dr. W. T. Ingram, Dr. R. J. Thompson, and Dr. R. S. Saltzman, the present chairman. Successive editors have been Dr. M. D. Thomas, Dr. Morris Katz, and Dr. James P. Lodge, Jr., the editor for this edition. Over the entire history of the Committee, Dr. George J. Kupchik has been Executive Secretary. All correspondence concerning these methods should be directed to him at the School of Health Sciences, BC 612, Hunter College, 425 East 25th Street, New York, NY 10010.

The Intersociety Committee gratefully acknowledges the past support of the U.S. Public Health Service under Grant AP-00256; of the Environmental Protection Agency under Contract Grant 68-02-0004, and of the National Institute for Occupational Safety and Health under Contract number HSM 99-73-89. Acknowledgments of permission to reprint, with editorial modifications, methods from other societies and agencies are duly noted below.

Robert S. Saltzman, ISA
Chairman
Willard A. Crandall, ASME
Andrew P. Hull, HPS
William T. Ingram, APWA
Theodore J. Kneip, APCA
Bernard E. Saltzman, AOAC
Richard J. Thompson, ACS
Richard F. Toro, AIChE

A KEY TO A CLEANER ENVIRONMENT

Other books have been written about air sampling and analysis, but there are none like this one.

To do any job well, you must use the proper tools, and this book, as with its two predecessor editions, gives you the know-how you need to sample and analyze air. Updated, with new methods added and in a streamlined form for ease of reading and understanding, this new Third Edition is in itself a tool for today's use.

Though full of changes from the edition of 11 years ago, this all-new work leans heavily on the two previous editions. Without the farsightedness of the leaders who gave us the first and second books through the Intersociety Committee for a Manual on Methods of Air Sampling and Analysis (beginning in 1960) this new work could have been created only by starting from "scratch." Even so, it was no mean task for the scores of committees and their editor to come forth with this matchless volume that will guide today's professionals who work daily to understand and solve the problems of both indoor and outdoor air pollution.

Our hat is off to the experts who labored long and hard — with intelligence and integrity — to make this book a reality. (These pioneers are noted in the front pages of this book and in the committee rosters appended to each method.) We are extremely pleased to be associated with the people who created this worthwhile and useful venture.

Methods of Air Sampling and Analysis, Third Edition is both a timely volume and a book destined for a long life. It describes and evaluates new methods along with the classic. It is a current and up-to-date reference to be regarded as a principal tool in the continuing, expanding battle to help solve the never-ending problem of maintaining clean air and keeping our environment workably pure.

— The Publisher

INTERSOCIETY COMMITTEE METHODS OF AIR SAMPLING AND ANALYSIS

Representatives of Participating Societies

Organization*	Representative	Dates of Service
ACS	Henry Freiser	1971–80
	Richard J. Thompson	1982–
AIChE	Robert Coughlin	1972–74
	Richard F. Toro	1974–
APWA	James V. Fitzpatrick	1972–79
	William T. Ingram	1981–
ASME	Arthur C. Stern	1963–74
	Wesley C. L. Hemeon	1976–79
	Willard A. Crandall	1982–
AOAC	Bernard E. Saltzman	1966–
AWMA	E. R. Hendrickson	1963–86
	Theo J. Kneip	1987–
HPS	Lysle C. Schwendiman	1972–83
	Andrew Hull	1984–
ISA	Robert L. Chapman	1972–74
	Robert S. Saltzman	1974–

*Participating Organizations
   ACS – American Chemical Society
   AIChE – American Institute of Chemical Engineers
   APWA – American Public Works Association
   ASME – American Society of Mechanical Engineers
   AOAC – Association of Official Analytical Chemists
   AWMA – Air and Waste Management Association
   HPS – Health Physics Society
   ISA – Instrument Society of America

ISC CHAIRMEN AND STAFF

	Name	Dates of Service
Chairman	Leonard Greenburg	1963–66
	E. R. Hendrickson	1966–69
	Arthur C. Stern	1969–72
	Bernard E. Saltzman	1972–76
	Richard F. Toro	1976–80
	William T. Ingram	1980–83
	Richard J. Thompson	1983–88
	Robert S. Saltzman	1988–
Editor	Moyer Thomas	1963–69
	Morris Katz	1969–83
	James P. Lodge	1983–
Executive Secretary	George J. Kupchik	1963–

SUBCOMMITTEE CHAIRMEN

Name	Dates of Service
#1 (Sulfur Compounds)	
Donald F. Adams	1966–83
Purnendu K. Dasgupta	1983–
#2 (Halogen Compounds)	
Lester V. Cralley	1966–70
C. Ray Thompson	1970–81
Richard A. Mandl	1981–
#3 (Oxidants and Nitrogen Compounds)	
Paul M. Giever	1966–67
Bernard E. Saltzman	1967–71
Evaldo L. Kothny	1971–74, 1976–79
Basil Dimitriades	1974–75
Dario Levaggi	1980–
#4 (Carbon Compounds)*	
Ralph Smith	1966–79
Milton Feldstein	1979–80
#5 (Hydrocarbon Compounds)*	
Eugene Sawicki	1966–79
#4/5 (Carbon and Hydrocarbon Compounds)	
Milton Feldstein	1981–
#6 (Metals I)	
William G. Fredrick	1966–70
Theo J. Kneip	1971–79
Richard Thompson	1980–82
Michael Kleinman	1983–

* Terminated 1980

#7 (Metals II)*

Elbert C. Tabor	1966–70
Robert E. Kupel	1970–72

#8 (Radioactive Compounds)

Bernard Shleien	1966–72
Andrew P. Hull	1972–83, 1987–
Donald C. Bogen	1984–87

#9 (General Precautions)

Morris Katz	1967–69
A. L. Linch	1969–79
Alvin L. Vander Kolk	1980–85
John N. Harman III	1986–

#10 (Particulate Matter)

Robert S. Sholtes	1968–71
Robert A. Herrick	1971–74
Edward Stein	1975–79
Dale A. Lundgren	1980–83
John Watson	1984–

#11 (Stationary Source Sampling)

John S. Nader	1971–73
M. Dean High	1974–80
Knowlton Caplan	1981–83
Walter S. Smith	1984–

#12 (Standardization Coordination)**

William D. Kelley	1971–72
Robert W. Garber	1972–74
Robert Spirtas	1975–80

* Terminated 1972
** Terminated 1980

PERMISSIONS

The Intersociety Committee expresses its appreciation for permission to use data as indicated from the following:

Academic Press, Orlando, Florida 32887, for use in Part I, Section 14, of data from "Detection of Volatile Organic Compounds and Toxic Gases in Humans by Rapid Infrared Techniques," by R. D. Stewart and D. S. Erley, in *Progress in Chemical Toxicology*, A. Stolman, Ed., Volume 2, 1965, pages 183–200.

The American Conference of Governmental Industrial Hygienists (ACGIH), 6500 Glenway Avenue, Cincinnati, Ohio 45211 for use in Part I, Section 27, of text from "Detector Tubes, Direct Reading Passive Badges and Dosimeter Tubes," by B. E. Saltzman and P. E. Caplan; and for use in Part I, Section 29, of text from "Filter Media for Air Sampling," by Morton Lippmann in *Air Sampling Instruments for Evaluation of Atmosphere Contaminants*, 7th ed., American Conference of Governmental Industrial Hygienists, Cincinnati, 1988, in press.

The American Industrial Hygiene Association, 475 Wolf Ledges Parkway, Akron, Ohio 44311–1087 for data in Part I, Section 13, from "U-V Absorptivity Curves"; and of material from AIHA Analytical Abstracts, 1965 (UV) for use in Part I, Section 14, in the *American Industrial Hygiene Association Journal*, Volume 32, 412, June 1971.

The American Public Health Association, 1015 15th Street, N.W., Washington, D.C. 20005 for use in Part I, Section 19, of text from Part 11, Section 104A of *Standard Methods for the Examination of Water and Waste Water*, 14th ed., 1976.

The American Society for Testing and Materials, 1916 Race Street, Philadelphia, Pennsylvania 19103 for use in Part I, Section 11, of text from E:131; for use in Part I, Section 12, of text ASTM E60; and for use on Part V of a table from ASTM E380.

Preston Publications, a Division of Preston Industries, Inc., 7800 Merrimar Avenue, Niles, Illinois 60048 for use in Part II, Method 101, of "Adsorption Chromatography on PLOT Columns: A New Look at the Future of Capillary GC," by J. de Zeeuw, R. C. M. de Nijs and L. T. Henrich, in *Journal of Chromatographic Science*, Volume 25, 76, 1987, Figure 6.

John Wiley & Sons, Inc., 605 Third Avenue, New York, New York 10158, for permission to use material in Part I, Section 20, from "Safety in the Analytical Laboratory," by R. F. Stalzer, J. R. Martin, and W. E. Railing, in I. M. Kolthoff, P. J. Elving, and E. B. Sandell, eds., *Treatise on Analytical Chemistry*, Part I, Volume 11, pp. 6743–6752, 1975.

James P. Lodge, Jr. attended the University of Illinois, with his undergraduate training there interrupted by two years of service in the U.S. Navy during World War II. After graduation, he studied at the University of Rochester with Professor V. Boeckelheide, receiving his doctorate in 1951. He was Assistant Professor of Chemistry at Keuka College in New York from 1950 to 1952. During the summer of 1952, he joined the Meteorology Department at the University of Chicago, and has worked in the field of Atmospheric Chemistry ever since. From 1955 to 1961, he was Section Head in Chemistry in the Community Air Pollution Program, U.S. Public Health Service, in Cincinnati; this was one of the predecessors of the present U.S. Environmental Protection Agency.

From 1961 to 1974, Dr. Lodge was a Program Scientist at the National Center for Atmospheric Research in Boulder, Colorado, and since then, has been a Consultant in Atmospheric Chemistry. He was chairman of the section on air in the American Chemical Society report, "Cleaning Our Environment: The Chemical Basis for Action," for which the Board of Directors of ACS awarded him a Special Certificate. He held the same position for the revision, "Cleaning Our Environment, A Chemical Perspective." Dr. Lodge has also received the Distinguished Service Award of the Division of Environmental Chemistry, ACS, and the Frank A. Chambers Award of the Air Pollution Control Association.

As Chairman of the Colorado State Air Pollution Control Commission for six years, Dr. Lodge received several local awards. He has contributed to the fields of air pollution analysis, atmospheric kinetics and photochemistry, atmospheric electricity, and theory of filtration, publishing over 200 papers. In addition to serving as Western Hemisphere Editor of *Atmospheric Environment*, he is its principal book reviewer. Dr. Lodge has also had several papers published in the field of his hobby, the study of theology.

TABLE OF CONTENTS

PART I   GENERAL TECHNIQUES

PART III METHODS FOR CHEMICALS IN AIR OF THE WORKPLACE AND IN BIOLOGICAL SAMPLES

PART IV STATE-OF-THE-ART REVIEWS

PART I
GENERAL TECHNIQUES

PART I

GENERAL TECHNIQUES

1. Physical Precautions

1.1 HOMOGENEITY OF THE SAMPLE. No analytical result regardless of the accuracy and precision of the procedure can be any better than the quality of the sample submitted for analysis. Therefore, the primary concern of the investigator must be directed to the collection of representative samples and the homogeneity of the air mixtures employed to calibrate both the collection and analytical systems. Human sensory perceptions cannot guide the uninitiated with respect to variations of contaminant concentrations in either space or time. Almost without exception gaseous components are either colorless or are present at such low concentrations as to be effectively colorless. The human sense of smell is notoriously deceptive with respect to concentrations of even highly odoriferous components, and therefore cannot be relied upon for estimating relative quality or uniformity of the ambient atmosphere (**2,4-Part 3,7,43,44**).

The source of the contaminant, air flow direction and velocity whether due to wind or thermal gradients, density of the contaminant, intensity of sunlight, time of day, presence of obstructions such as trees, buildings, partitions, machinery, etc., which act as baffles to produce turbulence, humidity, and half-life of the contaminant together determine the concentration at any given location. That the concentration can vary by several orders of magnitude within a relatively short radius from the point of reference has been amply confirmed by many investigators. Air may flow either in a stream line or turbulent flow and this factor alone can determine the dispersal pattern of the pollutant (**2,3,5,7,46**).

The density of the pollutant in many cases will counteract diffusion processes to an extent which will establish stratification. This phenomenon prevails in the case of many highly toxic and irritant gases such as phosgene, mustard gas, lewisite, chlorine, etc. (**5**). Natural convective circulation and diffusion in confined spaces such as silos, coal mines, wells, caves, tanks, etc., is not sufficient to maintain a normal atmosphere containing approximately 21% oxygen.

The location of sampling sites whether on a temporary or "grab" sample basis or on a fixed station basis for continuous monitoring over long periods of time is often critical and will determine the validity of the conclusions drawn from the results (**1,2,3,7,46**). In the vicinity of industrial atmospheres the variability may be sufficiently extreme to vitiate the results from fixed station monitors regardless of the care taken in locating the samplers. In an alkyl lead manufacturing facility, for example, no relationship could be established between exposure as indicated by urinary lead excretion and numerous fixed station samplers. In this case, correlation was not established until the monitors in the form of a small filter-microimpinger assembly were actually worn during the entire 8-h shift by the workmen assigned to these areas (**1**).

Several excellent treatises have been published on planning the air pollution survey and the effects of meteorological and geographical conditions on the validity of the results. This subject is outside of the field of General Techniques and Precautions; therefore, the reader is advised to consult one or more of these publications for full details (**2,3,6,7,32,33,34,47**).

On a smaller scale, even greater care is required to establish uniform mixtures of pollutants in air for use as calibration mixtures whether the system be dynamic or static in principle (**9**). Passage through lengths of small-diameter tubing to ensure turbulent mixing is one relatively simple expedient for dynamic mixing (**3**). Other systems employ some form of baffled chamber inserted between the mixing ports and the delivery point (**10**). Rotating paddles or even loose pieces of sheet metal or inert plastic to be activated by shaking have been inserted in containers for static mixtures for mixing by turbulence (**11**). Probably a more effective alternative is to evacuate the container, then mix the contaminant with the air at a uniform rate as it is released into the void (**11**). (Inject from a hypodermic syringe through a needle inserted through a septum.) Diffusion alone cannot be relied upon to produce a homogeneous mixture in most cases.

Other effects such as container wall absorption, humidity, volumetric errors, etc., discussed elsewhere in this section must be recognized and appropriate precautions taken to avoid concentration errors derived from these seemingly extraneous sources when attempts are made to produce homogeneous, known concentrations for calibration purposes.

1.2 ABSORPTION EFFECTS ON CONTAINER WALLS AND CONNECTING TUBES. Failure to recognize absorption effects can lead to serious errors both in collection of samples and in preparation of calibration mixtures, especially in static systems. Losses of analyte can occur either by primary adsorption, which may be reversible in response to pressure or temperature changes, or by secondary adsorption, which may involve chemical reaction with another material previously adsorbed (2,4,7). Long-term stability and accuracy of calibration blends have been studied as a function of the container material by several investigators. (48,49,50,51,52). The affinity of glass for water, nitrogen dioxide, benzene, aniline, the isocyanates, etc., has been recognized (8,12,13,31). The interior surface of a 5-gallon (19-L) borosilicate glass carboy can adsorb as much as 50% of the benzene in the 100 to 200 ppm range from a calibration mixture in air within 30 min after preparation. These losses can be reduced to some extent by preconditioning at a concentration higher than the working level but desorption may contribute positive errors (Table 1:I). Losses by adsorption of polynuclear aromatic compounds on glass have also been reported (14).

This adsorption effect can be minimized in the case of collection of air samples in evacuated glass flasks (Shepherd flasks) or sampling tubes by the introduction of a suitable solvent for the vapor, such as isooctane for benzene and xylidine, directly into the container with thorough rinsing of the inner surfaces (31). In some cases, such as the analysis for nitrogen dioxide, the solvent or reagent is introduced before the sample is taken; *e.g.*, 100 mL sample drawn into a syringe containing 10 mL of Saltzman's reagent (15).

Nitrogen oxides are so tenaciously adsorbed on glass surfaces that rinsing with Saltzman's reagent has been found necessary to obtain reliable results. Adsorption of isocyanates on glass, metal or plastic surfaces occurs to such a degree as to preclude preparation of static calibration mixtures even under strictly anhydrous conditions (8).

Positive errors can be included when evacuated double entry gas sampling tubes (fitted with 2 stopcocks) are used for sample collection (7,16). In this case, after release of the vacuum, additional sample is drawn through the tube for 3 min before closing the stopcocks. Recovery of xylidine increased to 120% in the 50 to 70 ppm range under these conditions. This error was increased to 40% by adding glass beads to the sampling flask to increase the surface area (31). These conditions can be avoided by collection of the pollutant in an impinger assembly (Greenburg-Smith, midget or micro).

In dynamic systems, prolonged equilibration is required before dependable calibration can be assured. Calibration of toluene diisocyanate (TDI) in the TLV range required passage of the air-TDI vapor mixture through a mixing chamber composed of a 6' (1.8 m) length of $1/4$" (6 mm) copper tubing under strictly anhydrous conditions for 60 to 72 h to reach equilibrium. This phenomenon was not entirely caused by the film of water normally adsorbed on the inner walls of the tubing, as the system had been carefully heated under vacuum to remove volatile impurities before use (8).

Adsorption is a critical factor in making up standard mixtures for phosgene calibration at concentrations below 1 ppm even after meticulous removal of the adsorbed water film. The use of metal cylinders especially must be avoided. In the development of a continuous monitor for phosgene concentrations below 5 ppm, decomposition in metal (steel, stainless steel, copper, aluminum, etc.) or glass tubing was found to be so extensive that their use in the sampling system could not be considered. In this case polyethylene tubing was acceptable. Water adsorbed to the contact surfaces was probably responsible for the losses as HCl was observed to be a by-product. Catalytic de-

Table 1:I Static Calibration Mixture
Benzene Loss by Absorption on Container Walls
Borosilicate Glass   Temperature 73 ± 5°F (24 ± 3°C)

Time Elapsed Hours	Benzene PPM Added	Benzene PPM Found	Percent Loss	Container Conditioned Time Hours	Container Conditioned Benzene PPM	Remarks Inside Surface Approx. 800 Sq. In. (0.52 m²)
0.5	100	56	44	0	none	
0.5	100	67	33	0	none	
16	200	100	50	0	none	
0.5	100	84	16	16	100	
1	–	78	22	–	–	
0.5	100	87	13	16	200	
0.5	100	107	–7	–	–	Swept, 2nd 100 ppm added
0.5	200	128	36	–	–	3rd 100 ppm added
0.5	50	56	–12	–	–	Swept, 50 ppm added
0.1	100	75	25	0	none	Idle 30 days
0.25	–	72	28	–	–	
0.5	–	55	45	–	–	
1.0	–	55	45	–	–	
4.5	–	48	52	–	–	
24	–	42	58	–	–	
192	–	0	100	–	–	

Note: Five-gallon (19-L) borosilicate glass carboy jacketed with a canvas cover. Calculated quantity of benzene was added and mixed by rotating a polytetrafluoroethylene (PTFE) paddle and sampled at the indicated time after addition. In two cases indicated by 16 h (column 5), equilibrated by standing overnight, then refilled.

composition on anhydrous surfaces cannot be ruled out, however (17).

In a study of the efficiency of midget and micro-impingers by collection of aniline from air, loss by adsorption in the glass batch-type generator (5 to 15% in the 100 μg range) was encountered unless the system was heated to 60°C in a hot water bath. Even greater were losses due to absorption by gum rubber tubing. Even 3-inch (8-cm) connections introduced significant errors (10% loss) but the losses were not entirely proportioned to hose length. These losses increased as the age and number of uses increased; e.g., recoveries decreased from 83 to 57% over a period of 8 days (18).

Failure to recognize the effect of adsorption on container walls delayed the determination of the half-life of SO_2 during a study of the fate of SO_2 in the atmosphere at Syracuse University. The half-life was found to be unexpectedly short even after correction for adsorption effects (19).

Tubing through which high concentrations of tetraethyl lead vapor have been drawn will become a source of contamination when "clean" air is passed through the system. These effects are eliminated by use of all-glass systems connected by standard taper, ball joints, "O" ring seals or polytetrafluoroethylene (Teflon) compression connectors. Chemical transformation occurred when sampling oxides of nitrogen through sampling probes from a simulated combustion source. The species interconversion or loss was a function of temperature and sample probe material and was found to be a function of the air-fuel ratio of the input sample. Glass sampling probes were found to be inert, while the stainless steel probes were found to be reactive (53).

The adsorption of traces of analytes from liquid solution onto container walls is probably somewhat more widely recognized but certainly requires review to reemphasize the importance of this source of errors. Heavy metal ions, especially lead, mercury (22), silver (20,21), cadmium, and antimony, are readily adsorbed by glass and in some cases by plastic surfaces. Lead in urine specimens, for example, is adsorbed to an extent sufficient to introduce significant errors after storage for more than 3 days in borosilicate glass specimen bottles

(23). This effect probably is produced by bacterial fermentation which elevates the pH by ammonia release, and coprecipitates lead with calcium and magnesium phosphates on the glass. In the early stages the precipitate is invisible but nevertheless may contain a significant fraction of the total lead in the sample. In many cases these losses can be held within tolerable limits by reducing the pH to 2 or less with nitric acid. This adsorption loss can be a critical factor in the storage of standard reference solutions and must be taken into account when collaborative testing programs are set up. Efforts to prepare stable aqueous standardizing solutions of metallic mercury met with failure unless excess mercury was maintained in contact with the liquid phase. Furthermore, centrifugation is necessary to remove dispersed microdroplets (or HgO) if shaking is employed to accelerate saturation, and a relatively large temperature coefficient of solubility must be taken into account. In spite of these limitations, this saturated aqueous mercury solution can be employed as a convenient source of mercury for calibration by comparing recovery from a collecting system with direct analysis of the standard (22).

Filtration of samples collected in a liquid medium should in all cases be avoided if the desired contaminant is to be recovered quantitatively in the liquid phase. Not only will the contact surfaces of the filter funnel and the filtrate receiver adsorb trace materials, but the filter medium (paper, membrane, asbestos mat, glass fiber, micro metallic, etc.) may be highly adsorbent (14). Also, the insoluble fraction removed may retain the analyte as in the case when calcium phosphate is separated from a mixture containing traces of soluble lead compounds (24).

The problem of adsorption on glass surfaces previously discussed requires reemphasis in the case of porous glass used either as a filter for particulate matter or as a diffuser for collection in liquid absorption systems. Filtration of aqueous solutions of phenanthrene, naphthalene, pyrene or anthracene through a porous glass plate reduced the hydrocarbon concentration as much as 40% (14). When ultraviolet (UV) absorption is employed for analysis, errors from background absorption produced from UV-absorbing materials leached from the filters and light scattering from suspended particles must be taken into account (23). A base-line technique for correction is often required as the solute often alters the light scattering and sedimentation properties with the result that the reference blank and the sample are no longer exactly comparable. If removal of sediment is necessary, filtration through polytetrafluoroethylene (PTFE) mat or decantation after centrifugation will minimize losses. In any case, potential losses at each step must be recognized and determined.

Rust, scale and other metallic corrosion products can introduce serious errors through adsorption of air pollution components and should be meticulously scoured out of systems used for air sampling, storage or analysis. Many of these oxidation and corrosion products are highly catalytic, as illustrated by the rate at which allkyl lead compounds are oxidized by hydrated iron and bismuth oxides (25). Pipe dope used on threaded connections and oils used in cutting threads or lubricating moving parts also can present significant reversible absorption effect. Polytetrafluoroethylene (PTFE) ribbon thread sealant, which has a very low absorption coefficient, is recommended for threaded joints that require lubrication. Moving parts machined from PTFE are self-lubricating and can be sealed with modified PTFE "O" rings.

The reverse effect has been encountered also. The diffusion of HF in Conway dishes is enhanced ten-fold when silicone grease is used as a sealant for the dual concentric chambered dish cover. Fluoride can be removed quantitatively from aqueous HCl by gentle boiling in the presence of silicone fluid (26).

Adsorption of a secondary layer on container walls may completely alter the character of the surface presented to the collection and analytical system. Soaps and synthetic detergents are especially troublesome in this respect. The film of soap on glass is not removed by 5% trisodium phosphate, chromic acid cleaning mixture, or concentrated nitric acid. Only by baking out in an annealing furnace can "chemical sterility" be achieved. In the colorimetric determination of aromatic nitro and amino compounds in air, this soap film reacts with the color-generating reagents. These adsorbed layers appear to retain nitrous acid or nitrogen oxides in a state that does not react with starch-iodide indicator or sulfamic acid but does react with the coupling agent (1-naphthyl ethylenediamine or Chicago Acid) to produce a deep orange color. This spurious color reaction produces high results, which may be sufficiently uniform to produce false calibration curves (zero intercept at some light transmittance value below 100%). To avoid this nitrite-absorbing contamination and remove residual azo dyes which may be adsorbed on the coupling flasks and spectrophotometer cuettes, soak all glassware in 5% aqueous trisodium phosphate for several hours and rinse thoroughly with distilled water. Soak Corex cuvettes only in 37% nitric acid wash (27).

The nonabsorption of heavy metal ions on hydrophobic surfaces of PTFE and polyethylene (PE) may be reversed by the presence of certain organic anions. In trace quantities, cesium is adsorbed on glass surfaces by interaction with the layer of adsorbed hydroxyl ions that account for the hydrophilic character of glass surfaces. Addition of a salt of high ionic activity such as sodium nitrate reduces this adsorption effect. The ion exchange capacity of the glass surface can also become a significant factor at very low concentrations. The overall effect is orders of magnitude less (adsorption coefficient increases in the order PTFE-PE-glass) for the nonionic hydrophobic plastic surfaces. However, the absorption coefficients reverse when the tetraphenyl borate (TPB) ion is introduced. This reversal is due to adsorption of TPB ions by van der Waals forces on the hydrophobic PTFE and PE which then sequester cesium ions. On the hydrophilic glass surface the adsorption is ionic in character, which reduces the binding sites for the cesium ions (28).

Lubricants in stopcocks and ground glass joints must be avoided as these materials are a source of contamination and/or absorb those constituents of interest either directly from the air sample or from the liquid reagent system after collection. Teflon plug cocks for stopcocks and sleeves for joints provide seals that require no lubrication or liquid-phase sealant for air-tight joints.

Fungus growth within an atmospheric analyzer can be a particularly vexing problem. Fungus within an SO_2 analyzer that uses liquid reagents results in an indicated concentration lower than that actually present. Fungus effect can be minimized in an analyzer by: (a) Keeping the instrument running continually, (b) cleaning periodically, and (c) using a small amount of fungicide in the reagent. A fungus problem will usually be found on dynamic calibration.

1.3 SOLID ADSORBENT COLLECTING SYSTEMS. Extended surface adsorbents such as activated silica gels, alumina gels, charcoals and carbons, glass beads, and spheres offer a very attractive solution to the collection and concentration of gas phase pollutants from atmospheric samples. However, the tenacity of the forces that provide quantitative extraction of the components sought may prove to be self-defeating when quantitative removal of these same components for analysis is attempted (54). Failure to recognize and equate these factors may, and has, produced serious errors. Mercury, either as metallic vapor or as volatile organic derivatives, is not removed quantitatively either by heating or elution with liquid reagents from silica gel. The same deficiency has been encountered in attempts to use activated charcoal for collection of alkyl lead compounds.

Since the polarity of the adsorbed element or compound determines the binding strength on silica and alumina gels, components of high polarity will displace components of lesser polarity—the principle of chromatography. Therefore, in attempting to collect relatively nonpolar compounds such as benzene, the presence of coexisting polar compounds such as phenol, acetic acid or ammonium chloride must be recognized as a source of interference in the use of solid adsorbents for sample collection. Under high humidity conditions the absorbent may be deactivated by saturation with water vapor (23). Activated charcoal exhibits this effect only to a minor degree. Aromatic hydrocarbons are not removed quantitatively from charcoal (32).

Selective displacement may be taken advantage of in eluting the sample from the adsorbent. An inert carrier gas with or without heat, for example, may be used to separate a mixture of aromatic hydrocarbons from a silica gel column. Proper choice of eluting solvent may well serve to separate components by liquid extraction also. An example would be separation of nitrobenzene from aromatic hydrocarbons for analysis by UV spectrophotometry.

Liquid eluents must be chosen and used with care however as decrepitation of the gel may occur. The finely dispersed fragments can produce difficulties in the determination of optical density in the final step of the analysis. This problem is especially acute in UV spectroscopy. Usually this effect is caused by the heat generated when a polar solvent such as ethanol is introduced. Dilution with a nonpolar solvent such as isooctane will usually avoid this difficulty (23). Elution from charcoal is dependent on molecular size rather than polarity (CS_2, CCl_4, etc.)

Change in air flow resistance due to either swelling, shrinking or channeling of the adsorbent is sometimes encountered. Collection of alkyl lead vapor in crystals of iodine was plagued with channeling and shrinkage due to iodine sublimation. Addition of a head of fine inert sea sand eliminated the problem by providing enough pressure to keep the iodine crystals compacted without appreciably altering the pressure drop across the column (29).

Many dusts are highly adsorbent and often carry a significant gaseous component present in trace amounts such as SO_2 at concentrations below 0.5 ppm. If a prefilter is installed in the collection of trace contaminants, particles collected on the filter must be analyzed also if a quantitative accounting for the total contaminant present is desired (30). In some cases such as in the analysis for lead in air, the filter would be analyzed separately from the fraction collected in the liquid reagent to provide a differential analysis for organic and inorganic lead. This is especially important in cases where the TLVs for the two forms are different and coefficients must be used for hazard evaluation (1).

The use of pretreatment or guard systems such as solid dehydrating agents to remove moisture, soda lime to remove CO_2, or sulfuric acid on silica gel to remove NH_3 interference, and liquid scrubbers such as an oxidizing solution to remove reducing interferences such as SO_2 and H_2S before adsorption of the desired pollutant can introduce serious errors (23,35). Silica based and anhydrous $CaSO_4$ granules may partially or completely adsorb the subject element or compound as well as eliminate the interference in the air sample. Careful calibration of the collection system under use conditions is critical when interferences must be eliminated from the air sample by pretreatment before collection of the desired component.

1.4 DIFFUSION EFFECTS OF PLASTICS. Diffusion outward from the air sample, into, or through plastic containers or tubing probably is one of the most frequent sources of error encountered in air analysis. These irreversible losses are in addition to the reversible surface adsorption effects noted under Section 1.2. Elastomeric materials, which include natural rubber, neoprene and plasticized polyvinyl chloride (PVC) are especially troublesome. (See section 3.1 on Permeation Tubes—Dynamic Calibration.)

The plasticizers employed in the manufacture of flexible tubing from PVC type polymers may act as cosolvents to remove organic components from air samples by absorption. This mechanism is often temperature sensitive and to some extent may be reversible in a fashion resembling gas chromatography column packings. In a study of the recovery of aniline vapor from air, losses as high as 1.7%/cm of exposed gum rubber tubing were encountered (18). Similar experience has been obtained with alkyl lead and mercury compounds, aromatic hydrocarbons (benzene, toluene, etc.), chlorinated hydrocarbons, and aromatic nitro compounds. An all-glass system with joints and connections fitted with PTFE gaskets and the shortest possible path between air sample intake and the collecting medium to a great extent eliminates these losses. However, adsorption effects must not be neglected even in all-glass systems.

The permeability of even the inert plastics exemplified by PTFE has been employed for the constant rate generation of calibration mixtures of certain trace components such as SO_2, HF, $COCl_2$, aromatic hydrocarbons, etc. Although this property would be undesirable in the collection of air samples, the magnitude of loss by diffusion through short inert plastic sampling lines

(PTFE, polyethylene, polypropylene) can be neglected. However, when samples are stored in plastic bottles or bags, diffusion may become a controlling factor. Indication of contamination from fluoropolymer bags used in air pollution studies has also been noted (45). Plastic containers must be used to facilitate collection or transportation; then the diffusion rate factor must be determined for each component sought over the concentration and temperature range to be encountered and a correction factor applied to correct for the time interval between collection and analysis. Aluminum foil-lined polyester bags (Mylar), which materially reduce diffusion losses, are available for CO_2, CO, water vapor and some other gases but are not recommended for the highly polar gases (SO_2, NO_2 and O_3) (36). Polyester film in 2-mil (50 μm) thickness is considered the best all-around material for sample-bag fabrication. However, polyvinylidene chloride (Saran) film (1-mil [25 μm] thickness) which has been pre-aged to reduce outgassing of compounds from the plastic into the contents of the bag has been widely accepted on the basis of inertness to many common gases and vapors, typical high tensile strength and heat sealability. Data relative to gas transmission is summarized in the following table.

Gas Transmission: cm^3/m^2 μm d atm at 23°C

O_2	32–43
CO_2	150–236
N_2	4.8–6.3
Air	8.3–17.3

Water Vapor Transmission: g/m^2 d

at 24°C — 0.31
at 38°C — 3.05

For ultimate chemical inertness fluoroplastic film bags are available. Laminated plastics—polyethylene on polyester, polyethylene on cellulose acetate and polypropylene on either base—used in packaging foods offer interesting alternatives for gas-impervious containers and a choice of surface for contact with the air sample. Semiflexible 10-L polyethylene bottles with greater wall thickness also offer a solution

to the diffusion problem and are convenient for preparation of calibration mixtures. (See Static Calibration, section 2.1.)

Inward diffusion also may be encountered and may create problems in shipment or in prolonged storage. Samples taken in a low humidity environment may be altered by inward diffusion of water vapor when the plastic container is stored in a high-humidity climate. Solvent vapors, gasoline vapors, and gasoline and jet engine exhaust components may contaminate the sample in shipment and show up as unexplained peaks in gas chromatographic analysis, or introduce interferences in wet chemical procedures. Shipment in air-tight containers would be a reasonable precaution to avoid contamination from this source.

Although obvious, the testing of all plastic containers, especially thin film bags, for gross leaks before use must be emphasized. The absence of pin holes, gaps in the seams, and leakage around the sampling port and in the retaining valve or septum assembly must be verified before each use. This examination may consist of full inflation under slight positive pressure (2 to 3" water, 0.5–0.8 kPa) under water or in a rigid sealed container with the air line introduced through a bulkhead fitting and connected to a water manometer. Soap solution stabilized with glycerol may be painted or sprayed on for leak detection to pinpoint "micro" leaks.

1.5 MECHANICAL DEFECTS OF SAMPLING EQUIPMENT. As well recognized as this source of error has become, leaks and faulty seals remain one of the major contributors to unreliable results. Any system used to collect air samples, to remove contamination components by absorption, or to calibrate such systems must be checked for leakage, either inward or outward depending upon whether the air stream is pushed or pulled. Perhaps the most expedient procedure involves either an increase or decrease of the pressure in the system within limits of the maximum operational range of the system, by closing off tightly and measuring the rate of loss of either pressure or vacuum with a water manometer. Pinholes in glass seals, especially ring seals; plastic or metal tubing; joints that do not seal perfectly or are insufficiently gasketed; "O" rings that may be cracked, loose or hardened or lack a sealing lubricant; loose glass-to-rubber or plastic tubing connections; dirt in glass-to-glass joints; glass plugcocks that do not perfectly mate with the stopcock barrels; poor valve stem packing or insufficient sealing pressure; and poorly sealed threaded fittings provide the most frequent leak sources. Pinholes in stainless steel valves, reducing valve diaphragms, and two stage regulators also have contributed their share of difficulties **(10)**. A review of the requirements for gases to be used with gas chromatographic systems and the overall gas sampling requirements is valuable in determining how pollutant standards should be handled **(55)**.

Fouling of rate meters and total volume meters by particulate matter or corrosion products often is not so obvious and requires eternal vigilance to minimize. Orifice type and positive displacement meters and rotameters should be protected by an efficient filter to prevent reduction in orifice opening, erosion of moving parts, or complete plugging. Rotameter calibration can be shifted significantly by dust accumulation, and plastic tube models require protection from solvents which soften or render the bore tacky. Operating characteristics of flow meters were discussed **(56)**. Metal floats are attacked by corrosive gases and some reagents such as iodine may in time render the tube body opaque. Corrosive gases and aerosols must be rigidly excluded from total volume and positive displacement meters to maintain calibration validity and integrity of moving vanes or pistons. Liquid aerosols may be especially troublesome by forming a liquid film that by surface tension and capillary effects reduces air flow in small orifice throats, interferes with the movement of rotameter floats, and increases the resistance to movement of critical components in positive displacement meters. Galvanic corrosion from airborne electrolytes under high-humidity conditions can be a problem with metal components.

Flow meters based on thermal conductivity (hot filament type) also are affected by dust and reactive gases, but in a different fashion. Corrosion will alter the electrical conductivity and leads ultimately to failure of the element, and coating by inert material will reduce heat transfer and ultimately reduce the response of the instrument. Chlorides and lead are especially destructive. The volumetric meter and the pump or aspirator sub-assembly of any air sampling system should be adequately protected by appropriate traps that are renewed frequently. Activated charcoal granules (5 to 10 mesh) probably offer the most efficient protection for all but a few destructive gases and vapors. Cellulose acetate membrane filters (0.8 μm pore size) will effectively remove over 99% of the dust particles that might affect instrument performance without introducing prohibitive pressure drop **(37)**.

The orifice in impinger-type collecting devices, and the interstices in porous glass diffusers are also vulnerable to partial or complete closure by particulate matter and should be protected by a filter when the solid fraction is not an integral part of the sample. Porous glass diffusers may accumulate particles that are not removed by conventional cleaning procedures after repeated use **(10)**. Also irreversible retention of analyte sometimes occurs. In the collection of alkyl lead vapors from 5 to 10% of known quantities added to calibration mixtures were retained in a pencil-type diffuser substituted for the impinger tube in the standard midget impinger sampler **(31)**. The increase in pressure drop produced by the decrease in free cross-sectional area will contribute to volumetric sampling errors, and may significantly reduce the collection efficiency. In dust collection by impingement, this factor becomes critical.

Pressure drop must be taken into account when a sampling system is calibrated volumetrically. Rotameters and orifice type meters should be calibrated with the pressure reducing or elevating components in place or volume correction made to compensate for the deviations from ambient conditions **(38)**. In systems based on filter collection, the pressure drop across the system may steadily increase as the filter medium loads up with retained particles. Under these conditions an averaged correction may be within tolerable limits of volumetric error. Changes in barometric pressure usually introduce only marginal error if ignored. (Deviation of 2 from 101.3 kPa will introduce a 2% volumetric error).

In the use of cold traps to collect contaminated air components (freeze-out technique), consideration must be given to the possible accumulation of ice crystals in the delivery tube. This possibility can be minimized by introducing the sample through a side arm into the chamber and exhausting through the inner tube, i.e., reverse flow (11).

1.6 PARTIAL VAPOR PRESSURE EFFECTS. In collection systems that depend exclusively upon solubility of the component sought in a nonreactive solvent rather than retention by chemical reaction to a nonvolatile derivative, the amount absorbable is determined by the partial pressure of absorbate. The ratio of the air sample volume to the volume of absorbant is a critical factor in collection efficiency and there is a limiting value of the ratio for any system that cannot be exceeded if a given degree of efficiency is to be attained. In the case of acetone in water the limit would be approximately 5 L of air through 20 mL of water aliquoted between two bubblers (39). In the case of mesityl oxide, which has poor solubility in water to begin with, absorption efficiency would be very poor although relative volatility is low. This is especially critical where a low TLV is involved, as large volumes of air would be required. The formation of lower-boiling binary and ternary mixtures also can limit absorption. Failure to collect benzene vapor in ethanol even when chilled undoubtedly can be attributed to revaporization of the higher vapor-pressure ternary alcohol-benzene-water mixture.

In general, the vapor pressure of the volatile analyte should be reduced to the lowest practical limit by refrigeration to promote efficient collection. In many cases a freeze-out trap, without solvent, in solid carbon dioxide, liquid air, or a mixture of liquid air and ethanol proves to be a more effective procedure (11).

Sampling in the region at or close to the vapor saturation limit can produce serious errors by condensation in sampling lines. This condition is most likely to be encountered when collecting samples of air at temperatures above ambient and may involve condensation of water vapor which in turn will entrain or dissolve a significant portion of the components present in the sample under high humidity conditions.

1.7 SOLUBILITY EFFECTS. Limitation may occur in cases where chemical reaction should but actually does not occur readily. The classical example is the failure for SO_3 to be absorbed in pure water. The collection of aniline vapor in dilute aqueous HCl is not quantitative; therefore, chilled alcohol is the medium of choice (13). Until the reaction is initiated by a catalyst, fluorine gas will bubble through distilled water without reacting (to produce F_2O and HF) although once initiated, the reaction may be violent. Other examples that can be cited are failure to obtain absorption of alkyl mercury vapor in acid permanganate, and the less-than-quantitative absorption of alkyl mercury and lead compounds in aqueous iodine reagent (22). The assumption that a given contaminant will be collected based on its reactivity with the reagents can be disastrous if calibration of the system under use conditions is omitted.

If the product from the reaction of the analyte with the collecting reagent has a very low solubility product, losses are to be expected unless care is exercised to avoid the saturation region. The precipitate may adsorb to the container walls, orifice tube, or diffuser surfaces or fail to react in subsequent color development. If the solid phase is finely dispersed and has a refractive index approximately the same as the liquid system, the losses may not be observed; e.g., silica in chloroform.

Conversely, if the collecting train does not include a prefilter, particulate fractions may go into solution and produce false positive results or interferences; e.g., lead dust or fume in the collection of tetra-alkyl lead vapor (30).

1.8 TEMPERATURE EFFECTS. The physical effects of heat on the collection system under normal operating conditions is usually ignored without introducing appreciable difficulties. However, at the design and development stages, attention must be given to the upper and lower temperature limits to be anticipated. At low temperatures components of the liquid reagent may crystallize out or the solvent may freeze. Frequently, a change of solvent will solve this problem; e.g., aqueous KI_3 versus methanolic iodine reagent (22,29). Close attention must be given to possible formation of ice or reagent crystals in orifices and porous glass diffusion beds when operating near the freezing range.

On the other end of the temperature scale, excessive evaporation may limit the choice of reagents, or sampling time may be reduced below practical limits (again alcoholic versus aqueous iodine reagents). Refrigeration usually can be applied in fixed station or mobile systems to alleviate this problem. A portable assembly that derives its sampling suction from the expansion of difluorodichloromethane vapor through an aspirator also provides useful refrigeration by forming the vaporizer coil into a cavity which will receive a midget impinger (11). In some cases a colorimetric system can be adapted to a nonvolatile solvent, as illustrated by the use of diethyl phthalate as a solvent for the detection of phosgene in trace ranges over extended periods of time—4 to 6 h (40).

As noted in section 1.6 (Partial Vapor Pressure Effects) the temperature coefficient may be quite large and preclude collection by solution in a solvent above a critical temperature, above which no collection will be obtained, and which may be only slightly above room temperature (39). In general, the adsorption capacity of solid adsorbents is decreased exponentially with rising temperature, which may limit the thermal operating range. Temperature coefficients should be determined before sampling is attempted (23).

Excessive evaporation at elevated temperature aggravates the scavenging problem for the metering devices and pump downstream from the collecting unit. Larger activated charcoal cartridges and more frequent replacement must be provided. Generation of volatile components from the reagent will create an even greater problems, e.g., HCl from aqueous iodine monochloride solution (1).

A word of caution with respect to liquid-air cooled freeze-out traps—liquid oxygen from the air sample may condense and create an explosion hazard that may not arise until the trap is removed from the cold bath. In other subambient systems, moisture may be a mixed blessing. Condensation of water vapor often assists with the retention of contaminants, but in the end interferes with the final analysis. Analytical systems that require anhydrous conditions cannot be employed for air analysis unless dehydration can be employed before the sample enters the collecting system without loss of the analyte; e.g., CO by infrared.

Although temperature extremes sufficient to alter the mechanical performance of collecting systems are not often encountered, this factor cannot be ignored when arctic or tropical desert conditions are encountered. Batteries for power supplies must be chosen with these conditions in mind. Piston-type pumps may bind or blow by; check valves may become inoperable; volumetric factors will require recalibration; metering valve and plug cock tolerances may shift significantly; elastomeric components lose elasticity beyond tolerable limits (tubing, "O"-ring seals, check valves, gaskets, etc.); and diffusion through plastic containers, especially sample collection and storage bags may become excessive.

Instruments that require electrical circuitry or electronic components for generating analytical data will be affected by temperature extremes and must either be provided with thermal insulation or compensated for altered performance. Extrapolation of response obtained in the range of normal ambient temperature is not a reliable substitute for actual testing under use conditions.

1.9 VOLUMETRIC ERRORS. Rate meters of all types require frequent calibration to detect and correct drift or shifts in reference points, (see Section 1.5). Portable instruments should be checked every 40 to 50 h of operation with at least a secondary standard (wet or dry test meter or glass rotameter) which in turn is compared frequently with a primary standard of the positive displacement total volume type (spirometer).

In volumetric calibration procedures, the pressure drop across the instrument being calibrated and across the system in which it will operate must be determined and appropriate corrections applied (see Section 1.5 for effects of fouling). When filter systems are used, the increasing pressure drop and decreasing sampling rate that develop as the filter loads must also receive consideration. Unnecessary restrictions to free sample flow also produce undesirable pressure drop increments that may limit the range of the sampling system.

Temperature effects do alter volumetric calibration significantly if the system is operated under ambient conditions more than ±20°C from the original calibration temperature conditions. A high degree of volumetric accuracy requires determination of the instrumental temperature coefficient as well as correction of the sample volume to standard temperature and pressure (25°C and 101.3 kPa).

Variability in certain types of pumping systems, both between units of the same model or within any given unit, may prevent constant rate sampling within tolerable limits over extended time intervals. Direct current motors are especially vulnerable to rate fluctuations unless effective control circuits are incorporated in the design. Manually operated squeeze-bulb devices seldom deliver reproducible volumes and should be avoided if better than semi-quantitative accuracy is required. Positive displacement syringe-type hand pumps are capable of a high degree of volumetric accuracy and are to be preferred (41).

1.10 OBSERVATIONAL ERRORS. The optical judgment of color shade and optical density is one, if not the most, variable factor to be considered in air analysis. Reasonably accurate results can be attained if the individual using a color comparator is willing to calibrate the instrument for his own personal use, but he must take into account the quality and intensity of the light available for illumination of the comparator, day-to-day variations in his own optical acuity, shade changes introduced by interferences, and stability of the color standards. Field kits based on optical color standards provide a definite, valuable service not available by any other means, but the unavoidable sacrifice of accuracy and precision must be carefully weighed when planning and evaluating investigations based on such equipment (17,29,42).

Optical acuity and judgment also enters as a factor in the detection of air contaminants with the length-of-stain type of detector tubes. Diffusion, migration and trailing may so obscure the stain front as to render accurate reading very questionable. Again individual calibration with knowns can be quite helpful in providing a basis for judgment.

Errors in the application of volumetric glassware are not unique to air sampling and are recognized by those familiar with analytical laboratory techniques. Such items as "to deliver" versus "to contain," delivery time for pipets, whether or not to "touch off" or "blow out" the drop retained in the pipet tip, capillary effects in miniscus reading, error introduced by a water-repellant film on the glass, and chipped buret and pipet tips should be too well recognized to require further elaboration. The relative accuracy of the original glassware calibration may be overlooked. Most laboratory-ware manufacturers adhere to two standards: precision-grade (within NBS tolerances) and laboratory grade (twice precision-grade tolerances). The choice of buret, volumetric or Mohr pipet or graduated cylinder will be determined by the relative accuracy required in the measurement. That is, volumetric pipets or a buret would be used to measure out standardizing solutions whereas a graduated cylinder is suitable for making up most reagents.

Reading instruments that indicate by mechanical movement, such as a rotameter float, voltmeter needles, etc., involves critical judgment also. Whether or not to read the top, bottom, or center of a rotameter float, parallax effects in observing meter needles (some of the better precision meters provide a mirror background to eliminate this problem), capillary effects on the meniscus of liquid-filled manometers, and scale units, illustrate several of the more common observational errors. The increasing relative error inherent in any instrument as either the minimum or maximum range limits are approached may introduce serious observational error. The upper and lower 10% of any instrument's range should be avoided, and if possible the capacity should be so chosen as to permit operating only in the center 50% of the overall range.

1.11 OPTIMUM SAMPLING RATE. Calibration of any sample collection device should include a sufficient range of rates to delineate the optimum range consistent with the efficiency demanded of the system and to establish the penalty incurred when the rate exceeds or falls below these limits. Failure to include this factor has contributed serious errors (dust collection by impinger) which are inexcusable. Collection efficiency cannot be assumed; it must be determined by calibration.

The design of the collector in liquid reagent collection systems for gases and vapors is, in most cases, not critical if in addition to acceptable collection efficiency the design provides retention of entrainment (baffles), sufficient freeboard to retain bubbles and foam, and interchangeability of component parts, and the device is easily assembled, dismantled and cleaned **(29)**. Only in those cases that involve surface reactions and adsorption will a "fritted" or porous glass diffusing element perform more efficiently than a standard single-orifice impinger **(10,12)**. Frequently these diffusers create problems such as retention of analyte, excessive frothing and foaming, variable pressure drop, fragility, and particle retention that are not encountered with the use of impinger-type absorbers.

REFERENCES

1. LINCH, A.L., E.G. WIEST, AND M.D. CARTER, 1970. Evaluation of Tetraalkyl Lead Exposure by Personnel Monitor Surveys, Amer. Ind. Hyg. Assoc. J. 31:170.
2. AMERICAN SOCIETY FOR TESTING AND MATERIALS. 1982. *Annual Book of ASTM Standards*, Vol. 11.03, ASTM Designation: D1357-82, 1982, ASTM, Philadelphia.
3. STERN, A.C., 1968. *Air Pollution and Its Effects*, Vol. I:Part II (2nd Edition) Academic Press, New York.
4. STERN, A.C., 1976. *Air Pollution*, Vol. 1 "Air Pollutants, Their Transformation & Transport," Vol. 2 "The Effects of Air Pollution," Vol. 3 "Measuring, Monitoring and Surveillance of Air Pollution," Vol. 4 "Engineering Control of Air Pollution," Vol. 5 "Air Quality Management": Supplementary volumes published in 1986, Vol. 6 "Air Pollutants, Transformation, Transport and Effects," Vol. 7 "Measurement, Monitoring, Surveillance and Engineering Control of Air Pollution," and Vol. 8 "Management of Air Quality," Academic Press, New York.
5. CRALLEY, L.V., L.J. CRALLEY, AND G.D. CLAYTON, 1968. *Industrial Hygiene Highlights* 1:297. Industrial Hygiene Foundation of America, Inc., Pittsburgh, PA.
6. WEISBURD, M.I. (ed.), 1962. *Air Pollution Control Field Operations Manual*, U.S. Department of Health, Education and Welfare, Public Health Service.
7. AMERICAN INDUSTRIAL HYGIENE ASSOCIATION, 1960. *Air Pollution Manual*, Part I-Evaluation, Amer. Ind. Hyg. Assoc.
8. MARCALI, KALLMAN, 1957. Microdetermination of Toluenedisocyamates in Atmosphere, Anal. Chem. 29:552.
9. COTABISH, H.N., P.W. McCONNAUGHEY, AND H.C. MESSER, 1961. Making Known Concentrations for Instrument Calibration. Amer. Ind. Hyg. Assoc. J. 22:392.
10. SALTZMAN, B.E., 1961. Preparation and Analysis of Calibrated Low Concentrations of Sixteen Toxic Gases. Anal. Chem. 33:1100.
11. LINCH, A.L., R.C. CHARSHA, 1960. Development of a Freeze-Out Technique and Constant Sampling Rate for the Portable Uni-Jet Air Sampler. Amer. Ind. Hyg. Assoc. J. 21:325.
12. SALTZMAN, B.E., AND A.F. WARTBURG, JR., 1965. A Precision Flow Dilution System for Standard Low Concentrations of Nitrogen Dioxide. Anal. Chem. 37:1261.
13. LINCH, A.L., AND M. CORN, 1965. The Standard Midget Impinger-Design Improvement and Miniaturization. Amer. Ind. Hyg. Assoc. J. 26:601.
14. INSCOE, M.N., 1966. Losses Due to Adsorption During Filtration of Aqueous Solutions of Polycyclic Aromatic Hydrocarbons. Nature 211:1083.
15. SALTZMAN, B.E., 1954. Colorimetric Microdetermination of Nitrogen Dioxide in the Atmosphere. Anal. Chem., 26:1949-1955.
16. YAFFE. C.D., D.H. BYERS, AND A.D. HOSEY, 1956. *Encyclopedia of Instrumentation for Industrial Hygiene*. Ann Arbor, Michigan.
17. LINCH, A.L., S.S. LORD, K.A. KUBITZ, AND M.R. DE BRUNNER, 1965. Phosgene in Air-Development of Improved Detection Procedures. Amer. Ind. Hyg. Assoc. J. 26:465.
18. LINCH, A.L., AND M. CORN, 1965. The Standard Midget Impinger-Design Improvement and Miniaturization. Amer. Ind. Hyg. Assoc. J 26:601.
19. KATZ, M., 1970. Photochemical Reactions of Atmospheric Pollutants. Canad. J. of Chem. Eng. 48:3-11.
20. WEST, F.K., P.W. WEST, AND F.A. IDDINGS, 1966. Adsorption of Traces of Silver on Container Surfaces. Anal. Chem. 38:1566.
21. DYCK, W. 1968. Adsorption of Silver on Borosilicate Glass. Anal. Chem. 40:454.
22. LINCH, A.L., R.F. STALZER, AND D.T. LAFFERTS, 1968. Methyl and Ethyl Mercury Compounds-Recovery from Air and Analysis. Amer. Ind. Hyg. Assoc. J. 29:79.
23. KEENAN, R.G., 1963. Determination of Lead in Air and in Biological Materials, *Manual of Analytical Methods*, American Conference in Governmental Industrial Hygienists.
24. WOESSNER, W.W., AND J. CHOLAK, 1953. Improvements in the Rapid Screening Method for Lead in Urine. A.M.A. Arch. Ind. Hyg. & Occup. Med. 7:249.
25. DOWNING, F.B., AND A.L. LINCH, 1946. Process for Stabilizing of Deactivating Sludges, Precipitates and Residues Occurring or Used in the Manufacture of Tetraalkyl Leads. U.S. Patent 2,407,261.
26. TARES, D.R., 1968. Effect of Silicone Grease on Diffusion of Fluoride. Anal. Chem. 40:204.
27. KONIECKI, W.B., AND A.L. LINCH, 1958. Determination of Aromatic Nitro-Compounds. Anal. Chem. 30:1134.
28. SKULSKII, I.A., AND V.V. GLASMO, 1966. Absorption of Caesium Tetraphenylborate from Aqueous Solutions on Glass, Polyethylene and Polytetrafluoroethylene. Nature 211:631.
29. LINCH, A.L., R.B. DAVIS, R.F. STALZER, AND W.F. ANZILOTTI, 1964. Studies of Analytical Methods for Lead-in-Air Determination and Use

with an Improved Self-Powered Portable Sampler. Amer. Ind. Hyg. Assoc. J. 25:81.

30. PILAT, M.J., 1968. Application of Gas-Aerosol Adsorption Data to the Solution of Air Quality Standards. J. Air Poll. Control Assoc., 18:751.

31. ANDREWS, M.L., AND D.C. PETERSON, 1947. A Study of the Efficiency of Methods for Obtaining Vapor Samples in Air. J. Ind. Hyg. Toxic. 29:403.

32. CAMPBELL, E.E., AND H.M. IDE, 1966. Air Sampling and Analysis with Microcolumns of Silica Gel. Amer. Ind. Hyg. Assoc. J. 27:323.

33. SALTZMAN, B.E., 1970. Factors in Air Monitoring Network Design. Proceedings of the Eleventh Conference on Methods in Air Pollution and Industrial Hygiene Studies, California Air and Industrial Hygiene Laboratory. Berkeley, California. March 30.

34. SALTZMAN, B.E., 1970. Significance of Sampling Time in Air Monitoring. J. Air Poll. Control Assoc., 10:660.

35. NICHOLS, R., AND A. TOPPING, 1966. Absorption of Ethylene by Self-Indicating Soda-Lime. Nature 211:217.

36. CALIBRATED INSTRUMENTS, INC., 729 Sawmill River Road, Ardsley, NY 10502.

37. MILLIPORE FILTER CORP., 80 Ashby Road, Bedford, Mass. 01730 Type Aerosol (MAW G037A0).

38. VEILLON, C., AND J.Y. PARK, 1970. Correct Procedures for Calibration and Use of Rotameter-Type Gas Flow Measuring Devices. Anal. Chem., 42: 684.

39. ELKINS, H.B., 1959. *The Chemistry of Industrial Toxicology*, (2nd Ed.) pp. 284–288. John Wiley & Sons, New York

40. NOWEIR, M.H., AND E.A. PFITZER, 1971. An Improved Method for Determination of Phosgene in Air. Amer. Ind. Hyg. Assoc. J. 32:163.

41. KITAGAWA, T., 1960. The Rapid Measurement of Toxic Gases and Vapors. Proceedings of the 13th International Congress on Occupational Health. New York.

42. GRIMM, K.E., AND A.L. LINCH, 1964. Recent Isocyanate-In-Air Analysis Studies. Amer. Ind. Hyg. Assoc. J. 25:285.

43. TURK, A. AND J.W. JOHNSTON, JR., 1974. *Human Responses to Environmental Odors*. Academic Press, New York.

44. FOX, D.L., AND H.L. JEFFERIES, 1983. Air Pollution. Anal. Chem., 55:233R-245R.

45. LONNEMAN, W.A., AND J.J. BUFFALINI, 1981. Contamination from Fluorocarbon Films. Environ. Sci. & Technol. 15:99–103.

46. MEYER, B., 1983. *Indoor Air Quality*. Addison Wesley Publishing Co., New York, p. 56ff.

47. LUDWIG, F.A., 1978. Air Monitoring. Environ. Sci. & Technol. 12:774–778.

48. BENNETT, B.I., 1979. Stability Evaluation of Ambient Concentrations of Sulfur Dioxide, Nitric Oxide, and Nitrogen Dioxide Contained in Compressed Gas Cylinders. EPA/600/4–79/006, U.S. NTIS PB-292749.

49. KEBBEKUS, E. AND F. CORNAVACCA, 1977. Factors in the Selection of Calibration Gas Standards. American Laboratory, 9(7):51–57.

50. HUGHES, E.E., AND W.D. DROKO, 1977. Long Term Investigation of the Stability of Gaseous Standard Reference Materials, NBS Spec. Publ. 464:535–539.

51. CADOFF, B.C., 1977. Standard Reference Materials for the Analysis of Organic Vapors in the Air. NBS Spec. Publ. 464:541–543.

52. WRIGHT, R.S., W.C. EATON, C.E. DECKER, AND D.J. VON LEHMDEN, 1983. The Performance Audit Program for Gaseous Certified Reference Materials. Proc. Annual Meeting, Air Pollution Control Assoc. 76th (Vol.2). pp. 83–98.

53. SAMUELSEN, G.S., AND J.N. HARMAN, III, 1977. Chemical Transformation of Nitrogen Oxides While Sampling Combustion Products. J. Air Poll. Control Assoc. 27:648–655.

54. CRECELIUS, H.J., AND W. FORWEG, 1975. *Staub Reinhalt Luft*, 35:330, in German. APA 576017.

55. SCOTT RESEARCH LABORATORIES, 1985. "Gas Chromatography Newsletter," Fall 1985. Scott Research Laboratories, San Bernadino, CA.

56. BARKER, W.C., AND J.F. PONDROT, 1983. The Measurement of Gas Flow, Part I. J. Air Poll. Control Assoc. 33:66–72.

Approved with modifications from
2nd edition
Subcommittee 9
J. N. HARMAN, *Chairman*

2. Calibration Procedures

2.1 PREPARATION OF STATIC CALIBRATION MIXTURES.

2.1.1 *Introduction.* Static calibration mixtures are prepared by introducing a known weight or volume of contaminant into a given volume of air (1). Generally the mixture is held in a container of fixed dimensions, but flexible chambers may be used.

2.1.2 *Measuring the Contaminants.* Liquid contaminants are commonly introduced into the calibration system with a microsyringe or micropipet. Gaseous materials are handled with a gas-tight syringe.

A pure gas sample may be obtained from a lecture bottle by either of the methods shown in Figure 2:1. The first technique involves attaching a lecture bottle septum directly to a regulated cylinder (2). Just enough gas to purge the system is bled through the pressure relief valve. The sampling chamber can then be maintained at about two atmospheres with no additional loss of gas to the atmosphere. Contaminant gas is then obtained through the septum using a gas-tight syringe.

In the second method, the pure gas is passed slowly through a T with a septum over one arm. The gas then flows from a dry midget impinger into one containing 10 mL of water. The pure gas is then with-

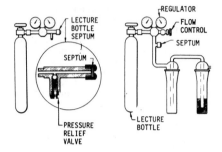

Figure 2:1 — Two methods for removing small amounts of contaminant gas from a lecture bottle.

drawn through the septum after the gas lines have been purged.

2.1.3 *Rigid Chambers.* Generally glass bottles and flasks make the best containers. However, plastic and metal chambers may be used if wall interactions with the contaminant gases are negligible.

The volume of the container is obtained by direct measurement of the chamber boundaries or volumetrically by filling with water. The usual size is on the order of 20 to 40 L, which allows the removal of enough useful gas without causing excessive dilution by the replacement gas.

A typical static calibration system is shown in Figure 2:2. The vessel is first purged with fresh air to remove any residual contamination, then the contaminant is injected directly into the vessel through the rubber septum. The gases and vapors are evaporated and mixed by an externally driven stirrer or magnetic stirrer or by agitating with internally placed aluminum or Teflon strips. The gas mixture is then withdrawn after opening both valves. Placement of two or three of these vessels in series as shown in Figure 2:3 greatly reduces the dilution effect of the air as the sample gas mixture is removed (1).

Sometimes it is desirable to sample from a rigid container without suffering makeup gas dilution or a significant internal pressure decrease. This can be done by attaching a deflated plastic bag to the dilution-gas inlet as shown in Figure 2:4. As the mixture is sampled, dilution air fills the plastic bag rather than diluting the mixture. A bag which does not absorb appreciable quantities of the contaminant gas or vapor must be selected for this purpose.

2.1.4 *Piston Type Chambers.* Figure 2:5 illustrates a piston type container (3). A calibration mixture is prepared by introducing the contaminant through the inlet as the piston is raised. The inlet is then closed and 15 min are allowed for mixing.

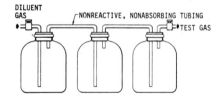

Figure 2:3 — Rigid chambers connected in series.

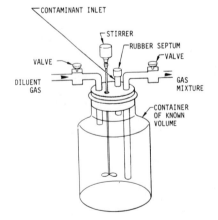

Figure 2:2 — Diagram of a typical state calibration system.

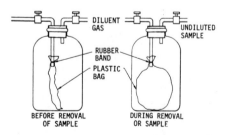
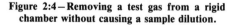

Figure 2:4 — Removing a test gas from a rigid chamber without causing a sample dilution.

2.1.5 *Nonrigid Chambers.* Mixtures may be prepared in plastic bags, generally made from Teflon, Mylar and aluminized Mylar. The bag is alternately filled and deflated with air to remove residual contamination and final purge is accomplished by applying a vacuum to the system. The container is connected to a flow meter or wet or dry gas meter. As the bag fills, the contaminant is introduced by one of the methods shown in Figure 2:6.

The test mixture should be used as soon as possible because the initial concentration often decays with time. Such decay can be lessened if the container is first preconditioned to the test substance however.

2.2 CALCULATIONS. The concentration by volume of a contaminant gas or vapor can be calculated either in per cent or in parts per million from the following equations.

For gas mixtures:

$$C_{\%} = \frac{100V_c}{V_c + V} \text{ and} \quad (1)$$

$$C_{ppm} = \frac{10^6 V_c}{V} \quad (2)$$

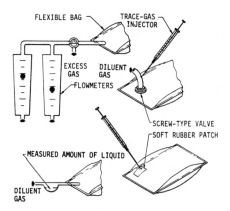

Figure 2:6 — Four methods of introducing a sample into a nonrigid chamber.

where V_c and V are the contaminant and system volumes respectively. If a liquid is added to a closed system, the resulting concentration is

$$C_{ppm} = 24.45 \times 10^6 \left(\frac{\rho V_L}{MV}\right)\left(\frac{T}{298}\right)\left(\frac{101.3}{P}\right) \quad (3)$$

where ρ is liquid density (g/mL), V_L is the volume of liquid (mL), T is the temperature (K), P is the pressure (kPa), M is the molecular weight (g/mol) and V is the system volume (L).

2.3 SOURCES OF ERROR. Lubricant greases and sealants should be avoided since they tend to absorb the trace contaminants. The use of rubber and plastic tubing, other than Teflon, should be avoided if possible. The compressed air should be of the highest quality available and a purification train should be used to remove unwanted contaminants.

For additional sources of error consult principles stated in the chapter on "Physical Precautions" which precedes this section.

REFERENCES

1. NELSON, G.O. 1971. *Controlled Test Atmospheres — Principles and Techniques.* Ann Arbor Science Publishers. Inc., Ann Arbor, Mich.
2. HAMILTON CO., P.O. Box 10030, Reno, Nevada 89520
3. HOUSTON ATLAS CO., P.O. Box 40052, Houston, Texas 77041.

Approved with Modifications
from 2nd edition
Subcommittee 9
J.N. HARMAN, *Chairman*

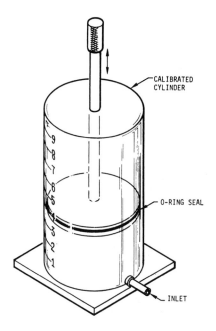

Figure 2:5 — A piston-type container.

3. Dynamic Calibration of Air Analysis Systems With Permeation Tubes

3.1 PERMEATION TUBES. The use of permeation tubes as primary standards for trace gas analysis was first documented by O'Keeffe and Ortman in 1966 (1). The principle of this device is based on diffusion of gas or vapor through a plastic membrane at very slow rates. A liquefied gas or volatile liquid sealed in a section of Teflon FEP tubing placed in a metered air stream can be used as a dynamic calibration standard. The diffusion rate is a nonlinear function of temperature; therefore, constant temperature conditions must be maintained for the permeation tube during gravimetric standardization and use as a calibration source. Also, the diffusion rate is influenced by the molecular weight of the carrier gas employed for dilution. The difference between air and nitrogen is negligible; but if a lighter gas such as helium or a heavier gas such as argon is used, corrections or recalibration would be required.

Certain compounds react with oxygen or water vapor at the surface of the plastic tubing; e.g., sulfur dioxide and hydrogen sulfide. However, such compounds can be used if air is excluded from contact with the loaded tube by storage and standardization in a dry nitrogen atmosphere. A relatively high concentration in dry nitrogen carrier gas then is diluted stepwise with air to provide calibration mixtures.

In general, Teflon fluorocarbon resins are chemically inert. However, certain qualifications that govern their application to ambient air analysis problems must be considered. A clear distinction between chemical compatibility and purely physical properties will best serve to delineate the application limits. Very few chemicals produce chemical degradation of fluorocarbon resins under normal conditions. Reaction will occur with molten alkali metals, fluorine, strong fluorinating agents, and sodium hydroxide above 300°C. If ignition is provided finely divided fluorocarbon resins also will react with 100% oxygen, aluminum or magnesium dust. Otherwise, Teflon fluorocarbons may be considered nonreactive. Examples of chemicals which Teflon

resists are listed in Table 3:1 (2). In most cases the tests included the boiling range.

Nearly all plastics absorb small quantities of certain materials on contact. Since no chemical reaction or solubility occurs in Teflon, the absorption detectable by weight increase is a result of the material filling submicroscopic voids between the polymer molecules (Table 3:II). The amount is proportional to contact time, pressure and temperature and the effect is reversible. If temperature or pressure cycling includes the boiling range, liquid inclusions within the submicroscopic pores will vaporize and thereby create internal pressure which may enlarge the pores and eventually mechanically damage the plastic. Surface blisters containing liquid provide visible evidence of this phenomenon. The absorption of aqueous solutions is minimal due in part to the low degree of wettability exhibited by fluorocarbon resins (3). Teflon is wetted by most organic solvents.

Gases and vapors permeate through fluorocarbon resins at a considerably lower rate than through most other plastics. In general, permeation increases with temperature, pressure and surface contact area, and decreases with increased film thickness and polymer density (Figure 3:1, 2, 3). The permeability coefficient P is expressed as:

$$P = \frac{Wt \times Th}{A \times Tc \times \Delta P}$$

Where:

Wt = Weight of permeant.
Th = Film thickness.
A = Area of film.
Tc = Contact time.
ΔP = Pressure difference.

Film thickness and ΔP can be combined into pressure gradient, $G = \Delta P/Th$ = atmospheres per mil (0.001 in.)

$$\text{Then: } P = \frac{Wt}{A \times Tc \times G}$$

Gases and liquids permeate plastics by molecular vibrations and motion between the plastic molecules. Therefore, an increase in plastic density for a well-molded resin will reduce the diffusion process due to reduced intermolecular spacing (Figure 3:1). Since Teflon FEP resins are processed by melt extrusion, voids larger than

Table 3:I Typical Chemicals with Which Teflon Resins are Compatible

Abietic acid	Cyclohexanone	Hydrogen peroxide	Phthalic acid
Acetic acid	Dibutyl phthalate	Lead	Pinene
Acetic anhydride	Dibutyl sebacate	Magnesium chloride	Piperidine
Acetone	Diethyl carbonate	Mercury	Polyacrylonitrile
Acetophenone	Diethyl ether	Methyl ethyl ketone	Potassium acetate
Acrylic anhydride	Dimethyl formamide	Methacrylic acid	Potassium hydroxide
Allyl acetate	Di-isobutyl adipate	Methanol	Potassium permanganate
Allyl methacrylate	Dimethylformamide	Methyl methacrylate	Pyridine
Aluminum chloride	Dimethyl hydrazine	Naphthalene	Soap and detergents
Ammonia, liquid	unsymmetrical	Naphthols	Sodium hydroxide
Ammonium chloride	Dioxane	Nitric acid	Sodium hypochlorite
Aniline	Ethyl acetate	Nitrobenzene	Sodium peroxide
Benzonitrile	Ethyl alcohol	2-Nitro-butanol	Solvents, aliphatic and
Benzoyl chloride	Ethyl ether	Nitromethane	aromatic*
Benzyl alcohol	Ethyl hexanoate	Nitrogen tetroxide	Stannous chloride
Borax	Ethylene bromide	2-nitro-2-methyl	Sulfur
Boric acid	Ethylene glycol	propanol	Sulfuric acid
Bromine	Ferric chloride	n-Octadecyl alcohol	Tetrabromoethane
n-Butylamine	Ferric phosphate	Oils animal and	Tetrachloroethylene
Butyl acetate	Fluoronaphthalene	vegetable	Trichloroacetic acid
Butyl methacrylate	Fluoronitrobenzene	Ozone	Trichloroethylene
Calcium chloride	Formaldehyde	Pentachlorobenzamide	Tricresyl phosphate
Carbon disulfide	Formic acid	Perfluoroxylene	Triethanolamine
Cetane	Furan	Phenol	Vinyl methacrylate
Chlorine	Gasoline	Phosphoric acid	Water
Chloroform	Hexachloroethane	Phosphorus	Xylene
Chlorosulfonic acid	Hexane	pentachloride	Zinc chloride
Chromic acid	Hydrazine		
Cyclohexane	Hydrochloric acid		

Based on experiments conducted up to the boiling points of the liquids listed. Teflon resins have normal service temperatures up to 500°F, (260°C) for TFE resins, 400°F (205°C) for FEP resins.
*Some halogenated solvents may cause moderate swelling.

intermolecular are not normally present and permeation is basically molecular diffusion only (**2**). See Figure 3:2 and Table 3:III.

Surface attractive forces between permeant and plastic barrier influence both absorption quantity and permeability coefficient P. Gases or vapors chemically related to the plastic normally show a higher permeation rate than dissimilar materials. High solubility of a material in a plastic is indicative of high permeation rate (molecular size and vapor pressure are also influential). Swelling produced by liquid absorption which increases molecular spacing of the plastic, and permeation is not a problem with fluorocarbon resins (**2**).

Increased temperature produces higher molecular activity of the diffusing material and hence permeation (Figure 3:2). This increase in not only due to increased solvent molecule activity but also due to the in-creased vapor pressure. Since the permeation rate is temperature dependent (as much as 10% per degree C) precise temperature control is critical.

Permeation tubes are available from several commercial sources and a wide range of permeant gas choices is available. Table 3:IV illustrates a representative selection of the currently available permeation tubes. The vendors of permeation tubes also offer custom-built permeation tubes, with the material designated by the customer, as special items. Also available are alternative permeation-based sources such as those employing a wafer of a diffusion-limiting plastic which may be connected to a lecture bottle of the desired permeant to afford a very long-life source.

Perhaps the most important parameter in the standardization of permeation tubes is the time factor required for diffusion equilibrium to be reached. Two to 5 days should

Table 3:II Absorption of Common Solvents in Teflon Resins

Solvent	Exposure °C	Temp* (°F)	Exposure Time	Weight Increase %
Acetone	25	(77)	12 mo	0.3
	50	(122)	12 mo	0.4
	70	(158)	2 wk	0
Benzene	78	(172)	96 hr	0.5
	100	(212)	8 hr	0.6
	200	(392)	8 hr	1.0
Carbon	25	(77)	12 mo	0.6
tetrachloride	50	(122)	12 mo	1.6
	70	(158)	2 wk	1.9
	100	(212)	8 hr	2.5
	200	(392)	8 hr	3.7
Ethyl	25	(77)	12 mo	0
alcohol (95%)	50	(122)	12 mo	0
	70	(158)	2 wk	0
	100	(212)	8 hr	0.1
	200	(392)	8 hr	0.3
Ethylacetate	25	(77)	12 mo	0.5
	50	(122)	12 mo	0.70
	70	(158)	2 wk	0.7
Toluene	25	(77)	12 mo	0.3
	50	(122)	12 mo	0.6
	70	(158)	2 wk	0.6

*Exposure at over the boiling point of the solvent are at its vapor pressure.
Note: Values are averages only and not for specification purposes.

be allowed in a constant temperature oven (30°C ± 0.02°C) under a flowing dry air stream for accurate standardization. The time is dependent upon diffusion rate and accuracy of weighing. In use, constant temperature should be maintained to ± 0.1°C for precision within ± 1% (3). Gas chromatography ovens are ideal for this purpose and internal standards for calibration of field gas chromatography apparatus can be conveniently installed inside the calibration assembly (4).

Care must be taken to assure tight seals at the tube ends; this is most important with low-boiling compounds. It is doubtful if tubes can be fabricated using materials that exert pressures of greater than 50 psig at normal temperatures. The method used is that described by O'Keeffe and Ortman (1). Others are:

a. Swagelok–type plugs at each end (may be cumbersome).

b. Tight fitting caps sealed with Apiezon W—for low pressure applications.

3.2 CALCULATION OF LENGTH REQUIRED. To a very good approximation, the output of a permeation tube is proportional to its active length (the length of tube in contact with the contained fluid). If the output of any given tube is known, then a rate per unit length can be calculated. Table 3:V shows rates for a particular size and material. To achieve any desired output, the length is then

1)
$$L = \frac{Pr}{R}$$

where L = length (cm) of permeation tube required.
Pr = output required (ng/min).
R = permeation rate in ng/min cm.

Thus, if an output of 4500 ng/min of SO_2 is needed, using the tubes considered in Table 3:V, then the required length is 4500/290 = 15.5 cm.

3.3 CONVERSION OF NG/MIN TO PPMV AT 30°C. Because many instruments are cali-

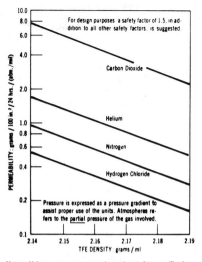

Notes: Values are averages only and not for specification purposes. To convert the permeation values for 100 sq. in. to those for 1 sq. cm. multiply by 0.00155. One mil = 25.4 μ.

Figure 3:1 – Effect of density of "Teflon" TFE resins on their permeability to gases at 30°C (86°F).

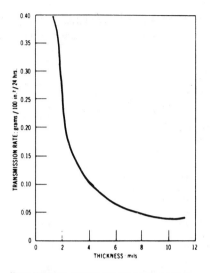

Notes: Values are averages only and not for specification purposes. To convert the permeation values for 100 sq. in. to those for 1 sq. cm. multiply by 0.00155. One mil = 25.4 μ.

Figure 3:3 – Water vapor transmission rate of "Teflon" FEP resins at 40°C (104°F).

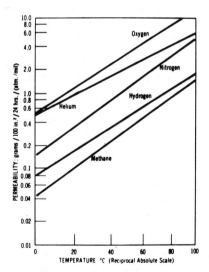

Notes: Values are averages only and not for specification purposes. To convert the permeation values for 100 sq. in. to those for 1 sq. cm. multiply by 0.00155. One mil = 25.4 μ.

Figure 3:2 – Permeability of "Teflon" FEP resins to gases, at various temperatures.

brated in ppmv and outputs are given in ng/min, a conversion formula is required. To convert from ng/min to ppm the required formula is:

2)
$$C = \frac{Pr \cdot K}{F}$$

where
C = concentration in ppm (vol).
Pr = output in ng/min.
F = flow rate in cm³/min.
K = ratio of molar volume (L/mol) to molecular weight (Table 3:VI).

Example: A permeation tube containing propane and having an output of 950 ng/min is to be used at 30°C and a flow rate of 500 cm³/min. What is the concentration of propane (in ppm) in the gas mixture?

$$C = \frac{950 \times 0.565}{500} = 1.07$$

3.4 Permeation tubes that contain liquefied gases or volatile liquids are calibrated gravimetrically and used to prepare standard concentrations of pollutants in air as follows (1,3,4,5). Analyses of these known concentrations give calibration curves

Table 3:III Permeability of Teflon Fluorocarbon Resins to Vapors (g/100 in² 24 h)*

	TFE		FEP		
	23°C. (73°F.)	30°C. (86°F.)	23°C. (73°F.)	35°C. (95°F.)	50°C. (122°F.)
Acetic acid	–	–	–	0.42	–
Acetone	–	–	0.13	0.95	3.29
Acetophenone	0.56	–	0.47	–	–
Benzene	0.36	0.80	0.15	0.64	–
n-Butyl ether	–	–	0.08	–	0.65
Carbon tetrachloride	0.06	–	0.11	0.31	–
Decane	–	–	0.72	–	1.03
Dipentene	–	–	0.17	–	0.35
Ethylacetate	–	–	0.06	0.77	2.9
Ethyl alcohol	0.13	–	0.11	0.69	–
Hexane	–	–	–	0.57	–
HCl. 20%	<0.01	–	<0.01	–	–
Methanol	–	–	–	–	5.61
Piperidine	0.07	–	0.04	–	–
"Skydrol" hydraulic fluid	0.06	–	0.05	–	–
NaOH. 50%	5×10^{-5}	–	4×10^{-5}	–	–
H_2SO_4. 98%	1.8×10^{-5}	–	8×10^{-6}	–	–
Toluene	–	–	0.37	–	2.93
Water	–	0.35	0.09	0.45	0.89

*Test Method: ASTM E-96-53T (at vapor pressure; for 0.001″ film thickness; one mil = 0.001″ = 25.4 μm).

Notes: Values are averages only and not for specification purposes. To convert the permeation values for 100 sq cm, multiply by 0.00155.

which simulate all of the operational conditions performed during the sampling and chemical procedure. The calibration curves include the important corrections for collection efficiency at various concentrations of pollutants.

Prepare or obtain (3,8) a Teflon FEP permeation tube that emits gas or vapor at a rate consistent with the desired concentration level. A permeation tube with an effective length of 1 to 2 cm and a wall thickness of 0.76 mm. (0.030") will usually yield the desired permeation rate if held at a constant temperature of 20°C for SO_2.

Permeation tubes are calibrated under a stream of dry nitrogen when reactions occur with atmospheric oxygen or moisture on or within the walls of the permeation tube (H_2S, SO_2, etc.).

3.5 To prepare standard concentrations of pollutant, select either a system designed for laboratory use or one for field use. (Figure 3:4 and 3:5, respectively). Assemble the apparatus as shown in one of these systems, consisting of a water-cooled condenser;

constant-temperature water bath maintained at the desired temperature; cylinder containing pure, dry air with appropriate pressure regulators; needle valves and flow meters for the nitrogen, if necessary; and dry air, diluent gas streams. The diluent gases are brought to temperature by passage through a 2-m-long copper coil immersed in the water bath. Insert a calibrated permeation tube (7) into the central tube of the condenser (maintained at the desired temperature by circulating water from the constant-temperature bath) and pass a stream of nitrogen or air over the tube at a fixed rate of approximately 50 cm³ min. Dilute this gas stream to the desired concentration by varying the flow rate of the clean, dry air. Clean, dry air may also be prepared by passing ambient air from a relatively uncontaminated outside source first through absorption tubes packed with activated carbon and soda-lime, then through an efficient fiber glass filter in series. This flow rate can normally be varied from 0.2 to 15 L/minute. The flow rate of

Table 3:IV Commercially Available Permeation Tubes

Chemical Permeant Material Gas	Formula	Chemical Permeant Material Gas	Formula
Acetaldehyde	CH_3CHO	Heptane (n)	C_7H_{16}
Acetic acid	CH_3COOH	Hexane (n)	C_6H_{14}
Acetone	CH_3COCH_3	Hydrazine	$NH_2NH_2 \cdot H_2O$
Acetonitrile	CH_3CN	Hydrogen chloride	HCl
Acrylonitrile	CH_2CHCN	Hydrogen cyanide	HCN
Ammonia	NH_3	Hydrogen fluoride	HF
Benzaldehyde	C_6H_5CHO	Hydrogen sulfide	H_2S
Benzene	C_6H_6	Iodine	I_2
Bromine	Br_2	Isobutylene	$(CH_3)_2CCH_2$
Butane	C_4H_{10}	Isopropyl alcohol	$(CH_3)_2CHOH$
Butanol	C_4H_9OH	Isopropylbenzene	C_9H_{12}
Butene (1)	C_4H_8	Isopropyl mercaptan	$(CH_3)_2CHSH$
Butyl acetate	$CH_3COOC_4H_9$	Mercury	Hg
Butyl mercaptan (n)	C_4H_9SH	Methanol	CH_3OH
Butyl mercaptan (Tert)	C_4H_9SH	Methyl acetate	$CH_3CO_2CH_3$
Carbon disulfide	CS_2	Methyl amine	CH_3NH_2
Carbon tetrachloride	CCl_4	Methyl bromide	CH_3Br
Carbonyl sulfide	COS	Methyl ethyl ketone	$CH_3COC_2H_5$
Chlorine	Cl_2	Methyl ethyl sulfide	$CH_3SC_2H_5$
Chlorobenzene	C_6H_5Cl	Methyl iodide	CH_3I
Chloroethane	C_2H_5Cl	Methyl isocyanate	CH_3NCO
Chloroform	$CHCl_3$	Methyl isopropyl sulfide	$CH_3SCH(CH_3)_2$
Chloromethane	CH_3Cl	Methyl mercaptan	CH_3SH
Chlorotoluene (o)	$CH_3C_6H_4Cl$	Napthalene	$C_{10}H_8$
Cyclohexane	C_6H_{12}	Nitric acid	HNO_3
Cyclopropane	$CH_2CH_2CH_2$	Nitrobenzene	$C_6H_5NO_2$
Decane	$C_{10}H_{22}$	Nitrogen dioxide	NO_2
Dichlorobenzene (p)	$C_6H_4Cl_2$	Nonane (n)	C_9H_{20}
Dichloroethane (1,1)	Cl_2CHCH_3	Octane (n)	C_8H_{18}
Dichloroethane (1,2)	$ClCH_2CH_2Cl$	Octanethiol (n)	$C_8H_{17}SH$
Dichloromethane	CH_2Cl_2	Pentane (n)	C_5H_{12}
Diethyl disulfide	$(C_2H_5)_2S_2$	Phenol	C_6H_5OH
Diethyl sulfide	$(C_2H_5)_2S$	Phosgene	$COCl_2$
Dimethyl amine	$(CH_3)_2NH$	Propane	C_3H_8
Dimethyl disulfide	$(CH_3)_2S_2$	Propyl mercaptan (n)	C_3H_7SH
Dimethyl ether	CH_3OCH_3	Propylene	C_3H_6
Dimethyl mercury	$Hg(CH_3)_2$	Propylene oxide	C_3H_6O
Dimethyl sulfide	$(CH_3)_2S$	Pyridine	C_5H_5N
Dimethyl sulfoxide	$(CH_3)_2SO$	Pyrrole	C_4H_5N
Dinitrotoluene (2,4)	$C_7H_6(NO_2)_2$	Sulfur dioxide	SO_2
Dipropyl sulfide	$(C_3H_7)_2S$	Sulfur hexafluoride	SF_6
Epichlorohydrin	C_3H_5OCl	Tetrachloroethylene	C_2Cl_4
Ethanol	C_2H_5OH	Thionyl chloride	$SOCl_2$
Ethyl acetate	$CH_3CO_2C_2H_5$	Thiophane	C_4H_8S
Ethyl amine	$C_2H_5NH_2$	Toluene	$C_6H_5CH_3$
Ethyl benzene	$C_6H_5C_2H_5$	Toluene diisocyanate	$H_3CC_6H_3(NCO)_2$
Ethyl mercaptan	C_2H_5SH	Trichloroethane (1,1,1)	CH_3CCl_3
Ethylene dibromide	$C_2H_4Br_2$	Trichloroethylene	$CHClCCl_2$
Ethylene oxide	$(CH_2)_2O$	Trimethylamine	$(CH_3)_3N$
Formaldehyde	CH_2O	Trioxane	$(CH_2O)_3$
Formic acid	$HCOOH$	Vinyl acetate	$CH_3COOC_2H_3$
Freon-11	CCl_3F	Water	H_2O
Freon-113	$C_2Cl_3F_3$	Xylene (m)	$C_6H_4(CH_3)_2$
Freon-12	CCl_2F_2	Xylene (o)	$C_6H_4(CH_3)_2$
Freon-21	$CHCl_2F$	Xylene (p)	$C_6H_4(CH_3)_2$
Freon-22	$CHClF_2$		

Table 3:V Typical Permeation Rates

Chemical	Permeation Rate (ng/min cm) at 30°C*
SO$_2$	290
NO$_2$	1200
H$_2$S	250
Propane	80
n-Butane	2
Chlorine	1500
Ammonia	170
Methyl Mercaptan	30

*Minimum Tube Length is 2.5 cm.

the sampling system determines the lower limit for the flow rate of the diluent gases. The flow rates of the nitrogen and the diluent air must be measured to an accuracy of 1 to 2%. With a tube emitting SO$_2$ at a rate of 0.26 μg/min, the range of concentration will be between 0.007 to 0.5 ppm (17 to 1300 μg m^3), a generally satisfactory range for ambient air conditions. When higher concentrations are desired, calibrate and use longer permeation tubes.

3.6 PROCEDURE FOR PREPARING SIMULATED CALIBRATION CURVES. Obviously one can prepare a multitude of curves by selecting different combinations of sampling rate and sampling time. The following description for SO$_2$ represents a typical procedure for ambient air sampling of short duration with a brief mention of a modification for 24-h sampling. The system is designed to provide an accurate measure in the 0.01 to 0.5 ppm range. It can be easily modified to meet special needs.

The dynamic range of the colorimetric procedure fixes the total volume of the sample at 30 L; then, to obtain linearity between the absorbance of the solution and the concentration in ppm, select a constant

Table 3:VI K Values.

	*K Values at 30°C
SO$_2$	0.389
NO$_2$	0.541
H$_2$S	0.732
Propane	0.565
N-Butane	0.429
Chlorine	0.351
Ammonia	1.463
Methyl Mercaptan	0.518

*Diverse density, L/g.

sampling time. This fixing of sampling time is desirable also from a practical standpoint. In this case, select a sampling time of 30 min. Then to obtain a 30-L sample of air requires a flow rate of 1 L/min. A 22-gauge hypodermic needle operating as a critical orifice will control air flow at this approximate desired rate (10). The concentration in air is computed as follows:

$$C = \frac{Pr \times K}{r_1 + r_2}$$

Where   C = Concentration in ppm.
    Pr = Output of tube in μg/min.
    K = Constant from Table 3:VI.
    r_1 = Flow rate of diluent air, liter/min.
    r_2 = Flow rate of diluent nitrogen, liter/min.

Data for a typical calibration curve are listed in Table 3:VII.

A plot of the concentration of sulfur dioxide in ppm (x–axis) against absorbance of the final solution (y–axis) will yield a straight line, the slope of which is the factor for conversion of absorbance to ppm. This factor includes the correction for collection efficiency. Any deviation from linearity at the lower concentration range indicates a change in collection efficiency of the sampling system. Actually, the standard concentration of 0.01 ppm and below of sulfur dioxide is slightly below the dynamic range of the method. If this is the range of interest, the total volume of air collected should be increased to obtain sufficient color within the dynamic range of the colorimetric procedure. Also, once the calibration factor has been established under simulated conditions the conditions can be modified so that the concentration of SO$_2$ is a simple multiple of the absorbance of the colored solution.

For long-term sampling of 24-h duration the conditions can be fixed to collect 300 L of sample in a larger volume of tetrachloromercurate. For example, for 24 h at 0.2 L/min, approximately 288 L of air are collected. An aliquot representing 0.1 of the entire amount of sample is taken for the analysis.

3.7 EQUILIBRIUM. Each tube's permeation rate changes with *its* temperature. Hence, time must always be allowed for the

PERMEATION TUBE SCHEMATIC FOR LABORATORY USE

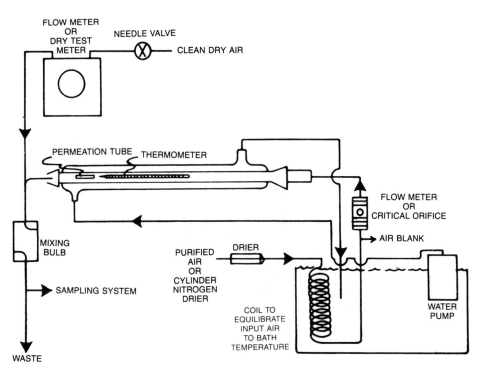

Figure 3:4 — Gas dilution system for preparation of standard concentrations of sulfur dioxide for laboratory use by the permeation tube method.

tube to reach an equilibrium condition. This is particularly true for tubes stored in a cool place and then inserted in a warm chamber for gas generation control. In such cases the tube will accurately provide a gas source based on its actual temperature, rather than the temperature of the warm air flowing through the chamber. Output concentration will be lower than that predicted

Table 3:VII Typical Calibration Data

Concentrations of SO_2 ppm	Amount of SO_2 in μL for 30 L	Absorbance of sample
0.005	0.15	0.01
0.001	0.30	0.02
0.05	1.50	0.117
0.10	3.00	0.234
0.20	6.00	0.468
0.30	9.00	0.703
0.40	12.00	0.937

by a permeation rate vs. air temperature calculation (9).

3.8 VENTILATION. Each tube permeates continuously. Thus if a tube is enclosed in a chamber without ventilation, concentration buildup will occur. This is particularly true for tubes stored in a warm, closed chamber environment. In such cases the actual concentration coming out of the chamber upon the first flush of ventilation air will be higher than that predicted by a permeation rate *vs* air temperature calculation. Often, considerable time must be allowed to "degas" a saturated system (9). An attractive alternative is to maintain a slow purge on the tube during periods of storage so that evolution of a high concentration plug of sample is avoided when the permeation device is first connected.

3.9 CONDITIONING. When working with low concentrations of gas (*i.e.*, concentra-

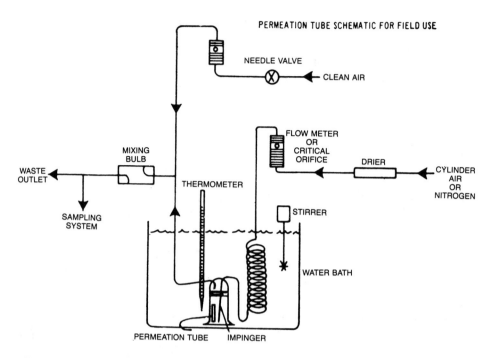

PERMEATION TUBE SCHEMATIC FOR FIELD USE

NEEDLE VALVE

CLEAN AIR

FLOW METER
OR
CRITICAL
ORIFICE

DRIER

CYLINDER
AIR
OR
NITROGEN

MIXING
BULB

WASTE
OUTLET

THERMOMETER

STIRRER

SAMPLING
SYSTEM

WATER BATH

PERMEATION TUBE   IMPINGER

Figure 3:5 — Gas dilution system for preparation of standard concentrations of sulfur dioxide for field use by the permeation tube method.

tions near nominal ambient levels) it is important to condition the system by allowing the gas to pass through the system for a period of day(s) before conducting any experiments. Conditioning is necessary even if nominally inactive materials such as glass or stainless steel are used. Active materials such as rubber should, of course, be avoided (9).

3.10 TESTING *vs* CALIBRATION. Dynacal permeation tubes can be used for testing atmospheric analyzers without temperature control. However, the existing temperature needs to be measured and taken into account in calculating the approximate permeation rate. Testing is considered to be an operation to establish whether an analyzer is operating "about right." It should be performed frequently. Testing, on the other hand, is not a substitute for less frequent, periodic calibrations in which known concentrations of gas are delivered to the analyzer with precision (9).

Acknowledgment. Mr. Arthur Johnston assisted with this section as Chairman of the

Analytical Chemistry Committee Standardization Subcommittee.

REFERENCES

1. O'KEEFFE, A.E., AND G.C. ORTMAN, 1966. Primary Standards for Trace Gas Analysis. Anal. Chem. 38:760.
2. DIGEL, W.A., (ed.), 1970. The Journal of "Teflon," Reprint No. 41. E.I. Dupont de Nemours Company.
3. ANALYTICAL INSTRUMENT DEVELOPMENT, INC., 250 S. Franklin Street, West Chester, Pa., 19380.
4. PATTY, F.A., 1963. *Industrial Hygiene and Toxicology*. 2nd Edition, Vol. II. Interscience Publishers. New York.
5. SCARINGELLI, F.P., A.A. FREY, AND B.E. SALTZMAN, 1967. Evaluation of Teflon Permeation Tubes for Use with Sulfur Dioxide. Amer. Ind. Hyg. Assoc. J. 28:260.
6. THOMAS, M.D., AND R.E. AMTOWER, 1966. Gas Dilution Apparatus for Preparing Reproducible Dynamic Gas Mixtures in any Desired Concentration and Complexity. J. Air Pollut. Contr. Assoc. 16:618.
7. SCARINGELLI, F.P., A.E. O'KEEFFE, E. ROSENBERG, AND J.P. BELL, 1970. Preparation of Known Concentrations of Gases and Vapors with Permeation Devices Calibrated Gravimetrically. Anal. Chem. 42:871.

8. VICI METRONICS, 2991 Corvin Drive, Santa Clara, Calif. 95061.
9. METRONICS ASSOCIATES, 3201 Porter Drive, Stanford Industrial Park, Palo Alto, California.
10. LODGE, J.P., JR., J.B. PATE, B.E. AMMONS, AND G.A. SWANSON, 1966. The Use of Hypodermic Needles as Critical Orifices in Air Sampling. J. Air Pollut. Control Assoc. 16:197.

Approved with modifications
from the 2nd edition
Subcommittee 9
J. N. HARMAN, *Chairman*

4. Sampling and Storage of Particles

4.1 INTRODUCTION. The collection of a particle sample that is not too different from the population of interest is often difficult and requires attention to details. This section deals with general information concerning the collection of such samples. Specific methods of analysis are covered in other sections.

Particles are sampled for many reasons such as:

a. to determine whether there are hazardous concentrations of pollutants in the atmosphere or if ambient air standards have been exceeded;
b. to determine the effectiveness of control programs in reducing ambient concentrations of pollutants;
c. to determine the emission levels from a source;
d. to determine the effectiveness of control equipment;
e. to determine the sources contributing to pollution at a receptor; or
f. to identify pollutants in the atmosphere.

Aerosols are defined here to mean dispersions of any material in the solid or liquid phase in a gas stream or the atmosphere. Dusts, smoke, soot, mist, fumes, and fog are terms used on occasion to describe certain types of aerosols. Some of the more useful terms may be defined as follows:

a. *Settleable Particles*—larger particles that settle out of the air fairly fast, such as those caught in an open jar. They are usually larger than 30 μm diameter, but in still air some particles 10 μm or smaller may settle.

b. *Suspended Particles*—particles that tend to remain suspended in the atmosphere for long periods of time. These are generally smaller than 30 μm in diameter. Commonly, the material collected by a High Volume Sampler (Hi-Vol) is limited by the shelter design and filter to particles less than 40 μm.

c. *Condensation Nuclei*—sometimes called *Aitken nuclei*, are small particles that act as condensation sites for supersaturated vapors in the atmosphere. They are predominantly 0.01 to 0.1 μm diameter.

d. *Agglomerates*—particles that are composed of several smaller particles that are attracted to a large particle or to each other and travel together in the atmosphere as a single particle.

e. *Modes*—the distribution of particle sizes in the atmosphere is usually bimodal or trimodal. The size range defined above as *condensation nuclei* are now termed the *nuclei mode*. Particles 0.1–2.5 μm are called the *accumulation mode*, being most persistent in the atmosphere. These two classes are collectively called *fine particles*. Particles larger than 2.5 μm are called *coarse particles*.

It should be kept in mind that there are no sharp cutoffs between classifications and the particle sizes in each class overlap the neighboring classes. This may be due to several factors that tend to make the proper collection and classification of particles difficult and somewhat inexact. These include the shape of the particles (which affects their aerodynamic properties), their density and velocity (which affects their inertia), and electrical charge. These will be discussed in more detail later in this section.

There is a great variety of methods available for the collection of particles, and the method selected will often depend on the purpose for which the sample is being taken. For instance, if information on individual particles is desired, a few hundred particles may be collected for microscopic examination. This can determine particle shape and size distribution but does not tell all about aerodynamic properties, weight, or chemical composition. A larger amount may be collected in an inertial collector in which the particle size distribution can be determined. This does not provide separate particles that can be examined nor does it usually provide enough samples for chemical analysis. For chemical analysis, a Hi-Vol sampler is usually employed. The average composition of all of the particles can be determined, but this does not provide information on the composition of individual particles or their sizes or shapes, or on the compounds present. Sometimes optical properties such as visibility reductions are of interest and none of the above methods is entirely suitable.

4.2 PRECAUTIONS IN COLLECTING AEROSOL SAMPLES.

4.2.1 *The Problem.* The most bothersome aspect of aerosol sampling is a result of the momentum of the particles, which is a product of its mass times its velocity (MV). Thus, this problem becomes more severe with larger particles or fast moving streams. Since particles are much larger than gas molecules, each time the flow direction of the gas stream changes, as at a bend in a pipe or in flow around an object, the larger particles tend to continue on their original line and are displaced somewhat from the original part of the gas stream they were with, as illustrated in Figure 4:1. This often leads to deposition on a nearby surface or unevenly dispersed streams where more particles will be found in one side of a pipe than another. Different sized particles will be displaced by different amounts. This makes it necessary for all parts of the stream to be sampled and properly weighted to be representative of the whole stream.

4.2.2 *Isokinetic sampling* refers to taking a sample under such conditions that there is no change in momentum, so that the sample will be representative of the gases and aerosols in that portion of the stream being sampled. This is accomplished by using a thin-walled tube aligned with the stream flow and drawing sample into it at the same linear velocity as the stream flow at that point. Even when these precautions are taken, there can be several reasons why the sample is not truly representative:

a. The probe always has a finite wall thickness that disturbs the flow;

b. the point sampled may not be representative of the whole stream;

c. if the flow is turbulent at the point of sampling there is no way to get a truly isokinetic sample; the sample will contain smaller portions of the larger particles; or

d. sample may be lost by deposition or changed by agglomeration or deagglomeration in the sample line after it enters the probe and before it enters the main sampling device.

Fortunately, it is not always necessary to use isokinetic sampling to obtain a reasonably good sample. Figure 4:2 shows the collection efficiency for particles of various sizes as a function of stream velocity. Figure 4:3 shows the effect of not aligning the sample tube in the same direction as the stream flow.

These data show that:

a. Sample velocity and alignment are both important conditions of isokinetic sampling;

b. under isokinetic conditions all particle sizes are collected efficiently;

c. small particles (under about 3 μm diameter) do not require isokinetic sampling for efficient collection – their small mass minimizes inertial effects;

d. stagnant gases, such as calm ambient air, do not require isokinetic conditions for efficient sampling of all particle sizes because slow moving particles have little or no momentum.

4.2.3 *Gravitational Effect.* In the case of quiescent air masses, as discussed immediately above, there will be a small effect because of gravitational settling of the particles. If the probe points up, the concentration of particles collected will be increased by a factor of $(1 + V_s/U)$ where V_s is the settling velocity of a particle and U is the sampling velocity. If the probe aims down (such as the shed opening of a Hi-Vol), the sample will contain less particles by the factor $(1-V_s/U)$. Here again, small particles are not much of a problem. The correction factor would predict a reduction of about 1% for 11 μm particles entering a Hi-Vol (assuming 645 cm^2 of opening and 1.42 m^3/min) but for 36 μm particles there would be 10% less, for 85μ particles about 50% less, and no particles larger than 135μm would enter. Of course, turbulent winds could

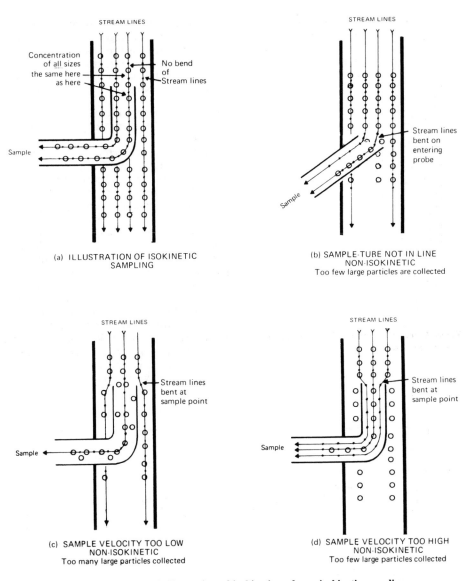

(a) ILLUSTRATION OF ISOKINETIC
SAMPLING

(b) SAMPLE-TUBE NOT IN LINE
NON-ISOKINETIC
Too few large particles are collected

(c) SAMPLE VELOCITY TOO LOW
NON-ISOKINETIC
Too many large particles collected

(d) SAMPLE VELOCITY TOO HIGH
NON-ISOKINETIC
Too few large particles collected

Figure 4:1 — Schematic illustration of isokinetic and non-isokinetic sampling.

help a few larger particles to reach the filter. (Obviously in the real system other factors enter, since experimentally the Hi-Vol sampler collects 50% of particles of diameters 30–45 μm depending on wind speed.)

4.2.4 *Sampling Rate.* The linear sampling rate can have an effect on several other types of aerosol collectors. For instance, the efficiency of impingement collectors such as cascade impactors, Anderson samplers, Rotorods, and jars with sticky paper all are dependent on the Stokes number. For impactors,

$$\text{Stk} = \frac{I_i}{R} = \frac{2V_o r^2 \rho}{9 \mu R}$$

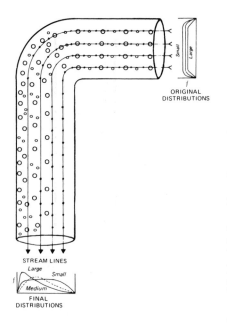

STREAM LINES

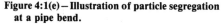

FINAL
DISTRIBUTIONS

**Figure 4:1(e) — Illustration of particle segregation
at a pipe bend.**

where I_i is the stopping distance, R is the radius of the impingement orifice, r the radius of the particle, V_o the velocity of the stream as it leaves the orifice, ρ the density of the particle, and μ the viscosity of the medium.

The mass of the particle, its linear velocity and the size of the jet or object are important factors.

Other types of collectors such as electrostatic precipitators and filters also depend on velocity, but to a much smaller extent.

4.2.5 *Miscellaneous Factors.* Previous mention was made of particles being lost from a sample by impingement on surfaces or bends in the sampling line, which can cause errors even after a good sample has been taken at the probe. In addition to this, particles can settle out in tubes or other components of the sampler or can be lost by electrostatic attraction to various surfaces in the sampler — especially nonconducting surfaces such as glass or plastics. Particles can also be lost due to adhesion on wet, oily, or sticky surfaces. As shown in Figure 4:4 and Table 4:I this can vary with relative humidity and surface roughness.

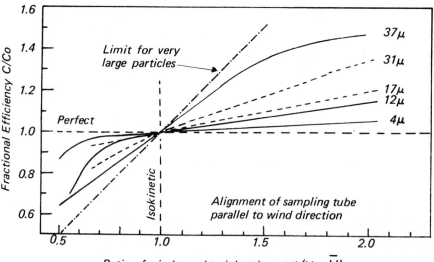

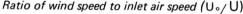

**Figure 4:2 — Diagram showing change in ratio of observed concentration to true concentration with
departure from isokinetic conditions.**
(Watson, H. 1954, Amer. Ind. Hyg. Assoc. Quart. 15:21.)

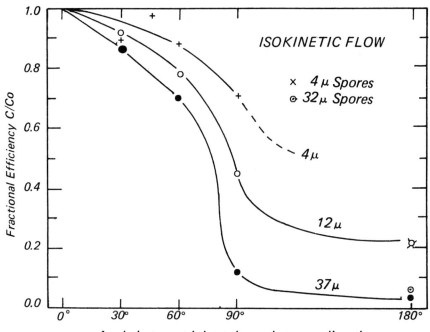

Figure 4:3 — Variation of the ratio of observed concentration to true concentration with angle of yaw of inlet tube for 4, 12 and 37 micrometer mass median diameter diethyl phthalate droplets and for 4 and 32 micrometer spores.
(Watson. H. 1954. Amer. Ind. Hyg. Assoc. Quart. 15:21)

Several factors can result in changes in size of particles, which may change the collection efficiency as discussed above. During periods of high relative humidity some particles such as sulfuric acid can absorb water vapor and get larger as illustrated in Figure 4:5. Most particles will adsorb water or other vapors; therefore, collected samples are generally equilibrated to a standard relative humidity, such as 40%, before weighing. Water adsorption is generally minimal up to about 50% R.H. but can cause considerable variation and error above this value.

There is generally a tendency for small particles to agglomerate to larger particles. This can occur in the sampling system but is not fast enough to be of concern for particles larger than about 0.2μm. Even for this size, particles only agglomerate at a rate of about 1%/h. In impingement collection methods, particles that are normally agglomerated in the atmosphere may be fractured to smaller particles when they strike the collector. Some of these smaller particles may then be redispersed and either collected as smaller particles or lost from the collector.

The following precautions should be mentioned as possibilities but seldom create a serious problem. Collected particles may react after collection so that the material analyzed may not represent the composition of the original aerosol. For instance, sulfuric acid particles may react with cement kiln dust (CaO) to form calcium sulfate. Particles may also react with the collection media such as Millipore filters, or in wet collectors they may react with the solvent or partially dissolve in it. Gases may adsorb on solid or liquid aerosols. Collected aerosols may also sublime or evaporate after collection. It is not uncommon for the concentration of pollutants to vary throughout the day or for the temperature to vary so that a pollutant collected at night

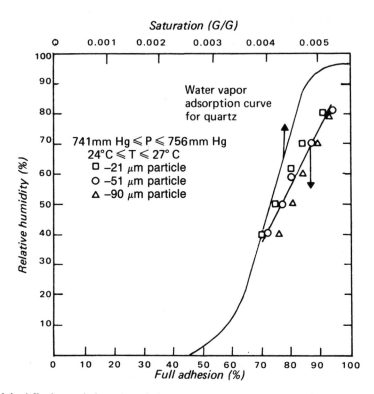

Figure 4:4 — Adhesion variation with relative humidity (quartz particles to pyrex flat). (Corn, M. 1961. J. Air Poll. Control Assoc. 11:566.)

or in the morning may be lost from the sampler after a warm afternoon. Although it is common for aerosol sampling to extend for 24 h or more, most solids have low vapor

Table 4:I Effect of Surface Irregularities on the Adhesion of Quartz Particles to Pyrex Glass

Surface	Range of profilometer reading (nm)	Root-mean-square surface irregularity (nm)	Adhesion (%)*
Pyrex flat 1	203.2–228.6	215.9 ± 12.7	100
Pyrex flat 2	203.2–381.0	292.1 ± 12.7	67
Pyrex microscope slide	304.8–381.0	342.9 ± 12.7	59
Pyrex flat 3	457.2–508.0	482.6 ± 12.7	45

*As percentage of adhesion to pyrex flat 1.

(Corn, M. 1961. J. Air Poll. Control Assoc. 11:566)

pressures and evaporative loss is small. For many purposes, particulate samples tend to keep well for long periods of time.

4.3 FACTORS AFFECTING SELECTION OF SITES FOR AEROSOL SAMPING OF THE ATMO-SPHERE. Site selection is important in any air sampling but especially so for particles because they are much less uniformly dispersed in the ambient air as well as in process equipment. (Factors involved in selecting source sampling locations are discussed in the next section, 4.4). Special consideration must be given to sources, obstacles, meteorology, and time. Particles of all sizes are being continually emitted into the atmosphere. The larger particles fall out rapidly and moderate sized ones more slowly, as listed in Table 4:II. At the same time the very small particles (Aitken nuclei) tend to become attached to larger particles. This process is fastest for the very small particles (80%/min. for 0.02 μm particles) becoming

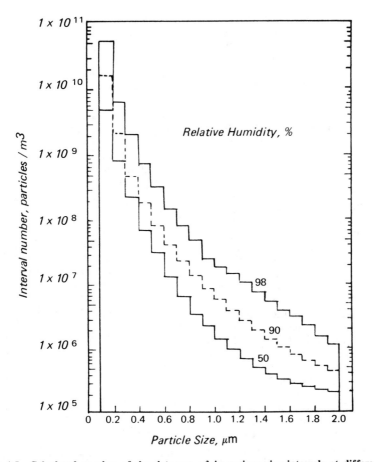

Figure 4:5—**Calculated number of droplets per m³ in various size intervals at different relative humidities in a sulfuric acid mist sample having a concentration of 39 μg/m³.**
(National Air Pollution Control Administration, 1969. U.S. Dept. Health, Education and Welfare, *Air Quality Criteria for Sulfur Oxides.* AP-50, pp. 1–22.)

Table 4:II Settling Velocity of Unit Density Spheres in Still Air

Particle Diameter, μm	Settling Velocity, V_s, cm/s
1000	385
500	200
220	76
100	25.1
50	7.2
40	4.8
30	2.7
10	0.30
1	0.0035

ticles) becoming attached to larger particles, and much slower for larger particles (14%/day for 0.2μm size). These two processes then tend to produce an equilibrium particle size distribution in the atmosphere, but because of the variable speeds of these processes there will be significant gradients near major sources, especially if they emit large particles. The height of the source, the wind velocity and turbulence, and particle size distribution will determine how fast the particles settle out.

Point sources do not generally distribute

their particles evenly in all directions. The particles travel with the wind which is quite variable. A sampler located directly in the path of a plume may show 10 or 100 times as much aerosol as one located a few meters away. In the lee of buildings, trees, hills or other obstacles, the air is often quiescent and will contain fewer particles. Strong gradients usually exist in the vertical direction also. Because most sources are near the ground and vertical mixing is frequently limited, particle loading usually decreases with height.

Time is also an important variable. In most cities the mixing height varies greatly throughout the day. Often there is an overnight inversion which keeps all of the pollution close to the ground and produces relatively high concentrations. When the sun comes up the inversion gradually lifts and generally breaks in the morning or early afternoon, at which time there is good dispersion. Particle concentrations at a given location can easily vary by a factor of 10 or more during a 24-h period. Thus, a sample taken in the afternoon may give very little information on the air pollution problem. On the other hand, a 24-h sample might average a few hours of critically severe pollution over a 24-h period and cause little alarm although a serious problem may exist. In spite of this, 24-h samples have become a standard practice, and samplers are generally started at midnight by an automatic timer and run until the following midnight. From the point of view of record-keeping this keeps the entire sample on the same date, but from an air pollution point of view it produces a sample that has one day plus parts of two different nights. It would be preferable to include a single night by starting 24-h samples at 6.00 P.M. or 6.00 A.M. or to sample for a shorter period, or for the critical period only.

To summarize:

a. A sample should be taken at the point of major interest, if possible, but care should be exercised in extrapolating its relevance to the ambient air of a whole city or even a nearby part of it; and

b. if it is desired to get a general measure of atmospheric aerosol as it affects man, plants and animals, one should:

(1) not be directly downwind from a major point source;

(2) locate the sampler about 1.5 m above ground level;

(3) locate downwind from major obstacles a distance of about 10 times their height;

(4) take several such samples at different locations in the area of interest;

(5) sample the time of day of greatest interest but do not consider this representative of all day or of any small part of the sampling period.

4.4 FACTORS INVOLVED IN SOURCE SAMPLING.

4.4.1 *General Considerations.* Source sampling refers to the collection of pollutants at their source as they are being released into, and before they are diluted by the ambient atmosphere, or for design purposes at a point where controls may be installed. The objective is to collect a sample that is representative of the material that is emitted or to be controlled. The material may not always be confined in a stack or duct, and the composition and amount may vary with times, as in batch processes. It may be coming out of vents, windows, etc. Collection of such samples is usually difficult and costly and requires advanced planning. Because of this, source testing is seldom done unless there is reason to suspect pollution or to determine the efficiency of control equipment or to get data for design purposes.

4.4.2 *Planning and Preparation for the Test.* The specific points of sampling are generally determined by examination of drawings or flow diagrams, and discussions with plant engineers or others who understand the process or source of emissions and its variation with time. A site visit is generally required to make the final selection. Criteria for selection of the sampling location are:

a. At the point of greatest interest (such as stack outlet);

b. in a straight section of pipe 5 or more diameters downstream and 3 or more diameters upstream from any bends or flow disturbances;

c. accessible to sampling personnel and equipment;

d. utilities such as electricity, water, and air available if needed; and

e. the point of sampling is a safe location for testing.

Often an ideal site cannot be found and a compromise must be reached.

Generally the selected site must then be prepared for testing by making access holes (confined source) or a collection system (unconfined source). A hole to which a 3" (8 cm) pipe coupling can be welded is generally about the right size but the particular sampling probes that are to be inserted must be considered before this decision is reached. A scaffold accessible to these holes may have to be constructed and utilities supplied. Some estimate of flow rate, temperature, pressure, dew point, particle size and gas composition must be made in advance to plan the sample collection. After the detailed procedure for collection has been worked out, the sampling and analysis equipment must be assembled and prepared for transportation to the site. The process variables must be considered in determining the proper time or times for sampling. The management or persons responsible for the source must be contacted and arrangements made to have access at the desired times. The sampling crew must be organized and briefed and the analysts alerted.

4.4.3 *Sampling*. The equipment and crew are transported to the site. Source flow conditions must first be measured. Although a number of methods are available, pitot tubes are generally preferred. In essence a pitot tube consists of two concentric tubes (Figure 4:6) one of which has an opening facing upstream and the other has openings perpendicular to the flow. The pressure difference between these is called the velocity head and may be used to determine the flow rate from handbook tables or the following equation:

$$V = 953\sqrt{\frac{P_v \cdot T}{S \cdot P}}$$

where V = flow velocity in feet per minute at measured conditions.

P_v = velocity head in inches of water.
S = specific gravity of the gas stream compared to air.

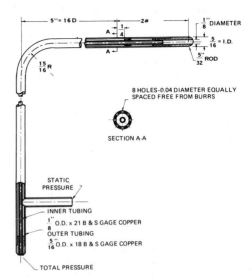

Figure 4:6 — Standard pitot tube.

P = absolute pressure of gas in duct, inches of Hg.
T = absolute temperature of gas in duct, ° R.

If measurements are made in SI units (velocity in m/s, pressures in kPa, temperature in K), the constant in the above equation becomes 23.9. The specific gravity is the ratio of mean molecular weight of the gas in the stream to that of air, 28.95. From this point onward, this discussion will use English units; operation in SI units is reasonably obvious.

A number of measurements must be made at various points in the stream. The cross-section should be divided into equal areas of about one square foot. In a rectangular duct, a measurement should be made in the center of such rectangle with a minimum of nine measurements.

X	X	X
X	X	X
X	X	X

In the case of a circular duct, annular rings of equal area (about 1 ft.²) should be visualized and measurements should be made at the center of each such ring on each of two diameters which are 90° from each other (with a minimum of 8) as shown below in Figure 4:7.

After calculating velocities, each velocity is multiplied by the area it represents and these flow rates may be added to obtain the total flow rate. Samples should be taken at each of the flow measuring points and the amounts collected and weighted together in proportion to the individual flow rates found at the corresponding points.

Particulate sampling should be carried out with probes inserted in the duct at each of the points of flow measurement. The probes must be pointed directly upstream and the sampling rate adjusted to provide isokinetic conditions at each point.

It is often more convenient to make flow measurements at the same time as the sample is collected by using a combined probe. This is especially useful in cases where the flow rate is varying and the sampling rate can be correspondingly varied to maintain isokinetic conditions.

Lack of equipment and personnel generally dictates that the points are sampled successively rather than simultaneously. However, in determining the efficiency of control equipment, it is highly desirable to take simultaneous samples before and after the control. Greater variability ($< \pm 50\%$) for flow rate or sample concentration from point to point or time to time make it advisable to increase the number of samples recommended above and place more emphasis on simultaneous sampling, whereas a somewhat smaller number can be taken if the stream is more consistent ($< \pm 20\%$). Sample concentrations determined at various cross-sectional points at a given location should be weighted together in proportion to the flow rates found at each point and the areas represented as follows:

$$M = 60(C_1 V_1 A_1 + C_2 V_2 A_2 + \ldots C_n V_n A_n)$$

Where $\quad$ $M =$ the mass of pollutant in pounds per hour.

$C_1, C_2, \ldots C_n =$ the concentration of the pollutant found at sample points 1, 2, . . . n, respectively. In pounds per cubic foot at measured conditions.

$V_1, V_2, \ldots V_n =$ the flow velocity at sample points 1, 2, . . . n respectively in feet per minute at the same conditions used for values of C.

$A_1, A_2, \ldots A_n =$ the square feet of cross-sectional areas represented by sample points 1, 2, . . . n respectively.

Separate calculations will need to be made for each pollutant analyzed.

Diffuse emissions such as wafts out doors or windows, or dust from an open quarry, or construction site, generally require individual ingenuity in sampling. A useful technique applicable to some of these is referred to as variable dilution sampling. In this method a blower is operated at the site and ducted so that all of the emissions from the location of interest enter this duct. Although the emissions rate may vary considerably, the fan mass-flow rate is kept

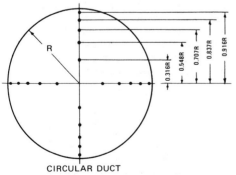

REGULAR DUCT

CIRCULAR DUCT

Figure 4:7 — Location of pitot traverse points. Note: • Indicates points of locations of Pitot Tube.

constant at a value exceeding the emission rate. A tight seal is not made so that when the effluent rate is below the blower rate, ambient air can enter freely to make up the difference so the total flow remains constant. This approach prevents creation of any appreciable unnatural suction at the source that might cause excessive emissions. At the output of the blower or some other point where the emissions and makeup air are thoroughly mixed, a sample can be taken for analysis. The sample concentration (C) times the total fan rate (V) will give the emission rate. If a constant fan mass-flow rate isused, a sample can be collected at any smaller constant mass-flow-rate and combined over any desired period of time and the combined sample concentration will be a true measure of the weighted pollution concentration over that period. This can be multiplied by the total blower flow over that period to obtain the mass of emissions for the period.

Care must be exercised to make sure that particles do not settle out in the duct prior to the sample point and that the sample point is operated isokinetically, if necessary. Another precaution of importance is to separate the particulate sample at the temperature of interest. If the temperature is too hot, liquid aerosols of interest may be vaporized and not be collected, whereas if the temperature is too low, water or other condensable vapors may form mists or aerosols which will collect with the solids and plug up the filter or give unrealistic results. The variable dilution technique tends to minimize condensation of vapors, which may be beneficial or undesirable depending on the particular case.

While the sample is being taken, the following data should generally be taken simultaneously:

Time and Date
Sample Location
Sample Flow Rate
Sample Pressure
Sample Temperature
Dew Point
Plant Operating Conditions
    Flow Rates and Variability
    Production Rates and Variability
    Pressures
    Temperatures

Persons Involved in Sampling

Approved with modifications from
2nd edition

Subcommittee 9
J.N. HARMAN, *Chairman*

5. Sampling and Storage of Gases and Vapors

Gases and vapors follow the normal laws of diffusion and mix freely with the surrounding atmosphere. They are not affected by inertial or electrostatic forces which may disturb particles. These characteristics make gases and vapors easier to sample than dusts or fumes.

5.1 GENERAL CONSIDERATIONS. In the selection of an appropriate sampling procedure, some important factors should be considered:

a. The integrity of the stored gases or vapors collected; is it maintained prior to analysis, i.e. will degradation or reactions take place in the sample probes or collection media?

b. The sampling procedure; is it optimal for the analytical procedure to be used?

c. The stability of the gas or vapor of interest; is it capable of being stored without wall losses in the container chosen for sampling?

d. That all precautions to prevent leaks and sample contamination have been addressed by appropriate quality control measures.

In general, one should always attempt to reduce time intervals between sample collection and analysis—good technique calls for protection of collected samples against sunlight and temperature extremes during transport.

5.2 GAS PUMPS (AIR MOVERS). Appropriate gas pumps are available in several different types and sizes. They should have adequate capacity and perform uniformly. Mechanical pumps, nearly always driven with electric motors, are necessary for prolonged periods of operation. Induction motors are preferred, since this type operates uniformly in spite of variations in line load.

Pumps that are Teflon* or stainless steel lined are available from many commercial sources. These pumps with inert surfaces are necessary when it is required to have a sample stream pass through the pump prior to the sample collector. This type of pump is essential when sampling for reactive gases such as ozone, or collecting trace quantities of hydrocarbons and halocarbons.

Other air movers such as hand pumps, air aspirators or siphons are rarely, if ever, used. Use of these devices as air movers is not recommended, as they are not able to control flow rates accurately or be used for prolonged sampling periods.

5.3 GAS METERING DEVICES. It is usually necessary to measure the total volume of air sampled in an experiment. For this purpose two types of meters are available, volume meters and rate meters. Volume meters record the total volume of sample which is passed through the sampling train. Rate meters indicate the flow rate, and this rate multiplied by the sampling time gives the total volume sampled.

Rate meters consist of rotameters, electronic flow controllers, and critical orifices. Rotameters are made to cover a large range of flow rate, from cubic centimeters per minute to hundred of liters per minute, and are made of glass or plastic. Glass rotameters are preferred when sampling for reactive gases or vapors.

Electronic mass flow meters are a recent development that are read out electronically. These units are considered more accurate than rotameters, as dependable, but are more expensive.

Use of critical orifices offers a convenient and inexpensive way to produce constant flow rates in sampling trains. Use of hypodermic needles is particularly advantageous (1). To maintain constant flow it is necessary that the pressure drop across the critical orifice be equal to or greater than 0.55 times the upstream pressure. A sampling pump of sufficient capability to maintain this pressure drop must be used. The critical orifice must also be kept free of dust to prevent plugging. A typical sampling train containing a critical orifice for flow control is shown in Figure 5:1.

---

*Teflon is a registered trademark of E.I. DuPont de Nemours Co., Inc., Wilmington, Delaware

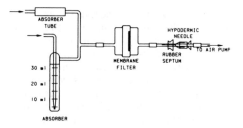

Figure 5:1 — Use of critical orifices for sample collections

5.4 SAMPLE PROBES. It is essential that an air sample stream be able to reach the collection device in an unaltered state. Sampling probes used for sample transport are sometimes of substantial length; e.g., sampling the outside air from within a building. To insure no losses of gases or vapors the sampling probe must be constructed of inert materials such as glass or Teflon (2). Use of materials such as rubber or nylon is not advised even when sampling for relatively inert gases such as carbon monoxide or nitric oxide. Sampling probes should be kept to the shortest possible length, and never allowed to accumulate particulate matter or moisture.

5.5 SAMPLING TECHNIQUES FOR COLLECTION OF THE WHOLE SAMPLE WITHOUT CONCENTRATION OF GASES AND VAPORS. These techniques include the use of plastic bags, glass containers and metal containers. For many low-boiling compounds such as methane, ethane, propane, and the fixed gases (N_2, O_2 and CO_2), this type of sampling is the preferred and/or sole method of collection.

5.5.1 *Plastic Bags.* The typical materials used in plastic bags are Aluminized Scotch-pak*; Scotch-pak, Mylar**, Saran***, and Tedlar****. Entry into the bags is by a fitting seated in and connected to the bag to form an integral part of the

---

*Scotch-pak is a registered trademark of Minnesota Minning and Manufacturing Co., St. Paul, Minnesota
**Mylar is a registered trademark of E.I. DuPont de Nemours Co. Inc., Wilmingon, Delaware
***Saran is a registered trademark of Dow Chemical, USA, Midland, Michigan
****Tedlar is a registered trademark of E.I. DuPont de Nemours Co. Inc., Wilmingon, Delaware

bag. Some bags have two fittings to facilitate sample taking by simply passing sample air through the bag. These plastics are appropriate for inert gas collection and storage. Bags are particularly well suited for integrated air sampling. An integrated bag sampling train is depicted in Figure 5:2. Tedlar bags are especially resistant to wall losses for many reactive gases, and have been in use in a large toxics air monitoring network (3).

Whether a substance can be sampled and stored in a plastic bag should be determined in the laboratory prior to field use. Literature references may also be used to determine proper usage of bag sampling. Bags can be reused after purging with clean air and checking for any residual gaseous components. Even "inert" plastics display outgassing properties (4).

5.5.2 *Glass Containers.* Glass containers can be used both in an evacuated or pass-through displacement mode. In the displacement mode gas sample tubes such as those in Figure 5:3 are opened on both ends. A pump is attached to one end and a sampling probe is attached on the other end, if needed. The pump is activated and approximately 10 times the sampling tube volume is passed through the system. The stopcocks are immediately shut and the sample is then secured. Alternatively, the tubes may be evacuated using a vacuum pump, then the sample collected by simply opening one of the stopcocks. Glass containers, by nature of their small volumes, provide basically instantaneous "grab" samples.

Glass containers are also convenient for

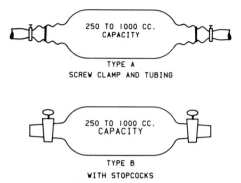

TYPE A
SCREW CLAMP AND TUBING

TYPE B
WITH STOPCOCKS

Figure 5:3 — Gas sample tubes for collection of sample by displacement or vacuum

collecting samples directly in a chemical absorbent in which they are later analyzed. When used in this manner, rather than sample tubes, 1-3 L flasks are better suited, with only one stopcock used for both evacuation and later sample entry. This is a proven and frequently used technique (5,6).

In general, glass containers are excellent for inert gases such as oxygen, nitrogen, methane, carbon monoxide and carbon dioxide. Reactive gases such as hydrogen sulfide, oxides of nitrogen and sulfur dioxide are not recommended for direct collection and storage. The reactive gases are, however, amenable to the chemical absorbent technique referred to earlier, which fixes and stabilizes the reactive gas prior to analysis.

5.5.3 *Metal Containers.* Stainless steel containers are also widely used for the collection of the inert gases mentioned in Section 5.5.2. They are not suitable for a majority of the reactive gases commonly encountered. In metal containers, wall losses are extremely rapid for the reactive gases. Metal containers come in a variety of sizes from 1 to 34 L. An advantage of using the large sizes is that a longer, more integrated, sampling time is possible when compared to the glass containers.

An important improvement in stainless steel containers has recently taken place. A treatment process for the containers has been introduced which reduces the surface adsorption of gases by reduction of the internal exposed metal surface area. This pro-

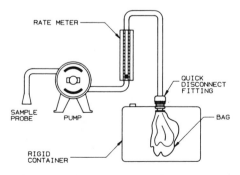

Figure 5:2 — Integrated bag sampling train

prietary process of SUMMA* polishing is available in various container sizes up to 6 L from two manufacturers (**7,8**). Extensive studies have shown excellent stability, in the order of weeks, for hydrocarbons and various halocarbons even in the low and sub parts per billion (**9**). Extensive studies on inorganic reactive compounds have not been performed to date. Figure 5.4 depicts a typical sampling setup for long term (up to 24 h) sampling for ambient hydrocarbons and halocarbons. This technique is considered state-of-the-art for the sampling of low level toxics in air.

5.6 SAMPLING TECHNIQUES WHICH EXTRACT AND CONCENTRATE GASES AND VAPORS FROM AIR SAMPLES.

5.6.1 *Wet Collection Systems.* Sampling techniques employing simple bubblers or bubblers with diffusers are widely used, and a large diversity of applications may be found. Bubblers, depending on design and use, may contain from 5 to 200 mL of absorbent. Bubblers with diffusers to provide more efficient air-to-absorbent contact are especially needed for acceptable collection of gases that are difficult to absorb; e.g., nitrogen dioxide (**5**). Figure 5:5 shows the basic design differences among bubbler types.

Specific use of a bubbler sampling train must be based on prior knowledge that a bubbler-absorbent-flow rate combination will result in an acceptable collection efficiency, usually 90% or greater. A typical complete sampling train using bubblers is shown in Figure 5:6.

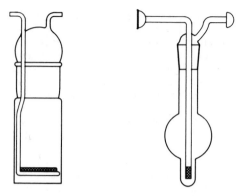

Figure 5:5A — Bubbler absorbers with diffusers

5.6.2 *Dry Collection Systems.* Any gas or vapor will, to some degree, adhere to any solid surface at ordinary or low temperatures. This phenomenon is called adsorption. Porous solids expose not only their exterior surface, but their interior surfaces as well, and some such solids indeed possess a vast network of extremely minute channels and submicroscopic pores within their bodies. Such materials have practical value as adsorbents. These solids include acti-

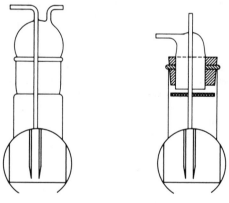

Figure 5:5B — Simple bubble absorbers

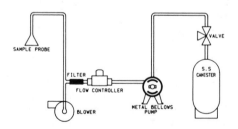

Figure 5:4 — Use of stainless steel canisters for integrated air samples

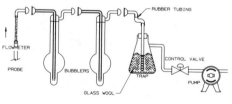

Figure 5:6 — Typical sampling train for bubbler absorption of gases

*SUMMA — Proprietary process of Molectrics, Inc., Carson, California

vated carbon, silica gel, activated alumina, and various active earths. Synthetic polymeric materials have also been used for adsorption of gases, particularly volatile organic gases. Among the most commonly utilized of the synthetics has been Tenax*. All these differ widely in the number and kinds of substances they absorb, as well as in the amount of sorbed substances they will retain.

Silica gel, charcoal and Tenax are the most widely used adsorbents, and will be individually addressed as to their sampling characteristics. These materials adsorb a variety of volatile organic gases, and have excellent storage properties prior to analysis. Rarely are the materials used for the quantitative collection and analysis of inorganic gases. In the use of the materials it is important after the sampling is completed that the adsorbent tubes be immediately sealed and properly stored. Storage in a sealed plastic or glass vial is most satisfactory.

Silica gel is a highly polar substance, and therefore will preferentially adsorb water more strongly than organic gases and vapors. Caution must be exercised when sampling high relative humidity gas streams to insure high collection efficiency for gases of interest. Silica gel is an excellent adsorbent for a variety of organic substances in the part per million concentration range. The adsorbed gases must be solvent extracted, for subsequent GC or IR analysis. Extractants are usually polar substances such as alcohols or dimethylsulfoxide or water in combination with carbon disulfide **(10)**. Experimental data show excellent recoveries from standard gas streams for a large variety of gases **(10)**.

Activated charcoal is the most widely used adsorbent for the concentration and storage of organic vapors from air streams. Many NIOSH/OSHA and ACGIH methods are based on the use of activated charcoal. Since charcoal is nonpolar, sample streams of high relative humidity do not present the difficulties mentioned for silica gel. Larger sample volumes may be taken, resulting in procedures which can provide sensitivity capabilities for analysis of subpart

*Tenax—Developed by AKZO Research Laboratories, marketed by Enka N.V., the Netherlands.

Table 5:I Retention Volume Estimates for Compounds on Tenax

Compound	Estimated Retention Volume at 100° F (38° C), L/g.
Benzene	19
Toluene	97
Ethyl Benzene	200
Xylene(s)	200
Cumene	440
n-Heptane	20
1-Heptene	40
Chloroform	8
Carbon Tetrachloride	8
1,2-Dichloroethane	10
1,1,1-Trichloroethane	6
Tetrachloroethylene	80
Trichloroethylene	20
1,2-Dichloropropane	30
1,3-Dichloropropane	90
Chlorobenzene	150
Bromoform	100
Ethylene Dibromide	60
Bromobenzene	300

per billion gas concentrations. Charcoal has been used for a wide spectrum of volatile organic gases and vapors. Solvent extraction, usually with carbon disulfide, prepares the samples for GC or IR analysis. When using charcoal for specific applications, refer to the literature to ascertain the quantity of charcoal to use, mesh size, and the sampling parameters; i.e., flow rate and volume. This is essential to insure acceptable collection efficiencies for a particular gas or vapor.

Tenax, a polymeric material (poly-2,6-diphenyl phenylene oxide) is used for air sampling at part per billion and lower levels of volatile organic compounds **(11)**. Tenax is nonpolar, and water vapor does not affect sampling in any way. Unlike silica gel and activated charcoal, the adsorbed gases are not solvent extracted after collection, but are thermally desorbed. The thermal desorption takes place over 10–15 seconds at a temperature of approximately 240° C. This rapid desorption is accomplished by heating to 240° C while purging the absorbent with helium gas, allowing all the collected gases to be introduced into a gas chromatograph as a single injection. This accounts for the extreme sensitivities possible when using Tenax. Table 5:I shows the

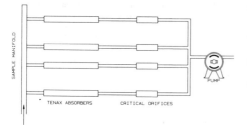

Figure 5:7 — Multiple and variable flow setup for Tenax sampling

sampling limitations of specific compounds, expressed as retention volumes. It is recommended that, prior to using Tenax for sampling, Reference 10 be reviewed. As with any methodology dealing with parts-per-billion detections, stringent quality control procedures must be followed. Figure 5:7 depicts a multiple and variable sampling system recommended when using Tenax absorbers (12). Flow rates normally used are in the 5 to 50 L/min range. The flow rate and time of sampling are determined primarily by the retention volume of the compound(s) of interest (see Table 5:I). Multiple parallel sampling using variable flow rates allows specific indications of data quality. This sampling approach adds an analytical burden, but provides important data regarding adequate retention, artifact formation, and hence, applicability to specific compounds.

REFERENCES

1. LODGE J.P., J.B. PATE, B.E. AMMONS AND G.A. SWANSON. 1966. The Use of Hypodermic Needles as Critical Orifices in Air Sampling. J. Air Pollution Control Assoc. 16:197-200
2. FEDERAL REGISTER. Vol. 44 No. 92, Thursday, May 10, 1979 p 27.
3. OSLUND W.E. 1986. Implementing a Quality Assurance Program for Sampling and Analysis of Ambient Air Toxics Compounds. Proceedings of the 1986 EPA/APCA Symposium on Measurement of Toxic Air Pollutants. pp 364-374.
4. LONNEMAN W.A., J.J. BUFALINI, R.L. KUNTZ AND S.A. MEEKS. 1981. Contamination from Fluorocarbon Films. Anal. Chem. 15:99-103.
5. ASTM COMMITTEE D-22. 1977. Method of Test for Oxides of Nitrogen in Gaseous Combustion Products (Phenoldisulfonic Acid Procedures) ASTM Designation D 1608-77 (Reapproved 1985). Book of ASTM Standards, Vol. 11.03.
6. SALTZMAN B.E. 1954. Colorimetric Determination of Nitrogen Dioxide in the Atmosphere. Anal. Chem. 26:1949-1955.
7. DEMARY SCIENTIFIC INSTRUMENT LTD., S.E. 1122 Latah Street, Pullman, Washington 99163.
8. BIOSPHERIC RESEARCH CORP., 1121 N.W. Donelson Road, Hillsboro, OR 97123.
9. OLIVER K.D., J.D. PLEIL AND W.A. McCLENNY. 1986. Sample Integrity of Trace Level Volatile Organic Compounds in Ambient Air Stored in SUMMA Polished Cannisters. Atmos. Environ 20:1403-1411.
10. FELDSTEIN M., S. BALESTRIERI AND D.A. LEVAGGI. 1967. The Use of Silica Gel in Source Testing. Am. Ind. Hyg. Assoc. J. 28:381-385.
11. RIGGIN R.M. 1984. Compendium of Methods for the determination of toxic organic compounds in ambient air. EPA-600/4-84-041. April 1984, USEPA, Research Triangle Park, N. C.
12. WALLING J.F. 1984. The Utility of Distributed Air Volume Sets When Sampling Ambient Air Using Solid Absorbents. Atmos. Environ. 18:855-859.

Subcommittee 9
D.A. LEVAGGI
J.N. HARMAN, *Chairman*

6. Use and Care of Volumetric Glassware

6.1 GENERAL CONSIDERATIONS. The National Bureau of Standards has specified certain minimum requirements which must be adhered to in the manufacture of volumetric glassware.

6.1.1 The instrument must bear, in legible characters, the capacity, the temperature at which it is to be used, and whether it is to contain or to deliver the specified volume.

6.1.2 The instrument must bear the name or trade-mark of the manufacturer and an identification number. If the instrument has detachable parts, such as stoppers or stopcocks, such parts must also bear the same numbers.

6.1.3 The inscription and capacity mark must be etched or engraved but not scratched on the instrument.

6.2 VOLUMETRIC FLASKS. Volumetric flasks most generally employed are those which are calibrated to "contain" a definite volume and have only one mark etched on the neck of the flask. There are also flasks available which have two etched marks. The lower mark indicates the volume of solution the instrument "contains," while the upper mark indicates the volume of solution the instrument will "deliver." The flask may be provided with a ground glass or plastic stopper to prevent any spillage when inverted for mixing solutions.

The sizes of volumetric flasks generally employed in the laboratory "to contain" (TC) are 1000-mL, 500-mL, 250-mL, 100-mL, 50-mL, 25-mL, and 10-mL. Flasks calibrated "to deliver" (TD) are not sufficiently accurate since the amount of liquid adhering to the sides of the flask after emptying can only be approximated.

6.3 THE USE OF VOLUMETRIC FLASKS.

6.3.1 *Preparation of a Solution of Definite Volume.* Transfer the solution, which has been previously prepared in a beaker, by means of a funnel, to the volumetric flask. Pour the solution with the aid of a glass rod held across the lip of the beaker. Wash the beaker, when empty, thoroughly into the flask by means of a wash bottle, holding the beaker with its lip in the funnel and then directing a stream of distilled water from the wash bottle around the inside walls of the beaker. Do not use an excessive amount of wash water, so that the capacity of the flask is not exceeded. Rinse the funnel into the flask and then remove. Stopper the flask and mix the solution by shaking and inverting the flask several times. Return the flask to the upright position and add water until the level of the liquid is slightly below the graduation mark. Repeat the mixing procedure. Return again the flask to the upright position. With the eye on the level of the graduation mark, add water slowly by means of a pipet until the level of the liquid reaches exactly the graduation mark. Stopper and shake well by inverting many times, so that the contents are completely mixed.

It should be kept in mind that the solution and the diluting water must both be at room temperature.

At times, it is necessary to use a portion of an "aliquot" of a prepared solution of definite volume. The "aliquot" portion, taken from the volumetric flask, is measured out by means of a pipet, buret, or a smaller volumetric flask calibrated to "deliver."

Volumetric flasks calibrated to "deliver," should be emptied by inclining them gradually until they are almost in a vertical position. Allow to drain in this position for about 30 s and then bring the mouth of the flask in contact with the receiving vessel so that the last adhering drop is removed from the flask.

6.3.2 *Calibration of Volumetric Flasks.* Calibrated volumetric flasks purchased from reliable manufacturers are sufficiently accurate. However, volumetric flasks should be calibrated when the accuracy of the capacity of the flask is in doubt.

Circular No. 602 of the National Bureau of Standards gives capacity tolerances for flasks. Determine from the weight of the water contained the capacity of the flask at 20°C using Table 6:I. For example, if the weight of water contained in a 50-mL volumetric flask at 23°C is 49.8310 g, the capacity of the flask calculated from the table is 49.8310 × 1.0034 or 50.000 mL.

6.4 BURETS. The most common size buret employed in the laboratory is the 50-mL capacity buret, graduated to 0.2 mL in its smallest subdivision. The buret has a capillary tip of definite size so that the free outflow time of a liquid does not exceed 3 min nor is it less than 90 s. The optimum outflow time for a 50-mL buret is about 100 s. Other sizes of burets are of 10-, 25-, and 100-mL capacity.

The Friedman and La Mer weighing buret of somewhat different construction allows increased accuracy in titration. The amount of liquid necessary for a titration is weighed rather than measured by volume. The buret is short, of light weight, and equipped with lugs so that it can be suspended from hooks on the stirrups of a balance. It is also provided with a glass stopper, and a cap that fits over the delivery tip to prevent evaporation.

There are also microburets available which have a capacity of 5 mL or less, with the error of delivery not exceeding 0.1%. The 5-mL microburet is graduated to 0.01 mL in its smallest subdivision.

The microburet is provided with a delivery tip drawn out to a fine point. Thus, the liquid is delivered in minute drops and the amount adhering to the tip is negligible. These burets are available with a variety of detachable tips that are attached to the bu-

Table 6:I Apparent Weights and Volumes of Water Weighed in Air for Calibration of Volumetric Glass Apparatus, Coefficient of Cubical Expansion, 0.000025 per °C

Temperature °C	Weight of 1 ml, g	Volume of 1g, mL	Temperature °C	Weight of 1 mL, g	Volume of 1g, mL
15	0.9979	1.0021	23	0.9966	1.0034
16	0.9978	1.0022	24	0.9964	1.0036
17	0.9977	1.0023	25	0.9961	1.0039
18	0.9975	1.0025	26	0.9959	1.0041
19	0.9973	1.0027	27	0.9956	1.0044
20	0.9972	1.0028	28	0.9954	1.0046
21	0.9970	1.0030	29	0.9951	1.0049
22	0.9968	1.0032	30	0.9948	1.0052

ret by means of an adapter and can be easily removed for cleaning.

There are available microburets which discharge very small volumes of liquids with high precision and reproducibility. The Rehberg microburet does not have a stopcock but uses a mechanical device which discharges very small volumes of liquid by means of a micrometer screw pushing against a column of mercury which, in turn, is in contact with the solution to be measured. The Rehberg microburet is of 0.125- or 0.225-mL capacity and, when used, the tip is dipped below the surface of the solution to be titrated.

6.4.1 *Use of the Buret.* The buret is thoroughly cleaned before use. If the buret has a glass stopcock, it should be greased with stopcock grease, not petrolatum, so that it is well lubricated and the capillary tip allows the liquid to flow freely. The temperature of the solution withdrawn from the buret should be near 20°C.

When measuring liquids by means of a buret, the readings are taken at the lowest part of the meniscus since this area is most clearly defined. Exceptions are made for dark liquids, such as permanganate and iodine solutions, where the bottom of the meniscus cannot be seen. In this case, the top of the meniscus is taken as the reading. It is essential that the point chosen for reading the meniscus should be the same for all readings.

In observing the bottom of the meniscus, the eye should be in the same plane as the meniscus as otherwise there will be an error due to parallax. After recording the buret reading, allow about 30 s for drainage before taking the final reading.

6.4.2 *Calibration of Burets.* Burets are generally purchased from reliable firms that guarantee their burets to meet the tolerances specified by the National Bureau of Standards. Burets are also available certified for accuracy by the National Bureau of Standards, with a Certificate of Accuracy.

Circular No. 602 of the National Bureau of Standards gives capacity tolerances for burets. The capacities of burets (even the certified ones) may change after a period of use, especially if caustic solutions are used. Therefore, burets should be calibrated when the need arises.

6.4.3 *Titrating with a Buret.* Titration is a method by which the quantity of a sample dissolved in a liquid is determined by adding a volume or weight of a reagent solution that just neutralizes or completely reacts with this sample.

A standard solution is one of definite and accurately known concentration, usually expressed in terms of normality.

The standard solution is prepared by dissolving a definite weight of a pure substance (primary standard) in a specific volume of liquid. The standard solution may also be prepared by dissolving a definite weight of a known substance (secondary standard) in a specific volume of liquid and then standardizing it, *i.e.*, determining its concentration by titrating against a primary standard.

The stoichiometric point is defined as the point at which the addition of a definite amount of substance is exactly equivalent to the dissolved substance being titrated. Actually, this is the theoretical end point which is determined from the equation of the reacting substances. However, the stoichiometric point will not always coincide with the end point of the titration.

The end point is the stage in a titration where the reaction is complete. This point is determined by means of indicators or other devices and should coincide with or be as close as possible to the stoichiometric point.

6.5 VARIATIONS OF VOLUMES WITH TEMPERATURES. The National Bureau of Standards has designated 20°C as the standard temperature at which volumetric apparatus will "contain" or "deliver" the stated volume.

Errors, although very slight, will occur if the volumetric apparatus is employed at other temperatures. Glass will expand slightly above 20°C and contract slightly at lower temperatures. The average coefficient of cubical expansion of soft glass is 0.000025 and of borosilicate glass, 0.00001/°C.

The variation in volume, due to the expansion of glass, over a small temperature range from the standard 20°C is very slight and may be disregarded for ordinary work. For example, for a 50-mL buret, a variation in volume due to a decrease or increase in temperature of 5°C from the standard 20°C will be only plus or minus 0.007 mL. For accurate work, however, the volume of the vessel may be corrected to 20°C by the following formula:

$$V_{20} = V_t[1 + a(20 - t)]$$

where V_{20} = volume at 20°C
   V_t = recorded volume at observed temperature
   t = observed temperature
   a = coefficient of cubical expansion of the material of the apparatus (for soft glass 0.000025/°C)

The volume of a liquid is also affected by temperature. Liquids generally expand at higher temperatures. The more concentrated the solution, the greater will be the variation. Fortunately, most standard solutions are not too concentrated and usually vary from 0.5 normal to less than 0.1 normal. Although the error involved in the variation in the volume of a 0.5 normal solution due to a 5°C change in temperature is small (about 0.05 mL for a 50-mL soft glass

buret), nevertheless, it is important, when employing volumetric apparatus or titrating, that the temperature of the solutions used is as close as possible to the standard temperature of 20°C.

6.6 PIPETS. Pipets of different types and capacities are available. The transfer pipets, calibrated "to deliver" at 20°C, are generally used to measure a single fixed volume.

The orifice of the tip of the pipet is of definite size so that the free outflow time should not be more than 1 min nor less than 15 s for a 5-mL pipet; 20 s for a 10-mL pipet; 30 s for a 50-mL pipet; 40 s for a 100-mL pipet and 50 s for a 200-mL pipet.

Pipets of a capacity of 0.1 mL or less are calibrated to contain a definite volume. When a liquid has been delivered from this pipet into a receiving vessel, the pipet must be rinsed with water at least three times. These rinsings are added to the receiving vessel.

Measuring or Mohr pipets are narrow straight tubes which are graduated to deliver variable amounts of liquids. They are not ordinarily used for precise work.

6.6.1 *Calibration of Transfer Pipets.* The accuracy of a transfer pipet may be checked when necessary. The maximum tolerances allowed for pipets by the National Bureau of Standards are listed in Table 6:II.

The procedure used in calibrating the pipet should be the same as that employed in the subsequent use of the pipet. A glass-stoppered weighing bottle of a capacity of about five times the volume of water to be received is employed as the receiving vessel. The bottle should be twice as tall as wide so that the evaporation of the water during delivery will be negligible.

Table 6:II Capacity Tolerances for Transfer Pipets

Capacity (mL) Less Than and Including	Limit of Error (mL)
2	0.006
5	0.01
10	0.02
30	0.03
50	0.05
100	0.08
200	0.10

The bottle is weighed to the nearest milligram by substitution if a two-pan balance is used.

6.7 GRADUATED CYLINDERS. Graduated cylinders are calibrated "to deliver" or "to contain" their stated volumes at 20°C. These cylinders are not as accurate as burets, pipets, or volumetric flasks since they are graduated to a tolerance of 1%.

Graduated cylinders are generally used to deliver approximate volumes and where exact volumes are not essential. Therefore, for accurate measurements, use the other volumetric apparatus described.

When using a graduated cylinder, select a size which will be nearly filled by the volume of liquid to be measured.

6.8 CLEANING GLASSWARE. Keeping laboratory glassware clean is a necessary but tedious chore. Much of the glassware may be cleaned by rinsing with distilled water if the rinsed glassware shows satisfactory drainage of water without the appearance of droplets adhering to its walls.

Glassware can be cleaned satisfactorily by soaking directly after use in a hot soapy solution and then scrubbing with a brush. The glassware is then rinsed thoroughly in hot tap water and, finally, in distilled water.

If the glassware is greasy and dirty and resists ordinary cleaning procedures, the following general chemical methods may be used.

6.8.1 *Sulfuric Acid–Sodium Dichromate Mixture.* A solution of sodium dichromate in concentrated sulfuric acid (cleaning solution) is widely used in laboratories to clean volumetric glassware thoroughly. The oxidizing potential of this mixture, when hot, will cut most greases adhering to glass walls and remove them completely. The cleaning solution is very corrosive and must be used with great care.

To prepare the solution, moisten about 20 g of powdered sodium or potassium dichromate with water to form a thick paste in a glass container. Add about 500 mL of concentrated sulfuric acid to this paste and stir. Store the mixture in a glass-stoppered bottle. Some of the salt may be left undissolved in the solution. Do not remove this excess salt but use the supernatant liquid. The solution can be repeatedly used by pouring back the unused portions into the stock bottle. Continue using this red solution until it turns green, indicating that the solution is spent. The color change is caused by the chromium being reduced to the chromic state.

Use this solution with the greatest care. Never allow it to come into contact with clothing or skin since sulfuric acid is very corrosive and bad burns will result. If some of the solution does spill, wash off immediately with running water. Never store or use this solution in metal or enameled containers. Finally, do not allow the solution to spill on floors or desks.

6.8.2 *The Use of the Cleaning Solution.* Rinse the glass vessel with the cleaning solution or immerse it in this solution. The immersion time is dependent on the stubbornness of the dirt, and may vary from a few minutes to 24 h. A hot cleaning solution is more effective than a cold one. After cleaning, rinse the glass vessel thoroughly with tap water until all traces of the solution are removed. Finally, rinse with distilled water and allow to drain and dry. A good test for cleanliness of a glass vessel is to fill it with water and then empty. An unbroken film of water indicates that the vessel is clean.

6.8.3 *Alcoholic Sodium Hydroxide.* An alcoholic solution of sodium hydroxide is very effective for removing greasy films from glassware.

Prepare a solution by dissolving 120 g of sodium hydroxide in 120 mL of water. Allow to cool and add sufficient isopropyl alcohol to make 1 L.

Immerse the glassware in this solution for not more than 30 min since the caustic may attack the glass. Remove from the solution and scrub with a stiff brush. Rinse thoroughly with tap water and finally with distilled water. Allow to drain and dry.

6.8.4 *Synthetic Detergents.* Anionic or nonionic synthetic detergents are available under proprietary names. Their cleaning properties are due to their ability to reduce surface tension and, at the same time, to wet objects thoroughly. Synthetic detergents are compounded with alkaline salts to increase their detergency. They are efficient in that they do not form scum on glassware when used in hard water as soaps do. Dilute solutions of about 0.5% concentration are sufficient to clean glassware.

Immerse the glassware in a hot or cold

solution for about 30 min, remove from the solution, and scrub with a stiff brush. Rinse thoroughly with tap water and finally with distilled water. Allow to drain and dry.

6.8.5 *Cleaning Burets.* Fill the buret with the "cleaning solution" and allow to remain overnight. Drain the solution into a beaker and return to the stock bottles. Rinse the buret thoroughly with tap water and finally with distilled water. Grease the stopcock with stopcock grease, if necessary. Invert the buret and allow to drain. A test for the cleanliness of a buret is to fill it with water and then allow to drain. An unbroken film of water, without droplets, should remain on the walls of the buret.

6.8.6 *Cleaning Pipets.*

a. Routine Procedure—Secure the pipet in a clamp in a vertical position with its tip inserted into a beaker containing water or a detergent. Attach, by means of rubber tubing, the mouth of the pipet to a safety bottle which, in turn, is connected to a vacuum source. Draw up the solution by suction until it almost fills the pipet. Release the suction and allow the solution to drain. Repeat this procedure several times. Change the beaker resting under the tip of the pipet and repeat the procedure with tap water and finally with distilled water. Remove the pipet from the clamp and allow to drain and dry.

b. Cleaning Greasy Pipets—Secure the pipet in a clamp in a vertical position with its tip inserted in a beaker containing warm sodium dichromate-sulfuric acid cleaning solution. Draw up the warm solution until it fills the pipet within an inch from the top, using the procedure described earlier.

Pinch the rubber tubing with a clamp and allow the solution to remain in the pipet for about 10 min. Then remove the pinch-clamp and discharge the cleaning solution. Rinse the pipet thoroughly with tap water to remove any trace of the solution and finally rinse with distilled water. Allow the pipet to drain and dry. To dry the pipet quickly, rinse it several times with ethyl alcohol, after draining the distilled water, and then rinse once with ether. Draw dry air through the pipet until dry.

Approved with modifications from 2nd edition
Subcommittee 9
J.N. HARMAN, *Chairman*

7. Reagent Water

Wet chemical analysis implies that water is used in the procedure. The water used in the reagents and for effecting solubilization, dilution, and transfer is called reagent water. Assumption that the reagent water contributes insignificantly to the bias of a determination is warranted only when the method of test is not sensitive to the impurity level in the water.

7.1 PREPARATION. Preferred practice is to minimize the probability of error by preparing and storing high purity water. The preparation frequently begins with single distillation from conventional apparatus. Condensate from a conventional steam boiler is suitable if the conductivity of this condensed steam is no greater than 20 $\mu S/$ cm. A condensate of adequate quality can almost always be obtained by partial condensation. The condensing train for this purpose is arranged in steps, to (*a*) effect partial condensation in the steam pipe, (*b*) dry the steam in a separator (retains the bulk of the more soluble gases and the entrained boiler water), (*c*) effect additional partial condensation of the steam, and (*d*) separate and store this condensate (the bulk of the less soluble gases like carbon dioxide remains in the uncondensed steam).

7.2 DEMINERALIZATION. Additional processing of the distillate or condensate is now usually accomplished by demineralization, the water being passed through a commercially available mixture of anion and cation exchange materials. Assumption of adequate upgrading by this process is unwarranted, especially as ion exchange materials vary from lot to lot, and the effluent in each instance should be discarded until the desired quality is obtained.

The successive steps of distillation and demineralization are both important. The former removes nonvolatile electrolytes and colloidal material, and the latter removes ionizable volatiles such as carbon dioxide and ammonia (which together impart the greater part of the 20 $\mu S/cm$ conductivity of the initial condensate).

7.3 FILTRATION. High purity water prepared in this way is usually satisfactory for the preparation of reagents, for the laboratory preparation and dilution of samples, and for the final rinsing of glassware and the other apparatus. It may not be adequate, however, for the determination of trace concentrations of some of the materials extracted from air samples. Organic contaminants, in a range from 1 to 5 ppb, are usually imparted to the water by its contact with the ion-exchange materials. For some determinations, such as tests for airborne micro-organisms, the "pure" water is further purified by special low porosity filtration. Examples of special purpose "purifiers" are the sized Gelman or Millipore membrane filters and the Barnstead Organic Removal Cartridge.

7.4 PROTECTION AGAINST CONTAMINATION. Reagent water should at all times be protected from atmospheric contamination. Replacement air should be passed through a vent guard which removes undesired species such as carbon dioxide and oxidant or reductant species from the makeup air (for example, a drying tube filled with equal parts of 8- to 20-mesh soda lime, oxalic acid, and 4- to 8-mesh calcium chloride, each product being separated from the other by a glass wool plug). Sample containers and tubing should be made of material that has been proven to be resistant to even minor solvation by, or reaction with, the water. Such materials usually are TFE-fluorocarbon, block tin, quartz, or polyethylene—in that order of preference. It is general practice to filter the output of a reagent water source with a 10-μm filter to remove particulate contamination. For specific grades of reagent water, finer filters are required.

The stored water is usually considered suitable until it fails to pass the maximum electrical conductivity test (0.1 μS/cm at 25°C). It may also be checked occasionally for its consumption of potassium permanganate, the requirement being that the permanganate color must persist for at least one hour after 0.20 mL of $KMnO_4$ solution (0.316 g/L) is added to a mixture of 500 mL of the reagent water and 1 mL of reagent grade sulfuric acid (conc) in a stoppered bottle of chemically resistant glass.

7.5 BLANK DETERMINATIONS. Despite all of these preparations and precautions, minute concentrations of impurities in reagent water may still affect the precision and accuracy of some determinations. Pretesting is a reasonable precaution. The usual practice is to run a "blank" to disclose whether or not the water contains a detectable quantity of the material for which the analysis is to be made. This "blank" is often designed to include the reactive impurities in the reagents as well as those in the water.

Occasionally, the analyst will purposely contaminate "pure" reagent water before it is used. For example, water for the preparation of the reagents used to calibrate conductivity cells is deliberately equilibrated with the atmosphere in which the calibration will be made. This tends to stabilize the content of dissolved gases, such as carbon dioxide, that affect conductivity.

When the analyst is seeking especially small concentrations of material in samples of air, he must necessarily be especially cautious in determining and correcting for any bias introduced because the water is not perfectly pure. He will be concerned, for example, with the magnitude of the "blank" in relation to the lower limit of detection of the method. He will determine the range of deviation in the "blank" to be expected from random error, and design his analysis so that the "blank" is a proper correction on the quantity he finds in the air sample. Above all, he will start with reagent water that is of a purity consistent with the projected testing and with its practical utility.*

Approved with modifications from 2nd edition

Subcommittee 9
J.N. HARMAN, *Chairman*

8. Common Acid, Alkali, and Other Standard Solutions

8.1 INTRODUCTION. Reagent grade chemicals shall be used in preparing and standardizing all solutions. Reagents shall conform to the current specifications of the Committee on Analytical Reagents of the American Chemical Society, where such specifications are available.** Other grades may be used, provided it is first ascertained that the reagent is of sufficiently high pu-

---

*For a detailed discussion of the various grades of reagent water, see ASTM D-1193.

**"Reagent Chemicals. American Chemical Society Specifications," Amer. Chemical Soc., Washington, D.C.

rity to permit its use without lessening the accuracy of the determination.

The National Bureau of Standards offers for sale certified standard samples of arsenic trioxide, benzoic acid, potassium hydrogen phthalate, potassium dichromate, and sodium oxalate. These samples of commercially available primary standards are to be used in standardizing the volumetric solutions.

Directions are given for the preparation of the most commonly used concentrations of the standard volumetric solutions. Stronger or weaker solutions are prepared and standardized in the same general manner as described, using proportionate amounts of the reagents. Similarly, if quantities larger than 1 liter are to be prepared, proportionate amounts of the reagents should be used.

When quantities of solution larger than 1 or 2 liters are prepared, special problems are encountered in being sure that they are well mixed before being standardized. While blade stirrers with glass or metal shafts are suitable for many solutions, they are not suitable in every case. In those cases where contact of a glass or metal stirrer with the solution would be undesirable it may be possible to use a sealed polyolefin coated stirrer. In those cases where only contact of the solution with metal must be avoided, the solution can be mixed by inserting a fritted glass gas dispersion tube to the bottom of the container and bubbling nitrogen through the solution for 1 or 2 h.

In order to make a solution of exact normality from a chemical that cannot be measured as a primary standard, a relatively concentrated stock solution may first be prepared and standardized, and then an exact dilution of this may be made to the desired strength. Another method is to make a solution of slightly higher concentration than desired, standardize, and then make suitable adjustments in the concentration. Alternatively, the solution may be used as first standardized, with appropriate modification of the factor used in the calculation. This alternative procedure is especially useful in the case of a solution that slowly changes strength—for example, thiosulfate solution, which must be restandardized at frequent intervals. Often, however, adjustment to the exact normality

specified is desirable when a laboratory runs a large number of determinations with one standard solution, since this simplified the final calculation.

As long as the normality of a standard solution does not result in a titration volume so small as to preclude accurate measurement or so large as to cause abnormal dilution of the reaction mixture, and as long as the solution is properly standardized and the calculations are properly made, the determinations can be considered to be in accord with the instructions in this manual.

If a solution of exact normality is to be prepared by dissolving a weighed amount of a primary standard or by dilution of a stronger solution, it is necessary that the solution be brought to exact volume in a volumetric flask.

The stock and standard solutions prescribed for the colorimetric determinations in the chemical sections of this manual should also be accurately prepared in volumetric flasks. Where the concentration does not need to be exact, it is often easier to mix the concentrated solution or the solid with measured amounts of water, using graduated cylinders for these measurements. There is usually a significant change of volume when strong solutions are mixed, so that the total volume is less than the sum of the volumes used. For approximate dilutions, the volume changes are negligible when concentrations of 6N or less are diluted.

Very thorough and complete mixing is essential when making dilutions. One of the commonest sources of error in analyses using standard solutions diluted in volumetric flasks is failure to attain complete mixing.

Glass containers are suitable for the storage of most of the standard solutions, although the use of polyolefin containers is recommended for alkali solutions.

When large quantities of solutions are prepared and standardized, it is necessary to provide protection against changes in normality due to absorption of gases or water vapor from the laboratory air. As volumes of solution are withdrawn from the container, the replacement air should be passed through a drying tube filled with equal parts of 8 to 20-mesh soda lime, oxalic acid, and 4 to 8-mesh anhydrous calcium chloride, each product being separated from the other by a glass wool plug.

8.2 SODIUM HYDROXIDE, 0.02 TO 1.0 N.

8.2.1 *Preparation of 50% NaOH Solution and of other Standard NaOH Solutions.* Dissolve 162 g of sodium hydroxide (NaOH) in 150 mL of carbon dioxide-free water. Cool the solution to 25°C and filter through a Gooch crucible, hardened filter paper, or other suitable medium. Alternatively, commercial 50% NaOH solution may be used.

To prepare a 0.1 N solution, dilute 5.45 mL of the clear solution to 1 L with carbon dioxide-free water, mix well, and store in a tight polyolefin container.

For other normalities of NaOH solution, use the requirements given in Table 8:I.

8.2.2 *Standardization.* Crush 10 to 20 g of primary standard potassium hydrogen phthalate† ($KHC_8H_4O_4$) to 100-mesh fineness, and dry in an open glass container at 120°C for 2 h. Stopper the container and cool in a desiccator.

To standardize a 0.1 N solution, weigh accurately 0.95 ± 0.05 g of the dried $KHC_8H_4O_4$, and transfer to a 500-mL conical flask. Add 100 mL of carbon dioxide-free water, stir gently to dissolve the sample, add 3 drops of a 1.0% solution of phenolphthalein in 95% ethanol and titrate with NaOH solution to the appearance of a very faint pink color that persists.

Table 8:I Sodium Hydroxide Dilution Requirements

Desired normality	Grams of NaOH required per L of solution	Volume of 50% NaOH Solution (25°C) required per L of Solution, mL
0.02	0.8	1.09
0.04	1.6	2.18
0.05	2.0	2.73
0.1	4.0	5.45
0.2	8.0	10.90
0.25	10.0	13.63
0.5	20.0	27.25
1.0	40.0	54.54

†A primary standard grade of potassium hydrogen phthalate ($KHC_8H_4O_4$) is available from the Office of Standard Reference Materials, National Bureau of Standards, Washington, D.C. 20234.

The weights of dried $KHC_8H_4O_4$ suitable for other normalities of NaOH solution are given in Table 8:II.

8.2.3 *Calculation.* Calculate the normality of the NaOH solution, as follows:

$$A = \frac{B}{0.20423 \times C}$$

where A = normality of the NaOH solution,

B = weight of $KHC_8H_4O_4$ used, g, and,

C = volume of NaOH solution consumed, mL.

8.2.4 *Stability.* The use of polyolefin containers eliminates some of the difficulties attendant upon the use of glass containers, and their use is recommended. Should glass containers be used, the solution must be standardized frequently if there is evidence of action on the glass container, or if insoluble matter appears in the solution.

8.3 HYDROCHLORIC ACID. 0.02 TO 1.0 N.

8.3.1 *Preparation.* To prepare a 0.1 N solution, measure 8.3 mL of concentrated hydrochloric acid (HCl, sp gr 1.19) into a graduated cylinder and transfer it to a 1-L volumetric flask. Dilute to the mark with water, mix well, and store in a tightly closed glass container.

For other normalities of HCl solution, use the requirements given in Table 8:III.

8.3.2 *Standardization.* Transfer 2 to 4 g of anhydrous sodium carbonate (Na_2CO_3) to a platinum dish or crucible, and dry at 250 C for 4 h. Cool in a desiccator.

To standardize a 0.1 N solution, weigh accurately 0.22 ± 0.01 g of the dried

Table 8:II Weights of Dried Potassium Hydrogen Phthalte

Normality of Solution	Weight of dried $KHC_8H_4O_4$ to be used, g*
0.02	0.19 ± 0.005
0.04	0.38 ± 0.005
0.05	0.47 ± 0.005
0.1	0.95 ± 0.005
0.2	1.90 ± 0.05
0.25	2.35 ± 0.05
0.5	4.75 ± 0.05
1.0	9.00 ± 0.05

Table 8:III Hydrochloric Acid Dilution Requirements

Desired normality	Volume of HCl to be diluted to 1 L, mL
0.02	1.66
0.04	3.32
0.1	8.3
0.2	16.6
0.5	41.5
1.0	83.0

Na_2CO_3, and transfer to a 500-mL conical flask. Add 50 mL of water, swirl to dissolve the carbonate, and add 2 drops of a 0.1% solution of methyl red in 95% ethanol. Titrate with the HCl solution to the first appearance of a red color, and boil the solution carefully, to avoid loss, until the color is discharged. Cool to room temperature, and continue the titration, alternating the addition of HCl solution and the boiling and cooling to the first appearance of a faint red color that is not discharged on further heating.

The weights of dried Na_2CO_3 suitable for other normalities of HCl solution are given in Table 8:IV.

8.3.3 *Calculation.* Calculate the normality of the HCl solution, as follows:

$$A = \frac{B}{0.0530 \times C}$$

where A = normality of the HCl solution,
       B = weight of Na_2CO_3 used, g, and

Table 8:IV Weights of Dried Sodium Carbonate

Normality of Solution	Weight of dried Na_2CO_3 to be used, g
0.02	0.088 ± 0.001*
0.04	0.176 ± 0.001*
0.1	0.22 ± 0.01†
0.2	0.44 ± 0.01†
0.5	1.10 ± 0.01†
1.0	2.20 ± 0.01†

*A 100-mL buret should be used for this standardization.

†The listed weights are for use when a 50-mL buret is used. If a 100-mL buret is to be used, the weights should be doubled.

C = volume of HCl solution consumed, mL.

8.3.4   *Stability.*   Restandardize monthly.

8.4 SULFURIC ACID. 0.02 TO 1.0 N.

8.4.1 *Preparation.* To prepare a 0.1 N solution, measure 3.0 mL of concentrated sulfuric acid (H_2SO_4, sp gr 1.84) into a graduated cylinder and slowly add it to one half the desired volume of water in a 600-mL beaker. Rinse the cylinder with water. Mix the acid-water mixture, allow it to cool, and transfer to a 1-L volumetric flask. Dilute to the mark with water, mix well, and store in a tightly closed glass container.

For other normalities of the H_2SO_4 solution, use the requirements given in Table 8:V.

8.4.2 *Standardization.* Transfer 2 to 4 g of anhydrous sodium carbonate (Na_2CO_3) to a platinum dish or crucible, and dry at 250° C for 4 h. Cool in a desiccator.

For standardization of a 0.1 N solution, weigh accurately 0.22 ± 0.01 g of the dried Na_2CO_3 and transfer to a 500-mL conical flask. Add 50 mL of water, swirl to dissolve the Na_2CO_3, and add 2 drops of a 0.1% solution of methyl red in 95% ethanol. Titrate with the H_2SO_4 solution to the first appearance of a red color, and boil the solution carefully, to avoid loss, until the color is discharged. Cool to room temperature and continue the titration, alternating the addition of H_2SO_4 solution and the boiling and cooling, to the first appearance of a faint red color that is not discharged on further heating.

The weights of dried Na_2CO_3 suitable for other normalities of H_2SO_4 solution are given in Table 8:IV.

Table 8:V Sulfuric Acid Dilution Requirements

Desired normality	Volume of H_2SO_4 to be diluted to 1 L, mL
0.02	0.60
0.1	3.0
0.2	6.0
0.5	15.0
1.0	30.0

8.4.3 *Calculation.* Calculate the normality of the H_2SO_4 solution, as follows:

$$A = \frac{B}{0.0530 \times C}$$

where
A = normality of the H_2SO_4 solution,
B = weight of Na_2CO_3, g, and
C = volume of H_2SO_4 solution consumed, mL.

8.4.4   Stability.   Restandardize monthly.

8.5 Iodine (0.1 N).

8.5.1 *Preparation.* Transfer 12.7 g of iodine and 60 g of potassium iodide (KI) to an 800-mL beaker, add 30 mL of water, and stir until solution is complete. Dilute with water to 500 mL, and filter through a sintered-glass filter. Wash the filter with about 15 mL of water, transfer the combined filtrate and washing to a 1-L volumetric flask, dilute to the mark with water, and mix. Store the solution in a glass-stoppered, amber-glass bottle in a cool place.

8.5.2 *Standardization.* Transfer 1 g of arsenic trioxide (As_2O_3) to a platinum dish, and dry at 105°C for 1 h. Cool in a desiccator. Weigh accurately 0.20 ± 0.01 g of the dried As_2O_3 and transfer to a 500-mL conical flask. Add 10 mL of sodium hydroxide solution (NaOH, 40 g/L), and swirl to dissolve. When solution is complete, add 100 mL of water and 10 mL of sulfuric acid (H_2SO_4, 1:35), and mix. Slowly add sodium bicarbonate ($NaHCO_3$) until effervescence ceases, add 2 g of $NaHCO_3$ in excess, and stir until dissolved. Add 2 mL of starch solution (10 g/L) and titrate with the iodine solution to the first permanent blue color.

8.5.3 *Calculation.* Calculate the normality of the iodine solution, as follows:

$$A = \frac{B}{0.040455 \times C}$$

where
A = normality of the iodine solution,
B = weight of As_2O_3 used, g, and
C = volume of iodine solution required for titration of the solution, mL.

8.5.4 *Stability.* Restandardize sealed bottles monthly. Restandardize open bottles weekly.

8.6 Sodium Thiosulfate (0.1 N).

8.6.1 *Preparation.* Dissolve 25 g of sodium thiosulfate pentahydrate ($Na_2S_2O_3 \cdot 5H_2O$) in 500 mL of freshly boiled and cooled water, and add 0.11 g of sodium carbonate (Na_2CO_3). Dilute to 1 L with freshly boiled and cooled water, and let stand for 24 h. Store the solution in a tightly-closed glass bottle.

8.6.2 *Standardization.* Pulverize 2 g of potassium dichromate ($K_2Cr_2O_7$), transfer to a platinum dish, and dry at 120°C for 4 h. Cool in a desiccator. Weigh accurately 0.21 ± 0.01 g of the dried $K_2Cr_2O_7$, and transfer to a 500-mL glass-stoppered conical flask. Add 100 mL of water, swirl to dissolve, remove the stopper, and quickly add 3 g of potassium iodide (KI), 2 g of sodium bicarbonate ($NaHCO_3$), and 5 mL of hydrochloric acid (HCl). Stopper the flask quickly, swirl to ensure mixing, and let stand in the dark for 10 min. Rinse the stopper and inner walls of the flask with water and titrate with the $Na_2S_2O_3$ solution until the solution has only a faint yellow color. Add 2 mL of starch solution (10 g/L), and continue the titration to the disappearance of the blue color.

8.6.3 *Calculation.* Calculate the normality of the $Na_2S_2O_3$ solution, as follows:

$$A = \frac{B}{0.04904 \times C}$$

where
A = normality of the $Na_2S_2O_3$ solution,
B = weight of $K_2Cr_2O_7$ used, g, and
C = volume of $Na_2S_2O_3$ solution required for titration of the solution, mL.

8.6.4 *Stability.* Restandardize weekly.

8.7 Ammonium Thiocyanate (0.1 N).

8.7.1 *Preparation.* Transfer 7.8 g of ammonium thiocyanate (NH_4SCN) to a flask, add 100 mL of water, and swirl to dissolve the NH_4SCN. When solution is complete, filter through a Cooch crucible, hardened filter paper, or other suitable medium. Dilute the clear filtrate to 1 L with water and mix. Store the solution in a tightly-stoppered glass bottle.

8.7.2 *Standardization.* Measure accurately about 40 mL of freshly standardized 0.1 N silver nitrate ($AgNO_3$), and transfer to a 250-mL conical flask. Add 50 mL of water, swirl to mix the solution, and add 2 mL of nitric acid (HNO_3) and 1 mL of ferric ammonium sulfate solution ($FeNH_4$ $(SO_4)_2 \cdot 12H_2O$, 80 g/L). Titrate the $AgNO_3$ solution with the NH_4SCN solution until the first permanent reddish-brown color appears and persists after vigorous shaking for 1 min.

8.7.3 *Calculation.* Calculate the normality of the NH_4SCN solution, as follows:

$$A = \frac{B \times C}{D}$$

where A = normality of the NH_4SCN solution,
   B = volume of Ag NO_3 solution used, mL
   C = normality of the $AgNO_3$ solution, and
   D = volume of NH_4SCN solution required for titration of the solution, mL.

8.7.4 *Stability.*   Restandardize monthly.

8.8 POTASSIUM DICHROMATE (0.1 N).

8.8.1 *Preparation.* Transfer 6 g of potassium dichromate ($K_2Cr_2O_7$) to a platinum dish and dry at 120°C for 4 h. Cool in a desiccator. Place 4.9 g of the dried $K_2Cr_2O_7$ in a 1-L volumetric flask, and add 100 mL of water. Swirl to dissolve and when solution is complete, dilute to the mark with water and mix. Store the solution in a glass-stoppered bottle.

8.8.2 *Standardization.* Place 40 mL of water in a 250-mL glass-stoppered conical flask, and add 40 mL, accurately measured, of the $K_2Cr_2O_7$ solution. Stopper the flask, swirl to mix, remove the stopper, and add 3 g of potassium iodide (KI), 2 g of sodium bicarbonate ($NaHCO_3$) and 5 mL of concentrated hydrochloric acid (HCl). Stopper the flask quickly, swirl to ensure mixing, and let stand in the dark for 10 min. Rinse the stopper and inner walls of the flask with water and titrate with freshly standardized sodium thiosulfate solution ($Na_2S_2O_3$) until the solution has only a faint yellow color. Add 2 mL of starch solution (10 g/L), and continue the titration to the disappearance of the blue color.

8.8.3 *Calculation.* Calculate the normality of the $K_2Cr_2O_7$ solution, as follows:

$$A = \frac{B \times C}{D}$$

where A = normality of the $K_2Cr_2O_7$ solution,
   B = volume of $Na_2S_2O_3$ solution required for titration of the solution, mL
   C = normality of the $Na_2S_2O_3$ solution, and
   D = volume of $K_2Cr_2O_7$ solution used, mL.

8.8.4   *Stability.*   Restandardize monthly.

8.9 SILVER NITRATE (0.1 N).

8.9.1 *Preparation.* Dry 17.5 g of silver nitrate ($AgNO_3$) at 105°C for 1 h. Cool in a desiccator. Transfer 16.99 g of the dried $AgNO_3$ to a 1-L volumetric flask. Add 500 mL of water, swirl to dissolve the $AgNO_3$, dilute to the mark with water, and mix. Store the solution in a tightly-stoppered amber-glass bottle.

8.9.2 *Standardization.* Transfer 0.3 g of sodium chloride (NaCl) to a platinum dish and dry at 105°C for 2 h. Cool in a desiccator. Weigh accurately 0.28 ± 0.01 g of the dried NaCl and transfer to a 250-mL glass-stoppered conical flask. Add 25 mL of water, swirl to dissolve the NaCl, and add 2 mL of nitric acid (HNO_3). Add, from a volumetric pipet, 50 mL of the $AgNO_3$ solution, mix, and add 1 mL of ferric ammonium sulfate solution ($FeNH_4(SO_4)_2$ $\cdot 12H_2O$, 80 g/L) and 5 mL of nitrobenzene (CAUTION‡). Stopper the flask and shake vigorously to coagulate the precipitate. Rinse the stopper into the flask with a few milliliters of water and titrate the excess of $AgNO_3$ with ammonium thiocyanate solution (NH_4SCN) until the first permanent reddish-brown color appears and persists after vigorous shaking for 1 min. Designate the volume of NH_4SCN solution required for the titration as Volume I.

---

‡**Caution**—nitrobenzene, used in this section, is extremely hazardous when absorbed through the skin or when its vapor is inhaled and is currently considered a carcinogen. Such exposure may cause cyanosis; prolonged exposure may cause anemia. Do not get in eyes, on skin, or on clothing. Avoid breathing vapor. Use only with adequate ventilation.

Transfer 50 mL of the $AgNO_3$ solution to a clean, dry, 250-mL glass-stoppered conical flask. Add 25 mL of water, 2 mL of HNO_3, 1 mL of $FeNH_4 (SO_4)_2$.$12H_2O$ solution (80g/L), stopper the flask, and shake vigorously. Rinse the stopper into the flask with a few milliliters of water and titrate the $AgNO_3$ solution with NH_4SCN solution until the first permanent reddish-brown color appears and persists after vigorous shaking for 1 min. Designate the volume of NH_4SCN solution consumed as Volume II.

Measure accurately, from either a buret or a volumetric pipet, 2.0 mL of the $AgNO_3$ solution, designate the exact volume as Volume III, and transfer to a 100-mL glass-stoppered conical flask. Add 25 mL of water, 2 mL of NHO_3, 1 mL of $FeNH_4 (SO_4)_2$.$12 H_2O$ solution (80 g/L), and 5 mL of nitrobenzene, stopper the flask, and shake vigorously. Rinse the stopper into the flask with a few milliliters of water and titrate the $AgNO_3$ solution with NH_4SCN solution until the first permanent reddish-brown color appears and persists after vigorous shaking for 1 min. Designate the volume of NH_4SCN solution consumed as Volume IV.§

8.9.3 *Calculation.* Calculate the normality of the $AgNO_3$ solution, as follows:

$$A = \frac{B}{0.05845 \times (C - D)}$$

where   A = normality of the $AgNO_3$ solution,

B = weight of NaCl used,

C = volume of $AgNO_3$ solution consumed by the total chloride, mL = 50 - (Volume I × 50/Volume II), and

D = volume of $AgNO_3$ solution consumed by any chloride ion in the nitrobenzene, mL = Volume III - (Volume IV × 50/Volume II)

8.9.4 *Stability.* Restandardize monthly.

_____

§The ammonium thiocyanate titrant used in the three titrations must be from the same, well-mixed solution. The nitrobenzene used in each titration must also be from the same, well-mixed container.

8.10 POTASSIUM PERMANGANATE (0.1 N).

8.10.1 *Preparation.* Dissolve 3.2 g of potassium permanganate ($KMnO_4$) in 100 mL of water, and dilute the solution with water to 1 L. Allow the solution to stand in the dark for two weeks and then filter through a fine-porosity sintered-glass crucible. **Do not wash the filter.** Store the solution in glass-stoppered, amber-colored glass bottles.

NOTE 1. Do not permit the filtered solution to come into contact with paper, rubber, or other organic material.

8.10.2 *Standardization.* Transfer 2 g of sodium oxalate ($Na_2C_2O_4$) to a platinum dish and dry at 105°C for 1 h. Cool in a desiccator. Weigh accurately 0.30 ± 0.01 g of the dried $Na_2C_2O_4$ and transfer to a 500-mL glass container. Add 250 mL of sulfuric acid (H_2SO_4, 1:19) that was previously boiled for 10 to 15 min and then cooled to 27 ± 3°C, and stir until the sample is dissolved. Add 39 mL of the $KMnO_4$ solution at a rate of 30 ± 5 mL per min. while stirring slowly, and let stand for about 45 s until the pink color disappears. Heat the solution to 60°C, and complete the titration by adding $KMnO_4$ solution until a faint pink color persists for 30 s. Add the final 0.5 to 1.0 mL dropwise, and give the solution time to decolorize before adding the next drop.

Carry out a blank determination on a second 250-mL portion of the H_2SO_4 (1:19), and make sure that the pink color at the end point matches that of the standardization solution. Correct the sample titration volume as shown to be necessary.

NOTE 2. The specified 0.30 g sample of $Na_2C_2O_4$ should consume about 44.8 mL of 0.1 N $KMnO_4$.

NOTE 3. If the pink color of the solution persists more than 45 s after the addition of the first 39 mL of $KMnO_4$ solution is complete, discard the solution and start over with a fresh solution of the $Na_2C_2O_4$, but add less of the $KMnO_4$ solution.

NOTE 4. The blank correction usually amounts to 0.03 to 0.05 mL.

8.10.3 *Calculation.* Calculate the normality of the $KMnO_4$ solution as follows:

$$A = \frac{B}{0.06701 \times C}$$

where A = normality of the $KMnO_4$ solution,

B = weight of $Na_2C_2O_4$ used, g, and

C = volume of $KMnO_4$ solution required for titration of the solution, mL.

8.10.4 *Stability.* Restandardize weekly.

Approved with modifications from 2nd edition

Subcommittee 9
J.N. HARMAN, *Chairman*

9. Recovery and Internal Standard Procedures

9.1 RECOVERY PROCEDURE. A recovery procedure does not enable the analyst to apply any correction factor to the results of an analysis; however, it does give him some basis for judging the applicability of a particular method of analysis to a particular sample.

A recovery determination may be performed at the same time as the determination itself. Of course, recoveries would not be run on a routine basis with samples whose general composition is known or when using a method whose applicability to the sample is well established. Recovery methods are to be regarded as tools to remove doubt about the applicability of a method to a sample.

In brief, the recovery procedure involves applying the analytical method to a reagent blank, to a series of known standards covering the expected range of concentration of the sample, to the sample itself, in at least duplicate, and to the recovery samples, prepared by adding known quantities of the substance sought to separate portions of the sample itself, each portion equal to the size of sample taken for the run. The substance sought should be added in sufficient quantity to overcome the limits of error of the analytical method, but without causing the total in the sample to exceed the range of the known standards used.

The results are first corrected by subtracting the reagent blank from each of the other determined values. The resulting known standards are then graphically represented. From this graph the amount of sought substance in the sample alone is determined. This value is then subtracted from each of the determinations consisting of sample plus known added substance. The resulting amount of substance divided by the known amount originally added and multiplied by 100 gives the percentage recovery.

The procedure outlined above may be applied to a colorimetric, flame photometric, or fluorimetric analysis. It may also be applied in a little more simple form to titrimetric, gravimetric, and other types of analyses.

Rigid rules concerning the percentage recoveries required for acceptance of results of analyses for a given sample and method cannot be stipulated. Recoveries of substances in the range of the sensitivity of the method may, of course, be very high or very low and approach a value nearer to 100 per cent recovery as the error of the method becomes small with respect to the magnitude of the amount of substance added. In general, intricate and exacting procedures for trace substances that have inherent errors due to their complexity may give recoveries that would be considered very poor and yet, from the practical viewpoint of usefulness of the result, may be quite acceptable. Poor results may reflect either interferences present in the sample or real inadequacy of the method of analysis in the range in which it is being used.

It must be stressed, however, that the judicious use of recovery determinations for the evaluation of analytical procedures and their applicability to particular samples is an invaluable aid to the analyst in both routine and research investigations.

9.2 INTERNAL STANDARD PROCEDURE. The internal standard technique is used primarily for emission spectrograph and polarographic procedures. This procedure enables the analyst to compensate for electronic and mechanical fluctuations within the instrument.

In brief, the internal standard method involves the addition to the sample of known amounts of a substance that will respond to the instrument in a manner similar to the contaminant being analyzed. The ratio of the measurement of the internal standard to the measurement of the contaminant is the value used to determine concentration of contaminant present in the sample. Any changes in conditions during analyses will affect the internal standard and the contaminant the same and so will compensate for such changes. The internal standard should be of similar chemical reactivity to the contaminant, approximately the same concentration as anticipated for the contaminant, and as pure a substance as possible.

Approved with modifications from 2nd edition

Subcommittee 9
J.N. HARMAN, *Chairman*

10. Interferences

10.1 INTRODUCTION. The term interference covers various effects of dissolved or suspended substances upon analytical procedures. The known interferences and related information have been included in the specific methods of analysis. Many of the methods in this manual have been developed specifically to minimize common interferences. The purpose of this section is to offer guidelines for detecting the presence of interfering substances and overcoming their effects when encountered.

In analytical chemistry, one must be prepared to use an alternate procedure if one method would be less affected than another due to the presence of the interfering substance. The most expedient method, however, for overcoming a suspected interference is often the use of a smaller initial aliquot. Thus, the effect of the interfering element is diminished or eliminated through dilution of the original sample. The level of the parameter being measured is likewise reduced, so care must be taken to insure an aliquot containing more than the minimum detectable concentration. If the data display a consistently increasing or decreasing pattern with dilution, interference is indicated. Dilution of final reaction volume must never be used in overcoming interferences. Certain reagents are prepared to react with a limited amount of the test element. If this limit is exceeded, a portion of the test element goes undetected. Then

the analyst must repeat the entire test procedure with a smaller initial aliquot.

10.2 INTERFERENCE REACTIONS. Essentially, three reactions may occur to produce an interference:

a. An interfering substance may react with the reagents in the same manner as the element being sought. (Positive interference);

b. an interfering substance may react with the element being sought to prevent complete isolation. (Negative interference); or

c. an interfering substance may combine with the reagents to prevent further reaction with the element being sought. (Negative interference).

These reactions will produce either high or low results. An estimate of the magnitude of an interference may be obtained by the internal standard procedure. After establishing a calibration graph for the appropriate range and performing an initial analysis of the sample, this procedure may be employed.

10.3 INTERNAL STANDARD PROCEDURE.

10.3.1 Select a standard which, when added to the unknown sample, will produce a final result approximately midrange of the standard calibration graph. Example: Calibration graph: 0 to 100 ppb.

Sample analysis: 10 ppb.
Select a standard concentration, which when added to 10 ppb, will produce a total concentration of 50 ppb; *i.e.*, 40 ppb.

10.3.2 Analyze the prepared solution (sample + standard) and read the result from the graph.

10.3.3 Subtract from this result the original sample analysis.

10.3.4 The difference divided by the amount originally added and multiplied by 100 gives the percentage recovery. Example:

Sample: 10 ppb

Sample + Standard:

10 ppb + 40 ppb = 50 ppb

Sample + Standard as determined:

56 ppb

Subtract original determination:

$$56 - 10 = 46 \text{ ppb}$$

Recovery equals:

$$\frac{46}{40} \times 100 = 115 \text{ percent}$$

Thus, a positive interference is present. If the result is below 100%, a negative interference is indicated. By establishing the direction of interference, the nature of the reaction taking place may be determined and appropriate steps taken for eliminating the effect. A positive error indicates the interference is reacting like the test element, and the possibility of isolation by a pH adjustment should be considered. This procedure should be applied in triplicate to both the sample and the sample plus standard and must include a reagent blank.

10.4 REMOVAL OF INTERFERENCE. If further treatment of the sample for removal of the interfering substance is necessary, several approaches are available to the analyst. These may be physical or chemical or a combination of both, such as:

a. Distillation of the sample leaving the interference behind;

b. removal of the interference by ion exchange resins;

c. addition of complexing agents;

d. extraction into organic solvents;

e. ashing;

f. pH adjustment;

g. different reaction rate;

h. different temperature.

These treatment methods serve as examples only, since a thorough review of the literature must be undertaken before proceeding to undertake a change in an established method.

10.5 SOURCES OF INTERFERENCE. In the preceding discussion, the analyst was given instructions in detecting an interference in quantitative analysis. The question of the source of the interference must be identified and corrected, if possible. The origin of an interference may be found in the follow situations:

a. Present at the sampling site;

b. imparted during sample collection;

c. developed in sample storage; or

d. imparted or developed in laboratory analysis.

These four source situations may normally be avoided in applying proper sampling and laboratory techniques. These techniques are discussed in this section.

The sources may be a result of physical, chemical, or biological phenomena. The possibility of interaction (physical upon chemical, etc.) should not be overlooked. These three classifications of sources of errors are further subdivided to assist the analyst in search of the action producing the effect.

10.6 PHYSICAL SOURCES.

10.6.1 *Heat.*

a. Effect on chemical equilibria.

b. Promote side reactions at elevated temperatures — decomposition.

c. Affect rate of reaction — temperature coefficient correction. Serious errors may be encountered at low temperatures.

10.6.2 *Light.*

a. Effect of visible spectrum — accelerate photooxidation.

b. Effect of UV — yellowing of KI reagent — positive error.

c. Fading of detector reagent colors (lead and mercury dithizonates) — negative error.

10.6.3 *Humidity.*

a. Reaction of H_2O with contaminant — toluene diisocyanate, phosgene.

b. Reaction of H_2O with the collecting substrate.

c. Dilution of the collecting substrate.

10.6.4 *Time.*

a. Deterioration rate after sample collection: half-life must be determined.

b. Fading of the developed color after maximum density attainment.

c. Reagents often deteriorate with time.

10.6.5 *Chemical Contamination.*

a. Contamination of sampling equipment during use.

(1) Collection in high concentration locations — tetraethyl lead.

(2) Extraneous dust and dirt.

(3) Failure to keep collection equipment and reagents stoppered.

(4) Cross contamination in recharging collector or changing filters, especially in the field.

10.7 CHEMICAL SOURCES.
10.7.1 *pH Control in Aqueous Systems.*

a. pH sensitive collection.
b. pH sensitive color responses.
c. Buffered versus unbuffered solutions.

10.7.2 *Chemical Contamination.*

a. Impure reagents — Sulfates in gum arabic used to stabilize $BaSO_4$ colloid.
b. Failure to run blanks and controls.

10.7.3 *Interferences.*

a. Negative — NH_3 in Hg detector tube — redox cancellations.
b. Positive
(1) Different shade or color produced — Dithizone + oxidizing agents.
(2) Reaction the same, increased by interference — Ozone — nitrogen oxides system.
(3) Electrolytes in conductivity and pH meters.

10.7.4 *Sensitivity.*

a. Concentration effects.
(1) Adjust to obtain maximum effect — maximum color density. Reactant ratios often nonstoichiometric.
(2) Beer's law may not be followed at low concentrations.
(3) Background "noise" in electrolytic instruments.
(4) Effect of reagent concentrations on reaction rate.
(5) Effect of reagent concentrations on reaction equilibrium — CH_2O + fuchsin reagent.
(6) Attempting to operate outside the optimum range for which the method was developed.
b. Color former structure versus extinction coefficient — Phenanthroline series for iron detection.
c. Length-of-stain tubes.
(1) Attempts to increase sensitivity by repeated sample aliquots — benzene tube requires recalibration.
(2) Change in sampling rate without recalibration may produce serious errors.

10.7.5 *Catalytic Effects.*

a. Decomposition on contact with tubes and containers — ozone on metals — H_2O_2 + heavy metal.
b. Promotion of side reactions.

c. Failure to react in absence of catalyst.
d. Surface reactions — porous glass bubblers.
e. Inhibition of reactions.

10.7.6 *Half-life in the Ambient Atmosphere.*

a. Effect of light, H_2O vapor, and coexisting contaminants.
(1) tetraethyl lead
(2) toluene diisocyanate
(3) $COCl_2$, phosgene
(4) Nitrogen Oxides
(5) Carcinogenic Hydrocarbons*
(6) Polymerization
(7) Carbon Monoxide

10.8 BIOLOGICAL SOURCES.
10.8.1 *Algal Growth.*
10.8.2 *Insects.*
10.8.3 *Waste Products (Animal and Insects).*

10.9 Procedures used herein are usually well established and, when applied to the selected sample, give predictable results. If a result is suspected to be in error, samples of known composition should be analyzed by the method in use and an alternate method. If a method fails on the unknown sample while giving correct results on a known, an interference is indicated. The best way to solve a problem of this kind is by reviewing the literature. Many hours may be saved through a few hours of diligent library searching. Solving such problems depends on the skill and ingenuity of the investigator.

Approved with modifications from 2nd edition

Subcommittee 9
J.N. HARMAN, *Chairman*

---

*Anthracene and naphthalene (polynuclear aromatic hydrocarbons), benzene, nitrobenzene.

11. Terms and Symbols Relating to Molecular Spectroscopy*

Absorbance, A — the logarithm to the base 10 of the reciprocal of the transmittance, (T).

$$A = \log_{10}(1/T) = -\log_{10}T$$

In practice the observed transmittance must be substituted for T. Absorbance expresses the excess absorption over that of a specified reference or standard. It is implied that compensation has been effected for reflectance losses, solvent absorption losses, and refractive effects, if present, and that attenuation by scattering is small compared with attenuation by absorption. Apparent deviations from the absorption laws (see *Absorptivity*) are due to inability to measure exactly the true transmittance or to know the exact concentration of an absorbing substance.

Absorption Band — a region of the absorption spectrum in which the absorbance passes through a maximum.

Absorption Coefficient, α — a measure of absorption of radiant energy from an incident beam as it traverses an absorbing medium according to Bouguer's law, $P/P_o = e^{\alpha}$.

In IRS, α is a measure of the rate of absorption of energy from the evanescent wave.

Absorption Parameter, a — the relative reflection loss per reflection that results from the absorption of radiant energy at a reflecting surface: $a = 1 - R$, and R = the reflected fraction of incident radiant power.

Absorption Spectrum — a plot of absorbance, or any function of absorbance, against wavelength or any function of wavelength.

Absorptivity, a — the absorbance divided by the product of the concentration of the substance and the sample pathlength, $a = A/bc$. The units of b and c shall be specified.

The recommended unit for b is the centi-

metre. The recommended unit for c is kilogram per cubic metre. Equivalent units are g/dm³, g/L, or mg/cm³. The equivalent IUPAC term is "specific absorption coefficient."

Absorptivity, Molar, ϵ — the product of the absorptivity, a, and the molecular weight of the substance. The equivalent IUPAC term is "molar absorption coefficient."

Aliasing — the appearance of features at wave-numbers other than their true value caused by using a sampling frequency less than twice the highest modulation frequency in the interferogram; also known as "folding."

Analytical Curve — the graphical presentation of a relation between some function of radiant power and the concentration or mass of the substance emitting or absorbing it.

Analytical Wavelength — any wavelength at which an absorbance measurement is made for the purpose of the determination of a constituent of a sample.

Angle of Incidence, θ — the angle between an incident radiant beam and a perpendicular to the interface between two media.

Anti-Stokes Line — a Raman line that has a frequency higher than that of the incident monochromatic beam.

Aperture of an IRE, A' — that portion of the beveled area that can be utilized to conduct light into the IRE at the desired angle of incidence.

Provided the beam length is equal to or greater than the IRE width, A' is equal to the total area of the bevel for $\theta \geq 45°$, and is given by: $A' = w(t/\sin \theta)$, where w = width of IRE or length of aperture, t = thickness of IRE, and θ = angle of incidence. For θ less than 45°, only part of the bevel contributes to the useful aperture and is given by

$$A' = w(2t \sin \theta)$$

The projected aperture width should be at least equal to the maximum width of the spectrometer slits to be employed. The length of the aperture is equal to the width of the IRE (provided the beam length is equal to or greater than the IRE width), and should be at least equal to the spectrometer slit length.

Apodization — modification of the ILS function by multiplying the interferogram

---

*Adapted, with permission, from the *Annual Book of ASTM Standards*, © American Society for Testing and Materials, 1916 Race St., Philadelphia, PA 19103. ASTM is not responsible for any modifications made by ISC.

by a weighting function the magnitude of which varies with retardation.

This term should strictly be used with reference to a weighting function whose magnitude is greatest at the centerburst and decreases with retardation.

Attenuated Total Reflection (ATR) — reflection that occurs when an absorbing coupling mechanism acts in the process of total internal reflection to make the reflectance less than unity.

In this process, if an absorbing sample is placed in contact with the reflecting surface, the reflectance for total internal reflection will be attenuated to some value between zero and unity ($O < R < I$) in regions of the spectrum where absorption of the radiant power can take place.

Attenuation Index, K — a measure of the absorption of radiant energy by an absorbing material. K is related to the absorption coefficient by: $nK = ac_o/4\pi\nu$, where $c_o =$ the speed of light in vacuo, $\nu =$ the frequency of radiant energy, and $n =$ the refractive index of the absorbing medium.

Background — apparent absorption caused by anything other than the substance for which the analysis is being made.

Baseline — any line drawn on an absorption spectrum to establish a reference point representing a function of the radiant power incident on a sample at a given wavelength.

Basic NMR Frequency, ν_o — the frequency, measured in hertz (Hz), of the oscillating magnetic field applied to induce transitions between nuclear magnetic energy levels.

Beamsplitter — a semireflecting device used to create, and often to recombine, spatially separate beams. Beamsplitters are often made by depositing a film of a high refractive index material onto a flat transmitting substrate with an identical compensator plate being held on the other side of the film.

Beamsplitter Efficiency — the product $4RT$, where R is the reflectance and T is the transmittance of the beamsplitter.

Beer's Law — the absorbance of a homogeneous sample containing an absorbing substance is directly proportional to the concentration of the absorbing substance. See also *Absorptivity*.

Bouguer's Law — the absorbance of a ho-

mogeneous sample is directly proportional to the thickness of the sample in the optical path. Bouguer's law is sometimes also known as Lambert's law.

Boxcar Truncation — identical effective weighting of all points in the measured interferogram prior to the Fourier transform; all points outside of the range of the measured interferogram take a value of zero.

Centerburst — the region of greatest amplitude in an interferogram. For unchirped or only slightly chirped interferograms, this region includes the "zero path difference point" and the "zero retardation point."

Chemical Shift (NMR), δ — the defining equation for δ is the following:

$$\delta = \frac{\Delta\nu}{\nu_R} \times 10^6$$

where ν_R is the frequency with which the reference substance is in resonance at the magnetic field used in the experiment and $\Delta\nu$ is the frequency difference between the reference substance and the substance whose chemical shift is being determined, at constant field. The sign of $\Delta\nu$ is to be chosen such that shifts to the high frequency side of the reference shall be positive. If the experiment is done at constant frequency (field sweep) the defining equation becomes

$$\delta = \frac{\Delta\nu}{\nu_R} \times (1 - \frac{\Delta\nu}{\nu_R}) \times 10^6$$

Chirping — the process of dispersing the zero phase difference points for different wavelengths across the interferogram, so that the magnitude of the signal is reduced in the short region of the interferogram where all wavelengths would otherwise constructively interfere.

Concentration, c — the quantity of the substance contained in a unit quantity of sample. For solution work, the recommended unit of concentration is grams of solute per litre of solution.

Critical Angle, θ_c — the angle whose sine is equal to the relative refractive index for light striking an interface from the greater to the lesser refractive medium: $\theta_c = \sin^{-1} n_{21}$, where $n_{21} =$ the ratio of the refractive indices of the two media. Total reflection occurs when light is reflected in the more refractive of two media from the interface between them at any angle of incidence exceeding the critical angle.

Depth of Penetration, d_p — in internal reflection spectroscopy, the distance into the less refractive medium at which the amplitude of the evanescent wave is e^{-1} (that is, 36.8%) of its value at the surface:

$$d_p = \frac{\lambda_1}{2\pi(\sin^2\theta - n_{21}{}^2)^{1/2}}$$

where: $n_{21} = n_2/n_1$ = refractive index of sample divided by that of the IRE; $\lambda_1 = \lambda/n_1$ = wavelength of radiant energy in the sample; and θ = angle of incidence.

Derivative Absorption Spectrum — a plot of rate of change of absorbance or of any function of absorbance with respect to wavelength or any function of wavelength, against wavelength or any function of wavelength.

Digitization — the conversion of an analog signal to digital values using an analog-to-digital converter "sampling" or "digital sampling."

Digitization Noise — the noise generated in an interferogram through the use of an analog-to-digital converter whose least significant bit represents a value comparable to, or greater than, the peak-to-peak noise level in the analog data.

Dilution Factor — the ratio of the volume of a diluted solution to the volume of original solution containing the same quantity of solute as the diluted solution.

Double Modulation — a technique in which an interferometer generates a conventional signal, while at the same time the signal is also modulated by an external high frequency modulator; on detection, the conventional interferometric signal is filtered out so that only the high frequency signal is recorded.

Double-Pass Internal Reflection Element — an internal reflection element in which the radiant power tranverses the length of the optical element twice, entering and leaving via the same end.

Effective Pathlength (or Effective Thickness), d_e — in internal reflection spectroscopy, the analog of the sample thickness in transmission spectroscopy that represents the distance of propagation of the evanescent wave within an absorbing sample in IRS. It is defined from the relationship:

$R = 1 - \alpha d_e$, and is related to the absorption parameter by: a $= \alpha d_e$.

Evanescent Wave — the standing wave that exists in the less refractive medium, normal to the reflecting surface of the IRE during internal reflection.

Fast Fourier Transform (FFT) — a method for speeding up the computation of a discrete FT by factoring the data into sparse matrices containing mostly zeroes.

Filter — a substance that attenuates the radiant power reaching the detector in a definite manner with respect to spectral distribution.

Filter, Neutral — a filter that attenuates the radiant power reaching the detector by the same factor at all wavelengths within a prescribed wavelength region.

Fixed-Angle Internal Reflection Element — an internal reflection element which is designed to be operated at a fixed angle of incidence.

Fluorescence — the emission of radiant energy from an atom, molecule, or ion resulting from absorption of a photon and a subsequent transition to the ground state without a change in total spin quantum number.

The initial and final states of the transition are usually both singlet states. The average time interval between absorption and fluorescence is usually less than 10^{-6} s.

Folding — see *Aliasing.*

Fourier Transform (FT) — the mathematical process used to convert an amplitude-time spectrum to an amplitude-frequency spectrum, or vice versa. In FT-IR spectrometry, retardation is directly proportional to time; therefore the FT is commonly used to convert an amplitude-retardation spectrum to an amplitude-wavenumber spectrum, and vice versa.

Fourier Transform Infrared (FT-IR) Spectrometry — a form of infrared spectrometry in which an interferogram is obtained; this interferogram is then subjected to a Fourier transform to obtain an amplitude-wavenumber (or wavelength) spectrum. The abbreviation FTIR is not recommended. When FT-IR spectrometers are interfaced with other instruments, a slash should be used to denote the interface; e.g. GC/FT-IR; HPLC/FT-IR, and the use of FT-IR should be explicit; i.e. FT-IR, not IR.

Frequency, ν – the number of cycles per unit time. The recommended unit is the hertz (Hz) (one cycle per second).

Frustrated Total Reflection (FTR) – the reflection which occurs when a nonabsorbing coupling mechanism acts in the process of total internal reflection to make the reflectance less than unity.

In the process the reflectance can vary continuously between zero and unity if: (*1*) An optically transparent medium is within a fraction of a wavelength of the reflecting surface and its distance from the reflecting surface is changed, or (*2*) both the angle of incidence and the refractive index of one of the media vary in an appropriate manner.

In these cases part of the radiant power may be transmitted through the interface into the second medium without loss at the reflecting surface such that transmittance plus reflectance equals unity. It is possible, therefore, to have this process taking place in some spectral regions even when a sample having absorption bands is placed in contact with the reflecting surface.

High-Resolution NMR Spectrometer – an NMR apparatus that is capable of producing, for a given isotope, line widths that are less than the majority of the chemical shifts and coupling constants for that isotope.

By this definition, a given spectrometer may be classed as a high-resolution instrument for isotopes with large chemical shifts, but may not be classed as a high-resolution instrument for isotopes with smaller chemical shifts.

Infrared – pertaining to the region of the electromagnetic spectrum from approximately 0.78 to 300 μm.

Instrument Line Shape (ILS) Function – the FT of the function by which an interferogram is weighted. This weighting may be performed optically, due to the finite optical throughput, or digitally, through multiplication by an apodization function, or both. The ILS function is the profile of the spectrum of a monochromatic source producing a beam with the same throughput as the beam in the actual measurement being performed.

Instrument Response Time – the time required for an indicating or detecting device to undergo a defined displacement following an abrupt change in the quantity being measured.

Integration Period, π – the time, in seconds, required for the pen or other indicator to move 98.6% of its maximum travel in response to a step function. For instruments with a first-order response, the integration period will be approximately equal to four times the exponential time constant. It is equal to the period, classically defined, for a second order, critically damped response system.

Interferogram, $I(\delta)$ – record of the modulated component of the interference signal measured as a function of retardation of the detector.

An alternate symbol is $I(x)$. The recommended symbol for the spectrum computed from $I(\delta)$ is $B(\nu)$. An alternate symbol is $B(\sigma)$.

Interferogram, Double-Sided – interferogram measured with approximately equal retardation on either side of the centerburst.

Interferogram, Laser Reference – sinusoidal interferogram of a laser source measured at the same time as the signal interferogram. The zero crossings of this interferogram are used to control sampling of the signal interferogram. It may also be noted that other effectively monochromatic sources can be used in place of the laser.

Interferogram, Signal – interferogram of the beam of radiant energy whose spectrum is desired.

Interferogram, Single-Sided – interferogram in which sampling is initiated close to the centerburst and continues through that point to the maximum retardation desired.

Interferogram, White Light – reference interferogram of a broadband light source measured at the same time as the signal interferogram and used to initiate data acquisition of consecutive scans for signal-averaging.

Interferometer – device used to divide a beam of radiant energy into two or more paths, generate an optical path difference between the beams, and recombine them in order to produce repetitive interference maxima and minima as the optical retardation is varied.

Interferometer, Genzel – interferometer in which the beam is focused in the plane of the beamsplitter and collimated before the moving mirror(s).

Interferometer, Lamellar Grating — interferometer in which the beam is reflected from two interleaved mirrors, one of which is stationary while the other is movable. This type of interferometer is generally used only for far infrared spectrometry.

Interferometer, Michelson — interferometer in which an approximately collimated beam of radiant energy is divided into two paths by a beamsplitter; one beam is reflected from a movable mirror and the other from a stationary mirror, and they are then recombined at the beamsplitter.

Interferometer, Rapid-Scanning — interferometer in which the retardation is varied rapidly enough that the modulation frequencies in the interferogram are sufficiently high that the interferogram signal can be amplified directly without additional modulation by an external chopper.

Interferometer, Refractively Scanned — interferometer in which the retardation between two beams is generated by the movement of a wedged optical element.

Interferometer, Slow-Scanning — interferometer in which the retardation is continuously varied, but so slowly that an external chopper is needed to modulate the beam at a frequency which is high enough for a-c signal amplification.

Interferometer, Stepped-Scanning — interferometer in which the movable element is held stationary for the length of time required for signal integration and digitization of each sample point, and then translated to the next sample point.

Internal Reflection Attachment, IRA — the transfer optical system which supports the IRE, directs the energy of the radiant beam into the IRE, and then redirects the energy into the spectrometer or onto the detector. The IRA may be part of an internal reflection spectrometer or it may be placed into the sampling space of a spectrometer.

Internal Reflection Element (IRE) — the transparent optical element used in internal reflection spectroscopy for establishing the conditions necessary to obtain the internal reflection spectra of materials.

Radiant power is propagated through it by means of internal reflection. The sample material is placed in contact with the reflecting surface or it may be the reflecting surface itself. If only a single reflection takes place from the internal reflection element, the element is said to be a single reflection element; if more than one reflection takes place, the element is said to be a multiple reflection element. When the element has a recognized shape it is identified according to each shape; for example, internal reflection prism, internal reflection hemicylinder, internal reflection plate, internal reflection rod, internal reflection fiber, etc.

Internal Reflection Spectroscopy (IRS) — the technique of recording optical spectra by placing a sample material in contact with a transparent medium of greater refractive index and measuring the reflectance (single or multiple) from the interface, generally at angles of incidence greater than the critical angle.

Intersystem Crossing — a transition between electronic states that differ in total spin quantum number (multiplicity). Current experimental evidence indicates this process is nonradiative.

Isoabsorptive Point — a wavelength at which the absorptivities of two or more substances are equal.

Isosbestic Point — the wavelength at which the absorptivities of two substances, one of which can be converted into the other, are equal.

Linear Dispersion — the derivative, $dx/d\lambda$, where x is the distance along the spectrum, in the plane of the exit slit, and λ is the wavelength.

Lock Signal (NMR) — the NMR signal used to control the field-frequency ratio of the spectrometer. It may or may not be the same as the reference signal.

Luminescence — the emission of radiant energy during a transition from an excited electronic state of an atom, molecule, or ion to a lower electronic state.

Modulation Frequency, f_v — the frequency, in Hz, at which radiant energy of a given wavenumber is modulated by a rapid-scanning interferometer. This is given by the product of the wavenumber (cm^{-1}) and the rate of change of retardation ($cm.s^{-1}$). An alternate symbol is f_o.

Molar Absorptivity, ϵ — see *Absorptivity, Molar*.

Monochromator — a device or instrument that, with an appropriate energy source, may be used to provide a continuous calibrated series of electromagnetic energy bands of determinable wavelength or frequency range.

Neutral Filter — see *Filter, Neutral.*

NMR Absorption Band; NMR Band — a region of the spectrum in which a detectable signal exists and passes through one or more maxima.

NMR Absorption Line — a single transition or a set of degenerate transitions is referred to as a line.

NMR Apparatus; NMR Equipment — an instrument comprising a magnet, radio-frequency oscillator, sample holder, and a detector that is capable of producing an electrical signal suitable for display on a recorder or an oscilloscope, or which is suitable for input to a computer.

Nuclear Magnetic Resonance (NMR) Spectroscopy — that form of spectroscopy concerned with radio-frequency-induced transitions between magnetic energy levels of atomic nuclei.

Nyquist Frequency — modulation frequency or wavenumber above which aliasing occurs. The Nyquist frequency is one half of the sampling frequency.

Optical Path Difference — see *Retardation.*

Optical Retardation — see *Retardation.*

Phase Correction — the operation in which the effects of an asymmetrical or chirped interferogram are corrected to eliminate instrumental phase contributions.

Phase Modulation — modulation produced by rapid oscillation of one mirror of a scanning interferometer through an amplitude which is smaller than the shortest wavelength in the spectrum to produce an interferogram which is, to a good approximation, the first derivative of the conventional interferogram.

Phosphorescence — the emission of radiant energy from an atom, molecule, or ion resulting from absorption of a photon and a subsequent transition to the ground state with a change in total spin quantum number (see also *Intersystem Crossing*). The initial state of the transition is usually a triplet state. The average time interval between absorption and phosphorescence is usually greater than 10^{-6} s.

Photometer — a device so designed that it furnishes the ratio, or a function of the ratio, of the radiant power of two electromagnetic beams. These two beams may be separated in time, space, or both.

Photometric Linearity — the ability of a photometric system to yield a linear relationship between the radiant power incident on its detector and some measurable quantity provided by the system. In the case of a simple detector-amplifier combination, the relationship is a direct proportionality between incident radiant power and the deflection of a meter needle or recorder pen.

Pulse Fourier Transform NMR — a form of NMR in which the sample is irradiated with one or more pulse sequences of radio-frequency power spaced at uniform time intervals, and the averaged free induction decay following the pulse sequences is converted to a frequency domain spectrum by a Fourier transformation.

Radiant Energy — energy transmitted as electromagnetic waves.

Radiant Power, P — the rate at which energy is transported in a beam of radiant energy.

Raman Line (band) — a line (band) that is part of a Raman spectrum.

Raman Shift — the displacement in wavenumber of a Raman line (band) from the wavenumber of the incident monochromatic beam. Raman shifts are usually expressed in units of cm^{-1}. They correspond to differences between molecular vibrational, rotational, or electronic energy levels.

Raman Spectrum — the spectrum of the modified frequencies resulting from inelastic scattering when matter is irradiated by a monochromatic beam of radiant energy.

Raman spectra normally consist of lines or bands at frequencies higher and lower than that of the incident monochromatic beam.

Reference Compound (NMR) — a selected material to whose signal the spectrum of a sample may be referred for the measurement of chemical shift (see also *Chemical Shift*).

Reference Material — a material of definite composition that closely resembles in chemical and physical nature the material with which an analyst expects to deal, and that is used for calibration or standardization.

Reflectance, R — the ratio of the radiant power reflected by the sample to the radiant power incident on the sample.

Refractive Index, n — the phase velocity of radiant power in a vacuum divided by the phase velocity of the same radiant power in a specified medium. When one medium is a vacuum, n is the ratio of the sine of the angle of incidence to the sine of the angle of refraction.

Retardation, δ — optical path difference between two beams in an interferometer; also known as "optical path difference" or "optical retardation." The recommended unit for retardation is cm. An alternate symbol is x.

Retardation, Maximum, Δ — the greatest retardation generated by an interferometer in a given scan. The nominal resolution of the spectrum is $1/\Delta$ cm^{-1}. An alternate symbol is X.

Sample Pathlength, b — *in a spectrophotometer*, the distance, measured in the direction of propagation of the beam of radiant energy, between the surface of the specimen on which the radiant energy is incident and the surface of the specimen from which it is emergent. The recommended unit for "sample pathlength" is centimetres. This distance does not include the thickness of the walls of any absorption cell in which the specimen is contained. In strict usage, a more appropriate term would be "specimen pathlength."

Sampling — see *Digitization.*

Sampling Frequency — number of interferogram data points digitized per second in a single scan.

Sampling Interval — difference in retardation between successive sample points in an interferogram.

Scattering, 90° (or 180°) — scattering which is observed at an angle of 90° (or 180°) to the direction of the incident beam. These are the usual scattering angles for Raman spectroscopy.

Sequential Excitation NMR; Continuous Wave (CW) NMR — a form of high-resolution NMR in which nuclei of different field-frequency ratio at resonance are successively excited by sweeping the magnetic field or the radio frequency.

Signal-to-Noise Ratio, S/N — the ratio of the signal, S, to the noise, N, as indicated by the instrumental read-out indicator. Noise as used here is the random variation of signal with time. The recommended measure of noise is the maximum peak-to-peak excursion of the indicator averaged over a series of five successive intervals, each of duration ten times the integration period. In some instruments signal-to-noise ratio varies with the signal.

Single-Pass Internal Reflection Element — *in internal reflection spectroscopy*, an internal reflection element in which the radiant power traverses the length of the element only once; that is, the radiant power enters at one end of the optical element and leaves via the other end.

Singlet State — an electronic state with a total spin quantum number of zero.

Specimen Pathlength — see *Sample Pathlength.*

Spectral Bandwidth, $\Delta\lambda$ — the wavelength interval, $\Delta\lambda$, or wavenumber interval $\lambda\nu$, of radiant energy leaving the exit slit of a monochromator measured at half the peak detected radiant power. The spectral bandwidth is commonly used as a measure of resolution. For further clarification, the conditions for measurement of spectral bandwidth should be given. The term "practical spectral bandwidth," $(\Delta\lambda)^\pi S/N$, is the spectral bandwidth applicable to an instrument operated at a given integration period, π, and a given signal-to-noise ratio, S/N, measured at or near 100% on a transmittance scale. The term "limiting spectral bandwidth," $(\Delta\lambda)_L$, is the minimum spectral bandwidth achievable under optimum experimental conditions. The term "theoretical spectral bandwidth," $(\Delta\lambda)_o$, is the computed spectral bandwidth. The term, $(\Delta\lambda)_o$ should be used sparingly and only when all the factors in the computation of spectral bandwidth are given.

Spectral Position — the effective wavelength or wavenumber of an essentially monochromatic beam of radiant energy.

Spectral Resolution, R — the ratio $\lambda/\Delta\lambda$ where λ is the wavelength of radiant energy being examined and $\Delta\lambda$ is the spectral bandwidth expressed in wavelength units; or, alternatively, the ratio $\bar{\nu}/\Delta\bar{\nu}$ where $\bar{\nu}$ is the wavenumber of radiant energy being examined and $\Delta\bar{\nu}$ is the spectral bandwidth expressed in wavenumber units.

Spectral Slit Width — the mechanical width of the exit slit, divided by the linear dispersion in the exit slit plane.

Spectrograph — an instrument with one slit that uses photography to obtain a record of a spectral range simultaneously. The radiant power passing through the optical system is integrated over time, and the quantity recorded is a function of radiant energy.

Spectrometer — an instrument with an entrance slit and one or more exit slits, with which measurements are made either by scanning the spectral range, point by point, or by simultaneous measurements at several spectral positions. The quantity measured is a function of radiant power.

Spectrophotometer — a spectrometer with associated equipment, so designed that it furnishes the ratio, or a function of the ratio, of the radiant power of two beams as a function of spectral position. The two beams may be separated in time, space, or both.

Spectrum, Internal Reflection — the spectrum obtained by the technique of internal reflection spectroscopy. Depending on the angle of incidence, the spectrum recorded may qualitatively resemble that obtained by conventional transmission measurements, may resemble the mirror image of the dispersion in the index of refraction, or may resemble some composite of the two.

Spin-Spin Coupling Constant (NMR), J — a measure, expressed in hertz (Hz), of the indirect spin-spin interaction of different magnetic nuclei in a given molecule. The notation $^nJ_{AB}$ is used to represent a coupling over n bonds between nuclei A and B. When it is necessary to specify a particular isotope, a modified notation may be used, such as $^3J(^{15}NH)$.

Standard Reference Material — a reference material, the composition or properties of which are certified by a recognized standardizing agency or group. The equivalent ISO term is "certified reference material."

Stray Radiant Energy — all radiant energy that reaches the detector at wavelengths that do not correspond to the spectral position under consideration.

Stray Radiant Power, P_s — the total detected radiant power outside a specified wavelength (wave number) interval each side of the center of the spectral band passed by the monochromator under stated conditions for wavelength (wave number), slit dimensions, light source, and detector.

Stray Radiant Power Ratio, P_s/P_t — the ratio of stray radiant power to the total detected radiant power. $P_t = P_d + P_s$ where P_d is the power detected within the specified wavelength (wavenumber) interval each side of the center of the spectral band passed by the monochromator.

Stokes Line — a Raman line that has a frequency lower than that of the incident monochromatic beam.

Throughput — the vector product of the area and solid angle of a beam at its focus and the square of the refractive index of the medium in which the beam is focused.

Transmittance, T — the ratio of radiant power transmitted by the sample to the radiant power incident on the sample. In practice the sample is often a liquid or a gas contained in an absorption cell. In this case, the observed transmittance is the ratio of the radiant power transmitted by the sample in its cell to the radiant power transmitted by some clearly specified reference material in its cell, when both are measured under the same instrument conditions such as spectral position and slit width. In the case of solids not contained in a cell, the radiant power transmitted by the sample is also measured relative to that transmitted by a clearly specified reference material. The observed transmittance is seldom equal to the true transmittance.

Triplet State — an electronic state with a total spin quantum number of one.

Ultraviolet — pertaining to the region of the electromagnetic spectrum from approximately 10 to 380 nm. The term ultraviolet without further qualification usually refers to the region from 200 to 380 nm.

Variable-Angle Internal Reflection Element — an internal reflection element which can be operated over a range of angles of incidence.

Visible — pertaining to radiant energy in the electromagnetic spectral range visible to the normal human eye (approximately 380 to 780 nm).

Wavelength, λ — the distance, measured along the line of propagation, between two points that are in phase on adjacent waves. The recommended unit of wavelength in the infrared region of the electromagnetic spectrum is the micrometre. The recommended unit in the ultraviolet and visible region of the electromagnetic spectrum is the nanometre.

Wavenumber, ν — the number of waves per unit length. The usual unit of wavenumber is the reciprocal metre, m^{-1}. In terms of this unit the wavenumber is the reciprocal of the wavelength, λ, when λ is expressed in metres.

Zero-Filling — addition of zero-valued points to the end of a measured interferogram. The result of performing the FT of a zero-filled interferogram is to produce correctly interpolated points in the computed spectrum.

Based on ASTM Method E131–84.
Subcommittee 9
J.N. HARMAN, *Chairman*

12. Photometric Methods for Chemical Analysis*

12.1 INTRODUCTION.

12.1.1 This section covers general recommendations for photoelectric photometers and for photometric practice prescribed in ISC photometric methods for chemical analysis. A summary of the fundamental theory and practice of photoelectric photometry is given. No attempt has been made, however, to include in this section a description of every apparatus or to represent recommendations on every detail of practice in photometric methods of chemical analysis.

12.1.2 The inclusion of the following paragraph, or a suitable equivalent, in any ISC photometric method shall constitute due notification that the photometers and photometric practice prescribed in that method are subject to the recommendations set forth in this section.

"Photometers and Photometric Practice — Photometers and photometric practice prescribed in this method shall conform to the recommendations presented in **Section 12** of Part I of the ISC Manual."

12.2 DEFINITIONS AND SYMBOLS.

12.2.1 For definitions of terms relat-

*Adapted, with permission, from the *Annual Book of ASTM Standards*, © American Society for Testing and Materials, 1916 Race St., Philadelphia, PA 19103. ASTM is not responsible for any modifications made by ISC.

ing to absorption spectroscopy, refer to **Section 11, Terms and Symbols Relating to Molecular Spectroscopy.**

12.2.2 *Concentration Range.* The recommended concentration range shall be designated on the basis of the optical path of the cell, in centimeters, and the final volume of solution as recommended in a photometric procedure. In general, the concentration range in any photometric method shall be specified as that which will produce transmittance readings between 90 and 15% under the conditions of the method.

12.2.3 *Photometric Reading.* The term "photometric reading" refers to the scale reading of the photometric instrument being used. Available instruments have scales calibrated in transmittance, T, (6) or absorbance, A, (8), or even arbitrary units proportional to transmittance or absorbance.

12.2.4 *Reference Solution.* Photometric readings consist of a comparison of the intensities of the radiant energy transmitted by the absorbing solution and the radiant energy transmitted by the solvent. Any solution to which the transmittance of the absorbing solution of the substance being measured is compared shall be known as the reference solution.

12.2.5 *Initial Setting.* The initial setting is the photometric reading (usually 100 on the transmittance scale or zero on the absorbance scale) to which the photometer is adjusted with the reference solution in the absorption cell. The scale will then read directly in percentage transmittance or in absorbance.

12.2.6 *Background Absorption.* Any absorption in the solution due to the presence of absorbing ions, molecules, or complexes of elements other than that being determined is called background absorption.

12.2.7 *Reagent Blank.* The reagent blank determination is a determination of the amount of the element sought that is present as an impurity in the reagents used.

12.3 THEORY.

12.3.1 Photoelectric photometry is based on Bouguer's and Beer's (or the Lambert-Beer) laws which are combined in the following expression:

$$P = P_0 10^{-abc}$$

where P = transmitted radiant power.

P$_0$ = incident radiant power, or a quantity proportional to it, as measured with pure solvent in the beam.

a = absorptivity, a constant characteristic of the solution and the frequency of the incident radiant energy.

b = internal cell length (usually in centimeters) of the column of absorbing material, and

c = concentration of the absorbing substance in grams per liter.

12.3.2 Transmittance, T, and absorbance, A, have the following values:

$$T = \frac{P}{P_0}$$

$$A = \log_{10} \frac{1}{T} = \log_{10} \frac{P_0}{P}$$

where P and P$_0$ have the values given in Paragraph 12.3.1.

12.3.3 From the transposed form of the Bouguer-Beer equation, $A = abc$, it is evident that at constant b, a plot of A versus c gives a straight line if Beer's law is followed. This line will pass through the origin if the usual practice of cancelling out solvent reflections and absorption and other blanks is employed.

12.3.4 In photoelectric photometry it is customary to make indirect comparisons with solutions of known concentration by means of calibration curves or charts. When Beer's law is obeyed and when a satisfactory photometer is employed, it is possible to dispense with the curve or chart. Thus from the transposed form of the Bouguer-Beer law, $c = A/ab$, it is evident that once a has been determined for any system, c can be obtained, since b is known and A can be measured.

12.3.5 The value for a can be obtained from the equation $a = A/cb$ by substituting the measured value of A for a given b and c. Theoretically, in the determination of a for an absorbing system, a single measurement at a given wavelength on a solution of known concentration will suffice. Actually, however, it is safer to use the average value obtained with three or more concentrations, covering the range over which the determinations are likely to be made and making several readings at each concentration. The validity of the Bouguer-Beer law for a particular system can be tested by showing that a remains constant when b and c are changed.

12.4 APPARATUS.

12.4.1 *General Requirements for Photometers.*

A photoelectric photometer consists essentially of the following†:

a. An illuminant (radiant energy source);

b. a device for selecting relatively monochromatic radiant energy (consisting of a diffraction grating or a prism with selection slit, or a filter);

c. one or more absorption cells to hold the sample, standards, reagent blank or reference solution;

d. an arrangement for photoelectric measurement of the intensity of the transmitted radiant energy, consisting of one or more photocells and suitable devices for measuring current or potential.

Precision photometers that employ monochromators capable of supplying radiant energy of high purity at any chosen wavelength within their range are usually referred to as spectrophotometers. Photometers employing filters are known as filter photometers or abridged spectrophotometers, and usually isolate relatively broad bands of radiant energy. In most cases the absorption peak of the compound being measured is relatively broad, and sufficient accuracy can be obtained using a fairly broad band (10 to 75 nm) of radiant energy for the measurement. In other cases the absorption peaks are narrow, and radiant energy of high purity (1 to 10 nm) is required. This applies particularly if accu-

---

†The choice of an instrument may naturally be based on price considerations, since there is no point in using a more elaborate (and, incidentally, more expensive) photometer than is necessary. In addition to satisfactory performance from the purely physical standpoint, the instrument should be compact, be rugged enough to stand routine use, and not require too much manipulation. The scales should be easily read, and the absorption cells should be easily removed and replaced, as the cleaning, refilling, and placing of the cells in the instrument consume a major portion of the time required. It is advantageous to have a photometer that permits the use of cells of different depth.

rate values are to be obtained in those systems of measurement based on the additive nature of absorbance values.

12.4.2 Types of Photometers.

a. Single-Photocell Photometers. In most single-photocell photometers, the radiant energy passes from the monochromator or filter through the reference solution to a photocell. Many commercially available single cell photometers employ stabilized optics. The photocurrent is measured by a galvanometer or a microammeter and its magnitude is a measure of the incident radiant power, P_0. An identical absorption cell containing the solution of the absorbing component is now substituted for the cell containing the reference solution and the power of the transmitted radiant energy, P, is measured. The ratio of the current corresponding to P to that of P_0 gives the transmittance, T, of the absorbing solution, provided the illuminant and photocell are constant during the interval in which the two photocurrents are measured. It is customary to adjust the photocell output so that the galvanometer or microammeter reads 100 on the percentage scale or zero on the logarithmic scale when the incident radiant power is P_0, in order that the scale will read directly in percentage transmittance or in absorbance. This adjustment is usually made in one of three ways. In the first method, the position of the cross-hair or pointer is adjusted electrically by means of a resistance in the photocell-galvanometer circuit. In the second method, adjustment is made with the aid of a rheostat in the source circuit. Kortum (5) has pointed out that on theoretical grounds this method of control is faulty, since the change in voltage applied to the lamp not only changes the radiant energy emitted but also alters its chromaticity. Actually, however, instruments employing this principle are giving good service in industry, so the errors involved evidently are not too great. The third method of adjustment is to control the quantity of radiant energy striking the photocell with the aid of a diaphragm somewhere in the path of radiant energy.

b. Two-Photocell Photometers. In order to eliminate the effect of fluctuation of the source, a great many types of two-photocell photometers have been proposed. Most of these are good, but some have poorly designed circuits and do not accomplish the purpose for which they are designed. Many improvements have occurred in the area of two-cell photometry over the recent past few years. The most significant have been: 1) the incorporation of light interference devices such as optical interference filters and spectrally selective mirrors; 2) incorporation of a great variety of solid state detectors, such as the thermopile and pyroelectric detectors, which offer the advantages of high sensitivity and low cost; 3) incorporation of signal processing electronics into the detector package itself and, 4) the utilization of a dual wavelength approach to minimize source-related noise by subtracting out common mode sources. Following is a brief description of two types of two-photocell photometers that are in general use and that have been found satisfactory:

(1) In the first type of two-photocell photometer, beams of radiant energy from the same source are passed through the reference solution and the sample solution and are focused on their respective photocells. Prior to insertion of the sample, the reference solution is placed in both absorption cells, and the photocells are balanced with the aid of a potentiometric bridge circuit. The reference solution and sample are then inserted and the balance reestablished by manipulation of the potentiometer until the galvanometer again reads zero. By choosing suitable resistances and by using a graduated slide wire, the scale of the latter can be made to read directly in transmittance. It is important that both photocells show linear response, and that they have identical radiation sensitivity if the light is not monochromatic.

(2) The second type of two-photocell photometer is similar to the first, except that part of the radiant energy from the source is passed through an absorption cell to the first photocell; the remainder is impinged on the second photocell without, however, passing through an absorption cell. Adjustment of the calibrated slide wire to read 100 on the percentage scale, with the reference solution in the cell, is accomplished by rotating the second photocell. The reference solution is then replaced by

the sample and the slide wire is turned until the galvanometer again reads zero.

12.4.3 *Radiation Source.*

a. In most of the commercially available photoelectric photometers the illuminant is an incandescent lamp with a tungsten filament. This type of illuminant is not ideal for all work. For example, when an analysis calls for the use of radiant energy of wavelengths below 400 nm, it is necessary to maintain the filament at as high a temperature as possible in order to obtain sufficient radiant energy to ensure the necessary sensitivity for the measurements. This is especially true when operating with a photovoltaic cell, for the response of the latter falls off quickly in the near ultraviolet. The use of high-temperature filament sources may lead to serious errors in photometric work if adequate ventilation is not provided in the photometer in order to dissipate the heat. Another important source of error results from the change of the shape of the energy distribution curve with age. As a lamp is used, tungsten will be vaporized and deposited on the walls. As this condensation proceeds, there is a decrease in the radiation power emitted and, in some instances, a change in the composition of the radiant energy. This change is especially noticeable when working in the near ultraviolet range and will lead to error (unless frequent standardization is resorted to) in all except those cases where essentially monochromatic radiant energy is used. These errors have been successfully overcome in commercially available photometers. One instrument has been so designed that a very low-current lamp (or the order of 200 ma) is employed as the source. This provides for long lamp life, freedom from line fluctuation (since a storage battery is employed), stability of energy distribution, reproducibility, and low-cost operation. In addition, the stable illuminant permits operation for long periods of time without need for restandardization against known solutions. Significant improvements have been made in the technology of light sources by the introduction of deuterium sources and tungsten/halogen lamps.

b. In most of the commercially available photometers where relatively high-wattage lamps are used, the power is derived from the ordinary electric mains with the aid of a constant-voltage transformer. Where the line voltages vary markedly, it is necessary to resort to the use of batteries that are under continuous charge, or to a very good voltage regulator.

12.4.4 *Filters and Monochromators.*

a. Filters. Spectrophotometric methods call for the isolation of more or less narrow wavebands of radiant energy. Relatively inexpensive instruments employing filters are adequate for a large proportion of the methods used, since most of the absorbing systems show broad absorption bands. In general, filters are designed to isolate as narrow a band of the spectrum as possible. Actually, it is usually necessary, especially when the filters are to be used in conjunction with an instrument employing photovoltaic cells, to sacrifice spectral purity in order to obtain sufficient sensitivity for measurement with a rugged galvanometer or a microammeter. Glass filters are most often used because of their stability to light and heat, but gelatin filters and even aqueous solutions are sometimes used.

b. Monochromators. Two types of monochromators are in common use: the prism and the diffraction grating. Modern monochromators are predominantly of the diffraction grating type. Prisms have the disadvantage of exhibiting a dependence of dispersion upon wavelength. On the other hand, the elimination of stray radiation energy is less difficult when a prism is used. In a well-designed monochromator, stray radiant energy resulting from reflections from optical and mechanical members is reduced to a minimum, but some radiant energy, caused by nonspecular scatterings by the optical elements, will remain. This unwanted radiant energy can be reduced through the use of a second monochromator or a filter in combination with a monochromator. Unfortunately, any process of monochromatization is accompanied by a reduction of the radiant power, and the more complex the monochromator the greater the burden upon the measuring system.

12.4.5 *Absorption Cells.* Some photometers provide for the use of several sizes and shapes of absorption cells. Others are designed for a single type of cell. It is ad-

vantageous to have a photometer that permits the use of cells of different path lengths. In some single-photocell instruments there is only one receptable for the cell; in others (and this is especially desirable in those instruments where the illuminant is unstable) a sliding carriage is provided so that two cells can be interchangeably inserted in the beam of radiant energy coming from the monochromator.

12.4.6 *Photocells.* In photoelectric photometry, the measurement of radiant energy is usually accomplished with the aid of either photoemission or photovoltaic cells.

a. The spectral response of a photoemission cell will depend upon the alkali metal employed and upon its treatment during manufacture. The spectral response of a photovoltaic (or barrier-layer) cell is crudely similar to that of the human eye, except that it extends from about 300 to 700 nm. In general, neither the voltage nor the current response of a photovoltaic cell is a linear function of the flux incident on the cell, but the current response is more nearly linear than the voltage response. Thus, current-measuring devices should be used with photovoltaic-cell photometers. The degree to which the response of these cells departs from linearity depends on the individual cell, its temperature, its level of illumination, the geometric distribution of this illumination on its face, and the resistance of the current measuring circuit.

b. In order for a photocell to be useful it must exhibit a constancy of current with time of exposure. Most commercial alkali cells in use at the present time produce a constant current after an exposure of a few minutes. The photovoltaic cells, on the other hand, frequently exhibit enough reversible fatigue to interfere with their use. The measures which improve linearity of response also tend to reduce fatigue. With most commercial photoelectric photometers, the errors due to reversible fatigue are usually less than 1%.

12.4.7 *Current Measuring Devices.*

a. The usual types of photoelectric photometers employ photovoltaic cells in conjunction with a microammeter or a moderately high-sensitivity galvanometer as may

be appropriate for the illumination level employed. Modern electrometer amplifiers have been almost universally applied to the production of new instrumentation for both photoelectric and photoemission devices. The scales for the galvanometers are sometimes designed to permit reading of absorbance values but more often yield only the more conveniently read T or percentage T values. Some photometers are designed so that the current is measured potentiometrically, using the galvanometer as a null instrument. It is stated that the error due to nonlinearity of the galvanometer under load is eliminated. Actually, however, this error is usually small and, moreover, many photometers provide individual calibration of the galvanometer.

b. Where photoemission cells are used, current amplification is usually resorted to before the galvanometer or meter is used.

12.5 PHOTOMETRIC PRACTICE.

12.5.1 *Principle of Method.* Photometric methods are generally based on the measurement of the transmittance or absorbance of a solution of an absorbing salt, compound, or reaction product of the substance to be determined. It is usually desirable to perform a rather complete spectrophotometric investigation of the reaction before attempting to employ it in quantitative analysis. The investigation should include a study of the following:

a. The specificity of any reagent employed to produce absorption;

b. the validity of Beer's law;

c. the effect of salts, solvent, pH, temperature, concentration of reagents, and the order of adding the reagents;

d. the time required for absorption development and the stability of the absorption;

e. the absorption curve of the reagent and the absorbing substances; and

f. the optimum concentration range for quantitative analysis.

In photometry it is necessary to decide upon the spectral region to be used in the determination. In general it is desirable to use a filter or monochromator setting such that the isolated spectral portion is in the region of the absorption maximum. In the ideal case (and, fortunately, this is true of most of the absorbing systems encountered in quantitative inorganic analysis) the absorption maximum is quite broad and flat so that deviations from Beer's law resulting from the use of relatively heterogeneous radiant energy will be negligible. Sometimes it will not be possible or desirable to work at the point of maximum absorption. In those cases where there is interference from other absorbing substances in the solution or where the absorption maximum is sharp, it is sometimes possible to find another flat portion of the curve where the measurements will be free from interference. When no flat portion free from interference can be found, it may be necessary to work on a steep portion of the curve. In this case Beer's law will not hold unless the isolated spectral band is quite narrow. There is no real objection to operation on a steep part of an absorption curve, provided the usual standard calibration curve is obtained, except that with most instruments the reproducibility of the absorbance readings will be poor unless a fixed wavelength setting of the monochromator is maintained or unless filters are used. A small change in any of a large number of conditions will decrease the accuracy by a larger amount than when observations are made where the change in absorption is more gradual. With the advent of microprocessor-based instrumentation and the mathematical manipulations that are possible employing this technology, mathematical methods to nullify the effects of cross interferences are often applied to spectral data before presentation to the operator.

In most photometric work it is best to prepare a calibration curve or chart rather than to rely on the assumption of linearity, since it is not at all uncommon to obtain curved lines in the calibration of solutions that are known to obey Beer's law. The two most common causes of this are the presence of stray radiant energy, and the use of filters or monochromators that isolate too broad a spectral region for the analysis. Nonlinearity will generally be more pronounced the greater the heterogeneity of the radiant energy employed. Thus, one is more likely to obtain linearity with a spectrophotometer having a prism or grating with a high resolving power than with a photometer employing rather broad-banded filters. On the other hand, high resolving power or a narrow slit width is no guarantee of linearity unless stray radiant energy is rigorously excluded. When nonlinearity is encountered at one wavelength setting, it is sometimes possible to eliminate it by changing to another wavelength (where stray radiant energy is negligible) even though the latter is less favorable from the standpoint of flatness and sensitivity. A filter photometer employing a good filter will sometimes yield a more nearly straight calibration curve than can be obtained with certain spectrophotometers. This is especially true in the violet and near ultraviolet regions where stray radiant energy is likely to be encountered in grating monochromators.

12.5.2 *Concentration Range.* The concentration of the species being determined should be adjusted, preferably so that the transmittance readings fall within the range that yields the minimum error for the amount of constituent being determined. There are several sources of error in photometric analysis, including instrumental and sample manipulative errors, which must be considered when selecting the optimum transmission region. These sources of error have been discussed in detail by Rothman and co-workers (7). These workers suggest that the optimum absorbance range for a photometric analysis be determined by preparing a working curve with enough measurements to get standard deviations on each absorbance value. However, for practical purposes, a simple test using a Ringbom-type plot may be useful. The Ringbom method has been discussed by Ayres (1).

The Ringbom test for optimum concentration range for minimum photometric error involves the plotting of experimental calibration data. A plot of the appropriate Ringbom parameter versus logarithm of concentration should exhibit a point of inflection where the relative error in concentration will be a minimum. If this curve is fairly straight over an interval surrounding the point of inflection, all values corresponding to that interval will be approximately as good as the best. The appropriate Ringbom parameter to be used will depend on the relationship between the error in transmittance measurement and transmittance for the specific instrument employed in the analysis. Three such relationships proposed for spectrophotometric instruments (3) are tabulated in Table 12:I. The corresponding Ringbom parameter to be plotted against logarithm of concentration is also given. The parameter to be used depends on the dominant error characteristic of the specific instrument involved in the analysis. The extended Ringbom method cannot determine this error characteristic; it does, however, provide a simple test for determining the optimum analytical range for any assumed dependence of transmittance error on transmittance.

If the dominant error source for an instrument is not known, the following guidelines are suggested. For any noise-limited instrument with a photovoltaic or thermal detector, error in intensity is independent of intensity and the appropriate Ringbom parameter is transmittance, as in the original Ringbom method. The optimum transmittance for this case will typically be in the 20 to 60% range. For modern instruments employing photomultiplier detectors and advanced read-out systems and operating under noise-limited conditions, the $T^{1/2}$ parameter should be applicable. In this case the optimum transmittance is typically found to be in the 5 to 40% range. The log T parameter may be appropriate for some specific instrument or sample systems, or both, but its use cannot be generalized. The effect of plotting log T is to move the optimum range to even lower transmission.

12.5.3 *Stability of Absorption.* The absorbing compounds on which photometric methods are based vary greatly in stability. In some instances the absorption is stable indefinitely, but in the majority of methods the absorption either increases or decreases on standing. In some cases a completely (or relatively) stable absorption is obtained on standing; in other cases it is stable for a time and then changes; in still other cases it never reaches a stable intensity. In all photometric work it is desirable to measure both standards and samples during the time interval of maximum stability of the absorption, provided that this occurs reasonably soon after development of the absorption. In those cases where the absorption changes continuously, it is necessary to control rigidly the time of standing.

12.5.4 *Interfering Elements.* In photometry there are two basic types of methods to be considered: one type in which the photometric measurement is made without previous separations, and a second type in which the element to be determined is partially or completely isolated from the other elements in the sample.

a. In the first type of method it usually happens that one or more of the elements or reagents present may cause interference with an absorbing reaction. Such interference may be due to the presence of a colored substance, to a suppressive or enhancing effect on the absorption of the substance being measured, or to the destruction or formation of a complex with the reagents thus preventing formation of the absorbing substance. The most important methods, not involving separations, used to eliminate such interferences are as follows:

(1) The use of standards whose composition matches the sample being analyzed as closely as possible;

(2) performing the measurement at a

Table 12:I Relationship Between Error in Transmittance (E_T) and Transmittance (T)

Error Relationship	Type of Error	Ringbom Parameter (T = Transmittance)
E_T independent of T.	scale reading errors, dark current drift (noise-limited instruments with photovoltaic or thermocouple detectors)	T
$E_T \propto T^{1/2}$	detector shot noise error (photoemissive detectors)	$T^{1/2}$
$E_T \propto T$	cell and sample preparation errors, wavelength error, source change errors	log T

wavelength where interference is at a minimum;

(3) the use of reagents that form complexes with the interfering elements.

The question as to how much interference can be tolerated in a given method will depend upon many factors, including the degree of accuracy required in the determination. In general, it is desirable to avoid using a method where the error to be "blanked out" is appreciable. The methods involving no separations suffer from the distinct disadvantage that the analyst must often know the matrix of the sample to be analyzed and, what is more important, must be able to prepare a standard to duplicate it. This is not always easy to do, for it often happens, especially in the determination of traces, that the so-called pure substances used for preparing the synthetic standards contain more of the element to be measured than does the sample being analyzed.

b. In the second type of method, the separations may involve removal of one or more interfering elements or may provide for complete isolation of the element in question before its photometric estimation. In this type of method there is usually no attempt made to adjust the matrix of the standard solution to fit that of the sample being analyzed, since presumably all extraneous interference has been removed. The standard in this case is a standard solution of the element in question. In any photometric determination it is desirable to keep the manipulation and separations as simple as possible, for the greater the number of reagents and the more manipulation involved the greater the blank and hence the more chance of error. Very useful tabulations have been compiled of methods used to eliminate interference in photometric analysis **(2,8)**.

12.5.5 *Concentrations of Standard Solutions.* The concentrations of standard solutions shall be expressed in milligrams or micrograms of the element per milliliter of solution, or other units of concentration appropriate to the task.

12.5.6 *Cell Corrections.* To correct for differences in cell paths in photometric measurements using instruments provided with multiple absorption cells, cell corrections should be determined according to the following procedure: Transfer suitable portions of the reference solution prepared in a specific method to two absorption cells (reference and "test") of approximately identical light paths. Using the reference cell, adjust the photometer to the initial setting using a light band centered at the appropriate wavelength. While maintaining this adjustment, take the absorbance reading of the "test" cell and record as the cell correction. Make certain that a positive absorbance reading is obtained. If it is negative, reverse the positions of the cells. ("Matched" cells frequently show no reading). Subtract this cell correction (as absorbance) from each absorbance value obtained in the specific method. Keep the cells in the same relative positions for all photometric measurements to which the cell correction applies.

12.5.7 *Calibration Curve or Chart.*

a. Linear relation between absorbance and concentration is not always obtained with commercially available photometers, even though the absorbing system is known to obey Beer's law. Because of this, it is evident that the use of calibration curves or charts will be necessary with such photometers. Moreover, it is not safe, with most photometers on the market today, to use calibration curves or charts interchangeably, even though the photometers may be of the same make and model. A separate calibration curve or chart must be prepared for each photometer.

b. The use of a calibration curve or chart in photometric analysis ensures correct measurement of concentration only when the composition of the radiant energy used in the work does not change. In most cases it is necessary to restandardize from time to time to guard against change in the photocell, filters, measuring circuit, and illuminant. In addition to this, in certain photometric methods, the temperature coefficient of the reactions employed to produce absorption is large enough to require winter and summer standard curves or charts.

c. When a calibration curve is used, the usual procedure is to plot the values of A, obtained from a series of standard solutions whose concentrations adequately cover the range of the subsequent determinations, against the respective concentrations, on ordinary graph paper. When the scale of the photometer being used does not read directly in absorbance, it is then convenient to plot concentration, c, against percent transmittance on semilogarithmic paper, using the logarithmic scale for the percent T values. In some cases it is more convenient to prepare a table of c versus A, or percent T values. In plotting, a straight line should be obtained if a good photometer is employed and if the solution obeys Beer's law. If all blanks and interference have been eliminated, the lines should pass through the origin (the point of zero concentration and zero absorbance or 100% transmittance). The use of A in the plotting is advantageous because it is directly proportional to the concentration. On the other hand, while percentage transmittance has the disadvantage of decreasing in magnitude as the concentration increases, it is more convenient to use when the photometer employed does not have a scale calibrated in absorbance, though the resulting plot will not be a straight line.

12.5.8 *Procedure.* Detailed instructions for the procedure to be followed will be found in each of the ISC photometric methods.

12.5.9 *Blanks.* In taking the photometric reading of the absorption in any solution, all the components present that absorb radiant energy in the region of interest must be taken into account. These sources of absorption are:

a. Absorption of the element sought;
b. absorption of the element sought, present as an impurity in the reagents used;
c. background;
d. absorption of all reagents used;
e. absorption produced by reaction of reagents with other elements present;
f. turbidities; and
g. absorption of cells.

These absorptions are additive and all or some will be included in the photometric reading, depending upon the method of preparing the calibration curve and the reference solution. Items **(e)** and **(f)** are interferences and are assumed to be eliminated by preliminary conditioning operations.

Items **(b)** to **(d)** have been loosely designated as "blanks". It is less confusing to restrict the usage of the word "blank" to reagent blank, Item **(b)** in the above list. Item **(c)** is defined in Paragraph 12.2.6 and Item **(d)** is usually taken care of by the "reference solution" (Paragraph 12.2.4). Item **(g)** should not be encountered if cells are properly matched and adequately clean. Cells should never be touched with fingers on the "window" surfaces.

The general case is stated in the preceding paragraphs, and it is desirable that all these factors be considered in the development of a photometric method. However, it is often possible to combine some or all of these factors into the reference solution. Thus, the reference solution may, in some cases, include the reagent blank, the background, and any absorption due to the reagents used. In other cases it may be desirable to measure the reagent blank alone in order that a check may be had on the purity of the reagents. It should be noted, however, that in the case of absorbing systems that do not obey Beer's law, it may be dangerous to use the reagent blank for the reference solution, particularly if the magnitude of the absorption due to the reagent blank becomes appreciable. If the values of the blank are high, it should alert the operator to try to clean up the problem. Looking for small absorption changes on top of a large blank correction does not afford an optimum accuracy or low detectable limit situation.

The requirements for the preparation and measurement or application of these various corrections, both in the preparation of the calibration curve and in the procedure, will be found in each of the ISC photometric methods.

12.6 PRECISION AND ACCURACY.

12.6.1 The primary advantages of photometric methods are those of speed, convenience, and relatively high precision and accuracy in the determination of micro- and semi-micro-quantities of constituents. For the determination of macro-quantities, differential spectrophotometric techniques **(4)** or other analytical techniques are often preferable, since they are generally more accurate when larger quantities are involved. It should be remembered that even under the most favorable circumstances it is difficult to obtain an accuracy better than about 1% of the amount present in most photometric determinations. This does not mean that it is not practicable to analyze macro-samples photometrically. Instrumental methods are now often more repeatable and accurate than wet chemical methods. In evaluating the precision and accuracy of any photometric method, the quality of the apparatus and the chemical procedure involved must be considered.

REFERENCES

1. AYRES G.H. 1949. Evaluation of Accuracy in Photometric Analysis. Anal. Chem. 21:652.
2. BOLZ D.F. 1958. *Colorimetric Determination of Nonmetals.* John Wiley and Sons, Inc., New York, NY.
3. L. CAHN. 1955. Some Observations Regarding Photometric Reproducibility between Ultraviolet Spectrophotometers. J. of the Optical Society of America *45*: 953
4. JONES A.G. 1959. *Analytical Chemistry, Some New Techniques* Academic Press, Inc., New York, NY.
5. KORTUM G. 1937. Photoelectric Spectrophotometry. Angewandte Chemie, 50:193.
6. MELLON M.G. 1950. *Analytical Absorption Spectroscopy.* John Wiley and Sons, Inc., New York, NY.
7. ROTHMAN L.D., S.R. CROUCH AND J.D. INGLE, JR. 1974. Theoretical and Experimental Investigation of Factors Affecting Precision in Molecular Absorption Spectrophotometry. Anal. Chem. 47:1226.
8. SANDELL E.B. 1959. *Colorimetric Determination of Traces of Metals* (3rd ed.). John Wiley and Sons, Inc., New York, NY.

Based on ASTM Method E60-80.
Subcommittee 9
J.N. HARMAN, *Chairman*

13. Ultraviolet Absorption Spectroscopy

13.1 INTRODUCTION.

13.1.1 The property of ultraviolet absorption can be utilized for quantitative determination of a number of materials (1,2,3). It is particularly well suited for the analysis of aromatic compounds but any molecule containing a chromophoric group would be expected to absorb light in the ultraviolet region of the spectrum. The amount of absorbance is directly proportional to the concentration of the material present in solution.

13.2 SAMPLING.

13.2.1 Ultraviolet absorption is a convenient method for air samples which can be collected directly in an appropriate solvent (4,5) or adsorbed on silica gel for subsequent leaching with the solvent. Certain precautions must be observed, however, in sampling and analysis.

13.2.2 All-glass sampling equipment must be used for samples collected in organic liquids. If two samplers are used in series, the connection should also be glass. Samples are readily contaminated by rubber or plastic connections, lubricants, and by rubber or cork stoppers; in essence, contact with any extraneous organic material should be avoided. Although samples collected in aqueous solutions are not so easily contaminated, it is advisable to prevent contact with organic substances. If a dusty atmosphere is being sampled, a prefilter should be used; suspended dust cannot be removed satisfactorily by centrifuging. Dust causes an apparent increase in absorbance due to scattering, particularly in the region below 300 nm.

13.2.3 For transport of liquid samples to the laboratory, the samples should be retained in the sampling vessel. If transfer is required, it should be either to all-glass bottles or to glass bottles with aluminum liners in the caps (1).

13.2.4 Adsorption of vapors on silica gel is a convenient method of collection and such samples are easy to transfer for transport. However, interferences introduced by leaching with solvents for analysis must be anticipated.

13.3 ANALYSIS.

13.3.1 Absorption maxima of specific materials will vary from one instrument to another. Thus each laboratory must determine the peaks on its own instrument. If the maximum is broad, as in the case of acetone, a precise setting is not so important. A precise setting is needed, however, for materials such as benzene and toluene that have sharp maxima; a difference of a nanometer will result in considerable error in the absorbance reading obtained.

13.3.2 Determinations are usually made at the peak wavelengths to utilize the greatest sensitivity, but the minima of the curves can also be used. When calibration curves are drawn, it is advisable to prepare curves for the various maxima and minima. With samples where the concentration is too high to use the peak wavelength, it is frequently possible to use a curve based on another maximum or a minimum without need for dilution of the sample.

13.3.3 In addition to proper wavelength settings, an important factor in analysis is the condition of the silica cuvettes. Not only must they be thoroughly clean but any pair used for analysis should, when filled with solvent blanks, have absorbance readings within ± 0.005 at the wavelength at which they are to be used. If there is a difference greater than ± 0.005, a correction factor will have to be applied to the readings obtained with such cuvettes.

13.3.4 The reference blank should always be prepared from the same bottle of solvent that is used for sampling, as reagent-grade solvents vary widely from one bottle to another.

13.4 PLOTTING OF CURVES.

13.4.1 The curves are plotted as the logarithm of molar absorptivity versus the wavelength in nanometers. The absorptivity and molar absorptivity values are given with each curve.

13.4.2 As indicated previously, the quantitative determination of materials by ultraviolet absorption is possible because the amount of absorbance is directly proportional to the concentration of the substance in solution. According to the Beer-Lambert law:

$$A = abc = (E/M) \, bc$$

where   A = absorbance (spectrophotometer reading)
     a = absorptivity in liters per gram-centimeter
     b = cell path length in centimeters
     c = concentration in grams per liter
     E = molar absorptivity = aM
     M = molecular weight

The terms and symbols used are those proposed by ASTM Committee E-13, (**6**, see section 11 above) and are also the preferred usages of *Analytical Chemistry* (**7**).

13.4.3 By means of the foregoing equation, calibration curves for a particular substance in a designated solvent can be drawn. The values "*a*" and "*b*" are known; values may be assumed for "*A*" and the "*c*" values calculated. When the absorbance values are plotted against the calculated concentrations, a straight line calibration curve will be obtained. With a fluorescent solution the true calibration curve would have a slight bend caused by the effect of the emitted light on the phototube. In the use of these calculated calibration curves, a blank must be employed so that there will be zero absorbance with zero concentration.

13.4.4 As an illustration of what results might be expected from the above procedure, the case of acetone collected in water may be considered. Acetone in water is not fluorescent—hence the curve of absorbance versus concentration will be a straight line. With the value of "*a*" = 0.30 (given on the absorptivity curve) and a 1-cm cell, values for "*A*" at the maximum absorptivity, 265 nm, are selected and each "*c*" is calculated with the following results:

A	c grams per liter
0.100	0.334
0.200	0.667
0.400	1.330
0.600	2.000
0.800	2.660

13.4.5 In the data from which the molar absorptivity curve was drawn, the value of "*a*" was actually 0.304. At an absorbance of 0.600, use of this figure in the equation $A = abc$ yields a concentration of 1.975 g/L instead of 2.000 g found by using the absorptivity value given on the absorptivity curve.

13.4.6 As an example of what might happen in the case of a fluorescent solution, hydroquinone in isopropanol is representative. Although the true curve would be bent, a straight line calibration curve at the maximum of 294 nm can be developed from calculations based on the value given for "*a*", namely, 32.4. From it, an absorbance reading of 0.800 with a 1-cm cell would indicate a concentration of 24.7 µg/mL. The actual concentration, however, determined from standard solutions, would be about 27.4 µg/mL.

13.4.7 Because of variations in spectrophotometers the greatest accuracy requires that standards be run in individual laboratories. This is especially true for fluorescent solutions.

13.5 DETERMINATION OF MIXTURES.

13.5.1 In many cases it is possible to determine two ultraviolet or visible absorbant materials in the same solution **(8)**. Each component contributes to the absorbance at a specific wavelength in this manner:

$$A_0 = A_x + A_y = a_x bc_x + a_y bc_y \quad (1)$$

where A_0 = absorbance observed
  x = one material
  y = second material
  b = cell length, assumed constant

The absorbance of the mixture is observed at two different wavelengths and the absorptivity of each material is determined at the same wavelengths. These data are used to form two simultaneous equations that are solved to give c_x and c_y.

13.5.2 A mixture of benzene and toluene in cyclohexane will serve to illustrate this procedure. Benzene has a maximum of 249 nm with an absorptivity of 2.3; toluene has a maximum at 270 nm with an absorptivity of 3.1. At 270 nm the absorptivity of benzene is 0.11, at 249 nm, that of toluene is 2.6. A sample of a mixture containing 0.100 g/L both of benzene and toluene in cyclohexane was found to have an absorbance of 0.319 at 270 nm and of 0.493 at 249 nm. A 1-cm cell was used.

13.5.3 The absorbance readings and the respective absorptivities are substituted in equation (1) to obtain two simultaneous equations:

$$(\text{at } 270\text{nm}) \; 0.319 = 3.1c_t + 0.11c_b \quad (2)$$

$$(\text{at } 249\text{nm}) \; 0.493 = 2.6c_t + 2.3c_b \quad (3)$$

In order to eliminate one unknown, equation (2) is multiplied by 0.839 to give:

$$0.268 = 2.6c_t + 0.092c_b \quad (4)$$

or

$$2.6c_t = 0.268 - 0.092c_b \quad (5)$$

If (5) is substituted in (3), it yields

$$0.225 = 2.2c_b \quad (6)$$

$$c_b = 0.102 \text{ g/L}$$

This value for c_b is substituted in either (2) or (3) to give $c_t = 0.099$ g/L.

Both concentrations are very close to the theoretical values of 0.100 g/L.

13.6 ADVANTAGES OF ULTRAVIOLET ABSORPTION.

13.6.1 Although there are many difficulties and interferences in carrying out ultraviolet absorption, it is an easy and convenient technique. As such, it is useful as a standard procedure for many routine samples. It is especially valuable in emergency situations when it is necessary to sample for a material for which there has not been time to establish a suitable chemical procedure **(1)**.

REFERENCES

1. HOUGHTON, J.A., G. LEE, M.A. SHOBAKEN AND A. FOX. 1964. Practical Applications of Analysis by the Ultraviolet Absorbance Method. Amer. Ind. Hyg. Assoc. J. 25:381.
2. MAFFETT, P.A., T.F. DOHERTY AND J.L. MONKMAN. 1956. Collection and Determination of Micro Amounts of Benzene or Toluene in Air. Amer. Ind. Hyg. Assoc. Quart. 17:186.
3. HIRT, R.C. AND J. GISCLARD. 1951. Determination of Parathion in Air Samples by Ultraviolet Absorption Spectroscopy. Anal. Chem. 23:185.
4. HOUGHTON, J.A. AND G. LEE. 1960. Data on Ultraviolet Absorption and Fluorescence Emission. Amer. Ind. Hyg. Assoc. J. 21:219.
5. HOUGHTON, J.A. AND G. LEE. 1961. Ultraviolet Spectrophotometric and Fluorescence Data. Amer. Ind. Hyg. Assoc. J. 22:296.
6. DEFINITIONS OF TERMS AND SYMBOLS RELATING TO MOLECULAR SPECTROSCOPY. 1984. E131-84. ASTM Committee E-13. American Society for Testing and Materials. Philadelphia, PA.
7. AMERICAN CHEMICAL SOCIETY. 1963. Spectrophotometry Nomenclature. Anal. Chem. 35:2262.
8. HIRT, R.C., F.T. KING AND R.G. SCHMITT. 1954. Graphical Absorbance–Ratio Method for Rapid Two-Component Spectrophotometric Analysis. Anal. Chem. 26:1270.

Based on data from AIHA Journal
Subcommittee 9
J.N. HARMAN, *Chairman*

14. Infrared Absorption Spectroscopy

14.1 INTRODUCTION.

14.1.1 Infrared is a powerful analytical tool because of its specificity, sensitivity, versatility, speed and simplicity. Solids, liquids and gases may all be analyzed using a few general sampling techniques; the spectra provide the necessary information for identifying and quantitating the components present.

14.2 PRINCIPLE.

14.2.1 Infrared radiation is passed through a chamber containing the sample, after which the radiation is dispersed and detected. Because each compound absorbs the radiation in a characteristic pattern, the graph of absorption versus wavelength (or spectrum) produced by the spectrometer may be used to identify each component. Furthermore, the amount of light absorbed is proportional to the concentration of the component; therefore, a quantitative analysis may also be obtained. A file of standard spectra of pure materials at known concentrations is required for the analysis. Fourier transform infrared spectrophotometers have recently been applied to the analytical tasks previously addressed by conventional infrared spectrophotometry. This Fourier transform infrared (FT–IR) technique offers the advantages of low detection limits, excellent accuracy, the use of mathematical interference correction techniques and good repeatability.

14.3 SAMPLING.

14.3.1 The analysis of solids and liquids by infrared techniques is a well known field; the reader is referred to Potts **(1)** for further detail. Less common, and of more direct concern to the industrial hygienist, is the use of infrared to analyze gases and vapors, specifically the analysis of air for trace contaminants (100 ppm or less). Because oxygen and nitrogen are transparent to infrared radiation, multiple-reflection sampling cells with path lengths from 10 to 40 m may be utilized to achieve high sensitivity to many atmospheric contaminants.

14.3.2 The sampling cell most commonly employed is the 10-m path length cell shown in Figure 14:1.

14.3.3 Air may be sampled in a plastic bag, preferably of Saran or Mylar; the cell is then evacuated, the sample drawn in and the spectrum scanned. About 5 L of air are required. Some vapors may be adsorbed on silica gel and quantitatively desorbed in the laboratory **(2)**.

14.3.4 It is also possible to monitor air flowing continuously through the cell if

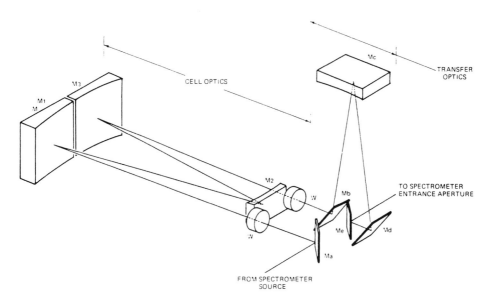

Figure 14:1 – Optical diagram of multiple reflection gas cell set for 1-meter path length. The number of beam traversals can be increased by resetting M_1 and M_3 to give path lengths up to 10 meters. (Courtesy of Perkin-Elmer Corporation, Norwalk, Connecticut.)

the spectrometer is set to detect a wavelength where the contaminant being monitored has a unique absorption band.

14.3.5 Many compounds can similarly be determined in the expired breath of workers following exposure to them. In most cases the magnitude of a chemical exposure can be established as well as the identity of the toxins (3).

14.4 ANALYSIS.

14.4.1 The analysis is performed by comparing the unknown spectrum with standard spectra of the compounds suspected to be present. Quantitative analysis requires the measurement of the absorbances of bands unique to each component. The following equation is used:

$$C_x = \frac{A_x}{A_s} \times \frac{b_s}{b_x} \cdot C_s$$

where A = absorbance
b = cell length
C = concentration
x = unknown
s = standard

14.4.2 A rough estimate of concentrations may be obtained from comparisons to published spectra; however, for accurate analysis the same instrument must be used to obtain both the sample and standard spectra. Several methods for preparing gaseous standards have been described (3). Catalogs of vapor spectra are available privately (4,5) or from commercial publishers (6).

14.4.3 Table 14:I lists the sensitivity of infrared analysis to a variety of gases and vapors using a 10-m path length cell at atmospheric pressure.

Table 14:I Infrared Analysis of Gases and Vapors in Expired Air

Compound	1986–7 Threshold Limit Value (ppm)	Infrared detection limit (ppm)	Analytical wavelength (μm)
Acetaldehyde	100	30	8.90
Acetic Acid	10	5	8.50
Acetone	750	5	8.20
Acetonitrile	40	100	9.58
Acetylene Tetrabromide	1	—†	—
Acrolein	0.1	10	10.43
Acrylonitrile	2	5	10.49
Allyl Chloride	1	5	13.22
Ammonia	25	20	10.77
n-Amyl Acetate	100	1	8.05
Amyl Alcohol	150 (1969)	10	9.47
Benzene	10	20	9.62
Benzyl Chloride	1	15	7.88
Bromine	0.1	—‡	—
Bromobenzene	—	10	9.30
1.3-Butadiene	10	5	11.02
2-Butanone (MEK)	200	10	8.52
sec-Butyl Acetate	200	1	8.05
n-Butyl Alcohol	50	10	9.35
tert-Butyl Alcohol	100	5	10.88
Butylamine	5	10	12.85
n-Butyl Glycidyl Ether	25	10	8.80
Butyl Mercaptan	0.5	5§	8.52
p-tert-Butyltoluene	10	20	12.25
Carbon Dioxide	5000	5	4.27
Carbon Disulfide	10	20	4.57
Carbon Monoxide	50	20	4.58
Carbon Tetrachloride	5	0.5	12.60
Carbonyl Sulfide	—	1	4.82
Chlorine	1	—‡	—

Table 14:I — (continued)

Compound	1986–7 Threshold Limit Value (ppm)	Infrared detection limit (ppm)	Analytical wavelength (μm)
Chlorobenzene	75	10	9.16
Chlorobromomethane	200	10	13.35
Chloroform	10	1	12.95
Chloropicrin	0.1	2	11.50
Cyclohexane	300	40	11.60
Cyclohexanol	50	10	9.32
Cyclohexanone	25	25	8.88
Cyclohexene	300	25	10.90
Diacetone Alcohol	50	5	8.50
1,2-Dibromoethane	10	5	8.38
o-Dichlorobenzene	50	5	13.37
m-Dichlorobenzene	—	5	12.75
p-Dichlorobenzene	75	2	9.10
Dichlorodifluoromethane	1000	1	10.85
1,1-Dichloroethane	200	5	9.42
1,2-Dichloroethane	10	10	8.18
cis-1,2-Dichloroethylene	5	5	11.58
trans-1,2-Dichloroethylene	5	2	12.05
Dichloroethylether	5	5	8.78
Dichlorofluoromethane	100	1§	12.50
1,1-Dichloro-1-nitroethane	2	2	9.07
1,2-Dichloropropane	75	(See Propylene Dichloride)	
Dichlorotetrafluoroethane	1000	0.5§	8.40
Diethylamine	25	10	8.70
Difluorodibromomethane	100	0.5§	12.10
Dimethyl Ether	—	2§	8.51
Dimethylformamide	10	5	9.22
Dioxane	25	2	8.80
Epibromohydrin	—	20	11.80
Epichlorohydrin	2	20	13.36
Ethanolamine	3	50	12.73
2-Ethoxyethyl Acetate	5	1	8.05
Ethyl Acetate	400	1	8.02
Ethyl Acrylate	5	1	8.35
Ethyl Alcohol	1000	5	9.37
Ethylamine	10	5§	12.95
Ethyl Benzene	100	50	9.70
Ethyl Bromide	200	10	7.98
Ethyl Chloride	200	10	10.18
Ethyl Ether	400	5	8.75
Ethyl Formate	100	1	8.43
Ethyl Mercaptan	0.5	40§	10.20
Ethyl Nitrate	—	5§	11.75
Ethylene	—	5	10.55
Ethylene Chlorohydrin	1	10	9.33
Ethylenediamine	10	15	12.85
Ethylene Dibromide		(See 1,2-Dibromoethane)	
Ethylene Dichloride		(See 1,2-Dichloroethane)	
Ethylene Glycol Ethyl Ether	—	5	8.78
Ethylene Imine	0.5	15	11.75
Ethylene Oxide	1	5	11.48
Fluorotrichloromethane	1000	1	11.82
FREON 11 Fluorocarbon	1000	(See Fluorotrichloromethane)	
FREON 12 Fluorocarbon	100	(See Dichlorodifluoromethane)	

Table 14:I — (continued)

Compound	1985 Threshold Limit Value (ppm)	Infrared detection limit (ppm)	Analytical wavelength (μm)
FREON 21 Fluorocarbon	100	(See Dichlorofluoromethane)	
FREON 112 Fluorocarbon	500	(See Tetrachlorodifluoroethane)	
Furfural	2	5	13.27
Furfuryl Alcohol	10	10	9.80
Gasoline	300	5‖	3.40
Heptane	400	5‖	3.40
Hexane	50	5‖	3.40
2-Hexanone	100 (1969)	10	8.57
Hexone (methyl isobutyl ketone)	50	10	8.52
Hydrogen cyanide	10	50§	3.00
4-Hydroxy-4-methylpentanone		(See Diacetone alcohol)	
Isophorone	5	— †	—
Isopropyl Ether	250	5	8.92
Mesityl Oxide	15	5	8.57
Methylacetylene	1000	20§	8.05
Methylal	1000	5	8.72
Methyl Alcohol	200	5	9.45
Methylamine	10	10§	12.80
Methyl Bromide	5	50	3.36
Methyl Chloride	50	30	3.35
Methyl Chloroform	350	2	9.20
Methylcyclohexane	400	5‖	3.40
2-Methylcyclohexanol	50	15	9.50
3-Methylcyclohexanol	50	20	9.50
2-Methylcyclohexanone	50	25	8.42
Methyl Ethyl Ketone	200	(See 2-Butanone)	
Methyl Formate	100	5	8.53
Methyl Isobutyl Ketone	50	(See Hexone)	
Methyl Mercaptan	0.5	100§	9.48
Methyl Methacrylate	100	1	8.55
alpha Methyl styrene	50	10	11.18
Methyl Trimethoxystyrene	—	1	9.06
Methylene Bromide	—	5	8.38
Methylene Chloride	100	2	13.10
Methylene Chlorobromide		(See chlorobromethane)	
Naphtha (coal tar)	100 (1969)	5§‖	3.40‖
Naphtha (petroleum)	—	5‖	3.40‖
Nitric Oxide	25	100§	5.25
Nitroethane	100	30	11.43
Nitrogen Dioxide	3	5§	7.90
Nitromethane	100	40	10.90
1-Nitropropane	25	25	12.45
2-Nitropropane	25	10	11.75
Nitrosylchloride	—	50§	10.75
Nitrous Oxide	—	25	7.68
Octane	300	5‖	3.40‖
Pentane	600	5‖	3.40‖
2-Pentanone	200	10	8.52
2-Pentene	—	10	3.37
Perchloroethylene	50	2	10.92
Phosgene	0.1	1	11.68
Propargyl Bromide	—	5	8.18
Propyl Acetate	200	1	8.05
i-Propyl Alcohol	200	15	10.47
n-Propyl Ether	500	5	8.81

Table 14:I — (continued)

Compound	1985 Threshold Limit Value (ppm)	Infrared detection limit (ppm)	Analytical wavelength (μm)
Propylene Dichloride	75	15	9.80
Propylene Oxide	100	10	11.96
Pyridine	5	30	9.60
Stoddard Solvent	500	5	3.40
Styrene Monomer	50	10	12.90
Sulfur Dioxide	5	15§	8.55
Sulfuryl Fluoride	2	1	11.32
1,1,2,2-Tetrachloro-1,2-difluoroethane	500	2	11.90
Tetrachloroethylene		(See perchloroethylene)	
Tetrahydrofuran	200	10	9.22
Toluene	100	5	13.75
1,1,2-Trichloroethane	10	10	10.60
Trichloroethylene	100	2	11.78
1,2,3-Trichloropropane	50		12.44
Trimethylamine	–	5§	12.10
Turpentine	100	10§	3.40
VIKANE Fumigant		(See Sulfuryl fluoride)	
Vinyl Chloride	5	10	10.63
Vinylidine Chloride	–	5	12.60
o-Xylene	100	10	13.51
m-Xylene	100	10	13.02
p-Xylene	100	10	12.58

*Materials marked * have been detected in postexposure expired air.
†no spectrum obtained from saturated vapor at 23°C, 1 atm pressure.
‡material produces no infrared spectrum.
§sensitivity estimated by extrapolating absorbance of vapor observed in 10-cm cell.
||Aliphatic hydrocarbons can be detected at 3.40 μm but cannot usually be identified specifically at low concentration.
Note: The above is a condensation of material found in Detection of Volatile Organic Compounds and Toxic Gases in Humans by Rapid Infrared Techniques, by R.D. Stewart and D.S. Erley, in "Progress in Chemical Toxicology," A. Stolman, ed., Vol. II, pp. 183–200. Academic Press, NY, 1965. The permission of the publisher to reprint this material is gratefully acknowledged.

REFERENCES

1. POTTS, W.J., JR. 1963. *Chemical Infrared Spectroscopy*, Vol. I: Techniques. John Wiley and Sons, New York, NY.
2. ROBERTSON, D.N. AND D.S. ERLEY. 1961. Anal. Biochem 2:45.
3. STOLMAN, *et al.* 1965. *Progress in Chemical Toxicology*. Vol. II. Academic Press, New York, NY.
4. ERLEY, D.S., AND B.H. BLAKE. *Infrared Spectra of Gases and Vapors*. The Dow Chemical Company, Midland, MI.
5. PIERSON, R.H., A.N. FLETCHER AND E.ST.C. GANTZ. 1956. Anal. Chem. 28:1218.
6. SADTLER RESEARCH LABORATORIES, Division of Bio-Rad Laboratories, 3316 Spring Garden St., Philadelphia, PA 19130.

Based on data from AIHA Journal
Subcommittee 9
J.N. HARMAN, *Chairman*

15. Atomic Absorption Spectroscopy

15.1 INTRODUCTION. Atomic absorption (AA) has rapidly become the most widely used analytical technique for quantitative trace metal analysis. The technique is directly applicable to some 68 elements, including most of the metals and semi-metals, but excluding sulfur, carbon, halogens and inert gases. Atomic absorption is essentially a solution-based technique, with only very limited applications for the direct analysis of solids or gases. Advantages include specificity, speed, ease of use and excellent sensitivity. Detection limits are typically in the sub-ppb to low ppm range by weight in the sample solution. A vast amount of published AA methodology is now available, covering a wide range of application areas.

15.2 PRINCIPLE OF THE METHOD.

15.2.1 If light of the proper wavelength strikes a free, ground-state atom, the atom may absorb the light in a process known as atomic absorption. By measuring the amount of light absorbed, a quantitative measurement of the analyte concentration can be made. Ground state atoms are typically produced by dissociating the sample in a flame or graphite tube furnace. A hollow-cathode lamp, with a cathode containing the element of interest, typically provides the narrow emission lines of characteristic radiation.

15.2.2 The functional components of an atomic absorption spectrophotometer are shown in Figure 15:1. The light source emits the narrow-line spectrum of the analyte element. In addition to the more commonly used hollow cathode lamps, electrodeless discharge lamps are used for several of the more volatile elements. Commercial instruments employing both single and double beam optical designs are available. In a double beam system (shown in Figure 15:1), the light output is split into separate sample and reference beams. Ratioing these two beams compensates for lamp drift and eliminates the lamp warmup typically required for single beam systems. Single beam designs employ no separate reference beam and are usually less expensive as a result of fewer optical components.

15.2.3 The sample cell, typically a flame or graphite furnace, produces the atomic vapor. A monochromator disperses the various wavelengths of light that are emitted from the source and isolates the specific analyte wavelength of interest. This combination of a specific source and the isolation of a particular wavelength is what allows the determination of a selected element in the presence of others. The particular wavelength of light isolated by the monochromator is directed to a photomultiplier detector. The resultant signal is then amplified and processed by the instrument electronics, producing a signal directly related to the amount of light absorbed in the sample cell.

15.2.4 Recent advances in AA instrumentation have focused on automation and data handling. Microprocessor-based electronics provide features such as direct concentration readout and curvature correction. Modern AA instruments are now often fully controlled by stand-alone computers. Automatic control of wavelength, slit position, lamp position and gas controls is now common. Multielement, sequential instruments featuring 6 to 8 lamps in a rotating turret are available. Other features available on current instrumentation include: programmable autosamplers, graphics display of analytical signals, storage and recall of complete analytical methods and comprehensive data analysis/storage and report generation software.

15.3 SAMPLE ATOMIZATION SYSTEMS.

15.3.1 The function of the sample atomization device is to generate ground-state atoms in the optical path of the spectrophotometer. The most commonly used sample atomization device remains a flame. However, in recent years the use of higher sensitivity sampling devices such as graphite furnaces has become widespread.

15.3.2 *Flame AA* systems (Figure 15:2) employ a pneumatic nebulizer through which sample solution is aspirated and sprayed as a fine aerosol into the burner mixing chamber. Here, fuel and oxidant gases are mixed with the sample aerosol and carried to the burner head, where combustion and sample atomization occurs. Two types of flame mixtures are commonly used, air-acetylene and nitrous oxide-acetylene. The air-acetylene flame provides a temperature around 2300°C while the nitrous oxide-acetylene flame provides a temperature around 2700°C. The hotter nitrous oxide-acetylene flame is required for the more refractory elements such as B, V, Ti and Si. In addition to aqueous solutions, a variety of organic solvents (ketones, xylene, alcohols, etc.) are commonly used in flame AA applications. To improve sensitivity, sample concentration using chelation and solvent extraction may be employed.

15.3.3 *Graphite Furnace AA* systems provide a 50- to 500-fold improvement in sensitivity relative to flame AA. This higher sensitivity is achieved by atomizing samples in a small, electrically heated graphite tube furnace (Figure 15:3). A small sample aliquot of 10 to 100 μL is pipetted into the

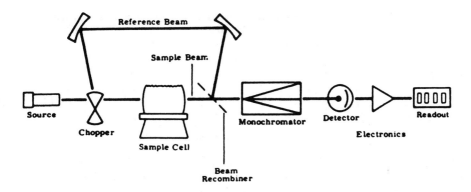

Figure 15:1 – Atomic absorption spectrophotometer (double-beam design).

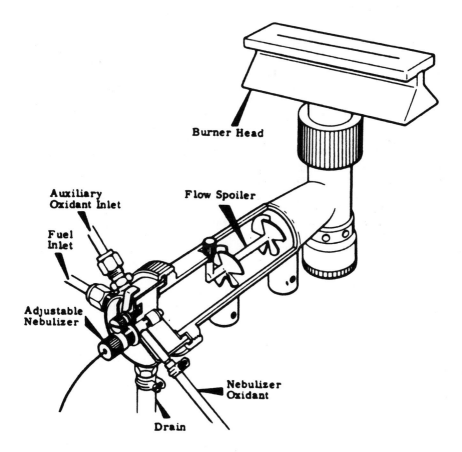

Figure 15:2 – Atomic absorption burner system.

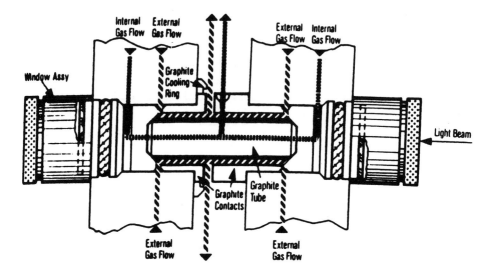

Figure 15:3 – Graphite furnace sampling system (cross-section view).

graphite tube and the temperature of the tube is then increased in a controlled manner to dry, ash and atomize the sample. Most of the sample matrix is removed during the ashing period. Atomization temperatures up to 2800°C can be achieved, allowing measurement of most refractory elements. The residence time of sample atoms in the graphite tube is several orders of magnitude longer than the atom residence time in flame applications. This longer atom residence time in the optical path accounts for the dramatically improved sensitivity achieved with the graphite furnace. While providing the advantage of very high sensitivity, the furnace technique is noticeably slower than the flame technique and also requires automatic correction for nonspecific background absorption generated by the sample matrix. This automatic background correction is typically achieved by using continuum sources (1,2) or by use of the Zeeman effect, (3,4) which is based on magnetic splitting of the absorption line. These automatic background correction systems are available on commercial instrumentation.

15.3.4 *Hydride AA* applications involve the generation of a volatile hydride of the analyte by means of a reducing agent added to a reaction vessel containing an acidic solution of the sample. The liberated hydride is transported to a heated quartz cell atomizer by an inert carrier gas. Sodium borohydride is most commonly used as the reducing agent. Since the hydride is separated from the matrix, advantages include high sensitivity and reduced interferences. However, this technique is only applicable to a limited number of elements. Those elements that form volatile hydrides include As, Se, Bi, Sb, and Te.

15.3.5 *Cold Vapor AA* analysis is specific for the determination of mercury, which can exist in an atomic state at ambient temperature due to its high vapor pressure. In a manner similar to hydride analysis, a reducing agent is added to a reaction vessel containing a sample with trace levels of ionic mercury. Stannous chloride or sodium borohydride are the most commonly used reducing agents. The atomic mercury vapor is then transported to a photometer. While extremely sensitive, this technique is only applicable for the determination of mercury.

15.4 DETECTION LIMITS.

15.4.1 Table 15:I lists detection limits for 40 of the most common elements determined by flame and graphite furnace AA. Detection limits obtained using hydride and cold vapor sampling techniques are listed in

Table 15:I Typical AA Detection Limits (μg/L) Using Flame and Graphite Furnace Sampling (6)

Element	Flame	Furnace	Element	Flame	Furnace
Ag	0.1	0.005	Na	0.2	< 0.5
Al*	30	0.01	Nb*	1000	–
As	100	0.2	Ni	4	0.1
Au	6	0.1	P*	50,000	40
B*	700	20	Pb	10	0.05
Ba*	8	0.04	Pt	40	0.2
Be*	2	0.05	Sb	30	0.2
Bi	20	0.1	Se	70	0.5
Ca	1	0.05	Si*	60	0.1
Cd	0.5	0.003	Sn*	100	0.2
Co	6	0.02	Sr*	1	0.2
Cr	2	0.01	Ta*	1,000	–
Cu	1	0.02	Te	20	0.1
Fe	3	0.02	Ti*	50	–
Hg	200	20	Tl	9	0.1
K	2	0.02	U*	11,000	–
Li	0.5	0.3	V*	40	0.2
Mg	0.01	0.004	W*	1,000	–
Mn	1	0.01	Zn	0.8	0.001
Mo*	30	0.02	Zr*	400	–

*N_2O-C_2H_2 flame used, all other values obtained using air-C_2H_2 flame.
Note: All furnace values based on 100 μL sample size.

Table 15:II. Detection limits are defined as the analyte concentration producing a signal twice the standard deviation of the blank solution.

15.5 ADVANTAGES.

15.5.1 For trace metal determinations, AA offers a number of important advantages, including specificity, speed, and moderate equipment cost. While various interferences exist, they are usually moderate in nature with documented methods of control. Relative to electrochemical and ion chromatography techniques, AA is applicable to many more metals and is generally more specific and faster. From a detection limit standpoint, the extremely low

Table 15:II Typical AA Detection Limits (μg/L) Using Hydride and Cold Vapor Mercury Sampling (6)

Element	Value
As	0.02
Bi	0.02
Hg*	0.01
Sb	0.1
Se	0.02
Te	0.02

*Value determined by cold vapor method using amalgamation. All other values determined by hydride generation.

levels measured by graphite furnace AA are surpassed only by neutron activation or ICP/Mass Spectrometry techniques.

15.6 DISADVANTAGES.

15.6.1 If a large number of elements are to be determined, the cost of lamps can be significant. Also, lamps have a finite life and generally need to be replaced after approximately 2–3 years of normal use. While the graphite furnace is extremely sensitive, it is considerably slower than flame sampling. A single furnace determination requires approximately 2 min, versus 5 to 10 s using a flame. The equipment cost of a graphite furnace and associated autosampler is also significant. Atomic absorption is a single-element technique, which can be a limitation if the number of elements to be determined is very large. Sample throughput using multi-element Inductively Coupled Plasma (ICP) systems will be much greater, although the cost of such systems is considerably higher (see §16).

15.7 CALIBRATION.

15.7.1 Regardless of the sampling device employed (flame, furnace, etc.), calibration curves must be constructed for each element to be determined. Suitably prepared reference standards and blanks are required for calibration. Commercially prepared standards are now available from many sources. In most cases, commercially available AA instruments are supplied with applications manuals that provide recommended operating parameters plus information on recommended standard concentrations and calibration ranges. Operator training courses are also available from most manufacturers.

15.8 INTERFERENCES.

15.8.1 While AA is virtually free from spectral interferences, chemical interferences due to other matrix constituents may occur. Samples containing high salt concentrations are more likely to exhibit such interferences. Fortunately, most of these interference effects are known, with documented methodology available on controlling such effects. Many flame interferences can be eliminated by adding a suitable releasing agent that forms a thermally stable compound with the interferent. The use of lanthanum to control an interference from phosphate when determining calcium **(5)** is one such example. Ionization interferences are controlled by adding an excess of an ionization suppressant such as potassium. In graphite furnace applications, background absorption interferences are usually present, necessitating automatic background correction.

15.8.2 Graphite furnace applications involve measurements at extremely low concentration levels. Hence, special attention must be paid to eliminating sources of contamination from the lab environment as well as the reagents and sample collection equipment. Information regarding specific interferences and means of control is generally provided in the applications manuals supplied with commercial instrumentation.

15.9 OPERATION/MAINTENANCE.

15.9.1 Atomic absorption instruments and accessories are relatively easy to operate and maintain. For flame operation, sources of gases and gas regulators are required, as is a suitable vent to remove heat and combustion products from the flame. Periodic cleaning of the drain and burner system is necessary, with occasional replacement of the nebulizer and burner system O-rings required. Corrosion-resistant nebulizers are commercially available and recommended for extensive use with acidic solutions. Operation of the graphite furnace requires a source of argon purge gas plus a source of cooling water. Graphite tubes are a consumable item and must be replaced after about 200 to 300 determinations.

15.10 APPLICATIONS.

15.10.1 A vast amount of published AA literature is available, covering a wide range of application areas and sample types. Of particular interest are several recent review articles **(7–9)** and publications describing sample collection/preparation procedures **(10–11)** and element-specific methodology **(12–16)**. Much of the recent AA literature has focused on graphite furnace applications. The manuals supplied by most manufacturers are often good sources of applications references.

REFERENCES

1. KOIRTYOHANN, S.R. 1965. Background Corrections in Long Path Atomic Absorption Spectrometry. Anal. Chem. 37:601.

2. KAHN, H.L. 1968. A Background Compensation System for Atomic Absorption. At. Absorption Newslett. 7:40.

3. HADEISHI, T. AND R.D. MCLAUGHLIN. 1971. Hyperfine Zeeman Effect AA Spectrometer for Mercury. Science 174:404.

4. DELOOS-VOLLEBREGT, M.T.C. AND L. DE GALAN. 1978. Theory of Zeeman Atomic Absorption Spectrometry. Spectrochim Acta, Part B, 33:495.

5. TRUDEAU, D.L. AND E.F. FREIER. 1967. Determination of Calcium in Urine and Serum by Atomic Absorption Spectrometry. Clin. Chem. 13:101.

6. Analytical Methods for Atomic Absorption Spectrophotometry, Perkin-Elmer Corp. 1982.

7. DE JONGHE, W.R. AND F.C. ADAMS. 1982. Measurements of Organic Lead in Air. A Review. Talanta 29:1057.

8. FOX, D.L. 1985. Air Pollution. Anal. Chem. 57:223R.

9. CRESSER, S.M. AND L.C. EBDON. 1986. Atomic Spectrometry Update – Environmental Analysis. J. Anal. At. Spectrom. 1:1R.

10. HRSAK, J. AND M. FUGAS. 1981. Preparation of Samples of Airborne Particles For Determination of Metals. Mikrochim. Acta 2:111.

11. SNEDDON, J. 1983. Collection and Atomic Spectroscopic Measurement of Metal Compounds in the Atmosphere: A Review. Talanta 30:631.

12. LONG, S.J., J.C. SUGGS AND J.F. WALLING. 1979. Lead Analysis of Ambient Air Particulates: Interlaboratory Evaluation of EPA Lead Reference Method. J. Air Poll. Control Assoc. 29:28.

13. COLOVOS, G., W.S. EATON, G.R. RICCI, L.S. SHEPARD AND H WANG. 1979. Collaborative Testing of NIOSH Atomic Absorption Method. U.S. DHEW (NIOSH) Publ. 79–144, Washington, DC.

14. BRYCE-SMITH, D. 1982. Environmental Lead and the Analyst. TrAC 1:199.
15. BYRNE, R.E. 1983. Rapid Method for Determination of Arsenic, Cadmium, Copper, Lead and Zinc in Airborne Particulates by Flame Atomic Absorption Spectrometry. Anal. Chim. Acta 151:187.
16. RAPTIS, S.E., G. KAISER AND G. TOELG. 1983. Survey of Selenium in the Environment and Critical Review of its Determination at Trace Levels. Z. Anal. Chem. 316:105.

Subcommittee 9
F.J. FERNANDEZ
J.N. HARMAN, *Chairman*

16. Inductively Coupled Plasma Emission Spectroscopy

16.1 INTRODUCTION. The inductively coupled plasma (ICP) emission spectroscopic technique has only been available commercially since the mid-1970's but has become widely accepted for the analysis of most elements because of its simplicity of use, wide linear dynamic range, speed and accuracy of analysis. It has become established as the accepted method for the analysis of the majority of the elemental priority pollutants in waste water.

16.2 PRINCIPLE OF THE TECHNIQUE. The technique is an emission spectroscopic method in which the sample is dissociated into its atomic form and excited to high energy levels including the ionic form by introducing the sample into the center of a gaseous plasma sustained inside an induction coil energized with a high frequency alternating current (Figure 16:1). The excited species then emit characteristic radiation as they relax back to the atomic and ionic ground states.

16.3 THEORY. Unlike the majority of emission sources, the ICP is an optically thin source and consequently exhibits linearity over five or more orders of magnitude. The measured light intensity at any wavelength is the sum of the background emission from the source and the emission from the line less the intensity absorbed by the ground state atoms or ions of the element:

$$I_{obs} = I_{bkg} + I_{em} - I_{abs}$$

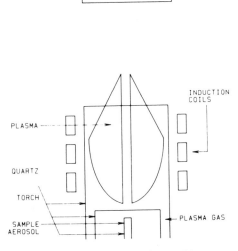

Figure 16:1 — Plasma torch assembly.

The emitted intensity is defined by the Einstein-Boltzmann equation:

$$I_{em} = k_{em}N \exp(-E_{em}/kT)$$

where k_{em} is a constant of the element in question, N is the number of atoms (or ions) of the element, k is the Boltzmann constant, T is the temperature and E_{em} is the energy of the excited state above the ground state. The absorbed intensity is defined by the equation:

$$I_{abs} = I_{in} \exp(-K_v L)$$

where L is the path length, I_{in} is $I_{bkg} + I_{em}$ and the absorption coefficient is defined by:

$$K_v = k_{abs}N$$

In the optically thin ICP source, the path length is small and the temperature is high so that the degree of self-absorption is low and may often be ignored. Thus the intensity of measured emission (I) may, to a first approximation, be calculated from the concentration (C) of the element via:

$$I = a_o + a_1 C$$

where the a's are obtained by statistical least-squares fit to a series of calibration standards. In practice, additional coefficients are sometimes added to the basic equation shown above to extend the analytical capability of the line. The typical equations used in data processing are of the general form:

$$I := a_0 + a_1C + a_2C^2 + a_3C^3$$

16.4 FEATURES OF THE ICP. The plasma temperatures are sufficiently high that chemical interferences often observed in flames are not observed. There is, however, background emission from stable molecular species formed during the decomposition of the sample (e.g. NO, CO, CN, C_2) as well as recombination spectra from ionic species in the sample. Additionally, all instruments exhibit some degree of stray light which appears as a background increase.

Spectral interferences are observed with the ICP. These interferences are a result of the spectrometer being unable to resolve completely the analytical wavelength from wavelengths of interfering elements in the same sample and are to an extent dependent upon the instrumentation employed. As the ICP, however, is a very high temperature source, the number and degree of such interferences are generally greater than with other sources and they require correction during the data processing in order to maintain accuracy.

16.5 INSTRUMENTATION. The typical instrument consists of a high-powered radiofrequency generator, sample introduction system, plasma torch, wavelength isolation, detection device and computerized data processing module, as shown in Figure 16:2. Radio-frequency generators typically deliver 1–3 kW at frequencies of 27–55 MHz with 27 MHz being the most commonly used. The plasma itself is usually argon gas at a flow of 5–20 L/min passing through the outermost tube of the quartz torch. The most common sample introduction system is a pneumatic nebulizer similar to those employed in flame atomic absorption that converts the sample solution to an aerosol in the carrier gas (usually argon flowing at 0.5–1.0 L/min) at typical sample flow rates of 1–2 mL/min. This aerosol then passes through an expansion chamber where large droplets are removed. The aerosol is then injected into the central portion of the plasma torch as shown in Fig-

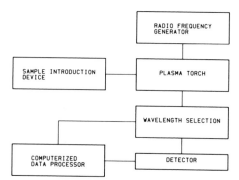

Figure 16:2 – Schematic of ICP instrumentation.

ure 16:1. The light emitted from the plasma is then collected and transferred by either a mirror or lens optical system to the wavelength selection device.

The instrumentation can either measure each element in the sample sequentially by moving the wavelength isolation device to capture the analytical line onto a single detector (Figure 16:3) or simultaneously by positioning multiple detectors to capture the analytical lines of all required elements (Figure 16:4). In the normal sequential instrument, a small wavelength scan around the analytical wavelength is performed because of the mechanical difficulties of accurately positioning the grating.

Each of these two instrumental approaches has unique advantages and limitations. The simultaneous approach obvi-

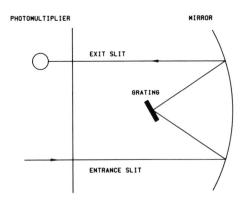

Figure 16:3 – Schematic of ICP sequential instrumentation.

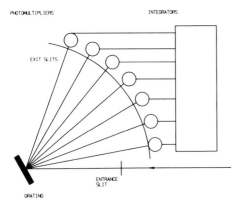

PHOTOMULTIPLIERS   INTEGRATORS

EXIT SLITS

ENTRANCE
SLIT

GRATING

Figure 16:4—Schematic of ICP simultaneous instrumentation.

ously provides a more rapid analysis and requires a smaller sample size than the sequential approach, but it does not permit rapid changing of analytical requirements. In general terms, the simultaneous approach is more suited to routine analysis and the sequential to methods development.

Once the wavelength(s) have been isolated, light is converted to electrical current via a photomultipler. In simultaneous instruments these currents are usually stored on integrating capacitors during the sample analysis time and then read by the computer at the end of the analysis. In sequential instruments, the photomultiplier is usually read directly by the computer as part of a small wavelength scan around the line of interest. The data processing unit then converts the measured line intensities into concentrations using coefficients (a_i) determined during the instrumental calibration process.

With ICP emission spectroscopy, background correction is performed by measuring the background emission adjacent to the analytical wavelength. In sequential instruments, such a measurement is generally available as part of the small wavelength scan performed to isolate the analytical line. In simultaneous instruments, the background is measured on all elements simultaneously by moving the spectrum across the analytical detectors. This latter effect is achieved by either refracting

through a quartz plate mounted inside the entrance slit of the polychromator or by moving the entrance slit of the polychromator.

Interference caused by line overlap is typically corrected by determining the interfering element at a wavelength free from interference, then using its concentration and the known line-intensity ratio to subtract the interference and arrive at a corrected concentration of the element sought.

16.6 ADVANTAGES OF THE ICP. The ICP is capable of determining most elements including the non-metals down to the part per billion level in solution. Table 16.I shows a comparison of the detection limits by ICP compared to atomic absorption. The dynamic range of the ICP is such that the technique can also measure up to the hundred part per million level or greater with precisions typically about 1% or better. Thus the ICP does not require several dilutions of the sample to be made in order completely to analyze a given sample.

The freedom from chemical interferences is also a distinct advantage of the ICP over atomic absorption. This, coupled with the dynamic range, permits sample preparation time to be reduced for ICP analysis compared to atomic absorption.

The simultaneous ICP instrumentation is capable of determining thirty or more elements from a single exposure which is typically about ten to twenty seconds in duration. The sequential ICP instrumentation takes longer to collect the data than the simultaneous instrument, but it also is capable of determining over ten elements per minute in the same sample solution. The ICP is thus a very efficient method of analyzing samples. Despite the high consumption of costly argon gas, the increase in sample throughput and reduction in sample preparation time compared to atomic absorption means that the cost per analysis of ICP is considerably less than atomic absorption.

The ICP is an electrical discharge and as such can be safely left unattended during operation; the use of autosamplers with ICP instrumentation is commonplace.

16.7 DISADVANTAGES OF THE ICP. The ICP suffers from spectral interferences which need correction via computerized data processing. The determination of these

Table 16:I Comparison of Detection Limits for Inductively Coupled Plasma Emission Spectroscopy with Atomic Absorption Spectrometry

Element	ICP Wavelength nm	Detection Limits ICP(1)	(mg/L) AA(2)	Element	ICP Wavelength nm	Detection Limits ICP(1)	(mg/L) AA(2)
Ag	328.0	0.001	0.004	Nd	430.3	0.001	ND
Al	308.2	0.02	0.03	Ni	231.6	0.002	0.005
As	193.7	0.05	0.1	Os	225.6	0.0004	0.1
Au	267.6	0.002	0.02	P	178.2	0.05	100
B	249.7	0.004	2	Pb	220.3	0.01	0.01
Ba	455.4	0.00005	0.005	Pd	340.4	0.04	0.01
Be	234.8	0.00006	0.001	Pr	390.8	0.04	ND
Bi	223.0	0.025	0.05	Pt	265.9	0.004	0.05
C	193.1	0.05	ND*	Re	227.5	0.001	0.6
Ca	393.4	0.00004	0.0005	Rh	343.4	0.007	0.02
Cd	228.8	0.0004	0.001	Ru	240.2	0.003	0.06
Ce	418.6	0.05	ND	S	180.7	0.01	ND
Co	228.6	0.001	0.002	Sb	206.8	0.03	0.03
Cr	267.7	0.002	0.002	Sc	361.3	0.00006	0.05
Cu	324.7	0.0003	0.004	Se	196.0	0.025	0.1
Dy	353.2	0.0003	0.2	Si	251.6	0.003	0.08
Er	390.6	0.0006	0.1	Sm	442.4	0.05	2
Eu	382.0	0.00008	0.02	Sn	189.9	0.02	0.03
Fe	259.9	0.0005	0.004	Sr	407.7	0.00003	0.004
Ga	294.4	0.04	0.05	Ta	226.2	0.025	2
Ge	209.4	0.04	0.2	Tb	350.9	0.001	ND
Hf	277.3	0.015	2	Te	214.2	0.04	0.05
Hg	194.2	0.025	0.5	Th	283.7	0.06	ND
Ho	345.6	0.006	ND	Ti	334.9	0.0003	0.09
In	230.6	0.06	0.03	Tl	190.8	0.04	0.02
Ir	224.3	0.025	1	Tm	346.2	0.0004	0.02
La	379.5	0.01	2	U	409.0	0.02	23
Li	670.7	0.0003	0.001	V	292.4	0.0007	0.05
Lu	261.5	0.0002	ND	W	207.9	0.03	1
Mg	279.5	0.00002	0.0003	Y	371.0	0.0001	0.3
Mn	257.6	0.0001	0.001	Yb	328.9	0.0001	0.01
Mo	202.0	0.005	0.03	Zn	213.8	0.001	0.001
Nb	316.3	0.003	1	Zr	339.2	0.0006	1

*ND = not detected

interferences increases the calibration time of the instrument and the presence of these interferences can reduce the accuracy and precision of trace elements in complex matrices compared to pure solutions.

Capital costs of ICP instrumentation are typically over twice the cost of an atomic absorption instrument.

16.8 APPLICATION OF ICP. The ICP emission spectroscopic technique can be applied to the determination of most metallic elements in solution over the concentration range 0.01–100 mg/L and most non-metallic elements in solution over the range 0.1–1000 mg/L. The operating conditions of the ICP need to be controlled carefully, but the same conditions are applied to all elements. Standard curves and interference factor determinations must be made for each instrument, but once these have been recorded a simple procedure to reset the operating conditions and a two-point standardization are all that are required to prepare the instrument for analysis.

REFERENCES

1. BARNES, R. ICP Information Newsletter, 5:416 (1980).
2. PARSONS M.L. AND P.M. McEIFERSH. Applied Spectroscopy, 26:472 (1972).

Subcommittee 9
J.N. HARMAN, *Chairman*

17. Gas Chromatography

17.1 INTRODUCTION.

17.1.1 Chromatography is a technique for the separation of closely related compounds. The technique has been known for at least a century, but only since 1952 have its applications been investigated. In approximately 1956, U.S. scientists began using chromatographic methods for the separation and analysis of mixtures of gases and volatile materials. Today the field of chromatography has grown to amazing proportions with almost countless applications (1–4).

In gas chromatography, the sample is vaporized (if not already in this form) and the mixture is passed by a stream of inert gas carrier through a rigid container (column) containing a packing material in the case of packed column chromatography, or coated with a liquid phase in the case of capillary chromatography. The packing or liquid coating has different affinities for each particular component in the mixture and lets each component pass through at a different rate. As each component emerges from the column it is observed by a sensitive detection device.

17.2 PRINCIPLE OF THE METHOD.

17.2.1 Basically, chromatography requires two phases. One phase is the stationary phase or fixed phase. This phase may be either a solid, as in adsorption chromatography, or a liquid, held by a solid or coated on tubing walls, as is in partition chromatography. The other phase is moving and is generally referred to as the mobile phase. When phase equilibria occur between the sample components, the moving phase and the stationary phase, the sample components are distributed between the phases.

17.2.2 The gas chromatograph can best be compared to a fractional distillation apparatus. Chromatography is far more efficient than fractional distillation. A good distillation column may have 100 to 200 theoretical plates, while a chromatographic column operates with the range of theoretical plates from approximately 1000 to as many as 500,000 theoretical plates. Each of the above separates the mixture into its component parts, which can then be analyzed separately. A number of reviews de-

scribe the technique and its applications (e.g., **6–9**).

17.2.3 There are two basic types of chromatography, gas adsorption and gas-liquid partition. For better understanding of the field of chromatography, a few definitions are in order:

a. gas-liquid chromatography—the mobile phase is a gas and the stationary phase is a liquid distributed on an inert solid support or on tubing walls.

b. gas-solid chromatography—the mobile phase is a gas and the stationary phase is an active solid such as alumina, activated charcoal, silica gel, Molecular Sieve, or a porous polymer.

c. gas chromatography—all methods in which the mobile phase is a gas (or vapor).

17.3 EQUIPMENT.

17.3.1 The three main components of a gas chromatograph are a sample injection system, a column and a detector. The sample injection system must be capable of introducing the sample *unchanged*, instantly and reproducibly at the head of the column. The column is a rigid container made of metal, glass, or other inert material which contains the stationary phase. This column will have various shapes depending upon the space available for housing the column (the oven). The detector is an apparatus which measures the compositional changes in the mobile phase (Figures 17:1 and 17:2).

17.4 REAGENTS.

17.4.1 *Carrier Gas.* The mobile phase used to move the sample through the column. Gases generally used are helium, hydrogen and nitrogen. Other gases may be

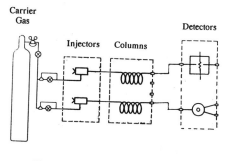

Figure 17:1 – **Basic diagram of a gas chromatograph.**

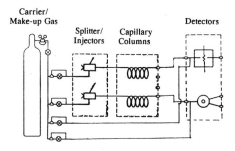

Figure 17:2—Basic diagram of a capillary gas chromatograph.

used. For example, SF_6 is used when detecting permanent gases using a gas density balance detector.

17.4.2 *Stationary Phase.* An essentially non-volatile (at the operating temperature used) liquid which is capable of dissolving the sample components and releasing them, preferentially, by difference in their volatility, from the solution.

17.4.3 *Solid Support.* Usually an inert, porous solid either inorganic or organic, which is of a known and standard size. Active solids may be used if they are chemically inactivated. The walls of a capillary tubing may be used as a solid support.

17.4.4 *Active Solid.* A solid capable of adsorption of gaseous sample components and their preferential release; e.g., the Molecular Sieves.

17.5 PARTS OF THE CHROMATOGRAPH.

17.5.1 *Sample Injection System.* Most samples are introduced to the column with a small calibrated syringe (microsyringe) such as those commercially available from several suppliers (Figure 17:3). These syringes are made for delivering either liquids or gases. Liquid syringes are available in sizes from 1 to 500 μL. Gas tight syringes are available from 50 to 2000 mL. For calibration with larger volumes of gases, plastic syringes are available in sizes 500 mL to 1500 mL (Figure 17:4).

Samples may also be introduced to the column using a sample injection valve as shown in Figure 17:5. Sample injection valve internal volumes and sample loops range in size from 0.06 to 1.0 mL for use with capillary columns and from 1.0 to 10.0 mL for use with packed columns.

The gas samples may be introduced to the gas sampling loop either by pressure or

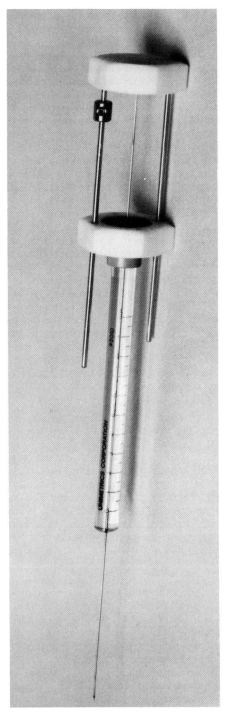

Figure 17:3—Syringe for liquids.

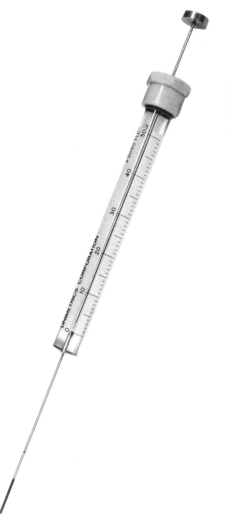

Figure 17:4 — Syringe for gases.

Figure 17:5 — Sample injection valve (Valco).

nal diameter for capillary columns. The tubing is generally coiled but can take other shapes depending on the oven designed to accommodate it, as seen in Figure 17:6. The selection of column packing material is discussed in section 17.6.

17.5.3 *Detectors.* There are numerous types of detectors:

a. Thermal conductivity (katharometer) — measures changes in heat capacity;

b. gas density balance — measures change of density;

c. flame ionization — measures difference in flame ionization due to combustion of the sample;

d. electron capture — measures current flow between two electrodes caused by ionization of the gas by radioactivity;

e. photoionization — measures current flow between two electrodes caused by ionization of the gas by ultraviolet radiation;

f. glow discharge — measures the voltage change between two electrodes caused by the change in discharge by different compositions;

g. alkali flame ionization — measures differences in flame ionization due to combustion of phosphorus or halogen compounds in the presence of an alkali salt at the flame tip;

h. flame photometric — measures light emitted from the excited-state sulfur and phosphorus compounds in the hydrogen flame;

i. helium ionization — measures current flow between electrodes caused by ionization of gas from metastable helium atoms formed by a radioactive source.

by vacuum. The gas sample can be delivered to the loop from a pressure container, sampling pump, or plastic or rubber bag or by means of a large syringe. The only requirement is that sufficient sample is available to thoroughly purge the sample loop.

17.5.2 The column is the rigid container made of glass, metal or Teflon tubing which contains a packing material or is coated with a liquid phase to effect the component separation. Column tubing comes in various sizes from 1/4" to 1/16" (6–1.6 mm) outer diameter for packed columns and from 0.8 mm to 0.1 mm inter-

PACKED COLUMNS

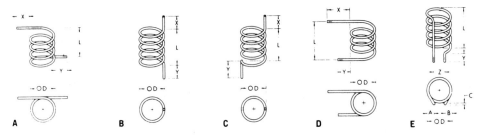

Figure 17:6 — Examples of columns used for gas chromatography.

The detector used depends upon the type of sample to be analyzed and the sensitivity needed for the analysis. Table 17:I lists the sensitivity obtainable with various detector types.

The detector output is generally presented on a recorder as a chromatogram. This chromatogram is a plot of detector response versus time. The area of each peak is a measure of the mass of the sample component, and the time for the peak to appear is used to identify the nature of the component. A recorder plot is shown in Figure 17:7.

17.6 GENERAL CONSIDERATIONS IN SELECTING THE PROPER COLUMN PACKING MATERIAL.

Table 17:I Sensitivity Obtained with Various Detector Types

Detector Type	Sensitivity in g/s
Thermal conductivity	10^{-7}
Ionization detector	
ion cross section	10^{-3}
Argon diode	10^{-13}
Electron affinity	10^{-14}
Flame ionization	10^{-12}
Thermionic emission	10^{-10}

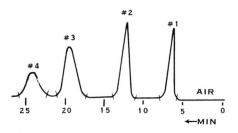

Figure 17:7 — Detector response versus time.

17.6.1 *Support Mesh Size and Liquid Phase Concentration.* The support mesh size refers to the size of the particles used to fill the column. Closely sized, small particles result in the best column effeciencies. However, pressure drop, available screen sizes and analysis time requirements are also considerations in determining mesh size. In general, the column diameter can be used as a guideline for determining the proper mesh as follows:

1/4" (6 mm) diameter column — 60/80 mesh
1/8" (3 mm) diameter column — 80/100 mesh

The liquid phase concentration is determined by the demands of the analysis and the type of support material. High phase loadings (> 20 wt%) tend to cover most active sites on the support, separate low boiling point compounds well, and can accommodate larger sample sizes. Low phase loadings give higher column efficiencies, are easier to pack in the column and give less column bleed. Maximum recommended column loadings for diatomaceous earth supports is as follows;

Pink (e.g. Chromosorb P) 30 wt%
White (e.g. Chromosorb W) 25 wt%
Oyster (e.g. Chromosorb G) 5 wt%

Teflon supports are difficult to coat evenly. Loadings as low as possible should be used. Glass beads can only be coated with 1.5% or less of a liquid phase.

17.6.2 *Selection of Proper Support Material.* The most generally used support material is a diatomaceous earth. The pink supports are harder and less friable than the white supports. The surface of the pink supports, however, is active and should be used for the analysis of nonpolar compounds. The white supports are more inert than the pink but they are extremely fragile. They are mainly used for the analysis of medium polarity compounds. The oyster supports are the most inert of the diatomaceous earth supports. They are most useful for the analysis of polar compounds. Additional treatment of these supports (acid washing, base washing, DMCS (dimethyldichlorosilane) or HMDS (hexamethyldisilazane) treatment) further deactivates their surfaces. Teflon supports are used where a highly inert surface is required to avoid tailing of such compounds as water, hydrazine, sulfur dioxide and ammonia.

17.6.3 *Selection of the Proper Column Length.* The column lengths, of course, must be determined by the analytical requirement. However, generally 6 to 12 ft (2-4 m) packed columns are used. Capillary columns vary in length from 10 to 100 m. The column efficiency is increased by increasing the column length; however, the analysis time is also increased with column length. The proper column must, therefore, take both efficiency and time of analysis into account.

17.6.4 *Gas-Solid Chromatography.* The most widely used adsorbents are:

a. Silica gel–used in the analysis of inorganic gases and light hydrocarbons;

b. Molecular Sieves — for the separation of permanent gases such as hydrogen, oxygen, nitrogen, carbon monoxide, methane and ethane. Carbon dioxide and higher hydrocarbons are irreversibly adsorbed on Molecular Sieves at low temperatures. Sieves are available in types: e.g., 4A, 5A and 13X;

c. activated charcoal — for the separation of air, CO, CH_4, CO_2, C_2H_2, C_2H_4, C_2H_6 and nitrous oxide;

d. activated alumina — for the separation of air, CO, CH_4, CO_2, C_2H_2, C_3H_6 and C_3H_8;

e. porous polymers–used for the analysis of gases and small, short-chain, polar molecules such as glycols, acids and amines. They are used extensively for analysis of aqueous samples and for determination of trace components in water. Porapak Q and Chromosorb 102 are similar and are divi-

nylbenzene polymers. By changing the chemical composition of the polymers to produce materials such as Porapak S or Chromosorb 103, the selectivity is changed. All porous polymers require pretreatment before use. The columns should be purged with carrier gas while heating to rid the column of residual solvent and volatile monomer and low polymer.

Conditioning time and temperature for adsorbents are as follows:

a. alumina, silica gel and activated charcoal — 4 h at 150°C.

b. Molecular Sieves — 4 h at 300°C

c. Porapak N and T — 4 h at 180°C

d. Porapak P, Q, R, S, PS and QS-4 h at 230°C

e. Chromosorb 101 and 103 — 4 h at 275°C; and

f. Chromosorb 102, 104, 105, 106, 107 and 108 — 4 h at 250°C

17.6.5 *Solid Supports.* The purpose of the solid support is to provide a large surface area for holding a fairly thin film of liquid phase. The main requirements for adequate support material are: chemical inertness and stability, large surface area, relatively low pressure drop and mechanical strength. The most generally used materials consist of diatomaceous earth processed or modified in various ways. An example of diatomaceous earth supports is the Chromosorb series (Johns-Manville Corporation):

Chromosorb P — calcined diatomaceous earth processed from firebrick (C-22)

Chromosorb W — flux calcined diatomite prepared from Celite filter aids

Chromosorb G — developed especially for use in gas chromatographic analysis

Since diatomaceous earth supports are not completely inert, they are often treated chemically to inactivate them. Types of treatment include:

a. Non-acid washed (NAW) — untreated form;

b. Acid washed — treatment with hydrochloric acid to remove mineral impurities from the support surface and to reduce surface activity;

c. Hexamethyldisilizane treated (HMDS) —

treated with hexamethyldisilizane to decrease surface activity;

d. High performance (HP) — acid washed and dimethyldichlorosilane treated to yield the most efficient inert grade of support.

Besides the various diatomaceous earth supports, porous polymer beads, Teflon and glass beads are used as solid supports.

The primary requirement of a stationary or liquid phase is to provide sample separation while maintaining phase integrity over a reasonable period of time. In other words, the phase must have a reasonable chemical and thermal stability in addition to having suitable selectivity. The choice of a stationary phase for a specific separation with a given selectivity can be based on experience, literature-reported results and/or experimentation. A systematic method for expressing selectivity of a stationary phase for a particular compound is the degree to which the retention index of that compound on that phase differs from the retention index of the same compound on a non-polar column. Considerations when choosing a phase, in addition to selectivity, include peak symmetry, analysis time, and degree of separation. If peaks tail severely, the column is incompatible with the sample. When the sample is strongly acidic, the packing should be acidic; conversely, strongly basic samples require basic packings. If the sample contains hydroxyl groups, as in the case of alcohols, a stationary phase or a tail reducer that also contains hydroxyl groups is needed. Some stationary phases that have been used for separation of different classes of compounds are:

a. Acids — Chromosorb 102, Porapak, FFAP;

b. Alcohols — Porapak Q, Chromosorb 102, OV-101, Carbowax 20M;

c. Aldehydes — Carbowax 1540, Carbowax 20M, SE-30;

d. Alkaloids — SE-30, OV-1, OV-210;

e. Amines — Versamid 900, Apiezon L;

f. Amides — Chromosorb 103, Carbowax 20M + 2% KOH;

g. Amino acid derivatives — EGA, OV-17;

h. Carbohydrate derivatives — OV-101, SE-30, OV-225;

i. Drugs — OV-1, SE-30, OV-17, Dexsil 300 GC, SE-30;

j. Esters — FFAP, Silar 10C, Dexsil 300 GC, OV-101;

k. Ethers — Carbowax 20M, SE-30;

l. Gases — Molecular Sieves 5A or 13X, Porapak Q, N or T, Chromosorb 102, 104 or 105;

m. Glycols — Porapak P, Chromosorb 102;

n. Halogenated compounds — Carbowax 20M, OV-210, SE-30

o. Hydrocarbons — Squalene, Chromosorb 102, Porasil B, Porapak Q, Apiezon L, N,N-bis (cyanoethyl) formamide, DC-200, Bentone 34, Dexsil 300 GC;

p. Ketones — FFAP, DC-550, Porapak Q, Carbowax 20M;

q. Nitriles — FFAP, OV-225;

r. Pesticides — DC-200, OV-101, SE-30, OV-17, OV-225, OV-210, Carbowax 20M;

s. Phenols — OV-1, OV-101, SE-30, OV-17;

t. Steroids — OV-1, OV-101, SE-30, DC-200, Dexsil 300 GC;

u. Sugar derivatives — Dexsil 300, OV-17, OV-210;

v. Sulfur compounds — Porapak QS, Chromosorb 102, Chromosorb 104;

w. Terpenoids — Apiezon L, SE-30, Carbowax 20M.

The following compilations are useful to the determination of proper column and operation conditions for analysis:

Gas Chromatographic Data, ASTM Special Technical Publication No. 343.
Gas Chromatographic Data Compilation, ASTM Special Technical Publication No. DS-25A.
Gas Chromatographic Data Compilation, First Supplement, ASTM Special Technical Publication No. AMD 25A SM1

The above mentioned publications are available from the American Society for Testing and Materials, 1916 Race St., Philadelphia, PA., 19103.

Another helpful resource is *Gas Chromatography Literature, Abstracts and Index*, Preston Technical Abstracts Co., PO Box 312, Niles, IL 60648.

Major technical journals dealing with chromatography are: *Journal of Chromatographic Science, Journal of Chromatography, Chromatographia, Journal of High Resolution Chromatography, Chromatography Communications, Analytical Chemistry, Analyst, Analytical Letters,* and

Journal of the Association of Official Analytical Chemists.

17.7 QUALITATIVE ANALYSIS.

17.7.1 From the foregoing information, it is obvious that practically any mixture can be separated by gas chromatographic techniques. The only problem is the qualitative determination of the mixture components. The separation achieved depends on the column, the temperature, the detector, and the flow rate. Therefore, it is imperative that all these parameters be kept the same for both the sample and the standard used for the determination of the component peak location. It is important to know that the unknown component is eluted in the same time as the known compound. It is also important to know that no other compound can appear at this location with the parameters used. Figure 17:8 illustrates the use of a known sample to determine the unknown components of the sample.

17.7.2 A more specific method of qualitative analysis is the use of auxiliary instrumentation such as infrared, ultraviolet or visible spectrophotometry or mass spectrometric analysis. The sample components are trapped at the outlet from the gas chromatograph and then transferred to the appropriate instrument and a qualitative analysis is performed; alternatively, systems are available in which a dedicated spectrophotometer or mass spectrometer is appropriately coupled directly to the column outlet.

17.8 QUANTITATIVE ANALYSIS

17.8.1 The primary application of gas chromatography is quantitative analysis. For most detectors the area under a chromatographic peak is proportional to the amount of sample component in the carrier gas stream. This means that in order to be able to use gas chromatography for quantitative analysis, one must know, first, the area of the peak and second, the proportionality factor to convert this area to concentration.

Areas can determined by many techniques; the most frequently used techniques in modern instrumentation utilize electronic integration of the area under the detector response versus time curve. Other approaches utilize geometric approximations or graphical techniques, but such approaches are rarely used.

17.8.2 The most commonly used methods in current instrumentation are electronic and display integrated area in units of square centimeters. In the ideal case, where detector response is the same for all components in a mixture, a simple relationship is used to calculate percentage. As an example, assume that we have a four-component mixture of methane, ethane, propane and butane. The areas are 2.50, 1.25, 5.00 and 0.625 square centimeters respectively. The total area is 9.375 square centimeters and thus the percentage distribution is Methane = 26.67, Ethane = 13.33, Propane = 53.33 and Butane = 6.67 %.

The ideal case does not always apply, as the response factors are different for individual components and those must be taken into account before calculation of percentage composition. Microprocessor-based instruments will calculate and store response factors and report concentrations of components in either normalized or unnormalized form.

17.8.3 Standard Calibration methods are numerous and only one will be discussed here. In this method, standards are run and the detector response is plotted versus concentration as shown in Figure 17:9.

17.9 CALIBRATION.

17.9.1 For accurate quantitative analysis, the gas chromatographic system must be calibrated against known concentrations of the component of interest. Several methods of preparing known concentration of gases and vapors for use in calibration are available and reported in the literature.

Known concentrations of gases are generally prepared by dynamic means in which known amounts of the gas are mixed with diluting gas to yield the required concentration. A simple system is shown in Figure 17:10.

17.9.2 A second method which may be used either for the preparation of known concentrations of gases or vapors in air is the static method where a known volume of gas or volatile liquid is introduced (with an accurate volume measuring device) through a septum into a rigid container (which has been previously evacuated) of known volume. The gas or vapor (from the liquid) is then mixed with the diluting gas and stirred

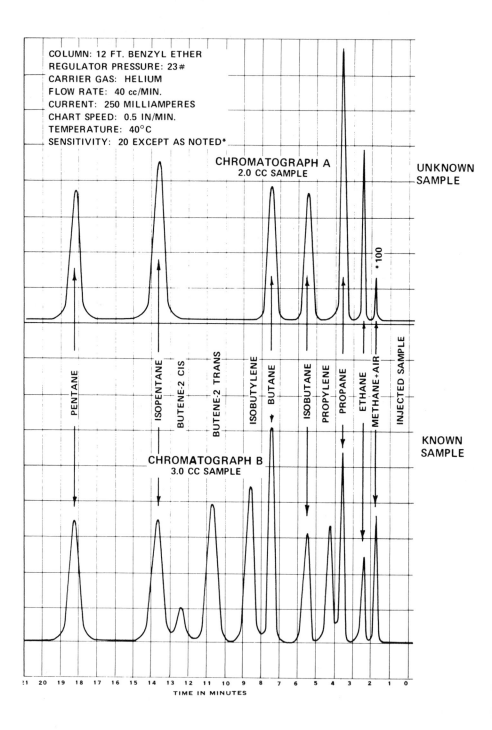

COLUMN: 12 FT. BENZYL ETHER
REGULATOR PRESSURE: 23#
CARRIER GAS: HELIUM
FLOW RATE: 40 cc/MIN.
CURRENT: 250 MILLIAMPERES
CHART SPEED: 0.5 IN/MIN.
TEMPERATURE: 40°C
SENSITIVITY: 20 EXCEPT AS NOTED*

CHROMATOGRAPH A
2.0 CC SAMPLE

UNKNOWN SAMPLE

* 100

PENTANE
ISOPENTANE
BUTENE-2 CIS
BUTENE-2 TRANS
ISOBUTYLENE
BUTANE
ISOBUTANE
PROPYLENE
PROPANE
ETHANE
METHANE+AIR
INJECTED SAMPLE

KNOWN SAMPLE

CHROMATOGRAPH B
3.0 CC SAMPLE

21 20 19 18 17 16 15 14 13 12 11 10 9 8 7 6 5 4 3 2 1 0
TIME IN MINUTES

Figure 17:8 — Use of known sample to determine unknown components of sample.

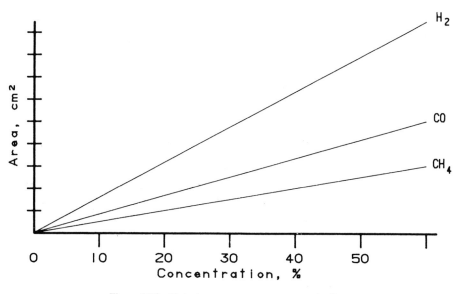

Figure 17:9 – Detector response versus concentration.

either by mechanical or thermal methods and samples are withdrawn from the vessel for use in calibration. This static system is shown in Figure 17:11.

17.9.3 Samples for either calibration system can be introduced directly to the gas sampling valve of the chromatograph (if available) or may be contained in a plastic or rubber bag and samples taken from the bag with gas tight syringes and introduced into the chromatograph.

17.9.4 For accurate calibration, the concentration of the known mixture must be determined by standard chemical methods. A convenient method for the determi-

nation of the concentration of many components is by infrared analysis.

17.9.5 There are several companies that market calibration gas standards contained in high pressure cylinders suitable for use with gas chromatogaphs. These calibration standards are very convenient to use, but should have the internal concentration verified before use, due to possible inaccuracies in the claimed concentration value or target value as received from the

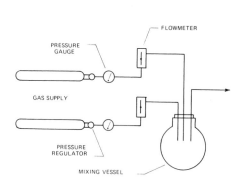

Figure 17:10 – A simple mixing system.

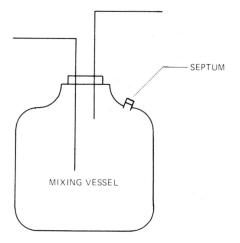

Figure 17:11 – Static system.

vendor or because of possible reaction of the calibration standard with the materials of cylinder construction.

17.10 METHOD FOR THE PRESENTATION OF CHROMATOGRAPHIC DATA.

17.10.1 To be able to reproduce your own work, or that of other investigators, the pertinent information concerning the analyses must be made available. The following information is recommended in chromatographic reports (5):

Apparatus
  Instrument.
  Detector type — Thermal conductivity, flame ionization, electron capture, etc.
    Recorder range.
    Detector voltage.
    Bridge current.
Column
  Length and diameter — Inside diameter is preferred.
  Material — Glass, copper, stainless steel, etc.
  Packing — Weight % of liquid phase on support (give mesh size and pretreatment of support, e.g. silanization).
  Capillary — No packing, only liquid phase — state liquid phase and film thickness coated on the column.
Temperature
  Injection port.
  Detector.
  Column (oven) — Isothermal or temperature programmed — give temperature or initial and final temperature and rate of temperature rise.
Flow rate of Gases
  In milliliters per minute at the exit port.
  Carrier gas.
  Other gases.
Sample
  Volume injected in microliters or milliliters if a gas, or if a solution, the concentration of the injected solution.
  Solvent used.
  Retention time of compounds expressed in minutes.
  Relative retention time.
Quantitation
  Methods. Peak area, peak height, integrator, planimeter etc.
  Precision. Repeatability.
  Recovery of added amounts.

Weight. Expressed as weight, volume or mole percent.
  Minimum detectable limits; basis, 2x noise level or 2x interference.
  Standards. Internal standard used for calibration.
  Interference of substrate. Baseline correction.
Chromatogram
  Typical analysis.
  Retention time (horizontal) vs. recorder response (vertical).
  Labels, peaks, attenuation, temperature program, title.
  Gas holdup. Air peak, solvent front.
Special
  Column conditioning procedures.
  Adequate extraction (exhaustive) of compounds from sample.
  Specificity of analysis.
  Speed of analysis.
  Interfering compounds related to those analyzed.
  Dual column operation, background subtraction.
  Reaction gas chromatography — e.g., hydrogenation, pyrolysis, etc.
  Backflush technique.
  Modifications. Valves, injection devices, stream splitters, trapping, flow controllers, equipment for transfer to other instruments (e.g., spectrometers). Full description of material or apparatus not available commercially.
  Technique for selecting representative samples.
  Collaborative results.
  Errors to be avoided.
  How to keep the apparatus clean and functional.

REFERENCES

1. ETTRE, L. S. AND A. ZLATKIS. 1967. *The Practice of Gas Chromatography*. Interscience Publishers, New York, NY.
2. LITTLEWOOD, A. B. 1966. *Gas Chromatography-Principles, Techniques and Applications*. Academic Press, New York, NY.
3. KAISER, R. 1963. *Gas Phase Chromatography*-volumes I, II and III. Butterworths, Great Britain.
4. LODDING, W. 1967. *Gas Effluent Analysis*. Marcel Dekker, Inc., New York, NY.
5. ANON. 1970. Reporting of Gas Chromatographic Methods, J. Agr. Food Chem., 18:552ff.

6. MCNAIR, H. AND E. BONELLI. 1968. *Basic Gas Chromatography*, Consolidated Printers, Berkeley, Califonia.
7. SUPINA, W. 1979. The Packed Column in Gas Chromnatography, Supelco, Inc., Bellefonte, Pa.
8. WALKER, J. ET AL. 1977. *Chromatographic Systems*, Academic Press, New York, NY.
9. GROBB, R., ed. 1977. Modern Practice of Gas Chromatography, John Wiley & Sons, New York, NY.

Approved with modifications for 2nd edition

Subcommittee 9
DIANE SANDERS
J.N. HARMAN, *Chairman*

18. Radioactivity Analysis*

18.1 INTRODUCTION.

18.1.1 Radioactivity in samples of airborne particles or gases is measured by counting either *alpha, beta* or *gamma* emissions. The methods for detecting radiations vary, but the basic concept is to capture part or all of the energy released through decay, and to convert it into a current flow or pulse which can be recorded. A variety of sensitive instruments of the pulse recording type are available.

18.1.2 The gross count can be taken by totaling the pulses recorded on scalers. For spectral analysis, the pulses are sorted out by amplitude, a function of energy, and the number of pulses counted in each energy band is stored separately. A pulse height analyzer is used for this spectral analysis.

18.1.3 Most airborne radioactivity samples have low activity. Therefore, background radiation seen by the detector, such as natural activity in the earth and cosmic rays, must be minimized. Massive shielding and special electronic circuitry are used for this purpose.

18.1.4 Some of the important considerations in the selection of the proper instrument are:

a. The type of radiation emitted—*alpha, beta, gamma*;

b. the energy of the emission;

c. the condition of the sample when prepared for counting;

d. the activity level of the sample;

e. the radionuclide composition of the sample; and

f. the accuracy needed or desired.

18.1.5 The preparation of samples and standards for calibration of the selected instrument also depends upon the points mentioned above as well as on the type of data needed. Qualitative data identify the radionuclides present, whereas quantitative data determine how much of each is present. Spectrometry can provide both types of data depending on the extent to which the spectrum is analyzed. Without spectrometry, radiochemistry quantitatively isolates the element (a quantitative separation of the nuclide) so that the quantity of radionuclide present can be counted.

18.2 EQUIPMENT.

18.2.1 Instruments for measuring radioactivity make use of the ionizing properties of nuclear radiation. Basically, they provide a medium with which the radiation can interact and a means of detecting and identifying this interaction. All instruments can be classified by their interaction medium under three general categories: gas ionization detectors, scintillation detectors or phosphors, and solid state detectors.

18.2.2 When a charged particle such as an *alpha* or *beta* particle moves through a gas at high velocity it may act on a gas molecule or atom to remove an electron and produce an ion pair. Gas ionization detectors contain an enclosed volume of gas in which two collecting electrodes are placed. When a direct voltage is applied across these electrodes in the presence of ionizing radiation, five regions of response will be observed as the voltage is increased.

a. In the *ion recombination region* of response many of the ion pairs recombine with each other before reaching the collecting electrode. Gas ionization instruments are not operated in this region of response.

b. In the *ionization chamber region* of response the applied voltage is sufficient to prevent significant recombination of ion pairs and the ions are collected in a one-to-one ratio with those formed. Ionization chamber instruments operate in this region of response.

*The material in this section is based on "Radioassay Procedures for Environmental Samples," January 1967. U.S. Department of Health, Education, and Welfare, Public Health Service Publication No. 999-RH-27.

c. At voltages in the *proportional region* of response the ions are accelerated to achieve enough energy to produce secondary ionization in the gas medium, thus amplifying the primary ion current. Although the ion current or pulse size is amplified to an extent described by the "gas amplification factor," it is still proportional to the energy of the incident particle *beta* (or *gamma*) radiation. *Alpha* and *beta* proportional counters with or without windows to separate samples from the gas ionization medium operate at voltages in the proportional region of response.

d. Geiger counters, or GM counters, operate in the *Geiger-Mueller region* of response where the gas amplification factor is so large that an avalanche of electrons spreads along the entire length of the anode. Here the pulse size is independent of the number of primary ions or energy of the incident radiation. This accounts for the high sensitivity of GM detectors and their inability to distinguish between the various types of radiation.

e. In the *continuous discharge region* of response the gas amplification factor is so large that the gas begins to arc, generating a series of self-perpetuating discharges. Operations of any gas ionization instrument at voltages in this region even for a few seconds will seriously damage the electrodes.

18.2.3 Scintillation detectors make use of the long-known ability of ionizing radiation to produce short-lived flashes of light (scintillations) in phosphors. The light flashes can produce photoelectrons from a photosensitive cathode in a photomultiplier tube and these photoelectrons are further amplified into a pulse that can be counted. When the scintillator is large enough to absorb all the energy of the radiation in the excitation and ionization interaction, the output pulse will be proportional to the energy of the incident radiation. This proportional response is necessary for gamma scintillation spectrometry. Scintillators are of many materials including organic crystals such as anthracene, liquid solutions of an organic scintillator such as p-terphenyl in an organic solvent, solid solutions of an organic scintillator in a plastic solvent (plastic scintillation detectors), inorganic

crystals such as sodium iodide or zinc sulfide, and noble gases.

18.2.4 Solid state detectors make use of the property of some insulators such as diamond or silver chloride crystals, or semiconductors such as germanium or silicon, to show an instantaneous conductivity when high-energy particles or radiation interact with the material of which they are composed.

18.2.5 A useful description of the more common type of laboratory instruments used in radioactivity determinations may be found in "Common Laboratory Instruments for Measurement of Radioactivity." June 1968, U.S. Department of Health, Education, and Welfare, Public Health Service Publication No. 999-RH-32.

18.3 CALCULATION OF RESULTS AND COUNTING ERROR.

18.3.1 *Integral Counting.* Any of the integral counters provides a gross count rate that consists of counts from both sample and background activity. To determine that activity from the sample alone, A_s, the background count rate of the instrument with no sample present, but otherwise identical, B, must be obtained. A clean sample container, similar to that used for the sample, should be in place for the background determination. The background count rate is subtracted from the gross count rate, S, to obtain a net count rate. This net count rate is then divided by the counting efficiency, E, previously determined for the particular sample type in this instrument, to obtain the disintegration rate, D, of the sample. Thus

$$D = \frac{S - B}{E}$$

where D is in disintegrations per minute (Bq × 60, S and B in counts per minute, and E in counts per minute/disintegration per minute. See 18.6 for units.)

To express the resulting activity in more customary concentration units of pCi/L,

$$A_s = \frac{D}{2.22\ V}$$

where V is sample volume in liters, and the constant is the conversion factor from disintegrations per minute to picocuries.

A large enough sample must be taken to insure that the measured disintegration rate is large compared with the counting error, and hence statistically significant. The counting error, E, is given by

$$E = Z(\frac{S}{t_s} + \frac{B}{t_b})^{1/2}$$

where t_s and t_b are the duration of counting (in minutes) for sample and background, respectively, and Z is the normal deviate corresponding to the confidence level desired (two-tailed).

As an example, assume the gross count rate of the sample to be 120 counts/min and the background count rate to be 40 counts/min. If the counting time of the sample is 30 min. and the counting time of the background is 100 min. the counting error at the 95% confidence level will be

$$E = \pm 1.96 \left(\frac{120}{30} + \frac{40}{100}\right)^{1/2}$$

$$= \pm 1.96 (4 + 0.4)^{1/2}$$

$$= \pm 4.1$$

The sample then has a net count rate of 80 ± 4.1 counts/min. or slightly more than a 5% counting error at the 95% confidence level. This means that we are 95% certain that the true count rate for sample lies between 75.9 and 84.1 counts/min.

When the conversion is made from counts to disintegrations, care must be taken to obtain the proper efficiency value. For *alpha* or *beta* counting, self-absorption in the sample can be very significant, and appropriate efficiency values must be determined and applied. For particle counting, either *alpha* or *beta*, backscatter from the sample container contributes to the count rate, and the efficiency values must be obtained for the proper container, which is the same container in which the samples will be counted.

18.3.2 *Spectral Analysis.* It should be noted that the counting error for each radionuclide in a complex spectrum is difficult to determine mathematically, and that the method used for integral counting does not apply. There are errors associated with each radionuclide in its own and other photopeak areas. In general, the more radionuclides present in the spectrum, the larger the error becomes for any single nuclide, but the percentage of interference is of primary importance. It should also be noted that the low-energy gamma photons do not add to the counting error of high-energy gamma photons.

Descriptive material on the calculation of results of gamma analyses can be found in "Radioassay Procedures for Environmental Samples," January 1967. U.S. Department of Health, Education, and Welfare, 999-RH-27. Computer calculation of results is described in a "Computer Program for the Analyses of Gamma Ray Spectra by the Method of Least Squares." August 1966. U.S. Department of Health, Education, and Welfare, Public Health Service Publication No. 999-RH-21.

18.4 RADIOACTIVE STANDARDS.

18.4.1 The proper selection and proper use of standards for calibrating any detection system are both essential to obtain reliable quantitative results.

18.4.2 The decay scheme used in standardization should be identified by the supplier, since there are discrepancies in reported decay schemes for most nuclides, and for some methods of standardization a knowledge of the decay scheme is essential. The amount and the chemical form of carrier added to the standard can also be of importance in minimizing adsorption on the container walls or volatilization of the solution. Such information is needed to make accurate dilutions for calibrating, since most standards have too high a disintegration rate to be counted on sensitive laboratory instruments without prior dilution.

18.4.3 It is essential that any standard prepared for instrument calibration be counted in the same geometrical configuration as the samples.

18.5 QUALITY CONTROL.

18.5.1 Application of quality control principles to the analysis of airborne radioactivity measurements can minimize the amount of disparity within and among laboratories to a considerable extent. This is done by isolating discrepancies in analytical results and taking appropriate action to remedy the causative factor. While quality control procedures can greatly assist in obtaining consistent data, they obviously cannot guarantee accuracy. For maintaining production of precise, unbiased data, quality control is a continuous task.

18.5.2 Accurate data are data that are both precise and unbiased. Complete control over analytical measurements requires day-to-day control over instrumentation, chemical steps, and associated factors at each individual laboratory. The application of three allied but independent procedures has been found most useful for insuring accurate data. The procedures are:

a. "Blind" duplicate analysis of actual samples within one laboratory. This procedure allows a laboratory to evaluate its internal precision or reproducibility;
b. cross-check analysis of samples among several laboratories. This intercomparison procedure allows a laboratory to determine its agreement with other laboratories doing similar work;
c. special standard sample analysis. This procedure allows a laboratory to study its accuracy, *i.e.,* the agreement of its results with a known value.

18.5.3 Each of the three quality control procedures supplies the observer with different but complementary information. Depending on the number or scope of the analysis, one or more can be employed.
18.6 UNITS OF RADIOACTIVITY. The original unit of radioactivity was the *curie*, (Ci) defined as that activity resulting from the decay of 1 g of pure radium. This is equivalent to 3.7×10^{10} disintegrations/s, or $2.22 \cdot 10^{12}$ disintegrations per min. The SI unit, just coming into use, is the bequerel (Bq), equal to 1 disintegration per second. Because of its wide use, the curie is "temporarily" still approved for use, and it takes the same prefixes as other SI units. For most environmental monitoring the convenient unit is the picocurie (pCi), 10^{-12} Ci, equal to 37 mBq.
18.7 SAFETY.
18.7.1 Analyses of airborne radioactivity samples require no unusual safety measures. Although the procedures are for radioassay, the samples themselves constitute no hazard greater than that of the environment from which they were taken. The most likely source of potential radiation hazard is the undiluted radioactive solution obtained for purposes of preparing spiked samples for tracer studies or standards for calibration, performance references, or quality control.

18.7.2 People who are to work with radioactive materials should be thoroughly trained in their proper use. Basic radiological safety regulations appropriate to the license held by the laboratory for use of such materials and machines are set forth by the Atomic Energy Commission in the Code of Federal Regulations. Title 10 — Atomic Energy. (United States Government. "Standards for protection against radiation," Code of Federal Regulations, Title 10 — Atomic Energy, Chapter II, Section II, Part 20. U.S. Atomic Energy Commission, Revised 1966, Washington, Government Printing Office (1963).

Approved with modifications from the 2nd edition

Subcommittee 9
J.N. HARMAN, *Chairman*

19. Precision and Accuracy*

19.1 INTRODUCTION. A clear distinction should be made between the terms "precision" and "accuracy" when they are applied to methods of analysis. *Precision* refers to the reproducibility of a method when it is repeated on a homogeneous sample under controlled conditions, regardless of whether or not the observed values are widely displaced from the true value as a result of systematic or constant errors present throughout the measurements. Precision can be expressed by the standard deviation. *Accuracy* refers to the agreement between the amount of a component measured by the test method and the amount actually present. *Relative error* expresses the difference between the measured and the actual amounts, as a percentage of the actual amount. A method may have very high precision but recover only a part of the element being determined, or an analysis, although precise, may be in error because of poorly standardized solutions, inaccurate dilution techniques, inaccurate balance weights, or improperly calibrated equipment. On the other hand, a method may be accurate but lack precision because of low instrument sensitivity, variable rate of bio-

---

*Modified and adapted from Standard Methods for the Examination of Water and Waste water. (14th Ed.). 1976. Part 104A. pp. 20–26.

logical activity, or other factors beyond the control of the analyst.

It is possible to determine both the precision and the accuracy of a test method by analyzing samples to which known quantities of standard substances have been added. It is possible to determine the precision, but not the accuracy, of such methods as those for suspended particulate matter, polycyclic aromatic hydrocarbons and many other contaminants because of the unavailability of standard substances that can be added in known quantities on which percentage recovery can be based.

19.2 STATISTICAL APPROACH.

19.2.1 *Standard Deviation (σ).* Experience has shown that if a determination is repeated a large number of times under essentially the same conditions, the observed values, x, will be distributed at random about an average as a result of uncontrollable or experimental errors. If there is an infinite number of observations from a common universe of causes, a plot of the relative frequency against magnitude will produce a symmetrical bell-shaped curve known as the Gaussian or normal curve (Figure 19:1). The shape of this curve is completely defined by two statistical parameters: (1) the mean or average $\bar{x}$, of n observations; and (2) the standard deviation, σ, which fixes the width or spread of the curve on each side of the mean. The formula is:

$$\sigma = \sqrt{\frac{\Sigma(x - \bar{x})^2}{n - 1}}$$

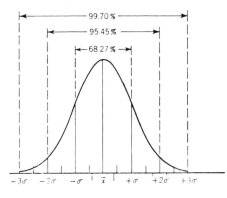

The proportion of the total observations lying within any given range about the mean is related to the standard deviation. For example, 68.27% of the observations lie between $\bar{x} \pm 1\,\sigma$, 95.45% between $\bar{x} \pm 2\,\sigma$, and 99.70% between $\bar{x} \pm 3\,\sigma$. These limits do not apply exactly for any finite sample from a normal population; the agreement with them may be expected to be better as the number of observations, n, increases.

19.2.2 *Application of Standard Deviation.* If the standard deviation, σ, for a particular analytical procedure has been determined from a large number of samples, and a set of n replicates on a sample gives a mean result $\bar{x}$, there is a 95% chance that the true value of the mean for this sample lies within the values $\bar{x} \pm 1.96\sigma\sqrt{n}$. This range is known as the 95% confidence interval. It provides an estimate of the reliability of the mean, and may be used to forecast the number of replicates needed to secure suitable precision.

If the standard deviation is not known and is estimated from a single small sample†, or a few small samples, the 95% confidence interval of the mean of n observations is given by the equation $\bar{x} \pm t\sigma\sqrt{n}$, where t has the following values:

n	t
2	12.71
3	4.30
4	3.18
5	2.78
10	2.26
∞	1.96

The use of t compensates for the tendency of small samples to underestimate the variability.

19.2.3 *Range (R).* The difference between the smallest and largest of n observations is also closely related to the standard deviation. When the distribution of errors is normal in form, the range, R, of n observations exceeds the standard deviation times a factor d_n only in 5% of the cases. Values for the factor d_n are:

Figure 19:1 — Gaussian or normal curve of frequencies.

†A "small sample" in statistical discussions means a small number of replicate determinations, n, and does not refer to the quantity used for a determination.

n	d_n
2	2.77
3	3.32
4	3.63
5	3.86
6	4.03

As it is rather general practice to run replicate analyses, use of these limits is very convenient for detecting faulty technique, large sampling errors, or other assignable causes of variation.

19.2.4 *Rejection of Experimental Data.* Quite often in a series of observations, one or more of the results deviates greatly from the mean whereas the other values are in close agreement with the mean value. At this point, one must decide whether to reject disagreeing values. Theoretically, no results should be rejected, since the presence of disagreeing results shows faulty techniques and therefore casts doubt on all the results. Of course the result of any test in which a known error has occurred is rejected immediately. For methods for the rejection of other experimental data, standard texts on analytical chemistry or statistical measurement should be consulted.

19.3 GRAPHICAL REPRESENTATION OF DATA. Graphical representation of data is one of the simplest methods for showing the influence of one variable on another. Graphs frequently are desirable and advantageous in colorimetric analysis because they show any variation of one variable with respect to the other within specified limits.

19.3.1 *General.* Ordinary rectangular-coordinate paper is satisfactory for most purposes. Twenty lines per inch or one line per millimeter is often convenient. For some graphs, semilogarithmic paper is preferable.

The five rules listed by Worthing and Geffner for choosing the coordinate scales are useful. Although these rules are not inflexible, they are satisfactory. When doubt arises, common sense should prevail. The rules are:

a. The independent and dependent variables should be plotted on abscissa and ordinate in a manner that can be comprehended easily.

b. The scales should be chosen so that the value of either coordinate can be found quickly and easily.

c. The curve should cover as much of the graph paper as possible.

d. The scales should be chosen so that the slope of the curve approaches unity as nearly as possible.

e. Other things being equal, the variables should be chosen to give a plot that will be as nearly a straight line as possible.

The title of a graph should describe adequately what the plot is intended to show. Legends should be presented on the graph to clarify possible ambiguities. Complete information on the conditions under which the data were obtained should be included in the legend.

19.3.2 *Method of Least Squares.* If sufficient points are available and the functional relationship between the two variables is well defined, a smooth curve can be drawn through the points. If the function is not well defined, as is frequently the case when experimental data are used, the method of least squares may be used to fit a straight line to the pattern.

Any straight line can be represented by the equation $x = my + b$. The slope of the line is represented by the constant m and the intercept (on the x axis) is represented by the constant b. The method of least squares has the advantage of giving a set of values for these constants not dependent upon the judgment of the investigator. Two equations in addition to the one for a straight line are involved in these calculations:

$$m = \frac{n\Sigma xy - \Sigma x\Sigma y}{n\Sigma y^2 - (\Sigma y)^2}$$

$$b = \frac{\Sigma y^2\Sigma x - \Sigma y\Sigma xy}{n\Sigma y^2 - (\Sigma y)^2}$$

n being the number of observations (sets of x and y values) to be summed. In order to compute the constants by this method, it is necessary first to calculate Σx, Σy, Σy^2, and Σxy. These operations are carried out to more places than the number of significant figures in the experimental data because the experimental values are assumed to be exact for the purposes of the calculations.

Example: Given the following data to be graphed, find the best line to fit the points:

Absorbance	Solute Concentration mg/L
0.10	29.8
0.20	32.6
0.30	38.1
0.40	39.2
0.50	41.3
0.60	44.1
0.70	48.7

Let y equal the absorbance values, which are subject to error, and x the accurately known concentration of solute. The first step is to find the summations (Σ) of x, y, y^2, and xy:

x	y	y^2	xy
29.8	0.10	0.01	2.98
32.6	0.20	0.04	6.52
38.1	0.30	0.09	11.43
39.2	0.40	0.16	15.68
41.3	0.50	0.25	20.65
44.1	0.60	0.36	26.46
48.7	0.70	0.49	34.09
$\Sigma = 273.8$	2.80	1.40	117.81

Next substitute the summations in the equations for m and b; $n = 7$ because there are seven sets of x and y values:

$$m = \frac{7(117.81) - 2.80(273.8)}{7(1.40) - (2.80)^2} = 29.6$$

$$b = \frac{1.4(273.8) - 2.80(117.81)}{7(1.40) - (2.80)^2} = 27.27$$

To plot the line, select three convenient values of y — say, 0, 0.20, 0.60 — and calculate the corresponding values of x:

$$x_0 = 29.6(0) + 27.27 = 27.27$$
$$x_1 = 29.6(0.20) + 27.27 = 33.19$$
$$x_2 = 29.6(0.60) + 27.27 = 45.04$$

When the points representing these values are plotted on the graph, they will lie in a straight line (unless an error in calculation has been made) that is the line of best fit for the given data. The points representing the latter are also plotted on the graph, as in Figure 19:2. Today many computers and programmable calculators have built-in programs for curve-fitting, removing the necessity of doing more than entering the data pairs.

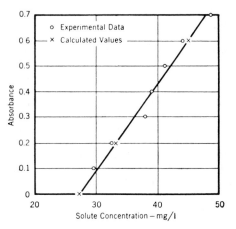

Figure 19:2 — Example of least-squares method.

19.4 SELF-EVALUATION (DESIRABLE PHILOSOPHY FOR THE ANALYST). A good analyst continually tempers his confidence with doubt. Such doubt stimulates a search for new and different methods of confirmation for his reassurance. Frequent self-appraisals should embrace every step — from collecting samples to reporting results.

The analyst's first critical scrutiny should be directed at the entire sample collection process in order to guarantee a representative sample for the purpose of the analysis and to avoid any possible losses or contamination during the act of collection. Attention should also be given to the type of container and to the manner of transport and storage, as discussed elsewhere in this volume.

A periodic reassessment should be made of the available analytical methods, with an eye to applicability for the purpose and the situation. In addition, each method selected must be evaluated by the analyst himself for sensitivity, precision, and accuracy, because only in this way can he determine whether his technique is satisfactory and whether he has interpreted the directions properly. Self-evaluation on these points can give the analyst confidence in the value and significance of his reported results.

The benefits of less rigid intralaboratory as well as interlaboratory evaluations deserve serious consideration. The analyst can regularly check standard or unknown concentrations with and without interfering elements and compare results on the same sample with results obtained by other workers in the laboratory. Such programs can uncover weaknesses in the analytical chain and permit improvements to be instituted without delay. The results can disclose whether the trouble stems from faulty sample treatment, improper elimination of interference, poor calibration practices, sloppy experimental technique, impure or incorrectly standardized reagents, defective instrumentation, or even inadvertent mistakes in arithmetic.

Approved with modifications from the 2nd edition

Subcommittee 9
J.N. HARMAN, *Chairman*

20. General Safety Practices

20.1 INTRODUCTION. The safety practices described below are the result of many decades of experience in academic and industrial laboratories. Some of them appear too obvious to be mentioned, but at least one unwary person has suffered at the hands of each of them.

20.2 PERSONAL PROTECTION. The corrosive nature of most chemicals on the eyes and the consequences of blindness are so severe that some form of eye protection is essential. This can be spectacles with specially armored lenses, which are available from most optical suppliers, either plain or with prescription lenses and with or without side shields. These glasses would offer protection against the unexpected. Where a splash or flying particles are expected, a face shield should be worn.

There are many types of gloves available: rubber for hazardous liquids, leather for physical hazards such as glassware and dry ice, mineral fiber for very hot objects, and even polyethylene hand guards to keep loose or dusty chemicals off the hands. Experience has shown that the majority of laboratory injuries is to the hands; proper use of gloves could have prevented most of them. All the special equipment mentioned above can be obtained from most laboratory supply houses.

It is essential that contact of the body with chemicals be kept to a minimum. The use of laboratory coats or aprons, the use of gloves, frequent showers and change of clothing will help. If a chemical does touch the body, it should be removed promptly by thorough washing with water. If the chemical is known to be toxic, or if damage to the skin or eye results, a qualified physician should be consulted, preferably one who is familiar with the action of chemicals on the body.

For the same reason, the consumption of food or beverage in a chemical laboratory should be avoided wherever possible. Hands and face should be washed before eating. Chemicals should not be stored in refrigerators containing food.

Smoking in a laboratory not only creates a fire hazard, but offers one more possibility for ingesting chemicals. It must not be done near any flammable solvents, and the part that contacts the mouth must be protected from contamination by chemicals.

No one should work alone in a laboratory without arranging in advance for frequent contacts with some other person who can summon aid in case he is overcome.

20:3 ELECTRICAL EQUIPMENT. Electrical equipment (even 110 V) has been known to kill people. Whenever possible, the chassis should be grounded so that any short will blow a fuse and not subject someone to electrocution. Electrical repairs should not be attempted by those unfamiliar with the equipment, and should be done with the equipment disconnected. Working on "hot" circuits is for professionals only. Some electrical equipment can cause a spark, and should not be used near flammable solvents.

20:4 LABORATORY MATERIALS. Many accidents have been caused by use of unlabeled or poorly labeled chemicals. Keep all chemicals clearly labeled and discard any that are not so.

Recent legislation has been enacted in an effort to reduce the incidence of illness and injuries due to chemical sources. The purpose of the enacted standard is to assure that chemical hazards are evaluated and that hazard information is transmitted to affected individuals. This hazard information is communicated by means of a material safety data sheet (MSDS) form, which must be readily available to the affected worker at the point of use **(16)**. This document is a useful resource for worker training programs and provides useful emergency medical treatment instructions.

It is the legal responsibility of all chemical manufacturers and importers to provide an MSDS to all downstream users of toxic chemicals. A strategy to ensure that MSDS sheets are supplied is to require them as an integral part of any purchase order for a toxic chemical. An additional benefit to the user of the chemical is that the MSDS documentation gives a recommendation for the disposal of that material when it is a waste.

Because of the fire hazard, the amount of flammable solvents should be kept at a minimum. Quantities in excess of 5 gal. (19 L) should be stored outside the main building. Metal safety cans having flame arresters in the pouring spouts and a spring-loaded lid on the spout should be used for quantities from 1 L to 5 gal (19 L). Smaller amounts may be stored in glass, but only one bottle of each solvent should be in the laboratory.

Acid containers should be stored in an acid-proof tray that will hold the entire contents should the bottle break. Do not store chemicals that may react with each other in the same tray, e.g., acids and ammonium hydroxide. To dispense small amounts more easily, 500-mL reagent bottles may be filled from the larger bottles.

Flammable solvents constitute a severe fire hazard and must be handled with care. Heat only on a steam or water bath, or spark-proof hot plates. Ground all metal containers when pouring from one to another.

Strong oxidizing agents (such as chlorates, perchlorates, permanganates, hypochlorites, peroxides, etc.) can cause a fire or explosion in contact with organic material. Clean up any spilled material promptly. Do not throw waste material in the trash can, but flush it down the sink with plenty of water. Do not use perchlorate drying agents (Dehydrite, Desichloride or Anhydrone) for drying organic materials.

The free acids of such materials as potassium permanganate and potassium chlorate are very unstable. Keep strong acids away from these materials. Strong hydrogen peroxide solutions can cause burns. Handle them with rubber gloves.

It is frequently desirable to use flammable solvents in wash bottles. Plastic or plastic-coated bottles are much safer than glass for this purpose. Flexible plastic bottles have the additional advantage that the liquid may be dispensed by squeezing the bottle.

The use of open flames for heating in the analytical laboratory is decreasing due to the hazards involved. Hot plates, particularly spark-proof units, are more convenient and much safer.

Heating of solvents in open vessels is dangerous and should be done only in a hood. Large quantities (250 mL or more) should be heated with a metal pan under the glass vessel. Certain solvents, such as ethyl ether, dioxane and Cellosolve can form peroxides. These solvents should never be evaporated to dryness. Empty glass vessels should not be heated. Severe strains can be set up, causing the vessel to fail unexpectedly.

20.5 GLASSWARE. Glassware represents one of the chief hazards in the laboratory due to the ease with which it is broken and the razor-sharp edges which broken glass presents. Broken or chipped glassware should be repaired, including reannealing, or discarded immediately. A special container should be provided for broken glassware to prevent injury to janitorial employees.

Beakers should not be carried by the top edge. Small beakers should be carried by gripping around the side. Large beakers should be carried with one hand supporting the bottom.

Pipets obviously were designed for filling by mouth suction, but this is a very dangerous practice. Apply vacuum only by a rubber bulb or a vacuum hose.

The neck of a volumetric flask is its weakest point because of the calibration ring. Flasks over 100-mL capacity should be handled with both hands. Avoid vigorous motion which might snap the neck from the base.

Graduated cylinders tip very easily. Place them at the back of the work bench.

20.6 LABORATORY OPERATIONS. When using Buchner funnels, apply vacuum *before* pouring material on the funnel. Accidents have occurred through connecting the hose to an air or gas outlet by mistake and blowing the material from the funnel.

When heating material in test tubes, do not point the mouth of the test tube at anyone. The material may spurt out. Apply heat only to test tubes of heat-resistant glass such as pyrex, and use a test tube holder while heating. Pour out the contents before leaving the test tubes for final cleaning.

Protect the hands with leather gloves when cutting glass tubing or inserting tubing or thermometers in rubber stoppers or tubing. Fire polish the sharp edges of glass tubing before using or storing. Remove frozen rubber stoppers or tubing from glass tubing by cutting with a knife. Wear leather gloves. An alternate method is to work a cork borer of the proper size between the glass and the stopper until the glass is free.

Wherever possible, substitute plastic for glass tubing and fittings. Plastic or metal "Y's" and "T's" are now available.

Patience is essential for the safe freeing of frozen stopcocks. Use a stopcock puller if available. Wear leather gloves. Tap the plug lightly with the wooden handle of a spatula and/or apply heat gently.

When glassware assemblies are erected, all glassware should be adequately supported by metal rings, tripods, or clamps. If heat is to be applied, arrange for rapid removal of the heat source if needed. When adjusting clamps, be careful not to put any lateral tension on the glass when the clamp is tightened. When arranging clamps in a vertical assembly, keep all the clamps on the same side of the vertical supporting rod. This makes backward and forward movement of the whole assembly possible. If the assembly is to contain dangerous solvents or mercury, a catch pan should be placed under it to contain the liquids in case of breakage.

Rubber tubing attached to glassware or plastic should be secured with wire or several wraps of a rubber band to keep it from coming off.

20.7 DEWAR FLASKS. If a Dewar flask should break, the resulting implosion can scatter sharp glass fragments over a wide area. To guard against this hazard, Dewar flasks should be encased in metal jackets or covered with tape.

The tape to be used should have a cloth body (medical type) to give it sufficient strength. Masking tape or plastic electrical tape is not recommended. A 25% minimum overlap between rows should be used. The flask should be completely taped over the outside wall whenever possible. If it is necessary to view through the wall, the viewing area should be minimized and be shielded by taping a piece of clear plastic in place under the tape. The tape should be cut carefully to expose only as much plastic as necessary.

20.8 CHEMICAL EFFECTS. Chemicals can affect the human body in a variety of ways. Some of these are:

Dermatitis. This manifests itself as a reddening of the skin or even eruptions. This has been known to result in death in extreme cases. Immediate treatment by a physician familiar with chemical dermatitis is essential.

Cyanosis. This causes a bluish cast to the lips and fingernails, dizziness and headache, and may not appear until several hours after exposure. The consumption of alcoholic beverages will often aggravate the condition. The patient should be kept quiet and a physician summoned. Oxygen should be administered to help breathing. Cyanosis can be fatal; prompt treatment is essential.

20.9 HOODS. Laboratory fume hoods provide not only efficient ventilation, but also shielding for potentially dangerous reactions. However, their safe use requires proper attention to several factors.

Hoods should not be used for disposal of large amounts of noxious gases. Use a scrubbing or absorption system. Most analytical operations are conducted at benchtop level. Therefore, the hood should be equipped with a baffle in the rear to allow most of the draft to be at bench top. If manipulations require the hood window to be open during use, place all equipment at least 10″ in from the edge of the bench to make sure that no fumes are carried out of the hood by stray air currents.

Hoods should be tested at least once a year to insure their effectiveness. This is best done by producing a smoke (titanium tetrachloride ampoules are very convenient) and observing the hood's action visually. A satisfactory hood will not release any smoke into the room when the smoke is released 10″ in from the hood face. Even cross currents from windows or people walking should not cause smoke to escape from the hood.

20.10 PRESSURE PRECAUTIONS. Any reactions that may endanger personnel through splashes or explosion should be shielded. The shield should completely surround the reaction if it is set up on an open bench. All shields should be well supported to contain flying debris in case of an explosion.

Pressure vessels should be equipped with suitable relief devices, and should be tested hydrostatically at least once a year to 2½ times their working pressure.

Large cylinders (over 5 L) should be stored outside the laboratory and their contents piped in. Cylinders should be anchored at all times to prevent tipping. They should not be stored near radiators or other sources of heat. If it is necessary to heat a cylinder, first connect it to a suitable pressure relief device. Use water or steam baths or infrared lamps for heating. If gas from a cylinder is being released below the surface of a liquid, be sure to have a trap in the line to prevent suck-back into the cylinder. Systems to carry oxygen under pressure should be carefully cleaned of any oxidizable substance such as oil.

20.11 NEW EQUIPMENT. New equipment received in the laboratory can constitute a hazard because of the unfamiliarity of personnel with it. Also, the manufacturers do not engineer their products for safety as well as they might. Therefore, new equipment merits special attention.

Electrical equipment should have its chassis grounded by a three-wire cord. Hot areas, which might cause burns, or sharp edges which might cause cuts should be adequately guarded. Electrical connections should not be exposed. Any pressure equipment should have a suitable pressure relief device in working order.

After the preliminary examination is complete, a "dummy" run should be made, with special attention to any possible hazards. This, also, familiarizes personnel with the equipment. After all safety precautions have been taken, a run with a sample may be made.

20.12 SAFETY IN UNIT OPERATIONS OF ANALYTICAL CHEMISTRY.

20.12.1 *Introduction.* Fortunately, the hazards associated with analytical chemistry are minimized by the use of small samples and known reactions. Experience indicates, however, that analysts are subject to many hazards that can cause severe consequences if precautions are not taken. Problems arise from two major sources—the chemical nature of the sample, and hazards associated with the procedure (reagents, reactions and equipment).

Effective communication of safety information is very important for safe operation. In many cases, an analyst is called upon to analyze a sample prepared by someone else using a procedure developed by another analytical chemist. Accordingly, the analyst may have no first-hand knowledge of the hazards associated with either the sample or the procedure. This problem is most acute in an analytical laboratory providing non-routine analyses on samples from research groups. In order to ensure communication of safety information, personal contacts between the submitter and analyst are encouraged. It is also advisable to include safety information in formal written procedures or publications and on request-for-analysis forms.

Analytical procedures that are written in a standard format should include a definite safety section that reminds the writer to consider these problems and enables the reader to find the information conveniently. This section should include, when appropriate, the type and degree of hazard, routes for exposure, need for special medical department clearance, safety equipment to be used, instructions for special disposal of sample, decontamination of equipment, and procedure to be followed in case of an accident. Brief safety instructions should also be included where pertinent in the procedure section of the analytical method. If an analytical procedure is used to analyze many different kinds of samples, it is cumbersome to include sufficient information on sample hazards in the procedure and this information is then indicated on request-for-analysis forms. These forms should include a check list that reminds the submitter to make recommendations regarding precautions to be taken in handling the sample.

When an analytical research chemist is studying a new technique or an unknown sample, he may have no source of first-hand pertinent safety information. In this case, he should give considerable thought to the hazards that might be foreseen from general safety and chemical knowledge.

20.12.2 *Preparation of Sample.* Hazards associated with preparing liquid or solid samples for analysis are usually derived from the chemical nature of the sample. Before drying or grinding a sample, it is well to make sure that it is not sensitive to impact, friction, or heat. Many samples are corrosive or can cause dermatitis; contact with these chemicals must be avoided.

Materials that are gases at room temperature provide additional handling problems. Small samples at or below atmospheric pressure are usually taken in glass sampling bulbs. Hazards associated with the use of these containers are (*a*) the possibility of implosion during evacuation, (*b*) explosion from too great an internal pressure, and (*c*) contamination of laboratory air by leakage of toxic contents. Bulbs should be covered with a metal screen and evacuated behind a shield. Filling should be monitored with a manometer. Stopcock holders should be used and bulbs containing toxic materials should be stored in a hood.

Large gas samples at pressures exceeding atmospheric pressure are best handled in metal cylinders. The common method of transferring gaseous materials to cylinders involves their condensation at atmospheric pressure with a carbon dioxide ice-solvent or liquid nitrogen bath for cooling followed by their transferral into suitable cylinders. In addition to the possible toxicity of the gas, the following dangers are inherent in these procedures:

a. If the vapor pressure of the compound at ambient temperatures exceeds the service pressure of the cylinder, rupture of the cylinder may result. Interstate Commerce Commission (ICC) regulations require that a product's vapor pressure at 21°C (70°F) not exceed the service pressure of the cylinder and that the vapor pressure P at 55°C (130°F) not exceed 1.25 times the cylinder's service pressure (**3**). These regulations also require periodic hydrostatic testing of the cylinders;

b. if too much sample is loaded into a cylinder at a low temperature, expansion of the liquid during warming to room temper-

ature may cause the cylinder to become liquid-full and rupture explosively. This hazard is avoided by determining the maximum weight of sample that the cylinder will hold safely and weighing the cylinder before and immediately after filling to make sure that this weight has not been exceeded;

c. oxygen (boiling point, -183°C) from the air tends to condense in cylinders cooled with liquid nitrogen (bp -196°C) and creates a possible explosion hazard due to excess pressure build-up or explosive reaction with organic compounds. For this reason, a carbon dioxide ice-solvent bath is used where possible (*i.e.*, for compounds boiling above -50°C). When liquid nitrogen is used, precautions must be taken to avoid introduction of air into the system. Liquid air should not be used as a coolant since the less volatile oxygen is concentrated during evaporation and has been known to react violently with organic material which leaked from a broken glass trap;

d. common high-pressure cylinders are constructed of steel and may be weakened by crystallization at liquid nitrogen temperatures. Suitable stainless steel cylinders are available commercially;

e. during the transfer of a liquefied gas from a glass cold trap to a sampling cylinder, pressure buildup may occur due to a restriction in the lines or insufficient cooling. An open-end manometer should be included in the system to indicate such a problem so that corrective measures can be taken.

20.12.3 *Decomposition and Dissolution of Samples.*

a. Inorganic Samples — Dissolution of inorganic compounds by heating with acids or fusion fluxes requires careful handling of very corrosive reagents. Use of a hood is required if toxic gases are evolved. Perchloric acid must be used cautiously. Use of this acid with metallic bismuth and antimony (**1**), strong reducing agents (**12**) and finely divided metals requires special precautions. Routine use of perchloric acid requires a special hood. In general, dilution of the perchloric acid with water or nitric acid tempers the reaction (**1**), but specific information should be obtained for each system. Shields are recommended for perchloric acid dissolutions and fusions with strong oxidants. Leather gloves and face shields are sometimes appropriate.

Hydrofluoric acid is sometimes used for dissolution of siliceous materials and volatilization of silicon tetrafluoride. Both concentrated and dilute hydrofluoric acid will cause painful skin burns which heal slowly. Burns caused by dilute acid may not appear until many hours after exposure. Vapors of hydrofluoric acid will cause extreme irritation of eyes, skin, mucous membranes and lungs. Moderate concentrations of the vapor will cause first degree skin burns; large concentrations will cause pulmonary edema and death. Containers of the acid should be stored in a lead tray in a cool and well-protected location. Analysts should wear rubber gloves and a rubber apron when handling the reagent. Work should be carried out in a fume hood and the hood window or a safety shield should be used for protection of the analyst when the reagent is being used.

b. Organic Samples—The initial step in analysis of organic compounds is often destruction of the organic matter. It is usually at this step that the most concentrated reagents and most vigorous reactions are employed. As a result, this is often the most hazardous operation in the procedure. Prior knowledge of the physical and chemical nature of the sample and information on hazards of the reagents and reactions are very important for safe operation.

c. Dry Ashing—Dry ashing is often the least hazardous way of destroying organic materials. Heat-sensitive explosive compounds, *e.g.,* nitro compounds, must not be ashed until it is established that this can be done safely. Experiments with questionable materials should be carried out on a small scale and behind a barricade. Preliminary ashing of nitrogenous materials should be carried out in a hood since hydrocyanic acid is sometimes evolved.

Use of an oxygen bomb for combustion of organic materials with oxygen under pressure involves the potential hazard of explosive rupture of the bomb. However, if the equipment is kept in good condition and operating instructions are followed carefully, the technique can be utilized quite safely. Bombs should be hydrostatically tested periodically and the sample size should never exceed the amount specified by the manufacturers of the bomb. A barricade and a normally open ignition switch should be used.

The oxygen flask technique for combusting organic samples has been reported to be perfectly safe by some investigators; however, a few isolated explosions have occurred with the hazard of flying glass from the flask. Two systems have been proposed for igniting the sample while the flask is behind a shield. One involves ignition with an electrically heated platinum wire (7), while the other utilizes the heat from a projector bulb for the ignition (8). Both procedures are simple and convenient, and certainly reduce the hazard when burning thermally unstable samples. If flame ignition is used, the flask should be held behind a $1/4''$ plastic shield and a heavy leather glove should be worn. Not more than 100 mg of sample should be burned in a 500-mL flask. The flask should be made of heavy-walled heat-resistant glass.

d. Wet Ashing—Wet ashing of organic samples may involve the possibility of explosions due to too rapid oxidation, formation of explosive intermediates (*e.g.,* nitro compounds or organic perchlorates), or rapid depolymerization of polymers. In general, digestions should be carried out in fume hoods in order to remove the acid fumes and confine any vigorous reactions or explosions that may occur. Safety spectacles, rubber gloves and a rubber apron should be worn by the analyst. A shield should be used for experiments with untested reactions. Use of micro methods reduces the hazards considerably.

The choice of reagents for wet combustion has an important influence on the hazards involved. Nitric acid is hazardous when the sample may be nitrated. Carius digestions with nitric acid also involve the difficulty of obtaining a good strain-free seal in the tube. Several procedures for sealing have been recommended (2,9). Again, the amounts of sample and reagent should not exceed amounts known to be safe. Tubes should be enclosed in proper steel sheaths and should not be removed until the tubes have been cooled and opened.

Most organic materials can be mineralized conveniently by charring with sulfuric acid followed by heating with small incremental additions of nitric acid and finally heating with 30% hydrogen peroxide. The order of addition of reagents must be strictly followed and each reaction should be complete before the following reagent is added.

Perchloric acid is a very useful reagent for digestions, but its hazards must be understood and precautions followed exactly if explosions are to be avoided. Most explosions are probably attributable to formation of anhydrous perchloric acid or organic perchlorates. Ethyl alcohol, glycerol, cellulose, sugars, carbohydrates and pyridine form explosive compounds (10,12,15). Precautions for the use of perchloric acid are beyond the scope of this chapter; special storage, hoods and handling procedures are required. In spite of the hazards, perchloric acid is being used without incident in many laboratories and has been accepted for use in recommended standard procedures (1,4). Safe use of this material demands an understanding of the hazards and exact compliance with operating instructions.

e. Special Reagents — Fusion of organic materials with sodium peroxide in a closed bomb presents an explosion hazard if too much organic material is charged into the bomb, the bomb has been weakened by previous use, the gasket is worn, or the heating is not done properly. Use of a shield (at least 3/16″ thick steel) is imperative. Heat should be applied locally; it should be kept away from the gasket and should be removed as soon as the reaction occurs. Liquid samples which attack gelatin capsules should be handled in glass weighing bulbs. Double gelatin capsules may be satisfactory if the charging operation is carried out promptly. Hazards are reduced by operation on a micro scale. However, the macro procedure is quite safe if precautions are observed.

Fusions of organic material with metallic sodium have been performed in glass tubes, but nickel bombs are preferable — particularly for high-temperature fusions. Potassium should never be melted in glass tubes since it reacts with glass with explosive violence. Properly designed bombs and shields eliminate the danger from explosions. The remaining hazards are those associated with the use of the alkali metals:

(1) Fire and explosion may result if alkali metals contact moisture;

(2) eye injuries and flesh burns may be caused by small pieces of metals or their oxides;

(3) irritation of eyes and nasal tissues can be caused by exposure to products from burning alkali metals. Fumes from burning potassium are toxic;

(4) sodium detonates in carbon tetrachloride. Carbon dioxide-propelled sodium chloride or soda ash extinguishers or dry sand may be used on sodium fires. Potassium reacts explosively with sand and forms explosive carbonyl with carbon dioxide. The only recommended extinguisher for potassium fires is sodium chloride powder or a mixture of sodium chloride and sodium carbonate. Potassium is also more hazardous than sodium since it forms an explosive superoxide KO_2 which can cause explosions even when the metal is stored under "inert" solvents. Scrap sodium should be disposed of by covering it with n-butyl alcohol. The safest means of disposal for potassium is removal to a safe open area where the potassium is allowed to react with atmospheric oxygen and moisture, then washed with a large excess of water.

20.12.4 *Separations.*

a. Distillation — The hazards associated with laboratory distillation at atmospheric pressure or under vacuum are as follows:

(1) The flask may implode due to a weakness or a blow. If the contents of the flask are flammable a fire may result if a source of sparks or a hot surface is nearby;

(2) toxic substances may be released into the atmosphere if the condenser is inadequate or the coolant flow is interrupted;

(3) a pressure buildup and explosion may result if the column or head is plugged with polymeric material or high-melting solids; and

(4) an explosion may result if thermally unstable compounds are distilled.

Distillation flasks containing flammable materials should never be heated with open flames or electric heaters with exposed elements. A heating mantle controlled with a variable transformer is the preferred equipment. The flask and heater should be placed in a metal container large enough to confine any spills from a broken flask.

Any large (over 250-mL) glass distilling flask being used under vacuum should either be completely enclosed or placed behind a shield. This rule also applies to distillations at atmospheric pressure if thermally sensitive compounds are being distilled. In this case, the addition of a high-boiling chaser is advisable.

Solvents that may peroxidize (especially hydrocarbons, aldehydes and ethers) should always be analyzed for active oxygen prior to distillation (6). If the active oxygen content exceeds 0.02%, the peroxide should be removed (14,11) before distillation. In distillation of materials which have a tendency to polymerize, it is well to add an inhibitor. In some cases, it is advisable to add a solution of the inhibitor dissolved in part of the distillate dropwise through the reflux condenser during the entire distillation.

b. Centrifugation — Use of a centrifuge involves the hazards of injury from whirling tubes, breakage of tubes due to imbalance and fire or explosion if a flammable solvent is spilled in a centrifuge which is not of explosion-proof construction. Centrifuges are built with a shield over the moving parts, but care must be taken to stop the equipment completely before opening the cover. Tubes with their contents should be carefully balanced. The safe speed recommended by the centrifuge manufacturer for a given type of tube should not be exceeded.

c. Extractions — If volatile solvents or high temperatures are used in extractions, care should be used to relieve the pressure which is generated by shaking; otherwise the analyst may be exposed to toxic solvents and sample may be lost by spurts that accompany the sudden release of pressure by lifting of the glass stopper. The preferred technique is to invert the extraction funnel, then cautiously open the stopcock to relieve the pressure. This should be done frequently, especially at the start of the extraction.

d. Ion Exchange — The volume of ion exchange resin in a column will change, depending on the ion associated with it and/or the solvent being used. If glass columns are used, plenty of room should be allowed for expansion or broken tubes may result.

e. Liquid-Solid Chromatography — The use of fairly large quantities of flammable solvents requires that this procedure be performed in an area which is free from flames, hot surfaces or sparking electrical equipment. Glass joints, tubing connections and stopcocks should be clamped. The column and reservoir should be placed over a tray which will confine any spill that does occur. Explosion-proof fraction collectors should be operated in a hood with sufficient air flow to prevent the concentration of solvent vapor from reaching the explosive range.

f. Electrophoresis and Electrochromatography — The use of relatively high voltages requires shielding of electrodes and care to avoid physical contact.

20.12.5 *Miscellaneous.*

a. Hydrogenation — Analytical hydrogenations are usually carried out in glass equipment at pressures not exceeding 30 psia (210 kPa). The possibility of an unsuspected leak or a break in the equipment makes it advisable to ban any smoking in the vicinity of the experiment. Some hydrogenation catalysts are pyrophoric and must be kept wet or fires may result. If mercury is used as a confining liquid, the whole apparatus should be over a large tray that will confine any mercury spills.

b. Pressure Bottle Reactions — Saponifications, acetylations, hydrolyses and similar reactions are often carried out in citrate bottles or other pressure bottles. The use of volumetric flasks as pressure bottles is not recommended since they occasionally have areas of thin glass in their bulbs and they may lose their calibration during heating. Heavy-walled pyrex pressure bottles are recommended. They should be well annealed and inspected for strains with polarized light. If facilities are available, it is advisable to perform a hydrostatic bursting test on several bottles and then use the remaining bottles at pressures not exceeding 40% of the minimum bursting pressure. The bottles should be as small as possible, but not filled more than half full. They should be wrapped with strong adhesive tape or enclosed in metal screening. The heating bath should have a safety high-temperature cut off that will not allow the temperature to rise sufficiently to produce dangerous pressures if the primary temperature control sticks in the "on" position. An enclosed constant temperature bath or a shield should be used. The bottle closure should be well designed, sturdy and positive in action.

20.13 INSTRUMENTAL METHODS.

20.13.1 *Gas Chromatography.* The principal hazards associated with gas chromatography are connected with injection of liquid samples. At least one serious injury has resulted from accidental injection of a chemical when a technician's finger was inadvertently jabbed with a hypodermic syringe needle. A severe facial chemical burn resulted in another instance when a corrosive chemical was being injected into a poorly designed injection port and the pressure in the gas chromatograph forced the plunger and the sample out of the back of the syringe onto the face of the chemist.

Injection ports should be designed and located so injections can be made without having the analyst's face close to the syringe. Unnecessarily high operating pressure in the instrument should be avoided where possible. Permanently attached needles eliminate the hazard of the needle becoming detached from the syringe with possible exposure to the sample. Short needles reduce the hazard of breaking needles. Unguarded needles should not be left on a bench top or in a drawer, but should be kept in a tray or box. The tips should be covered with tubing or inserted into cork or rubber stoppers. Needles should not be discarded in waste baskets. Syringes must be handled carefully; they should never be pointed at the operator or other workers. When making an injection into a gas chromatograph, the analyst should keep his thumb or the palm of his hand firmly against the back of the plunger so that it will not be dislodged from the barrel by the pressure in the instrument. If the sample is corrosive, rubber gloves should be worn by the analyst. Long needles should be guided with the fingers to prevent bending or breakage. Bent needles should not be used. They are more difficult to control and are prone to break. Sharp needles decrease the danger of breakage.

In case a chemical is injected into the body accidentally, the following steps should be taken as quickly as possible:

a. Squeeze or suck out (slight vacuum) the injected material;

b. wash the area with a large amount of water;

c. where possible, apply a light tourniquet above the puncture wound, and

d. go to a physician promptly (with an escort) and report the nature and amount of material injected and the depth of needle penetration.

20.13.2 *X-Ray Diffraction and Fluorescence Spectrography.* The use of X-ray analytical equipment involves possible exposure to hazardous X-rays and high voltages. Care must be taken to avoid exposure of any part of the body to a direct X-ray beam or to the secondary (emitted or scattered) radiation that occurs when the primary beam strikes or passes through any material. Precautions should be taken to make it impossible for unqualified personnel to operate the instrument. The available range of potential and current provided by the instrument and specified in the manufacturer's instructions should not be exceeded, or the provided shielding may not be sufficient for prevention of penetration or emergence of unsafe amounts of radiation.

Operating personnel should stay as far away from the instrument as consistent with proper manipulation. Radiation badges or dosimeters should be worn and checked at monthly or more frequent intervals. Persons not involved in the operation of the equipment should not be permitted to remain in the area.

Primary and secondary radiation around the equipment should be checked periodically with a monitoring instrument. The cause of any unusual amount of radiation should be ascertained and removed. If there is a radiation protection officer at the location of the laboratory, he should also make periodic inspections of the X-ray facilities.

When a diffraction apparatus is used, care should be taken to see that all ports to the X-ray tube are covered by a camera, a shutter or a cap. If the instructions call for radiation baffles on the cameras, they must be put in place to minimize side leakage. Shields designed to prevent accidental exposure to X-rays should not be removed. Instruments should be provided with automatic shutters which will prevent accidental exposure of the analyst's hands to the X-ray beam while changing samples.

During alignment of the goniometer in a diffraction instrument or adjustment of the counter arms and crystal holders in a spectrograph, it may be necessary to operate without all the shields and guards in position. These are hazardous operations. Great care should be taken to avoid placing hands or other parts of the body in the direct path of primary or secondary radiation. It is often feasible to reduce power to the X-ray tube during these adjustments to minimize the effects of any accidental exposures. In some spectrographs, removal of the specimen chamber causes an extremely dangerous concentration of X-rays to stream out of the opening. The power to the X-ray tube must be shut off when the specimen chamber or turret head is removed (5).

Before changing tubes or making internal adjustments in the instrument, the power must be shut off. Access door interlocks designed to prevent exposure to high voltages should not be by-passed or wired out. Certain parts of the equipment will store high voltages even after the power is shut off. The charge must be removed with an insulated wire which has previously been connected to ground before service work is attempted.

20.13.3 *Electron Microscopy.* High voltages (50 to 100 kV or more) and secondary emission of X-rays are the hazards associated with electron diffraction and electron microscopy. The precautions in the use of high-voltage equipment specified in the section on X-ray diffraction and spectrometry are equally applicable in electron microscopy. Circuit checking and trouble shooting should be performed by no fewer than two technicians working together. A grounding rod or wire must be used to remove high voltages which are stored in capacitors even after the power is turned off. In addition to grounding, it is well to check the voltage in each circuit with a voltmeter and to leave the grounding rod touching a conducting surface in the circuit under investigation before any physical contact is made. Interlocks and automatic grounding devices should not be tampered with.

Under certain conditions, significant amounts of X-rays may be emitted while operating continuously at 100 kV. Each instrument should be checked for X-ray emission after installation and at periodic intervals thereafter. If significant emission of X-rays is observed when high voltages are in use, it may be necessary to add additional shielding and additional lead glass in front of the viewing windows.

When metal evaporation under vacuum is used for shadow casting, the glass bell jar should be covered with a plastic shield and welder's goggles should be used to observe the metal filament.

20.13.4 *Flame Spectrophotometry and Atomic Absorption.* In the use of hydrogen or acetylene as fuel, it is important that the system be free from leaks and that the fuel be ignited promptly after the valve is opened. The oxygen should be turned on before the fuel when igniting the flame and it should be turned off after the fuel when extinguishing the flame. Oxygen lines, gauges and fittings must be free of oil, grease and pipe dope. When many samples containing large amounts of mineral acids or significant amounts of toxic metals are to be analyzed, a fume exhaust line should be provided over the burner. If samples are dissolved in volatile flammable solvents, they should be prepared and stored at a location away from the flame. Atomic absorption burners that have sample-air mixing chambers should always be shielded and the mixing chamber drain should have a water seal.

20.13.5 *Polarography.* The major hazard associated with polarography is the use of fairly large quantities of mercury. The entire dropping mercury electrode assembly should be in a tray which will confine any mercury spills, and it is advisable to place the dropping mercury electrode assembly in a hood to minimize the danger of inhalation of mercury vapor.

20.13.6 *Infrared Spectrography.* The use of a high-pressure press for forming potassium bromide pellets has potential hazards for finger injuries, but no incidents should occur if reasonable care is observed.

Carbon disulfide, which is a popular solvent for infrared work, is extremely flammable. Preparation of solutions in carbon disulfide and other flammable solvents should be carried out in a fume hood and not in the vicinity of the spectrometer. Some spectrometers have no panel enclosing the underside of the instrument and contact with 110 volts can result if an operator's fingers extend under the chassis while lifting or tilting the instrument with the power on. The power should be turned off before carrying out this operation, or leather gloves should be worn. The chart drive gear shield should not be removed during operation of the spectrometer.

The use of prisms or other optical parts made of thallium salts involves a hazard, since thallium is a cumulative poison. Any broken pieces should be collected and carefully discarded; they should not be allowed to lie on the laboratory bench or floor where they may become pulverized to a dust and, thus, present an inhalation hazard.

20.13.7 *Emission Spectrography.* The use of arc and spark sources provides the major hazards associated with emission spectrography. The electrodes should be protected by shields as much as possible in order to avoid inadvertent contact with high voltages. If many samples of toxic metals are to be analyzed, a fume vent should be provided over the source. Proper eye protection in the form of a shield should be provided in front of the arc source.

REFERENCES

1. A.S.T.M. COMMITTEE. 1956. *Recommended Practices for Apparatus and Reagents for Chemical Analysis of Metals.* A.S.T.M. Designation E50–53, American Society for Testing Materials. Philadelphia, Pennsylvania.
2. GORDON, C.L. 1943. *J. of Research, Natl. Bur. of Standards,* 30:107–111.
3. *Interstate Commerce Commission Regulations.* 1963. Sections 73.301e, 73.304e, and 73.308. T.C. George, New York.
4. JOLLY, S.C. 1963. *Official, Standardized and Recommended Methods of Analysis,* pp. 3–19 and 42–44, The Society for Analytical Chemistry, Heffer, Cambridge, England.
5. MANUFACTURING CHEMISTS' ASSOCIATION, SAFETY AND FIRE PROTECTION COMMITTEE, 1962. *Case Histories of Accidents in the Chemical Industry* Vol. 1, Case Histories No. 547, 590, 592. Manufacturing Chemists' Association, Inc.
6. MARTIN, A.J., 1960. IN J. MITCHELL, JR., ET AL. *Organic Analysis,* p. 1–64. Interscience, New York.
7. MARTIN, A.J. AND H. DEVERAUX. 1959. *Anal. Chem.* 31:1932.
8. OGG, C.L., R.B. KELLY, AND J.A. CONNELLY. 1961. Design of Apparatus for Safe Oxygen Filled Combustion. *Abstracts of Papers from International Symposium on Microchemical Techniques,* Paper No. 51, p. 34. The Pennsylvania State University, University Park, Pa.
9. PARR OXYGEN COMBUSTION BOMBS. Specifications No 1100, Parr Instrument Company, Moline, Illinois.
10. NATIONAL SAFETY COUNCIL, Perchloric Acid. Data Sheet D-311 (D-Chem. 44). Chicago, Illinois.
11. RAMSEY, J.B., AND F.T. ALDRICH. 1955. *J. Am. Chem. Soc.* 77:2561.
12. SMITH, G.F. 1953. *Anal. Chim. Acta* 8:397–421.
13. SMITH, G.F. 1942. *Mixed Perchloric, Sulfuric and Phosphoric Acids and their Applications in Analysis* (pamphlet). 2nd ed., The G. Fredrick Smith Chemical Company, Columbus, Ohio.
14. WILLIAMS, F.E. 1951. Distillation in Weissbergers' Technique of Organic Chemistry, Vol. IV, pp. 300–302. Interscience, New York.
15. ZACKERL, M.K. 1948. *Mikrochemie ver. Mikrochim. Acta* 33:387–388, (Through Chem. Abstracts. 42, 6538b 1948).
16. 29 CFR Part 1910, Federal Register *48*, no. 228, Friday November 25, 1983 p. 53280ff.

Approved with modifications from 2nd edition
Subcommittee 9
J.N. HARMAN, *Chairman*

21. Air Purification*

21.1 INTRODUCTION. Laboratory compressed air is the most common source of diluent gas for low-concentration, high-volume standard gas mixtures. It is continuously supplied as needed, usually by diesel or electric compressors at pressures of 80 to 125 psi (550–860 kPa), and it is stored in holding tanks. Several undesirable contaminants can be introduced during compression and storage. Oil mists are a common by-product, as are substantial amounts of carbon dioxide, nitrogen dioxide, aldehydes, carbon monoxide, and unburned hydrocarbons. Acid gases as well as dust particles and pipe scales of all sizes are also a problem. Even if the air is known to be 99.9% pure it can still contain up to 1000 ppm of undesirable materials. Before any quality low-concentration work can be

---

*Abstracted from G. O. Nelson, "Controlled Test Atmospheres," Chap. 2, Air Purification. Ann Arbor Science Publishers, Inc., P.O. Box 1425, Ann Arbor, Michigan 48106.

done, the air-supply system must be scrupulously cleaned to prevent contamination and possible chemical reaction. The composition of clean, dry air is given in Table 21:I.

21.1.1 This section describes the basic methods of removing contaminants from flowing air streams. General multipurpose filtering devices are discussed, as are methods for removing excess water vapor, oil mists, extraneous gases, and particulate matter. The air-purification procedures that are described can also be applied to such relatively stable gases as nitrogen and oxygen as well as to inert gases.

21.2 REMOVAL OF WATER VAPOR. Moisture can be removed from gases by a variety of methods. Chief among these are adsorption, absorption, cooling, compression and combined compression and cooling. Usually, only the first three methods are used in the laboratory.

21.3 SOLID DESICCANTS.

21.3.1 Solid desiccants constitute the most conventional method of removing water vapor in the laboratory. They remove moisture either by chemical reaction (absorption) or by capillary condensation (adsorption) (1). Solid absorbing agents include calcium chloride, calcium sulfate, and magnesium perchlorate; solid adsorbing agents include activated alumina and silica gel. Solid desiccants are one of the most practical tools for drying gases because they are commercially available, they are easy to store, they can be regenerated by heating, and they often indicate their condition by their color.

21.3.2 Solid desiccants are generally evaluated by comparing their drying efficiencies and capacities. The efficiencies of drying agents (*i.e.*, the degree of dryness

Table 21:I Composition of Clean, Dry Air

Nitrogen	78.08%
Oxygen	20.95%
Argon	0.934%
Carbon dioxide	0.033%
Neon	18.2 ppm
Helium	5.24 ppm
Methane	2.0 ppm
Krypton	1.14 ppm
Hydrogen	0.5 ppm
Nitrous oxide	0.5 ppm
Xenon	0.087 ppm

achieved) can be compared by measuring the water vapor remaining in a gas after it passes through the desiccant at the equilibrium velocity (2). The drying efficiencies of several solid desiccants are compared in Table 21:II. Barium oxide and magnesium perchlorate are the most efficient desiccants of those compared, whereas copper sulfate and granular calcium chloride are the least efficient. An extensive investigation by Trusell and Diehl evaluated the efficiencies of 21 desiccants in drying a stream of nitrogen (3). Their results, which are listed in Table 21:III, show that the most efficient desiccant is anhydrous magnesium perchlorate.

21.3.3 The capacity of a desiccant is the amount of water it is able to remove per unit of the desiccant's dry weight. Often, a drying agent is efficient but is unsuitable for drying large quantities of gas because of its limited capacity. Anhydrous calcium chloride is an example of such a desiccant. The capacity of a desiccant depends not only on the kind of material of which it is composed, but also on the size of the grains, the amount of surface area exposed to the gas, and the thickness through which the gas flows. Additional factors include the type of gas being dried as well as its velocity, temperature, pressure, and moisture content (7). The capacities of several desiccants as a function of relative humidity are shown in Figure 21.1. The relative capacities of the most common drying agents can be determined from Table 21:III by comparing the volumes of gas each desiccant is able to dry. The materials with the highest capacities are anhydrous magnesium perchlorate, calcium sulfate, and phosphorus pentoxide.

21.3.4 One should not indiscriminately choose a desiccant simply because it fits the proper efficiency and capacity. The geometry and size of the drying train must be considered to allow enough residence time to achieve equilibrium. In addition, many drying agents heat up violently when they are exposed to too much moisture over too short a time, and the pressure drop through the desiccant can be a problem at high flow rates. The most acceptable desiccants are anhydrous magnesium perchlorate, calcium sulfate, silica gel, and activated alumina. These are described below.

Table 21:II Comparative Efficiencies of Various Solid Desiccants Used in Drying Air*

Desiccant	Granular Form	Residual Water† (mg/L)
BaO	—	0.00065
$Mg(ClO_4)_2$	—	0.002
CaO	—	0.003
$CaSO_4$	Anhydrous	0.005
Al_2O_4	—	0.005
KOH	Sticks	0.014
Silica gel	—	0.030
$Mg(ClO_4)_2 \cdot 3H_2O$	—	0.031
$CaCl_2$	Dehydrated	0.36
NaOH	Sticks	0.80
$Ba(ClO_4)_2$	—	0.82
$ZnCl_2$	Sticks	0.98
$CaCl_2$	Technical anhydrous	1.25
$CaCl_2$	Granular	1.5
$CuSO_4$	Anhydrous	2.8

*Data taken from Reference 2.
†After drying to equilibrium.

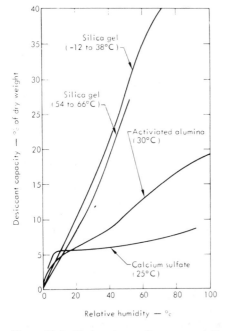

Figure 21:1 — Desiccant capacity versus relative humidity for silica gel, activated alumina, and calcium sulfate (Drierite). The calcium sulfate temperature is estimated. (Taken from Reference 5.)

a. Anhydrous magnesium perchlorate (Anhydrone or Dehydrite) has the highest efficiency as well as the greatest capacity (**3**). It is hygroscopic but not deliquescent, and it can absorb up to 35% of its own weight without evolving corrosive fumes as phosphorus pentoxide does. Since the monohydrate does not dissociate to liberate water until 134°C, it can be used to dry gases at high temperatures. Hydration continues until the hexahydrate, which has a theoretical capacity of 48.4%, is formed. Magnesium perchlorate is available in either the regular or the indicating form, the latter containing about 1% potassium permanganate.

The chief disadvantages of anhydrous magnesium perchlorate are its relatively high cost (roughly four times the cost of the other three desiccants), and the difficulty of regeneration. The temperature must be raised slowly while the perchlorate is dried in a vacuum in order to prevent the crystals from fusing. A final temperature of about 245°C is recommended to return the perchlorate to its anhydrous state (**3**). A further disadvantage of this and other perchlorate desiccants is the *tendency to form explosive compounds in the presence of or-*

Table 21:III Comparative Efficiencies and Capacities of Various Solid Desiccants in Drying a Stream of Nitrogen*

Desiccant	Initial Composition	Regeneration Requirements		Average Efficiency† (mg/L)	Relative Capacity‡ (L)
		Drying Time (h)	Temperature (°C)		
Anydrous magnesium perchlorate§	$Mg(ClO_4)_2 \cdot 0.12H_2O$	48||	245||	0.0002	1168
Anhydrone§#	$Mg(ClO_4)_2 \cdot 1.48H_2O$	—	240**	0.0015	1157
Barium oxide	96.2% BaO	—	1000‡‡	0.0028	244
Activated alumina	Al_2O_3	6–8‡‡	175, 400‡‡	0.0029	263
Phosphorus pentoxide§§	P_2O_5	—	—	0.0035	566
Molecular sieve 5A#	Calcium aluminum silicate	—	—	0.0039	215
Indicating anhydrous magnesium perchlorate§	88% $Mg(ClO_4)_2$ and 0.86% $KMnO_4$	48||	240||,**	0.0044	435
Anhydrous lithium perchlorate§§	$LiClO_4$	12||, 12	70||, 110	0.013	267
Anhydrous calcium chloride§§	$CaCl_2 \cdot 0.18H_2O$	16||	127||	0.067	33
Drierite#	$CaSO_4 \cdot 0.02H_2O$	1–2	200–225||||	0.067	232
Silica gel	....	12	118–127|††	0.070	317
Ascarite#	91.0% NaOH	—	—	0.093	44
Calcium chloride§§	$CaCl_2 \cdot 0.28H_2O$	—	200||	0.099	57
Anhydrous calcium chloride§§	$CaCl_2$	16||	245||	0.137	31
Sodium hydroxide§§	$NaOH \cdot 0.03H_2O$	—	—	0.513	178
Anhydrous barium perchlorate	$Ba(ClO_4)_2$	16	127	0.599	28
Calcium oxide	CaO	6	500, 900**	0.656	51
Magnesium oxide	MgO	6	800	0.753	22
Potassium hydroxide§§	$KOH \cdot 0.52H_2O$	—	—	0.939	18.4

*Nitrogen at an average flow rate of 225 mL/min was passed through a drying train consisting of three Swartz drying tubes (14 mm i.d. × 150 deep) maintained at 25°C. Except as noted in columns 3 and 4, the data in this table are taken from Reference 3.
†The average amount of water remaining in the nitrogen after it was dried to equilibrium.
‡The average maximum volume of nitrogen dried at the specified efficiency for a given volume of desiccant.
§Hygroscopic
||Dried in a vacuum.
#Trade name.
**Taken from Reference 4.
††Taken from Reference 5.
‡‡Taken from Reference 6.
§§Deliquescent.
||||Taken from Reference 2.

ganic materials, especially when they are heated **(6)**. Oil mists and other organic vapors must therefore be removed before such desiccants are used.

b. Calcium sulfate (Drierite) has an average efficiency of about 0.1 mg/L and a capacity of 7 to 14% at 25°C. It is stable, inert and does not deliquesce even at peak capacity. It is easily regenerated (1 to 2 h at 200°C), and it operates at an almost constant efficiency over a wide range of temperatures. However, continued regeneration is difficult because the constant formation and destruction of the hemihydrate breaks down the grains. Calcium sulfate is available in sizes from 4 to 20 mesh and in either the regular or the indicating form.

c. Silica gel has a moderately high efficiency and capacity because of its large number of capillary pores, which occupy about 50% of the gel's specific volume **(1)**. The capacity of the gel varies from batch to batch because of differences in the size and shape of the pores. The gel maintains its efficiency until it has absorbed 20% of its weight, and it can be regenerated indefinitely at 120°C. At this relatively low regeneration temperature, however, the gel cannot be used for high-temperature drying. If it is regenerated above 260°C, it loses some of its capacity. Silica gel is available in sizes from 2 to 300 mesh. The addition of cobalt chloride to the surface of the gel provides an indicating ability.

d. Activated alumina has a higher efficiency than silica gel but offers less capacity (12 to 14%), especially at high humidities. It can be regenerated between 180° and 400°C without losing much of its capacity. It is available in sizes from 14 mesh to 2.5 cm and in either the regular or the indicating form.

21.4 LIQUID DESICCANTS.

21.4.1 If solid desiccants are not practical, then liquid desiccants can be used. Liquid desiccants have a much higher capacity than their solid counterparts (see Figure 21:2), and they can be continuously regenerated via spraying, pumping, or recirculating. On the other hand, their efficiencies are normally very low unless the anhydrous forms are used, and they usually cannot produce relative humidities below about 20% **(5)**. Some of the more common

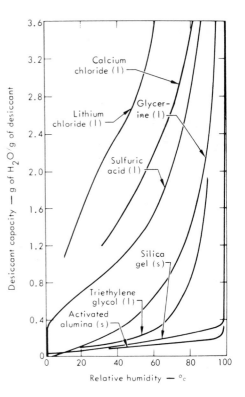

Figure 21:2 — Desiccant capacity versus relative humidity for two solid desiccants and five liquid desiccants. (Taken from Reference 5.)

liquid desiccants are described in Table 21:IV. Note that although the strong acids and bases achieve the best efficiencies, they also emit corrosive vapors.

21.5 COOLING.

21.5.1 Cooling is the most efficient laboratory method of removing water from a stream of gas. The gas is directed through a vessel in a low-temperature bath, and the excess water condenses on the cold walls of the vessel. As an example, a bath of dry ice and acetone at –70 C removes all but about 0.01 mg/L of water in air at equilibrium. A liquid-nitrogen bath at –194°C removes all but about 1×10^{-23} mg/L of water in air at equilibrium. This is about 19 orders of magnitude more efficient than anhydrous magnesium perchlorate, the best solid desiccant.

21.6 REMOVAL OF PARTICLES.

21.6.1 There are a number of inline filters that remove micrometer-sized parti-

Table 21:IV Comparative Properties of Eight Liquid Desiccants*

Desiccant	Relative Humidity Achieved at 21°C (%)	Solution Concentration (%)	Operating Temperature Range (°C)	Remarks
Calcium chloride	20–25	40–50	32–49	None.
Diethylene glycol	5–10	70–95	16–43	Can be regenerated with heating to 150°C.
Glycerol	30–40	70–80	21–38	Oxidizes and decomposes at high temperatures; can be regenerated with vacuum evaporation.
Lithium chloride	10–20	30–45	21–38	None.
Phosphoric acid	5–20	80–95	16–38	Corrosive; fumes carried over during the drying process; does not fume during regeneration.
Sodium and potassium hydroxides	10–20	Saturated	29–49	Corrosive; frequently used to remove CO_2 and water simultaneously.
Sulfuric acid	5–20	60–70	21–49	Corrosive; most efficient liquid desiccant.
Triethylene glycol	5–10	70–96	16–43	None.

*Data taken from Reference 1.
†5% with anhydrous diethylene glycol.

cles. One example is a sintered-bronze mesh 1″ in diam and 2.5″ (6.35 cm) long that can filter out 2-μm particles at a flow rate of 1 ft³/min (28 L/min) and an operating pressure of 110 psi (760 kPa) (8). Filters made of metal fibers (pore size = 5 to 750 μm) (9), sintered stainless steel (pore size = 2 to 150μm), and foamed metals (pore size = 5 μm to 2.5 mm) are available. Porous Teflon and Kel-F filters that can remove particles as small as 2 μm are also available (12).

21.6.2 If removal down to certain precise particle sizes is required, membrane filters are often useful. These are available in sizes from 13 to 293 mm with pore sizes from 7.5 nm to 8μm (10–12). These filters are constructed from a wide selection of microporous materials (e.g., regenerated cellulose, poly(vinyl chloride), glass fibers, and polypropylene) whose capacities are accurately known (10). The flow rate per unit of filter area at a given temperature is a function of the pore size and the upstream pressure. The relationship between pressure, pore size, and flow rate for a typical membrane filter is shown in Figure 21:3.

21.7 REMOVAL OF ORGANIC VAPORS.

21.7.1 Organic vapors such as unburned hydrocarbons are sometimes present in a compressor-generated gas even after it has been passed through an in-line filter. The concentrations of such organic vapors can be further reduced either by passing the gas through an activated-charcoal filter or by continuously burning the vapors in a combustion furnace.

21.7.2 Activated coconut charcoal has long been the most popular material for removing organic vapors. Instead of water vapor being adsorbed on the filter, the organic vapors tend to displace any water that may be present. A filter containing fine grains of coconut charcoal uniformly dispersed throughout a web matrix is also available (13). Not all organic vapors are completely adsorbed by charcoal filters, even when large filter areas and low flow rates are used. For example, such low molecular weight compounds as acetylene, ethane, ethylene, methane, hydrogen, carbon monoxide, and carbon dioxide have almost no affinity for activated charcoal (14).

21.7.3 As for the second method of removing organic vapors, Kusnetz et al. recommend passing the gas through a 2″-diam (5-cm) Mullite tube that is filled with copper shavings and surrounded by a combustion furnace maintained at 675°C (15). The exiting gas is cooled to 425°C by a

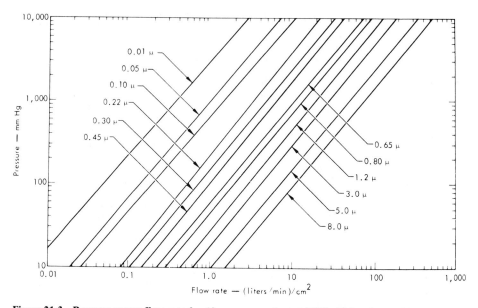

Figure 21:3 – Pressure versus flow rate for 12 mean pore sizes at 25°C. (Taken from Reference 12.)

finned brass tube and a water-cooled condenser. To remove the combustion products, the gas is bubbled through a sulfuric acid-dichromate solution and passed through filters of Ascarite, activated charcoal, and glass wool, in that order.

21.8 REMOVAL OF MISCELLANEOUS CONTAMINANTS.

21.8.1 A host of in-line filters are available for treating compressed gases in the laboratory. These filters can remove both water and oil mists as well as particles. Drain plugs are provided for periodically removing the collected liquid, and the filters themselves can be exchanged when they become clogged or excessively loaded. Automatic drain traps are also available (16). Most compressed-gas filters operate via a two-stage separation that involves centrifugal or inertial separation followed by diffusion through a filter (17–18) or a special absorption material (19).

21.8.2 Compressed gases may also contain acid gases, carbon monoxide, and carbon dioxide. The acid gases are those that produce hydrogen ions either by direct dissociation or by hydrolysis and include hydrogen cyanide, hydrogen chloride, and sulfur dioxide. Both acid gases and carbon dioxide can be filtered out by soda lime, a

mixture of calcium and sodium hydroxides. Soda lime with various moisture contents is available in sizes from 4 to 14 mesh, in either the regular or the indicating form, and can absorb up to 25% of its weight of carbon dioxide. The rate at which acid gases are absorbed by soda lime depends on the condition of the lime. As the lime becomes spent, a thin film of calcium carbonate covers the surface of the soda lime particles and cannot be removed by regeneration.

21.8.3 Both carbon dioxide and water vapor can be absorbed by Ascarite, which consists of sodium hydroxide in a woven asbestos matrix. Ascarite is available in sizes from 8 to 30 mesh.

21.8.4 Carbon monoxide is relatively unaffected by its passage through soda lime, activated carbon, or the previously described desiccants, so special provisions for its removal must be made. Hopcalite, usually a mixture of copper and manganese oxides, has traditionally been the most practical agent for removing carbon monoxide. It operates as a catalytic oxidizing agent, converting the carbon monoxide to carbon dioxide which can ultimately be absorbed by Ascarite or soda lime. The main requirement for using Hopcalite is that it be kept scrupulously dry, for it loses its catalytic ability in the presence of water. Other carbon monoxide filters are also available commercially **(19)**.

21.8.5 For the removal of specific substances, the appropriate sorbents are listed in Reference **20**.

REFERENCES

1. PERRY, J.H. 1950. "Chemical Engineers Handbook," McGraw-Hill Book Company, Inc., New York, p. 879.
2. HAMMOND, W.A. 1958. Drierite, the Versatile Desiccant, and its Applications in the Drying of Solids, Liquids, and Gases. The Stoneman Press, Columbus, OH.
3. TRUSELL, F. AND H. DIEHL. 1963. Efficiency of Chemical Desiccants, Anal Chem, *35*:674–77.
4. SKOOG, D.A. AND D.M. WEST. 1963. "Fundamentals of Analytical Chemistry," Holt, Rinehart & Winston, Inc., NY.
5. HOUGEN, O.A. AND F.W. DODGE. 1947. "The Drying of Gases," Edwards Brothers, Inc., Ann Arbor, MI.
6. MORTON, A.A. 1938. "Laboratory Techniques in Organic Chemistry," McGraw-Hill Book Company, Inc., NY.
7. NONHEBEL, G. 1964. "Gas Purification Processes," George Newnes, Ltd., London.
8. MOTT METALLURGICAL CORPORATION, Farmington Industrial Park, Farmington, CT 06032.
9. Brochure. Pall Corporation. Glen Cove, NY.
10. 1969 catalog. Pure Aire Corporation, 8441 Canoga Ave., Canoga Park, CA 91304.
11. Brochure. Arthur H. Thomas Company, P.O. Box 779, Philadelphia, PA 19105.
12. MILLIPORE CORPORATION, 80 Ashby Road, Bedford, MA 01730.
13. THE DEXTER CORPORATION, One Elm St., Windsor Locks, CT 06096.
14. BARNEBEY, H.L. 1958. Heating, Piping, Air Conditioning, *30*:153.
15. KUSNETZ, H.L., B.E. SALTZMAN AND M.E. LANIER. 1960. Amer Ind Hyg Assoc J, *21*:361.
16. KING ENGINEERING. 3201 S. State St., Ann Arbor, MI 98106
17. R.P. ADAMS CORPORATION, 225 E. Park Dr., Buffalo, NY 14240
18. WILKERSON CORPORATION, 1201 W. Mansfield Ave., Englewood, CO 80110.
19. DOLLINGER CORPORATION, One Town Line Circle, Rochester, NY 14692.
20. Respiratory protective Devices Manual, 1963. Chapter 5. Braun and Brumfield, Inc., 100 N. Staebler Road, Ann Arbor, MI 48106.

Approved with modifications from 2nd edition
Subcommittee 9
J.N. HARMAN, *Chairman*

22. Liquid Chromatography

22.1 INTRODUCTION. In comparison to gas chromatography (GC), liquid chromatography (LC) has several advantages. Foremost is the fact that roughly 85% of all known chemical compounds are not sufficiently volatile or stable to be separated by GC. Thus, liquid chromatography is applicable to a much wider range of materials than GC — particularly to those materials with low volatility and to labile or unstable compounds. Liquid chromatography is especially useful to separate mixtures of compounds with identification by supplementary techniques, such as infrared and mass spectroscopy. The text by Snyder and Kirkland **(1)** is recommended as a general reference.

22.1.1 Separations involve a mobile liquid traveling through a packed column containing a stationary phase. The sample mixture, introduced into the mobile liquid, undergoes a series of interactions with the stationary phase as it moves through the column. Separated components emerge in the order of increasing interaction with the stationary material — the least retarded component elutes first, the most strongly retained material elutes last.

22.1.2 With presently available columns and instrumentation, it is possible to perform liquid chromatography where the time necessary to achieve a given separation with the desired resolution rivals that attainable by packed column GC. The technique is commonly referred to as High Performance Liquid Chromatography (HPLC) to distinguish it from less efficient liquid chromatographic systems of the past.

22.1.3 One of the most important considerations in obtaining faster, more efficient separations involves the particles used in the column packing. For HPLC use, the packing should be of spherical shape, displaying a highly uniform particle size distribution, and possess high enough mechanical strength to maintain its integrity under the operating and packing pressures (as high as 69 MPa, ~ 10,000 psi). For most applications, particles of high surface area are desirable in that they provide larger column capacities and better column efficiencies.

22.2 STATIONARY PHASES.

22.2.1 Stationary phases and operational modes used in the current practice of HPLC can be divided in three broad categories: (a) normal phase, which utilizes a polar stationary phase (commonly underivatized silica with adsorbed water actually functioning as the chromatographic phase) in conjunction with nonpolar eluents (e.g., n-hexane) for the separation of primarily polar solutes; (b) reverse phase, which utilizes a non-polar stationary phase (most commonly a long-chain hydrocarbon bonded to silica or a porous polymer support, or a hydrophobic macroporous polymer support itself) in conjunction with polar eluents (e.g., methanol/acetonitrile/water) for the separation of primarily nonpolar solutes (it is estimated that 85% of present HPLC applications involve the reverse-phase mode); (c) ion exchange, which involves cation or anion exchange functionalities bonded to silica or porous polymer supports, used in conjunction with saline aqueous eluents for separation of ionic solutes.

Silica is unsurpassed in mechanical strength, and it is easily available in the form of tightly size-controlled spherical particles. However, silica-based support can only be used with eluents having a pH between 2 and 8. Polymer-based supports (typically macroreticular poly(styrenedivinylbenzene)) allow a much greater pH latitude (typically 1–13) but cannot be used at pressures higher than 28 MPa (~4000 psi) and are generally not as efficient as silica-based packings.

22.2.2 Typical HPLC packing in use today ranges in diameter from 1.5 to 15 μm with 5μm being the most common. Column packings of various types are available from several dozen independent vendors, both as packed columns and as bulk packing, intended for packing by the user.

22.2.3 Columns with internal diameters ranging from 1 to 4.6 mm are used in analytical HPLC. Columns up to 30 cm in diameter are available for industrial-scale separations. Many manufacturers guarantee lot-to-lot reproducibility of their packed columns, with supporting chromatographic data supplied with each column. Reproducible user packing of the microparticles used in the current practice of HPLC requires specialized packing equipment and even then is far from facile. Unless one is buying in excess of 50 columns annually, it is rarely worthwhile to pack columns in-house.

22.2.4 Smaller-diameter particles lead to better column efficiencies, but also result in higher operating pressures. Smaller-volume detectors, injectors, and connecting hardware are also needed with columns packed with smaller particles.

Smaller-diameter columns lead to savings in solvent usage but also require smaller-volume hardware as indicated above. For the vast majority of applications, 4–4.6 mm i.d., 6.4 mm o.d. (¼ inch), 250 mm length columns packed with 5 μm spherical supports represent the optimum choice and are most frequently used.

22.3 HPLC HARDWARE.

22.3.1 Most HPLC applications involve operating pressures less than 21 MPa (~3000 psi) and flow rates ranging from 0.5 to 2.0 mL/min. Operating pressure depends on the flow rate, viscosity of the eluent, column length and diameter and particle diameter of the stationary phase. All other factors remaining the same, the operating pressure varies inversely as the square of the particle diameter. When eluent composition is deliberately changed during the course of a chromatographic separation (gradient elution), at constant flow rate the pressure drop changes due to a change in viscosity. It should be noted that mixtures of two pure solvents A and B often display a higher viscosity than either pure A or pure B.

Pumps capable of maintaining precise flow rates at pressures up to 69 MPa (10,000 psi) are commercially available. The majority of pumping systems permit, however, operating pressures up to 35–42 MPa (5000–6000 psi) and are adequate for most purposes. Pump seal-life decreases with increasing operating pressures for all pumps. Many available pumping systems can function either in a constant pressure or a constant flow mode.

There is presently a multitude of manufacturers that offer HPLC equipment; it is not practical to provide a listing here. Trade magazines devoted to the field, e.g., LC·GC (Asher Publishing Corp., Eugene, OR 97440-1046) and Chromatography Forum (Technical Publishing, 1301 S. Grove

Ave., Barrington, IL 60010) are recommended as sources of commercial information.

22.3.2 In some applications, especially in trace metal analysis or analysis of biological material prone to metal-catalyzed degradation, completely nonmetallic chromatographic systems (pump, injector, column, detector, connecting tubing and connecting hardware) may be desirable. Such systems, capable of operation up to 28 MPa ($\sim$4000 psi) are available from Dionex Corp. (1228 Titan Way, Sunnyvale, CA 94086) and Wescan Instruments (3018 Scott Blvd., Santa Clara, CA 95050).

22.3.3 The pump design in most common use involves a reciprocating piston pump with a sapphire piston operating in a precisely machined stainless steel cavity, equipped with one or more spring loaded ring-seals made of synthetic polymers. Unidirectional flow is provided by inlet and outlet check valves made from ruby balls on sapphire seats; frequently these are spring-loaded and two in a row are utilized.

The reciprocating single piston pump tends to produce a relatively high level of pulsations; this is particularly problematic with detectors that are especially flow-sensitive, e.g., electrochemical or refractive index detectors. Pulsation can be minimized by design within the basic reciprocating single piston genre, for example by using a very rapid fill stroke and an accelerated piston movement during the initial portion of the pressurization stroke until the operating pressure is reached. Another variation is to use small stroke volumes and high stroke frequencies, the column itself acting as an effective dampener for stroke frequencies exceeding 10 Hz. External pressure dampeners, including Bourdon gauges and other similar devices, are often used for pulse dampening; some of the better designs are highly effective. However, these add undesirable hold-up volume on the high pressure side of the pumping system, creating a significant difference between the programmed solvent composition and the solvent composition actually experienced by the column, during gradient elution.

Dual and triple independent pumping heads operating 180° to 120° out of phase provide more pulsation-free pumping systems but are more expensive.

Nearly pulseless pumping is attainable with high pressure syringe pumps; the time spent in refilling the pump is wasted, however, and represents a significant disadvantage.

22.3.4 Injectors used in HPLC are loop-type; the loop may be external or internal. The internal-loop design is used for loop volumes of ≤ 5 μL; loop valves with injection volumes as small as 0.2 μL are available. The injector has two positions. In the load position, sample is loaded into the loop with a syringe (or from an autosampler). Any excess sample goes to waste, while the eluent is pumped directly to the column. As the valve is switched to the inject position, contents of the sample loop are flushed by the eluent to the top of the column. Electrically or electropneumatically actuated loop injectors are available.

Typical sample volumes used with columns containing 5-μm packing range from 5 to 20 μL. Larger sample volumes can occasionally be used to advantage if the component of interest is late-eluting and gradient elution is being used; the detection limit then improves with a larger sample volume. In routine use, however, large sample volumes are undesirable; chromatographic efficiency decreases and resolution is lost.

22.3.5 The need for minimizing connection and detection volumes cannot be overemphasized for the proper practice of present-day HPLC, else all advances made in the past decade to produce efficient stationary phases are wasted. With columns containing 5-μm packing, connections should be made with minimum lengths of 0.25 mm i.d. tubes from the injector to the column and from the column to the detector. The fittings used to make such connections must be of the "zero dead volume" type. With 3 μm and 1.5 μm packing, the connection requirements are more stringent.

22.3.6 Detectors used to monitor the column effluent must exhibit a small detection volume. With 5 μm packing, detector cell volume should be ≤ 8 μL. A variety of flow-through detectors are now available specifically for HPLC applications. A volume devoted to liquid chromatography detectors is available (2). Considerable advances have also been made towards adapting proven GC detectors, such as the electron capture detector, thermionic ionization detector and the flame photometric detector for HPLC use; a review by Brinkmann and Maris is recommended for further details (3). Although commercial instrumentation is available that interfaces an HPLC to a mass spectrometer or a flame ionization detector, considerable improvements are needed before these can be recommended for routine analysis.

The mainstay of HPLC detection remains with the measurement of optical absorbance in the UV-visible range. Detectors with noise levels down to 1×10^{-5} absorbance units are commercially available. The degree of sophistication ranges from fixed wavelength (254 nm) instruments to filter photometers to programmable, monochromator-equipped variable-wavelength devices to diode array detectors. The latter can measure the entire spectral absorption profile (190–800 nm) simultaneously on a subsecond time scale. Fluorescence and electrochemical (amperometric/coulometric) detectors are among the most sensitive and selective. Conductivity detectors are particularly useful for ionic analysis and are widely used in the subdiscipline of Ion Chromatography (4). Radioactivity detectors are available for the detection of radiolabelled compounds. The refractive index detector approaches the closest to an universal detector; however, good sensitivity with such a detector can only be attained with very precise temperature control.

22.4 APPLICATIONS.

22.4.1 The applications of HPLC are varied in scope and include both qualitative and quantitative analyses. The retention times of peaks are frequently used as the principal means of qualitative identification. Components, even in trace quantities, can be trapped and submitted for further characterization by mass spectrometry or by other techniques.

22.4.2 Quantitative analyses by HPLC can be accomplished with good precision and accuracy. When an electronic integrator is used to measure peak areas of separated components, LC analyses with a standard deviation of less than 0.5% are not uncommon.

22.4.3 Techniques such as temperature, flow, and solvent programing, whereby one of these parameters is varied during the run, are useful in HPLC. For separations of particularly complex mixtures, gradient elution appears to be the most valuable approach. In this process, the composition of the mobile phase is changed stepwise or continuously throughout the chromatogram by increasing polarity, ionic strength, pH, and so forth. The overall effect is to resolve the early-eluting compounds as well as the more strongly retained ones and to complete the elution of all components in a reasonably short time. Gradient elution or solvent programming is thus used to attack the "General Elution Problem" and is completely equivalent to temperature programming used in GC.

22.4.4 Preparative scale HPLC can be carried out with larger diameter columns. Column efficiencies are lower than those attained in analytical scale operation.

22.4.5 Many HPLC packings are available coated on thin layer plates (HPTLC plates). Scouting or trial separation of the analyte mixture on such plates often provides a quick way to establish the feasibility of the intended separation using the same stationary phase and eluent in the HPLC mode.

REFERENCES

1. SNYDER, L.R. AND J.J. KIRKLAND, 1986. "An Introduction to Modern Liquid Chromatography", 3rd ed., Wiley, New York.
2. VICKREY, T.M. ed. 1982. "Liquid Chromatography Detectors", Chromatographic Science Series, Vol. 23, Marcel Dekker, New York.
3. BRINKMANN, U. TH, AND F.A. MARIS, 1987. LC·GC, 5: 476–484.
4. TARTER, J.G. ed., 1986. "Ion Chromatography", Chromatographic Science Series, Vol. 37, Marcel Dekker, New York.

Subcommittee 9
P.K. DASGUPTA
J.N. HARMAN, *Chairman*

23. Thin-Layer Chromatography

23.1 INTRODUCTION.

23.1.1 One of the basic problems of environmental analytical chemistry is the separation of complex mixtures. Analytical methods usually are designed for relatively pure materials and may be confounded by the complexity of many environmental samples. Proper separation of sample mixtures into their components can usually aid the analytical process by removing interferences, increasing reproducibility and pushing back the limits of detection. Most analytical schemes, therefore, usually consist of one or more separation/purification steps prior to analysis. The separation step(s) is often the most challenging.

23.1.2 Chromatographic techniques provide the most practical, versatile, and hence widely used separatory tools. There are a variety of chromatographic techniques, each with its unique advantages and limitations. Gas chromatography is useful in separating volatile, relatively nonpolar materials or members of a homologous series (e.g., individual PCB congeners). Ion exchange chromatography can readily separate inorganic ions, while ion chromatography coupled with high performance liquid chromatography(HPLC) is often employed to separate both inorganic and organic ions. Size-specific separations can often be accomplished using gel-permeation or size-exclusion liquid chromatographies. Under certain circumstances more than one separatory principle can be combined, as is the case with gel-electrophoresis where both size and charge provide the basis for separation. Thin-layer chromatography (TLC) seems to provide the analyst with a greater degree of versatility and adaptability than any of the techniques previously mentioned.

23.1.3 Thin-layer Chromatography has been an accepted separation technique for over 20 years. Although the 2nd Edition of this book stated that TLC had "reached the point of maturity," advances in materials science have continued to augment the strength and versatility of TLC. In the early '60's the literature abounded with accounts of TLC separations involving materials, matrices, and quantities not previously pos-

sible. The advent of High Performance Liquid Chromatography (HPLC) seemed to supersede the need for TLC. However, it was the theoretical considerations of optimization in HPLC and the concomitant advances in materials science (providing better packing materials for HPLC columns and subsequently coating materials for TLC plates) that actually ushered in the current explosive growth in TLC and High Performance TLC (HPTLC)(1). Although most references to TLC and HPTLC emphasize the practical rather than the theoretical basis of application, Fried and Sherma (2) treat the theory behind HPTLC optimization and briefly review the theoretical literature.

23.1.4 Thin layer chromatography is one of the few modern analytical techniques that does not require the use of sophisticated and expensive instrumentation. Equipment and materials suitable for a wide variety of separations can be acquired for between several hundred and less than a few thousand dollars. A variety of commercial or hand-built equipment is available to make the manipulations easier. Although most early workers hand-coated their plates, the development of pre-coated plates (with an ever-increasing variety of coatings) offers not only convenience, but also a level of reproducibility and purity not previously possible.

23.1.5 Thin layer chromatography offers many other advantages. It is both sensitive and selective. Separations of picogram quantities of materials have been made; yet, when necessary, the same separation can usually be scaled up to handle masses of 100 mg or more. When HPTLC techniques are used even nanogram separations are practical. Many factors of TLC separations can be varied to produce a high degree of selectivity; these include type of sorbent, binder, layer thickness, particle size of the sorbent, activation of the layer, solvent selection, chamber saturation, visualization, and recovery procedure. One might think that this list of variables could make TLC an inconsistent or variable technique. On the contrary, this diversity is the prime reason for the versatility and selectivity of HPTLC.

23.1.6 As a practical laboratory technique, TLC is relatively rapid. An adequate separation can often be obtained within a few hours, even when a completely new problem is being studied. Each separation requires 10 to 45 min to complete, and once the experimenter has become familiar with the basic principles and the nature of the system, adjustments in the experimental variables can usually be made in a short time. While other forms of chromatography can offer separations in as little as 10 to 45 min they are limited to sequential analysis. The geometry of the TLC plate allows simultaneous separations to be made, thus increasing the through-put of samples and decreasing the per-sample analysis time. This simultaneous aspect of TLC points up its advantage as a time saver compared to other suitable separation techniques.

23.1.7   TLC is a versatile technique that can be used to separate inorganic or organic materials, and low-molecular weight species up to high chain-length polymers. There are few volatility restrictions, as opposed to gas chromatography; however, highly volatile materials may require special adaptations as they are difficult to migrate and recover without loss of sample. There are virtually no polarity restrictions on materials to be separated. If a solvent is available for the material, generally a TLC separation system can be found to resolve it. The extensive use of relatively stable sorbents makes TLC separations feasible for many reactive species that are difficult to separate by other means. The inexpensiveness of HPTLC plates when compared to the cost of capillary columns for GC or HPLC columns provides a medium that is forgiving in terms of incomplete sample clean-up. Injection of a dirty sample may render a column valued at hundreds of dollars useless for further analysis, while application of the same sample onto a TLC plate may provide adequate separation (thereby ending the process). At worst, such an error requires disposal of a plate worth several dollars.

23.1.8   A variety of reference books is available that describe the variables of TLC. The books by Bobbitt (**3**) and Randerath (**4**) offer concise information for those who have little experience with the technique. The compilations of Kirchner (**5**) and Mikes (**6**), plus the comprehensive manual by Stahl (**7**), not only describe the

general techniques, but contain many literature references on detailed separations of a variety of materials. The books by Truter (**8**) and Gordon and Eastoe (**9**) also furnish much useful information. Updated techniques are reviewed by Fried and Sherma (**2**) and Touchstone and Sherma (**10**). Quantitative aspects of modern TLC techniques are collected in the book by Treiber (**11**). Lastly, no TLC laboratory should be considered complete without a copy of the CRC *Handbook of Chromatography* (**12**), which provides the particulars of TLC separations listed by compound separated, with literature citations. Keeping up with new TLC and HPTLC developments is readily facilitated, in that *Analytical Chemistry* provides reviews of current techniques and literature every other year in its annual review (e.g., ref. **1**).

23.2   MEDIUM AND SOLVENT SELECTION.

23.2.1   The two most important variables in TLC are media and solvents. Although it is possible to approach their selection from a purely theoretical view (**13,14,15**) it is often advantageous to approach a separation from a practical aspect, i.e., asking just what is to be separated.

23.2.2   In devising a new separation procedure for an unknown system, the physical or chemical differences between the materials to be separated should be understood. For separation problems where even the functional groups are unknown, an evaluation of solubility is highly desirable. The old rule "like dissolves like" is helpful. Learning what will dissolve the sample is a good starting point for selecting the basic thin layer separation process and the type of medium. As an example, one would take a completely different approach if the compounds to be separated had maximum solubility in benzene or ether as opposed to solubility in water.

23.2.3   With the supply of TLC media on the market today, special separation techniques such as ion-exchange (**16**), ion-exclusion, electrophoresis, and even the separation of chiral compounds into their pure enantiomeric forms (**17**) are possible. The scope of this summary prevents a complete discussion of these special separation techniques; however, with a basic understanding of TLC as presented here, the reader can easily branch out.

23.2.4 Most TLC separations involve either adsorption or partition processes. A continuous spectrum of conditions can apply between the two extremes; some systems cannot be defined as applying completely to either partition or adsorption.

23.3 ADSORBENTS AND SOLVENTS.

23.3.1 Adsorption processes are generally associated with the separation of relatively nonpolar, hydrophobic materials, in which case a strong adsorbent such as alumina or silica gel with a moderately nonpolar (low dielectric constant) organic solvent or mixture is used. The activity of the layer and the polarity of the solvent are the prime variables of these systems. The adsorbent serves as the stationary phase; the organic solvent is the mobile phase of the two-phase system. As materials in the liquid phase are brought into contact with the solid phase, a concentration gradient is established at the interface. Components in the mixture separate because of differences in adsorption coefficients. The benzene-soluble material mentioned previously is relatively hydrophobic; thus separation involving adsorption chromatography should be considered. Other media are now available to expand the range of separations possible with this system of separations. These plates, called reversed-phase, have an organic moiety bound to the silica or alumina in such a manner that materials associated with the solid phase are actually dissolved in the organic moiety. Thus the principal of solvent-solvent interaction is actually invoked.

23.3.2 Silica gel is probably the most widely used adsorbent in TLC. It has high capacity and is useful for separating a variety of acidic or neutral materials. For basic materials, such as amines, alumina is often used; however, alumina has the disadvantage that it may catalyze the decomposition of many organic compounds, especially if post-separation derivativization is to be employed.

23.3.3 Other specialized TLC adsorbents include: kieselguhr, magnesium silicate, talc, calcium sulfate (also used as a binder for stronger adsorbants) magnesium oxide, calcium carbonate, cellulose, agarose gels, and even charcoal. As stated earlier, these basic adsorbents can be fur-

ther modified for a variety of purposes. Organic molecules can be attached to either aid in specialized separations (e.g., ion-chromatographies, chiral separations etc.) or act as a generalized hydrophobic region (e.g., reversed phase TLC).

23.3.4 The activity of the adsorbent layer is an important variable. Adsorbent layers should generally be activated in an oven before use. After treatment, precautions must be taken to prevent the uncontrolled adsorption of water vapor from the air. This occurs rapidly with both silica gels and aluminas. Separations on a highly active layer (silica used directly after treatment in an oven at 120°C) will be markedly different from those where the layer has been allowed to equilibrate with the water vapor in normal room atmospheres. Depending upon the specific separation, a highly active adsorbent may not be required, or desired. To obtain reproducible results the adsorbent activity must be controlled and rigidly specified. Complete saturation of silica gel with water produces a relatively inert adsorbent suitable for partition chromatography.

23.3.5 In practice it is best to pick a specific adsorbent and then vary the solvent system to maximize the separation. Solvent polarity is the dominant variable in adsorption chromatography. An eluotropic series is a good aid in selecting a solvent (see Table 23:I). More refined eluotropic series as well as a discussion of them are given by Fried and Sherma (2).

Table 23:I Eluotropic Solvent Series

Solvent	Dielectric Constant
n-Hexane	1.9 (20°C)
Cyclohexane	2.0 (20°C)
Carbon tetrachloride	2.2 (20°C)
Benzene	2.3 (20°C)
Toluene	2.4 (25°C)
Trichloroethylene	3.4 (16°C)
Diethyl ether	4.3 (20°C)
Chloroform	4.8 (20°C)
Ethyl acetate	6.0 (25°C)
1-Butanol	17. (25°C)
1-Propanol	20. (25°C)
Acetone	21. (25°C)
Ethanol	24. (25°C)
Methanol	33. (25°C)
Water	80. (20°C)

23.3.6 Generally the more polar the solvent, the lower is the adsorption affinity of a given material and the greater its migration distance. A rapid evaluation of solvents can often be made by spotting the unknown on a series of small strips and effecting the migration in a series of six or so solvents contained in small jars. In fact, at least one manufacturer provides a specialized development chamber that allows development of a single plate with several solvents simultaneously, so that this type of initial evaluation can be facilitated. Adjustments involving solvent mixtures to obtain more subtle separations can often be made in the same way. If a separation can be accomplished with a single solvent, the use of mixed solvent systems (particularly those consisting of components of widely varying volatility) should be avoided because of "second-order" factors. If a mixed solvent system is unavoidable the mixed solvent must be prepared only as needed in order to avoid changes in the mixture due to differential evaporation.

23.3.7 Plates are available in many sizes, with a variety of backing materials as well as a wide selection of coating materials. Proper selection of backing material or dimensions for the TLC plates is largely a matter of convenience and expense. Plastic and foil backed plates are usually inexpensive and useful in methods development as long as the solvent system being investigated is not a solvent for the plastic backing or any adhesives that may have been used in preparing the plates. Likewise the size and dimensions of the plate may be selected for economy or ease of use. The industry standard for clinical chemistries seems to be the 20 × 20 cm plate, which is also very useful for other separations and exploratory work. HPTLC plates are generally available in 10 × 10 cm, 10 × 20 cm, or 20 × 20 cm formats. Some suppliers provide special adaptations such as plates that are pre-ruled into lanes, or pre-scored so as to separate into individual lanes by breaking the plate along the score-marks.

23.3.8 Early workers had to hand-coat TLC plates by pouring, dipping, or using a knife-edge coater. Not only was this very time-consuming, but it was difficult for the novice to produce defect-free smooth-surfaced plates. Today, hand-coating of plates is no longer a necessity because a variety of high-purity, uniform commercial coatings are available. In fact virtually any packing material available for HPLC columns can be obtained in a plate format, and manufacturers take great pride in providing custom-coating of TLC plates at reasonable cost. The normal layer thicknesses range from 100 to 250 μm, although layers 1 to 2 mm thick may be used for preparative work. Generally the phrase HPTLC refers to a small monodisperse particle size in the coating material coupled with a thickness of 100 to 150 μm. The number of theoretical plates achievable by this optimization of materials is an astonishing 5000 theoretical plates **(18)**, compared to an average 600 theoretical plates achievable by conventional TLC.

23.4 SAMPLE APPLICATION.

23.4.1 Many devices are useful for applying sample solutions to the TLC sorbant layer. These include capillary tubes (Figure 23:1), syringes, and micropipettes of all types, as well as an ever-increasing number of mechanical or automated spotting devices. The zone of application, or spot, must be very compact. It is helpful to

Figure 23:1 – Sheets can be spotted most easily and conveniently using capillary tube applicators.

use as non-polar a solvent as possible in order to minimize sample diffusion during the spotting operation. Optimum chromatographic resolution is affected by both the amount of material applied per spot, and the concentration of the analyte in the working solution. Generally 5 μl of a 1% solution of the sample mixture is optimal; however, differences in solubility of the analyte etc. may necessitate sequential spotting of the mixture until enough analyte is deposited at the starting point to enable adequate detection later. There are two main types of commercially available sample applicators. The first simply is an adaptation of the hand held capillary tubes but provides much more uniformity in spot delivery than hand spotting can provide. A variety of these devices are described by Fried and Sherma (2, Chapter 5). The second type consists of a vacuum-actuated teflon strip that is drawn into a template. The sample is applied to the resulting depression and the solvent is evaporated. The low surface tension of the teflon strip allows the analyte to form a tight spot at the bottom of the well. A chromatographic plate is then placed over the teflon strip, the vacuum reversed, and the spot is then transferred to the TLC plate. This latter device is recommended if the analyte is of low solubility, and several applications need to be made to the same spot (19). The use of a volatile solvent is advisable with any spotting technique to ensure complete removal of the solvent prior to development. The solubility of analyte in the chosen solvent must, however, allow complete removal from the walls of any application device.

23.4.2 If a compound migrates with the solvent front, even with a relatively nonpolar solvent such as hexane, a stronger adsorbent should be considered. Conversely, compounds that fail to migrate from the point of application with very polar solvents such as methanol should be separated on a weaker adsorbent or perhaps by partition chromatography. Non-polar sample mixtures are best separated with nonpolar solvents on strong adsorbents. More polar materials may require more polar solvents on weaker or less highly activated adsorbents or may be considered for separation by partition chromatography.

23.5 PARTITION CHROMATOGRAPHY.

23.5.1 Partition chromatography is used primarily to separate relatively polar organic compounds or inorganic salts. The layer, as such, plays no active part in the separation. Two immiscible liquid phases are present. The mobile phase is organic and hydrophobic; the stationary phase often consists merely of water held on cellulose fibers (as in paper chromatography) or an organic moiety covalently bonded to the coating material. The components of the mixture to be separated distribute in different ratios between the two liquid phases. Although partition chromatography has been practiced for many years in the form of paper chromatography, the experimental variables are harder to define because the adsorption activity rules, the solubility rules, and the eluotropic series are not always applicable. Most existing separations originally designed for paper chromatography are easily adaptable to thin cellulose layers without extensive modification, and usually with greater reproducibility and lower limits of detection. The separation on TLC and HPTLC plates produces sharper zones (more concentrated areas of analyte), owing to a reduction in diffusion. Also the migration time is markedly reduced (30 to 90 min instead of up to 16 h) compared to classical paper chromatography. If large amounts of material are to be separated (preparatory chromatography) application of the sample can take the form of a narrow band rather than a small spot. Under ideal conditions 1 mg or more may be separated. The use of special preparative plates with thicker coatings, or coatings that vary from one thickness to another (taper) can further increase the mass of material that may be manageably separated.

23.5.2 A word should also be included about the new plates on the market that include a preconcentrating strip or zone. Generally these plates consist of an additional strip of adsorbent or coating near the origin of the plate. This strip serves to absorb the sample, allowing more material to be placed in the small spot. The discontinuity between the thickness of the applicating strip and the actual chromatographic coating causes the migrating spot to bunch up (become more concentrated, resulting in a smaller spot diameter) just before the chromatographic separation actually begins. These plates are very useful when spots are to be applied by hand held devices, or when the solvent used has shown a tendency to spread on the given coating material. The advent of automatic spotting devices should diminish the necessity of using plates with pre-adsorbent strips.

23.5.3 For comparison, standards should always be run alongside the sample on the same sheet or plate. It is important that sample and standard be applied in the same solvent and in the same manner.

23.6 DEVELOPMENT.

23.6.1 Non-reproducible results and difficulties in duplicating separations can usually be traced to differences in developing conditions. The type of chamber, the initial degree of saturation and changes in conditions during the course of development significantly influence the results.

23.6.2 The ascending method of development is by far the most common in TLC; however, it is rapidly being replaced by linear development. The two most common types of development chamber in use are the jar (usually lined with filter paper or containing a saturation pad or wick to promote saturation of the air in the chamber with solvent vapors), and the sandwich chamber (designed so that minimal volume remains after the plate is inserted, and also promoting saturation of solvent vapors; see Figure 23:2). While it is difficult to predict in advance which chamber will yield the best separations, it seems that the sandwich chamber is enjoying more widespread usage lately. Since both the sample mixture and the solvent system behave differently in each type of chamber, trial and error are still necessary; and time spent experimenting with various chamber environments is well invested.

23.6.3 The time required for a development distance of 10 cm above spot origin (the most commonly reported and the most practical development distance for 20 × 20 cm plates) varies between 10 min and 2 h, depending on the type of sorbent, solvent system, and developing conditions. If more than one sample is applied to a plate (separated into lanes by scoring the plate) than the separation time per sample is greatly reduced.

Figure 23:2 — The sandwich chamber is a simple and efficient piece of laboratory apparatus in which to develop a chromatogram. After the sheet has been spotted, it is clipped into the glass plate sandwich and immersed in a trough in which the solvent level is adjusted to immerse just the bottom of the sheet. As the solvent rises through the sheet by capillary action, components of the mixture to be separated are carried in a differential manner to produce discrete zones.

23.6.4 Special techniques of stepwise, continuous, multiple, two-dimensional, forced flow, controlled migration, over pressure, and radial development, as well as a combination of TLC and thin-layer electrophoresis (TLE) have also been used to advantage in the separation of the more complex mixtures.

23.6.5 Stepwise development consists of repeated migrations on a single chromatographic plate using a series of solvents (usually of increasing polarity). It is a useful technique for complex mixtures that do not separate with a single migration. Not only are recovery and respotting steps eliminated, but it may be possible to avoid a time-consuming evaluation of the solvents to find the best one. An added benefit of this method accrues from the fact that the spots tend to condense (by the same mechanism that concentrates the spots in pre-adsorbent strips) as the solvent front progresses through their location after the previous migration. Although specialized TLC equipment exists, the simple addition of a gradient pump system to deliver a changing solvent mixture to the development solvent reservoir (35) can easily provide gradient TLC separations. Chromatograms can thus be developed with, against, or across the stationary-phase gradient (20,21,22).

23.6.6 In continuous or overrun development the solvent is allowed to evaporate from the solvent front to a wick or the chamber air. This enhances resolution of materials that migrate poorly. This almost always results in a faster separation than conventional techniques (23).

23.6.7 Multiple development involves repeated but discrete migrations with a single solvent. It can be used to remove a small amount of material from a relatively insoluble zone, and also compresses the zones as in stepwise development. Its use has been reviewed by Perry (24).

23.6.8 Complex mixtures such as amino acids have been separated by two-dimensional development. A sample is spotted in one corner and caused to migrate with one solvent. The chromatogram is then rotated 90° and a second solvent is used for the second migration. The general result is a band of discrete spots placed diagonally across the plate. Special considerations in two-dimensional TLC are discussed by Giddings (25).

23.6.9 Forced flow and controlled migration both try to overcome the same problem by different strategies. A problem with conventional TLC is the propensity for the solvent front speed to decrease with increasing distance from the solvent reservoir. In forced flow TLC the developing solvent is applied by pumping at specified flow rates (26). Controlled migration TLC delivers the mobile phase solvent from a syringe that actually moves along the plate in the direction of development (27).

23.6.10 Overpressure chromatography very closely resembles HPLC. A pressurized plenum placed over the TLC plate eliminates the vapor phase in the development process, and allows moderate pressures to be applied. The methodology shows great promise for the TLC separation of volatile compounds. Plates are newly available on the market that allow 30 samples to be separated under 25 Atm. Newman (28) reviews the literature concerning this technique.

23.6.11 The advent of HPTLC has much diminished the migration distance re-

quired to effect total separation, so much so that equipment now exists to allow radial development of samples. Here samples are spotted around a circle ("bull's eye") in the center of a 10 × 10 cm HPTLC plate. Solvent is applied to the plate by the use of a tiny wick. Migration then proceeds outwards in a radial manner. The advantages here are two-fold: Increased speed of development (due to multiple sample application and shorter migration distances); and independence of gravitational effect (29). A modification of this process utilizes a rotating disc, applying centrifugal force to the separation (30).

23.7 VISUALIZATION TECHNIQUES.

23.7.1 Separations involving dyes and pigments can be viewed directly. However, when compounds are not detectable by their visible color, one must search for a method of locating these materials on the chromatogram. The immediate tendency is to spray the plate with a non-selective color forming reagent (see Figure 23:3). This practice should be resisted at least until some thought is given to the many nondestructive visualization methods available. Often, compounds can be located by their fluorescence under either long (366 nm) or short (254 nm) wavelength ultraviolet light, or by their ability to quench the fluorescence of an inorganic phosphor. Plates can be purchased with the fluorescent phosphor impregnated into the plate coating, or it can be added as a spray reagent after migration. A light spray of water is often repelled by

Figure 23:3 — Materials that are not colored can be visualized by spraying with a reagent that reacts to form a colored product.

hydrophobic compounds on the chromatogram so that white spots appear on a translucent background. Iodine vapor is reversibly absorbed by many organic compounds and thus may be used as a nondestructive visualization technique. Autoradiography is a highly sensitive method for monitoring of separations of radioactive or radiolabelled compounds.

23.7.2 If all the previous techniques fail, then it may be necessary to turn to one of the many destructive methods of visualization. These range in sophistication from highly sensitive and specific sprays for forming fluorescent derivatives with picogram quantities of materials to sprays of sulfuric acid or other reagents that, because of their strong oxidizing capabilities, nonselectively turn all organic matter present to black carbon. Chemical spot tests are known for scores of functional groups and hundreds of different compounds (31,32). Most of these are directly applicable to use on thin layers. These color reactions are adequately described in the literature (5,7). If starting material is at a premium, but no indication of the compound's presence can be found without the formation of derivatives, the use of scored plates is in order. Develop the chromatogram, break off one of the pre-scored lanes, derivatize it, and hold it in close proximity to the original plate to determine the area that holds the compound of interest.

23.7.3 One must not overlook the new scanning densitometers on the market, which may be able to detect the presence of a compound by scanning with a uv/visible spectrometer. These TLC scanning densitometers are capable of producing a time dependent chromatogram that appears like that produced by HPLC.

23.8 CHEMISTRY OF THE LAYER.

23.8.1 Carrying out chemistry on the sorbent layer can be extended beyond visualization reactions. Derivatives can be formed after spot application and before development. Other chemical conversions can be carried out to enhance multiple, step-wise, or two-dimensional development. Layers can be chemically altered either during preparation or by modifying the solvent system to control ionization or to complex certain species under study.

23.9 ELUTION OF SEPARATED ZONES.

23.9.1 After the TLC separation is complete, it may be desired to remove and isolate the pure material contained in a single zone. One may then wish to quantitate this material, confirm its identity by some additional instrumental or chemical technique, use it in a subsequent experiment as a pure reference compound, determine one of its physical constants, or even isolate milligram quantities for special studies.

23.9.2 TLC zones are conveniently removed from the adsorbent by a variety of methods. These methods include: scraping the zone to be harvested and desorbing the analyte in a suitable solvent; cutting out the zone to be harvested (flexible plate backing material may be required) and soaking it in a suitable solvent; and the use of a special automatic eluter. The automatic eluter requires that each zone to be harvested be circumscribed to separate it from other zones and the media. Then a special elution head is clamped down over the zone and a suitable solvent is passed from a reservoir over the spot and to a second reservoir. This device is largely a labor-saving device in that the principle of extraction is the same as manual methods, but up to six zones can be eluted simultaneously. Regardless of the method chosen to harvest the zone, some procedural guidelines are needed:

a. Prewash the TLC plate or precoated sheet with the eluting solvent prior to activation and sample application. This removes impurities that would be eluted with the separated zones after chromatography;

b. use either flexible TLC media or prescored plates so that a sample of the zone may be removed from the rest if destructive visualization techniques are needed to locate the material to be harvested;

c. when scraping separated zones for the removal of adsorbent by filtration, apply a cake of calcium sulfate to the filter from a slurry of the finely divided powder before carrying out the filtration. This procedure aids in removing the last traces of finely divided sorbent, which can often interfere with subsequent procedures, particularly if an infrared spectrum is being used for characterization of unknown zones;

d. when attempting to elute very polar

compounds from a strong sorbent, add a material to the eluting solvent that will be more strongly adsorbed to the solid phase than is the compound under study;

e. obtain useful concentrations of the materials being separated in the horizontal bands described earlier by carrying out the chromatography a second time at a 90° angle to the first development. A very polar solvent is used that will carry the compound of interest with the moving front. Scraping channels in the sorbent layer or cutting the flexible TLC sheet can be advantageous. At the end of the second development the ratio of separated sample to sorbent should be favorable.

23.10 RECORDING RESULTS.

23.10.1 Keeping records in chromatography may present some special problems, although at first it appears to be quite simple. Often, sufficient attention is not paid to this subject until serious problems arise. At that point, the data needed may be irretrievable.

23.10.2 The most important and perhaps the most often neglected aspect of TLC record keeping is recording the EXACT chromatographic conditions. If work is to be reproduced at a later date it is essential to know all the system variables; the type of adsorbent, activation conditions, sample application, developing solvent and technique, type of developing chamber, degree and method of chamber saturation, type and distance of development, and visualization details. Temperature and humidity conditions (for critical work or in laboratories where seasonal variations are common) are two variables that are critical and often overlooked in TLC literature. Devices are available on the market in which both temperature and humidity can be tightly controlled in the development chamber. Other special conditions, particularly those unique to the sample being analyzed, should also be recorded. The best way to proceed for routine work is to prepare a form that the analyst fills out and files for each chromatogram.

23.10.3 Abrasion-resistant plates and precoated sheets lend themselves well to file storage. Although at first this may well seem like the ultimate in record retention, it must be remembered that most zones fade or change with time, or worst of all (chemi-

cal safety should be considered) the zone may desorb to a neighboring plate or sheet of paper. If the chromatograms themselves are to be kept for permanent records, outlining the separated zones with a pencil to avoid later confusion is recommended. If permanent record TLC plates will not to be reextracted to harvest a zone the researcher should consider coating the TLC plate with a thin layer of paraffin, as a preservative.

23.10.4 Two methods of chromatographic data logging should be considered for permanent records. A photograph of the chromatogram using either visible or ultraviolet light can be made easily with equipment on the market designed especially for this purpose. If quantitative TLC is the goal, then specially designed scanning densitometers can be used to produce a permanent strip-chart type recording or computer data set from a developed chromatogram.

23.11 RF VALUES AND RM VALUES.

23.11.1 The results of a separation of a mixture of compounds are often described by calculating their Rf values, defined as the ratio of the distance traveled by the sample component to the distance traveled by the solvent. Materials that do not move from the origin have a Rf value of 0; migration with the solvent front means an Rf value of 1.0. Most useful separations are those that produce widely spaced Rf values between 0.3 and 0.7. The origin is defined as the spot where the sample was applied to the TLC plate, unless plates with pre-adsorbent strips are used, in which case the beginning of the TLC adsorbent (the end of the pre-adsorbent strip) is considered the origin.

23.11.2 Although Rf values are useful in describing results, they are quite inadequate as absolute standards and their importance is often overemphasized. When they are used to communicate observed effects, they should be compared to actual standards run on the same plate in a different lane, and they should be heavily qualified, because they are affected by so many variables. The Rm value, though equally constrained by the same variables, is now more widely used **(33)**. The Rm is related to the Rf by the following: $Rm = \log_{10} (1/Rf - 1)$. This normalization allows migration distance to be much more useful **(33)** and has even been used in qualitative analytical identifications of unknowns and structural analysis (34).

23.12 QUANTITATION.

23.12.1 Great strides have been made in quantitative instrumentation for TLC in the past decade. At that time it was generally considered that analytical accuracy required removal of the material from the layer and subsequent analysis and quantification by conventional analytical techniques. These and other quantitative techniques are described by Shellard (36). A collection of newer techniques including a consideration of scanning densitometry is available as edited by Trieber (11). Many forms of quantification have been utilized, from the semi-quantitative method of measuring zone diameters to highly technical approaches including: photothermal deflection densitometry using lasers (36); fast atom bombardment (38), secondary ion (40) and laser mass spectrometry (39); FTIR (41), and video densitometry (16).

23.12.2 A variety of scanning densitometer designs is on the market. These include densitometers with fixed wavelength scanning that are widely used in scanning gels from electrophoresis, etc. Due to their limited range of response, they are unacceptable for TLC work as are those that scan a photograph of the developed plate. However, there are several manufacturers with instruments that have been designed specifically for TLC work. They usually have a selection of operating wavelengths so that the one with greatest sensitivity for absorption by the sample can be used. Additionally, most of these systems use a combination of reflected and transmitted light to allow for differences in plate or coating thickness. These on-plate scanners are programable as to scan speed and distance, and can generally be adapted to existing laboratory information management systems. An added benefit from these scanning densitometers is that they can be used at a variety of wavelengths sequentially on the same plate in order to quantify different classes of compounds, or to give information as to the identity of a given zone. A permanent record of these nondestructive method results can be saved in the data format of choice, and destructive visualization techniques can be used subsequently if needed.

23.12.3 The new instrumental trend in TLC does much to aid in data aquisition. Both hardware and software are readily available (42), as are identification retrieval systems (43), and the potential for automation (44).

23.12.4 This points out another advantage of TLC. The semi-permanent nature of the TLC plate means that it can be sequentially processed so long as a given visualization technique does not preclude subsequent techniques. Each of these visualization techniques can be optimized independently of the separation step and of previous nondestructive techniques. HPLC or GC quantification techniques often have to be compromised by column flow rate, etc., that must be chosen to maximize the resolution of separation.

23.13 LIMITATIONS.

23.13.1 TLC is also subject to limitations. It is difficult to separate high molecular weight polymers and closely related materials, such as isomers and homologs. Volatile materials require special equipment for handling. Quantities greater than several grams are unwieldy to handle. Characterization of materials requires the use of reference materials or other visualization techniques.

23.14 APPLICATIONS.

23.14.1 Thin layer chromatography is used routinely in industrial, biomedical, clinical, and environmental laboratories. The technique has found use in air-pollution monitoring, especially with regard to detection and determination of polynuclear hydrocarbons. Some representative compounds that have been detected by this technique include:

7-H-Benz[de]anthracen-7-one   (45,46)
Phenalen-1-one   (45,46)
Benzo[a]pyrene   (46)
Benzo[c]acridine   (47)
Benzo[h]quinoline   (47)

Fatiadi (48) has studied the oxidation products of pyrene using TLC. The direct reaction of azulene on a TLC chromatogram has been applied to identify 5-hydroxymethylfurfural in the effluent from a coffee roasting plant (49). Components in automobile exhaust, including benzaldehyde, salicylaldehyde, and acetophenone, have been separated and identified by paper and thin layer chromatography **(50)**.

23.14.2 In the study of airborne particles, other techniques have been combined with TLC to enhance the detection limits or improve the quantitative aspects. Complex mixtures of aromatic hydrocarbons have been separated by TLC and subsequently identified by mass spectrometry **(51)**. Minute quantities of benzo[a]pyrene are detected by combining TLC and spectrofluorometry **(52)**. A method for the direct spectrofluorometric determinations of polycyclic hydrocarbons and azaheterocyclics on chromatograms has been devised **(53)**. Two-dimensional TLC and fluorometry have been combined to determine 9-acridanone in urban air particulate matter **(54)**. There are numerous other applications of the use of TLC in the field of air pollution.

23.14.3 The analysis of pesticide residues is another area where TLC can be used to practical advantage. TLC separation data of 60 different pesticides using 19 different solvents have been summarized by Walker and Beroza **(55)**. TLC is an ideal technique for separating a diversity of moderately nonpolar materials encountered in this field. Kirchner **(5)** presents a summary of references for the TLC separation of various fungicides, herbicides, and insecticides such as pyrethrins and thiocarbamates, in addition to those mentioned above.

23.14.4 TLC has been combined with other techniques in the environmental field. Bjorseth **(56)** proposed combining TLC with the Ames Salmonella/mammalian microsome mutagenicity assay **(57)**. Butler **(58)** took this approach one step further by performing sequential extractions (with solvents of increasing polarity) of particulate samples collected on high-purity glass-fiber or quartz filters. The resultant extracts were each developed with different mobile phases before being used in the mutagenicity assay. Reilly **(59)** showed the applicability of these techniques to a wide variety of environmentally important compounds including aza-arenes, PCB's, and other compounds identified by their functional groups.

23.15 SUMMARY.

23.15.1 Rapid and efficient separation and identification of chemical compound classes in environmental samples is best accomplished with TLC **(60,61)**, especially if the sample load is large or if screening of the samples is the objective. TLC in recent years has experienced a resurgence of interest and application, in that it provides many advantages over classical GC and LC analyses. Modern, high-precision, thin-layer chromatography (HPTLC) saves time and sample clean-up, avoids problems associated with column "poisoning" and presents no limitation on the type of solid phase used, the type of solvent system used, or time for development **(62)**. Perhaps more importantly, HPTLC provides a "hardcopy" of the analysis performed, in the form of a plate, which allows frequent re-inspection, re-analysis, if necessary, and the application of detection procedures not possible with classical, column-based GC or LC analyses **(62,60,58)**.

REFERENCES

1. SHERMA, J. 1986. Anal. Chem. 58:69–81R.
2. FRIED, B. AND J. SHERMA. 1986. *Thin-Layer Chromatography Techniques and Applications*, 2nd Ed. Marcel Dekker Inc., New York.
3. BOBBITT, J.M. 1963. *Thin-Layer Chromatography*, Reinhold Publishing Company, New York.
4. RANDERATH, K. 1966. *Thin-Layer Chromatography*, 2nd Ed. Academic Press Inc., New York.
5. KIRCHNER, J.G. 1967. *Thin-Layer Chromatography*, Vol. XII of *Technique of Organic Chemistry*, Interscience Publishers, New York.
6. MIKES, O. 1966. *Laboratory Handbook of Chromatographic Methods*, D. Van Nostrand Co., Inc., New York.
7. STAHL, E. 1969. *Thin-Layer Chromatography—A Laboratory Handbook*, 2nd Ed. Academic Press Inc., New York.
8. TRUTER, E.V. 1963. *Thin-Film Chromatography*, Cleaver-Hume Press, Ltd., London.
9. GORDON, A.H., AND J.E. EASTOE. 1964. *Practical Chromatographic Techniques*, D. Van Nostrand Co., Inc., New York.
10. TOUCHSTONE, J.C., AND J. SHERMA. 1985. *Techniques and Applications of Thin Layer Chromatography*. John Wiley & Sons, Inc., New York.
11. TREIBER, L.R., 1987. *Quantitative Thin-Layer Chromatography and its Industrial Applications*. Marcel Dekker, Inc., New York.

12. ZWEIG, G., AND J. SHERMA. 1972. *Handbook of Chromatography*, Vol. I. CRC Press Inc., Cleveland.
13. ROZYLO, J.K., J. OSCIK, B. OSCIK-MENDYK, AND M. JARONIEC. 1981. J. High Resol. Chromatog. 4:17–23.
14. PETROVIC, S.M., E. LOUCAR, L. KOLAROV. 1984. Chromatographia. 18:683–691.
15. SOCZEWINSKI, E. 1980. J. Liquid Chrom. 3:1781–1806.
16. ELODI, R., AND T. KARSAI. 1980. J. Liquid Chrom. 3:809–831.
17. GUNTHER, K., M. SCHICKEDANZ, K. DRAUTZ, AND J. MARTENS, 1986. Z. Anal. Chem. 325:513–514.
18. POOLE, C.F., AND S.A. SCHUETTE. 1984. *Contemporary Practice of Chromatography*. Elsevier, New York.
19. FENIMORE, D.C., AND C.J. MEYER. 1979. J. Chromatogr. 366:71–77.
20. STAHL, E., 1959. Arch. Pharm. 292:411–416.
21. STAHL, E., 1965. J. Pharm. Belg. 20:159–168.
22. STAHL, E., AND E. DUMONT. 1969. J. Chromatogr. Sci. 7:517–525.
23. NUROK, D., R.E. TECKLENBURG, AND B.L. MAIDAK. 1984. Anal. Chem. 56:293–297.
24. PERRY, J.A., T.H. JUPILLE, AND L.H. GLUNZ. 1975. Anal. Chem. 47(1):65A–74A.
25. GIDDINGS, J.C. 1984. Anal. Chem. 56:1258A–1270A.
26. KALASZ, H. 1985. V. Danube Symposium on Chromatography Proc.
27. NUROK, D. 1981. Anal. Chem. 53:714–716.
28. NEWMAN, J.M. 1985. International Lab. 15(5):22–33.
29. PO, A. LI WAN, AND W.J. IRWIN. 1979. High Resolut. Chromatogr. & Chromatogr. Commun. 2:623–627.
30. ZINK, D.L. 1985. Ch. 4 in *Techniques and Applications of Thin Layer Chromatography*. J.C. Touchstone and J. Sherma Eds. John Wiley & Sons, Inc., New York.
31. FEIGL. F. 1966. *Spot Tests in Inorganic Analysis*, 7th ENGLISH Ed. American Elsevier Publishing Co. Inc. New York.
32. FEIGL. F. 1958. *Spot Tests in Inorganic Analysis*, 5th ENGLISH Ed. American Elsevier Publishing Co. Inc., New York.
33. OSCIK, J., J.K. ROZYLO, B. OSCIK-MENDYK, M. JARONIEC. 1981. Chromatographia. 14:95–99.
34. BARK, L.S. 1979. Ch 1. in *Progress in Thin-Layer Chromatography and Related Methods*, Vol. I. A. Neiderwieser and G. Pataki Eds. Ann Arbor-Humphrey Science Publishers. Ann Arbor.
35. SANDER, L.C., AND L.R. FIELD, J. Chromatogr. Sci. 18:133.
36. SHELLARD, E.J. 1968. *Quantitative Paper and Thin-Layer Chromatography*. Academic Press Inc., New York.
37. PECK, K., F.K. FOTIOU, AND M.D. MORRIS. 1985. Anal. Chem. 57:1359–1362.
38. CHANG, T.T., J.O. LAY, JR., AND R.J. FRANCEL. 1984. Anal. Chem. 56:109–111.
39. NOVAK, F.P., AND D.M. HERCULES. 1985. Anal. Letters 18(A4):503–518.
40. KUSHI, Y., AND S. HANDA. 1985. J. Biochem. (Tokyo) 98:265–268.
41. WHITE, R.L., 1985. Anal. Chem. 57:1319–1822.
42. FOSS, R.G., C.W. SIGEL, R.J. HARVEY, AND R.L. DEANGELIS. 1980. J. Liquid Chrom. 3:1843–1852.
43. GILL, R., B. LAW, C. BROWN, 1985. Analyst 110:1059–1065.
44. ROGERS, D., 1985. Intern. Lab. 15:12–20.
45. ENGEL, C.R., AND E.J. SAWICKI. 1968. Chromatog. 37:508.
46. STANLEY, T.W., M.J. MORGAN, AND E.M. GRISBY. 1967. Environ. Sci. & Technol. 1:927.
47. SAWICKI, E.J., T.W. STANLEY, AND W.C. ELBERT. 1967. J. Chromatogr. 24:72.
48. FATIADI, A.J. 1968. Environ. Sci. & Technol. 2:464.
49. ENGEL, C.R., AND E.J. SAWICKI. 1968. Microchem. J. 13:202.
50. BARBER, E.D., AND E.J. SAWICKI. 1968. Anal. Chem. 40:984.
51. WALLCAVE, L. 1969. Environ. Sci. & Technol. 3:948.
52. BENDER, D.F. 1968. Environ. Sci. & Technol. 2:204.
53. SAWICKI, E.J., T.W. STANLEY, AND H. JOHNSON. 1964. Microchem. J. 8:257.
54. SAWICKI, E.J., T.W. STANLEY, AND W.C. ELBERT. 1967. J. Chromatogr. 24:431.
55. WALKER, K.C., AND M. BEROZA. 1963. J. Assoc. Official Analyt. Chemists. 46:250.
56. BJORSETH, A., G. EIDSO, J. GETHER, L. LANDMARK, AND M. MOLLER. 1982. Science. 215:87–89.
57. AMES, B.N., J. MCCANN AND E. YAMASAKI. 1975. Mutat. Res. 31:347–364.
58. BUTLER, J.P., T.J. KNEIP AND J.M. DAISEY. 1987 Atmos. Environ. 21(4):883–892.
59. REILLY, F.J., JR., J.M. O'CONNOR AND P.M. BOONE. Oceans '86. Vol. 3 Monitoring Strategies. 797–802.
60. TOUCHSTONE, J.C. AND M.F. DOBBINS. 1983. *Practice of Thin-Layer Chromatography*. 2nd Ed. John Wiley & Sons, New York.
61. BORMAN, S.A. 1982. Anal. Chem. 54:790A–794A.
62. FENIMORE, D.C. AND C.M. DAVIS. 1981. Anal. Chem. 53(2):252A–266A.

Subcommittee 9
A.L. LINCH, *Chairman*
E.R. HENDRICKSON
M. KATZ
J.R. MARTIN
G.O. NELSON
J.N. PATTISON
A.L. VANDER KOLK
R.B. WEIDNER

Approved with modifications from 2nd edition
F.J. REILLY, JR.
T.J. KNEIP
J.M. O'CONNOR
J.N. HARMAN, *Chairman*

24. Instrumental Neutron Activation Analysis of Atmospheric Particulate Matter

24.1 INTRODUCTION.

One of the penalties of progress in our society has been the unwitting alteration of the natural balance of elements in our environment. By using extremely small amounts of "marker elements," researchers can use instrumental neutron activation analysis (INAA) to locate atmospheric pollution sources and pathways. Already there have been a number of successful applications in such areas as the detection and identification of local, regional, and global sources of atmospheric particles (1–4).

24.1.1 In studies of the origin and fates of trace elements, it is advantageous to be able to analyze samples for a wide spectrum of elements with a high sensitivity and accuracy. This condition is best satisfied with INAA, since it is one of the most sensitive, selective, and reliable multielement analysis techniques available. In spite of these advantages, use of the technique has been limited by the expense and sophistication of the equipment required, including a nuclear research reactor and counting equipment. Another limitation has been the availability of people with the special expertise needed to perform the analyses. But recently the technique became more accessible, as several laboratories, including the Nuclear Reactor Laboratory at MIT, have made their facilities available to the research community.

24.2 METHODOLOGY.

24.2.1 The method of activation analysis takes advantage of the fact that when a medium is irradiated with subatomic particles, such as neutrons, some of the nuclei in the target material will be converted into radioactive isotopes. The isotopes so produced will decay at a characteristic rate, in most cases emitting γ-rays of characteristic energies and intensities. These specific γ-rays are used to identify and also to determine the concentrations of elements present (5–7). Samples of atmospheric particulate matter are usually subjected to two different irradiations, the first for a few minutes to produce radioisotopes with short half-lives, up to several hours. For this irradiation a pneumatic tube facility with a short

sample transfer time should be used. Following the irradiation, samples are transferred into unirradiated containers such as previously cleaned polyethylene bags and heat sealed. Two γ-ray spectra are taken in succession to observe radioisotopes with half-lives ranging from 2.2-min Al-28 to 15–h Na-24.

After a few days, the activity due to short-lived radionuclides will be gone and the samples, elemental monitors and standard reference materials are placed in new sample carriers and irradiated for several hours to produce substantial amounts of long-lived activities. They are transferred into unirradiated containers again after the irradiation. Gamma-ray spectra of each sample are taken twice, for a few hours and for several hours, following 4–5 days and 2–3 weeks of cooling, respectively. Although this procedure has proved to be very appropriate for matrices such as atmospheric particulate matter and most of the other environmental samples (8), it can be modified easily for special applications. The general irradiation, cooling, and counting procedures, along with the elements that can possibly be observed in each spectrum, are given in Tables 24:I–IV.

24.2.2 The γ-ray spectrum is analyzed using computer-directed programs to search for the best peak of each element which is interference free, or of those for which interferences are well documented for necessary corrections to be made. There are certain energy regions where even very high-resolution detectors have difficulty resolving different γ-rays from each other, depending upon their relative intensities. One such region is the portion of spectra where γ-rays of 554.3 keV (Br-82), 559.5 keV (As-76) and 564.1 keV (Sb-122) are located. In this particular region one has to be very careful in integrating the peak area of 559.5-keV line, especially if the other line, 657.2 keV of As-76, is not observed. Bromine and Sb concentrations can be determined very accurately from other isotopes or γ-lines.

Another region that poses problems is the 1115.5-keV of Zn-65 which is covered by the 1120.5-keV peak of Sc-46, especially when there are high concentrations of Sc in the matrix relative to Zn. For an accurate Zn determination, in most cases,

Table 24:I

Thermal neutron flux:	$1 - 5 \times 10^{13}$ n/s cm^2		
Irradiation time:	1–5 min		
Cooling time:	3–5 min		
Counting time:	5 min		

Isotopes Measured in Count 1 Following Short Irradiation

Isotope	$T_{1/2}$ (min)	Energy (keV)	Interferences w/ Reference Peak	Special Notes
1. Mg–27	9.45	843.8 1014.4	Mn–56 (846.6/1811.2)	
2. Al–28	2.24	1778.9		
3. S–37	5.06	3103.3		
4. Ca–49	8.72	3084.4		
5. Ti–51	5.76	320.1	Cr–51 (320.1)	
6. V–52	3.76	1434.2		
7. Cu–66	5.10	1039.0	Br–82 (1044.0/776.5)	

hand integration of the 1115.5-keV line is required.

24.2.3 Although INAA is generally considered as an interference-free technique, there is some possibility of error due to interferences when the experimental conditions are not properly adjusted or neutron flux is not characterized well enough. There are two possible kinds of interference.

a. Primary Interference — where an isotope used for the determination of a specific element is also produced from a different element present in the matrix. For example, in the determination of Al, the Al-28 isotope produced through the ^{27}Al $(n,\gamma)^{28}$Al reaction is used. If the matrix contains high concentrations of Si and/or if the irradiation location has a relatively high fast neutron/thermal neutron ratio, Al-28 will also be produced through the ^{28}Si $(n,p)^{28}$Al reaction. This kind of interference can only be corrected by irradiation of a known amount of Si along with the sample and subtracting the corresponding activity from that of the 1778.9 keV line of Al-28.

b. Secondary Interference — which is spectral interference, is much easier to handle, especially if we use a high resolution detector and multi-channel analyzers

containing 4096 or more channels. One of the classical examples is the interference of Ta with Se determinations. The major γ-ray of Se-75 at 264.6 keV is inseparable from a line at 264.1 keV from Ta-182. However, the latter has some other strong lines which can be used to make the necessary correction using the very same spectrum along with a spectrum of pure Ta-182 taken on the same detector.

24.3 SOME OTHER NEUTRON ACTIVATION ANALYSIS TECHNIQUES.

24.3.1 *Fast Neutron Activation Analysis.* Fast neutron activation analysis, which is not used as extensively as thermal neutron activation analysis, became a promising technique in accordance with the developments in high yield 14-MeV neutron sources. Neutrons of about 14 MeV are produced by the ^{3}H(d,n)^{4}He reaction. Also, there were several attempts to use cyclotron-produced fast neutrons for analytical purposes, but their usage is rather limited **(9)**.

Irradiation of a sample with 14-MeV neutrons could produce a variety of reactions, such as (n,2n), (n,p), etc. This provides certain advantages (production of different radioisotopes than that of thermal neutrons, which enlarges the number of iso-

Table 24:II

Cooling time:	20–30 min
Counting time:	30 min

Isotopes Measured in Count 2 Following Short Irradiation

	Isotope	$T_{1/2}$ (h)	Energy (keV)	Interferences w/ Reference Peak	Special Notes
1.	Na–24	15.02	1368.5		
			2753.9		
2.	Cl–38	0.620	1642.4		
			2167.5		
3.	K–42	12.36	1524.7		
4.	Mn–56	2.576	846.6	Mg–27 (843.7/1014.4)	
				Eu–152m (841.5/121.8)	
			1811.2		
			2112.6	In–116m (2112.1/417.0)	
5.	Zn–69m	13.90	438.7	W–187 (437.5/479.5)	
				I–128 (442.9)	
6.	Ga–72	14.10	834.0		
			2201.7		
7.	Br–80	0.290	616.2	Br–82 (619.1/776.5)	
	+ Br–80m	4.42		Cl–38 (d.e.* 620.4/1642.4)	
8.	Sr	0.291	388.4		
9.	In–116m	0.903	417.0		
			1097.3	Fe–59 (1099.2/1291.6)	Ar–41 - 1293.6
					In–116m - 1293.5
10.	I–128	0.417	442.9	Eu–152m (444.0/121.8)	
				Zn–69m (438.7)	
11.	Ba–139	1.388	165.8		

*Double escape peak of 1642.4 keV

topes available for analytical purposes) and disadvantages (increase in the interferences due to the other elements present in the matrix). Also, one has to be very critical because of the large variations in the neutron flux from one experimental condition to another. Therefore, the use of flux monitors simultaneously irradiated with each sample is required.

Use of 14-MeV neutron activation analysis for trace element determinations is limited due to:

a) the small activation cross sections of most elements for fast neutrons,
b) the low neutron flux available,
c) other factors such as large variations in the neutron flux, shorter irradiation time, etc.

In spite of these drawbacks 14-MeV neutron activation is applied to on-line analysis in industry and used as a complementary analytical tool to other techniques.

24.3.2 *Prompt Gamma Activation Analysis (PGAA).* The PGAA is similar to INAA, as usually thermal neutrons are utilized for irradiations and emitted γ-rays for the measurements. But there are some fundamental differences. Determination of elemental composition of a sample by INAA depends upon the measurement of delayed radiations following neutron capture. Therefore, for an element to be measured by INAA, at least one of its isotopes must have a fairly large product of its isotopic abundance and thermal neutron cross section. Also the reaction products must not have very short or very long half-lives and should emit γ-rays of measurable intensity in their decays. If we consider that the above parameters are equal, the success of PGAA is based on the fact that it is inde-

Table 24:III

Irradiation time:	8–12 h
Cooling time:	3–4 d
Counting time:	2–4 h

Isotopes Measured in Count 1 Following Long Irradiation

	Isotope	Energy (keV)	$T_{1/2}$		Interferences w/Reference Peak	Special Notes
1.	Hg–197	77.4	2.671	d		
2.	Nd–147	91.0	11.06	d	Ba–131 (92.3/216.0)	
3.	Sm–153	103.2	1.948	d	Pa–233 (103.9/311.9) Np–239 (103.7/106.4)	Possible Gd–153 interference
4.	Np–239	106.4	2.355	d		
5.	Mo–99	140.5	2.758	d		
6.	Te–131m	149.8	1.25	d		
7.	Lu–177	208.4	6.71	d	Np–239 (209.7/106.4)	Possible Lu–177m interference
8.	Yb–175	282.5	4.19	d		
9.	Te–131m	334.3	1.25	d	Fe–59 (334.9/1099.2)	
10.	Cd–115	336.3	2.224	d	Fe–59 (334.9/1099.2) Tb–160 (337.3/879.4)	
11.	Yb–175	396.3	4.19	d	Pa–233 (398.5/311.9)	Possible Nd–147 interference
12.	Au–198	411.8	2.697	d	Eu–152 (411.1/1408.0) Nd–147 (410.4/531.0)	
13.	Zn–69m	438.7	13.8	h	Nd–147 (439.8/531.0)	
14.	W–187	479.5	23.9	h	Ba–131 (480.6/216.0)	
15.	La–140	487.0	40.27	h		Possible radium background
16.	Cd–115	527.9	2.224	d		
17.	Br–82	554.3	1.47	d		
18.	As–76	559.5	1.096	d		
19.	Sb–122	564.1	2.70	d	Eu–152 (564.0/1408.0) Eu–152 (566.0/1408.0) As–76 (562.8/559.5) Cs–134 (563.2/795.8)	
20.	Br–82	619.1	1.47	d	W–187 (618.2/479.5) Ag–110m (620.3/884.6) Ba–131 (620.0/216.0)	
21.	Ga–72	629.9	14.1	h		
22.	As–76	657.2	1.096	d	Ag–110m (657.7/884.6) Eu–152 (656.5/1408.0)	
23.	W–187	685.7	23.9	h	Nd–147 (685.9/531.0) Ag–110m (686.8/884.8)	
24.	Mo–99	739.4	2.758	d		
25.	Te–131	773.7	1.25	d	W–187 (772.9/479.5)	
26.	Br–82	776.5	1.47	d	Mo–99 (777.8/739.4) Eu–152 (778.9/1408.0)	
27.	La–140	815.8	40.27	h	Ag–110m (818.0/884.7)	Possible Eu–154 interference
28.	Ga–72	834.0	14.1	h		Possible Mn–54 interference
29.	Br–82	1044.0	1.47	d	Sb–124 (1045.1/1690.9)	
30.	Na–24	1368.5	15.03	h	Sb–124 (1368.1/1690.9)	
31.	K–42	1524.7	12.36	h		
32.	La–140	1596.4	40.27	h	Ga–72 (1596.7/834.0)	Possible Eu–154 interference

Table 24:IV

Cooling time: **2 - 3 weeks**
Counting time: **8 - 12 h**

Isotopes Measured in Count 2 Following Long Irradiation

	Isotope	Energy (keV)	$T_{1/2}$ (days)	Interferences w/Reference Peak	Special Notes
1.	Nd-147	91.0	10.98	Ba-131 (92.3/496.2)	
2.	Gd-153	97.4	241.6	Se-75 (96.7/264.5) Pa-223 (98.4/311.9)	
3.	Hf-181	133.0	42.5	Ba-131 (133.5/496.2)	
4.	Se-75	135.9	118.45	Hf-181 (136.3/482.2) Ir-192 (136.4/316.5) Lu-177 (136.7/208.4)	
5.	Ce-141	145.4	32.55		
6.	Yb-169	177.2	32.02		Possible Ta-182 interference
7.	Yb-169	197.9	32.02	Tb-160 (197.0/879.4) Ta-182 (198.3/1189.0) Se-75 (198.4/264.5)	
8.	Lu-177	208.4	6.71		Possible Ir-192 interference
9.	Ba-131	216.0	11.8	Tb-160 (215.6/879.4)	Possible Lu-177m interference
10.	Eu-154	248.0	3102.5	Pa-233 (248.3/311.9)	
11.	Se-75	264.5	118.45	Ta-182 (264.1/1189.0)	
12.	Hg-203	279.2	46.76	Se-75 (279.5/264.5)	
13.	Tb-160	298.6	72.1	Pa-233 (300.1/311.9)	
14.	Pa-233	311.9	26.95		
15.	Ir-192	316.5	74.17		
16.	Cr-51	320.0	27.7	Nd-147 (319.7/531.0) Lu-177 (319.8/208.4) Lu-177 (321.0/208.4)	
17.	Ir-192	468.0	74.17		
18.	Hf-181	482.2	42.5	Ba-131 (480.5/496.2)	
19.	Ba-131	496.2	11.8		Possible Ru-103 interference
20.	Sr-85	514.0	64.85		
21.	Nd-147	531.0	11.06		
22.	Ag-110m	657.7	252.2	Eu-152 (656.5/1408.0)	
23.	Zr-95	724.2	63.98	Eu-154 (723.3/248.0) Sb-124 (722.8/1691.0)	
24.	Zr-95	756.7	63.98	Eu-154 (756.7/248.0)	
25.	Nb-95	765.8	34.97	Ag-110m (763.8/884.7) Tb-160 (765.2/879.4)	
26.	Cs-134	795.8	752.63		
27.	Co-58	810.8	70.78	Eu-152 (810.5/1408.0)	
28.	Tb-160	879.4	72.1		
29.	Ag-110m	884.7	252.2	Ir-192 (884.5/316.5)	
30.	Sc-46	889.3	83.8		
31.	Ag-110m	937.3	252.2		
32.	Eu-152	964.0	4818.0	Tb-160 (962.1, 966.1/879.4)	
33.	Rb-86	1076.6	18.82		
34.	Fe-59	1099.2	44.56	Tb-160 (1102.9/879.4)	
35.	Zn-65	1115.5	244.0	Tb-160 (1115.2/879.4) Ta-182 (1113.2/1189.0)	Possible Eu-152 interference
36.	Sc-46	1120.5	83.8	Ta-182 (1121.3/1189.0)	

Table 24:IV   (Continued)

Isotope	Energy (keV)	$T_{1/2}$ (days)	Interferences w/Reference Peak	Special Notes
37. Co–60	1173.2	1924.2		
38. Ta–182	1189.0	115.0		
39. Ta–182	1221.5	115.0		
40. Fe–59	1291.6	44.56	Eu-152 (1292.0/1408.0)	
			Ta-182 (1289.1/1189.0)	
41. Co–60	1332.5	1924.2		
42. Eu–152	1408.0	4818.0	Ta-182 (1410.0/1189.0)	
43. Sb–124	1691.0	60.2		

pendent of the stability or radioactive properties of the product nuclide. It utilizes the large number of γ-rays emitted immediately (10^{-12}sec or less) following the decay of the excited nucleus formed by neutron capture. Since the technique does not depend on the measurement of delayed radiation, one can get the results immediately following the irradiation. There is no need for cooling samples to permit short-lived isotopes to decay for several weeks as required with INAA. This property of the method not only shortens the turn-out times of analysis but also allows the measurement of some elements, such as H, B, C, N, Si, that are not detectable with conventional INAA **(10)**.

24.4 ADVANTAGES OF NUCLEAR ANALYTICAL TECHNIQUES OVER THE OTHER COMPETING METHODS.

24.4.1 *Highly Sensitive for a Wide Range of Elements.* If "interference-free" detection limits are calculated for each element, we would obtain meaninglessly small numbers, $\cong 10^{-12}$g or less for many elements. In real samples, the limit of detectability of a particular element depends on the presence and amounts of other elements in the matrix, which increase the general background. In Table 24:V, as an example, minimum detection limits, ng/m³, are given for the atmospheric concentrations of elements observed in urban air from an actual experiment in which about 17 m³ of air was pumped through a filter paper and analyzed by instrumental neutron activation analysis with the procedure explained previously.

24.4.2 *There are Virtually No Matrix Effects from Self-Absorption or Enhancement.* There are some other instrumental analysis techniques that may not require chemical separations, e.g. x-ray fluores-

cence (XRF), scanning electron microscopy coupled with XRF, secondary-ion mass spectrometry, Auger spectroscopy, etc. These techniques have many powerful applications, but they cannot accurately handle samples that are thick or inhomogeneous. For example, in XRF, the K x-rays of Si have 1.74 keV energies, so low that 50% of the x-rays would be absorbed upon passage through a 1-μm thickness of Al. By contrast, in nuclear methods of analyses long range projectiles and emitted radiations are utilized. Self-absorption of thermal neutrons is negligible unless the sample used has an unusual composition, e.g., > 10% Cd.

24.4.3 *Methods are Fairly Nondestructive.* In most cases there is no need to dissolve or change the physical forms of the samples. Also, after the activities have decayed to low enough levels, samples can be re-analyzed using other analytical techniques. However, there is a possibility of heating (usually below 40°C) and radiation damage during long irradiations, which could be important for some samples.

24.4.4 *A Large Number of Elements Are Measurable in One Small Sample.* It is possible to determine concentrations of 20 to 40 elements in individual samples weighing 10 to 100 mg. This is rather important in cases where the amount of sample is limited or large sample-to-sample variation does not allow duplicate sampling.

24.4.5 *Most Analyses Are Completely Instrumental, Requiring Minimal or No Sample Handling.* Aside from some specific applications, one does not need to do radiochemical separations. Therefore, it is easy to handle samples that would be very difficult to dissolve, such as coal, rocks, or minerals. This eliminates the possibility of

Table 24:V   Sensitivity and Detection Limits for Elements on Air Filter Samples from Urban Air (Washington, D.C.) Observed by Instrumental Neutron Activation Analysis.

ELEMENTS	Sensitivity Counts/s / ng/m³ ($\times 10^3$)	Minimum Detection Limit (ng/m³)	Typical Urban Conc. (ng/m³)	Absolute Minimum Detection Limit Mass (ng)
Na	4	4	400	70
Mg	0.4	520	650	9000
Al	3.5	40	1700	700
S	0.002	9800	7200	1.7×10^5
Cl	3.5	8	180	140
K	0.2	40	600	700
Ca	0.1	160	1600	2700
Sc	1600	2×10^{-3}	0.5	0.03
Ti	35	110	100	1900
V	210	1	100	17
Cr	40	0.3	30	5
Mn	720	0.2	50	3
Fe	0.5	6	1600	100
Co	95	0.03	1.4	0.5
Cu	5	50	25	850
Zn	3	5	360	85
Ga	30	0.8	3	14
As	130	0.3	15	5
Se	70	0.1	6	2
Br	30	0.6	200	10
Rb	0.3	10	30	170
Sr	5	30	9	140
Ag	40	0.2	0.3	3
Cd	6	6	4	100
In	5800	0.01	0.1	0.2
Sb	10	0.1	20	2
I	80	2	10	35
Cs	160	0.05	0.2	0.9
Ba	0.6	10	30	170
La	150	0.08	2	1.4
Ce	80	0.1	4	2
Sm	2600	0.02	0.3	0.3
Eu	160	0.01	0.05	0.2
Hf	700	0.02	0.1	0.3
Ta	75	0.04	0.06	0.7
W	100	0.3	0.4	5
Th	350	0.02	0.3	0.3

contaminating the sample, losing volatile species or leaving behind insoluble residues and sample material on container walls. These effects can cause serious errors, especially when analyzing for ultra-trace species.

24.4.6 *They are Virtually Interference-free.* There are very few inter-element interferences, but when they occur they can be unraveled by looking at the other characteristic γ lines of interfering species.

In many other types of trace element analysis (e.g., atomic absorption spectrophotometry, in which one does not normally see the spectrum), one is often not aware of interferences unless the sample is also analyzed for the interfering species, which frequently is not the case for complex samples.

24.5 DISADVANTAGES OF NUCLEAR METHODS.

24.5.1 The major disadvantage of nuclear methods is that the equipment required is expensive and the number of people skilled in its use is so small that most potential users do not have access to these methods. To pursue work of this sort one needs access to a major nuclear facility, plus some $50,000–$100,000 worth of counting equipment. If these are available it is possible to handle very large numbers of samples with a reasonable cost. But the initial cost is so high that it requires a strong commitment to get into the field. Fortunately there are centers such as MIT where these capabilities are already available to users.

24.5.2 Another disadvantage is that turn-around times for the analysis of the long-lived isotopes are around 2–3 weeks because of the wait for the decay of the short-lived isotopes.

This waiting period greatly improves the signal to background for the long-lived isotopes.

24.5.3 They are elemental techniques, which therefore do not give information about the chemical composition of the samples.

24.5.4 Although the number of elements that can be determined by INAA as given in Table V is impressive, several environmentally important elements such as Si, Ni, Pb, and all elements lighter than Na except F can be determined marginally or are impossible to observe. Therefore it should be supplemented by other analytical techniques for the determination of these elements. For example, by the use of fast neutron activation analysis or PGAA as complementary techniques to INAA, these and most of the low-Z elements can be determined in atmospheric particulate material.

REFERENCES

1. R. DAMS, J.A. ROBBINS, K.A. RAHN, AND J.W. WINCHESTER (1970), Nondestructive Neutron Activation Analysis of Air Pollution Particulates, Anal. Chem. *42*, pp. 861–867.
2. S.G. TUNCEL, I. OLMEZ, J.R. PARRINGTON, G.E. GORDON, AND R.K. STEVENS (1985), Composition of Fine Particle Regional Sulfate Component in Shenandoah Valley, Environ. Sci. Technol., *19*, pp. 529–537.
3. K.A. RAHN AND D.H. LOWENTHAL (1985), Pollution Aerosol in the Northeast: Northeastern-Midwestern Contributions, Science, *228*, pp. 275–284.
4. I. OLMEZ AND G.E. GORDON (1985), Rare Earths:

5. Atmospheric Signatures for Oil-Fired Power Plants and Refineries, Science, *229*, pp. 966–968.
5. P. KRUGER (1971), *Principles of Activation Analysis*, Wiley-Interscience, New York.
6. D.D. SOETE, R. GIJBELS, AND J. HOSTE (1972), *Neutron Activation Analysis*, Wiley-Interscience, New York.
7. J.M.A. LENIHAN, S.J. THOMSON, AND V.P. GUINN (editors) – (1972), *Advances in Activation Analysis*, Vol. 2, Academic Press, London and New York.
8. M.S. GERMANI, I. GOKMEN, A.C. SIGLEO, G.S. KOWALCZYK, I. OLMEZ, A.M. SMALL, D.L. ANDERSON, M.P. FAILEY, M.C. GULOVALI, C.E. CHOQUETTE, E.A. LEPEL, G.E. GORDON, AND W.H. ZOLLER (1980), Concentration of Elements in the National Bureau of Standards' Bituminous and Subbituminous Coal Standard Reference Materials, Anal. Chem. *52*, pp. 240–245.
9. I. OLMEZ, G.S. KOWALCZYK AND G.H. HARRISON (1985), Use of Cyclotron Produced Fast Neutrons in Activation Analysis, Radioanal. Nucl. Chem. Letters, *94*, pp. 391–398.
10. M.P. FAILEY, D.L. ANDERSON, W.H. ZOLLER, G.E. GORDON, AND R.M. LINDSTROM (1979), Neutron Capture Prompt γ-Ray Activation Analysis for Multi-Element Determinations in Complex Samples, Anal. Chem. *51*, pp. 2209–2221.

Subcommittee 9
I. OLMEZ
J.N. HARMAN, *Chairman*

25. Use of Selective Ion Electrodes to Determine Air Pollutant Species (Ambient and Source Level)

25.1 INTRODUCTION.

25.1.1 *Definition.* Selective ion electrodes measure the activity of an ionic species in solution by developing an absolute potential which is related to the ionic activity; it may be determined experimentally by measuring the difference in potential of an electrode pair consisting of the selective ion electrode and a suitable reference electrode (such as a calomel or silver-silver chloride reference electrode) with a high input impedance voltmeter. The potential response of a selective ion electrode to an ionic species A is defined by the Nernst equation

$$E = E^0 + \frac{RT}{nF} \ln A$$

or

$$E = E^0 + \frac{2.303\ RT}{nF} \log A$$

where
- E = equilibrium electrode potential
- R = the gas constant
- T = the temperature, K
- F = the faraday
- A = ionic activity of species of interest
- n = charge on ion of interest
- E^0 = constant dependent on choice of working electrode — reference electrode system

Selective ion electrode systems are usually calibrated against known standards to account for behavioral anomalies and to insure that the system is functional for a given determination.

The first selective ion electrode to find widespread use was the glass pH electrode; it responds to changes in hydrogen ion activity according to the Nernst equation, in that for each tenfold change in hydrogen ion activity at 25°C, the change of the electrode's absolute potential will be 59.15 mV (the value of 2.303 RT/F at 25°C). This is the logical point to distinguish the selective ion electrode from the only theoretically achievable "specific" ion electrode which does not suffer from interference by other ionic species; the pH electrode is predominantly sensitive to hydrogen ion, but in highly alkaline solution, where the activity of $(H_3O)^+$ is significantly less than such other monovalent cations as sodium and potassium, it will respond to these ions. This phenomenon is known as the "sodium ion error" and when capitalized upon led to the development of the technique to determine NH_4^+, K^+ and Na^+ in alkaline aqueous media by the use of the pH electrode.

25.1.2 *Construction — Mechanism.* Construction is based on sealing a membrane of a substance such as glass, an insoluble inorganic salt or an immobilized organic ion exchanger or ion exchange membrane across the end of a tube that serves to contain a filling solution and an internal half cell assembly. The membrane serves as a filter, allowing only ions of the test species (in the ideal case) to diffuse across the membrane separating the test solution and the inner solution. Such diffusion derives from the chemical potential difference of the test ion existing between the two solutions. Diffusional transport of the ions continues until a back EMF sufficient to retard further diffusion is produced — this is the variable which is normally measured when the equilibrium potential of a test electrode is determined and is related to the activity of the ionic species being determined.

25.1.3 *Problems.* Selective ion electrodes exhibit the advantage or disadvantage, depending on one's point of view, of determining ionic activities, not ionic concentrations. Activities are generally most meaningful in kinetic determinations and in equilibrium situations, but are of limited value to the determination of ionic concentrations — for assessment of concentrations, a conventional titration technique should be employed (selective ion electrodes may serve as indicator electrodes). The addition of a supporting electrolyte solution containing decomplexing agents to free the desired species of the analysis is often employed to offer a fixed background medium — these solutions are known by the acronym "TISAB" (an abbreviation of "total ionic strength adjustment buffer"). Techniques involving the addition of known amounts of the analyte, the "standard addition" method, are often of value in the analysis of air pollutant samples. Along with the inherent problems of activity measurement are the difficulties caused because the ionic activity coefficient of a particular species of interest is a function of the total ionic strength of the solution and because the selective ion electrode may exhibit anomalous values due to the effects of ionic interferents or its inability to determine species present in a complexed form. The electrodes are more suited for use in the laboratory environment than the continuous process analytical utilization where limited life of liquid ion exchanger-based systems, the problems of interferences, and the lack of long term stability contribute to lessened utility.

25.1.4 *References and Scope.* Thorough descriptive treatments of selective ion electrode principles are given in the articles by Weber, (1) Frant (2) and the publications of Orion Research Inc. (3,4,5). The reader is referred to these publications for a more fundamental and general description of selective ion electrodes and for the emphasis that these articles place on the more conventional aqueous sample analysis role of this device. The use of selective ion electrodes has gained widespread acceptance for analytical determinations in chemical, industrial, pharmaceutical and general analytical chemistry areas. The bulk of such usage has been for the determination of ionic species already present in the aqueous sample.

The success of selective ion electrodes in providing accurate and sensitive determinations over a wide range of analyte concentrations in a wide variety of sample matrices has lead to the extension of this basic detection and measurement device to the determination of gaseous species that are soluble in a collection medium.

A listing of commercially available selective ion electrodes by detector type, along with detection limits, where available, is given in Tables 25:I and 25:II.

25.2 DETECTION OF GASEOUS SPECIES WITH SELECTIVE ION ELECTRODES. BACKGROUND. Any gaseous species which is readily soluble in an aqueous collection medium may be determined by a selective ion electrode if a selective ion electrode exists for the ionic hydrolysis product of that species. The determination may be made on

Table 25:I Classification of Commercially Available Selective Ion Electrodes by Analyte and Type

Glass	Solid State	Liquid Filled	Gas Sensing
H^+	Br^-	Cu^{++}	NH_3
Na^+	Cd^{++}	Cl^-	CO_2
K^+	Cl^-	NO_3^-	NO_2
NH_4^+	F^-	ClO_4^-	SO_2
Ag^+	I^-	K^+	
	Pb^{++}	Ca^{++}	
	Ag^+	$Ca^{++} + Mg^{++}$	
	CNS^-	BF_4^-	
	Cu^{++}		
	CN^-		
	S^{--}		
	Na^+		

Principal sources:
Beckman Instruments, Corning Glass Works, H-Nu Systems, Lazar Research Laboratories, Leeds & Northrop Co., London Co., Orion Research Laboratories, Philips Electronics, Radiometer Corp., Sensorex

either a batch or continuous basis and the detection limit of the gaseous species will be an inverse function of the collection time in the case of batch analyses or a function of the gaseous sample flow rate/collection medium volume for a continous analysis. Orion Research (6) has detailed the demonstrated and possible systems suitable for analysis by selective ion electrode; these are noted in Table 25:III.

25.2.1 *Applications of Solid State and Glass Electrodes to the Measurement of Gaseous Species.*

a. Fluoride. The fluoride selective ion electrode was one of the first reliable selective ion electrodes that sensed a species of particular importance to the analysis of source level and ambient level air pollution. Several writers (7,8,9) have compared the accuracy of fluoride ion determination by selective ion electrode to that of the SPADNS distillation technique, where ambient or source levels of HF were determined. Purcell (10) has determined HF and SiF_4 being emitted from a combustion source with a fluoride ion electrode-based continuous monitoring system. Thompson and Johnston (11) have developed a continuous fluoride monitor employing a counter-current absorber and selective ion electrode detection system and used it to determine ambient levels of fluorides.

b. NO_x (NO + NO_2). NO_2 is a product of air oxidation of NO liberated by most combustion processes. It has been determined by a selective ion electrode in several fashions. Driscoll (12) found that by collecting the NO_x sample in a $\sim 10^{-3}$M H_2SO_4 + 3 × 10^{-2}M H_2O_2 solution, it was possible to obtain a correlation coefficient of 0.987 over a wide range of input concentrations between the selective ion electrode and the conventional Saltzman analyses. L. A. Dee (13,14) employed the NO_3^- solid state electrode to determine NO_2 concentrations in stack gases by allowing this pollutant to react with PbO_2 in the collection medium to form Pb $(NO_3)_2$, which is fully ionized and readily detectable. Di Martini (15) has used gas phase oxidation of NO to NO_2 with O_3 and subsequent bubbling through a collection medium to determine NO_x with an NO_3^- selective ion electrode.

25.2.2 *Application of Gas Sensing Selective Ion Electrodes to the Measurement*

Table 25:II Nominal Useful Ranges of Selective Ion Electrodes for Various Analytes

Ammonia	0.0085 to 17,000 ppm	$(5 \times 10^{-7}M$ to $10^{0}M)$
Bromide	0.040 to 79,900 ppm	$(5 \times 10^{-7}$ to $10^{0}M)$
Calcium	0.201 to 40,100 ppm	$(5 \times 10^{-6}$ to $10^{0}M)$
Carbon dioxide	0.440 to 1320 ppm	$(10^{-5}$ to $3 \times 10^{-2}M)$
Chloride	0.178 to 35,500 ppm	$(5 \times 10^{-6}$ to $10^{0}M)$
Copper (II)	0.032 to 63,500 ppm	$(5 \times 10^{-7}$ to $10^{0}M)$
Cyanide	0.013 to 260 ppm	$(5 \times 10^{-7}$ to $10^{-2}M)$
Calcium + Magnesium	0.080 to 40,000 ppm as Ca	$(2 \times 10^{-6}$ to $10^{0}M)$
Fluoride	0.0095 to 19,000 ppm	$(5 \times 10^{-7}$ to $10^{0}M)$
Hydrogen		$(10^{-14}$ to $10^{0}M)$
Iodide	0.635 to 127,000 ppm	$(5 \times 10^{-6}$ to $10^{0}M)$
Lead	0.0095 to 207,000 ppm	$(10^{-6}$ to $10^{0}M)$
Nitrate	0.310 to 310,000 ppm	$(5 \times 10^{-6}$ to $10^{0}M)$
Nitrite	0.002 to 920 ppm	$(5 \times 10^{-6}$ to $2 \times 10^{-2}M)$
Silver	0.011 to 107,900 ppm	$(10^{-7}$ to $10^{0}M)$
Perchlorate	0.7 to 99,500 ppm	$(7 \times 10^{-6}$ to $10^{0}M)$
Sodium	0.023 ppm to saturated	$(10^{-6}M$ to sat.)
Sulfide	0.016 to 32,100 ppm	$(10^{-7}$ to $10^{0}M)$
Sulfur Dioxide	0.064 to 640 ppm	$(10^{-6}M$ to $10^{-2}M)$
Thiocyanate	0.29 to 58,100 ppm	$(5 \times 10^{-6}$ to $10^{0}M)$

Table 25:III Systems Suitable for Analysis by Selective Ion Electrode

Gaseous Species	Electrolyte Equilibrium		Sensing Electrode
NH_3	$NH_3 + H_2O$	$\rightleftharpoons NH_4^+ + OH^-$	$H+$
SO_2	$SO_2 + H_2O$	$\rightleftharpoons H^+ + HSO_3^-$	$H+$
NO_2	$2NO_2 + H_2O$	$\rightleftharpoons NO_3^- + NO_2^- + 2H^+$	H^+, NO_3^-
H_2S	$H_2S + H_2O$	$\rightleftharpoons 2H^+ + S^{--}$	S^{--}
HCN	$Ag^+ + 2CN^-$	$\rightleftharpoons Ag(CN_2)^-$	$Ag+$
CO_2	$CO_2 + H_2O$	$\rightleftharpoons H^+ + HCO_3^-$	H^+
HF	HF	$\rightleftharpoons H^+ + F^-$	F^-
Cl_2	$Cl_2 + H_2O$	$\rightleftharpoons 2H^+ + OCl^- + Cl^-$	H^+, Cl^-

of Gaseous Species. Gas sensing electrodes consist of a specific ion electrode and companion reference electrode immersed in an appropriate internal electrolyte, isolated from the gaseous test sample by a gas-permeable membrane. The species to be measured diffuses across the membrane and across a thin layer of electrolyte to the selective ion electrode; as the species of interest dissolves in this thin film of electrolyte, the concentration of the hydrolyzed product species changes in a fashion directly related to the gas phase concentration. The potential change of the electrode with changes in gas phase concentration approximates Nernstian behavior very closely.

Advantages of the membrane-isolated electrode system over a selective ion electrode-conventional reference electrode system include virtual elimination of liquid junction potentials, isolation of the electrode from interferences that do not exhibit high vapor pressures in aqueous solution, and elimination of interferences from ionic species that enter into redox couples at the electrode surface (since there is no ionic transport).

a. The Nitrogen Oxide Gas Sensing Electrode. This electrode responds to dissolved nitrous acid or nitrite and is subject to few interferences except formic and acetic acids, when present in great excess. SO_2 and

CO_2 do not interfere when the procedures specified by the manufacturer are followed. NO_2 in air samples can be determined by collecting the NO_2 in an appropriate collection medium in which the NO_2 hydrolyzes to nitrite, which is measured by the electrode. The dissolved NO_2 and NO in equilibrium with HNO_2 in any acidified sample diffuse through the gas-permeable membrane of the electrode until equilibrium is reached between the level of HNO_2 in the test solution and in the inner filling solution of the electrode. The hydrogen ion response is measured and exhibits a 59 mV/decade span over the concentration range of 5×10^{-7} M NO_2 to 10^{-2} M NO_2 (2×10^{-2} to 460 ppm NO_2). Response time is faster at the higher concentrations and an upper temperature limitation of 50°C is to be observed.

This electrode would serve as the basis of assessing NO_X concentrations with the use of a suitable prior oxidation medium, such as acidified $KMnO_4$, $K_2Cr_2O_7/H_2SO_4$ or O_3 to oxidize NO to NO_2, which could be collected in an appropriate collection medium with subsequent determination by the NO_2 gas sensing electrode.

b. Sulfur Dioxide Gas Sensing Electrode. The sulfur dioxide gas sensing electrode responds to dissolved SO_2 and is subject to interference only from large excess concentrations (> 30 fold) of weak acids such as acetic acid and HF. Cl_2, NO_2 and CO_2 do not interfere, if the recommendations about collection media given by the manufacturer are followed. SO_2 at ambient and source levels can be determined by collection in an appropriate medium such as sodium hydroxide, which is subsequently acidified to release the gaseous SO_2 to be measured by the selective ion electrode. Sensitivities as high as 0.02 ppm SO_2 are claimed (based on the sampling of 30 L of air through the basic collection medium and a final test solution volume of 25 mL). The SO_2 gas measured by the electrode creates hydrogen ions in the inner filling solution of the electrode; these H^+ ions are measured and a 59 mV/decade span for each tenfold SO_2 concentration change is observed. The accurate concentration range for correct electrode behavior is $\sim 10^{-6}$M to 10^{-1}M (0.1 to 1000 ppm SO_2). Response time is faster in more concentrated solu-

tions and an upper temperature limitation of 50°C is to be observed.

c. Ammonia Gas Sensing Electrode. The ammonia gas sensing electrode responds to dissolved ammonia and is subject to few interferences other than volatile amines. Dissolved ammonia in a collection medium diffuses through the gas-permeable membrane of the electrode until the concentration in the inner filling solution of the electrode is equal to that in the external solution. The (OH^-) level in this filling solution is directly proportional to the external ammonia concentration and a response of 59 mV/decade ammonia change is observed. The accurate concentration range for correct electrode behavior is 10^{-6}M to 10^0M ammonia (0.017 to 17,000 ppm NH_3). Response time is faster in more concentrated solutions and an upper temperature limitation of 50°C is to be observed.

25.3 PRECAUTIONS.

25.3.1 *Calibration Standards and Techniques.* Calibration of selective ion electrodes used to monitor the concentration of gaseous species should be conducted according to the recommendations of the manufacturer or based upon the theoretical concepts and good laboratory practice as detailed in the articles by both Frant and Weber **(1,2)** and Eynon **(16)**. Attempts should be made to match the composition and ionic strength of the analytical standards to those of the collection medium employed. The best approach to this is to make incremental additions of the test ion of known value to the collection medium in order to develop an electrode calibration curve. This technique is detailed in most manufacturers' literature. One should observe the sequence of immersion in low concentration standards, working toward high concentration standards to minimize gross contamination and absorption/desorption effects. Calibration of gas sensing electrodes may be conducted by the above-mentioned techniques or may be accomplished by exposure to a gas stream containing a known concentration of the gas being determined. Such a calibration gas standard may be prepared by dynamic dilution, the use of permeation tubes or by the use of analyzed standards or mixtures. (See section 3.)

25.3.2 *Measurement Techniques.*
a. Calibration.
b. Electrical Considerations. Potential measurements of the test electrode must be made with the ultimate accuracy possible; at 25°C a 0.5 mV error will account for a 1.9% error for monovalent ions, 3.8% for divalent ions and 5.7% for trivalent ions. Care should be taken to provide adequate shielding of the sensing electrode so that spurious potentials are not measured.

c. These potential measurements should be taken with a high input impedance device (typically $10^{13}\,\Omega$) in order to minimize loading of the selective ion electrode-reference electrode pair and the observation of erroneous open circuit potentials due to current flow in the selective ion electrode-reference electrode-measuring instrument-solution loop. Care should be exercised in grounding the electrode measurement system to avoid ground loop conditions and to observe proper shielding techniques. Such precautions are explained in detail in pertinent references **(17,18,19,20)**. Generally, reference electrodes which exhibit impedances on the order of 1 to 20kΩ without significant bleed rates of internal solution into the test solution (which may be variable and contribute to higher and variable ionic strength of the solution being measured) are most suitable for use in such systems.

d. Temperature Effects and Temperature Compensation. This subject has been treated in a significant article by Negus *et al.* **(21)**. They classify temperature effects on selective ion electrodes into three categories:

1) The change in the potential of a selective ion electrode system due to non-uniform response of the selective ion electrode and reference electrode to a temperature excursion.

2) the temperature coefficient slope term of the Nernst equation, RT/nF. This is the parameter usually compensated for by the use of a thermal compensator in a process selective ion electrode monitoring system; the thermocompensator corrects for the slope change in the potential difference versus ion activity relationship due to the change in the slope factor RT/nF with T. This

is the major thermal compensation characteristic to be corrected.

3) the solution temperature coefficient term that arises from temperature-dependent equilibria and activities existing in the test solution. The magnitude of these effects is usually so small that it is neglected.

In order to minimize the effects of temperature or temperature variations on the accuracy of results obtained from a given selective ion electrode-based system, calibration of the system should be conducted with standards maintained at the same temperature as the test sample which is measured. Sufficient time should be allowed when changing from calibration standard to calibration standard so that thermal equilibrium is achieved. Where practicable, it is recommended that the sample temperature be held at a constant value by thermostatic control and the electrode pair and sensing cell be maintained in a temperature-controlled environment held at the same temperature as the input aqueous sample. If this is not possible, and reduced readout accuracy is allowable, it is sufficient either to compensate automatically for temperature excursions with a thermal compensator or to use manual thermal compensation or some mean sample temperature when temperature excursions are small and of long diurnal nature.

25.4 SUMMARY AND CONCLUSIONS. Selective ion electrodes show great promise in the determination of certain air pollutant species. This technique has not yet been implemented into commercially available continuous air pollutant monitoring instrumentation, but may be in the near future. The use of these electrodes for the determination of air pollutant species is currently primarily oriented toward batch analysis of collected samples in a laboratory setting, but reliable continuous monitoring instrumentation based on such electrodes is under development. If the limitations exhibited by the electrodes are understood and techniques for compensating for these limitations are employed, these electrodes offer a powerful, inexpensive and reliable technique for the determination of ambient and source level air pollutant species over a wide dynamic range. For specific applications, when the user is knowledgeable in the art of application of the electrode to the system chosen, selective ion electrodes offer a viable alternative technique for the determination of ambient and source level air pollutant species.

REFERENCES

1. WEBER, S.J. 1970. Specific Ion Electrodes in Pollution Control, American Laboratory, p. 15–23, July.
2. FRANT, M.S. 1974. Detecting Pollutants with Chemical Sensing electrodes, Environmental Science and Technology, 8:224–228.
3. ANON. 1978. Orion Research Analytical Methods Guide, Orion Research, Inc., Cambridge, Mass.
4. ANON. 1970–1985, Orion Research Newsletter, Orion Research, Inc., Cambridge, Mass.
5. ANON. 1987. U.S. Ion Selective Electrode Catalgue and Guide to Ion Analysis, Orion Research, Cambridge, Mass.
6. ANON. 1973. Orion Newsletter, Vol. V, No. 2 Orion Research, Cambridge, Massachusetts.
7. ELFERS, L.A. AND C.E. DECKER, 1968. Determination of Fluoride in Air and Stack Gas Samples by Use of an Ion Specific Electrode. Analytical Chemistry, 40:1658–1661.
8. MACLEOD, K.E. AND H.L. CHRIST. 1973. Comparison of the SPADNS-Zirconium Lake and Specific Ion Electrode Methods of Fluoride Determination in Stack Emission Samples. Analytical Chemistry, 45:1272–1273.
9. THOMPSON, R.J. 1971. Fluoride Concentrations in the Ambient Air. Journal of the Air Pollution Control Association, 21:484–487.
10. POWELL, R.A. AND M.C. STOKES. 1973. A Method for the Continuous On-Site Measurement of Fluorides in Stack Gases and Emissions for Periods of up to Five Hours, Atmospheric Environment, 7:169–176.
11. THOMSON, C.R. 1969. University of California. Riverside, California and G.W. Johnston, Jr., Beckman Instruments, Inc. – Private Communication.
12. DRISCOLL, J.N. 1971. Determination of Oxides of Nitrogen in Combustion Effluents with a Nitrate Selective Ion Electrode., #71–49, Walden Research Corporation, Walden, Massachusetts.
13. DEE, L.A. 1972. An Improved Manual Method for NO_x Emission Measurement. EPA-R2-72-067.
14. DEE, L.A., H.H. MARTENS, C.I. MERRILL, J.T. NAKAMURA AND F.C. JAYE. 1973. A New Manual Method for Nitrogen Oxides Emission Measurement, Analytical Chemistry, 45:1477–1481.
15. DI MARTINI, R. 1970. Determination of Nitrogen Dioxide and Nitric Oxide in the Parts per Million Range in Flowing Gaseous Mixtures by Means of the Nitrate Specific Ion Electrode. Analytical Chemistry, 42:1102–1105.
16. ENYON, J.U. 1970. Known Increment and Known Decrement Methods of Measurement with Selective Ion Electrodes. American Laboratory, 59–66, September.
17. MATTOCH, G. 1961. pH Measurement and Titration, New York, The MacMillan Co.
18. BLEAK, T.M. AND K.B. SAWA. 1971. A New Process pH Analyzer Design. 17th National Symposium, Analysis Instrument Division, Instrument Society of America. Houston, Texas. paper I-3.
19. MORRISON, R. 1970. Grounding and Shielding Techniques. John Wiley & Sons. New York.
20. MOORE, F. 1969. Taking the Errors out of pH Measurements by Grounding and Shielding. Instrument Society of America Journal, 13:60–66.
21. NEGUS, L.E. AND T.S. LIGHT. 1972. Temperature Coefficients and Their Compensation in Ion Selective Systems, Instrumentation Technology, 19:23–29.

Subcommittee 9
A. L. LINCH, *Chairman*
J. N. HARMAN
E. R. HENDRICKSON
M. KATZ
R. J. MARTIN
G. O. NELSON
J. N. PATTISON
A. L. VANDER KOLK

Approved with modifications from the 2nd edition
Subcommittee 9
J. N. HARMAN, *Chairman*

26. Quality Control For Sampling and Analysis

26.1 INTRODUCTION. The measurement of physical entities such as length, volume, weight, electromagnetic radiation, and time involves uncertainties that cannot be eliminated entirely, but when recognized can be reduced to tolerable limits by meticulous attention to detail and close control of the significant variables. In addition errors, often unrecognized, are introduced by undesirable physical or chemical effects and by interferences in chemical reaction systems. In many cases, absolute values are not directly attainable, and therefore standards from which the desired result can be derived by comparison must be established. Errors are inherent in the measurement system. Although the uncertainties cannot be reduced to zero, methods are available by which reliable estimates of the probable true value and of the range of measurement error can be made.

In this section the fundamental procedures for the administration of an effective quality control program are presented with sufficient explanation to enable the investigator both to understand the principles and to apply the techniques. First, the detection and control of systematic and random error will be considered. Based on this foundation, the types of errors and meanings of the common terms used to define an accurate method are discussed as a basis for application of quality control to sampling and analysis. The theory, construction, applications, and limitations of control charts are developed in sufficient depth to provide practical solutions to actual quality control problems. Additional statistical approaches are included to support those systems, which may require further refinement of precision and accuracy to evaluate and control sampling and analysis reliability. A valuable discussion of quality control procedures utilized in determination of standard values of standard reference materials is found in the "Handbook for SRM users," (27). Taylor has produced a comprehensive review of modern quality assurance practices (28).

Finally, a discussion of collaborative testing projects and intra-laboratory quality control programs designed to improve and test the integrity of the laboratory's performance completes the survey of quality control principles and practices.

26.2 QUALITY CONTROL PRINCIPLES.

26.2.1 *Total Quality Control.* A quality control program concerned with sampling and laboratory analysis is a systematic attempt to assure the precision and accuracy of future analyses by detecting systematic errors in analysis and preventing their recurrence. Confidence in the accuracy of analytical results and improvements in analysis precision are established by identification of the systematic sources of error. The precision will be governed by the random error inherent in the procedure, and can be estimated by statistical techniques. For a result to be acceptable, the procedure must not only be precise, but must also be without bias. Techniques have been developed for the elimination of bias. The quality control program should cover instrumental control as well as total analysis control. The use of replicates submitted in support of the quality control program provides assurance that the procedure will remain in statistical control.

Quality must be defined in terms of the characteristic being measured. Control must be related to the source of variation, which may be either systematic or random. Usually the basic variable is continuous (any value within some limit is possible). A numerical value of an analysis for which

the range of uncertainty inherent in the method has not been established cannot be reliably considered a reasonable estimate of the true or actual value. The basic quality control program incorporates the concepts of:

a. Calibration to attain accuracy;

b. replication to establish precision limits, and

c. correlation of quantitatively related tests to confirm accuracy, where appropriate.

Evaluation of the overall effectiveness of the quality control program encompasses a number of parameters:

a. Equipment and instruments;

b. the current state of the art;

c. expected ranges of analytical results;

d. precision of the analytical method itself;

e. control charts to determine trends as well as gross errors;

f. data sheets and procedures adopted for control of sample integrity in the laboratory, and

g. quality control results on a short term basis (daily if appropriate) as well as on an accumulated basis.

The manipulative operations which are directly influenced by quality control include:

a. Sampling techniques;

b. preservation of sample integrity (identification, shipping and storage conditions, contamination, desired component losses, etc.);

c. aliquoting procedures;

d. dilution procedures;

e. chemical or physical concentration, separation and purification, and

f. instrument operation.

26.2.2 *Statistical Quality Control.* Statistical quality control involves application of the laws of probability to systems where chance causes operate. The technique is employed to detect and separate systematic (determinate) from random (indeterminate) causes of variation. Uncertainty can be quantified in terms of well-defined statistical probability distributions, which can be applied directly to quality control. The application of statistical quality control can most efficiently indicate when a given procedure is in statistical control, and a continuing program that covers sampling, instrumentation and overall analysis quality will assure the validity of the analytical program.

26.2.3 *Quality Control Charts* **(1).** The Shewhart Control Chart **(2)** is one of the most generally applicable and easily adapted statistical techniques, which can be applied to almost any phase of production, research or analysis. Control charts originally were developed to monitor production lines where large numbers of manufactured articles were inspected on a continuous basis. Since analyses frequently are produced on an intermittent basis, or on a greatly reduced scale, fewer data are available to work with. Therefore, certain concessions must be made in order to respond quickly to objectionable changes in the analytical procedure.

This control chart may serve several functions:

a. To determine empirically and to define acceptable levels of quality;

b. to achieve the acceptable level established, and

c. to maintain performance at the established quality level.

Certain assumptions reside in this technique. The first and major assumption is that there will be variation. No process or procedure has been so well perfected, or is so unaffected by its environment, that exactly the same result will always be produced. Either the device used for measurement is not sufficiently controlled or the operator performing the measurement is not sufficiently skilled. The sources of variation present in analytical work include:

a. Differences among analysts;

b. instrumental differences;

c. variations in reagents and related supplies;

d. effect of time on the differences found in Items **a, b,** and **c;** and

e. variations in the interrelationship of Items **a, b** and **c** with each other and with time.

A "system of chance causes" is inherent in the nature of processes and procedures and will produce a pattern of variation. When this pattern is stable, the process or procedure is considered to be "in statistical control," or just "in control." Any result that falls outside of this pattern will have an assignable cause that can be determined and corrected.

The control chart technique provides a means for separating the systematic cause variant from the stable pattern. The chart is a graphical presentation of the process or procedure test data that compares the variability of all results with the average or expected variability from small arbitrarily defined groups of the data. The technique in effect is a graphical analysis of variance.

The data from such a system can be plotted with vertical scale in test result units and the horizontal scale in units of time or sequence of results. The average value, or mean, and the limits of the dispersion (spread, or range of results) can be calculated.

26.3 ERRORS.

26.3.1 *Introduction.* Numbers are employed either to enumerate objects or to delineate quantities. If sixteen air samples are taken simultaneously at different locations in a warehouse where gasoline-powered fork lift trucks are in motion, the number, *i.e.,* the count, would be the same regardless of who counted them, when the count was made or how the count was made. However, if each individual sample is analyzed for carbon monoxide, sixteen different numbers, *i.e.,* the concentration, undoubtedly would be obtained. Furthermore, when replicate determinations are made on each sample, a range of carbon monoxide concentrations would be found **(3).**

Experimental errors are classified as systematic or random. A fifteen count of the warehouse samples would be a systematic error quickly disclosed by recount. A random error would be encountered due to the inherent variability in repetitive determinations of carbon monoxide by gas chromatography, infrared or a colorimetric technique.

If the estimation of carbon monoxide concentration is made with a length of stain detector tube and a 6.5-mm stain length, equivalent to 57 ppm, is recorded by the observer, whereas the true stain length is 6.0 mm and equivalent to 50 ppm, the observational error would be $(57-50) \times 100 \div 50 = 14\%$.

All analytical methods are subject to errors. The systematic ones contribute constant error or bias, while the random ones produce random fluctuations in the data. The concepts of accuracy and precision as applied to the detection and control of error have been clearly defined and should be used exactly.

Accuracy. Accuracy relates the amount of an element or compound recovered by the analytical procedure to the amount actually present. For results to be accurate, the analysis must yield values close to the true value.

Precision. Precision is a measure of the method's variability when repeatedly applied to a homogeneous sample under controlled conditions, without regard to the magnitude of displacement from the true value as the result of systematic errors that are present during the entire series of measurements. Stated conversely, precision is the degree of agreement among results obtained by repeated measurements or "checks" on a single sample under a given set of conditions (4,5).

26.3.2 *Detection and Elimination of Determinate Error.* The terms "determinate" error, "assignable" error, and "systematic" error are synonymous. A systematic error contributes constant error or bias to results that may agree precisely among themselves.

Sources of Systematic Error. A method may be capable of reproducing results to a high degree of precision, but only a fraction of the component sought is recovered. A precise analysis may be in error due to inadequate standardization of solutions, inaccurate volumetric measurements, inaccurate balance weights, improperly calibrated instruments, or personal bias (color estimation). Method errors that are inherent in the procedure are the most serious and most difficult to detect and correct. The contribution from interferences is discussed later.

Personal errors other than inherent physical visual acuity deficiencies (color judgment) include consistent carelessness, lack of knowledge and personal bias which are exemplified by calculation errors, use of contaminated or improper reagents, nonrepresentative sampling or poorly calibrated standards and instruments.

Types of Systematic Error. *Additive*: An additive error occurs when the mean error has a constant value regardless of the amount of the constituent sought in the sample. A plot of the analytical value versus the theoretical value will disclose an intercept somewhere other than zero. *Proportional*: A proportional error is a systematic error in which magnitude is changed according to the amount of constituent present in the sample. A plot of the analytical value versus the theoretical value not only fails to pass through zero, but discloses a curvilinear rather than a linear function, or has a slope significantly different from 1.0.

Recovery or "Spiked" Sample Procedures. A recovery procedure in which spiked samples are used provides a technique for the detection of systematic errors. The technique provides a basis for evaluating the applicability of a particular method to any given sample. It allows derivation of analytical quality control from the results, thus providing the basis for an excellent quality control program.

The recovery technique applies the analytical method to a reagent blank, to the sample itself (in at least duplicate), and to "spiked" samples prepared by adding known quantities of the substance sought to separate aliquots of the sample that are equal in size to the unspiked sample taken for analysis. The substance sought should be added in sufficient quantity to exceed in magnitude the limits of analytical error, but the total should not exceed the range of the standards selected.

The results are first corrected for reagent influence by subtracting the reagent blank from each standard, sample, and "spiked" sample result. The average unspiked sample result is then subtracted from each of the "spiked" determinations, the remainder divided by the known amount originally added, and expressed as percentage recovery. Table 26:I illustrates an application of this technique to the analysis of blood for lead content.

Specifications for acceptance of analytical results usually are limited by the state of the art and/or the final disposition of the results. Recoveries of substances within the range of the method may be very high or very low and approach 100% as the errors diminish and as the upper limit of the cali-

Table 26:I Lead in Blood Analysis

Basis: 10.0 g blood from blood bank pool, ashed and lead determined by double extraction, mixed color, dithizone procedure.

		Analyst: DJM		
µg Pb Added	Optical Density	µg Pb Found		Recovery, %
		Total	Recovered	
None-blank	0.0969	–	–	–
None	0.1427	1.6	–	–
None	0.1337	1.3	–	–
None	0.1397	1.4	–	–
Average	0.1389	1.4	–	–
2.0	0.1805	2.9	1.5	75
4.0	0.2636	5.4	4.0	100
6.0	0.3372	7.8	6.4	107
8.0	0.3925	9.4	8.0	100
10.0	0.4437	11.4	10.0	100
30.0 Total	–	36.9	29.9	96

Calculation of mean error (5)
Mean error = $36.9 - (30.0 + 5 \times 1.4) = 0.1$ µg for entire set
$= 2.9 - (2.0 + 1.4) = 0.5$ µg for 2 µg spike
Calculation of relative error
Relative error = $(0.1 \times 100)/37.0 = 0.27\%$ for entire set
$= (0.5 \times 100)/3.4 = 14.7\%$ for 2 µg spike

bration range is approached. Trace analysis procedures that inherently have relatively large errors when operated near the limits of sensitivity deliver poor recoveries based on classical analytical criteria and yet, from a practical viewpoint of usefulness, may be quite acceptable. Poor recovery may reflect excessive manipulative losses, or the method's technical inadequacy in the range of application. The limit of detection may be considered the point beyond which random error exceeds the measured quantities.

Control Charts. Trends and shifts in control chart responses also may indicate systematic error. The standard deviation is calculated from spiked samples, and control limits (usually ± 3 standard deviations) for the analysis are established. In some cases, such as BOD and some pesticide samples, spiking to resemble actual conditions is not possible. However, techniques for detecting bias under these conditions have been developed (6).

Control charts may be prepared even for samples that cannot be spiked or for which the recovery technique is impractical. A reference value is obtained from the average of a series of replicate determinations performed on a composite or pooled sample that has been stabilized to maintain a constant concentration during the control period.

Change in Methodology. Analysis of a sample for a particular constituent by two or more methods that are entirely unrelated in principle may aid in the resolution of systematic error.

In Table 26:II, an interlaboratory evaluation of three different methods for the determination of lead concentration in ashed urine specimens (mixed color dithizone, atomic absorption and polarography) is summarized. If the highly specific polarographic method was selected as the primary standard, then the dithizone procedure is subject to a + 7.4µg/L bias as compared with a + 3.6 µg/L bias in the atomic absorption method for lead.

Quantity of Sample. If the systematic error is additive, the magnitude may be estimated by plotting the analytical results versus a range of sample volumes or weights.

Table 26:II An Interlaboratory Study of the Systematic Error in the Dithizone Procedure for the Determination of Lead in Nine Urine Specimens, μg/L

Polarographic Method	Mixed Color	Dithizone	Atomic	Absorption
	Found	Difference	Found	Difference
10	25	15	10	0
14	28	14	22	8
12	12	0	16	4
15	20	5	16	1
21	20	−1	22	1
22	30	8	24	2
27	40	13	36	9
19	22	3	22	3
12	22	10	16	4
Mean	−	+ 7.4	−	+ 3.6

If the error has a constant value regardless of the amount of the component sought, then a straight line fitted to the plotted points will not pass through the origin.

Elimination. a. *Physical.* In many cases error can be reduced to tolerable levels by quantitating the magnitude over the operating range and developing either a corrective manipulation directly in the procedure or a mathematical correction in the final calculation. Temperature coefficients (parameter change per degree) are widely applied to both physical and chemical measurements. For example, the stain length produced by carbon monoxide in a detector tube is dependent on the temperature as well as the air sampling rate and CO concentration. Therefore, when these tubes are used outside the median temperature range, a correction must be applied to the observed stain length (Table 26:III) (7).

As a general rule, most instruments exhibit maximum reliability over the center 70% of their range (midpoint ± 35%). As the extreme to either side is approached the response and reading errors become increasingly greater. Optical density measurements, for example, should be confined to the range 0.045 to 0.80 by concentration adjustment or cell path choice. Extrapolation to limits outside the range of response established for the analytical method or instrumental measurement may introduce large errors as many chemical and physical responses are linear only over a relatively

narrow band in their total response capability. In absorption spectrophotometric measurements, Beer's law relating optical density to concentration may not hold outside of rather narrow limits in some instances (colorimetric determination of formaldehyde at high dilution by the chromotropic acid method).

b. *Internal Standard.* The internal standard technique is used primarily for emission spectrograph, polarographic, and chromatographic (liquid or vapor phase) procedures. This technique enables the analyst to compensate for electronic and me-

Table 26:III Kitagawa Carbon Monoxide Detector Tube No. 100(7)
Temperature Correction Table

Chart Readings (ppm)	Corrected Concentration (ppm)				
	0°C	10°C	20°C	30°C	40°C
1,000	800	900	1,000	1,060	1,140
900	720	810	900	950	1,030
800	640	720	800	840	910
700	570	640	700	740	790
600	490	550	600	630	680
500	410	470	500	520	560
400	340	380	400	420	440
300	260	290	300	310	320
200	180	200	200	200	210
100	100	100	100	100	100

chanical fluctuations within the instrument.

In brief, the internal standard method involves the addition to the sample of known amounts of a substance to which the instrument will respond in a manner similar to the analyte measured. The ratio of the internal standard response to the analyte response determines the concentration of analyte in the sample. Conditions during analyses will affect the internal standard and the analyte identically, and thereby compensate for any changes. The internal standard should be of similar chemical composition to the analyte, of approximately the same concentration anticipated for the analyte, and of the purest attainable quality.

c. *Chemical Interference.* The term "interference" relates to the response of the analytical system to dissolved or suspended materials other than the analyte of interest. A reliable analytical procedure must anticipate and minimize interferences.

The investigator must be aware of possible interferences and be prepared to use an alternate or modified procedure to avoid errors. Analyzing a smaller initial aliquot may suppress or eliminate the effect of the interfering element through dilution. The concentration of the substance sought is likewise reduced; therefore, the aliquot must contain more than the minimum detectable amount. When the results display a consistently increasing or decreasing pattern by dilution, then interference is indicated.

An interfering substance may produce one of three effects: (a) React with the reagents in the same manner as the component being sought (positive interference); (b) react with the component being sought to prevent complete isolation (negative interference); (c) combine with the reagents to prevent further reaction with the component being sought (negative interference).

The sampling and analytical technique employed for the surveillance of airborne toluene diisocyanate (TDI) **(8)** in the manufacturing environment furnishes a good example in which all three factors can be encountered. The TDI vapor is absorbed and quantitatively hydrolyzed in an aqueous acetic acid-hydrogen chloride mixture to toluene diamine (MTD) which then is diazo-

tized by the addition of sodium nitrite. The excess nitrous acid is destroyed with sulfamic acid and the diazotized MTD coupled with N-1-naphthylethylenediamine to produce a bluish-red azo dye **(9)**. In the phosgenation section of the operations, the starting material (MTD) may coexist with TDI in the atmosphere sampled. If so, then a positive interference will occur as the method cannot distinguish between free MTD and MTD from the hydrolyzed TDI.

This problem can be resolved by collecting simultaneously a second sample in ethanol. The TDI reacts with ethanol to produce urethane derivatives that do not produce color in the coupling stage of the analytical procedure. The MTD is determined by the same diazotization and coupling procedure after boiling off the ethanol from the acidified scrubber solution. Then the difference represents the TDI fraction in the air sampled.

On the other hand, if the relative humidity is high or alcohol vapors are present, negative interference will reduce the TDI recovered by formation of the carbanilide (dimer) or the urethane derivative which will not produce color in the final coupling stage. Alternative methods have not been developed for these conditions. If high concentrations of phenol are absorbed, then a negative interference will arise from side reactions with the nitrous acid required to diazotize the MTD. This loss can be avoided by testing for excess nitrous acid in the diazotization stage and adding additional sodium nitrite reagent if a deficiency is indicated.

An estimate of the magnitude of an interference may be obtained by the recovery procedure.

If recoveries of known quantities exceed 100%, a positive interference is present (Condition a). If the results are below 100%, a negative interference is indicated (Condition b or c.)

26.3.3 *Random Error and Its Control.* **Nature.** Even though all systematic errors are removed from a sampling or analytical procedure, replicate analyses will not produce identical results. This erratic variation arises from random error. Examples of this type of variation would be variation in reagent addition, instrument response, line voltage transients and physical measurement of volume and mass. In environmental analysis the sample itself is subject to a great variety of variability. Random errors conform to the laws of chance, therefore statistical measures of precision can be employed to quantitate their effects.

A measure of the degree of agreement (precision) among results can be ascertained by analyzing a given sample repeatedly under conditions controlled as closely as conditions permit. The standard deviation of these replicate results provides a measure of the random variations.

Quantification. a. *Distribution of Results.* Random error can be estimated by calculation of the standard deviation (σ). When experimental errors occur in a random fashion, the observed results (x) will be distributed at random around the average or arithmetic mean ($\bar{x}$) for a normal distribution.

Given an infinite number of observations, a graph of the relative frequency of occurrence plotted against magnitude will describe a bell-shaped curve known as the Gaussian or normal curve. However, if the results are not from a normal distribution, the curve may be flattened (no peak), skewed (unsymmetrical), narrowed, or exhibit more than one peak (multi-modal). In these cases the arithmetic mean will be misleading, and unreliable conclusions with respect to deviation ranges (σ) will be drawn from the data.

If there is any doubt, the investigator should confirm the normalcy of the data at hand to the extent possible. Various procedures are available to test this assumption. However, all require rather large data sets. One method is to construct a histogram, if the sample is large enough, and then to plot a normal curve having the same mean and standard deviation with the histogram to see how well the normal curve fits. This is an imprecise method at best and, unless there is an extremely good fit of a normal curve laid over the resulting histogram or polygon, the cumulative distribution should be plotted on normal probability paper before proceeding. Another way of testing for normality is to use the χ^2 test for normality of data. Standard χ^2 computer programs are commonly available, but judgment must be used to weigh the cost of

getting an accurate determination against the value of the information.

The distribution of results within any given range about the mean is a multiple of σ. The proportion of the total observations which reside within $\bar{x} \pm 1\,\sigma$, $\bar{x} \pm 2\,\sigma$ and $\bar{x} \pm 3\,\sigma$ have been thoroughly established. Although these limits do not define exactly any finite sample collected from a normal group, the agreement with the normal limits improves as n increases. As an example, suppose an analyst were to analyze a composite urine specimen 1000 times for lead content. He could reasonably expect 50 results would exceed $\bar{x} \pm 2\,\sigma$ and only 3 results would exceed $\bar{x} \pm 3\,\sigma$. However, the corollary condition presents a more useful application. In the preceding example, the analyst has found $\bar{x}$ to be 0.045 mg/L with $\sigma = \pm 0.005$ mg/L. Any result which fell outside the range 0.035–0.055 mg/L (0.045 $\pm 2\,\sigma$) would be questionable as the normal distribution curve indicates this should occur only 5 times in 100 determinations. This concept provides the basis for tests of significance, a concept that is discussed in detail in any good statistical reference such as those cited in this section.

b. *Range of Results.* The difference between the maximum and minimum of n results (range) also is related closely to σ. The range (R) for n results will exceed σ multiplied by a factor d_n only 5% of the time when a normal distribution of errors prevails.

Values for d_n:

n	d_n
2	2.77
3	3.32
4	3.63
5	3.86
6	4.03

Since the custom of analyzing replicate (usually duplicate) samples is a general practice, application of these estimated limits can provide detection of faulty technique, large sampling errors, inaccurate standardization and calibration, personal judgment and other determinate errors. However, resolution of the question whether the error occurred in sampling or in analysis can be answered more confidently when single determinations on each of three samples rather than duplicate determinations on each of two samples are made. This approach also reduces the amount of analytical work required (6). Additional information relative to the evaluation of the precision of analytical methods will be found in ASTM Standards (9).

  c. *Collaborative Studies or "Round Robins."* After an analytical method has been evaluated fully for precision and accuracy, collaborative testing should be initiated. The values for precision and accuracy as determined by the results from a number of laboratories can be expected to be somewhat larger than the performance of the originating laboratory. Because technicians in different laboratories apply to the procedure their own characteristic systematic and random errors, which may differ significantly from the original technique, the values for precision and accuracy will disclose the true reliability (ruggedness, or immunity to minor changes) of the method. Participation in collaborative programs will aid the investigator in evaluating his laboratory's performance in relation to other similar facilities and in locating sources of error.

  Duplicate analyses are employed for the determination and control of precision within the laboratory and between laboratories. Initially, approximately 20% of the routine samples, with a minimum of 20 samples, should be analyzed in duplicate to establish internal reproducibility. A standard or a repeatedly analyzed control, if available, should be included periodically for long-term accuracy control. The control chart technique is directly applicable, and appropriate control limits can be established by arbitrarily subgrouping the accumulated results or by using appropriate estimates of precision from an evaluation of the procedure.

26.4 CONTROL CHARTS.

  26.4.1 *Description and Theory.* The control chart provides a tool for distinguishing the pattern of random variation from the systematic variation. This technique displays the test data from a process or method in a form that graphically compares the variability of all test results with the average or expected variability of small groups of data — in effect, a graphical analysis of variance, and a comparison of the "within groups" variability versus the "between groups" variability.

  The data from a series of analytical trials can be plotted with the vertical scale in units of the test result and the horizontal scale in units of time or sequence of analyses. The average or mean value can be calculated and the spread (dispersion or range) can be established.

  The determination of appropriate control limits can be based on the capability of the procedure itself or can be arbitrarily established at any desirable level. Common practice sets the limits at 3 σ on each side of the mean. If the distribution of the basic data exhibits a normal form, the probability of results falling outside of the control limits can be readily calculated.

  The control chart is actually a graphical presentation of the degree of statistical control achieved. If the procedure is "in control," the results will fall within the established control limits. Further, the chart can disclose trends and cycles from assignable causes that can be corrected promptly. Chances of detecting small changes in the process average are improved when several values for a single control point (an x chart) are used. As the sample statistical size increases, the chance that small changes in the average will not be detected is decreased.

  The basic procedure of the control chart is to compare "within group" variability to "between group" variability. For a single analyst running a procedure, the "within group" may well represent one day's output and the "between group" represents between days, or day-to-day variability. When several analysts or several instruments or laboratories are involved, the selection of the subgroup unit is critical. Assignable causes of variation should show up as "between group" and not "within group" variability. Thus, if the differences between analysts should provide systematic causes of variation, their results may not be lumped together in a "within group" subgrouping.

  26.4.2 *Application and Limitations.* In order for quality control to provide a means for separating the systematic from random sources of variation, the analytical method must clearly emphasize those details that should be controlled to minimize variability. A check list would include:

a. Sampling procedures;
b. preservation of the sample;
c. aliquoting methods;
d. dilution techniques;
e. chemical or physical separations and purifications;
f. instrumental procedures; and
g. calculation and reporting results.

The next step to be considered is the application of control charts for evaluations and control of these unit operations. Decisions relative to the basis for construction of a chart are required:

a. Choose method of measurement;
b. select the objective
(1) precision or accuracy evaluation
(2) observe test results, or the range of results
(3) measurable quality characteristics;
c. select the variable to be measured (from the check list above);
d. basis of subgroup, if used
(1) Size – A minimum subgroup size of n = 4 is frequently recommended. The chance that small changes in the process average remain undetected decreases as the statistical sample size increases.
(2) Frequency of subgroup sampling – Changes are detected more quickly as the sampling frequency is increased;
e. Control Limits – Control limits (CL) can be calculated, but judgment must be exercised in determining whether or not the values obtained satisfy criteria established for the method; *i.e.*, does the deviation range fall within limits consistent with the solution or control of the problem. After the mean $(\bar{X})$ of the individual results (X), and the mean of the range $(\bar{R})$ of the replicate result differences (R) have been calculated, then CL can be calculated from data established for this purpose (Table 26:IV)(5).

Mean of the Means $(\bar{\bar{X}}) = \Sigma \bar{X}/k$
CL's on Mean = $\bar{\bar{X}} \pm A_2 \bar{R}$
Mean of the Range $(\bar{R}) = \Sigma R/k$, or $d_2 \sigma$
Upper Control Limit (UCL) on
    Range = $D_4 \bar{R}$
Lower Control Limit (LCL) on
    Range = $D_3 \bar{R}$

Table 26:IV Factors for Computing Control Chart Lines*

Observations in Subgroup (n)	Factor A_2	Factor d_2	Factor D_4	Factor D_3
2	1.88	1.13	3.27	0
3	1.02	1.69	2.58	0
4	0.73	2.06	2.28	0
5	0.58	2.33	2.12	0
6	0.48	2.53	2.00	0
7	0.42	2.70	1.92	0.08
8	0.37	2.85	1.86	0.14

*ASTM Manual on Quality Control of Materials (5)

Where: k = number of subgroups, A_2, D_4 and D_3 are obtained from Table 26:IV. R may be calculated directly from the data, or from the standard deviation (σ) using factor d_2. The lower control limit for R is zero when n ≤ 6.

The calculated CL's include approximately the entire data under "in control" conditions and, therefore, are equivalent to ±3 σ limits which are commonly used in place of the more laborious calculation. Warning limits (WL) set at ±2 σ limits (95%) of the normal distribution serve a very useful function in quality control. The upper warning limit (UWL) can be calculated by:

$$UWL = \bar{R} + 2\sigma_R$$
$$UWL = \bar{R} + 2/3 \, (D_4 \bar{R})$$

Where the subgrouping is n = 2, UWL reduces to

$$UWL = 2.51 \, \bar{R}.$$

26.5 CONSTRUCTION OF CONTROL CHARTS.

26.5.1 *Precision Control Charts.* The use of range (R) in place of standard deviation (σ) is justified for limited sets of data n ≤ 10 since R is approximately as efficient and is easier to calculate. The average range $(\bar{R})$ can be calculated from accumulated results, or from a known or selected σ ($d_2\sigma$). $LCL_R = 0$ when n ≤ 6. (LCL = lower control limit).

The steps employed in the construction of a precision control chart for an automatic analyzer illustrate the technique (Table 26:V):

a. Calculate R for each set of side-by-side duplicate analyses of identical aliquots;

b. calculate $\bar{R}$ from the sum of R values divided by the number (k) of sets of duplicates;

c. calculate the upper control limit (UCL_R) for the range:

$$UCL_R = D_4\bar{R}$$

Since the analyses are in duplicates, D_4 = 3.27 (from Table 26:IV);

d. calculate the upper warning limit (UWL): UWL_R =

$\bar{R} + 2\sigma_R = \bar{R} + 2/3\ (D_4\bar{R}) = 2.51\ \bar{R}$

26.6 ACCURACY OF CONTROL CHARTS— MEAN OR NOMINAL VALUE BASIS

$\bar{X}$ charts simplify and render more exact the calculation of CL since the distribution of data which conform to the normal curve can be completely specified by $\bar{X}$ and σ. Stepwise construction of an accuracy con-

Table 26:V Precision (Duplicates) Data

Date	Data			Range (R)
9/79	# 8	25.1	24.9	0.2
	#16	25.0	24.5	0.5
	#24	10.9	10.6	0.3
10/79	# 7	12.6	12.4	0.2
	#16	26.9	26.2	0.7
	#24	4.7	5.1	0.4
2/80	# 6	9.2	8.9	0.3
	#12	13.2	13.1	0.1
	#16	16.2	16.3	0.1
	#22	8.8	8.8	0.0
4/80	# 6	14.9	14.9	0.0
	#12	17.2	18.1	0.9
	#18	21.9	22.2	0.3
5/80	# 6	34.8	32.6	2.2
	#12	37.8	37.4	0.4
6/80	# 6	40.8	39.8	1.0
	#10	46.0	43.5	2.5
	#17	40.8	41.2	0.4
	#24	38.1	36.1	2.0
7/80	# 6	12.2	12.5	0.3
	#12	25.4	26.9	1.5
	#18	20.4	19.8	0.6

$\bar{R}$ = 14.9/22
= 0.68
UCL = 3.27 × 0.68 = 2.2
UWL = 2.51 × 0.68 = 1.7

trol chart for the automatic analyzer based on duplicate sets of results obtained from consecutive analysis of knowns serves as an example (Table 26:VI):

a. Calculate $\bar{X}$ for each duplicate set;

b. group the $\bar{X}$ values into a consistent reference scale (in groups by orders of magnitude for the full range of known concentrations);

c. calculate the UCL and lower control limit (LCL) by the equation: $CLs = \pm A_2\bar{R}$ (A_2 from Table 26:IV);

d. calculate the Warning Limits (WL) by the equation: $WLs = \pm 2/3\ A_2\bar{R}$;

e. chart CL's and WL's on each side of the standard which is set at zero, and

f. plot the difference between the nominal value and $\bar{X}$ (see Table 26:VI) and take action on points which fall outside of the control limits.

26.7 CONTROL CHARTS FOR INDIVIDUAL RESULTS. In many instances a rational basis for sub-grouping may not be available, or the analysis may be so infrequent as to require action on the basis of individual results. In such cases X charts are employed. However, the CL's must come from some subgrouping to obtain a measure of "within group" variability. This alternative has the advantage of displaying each result with respect to tolerance, or specification limits. The disadvantages must be recognized when considering this approach:

a. The chart does not respond to changes in the average;

b. changes in dispersion are not detected unless an R chart is included;

c. the distribution of results must approximate normal if the control limits remain valid.

Additional refinements, variations and control charts for other variables will be found in standard texts (1,12,13).

26.8 MOVING AVERAGES AND RANGES. The $\bar{X}$ control chart is more efficient for disclosing moderate changes in the average as the subgrouping size increases. A logical compromise between the X and $\bar{X}$ approach would be application of the moving average. For a given series of analyses, the moving average is plotted. (See data shown in Table 26:VII). The moving range serves well as a measure of acceptable variation when no rational basis for subgrouping is

Table 26:VI Accuracy Data

Date	Calibration Range	Nominal (N)	Values	X	N - X
9/79	10–400 ppm	100 ppm	22.9,21.5/	22.2	-0.7
	1.7–69.7 scale	22.9	22.7,22.3	22.5	-0.4
10/79	10–400	100	21.6,21.3/	21.5	0.0
	1.5–67.6	21.5			
2/80	10–400	100	23.6,24.1/	23.9	-0.6
	1.4–62.5	24.5			
4/80	10–400	100	25.8,26.5/	26.2	+ 0.2
	1.6–59.4	26.0	26.0,26.7/	26.4	+ 0.4
5/80	10–150	100	72.2,70.2/	71.2	+ 1.2
	6.3–83.0	70.0			
6/80	10–150	100	71.0,70.8/	70.9	-0.1
	6.6–85.0	71.0	71.0,71.3	71.2	+ 0.2
7/80	10–150	60	14.9,14.7/	14.8	-0.2
	1.8–33.5	15.0	15.1,14.4	14.8	-0.2

available or when results are infrequent or expensive to gather.

26.9 OTHER CONTROL CHARTS FOR VARIABLES. Although the standard $\bar{X}$ and R control chart for variables is the most common, it does not always do the best job. Several examples follow where other charts are more applicable.

26.9.1 *Variable Subgroup Size.* The standard $\bar{X}$ and R chart is applicable for a constant size subgroup of n = 2,3,4,5. In some cases such a situation does not exist. Control limit values must be calculated for each sample size. Plotting is done in the usual manner with the control size limits drawn in for each subgroup depending on its size.

26.9.2 *R or σ Charts.* In some situations the dispersion is equal over a range of assay values. In this case, a control chart

Table 26:VII Moving Average and Range Table (n = 2)

Sample No.	Assay Value	Sample Nos. Included	Moving Average	Moving Range
1	17.09	–	–	–
2	17.35	1–2	17.22	+ 0.26
3	17.40	2–3	17.38	+ 0.05
4	17.23	3–4	17.32	-0.17
5	17.00	4–5	17.12	-0.23
6	16.94	5–6	16.97	-0.06
7	16.68	6–7	16.81	-0.26
8	17.11	7–8	16.90	+ 0.43
9	18.47	8–9	17.79	+ 1.36
10	17.08	9–10	17.78	-1.39
11	17.08	10–11	17.08	0.00
12	16.92	11–12	17.00	-0.16
13	18.03	12–13	17.48	+ 1.11
14	16.81	13–14	17.42	-1.22
15	17.15	14–15	16.98	+ 0.34
16	17.34	15–16	17.25	+ 0.19
17	16.71	16–17	17.03	-0.63
18	17.28	17–18	17.00	+ 0.57
19	16.54	18–19	16.91	-0.74
20	17.30	19–20	16.92	+ 0.76

for either range or standard deviation is appropriate.

When the dispersion is a function of concentration, control limits can be expressed in terms of a percentage of the mean. In practice such control limits would be given as in the example below:

± 5 units/liter for 0–100 units/liter concentration

± 5% for > 100 units/liter concentration

An alternative procedure involves transformation of the data (13). For example, logarithms could be the appropriate transformation.

26.9.3 $\bar{X}$ *and* σ *Charts.* If the subgroup size exceeds 10, the Range Chart becomes inefficient. The use of a σ chart would then be appropriate. Where the cost of obtaining the test data is high, the increase in efficiency using σ rather than R may be worthwhile.

26.10 OTHER STATISTICAL TOOLS.

26.10.1 *Rejection of Questionable Results.* The question whether or not to reject results which deviate greatly from $\bar{X}$ in a series of otherwise normal (closely agreeing) results frequently arises. On a theoretical basis, no result should be rejected, as the one or more errors that render the entire series doubtful may be systematic errors that can be resolved. Tests that are known to involve mistakes, however, should not be reported exactly as analyzed. Mathematical basis for rejection of "outliers" from experimental data may be found in statistics text books (12,13).

26.10.2 *Correlated Variables-Regression Analysis.* A major objective in scientific investigations is the determination of the effect that one variable exerts on another. For example a quantity of sample (x) is reacted with a reagent to produce a result (y). The quantity x represents the independent variable over which the investigator can exert control.

The dependent variable (y) is the direct response to changes made in x, and varies in a random fashion about the true value. If the relationship is linear, the equation for a straight line will describe the effect of changes in x on the response y: $y = a + bx$, in which a is the intercept with the y axis and b is the slope of the line (the change in y per unit change in x). In chemical analysis a

is a measure of constant error arising from a colorimetric determination, trace impurity, blank, or other systematic source. The slope b may be controlled by reaction rate, equilibrium shift or the resolution of the method. The term "regression analysis" is applied to this statistical tool. Additional useful information can be obtained by certain transformations and shortcuts (13,14,15).

26.11 GRAPHIC ANALYSIS FOR CORRELATIONS. Useful shortcuts may be elected to determine whether a significant relationship exists between x and y factors in the equation for a straight line ($y = a + bx$). The data are plotted on linear cross section paper and a straight line drawn by inspection through the points with an equal number on each side or fitted by the least squares method. If the intercept a must be zero (a blank correction may produce such a situation), the fitting is greatly simplified. Then on each side equidistant from this line draw parallel lines corresponding to the established deviation (σ) of the analytical procedure, tally up the points falling inside of the band formed by the $\pm\ \sigma$ lines and calculate percent correlation (conformance = No. within band $\times$ 100/total points plotted). (See **16** for illustration of this technique).

Curvilinear functions can be accommodated, especially if a log normal function is involved and a plot of the data on semi-log paper yields a straight line (17). Log-Log paper also is available for plotting complex functions.

A combination of curvilinear and bar charts in some cases will reveal correlations not readily detected by mathematical processes. The data derived from an industrial cyanosis control program (18,19) illustrate an application that revealed a rather significant relationship between abnormal blood specimens and the frequency of cyanosis cases on a long-term basis.

Grouping data on a graph and approximating relationships by the quadrant sum test (rapid corner test for association) can provide useful results with a minimum expenditure of time (20).

In those cases where application of mathematical tools is tedious or completely impractical, a system of ranking is sometimes applicable to the restoration of order out of chaos. Again with reference to the cyanosis control program, a relationship between causative agent structure and biochemical potential for producing cyanosis and anemia was needed. Ten factors (categories) common to some degree for each of the 13 compounds under study had been recognized. The 13 compounds were ranked in each category in reverse order of activity (No. 1 most, No. 13 least active) and the sum of the rankings obtained for each compound. These sums then were divided by the number of categories used in the total ranking to obtain the "score." The scores were then arranged in increasing numerical order in columnar form. The most potent cyanogenic and anemiagenic compounds then appeared at the top of the table and the least at the bottom (19).

26.12 CHI-SQUARE TEST. Control charts are a convenient tool for daily checking with reference standards, but the answers are not always as nearly quantitative as needed. Periodic checking of the accumulated daily reference results to determine more rigorously whether all of the data belong to the same normal distribution may become necessary. One approach to this question and to assign a probability to the answer is provided by the chi-square (χ^2) test.

The chi-square distribution describes the probability distribution of the sums of the squares of independent variables that are normally or approximately normally distributed. The general form of the expression provides a comparison of observed versus expected frequencies (21). The chi-square test has also been applied to variables that fall within the Poisson distribution (23).

26.13 THE ANALYSIS OF VARIANCE (ANOVA). The analysis of variance is one of the most useful statistical tools. Variation in a set of results may be analyzed in such a way as to disclose and evaluate the important sources of the variation. For a detailed description of this technique consult standard statistics textbooks.

26.14 YOUDEN'S GRAPHICAL TECHNIQUE (6,23). Dr. W.J. Youden has devised an approach to test for determinate errors with a minimum of effort on the part of the analyst.

Two different test samples (X and Y) are prepared and distributed for analysis to as many individuals or laboratories as possible. Each participant is asked to perform only one determination on each sample (**NOTE:** It is important that the samples are relatively similar in concentration of the constituent being measured.)

The results from each laboratory can then be plotted as a point on a graph, with the test value for sample X as the x-coordinate, and for Y as the y-coordinate.

A vertical line is drawn through the average of all the results obtained on sample X; a horizontal line is drawn through the average of all the results obtained on sample Y. If the ratio of the bias to standard deviation is close to zero for the determinations submitted by the participants, then one would expect the distribution of the paired values (or points) to be close to equal among the four quadrants. The fact that the majority of the points fall in the (+, +) and (-, -) quadrants indicates that the results have been influenced by some source of bias.

Furthermore, one can even learn something about a participant's precision. If all participants had perfect precision (no random error), then all the paired points would fall on a line passing through the origin. Consequently the distance from such a line to each participant's point provides an indication of that participant's precision.

26.15 INTRALABORATORY QUALITY CONTROL PROGRAM.

26.15.1 *Responsibilities.* The attainment and maintenance of a quality control program in the laboratory is the direct responsibility of the laboratory manager or supervisor. The fundamental quality control techniques are based on:

a. Calibration to ensure accuracy;

b. duplication to ensure precision; and

c. correlation of quantitatively related tests to confirm accuracy and continual scrutiny to maintain the integrity of the results reported.

The individual technician can contribute significant assistance in this effort by his desire to deliver the best possible answers within the inherent limits of the equipment and procedure. Part of supervision's responsibility is adequate instruction to provide the "man on the bench" with sufficient "know how" to apply the principles on a routine basis.

The guidelines established by the American Industrial Hygiene Association for Accreditation of Industrial Hygiene Analytical Laboratories (12) delineate the minimum requirements that must be satisfied in order to qualify for proficiency recognition.

26.15.2 *Precision Quality Control.* In addition to the use of internal standards, recovery procedures and statistical evaluation of routine results, the laboratory should subscribe to a reference sample service to confirm precision and accuracy within acceptable limits. Apparatus should be calibrated directly or by comparison with National Bureau of Standards (NBS) certified equipment or its equivalent, reagents should meet or exceed ACS standards, and, calibration standards should be prepared from AR (analytical reagent) grade chemicals, and standardized with NBS standards if available. To illustrate, in a laboratory engaged in an exposure control program based on biological monitoring by trace analysis of blood and urine for lead content, at least two calibration points, blanks and a recovery should be included in each batch analyzed by the dithizone procedure. In addition, the wavelength integrity and optical density response of the spectrophotometer should be checked and adjusted — if necessary by calibration with NBS cobalt acetate standard solution. Until the standard deviation for the analytical procedure has been established within acceptable limits, replicate determinations should be made on at least two samples in each batch (either aliquot each sample or take duplicate samples), and thereafter with a frequency sufficient to ensure continued operation within these limits.

Control charts are probably the most widely recognized application of statistics. They provide "instant" quality control status when plotted daily, or at other intervals sufficiently short to disclose trends without undue oscillations from over-refinement of the data.

26.15.3 *Accuracy Quality Control.* A standard or well defined control sample should be analyzed periodically to confirm accuracy of a procedure. The control chart technique is directly applicable to long-term evaluation of the reliability of the analyst as well as the accuracy of the procedure. To attain and maintain the high level of analytical integrity presented earlier in this section, the major sources of "assignable cause" errors must be reduced to a minimum level that is consistent with cost penalties and the objective of the study for which the analytical service is rendered.

26.15.4 *Interlaboratory Reference Systems.* Participation in interlaboratory studies, whether by subscription from a certified laboratory supplying such a service or from a voluntary program initiated by a group of laboratories in an attempt to improve analytical integrity (10), is highly recommended. Evaluation of the analytical method, as well as evaluation of the individual laboratory's performance, can be derived by specialized statistical methods applied to the data collected from such a study. However, inasmuch as most investigators will not be called upon to conduct or evaluate interlaboratory surveys, the reader is referred to the literature in the event such specialized information is needed (10,24, 25,26). In the absence of such programs, the investigator or laboratory supervisor should make every effort to locate colleagues engaged in similar sampling and analytical activity and arrange exchange of standards, techniques and samples to establish integrity and advance the art.

26.16 SUMMARY. Identification of the systematic sources of error of a procedure provides the information required to reduce such error to a minimum level. The remaining (residual) random errors then determine the precision of analyses produced by the procedure. Statistical techniques have been developed to estimate efficiently the precision. For a procedure to be accurate, the results must be not only precise, but bias must be absent. Several approaches are available to eliminate bias both within the laboratory and between laboratories by collaborative testing. Quality control programs based on appropriate control charts must be employed on a routine basis to assure adherence to established performance standards. The total analysis control program must include instrumental control, procedural control and elimination of personal errors. The use of replicate determinations, "spiked" sample techniques, reference samples, standard samples, and quality control charts will provide assurance that the procedure remains in control.

The guidelines established by the American Industrial Hygiene Association for Accreditation of Industrial Hygiene Analytical Laboratories (11) summarize in a succinct fashion the requirements for proficiency.

REFERENCES

1. KELLEY, W.D. 1968. Statistical Method — Evaluation and Quality Control for the Laboratory. Training Course Manual in Computational Analysis. U.S. Dept. of Health, Education and Welfare, Public Health Service.
2. SHEWHART, W.A. 1931. Economic Control of Quality of Manufactured Products. Bell Telephone Laboratories.
3. LINCH, A.L., H.V. PFAFF, 1971. Carbon Monoxide — Evaluation of Exposure Potential by Personnel Monitor Surveys. Am. Ind. Hyg. Assoc. J. 32:745.
4. AMERICAN CHEMICAL SOCIETY. 1963. Guide for Measures of Precision and Accuracy. Anal. Chem. 35:2262.
5. AMERICAN SOCIETY FOR TESTING AND MATERIALS. 1976. ASTM Manual on Quality Control of Materials — Special Technical Publication 15-D. Philadelphia, Pa.
6. YOUDEN, W.J. 1951. Statistical Methods for Chemists. John Wiley and Sons, New York, N.Y.
7. KITAGAWA, T. 1971. Carbon Monoxide Detector Tube No. 100. National Environmental Instruments, Inc., P.O. Box 590, Fall River, Mass.
8. GRIM, K., A.L. LINCH 1964. Recent Isocyanate-In-Air Analysis Studies. Am Ind. Hyg. Assoc. J. 25:285.
9. AMERICAN SOCIETY FOR TESTING AND MATERIALS. 1977. ASTM D2777. Standard Practices for Determination of Precision of Committee D-19 Methods.
10. KEPPLER, J.F., M.E. MAXFIELD, W.D. MOSS, G. TIETJEN AND A.L. LINCH. 1970. Interlaboratory Evaluation of the Reliability of Blood Lead Analysis. Am. Ind. Hyg. Assoc. J. 31:412.
11. CRALLEY, L.J., C.M. BERRY, E.D. PALMES, C.F. REINHARDT, AND T.L. SHIPMAN. 1970. Guidelines for Accreditation of Industrial Hygiene Analytical Laboratories. Am. Ind. Hyg. Assoc. J. 31:335.
12. COWDEN, D.J. 1957. Statistical Methods in Quality Control. Prentice Hall Inc., Englewood Cliffs, New Jersey.
13. BAUER, E.J. 1971. A Statistical Manual for Chemists. 2nd Edition. Academic Press, New York, N.Y.
14. AMERICAN PUBLIC HEALTH ASSOCIATION. 1976. Standard Methods for the Examination of Water and Wastewater. 14th edition, Washington, D.C. 20036.
15. HINCHEN, J.D. 1969. Practical Statistics for Chemical Research. Methuen and Co. Ltd., London.
16. LINCH, A.L., E.G. WIEST AND M.D. CARTER. 1970. Evaluation of Tetraalkyl Lead Exposure by Personnel Monitor Surveys. Am. Ind. Hyg. Assoc. J. 31:170.
17. LINCH, A.L. AND M. CORN. 1965. The Standard Midget Impinger-Design Improvement and Miniaturization. Am. Ind. Hyg. Assoc. J. 26:601.
18. WETHERHOLD, J.M., A.L. LINCH AND R.C. CHARSHA. 1959. Hemoglobin Analysis for Aromatic Nitro and Amino Compound Exposure Control. Am. Ind. Hyg. Assoc. J. 20:396.
19. STEERE, N.V. (Editor). 1971. Handbook of Laboratory Safety, 2nd edition. The Chemical Rubber Co., Cleveland, Ohio.
20. WILCOXON, F. 1949. Some Rapid Approximate Statistical Procedures. Insecticide and Fungicide Section — American Cyanamid Co., Agricultural Chemicals Division. New York. N.Y.
21. MAXWELL, A.E. 1946. Analyzing Qualitative Data. Chap. 1, pp. 11–37, John Wiley & Sons, Inc., New York.
22. DUNCAN, A.J. 1965. Quality Control and Industrial Statistics, 3rd edition. R.D. Irwin, Inc., Homewood, Illinois.
23. YOUDEN, W.J. 1960. The Sample, the Procedure and the Laboratory. Anal. Chem. 32(13):23A–37A.
24. AMERICAN SOCIETY FOR TESTING AND MATERIALS. 1963. ASTM Manual for Conducting an Interlaboratory Study of a Test Method. Technical Publication No. 335.
25. WEIL, C.S. 1971. Critique of Laboratory Evaluation of the Reliability of Blood-Lead Analyses. Am. Ind. Hyg. Assoc. J. 32:304.
26. SNEE, R.D. AND P.E. SMITH. 1971. Statistical Analysis of Interlaboratory Studies. Paper prepared for presentation to the Am. Ind. Hyg. Conference in San Francisco, Calif.
27. TAYLOR, J.K. 1985. Handbook for SRM Users. Special Publication 260–100, National Bureau of Standards, Gaithersburg, MD 20899.
28. TAYLOR, J.K. 1987. Quality Assurance of Chemical Measurements. Lewis Publishers, Inc. Chelsea, MI. 48118.

Approved with modifications from 2nd edition
Subcommittee 9
J.N. HARMAN, *Chairman*

27. Direct Reading Colorimetric Indicators

27.1 INTRODUCTION.

27.1.1 Three types of direct reading colorimetric indicators have been in use for the determination of contaminant concentrations in air: liquid reagents, chemically treated papers, and glass indicating tubes containing solid chemicals. A comprehensive bibliography in this area was prepared by Campbell and Miller (1).

27.1.2 Convenient laboratory procedures using liquid reagents have been simplified and packaged for field use. Reagents are supplied in sealed ampoules or tubes, frequently in concentrated or even solid form which is diluted or dissolved for use. Unstable mixtures may be freshly prepared when needed by breaking an ampoule containing one ingredient inside a plastic tube or bottle containing the other. Commercial apparatus of this type is available for tetraethyl lead and tetramethyl lead. Certain liquid reagents, such as the nitrogen dioxide sampling reagent, produce a direct color upon exposure without requiring additional chemicals or manipulations. These permit simplified sampling equipment. Thus, relatively high concentrations of nitrogen dioxide may be directly determined by drawing an air sample into a 50 or 100 mL glass syringe containing a measured quantity of absorbing liquid reagent, capping, and shaking. Liquids containing indicators have been used for determining acid or alkaline gases by measuring the volume of air required to produce a color change. These liquid methods are somewhat inconvenient and bulky to transport and require a degree of skill to use. However, they are capable of good accuracy, as measurement of color in liquids is inherently more reproducible and accurate than measurement of color on solids.

27.1.3   Chemically treated papers have been used to detect and determine gases because of their convenience and compactness. An early example of this is the Gutzeit method in which arsine blackens a paper strip impregnated previously with mercuric bromide. Such papers may be prepared freshly and used wet, or stored and used in the dry state. Special chemical chalks or crayons have been used (2) to sensitize ordinary paper for phosgene, hydrogen cyanide, and other war gases. Semiquantitative determinations may be made by hanging the paper in contaminated air. Inexpensive detector tabs are available commercially that darken upon exposure to carbon monoxide (3). The accuracy of such procedures is limited by the fact that the volume of the air sample is rather indefinite and the degree of color change in the paper is influenced by air currents and temperature. More quantitative results may be obtained by using a sampling device capable of passing a measured volume of air over or through a definite area of paper at a controlled rate, as is done in a commercial device for hydrogen fluoride. Particulate matter contaminants such as chromic acid and lead may be determined similarly, usually by addition of liquid reagents to the sample on a filter paper. Visual evaluation of the stains on the paper may be made by comparison with color charts or by photoelectric instruments. Recording photoelectric instruments utilizing sensitized paper tapes operate in this manner. Accuracy of these methods requires uniform sensitivity of the paper, stability of all chemicals used, and careful calibration. In the case of particulate matter analysis, it may be necessary to calibrate with the specific dust being sampled if the degree of chemical solubility is an important factor.

27.1.4   Glass indicating tubes containing solid chemicals are another type of convenient and compact direct reading device. The early detection tubes were made for carbon monoxide (4–6), hydrogen sulfide (7,8), and benzene (9,10). During the past decades, there has been a great expansion in the development and use of these tubes, (11–32) and more than four hundred different types are now available commercially. Several manuals provide comprehensive descriptions and listing (33–35). Because of the great popularity and wide use of glass detector tubes, the bulk of this introduction will deal with them, although much of the information will be applicable to the liquid and paper indicators as well.

27.1.5   There are many uses for detector tubes. They are convenient for qualitative (36) and quantitative evaluation of toxic hazards in industrial atmospheres. They are also useful for air pollution studies, although in most situations currently available tubes do not have the required sensitivity. Detector tubes may be used for detection of explosive hazards, as well as for process control of gas composition. Confirmation of carbon monoxide poisoning may be made by determining carbon monoxide in exhaled breath or in gas released from a sample of blood (after an appropriate procedure). Detector tubes may be used for law enforcement purposes, such as determining alcohol in the breath, or gasoline in soil in cases of suspected arson or of leakage from underground tanks.

Minute quantities of ions in aqueous solutions also may be determined, such as sulfide in waste water from pulp manufacturing, chromic acid in electrolytic plating waste water, and nickel ion in waste water of refineries.

27.1.6 Detector tubes have been widely advertised as being capable of use by unskilled personnel. While it is true that the operating procedures are simple, rapid, and convenient, many limitations and potential errors are inherent in this method. The results may be dangerously misleading unless the sampling procedure is supervised and the findings interpreted by an adequately trained occupational hygienist.

27.2 OPERATING PROCEDURES.

27.2.1 The use of detector tubes is extremely simple. After its two sealed ends are broken open, the glass tube is placed in the manufacturer's holder which is fitted with a calibrated squeeze bulb or piston pump. The recommended air volume is then drawn through the tube by the operator. Adequate time must be allowed for each stroke. Even if a squeeze bulb is fully expanded, it may still be under a partial vacuum and may not have drawn its full volume of air. The manufacturer's sampling instructions must be followed closely.

27.2.2 The observer then reads the concentration in the air by examining the exposed tube. Some of the earlier types of tube are provided with charts of color tints to be matched by the solid chemical in the indicating portion of the tube. This visual judgment depends, of course, upon the color vision of the observer and the lighting conditions. In an attempt to reduce the errors due to variations between observers, most recent types of tubes are based upon producing a variable length of stain on the indicator gel. Although in a few tubes a variable volume of sample is collected until a standard length of stain is obtained, in most cases a fixed volume of sample is passed through the tube and the stain length is measured against a calibration scale. The scale may be printed either directly upon the tube or on a provided chart. In a few tubes, such as those for arsine and stibine, a variable volume of sample is drawn through the tube until the first visible discoloration is noted. This is a very difficult judgment, which must be made

retrospectively. The range in the interpretation of results by different observers is large, since in many cases the end of a stain front is not sharp. Experience in sampling known concentrations is of great value in training an operator to know whether to measure the length up to the beginning or end of the stain front, or some other portion of an irregularly shaped stain. In some cases the stains change with time, and thus the reading should not be unduly delayed.

27.2.3 Care must be taken to see that leak-proof pump valves and connections are maintained. A leakage test may be made by inserting an unopened detector tube into the holder and squeezing the bulb; at the end of two minutes any appreciable bulb expansion is evidence of a leak. If the apparatus is fitted with a calibrated piston pump, the handle is pulled back and locked. Two minutes later, it is released cautiously and the piston allowed to pull back in; it should remain out no more than 5% of its original distance. Leakage indicates the necessity of replacing check valves, tube connections, or the squeeze bulb, or of greasing the piston.

27.2.4 At periodic intervals the flow rate of the apparatus should be checked and maintained within specifications for the tube calibrations (generally ± 10%). This may be done simply by timing the period of squeeze bulb expansion. A more accurate method is to place a used detector tube in the holder, and to draw an air sample through a calibrated rotameter. Alternatively, the air may be drawn from a burette in an inverted vertical position, which is sealed with a soap film, and the motion of the film past the graduations timed with a stop watch (37). The latter method also provides a check on the total volume of the sample which is drawn. In some devices, the major resistance to the air flow is in the chemical packing of the tube; thus, each batch might require checking. An incorrect flow rate indicates a partially clogged strainer or orifice which should be cleaned or replaced.

27.2.5 With most types of squeeze bulbs and hand pumps the sample air flow rate is variable, being high initially and low towards the end when the bulb or pump is almost filled. This has been claimed to be an advantage because the initially high rate gives a long stain and the final low rate sharpens the stain front. Flow patterns for six commonly used pumps were found to be different (38,39). When five popular brands of carbon monoxide tubes were used with pumps other than their own, grossly erroneous results were obtained, even with identical sample volumes. The stains may depend more on flow rate than on concentrations. It should be noted that accuracy requires a close reproduction of the flow rate pattern for the calibrations to be correct.

27.2.6 A number of special techniques may be used in appropriate cases. When sampling in inaccessible places, the indicator tube may be placed directly at the sampling point and the pump operated at some distance away. A rubber tube extension of the same inside diameter as the indicator tube may be inserted between the pump and indicator tube. Such tubes are available commercially as accessories. Lengths as great as 60 feet have been successfully used without appreciable error, provided that more time is allowed between strokes of the pump to compensate for the reservoir effect and to obtain the full volume of sample. This method has the disadvantage that the detector tube cannot be observed during the sampling.

27.2.7 A second arrangement may be used when sampling hot gases such as from a furnace stack or engine exhaust. Cooling the sample is essential in these cases, otherwise the calibration would be inaccurate and the volume of the gas sample uncertain. A probe of glass or metal, available commercially as an accessory, may be attached to the inlet end of the detector tube with a short piece of flexible tubing (40). If this tube is cold initially, as little as four inches of tubing outside of the furnace is sufficient to cool the gas sample from 250°C to about 30°C. Such a probe has to be employed with caution. In some cases serious adsorption errors occur either on the tube or in condensed moisture. The dead volume of the probe should be negligible in comparison to volume of sample taken. Solvent vapors should not be sampled with this method. When sampling air colder than 0°C, clasping the tube in the hand warms it sufficiently to eliminate any error (40). Critical studies (41,42) of appli-

cations to analysis of diesel exhaust showed serious errors for some tubes.

27.2.8 Other special techniques may also be employed. Some symmetrical tubes can be reversed in the holder and used for a second test. In certain special cases tubes may be re-used if a negative test was previously obtained, or after the color has faded. Two tubes also may be connected in series in special cases, such as passing crude gas through first a Kitagawa hydrogen sulfide tube and then a phosgene tube to get two simultaneous determinations and remove interferences. These techniques may be used only after testing to demonstrate that they do not impair the validity of the results.

27.2.9 Tubes also have been used at pressures as high as several atmospheres. This situation would exist, for example, in underwater stations. If both the tube and pump are in the chamber, the calibrations and sample volumes are altered. It has been reported (40,43) that only the latter occurs for the following Draeger tubes: ammonia 5/a, arsine 0.05/a, CO_2 0.1%/a, CO 5/c, 10/b, H_2S 1/c, 5/b. For these tubes, the corrected concentration is equal to the scale reading (ppm or vol.%) divided by the ambient pressure (in atmospheres) at the pump. When tube tips are broken in a pressure chamber, the tube filling should be checked for possible displacement.

27.3 SPECIFICITY AND SENSITIVITY.

27.3.1 The specificity of the tubes is a major consideration for determining applicability and interpreting results. Most tubes are not specific. Chromate reduction is a common reaction used in tubes for detection of organic compounds. In the presence of mixtures, the uncritical acceptance of such readings can be grossly misleading. Comprehensive listing of reactions, as well as a discussion of other major aspects, are available (33,34,44). Six common reactions and the associated tube types are listed in Table 27:I. It can be seen that the name of the compound listed on the tube often refers to its calibration scale rather than to its contents.

27.3.2 The lack of specificity of some tubes may be used to advantage for detection of substances other than those indicated by the manufacturer. In this respect, tubes using colorimetric reactions 1, 2, and

Table 27:I Common Colorimetric Reactions in Gas Detector Tubes

1. Reduction of chromate or dichromate to chromous ion:

    Draeger: Acetaldehyde 100/a, alcohol 100/a; aniline 0.5/a; cyclohexane 100/a, diethyl ether 100/a, ethyl acetate 200/a, ethyl glycol acetate 50/a; n-hexane 100/a; methanol 50/a; n-pentane 100/a.

    Gastec: Acetone 151; aniline 181; butane 104; butyl acetate 142; ethanol 112; ethyl acetate 141; ethyl ether 161; ethylene oxide 163; gasoline 101, 101L; hexane 102H, 102L; isopropanol 113; LP gas 100A; methanol 111; methyl ethyl ketone 152; methyl isobutyl ketone 153; propane 100B; sulfur dioxide 5H; vinyl chloride 131.

    Kitagawa: Acetone 102A, acrylonitrile 128A, 128B; butadiene 168A; butyl acetate 138; cyclohexane 115; dimethyl ether 123; dioxane 154; ether 107; ethyl acetate 111; ethyl alcohol 104A; ethylene oxide 122; furan 161; n-hexane 113; isobutyl acetate 153; isopropanol 150; isopropyl acetate 149; methyl acetate 148; methyl alcohol 119; methyl ethyl ketone 139B; methyl isobutyl ketone 155; propyl acetate 151; propylene oxide 163; sulfur dioxide 103A; tetrahydrofuran 162; vinyl chloride 132.

    MSA: *Part 95097* for n-amyl alcohol, iso-amyl alcohol, sec-amyl alcohol, tert-amyl alcohol, 2-butoxyethanol (butyl Cellosolve), n-butyl alcohol, isobutyl alcohol, sec-butyl alcohol, tert-butyl alcohol, cyclohexanol, 2-ethoxyethanol (Cellosolve), ethyl alcohol (ethanol), ethylene glycol monomethyl ether, furfuryl alcohol, 2-methyloxyethanol, methyl alcohol , 2-methlcyclohexanol, methyl isobutyl carbinol (methyl amyl alcohol), n-propyl alcohol, iso-propyl alcohol.
                Part 460423 for acetone, methyl methacrylate.

2. Reduction of iodine pentoxide plus fuming sulfuric acid to iodine:

    Draeger: Benzene 5/b; carbon disulfide 5/a; carbon monoxide 2/a, 5/c, 8/a, 10/a, 10/b, 0.001%/a, 0.1%/a, 0.3%/a, 0.3%/b; ethyl benzene 30/a; hydrocarbon 0.1%/b; natural gas*; perchloroethylene 0.1%/a; petroleum hydrocarbons 100/a; polytest; toluene 5/a, 25/a.

    Gastec: Acetylene 171; benzene 121, 121L; carbon monoxide 1H, 1M; Stoddard solvent 128; toluene 122; vinyl chloride 131; xylene 123.

    MSA: *Part 93074* for benzene (benzol), chlorobenzene, monobromobenzene, toluene (toluol), xylene (xylol).

3. Reduction of ammonium molybdate plus palladium sulfate to molybdenum blue:

    Draeger: Ethylene 0.5/a; 50/a; methyl acrylate 5/a; methyl methacrylate 50/a.

    Gastec: Butadiene 174; ethylene 172, 172L.

    Kitagawa: Acetylene 101; butadiene 168B; carbon monoxide 106A, 106B, 106C*; ethylene 108B; hydrogen sulfide and sulfur dioxide 120C.

    MSA: *Part 47134* for carbon monoxide (NBS color change).
                Part 85802 for acetylene, ethylene, propylene.

4. Reaction with potassium palladosulfite:

    Gastec: Carbon monoxide 1L, 1La, 1LL; hydrogen cyanide 12H.

    MSA: *Part 91229* for carbon monoxide (length of stain).

5. Color change of pH indicators (e.g., bromphenol blue, phenol red, thymol blue, methyl orange):

    Draeger: Acetic acid 5/a; acrylonitrile 5/a*, 5/b*; ammonia 2/a, 5/a, 0.5%/a; chlorobenzene 5/a*; cyanide 2/a*; cyclohexylamine 2/a; dimethyl acetamide 10/a*; dimethyl formamide 10/b*; formic acid 1/a; hydrazine 0.25/a; hydrochloric acid 1/a, 50/a; hydrogen cyanide 2/a (HgCl$_2$ + methyl red); methacrylonitrile 1/a*; nitric acid 1/a; sulfur dioxide 0.1/a*, 50/a; triethylamine 5/a; vinyl chloride 0.5/a*.

Table 27:I continued

Gastec: Acetaldehyde 92*; acetic acid 81; acrolein 93*; acrylonitrile 191*, 191L*; amines 180: ammonia 3H, 3M, 3L; tert-butyl mercaptan 75; carbon dioxide 2H, 2L; carbon disulfide 13*, 13M; carbonyl sulfide 21; dimethylacetamide 184*; dimethylformamide 183*; formaldehyde 91L*; hydrogen chloride 14L, 14M; hydrogen cyanide 12L*; methacrylonitrile 192; nitric acid 15L; perchloroethylene 133*; pyridine 182; sulfur dioxide 5M, 5L, 5La; trichloroethylene 132H*, 132L*; vinyl chloride 131La*, 131L*.

Kitagawa: Acetaldehyde 133; ammonia 105B; carbon dioxide 126A, 126B; hydrogen cyanide 112B.

MSA: Part 85976 for carbon dioxide. Part 91636 for hydrogen chloride.
           Part 92030** for 1-chloro-1,1-difluoroethane (Genetron 142B), chlorotrifluoromethane (Freon 13), 1,2-dichloroethane (ethylene dichloride), dichloroethylene (trans-1,2), ethyl chloride, fluorotrichloromethane (Freon 11), methyl chloride, methylene chloride (dichloromethane), propylene dichloride (1,2-dichloropropane), 1,1,2-trichloro-1,2,2-trifluoroethane (Freon 113), vinyl chloride (chloroethylene).
           Part 92115 for ammonia, n-butylamine, cyclohexylamine, diisopropylamine, di-n-propylamine, ethylamine, ethylene imine, n-ethylmorpholine, isopropylamine, methylamine, propylene imine, triethylamine, trimethylamine.
           Part 92623 for sulfur dioxide. Part 93865 for ozone. Part 95739** for dimethyl sulfoxide. Part 460021 for acetic acid. Parts 460103 and 460158 for ammonia.
           Part 460425 for hydrazine, monomethyl hydrazine, unsymmetrical dimethyl hydrazine.

6a. Reaction with o-tolidine:

Draeger: Chlorine 0.2/a, 0.3/b, 50/a; chloroform 2/a*, epichlorohydrin 5/b*; perchloroethylene 10/b; trichloroethylene 2/a*, 10/a*, 50/d*; vinyl chloride 1/a*.

Gastec: Chlorine 8H, 8La; chloroform 137*; methyl bromide 136*; methyl chloroform 135*; methylene chloride 138*; nitrogen dioxide 9L; nitrogen oxides 10*, 11*.

Kitagawa: bromine 114; chlorine 109; chlorine dioxide 116; nitrogen dioxide 117.

6b. Reaction with tetraphenylbenzidine:

MSA: Part 82399 for bromine, chlorine, chlorine dioxide.
           Part 83099 for nitrogen dioxide.
           Part 85833* for chlorobromomethane; 1,1-dichloroethane; dichloroethylene (cis-1,2 and trans-1,2); ethyl bromide; ethyl chloride; perchloroethylene (tetrachloroethylene); trichloroethylene; 1,2,3-trichloropropane; vinyl chloride (chloroethylene).
           Part 85834* for chlorobenzene (mono); 1,2-dibromoethane (ethylene dibromide); dichlorobenzene (ortho); 1,2-dichloroethane (ethylene dichloride); dichloroethyl ether; 1,1-dichloroethylene (vinylidine chloride); methyl bromide; methylene chloride (dichloromethane); propylene dichloride (1,2-dichloropropane); 1,1,2,2-tetrabromoethane; 1,1,2,2-tetrachloroethane; 1,1,3,3-tetrachloropropane; trichloroethane (beta 1,1,2); vinyl chloride (chloroethylene).
           Part 87042 for bromine, chlorine.
           Part 88536** for carbon tetrachloride; chlorobromomethane; 1-chloro-1,1-difluoroethane (Genetron 142B); chlorodifluoromethane (Freon 22); chloroform (trichloromethane); chloropentafluoroethane (Freon 115); chlorotrifluoromethane (Freon 13); 1,2-dibromoethane (ethylene dibromide); dichlorodifluoromethane (Freon 12); 1,1-dichloroethylene (vinylidine chloride); dichloroethylene (cis-1,2); dichlorotetrafluoroethane (Freon 114); fluorotrichloromethane (Freon 11); Freon 113; Freon 502; methyl bromide; methyl chloroform (1,1,1-trichloroethane); methylene chloride (dichloromethane); perchloroethylene (tetrachloroethylene); trichloroethane (beta 1,1,2); trichloroethylene; 1,1,2-trichloro-1,2,2-trifluoroethane (Freon 113); trifluorobromomethane (Freon 13B1).

Table 27:I (continued)

*Part 91624*** for acetonitrile; acrylonitrile; 1-chloro-1-nitropropane; cyanogen; 1,1-dichloro-1-nitroethane; dimethylacetamide; dimethylformamide; fumigants (Acritet, Insect-O-Fume, Fume-I-Gate, Termi-Gas, Termi-Nate); methacrylonitrile; nitroethane; nitromethane; 1-nitropropane; 2-nitropropane; n-propyl nitrate; pyridine; vinyl chloride.
Part 460225 for chlorine. *Part 460424*** for nitric oxide.

*Multiple reaction or multiple layer tube for improved specificity or preliminary reaction.
**Pyrolyzer required.

6 (Table 27:I) are widely applicable. Thus, the Draeger screening tube (Polytest) and ethyl acetate tube may be used for qualitative indications of reducing and organic materials, respectively (45). The Draeger trichloroethylene tube is also applicable to chloroform, o-dichlorobenzene, dichloroethylene, ethylene chloride, methylene chloride, and perchloroethylene. The methyl bromide tube may be used for chlorobromomethane and methyl chloroform. The chlorine tube may be used for bromine and chlorine dioxide. The toluene tube may be used for xylene. Such use requires specific knowledge of the identity of the reagent and of the proper corrections to the calibration scales.

27.3.3   In some brands of indicator tubes the units of the calibration scales are in milligrams per cubic meter. Although it has been said that this method of expression eliminates the necessity of making temperature and pressure corrections, such a claim is debatable since the scale calibrations themselves may be highly dependent upon these variables. Units of parts per million or percent by volume are most common for industrial hygiene purposes and are used on most of the newer tubes.

27.3.4   Although detector tubes are generally designed for detection of relatively high gas concentrations found in industrial workplaces, some have been applied to the much lower outdoor air pollutant concentration. Kitagawa (46) determined 0.01 to 2 ppm of NO_2 using two glass tubes in series, with the temperature controlled at 40°C. The first tube contained diatomaceous earth impregnated with a specific concentration of sulfuric acid to regulate the humidity to the air sample. The second tube, 120 mm long × 2.4 mm inside diameter, contained white silica gel impregnated with ortho-tolidine. (It is not clear whether or not this is identical with the commercial No. 117.) Air was drawn through the tubes for 30 min at 180 mL/min by an electric pump with a stainless steel orifice plate at its inlet. Accuracy was ± 10%; no comments on the specificity were given. Grosskopf (47,48) determined 0.007 to 0.5 ppm NO_2 by drawing air through a Draeger 0.5/a nitrous gas tube with a diaphragm pump for 10 to 40 min at the rate of 0.5 L/min. Readings were not affected by flow rates if the flow rates exceeded 0.5 L/min. No comments were given on the specificity, except that humidity from 30 L of air at 70% relative humidity did not impair the sensitivity. This tube responds to nitric oxide and to oxidants, both of which commonly may be present. Leichnitz (49) reported a new tube (Draeger SO_2 0.1/a) capable of measuring 0.1 to 3 ppm of sulfur dioxide. This tube requires 100 strokes of a hand bellows pump (each taking 7 to 14 s), or use of the Draeger Quantimeter electric pump, in which a motor-driven crank controlled by a timer or counter operates a bellows. This pump is described in the text of the manufacturer's listing.

27.3.5   Less success has been attained when carbon monoxide detector tubes were used for sampling periods of 4 h or longer with continuous pumps. It was found that at low concentrations, after an initial period, the stain lengths cease to increase (23). However, at higher concentrations a new calibration could be made (50) (for 3- to 5-h samples at 8 mL/min through a Kitagawa 100 tube in the range 30-100 ppm of carbon monoxide). The latter investigator hypothesized that the oxygen in air bleached the black palladium stain and caused the front produced by low concentrations to remain stationary after the first 20-30 min. Effects of water vapor and of other contaminants also must be considered in this application. A new calibration is essential under the flow conditions to be used. Studies confirmed that secondary reactions that bleached the indication and prevented long term sampling could be avoided with appropriate reagent systems **(51)**.

27.3.6 Recently a considerable number of indicator tubes have been developed **(52,53)** for long duration sampling (4–8 h). These appear to be very similar to the tubes designed for short duration sampling, and are effective within the same concentration ranges. They are calibrated for use with a continuous sampling pump, but operate at lower flow rates. The application of these tubes is to provide time-weighted average concentrations, rather than short-term (few minutes) values. In order to provide valid averages, the calibrations must be linear both with concentration and time, and should display uniformly spaced markings for uniform increments of contaminants. The scales on these tubes usually are in terms of microliters of test gas (ppm $\times$ L), rather than ppm, and the latter is calculated by dividing the scale reading by the liters of air sampled. Over 30 types of long-duration tubes are now available commercially. It should be noted that they must be used within the ranges of flow rate and total sampling time established during their calibration by the manufacturer, using the specified continuous sampling pump. Low flow MSA Accuhaler pumps have been utilized, with some loss of accuracy **(54)**. They generally are not suitable for analysis of concentrations in lower ranges than those of ordinary tubes designed for short duration sampling **(55)**, because of the previously mentioned problems of water vapor, oxygen, and other contaminants.

27.3.7 Greater accuracy can be obtained when several detector tubes are used for replicate sampling. A simplified statistical approach based on an assumed normal distribution of values is recommended for 3 to 10 samples **(56)**. However, subsequent work indicated that most of the variations were due to the environmental fluctuations rather than to the relatively small analytical errors, and that a lognormal distribution was more appropriate. A step by step procedure was presented **(57)**, which categorized the results into non-compliance (less than 5% chance of erroneously citing when

actually compliance exists), no decision, and compliance (less than 5% chance of failing to cite when actually noncompliance exists).

27.4 PROBLEMS IN THE MANUFACTURE OF INDICATOR TUBES.

27.4.1 The accuracy, limitations, and applications of indicator tubes are highly dependent upon the skill with which they were manufactured. Generally, the supporting material is silica gel, alumina, ground glass, pumice, or resin. This is impregnated with an indicating chemical, which should be stable, specific, and sensitive, and should produce a color that strongly contrasts with the unexposed color and is non-fading for at least an hour. If the reaction with the test gas is relatively slow, a color is produced throughout the length of the tube, since the gas is incompletely absorbed and the concentration at the exiting end is an appreciable fraction of that at the entrance. Such a color must be matched against a chart of standard tints. A rapidly reacting indicating chemical is much more desirable, and yields a length-of-stain type of tube in which the test gas is completely absorbed in the stained portion.

27.4.2 There is a very wide and unpredictable variation in the properties of different batches of indicating gel. The major portion of the chemical reaction probably occurs upon the surface. Therefore, the number of active centers, which is highly sensitive to trace impurities, affects the reaction rate. These problems are well known in the preparation of various catalysts. Close controls must be kept on the purity and quality of the materials, the method of preparation, the cleanliness of the air in the factory or glove box in which the tubes are assembled, the inside diameter of the glass tubes, and even upon the size analysis of the impregnated gel, which in some cases is important in controlling the flow rate. The manufacturer also must accurately calibrate each batch of indicating gel.

27.4.3 Some tube types are constructed with multiple layers of different impregnated gels with inert separators. Generally, the first layer is a precleansing chemical to remove interfering gases and improve the specificity of the indication. Thus, in the case of some carbon monoxide tubes, chemicals are provided to remove in-

terfering hydrocarbons and nitrogen oxides. In carbon disulfide tubes, hydrogen
sulfide is first removed. In hydrogen cyanide tubes, hydrogen chloride or sulfur dioxide are removed first. In other cases, the
entrance layer provides a preliminary reaction essential to the indicating reaction.
Thus, in some trichloroethylene tubes, the
first oxidation layer liberates halogen
which is indicated in the subsequent layer.
In some tubes for NO_x gases, a mixture of
chromium trioxide and concentrated sulfuric acid is used to oxidize nitric oxide to
nitrogen dioxide, which is the form to
which the sensitive indicating layer responds. While such multiple layer tubes are
advantageous when properly constructed,
they frequently have a shorter shelf life because of diffusion of chemicals between
layers and consequent deterioration.

27.4.4 A shelf-life of at least two
years is highly desirable for practical purposes. A great deal of disappointment with
various tube performances is no doubt due
to inadequate shelf life. Since some tubes
have only been on the market for a short
time, the manufacturer himself may have
inadequate experiences as to the shelf life of
his product. Small variations in impurities,
such as the moisture content, may have a
large effect upon the shelf lives of different
batches. The storage temperature, of
course, greatly affects the shelf life, and it
is highly desirable to store these tubes in a
refrigerator. In some cases, shelf life has
been estimated by accelerated tests at
higher temperatures. Such a variation of
shelf life (length of time within which the
calibration accuracy is maintained at ±
25%) is illustrated by the data listed in Table 27:II received in a personal communication from Dr. Karl Grosskopf of the
Draeger Company. These data plot as an

**Table 27:II Shelf Life of Draeger Carbon
Monoxide Tubes**

Temperature °C	Shelf Life
25	> 2 yr
50	> 1/2 yr
80	weeks
100	1 week
125	3 days
150	1 day

approximately straight line when the logarithm of the shelf lifetime is plotted against
a linear scale of the reciprocal of absolute
temperature. Such a plot is usual for the
reaction rate of a simple chemical reaction.
In other cases, relationships may be more
complex.

27.4.5 The shipping properties of
tubes must also be controlled carefully.
Loosely packed indicating gels may shift,
causing an error in the zero point of scales
printed directly upon the tube, as well as an
error in total stain length. When the size
analysis includes an appreciable range, the
fines may segregate to one side of the bore,
causing different flow resistances and rates
on each side of the tube. This may cause
oval stain fronts which are not perpendicular to the tube bore. If the indicating gel is
friable, the size analysis may change during
shipping.

27.4.6 Obviously, satisfactory results
can be obtained only if the manufacturers
take great pains in the design, production,
and calibration of tubes.

27.5 THEORY OF CALIBRATION SCALES.

27.5.1 Up to now, calibration scales
have been entirely empirical. The variables
that can affect the length of stain are: concentration of test gas, volume of air sample, sampling flow rate, temperature, and
pressure, as well as a number of factors
related to tube construction. There is a
striking similarity in the fact that most of
the calibration scales are logarithmic with
respect to concentration in spite of the
widely differing chemicals employed in different tube types. Although very few data
are available for these relationships, a basic
mathematical analysis was made by Saltzman (58). The theoretical formulae discussed below will, of course, have to be
modified as more data become available.
The relationships were also studied by
Grosskopf (48) and Leichnitz (59).

27.5.2 In the usual case, although the
test gas is sorbed completely, equilibrium is
not reached between the gas and the absorbing indicator gel because the sampling
period is relatively short and the flow rate
relatively high. The length of stain is determined by the kinetic rate at which the gas
either reacts with the indicating chemical or
is adsorbed on the silica gel. The theoretical
analysis shows that the stain length is pro-

portional to the logarithm of the product of gas concentration and sample volume:

$$L/H = \ln (CV) + \ln (K/H) \quad (1)$$

Where

L = the stain length, cm
C = the gas concentration, ppm
V = the air sample volume, cm^3
K = a constant for a given type of indicator tube and test gas
H = a mass transfer proportionality factor known as the height of a mass transfer unit, cm.

The factor H varies with the sampling flow rate raised to an exponent of between 0.5 and 1.0, depending upon the nature of the process which limits the kinetic rate of sorption. This process may be diffusion of the test gas through a stagnant gas film surrounding the gel particles, the rate of surface chemical reaction, or diffusion in the solid gel particles. If the indicator tube follows this mathematical model, a plot of stain length, L, on a linear scale, versus the logarithm of product CV (for a fixed constant flow rate) will be a straight line of slope H. The equation indicates the importance of controlling flow rate, as it may affect stain lengths more than gas concentration.

27.5.3 If larger samples are taken at low concentrations and the value of L/H exceeds 4, the gel approaches equilibrium saturation at the inlet end, and calibration relationships are modified. The solution to the equations for this case has been presented by Saltzman **(58)** graphically in a generalized chart. However, there is little advantage to be gained in greatly increasing the sample size, since the stain front is greatly broadened and various errors are increased.

27.5.4 For some types of tubes such as hydrogen sulfide and ammonia the reaction rate is fast enough so that equilibrium can be attained between the indicating gel and the test gas. Under these conditions there is a stoichiometric relationship between the volume of discolored indicating gel and the quantity of test gas absorbed. In the simplest case the stain length is propor-

tional to the product of concentration and volume sampled:

$$L = K' CV \quad (2)$$

If adsorption is important, the exponent of concentration may differ from unity:

$$L = K'' C^{(1-n)} V \quad (3)$$

The value of n is the same as that in the Freundlich isotherm equation for equilibrium adsorption, which states that the mass of gas adsorbed per unit mass of gel is proportional to the gas concentration raised to the power n. If the value of n is unity, which is not unusual, equation 4 indicates that stain length is proportional to sample volume, but is independent of concentration. The physical meaning of this is that all concentrations of gas are adsorbed completely by a fixed depth of gel. Such a tube is obviously of no practical value.

27.5.5 Equilibrium conditions may be assumed for a given type of indicator tube if stain lengths are directly proportional to the volume of air sampled (at a fixed concentration), and are not affected by air sampling flow rate. A log-log plot then may be made of stain length versus concentration for a fixed sample volume. A straight line with a slope of unity indicates equation 2 applies; if another value of slope is obtained, equation 3 applies.

27.5.6 In some of the narrower indicator tubes, manufacturing variation in tube diameters produces an appreciable percentage variation in tube cross-sectional areas. This results in an error in the calibration as high as 50%, because the volume of sample per unit cross-sectional area is different from that under standard test conditions. An additional complicating factor is the variation produced in flow rate per unit cross-sectional area. If an exactly equal quantity of indicating gel is put into each tube, variations in cross-sectional areas will be indicated by corresponding variations in the filled tube lengths. Correction charts are provided by one manufacturer on which the tube is positioned according to the filled length and a scale is given for reading stain lengths. Although the corrections are rather complex, practically linear corrections are very close approximations that can reduce the errors to 10%. In some tubes, the tube diameters are controlled closely enough so that no correction is necessary.

27.5.7 Temperature is another important variable for tube calibrations. The effect is different for different tubes. Since the color-tint type of tube depends upon the degree of reaction, it is most sensitive to temperature. For example, some types of carbon monoxide tubes require correction by a factor of two for each deviation of 10°C from the standard calibration conditions.

27.5.8 Errors in judging stain lengths produce equal percentage errors in concentration derived from the calibration scale. Errors in measuring sample volume and in flow rate may also result in errors in the final value, although the exact relationships might vary according to the tubes.

27.5.9 Many other complications can be expected in calibration relationships. Thus, for nitrogen dioxide the proportion of side reactions is changed at different flow rates. Changing sample volumes freely from calibration conditions is not recommended unless the tube is known to be thoroughly free from the effects of interfering gases and humidity in the air.

27.5.10 A crucial factor in the accuracy of the calibration is the apparatus used for preparing known low concentrations of the test gas. This subject is also discussed in the Sections 2 and 3 of this manual. Some manufacturers have used static methods. However, in our experience, losses of 50% or more by adsorption are not uncommon. Low concentrations of reactive gases and vapors are best prepared in a dynamic system. This has further advantages of compactness and ability to rapidly change concentrations as required. With either type of apparatus it is highly desirable to check the concentrations using chemical methods of known adequacy. Some successful systems have been described (60–66).

27.5.11 A simple and completely dynamic apparatus for accurately diluting tank gas (which may be either pure or a mixture) was developed by Saltzman (61,62) and Avera (63). The asbestos plug flowmeter measures and controls gas flows in the range of a few hundredths to a few milliliters per minute. Air vapor mixtures of volatile organic liquids may be prepared in a flow dilution apparatus using a motor-driven hypodermic syringe. High quality gears, bearings, and screws are needed in the motor drive to provide the uniform slow motion. Some commercial devices have been found unsatisfactory in this regard. Many types of permeation tubes now available also have proven useful.

27.5.12 It is highly desirable for the user as well as the manufacturer to have facilities available for checking calibrations. Only in this manner may the user be confident that the tubes and his technique are adequate for his purposes. Tubes also may be applied to gases other than those for which they have been calibrated by the manufacturer, in certain special cases, if the user can prepare his own calibration.

27.6 STAIN LENGTH PASSIVE DOSIMETERS.

27.6.1 An important new advance has been the development of direct reading passive dosimeters. Passive dosimetry utilizes diffusion of the test gas, and eliminates the need for a sampling pump and its calibration. These attractive devices are compact, convenient, and relatively inexpensive. In early work detector tubes for toluene, ethanol, and 2-propanol were cut open at the entrance of the chemical packing (67). Later glass adapters with a membrane (e.g., Millipore, or silicone rubber) were used (68–70) to provide a draft shield, in some cases a pretreatment chemical layer, and a diffusion resistance. Simpler commercial devices merely provided a score mark which permitted breaking the tube at a controlled point (71). Some allowed a controlled air space (e.g., 15 mm) upstream from the indicating gel to serve as the initial resistance to diffusion (72). In some devices, rather than an indicating gel, a strip of chemically impregnated paper is inserted in the glass tube.

27.6.2 The theoretical calibration relationships for these devices rest upon Fick's First Law of Diffusion, which can be expressed as:

$$W = 10^{-6} \, C \, t \, D \, A/X \qquad (4)$$

Where:

W = volume of test gas collected, cm^3

t = time, s

D = diffusion coefficient, cm^2/s

A = effective orifice cross section area, cm^2
X = orifice length, cm.

This equation assumes that the concentration is completely absorbed in the indicating gel and that there is no significant back pressure. A second common assumption is that the stain length is proportional to the amount absorbed (analogous to equation 2):

$$L = k W \qquad (5)$$

where k = a constant for a given test gas and tube.

27.6.3 The test gas diffuses through a membrane or air space, then through the stained length of indicating gel, and is finally absorbed at the stain front, which is assumed to be relatively narrow. It is convenient to express X in terms of L:

$$X = r + L \qquad (6)$$

where r = effective length corresponding to the diffusive resistance of the membrane or air space.

Combining equations 4–6 and rearranging yields:

$$r L + L^2 =$$
$$(10^{-6} k D A) C t = k' C t \qquad (7)$$

where k' = a constant equal to the bracketed expression

This equation has been shown to fit MSA tubes with a 15 mm air space (72,73). When L was expressed in mm and t in h, r was taken as 15, and k' was for CO 0.59, NH_3 11.0, NO_2 14.2, H_2S 22.6, SO_2 67.3, and CO_2 74.0. For Draeger tubes, which do not utilize an air space, the equation applied with a zero value for r (71). For membrane type devices the equation was modified by adding another constant (68–70):

$$C t = a + b L + c L^2 \qquad (8)$$

where a,b,c, = empirical constants

These constants may differ for each individual membrane. The inapplicability of a general calibration is a disadvantage of this type.

27.6.4 A more complete mathematical analysis (74) showed that for rapidly changing concentrations the errors would be small. This was experimentally confirmed (75) for both passive dosimeters and for long term tubes. Most of the published work on passive dosimeters has been by staff of manufacturers. Much larger errors were reported (76,77) by users. Some of the stain boundaries were very diffuse and difficult to read, and some calibrations inaccurate. Since these tubes are in an early state of development, the values should be checked as much as possible.

27.6.5 Another type of passive dosimeter is the direct reading colorimetric badge. These provide a color tint that is related to the product of time and concentration. All passive devices require a minimum air velocity at their entrance (0.008 m/s or 15 ft/min) to avoid "starvation" effects (depletion of the air concentration near the entrance).

27.7 PERFORMANCE EVALUATION AND CERTIFICATION.

27.7.1 Evaluations by users of some types of tubes have been reported (30,78–95). Temperature and humidity were found to be significant factors in some cases (96,97). Accuracy has been found highly variable. In some cases, the tubes were completely satisfactory; in others, completely unsatisfactory. Manufacturers, in their efforts to improve the range and sensitivity of their products, are rapidly changing the contents of their tubes, and these reports are frequently obsolete before they appear in print. Improved quality control, and perhaps greater self-policing of the industry, would greatly increase the value of the tubes, especially for the small consumer who is not in a position to check calibrations.

27.7.2 After reviewing this need, a joint ACGIH-AIHA Committee made the following recommendations (98):

1. Manufacturers should supply a calibration chart (ppm) for each batch of tubes.

2. Length-of-stain tubes are preferable to those exhibiting change in hue or intensity of color.

3. Tests of calibrations should be made at 0.5, 1, 2, and 5 times the Threshold Limit Value (ACGIH).

4. The manufacturer should specify the

methods of tests. Values should be checked by two independent methods.

5. Calibration at each test point should be accurate within ± 25% (95% confidence limit).

6. Allowable ranges and corrections should be listed for temperature, pressure, and relative humidity.

7. Each batch of tubes should be labelled with a number and an expiration date. Instructions for proper storage should be given.

8. Tolerable concentrations of interferents should be listed.

9. Pumping volumes should be accurate within ± 5%, and flow rates should be indicated.

10. Special calibrations should be provided for extended sampling for low concentrations, and flow rates should be specified.

27.7.3 A performance evaluation program was initiated by the National Institute for Occupational Safety and Health (NIOSH). Known concentrations of test substances were generated in flow systems, from sources such as cylinder mixtures, vapor pressure equilibration at known temperatures, or permeation tubes. Although few tubes achieved an accuracy of ± 25%, many types showed accuracies in the range ± 25 to 35%.

27.7.4 A formal certification program **(99,100)** was the next step. In addition to passing performance evaluation tests at the Morgantown, WV laboratory of NIOSH, manufacturers were required to provide information on the contents of the tubes, and to conduct a specified quality control program. Because of the dependence of the calibrations on the pumps used with the tubes, certifications were periodically updated and issued **(101)** for specified combinations of tubes and pumps. By 1981 tubes of four manufacturers for 23 contaminants had been certified. Unfortunately, the program was terminated in 1983 for lack of funding **(102)**.

27.7.5 The requirements for certification generally followed the recommendations of the joint committee. However, the accuracy requirement was modified to ± 35% at 0.5 TLV, and ± 25% at 1.0, 2, and 5 times TLV, to be maintained until the expiration date if the tubes were stored ac-

cording to the manufacturer's instructions. At the TLV concentration, either the stain length had to be 15 mm or greater, or the relative standard deviation of the readings of the same tube by three or more independent tube readers had to be less than 10%. If the stain front was not exactly perpendicular to the tube axis (because of channeling of the air flow), the difference between the longest and shortest stain length measurements to the front had to be less than 20% of the mean length. Color intensity tubes had to have sufficient charts and sampling volume combinations to provide scale values including at least the following multiples of the TLV: 0.5, 0.75, 1.0, 1.5, 2.0, 2.5, 3.0, 4.0, and 5.0; the relative standard deviation for readings of a tube by independent readers had to be < 10%. Tests were to be conducted generally at 65–85°F (18.3–29.5°C), and at relative humidities of 50%, unless the humidity had to be reduced to avoid disturbing the test system. The manufacturer had to file a quality control plan and keep records of his inspections of raw materials, finished tubes, and calibration and test equipment. Acceptable statistical quality levels for defects in finished tubes were as follows: critical 0% where tests were non-destructive, otherwise 1.0%, major 2.5%, minor 4.0%, and accuracy 6.5%. Typical statistical calculations have been described **(103)**. Certification seals were affixed to approved devices. NIOSH reserved the right to withdraw certification for cause.

27.7.6 Since important legal and economic consequences depend upon the accuracy of measurements of contaminant concentrations, enforcement agencies will most likely prefer certified equipment. Standards for detector tubes have been issued by 25 organizations **(104)**, including OSHA **(105)**, IUPAC **(106)**, The Council of Europe, Great Britain, France, Soviet Union, and a variety of private organizations in the U.S. and Europe. Requirements are mostly similar to those cited above. In 1986 the Safety Equipment Institute announced a voluntary program for third-party certification of detector tubes. Manufacturers submit tubes for testing as the schedule for each type is announced. Two AIHA-accredited laboratories were selected to evaluate the tubes according to the NIOSH protocol **(99)**. An-

Table 27:III Certifications of Detector Tubes by Safety Equipment Institute as of May 1987*

Substance	Manufacturer	Tube Type
Ammonia	Matheson/Kitagawa	8014-105 Sc
	Mine Safety Appliances Co.	460103
	National Draeger, Inc.	5/a CH20501
	Sensidyne/Gastec	3La
Carbon dioxide	Matheson/Kitagawa	8014-126Sa
	Mine Safety Appliances Co.	85976
	National Draeger, Inc.	0.1% CH 23501
	Sensidyne/Gastec	2L
Carbon monoxide	Mine Safety Appliances Co.	465519
		91229
	National Draeger, Inc.	5/c CH 25601
		10/b CH 20601
	Sensidyne/Gastec	1La
Chlorine	Matheson/Kitagawa	8014–10956
	Mine Safety Appliances Co.	460225
	National Draeger, Inc.	0.3/b 6728411
	Sensidyne/Gastec	8La
Hydrogen cyanide	Matheson/Kitagawa	8014-112Sb
	Mine Safety Appliances Co.	93262
	National Draeger, Inc.	2/a (CH 25701)
	Sensidyne/Gastec	12L
Hydrogen sulfide	Mine Safety Appliances Co.	460058
	National Draeger, Inc.	2/a 67 28821
		1/c 67 19001
	Sensidyne/Gastec	4LL
Sulfur dioxide	Mine Safety Appliances Co.	92623
	National Draeger, Inc.	0.5/a 67 28491
	Sensidyne/Gastec	5Lb

*Tubes are certified only when used with pump model of same manufacturer, as follows: Matheson/Kitagawa—8014-100A; Mine Safety Appliance Co.—Samplair Pump 464080; National Draeger, Inc.—Bellows Pump Model 31, 6726065; Sensidyne/Gastec—Model 800 Pump.

other contractor makes on-site quality assurance audits of manufacturing facilities every six months for three audits, and then annually. If the tubes meet all requirements, the manufacturer may apply the SEI certification mark. This program should provide a stimulus for greater acceptance and use and for further improvements in detector tube technology. Tubes will be retested every three years. Table 27:III gives the current listing of certified tubes (107). Types for eleven more substances are in process of testing.

27.8 CONCLUSIONS.

27.8.1 Use of indicating tubes for analysis of toxic gas and vapor concentrations in air is a very rapid, convenient, and inexpensive technique which can be performed by semi-skilled operators. These tubes are in various stages of development, and highly variable results have been ob-

tained. Accuracy is dependent upon a high degree of skill in the manufacture of the tubes. At present, results may be regarded as only range-finding and approximate in nature. The best accuracy that can be expected from indicator tube systems of the best types is of the order of ±25%. Since many of the tubes are far from specific, an accurate knowledge of the possible interfering gases present is very important. The quantitative effect of these interferences depends upon the volume sampled in an irregular way. In order to avoid dangerously misleading results, the operation and interpretation should be under the supervision of a skilled industrial hygienist.

REFERENCES

1. CAMPBELL, E.E., AND H.E. MILLER. 1961, 1964. Chemical Detectors. A Bibliography for the Industrial Hygienist with Abstracts and Annotations. Los

Alamos Scientific Laboratory. LAMS-2378 (Vol. I, II).

2. Individual Protective and Detection Equipment. 1953. pp. 56–80. Dept. of the Army Technical Manual, TM 3-290; Dept. of the Air Force Technical Order, TO 39C-10C-1 (September).

3. MCFEE, D.R., R.E. LAVINE, R.J. SULLIVAN. 1970. Carbon Monoxide, A Prevalent Hazard Indicated by Detector Tabs. Am. Ind. Hyg. Assoc. J. 31:749–53.

4. LAMB, A.B., W.C. BRAY, AND J.C. FRAZER. 1920. Ind. Eng. Chem. 12:213.

5. HOOVER, C.W. 1921. Ibid. 13: 770.

6. SHEPHERD, M. 1947. Rapid Determination of Small Amounts of Carbon Monoxide, Preliminary Report on the NBS Colorimetric Indicating Gel. Anal. Chem. 19:77–81.

7. LITTLEFIELD, J.B., W.P. YANT, AND L.B. BERGER. 1935. U.S. Bur. Min. Rep. Inv., No. 3276.

8. KITAGAWA, T. 1951. Rapid Analysis of Phosphine and Hydrogen Sulfide in Acetylene. J. Japan Chem. Ind. Soc. No. 33. (Feb.).

9. HUBBARD, B.R., L. SILVERMAN. 1950. Rapid Method for the Determination of Aromatic Hydrocarbons in Air. Arch. Ind. Hyg. and Occ. Med. 2:49–55.

10. GROSSKOPF, K. 1951. Technical Analysis of Gases and Liquids by Means of Chromometric Gas Analysis. Angew. Chem. 63:306–311.

11. KITAGAWA, T. 1952. Rapid Method of Quantitative Gas-Analysis by Means of Detector Tubes. Kagaku no Ruoiki 6:386.

12. SACKS, V. 1956. Carbon Monoxide Detection by Means of the Colorimetric Gas Analyzer (German). Deutsche Zeitschrift für gerichtliche Medizin 45:68–71.

13. KINOSIAN, J.R., B.R. HUBBARD. 1958. Nitrogen Dioxide Indicator. Amer. Ind. Hyg. Assoc. J. 19:453–460.

14. GROSSKOPF, K. 1958. Detector Tubes as Detectors in Gas Chromatography (German). Erdöhl und Kohle. 11:304–306.

15. GROSSKOPF, K. 1959. Vaporous Reagents in the Detector Tube Technique for Measurement of Vapors and Gases (German). Zeitschrift für analytische Chemie. 170:271–277.

16. IDEM. 1959. Systox Detection (German). Chemiker-Zeitung-Chemische Apparatus 83:115–117.

17. HETZEL, K.W. 1959. Poisonous Action and Detection of Injurious Gases and Vapors in Mining Operations (German). Brennstoff-Chemie 41:115–122.

18. KITAGAWA, T. 1960. The Rapid Measurement of Toxic Gases and Vapors. Presented at The International Congress on Occupational Health, New York, July 25–29.

19. BRETZKE, W. 1960. The Determination of Carbon Dioxide Content in the Atmosphere of Silos and Fermenters (German). Die Berufsgenossenschaft (May).

20. KETCHAM, N.H. 1964. Practical Air-Pollution Monitoring Devices. Am. Ind. Hyg. Assoc. J. 23:127–131.

21. SILVERMAN, L. 1962. Panel Discussion of Field Indicators in Industrial Hygiene. Ibid., 23:108–111.

22. SILVERMAN, L. AND G.R. GARDNER: 1965. Potassium Pallado Sulfite Method for Carbon Monoxide Detection. Ibid. 26:97–105.

23. INGRAM, W.T. 1964. Personal Air-Pollution Monitoring Devices. Ibid. 25:298–303.

24. LINCH, A.L., S.S. LORD, JR., K.A. KUBITZ, AND DEBRUNNER, M.R. 1965. Phosgene in Air—Development of Improved Detection Procedures. Ibid. 26:465–473.

25. LINCH, A.L. 1965. Oxygen in Air Analyses—Evaluation of a Length of Stain Detector. Ibid. 26:645.

26. LEICHNITZ, K. 1967. Determination of Arsine in Air in the Work Place (German). Die Berufsgenossenschaft (Sept.).

27. LEICHNITZ, K. 1968. Cross-Sensitivity of Detector Tube Procedures for the Investigation of Air in the Work Place (German). Zentralblatt für Arbeitsmedizin und Arbeitsschutz 18:97–101.

28. LINCH, A.L., R.F. STALZER, D.T. LEFFERTS. 1968. Methyl and Ethyl Mercury Compounds—Recovery from Air and Analysis. Am. Ind. Hyg. Assoc. J. 29:79–86.

29. PEURIFOY, P.V., L.A. WOODS AND G.A. MARTIN. 1968. A Detector Tube for Determination of Aromatics in Gasoline. Anal. Chem. 40:1002–1004.

30. KOLJKOWSKY, P. 1969. Indicator-tube Method for the Determination of Benzene in Air. Analyst 94:918–920.

31. GRUBNER, O., J.J. LYNCH, J.W. CARES, W.A. BURGESS, 1972. Collection of Nitrogen Dioxide by Porous Polymer Beads. Am. Ind. Hyg. Assoc. J. 33:201–206.

32. NEFF, J.E. AND N.H. KETCHAM. 1974. A Detector Tube for Analysis of Methyl Isocyanate in Air or Nitrogen Purge Gas. Ibid. 35:468–75.

33. LEICHNITZ, K. 1985. Detector Tube Handbook, 6th ed. Drägerwerk, AG, P.O. Box 1339, D-24 Lübeck 1, Federal Republic of Germany (May).

34. SENSIDYNE/GASTEC. 1985. Precision Gas Detector System Manual. Sensidyne, Inc., 12345 Starkey Road, Largo, FL 33543.

35. Direct Reading Colorimetric Tubes—A Manual of Recommended Practices, 1977. 1st ed. Am. Ind. Hyg. Assoc., 475 Wolf Ledges Pkway., Akron, OH 44311.

36. GROTE, A.A., W.S. KUN AND R.E. KUPEL. 1978. Establishing a Protocol from Laboratory Studies to be Used in Field Sampling Operations. Am. Ind. Hyg. Assoc. J. 39:880–884.

37. KUSNETZ, H.L. 1960. Air Flow Calibration of Direct Reading Colorimetric Gas Detecting Devices. Ibid. 21:340–341.

38. COLEN, F.H. 1973. A Study of the Interchangeability of Gas Detector Tubes and Pumps. Report TR-71. National Inst. for Occup. Safety and Health, Morgantown, WV 26505 (June 15).

39. COLEN, F.H. 1974. A Study of the Interchangeability of Gas Detector Tubes and Pumps. Am. Ind. Hyg. Assoc. J. 35:686–694.

40. LEICHNITZ, K. 1977. Use of Detector Tubes under Extreme Conditions (Humidity, Pressure, Temperature). Ibid. 38:707–711.

41. CARLSON, D.H., M.D. OSBORNE AND J.H. JOHNSON. 1982. The Development and Application to Detector Tubes of a Laboratory Method to Assess Accuracy of Occupational Diesel Pollutant Concentration Measurements. Ibid. 43:275–285.

42. DOUGLAS, K.E. AND H.J. BEAULIEU. 1983. Field Validation Study of Nitrogen Dioxide Passive Sam-

plers in a "Diesel" Haulage Underground Mine. Ibid. 44:774–778.

43. LEICHNITZ, K. 1973. Effect of Pressure and Temperature on the Indication of Draeger Tubes. Dräger Review 31:1–7. Drägerwerk AG, P.O. 1339, D-24 Lübeck 1, Federal Republic of Germany (Sept.).

44. LINCH, A.L. 1974. Evaluation of Ambient Air Quality by Personnel Monitoring. CRC Press, Inc., Cleveland, Ohio.

45. LEICHNITZ, K. 1980. Qualitative Detection of Substances by Means of Draeger Detector Tube Ethyl Acetate 200 A. Dräger Review 46:13–21. Drägerwerk AG, P.O. 1339, D-24 Lübeck 1, FRG (December).

46. KITAGAWA, T. 1965. Detector Tube Method for Rapid Determination of Minute Amounts of Nitrogen Dioxide in the Atmosphere. Yokohama National Univ. (July).

47. INFORMATION SHEET NO. 44: 0.5a Nitrous Gas/Detector Tube. 1960. Drägerwerk AG, P.O. Box 1339, D-24 Lübeck 1, FRG (November).

48. GROSSKOPF, K. 1963. A Tentative Systematic Description of Detector Tube Reactions (German). Chemiker Zeitung-Chemische Apparatus 87:270–275.

49. LEICHNITZ, K. 1973. Determination of Low SO_2 Concentrations by Means of Detector Tubes. Dräger Review 30:1–4. Drägerwerk AG, P.O. Box 1339, D-24 Lübeck 1, FRG (May).

50. LINCH, A.L., H.V. PFAFF. 1971. Carbon Monoxide – Evaluation of Exposure Potential by Personnel Monitor Surveys. Am. Ind. Hyg. Assoc. J. 32:745–752.

51. LEICHNITZ, K. 1973. The Detector Tube Method and its Development Tendencies (German). Chemiker-Zeitung 97:638–645.

52. LEICHNITZ, K. 1977. An Analysis by Means of Long-Term Detector Tubes. Dräger Review 40:9–17. Drägerwerk AG, P.O. Box 1339, D-24 Lübeck 1, FRG (Dec.).

53. LEICHNITZ, K. 1979. Some Information on the Long-Term Measuring System for Gases and Vapors. Ibid. 43:6–13 (June).

54. HEUBENER, D.J. 1980. Evaluation of a Carbon Monoxide Dosimeter. Am. Ind. Hyg. Assoc. J. 41:590–591.

55. DHARMARAJAN, V. AND R.J. RANDO. 1979. Clarification – re: A Recommendation for Modifying the Standard Analytical Method for Determination of Chlorine in Air. Ibid. 40:746.

56. Criteria for a Recommended Standard – Occupational Exposure to Carbon Monoxide. 1972. Pub. No. HSM 73-11000. Nat. Inst. Occup. Safety and Health, Dept. Health, Ed. and Welfare, Rockville, MD.

57. LEIDEL, N.A., K.A. BUSCH. 1975. Statistical Methods for Determination of Noncompliance with Occupational Health Standards. DHEW (NIOSH) Publ. No. 75-159. National Institute for Occupational Safety and Health, Cincinnati, OH. (April).

58. SALTZMAN, B.E. 1962. Basic Theory of Gas Indicator Tube Calibrations. Am. Ind. Hyg. Assoc. J. 23:112–126.

59. LEICHNITZ, K. 1967. Attempt at Explanation of Calibration Curves of Detector Tubes (German). Chemiker Ztg.-Chem. Apparatus 91:141–148.

60. SCHERBERGER, R.F., G.P. HAPP, F.A. MILLER AND D.W. FASSETT. 1958. A Dynamic Apparatus for Preparing Air-Vapor Mixtures of Known Concentrations. Am. Ind. Hyg. Assoc. J. 19:494–498.

61. SALTZMAN, B.E. 1961. Preparation and Analysis of Calibrated Low Concentrations of Sixteen Toxic Gases. Anal. Chem. 33:1100–1112.

62. SALTZMAN, B.E. 1973. The Industrial Environment – Its Evaluation and Control, Chapter 12, Preparation of Known Concentrations of Air Contaminants, pp. 123–137. Nat. Inst. of Occup. Safety and Health, Contract HSM-99-71-45, Cincinnati, OH.

63. AVERA, C.B., JR. 1961. Simple Flow Regulator for Extremely Low Gas Flows. Rev. Sci. Instru. 32:985–986.

64. COTABISH, H.N., P.W. MCCONNAUGHEY, AND H.C. MESSER, 1961. Making Known Concentrations for Instrument Calibration. Am. Ind. Hyg. Assoc. J. 22:392–402.

65. HERSCH, P.A. 1969. Controlled Addition of Experimental Pollutants to Air. J. Air Poll. Control Assoc. 19:164–172.

66. HUGHES, E.E. ET AL. 1973. Gas Generation Systems for the Evaluation of Gas Detecting Devices. NBSIR 73-292. National Bureau of Standards, Washington, DC (Oct.).

67. HILL, R.H. AND D.A. FRASER. 1980. Passive Dosimetry Using Detector Tubes. Am. Ind. Hyg. Assoc. J. 41:721–729.

68. SEFTON, M.V., A.V. KOSTAS AND C. LOMBARDI. 1982. Stain Length Passive Dosimeters. Ibid. 43:820–824.

69. GONZALEZ, L.A. AND M.V. SEFTON. 1983. Stain Length Passive Dosimeter for Monitoring Carbon Monoxide. Ibid. 44:514–520.

70. GONZALEZ, L.A. AND M.V. SEFTON. 1983. Laboratory Evaluation of Stain Length Passive Dosimeters for Monitoring of Vinyl Chloride and Ethylene Oxide. Ibid. 46:591–598.

71. PANNWITZ, K.-H. 1984. Direct-Reading Diffusion Tubes. Dräger Review 53:10–16. Drägerwerk AG, P.O. Box 1339, D-24 Lübeck 1, FRG (June).

72. McKEE, E.S. AND P.W. MCCONNAUGHEY. 1985. A Passive, Direct Reading, Length of Stain Dosimeter for Ammonia. Am. Ind. Hyg. Assoc. J. 46:407–410.

73. MCCONNAUGHEY, P.W., E.S. MCKEE AND I.M. PRETTS. 1985. Passive Colorimetric Dosimeter Tubes for Ammonia, Carbon Monoxide, Carbon Dioxide, Hydrogen Sulfide, Nitrogen Dioxide, and Sulfur Dioxide. Ibid. 46:357–362.

74. BARTLEY, D.L. 1986. Diffusive Samplers Using Longitudinal Sorbent Strips. Ibid. 47:571–577.

75. PANNWITZ, K.-H. 1986. The Direct-Reading Diffusion Tubes on the Test Bench. Dräger Review 57:2–12. Drägerwerk AG, P.O. Box 1339, D-24 Lübeck 1, FRG (June).

76. CASSINELLI, M.E., R.D. HULL AND P.A. CUENDET. 1985. Performance of Sulfur Dioxide Passive Monitors. Am. Ind. Hyg. Assoc. J. 46:599–608.

77. HOSSAIN, M.A. AND B.E. SALTZMAN. 1987. Laboratory Evaluation of Passive Colorimetric Dosimeter Tubes for Carbon Monoxide. Paper 239, American Industrial Hygiene Conference, Montreal, Canada (June 3).

78. DITTMAR, P., AND G. STRESE. 1959. The Suitabil-

ity of Detector Tubes for the Detection of Toxic Substances in the Air. I. Hydrogen Sulfide Detector Tubes (German). Arbeitsschutz 8:173–177.

79. HESELTINE, J.K. 1959. The Detection and Estimation of Low Concentrations of Methyl Bromide in Air. Pest Technology (England) (July/August).

80. KUSNETZ, J.L., B.E. SALTZMAN, AND M.E. LA-NIER., 1960. Calibration and Evaluation of Gas Detecting Tubes. Am. Ind. Hyg. Assoc. J. 21:361–373.

81. BANKS, O.M., AND K.R. NELSON. 1961. Evaluation of Commercial Detector Tubes. Presented at Am. Ind. Hyg. Conference, Detroit, MI (April 13).

82. LANIER, M.B. AND H.L. KUSNETZ. 1963. Practices in the Field Use of Detector Tubes. Arch. Env. Health 6:418–421.

83. HAY III, E.B. 1964. Exposure to Aromatic Hydrocarbons in a Coke Oven By-Product Plant. Am. Ind. Hyg. Assoc. J. 25:386–391.

84. LARSEN, L.B. AND R.H. HENDRICKS, 1969. An Evaluation of Certain Direct Reading Devices for the Determination of Ozone. Ibid. 30:620–623.

85. MORGANSTERN, A.S., R.A. ASH AND J.R. LYNCH. 1970. The Evaluation of Gas Detector Tube Systems. I: Carbon Monoxide. Ibid. 31:630–632.

86. ASH, R.M. AND J.R. LYNCH. 1971. The Evaluation of Gas Detector Tube Systems: Benzene. Ibid. 32:410–411.

87. ASH, R.M. AND J.R. LYNCH. 1971, 1972. The Evaluation of Detector Tube Systems: Sulfur Dioxide. Ibid. 32:490–491; Ibid. 33:11.

88. ASH, R.M. AND J.R. LYNCH. 1971. The Evaluation of Detector Tube Systems: Carbon Tetrachloride. Ibid. 32:552–553.

89. ROPER, C.P. 1971. An Evaluation of Perchloroethylene Detector Tube. Ibid. 32:847–849.

90. JOHNSTON, B.A. AND C.P. ROPER. 1972. The Evaluation of Gas Detector Tube Systems: Chlorine. Ibid. 33:533–534.

91. JOHNSTON, B.A. 1972. The Evaluation of Gas Detector Tube Systems: Hydrogen Sulfide. Ibid. 33:811–812.

92. JENTZSCH, D. AND D.A. FRASER. 1981. A Laboratory Evaluation of Long-Term Detector Tubes: Benzene, Toluene, Trichloroethylene. Ibid. 42:810–823.

93. SEPTON, J.C. AND T. WILCZEK, JR. 1986. Evaluation of Hydrogen Sulfide Detector Tubes. App. Ind. Hyg. 1:196–198.

94. LEICHNITZ, K. 1981. Survey of Draeger Long-Term Tubes with Special Consideration of the Long-Term Tubes Sulfur Dioxide 5/a-L. Dräger Review 48:16–18. Drägerwerk AG, P.O. Box 1339, D-24 Lübeck 1, FRG (November).

95. LEICHNITZ, K. 1984. Draeger Long-Term Tubes Meet IUPAC Standard. Ibid. 52:11–15 (January).

96. STOCK. T.H. 1986. The Use of Detector Tube Humidity Limits. Am. Ind. Hyg. Assoc. J. 47:241–244.

97. McCAMMON, C.S., JR., W.E. CROUSE AND H.B. CARROL, JR. 1982. The Effect of Extreme Humidity and Temperature on Gas Detector Tube Performance. Ibid. 43:18–25.

98. JOINT COMM. ON DIRECT READING GAS DETECTING SYSTEMS. ACGIH-AIHA. 1971. Direct Reading Gas Detecting Tube Systems. Ibid. 32:488–489.

99. NATIONAL INSTITUTE FOR OCCUPATIONAL SAFETY AND HEALTH. 1983. Certification of Gas Detector Tube Units. Federal Register 38:11458–11463 (May 8); also 43 CFR pt. 84.

100. ROPER, C.P. 1974. The NIOSH Detector Tube Certification Program. Am. Ind. Hyg. Assoc. J. 35:438–42.

101. NIOSH Certified Equipment List as of October 1, 1981. DHHS (NIOSH) Pub. No. 82-106. Nat. Inst. for Occup. Safety and Health, Cincinnati, OH (Oct), periodically updated and reissued).

102. CENTER FOR DISEASE CONTROL–NATIONAL INSTITUTE FOR OCCUPATIONAL SAFETY AND HEALTH. 1983. NIOSH Voluntary Testing and Certification Program. Fed. Reg. 48(191):44931-44932 (September 30).

103. LEICHNITZ, K. 1979. How Reliable are Detector Tubes? Dräger Review 43:21–26. Drägerwerk AG, P.O. Box 1339, D-24 Lübeck 1, FRG (June).

104. LEICHNITZ, K. 1982. Comments of Official Organizations Regarding Suitability of Detector Tubes. Ibid. 49:19–22 (May).

105. U.S. DEPARTMENT OF LABOR. 1973. Directive 73-4. Use of Detector Tubes. Washington, DC (March).

106. LEICHNITZ, K. 1983. IUPAC Performance Standard for Detector Tubes. Dräger Review 51:12–15. Drägerwerk AG, P.O. Box 1339, D-24 Lübeck 1, FRG (April).

107. SAFETY EQUIPMENT INSTITUTE. 1987. Certified Products List, May. Safety Equipment Institute, 1901 N. Moore Street, Arlington, VA 22209.

Adapted from "Air Sampling Instruments," 7th edition, by permission of American Council of Governmental Industrial Hygienists.

Subcommittee 9
B.E. SALTZMAN
J.N. HARMAN, Chairman

28. Fluorescence Spectrophotometry

28.1 BASIC PRINCIPLES. Fluorescence spectrophotometry is the measurement of "fluorescent" light emitted by certain molecules when excited by a radiation source of appropriate energy or wavelength. Since energy is lost in the transition, the fluorescent or secondary light is of lower energy, and consequently longer wavelength, than the exciting or primary light. For this reason, and because intense light sources are available in this region, the ultraviolet and lower visible wavelengths are most useful as a source of excitation. The exciting and fluorescing wavelengths for a given compound are characteristic and permit identification in many cases, particularly when the fluorimeter is of the scanning type.

Organic molecules containing conjugated double bonds (alternating single and double bonds) are the most commonly encountered fluorescing materials. Substitutions in the molecule may alter the excitation or emission spectrum or substantially enhance or reduce fluorescence. Although solids may be fluorescent, quantitative measurements are usually made in solution. Emission intensity is dependent on the total number of excited molecules, and is thus theoretically directly proportional to concentration. This holds true at very low concentrations; however, as concentration increases, absorption by the sample of both primary and secondary light becomes significant, finally resulting in a phenomenon called "concentration quenching." The fluorescence may be visually intense in the first portion irradiated by the primary light but it is absorbed by the sample to such an extent that the light reaching the detector is reduced. The linear plot of concentration *vs.* instrumental emission intensity may thus reach a plateau and return towards the base line. For this reason, it is good practice to observe samples visually, and dilute those that appear to be self-quenching. Alternatively each sample may be routinely analyzed at two different dilutions to check for self-quenching.

The method has been the subject of numerous monographs and reviews (1–7).

28.2 INSTRUMENTATION. The basic fluorimeter (Figure 28:1) consists of a primary light source which is directed through an optical filter or monochromator to isolate a specific wavelength for excitation of the sample. The secondary filter or monochromator allows only the wavelength of light due to the fluorescence from the sample to reach the phototube, while blocking out any stray primary light or fluorescence of a different wavelength due to other substances in the sample. Detectors are usually photomultiplier tubes placed at 90° to the primary light source to reduce the possibility of interference from primary light. Variable slits may be employed to control resolving power or sensitivity.

Instruments using only filters for light transmission are referred to as filter fluorimeters while those using monochromators are called spectrofluorimeters. The spectrofluorimeters, which are much higher priced than the filter fluorimeters, offer the advantage of more precise control of wavelength of both primary and secondary light.

This permits scanning of samples both for identification and for more precise determination of exciting and fluorescing wavelengths. The value of the more reasonably priced filter fluorimeters should not be overlooked, as they offer excellent sensitivity and reproducibility for routine quantitative work.

Monochromators for spectrofluorimeters are usually diffraction gratings. Quartz prisms may be used for greater dispersion of primary light to obtain better resolution, but at the expense of some loss of intensity.

Light sources for filter type instruments are commonly of the mercury vapor type, which emit line spectra. The most prominent useful lines are 254, 312, 334, 365, 405, and 436 nm. The shorter wavelengths are obtainable only when the mercury lamp is encased in clear quartz or synthetic silica, since glass absorbs over 40% of radiation at 320 nm, and virtually 100% at 254 nm. For the longer wavelengths, glass envelopes may be used, frequently coated or colored to filter out undesirable wavelengths. Mercury lamps require a relatively long warm-up time for stabilization. Their intensity decreases with age, so that continued reference to standards is of great importance. The xenon arc lamp is usually used in spectrofluorimeters, because it is more stable and gives an intense continuous spectrum over the ultraviolet and visible regions. The intensity varies with wavelength, and tends to diminish at short wavelengths; however, this may be corrected for instrumentally.

Where the excitation wavelength is above 320 nm, ordinary borosilicate glass tubes may be used as sample containers. At lower excitation wavelengths it is necessary to use synthetic silica cells (usually square) in order to transmit the primary light.

Under optimum conditions, fluorimetric methods are among the most sensitive and specific available. Detection limits of less than 1 ng/mL are common.

28.3 APPLICATION. Fluorimetry finds most use in detection or determination of organic compounds, although inorganic compounds may be determined by formation of a fluorescent organic complex.

In the field of environmental analysis, fluorimetric methods have been used for the determination of coproporphyrin in

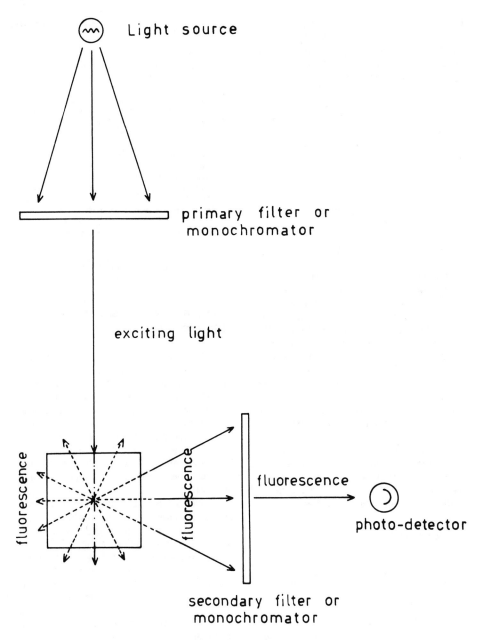

Figure 28:1 – Basic fluoremeter.

urine as an index of lead exposure; the determination of uranium in air, soil, or biological materials by fluorescence of a bead obtained by fluxing the sample with sodium-lithium fluoride; and for the identification and determination of polynuclear aromatic hydrocarbons. Due to the extreme sensitivity of detection of the fluorescent dyes rhodamine B and fluorescein, they have been used as tracers in water systems.

Uranine has been used as a tracer in air, specifically to determine the efficiency of filters for aerosols. Beryllium and selenium have been determined with great sensitivity by fluorimetric methods.

28.4 FACTORS AFFECTING RESULTS.
Intensity of primary light
Band width of primary and secondary light
Primary and secondary wavelengths used
Cell path
Efficiency of collection optics
pH of sample
Spectral sensitivity of phototube
Sample solvent
Self-absorption
Temperature of sample
Quenching

28.5 ANALYTICAL APPLICATIONS. Fluorimetry is used frequently in biochemistry, as many vitamins, hormones, etc. are fluorescent. It is extremely sensitive for determination of beryllium, selenium, boron, etc. It is widely used for determination of uranium after fluxing with solid NaF-LiF and is useful in filter efficiency studies with uranine aerosol. It is applied to tracer studies: rhodamine-B, fluorescein, in water; uranine in air. It is used for determination of polycyclic aromatic hydrocarbons in air pollution (8–17).

28.6 ADVANTAGES AND LIMITATIONS. Fluorimetry is one of the most sensitive methods of analysis. It is, however, relatively non-specific, and fluorescence of many compounds fades rapidly.

REFERENCES

1. RADLEY, J.A. AND J. GRANT. 1959. Fluorescence Analysis in Ultraviolet Light, Chapman and Hall, London, 4th Ed.
2. UDENFRIEND, S. 1962, 1969. Fluorescence Spectra in Biology and Medicine. Academic Press, New York, Vol. I, Vol. II.
3. PARKER, C.A., AND W.T. REES. 1962. Fluorescence Spectrometry, Analyst, 87:83.
4. HERCULES, D.M. (Ed.) 1960. Fluorescence and Phosphorescence Analysis, Interscience, N.Y.
5. HOUGHTON, J.A., AND G. LEE. 1960. Data on Ultraviolet Absorption and Fluorescence Emission. AIHAJ, 21:219.
6. BERLMAN, I.B. 1971. Handbook of Fluorescence Spectra of Aromatic Molecules, Academic Press, New York, 2nd Ed.
7. KALLMAN, H. AND G.M. SPRUCK. 1962. Luminescence of Organic and Inorganic Materials, Wiley, New York.
8. WELFORD, G.A., AND J.H. HARLEY. 1952. AIHAQ, 13:232.
9. WALKLEY, J. 1959. AIHAJ. 20:241.
10. SILL, C.W., AND C.P. WILLIS. 1959. Anal. Chem., 31:598.
11. WHITE, C.E., ET AL. 1947. Anal. Chem., 19:802.
12. WATKINSON, J.H. 1960. Anal. Chem., 32:981.
13. CENTANNI, F.A., A.M. ROSS, AND M.A. DESESA. 1956. Anal. Chem., 28:1651.
14. DUBOIS, L. AND J.L. MONKMAN. 1965. Int. J. Air and Water Pollution, 9:131.
15. SAWICKI, E., T.R. HAUSER, AND T.W. STANLEY. 1960. Int. J. Air Poll., 2:253.
16. CHAUDET, J.J. AND W.I. KAYE., 1961. Anal. Chem., 33:113.
17. See this volume, Part 2, Methods 102-A-D.

Approved with modifications from 2nd edition

Subcommittee 9
J.N. HARMAN, Chairman

29. Sampling Aerosols by Filtration

29.1 INTRODUCTION. Filtration is the most widely used technique for aerosol sampling, primarily because of its low cost and simplicity. The samples obtained usually occupy a relatively small volume, and may often be stored for subsequent analysis without deterioration. By appropriate choice of air mover, filter medium and filter size, almost any sample quantity desired can be collected in a given sampling interval.

Figure 29:1 is a schematic representation of the elements of a filter sampling system. It shows the arrangement of the component parts. These may include either all or some of the following: a sampling nozzle, filter holder, filter, flowmeter, air mover, and a means of regulating the flow. A nozzle is needed only when sampling from a moving stream; e.g., a duct or stack. For these applications, careful attention must be given to its shape, size, and orientation in order to obtain representative samples. The factors affecting the entry of particles into a sampling tube, i.e., particle inertia, gravity, flow convergence, and the inequality of ambient wind and suction velocity, have been critically evaluated by Davies (1). Errors can also arise from particle deposition between the probe inlet and the filter due to

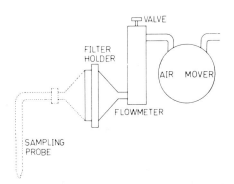

Figure 29:1—Elements of a filter sampling system.

impaction at the bends and turbulent diffusion (2).

The filter should be upstream of everything in the system but the nozzle, so that any dirt in the system, manometer liquid, or pump oil will not be carried accidentally onto it. The filter should be as close as possible to the sampling point, and all sampling lines must be free of contamination and obstructions.

The filter holder, designed for the specific filter size used, must provide a positive seal at the edge. A screen or other mechanical support is required to prevent rupture or displacement of the filter in service. With a properly designed holder, the air velocity will be uniform across the cross section of the filter holder. Uniform flow distribution is especially desirable when analyses are to be performed directly on the filter, or where only a portion of the filter will be analyzed.

The accurate measurement of flowrate and sampling time or sample volume is as important as the measurement of sample quantity, since aerosol concentration is determined by the ratio of sampled quantity to sampled volume. Unfortunately, air volume measurements are often inaccurate. When the volumetric capacity of the air mover is highly pressure-dependent, as it is for turbine blowers, ejectors, and some other types of air movers, the flow cannot be metered by techniques that introduce a significant pressure drop themselves. This precludes the use of most meters that re-

quire the passage of the full volume through them, and limits the choice to low resistance flowmeters. These include: bypass meters, which measure the flowrate of a small volume fraction of the sampled air; and meters utilizing very sensitive measurements of vane displacement or pressure drop. These can provide sufficiently accurate measurements, but often require more careful maintenance and more frequent calibration and adjustment than they are likely to receive in field use.

Most flowmeters are calibrated at atmospheric pressure, and many require pressure corrections when used at other pressures. Such corrections must be based on the static pressure measured at the inlet of the flowmeter. The flowmeter should be downstream of the filter, to preclude the possibility of sample losses in the flowmeter. It will therefore be metering air at a pressure below atmospheric, due to the pressure drop across the filter. Furthermore, if the filter resistance increases due to loading, as is often the case, the pressure correction will not be a constant factor.

If the sampling flowrate is to be controlled with a throttling valve, this valve should be downstream of the flowmeter to avoid adding to the pressure correction for the flowmeter. Flowrate adjustments can be made either with a throttling valve or by speed control of the air mover motor, and can be either manual or automatic. Automatic control requires pressure or flow transducers and appropriate feed-back and control circuitry.

The discussion which follows was designed to provide the background necessary for the proper selection of filters for particular applications. Filtration theory is outlined, the various kinds of commercial filter media used for air sampling are described, and the criteria which limit the selection in various sampling situations are discussed.

29.2 FILTRATION THEORY.

29.2.1 *Types and Structures of Filters.* All filters are porous structures with definable external dimensions such as thickness and cross section normal to fluid flow. They differ considerably in terms of flow pathways, flow rates, and residence times, and these factors are strongly influenced by their structure. One of the oldest and most common types for air sampling is the fibrous filter, which is composed of a mat of cellulose, glass, quartz, asbestos, or plastic fibers in random orientation within the plane of the filter sheet. Another classical type of filter is the granular bed, in which solid granules are packed into a definable sheet or bed. In granular bed filters used in air sampling, the granules are usually sintered to the point where they form a relatively rigid mechanical structure. Granules of glass and Alundum are frequently sintered in the form of a thimble for high temperature stack sampling. Thin sintered beds of silver granules are used in disc form for a variety of applications and are generally known as silver membranes.

The term membrane filter was originally applied to discs of a cellulose ester gel having interconnected pores of uniform size. Paulus *et al.* (3) described the various applications of such filters for air sampling in the 1950's. Gel-type membrane filters are now available in polyvinyl chloride (PVC), nylon, and other plastics. While the method of production is quite different from those used to make fibrous filters or granular beds, the flow pathways of all three types of structures are quite similar. The Nuclepore filter, a polycarbonate pore filter, is generally considered to be a membrane filter, but has a radically different structure; i.e., a series of parallel straight-through holes. It is made by exposing a thin sheet ($\sim 10\ \mu$m) of polycarbonate plastic to a flux of neutrons in a nuclear reactor, and then chemically etching the fission fragment tracks.

29.2.2 *Flow Fields and Collection Mechanisms.* Theoretical models of particle filtration have been developed using simplified flow field and particle motion in the vicinity of a single isolated cylindrical fiber. Extension of the theory to a filter mat depends upon taking proper account of the influence of adjacent fibers on the flow field (4–6).

Filters remove particles from a gas stream by a number of mechanisms. These include direct interception, inertial deposition, diffusional deposition, electrical attraction, and gravitational attraction. The mechanisms that predominate in a given case will depend on the flowrate, the nature of the filter, and the nature of the aerosol.

a. Interception occurs when the radius of a particle moving with a gas stream-line is greater than the distance from the streamline to the surface. This mechanism is important only when the ratio of the particle size to the void or pore size of the filter is relatively large.

b. Inertial Collection results from a change in direction of the gas flow. The particles, due to their relatively greater inertia, tend to remain on their original course and strike a surface. Capture is favored by high gas velocities and dense fiber packing. The operation of the inertial mechanism in a variety of commercially available fibrous filters was demonstrated experimentally by Ramskill and Anderson (7).

c. Diffusion is most effective for small particles at low flowrates. It depends on the existence of a concentration gradient. Particles diffuse from the gas stream to the surfaces of the fibers where the concentration is zero. Diffusion is favored by low gas velocities and high concentration gradients. The root mean square particle displacement of the particles, and hence the collection efficiency, increases with decreasing particle size.

Kirsch and Zhulanov (8) have tested the performance of high efficiency Whatman-type filters made of glass and polymeric polydisperse fibers and found good agreement with the theory proposed earlier by Kirsch, Stechkina and Fuchs. (6)

Gentry, Spurny and Schoermann (9) studied the diffusional deposition of ultrafine aerosols on Nuclepore filters. They found that particles $\leq 0.03\ \mu$m were collected by diffusion on the upstream surface of the filter and that the efficiency was only slightly higher than the values predicted by theory. On the other hand, for particles of 0.04 to 0.10 μm, the particles were collected primarily around the rims of the pores and the efficiencies were much higher than those predicted by theory.

d. Electrical Forces may contribute greatly to particle collection efficiency, if the filter or the aerosol has a static charge. The flow of air may induce charges on the filter.

Lundgren and Whitby (10) have shown that image forces, i.e., the forces between a charged particle and its electrical image in a neutral fiber, can strongly influence particle collection. The factors controlling particle deposition on a filter suspended in a uniform electric field, and the influence of such a field on the deposition of both charged and uncharged particles have been described by Zebel (11). Unfortunately, the data needed to predict the effect of electrostatic charges on the collection efficiency of sampling filters are seldom available.

e. **Gravitational forces** usually may be neglected when considering filter sampling. The settling velocities of airborne particles of hygienic significance are too low, and the horizontal components of the surface areas in the filters too small, for gravitational attraction to have any significant effect on particle collection efficiency, unless the face velocity through the filter is very low, e.g., <5 cm/s.

29.2.3 *Minimum efficiencies and most penetrating particle sizes*

Since a variety of collection mechanisms is involved in filtration, it is not surprising that, for a given aerosol and a given filter, the collection efficiency should vary with face velocity and particle size. The efficiency of a given filter for a given particle size could be high at low flows, due primarily to the effects of diffusion. With increasing velocity, it could first fall off and then, with still higher velocities, begin to rise due to increased inertial deposition. This pattern has been observed in several experimental penetration tests (12,13). At very high velocities, the retention could decrease because of re-entrainment. Additional data showing these effects are presented in Table 29:I.

Filter retention by the interception and diffusion mechanisms is also strongly influenced by particle size. Spurny et al. (14) and Liu and Lee (15) experimentally demonstrated the existence of large penetration maxima in Nuclepore filters with larger pore sizes (≥1 μm). These maxima are generally at particle diameter <1 μm, and are a function of face velocity, as expected. Polycarbonate pore filters have a very different structure than other types of filters, as has been discussed, and exhibit a more extreme size dependence. Rimberg (12) has demonstrated experimentally that there are sizes for maximum penetrations in IPC 1478 and H-V 5G fibrous filters, and that these sizes increase with decreasing face velocity.

The theoretical basis for predicting the minimum collection efficiency and most penetrating particle sizes for fibrous filters was addressed by Lee and Liu (16). They developed equations for such predictions which compared favorably with experimental filter efficiency data. Lee (17) extended his analysis of minimum efficiency and most penetrating particle size to granular bed filters. Predictive theories for deposition in such filters were also developed by Schmidt et al. (18) and Finchman et al. (19).

Spurny (20) investigated the collection efficiencies of membrane and polycarbonate pore filters for aerosols of chrysotile asbestos. For Millipore membrane filters with 8 μm pores the collection efficiency at a face velocity of 3.5 cm/s fell from 100% for fibers ≥ 5 μm in length to ~75% for fibers 2 μm in length, and to 25% for fibers ~0.5 μm in length. For Nuclepore filters with pore diameters of 0.2, 0.4, and 0.8 μm, collection efficiencies began to drop for fiber lengths < 3 μm and fiber diameters < 0.2 μm. For 0.2 μm pores, the efficiencies did not drop below ~80%, while for 0.8 μm pores, the efficiencies dropped to near zero for fiber lengths below 0.5 μm and diameters below 0.05 μm.

29.2.4 *Forces of Adhesion and Reentrainment.* The collection mechanisms discussed above act to arrest the motion of the particles in a gas stream as the gas flows through the voids of a filter. The particles removed from the gas stream are then subject to forces of adhesion. If the forces of adhesion on a particle are greater than the forces that tend to push the particle free, then that particle is "collected," and will be available for analysis. However, the forces exerted on the particle by the flowing gas stream may be greater than the forces of adhesion, resulting in reentrainment of the particle. At present, it is at least as difficult to predict forces of adhesion from theoretical considerations as it is to predict the effectiveness of the collection mechanisms. One reason is that it is usually not possible to determine whether particles that penetrate a filter were blown off after collection due to inadequate adhesion, or whether they underwent elastic rebound upon initial contact with the filter fibers. This question has been theoretically and experimentally investigated by Löffler (21). He concluded

Table 29:1

Flow Rate and Collection Efficiency Characteristics of Selected Air Filter Media

Filter Type	Filter	Characteristics at Indicated Face Velocities (cm/sec)								Flow Reduction Due to Loading **
		mm Hg Pressure Drop			Percent Penetration of 0.3 μm DOP					$\%/m^3/cm^2$
		53	106	211	26.7	53	106	211		
Cellulose	Whatman 1	86	175	350	7	0.95	0.061	0.001		17.9
	41	36	72	146	28	16	2	0.30		5.0
	541	30	61	123	56	40	22	9		10.4
	IPC 1478	1.5	3	5.5	90	90	90	85		<< 0.1
Cellulose-Asbestos†	H-V H-70 (9 mil)	64	127	254	1.8	0.8	0.20	0.05		1.7
Cellulose-Glass	H-V 5-G	5	10	21	32	32	26	16		0.20
Glass	MSA 1106BH	30	61	120	0.068	0.048	0.022	0.005		0.43
	Gelman A‡	33	65	129	0.019	0.018	0.011	0.001		0.50
	E‡	28	57	114	0.036	0.030	0.014	0.004		0.53
	Hurlbut 934AH	37	74	150	0.010	0.006	0.003	0.001		0.47
	Whatman GF/A	29	60	118	0.018	0.015	0.008	0.001		0.37
Polystyrene†	Delbag Microsorban	44	89	176	0.45	0.04	0.20	0.05		0.29
Membrane	Millipore AA (0.8 μm)	142	285	570	0.015	0.020	—	—		1.6
	Polypore AM-1 (5 μm)	23	46	95	12	8	5	2		2.4
	AM-3 (2 μm)	84	190	380	0.36	0.22	0.090	0.015		3.1

*Data extracted from NRL Report No. 6054 (13); 1 mm Hg + 0.133 kPa.

**Normalized to the dust loading in the atmosphere on an "average" summer day (Washington, DC–1964), using a positive-displacement pump without flow regulation.

†No longer available.

‡No longer available; replaced by type AE.

that the measured forces of adhesion were in good agreement with the Van der Waals forces calculated theoretically and that the flow velocity required for blowing collected particles off fibers is much higher than those normally used in air filtration. An increase in particle penetration with increasing velocity will usually be due to increased rebound or to the resuspension of particle flocs rather than individual particles.

29.3 COMMERICAL FILTER MEDIA. Filter media of many different types and with many different properties have been designed for or adapted to air sampling requirements. For purposes of discussion, they have been divided into groups determined by their composition. Air flow resistance and collection efficiency characteristics of some commonly used filters are tabulated in Table 29:I. Table 29:II summarizes the physical characteristics of commercially available filter media based on vendor-supplied or -approved data. These media have been subdivided on the basis of their composition and/or structure.

29.3.1 *Cellulose Filter Papers.* Most filters of this type are used primarily by analytical chemists for liquid-solid separations. They are made of purified cellulose pulp, are low in ash content, and are usually less than 0.25 mm thick. These papers are relatively inexpensive, are obtainable in an almost unlimited range of sizes, have excellent tensile strength, show little tendency to fray during handling, and are uni-versally obtainable. Their disadvantages include non-uniformity, resulting in variable flow resistance and collection efficiency, and hygroscopicity, which makes accurate gravimetric determinations very difficult.

For air sampling, Whatman No. 41 is the most widely used filter paper of this type. It has the advantages of low cost, high mechanical strength, and high purity typical of these papers and, in addition, has a moderate flow resistance. This group also includes hardened papers, such as Whatman No. 50, from which collected particles can be removed by washing.

29.3.2 *Glass Fiber Filters.* Glass fiber filters are, in most cases, more expensive and have poorer mechanical properties than cellulose papers. They also have many advantages, i.e., reduced hygroscopicity, ability to withstand higher temperatures, and higher collection efficiencies at a comparable pressure drop. These properties, combined with the ability to make benzene, water, and nitric acid extracts from particulate matter collected on them, led to the selection of a high efficiency glass fiber filter as the standard collection medium for high-volume samplers in air sampling networks.

As described by Pate and Tabor (22), a large number of tests were routinely performed on such filters. Nondestructive tests, e.g., weighing, gross β-activity, and reflectance were performed prior to the chemical extractions. Also, portions of the filters were stored untreated for possible

Table 29:II Summary of Air Sampling Filter Characteristics

A. Cellulose Fiber Filter Characteristics

	Filter	Void Size μm	Fiber Diam. μm	Thick-ness μm	Weight/ Area mg/cm^2	Ash Content %	Max. Oper. Temp. °C	Tensile Strength g/cm	ΔP_{100}* in H$_2$O	Source
Whatman	1	2+	NA	180	8.7	0.06	150	4700	40.5	WRA
	4	4+	NA	210	9.2	0.06	150	NA	11.5	WRA
	40	2	NA	210	9.5	0.01	150	4600	54	WRA
	41	4+	NA	220	8.5	0.01	150	4600	8.1	WRA
	42	>1	NA	200	10.0	0.01	150	NA	NA	WRA
	44	>1	NA	180	8.0	0.01	150	NA	NA	WRA
	50	1	NA	120	9.7	0.025	150	NA	NA	WRA
	541	4+	NA	160	7.8	0.008	150	NA	NA	WRA

*Pressure drop at face velocity of 100 ft/min (50.8 cm/s), 1 in H$_2$O = 0.25 kPa.
NA Information not available or not applicable.

Table 29:II Summary of Air Sampling Filter Characteristics

B. Glass Fiber Filter Characteristics

Filter	Void Size μm	Fiber Diam. μm	Thickness μm	Weight/Area mg/cm^2	Ash Content %	Max Oper. Temp. °C	Tensile Strength g/cm	ΔP_{100}* in H$_2$O	Source
MSA 1106B[a]	NA	NA	180–270	6.1	~95	540	625	19.8	MSA
1106BH[b]	NA	NA	180–460	5.8	~100	540	270	19.8	MSA
Gelman									
Type A/E[b]	1	NA	450	NA	100	550	NA	NA	G
Millipore									
AP 15[a]	NA	<1	380	8.0	95	500	625	70	M
AP 20[a]	NA	<1	330	7.3	95	500	625	16	M
AP 40[b]	NA	<1	410	6.9	100	500	450	18	M
Whatman									
GF/A[b]	<1	0.5–0.75	260	5.3	NA	540	500	NA	WRA
GF/B[b]	<1	0.5–0.75	680	14.3	NA	540	1000	NA	WRA
GF/C[b]	<1	0.2–0.5	260	5.3	NA	540	500	NA	WRA
934AH[b]	<1	NA	330	6.4	NA	540	180	24.4	WRA
EPM-2000[b]	NA	NA	430	8.0	NA	540	700	NA	WRA
QM-A Quartz	NA	NA	450	8.5	NA	540	250–300	15.3	WRA
H & V									
HB-5055[a]	NA	0.6	460–560	3.5	96–99	425	1070	14	H&V
BA-8045	NA	0.45	400–420	2.7	96–99	425	1100	19	H&V
HD-2142[a]	NA	1.0	380	2.4	96–99	425	890	4.5	H&V
HD-2025[a]	NA	1.6	460–560	3.5	96–99	425	1070	5.8	H&V
HA-8021[b]	NA	0.45	380	2.5	100	550	625	19.8	H&V
Pallflex									
600A	<0.4	0.4–.07	230	3	~95	315	1000	6	P
2500A	<0.4	0.4–0.7	500	6.5	~96	315	1500	15	P
2500 QAS Quartz	<0.2	NA	530	6	100	1000	NA	12	P
TX40H120WW[t]	<0.3	<0.5	175	5	~85	315	3500	20	P
T60A20[t]	<0.4	0.4–0.7	240	4	~80	315	1200	8	P
E70[c]	<0.4	0.4–0.8	175	3.5	~35	120–160	650	8	P
Nuclepore									
AA[a]	NA	NA	NA	NA	NA	NA	NA	NA	N
AAA[b]	NA	NA	NA	NA	NA	NA	NA	NA	N

[a]with organic binder
[b]without organic binder
*pressure drop at face velocity of 100 ft/min (50.8 cm/s); 1 in H$_2$O = 0.25 kPa
NA information not available or not applicable
[t]contains Teflon
c contains cellulose

use at later times to obtain background data on air concentrations whose need was not anticipated at the time of sample collection.

The types of chemical analyses that can be performed on extracts from the filters are determined by the sensitivity of the analyses and by the magnitude and variability of the extractable filter blank for the particular ion or molecule involved. One of the major problems in the operations of the National Air Sampling Network (NASN) was the variability in the composition and properties of the glass fiber filters used. The quality control requirements of NASN proved to be beyond the capability of the filter vendors. The filter characteristics are determined by the process variables in at least four production stages, i.e., the production of the glass, the production of glass fiber from the bulk glass, the production of the fiber mat from the glass fiber, and the packaging of the individual filters. In the case of the MSA 1106BH filter, the most widely used for NASN and related application, Pate and Tabor (22) describe four dif-

Table 29:II Summary of Air Sampling Filter Characteristics

C. Membrane Filter Characteristics

Filter	Composition	Pore Size μm	Thick-ness μm	Weight/ Area mg/cm^2	Ash Content %	Max Oper. Temp. °C	Tensile Strength psi†	Refractive Index	ΔP_{100}* in H$_2$O	Source
Millipore										
SC	Mixed Cellulose	8.0	130	5.2	<0.001	125	175	1.515	20	M
SM	Esters	5.0	130	2.8	<0.001	125	160	1.495	32	M
SS		3.0	150	3.0	<0.001	125	150	1.495	56	M
RA[a]		1.2	150	4.2	<0.001	125	300	1.510	75	M
AA[a,b]		0.80	150	4.7	<0.001	125	350	1.510	102	M
DA[a]		0.65	150	4.8	<0.001	125	400	1.510	112	M
HA[a]		0.45	150	4.9	<0.001	125	450	1.510	250	M
PH		0.30	150	5.3	<0.001	125	500	1.510	300	M
GS		0.22	135	5.5	<0.001	125	700	1.510	450	M
VC		0.10	130	5.6	<0.001	125	800	1.500	2290	M
VM		0.05	130	5.7	<0.001	125	1000	1.500	3610	M
VS		0.025	130	5.8	<0.001	125	1500	1.500	5100	M
LC	Teflon	10.0	125	8.0	NA	260	250	NA	125	M
LS		5.0	125	8.0	NA	260	150	NA	187	M
FA	PTFE-polyethylene	1.0	180	2.2	NA	130	NA	NA	NA	M
FH	reinforced	0.5	180	2.2	NA	130	NA	NA	NA	M
FG		0.2	180	2.2	NA	130	NA	NA	NA	M
Metricel										
GN 450	Mixed Cellulose	0.45	150	4.0	NA	74	NA	1.51	NA	G
GN 800	esters	0.8	150	4.0	NA	74	NA	1.51	NA	G
VM	Polyvinyl Chloride	5.0	150	1.0	NA	52	NA	1.55	NA	G
DM-800	PVC/Acrylonitrile	0.8	150	3.0	NA	66	NA	1.51	NA	G
DM-450		0.45	150	3.0	NA	66	NA	1.51	NA	G
Nylasorb	Nylon	1.0	NA	NA	NA	NA	NA	NA	NA	G
Zylon	PTFE	5.0	NA	NA	NA	NA	NA	NA	NA	G
TF	PTFE with poly-propylene support	1.0	NA	NA	NA	NA	NA	NA	NA	G
		0.45	NA	NA	NA	NA	NA	NA	NA	G
		0.20	NA	NA	NA	NA	NA	NA	NA	G

ferent types produced sequentially between 1956 and 1962, which differed in softening temperature, and chemical composition and extractability, and which resulted from manufacturing changes beyond MSA's control.

One of the determinations, made by NASN, was gross mass of particulate matter by gravimetric analysis. Many other investigators use the same types of high-volume samplers and filters for routine monitoring and analyze only for gross mass concentration. The validity of these determinations is suspect. The potential errors arising from inaccurate sample volume determinations, from inadequate temperature and humidity conditioning prior to weighing, and from the precision of the weighing procedure are well-known and have been discussed by Kramer and Mitchel (23). An additional serious source of error is the loss of filter fibers drawn through the support screen into the air mover during sample collection. Flash-fired binderfree filters are soft and friable and the loose fiber content is variable. Some NASN filters returned from the field had lower than tare weights, despite the presence of visible deposition on the filter face. If gross mass concentration analyses are to be performed, other non-hygroscopic filter media, which are both mechanically strong and efficient, should be used, or the filter should include a backing layer which will help prevent the loss of filter fibers.

The preceding discussion applied to glass

Table 29:II Summary of Air Sampling Filter Characteristics

D. Polycarbonate Pore Filters

Filter	Compo-sition	Pore Size μm	Thick-ness μm	Weight/Area mg/cm^2	Ash Content %	Max. Oper. Temp. °C	Tensile Strength psi†	Refractive Index	ΔP_{100}* in H$_2$O	Source
Nuclepore										
PC	Poly-	8.0	9.0	1.0	0.04	140	>3,000	1.58&1.614	3.0	N
PC, AP	carbonate	8.0	NA	NA	NA	NA	>3,000	NA	NA	N
PC		0.40	10.0	0.8	0.04	140	>3,000	1.58&1.614	83.0	N
PC		0.20	10.0	0.9	0.04	140	>3,000	1.58&1.614	208.0	N
Poretics										
	Poly-	8.0	10	1.0	0.01	140	>3,000	1.584+1.625	NA	POR
	carbonate	5.0	10	1.0	0.01	140	>3,000	1.584+1.625	NA	POR
		3.0	10	1.0	0.01	140	>3,000	1.584+1.625	NA	POR
		2.0	10	1.0	0.01	140	>3,000	1.584+1.625	NA	POR
		1.0	10	1.0	0.01	140	>3,000	1.584+1.625	NA	POR
		0.8	10	1.0	0.01	140	>3,000	1.584+1.625	NA	POR
		0.6	10	1.0	0.01	140	>3,000	1.584+1.625	NA	POR
		0.4	10	1.0	0.01	140	>3,000	1.584+1.625	NA	POR
		0.2	10	1.0	0.01	140	>3,000	1.584+1.625	NA	POR
		0.1	10	1.0	0.01	140	>3,000	1.584+1.625	NA	POR

AP   Apiezon coated
NA   Information not available or not applicable
*Pressure drop at face velocity of 100 ft/min (~50.8 cm/s)
†   1 psi = 6.9 kPa

fiber filters which are virtually 100% efficient for all particle sizes. For some applications, e.g., a filter-pack sampler designed to provide data on particle size distribution, less efficient glass fiber filters may be desirable. Shleien, Cochran and Friend (24) described the physical and collection efficiency characteristics of four less efficient glass fiber filters, produced for gas cleaning and air conditioning applications, which they selected for their filter pack.

29.3.3 *Mixed Fiber Papers.* This group includes cellulose-asbestos, cellulose-glass, and glass-asbestos mixtures. Filters of this type found extensive application in air cleaning, where their characteristics of high collection efficiency and low pressure drop were especially important. However, since it is extremely difficult to remove collected dust from these media, they have limited value for air sampling. The high ash content resulting from the mineral components of these filters often interferes with chemical analysis of the deposited material. Mixed fiber filters are used for sampling when simple gravimetric analyses are to be

performed and also when sampling radioactive particles where the activity can be counted without removing the sample from the filter. In any case, filters containing asbestos are no longer made.

29.3.4 *Plastic Fiber Filters.* Plastic fiber filters have also been used for air sampling applications. The most widely used of these has been the Microsorban (25) filter, made of mats of polystyrene fibers of submicrometer diameter. Its flow resistance is relatively low, being comparable to Whatman No. 41, while its efficiency of collection is relatively high, i.e., comparable to that of glass and cellulose-asbestos filters. It is no longer commercially available. Polystyrene filters are soluble in aromatic hydrocarbon solvents. Their mechanical strength is poor, and they must be well supported by a firm backup in the filter holder. Air sample filters composed of PVC fibers of micrometer size have been described by Berka (26).

29.3.5 *Membrane Filters.* Filters consisting of porous membranes can be used for many applications where fibrous filters cannot. Organic membranes are produced

Table 29:II Summary of Air Sampling Filter Characteristics

E. Filter Thimble Characteristics

Designation	Composition	Size	Void Size μm	Max. Oper. Temp. °C	Source	Remarks
D1013	Cellulose	43 × 123 mm	NA	120	A	Use with D1012 Paper Thimble Holder
D1016	Glass Cloth	2-3/16 × 14″	NA	400	A	Use with D1015 Glass Cloth Thimble Holder
RA-98	Alundum	NA	Standard	High	A	Use with D1021 Alundum Thimble Holder
RA-360	Alundum	NA	Fine	High	A	Use with D1021 Alundum Thimble Holder
RA-84	Alundum	NA	Extra Fine	High	A	Use with D1021 Alundum Thimble Holder
S&S 603 GV	Glass Fiber Heat Treated	from 19 × 90 mm to 90 × 200 mm	NA	510	S&S	
Whatman	Cellulose	from 10 × 50 mm to 90 × 200 mm	NA	120	WRA	

NA Information not available or not applicable

by the formation of a gel from an organic colloid, with the gel in the form of a thin ($\sim 150\ \mu$m) sheet with uniform pores. Membrane filters made from cellulose nitrate achieved widespread use for air sampling in the early 1950s. In recent years, membrane filters made of cellulose triacetate, regenerated cellulose, polyvinyl chloride, nylon, polypropylene, polyimide, polysulfone, a copolymer of vinyl chloride and acrylonitrile, Teflon, and silver have become available. Silver membranes are produced by a different technique and will be discussed separately at the end of this section.

Cellulose nitrate and cellulose triacetate membranes are the most widely used and, as indicated in Table 29:II, are available in the widest range of pore sizes. The mass of these filters is very low and their ash content is negligible. Some are completely soluble in organic solvents. Cellulose nitrate filters dissolve in methanol, acetone, and many other organic solvents. Cellulose triacetate, nylon, and PVC filters dissolve in fewer solvents, while filters composed of teflon and regenerated cellulose do not dissolve in common solvents. The ability to dissolve a filter completely in a solvent permits the concentration of the collected material within a small volume for subsequent chemical and/or physical analyses.

Collection efficiency increases with decreasing pore size, but even the large pore size filters have relatively high collection efficiencies for particles much smaller than their pores. Membrane filters do not behave at all like sieves. As in fibrous filters, particles are removed primarily by impaction and diffusion. Early investigators believed that electrostatic forces played a major role in particle deposition in membrane filters, but experimental studies by Spurny and Pich (27) and Megaw and Wiffen (28) demonstrate that diffusional and impaction

Table 29:II Summary of Air Sampling Filter Characteristics

F. Commercial Sources for Filters and Filter Holders

Symbol	Source	Symbol	Source
A	Andersen Samplers, Inc. 4215 Wendell Drive Atlanta, GA 30336	N	Nuclepore Corporation 7035 Commerce Circle Pleasanton, CA 94566
BGI	BGI, Incorporated 58 Guinam Street Waltham, MA 02154	P	Pallflex Products Corporation Kennedy Drive Putnam, CT 06260
G	Gelman Sciences, Inc. 600 South Wagner Road Ann Arbor, MI 48106	POR	Poretics Corporation 151-I Lindbergh Avenue Livermore, CA 94550–9925
GMW	General Metal Works, Inc. 145 S. Miami Ave. Cleves, OH 45002	RAC	Research Appliance Company Div. Anderson Samplers, Inc. 4215 Wendell Drive Atlanta, GA 30336
H&V	Hollingsworth and Vose Company East Walpole, MA 02032	RAD	SAI/RADēCO 10373 Roselle Street San Diego, CA 92121
HI-Q	HI-Q Environmental Products Co. 7386 Trade St. San Diego, CA 92109	S&S	Schleicher and Schuell, Inc. 10 Optical Ave. Keene, NH 03431
M	Millipore Corporation Bedford, MA 01730	ST	The Staplex Company 777 Fifth Avenue Brooklyn, NY 11232
MSA	Mine Safety Appliances Company 600 Penn Center Blvd. Pittsburgh, PA 15235	WRA	Whatman Reeve Angel 9 Bridewell Place Clifton, NJ 07014

deposition account for most of the observed collection and that the contributions of direct interception and electrostatic deposition, if present, are less important.

Membrane filters differ from fibrous filters in that a much greater proportion of the deposit is concentrated at or close to the front surface. Lindeken et al. (29) and Lossner (30) measured the penetration depth using test aerosols tagged with alpha emitters. Lindekin et al. were interested primarily in the use of the filters for measuring the concentration of α-emitters in air. If the deposit were truly at the surface, there would be no need for correcting for differences in distance from the detector face or for absorption of α-energy in the filter. They found that, on a microscopic scale, the filter surfaces were not smooth. The

surface roughness varied among different brands and, for Millipore filters, from the front surface to the back. They concluded that the smooth face of a SM Millipore was suitable for their application. Lossner (30) demonstrated the effect of pore size and face velocity on penetration depth for 0.55 μm SiO_2 particles.

The fact that particle collection takes place at or near the surface of the filter accounts for most of the advantages of membrane filters and also some of their disadvantages. The advantages arising from this property are:

1. It is possible to examine solid particles microscopically without going through a transfer step which might change the state or form of the particles. Examination can be by optical

microscopy using immersion oil having the same index of refraction as the filter. The oil renders the filter transparent to light rays. Transmission electron microscopy can be performed on a replica of the filter surface produced by vacuum evaporation techniques, while scanning electron microscopy can be performed directly on a segment of the filter.

2. Direct measurements of the deposit can be made on the surface without interference caused by absorption in the filter itself. This is advantageous in radiometric counting of air dust, and in soiling index measurements made by reflectance.

3. Autoradiographs of radioactive particles can be produced by a technique whereby photographic emulsion is placed in contact with the membrane filter sample (31).

The disadvantage arising from surface collection is that the amount of sample that can be collected is limited. When more than a single layer of dust particles is collected on a membrane filter, the resistance rapidly increases and there is a tendency for the deposit to slough off the filter.

Silver membranes for air sampling applications are made by sintering uniform metallic silver particles. These membranes possess a structure basically similar to that of the organic membranes previously described. They have a uniform pore size, and for a given pore size, about the same flow characteristics. For filters up to 47 mm in diameter, they are 50 μm thick. According to the manufacturer, the membrane is an integral structure of permanently interconnected particles of pure silver, contains no binding agent or fibers, and is resistant to chemical attack by all fluids which do not attack pure silver. Thermal stability extends from -130 to $+370°C$ (-200 to $+700°F$).

Richards, Donovan and Hall (32) described the use of silver membranes for sampling coal tar pitch volatiles. Other filter media evaluated were not suitable because of the high weight losses of blank filters in the benzene extraction step in the analysis, including 1106BH glass, cellulose acetate membrane, and Whatman 41 cellulose. The weight loss for the silver membrane was negligible. Another application of silver membranes is for sampling airborne quartz for X-ray diffraction analysis, as described by Knauber and VonderHeiden (33). Most instruments satisfying the ACGIH criteria for respirable dust samplers operate at low flowrates, and the sample masses on the backup filters are too small for conventional analyses. Using silver membranes, the X-ray diffraction background is very consistent, and quartz determinations can have a lower limit of sensitivity as low as 0.02 mg.

29.3.6 *Polycarbonate Pore Filters.* Polycarbonate pore filters are similar to membrane filters in that both contain uniform-sized pores in a solid matrix. However, they differ in structure and method of manufacture. They are made by placing polycarbonate sheets approximately 10 μm thick in contact with sheets of uranium into a nuclear reactor. The neutron flux causes U-235 fission, and the fission fragments bore holes in the plastic. Subsequent treatment in an etch solution enlarges the holes to a size determined by the temperature and strength of the bath, and the time within it. Commercial filters are available with pore diameters between 0.03 and 8 μm. Polycarbonate pore filters possess many of the attributes erroneously attributed to membrane filters in earlier days. They have a smooth filtering surface; the pores are cylindrical, almost all uniform in diameter, and essentially perpendicular to the filter surface. The filters also are transparent, even without immersion oil.

The structure and air paths are so simple that, as demonstrated by Spurny et al. (14), it is possible to predict their particle collection efficiency on the basis of measured dimensions and basic particle collection theory.

Although their pore volume is much lower, polycarbonate pore filters have about the same flowrate-pressure drop relations as membrane filters of comparable pore diameter. However, the filter penetrations at 5 cm/sec, reported by Spurny et al (14) are much greater than those of membrane filters with the same pore sizes. Pore filters have a lower and more uniform weight, and since they are non-hygroscopic, they can be used for sensitive gravimetric analyses.

The polycarbonate base is very strong and Nuclepore filter tapes do not require extra mechanical backing. They can be analyzed by light transmittance, or filter segments can be cut from discs or tapes for microscopy.

The very smooth surface makes polycarbonate pore filters good collectors for particles to be analyzed by electron microscopy and X-ray fluorescence analyses. Spurny et al. (14) show high resolution electron micrographs made from silicon monoxide replicas of the filter surface. The very smooth surface also permits good resolution of the collected particles by scanning electron microscopy. The very low collection efficiencies of pore filters under certain conditions permit their use in particle classifications which separate aerosols into size-graded fractions. Cahill et al. (34) and Parker et al. (35) proposed using two Nuclepore filters in series, with the first having a cut-characteristic approximating the ACGIH "respirable" dust criterion. Heidam (36) reviews the use of series pore filters for a variety of applications, but cautions that particle bounce may be a significant source of error. Particle bounce as a means of penetration of such filters also has been noted by John et al. (37), Buzzard and Bell (38), and Spurny (39). John et al. (37) found that the collection of solid particles was lower than that for liquid droplets of the same aerodynamic size, and this was attributed to the bouncing of solid particles off the collection surface.

29.3.7 *Plastic Foam Filters.* Gibson and Vincent (40) describe the use of porous filter media to simulate the collection characteristics of the MRE elutriator under a wide range of face velocities. They found that, for particles close to the respirable size, deposition by inertial impaction and gravitational sedimentation compete. As a result, the efficiency remains relatively constant over a substantial range of face velocities.

29.3.8 *Granular Beds.* Ground crystals of salicylic acid, sugar, and naphthalene have been used as aerosol filters. The particles are recovered either by volatilization or solution of the crystals. One difficulty with this type of filter is the high impurity level of most available crystals. The efficiency of these filters depends on the size of the crystals, the depth of the bed, etc. Usually, low flowrates are necessary in order to obtain high efficiencies through diffusional separation.

29.3.9 *Filters Occasionally Used for Air Sampling.*

a. Respirator filters

Respirator filters of felt and/or cellulose fiber can be, and have been, used for air sampling. In many of them, the filter is manufactured in a pleated form, which increases the surface area without increasing the overall diameter. Filters of this type have the same advantages and disadvantages as the mixed fiber filters previously discussed.

b. Thimbles

Filter thimbles are available in glass fiber, paper, and cloth. They are sometimes filled with loose cotton packing to reduce clogging. The advantage is that large samples can be collected.

Alundum thimbles and sintered glass filters are manufactured with a variety of porosities. They have considerably higher resistance to air flow than comparable paper filters, but can be used for very high temperature sampling.

29.4 FILTER SELECTION CRITERIA.

29.4.1 *General Considerations.* The selection of a particular filter type for a specific application is invariably the result of a compromise among many factors. These factors include cost, availability, collection efficiency, and requirements of the analytical procedures, and the ability of the filter to retain its filtering properties and physical integrity under the ambient sampling conditions. The increasing variety of commercially available filter media sometimes makes the choice seem somewhat more difficult, but more importantly, increases the possibility of a selection that satisfies all important criteria.

29.4.2 *Efficiency of Collection.* Before discussing experimental efficiency data, it is important that a distinction be made between particle collection efficiency and mass collection efficiency. The former refers to fractions of the total number of particles, while the latter refers to fractions of the total mass of the particles. These efficiencies will be numerically equivalent only when all the particles are the same size, as in some laboratory investigations of filter efficiency. In almost all other cases, the mass collection efficiency will be larger than the corresponding particle collection efficiency. When sampling for total mass concentration of particulate matter, or for the mass concentration of a component of an aerosol, the efficiency of interest is mass efficiency. Submicrometer particles often contribute only a small fraction of the total mass of an industrial dust, even when they represent the majority of the particles. Therefore, it is not always essential that an air sampling filter have a high efficiency for the smallest particles. Insistence on high efficiency for all particle sizes may restrict the selection to media with other limitations, such as high flow resistance, high cost, and fragility.

Collection efficiency data for a variety of filter media are given in Table 29:I for 0.3 μm diameter DOP droplets at various face velocities (13). This is a commonly used particle size for a test aerosol, since it is close to the size for maximum filter penetration for many commonly used sampling media operating at representative flowrates. On this basis, it is reasonable to assume that penetration of both smaller and larger particles would be lower, i.e., the collection efficiency would be higher. This was confirmed by Stafford and Ettinger (41), who showed that the collection efficiency of Whatman 41 is lowest for 0.264 μm particles at a face velocity of approximately 15 cm/s. It increases for both larger and smaller particles, and is approximately 95% or greater for all sizes at face velocities above 100 cm/s.

Liu and Lee (15) measured the collection efficiencies of Nuclepore and Teflon® membrane filters for particles in the 0.03 to 1.0 μm diameter range. For 10 μm Teflon filters (Type LC) the collection efficiencies for 0.003 to 0.1 μm particles at low face velocities were in the 60–65% range, while for 5 μm (Type LS) filters, they were in the 80–85% range. For higher velocities and/or large particles, the efficiencies were >99.99% under all conditions tested. For Nuclepore filters, the penetrations were much higher at comparable pore sizes, and reached 100% for small particles with 5 and 8 μm pore filters. The results were consis-

tent with predictions based on interception, impaction, and diffusion collection.

Liu (42) summarized the results of collection efficiency measurements at four particle sizes and four face velocities for 76 different air sampling filters. Key results of this extensive body of calibration data are summarized in Table 29:III.

The effect of particle shape on filter penetration was explored by Spurny (20) using aerosols of chrysotile asbestos, as discussed earlier. Collection efficiencies decreased substantially with fiber length for both membrane and polycarbonate pore filters of larger pore size. The orientation of the airborne fibers as they approach the filter pore entrances has an important effect on their ability to penetrate the filter.

Skocypec (43) measured the penetration of condensation nuclei in the 0.002 to 0.007 μm range through most of the commercially available membrane filters at a face velocity of 10 cm/s. Less than 1.0% of the particles penetrated through most of the filters. However, much higher penetrations were observed for some of them. Penetrations of 3.0% or more were found only for some of the large pore (≥ 3.0 μm) filters, Nuclepore filters with ≤ 0.08 or ≥ 0.6 μm pores, silver membranes with ≥ 0.8 μm pores, and Type FG-0.2 μm PTFE Fluoropore. Some of the large pore membranes, e.g., the cellulose ester filters of Millipore, cellulose triacetate filters of Gelman, and S & S nitrocellulose filters retained very high efficiencies for these very small particles.

John and Reischl (44) also determined the collection efficiency of various air sampling filters for condensation nuclei. Efficiencies of >99% were found for a variety of Teflon membranes including Ghia filters with 1–3 and 2–4 μm pores, Fluoropore filters with 1.0 and 3 μm pores, Gelman cellulose acetate with 5 μm pores (GA-1), and four glass fiber filters (Gelman A and Spectrograde, MSA 1106BH, and EPA). The Ghia Teflon membranes with 3–5 μm pores were almost as good, with efficiencies >98%. The Nuclepore (0.8 μm pore) filters had efficiencies of 72, 72, and 89% at face velocities of approximately 25, 50, and 150 cm/s, respectively, while the efficiencies for Whatman 41 were 64 and 83% at approximately 50 and 150 cm/s.

Lundgren and Gunderson (45) tested the

Table 29:III List of Filters Tested and Principal Results

Filter	Material	Pore Size, μm	Filter Permeability Velocity, cm/s (ΔP = 1.33 kPa)	Filter Efficiency Range, %*		
A. Cellulose Fiber Filter						
Whatman						
No. 1	Cellulose Fiber	—	6.1	49	–	99.96
No. 2		—	3.8	63	–	99.97
No. 3		—	2.9	89.3	–	99.98
No. 4		—	20.6	33	–	99.5
No. 5		—	0.86	93.1	–	99.99
No. 40		—	3.7	77	–	99.99
No. 41		—	16.9	43	–	99.5
No. 42		—	0.83	92.0	–	99.992
B. Glass Fiber Filter						
Gelman						
Type A	Glass Fiber	—	11.2	99.92	–>	99.99
Type A/E		—	15.5	99.6	–>	99.99
Spectrograde		—	15.8	99.5	–>	99.99
Microquartz		—	14.1	98.5	–>	99.99
MSA 1106B		—	15.8	99.5	–>	99.99
Pallflex						
2500 QAO	Quartz Fiber	—	41	84	–	99.9
E70/2075W		—	36.5	84	–	99.95
T60A20	Teflon Coated Glass Fiber	—	49.3	55	–	98.8
(another lot)		—	40.6	52	–	99.5
T60A25		—	36.5	65	–	99.3
TX40H12O		—	15.1	92.6	–	99.96
(another lot)		—	9.0	98.9	–>	99.99
Reeve Angel 934 AH	Glass Fiber	—	12.5	98.9	–>	99.99
(acid treated)		—	20	95.0	–	99.96
Whatman						
GF/A	Glass Fiber	—	14.5	99.0	–>	99.99
GF/B		—	5.5	> 99.99	–>	99.99
GF/C		—	12.8	99.6	–>	99.99
EPM 1000		—	13.9	99.0	–>	99.99
C. Plastic Fiber Filter						
Delbag[†]	Polystyrene					
Microsorban-98		—	13.4	98.2	–>	99.99
D. Membrane Filter						
Millipore						
MF-VS	Cellulose acetate/nitrate	0.025	0.028	99.999	–>	99.999
MF-VS		0.1	0.16	99.999	–>	99.999
MF-PH		0.3	0.86	99.999	–>	99.999
MF-HA		0.45	1.3	99.999	–>	99.999
MF-AA		0.8	4.2	99.999	–>	99.999
MF-RA		1.2	6.2	99.9	–>	99.999
MF-SS		3.0	7.5	98.5	–>	99.999
MF-SM		5.0	10.0	98.1	–>	99.99
MF-SC		8.0	14.1	92.0	–>	99.99

Table 29:III (continued)

Filter	Material	Pore Size, μm	Filter Permeability Velocity, cm/s ($\Delta P = 1.33$ kPa)	Filter Efficiency Range, %*	
Polyvic-BD	Polyvinyl chloride	0.6	0.86	99.94	-> 99.99
Polyvic-VS		2.0	5.07	88	-> 99.99
PVC-5		5.0	11	96.7	-> 99.99
Celotate-EG	Cellulose acetate	0.2	0.31	> 99.95	-> 99.99
Celotate-EH		0.5	1.07	99.989	-> 99.999
Celotate-EA		1.0	1.98	99.99	-> 99.99
Mitex-LS	Teflon	5.0	4.94	84	-> 99.99
Mitex-LC		10.0	7.4	62	-> 99.99
Fluoropore	PTFE-polyethylene reinforced				
FG		0.2	1.31	> 99.90	-> 99.99
FH		0.5	2.32	> 99.99	-> 99.99
FA		0.1	7.3	> 99.99	-> 99.99
FS		3.0	23.5	98.2	-> 99.98
Metricel					
GM-6	Cellulose acetate/nitrate	0.45	1.45	> 99.8	-> 99.99
VM-1	Polyvinyl chloride	5.0	51.0	49	- 98.8
DM-800	PVC/Acrylonitrile	0.8	2.7	99.96	-> 99.99
Gelman Teflon	Teflon	5.0	56.8	85	- 99.90
Ghia[†]					
S2 37PL 02	Teflon	1.0	12.9	> 99.97	-> 99.99
S2 37PJ 02		2.0	23.4	99.89	-> 99.99
S2 37PK 02		3.0	24.2	92	- 98.98
S2 37PF 02		10.0		95.4	-> 99.99
Zefluor	Teflon				
P5PJ 037 50		2.0	32.5	94.6	- 99.96
P5PI 037 50		3.0	31.6	88	- 99.9
Chemplast					
75-F	Teflon Filter	1.5	3	83	- 99.99
75-M		1.0	6.6	54	-> 99.99
75-C		1.0	32	26	- 99.8
Selas Flotronics					
FM0.45	Silver	0.45	1.8	93.6	- 99.98
FM0.8		0.8	6.2	90	- 99.96
FM1.2		1.2	9.2	73	- 99.7
FM5.0		5.0	19.0	25	- 99.2

E. Nuclepore Filter

Filter	Material	Pore Size, μm	Filter Permeability Velocity, cm/s ($\Delta P = 1.33$ kPa)	Filter Efficiency Range, %*	
Nuclepore					
N010	Polycarbonate	0.1	0.602	> 99.9	-> 99.9
N030		0.3	3.6	93.9	-> 99.99
N040		0.4	2.9	78	-> 99.99
N060		0.6	2.1	53	- 99.5
N100		1.0	8.8	28	- 98.1
N200		2.0	7.63	9	- 94.1
N300		3.0	12	9	- 90.4

Table 29:III (continued)

Filter	Material	Pore Size, μm	Filter Permeability Velocity, cm/s ($\Delta P = 1.33$ kPa)	Filter Efficiency Range, %*		
N500		5.0	30.7	6	–	90.7
N800		8.0	21.2	1	–	90.5
N1000		10.0	95	1	–	46
N1200		12.0	161.1	1	–	66

F. Miscellaneous Filter

Filter	Material	Pore Size, μm	Filter Permeability Velocity, cm/s	Filter Efficiency Range, %*		
MSA Personal Air Sampler		–	12	89	–	99.97

*The range of filter efficiency values given generally corresponds to a particular diameter range of 0.035 to 1 μm, a pressure drop range of 1 to 40 kPa and a face velocity range of 1 to 100 cm/s.
†No longer commercially available.

effects of temperature, face velocity and loading on the particle collection efficiency of glass fiber filters. At room temperatures, they found similar collection efficiencies for Gelman Type A, Gelman Type E, Gelman Spectrograde Type A, MSA 1106B, and the EPA Microquartz filters made of Johns-Manville "Microquartz" fibers by A.D. Little. The EPA filters had a low extractable background and are used for stack gas sampling at temperatures in excess of 500°C. All of the filters had similar pressure drop vs. flow rate characteristics and filter masses per unit area, and the high temperature comparisons were limited to the Gelman Type A and "Microquartz" filters.

In all tests, aerosol penetrations of nonvolatile particles were less than about 0.10%. The highest penetrations were for particles approximately 0.1 μm diameter at the highest face velocity tested, i.e., 51 cm/s. Penetrations dropped significantly with aerosol loadings of only several micrograms per square centimeter.

The effect of pinholes on filter efficiency was examined by punching two 0.75 mm pinholes through the filter mat. Although this produced higher initial penetrations by up to 30 times, the penetrations were never more than a few percent and fell rapidly with loading. Thus, their effect on sample collection would be essentially negligible.

Particle penetrations decreased with increasing temperature, except when the temperature was sufficient to volatilize the particles, or to contribute to mechanical leakage of the filter holder. For many fil-

ters, there is also a face velocity for maximum penetration.

The same filter can be inefficient at some face velocities and highly efficient at others. For example, Whatman 41 penetration below 10 cm/s exceeds 40%, while at 100 cm/s it is only about 4%, and at higher flowrates it is much less than that. This filter is often used in industrial hygiene surveys with both low and high volume samplers. When sampling with a 25 mm filter head at 25 L/min the face velocity (based on an effective filtration area of 3.68 cm²) is 113 cm/s. When sampling with a 102 mm (4 inch) filter head at 500 L/min (17.7 cfm), the face velocity (based on effective filtration area of 60 cm²) is 139 cm/s. On the other hand, when sampling at lower flowrates, as in personal air samplers, Whatman 41 would not be a good choice. With a 25 mm filter head and a flowrate of 2.5 L/min, the face velocity would only be 11.3 cm/s. For such an application, other filters more efficient at this flowrate would be preferred.

The necessity for caution in interpreting filter efficiency data in the literature is illustrated by the data of various authors for the penetration of Whatman 41 by 0.3 μm diameter particles. The most reliable data appear to be those of Rimberg (12), Stafford and Ettinger (41), and Lockhart et al. (13), which are in reasonably good agreement with one another. Lockhart's data are for both Whatman 41 and TFA-41, which is Whatman 41 packaged and sold by the Staplex Company. The differences between the two sets of data are presumably the differences to be expected from randomly selected batches. The Smith and Surprenant (46) data were based on the same techniques as the data of Stafford and Ettinger (41) and of Lockhart, i.e., light-scattering measurements of 0.3 μm DOP droplets, and the large discrepancy is inexplicable.

Rimberg (12) measured the penetration of charge-neutralized polystyrene latex spheres using a light scattering photometer. The 0.3 μm points are actually interpolated from the corresponding data for 0.365 and 0.264 μm particles. Lindekin et al. (47) used a similar technique except that they did not neutralize the electrical charge on their polystyrene test aerosols. Thus, their data appear to reflect the influence of particle charge on filter penetration.

Stafford and Ettinger (48) also compared the collection efficiencies of Whatman 41 and IPC 1478 filters for 0.3 μm DOP and latex spheres of similar sizes. The efficiencies were higher for the solid particles, especially at face velocities below 20 cm/s. They also showed that efficiency increased with loading of solid particles but not for liquid DOP droplets. Thus, some of the differences in their efficiency test results could have been due to the increase with loading during the test with the latex.

In interpreting filter efficiency data, it is also important to consider that the test data are usually based on the efficiency of a "clean" filter. For most filters, collection efficiency increases with the accumulation of solid particles on the filter surfaces. The resistance to flow also increases with increasing loading, but usually at a much slower rate. A theoretical basis for these phenomena has been developed by Davies (49). A practical implication is that even with reliable published filter efficiency and aerosol size distribution data it is not possible to know precisely what the collection efficiency of a filter will be for a given sampling interval. The filter efficiency data can only provide an estimate of the minimum collection efficiency. The actual collection efficiency will usually be higher.

Biles and Ellison (50) reported on the increase in collection efficiency for three types of cellulose fiber filters, i.e., Whatman 1, 4, and 451, for collecting lead and "black smoke" from the air of London, England. At a face velocity of 6.5 cm/s, the

clean paper efficiencies for lead were 50, 30, and 15%, respectively, and 70, 40, and 30% in terms of the light reflectance measurement for black smoke. As the percent soiling index approached 40%, the collection efficiencies of all three papers approached 100% for both lead and black smoke.

There have been reports in the literature that low concentrations of small atmospheric particles could have large penetration rates through filters like the Millipore HA or glass fiber filters (51,52). Since numerous careful investigations have shown such filters to have almost complete collection for all particle sizes and flowrates, as discussed earlier and illustrated in Table 29:III, it appears that such reports are most likely due to background or contamination problems associated with the analysis of the charcoal traps used by the investigators as back-up collectors. Kneip et al. (53) investigated the efficiency of Millipore AA and SC membrane filters and Gelman AE glass fiber filters for ambient air lead particles and laboratory generated dye aerosol particles ≤ 0.07 μm in diameter at very low loadings and face velocities as low as 1.0 cm/sec and found that all efficiencies were >99%.

29.5 REQUIREMENTS OF ANALYTICAL PROCEDURES.

29.5.1 *Sample Quantity.* In many instances the limited sensitivity of an analytical method, when combined with a low aerosol concentration, makes it necessary for large volumes of air to be sampled in order to collect sufficient material for an accurate analysis. In addition to the material being studied, background dust and co-contaminants must, unavoidably, also be collected. Therefore, it is highly desirable that the filter medium selected have the capacity to collect and retain large sample masses. Furthermore, it is usually desirable to have the sampling rate nearly uniform over the length of the sampling period. The flow resistance of all filters increases with increased loading, but some do so at much lower rates than others. Table 29:I shows the rates of resistance increase for a variety of filters when sampling the ambient air outside the Naval Research Laboratory. The loading rate would certainly differ for other aerosols and these data generally would not be applicable. However, they do indicate the relative loading characteristics of these filters. Those with low values load much more slowly than those with high values. Those filters with the lowest resistance build-up rate are most useful for collecting high-volume samples, especially when using the pressure-sensitive turbine-type blowers as air movers. In general, deep-bed fibrous filters have the lowest rates of resistance pressure increase.

29.5.2 *Sample Configuration.* Some analyses require that the sample be collected or mounted in a particular form. For example, microscopic particle size analysis can be performed only when the particles are on a flat surface. This is due to the limited depth of focus of the objective lens. In order to use fibrous filters for collecting samples for size analysis, it must be possible to remove the sample quantitatively and transfer it to a microscope stage without altering it. For such applications, the membrane and polycarbonate pore filters offer significant advantages over other filters. First, the samples can be analyzed directly on the filter surface. Second, since the sample does not have to be transferred, there is a greater likelihood that the sample observed is in the same form as when it was airborne.

Another situation in which the sample configuration may be important to the analysis is the determination of airborne radioactivity. Many radiation detectors such as Geiger-Mueller tubes and scintillation detectors are designed to view a limited surface area, usually a 2.5 cm diameter circle. Thus, to make efficient use of the detector, the effective filtering area should be limited to a similar size. An additional consideration in radiometric analysis is the depth of penetration of the particles into the filter, especially for alpha and beta emitters. The activity observed by the detector will be affected by the distance of the particles from the detector, and by absorption of radiation by intervening filter material.

Other characteristics influence the choice of filters when quantitative particulate analysis by X-ray diffraction is desired. Davis and Johnson **(54)** examined seven filter substrates of both fiber and membrane construction and found that the degree to which the filters were suitable for X-ray

diffraction analysis was primarily dependent on: 1) interfering background scatter, and 2) the mass per unit area of the particulate load collected. They found that Teflon filters were superior when mass loadings were less than 200 μg/cm^2. On the other hand, when mass loadings were greater than 300 μg/cm^2, quartz and glass fiber filters were more suitable because of their particle retention qualities and their lack of a substrate spectrum in the diffraction pattern.

29.5.3 *Sample Recovery from Filter.* High collection efficiency is valueless if all of the sample is not available for analysis. For most chemical analyses, it is necessary to either remove the sample from the filter, or to destroy the filter. Inorganic particles usually are recovered from cellulose paper filters by low temperature (plasma) ashing, wet ashing (digesting in concentrated acid), or muffling (incinerating) the filter. Samples collected on glass fiber filters can be recovered only by leaching or dissolving the sample from the filter. Samples can be recovered from membrane filters, polystyrene filters, and soluble granular beds by dissolving the filter in a suitable solvent.

Some of the membrane filters have a limited loading capacity in terms of the ability of the filter to retain the particles after they are collected. The material retained on the surface may have very poor adhesion to the surface or to itself and slough off the surface. The problem is especially severe for polycarbonate pore filters.

29.5.4 *Interferences Introduced by Filters.* Before selecting a filter for a particular application, the filter's blank count or background level of the material to be analyzed must be determined. All filters contain various elements as major, minor, and trace constituents, and the filter medium of choice for analyzing particular elements must be one with little or no background level for the elements being analyzed. The components of the filter medium itself may introduce undesirable or unacceptable background to the subsequent analyses. If the filter is dissolved or digested, then all of the material in the filter will be mixed with the sample. If it is oxidized, then the residual ash content of the filter will be mixed with the sample. On the other hand, if the sample is extracted from the filter by a solvent, the sample will contain only those components of the filter matrix which are soluble. Finally, if a non-destructive analysis, such as X-ray diffraction, is performed, the contribution of the components of the filter will depend on both the content of the filter, its distribution in space, and the amount of X-ray absorption by the matrix and sample.

Data on the composition and interference levels of some commonly used sampling filters have been presented by Zhang et al. (55), Gelman et al. (56) and Mark (57). Table 29:IV shows measured elemental impurity levels in some commonly used air sampling filters. Polycarbonate pore filters build up an electrical charge which can cause a serious weighing error when they are used in gravimetric analysis. A decrease in apparent weight of over 75 μg over 3 weeks due to the decay in charge on the filter was observed by Engelbrecht et al. (58). The charge effect was present despite a 30-s exposure to a [210]Po source prior to the weighing.

Another type of interference is inaccessibility of the sample to a measurement or sensing device. For instance, in determining reflectance of filtered particulate matter, the more the particles penetrate the surface the less they will be visible. In such an application, the sensitivity of measurement on a membrane filter surface would be greater than on a fibrous filter.

29.5.5 *Size or Mass of Filter.* The mass of the filter itself may be important in gravimetric determinations. In determining the mass of collected aerosol, the mass of the filter should be as small as possible, relative to the mass of the sample. Also, other things being equal, the less the filter weighs and/or the smaller it is, the simpler the sample handling and processing. Collecting the sample on a smaller filter may save a concentrating step in the analysis, and make it possible to use smaller analytical equipment and/or glassware.

29.6 LIMITATIONS INTRODUCED BY AMBIENT CONDITIONS.

29.6.1 *Temperature.* The temperature stability of a filter must be considered when sampling hot gases such as stack effluents. For such applications, combustible materials cannot be used, and a selection must be made from the several types of mineral, glass, or other refractory media. In order to

Table 29:IV Impurity Levels of Filter Media (ng-cm^{-2})

Element	W41	MFHA + W41	MFAA + W41	MFRA + W41	MFSS + W41
Na	150	800	740	700	250
Mg	<80	<400	<370	<340	<200
Al	12	30	17	36	38
Cl	100	1200	1400	520	540
K	15	145	62	18	5.9
Ca	140	810	560	450	150
Sc	<0.0005	0.06	0.008	0.0045	0.0049
Ti	10	25	<30	<16	<11
V	<0.03	<0.10	<0.21	<0.15	0.074
Cr	3	25	36	33	30
Mn	0.5	8	2.1	2.1	1.4
Fe	40	80	100	125	110
Co	0.1	0.3	0.25	0.14	0.22
Cu	<4	24	17	24	7.4
Zn	<25	50	<180	41	37
As	—	<0.4	0.13	<0.3	0.071
Se	—	—	1.0	0.34	0.28
Br	5	9	7.7	5	4.5
In	—	—	0.014	<0.017	0.0065
Sb	0.15	0.8	0.23	0.081	0.071
I	—	—	0.58	1.9	1.3
La	<0.2	<0.6	<0.5	0.074	0.013
Sm	—	—	0.010	0.013	0.013

—Not determined

select the appropriate medium, the peak temperature and duration of sampling must be known. Glass fiber filters are widely used for temperatures up to about 500°C.

29.6.2 *Moisture Content.* For sampling under conditions of high humidity, filter media that are relatively non-hygroscopic must be chosen. Some filters pick up moisture, and this may affect their filtering properties. If their efficiency is partially dependent on electrostatic effects, moisture may reduce it. Also when the filter picks up moisture it may become mechanically weaker and rupture more easily.

For some airborne dusts, the standards are based on gravimetric analyses without regard to dust composition. These include suspended particulate matter in the ambient air and coal mine dust. Mass concentrations are determined from the gain in weight of the filter during the sampling interval, divided by the sampled volume. Since the filter weighs much more than the sample collected on it, the accuracy of the analysis depends on the stability of the filter's weight. Serious errors can arise if some of the filter's mass is lost due to abrasion during handling between the tare and final filter weighings, or if there is a significant difference in atmospheric water vapor content at the time of analysis.

The highly variable water vapor retention characteristics of cellulose fiber filters usually rules out their selection for use when gravimetric analyses are to be performed **(59)**. However, even glass fiber and membrane filters, while much less affected by water vapor, may still have enough absorption to cause problems in gravimetric analyses. Charell and Hawley **(60)** examined the weight changes at various humidities for cellulose ester, polyvinyl chloride (PVC), and polycarbonate membrane filters. They found that all changed their weights reversibly in proportion to the water vapor concentration, that the minimum uptake was seen with polycarbonate and some PVC filters, that other PVC filters took up 6.6 times as much water, and that cellulose ester membranes took up 40 to 50 times more water vapor. Thus, pre-and post-sampling weighings should be done at the same humidity conditions. Mark **(57)** examined the weight changes associated with changes in humidity for a variety of PVC membranes,

some cellulose ester membranes, and a glass fiber filter. All showed weight changes that correlated with RH, with those of the cellulose esters being the largest. He also reported that the Sartorius PVC-type 12801 developed an electrical charge that repelled particles onto the filter holder during sampling, reducing the apparent collection efficiency. He was able to overcome this source of error by pre-treating the filters with a detergent solution.

29.6.3 *Artifact Formation.* Air sampling filters can collect gases and vapors as well as particles. When they are collected unintentionally by adsorption or absorption onto filter surfaces, or onto particles collected on those surfaces, their presence in the sample can constitute an artifact. For example, ordinary glass fiber filters are slightly alkaline and collect SO_2 while sampling ambient air. This led to overestimation of the ambient aerosol sulfate concentrations for many years.

As shown by Coutant **(61)**, Spicer and Schumacher **(62)**, and Appel et al. **(63)**, artifact particulate matter can be formed by oxidation of acidic gases (e.g., SO_2, NO_2) or by retention of gaseous nitric acid on the surface of alkaline (e.g., glass fiber) filters and other filter types. The effect is a surface-limited reaction and, depending on the concentration of the acidic gas, should be especially significant early in the sampling period. The magnitude of the resulting error depends upon such factors as the sampling period, filter composition and pH, and the relative humidity. The magnitude and the significance of artifact mass errors are variable and dependent on local conditions. Excluding the uncertainty associated with the collection and retention of organic particulate matter with appreciable vapor pressure, artifact mass primarily reflects the sum of the sulfates and nitrates formed by filter surface reactions with sulfur dioxide and nitric acid gas, respectively.

The study by Coutant **(61)** reported artifact sulfate for 24-hour samples from 0.3 to 3 $\mu g/m^3$. Stevens *et al* **(64)** found 2.5 $\mu g/m^3$ average artifact sulfate sampling at 8 sites around St. Louis, MO; and Rodes and Evans **(65)** noted 0.5 $\mu g/m^3$ artifact sulfate in West Los Angeles, CA.

Artifact sulfate formation can also occur on nylon filters. Chan et al. **(66)** examined the extent of conversion of SO_2 to sulfate on Membrana-Ghia (now Gelman) Nylasorb nylon filters used as nitric acid vapor collectors. The percent conversion was found to depend on both the concentration of SO_2 and the relative humidity.

Appel et al. (67) reported that artifact particulate nitrate on glass fiber filters is limited only by the gaseous nitric acid concentration. Such filters approximated total inorganic nitrate samplers, retaining both particulate nitrate and nitric acid even when the latter was present at very high atmospheric concentrations; e.g., 20 ppb. Nitric acid was found to represent from approximately 25 to 50% of the total inorganic nitrate at Pittsburgh, PA and Lennox and Claremont, CA. Based on an estimate of the most probable 24-hour artifact sulfate error, 3.0 $\mu g/m^3$, and of the most probable artifact particulate nitrate, 8.2 $\mu g/m^3$ in the Los Angeles, CA Basin and 3.8 $\mu g/m^3$ elsewhere, typical errors in mass due to sulfate plus nitrate artifacts are estimated at 11.2 $\mu g/m^3$ in the Los Angeles Basin and 6.8 $\mu g/m^3$ elsewhere.

Nitrate salts can be rapidly lost from inert filters (e.g., Teflon, quartz) by volatilization (65), and by reactions with acidic materials (68).

Sampling artifacts are also of serious concern for organic contaminants in air. Schwartz et al. (69) showed that the apparent concentration of extractable organics collected on glass fiber filters varied with the duration of the sampling period. They found that moderately polar organics extracted by dichloromethane were increasingly poorly recovered as the sampling period became progressively longer. This could have been due to volatilization of sampled material during continued sampling, or to their oxidation to a form not extracted by the solvent. For more polar organics extracted with cyclohexane, the apparent concentration increased with increasing sampling time, suggesting that the sampled material was behaving as a vapor adsorbent. Similar observations have been made by Appel et al. (70). Much more work is needed on the volatility of sampled material during further sampling, on chemical conversions that take place on filter substrates, and on adsorption of vapors by sampled materials before the extent and significance of those factors can be fully established.

29.7 LIMITATIONS INTRODUCED BY FILTER HOLDER.

29.7.1 *Filter Size.* In order to use any filter, it must be held securely and without leakage in an appropriate filter holder. This limits the diameter of a filter disc to a particular size, unless the filter holder is fabricated especially for the filter. Most filter media can be obtained in any desired size, but some, such as respirator filters, are preformed on molds and are available in only one size.

29.7.2 *Mechanical Properties of Filters.* Some filter holders can only be used with filters of high mechanical strength. A strong paper (e.g., Whatman #41) can be used in a simple head without a back-up screen, while soft papers (e.g., glass fiber) or brittle papers (e.g., the membrane filter) require a more elaborate holder with a firm back-up screen or mesh support to prevent rupture.

29.8 AVAILABILITY AND COST. There are great variations in the unit cost of filter media. For example, glass costs about twice as much as cellulose filter paper, while membrane filters may cost ten times as much. For large scale sampling programs, such price differentials can add up to significant annual cost increments. The less expensive paper should be chosen when the differences in performance are marginal. Ready availability is another factor to be considered. The cellulose and glass papers can be obtained from any chemical supply house, while other types may only be available from a limited number of suppliers.

29.9 SUMMARY AND CONCLUSIONS. The advantages of sampling by filtration have been discussed; filtration theory has been outlined; commercial filter media have been described; and the criteria for selecting appropriate filters for particular applications have been reviewed.

Of all the particle collection techniques, filter sampling is the most versatile. With appropriate filter media, samples can be collected in almost any form, quantity, and state. Sample handling problems are usually minimal, and many analyses can be performed directly on the filter. No single filter medium is appropriate to all problems, but a filter appropriate to any immediate problem can usually be found.

REFERENCES

1. DAVIES, C.N. 1968. The Entry of Aerosols into Sampling Tubes and Heads. Brit J Appl Physics, Ser 2, *1*:921–932.
2. IDEM. 1966. Deposition from Moving Aerosols, in "Aerosol Science," C.N. Davies, Ed., pp. 393–446. Academic Press, London.
3. PAULUS, H.J., N.A. TALVITIE, D.A. FRASER AND R.G. KEENAN. 1957. Use of Membrane Filters in Air Sampling. Am Ind Hyg Assoc Q, *18*:267–273.
4. STENHOUSE, J.I.T., J.A. HARROP AND D.C. FRESHWATER. 1970. The Mechanisms of Particle Capture in Gas Filters. Aerosol Sci, *1*:41–52.
5. EMI, H., K. OKUYAMA AND M. ADACHI, 1977. The Effect of Neighboring Fibers on the Single Fiber Interia-Interception Efficiency of Aerosols. J Chem Eng of Japan, *10*:148–153.
6. KIRSCH, A.A., I.B. STECHKINA AND N.A. FUCHS. 1975. Efficiency of Aerosol Filters Made of Ultrafine Polydisperse Fibers. J Aerosol Sci, 6119–124.
7. RAMSKILL, E.A. AND W.L. ANDERSON. 1951. The Inertial Mechanism in the Mechanical Filtration of Aerosols. J Coll Sci, *6*:415–428.
8. KIRSCH, A.A. AND U.V. ZHULANOV. 1978. Measurement of Aerosol Penetration Through High Efficiency Filters. J Aerosol Sci, *9*:291–298.
9. GENTRY, J.W., K.R. SPURNY AND J. SCHOERMANN. 1982. Diffusional Deposition of Ultrafine Aerosols on Nuclepore Filters. Atmos Environ, *16*:25–40.
10. LUNDGREN, D.A. AND K.T. WHITBY. 1965. Effect of Particle Electrostatic Charge on Filtration by Fibrous Filters. I&EC Process Des and Develop, *4*:345–350.
11. ZEBEL, G. 1965. Deposition of Aerosol Flowing Past a Cylindrical Fiber in a Uniform Electric Field. J Colloid Sci, *20*:522–543.
12. RIMBERG, D. 1969. Penetration of IPC 1478, Whatman 41, and Type 5G Filter Paper as a Function of Particle Size and Velocity. Am Ind Hyg Assoc J, *30*:394–401.
13. LOCKHART, L.B., JR., R.L. PATTERSON, JR. AND W.L. ANDERSON. 1964. Characteristics of Air Filter Used for Monitoring Airborne Radioactivity, NRL Report No. 6054, U.S. Naval Research Laboratory, Washington, DC (March 20).
14. SPURNY, K.R., J.P. LODGE, JR., E.R. FRANK AND D.C. SHEESLEY. 1969. Aerosol Filtration by Means of Nuclepore Filters: Structural and Filtration Properties. Environ Sci & Tech, *3*:453–464.
15. LIU, B.Y.H. AND K.W. LEE. 1976. Efficiency of Membrane and Nuclepore Filters for Submicrometer Aerosols. Environ Sci & Tech, *10*:345–350.
16. LEE, K.W. AND B.Y.H. LIU. 1980. On the Minimum Efficiency and the Most Penetrating Particle Size for Fibrous Filters. J Air Poll Control Assoc, *30*:377–381.
17. LEE, K.W. 1981. Maximum Penetration of Aerosol

18. Particles in Granular Bed Filters. J Aerosol Sci, *12*:79–87.
18. SCHMIDT, E.W., J.A. GIESEKE, P. GELFAND, T.W. LUGAR AND D.A. FURLONG, 1978. Filtration Theory for Granular Beds. J Air Poll Contr Assoc, *28*:143–146.
19. FICHMAN, M.C., C. GUTFINGER AND D. PNUELI. 1981. A Modified Model for the Deposition of Dust in a Granular Bed Filter. Atmos Environ, *15*:1669–1674.
20. SPURNY, K. 1986. On the Filtration of Fibrous Aerosols. J Aerosol Sci, *17*:450–455.
21. LÖFFLER, F. 1968. The Adhesion of Dust Particles to Fibrous and Particulate Surfaces. Staub (English trans.), 28:29–37.
22. PATE, J.B. AND E.C. TABOR. 1962. Analytical Aspects of the Use of Glass Fiber Filters for the Collection and Analysis of Atmospheric Particle Matter. Am Ind Hyg Assoc J, *23*:145–150.
23. KRAMER, D.N. AND P.W. MITCHEL. 1967. Evaluation of Filters for High-Volume Sampling of Atmospheric Particulates. Am Ind Hyg Assoc J, *28*:224–228.
24. SHLEIEN, B., J.A. COCHRAN AND A.G. FRIEND. 1966. Calibration of Glass Fiber Filters for Particle Size Studies. Am Ind Hyg Assoc J, *27*:253–359.
25. WINKEL, A. 1959. Uber neue Methode zur Staubmessung. Staub *19*:253.
26. BERKA, I. 1968. Organic Microfiber Filters for Sampling of Industrial Dusts. Staub (English trans.) *28*:27–28.
27. SPURNY, K. AND J. PICH. 1964. The Separation of Aerosol Particles by Means of Membrane Filters by Diffusion and Inertial Impaction. Int J Air Wat Poll, *8*:193–196.
28. MEGAW, W.J. AND R.D. WIFFEN. 1963. The Efficiency of Membrane Filters. Int J Air Wat Poll, *7*:501–509.
29. LINDEKEN, C.L., F.K. PETROCK, W.A. PHILLIPS AND R.D. TAYLOR. 1964. Surface Collection Efficiency of Large-Pore Membrane Filters. Health Physics, *10*:495–499.
30. LOSSNER, V. 1964. Die Bestimmung der Eindringtiefe von Aerosolen in Filtern. Staub, *24*:217–221.
31. GEORGE, L.A., II. 1961. Electron Microscopy and Autoradiography. Science, *133*:1423.
32. RICHARDS, R.T., D.T. DONOVAN AND J.R. HALL. 1967. A Preliminary Report on the Use of Silver Metal Membrane Filters in Sampling Coal Tar Pitch Volatiles. Am Ind Hyg Assoc J, *28*:590–594.
33. KNAUBER, J.W. AND F.H. VONDERHEIDEN. 1969. A Silver Membrane X-Ray Diffraction Technique for Quartz Samples. Presented at Amer Ind Hyg Conf, Denver, CO.
34. CAHILL, T.A., L.L. ASHBAUGH, J.B. BARONE, R.A. ELDRED, P.J. FEENEY, R.G. FLOCCHINI, C. GOODART, D.J. SHADOAN AND G.W. WOLFE. 1977. Analysis of Respirable Fractions in Atmospheric Particulates via Sequential Filtration. J Air Poll Contr Assoc, *27*:675–678.
35. PARKER, R.D., G.H. BUZZARD, T.G. DZUBAY AND J.P. BELL. 1977. A Two-Stage Respirable Aerosol Sampler Using Nuclepore Filters in Series. Atmos Environ, *11*:617–621.
36. HEIDAM, N.Z. 1981. Review: Aerosol Fractiona-

tion by Sequential Filtration with Nuclepore Filters. Atmos Environ, *15*:891–904.

37. JOHN, W., G. REISCHL, S. GOREN AND D. PLOTKIN. 1978. Anomalous Filtration of Solid Particles by Nuclepore Filters. Atmos Environ, 12:1555–1557.

38. BUZZARD, G.H. AND J.P. BELL. 1980. Experimental Filtration Efficiencies of Large Pore Nuclepore Filters. J Aerosol Sci, *11*:435–438.

39. SPURNY, K. 1977. Discussion: A Two-Stage Respirable Aerosol Sampler Using Nuclepore Filters in Series. Atmos Environ, *11*:1246.

40. GIBSON, H. AND J.H. VINCENT. 1981. The Penetration of Dust Through Porous Foam Filter Media. Ann Occup Hyg, *24*:205–215.

41. STAFFORD, R.G. AND H.J. ETTINGER. 1972. Filter Efficiency as a Function of Particle Size and Velocity. Atmos Environ, *6*:353–362.

42. LIU, B.Y.H., D.Y.H. PUI AND K.L. RUBOW. 1983. Characteristics of Air Sampling Filter Media, in "Aerosols in the Mining and Industrial Work Environment." V.A. Marple and B.Y.H. Liu, Eds. Vol. III, Ch. 70, Ann Arbor Science, pp. 989–1038.

43. SKOCYPEC, W.J. 1974. "The Efficiency of Membrane Filters for the Collection of Condensation Nuclei," M.S. Thesis. Univ. of North Carolina, School of Public Health, Chapel Hill, NC.

44. JOHN, W. AND G. REISCHL. 1978. Measurements of the Filtration Efficiencies of Selected Filter Types. Atmos Environ, *12*:2015–2019.

45. LUNDGREN, D.A. AND T.C. GUNDERSON. 1975. Efficiency and Loading Characteristics of EPA's High-Temperature Quartz Fiber Filter Media. Amer Ind Hyg Assoc J, *36*:806–872.

46. SMITH, W.J. AND N.F. SUPRENANT. 1963. Properties of Various Filtering Media for Atmospheric Dust Sampling. Presented at the Am Soc for Testing and Materials, Philadelphia, PA.

47. LINDEKEN, C.L., R.L. MORGIN AND K.F. PETROCK. 1963. Collection Efficiency of Whatman 41 Filter Paper for Submicron Aerosols. Health Physics, *9*:305–308.

48. STAFFORD, R.G. AND H.J. ETTINGER. 1971. Comparison of Filter Media against Liquid and Solid Aerosols. Amer Ind Hyg Assoc J, *32*:319–326.

49. DAVIES, C.N. 1970. The Clogging of Fibrous Aerosol Filters. Aerosol Sci, *1*:35–39.

50. BILES, B. AND J. McK. ELLISON. 1975. The Efficiency of Cellulose Fiber Filters with Respect to Lead and Black Smoke in Urban Aerosol. Atmos Environ, *9*:1030–1032.

51. ROBINSON, J.W. AND D.K. WOLCOTT. 1974. Simultaneous Determination of Particulate and Molecular Lead in the Atmosphere. Environ Lett, *6*:321–333.

52. SKOGERBOE, R.K., D.L. DICK AND P.J. LAMOTHE. 1977. Evaluation of Filter Inefficiencies for Particulate Collection Under Low Loading Conditions. Atmos Environ, *11*:243–349.

53. KNEIP, T.J., M.T. KLEINMAN, J. GORCZYNSKI AND M. LIPPMANN. 1981. A Study of Filter Penetration by Lead in New York City Air. in "Environmental Lead." D.R. Lynam, L.G. Piantanida and J.F. Cole, Eds. Academic Press, New York, pp. 291–308.

54. DAVIS B.L. AND L.R. JOHNSON. 1982. On the Use of Various Filter Substrates for Quantitative Partic-

ulate Analysis by X-ray Diffraction. Atmos Environ, *16*:273–282.

55. ZHANG, J., J. BILLIET AND R. DAMS. 1981. Stationary Sampling and Chemical Analysis of Suspended Particulate Matter in a Workplace. Staub-Reinhalt Luft, *41*:381–386.

56. GELMAN, C., D.V. MEHTA AND T.H. MELTZER. 1979. New Filter Compositions for the Analysis of Airborne Particulate and Trace Metals. Amer Ind Hyg Assoc J, *40*:926–932.

57. MARK, D. 1974. Problems Associated with the Use of Membrane Filters for Dust Sampling When Compositional Analysis is Required. Ann Occup Hyg, *17*:35–40.

58. ENGELBRECHT, D.R., T.A. CAHILL AND P.J. FEENEY. 1980. Electrostatic Effects on Gravimetric Analysis of Membrane Filters. J Air Pollut Control Assoc, *30*:391–392.

59. DEMUYNCK, M. 1975. Determination of Irreversible Absorption of Water by Cellulose Filters. Atmos Environ, *9*:523–528.

60. CHARELL, P.R. AND R.E. HAWLEY. 1981. Characteristics of Water Adsorption on Air Sampling Filters. Amer Ind Hyg Assoc J, *42*:353–360.

61. COUTANT, R.W. 1977. Effect of Environmental Variables on Collection of Atmospheric Sulfate. Environ Sci Tech, *11*:873–878.

62. SPICER, C.W. AND P.M. SCHUMACHER. 1979. Particulate Nitrate: Laboratory and Field Studies of Major Sampling Interferences. Atmos Environ, *13*:543–552.

63. APPEL, B.R., S.M. WALL, Y. TOKIWA AND M. HAIK. 1979. Interference Effects in Sampling Particulate Nitrate in Ambient Air. Atmos Environ, *13*:319–325.

64. STEVENS, R.K., T.G. DZUBAY, G. RUSSWURM AND D. RICKEL. 1978. Sampling and Analysis of Atmospheric Sulfates and Related Species. "Sulfur in the Atmosphere," Proceedings of the International Symposium, Dubrovnik, Yugoslavia, September, 1977, Atmos Environ, *12*:55–68.

65. RODES, C.E. AND G.F. EVANS. 1977. "Summary of LACS Integrated Measurements." EPA-600/4-77-034. U.S. Environmental Protection Agency, Research Triangle Park, NC.

66. CHAN, W.H., D.B. ORR AND D.H.S. CHUNG. 1986. An Evaluation of Artifact SO_4 Formation on Nylon Filters under Field Conditions. Atmos Environ, *20*:2397–2401.

67. APPEL, B.R. AND Y. TOKIWA. 1981. Atmospheric Particulate Nitrate Sampling Errors Due to Reactions with Particulate and Gaseous Strong Acids. Atmos Environ, *15*:1087–1089.

68. HARKER, A., L. RICHARDS AND W. CLARK. 1977. Effect of Atmospheric SO_2 Photochemistry Upon Observed Nitrate Concentrations. Atmos Environ, *11*:87–91.

69. SCHWARTZ, G.P., J.M. DAISEY AND P.J. LIOY. 1981. Effect of Sampling Duration on the Concentration of Particulate Organics Collected on Glass Fiber Filters. Amer Ind Hyg Assoc J, *42*:258–263.

70. APPEL, B.R., E.M. HOFFER, E.L. KOTHNY, S.M. WALL, M. HAIK AND R.L. KNIGHTS. 1979. Analysis of Carbonaceous Material in Southern California Atmospheric Aerosols, 2. Environ Sci Tech, *13*:98–104.

Adapted from "Air Sampling Instruments,"
7th edition, by permission of American Council of
Governmental Industrial Hygienists.

Subcommittee 9
M. LIPPMANN
J.N. HARMAN, *Chairman*

30. X-Ray Powder Diffraction

30.1 INTRODUCTION. X-ray diffraction (XRD) is the basis of a unique method of air monitoring; most of the airborne dusts to be monitored are crystalline compounds and therefore give distinctive X-ray diffraction patterns. X-ray powder diffraction is primarily used for inorganic compounds, an area in which elemental analysis has been the traditional method. But elemental analysis is insufficient when different compounds of a given element and even different crystalline forms of one compound may have different health effects. X-ray powder diffraction is not only an excellent tool for identifying any of these substances, but has been used for measuring airborne concentrations of some of them with relative success.

30.2 PRINCIPLE OF THE METHOD.

30.2.1 *Theory.* X-rays striking a substance are scattered by the electrons in that substance. For crystals, the scattered X-rays are reinforced at certain angles (θ), each of which depends on the wavelength (λ) of the radiation and the spacing (d) between the planes of atoms in the crystal causing reinforcement at that angle. This reinforcing phenomenon and the angle at which it occurs is often called a "reflection". The relationship between λ, θ, and d is the Bragg equation:

$$n \lambda = 2d \sin \theta \qquad (1)$$

Since higher order reflections can be ignored ($n = 1$), and the wavelength of radiation for a given instrumental setup is known, the quantity on the left is fixed. If one of the variables on the right can be determined by some method, the other can be calculated. The common method in X-ray diffraction is to use a mechanical device called a goniometer which moves the X-ray source and the detector on the circumference of a circle with the sample at the center of the circle (Figure 30:1). The goniometer precisely maintains equal angles (θ) between the plane of the flat powder specimen and the X-ray beams coming from the source and diffracting to the detector. A dial on the goniometer indicates the angle θ, or, more often, twice that angle (2θ). Since the angle θ at which a reflection (diffraction maximum) occurs is now known, a value of d for the interplanar spacing can be calculated. The units for d are the same as for λ, usually nanometers. The older data use the angstrom unit, which is 0.1 nm. A number of general descriptions of X-ray diffraction are available (1–4), and there are some articles describing its use for air monitoring (5,6).

30.2.2 *Qualitative Analysis.* Because each crystalline substance is a unique arrangement of various kinds of atoms, each has a different diffraction pattern (in several cases, different compounds exhibit indistinguishable diffraction patterns). About 30,000 diffraction patterns are published by the International Committee on Powder Diffraction Standards (7). The patterns are indexed in several ways so that the pattern of an unknown substance can be matched with one of the 30,000 patterns, and the substance identified. For example, the Hannawalt Method lists compounds in groups, each group covering a range of d spacings corresponding to those compounds' most intense peaks. Within a group, the listing is in order of decreasing secondary d spacing, and so on. Using this, or other manual methods, compounds can be identified. Mixtures are more difficult unless some of the compounds are known so that only the remaining peaks need be identified. In all cases, an elemental analysis is most helpful.

There are also available several computerized "search-match" programs; some of these are discussed in reference 8. In addition to the d values and intensities from the diffraction pattern, most programs will request any elemental identifications that can be made as well as an "error window" which will be used by the matching algorithm to allow for discrepancies between measured and true d values. Computerized searching of diffraction patterns is less reliable than for mass spectra because the data quality is not as good. Nevertheless, computer searching can be quite useful considering the relatively low cost of using some systems.

30.2.3 *Quantitative Analysis.* Because XRD sample holders are generally designed to hold bulk samples, some modifications to the holders may be necessary in order to analyze filter samples. Quantitative analysis of dust on filters is possible down to about 20 micrograms with a long counting time on the most intense peak of the analyte.

As always in quantitative analysis, standards must be used. The standards can be made by depositing known amounts of pure analyte on filters. Because the samples contain some matrix with the analyte, one method of analysis adapts the internal reference or internal "standard" method long used for bulk samples to nullify matrix effects **(9)**. In this method, a set amount (e.g.: 200 μg) of a non-interfering compound is co-deposited with all standards and samples. The intensity of a diffraction peak of this internal standard is divided into the analyte intensity for each filter analyzed and the results for standards and samples are compared in this form.

Another method of analysis leaves out the internal standard, taking advantage of the fact that matrix effects are less severe in the thin dust on filters than in bulk sample cakes. Analyte intensities for standards and samples are compared directly. One matrix effect, absorption of X-rays, can become significant with heavily loaded filters, however, and a method of compensating for the error is desirable. This is possible by using a silver filter to collect the dust. If the intensity of a silver reflection is measured before and after the dust is deposited, the ratio of these intensities (T) is a measure of the absorption of X-rays by the sample. A function of this T value:

$$f(T) = \frac{-R \ln T}{1 - T^R} \quad (2)$$

can be multiplied by the apparent mass of analyte to obtain a corrected mass **(10–12)**. In this function, R is the ratio of the sines of the θ angles of the analyte and silver peaks. This method has been referred to as the substrate standard method **(13)**.

30.2.4 *Interferences.* Mixtures of compounds very often have diffraction patterns with overlapping peaks. This will not usually be a problem in the qualitative analysis of bulk samples that give many peaks, but does complicate the quantitation of microgram quantities of an analyte whose primary peak is interfered with. The solution to this problem is simple if the analyte has another peak, less intense than the primary peak, but sufficiently intense for quantitation at the levels required. Another solution is possible if peak overlap is slight; longer wavelength radiation may be used to spread the peaks over a greater angular range (cf. equation 1). Still another solution is possible for mixtures of compounds that are separable by chemical means (e.g., differing solubilities) or physical means (e.g., differing magnetic properties). Lastly, the computer implementation of the mathematical procedure of deconvolution of diffraction peaks may be useful for assigning accurate peak areas to each of several overlapping peaks.

30.3 SAMPLE COLLECTION AND PREPARATION.

30.3.1 *Types of Samples.* Air monitoring for dusts implies the use of filter samples. For monitoring worker exposure to airborne dusts, portable battery–operated pumps can be attached to a small sampling device clipped to the collar. The sampling device includes a cassette that holds the filter and a cyclone that is a particle size selector. Dust–laden air enters the cyclone first, where sharp turns cause particles larger than 10 μm diameter to impact on the walls. Smaller particles, which are considered respirable, are passed to the filter. The amount of sample collected in a personal sampler is small, which often makes positive identification of dusts difficult. To solve this problem, and for measuring general air pollution, larger samples are employed. The components (pump, size selector, and filter) are the same for collecting these area respirable samples, but the capacity of each component is higher, particularly the pump, operated from the commercial power lines. Thus, much more dust is available for analysis. A greater variety of size selectors (cyclones, impactors, and elutriators) is available for such samplers. Another method, much less desirable, for collecting a large amount of dust is the collection of settled bulk dust. This is possible if dust has settled on surfaces above the points where dust has been generated; such "rafter samples" are much closer to respirable size than any bulk sample from the parent material.

30.3.2 *Filters for Personal Samples.* For the collection of personal samples, polymeric membrane filters, usually polyvinyl chloride (PVC) are employed. PVC filters of 5 μm pore size have good collection efficiency and have pressure drops compatible with the use of battery-operated pumps. The dust on the PVC collection filter, however, is not necessarily uniform because of the design of most collection devices. Therefore, the dust is redeposited on another filter for X-ray analysis. The filter most often used is the silver membrane filter. It allows the absorption correction already mentioned, but is sometimes chosen only for its very low scattered X-ray background.

30.3.3 *Redeposition.* Transferring all of the dust on a collection filter to a silver filter for analysis is best done by first destroying the collection filter. There are two methods in use for removal of the collection filter: dissolution and ashing. The dissolution of PVC filters in tetrahydrofuran is very fast, while ashing may be advantageous for the removal of organic dusts. Low-temperature R.F. plasma ashers are more convenient than muffle furnaces and avoid the unwanted reactions of sample matrix that sometimes occur at high temperatures. After ashing, a liquid such as isopropyl alcohol, or water with a wetting agent, is added to the sample containers, and from this point the ashed and dissolved filters are treated similarly. An internal standard is added at this point if desired. Thorough suspension of the sample dust in the liquid is attained by ultrasonic agitation. The suspension is then transferred to a filtering apparatus containing the silver filter. Washings of the sample vessel may be added to the liquid in the filtering apparatus, but after filtering, the silver filter should not be washed to avoid disturbing the deposited dust layer. After drying, the silver filter is ready for analysis.

30.4 STANDARDS AND CALIBRATION.

30.4.1 *Powder Standards.* Standard materials of high purity are desirable in X-ray diffraction, but properties other than purity are very important. A specimen may be pure elementally but may not be a suitable standard if it has low crystallinity. Particle size is also a consideration since X-ray diffraction intensities are sensitive to particle size. The particle size of respirable dust samples may vary somewhat, and if possible, one of several powder standards should be chosen by comparison under a microscope to match each set of dust samples. Although standard powders may be prepared in a laboratory by grinding large crystals, and sieving or otherwise separating the oversize particles, it is preferable to find a large supply of a standard that could be available to as many laboratories as perform the analysis. Presently, such standards exist for quartz and a few other common compounds **(14)**.

30.4.2 *Calibration Standards.* Once a standard powder has been chosen, there are several methods for preparing calibration standards on filters. The filters ultimately used for the measurement of standards and samples should be the same, usually silver. One method is to place the powder in a dust generator and disperse it in a cloud from which air samples are drawn on open-face filters. The weight of standard powder on each filter can be found by weighing the filters before and after sampling. A second method is to weigh out a given amount of standard powder and disperse it in the same liquid being used to disperse samples (e.g.: 10 mg in 1 L). Aliquots of this suspension are drawn by pipet immediately after thorough mixing and transferred to the filtering apparatus for deposition on a filter. A third method is similar to the second method in that deposition is from a liquid but each standard filter is made by weighing out the amount of powder to be deposited on it. Each of these methods has its advantages and disadvantages.

30.5 DATA COLLECTION.

30.5.1 *Instrument Calibration.* Quantitative analysis of air samples requires a stable diffractometer, optimized for intensity. The diffractometer usually has a "copper X-ray tube" which means the X-ray tube has a copper anode (the X-RAY SOURCE in Figure 30:1) which strongly emits 0.154 nm X-rays when struck by electrons. Near the X-ray source is the divergence slit, which should be chosen to illuminate as much of the sample as possible with

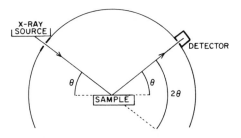

Figure 30:1 — Geometry of X-ray diffractometer

X-rays. The receiving slit in front of the detector is also chosen to have a large acceptance angle (e.g., 1 degree) for diffracted X-rays. Outfitting the diffractometer with an automatic sample changer and a computer to control the instrument and process the data is extremely helpful (15). In addition, a reference standard is very helpful for optimum quality control. A reference standard can be any stable specimen that is permanently affixed to a holder and that has a sharp diffraction peak. Arkansas stone (microcrystalline quartz) is commonly used but mica chips and other materials have also been used. The intensity of the reference standard should be checked whenever the instrument is used. With an automated instrument, this reference intensity can easily be checked as often as before every sample if desired. Additionally, the most recently obtained value of reference intensity can be divided into the intensity measured on every standard and sample to help eliminate errors due to drift in X-ray output or amplifier detector settings. The intensities for standards are, of course, plotted against mass to form a calibration curve, and if these intensities are "normalized" against the reference intensity as just described, then intensities for samples should be similarly normalized.

30.5.2 *Intensity Measurements.* The intensities discussed above may be either peak heights or peak areas. The use of peak heights is simple, and a set of standards would give an acceptable calibration line, but peak areas are preferred for good precision (1). For peak area measurements the goniometer is set to an angle at the side of the peak and moved in small steps (e.g.: $0.02°$ every 5 s) through the peak to the other side. X-ray counts are accumulated

throughout the period to give the peak area plus background. Backgrounds may be measured at either side of the peak for a period equal to half the time required for the step scan. The background counts are added together, then subtracted from step scan counts to obtain the peak area. One advantage of peak areas is that they are less susceptible to changes in peak shape that may occur with different powders of differing particle size. Another advantage is that there is no need to determine the exact peak position, which is made difficult by the small peaks encountered in air monitoring work. The measurement of reference specimen intensity may be done by peak height; the reference intensity will be high so that peak location is facilitated and counting precision is good for relatively short counting times, and, of course, peak shape is unimportant.

30.5.3 *Absorption Corrections.* As mentioned above, there are two commonly used methods for correcting matrix effects. One is the internal standard method, which can only be performed if the internal standard material was added during the sample and standard preparation steps. In this case, the correction involves only the extra step of measuring the intensity of an internal standard peak and dividing it into the analyte intensity for both standards and samples. The other, the substrate standard method, avoids the complication of adding the internal standard, but requires that a silver filter be used and that measurements be made on the silver peak both with and without the dust in place. This need not be done for smaller loadings of most analytes and matrices because the absorbance can be shown to be negligible. In the cases where absorbance may be significant, as for unknown matrices and/or higher loadings, one technique is to measure the silver intensity through the sample dust and then use the reverse face of that sample filter to measure the silver intensity without absorption. Measurement times are short since the silver peak is intense. The factor obtained from the equation above can then be used as an absorption correction to the apparent mass of an analyte. A table of correction factors for measured T values is convenient (as in method P&CAM 259, Volume 5 of Reference 16).

30.6 CONCLUSIONS. The technique of X-ray diffraction for the measurement of airborne contaminants presently finds limited application but has further potential. The greatest use for XRD has been for silica but it has been applied to asbestos, talc, oxides of zinc, zirconium, mercury, and others. The applicability of XRD is limited when, as in asbestos analysis, particle shape must also be known. Because XRD can determine the exact composition of crystalline dusts, it has an advantage over elemental analyses that cannot so speciate. Therefore, X-ray diffraction is likely to find increased use in air monitoring applications.

REFERENCES

1. KLUG, H.P. AND L.E. ALEXANDER. 1974. "X-Ray Diffraction Procedures for Polycrystalline and Amorphour Materials," Second Edition, John Wiley & Sons, NY.
2. CULLITY, B.D. 1978. "Elements of X-Ray Diffraction." 2nd edition. Addison Wesley Publishing Company, Inc., Reading, MA.
3. WARREN, B.E. 1969. "X-Ray Diffraction." Addison Wesley Publishing Company, Inc., Reading, MA.
4. JENKINS, R. AND J.L. DE VRIES. 1972. "Introduction to X-Ray Powder Diffractometry." N V Philips Gloeilampenfabrieken, Eindhoven, 7000.02.3770.11, 40 pp.
5. NENADIC, C.M., J.V. CRABLE. 1973. Applications of X-Ray Diffraction in Occupational Health Studies, in "Progress in Analytical Chemistry," T.L. Simmons and G.W. Ewing, Eds., Vol. 6, pp. 81–101, Plenum Press, NY.
6. DOLLBERG, D.D., M.T. ABELL AND B.A. LANGE. 1980. Occupational Health Analytical Chemistry: Quantitation Using X-Ray Powder Diffraction, in "Analytical Techniques in Occupational Health Chemistry," D.D. Dollberg and A.W. Verstuyft, Eds. ACS Symposium Series 120, pp. 43–66.
7. JCPDS—International Centre for Diffraction Data, Powder Diffraction File, published annually, Swarthmore, PA.
8. NICHOLS, M.C. AND Q. JOHNSON. 1980. The Search-Match Problem, in "Advances in X-Ray Analysis," Vol. 23, p. 273, Plenum Press, NY.
9. BUMSTED, H.E. 1973. Determination of Alpha-Quartz in the Respirable Portion of Airborne Particulates by X-Ray Diffraction. Amer Ind Hyg Assoc J, 34:150.
10. WILLIAMS, P.P. 1959. Direct Quantitative Diffractometric Analysis, Anal Chem, 31:1842.
11. LEROUX, J., A.B.C. DAVEY AND A. PAILLARD. 1973. Proposed Methodology for the Evaluation of Silicosis Hazards. Amer Ind Hyg Assoc J, 34:409.
12. ABELL, M.T., D.D. DOLLBERG, B.A. LANGE, R.W. HORNUNG AND J.C. HAARTZ. 1981. Absorption Corrections in X-Ray Diffraction Dust Analyses: Procedures Employing Silver Filters, in "Electron Microscopy and X-Ray Applications," P. Russel and A. Hutchings, Eds., Vol. II, Ann Arbor Science, Ann Arbor, MI.
13. LANGE, B.A. AND J.C. HAARTZ. 1979. Determination of Microgram Quantities of Asbestos by X-Ray Diffraction: Chrysotile in Thin Dust Layers of Matrix Material. Anal Chem, 51:520.
14. GRAF, J.L., P.K. ASE AND R.G. DRAFTZ. 1979. Preparation and Characterization of Analytical Reference Minerals. NIOSH Contract 210–75–0043, NTIS, Springfield, VA 22161, PB-80–148612.
15. ABELL, M.T., D.D. DOLLBERG AND J.V. CRABLE. 1981. Quantitative Analysis of Dust Samples from Occupational Environments Using Computer Automated X-Ray Diffraction, in "Advances in X-Ray Analysis," Vol. 24, pp. 37–48.
16. "NIOSH MANUAL OF ANALYTICAL METHODS," D.G. Taylor and J.V. Crable, Eds. NIOSH Publication Nos. 77–157 (Vols. 1–3), 78–175 (Vol. 4), 79–141 (Vol. 5), 80–125 (Vol. 6).

Subcommittee 9
M. ABELL
J.N. HARMAN, Chairman

31. X-Ray Fluorescence Analysis

31.1 INTRODUCTION.

31.1.1 X-ray fluorescence (XRF) is convenient for elemental analysis for several reasons: (1) lack of sample preparation; (2) speed of analysis; (3) indifference to the chemical state of the analyte element; and (4) non-destruction of the sample (so that the sample may be stored and analyzed subsequently). The XRF technique is applicable to every element past the first row of the periodic table, whether in bulk, liquid or filter samples.

31.1.2 Two significantly different types of XRF instrumentation are available: wavelength dispersive (WDXRF) and energy dispersive (EDXRF). Although quite different in their operation, they both have roughly comparable sensitivities. The analysis time is shorter with EDXRF, but more extensive computer data reduction is usually required for the quantitative analysis of mixtures.

31.1.3 Microanalysis instruments may use X-ray fluorescence to perform elemental analysis on very small areas ($\sim 1\mu m$) or of individual particles. Electron microprobe analysis instruments apply EDXRF or WDXRF to electron microscopy, while scanning electron microscopes usually employ EDXRF.

31.1.4 XRF is applicable for both qualitative and quantitative analysis. The latter requires exacting standardization procedures and computer reduction of the data. A number of techniques are available for quantitative analysis, which will be discussed below.

31.1.5 Difficulties with the XRF method include the influence of particle size, certain required sample characteristics, the overlap of the X-ray emission spectra of many elements, and the effects of interfering elements (X-ray absorption and enhancement).

31.2 THEORY AND OPERATIONAL PRINCIPLES (1).

31.2.1 X-rays are photons of electromagnetic radiation having a wavelength between those of ultraviolet and cosmic rays (10^{-6} to 10 nm). X-rays corresponding to the lower electronic levels of atoms, about 0.01 to 2 nm, are commonly employed by commercial X-ray spectrometers. Like other forms of electromagnetic radiation, X-rays may be absorbed, transmitted, reflected, polarized, or diffracted by matter. X-rays are designated by their wavelength, λ (usually in nanometers), or by energy (in electron volts, $eV = 1239.6/\lambda$).

31.2.2 Two processes contribute to the absorption of an X-ray by an atom: photoelectric absorption and scatter.

31.2.3 Photoelectric absorption may occur if the energy of the X-ray is sufficient to remove an inner electron from the atom. For example, iron will absorb X-rays of energies exceeding 7112.1 eV ($\lambda = 0.1743$ nm), the energy required to remove the 1s electron. This is the iron K absorption edge. L absorption, the removal of a second shell electron (2s or 2p), is of analytical importance for elements past cesium. Lastly, M absorption, the removal of a third shell electron (3s, 3p, or 3d) may be used for very heavy elements. (The K, L, and M absorptions can be further separated into the absorptions due to individual orbitals.) Because the 1s orbitals are least affected by the chemical state of the atom, and because higher energy X-rays are more penetrating, K absorption is usually the preferred method of X-ray excitation.

31.2.4 Scatter may involve a decrease in energy of the radiation (Compton or incoherent scatter), or no change in wavelength (Rayleigh or coherent scatter). Scattering of X-rays is important because the detection limit is defined in relation to the

background radiation, which is primarily scattered excitation radiation.

31.2.5 Following removal of an electron by an X-ray, one of two de-excitation processes can occur: (1) An electron in a higher orbital may fill the hole left by the emitted electron. An X-ray with an energy equal to the difference between the two energy levels (and thus less energetic than the excitation X-ray) will be emitted; this is X-ray fluorescence. (2) An electron in an upper orbital may drop down and fill the hole as above, but the excess energy is removed not by an X-ray but by the emission of another electron. This is denoted Auger emission, and is more common with very light elements.

31.2.6 The techniques of XRF, Auger, XPES (X-ray excited photoelectron spectroscopy), UV (Ultraviolet absorption and emission spectroscopy), and UPES (ultraviolet photoelectron spectroscopy) are closely related (Figure 31:1). The measurement of photoelectrons produced by ionization of core orbitals is called XPES, while the measurement of the radiation produced when these levels are filled by higher level electrons is XRF. Measurement of the energy of emitted valence orbital electrons excited by ultraviolet radiation is UPES. Ultraviolet absorption spectroscopy is the measurement of radiation absorbed by electrons excited from the valence level to unoccupied excited levels; the reverse processes result in optical emission (fluorescence and phosphorescence).

31.2.7 An XRF spectrometer must contain a source of X-ray excitation, the specimen to be analyzed, a disperser, a detector, and a display.

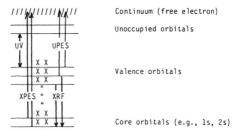

Continuum (free electron)

Unoccupied orbitals

Valence orbitals

Core orbitals (e.g., 1s, 2s)

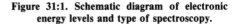

Figure 31:1. Schematic diagram of electronic energy levels and type of spectroscopy.

31.2.8 Several types of primary X-ray radiation sources are commercially used. The X-ray tube, producing X-rays by focussing a beam of accelerated electrons onto a metal target, is common for applications requiring high intensities, such as on WDXRF spectrometers. A particle or electron beam directed to the area of analysis is used in electron microprobe instruments. Radioactive nuclides are light, simple sources of low intensity X-rays, which are used in portable or light duty applications. Lastly, it is quite convenient in some applications to use secondary excitation, in which X-rays from a primary source such as an X-ray tube excite X-radiation from a metal foil; the latter are used to excite the sample.

31.2.9 Two methods of dispersion are in use. The WDXRF spectrometer utilizes a crystal that disperses the X-radiation emitted from the sample according to Bragg's law. A particular wavelength λ will be diffracted by the planes of a crystal lattice of dimension d, at an angle θ as follows

$$n \lambda = 2d \sin \theta$$

where: n = the order of the diffraction (always an integer)
 λ = the wavelength of the dispersed X-ray
 d = the distance between lattice planes
 θ = the angle of diffraction

A detector is placed at the appropriate angle to detect a particular wavelength. The detector may be a gas-flow meter or a scintillation counter. The X-ray source, sample, and detector are mounted on a goniometer which maintains the components at the proper angle. Several crystals may be used to cover the entire X-ray energy range. Quite good resolution is possible, minimizing the occurrence of overlapping fluorescence lines.

31.2.10 The EDXRF spectrometer, however, uses a different method. Here, dispersion and detection are combined into a single unit, usually a lithium-drifted silicon crystal detector with a multichannel analyzer. Both the intensity and the energy of the X-rays are measured simultaneously. Although simpler and cheaper than crystal dispersion, the resolution is not as good.

Also, the detector must be under vacuum and cooled in liquid nitrogen when used.

31.3 SAMPLING.

31.3.1 In general, air sampling usually employs the use of either personal or area sampling with filters. The first technique, personal sampling, is the most appropriate for worker exposure assessments in industrial hygiene. A small monitor cassette containing an appropriate filter is mounted on an individual. A battery-powered pump draws air through the monitor at a predetermined rate. After a suitable sampling time, the filter is removed from the cassettes and mounted directly on the XRF instrument for analysis. Area sampling, the second technique, is most appropriate for environmental assessments. In this technique, samples are obtained with a stationary unit consisting of a high-volume pump connected to a large diameter filter. As a result, much more analyte is collected than with a personal sampler.

31.3.2 A good sampling strategy should endeavor to attain the following optimal sample characteristics: (1) The sample size should match the X-ray beam size, and be compatible with the spectrometer. (2) The particles should be quite small ($<5\mu$m), or be of known size (*vide infra*). (3) The deposition should be homogeneous throughout the sample, both in gross deposition characteristics and in particle characteristics, as standards of similar composition will be required in most experimental regimens. (4) If the deposition is thin, the primary and emitted X-rays are not absorbed by the matrix, and subsequent conversion of X-ray intensities into concentrations is simplified considerably.

31.3.3 Filter material should have the following characteristics: (1) The filter should be of high purity, especially for the analytes expected to be present. For example, glass fiber filters have been reported to be high in metal impurities (2). (2) Filter material will absorb X-rays, especially low energy X-rays associated with lighter elements. A deposition on the surface of the filter, with minimum penetration, minimizes filter absorption. Consequently, a membrane filter is preferable to a depth filter such as a glass fiber filter (3) A filter of low mass will scatter the incident radiation minimally, resulting in low background radiation. (4) The filter material should not react chemically with the analyte or matrix.

31.4 ANALYSIS.

31.4.1 As a first approximation, the intensity of a particular analyte emission is proportional to the amount of analyte present. To perform a qualitative analysis, the detector output is presented as wavelength (or energy) vs intensity (counts/second); the higher the intensity above the background, the more analyte is present.

31.4.2 Quantitative analysis is complicated by a number of effects. Prominent among them are spectral line overlap, X-ray absorption and enhancement effects, and particle size effects. These are discussed individually below.

31.4.3 Line overlap occurs when the emissions of two elements are both detected simultaneously (i.e., at the same WDXRF goniometer setting or in the same EDXRF channel). In many cases, an analyte has another X-ray emission line that can be used. If not, corrections for line overlap can be computed by a suitable algorithm if the spectra of the pure elements are known. The poorer resolution of EDXRF instrumentation increases the necessity of these corrections. However, modern EDXRF instruments may obtain a rapid elemental analysis by computer-matching the spectra of the unknown sample to a library of pure element spectra (2).

31.4.4 An interelement interference occurs if (1) the primary or emitted radiation is absorbed by the matrix, resulting in an analyte emission line that is decreased relative to what would be found in a matrix of the pure analyte (absorption); or, (2) the emission of an interferent may be absorbed by the analyte, causing the line intensity of the latter to exceed that which would be found in pure analyte (enhancement). (Higher order interferences exist; see reference 1 for a full discussion.) These effects arise if the thickness of the deposit requires the X-rays to pass through substantial amounts of matrix material. The effect occurs in loadings exceeding a limit defined as m = 0.1 /(u/d), where m = the mass per unit area (g/cm²), and (u/d) is the sum of the mass-absorption coefficients (cm²/g) of the specimen for the primary and emitted X-rays (3). For iron oxide this maximum

thickness is about 50 μg/cm². The intensity of these interelement effects is dependent on the amount of the interferent in the sample, which is not known *a priori*. Consequently, the correction may be applied iteratively, and a number of successful algorithms exist. These corrections rely on published correction coefficients, or the coefficients may be generated from standards. Instrument product literature is the best source of information on which corrective algorithms are available on a particular instrument. Fortunately, these effects are not likely to be problematic in lightly deposited air filter samples.

31.4.5 A special manifestation of the absorption interference is the particle size effect. This effect arises because of the X-ray shadowing or shielding of one side of a particle by the opposite side, and is the spherical analog of the depth absorption described above. A mathematical correction for the effects of particle size of analytes in various matrices can be applied if the size is known (6), or the effect can be avoided by the use of standards containing particles of size and constitution similar to the samples.

31.4.6 A second type of particle size effect arises due to inhomogeneity of the particles. Since the matrix of an analyte will be different in the different kinds of particles, the absorption and emission effects will differ, and standardization may be difficult (7).

31.4.7 Penetration of the analyte into the filter, resulting in absorption of the primary and emitted X-rays, may be a significant source of error with light elements. The correction for this effect requires an estimation on the profile of deposition inside the filter; a number of correction routines exist (4,5).

31.5 STANDARDIZATION METHODS.

31.5.1 The mathematical expressions available for the conversion of XRF analyte line intensity into mass loadings will depend on the particular instrument used. Generally, these expressions take the form

$$C = K(I-B)F$$

where C = deposited mass, in g/cm²
I = intensity of analyte line radiation

K = conversion factor unique to the particular instrument and experimental arrangement. This is determined using appropriate standards.

F = correction factors appropriate to the system; e.g., inter-element, absorption, or particle size corrections.

31.5.2 Standards for XRF are of considerable importance. They should have the following characteristics: (1) They should closely mimic the physical and chemical characteristics of the samples; (2) they should be capable of withstanding exposure to X-rays; and (3) they should be capable of long-term storage without change.

31.5.3 Commercial standards are available from a variety of sources. Thin-film standards useful for air filter analysis are available from Micromatter Corporation, Rt. 1 Box 72-B, Eastsound, WA, 98245, and from Columbia Scientific Industries Corporation, P.O. Box 203190, Austin, TX, 78720. Bulk standards are available from Spex Industries, Inc., P.O. Box 798, Metuchen, N.J., 08840, or from the National Bureau of Standards, Washington, D.C.

31.5.4 Laboratory-made standards may be cheaper and more easily obtained (8). A number of techniques have been developed. For example, selected samples analyzed by an independent method are likely to have particle size and composition similar to the other samples (9). Simulated field samples, made by filter sampling a dust cloud of appropriate particles from a laboratory dust generator and analyzed by an independent method, have been used (10). Plastic standards can be made by dissolving an organometallic compound in a plastic forming solution, which is then dried or polymerized (11,12). An analyte solution may be sprayed onto a filter and dried; the deposition is then determined by weighing (8). Vacuum sublimation of the analyte onto a filter or a plate has been used (13). Analyte solutions may be pipetted onto substrates and dried (14), or solutions can be applied to a hydrophobic-edge filter and dried (15). A dispersion of an appropriate powder in a non-dissolving solution may be filtered and dried, and analyzed by an independent method or weighed to determine the amount of analyte present (16).

31.6 DETECTION LIMITS.

31.6.1 The detection limit in XRF is usually defined as the amount of analyte that gives a net line intensity equal to three times the square root of the background intensity for a specified counting time. The lowest limit of quantitation is defined as three times the detection limit (17).

31.6.2 For WDXRF, the detection limits for the different elements differ greatly; however, detection limits of 5–20 ng/cm^2 for deposits on Millipore filters have been reported (18). EDXRF detection limits approach 30–100 ng/cm^2, again with considerable variation among elements and with different instruments (19). In electron-probe microanalysis, as little as 10^{-15} grams may be qualitatively detected within the very small electron beam area ($< 1\mu m$) (20).

31.7 CONCLUSIONS.

31.7.1 The increased interest in atmospheric and industrial pollution is likely to result in a greater use of X-ray fluorescence analysis. The introduction of computer-operated instrumentation has mitigated the severity of a number of experimental difficulties such as particle size, overlapping X-ray fluorescence lines, and interelement effects, and has greatly shortened the analysis time.

REFERENCES

1. BERTIN, E.P. 1978. "Principles and Practice of X-Ray Spectrometric Analysis," 2nd Edition, Plenum, NY.
2. DZUBAY, T.G. AND D.G. RICKEL. 1978. X-ray Analysis of Filter-Collected Aerosol Particles, in "Electron Microscopy and X-ray Applications," P.A. Russell and A.E. Hutchings, Eds., Ann Arbor Science.
3. RHODES, R.J. 1973. Energy-Dispersive X-Ray Spectrometry for Multielement Pollution Analysis, Amer Lab, 5:57.
4. ADAMS, F.C. AND J. BILLIET. 1976. Experimental Verification of the X-Ray Absorption Correction in Aerosol Loaded Filters, X-Ray Spectrum 5(4):188. Adams, F.C. and R.E. Van Grieken. 1975. Absorption Correction for X-ray Fluorescence Analysis of Aerosol Loaded Filters. Anal Chem, 47:1767.
5. DAVIS, D.W., R.L. REYNOLDS, G.C. TSOU, AND L. ZAFONTE. 1977. Filter Attenuation Corrections for the X-Ray Fluorescence Analysis of Atmospheric Aerosols, Anal Chem, 49:1990.

6. CRISS, J.W. 1976. Particle Size and Composition Effects in X-Ray Fluorescence Analysis of Pollution Samples, Anal Chem, *48*:179.
7. BERTIN, E.P., (op. cit.) p. 747.
8. GIANQUE, R.D., R.B. GARRETT AND L.Y. GODA. 1977. Calibration of Energy-Dispersive X-Ray Spectrometers for Analysis of Thin Environmental Samples, in "X-Ray Fluorescence Analysis of Environmental Samples," T.G. Dzubay, Ed., Ann Arbor Science, p. 181.
9. BERTIN, E.P., (op. cit.) p. 713.
10. GRENNFELT, P., A AKERSTROM AND C. BROSSET. 1971. Determination of Filter-Collected Airborne Matter by X-Ray Fluorescence, Atmos Environ, *5*:1.
11. DZUBAY, T.G., N. MOROSOFF, G.L. WHITAKER AND H. YASUDA. 1980. Evaluation of Polymer Film Standards for X-ray Spectrometers, presented at the 4th Symposium on Electron Microscopy and X-ray Applications to Environmental Health and Occupational Health, State College, Penn.
12. BILLIET, J., R. DAMS AND J. HOSTE. 1980. Multielement Thin Film Standards for X-ray Fluorescence Analysis, X-Ray Spectrom., *9*:206.
13. HEAGNEY, J.M. AND J.S. HEAGNEY. 1979. "Thin Film X-Ray Fluorescence Calibration Standards." Nucl. Instrum. Methods, *167* (1):137-138.
14. BAUM, R.M., R.D. WILLIS, R.L. WALTER, W.F. GUTKNECHT AND A.R. STILES. 1977. Solution-Deposited Standards Using a Capillary Matrix and Lyophilization, in T. Dzubay (op. cit.) p. 165.
15. CARSEY, T.P., unpublished data.
16. SEMMLER, R.A., R.D. DRAFTZ AND J. PURETZ. 1977. Thin Layer Standards for the Calibration of X-ray Spectrometers, in T. Dzubay (op. cit.) p. 181.
17. BERTIN, E.P., (op. cit.) p. 531.
18. JAKLEVIC, J.M. AND R.L. WALTER. 1977. Comparison of Minimum Detectable Limits Among X-ray Spectrometers, in T. Dzubay (op. cit.) p. 63.
19. GILFRICH, J.V., P.G. BURKHALTER AND L.S. BIRKS. 1973. X-ray Spectrometry for Particulate Air Pollution—A Quantitative Comparison of Techniques, Anal Chem, *45*:2002.
20. BERTIN, E.P., (op. cit.) p. 903.

Subcommittee 9
T.P. CARSEY
J.N. HARMAN, *Chairman*

32. Chemiluminescent Analyzers

32.1 INTRODUCTION.

32.1.1 Chemiluminescent analyzers for gaseous analysis represent a significant advance over manual or automated wet chemical analytical techniques because no aqueous reagents are required, the instrumental determination of the species of interest is available in real time ($\sim$1–5 seconds response time), and the instrumental techniques show few interferences. Additional advantages are linearity, sensitivity and ease of use over a wide dynamic measurement range. Gas phase chemiluminescence is defined as the production of visible or infrared radiation by the reaction of two gaseous species to form an excited species product that decays to its ground state by the photoemissive act. Other possibilities exist for the decay of the excited species; instrumental design attempts to minimize deactivation by non-photoemissive events such as collision with a chamber wall or third body gaseous species.

32.1.2 Chemiluminescence has been studied on a fundamental laboratory basis (1,2,3) and these principles were practically implemented in 1970 to the analysis of nitric oxide by Fontijn (4) and adapted to the analysis of automotive exhaust emissions by Nikki (5). Further work on the application of this approach to the measurement of NO_2 by thermal decomposition of NO_2 to NO has been reported by Sigsby, et al. (6). Although these analyzers were developed for the analysis of high NOx levels from emissions sources, the methodology has been extended to the much lower levels that are typically encountered in ambient air (7,8). Commercially available chemiluminescent analyzers for oxides of nitrogen may cover typical ranges of 0.25 to 25 ppm for ambient monitoring purposes and 10 to 2500 ppm for source monitoring purposes.

32.1.3 Another significant species determined by chemiluminescence is ozone, and was first reported by Nederbragt, et al. (9). If ozone is reacted with ethylene, a photoexcited aldehyde product is formed that gives off blue light as it decays to its ground state, and this light emission is proportional to the ozone concentration. Other nitrogen-based gaseous species have been determined by the gas phase chemiluminescent techniques (10).

32.1.4 The two primary problems found in the application of the chemiluminescent techniques to the analysis of pollutant species are the integrity of the converter employed in NOx analyzers and the influence of changes in the background gas when determinations of a species in a background gas matrix other than ambient air are conducted, especially at source level concentrations of the species of interest (11).

32.2 CHEMILUMINESCENT ANALYZER DESIGN.

32.2.1 Commercially available chemiluminescent analyzers are of the same general design configuration. Incorporated in the analyzer are a pump to supply the sample to the reaction chamber, a source of the reactant gas to combine with the constituent of interest, a chamber to allow the reaction of the two gases, a photomultiplier tube to detect the chemiluminescent radiation, which is proportional to the concentration of the species of interest, and signal processing electronics. One significant difference in commercially available instruments is whether the reaction chamber is held at atmospheric pressure or is operated at subatmospheric pressure. Atmospheric pressure operation is effective in preventing possible dilution upstream of the reaction chamber by leaks in the upstream sample handling system, but suffers from the disadvantage of showing higher susceptibility to variations in the background gas sample composition. Subatmospheric pressure operation minimizes the effects of background gas composition changes but aggravates the possibility of sample dilution of unknown and variable amount if leaks are present ahead of the reaction chamber.

32.2.2   A typical block diagram of a chemiluminescent analyzer is shown in Figure 32:1. Some analyzers incorporate a thermoelectric cooler to maintain the photomultiplier tube at a low, constant temperature to reduce the magnitude of the dark current through the photomultiplier.

32.3 PRECAUTIONS.

32.3.1   Chemiluminescent NOx analyzers should be periodically characterized to assess the efficiency of the converter used to perform the NO_2-to-NO reduction process. Some converters employed in current instrumentation may become more active with time, and require periodic downward readjustment of the converter temperature as the catalyst becomes more active, to prevent the further reduction of the NO product from NO_2 to molecular nitrogen. Attention should be paid to this conversion efficiency test as the converters have limited life and are subject to poisoning from other gaseous species.

32.3.2   The sample handling system used to deliver the sample from the sampling point to the input of the analyzer should be composed of FEP or TFE Teflon or glass in order to prevent losses of the sample by adsorption or absorption by the tubing walls. These effects have been documented by many workers (12,13). Steps should be taken to minimize the intrusion of water from rain if outdoor sampling is being conducted, and care should be taken to minimize the condensation of ambient humidified air into liquid water by localized cold spots in the sampling system (liquid water present in the sampling system is capable of dissolving significant amounts of water-soluble gases like O_3 and NO_2).

32.3.3   Calibration of chemiluminescent analyzers to assure data validity and to check instrumental response should be conducted with permeation tube standards (14), standard O_3–producing UV lamps, standard gas mixtures or by gas-phase titration techniques (15,16). A good general introduction to various calibration techniques and hardware used in generating standard gases is given by Nelson (17).

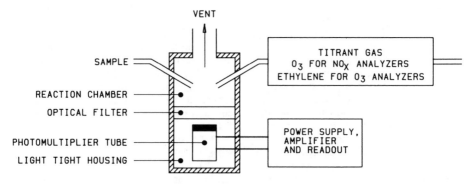

Figure 32:1 — Chemiluminescent measurement.

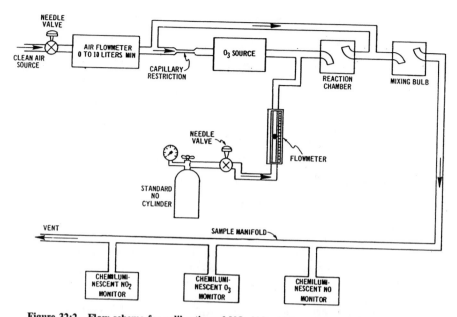

Figure 32:2 — Flow scheme for calibration of NO, NO₂, NOₓ, and O₃ monitors by GPT.

32.3.4 Commercially-obtained standards prepared in metal cylinders should have the concentration stated by the vendor determined separately and should be periodically monitored for changes in concentration. Time-related changes are generally due to the reaction of the desired species with cylinder walls; aluminum cylinders coated with proprietary inerting coatings are now available from gas standard manufacturers and have demonstrated much improved stability **(18)**. Components between the standard cylinder and the analyzer input should not be reactive to the species of interest, so the use of Teflon tubing and stainless steel pressure regulators as opposed to metal tubing or brass pressure regulators is recommended.

32.3.5 Permeation tube sources have been extensively used for the calibration of NO_2 analyzers **(19)**. They offer the disadvantages of the requirement of process temperature control, but allow for determinations of the emission rate by direct gravimetric determinations. Unfortunately, they are not available for NO. Performance characteristics of generalized permeation tube ppb-ppm level standard pollutants have been documented **(20,21)**.

32.3.6 The gas phase titration technique offers a convenient way of preparing dynamic NO_2 standards and of checking O_3 and NO by decrement as they react in the gas phase titration apparatus by the following:

$$O_3 + NO \rightarrow NO_2 + O_2$$

A typical gas phase titration system is shown in Figure 32:2. Complete information on the assembly and operation of the gas phase titration system is contained in the references.

32.4 SUMMARY. Chemiluminescent determinations offer the advantages of real time analysis, specificity and high sensitivity by means of a technique that has been widely accepted. Careful precautions must be exercised in the preparation of gas standards and in calibration of the analyzers, and attention must be paid to the use of inert materials in the sample handling system prior to the analyzers. Periodic checks should be performed on the condition of any converters employed in the analyzer.

REFERENCES

1. GREAVES, J.C. AND D. GARVIN. 1959. Chemically Induced Molecular Excitation: Excitation Spectrum

of the Nitric Oxide Ozone System. J Chem Phys, 30:348.

2. CLYNE, M.A.A., B.A. THRUSH AND R.P. WAYNE. 1964. Kinetics of the Chemiluminescent Reaction Between Nitric Oxide and Ozone. Trans Faraday Soc, 60:359.

3. CLOUGH, P.N. AND B.A. THRUSH. 1967. Mechanism of Chemiluminescent Reaction Between Nitric Oxide and Ozone. Trans Faraday Soc, 63:915.

4. FONTIJN, A., A.J. SABADELL AND R.J. RONCO. 1970. Homogeneous Chemiluminescent Measurement of Nitric Oxide with Ozone. Anal Chem, 42:575.

5. NIKI, H., A. WARNICK AND R.R. LORD. 1971. An Ozone-NO Chemiluminescent Method for NO Analysis in Piston and Turbine Engines. Presented at SAE meeting, Detroit, MI SAE paper #710072.

6. SIGSBY, J.E., JR., F.M. BLACK, T.A. BELLAR AND D.L. KLOSTERMAN. 1973. Chemiluminescent Method for Analysis of Nitrogen Compounds in Mobile Source Emissions (NO, NO_2, and NH_3), Environ Sci & Tech, 7:51–54.

7. STEVENS, R.K., T. CLARK, R. BAUMGARDNER AND J.A. HODGESON. 1974. Instrumentation for the Measurement of Nitrogen Dioxide. Instrumentation for Air Quality, p. 44 ff, ASTM STP 555, ASTM Philadelphia, PA.

8. HODGESON, J.A., K.A. REHME, B.E. MARTIN AND R.K. STEVENS. 1972. Measurement for Atmospheric Oxides of Nitrogen and Ammonia by Chemiluminescence. Presented at APCA meeting, Miami, FL.

9. NEDERBRAGT, L., A VAN DER HORST AND J. VANDUIJN. 1965. Rapid Ozone Determination Near an Accelerator. Nature, 206:87.

10. WINER, A., J.W. PETERS, J.P. SMITH AND J.N. PITTS, JR. 1974. Response of Commercial Chemiluminescent NO-NO_2 Analyzers to Other Nitrogen-Containing Compounds. Environ Sci & Tech, 8:1118.

11. MATTHEWS, R.D., R.F. SAWYER AND R.W. SCHEFER. 1977. Interferences in Chemiluminescent Measurement of NO and NO_2 Emissions from Combustion Systems. Environ Sci & Tech, 11: 1092.

12. WHITE, A. AND L.M. BEDDOWS. 1973. The Choice of Sampling Tube Material in the Determination of Nitrogen Oxide Concentrations in Products of Combustion. J Appl Chem Biotechnol, 23:759.

13. SAMUELSEN, G.S. AND J.N. HARMAN III. 1977. Chemical Transformations of Nitrogen Oxides While Sampling Combustion Products. J Air Poll Control Assoc, 27:648–655.

14. O'KEEFFE, A.E. AND G.C. ORTMAN. 1966. Primary Standards for Trace Gas Analysis. Anal Chem, 38:760.

15. REHME, K. 1976. Application of Gas Phase Titration in the Calibration of Nitric Oxide, Nitrogen Dioxide and Oxygen Analyzers. Calibration in Air Monitoring, ASTM STP 598, p. 198 ff, ASTM, Philadelphia, PA.

16. ELLIS, E.C. 1975. Technical Assistance Document for the Chemiluminescence Measurement of Nitrogen Dioxide. Final Report. EPA 600/4-75-003.

17. NELSON, G.O. 1971. "Controlled Test Atmospheres." Ann Arbor Science Pub. Ann Arbor, MI.

18. WECHTER, S.G. 1976. Preparation of Stable Pollution Gas Standards Using Treated Aluminum Cylinders, in "Calibration in Air Monitoring." ASTM STP 598, p. 40 ff, ASTM, Philadelphia, PA.

19. WILLIAMS, D. 1976. Permeation Tube Equilibration Times and Long-Term Stability, in "Calibration in Air Monitoring," ASTM STP 598, p. 183 ff, ASTM, Philadelphia, PA.

20. DEBBRECHT, F. AND E.M. NEEL. 1976. Application and Description of a Portable Calibration System, in "Calibration in Air Monitoring," ASTM STP 598, p. 55 ff, ASTM, Philadelphia, PA.

21. SCARINGELLI, F.P., A.E. O'KEEFFE, E. ROSENBERG AND J.P. BELL. 1970. Preparation of Known Concentrations of Gases and Vapors with Permeation Devices Calibrated Gravimetrically. Anal Chem, 42:871.

Subcommittee 9
J.N. HARMAN, Chairman

33. Anodic Stripping Voltammetry (ASV)

33.1 INTRODUCTION.

33.1.1 Anodic stripping voltammetry (ASV) is a sensitive and precise electroanalytical technique for the detection of a number of metals in trace concentrations in environmental samples. Many heavy metals are toxic to living organisms and can produce undesirable effects even when present in extremely low concentrations. Since such metals are not biodegradable, they may accumulate and persist in the environment for long periods of time. ASV involves the use of voltammetric technique to concentrate metals by reduction onto a micro-electrode, followed by reoxidation at the anode to produce a peak-shaped graph of the current response as a function of potential. The method has the advantages of preconcentration to increase sensitivity and also of selectivity due to deposition of the metal at a known, constant potential.

33.1.2 ASV is applicable principally to the ions of the following metals: Ag, As, Au, Ba, Bi, Cd, Cu, Ga, Ge, Hg, In, Mg, Ni, Pb, Pt, Sb, Sm, Sn, Tl, Zn. It is also applicable to halide species, such as Br^-, Cl^-, and I^-. The method requires dissolution of the sample but is essentially nondestructive, and provides both quantitative and qualitative information. It is suitable for the determination of any gaseous or particulate species that is soluble in an appropriate collection medium and is therefore readily applicable to samples collected from polluted air and water sources.

33.2 PRINCIPLE OF THE METHOD.

33.2.1 ASV measurements are conducted in an electrochemical apparatus consisting of a glass container or cell to hold the sample solution and a system of three electrodes. The electrode system consists of a working electrode, such as a mercury drop electrode or a thin film mercury electrode, a reference electrode, and an auxiliary electrode of platinum foil. A glass tube containing a frit is also placed in the cell for bubbling pure nitrogen through the sample solution to remove dissolved oxygen.

33.2.2 The method has been reviewed by Schieffer and Blaedel (1) and more recently by Wang (2). The first step in the ASV technique involves the electrolytic deposition of a small portion of the metal ions in solution into the mercury electrode to preconcentrate the metals. This is followed by the measurement or stripping step which involves the dissolution of the deposit. The preconcentration is usually carried out by cathodic deposition at a controlled potential and time. During this step the metal ions reach the electrode surface by diffusion and convection, facilitated by stirring the solution or rotating the working electrode. At this stage the metal ions are reduced and concentrated as amalgams in the mercury. The selected duration of this deposition step is dependent upon the concentration of the metal ions in solution and may vary from less than 1 min at a level of 0.1 mg/L (0.1 ppm) to about 10 min at the level of 1.0 μg/L (1 ppb).

33.2.3 After this deposition step, the potential is scanned anodically or toward more positive potentials in a potential-time waveform. During this scan, the metals are stripped out of the mercury electrode in an order that is a function of the standard potential of each metal and are reoxidized to yield the measured anodic peak currents.

33.2.4 The basis of the theory and diagnostic parameters employed in ASV analysis is illustrated in the following equations. For the mercury hanging drop electrode, the stripping peak current, i_p, is derived by the following equation:

$$i_p = kn^{3/2}D^{2/3}rv^{1/2} \, tmC_b \quad (1)$$

where k = constant, n = number of electrons, D = diffusion coefficient (cm²/s) of

the metal ion in solution, r = radius of the mercury drop, cm; v = potential scan rate, V/s; t = deposition time, s; m = mass transport coefficient, and C_b = concentration of the metal ions in the bulk solution, mol/cm³.

33.2.5 The stripping peak potential, E_p, is given by

$$E_p = E_{1/2} - \frac{1.1 \, RT}{nF} \quad (2)$$

where $E_{1/2}$ = polarographic half-wave potential, R = gas constant, T = absolute temperature, F = Faraday's constant.

33.2.6 If a thin film mercury electrode is employed, the stripping peak current, i_p, is governed by the following relation, if the film thickness is less than 10 μm, so that the diffusion in the mercury does not limit the stripping current. It is also assumed that the solution is stirred during the ASV measurement. This equation was derived by Roe and Toni (3).

$$i_p = \frac{(nF)^2}{RT} \cdot \frac{Av}{e} \cdot DtmC_b \quad (3)$$

where A = electrode area, cm²; e = base of Naperian logarithms and the other parameters are designated as above.

33.2.7 The stripping peak potential for the thin mercury film electrode is given by:

$$E_p = E_o + \frac{2.3 \, RT}{nF} \log \left(\frac{nF}{RT} \cdot \frac{\delta v \ell}{D} \right) \quad (4)$$

E_o is the formal (standard) redox potential of the metal cation species being determined; δ = the diffusion layer thickness (cm); ℓ = mercury film thickness.

33.2.8 Blaedel and Klatt (4) developed a relation for the limiting current in the case of tubular electrodes used for the deposition of a trace metal under conditions of laminar flow. This relation was used by Schieffer and Blaedel (1) to obtain the following expression for the anodic stripping peak current for tubular electrodes.

$$i_p = 2.023 \left(\frac{(nF)^2}{RT} \right) vD^{2/3}V_f^{1/3}X^{2/3}C_b t \quad (5)$$

V_f = volume flow rate, cm³/s; X = length of the tubular electrode, cm.

33.2.9 It is noted that the area, A, of a tubular electrode is given by $A = 2\pi rX$, where r is the radius of the tube, cm. The following expression can be derived for the diffusion layer thickness δ (cm).

$$\delta = 1.142 \, D^{1/3} \, X^{1/3} r V_f^{-1/3} \qquad (6)$$

Insertion of Equation 6 into Equation 4 for δ gives the following expression for E_p, the stripping peak potential for tubular electrodes.

$$E_p =$$

$$E^\circ + 2.3 \frac{RT}{nF} \log \left(1.142 \frac{nF}{RT} \cdot \frac{X^{1/3}rv\ell}{D^{2/3}V_f^{1/3}}\right) \qquad (7)$$

Equation 7 yields the stripping peak potential as a function of the electrode geometry, scan rate, mercury film thickness and flow rate.

33.2.10 The peak potential, E_p, is a characteristic of each metal and is related to the standard potential of its redox couple. It is useful, therefore, for qualitative identification. The peak current height, i_p, is proportional to the concentration of the corresponding metal ion in the test solution. For each metal under test, the concentration is derived by a calibration curve or by a standard addition.

33.3 RANGE, ACCURACY AND PRECISION.

33.3.1 The lower limit of the range of concentration depends upon the sample volume and the allowable deposition time. A sensitivity of the order of 10 ng is attainable and limits as low as 10^{-2}ng have been reached at $\pm 10\%$ relative standard deviation. The upper limit can be established by appropriate, accurate dilution of the sample which may consist of collected particles dissolved in an appropriate medium, or samples of lake or river water, sea water, leached soil, urine or blood.

33.3.2 Accuracy and precision may be determined by calibration against known standards or by investigation of the accuracy (%) as a function of deposition time coupled with comparisons against known standards. The following precision levels may be attained at corresponding metal concentration levels.

Concentration, ppb Level	% Relative Standard Deviation
100	2
10	4
1	8
0.01	12

A typical current-potential voltammogram is shown in Figure 33:1, as abstracted from Wang (2).

33.4 INTERFERENCES.

33.4.1 High purity nitrogen should be bubbled through the sample solution for about 10 minutes prior to the deposition step to remove dissolved oxygen which, if present, is electrolytically coreduced at the mercury electrode with the desired species, leading to lesser desired species deposition. Intermetallic compound formation and overlapping stripping peaks are the main types of interferences in ASV. The cause of overlapping peaks is similarity in oxidation potentials and this may cause interferences in the simultaneous determination of lead with tin, cadmium with thallium or bismuth

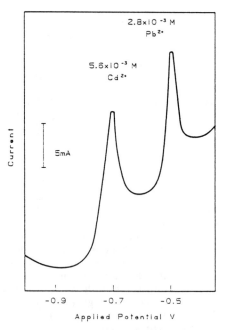

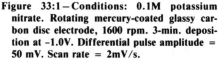

Figure 33:1 — Conditions: 0.1M potassium nitrate. Rotating mercury-coated glassy carbon disc electrode, 1600 rpm. 3-min. deposition at –1.0V. Differential pulse amplitude = 50 mV. Scan rate = 2mV/s.

with antimony. The problem can be minimized by the selective complexation of one of the two metals or the use of different supporting electrolytes to alter the peak potential.

33.4.2 The size or position of the peak current for a particular metal may be affected by the formation of intermetallic compounds in or near the working electrode. The deposition of zinc and copper in mercury results in such a mutual interference. This interference may be avoided or minimized by use of the hanging mercury drop electrode rather than a thin film mercury electrode. Other procedures for preventing the formation of intermetallic compounds are the addition of a third element that preferentially combines with one of the metals (for example, the combination of gallium with copper allows the determination of zinc); or the use of a deposition potential that prevents the plating of one of the metal ions.

33.4.3 Since ASV procedure involves the handling of extremely dilute solutions, adequate precautions must be taken to prevent contamination in trace analysis practice, such as sample preparation, storage and purity of reagents and standards, cleanliness of glassware, and prevention of losses due to adsorption or alterations of dissolved levels of trace metals.

33.4.4 To prevent electrical migration of the metal ion due to the influence of the electrical field, it is necessary to introduce into the sample solution a 100-fold excess of a supporting electrolyte such as an inorganic salt, base or mineral acid. Dissolved organic matter in the sample can cause interferences by adsorption of an organic layer on the electrode surface. This may decrease the rate of metal deposition or change the reversibility of the metal oxidation reaction, resulting in a broader and lower i_p peak or a shift of E_p to more positive values. The interferences by adsorption of organic matter during ASV analysis have been discussed in detail by Brezonik et al. **(5)**. Organic matter in the sample should be removed prior to ASV analysis or destroyed by ozone oxidation or UV irradiation in the case of water samples. Wang **(2)** has tabulated a number of selected measures for avoiding interferences in ASV analysis for trace metals in "real-life" samples.

33.5 APPLICATIONS OF ASV.

33.5.1 Airborne particulate matter can be collected on a suitable filter, such as Millipore, by use of a filtration pump or other sampling instrument. However, the filter material and organic matter content of the airborne particles must be destroyed by low temperature ashing, and the sample residue dissolved completely, prior to ASV analysis. A relatively small sample may be used for analysis because of the high sensitivity of ASV.

33.5.2 Many environmental, clinical and industrial applications of ASV for trace metal analysis have been reported, such as toxic metals in rain and snow **(6,7)**, soil samples **(8)**, blood **(9)**, urine **(10)**, hair **(11)** and food samples **(12)**. Measurements can be made with any polarographic instrument, especially one that operates in the differential-pulse mode. Several commercial ASV systems are available for laboratory analysis. For *in situ*, real-time analysis, continuous flow-through ASV systems have been developed to detect cases of unusual metal contamination in water environments **(13,14)**. Some of these instruments are based on the use of graphite coated with mercury in the form of tubular electrodes.

33.5.3 The measurement of toxic metals in airborne particles, water, soil and other environmental media by ASV has been proved to be a rapid, accurate and sensitive technique that is comparatively inexpensive. The results are comparable with those obtained by atomic absorption spectroscopy. The equipment is available from a considerable number of U.S. manufacturers, including Sargent-Welch, I.B.M. Instruments, Environmental Sciences Associates, McKee-Pedersen Instruments and others.

REFERENCES

1. SCHIEFFER, G.W. AND W.J. BLAEDEL. 1977. Study of Anodic Stripping Volammetry with Collection at Tubular Electrodes. Anal Chem, *49*:49.
2. WANG, J. 1982. Anodic Stripping Voltammetry as an Analytical Tool. Environ Sci Technol, *16*:104A.
3. ROE, D.K. AND J.E.A. TONI. 1965. An Equation for Anodic Stripping Curves of Thin Mercury-Film Electrodes. Anal Chem, *37*:1503.
4. BLAEDEL, W.J. AND L.N. KLATT. 1966. Revers-

ible Charge Transfer at the Tubular Platinum Electrode. Anal Chem, 38:879.

5. BREZONIK, P.L., P.A. BRAUNER AND W. STUMM. 1976. Trace Metal Analysis by Anodic Stripping Volammetry: Effect of Sorption by Natural and Model Organic Compounds. Water Research, 10:605.

6. EISNER, U. AND H.B. MARK, JR. 1970. The Anodic Stripping Volammetry of Trace Silver Solutions Employing Graphite Electrodes: Application to Silver Analysis of Rain and Snow Samples from Silver Iodide Seeded Coulds. J Electroanal Chem, 24:345.

7. LANDY, M.P. 1980. An Evaluation of Differential-Pulse Anodic-Stripping Voltammetry at a Rotating Glassy-Carbon Electrode for the Determination of Cadmium, Copper, Lead and Zinc in Antarctic Snow Samples. Anal Chim Acta, 121:39.

8. EDMONDS, T.E., P. GUOGANG AND T.S. WEST. 1980. The Differential Pulse Anodic Stripping Voltammetry of Copper and Lead and Their Determination in EDTA Extracts of Soils with the Mercury Film Glassy Carbon Electrode. Anal Chim Acta, 120:41.

9. OEHME, W. AND W. LUND. 1979. Comparison of Digestion Procedures for the Determination of Heavy Metals (Cadmium, Copper, Lead) in Blood by Anodic Stripping Voltammetry. Fresenius' Z. Anal Chem, 298:260.

10. LUND, W. AND R. ERICKSEN. 1979. The Determination of Cadmium, Lead and Copper in Urine by Differential Pulse Anodic Stripping Voltammetry. Anal Chim Acts, 107:37.

11. CHITTLEBOROUGH, G. AND B.J. STEEL. 1980. The Determination of Zinc, Cadmium, Lead and Copper in Human Hair by Differential Pulse Anodic Stripping Voltammetry at a Hanging Mercury Drop Electrode After Nitrate Fusion. Anal Chim Acta, 119:235.

12. FIORINO, J., R. MORRISON, A. WOODSON, R. GAJAN, G. HUSKEY AND R. SCHOLZ. 1973. Determination of Lead in Evaporated Milk by Atomic Absorption Spectrophotometry and Anodic Stripping Voltammetry. J Assoc Off Anal Chem, 56:1246.

13. WANG, J. AND M. ARIEL. 1977. Anodic Stripping Voltammetry in a Flow Through Cell with a Fixed Mercury Film Glassy Carbon Disk Electrode. J Electroanal Chem Interfacial Electrochem, 83:217.

14. ZIRINO, A., S. LIEBERMAN AND C. CLAVELL. 1978. Measurement of Copper and Zinc in San Diego Bay by Automated Anodic Stripping Voltammetry. Environ Sci & Technol, 12:73.

Subcommittee 9
M. KATZ
J.N. HARMAN, *Chairman*

34. Ion Chromatography

34.1 INTRODUCTION.

34.1.1 The method of ion exchange chromatography has been practiced for decades for selected applications. However, because of practical problems associated with detection methods, this technique did not achieve popularity until the mid-1970's.

In 1975, H. Small, T. S. Stevens and W. C. Bauman of the Dow Chemical Company published their paper **(1)** "Novel Ion Exchange Chromatographic Method Using Conductimetric Detection." The advances detailed in this paper set the stage for the wide-spread use of Ion Chromatography as a routine analytical technique.

34.1.2 The two significant aspects of this pioneering work were **(2)**:

(1) The development of a reproducible, low capacity, high efficiency anion exchange resin with selectivities identical to totally aminated materials but with better permeabilities, thus permitting the use of high flow rates.

(2) The development of post-column enhancement techniques (i.e., chemical suppression), which allowed the direct measurement of approximately 10 ng of an ionic species using an electrolytic conductivity detector.

Since that time, the practical application of Ion Chromatography has grown steadily and rapidly, due to the ongoing development of:

1. Improved separator column materials,
2. Alternate separation modes,
3. Improved post-column enhancement devices, and
4. A variety of additional detectors.

Presently, this method is routinely applied to a very broad range of inorganic (Table 34:I) and organic (Table 34:II) ions.

34.2 TYPICAL ION CHROMATOGRAPHIC SYSTEM.

34.2.1 A generalized schematic of an Ion Chromatograph is shown in Figure 34:1. The following is a description of how this apparatus is used for the determination of several anions in water solution (e.g. fluoride, chloride, nitrate, sulfate etc.).

34.2.2 With the sampling valve in the load position, the desired volume of an aqueous sample (e.g. 50 μL) is loaded into the sample loop. The valve is then switched to the inject position, where the eluent (e.g. sodium hydroxide in water), pumped by a high pressure pump, sweeps (e.g., 2 mL/min) the injected sample onto the separator (anion exchange) column. Here, the anions are retained momentarily by displacing hy-

Table 34:I Inorganic Ions by Ion Chromatography

Inorganic Ions		Inorganic Complexes
Aluminum	Mercury	Chromium EDTA
Ammonium	Molybdate	Cobalt EDTA
Arsenate	Monofluorophosphate	Copper EDTA
Arsenite	Nickel	Lead EDTA
Azide	Nitrate	Nickel EDTA
Barium	Nitrite	Cobalt Cyanide
Borate	Perchlorate	Gold (I, II) Cyanide
Bromide	Periodate	Iron (II, III) Cyanide
Cadmium	Phosphate	Palladium Cyanide
Calcium	Platinum	Platinum Cyanide
Carbonate	Potassium	Silver Cyanide
Cesium	Pyrophosphate	
Chlorate	Rhenate	
Chloride	Rubidium	
Chromate	Selenate	
Cobalt	Selenite	
Copper	Silicate	
Cyanide	Sodium	
Cyanate	Strontium	
Dithionate	Sulfate	
Fluoride	Sulfide	
Gold	Sulfite	
Hydrazine	Tetrafluoroborate	
Hypochlorite	Thiocyanate	
Hypophosphite	Thiosulfate	
Iodate	Tripolyphosphate	
Iodide	Tungstate	
Iridium	Uranium	
Iron (II, III)	Vanadate	
Lead	Zinc	
Lithium		
Magnesium		

Table 34:II Organic Ions by Ion Chromatography

Class	Examples
Amines	Methyl amine, diethanolamine
Amino acids	Alanine, threonine, tyrosine
Carboxylic acids	Acetate, oxalate, citrate, benzoate, trichloroacetate
Carbohydrates	Lactose, sucrose, xylitol, cellobiose, maltononose
Chelating agents	EDTA, NTA, DTPA
Quaternary ammonium compounds	Tetrabutylammonium ion, cetylpyridinium ion
Nucleosides	Adenosine monophosphate, guanidine monophosphate
Phenols	Phenol, chlorophenol
Phosphates	Dimethylphosphate
Phosphonates	Dequest 2000®, Dequest 2010®
Phosphonium compounds	Tetrabutylphosphonium ion
Sulfates	Lauryl sulfate
Sulfonates	Linear alkyl benzene sulfonate, hexane sulfonate
Sulfonium compounds	Trimethylsulfonium ion
Vitamins	Ascorbic acid

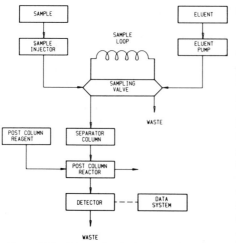

Figure 34:1 — Schematic of a basic ion chromatograph.

droxide from the anion resin. As the eluent continues to flow, the water and cations from the sample, as well as the displaced hydroxide, are swept through the column.

34.2.3 In time (e.g., 1 to 2 min), the retained anions are displaced (exchanged) by the hydroxide in the eluent and move down the column. Those anions (e.g., fluoride and chloride) that compete weakly with hydroxide for the anion exchange sites move faster down the column compared to nitrate and sulfate, which compete more strongly for the anion sites. This is the process by which the anions in the samples are separated from each other.

34.2.4 After the separator column, the eluent stream then passes through a post-column device (e.g., micromembrane suppressor) and then to a conductivity detector, where the separated anions are detected. In the suppressor, cations in both the eluent and the sample migrate across a membrane, while hydronium ions from a dilute sulfuric acid solution on the opposite side of the membrane replace the original cations in the eluent and combine with the hydroxide ions to form water. This exchange produces two beneficial effects. These are:

(1) The sodium hydroxide in the eluent is converted to water, which produces a low conductivity, low noise background and

(2) Each anion measured is associated with a hydronium ion which, on average, increases its conductivity at least tenfold.

This particular post–column enhancement process is called chemical suppression. Its benefit is to increase substantially the signal-to-noise ratio of the measurement. This produces an analytical method with both a high sensitivity (e.g., 10 ppb by direct injection, 10 ppt by preconcentration) and wide dynamic range (e.g., three decades).

34.2.5 The output of the conductivity detector is read by a strip chart recorder, a chromatographic integrator, or a computer system for data analysis and presentation. A typical analog anion chromatogram is shown in Figure 34:2. The anions are identified, based on their retention times, and compared to known standards. Quantitation is accomplished by measuring the height or area of the analyte peaks and comparing them to calibration curves generated from known standards.

34.2.6 Cations can be determined in a manner similar to anions, except the eluent is usually a dilute acid (e.g., hydrochloric acid) and the separator column is a cation exchange column. The postcolumn device is a cation micromembrane suppressor, where anions in the eluent are exchanged for hydroxide ions, producing a low conductivity, low noise background. A typical cation chromatogram is shown in Figure 34:3.

34.3 MODES OF SEPARATION.

34.3.1 There are three types or modes of separation currently used in ion chromatography (3). These are called High Performance Ion Chromatography (HPIC), High Performance Ion Chromatography by Exclusion (HPICE) and Mobil Phase Ion Chromatography (MPIC). These modes each employ different packing materials with different ion exchange capacities and each mode performs separations based on different mechanisms.

34.3.2 *HPIC Separation Mode.* The HPIC mode is characterized by direct ion exchange between ions in the sample solution and exchange sites on the separator column. A description of this mode has been given in the previous examples.

34.3.3 *HPICE Separation Mode.* The HPICE mode of separation uses a highly

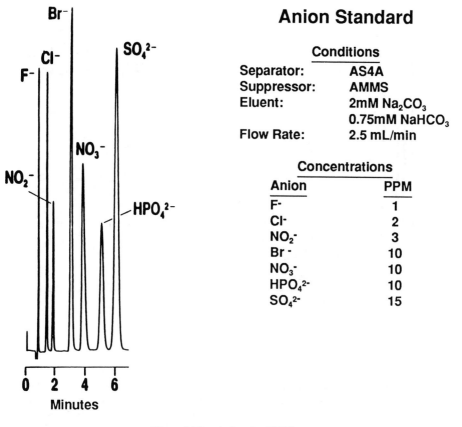

Anion Standard

Conditions

Separator:	AS4A
Suppressor:	AMMS
Eluent:	2mM Na_2CO_3
	0.75mM $NaHCO_3$
Flow Rate:	2.5 mL/min

Concentrations

Anion	PPM
F-	1
Cl-	2
NO_2^-	3
Br -	10
NO_3^-	10
HPO_4^{2-}	10
SO_4^{2-}	15

Figure 34.2. – Anions by HPIC.

sulfonated microporous cation exchange resin and a strong alkylsulfonic acid (RSO_3H) eluent. Under these conditions, a semipermeable charge layer builds up on the surface of the resin. Under these conditions, highly ionized species (e.g., chloride, nitrate, etc.) are rejected or excluded from interacting with the resin and are thus swept quickly through the separator column. On the other hand, species that are more neutral in the acid environment of the eluent (e.g., organic acids) can diffuse through the charged layer and interact with the resin.

In this separation mode, species that tend to ionize poorly (i.e., high pKa values) will interact more strongly and thus will be retained longer. Conversely, species that have a greater tendency to ionize (i.e., low pKa values) will interact less and will be eluted early.

As in HPIC, the HPICE method uses a micromembrane chemical suppressor after the separator column to reduce the eluent background conductance before entering the conductivity detector. In this case, tetrabutylammonium hydroxide (TBA-OH) is used as the regenerant. The eluent alkyl sulfonic acid conductivity is suppressed by the TBA^+ cation in the regenerant, exchanging into the eluent stream and combining with the eluent alkyl sulfonate, while the eluent hydronium ion exchanges into the regenerant stream to form water.

This method allows for the selective determination of weak organic acids (eg. acetate, formate, etc.) in the presence of large

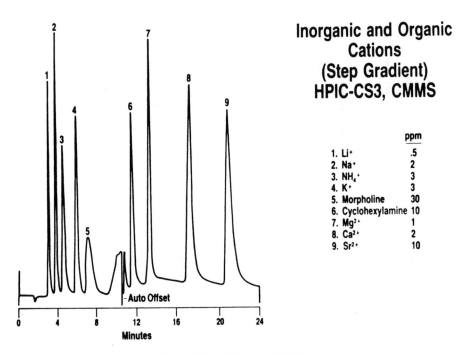

Inorganic and Organic Cations (Step Gradient) HPIC-CS3, CMMS

		ppm
1.	Li$^+$	.5
2.	Na$^+$	2
3.	NH$_4$$^+$	3
4.	K$^+$	3
5.	Morpholine	30
6.	Cyclohexylamine	10
7.	Mg^{2+}	1
8.	Ca^{2+}	2
9.	Sr^{2+}	10

Figure 34.3.—Cations by HPIC.

amounts of highly ionized species. A typical HPICE chromatogram is shown in Figure 34:4.

34.3.4 *MPIC Separation Mode.* This mode is better suited to the separation of more hydrophobic ions. Examples of these ions include both organics (e.g., alkyl sulfonates and amines) and inorganics (e.g., thiocyanate and metal cyanide complexes). This mode couples ion-pair separation on a neutral resin surface with chemically suppressed conductivity detection.

A practical illustration of the MPIC mode is the separation of ammonia and ethanolamine (HOCH$_2$CH$_2$NH$_2$). This separation is not possible using the HPIC cation exchange resins. If a hydrophobic acid is added to a solution of these two amines, the formation of "ion pairs" will occur, as shown in Equations 1 and 2.

$$\text{RSO}_3\text{H} + \text{NH}_3 \rightleftarrows [\text{RSO}_3^- \text{ NH}_4^+](\text{``ion pair''}) \quad (1)$$

$$\text{RSO}_3\text{H} + \text{HOCH}_2\text{CH}_2\text{NH}_2 \rightleftarrows$$
$$[\text{RSO}_3^- \text{ }^+\text{NH}_3\text{CH}_2\text{CH}_2\text{OH}] \text{ (``ion pair'')} \quad (2)$$

When these "ion pairs" approach the surface of the neutral MPIC resin, adsorption, and therefore retention, tends to occur. The degree of this interaction is dependent upon the total hydrophobicity of the ion pair. The ethane functionality of ethanolamine tends to make its ion pair more hydrophobic than ammonia and thus separation is accomplished.

The MPIC separator column is capable of separating either anions or cations. For the separation of cations by MPIC, an aliphatic sulfonic acid, (e.g., hexanesulfonic acid [CH$_3$(CH$_2$)$_5$SO$_3$H]) is commonly used as the ion pair reagent, with varying amounts of acetonitrile as an organic modifier. For the separation of anions, one commonly employs either quaternary ammonium hydroxides or ammonium hydroxide as the ion pair reagent. The proper choice of the ion pair reagent is dependent upon the hydrophobicity of the ion to be determined, as well as other ions in the matrix. In general, the more hydrophobic the analyte ions, the smaller the ion pair

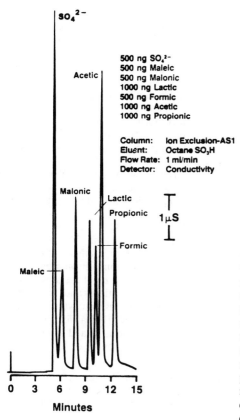

Figure 34.4. — **Organic acids by HPICE.**

reagent should be. Table 34:III illustrates suggested reagents for several hydrophobic anions. A typical chromatogram is shown in Figure 34:5.

After the MPIC separator column, chemical suppression of the eluent is accomplished in a manner similar to HPIC and MPIC. Thus, the MPIC mode combines the versatility of ion-pair separations with the sensitivity and wide dynamic range of chemically suppressed conductivity detection.

34.4 OTHER POST-COLUMN ENHANCEMENT OPTIONS.

34.4.1 So far, the only example of post-column enhancement discussed has been chemical suppression of eluent conductivity. Another enhancement option is to form a light-absorbing or fluorescing derivative of selected analytes as they exit the separator column. In these cases, a photometric or fluorescence detector would be used in place of the usual conductivity detector. Many transition metals can be determined in this manner by combining a metallochromic reagent containing 4-(2-pyridylazo)-resorcinol (PAR) with the separator column effluent. This reagent forms colored complexes with many metals that can be detected by measuring light absorption at 520 nm with a photometric detector. A typical chromatogram is shown in Figure 34:6.

Table 34:III Reagents for MPIC of Anions

Analyte	Reagent[1]	Acetonitrile Concentration
F^-, Cl^-, NO_2^-, SO_4^{2-}, etc.	$N(C_4H_9)_4^+$	8%
I^-, SCN^-	$N(C_4H_9)_4^+$	15%
Aromatic Sulfonates	$N(C_3H_7)_4^+$	20%
Alkyl sulfonates, sulfates $<C_8$	$N(C_4H_9)_4^+$	28%
$Fe(CN)_6^{3-}$, $Au(CN)_2^-$, etc.	$N(C_4H_9)_4^+$	30%
Alkyl sulfonate, sulfates $>C_8$	NH_4^+	30%

[1] All are used as the hydroxides.

Eluent: 10mM NH₄OH
in 12% Acetonitrile
Detector Sensitivity: 30µS

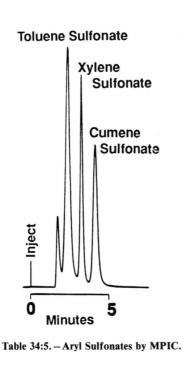

Toluene Sulfonate

Xylene
Sulfonate

Cumene
Sulfonate

Inject

0   5
Minutes

Table 34:5. — Aryl Sulfonates by MPIC.

34.4.2 Amino acids may also be determined using post-column reaction with an OPA reagent to form a fluorescent derivative that may be measured using a fluorescence detector.

34.5 DETECTOR OPTIONS. The electrolytic conductivity detector is used in the majority of Ion Chromatographic applications. However, any detector used in the field of liquid chromatography may be used if it will respond to the ions of interest. Photometric detectors operated in the ultraviolet (UV) region may be used to detect aromatic ions (e.g., benzenesulfonate). Photometric detectors in the visible (VIS) region may be used to detect organometallic complexes. Fluorescence detectors may be used to detect fluorescent derivatives of amino acids. Amperometric detectors may be used to detect electroactive ions, such as sulfide and cyanide.

34.6 SUMMARY OF SEPARATION AND DETECTION MODES. With so many separation and detection modes, the decision to determine which combination of capabilities will best suit his application needs is sometimes confusing to the non-user of Ion Chromatography. Certainly, manufacturers' training manuals and application notes, as well as technical journal publications, are excellent sources of information on the best selection of components and operating conditions. In addition to these information sources, Tables 34:IV and 34:V offer an overview as to the most useful combinations of separation and detection modes (4).

34.7 APPLICATIONS.

34.7.1 Tables 34:I and 34:II present a fairly complete list of analytes that have been measured by Ion Chromatography. Tables 34:IV & 34:V present the most likely selections of separation and detection modes to measure these analytes. Beyond this, the technical journal articles that are unique to a given industry and sample matrix are an excellent source of information on such things as:

1. Sample preservation and pretreatment
2. Proper calibrations procedures
3. Outlines of potential interference problems
4. Comparison of Ion Chromatography results with other analytical methods
5. Estimate of dynamic range and limit of detection of recommended methods.

34.7.2 In the case of the application of Ion Chromatography to air analysis, the two volume series entitled *Ion Chromatographic Analysis of Environmental Pollutants* (5) contains many of the fundamental papers written by pioneers in this vital application area. The following is a listing of titles of papers that deal specifically with the application of Ion Chromatography to air pollution analysis, from these two publications.

Volume 1

1. Ion Chromatographic Determination of Atmospheric Sulfur Dioxide
2. Ion Chromatographic Analysis of Ammonium Ion in Ambient Aerosols
3. Practical Experience on the Use of Ion Chromatography for Determination of Anions in Filter Catch Samples

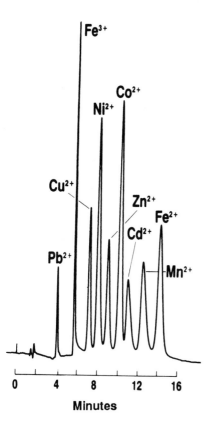

Determination of Nine Transition Metal Ions

Metal	Conc. (μg/L)
Lead(Pb^{2+})	10
IronIII(Fe^{3+})	1
Copper(Cu^{2+})	1
Nickel(Ni^{2+})	1
Zinc(Zn^{2+})	1
Cobalt(Co^{2+})	1
Cadmium(Cd^{2+})	3
Manganese(Mn^{2+})	1
Iron II(Fe^{2+})	1

Figure 34:6. – **Transition metal by HPIC with post column derivatization.**

Table 34:IV Selection of Separation and Detection Modes for Anions

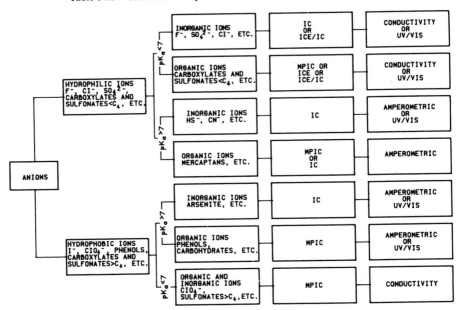

Table 34:V Selection of Separation and Detection Modes for Cations

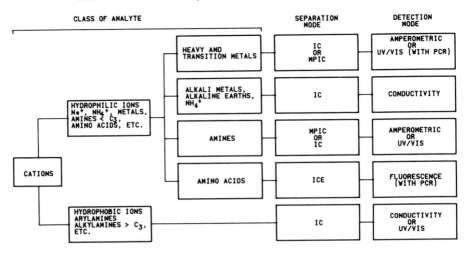

Desulfurization Systems by Ion Chromatography

4. Application of Ion Chromatography to the Analysis of Source Assessment Samples

5. Simultaneous Analysis of Anions and Cations in Diesel Exhaust Using Ion Chromatography

6. Ion Chromatographic Determination of Sulfur in Fuels

7. Determination of Azide in Environmental Samples by Ion Chromatography

8. Collection of Formic Acid Vapor and Analysis by Ion Chromatography

9. Sampling and Analysis of Formaldehyde in the Industrial Atmosphere

10. Analysis of Fuels by Ion Chromatography: Comparison with ASTM Methods

11. Measurement of Ambient Sulfuric Acid Aerosol with Analysis by Ion Chromatography

12. Comparison of Ion Chromatography and Automated Wet Chemical Methods for Analysis of Sulfate and Nitrate in Ambient Particulate Filter Samples

13. Evaluation of Ion Chromatography as an Equivalent Method for Ambient Sulfur Dioxide Analysis

14. An Evaluation of Sulfate Analyses of Atmospheric Samples by Ion Chromatography

15. Intensity-Weighted Sequential Sampling of Precipitation: A Technique for Monitoring Changes in Storm Chemistry During a Storm

16. Ion Chromatographic Analysis of Simulated Rainwater

17. Automation of the Ion Chromatograph and Adaptation for Rainwater Analysis

18. Ion Chromatographic Analysis of Cations at Baseline Precipitation Stations

34.7.3 New developments in the field through 1986 are highlighted in the recent volume edited by Tarter **(6)**. Many of these are directly applicable to atmospheric analysis.

REFERENCES

1. SMALL, H., T.S. STEVENS AND W.C. BAUMAN. 1975. Novel Ion Exchange Chromatographic Method Using Conductimetric Detection. Anal Chem, 47:1801–1804.

2. POHL, C.A. AND E.L. JOHNSON. 1980. Ion Chromatography—The State-of-the-Art. J Chromatographic Sci, *18*:442–452.

3. JOHNSON, E.L. AND K.K. HAAK. 1983. Anion Analysis by Ion Chromatography (Recent Developments), Chapter 6, in "Liquid Chromatography in Environmental Analysis, J.F. Lawrence, Ed. The Human Press, Clifton, NJ.

4. WETZEL, R.A., C.A. POHL, J.M. RIVIELLO AND J. MACDONALD. 1985. Ion Chromatography, Chapter 9, in "Inorganic Chromatography," J. MacDonald, Ed. John Wiley and Sons, Inc.

5. MULIK, J. AND E. SAWICKI, EDS. 1978, 1979. "Ion Chromatographic Analysis of Environmental Pollutants." Vols. I and II. Ann Arbor Science Publishers, Ann Arbor, MI.

6. TARTER, J.G., ED. 1987. "Ion Chromatography." Marcel Dekker, NY.

Subcommittee 9
R.J. JOYCE
J.N. HARMAN, *Chairman*

PART II
METHODS FOR
AMBIENT AIR SAMPLING
AND ANALYSIS

101.

Determination of C_1 Through C_5 Atmospheric Hydrocarbons

1. Principle of the Method

1.1 Atmospheric hydrocarbons are determined by capillary column gas chromatography (GC) using a temperature-programmed, fused silica porous layer open tubular (PLOT) Al_2O_3/KCl column (1) and a flame ionization detector.

1.2 Quantitation is accomplished using an electronic integrator, laboratory data station, or the like, calibrated by injection of standards.

1.3 The method is rapid and applicable to routine sampling and analysis of both grab and integrated samples. The separation of the following 22 hydrocarbons can be achieved in 30 min.: methane, ethane, ethylene, propane, cyclopropane, propylene, acetylene, isobutane, propadiene, n-butane, *trans*-2-butene, 1-butene, isobutene, *cis*-2-butene, isopentane, 1,2-butadiene, propyne, n-pentane, 1,3-butadiene, 3-methyl-1-butene, vinylacetylene, ethylacetylene. Additional hydrocarbons can be determined as well.

2. Range and Sensitivity

2.1 The lower limit of measurement for the various hydrocarbons is 0.01 ppm by volume (*i.e.*, 0.01 $\mu l/L$). The lower limit of measurement of the C_3 through C_5 hydrocarbons can be extended to 0.1 parts/billion (ppb) by concentrating 100 mL of the gas sample in a freeze trap. [Liquid oxygen (bp $-181°C$) is used in the freeze trap rather than liquid nitrogen (bp $-195°C$) to avoid condensation of oxygen in freezing the air sample.] Methane (bp $-164°C$) is not collected in the freeze trap and the resolution of ethane and ethylene chromatographic peaks is reduced in the concentrated sample. The upper limit of

measurement can be considerably greater than the anticipated concentrations of the various hydrocarbons in the ambient air.

3. Interferences

3. At the present time there are no known common pollutants in the ambient atmosphere in sufficient concentrations to interfere with the listed hydrocarbons. However, in order to obtain representative atmospheric samples, the selection of the sampling site is most important. The selected site should be reasonably distant from local sources of hydrocarbon emissions, and in a locale where the surrounding topography is conducive to adequate mixing of the ambient atmosphere. Under such conditions, local source emissions will not bias the hydrocarbon data toward a specific individual hydrocarbon or hydrocarbon group.

4. Precision and Accuracy

4.1 Replicate analysis of aliquots of uniform air samples and known standard hydrocarbon mixtures should not deviate by more than 10% relative standard deviation.

5. Apparatus

5.1 A GC equipped for capillary columns with split injection, a hydrogen flame ionization detector, and a data handling system is required, such as a Hewlett-Packard 5980 or equivalent.

5.1.1 *Chromatography Column.* The column recommended is a 50 m × 0.32 mm ID fused silica PLOT Al_2O_3/KCl column such as supplied by Chrompak Inc (Catalog # 7515, Chrompak Inc., Raritan, NJ 08869).

5.1.2 *GC Conditions.* Injector and detector temperatures, 250°C; column temperature programmed at 3°C/min from 70°C to 200°C; carrier gas, N_2 at 26 cm/s.

5.1.3 *Sample Injection.* A gas-tight syringe can be used to inject 100 μL of gas sample, with a split ratio of 15/1.

5.1.4 The freeze trap for sample concentration consists of a 20-cm U-tube of 1-mm ID stainless steel tubing packed with Chromosorb P. Each end of the U-tube is fitted with a toggle valve.

5.2 Air samples are collected and calibration standards are prepared in bags made of the following material:

5.2.1 Dupont Tedlar poly(vinyl fluoride) film bags, 2 mil (50 μm) thick, or equivalent, optionally covered with black polyethylene film to exclude light, and fitted with Roberts valves or equivalent.

5.3 The assembly of equipment used for preparation of calibration standards is shown in Figure 101:1. The principal components are:

5.3.1 *Precision Wet Test Gas Meter.*

5.3.2 *Stainless Steel Injection Tee.* A Swagelok all-tube tee, 6 mm, with the center arm fitted with a silicone rubber injection gasket.

5.3.3 *Gas-Tight Syringes,* 50-ml and 10-ml.

5.3.4 *Two Sizes of Tedlar Bags Fitted with Roberts Valves,* 10 L and 40 L.

5.4 The following equipment is used for collecting grab samples of air:

5.4.1 *Tedlar Bags Fitted with Roberts Valves.* Bag size 7.5 L.

5.4.2 *One Atomizer Rubber Bulb Set or Automatic Buret Bulb.* Available from all laboratory supply houses.

5.5 The following equipment is used for collecting integrated samples of air:

5.5.1 *Air Sampling Pump.* A diaphragm-actuated pump with constant flow, such as an SKC Aircheck pump (SKC Inc., Eighty-Four, PA).

5.5.2 *Tedlar Bags Fitted with Roberts Valves.* Bag sizes, 15 L and 30 L.

5.5.3 *Micrometering valve and check valve.* This is needed for control of gas sample flow.

Limiting orifice needles. 30-gauge hypodermic needles, 1.9 cm long, of hyperchrome stainless steel are suitable for this purpose, instead of a micrometering valve as mentioned in **5.5.3**.

5.5.4 *A Timer.* A useful timer for this purpose is an electric timer which automati-

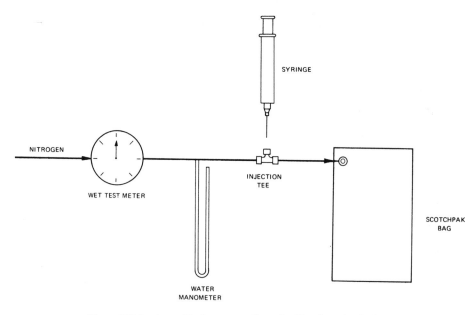

Figure 101:1 — Assembly for preparation of calibration standards.

cally shuts off the sampling pump at the end of the sampling interval. A single pole, double throw time switch may be employed.

5.5.5 *A Rubber Septum.* This is needed .when limiting orifice needles are used for flow control. A sleeve type serum bottle stopper may be used for this purpose.

5.5.6 *Flowmeter or Rotameter,* equipped with stainless steel ball.

5.5.7 A filter system upstream of the pump is recommended for the entrapment of particulate matter. An aerosol field monitor fitted with an all glass or a 0.8 μm pore membrane filter is useful for this purpose.

6. Reagents

6.1 PURITY. All chemicals should be analytical reagent grade.

6.2 CYLINDER GASES. All cylinder gases should be gas chromatography grade or equivalent.

6.3 REQUIRED CHEMICALS AND GASES. Compressed air, hydrogen, nitrogen, and hydrocarbon standards (99 mole % or better).

7. Procedure

7.1 COLLECTION OF FIELD SAMPLES.

7.1.1 As indicated in Section 3 the location of the field sampling site can have an important effect on the sample composition unless precautions are taken to avoid local sources of hydrocarbons.

a. Generally, for a homogeneous, urban air mass, hydrocarbon contaminants would be expected to fall within the ranges given as a guide in Table 101:I.

b. If there are large variances from the extremes of the range, a critical review of the data should be made. The review should consider such factors as: (a) Are there errors in the calculation of instrument response factors? (b) Are there errors in calculation of peak areas? (c) Was there possible contamination of the sample collection bags or sample valves and lines by atmosphere containing high hydrocarbon concentrations?

c. During collection of the sample, was there an intermittent source of hydrocarbon emissions such as an automobile start-

Table 101:I Ranges of Hydrocarbon Values Expected in Urban Air Masses

Component	Range, ppm in air	
	Minimum	Maximum
Methane	1.2	15
Ethane	0.005	0.5
Propane	0.003	0.3
Isobutane	0.001	0.1
n-Butane	0.004	0.4
Isopentane	0.002	0.2
n-Pentane	0.002	0.2
Ethylene	0.004	0.3
Propene	0.001	0.1
Butene-1	0.000	0.02
Isobutylene	0.000	0.02
Trans-2-butene	0.000	0.01
Cis-2-Butene	0.000	0.01
1,3 Butadiene	0.000	0.01
Acetylene	0.000	0.2

ing, stopping or passing near the sampling site; the filling of a gasoline fuel tank at a nearby service station? If the sample site is near a refinery or petroleum processing plant, was there a possible operational upset that resulted in venting unusual amounts of hydrocarbon compounds?

d. Depending upon the sampling site, other emission sources may be a possibility.

e. As a general "rule of thumb," if methane, ethane or propane concentrations are high relative to other identified hydrocarbons (See Table 101:I), look for local source emissions of natural gas. If acetylene and the butenes are high, look for a nearby automobile exhaust source. If n-butane, n-pentane or isopentane are high, look for a source of gasoline evaporation.

7.1.2 Grab samples are collected as follows:

a. Flush the bag out 3 times with the ambient air to be sampled. This is done with a rubber buret bulb connected to the valve of the bag or the air pump.

b. Fill the bag approximately three-fourths full, close the air valve securely and remove the atomizer bulb. Do not fill the bag to capacity. Some space must be allowed for expansion due to temperature and pressure variations.

c. Give the bag an identifying number and record appropriate field information, *e.g.*, date and time of collection, location of sample collection, weather and air pollution conditions, etc.

d. The samples should be sent to the laboratory and analyzed as soon as possible.

7.1.3 A diagram of an assembly of equipment for collecting integrated samples is given in Figure 101:2.

a. With the equipment as assembled in Figure 101:2, the air flow rate into the plastic bag is measured by attaching a flow meter to the upstream side of the diaphragm pump. If a micrometering valve is used instead of a limiting orifice, the valve is adjusted to the desired flow rate.

b. When the desired flow rate is obtained, disconnect the flowmeter. Set the electric timer for the time period over which the sample is to be collected.

c. Near the end of the sampling period, the flow rate is measured again.

d. The bag sample is identified and appropriate field data recorded.

e. It is not necessary to know the exact volume of the sample collected. A 100 μL aliquot of the sample is taken for gas chromatographic analysis and the concentration of hydrocarbons in the air is based on the 100 μL aliquot. A constant flow rate should be maintained during the sampling period or the change in flow rate during the period should be measured in order to insure a valid integrated sample.

7.2 ANALYSIS OF SAMPLES.

7.2.1 Air samples collected in the field and returned to the laboratory are analyzed by GC, using the procedure specified in the manufacturer's operating manual.

7.2.2 A chromatogram of a known mixture is given in Figure 101.3.

8. Calibration and Standards

8.1 A calibration standard should be prepared for each of the 17 hydrocarbons to be measured. Retention time is used for identification of the hydrocarbon and integrated peak area is used for quantitation of the hydrocarbon. Standard calibration mixtures using the assembly of equipment illustrated in Figure 101:1 are prepared as follows:

8.1.1 Purge wet test meter with nitrogen for one-half h.

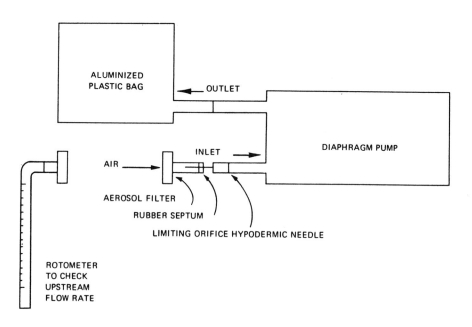

Figure 101:2 — Assembly for integrated hydrocarbon sampling.

Peak identification

1. methane
2. ethane
3. ethene
4. propane
5. cyclopropane
6. propene
7. ethyne
8. isobutane
9. propadiene
10. n-butane
11. trans-2-butene
12. 1-butene
13. isobutene
14. cis-2-butene
15. isopentane
16. 1,2-butadiene
17. propyne
18. n-pentane
19. 1,3-butadiene
20. 3-methyl-1-butene
21. vinylacetylene
22. ethylacetylene

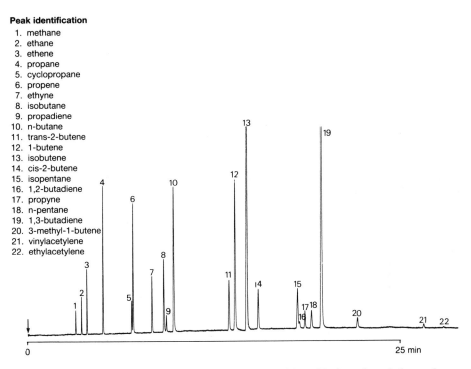

Figure 101:3 – Sample chromatogram showing relative position of hydrocarbon elution peaks.

8.1.2 After purge, connect the plastic bag to the system and start metering nitrogen into the bag. 40-L bags are used for the 40-L dilutions; 10 L are used for the 10-L dilutions.

8.1.3 Inject the predetermined amount of each hydrocarbon into the gas stream through the stainless steel tee. A 50-mL gas-tight glass syringe is used for quantities of gas greater than 10 mL; for quantities less than 10 mL, a 10-mL gas-tight syringe is used.

8.1.4 To make standards containing less than 10 ppm of a hydrocarbon, a double dilution is required. First a 1000 ppm standard is prepared. Aliquots of this standard are diluted to give standards with less than 10 ppm hydrocarbons (see Table 101:II).

8.2 CALIBRATION OF THE CHROMATOGRAPH.

8.2.1 100 μL of standard gas is injected into the gas chromatograph.

8.2.2 The response of the hydrogen flame ionization detector is linear from 0.001 to 10 ppm at least.

Table 101:II Dilutions for Preparation of Calibration Standards

Hydrocarbon Concentration in ppm	Aliquot of Hydrocarbon	Final Dilution Volume in Liters
1000	40 mL pure hydrocarbon	40
10	0.4 mL pure hydrocarbon	40
1	10 mL of 1000 ppm standard mixture	10
0.1	1 mL of 1000 ppm standard mixture	10

9. Calculation

9.1 Air samples collected in the field and returned to the laboratory are calculated according to the following procedure.

9.1.1 The response of the detector is printed on the integrator output for each hydrocarbon eluted from the column, along with the retention time of the peak maximum. The peak area is used for quantitation.

9.1.2 Determine the concentration of each hydrocarbon present in the sample using the calibration response factor:

$$\text{ppm hydrocarbon} = f \times \text{peak area}$$

where f is the calibration response factor for each hydrocarbon in units of ppm per peak area unit.

9.1.3 Calculate the response factor for each hydrocarbon at 10 ppm, 1.0 ppm and 0.1 ppm. The response factor equals the concentration of the standard in ppm divided by the detector response.

10. Effects of Storage

10.1 Storage experiments on samples in Tedlar bags showed average losses of 6% for C_1 through C_5 paraffinic and olefinic hydrocarbons after 24 h storage and 10% losses after 3 days storage.

11. References

1. DE ZEEUW, J., R.C.M. DE NIJS AND L.T. HENRICH. 1987. Adsorption Chromatography on PLOT Columns: A New Look at the Future of Capillary GC. J. Chromatogr. Sci., 25:71-83.

Subcommittee 4/5

D.C. LOCKE
M. FELDSTEIN, *Chairman*
R.J. BRYAN
D.L. HYDE
D.A. LEVAGGI
R.A. RASMUSSEN
P.O. WARNER

102.

Separation and Determination of Polynuclear Aromatic Hydrocarbons and Benzo[a]pyrene.

102A. Extraction and Clean-up Procedures for Polynuclear Aromatic Hydrocarbons in Atmospheric Particulate Matter

1. Principle of the Method

1.1 The extraction and clean-up procedures described here are used to prepare atmospheric particulate samples for analysis by one of the methods described in 102B, 102C or 102D. The polynuclear aromatic hydrocarbons (PAHs) are extracted from the particulate matter using a nonpolar solvent, cyclohexane, which extracts less extraneous material than a more polar solvent. The extracts can be further cleaned up by running them through an activated alumina column to separate the PAHs from interfering compounds (1–5).

2. Range and Sensitivity

2.1 The range and sensitivity are determined by the analytical method used. See Method 102B, C, D.

3. Interferences

3.1 The separation procedure will separate the PAHs from the other components in the sample such as aliphatics and more polar hydrocarbons. Alkyl-substituted PAHs will still be present and can interfere with subsequent analysis.

4. Precision and Accuracy

The precision and accuracy will be determined by the method used for analysis.

5. Equipment

5.1 MICROSOXHLET EXTRACTOR. With 30 mL boiling flask capacity (Fisher Scientific No. 20650 or equivalent).

5.2 KUDERNA-DANISH (K-D) APPARATUS. Apparatus consists of a 10 mL concentrator tube (Kontes K-570050–1025 or equivalent), 500 mL evaporation flask (Kontes K-57001–0500 or equivalent), three-ball macro Snyder column (Kontes K503000–0121 or equivalent).

5.3 CHROMATOGRAPHY COLUMN. Borosilicate glass column 10 mm I.D. and 30 cm in length. The column should be equipped with a small funnel reservoir at the top and have a coarse frit and Teflon stopcock at the bottom to control the flow.

5.4 WATER BATH. Heated, with concentric ring cover, capable of temperature control ($\pm 2°C$). Bath should be used in a hood.

6. Reagents

6.1 CYCLOHEXANE, METHYLENE CHLORIDE, PENTANE. HPLC grade, distilled in glass.

6.2 SILICA GEL. 100/120 mesh (Davidson Chemical Grade 923 or equivalent). Before use, activate for at least 16 hours at 130°C in foil-covered glass container.

6.3 SODIUM SULFATE. ACS Reagent Grade. Granular, anhydrous. Purify by heating at 400°C for 4 hours in a shallow tray.

6.4 BOILING CHIPS. 10/40 mesh, solvent extracted.

7. Procedure

7.1 The particulate matter collected on a filter sample is extracted in a soxhlet extractor. Using a clean metal punch or cork borer, four circles of 35.5 mm diameter are cut from the high-volume (20×25 cm) glass or silica fiber filter sample. These represent 1/10.5 of the total effective area of the filter. These are placed in a microsoxhlet extractor on top of a wad of glass wool, which prevents the carbon particles from being washed over into the extract and avoids subsequent filtration. Extract for 6 to 8 h with UV grade fluorescence-free cyclohexane. Carefully evaporate the solvent extract to 2 mL using a current of clean nitrogen.

7.2 Prepare a slurry of 10 g of activated silica gel in methylene chloride and pour this into the chromatography column. Gently tap the column to settle the silica gel, and elute the methylene chloride. Add 1–2 cm of anhydrous sodium sulfate to the top of the silica gel.

7.3 Adjust column elution rate to 2 mL/min. Pre-elute the column with 40 mL pentane. Just prior to exposure of the sodium sulfate layer, transfer the 2 mL cyclohexane sample extract onto the column, using an additional 2 mL of cyclohexane to complete the transfer. Add an additional 25 mL pentane and continue elution of the column. Discard the pentane eluate.

7.4 Elute the column with 25 mL of 40% methylene chloride/60% pentane. If the fluorescence method is used, 3-mL fractions should be collected and individually analyzed for the compounds of interest. All fractions containing those compounds should be combined and concentrated as described below. If the HPLC method is used, all of the eluate should be collected in a 500 mL K-D flask equipped with a 10 mL concentrator tube.

7.5 Add 1–2 clean boiling chips to the K-D flask and attach a three-ball Snyder column. Pre-wet the Snyder column by adding 1 mL of pentane to the top. Place the K-D apparatus in a hot water bath (60–65°C) so the concentrator tube is partially immersed in the hot water. Adjust the vertical position as required to complete the concentration in 15–20 min. At the proper rate of distillation the balls in the chamber will actively chatter, but the chambers will not flood. When the apparent volume reaches 1 mL remove the K-D apparatus and allow to drain for 10 min. Remove

Snyder column and rinse flask into concentrator with 1 mL pentane.

7.6 The sample is now ready for analysis by one of the described methods.

8. Comparison with Standards

8.1 Standards that are to be used in Method 102C or D should be run through the extraction and clean-up procedures to verify recovery.

9. Calculations

9.1 Calculations are described under the method used for analyzing the extracts.

10. Effect of Storage

10.1 Sample solutions should be analyzed the same day. If they cannot be analyzed until the next day, the solutions should be protected from light and stored in a cold box. The organic residues can be kept in a refrigerator in the dark for longer periods of time. Thus, the cyclohexane–soluble fraction, stored in this way, is stable for years as far as the BaP is concerned.

11. References

1. SAWICKI, E., T.R. HAUSER, AND T.W. STANLEY. 1960. Int. J. Air Poll. 2:253.
2. SAWICKI, E., T.W. STANLEY, W.C. ELBERT, J. MEEKER, AND S. MCPHERSON. 1967. Atmospheric Environment. 1:131.
3. SAWICKI, E., T.W. STANLEY, W.C. ELBERT, AND J.D. PFAFF. 1967. Anal. Chem. 36:497.
4. SAWICKI, E., T.W. STANLEY, J.D. PFAFF, AND W.C. ELBERT. 1964. Chemist-Analyst. 53:6.
5. "Sampling and Analysis Procedures for Screening of Industrial Effluents for Priority Pollutants." Method 610, Polynuclear Aromatic Hydrocarbons. USEPA, Environmental Monitoring and Support Laboratory, Cincinnati, Ohio (1977).

Subcommittee 4/5

P.O. WARNER
M. FELDSTEIN, *Chairman*
R.J. BRYAN
D.L. HYDE
D.A. LEVAGGI
D.C. LOCKE
R.A. RASMUSSEN

102B. Separation and Microanalysis of Airborne Particulate Matter for Benzo[a]pyrene Using Thin Layer Chromatography and Spectrofluorimetry

1. Principle of the Method

1.1 The particulate matter collected from the urban atmosphere is extracted with an organic solvent and then separated alongside pure benzo[a]pyrene (BaP) with alumina thin-layer chromatography. The unknown and standard spots are eluted, their solutions are evaporated, and the residues are dissolved in concentrated sulfuric acid. An intensely yellow fluorescent cationic salt is obtained. The concentration of benzo[a]pyrene is measured using the fluorescence method described in Method 102C.

2. Range and Sensitivity

2.1 The lower limit of determination of the spectrophotofluorimetric method is about 3 ng of BaP with an instrument capable of its full potential of sensitivity. The range of analysis is then 3 to 200 ng of BaP. This means that for moderately contaminated urban air samples (< 500 to 1000 μg BaP/g benzene-soluble fraction) more than 6 μg of benzene-soluble fraction or more than 0.6 m^3 of air is necessary for an analysis. For samples of low contamination ($<$ 50 μg BaP/g benzene-soluble fraction) at least 60 μg of benzene-soluble fraction or 6 m^3 of air are necessary for analysis.

3. Interferences

3.1 Because of the sensitivity of the method, the laboratory air must be clean. Cigarette smoking cannot be allowed since it contributes to the background and worsens the sensitivity of the method.

3.2 The method is highly selective for BaP. Hydrocarbons found with or near BaP in alumina chromatographic fractions do not interfere, e.g., benzo[k]-fluoranthene, benzo[g,h,i]perylene, benzo[e]pyrene, perylene, and anthanthrene (3). In fact, benzo[a]pyrene can be determined by the spectrophotofluorimetric method in the presence of 50 hydrocarbons (over 40 of which are polynuclear compounds ranging in size from 2 to 7 rings (1).

4. Precision and Accuracy

4.1 Eleven micromethods for the estimation of BaP have been compared in a composite benzene-soluble fraction of airborne particulate matter from over 100 communities (2). The methods gave values ranging from 720 to 1000 μg BaP/g samples. An average value of 870 was obtained. In comparison, the spectrophotofluorimetric method gave a range of 800 $\pm$ 50 in 8 determinations. Pure BaP was added to an airborne particulate sample, and recoveries of 90, 93, 94, and 100% were obtained, averaging 95 $\pm$ 5%.

4.2 The composite sample analyzed by the filter fluorimetric method gave a value of 950 $\pm$ 100 μg BaP/g sample in eight determinations. Recoveries of 80, 84, 84, 88, 88, 90, 100 and 100 (averaging 90 $\pm$ 10%) were obtained from the airborne particulate sample.

4.3 The fluorescence intensity is stable for at least 20 minutes (1,3). However, this stability is worsened if the solution is in a bright light or is repeatedly illuminated in the instrument.

4.4 The time necessary for separation and analysis has been compared for 11 methods (2). For the two instrumental methods in this paper about 1 1/2 h are necessary for the separation and assay of an organic fraction.

5. Apparatus

5.1 All laboratory ware must be cleaned and tested to insure absence of contamination by organic material.

5.2 Thin-layer plates are coated with alumina, 250 μm in thickness, and activated by heating at 100°C for 30 min. The activated plates are stored in a vacuum desiccator adjusted to 45% relative humidity with aqueous sulfuric acid. The plates should not be stored for more than 3 weeks and should be checked for fluorescent impurities before use (or use commercially prepared TLC plates).

5.3 Thin-Layer Chromatographic Chamber.

5.4 Chromato-Vue Cabinet or some other long wavelength ultraviolet source.

5.5 Microsoxhlet Extractor, with 30 mL boiling flask capacity (Fisher Scientific No. 20650 or similar apparatus from any other source).

5.6 Micropipets, a calibrated set to cover the range of 10 μL to 1.0 mL.

6. Reagents

6.1 METHYLENE CHLORIDE, PENTANE. HPLC grade, distilled in glass.

6.2 ETHER, ANHYDROUS. Grade suitable for HPLC, distilled in glass.

6.3 CONCENTRATED SULFURIC ACID.

6.4 BENZO[A]PYRENE. A standard of highest purity is required. Sources of zone-refined BaP of high purity are Eastman, Aldrich, K & K.

7. Procedure

7.1 The particulate matter collected from several m³ of urban atmosphere is extracted in a microsoxhlet extractor (boiling flask capacity 30 mL) as described in 102A. The combined extracts are evaporated to dryness in a *current of nitrogen* at room temperature or by vacuum, and the residue is dissolved in 0.2 mL of methylene chloride. An aliquot of this solution, anywhere from 10 μL to the entire sample, is placed on an alumina thin-layer plate 1.50 cm from the bottom. (The benzene-soluble fraction (0.02 to 1 mg) can also be used.)

7.2 Standards (0.02 μg BaP to a spot) are also placed on the plate at the origin. The plate is transferred to a thin-layer chromatographic chamber which contains 200 mL of pentane-ether (19:1). After development to 15 cm in dim light, the plate is quickly examined and marked under UV light. The absorbent in each spot is transferred quantitatively to a throwaway pipet containing a small wad of glass wool or to a small fine-porosity fritted glass funnel. The absorbent is eluted with 70 to 100 mL of ether. The eluent is collected in a 25 mL test tube fitted with a sidearm connected to a vacuum system. The ether is evaporated by vacuum and the residue is dissolved in 1 mL of concentrated sulfuric acid.

8. Comparison with Standards

8.1 Readings of standard and test spot solutions are taken using the fluorescence procedure described in Method 102C.

9. Calculations

9.1 Even though the relationship between the concentration and the product of the meter multiplier and transmittance readings is linear, it is advisable to run standards at the same time. On this basis the amount of benzo[a]pyrene in the test mixture is readily calculated for both procedures by the following equations (2),

where Cs = nanograms of Standard BaP analyzed;

Rs = product of the meter multiplier and transmittance readings of eluted BaP spot dissolved in 1 mL of sulfuric acid;

Rx = product of the meter multiplier and transmittance readings of the eluted unknown spot dissolved in 1 mL of sulfuric acid;

Wt = weight in mg of analyzed organic airborne particulate sample;

V = volume in m³ of analyzed air.

$$\text{ngBaP/g organic mixture} = \frac{Cs \cdot Rx \cdot 1000}{Wt \cdot Rs}$$

and

$$\text{ngBaP/g m}^3 \text{ air} = \frac{Cs \cdot Rx}{V \cdot Rs}$$

10. Effect of Storage

10.1 Sample solutions should be analyzed the same day. If they cannot be analyzed until the next day, the solutions should be protected from light and stored in a cold box. The organic residues can be kept in a refrigerator in the dark for longer periods of time. Thus, the benzene-soluble fraction, stored in this way, is stable for years as far as the BaP is concerned. The variations in value are what would be expected of the column chromatographic absorption spectral method.

11. References

1. SAWICKI, E., T.R. HAUSER AND T.W. STANLEY. 1960. Int. J. Air Poll. 2:253.
2. SAWICKI, E., T.W. STANLEY, W.C. ELBERT, J. MEEKER, AND S. MCPHERSON. 1967. Atmospheric Environment 1:131.
3. SAWICKI, E., T.W. STANLEY, W.C. ELBERT, AND J.D. PFAFF. 1967. Anal. Chem. 36:497.

Subcommittee 4/5

P.O. WARNER
M. FELDSTEIN, *Chairman*
R.J. BRYAN
D.L. HYDE
D.A. LEVAGGI
D.C. LOCKE
R.A. RASMUSSEN

102C. Measurement of Benzo[a]pyrene and Benzo[k]fluoranthene by Spectrofluorimetry

1. Principle of the Method

1.1 This is a rapid method for the measurement of benzo[a]pyrene (BaP) and benzo[k]fluroanthene (BkF) in the sample aliquots collected in either Method 102A or Method 102B. The concentrations of the PAHs are determined by measuring the fluorescence emission of the extracts and comparison with known standards (1). For more information see the discussion on fluorescence spectrophotometry in Part I, Section 28 of this manual.

2. Range and Sensitivity

2.1 The method can measure concentrations in a prepared air sample extract or fraction over the range of 0 to 0.25 μg of BaP or BkF/mL of solution.

3. Interferences

3.1 In the chromatographic separation used, BaP and BkF will be found in about 7 to 8 eluate fractions. These fractions are likely to contain both BaP and BkF, but unlikely to contain other strongly fluorescing compounds. It then becomes a question of ability to measure BaP (or BkF) accu-

rately in the presence of the other hydrocarbon.

4. Precision and Accuracy

4.1 Experiments have shown that 0.25 μg/mL of BaP can be measured with an accuracy of better than 0.002 μg, while 0.25 μg/mL of BkF can be measured with an accuracy of better than 0.001 μg. If the concentration of BaP found is more than twice the BkF concentration, the BkF results will be in error by a factor of 10%. In such a case, suitable mathematical corrections can be made.

5. Spectrophotofluorimeter

5.1 A fluorimeter equipped with motor-driven excitation and emission monochromators is required in the wavelength range of about 250 to 550 nm. The Aminco-Bowman spectrophotofluorimeter or any instrument with similar performance may be used.

6. Reagents

6.1 PENTANE. HPLC Grade, distilled in glass.

6.2 BENZO[A]PYRENE (BAP). This material is available from Eastman Kodak, Rochester, and Aldrich Chemical Co., Milwaukee. Material from these sources is zone refined and of high purity.

6.3 BENZO[K]FLUORANTHENE (BkF). Limited quantities for analytical purposes may be obtained from the Occupational Health Division, Environmental Health Centre, Ottawa, Canada.

6.4 STANDARD SOLUTIONS OF BAP. Solutions of BaP are prepared containing 0.005, 0.010, 0.015, 0.020, and 0.025 μg/mL in fluorescence-free toluene. Weigh accurately 1.25 mg of BaP on a micro balance and dissolve in 250 mL of spectrograde toluene. Measure accurately 1.0 mL of this stock solution and dilute to 1000 mL with spectrograde toluene. Repeat this with 2.0, 3.0, 4.0, and 5.0 mL portions of stock solution, diluted in each case to 1000 mL.

6.5 STANDARD SOLUTIONS OF BkF. These solutions are prepared in the same manner as that described in Section 6.4.

7. Procedures

7.1 Samples from one of the extraction procedures are placed in the spectrophotometer cell. Fluorescence emission measurements of peak heights are made using excitation wavelengths 307 and 384 nm. The method used to measure peak heights is shown in Figure 102C:2. These wavelengths had been established by running the excitation spectrum of both hydrocarbon standards at the 0.015 μg/mL concentration. Fluorescence spectra of typical eluted fractions 12–26 are shown in Figures 102C:1–5. A typical calibration curve for standards in toluene is shown in Figure 102C:6.

7.2 A blank determination is carried out on the glass fiber filter, glassware, and reagents.

8. Calibration and Standards

8.1 Standard curves of fluorescence emission, in arbitrary units, are prepared for the various concentrations of both BaP and BkF at the two exciting wavelengths 384 and 307 nm, i.e., the optimum excitation wavelengths for BaP and BkF. Four curves for the solutions in toluene are shown in Figure 102C:6. These are prepared by plotting the height of the peaks against the concentration. The fluorescence emission intensities of air sample extracts or fractions are also measured using the same two exciting wavelengths.

8.2 Similar calibration curves are obtained with standard solutions made up in cyclohexane. The sensitivity is noticeably less in toluene than in cyclohexane solution. However, though the sensitivity may be less, there is some advantage in making measurement in toluene. The emission of BaP with 384 excitation is higher than the emission of BkF with 384 excitation, so that the toluene-based BaP measurement is somewhat better than a measurement made in cyclohexane. The optimum excitation wavelengths for BaP differ somewhat with the solvent, being 382 nm in cyclohexane and 384 nm in toluene.

9. Calculation

9.1 Since the fluorescence emission intensity of BkF is much greater than that of BaP when a mixture of the two is excited at 307 nm, the reading at this wavelength is essentially due to BkF. Having determined the concentration of BkF, one can calculate the effect of this hydrocarbon when a mixture is excited at 384 nm after which the BaP concentration may be calculated. Thus:

$$\text{Conc. BkF } (\mu g/mL) = \frac{\text{Emission of sample at 307 nm excitation}}{\text{Slope of BkF Standard curve at 307 nm excitation}}$$

$$\text{Conc. BaP } (\mu g/mL) = \frac{\text{Emission of sample at 384 nm} - (\text{conc. BkF} \times \text{Slope BkF standard curve at 384 nm excitation})}{\text{Slope of BaP Standard curve at 384 nm excitation}}$$

The peak height at a given emission wavelength is measured by the base line technique illustrated in Figure 102C:2. The concentrations of the BaP and BkF in μg/g of particulate matter or per 1000 m^3 of air sample are calculated as follows:

To determine the concentration in μg/g, multiply the observed concentration in μg/mL (as above) by the total liquid volume (5 mL) and by the area factor of the glass fiber filter (10.5), then divide by the total weight of particulate matter collected on the filter (g).

$$\text{Conc. of hydrocarbon, } \mu g/g = \frac{\mu g/mL \times 5 \times 10.5}{\text{wt. of particles, g}}$$

$$\text{Conc. of hydrocarbon, } \mu g/1000m^3 = \frac{\mu g/mL \times 5 \times 10.5 \times 1000}{\text{Volume of air sample, m}^3}$$

The volume of the air sample should be corrected to standard conditions.

10. Effect of Storage

Polycyclic aromatic hydrocarbons in standard solutions or in sample extracts appear to remain constant for many months when such solutions are stored in the refrig-

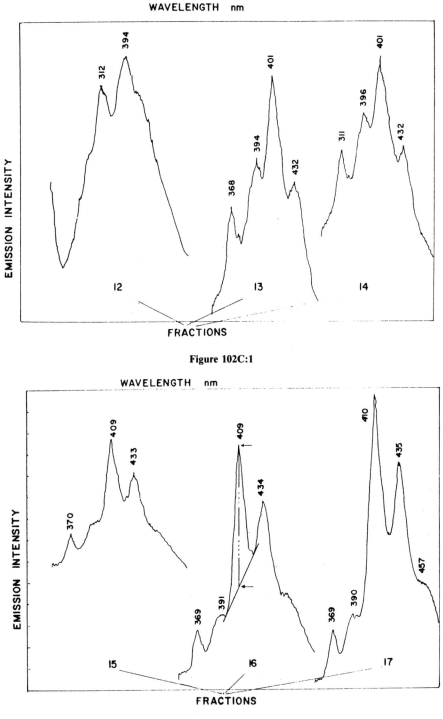

Figure 102C:1

Figure 102C:2

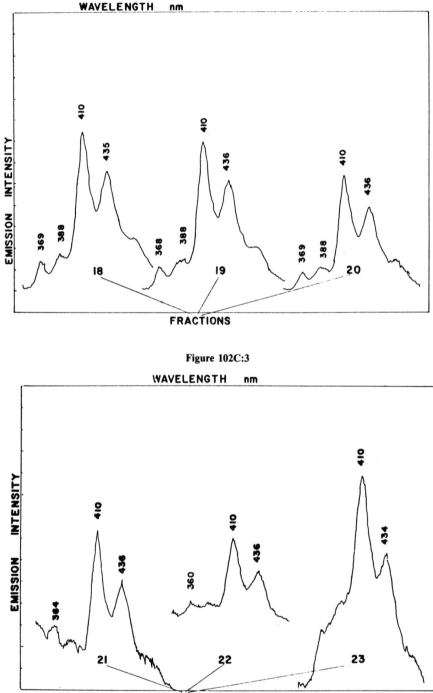

Figure 102C:3

Figure 102C:4

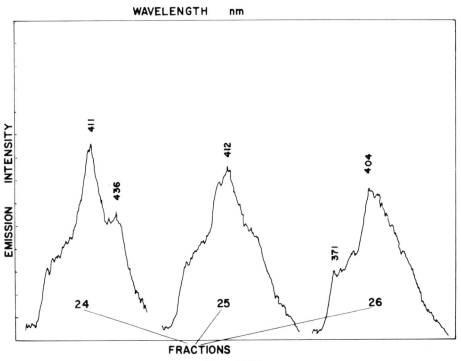

Figure 102C:5

erator in borosilicate glass bottles. The samples must be protected from light.

11. References

1. DuBois, L., A. Zdrojewski, C. Baker, and J.L. Monkman. 1967. Some Improvements in the Determination of Benzo[a]pyrene in Air Samples. J. Air Poll. Control. Assoc. 17:818–821.

Subcommittee 4/5

P.O. Warner
M. Feldstein, *Chairman*
R.J. Bryan
D.L. Hyde
D.A. Levaggi
D.C. Locke
R.A. Rasmussen

102D. Measurement of Polynuclear Aromatic Hydrocarbons Using Liquid Chromatography with Fluorescence Detection.

1. Principle of Method

1.1 The extract samples from method 102A are analyzed using high pressure liquid chromatography (HPLC) with a fluorescence detector **(1,2)**. Identification of peaks and measurement of the concentrations of individual PAHs is done by comparison with standard solutions. Since many compounds can be present, the user of the method should have some high expectation of finding the specific compounds of interest. This procedure is capable of resolving all 16 PAHs listed in Table 102D:I.

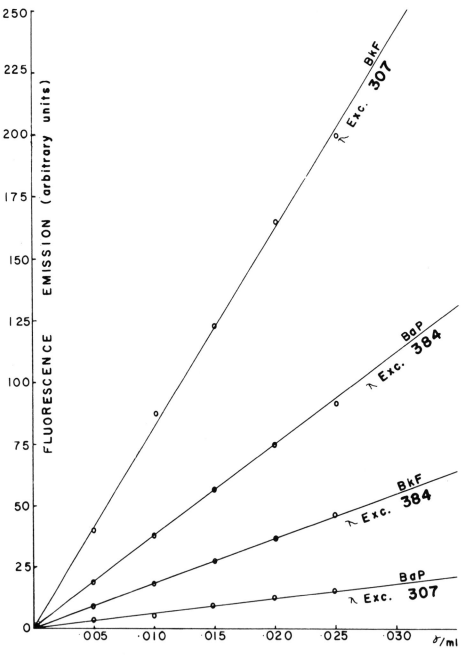

Figure 102C:6 — Standards in toluene.

Table 102D:I High Performance Liquid Chromatography of PAHs

Compound	Retention Time (min.)	Limit (μg/m^3) Fluorescence
Naphthalene	16.17	10
Acenaphthylene	18.10	50
Acenaphthene	20.14	2
Fluorene	20.89	1
Phenanthrene	22.32	0.1
Anthracene	23.78	0.8
Fluoranthene	25.00	0.02
Pyrene	25.94	0.02
Benzo(a)anthracene	29.26	0.02
Chrysene	30.14	0.02
Benzo(b)fluoranthene	32.44	0.02
Benzo(k)fluoranthene	33.91	0.02
Benzo(a)pyrene	34.95	0.02
Dibenzo(a,h)anthracene	37.06	0.04
Benzo(ghi)perylene	37.82	0.1
Indena(1,2,3-cd)pyrene	39.21	0.05

HPLC Conditions: Reverse phase HC-ODS Sil-X 2.6 × 250 mm Perkin-Elmer column; isocratic elution for 5 min using 40% acentonitrile/60% water, then linear gradient elution to 100% acetonitrile over 25 minutes; flow rate is 0.5 ml/min.

Detection limit is calculated from the minimum detectable HPLC response being equal to five times the background noise, assuming an equivalent of a 2 mL final volume of a 2 m^3 air sample, and assuming an HPLC injection of 2 μL.

2. Range and Sensitivity

2.1 The sensitivity of the method is usually dependent on the level of interferences rather than instrumental limitations. These values are listed in Table 102D:I.

3. Interferences

3.1 Solvents, reagents, glassware, and other sample processing hardware may yield discrete artifacts and/or elevated baselines causing misinterpretation of the chromatograms. All of these materials must be demonstrated to be free from interferences under the conditions of the analysis by running method blanks. Specific selection of reagents and purification of solvents by distillation in all-glass systems may be required.

3.2 Interferences coextracted from the samples will vary considerably from source to source, depending upon the diversity of the sample. Unique samples may require additional clean-up to achieve maximum sensitivity.

3.3 The extent of interferences that may be encountered using liquid chromatographic techniques has not been fully assessed. Although the chromatographic conditions described allow for a unique resolution of the specific PAH compounds covered by this method, other PAH compounds including alkyl substituted PAHs may interfere.

4. Precision and Accuracy

4.1 At this time the precision and accuracy of the method based on interlaboratory calibrations are not available.

5. Apparatus

5.1 HIGH PRESSURE LIQUID CHROMATOGRAPH (HPLC). With the following features. Constant flow gradient pumping system. Reverse phase, 5 μm HC-ODS Sil-X, 250 mm × 2.6 mm I.D. column (Perkin Elmer No. 809-0716 or equivalent). Fluorescence detector for excitation at 280 nm and emission at 389 nm. Strip chart recorder or integrator. Other wavelengths may be desirable to achieve more selectivity.

6. Reagents

6.1 ACETONITRILE. HPLC grade, distilled in glass.

6.2 STOCK STANDARDS. Prepare stock standard solutions at a concentration of 1.00 ng/μL by dissolving 0.100 grams of assayed reference material in 1000 mL pure isooctane or other appropriate solvent. Dilute 1 mL to 100 mL in a ground glass stoppered volumetric flask. The stock solution is transferred to a reagent bottle with Teflon-lined cap and stored in a refrigerator.

7. Procedures

7.1 To the extract in the concentrator tube, add 4 mL acetonitrile and a new boiling chip, then attach a micro-Snyder column. Increase the temperature of the hot water bath to 90–100°C. Concentrate the solvent as described in Method 102A. After cooling, remove the micro-Snyder column and rinse its lower joint into the concentrator tube with about 0.2 mL acetonitrile. Adjust the extract volume to 1.0 mL.

7.2 Good separation of the PAHs can be obtained using a reverse phase ODS column and a solvent gradient from 40% acetonitrile/60% water to 100% acetonitrile. Total time for gradient is 25 min. An example of the separation achieved by this column is shown in Figure 102D:1. Calibrate the system daily with a minimum of three injections of calibration standards.

7.3 Inject 2–5 μL of the sample extract with a high pressure syringe or sample injection loop. Record the volume injected to the nearest 0.05 μL, and the resulting peak size, in area units.

7.4 The fluorescence detector should be operated according to the manufacturer's instructions. If the peak area exceeds the linear range of the system, dilute the extract and re-analyze. If the peak area measurement is prevented by the presence of interferences, selected fractions may be chosen from the clean-up step.

8. Calibration and Standards

8.1 Prepare calibration standards that contain the compounds of interest, either singly or mixed together. The standards should be prepared at concentrations covering two or more orders of magnitude that will completely bracket the working range of the chromatographic system.

8.2 Assemble the necessary HPLC apparatus and establish operating parameters equivalent to those indicated in Table 102D:I. By injecting calibration standards, establish the sensitivity limit of the detectors and the linear range of the analytical systems for each compound.

8.3 The analyst should process a series of calibration standards through the cleanup procedure (Method 102A) to validate elution patterns and the absence of interferences from the reagents.

9. Calculations

9.1 The measurements of the standard PAHs in the chromatographic system and the compounds in the sample extract in area or peak height units are made by the baseline technique from the observed fluorescence spectra. The peak height above the baseline is divided by the concentration of the standard PAH to yield the slope S per μg/mL.

9.2 The concentrations are calculated as follows:

$$C_f \text{ } (\mu g/mL) = \frac{\text{observed peak height of sample}}{S/\mu g/mL}$$

where C_f = concentration of PAH compound in composite sample extract, μg/mL

C = concentration of the PAH compound in air, μg/m^3

V_f = volume of composite sample extract, mL

V = volume of air sampled in m^3 at standard conditions

a = area of aliquot of the particulate matter of extracted glass fiber filter, cm^2

A = total particle sample area of glass fiber filter, cm^2

$$C \text{ } (\mu g/m^3) = \frac{C_f \times V_f}{{}^a/_A \cdot V}$$

9.2.1 Concentration of PAH in μg/g of particulate loading = C/L where L = particulate loading of the glass fiber filter, g/m^3.

COLUMN: HC-ODS SIL-X
MOBILE PHASE   40% TO 100% ACETONITRILE IN WATER
DETECTOR: FLUORESCENCE

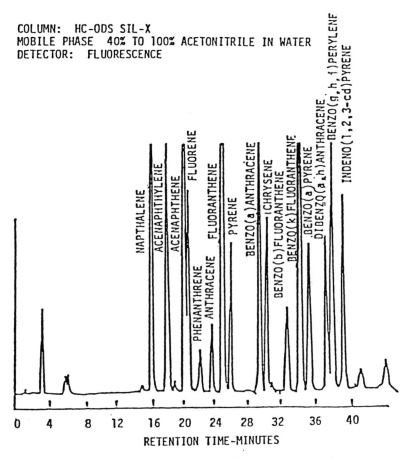

Figure 102D:1 – Liquid chromatogram of polynuclear aromatics.

10. Effect of Storage

Polynuclear aromatic hydrocarbons in standard solutions or in sample extracts appear to remain constant for many months when stored in a refrigerator in borosilicate glass bottles. The samples must be protected from light.

11. References

1. U.S. EPA Environmental Monitoring and Support Laboratory, Cincinnati, Ohio. 1977. Sampling and Analysis Procedures for Screening of Industrial Effluents for Priority Pollutants. Method 610. Polynuclear Aromatic Hydrocarbons.

2. DONG, M., D.C. LOCKE, E. FERRAND. 1976. High Pressure Liquid Chromatographic Method for Routine Analysis of Major Parent Polycyclic Aromatic Hydrocarbons in Suspended Particulate Matter. Anal. Chem., 48:368.

Subcommittee 4/5

P.O. WARNER
M. FELDSTEIN, *Chairman*
R.J. BRYAN
D.L. HYDE
D.A. LEVAGGI
D.C. LOCKE
R.A. RASMUSSEN

108.

Continuous Determination of Total Hydrocarbons in the Atmosphere (Flame Ionization Method)

1. Principle and Applicability

1.1 The sampling system is an integral part of most commercial total hydrocarbon analyzers designed to measure continuously the concentration of hydrocarbons (and other organic compounds) in the atmosphere. The sample line is attached to the inlet and the sample is pumped into a flame ionization detector. A sensitive electrometer coupled with a potentiometric recorder detects the increase in ion intensity resulting from the introduction into a hydrogen flame of sample air containing organic compounds (*e.g.*, hydrocarbons, aldehydes, alcohols). The response is approximately proportional to the number of carbon atoms in the sample. The analyzer is calibrated using methane and the results reported as methane equivalents (1,2).

1.2 The sample introduction pump may be by-passed for analysis of gases under pressure as is done with calibration gases.

2. Range and Sensitivity

2.1 The range of the analyzer may be varied so that full scale may be 2.6 mg/m³ (4 ppm) to 1960 mg/m³ (3000 ppm) hydrocarbon as methane by varying the attenuation and the sample flow rate to the detector. The 13 mg/m³ (20 ppm) range is normally used for atmospheric sampling.

2.2 Sensitivity is 1% of full scale recorder response.

3. Interferences

3.1 Carbon atoms bound to oxygen, nitrogen, or halogens give reduced or no response. There is no response to nitrogen, carbon monoxide, carbon dioxide, or water vapor.

4. Precision and Accuracy

4.1 Precision is approximately 0.5% full recorder scale on the 13 mg/m³ (20 ppm) scale.

4.2 Accuracy is dependent on instrument linearity and absolute concentration of the calibration gases used. Generally, accuracy is ± 1% of full scale on the 0 to 13 mg/m³ range.

4.3 Zero drift necessitates frequent calibration. The magnitude of the drift depends on the air flow rate, sample flow rate, fuel flow rate, ambient temperature changes, detector contamination, and electronic drift. Zero drift observations on various instruments indicate 2%/24 h on the 13 mg/m³ scale.

5. Apparatus

5.1 COMMERCIALLY AVAILABLE TOTAL HYDROCARBON ANALYZER. Instruments obtained must be installed on location and demonstrated by the manufacturer to meet or exceed manufacturer's specifications or those described in this method. Generally, hydrocarbon analyzers consist of a regulated fuel and air delivery system for the hydrocarbon burner, a regulated sample injection system, electrometer for measuring the flame ion current, meter readout with connections for a recorder or other data system, and a sample pump.

5.2 RECORDER. Potentiometric type, compatible with analyzer with an accuracy of 0.5% or better, or electronic data system with similar characteristics.

5.3 SAMPLE LINE. Any tubing that is not a source of interference or an absorbent of hydrocarbons. Inert materials such as glass, stainless steel, and Teflon are recommended. All tubing should be clean prior to

Table 108:I Suggested Minimum Performance Specifications for Total Hydrocarbon Analyzers

Range (minimum)	0–13mg/m^3 (0–20 ppm)
Output (minimum)	0–10,100,1000,5000 mv full scale
Minimum Detectable Sensitivity	0.1 mg/m^3 (0.16 ppm)
Lag Time (maximum)	30 s
Time to 90% response (maximum)	60 s
Rise Time (90% maximum)	30 s
Fall Time (90%)	30 s
Zero Drift (maximum)	2%/24 h
Span Drift (maximum)	2%/24 h
Precision (maximum)	± 1%
Operational Period (minimum)	3 d
Noise (maximum)	± 0.5%
Interference Equivalent (maximum)	1% of full scale
Operating Temperature Range (minimum)	5–40° C
Operating Humidity Range (minimum)	10–100%
Linearity (maximum)	1%

and during use. A particle filter should be installed in the sample line.

5.4 MINIMUM PERFORMANCE SPECIFICATIONS, see Table 108:I.

6. Reagents

6.1 COMBUSTION AIR. High purity air containing less than 1.3 mg/m^3 (2 ppm) hydrocarbon as methane, purified by passage over copper oxide maintained at 500°C and through a clean 55 μm filter.

6.2 FUEL. Hydrogen or a hydrogen-inert gas mixture; when ordering, specify hydrocarbon-free gas. A hydrogen generator is strongly advised for safety reasons. An automatic cut-off should be provided for the instrument as a safety measure in the event of a detector flameout.

6.3 ZERO GAS. Less than 0.05 mg/m^3 (0.1 ppm) hydrocarbon as methane in air.

6.4 SPAN GAS. Methane in air corresponding to 80% of full scale, 10.4 mg/m^3 (16.0 ppm) for 13 mg/m^3 range. A certified or guaranteed analysis is required.

7. Procedure

7.1 For specific operating instructions, refer to the manufacturer's manual (see also Section **5.1**).

8. Calibration and Efficiencies

8.1 Calibrate the instrument at the desired flow rate and attenuator setting. Introduce zero gas and set zero control to indicate proper value on the recorder. If a live zero recorder is not used, it is recommended that the zero setting be offset at least 5% of scale to allow for negative zero drift. In this case, the span setting must also be offset by an equal amount. Introduce span gas and adjust span control to indicate proper value on recorder scale (*e.g.*, 0 to 13 mg/m^3 (0 to 20 ppm) scale set at 10.4 mg/m^3 (16.0 ppm) standard to read 80% of recorder scale). Recheck zero and span until adjustments are no longer necessary. Since the scale is linear, the two-point calibration is valid. Make analogous adjustments to an electronic data system.

8.1.1 If attenuation is varied, some discrepancy between the true attenuation and the nominal attenuation may exist. The instrument should be calibrated using appropriate standards at each attenuator setting used.

9. Calculations

9.1 The recorder or data system is read directly in terms of concentration expressed as ppm methane by volume (1 ppm = 0.65 mg/m^3).

10. Effects of Storage

10.1 Not applicable.

11. References

1. NEWSOME, J.R., V. NORMAN, AND C.H. KEITH. 1965. *Tobacco Science*, IX, 102–110. Section of: Tobacco, 161 (4):24.
2. PIGLICUCCI, R., W. AVERILL, J.E. PURCELL, AND L.S.S. ETTRI. 1975. *Chromatographia*, 8:165–175.

Subcommittee 4/5

109.

Flame Ionization Detector

1. Principle of the Detector

1.1 In view of the use of this detector in Methods 101, 108, and 130, this expanded discussion is provided here. A flame ionization detector (FID) is a device which incorporates regulated fuel, air and sample delivery systems, an internal burner, and associated electronics for measuring the ion current produced by species introduced into the flame. The FID is used to sense and measure small amounts of gaseous organic-type components present in the carrier gas stream leaving the column of a gas chromatograph (G.C.) or to monitor methane and/or total hydrocarbon concentrations in ambient air samples.

1.2 Ideal characteristics of a detector may be defined as high sensitivity, low noise level, a wide linearity of response, ruggedness, insensitivity to flow and temperature changes, response to all types of compounds or an advantageous specificity for a particular type of compound. The FID is applicable to a wide range of compounds and has been demonstrated to be highly sensitive, reasonably stable, moderately insensitive to flow and temperature changes, rugged, and reliable **(9,10)**. The FID has been reported to have the best record for reliable performance among the ionization type detectors **(7)**.

1.3 A schematic diagram of a typical flame ionization burner (the sensing unit of the FID) is shown in Figure 109:1 **(10)**. The sample air or column effluent is mixed with fuel and burned at the tip of the metal jet in a diffusion flame with an excess of air.

1.4 The FID makes use of the principle that very few ions are present in the flame produced by burning pure hydrogen, or hydrogen diluted with an inert gas. However, the introduction of mere traces of organic matter into such a flame produces a large amount of ionization. The charged parti-

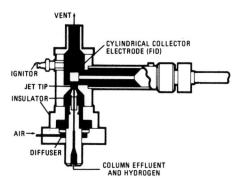

Figure 109:1 – Flame ionization detector.

cles generated in the flame are collected at two terminals resulting in a small flow of current which is amplified by an electrometer for output to a suitable recording device.

1.5 The response of the detector is roughly proportional to the carbon content of the solute. The response to most organic compounds on a molar basis increases with molecular weight. For example, propane has three times the response that an equimolar quantity of methane has. Although the FID does not respond to atoms other than carbon, other atoms change instrument sensitivity to carbon by altering the chemical environment of the carbon atom. The sensitivity of the FID varies with structure of the oxygenated or nitrogenated molecule and decreases with increasing heteroatom content **(10)**. The relative response or effective carbon number of any particular type of carbon atom is defined as the ratio between the instrument response caused by an atom of this type and the instrument response caused by an aliphatic carbon atom. Tables of effective carbon numbers have been prepared by various investigators which permit reasonable estimates to be made from chromatograms **(4,9,14)**. However, each detector must be calibrated for its response to each compound for accurate analysis.

2. Range and Sensitivity

2.1 The most extensive use of the FID in trace analysis has been in those cases where the matrix is selectively removed (traces concentrated) or where the detector does not respond to the matrix material. The FID does not respond generally to the compounds listed in Table 109:I (4,7), and has relatively low sensitivity to perfluoro compounds. The lack of response to air and water enables the FID to be especially suitable for the analysis of many air and water pollutants.

2.2 The FID has the widest linear range of any detector in common use. Linear dynamic range is about 6 to 7 orders of magnitude which allows quantitation of a broad range of sample sizes (8).

2.3 The wide linear range, lack of response to major atmospheric constituents, and excellent stability of the FID are important advantages for ambient air analysis in the 0.001 to 1 ppm concentration range. The detection limit for the FID is near 0.001 ppm of sample gas concentration (about 2 mg/m³ or 2 pg/mL) with optimized operating parameters (3,11). It is accepted practice to use twice the peak-to-peak noise level for establishing a limit of detection (7,10).

2.4 Concentration procedures have been described in the literature which can be utilized to estimate atmospheric concentrations far below the 1 ppb concentration level (2,10,15). The upper limit of measurement can be considerably greater than the concentrations normally encountered in the atmosphere.

3. Interferences

3.1 Any compound capable of ionizing in the burner flame is a potential interference. The FID is not specific for hydrocarbons, as other organic types such as alcohols, aldehydes, and ketones will respond to varying degrees. This should be taken into account in interpretation of data (9).

3.2 The FID is not a versatile qualitative tool since use of the FID without prior separation steps does not distinguish among individual compounds or hydrocarbon classes. Identification of components in G.C. atmospheric analysis usually is based on one or more retention times. However, the possibility always exists of more than one compound having the same retention time for any given set of instrumental parameters (1).

4. Precision and Accuracy

4.1 Precision is affected by instrumentation noise and drift and by variations in sample, air and fuel flow rates. Instrument stability will determine reproducibility and the need for and frequency of recalibrations.

4.2 Accuracy is affected primarily by sampling error, instrument linearity, the species being determined and the standardization or calibration employed in the analysis. Non-linearity may be overcome by calibration, which normally is easy and rapid if span gases or the pure compounds are available.

4.3 Replicate quantitative analyses of standards by the FID should not deviate by more than 2% relative standard deviation at optimum operating conditions. For the most accurate work, only the best commercially available instrumentation should be utilized.

Table 109:I Compounds Giving Little or No Response in the FID

Ar	Ne
HCHO	NH_3
CO	NO
CO_2	N_2
COS	NO_2
CS_2	O_2
He	$SiCl_4$
H_2O	SiF_4
H_2S	$SiHCl_3$
Kr	SO_2
N_2	Xe
HCOOH	

5. Apparatus

5.1 Commercially available instruments have varying modes of controlling the flow of sample air, carrier gas, fuel and combustion air streams.

5.1.1 Most of the continuous hydrocarbon analyzers utilize a pressure drop across a capillary to regulate the flow.

5.1.2 A constant differential-type low-flow controller in conjunction with a rotameter is used to regulate the carrier gas stream in most gas chromatographs. The controllers maintain a constant differential in the pressure across an external needle valve, making the flow a function of the needle valve. The use of a glass wool plug, sintered disk or porous bronze filter is recommended to protect capillaries or needle valves against particles that might change flow characteristics.

5.1.3 The rotameters are sized for the appropriate flows to give immediate visible readout on the front panel. Flow rates may appear to fluctuate on short rotameters even though an independent measurement will show a well-regulated constant flow.

In any type of flow control system, the flow rates should be verified periodically by a flow calibration device.

5.2 A burner system is recommended which is linear over a current range of at least 10^6 to allow wide versatility in application of the equipment.

5.2.1 After the ionization process, recombination of the electrons and positive ions occurs to a degree that is determined by the ion concentrations and the electrode voltage. At low voltages the ion current is proportional to applied voltage, *i.e.*, the electrode gap acts as a fixed resistance, while at high voltages a maximum number of ions are accelerated to the electrodes as determined by the equilibrium between the rates of ionization, recombination, and discharge at the electrodes.

If the jet is made the anode, higher voltages are required to obtain the saturation current. Presumably this is due to the low mobility and greater recombination rate of the positive ions, which must now move the increased distance from the inner zone of the flame to the electrode. Consequently, the best performance is obtained with the electrode over the flame as the electron collector **(10)**. In general, electrode spacing and size are not critical; spacing determines only the minimum voltage required to obtain a saturation current. A fixed-position, cylindrical collector electrode is recommended to assure a uniform spacing between the electrodes and provide long-term reproducibility.

5.2.2 The burner should be thermally isolated. Proportional temperature control up to at least 300°C is recommended to assure minimum baseline perturbations and long-term stability in G.C. isothermal and temperature programming applications. The FID must be maintained at a high enough temperature to prevent condensation of the sample, G.C. stationary phase, and the water and other by-products formed in the ionization process.

5.2.3 Corrosive samples may hasten the deterioration of the burner jet and electrodes. A FID with high-velocity air flow helps minimize the build-up of contaminants and causes SiO_2 dust, and other solids formed in the flame, to be blown out the detector's chimney. Suitable filters should be installed in the fuel, air and sample lines to protect the FID from particulate material.

5.2.4 A burner sensitivity of 0.01 C/g (5×10^{-12} g/s minimum detectability) or better is recommended. Sensitivity is a measure of the effectiveness of the detector as a transducer in converting the sample into a measurable electrical signal.

5.2.5 The burner should be equipped with a flame-sensing device to close the hydrogen supply solenoid valve in event of power failure or a flame out. This device is an important safety feature of methane and hydrocarbon analyzers that operate unattended for extended periods of time. The flame ionization detectors supplied on commercially available gas chromatographs are normally not equipped with such a device. Without exception, an exhaust hood must be provided above the instrument to remove all FID combustion products and any hydrogen which might be released to the atmosphere.

5.3 Electrometers furnished with FID's are usually of four types: (a) electron tube input, (b) vibrating capacitor input, (c) solid-state field effect transistor input, and (d) solid-state parametric oscillator amplifier. The solid state devices are greatly improved in terms of ruggedness, long instrument life and freedom from calibration drift.

5.3.1 The ion current produced in the flame is on the order of 4×10^{-11} A/10 ppm carbon. An extremely stable electrometer is required to measure currents of this magnitude. Criteria for such an electrometer include negligible noise (less than 10^{-13} A), low time constant, high input impedance circuitry, minimum zero shift due to mechanical shock or temperature variations (less than 0.5% drift per °C), overload protection, fast warm-up time, low background suppression current, and maximum reliability.

5.3.2 The current amplifier used to measure the signal should be capable of at least 5×10^{-12} A/mV for full scale output, have appropriate attenuation settings to reduce this output, and have a linear dynamic range of over 10^6. FID amplifiers are commercially available that provide output for electronic integrator and computer as well as strip chart recorders.

5.4 Recording devices include meters, servo-recorders, electronic integrators, and complete data handling systems. The device should meet user requirements and be compatible with the output of the FID amplifier selected.

6. Reagents

6.1 PURITY. All organic compounds utilized to make G.C. standards should be of ACS analytical grade or spectrograde quality.

6.2 FUEL. Fuels range from 40% hydrogen in nitrogen to pure hydrogen depending on FID design requirements. Hydrogen in compressed gas cylinders should be prepared from water pumped sources.

High quality electrolytic generators which produce ultrapure hydrogen under high pressure are recommended. These provide some degree of safety when compared to the explosive hazard of bottled hydrogen. All connections should be made with thoroughly cleaned stainless steel tubing and properly tested for leaks.

6.3 AIR. Filtered compressed air in cylinders, low in hydrocarbon content. Because hydrocarbon-free air is difficult to obtain, zero adjustment must be carried out for each new cylinder used if this air is also used as a calibration gas to set the zero point. Alternately, hydrocarbon-free synthetic air can be conveniently obtained by mixing oxygen and water-pumped nitrogen

or other equivalent diluent. Compressed commercial grade oxygen in cylinders is suitable for detectors requiring oxygen.

6.4 CARRIER GAS. Compressed nitrogen, argon, or helium should be prepared from water pumped sources or be of guaranteed high purity.

6.5 SPAN GASES. At least three different concentrations of hydrocarbons in air (methane and/or propane are normally used) are required for calibration purposes. Commercial gases having a certified analysis should be used. If possible, an independent G.C. analysis should be made to verify the hydrocarbon concentration in the tank and particularly if the span gas has been calculated and prepared in the laboratory **(12,13)**.

7. Operation

7.1 Instrument manuals are provided by manufacturers of commercially available flame ionization detectors and should be consulted for specific operating procedures.

7.2 FID performance depends on the proper choice of gas flow rates. Individual detectors should be calibrated to determine the hydrogen flow rate that gives maximum sensitivity and stability. At this point the signal is relatively insensitive to small variations in fuel rate **(8)**. Optimum performance is particularly dependent on the ratio of hydrogen to carrier or sample gas flow rate, since this ratio determines the flame temperature and therefore the efficiency of ionization. The optimum ratio of hydrogen to carrier gas has been determined to be about 1. The supply of air should be such that the ratio of air to hydrogen is about 10 **(4)**. Depending on the geometry of the detector chamber, an excessive supply of air may lead to turbulence in the flame zone with resultant noise and result in a loss of detectability of small sample concentrations.

7.3 Oxygen may be substituted for air in some FID's with a resulting increase in sensitivity but often with a decrease in the stability of the flame. All flow rates should be optimized for a particular instrument and maintained carefully.

7.4 When the burner is not operating, it is recommended that an inert gas (helium, argon, or nitrogen) be used to purge the FID. This will eliminate corrosion caused by the laboratory environment.

7.5 The FID is destructive in operation, since the sample stream is burned at the jet. However, this is not a serious handicap in G.C. applications, since the highly sensitive FID may be arranged in series with an effluent splitter which would allow a fraction of the components from the effluent stream to be collected for further analysis (mass spectrometry, nuclear magnetic resonance, etc.) **(5,10)**.

7.6 While the FID has generally been touted for its insensitivity to flow, exception must be taken to this claim from a consideration of liquid phase volatility (G.C. applications). The signal represents a rate at which solute is being ionized in the detector and any background contributed by liquid phase vapor will vary with flow rate. The concentration of liquid phase in the detector is a function of its vapor pressure which is established by the flow rate of the carrier gas and the column temperature. Consequently, both carrier gas flow rate and column temperature can alter the performance of the FID, particularly when temperature programming is employed **(6)**.

7.7 After continued operation of the FID, the detector may become contaminated resulting in increased noise and decreased sensitivity. The FID should be carefully disassembled and cleaned with a soft brush and a suitable solvent such as methylene chloride. Clean tweezers should be used to avoid contamination from fingers when replacing the jets. The electrodes should be accurately repositioned as the FID is reassembled. It is recommended that the FID be cleaned every 3 to 6 months to assure proper maintenance and optimum performance.

7.8 A malfunction in the FID normally manifests itself as an erratic reading on a strip chart recorder. Most instrument manuals include a section on troubleshooting to determine whether the malfunction is in the electronic system, the detector or the flow system.

8. Calibration and Standards

8.1 Hydrocarbon analyzers should be zeroed with a hydrocarbon-free air mixture and calibrated with span gases in the 25 to several hundred ppm range to be analyzed on a regular basis, at least once daily. At optimum operation conditions, FID's require only a one point, one component calibration since response is linear with the carbon content of the sample being admitted to the burner. However, a three or more point calibration is recommended to assure a more accurate analysis (confirm linearity) over the range of expected sample concentrations.

8.2 The instrument sample regulator is adjusted to cause the recorder reading to equal the value of the concentration of hydrocarbon as indicated on the cylinder of span gas. It is recommended that the use of tanks at gauge pressures lower than 3.45 MPa (500 psi) be discontinued since calibration gases in the lower ranges could change markedly in concentration (because of desorption).

8.3 Care must be exercised in selecting the sample flow rate to the burner, since at some maximum flow rate the flow versus response relationship becomes nonlinear because of saturation of the electronics. Since the output of the FID is directly dependent on the amount of hydrocarbons in the flame, the same setting of the flow regulator is used when the atmospheric sample is introduced, as the calibration (for hydrocarbon analyzers) is flow dependent **(9)**.

8.4 Quantitative analysis requires adequate calibrations (three or more points) for individual compounds. The same G.C. operating parameters that are used for analyzing the standards must be used when analyzing the samples. The samples and standards should be analyzed in sequence to minimize analytical, analyst, and instrumentation errors. Calibration standards for G.C. applications may be prepared by use of procedures recommended in this volume.

Note: Preparation of Hydrocarbon Mixtures: In the preparation of hydrocarbon mixtures in air in tanks, experience has shown that the lowest concentration feasible on a routine basis is 10 ppm. Even at this concentration, careful checks must be made by independent means to determine the change in concentration with time. To prepare stable hydrocarbon mixtures at lower levels of concentration in tanks under pressure, down to less than 1 ppm, extreme care and effort must be taken to account for sorption and desorption effects that are influenced by metal composition of the tank, previous "chemical history," temperature and other factors.

Calibration gas mixtures at levels normally used, such as 15 ppm propane, change markedly in concentration, because of desorption, as the tank gauge pressure drops below about 2 MPa. It is recommended that tank use be discontinued at gauge pressures below 3.45 MPa (500 psi).

9. Calculation

9.1 The meter or recorder recording of a calibrated FID (hydrocarbon analyzer) will normally represent the hydrocarbon concentration of the measured gas in parts per million carbon or methane.

9.2 The G.C. standard curves are drawn by plotting the concentration of the standards versus the respective peak area or integrator output. The concentration of any given organic contaminant in the sampled air is determined by referring its peak area or integrator output to the corresponding standard curve.

10. Effects of Storage

10.1 Not applicable.

11. References

1. ALTSHULLER, A.P. 1968. *Advances in Chromatography*, J.C. Giddings, and R.A. Keller, eds., Marcel Dekker, Inc., New York, N.Y., p. 236.
2. BELLAR, T.A., M.F. BROWN, AND J.E. SIGSBY, 1963. *Anal. Chem.* 35:1924.
3. BELLAR, T., J.E. SIGSBY, C.A. CLEMONS, AND A.P. ALTSHULLER, 1962. *Anal. Chem.* 34:763.
4. CONDON, R.D., P.R. SCHOLLY, AND W. AVERILL, 1960. *Gas Chromatography.* R.P.W. Scott, ed., Butterworths, Washington, D.C., p. 30.
5. COOPER, C.V., L.D. WHITE, AND R.E. KUPEL, 1974. Am. Ind. Hyg. Assoc. J. 32:383.
6. GERRARD, W., S.J. HAWKES, AND E.F. MOONEY, 1960. *Gas Chromatography.* R.P.W. Scott, ed., Butterworths, Washington, D.C., p. 199.
7. HARTMANN, C.H. 1971. Anal. Chem. 43:113A.
8. MCNAIR, H.M. AND E.J. BONELLI, 1969. *Basic Gas Chromatography.* Varian Aerograph, Walnut Creek, Calif., pp. 99–105.
9. MORRIS, R.A. AND R.L. CHAPMAN, 1961. Air Poll. Control Assoc. J. 11:467.
10. NOGARE, S.D. AND R.S. JUVET, 1966. *Gas-Liquid Chromatography, Theory and Practice.* Interscience Publishers, New York, N.Y., pp. 180–239.
11. RASMUSSEN, R.A. AND F.W. WENT, 1965. Proc. Natl. Acad. Sci. U.S. 53:215.
12. SALTZMAN, B.E. 1972. *The Industrial Environment . . . Its Evaluation and Control: Syllabus.* 3rd Ed., Chapter 12, Public Health Service Publication No. 614, U.S. Government Printing Office, Washington, D.C.
13. SALTZMAN, B.E., W.R. BURG, AND G. RAMASWAMY, 1971. *Environ. Sci. Technol.* 5:1121.
14. STENBERG, J.C., W.S. GALLAWAY, AND D.L.T. JONES. 1962. *Gas Chromatography.* N. Brenner, J.E. Callen, and M.D. Weiss, eds., Academic Press, New York, N.Y., p. 231.
15. WHITE, L.D., D.G. TAYLOR, P.A. MAUER AND R.E. KUPEL, 1970. Am. Ind. Hyg. Assoc. J. 31:225.

Subcommittee 5

E. SAWICKI, *Chairman*
P.R. ATKINS
T. BELSKY
R.A. FRIEDEL
D.L. HYDE
J.L. MONKMAN
R.A. RASMUSSEN
L.A. RIPPERTON
J.E. SIGSBY
L.D. WHITE

Approved with modifications from 2nd edition.

Subcommittee 4/5

M. FELDSTEIN, *Chairman*
R.J. BRYAN
D.L. HYDE
D.A. LEVAGGI
D.C. LOCKE
R.A. RASMUSSEN
P.O. WARNER

114.

Determination of Acrolein Content of the Atmosphere (Colorimetric)

1. Principle of Method

1.1 The reaction of acrolein with 4-hexylresorcinol in an ethyl alcohol-trichloroacetic acid solvent medium in the presence of mercuric chloride results in a blue-colored product with strong absorption maximum at 605 nm (1,2).

2. Range and Sensitivity

2.1 The absorbances at 605 nm are linear for at least 1 to 30 μg of acrolein in the 10 mL portions of mixed reagent (3).

A concentration of 0.01 ppm of acrolein can be determined in a 50-L air sample based on a difference of 0.05 absorbance unit from the blank using a 1-cm cell. Greater sensitivity could be obtained by use of a longer path length cell.

3. Interferences

3.1 There is no interference from ordinary quantities of sulfur dioxide, nitrogen dioxide, ozone and most organic air pollutants. A slight interference occurs from dienes: 1.5% for 1,3-butadiene and 2% for 1,3-pentadiene. The red color produced by some other aldehydes and undetermined materials does not interfere in spectrophotometric measurement (3).

4. Precision and Accuracy

4.1 Known standards can be determined to within ± 5% of the true value (4). No data are available on precision and accuracy of air samples.

5. Apparatus

5.1 ABSORBERS. All-glass standard midget impingers with fritted glass inlets are acceptable. The fritted end should have a porosity approximately equal to that of Corning EC (170 to 220 μm maximum pore diameter). A train of 2 bubblers is needed (3).

5.2 WATER BATH. Any bath capable of maintaining a temperature of 58° to 60°C is acceptable.

5.3 AIR PUMP. A pump capable of drawing at least 2 L of air/min for 60 min through the sampling train is required. A trap at the inlet to protect the pump from the corrosive reagent is recommended.

5.4 AIR-METERING DEVICE. Either a limiting orifice of approximately 1 or 2 L/min capacity or a glass flow meter can be used. If a limiting orifice is used, regular and frequent calibration is required.

5.5 SPECTROPHOTOMETER. This instrument should be capable of measuring the developed color at 605 nm. The absorption band is rather narrow, and thus a lower absorptivity may be expected in a broadband instrument (3).

6. Reagents

6.1 PURITY. All reagents must be ACS reagent grade. Distilled water must meet the specifications of ASTM reagent water, Type II.

6.2 ETHANOL (96%).

6.3 TRICHLOROACETIC ACID SOLUTION, SATURATED.* Dissolve 100 g of the acid (reagent grade) in 10 mL of water by heating on a water bath. The resulting solution has a volume of approximately 70 mL. Even reagent grade trichloroacetic acid has an impurity that affects product formation.

---

*CAUTION—*Both the solid acid and solution are corrosive to the skin. Breathing of the fumes evolved during preparation of the solution should be avoided.*

Every new batch of solution should be standardized with acrolein. It is convenient to prepare a large quantity of solution from a single batch of trichloroacetic acid to maintain a uniformity of response (1).

6.4 MERCURIC CHLORIDE SOLUTION† (3%). Dissolve 3 g of mercuric chloride in 100 mL of ethanol.

6.5 4-HEXYLRESORCINOL SOLUTION. Dissolve 5 g of 4-hexylresorcinol (MP 68 to 70°C) in 5.5 mL of ethanol. This makes about 10 mL of solution.

6.6 MIXED SAMPLING REAGENT. Mix, in order, reagents in the following proportions: 5 mL ethanol, 0.1 mL 4-hexylresorcinol solution, 0.2 mL mercuric chloride solution, and 5 mL saturated trichloroacetic acid solution. The mixed reagent may be stored for a day at room temperature. Prepare the needed quantity by selecting an appropriate multiple of these amounts. Protect from direct sunlight.

6.7 ACROLEIN, PURIFIED. Freshly prepare a small quantity (less than 1 mL is sufficient) by distilling 10 mL of the purest grade of acrolein commercially available. Reject the first 2 mL of distillate. (The acrolein should be stored in a refrigerator to retard polymerization.) The distillation should be done in a hood because the vapors are irritating to the eyes.

6.8 ACROLEIN, STANDARD SOLUTION "A" (1 MG/ML). Weigh 0.1 g (approximately 0.12 mL) of freshly prepared, purified acrolein into a 100-mL volumetric flask and dilute to volume with ethanol. This solution may be kept for as long as a month if properly refrigerated.

6.9 ACROLEIN, STANDARD SOLUTION "B" (10 µG/ML). Dilute 1 mL of standard solution "A" to 100 mL with ethanol. This solution may be kept for as long as a month if properly refrigerated.

7. Procedure

7.1 AIR SAMPLING. Draw measured volumes of the vapor laden air at a rate of either 1 L/min for 60 min or 2 L/min for 30 min through 2 bubblers in series, each containing 10 mL of mixed sampling reagent. An extra bubbler containing water may be

†CAUTION—*Mercuric chloride is highly toxic and direct contact should be avoided.*

added as a trap to protect the pump. A maximum of 60 L of air can be sampled before possible reagent decomposition may occur.

This sampling system collects 70 to 80% of the acrolein in the first bubbler and 95% of the acrolein in the first 2 bubblers (1), using absorbers with EC fritted glass inlets. The absorption efficiency might be increased by use of a C porosity frit (60 µm maximum pore diameter).

7.2 ANALYSIS.

7.2.1 If evaporation has occurred during sampling the sampling solution is diluted to its original 10 mL volume with ethanol.

7.2.2 Transfer the samples to glass-stoppered test tubes. Immerse the tubes in a 60°C water bath for 15 min to develop the colors. A test tube containing 10 mL of mixed sampling reagent must be run similarly and simultaneously. This serves as the reagent blank.

7.2.3 Cool the test tubes in running water immediately upon removal from the water bath.

7.2.4 After 15 min read the absorbances at 605 nm in a suitable spectrophotometer using 1-cm cells. There is no appreciable loss in accuracy if the samples are allowed to stand up to 2 h before reading the absorbances. Determine the acrolein content of the sampling solution from a curve previously prepared from the standard acrolein solution.

For very low acrolein concentrations it may be convenient to use a longer pathlength cell.

8. Calibration

8.1 PREPARATION OF STANDARD CURVE.

8.1.1 Pipet 0, 0.5, 1.0, 2.0, and 3.0 mL of standard solution "B" into glass stoppered test tubes.

8.1.2 Dilute each standard to exactly 5 mL with ethanol.

8.1.3 Add in order, to each tube, exactly 0.1 mL of 4-hexylresorcinol solution, 0.2 mL of mercuric chloride solution, and 5 mL of trichloroacetic acid solution.

8.1.4 Mix, develop and read the colors as described in the analytical procedure.

8.1.5 Plot absorbance against micrograms of acrolein in the color-developed solution.

9. Calculation

9.1 The concentration of acrolein in the sampled atmosphere may be calculated by using the following equation.

$$\text{PPM (Vol.)} = \frac{C \times 24.45}{V \times \text{M.W.} \times E}$$

where E = Correction factor for sampling efficiency

V = liters of air sampled, corrected to 25°C and 101.3 kPa

C = μg of acrolein in sampling solution

M.W. = molecular weight of acrolein (56.06)

24.45 = mL of acrolein vapor in one millimole at 101.3 kPa and 25°C

10. Effect of Storage

10.1 The color forms in the sampling solution and is fully developed in 2 h at room temperature. The solution starts fading after about 3 h. Therefore it is best to analyze the samples almost immediately after completion of sampling.

11. References

1. COHEN, I.R. AND A.P. ALTSHULLER. 1961. A New Spectrophotometric Method for the Determination of Acrolein in Combustion Gases and the Atmosphere. *Anal. Chem.* 33:726.
2. ALTSHULLER, A.P. AND S.P. MCPHERSON, 1963. Spectrophotometric Analysis of Aldehydes in the Los Angeles Atmosphere. *J. Air Poll. Control Assoc.* 13:109.
3. COHEN, I.R. AND B.E. SALTZMAN. 1965. Determination of Acrolein: 4-Hexylresorcinol Method. Selected Methods for the Measurement of Air Pollutants, Public Health Service Publication No. 999-AP-11, Page G-1.
4. LEVAGGI, D.A. AND M. FELDSTEIN, 1970. The Determination of Formaldehyde, Acrolein and Low Molecular Weight Aldehydes in Industrial Emissions on a Single Collected Sample. *J. Air Poll. Control Assoc.* 20:312.

Subcommittee 4

R.G. SMITH, *Chairman*
R.J. BRYAN
M. FELDSTEIN
B. LEVADIE
F.A. MILLER
E.R. STEPHENS
N.G. WHITE

Approved with modifications from 2nd edition.

Subcommittee 4/5

M. FELDSTEIN, *Chairman*
R.J. BRYAN
D.L. HYDE
D.A. LEVAGGI
D.C. LOCKE
R.A. RASMUSSEN
P.O. WARNER

116.

Determination of Formaldehyde Content of the Atmosphere (Colorimetric Method)

1. Principle of the Method

1.1 Formaldehyde reacts with chromotropic acid-sulfuric acid solution to form a purple monocationic chromogen. The absorbance of the colored solution is read in a spectrophotometer at 580 nm and is proportional to the quantity of formaldehyde in the solution **(2,6)**.

1.2 Feigl **(3)**, though stating "The chemistry of this color reaction is not known with certainty," proposes the reactions shown in Figure 116:1. Recent work by P. Georghiou (personal communication, 1988) involves attack at the β-carbon atoms, followed by ring closure to form a dibenzo [c, h] xanthylium structure.

2. Range and Sensitivity

2.1 From 0.1 μg/mL to 2.0 μg/mL of formaldehyde can be measured in the color-developed solution (10.1 mL).

2.2 A concentration of 0.1 ppm of formaldehyde can be determined in a 25-L air sample based on an aliquot of 4 mL from 20 mL of absorbing solution and a difference of 0.05 absorbance unit from the blank.

3. Interferences

3.1 The chromotropic acid procedure has very little interference from other aldehydes. Saturated aldehydes give less than 0.01% positive interference, and the unsaturated aldehyde acrolein results in a few per cent positive interference. Ethanol and higher molecular weight alcohols and olefins in mixtures with formaldehyde are negative interferences. However, concentrations of alcohols in air are usually much lower than formaldehyde concentrations and, therefore, are not a serious interference.

3.2 Phenols result in a 10 to 20% negative interference when present at an 8:1 excess over formaldehyde. They are, however, ordinarily present in the atmosphere at lesser concentrations than formaldehyde and, therefore, are not a serious interference.

3.3 Ethylene and propylene in a 10:1 excess over formaldehyde result in a 5 to 10% negative interference and 2-methyl-1,3-butadiene in a 15:1 excess over formaldehyde showed a 15% negative interference. Aromatic hydrocarbons also constitute a negative interference **(6)**. It has been found that cyclohexanone causes a bleaching of the final color **(4)**.

4. Precision and Accuracy

The method was checked for reproducibility by having three different analysts in three different laboratories analyze formaldehyde samples. The results, listed in Table 116:I, agreed within ± 5%.

5. Apparatus

5.1 ABSORBERS. All glass samplers with extra coarse fritted tube inlet. Figure 116:2 shows an acceptable absorber.

5.2 AIR PUMP. A pump capable of drawing at least 1 L of air per min for 24 h through the sampling train is required.

5.3 AIR METERING DEVICE. Either a limiting orifice of approximately 1 L/min capacity or a wet test meter can be used. If a limiting orifice is used, regular and frequent calibration is required.

5.4 SPECTROPHOTOMETER OR COLORIMETER. An instrument capable of measuring the absorbance of the color developed solution at 580 nm.

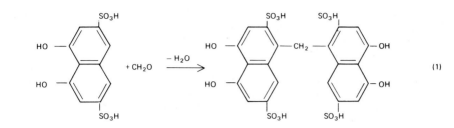

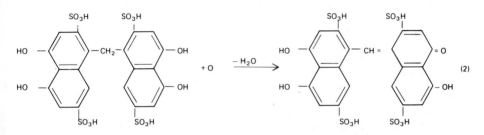

Figure 116:1 – A proposed mechanism of reaction of formaldehyde and chromotropic acid.

6. Reagents

6.1   Reagents must be ACS Reagent grade. Water is ASTM reagent water, Type II.

6.2 CHROMOTROPIC ACID REAGENT. Dissolve 0.10 g of 4,5-dihydroxy-2,7-naphthalenedisulfonic acid disodium salt (Eastman Kodak Company, Rochester, New York, Cat. No. P230) in water and dilute to 10 mL. Filter if necessary and store in a brown bottle. Make up fresh weekly.

6.3 CONCENTRATED SULFURIC ACID.

6.4 FORMALDEHYDE STANDARD SOLUTION "A" (1 MG/ML). Dilute 2.7 mL of 37% formalin solution to 1 L with distilled water. This solution must be standardized as described in section 8.1. The solution is stable for at least a 3-month period. Alternatively sodium formaldehyde bisulfite (Eastman Kodak Company, Cat. No. P6450) can be used as a primary standard **(4)**. Dissolve 4.4703 g in distilled water and dilute to 1 L.

6.5 FORMALDEHYDE STANDARD SOLUTION "B" (10 μG/ML). Dilute 1 mL of standard solution "A" to 100 mL with distilled water. Make up fresh daily.

6.6 IODINE, 0.1 N (APPROXIMATE). Dissolve 25 g of potassium iodide in about 25 mL of water, add 12.7 g of iodine and dilute to 1 L.

6.7 IODINE, 0.01 N. Dilute 100 mL of the 0.1 N iodine solution to 1 L. Standardize against sodium thiosulfate.

6.8 STARCH SOLUTION, 1%. Make a paste of 1 g of soluble starch and 2 mL of water and slowly add the paste to 100 mL of boiling water. Cook, add several mL of chloroform as a preservative, and store in a stoppered bottle. Discard when a mold growth is noticeable.

Table 116:1 Comparison of Formaldehyde Results From Three Laboratories

Micrograms Formaldehyde	Absorbance		
	Lab. 1	Lab. 2	Lab. 3
1	0.057	0.063	0.061
3	0.183	0.175	0.189
5	0.269	0.279	0.262
7	0.398	0.381	0.392
10	0.566	0.547	0.537
20	1.02	0.980	1.07

6.9 SODIUM CARBONATE BUFFER SOLUTION. Dissolve 80 g of anhydrous sodium carbonate in about 500 mL of water. Slowly add 20 mL of glacial acetic acid and dilute to 1 L.

6.10 SODIUM BISULFITE, 1%. Dissolve 1 g of sodium bisulfite in 100 mL of water. It is best to prepare a fresh solution weekly.

7. Procedure

7.1 AIR SAMPLING.

7.1.1 Draw measured volumes of ambient air at a rate of 1 L/min for 24 h through 20 mL of distilled water contained in the absorber (1,5). However, a shorter sampling time can be used providing enough formaldehyde is collected to be above the lower limit of sensitivity of the method. Two absorbers must be used in series because under conditions of sampling, collection efficiency of one absorber is approximately 80%. With two absorbers in series the total collection efficiency is approximately 95%.

7.1.2 Note that some loss of sampling solution due to evaporation will take place over a 24-h period so that it is either necessary to add water during the sampling period or else start out with a larger volume than 20 mL. Tests have shown that 35 mL in the first bubbler and 25 mL in the second bubbler is satisfactory.

7.2 ANALYSIS.

7.2.1 Transfer the sample from each absorber to either a 25-mL or 50-mL graduate. Note the volume of each solution.

7.2.2 Pipet a 4-mL aliquot from each of the sampling solutions into glass stoppered test tubes. A blank containing 4 mL of distilled water must also be run. If the formaldehyde content of the aliquot exceeds the limit of the method a smaller aliquot diluted to 4 mL with distilled water is used.

7.2.3 Add 0.1 mL of 1% chromotropic acid reagent to the solution and mix.

7.2.4 To the solution, pipet slowly and cautiously 6 mL of concentrated sulfuric acid. The solution becomes extremely hot during the addition of the sulfuric acid. If the acid is not added slowly, some loss of sample could occur due to spattering.

7.2.5 Allow to cool to room temperature. Read at 580 nm in a suitable spectrophotometer using a 1-cm cell. No change in absorbance was noted over a 3-hr period after color development. Determine the formaldehyde content of the sampling solution from a curve previously prepared from standard formaldehyde solutions.

8. Calibration and Standards

8.1 STANDARDIZATION OF FORMALDEHYDE SOLUTION.

8.1.1 Pipet 1 mL of formaldehyde standard solution "A" into an iodine flask. Into another flask pipet 1 mL of distilled water. This solution serves as the blank.

8.1.2 Add 10 mL of 1% sodium bisulfite and 1 mL of 1% starch solution.

8.1.3 Titrate with 0.1 N iodine to a dark blue color.

8.1.4 Destroy the excess iodine with 0.05 N sodium thiosulfate.

8.1.5 Add 0.01 N iodine until a faint blue end point is reached.

8.1.6 The excess inorganic bisulfite is now completely oxidized to sulfate, and the solution is ready for the assay of the formaldehyde bisulfite addition product.

8.1.7 Chill the flask in an ice bath and add 25 mL of chilled sodium carbonate buffer. Titrate the liberated sulfite with 0.01 N iodine, using a microburet, to a faint blue end point. The amount of iodine added in this step must be accurately measured and recorded.

8.1.8 One mL of 0.0100 N iodine is equivalent to 0.15 mg of formaldehyde. Therefore, since 1 mL of formaldehyde standard solution was titrated, the mL of 0.01 N iodine in the final titration multiplied by 0.15 mg gives the formaldehyde concentration of the standard solution in mg/mL.

8.2 PREPARATION OF STANDARD CURVE.

8.2.1 Pipet 0, 0.1, 0.3, 0.5, 0.7, 1.0, and 2.0 mL of standard solution "B" into glass stoppered test tubes.

8.2.2 Dilute each standard to 4 mL with distilled water.

8.2.3 Develop the color as described in the analytical procedure (2).

8.2.4 Plot absorbance against micrograms of formaldehyde in the color developed solution.

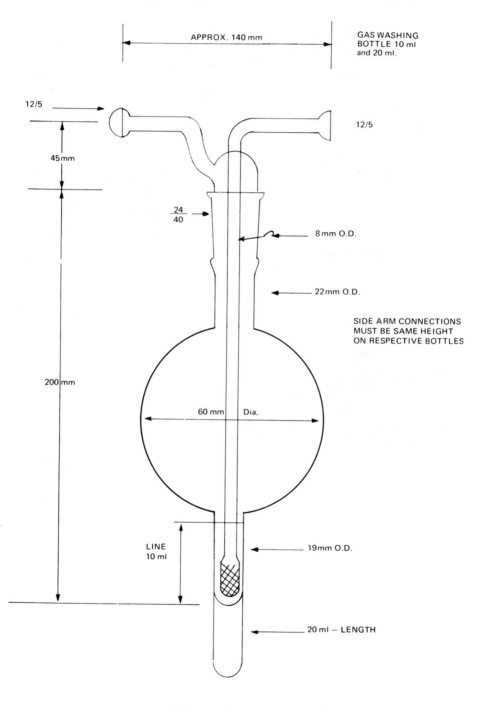

Figure 116:2 — Absorber.

9. Calculation

The concentration of formaldehyde in the sampled atmosphere may be calculated by using the following equation, assuming standard conditions are taken as 101.3 kPa and 25°C:

$$\text{ppm (volume)} = \frac{C \times A \times 24.47}{V \times M.W.}$$

where V = volume of air sampled, L
　　　C = formaldehyde in aliquot, μg
　　　A = aliquot factor (sampling solution volume in mL/mL in aliquot)
　M.W. = molecular weight of formaldehyde (30.03)
　24.47 = mL of formaldehyde gas in one millimole at 101.3 kPa and 25°C

10. Effect of Storage

10.1 The absorbance of the reaction product increases slowly on standing. An increase of 3% in absorbance was noted after 1 day standing and an increase of 10% after 8 days standing (6).

10.2 No information is available on the effect of storage on the collected air sample.

11. References

1. ALTSHULLER, A.P., L.J. LENG, AND A.F. WARTBURG. 1962. Source and Atmospheric Analyses for Formaldehyde by Chromotropic Acid Procedures. *Int. J. Air Wat. Poll.*, 6:381.
2. EEGRIWE, E. 1937. Reaktionen und Reagenzien zum Nachweis Organischer Verbindungen IV. *Z. Anal. Chem.*, 110:22.
3. FEIGL, F. 1966. Spot Tests in Organic Analysis. Seventh Edition, American Elsevier Publishing Company, New York.
4. FELDSTEIN, M. March, 1968. (Bay Area Air Pollution Control District) Personal communication.
5. MACDONALD, W.W. 1954. Formaldehyde in Air — A Specific Field Test. *Amer. Ind. Hyg. Assoc. Quarterly*, 15:217.
6. SLEVA, S.F. 1965. Determination of Formaldehyde: Chromotropic Acid Method, Selected Methods for the Measurement of Air Pollutants. Public Health Service Publication No. 999-AP-11, H-1.

Subcommittee 4

R.G. SMITH, *Chairman*
R.J. BRYAN
M. FELDSTEIN
B. LEVADIE
F.A. MILLER
E.R. STEPHENS
N.G. WHITE

Approved with modifications
from 2nd edition
Subcommittees 4 & 5
M. FELDSTEIN, *Chairman*
R.J. BRYAN
D.L. HYDE
D.A. LEVAGGI
D.C. LOCKE
R.A. RASMUSSEN
P.O. WARNER

117.

Determination of Formaldehyde Content of the Atmosphere (MBTH Colorimetric Method—Application to Other Aldehydes)

1. Principle of Method

1.1 The aldehydes in ambient air are collected in a 0.05% aqueous 3-methyl-2-benzothiazolinone hydrazone hydrochloride (MBTH) solution. The resulting azine is then oxidized by a ferric chloride-sulfamic acid solution to form a blue cationic dye in acid media, which can be measured at 628 nm (1,2,3).

1.2 The mechanism of the present procedure as applied to formaldehyde includes the following steps: reaction of the aldehyde with 3-methyl-2-benzothiazolinone hydrazone, A, to form the azine, B; conversion of A to a reactive cation, C; and formation of the blue cation, D (1).

2. Range and Sensitivity

2.1 From 0.03 μg/mL to 0.7 μg/mL of formaldehyde can be measured in the color-developed solution (12 mL). A concentration of 0.03 ppm of aldehyde (as formaldehyde) can be determined in a 25-L air sample based on an aliquot of 10 mL from 35 mL of absorbing solution and a difference of 0.05 absorbance unit from the blank.

3. Interferences

3.1 The following classes of compounds react with MBTH to produce colored products: aromatic amines, imino heterocyclics,

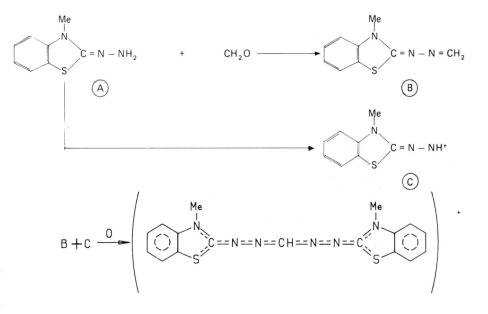

carbazoles, azo dyes, stilbenes, Schiff bases, the aliphatic aldehyde 2,4-dinitrophenyl hydrazones, and compounds containing the p-hydroxystyryl group. Most of these compounds are not gaseous or water soluble and, consequently, should not interfere with the analysis of water-soluble aliphatic aldehydes in the atmosphere (3).

4. Precision and Accuracy

4.1 The method was checked for reproducibility by having three different analysts in three different laboratories analyze standard formaldehyde samples. The results, listed in Table 117:I, agreed within ± 5%.

5. Apparatus

5.1 ABSORBERS. All-glass samplers with coarse fritted tube inlet. Figure 117:1 shows an acceptable absorber.

5.2 AIR METERING DEVICE. Either a limiting orifice of approximately 0.5 L/min capacity or a wet test meter can be used. If a limiting orifice is used, regular and frequent calibration is required.

5.3 AIR PUMP. A pump capable of drawing at least 0.5 L of air/min for 24 h through the sampling train is required.

5.4 SPECTROPHOTOMETER. An instrument capable of measuring accurately the developed color at the narrow absorption band of 628 nm.

6. Reagents

6.1 PURITY. All reagents must be ACS reagent grade; water must be ASTM reagent water, Type II.

6.2 3-METHYL-2-BENZOTHIAZOLINONE HYDRAZONE HYDROCHLORIDE ABSORBING SOLUTION (0.05%). Dissolve 0.5 g of

Table 117:I Comparison of Formaldehyde Results from Three Laboratories (Analysis of Standard Formaldehyde Samples)

Micrograms/mL Formaldehyde	Absorbance		
	Laboratory 1	Laboratory 2	Laboratory 3
0.05	0.078	0.077	0.082
0.10	0.151	0.156	0.146
0.30	0.430	0.457	0.445
0.50	0.720	0.700	0.728
0.70	0.990	1.04	1.02

MBTH in distilled water and dilute to 1 L. This colorless solution is filtered by gravity, if slightly turbid, and is stable for at least 1 week after which it becomes pale yellow. Stability may be increased by storing in a dark bottle in the cold.

6.3 OXIDIZING REAGENT. Dissolve 1.6 g of sulfamic acid and 1.0 g of ferric chloride in distilled water and dilute to 100 mL.

6.4 FORMALDEHYDE STANDARD SOLUTION "A" (1 MG/ML). Dilute 2.7 mL of 37% formalin solution to 1 L with distilled water. This solution must be standardized as described in "Calibration" section. This solution is stable for at least a 3-month period.

6.5 FORMALDEHYDE STANDARD SOLUTION "B" (10 μG/ML). Dilute 1 mL of standard solution "A" to 100 mL with 0.05% MBTH solution. Make up fresh daily.

6.6 IODINE 0.1 N (APPROXIMATE). Dissolve 25 g of potassium iodide in about 25 mL of water, add 12.7 g of iodine and dilute to 1 L.

6.7 IODINE 0.01 N. Dilute 100 mL of the 0.1 N iodine solution to 1 L. Standardize against sodium thiosulfate.

6.8 STARCH SOLUTION 1%. Make a paste of 1 g of soluble starch in 2 mL of water and slowly add the paste to 100 mL of boiling water. Cool, add several mL of chloroform as a preservative, and store in a stoppered bottle. Discard when a mold growth is noticeable.

6.9 SODIUM CARBONATE BUFFER SOLUTION. Dissolve 80 g of anhydrous sodium carbonate in about 500 mL of water. Slowly add 20 mL of glacial acetic acid and dilute to 1 L.

6.10 SODIUM BISULFITE 1%. Dissolve 1 g of sodium bisulfite in 100 mL of water. It is best to prepare a fresh solution weekly.

7. Procedure

7.1 AIR SAMPLING. Draw measured volumes of the vapor-laden air at a rate of 0.5 L/min for 24 h through 35 mL of MBTH absorbing solution contained in the absorber. A shorter sampling time can be used providing enough formaldehyde is collected to be above the lower limit of sensitivity of the method.

The average collection efficiency of formaldehyde in air has been determined to be

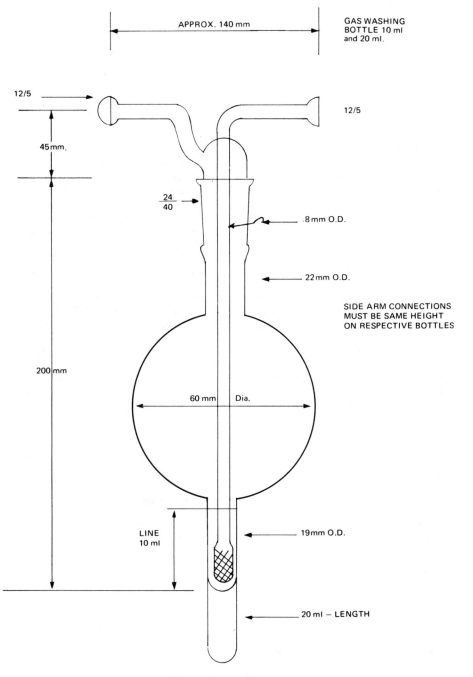

Figure 117:1 — Absorber

84% when air was sampled at a rate of 0.5 L/min over a 24-hr period in 35 mL of collecting reagent (3) in an absorber equipped with an extra coarse (EC) fritted tube inlet. Absorption efficiency may be improved by using a coarse (C) frit although data are lacking on this likelihood.

7.2 ANALYSIS.

7.2.1 Transfer the samples from the sampling bottles to 50-mL graduates, dilute to 35 mL with distilled water and allow to stand for 1 h.

7.2.2 Pipet a 10-mL aliquot of the sampling solution into a glass stoppered test tube. A blank containing 10 mL of MBTH solution must also be run. If the aldehyde content of the aliquot exceeds the limits of the method, a smaller aliquot diluted to 10 mL with MBTH solution is used.

7.2.3 Add 2 mL of oxidizing solution and mix thoroughly.

7.2.4 After at least 12 min, read at 628 nm on a suitable spectrophotometer using a 1-cm cell. No significant change in absorbance was noted over a 3-h period after color development. Determine the aldehyde content of the sampling solution from a curve previously prepared from the standard formaldehyde solution. This will give total aldehyde calculated as formaldehyde.*

8. Calibration

8.1 Pipet 1 mL of formaldehyde standard solution "A" into an iodine flask. Into another flask pipet 1 mL of distilled water. This solution serves as the blank.

8.2 Add 10 mL of 1% sodium bisulfite and 1 mL of 1% starch solution.

8.3 Titrate with 0.1 N iodine to a dark blue color.

8.4 Destroy the excess iodine with 0.05 N sodium thiosulfate.

8.5 Add 0.01 N iodine until a faint blue end point is reached.

*Note: The final colored solution tends to form bubbles that cling to the sides of the cuvettes. In order to eliminate this, the solution should be thoroughly shaken periodically during the 12-min standing time waiting for full color development. It has been found that this thorough shaking will eliminate bubble formation.

8.6 The excess inorganic bisulfite is now completely oxidized to sulfate and the solution is ready for the assay of the formaldehyde bisulfite addition product.

8.7 Chill the flask in an ice bath and add 25 mL of chilled sodium carbonate buffer. Titrate the liberated sulfite with 0.01 N iodine using a microburet, to a faint blue end point. The amount of iodine added in this step must be accurately measured and recorded.

8.8 One mL of 0.0100 N iodine is equivalent to 0.15 mg of formaldehyde. Therefore, since 1 mL of formaldehyde standard solution was titrated, the milliliters of 0.01 N iodine used in the final titration multiplied by 0.15 mg gives the formaldehyde concentration of the standard solution in mg/ml.

8.9 PREPARATION OF STANDARD CURVE.

8.9.1 Pipet 0, 0.5, 1.0, 3.0, 5.0, and 7.0 mL of standard formaldehyde solution "B" into 100-mL volumetric flasks. Dilute to volume with 0.05% MBTH solution. These solutions contain 0, 0.05, 0.1, 0.3, 0.5, and 0.7 µg of formaldehyde/mL.

8.9.2 After final dilution let stand for 1 h.

8.9.3 Transfer 10 mL of each solution to a glass stoppered test tube and add 2 mL of oxidizing reagent and mix.

8.9.4 After 12 min read the absorbance at 628 nm in a suitable spectrophotometer using 1-cm cells.

8.9.5 Plot absorbance against micrograms of formaldehyde/mL of solution.

9. Calculation

9.1 The concentration of total aliphatic aldehyde (as formaldehyde) in the sampled atmosphere may be calculated by using the following equation:

$$PPM \text{ (Vol.)} = \frac{C \times 35 \times 24.45}{V \times M.W. \times E}$$

where E = correction factor for sampling efficiency (0.84 may be used if absorber contains an EC frit)
V = liters of air sampled, corrected to 25°C and 101.3 kPa
C = µg/mL of formaldehyde in sampling solution. (Since each sample is diluted to 35 mL, this figure must be multiplied by 35

Table 117:II Acetaldehyde

μg/mL	Number of Samples	Average Absorbance	Range	% Range From Avg.
0.05	29	0.063	0.050–0.074	± 20
0.10	29	0.125	0.106–0.144	± 15
0.30	29	0.339	0.316–0.355	± 7
0.50	29	0.519	0.495–0.538	± 4
0.70	29	0.685	0.660–0.710	± 3
1.00	15	0.900	0.890–0.910	± 1

Table 117:III Propionaldehyde

μg/mL	Number of Samples	Average Absorbance	Range	% Range From Avg.
0.05	29	0.046	0.032–0.057	± 27
0.10	29	0.082	0.063–0.095	± 20
0.30	29	0.243	0.225–0.250	± 5
0.50	29	0.399	0.380–0.422	± 5
0.70	29	0.538	0.515–0.568	± 5
1.00	15	0.732	0.710–0.750	± 2

to give total micrograms in sampling solution.)

M.W. = molecular weight of formaldehyde (30.03)

24.45 = mL of formaldehyde gas in one millimole at 101.3 kPa and 25°C.

10. Effect of Storage

10.1 The time study of the reaction of microgram quantities of formaldehyde with 0.05% MBTH shows that the reaction is complete in approximately 45 min; therefore, a reaction time of 1 h is selected for this procedure. Formaldehyde is fairly stable in 0.05% MBTH since only approximately 5% of the formaldehyde is lost after standing in the MBTH for 13 days. The samples are, therefore, stable enough for later analysis (3).

11. References

1. SAWICKI, E., T.R. HAUSER, T.W. STANLEY AND W. ELBERT. 1961. The 3-Methyl-2-Benzothiazolone Hydrazone Test. *Anal. Chem.* 33:93.
2. HAUSER, T.R. AND R.L. CUMMINS. 1964. Increasing the Sensitivity of 3-Methyl-2-Benzothiazolone Hydrazone Test for Analysis of Aliphatic Aldehydes in Air. *Anal. Chem.* 37:679.
3. HAUSER, THOMAS R. 1965. Determination of Aliphatic Aldehydes: 3-Methyl-2-Benzothiazolone Hydrazone Hydrochloride (MBTH) Method. Selected Methods for the Measurement of Air Pollutants. *Public Health Service Publication No. 999-AP-11, Page F-1.*

ADDENDUM

Applications to Other Aldehydes

Acetaldehyde and propionaldehyde both yield a blue dye after reaction with 3-methyl-2-benzothiazolinone hydrazone hydrochloride and a ferric chloride-sulfamic acid solution. It has been found that as the length of chain increases, the sensitivity decreases. Therefore when measuring total aldehydes as formaldehyde this method would give low results if any aldehyde other than formaldehyde is present.

From 0.05 μg/mL to 1.0 μg/mL of both acetaldehyde and propionaldehyde can be measured in the color developed solution (12 mL). For the lower concentrations the method has poor reproducibility. However, at higher concentrations (0.30 μg/mL and above) reproducibility was very good. These data are summarized in Tables 117:II and III.

Acetaldehyde (Eastman Kodak Company, Cat. No. 468) and propionaldehyde (Eastman Kodak Company, Cat. No. 653) were considered to be primary standards when preparing solutions of known concentration. Exactly 1.28 mL of acetaldehyde was diluted to 1 L with distilled water and then 1 mL of this solution was diluted to 100 mL with MBTH solution giving a final concentration of 10 μg/mL. Exactly 1.24 mL of propionaldehyde was diluted to 1 L with distilled water and then 1 mL of this solution was diluted to 100 mL with MBTH solution giving a final concentration of 10 μg/mL. The strong standard solutions have a 2 month shelf life. The dilute standard solutions must be prepared fresh daily.

A series of 34 ambient air samples were collected in 35 mL of MBTH solution contained in each of two absorbers in series. The sampling time was 24 h and the sampling rate was 1 L/min. Collection efficiencies varied from 69% to 100% with the average for the 34 samples being 82%.

Subcommittee 4

R.G. SMITH, *Chairman*
R.J. BRYAN
M. FELDSTEIN
B. LEVADIE
F.A. MILLER
E.R. STEPHENS
N.G. WHITE

Approved with modifications
from 2nd edition
Subcommittee 4/5
M FELDSTEIN, *Chairman*
R.J. BRYAN
D.L. HYDE
D.A. LEVAGGI
D.C. LOCKE
R.A. RASMUSSEN
P.O. WARNER

118.

Determination of Mercaptan Content of the Atmosphere

1. Principle of the Method

1.1 Mercaptans (organic thiols) are collected by aspirating a measured volume of air through an aqueous solution of mercuric acetate-acetic acid. The collected mercaptans are subsequently determined by spectrophotometric measurement of the red complex produced by the reaction between mercaptans and a strongly acid solution of N,N-dimethyl-*p*-phenylenediamine and ferric chloride (1–3). The method determines total mercaptans and does not differentiate among individual mercaptans. Although the sensitivity for various mercaptans on a weight basis decreases with increasing molecular weight of the mercaptan, on a molar basis the sensitivity has been reported to be approximately the same (2). All concentration units in this description refer to methyl mercaptan, unless otherwise specified.

2. Sensitivity and Range

2.1 This method is intended to provide a measure of mercaptans in the range below 200 μg/m³ (102 ppb CH_3SH). For concentrations above 100 ppb, the sampling period can be reduced or the liquid volume increased either before or after aspirating. The minimum detectable amount of methyl mercaptan is 0.04 μg/mL (2) in a final liquid volume of 25 mL. When sampling air at the maximum recommended rate of 1 L/min for 2 h, the minimum detectable mercaptan concentration is 3.9 μg/m³ (2.0 ppb methyl mercaptan at 101.3 kPa and 25°C).

3. Interferences

3.1 The N,N-dimethyl-*p*-phenylenediamine reaction is also suitable for the determination of other sulfur-containing compounds including hydrogen sulfide (1) and dimethyl disulfide. The potential for interference from these latter compounds is especially important since all of these compounds commonly coexist in certain industrial emissions. Appropriate selection of the color formation conditions minimizes the interference from hydrogen sulfide and dimethyl disulfide.

3.2 Hydrogen sulfide, if present in the sampled air, may cause a turbidity in the sample absorbing solution. This precipitate must be filtered before proceeding with the analysis. One study showed that 100 μg H_2S gave a mercaptan color equivalent to 1.5 to 2.0 μg methyl mercaptan (4). Other studies reported no absorption at 500 nm in the presence of 150 μg of hydrogen sulfide (2,5).

3.3 An unexplained yellow tinge has been randomly observed in a few impingers after sampling. In these instances, the absorbing solution subsequently turned pink. A black precipitate then formed when the color developing reagent was added. Although the precipitate was removed by filtration just before the absorption was measured, it is not known whether this condition changed the apparent mercaptan concentration.

3.4 Approximately equimolar response is obtained from the hydrolysis products of dimethyl disulfide compared to methyl mercaptan. In practice, however, the collection of dimethyl disulfide in aqueous mercuric acetate is inefficient. Thus, the actual interference is negligible.

3.5 Sulfur dioxide up to 250 μg does not influence the color development even when sampling a test atmosphere containing 300 ppm SO_2 (2).

3.6 Nitrogen dioxide does not interfere up to 700 μg NO_2 when sampling a test atmosphere containing 6 ppm NO_2. Higher concentrations of NO_2 caused a positive interference when mercaptans were present but no interference in the absence of mercaptans (2).

3.7 The supply of mercuric acetate must be free of mercurous ion. If mercurous ion is present, turbidity will result when the chloride ion-containing reagents are added in the last step of the analytical procedure (2).

4. Precision and Accuracy

4.1 The relative standard deviation for the determination of ethyl, butyl, and hexyl mercaptan in the absorbing solution at a concentration of 8 mg/L ranged from 0.8 to 2.0% (2). The relative standard deviation increased with increasing molecular weight of the mercaptans. Field studies in which methyl mercaptan concentrations ranged from 0.7 to 9 mg/m³ showed a relative standard deviation of 2.6 to 7.1% in replicate determinations.

5. Apparatus

5.1 ABSORBER. Midget bubbler fitted with coarse porosity frit.

5.2 AIR PUMP with a flow meter and/or gas meter having a minimum capacity of drawing 2 L/min of air through a midget bubbler.

5.3 SPECTROPHOTOMETER. Capable of operation at 500 nm. A long path length cell (2–5 cm) may be used to improve sensitivity.

5.4 AIR VOLUME MEASUREMENT. The air meter must be capable of measuring the air flow within ±2%. Wet or dry gas meters, specially calibrated rotameters or critical orifices may be used.

6. Reagents

6.1 PURITY. Reagents must meet specifications of the American Chemical Society. Water should conform to the Standards for ASTM Reference Reagent Water, Type II.

6.2 Solutions should be refrigerated when not in use.

6.3 AMINE-HYDROCHLORIC ACID STOCK SOLUTION. Dissolve 5.0 g N,N-dimethyl-*p*-phenylenediamine hydrochloride (*p*-aminodimethylaniline hydrochloride) in 1 L

of concentrated hydrochloric acid. Refrigerate and protect from light. The solution is stable for at least 6 months.

6.4 REISSNER SOLUTION. Dissolve 67.6 g ferric chloride hexahydrate in water, dilute to 500 mL and mix with 500 mL nitric acid solution containing 72 mL boiled concentrated nitric acid (sp gr 1.42). This solution is stable.

6.5 COLOR DEVELOPING REAGENT. Mix 3 volumes of amine solution and 1 volume of Reissner solution. Prepare this solution fresh for each set of determinations.

6.6 ABSORBING SOLUTION. Dissolve 50 g of mercuric acetate in 400 mL water and add 25 mL glacial acetic acid. Dilute to 1 L. The mercuric acetate must be free of mercurous salts to prevent precipitation of mercurous chloride during color development. Reagent grade mercuric acetate sometimes contains mercurous mercury. Determine the acceptability of each new bottle of mercuric acetate by adding 3 mL of concentrated hydrochloric acid to 3 mL of the 5 per cent mercuric acetate. If the solution becomes cloudy, the mercuric acetate is not acceptable. (CAUTION: Mercuric acetate is highly toxic. If spilled on skin, wash off immediately with water.)

6.7 LEAD METHYL MERCAPTIDE. This material is available from Kodak Laboratory Chemicals as Lead (II) Methanethiolate in 99% purity. One mole of this compound is equivalent to 2 moles of methyl mercaptan.

6.8 CONCENTRATED STANDARD LEAD MERCAPTIDE SOLUTION. Weigh out 156.6 mg of the crystalline lead mercaptide and make up to 100 mL with the 5% mercuric acetate absorbing solution. This solution contains the equivalent of 500 µg of methyl mercaptan/mL.

6.9 DILUTED STANDARD MERCAPTAN SOLUTION. Dilute 2 mL of the concentrated standard solution to 100 mL with the 5% mercuric acetate absorbing solution. This solution contains the equivalent of 10 µg CH_3SH/mL.

6.10 MERCAPTAN PERMEATION SOURCE. Sources emitting ≤100 ng/min are needed for calibration at realistic levels. With an air flow rate of 2 L/min, emission rates of 4–100 ng/min are necessary to directly generate mercaptan concentrations in the range 2–50 µg/m³. Tubular permeation devices that emit a desired mercaptan at a rate ≥ 40 ng/min (at 30°C) are available (with certified emission rates if desired) from VICI/Metronics (2991 Corvin Dr., Santa Clara, CA 95051). From the same vendor, wafer-type permeation devices are available; these exhibit permeation rates even lower than tubular devices. However, the magnitude of the permeation rate makes gravimetric calibration commercially impractical. The wafer devices are not available calibrated and must be calibrated by the user. Methyl and ethyl mercaptan permeation sources are readily available and higher mercaptans up to octyl mercaptan are available on special order. The permeation sources must be stored in a wide-mouth glass bottle containing silica gel and solid sodium hydroxide to remove moisture and the emitted mercaptan. The storage bottle is immersed to two-thirds its height in a constant temperature water bath in which the water is controlled at the temperature of intended use, typically 25.0° or 30.0° ± 0.1°C.

7. Procedure

7.1 COLLECTION OF SAMPLE. Aspirate the air sample through 15 mL of the absorbing solution in a midget bubbler at 1.0 to 1.5 L/min for a selected period up to 2 h.

7.2 ANALYSIS. Quantitatively transfer the sample from the impinger to a 25-mL volumetric flask and dilute to approximately 22 mL with water that has been used to rinse the fritted bubbler. Add 2.0 mL of freshly prepared color developing reagent, dilute to volume with water and mix well. Prepare a reference blank in the same manner using 15 mL of 5% mercuric acetate and 2 mL color developing reagent and dilute to 25 mL. After 30 min, measure the absorbance at 500 nm with a spectrophotometer against the mercaptan-free reference blank.

7.3 For sampling of 24-h duration, the conditions can be fixed to collect 1200 L of sample in a larger volume of mercuric acetate-acetic acid. For example, for 24 h at 0.83 L/min, approximately 1200 L of air are collected. An aliquot representing 0.1 of the entire amount of sample is taken for analysis. The remainder of the analytical procedure is the same as described in the previous paragraph.

8. Calibration

8.1 AQUEOUS MERCAPTIDE. Prepare a calibration curve by pipetting appropriate aliquots of the diluted standard lead mercaptide into a series of 25-mL volumetric flasks, diluting each with 15 mL of 5% mercuric acetate absorbing solution and developing the color in the same way as the samples. Prepare a reference blank in the same manner without lead mercaptide. Determine the absorbance at 500 nm against the mercaptan-free reference blank. Prepare a standard curve of absorbance vs μg methyl mercaptan/mL.

8.2 GASEOUS MERCAPTAN. Refer to Section 3, Part I for the generation of standard gaseous atmospheres using permeation sources.

8.2.1 Procedure for preparing simulated calibration curves. Obviously one can prepare a multitude of curves by selecting different combinations of sampling rate and sampling time. The following description represents a typical procedure for ambient air sampling of short duration.

The system is designed to provide an accurate measure of methyl mercaptan in the 3 to 200 μg/m³ (approximately 1.5 to 100 ppb) range. It can be easily modified to meet special needs.

The dynamic range of the colorimetric procedure fixes the total volume of the sample at 120 L; then, to obtain linearity between the absorbance of the solution and the concentration of methyl mercaptan in ppb, select a constant sampling time. This fixing of sampling time is desirable also from a practical standpoint: in this case, select a sampling time of 120 min. Then to obtain a 120 L sample of air requires a flow rate of 1.0 L/min. The concentration of standard CH_3SH in air is computed as follows:

$$C = \frac{Pr \times M}{R}$$

where C = Concentration of CH_3SH in ppb,

Pr = Permeation rate in ng/min,

M = Reciprocal of vapor density, 0.51 nL/ng

Table 118:I. Typical Calibration Data

Concentration CH$_3$SH, ppb	Amount of CH$_3$SH in μL per 120 L air sample	Absorbance of sample
21	2.5	0.027
42	5.0	0.056
83	10.0	0.085
125	15.0	0.120
208	25.0	0.209
417	50.0	0.417

R = Total flow rate of air/ nitrogen through generation system in L/min.

Data for a typical calibration curve are listed in Table 118:I.

A plot of the concentration of methyl mercaptan in ppb (x-axis) against absorbance of the final solution (y-axis) will yield a straight line, the reciprocal of the slope of which is the factor for conversion of absorbance to ppb. This factor includes the correction for collection efficiency. Any deviation from linearity at the lower concentration range indicates a change in collection efficiency of the sampling system. If the range of interest is below the dynamic range of the method, the total volume of air collected should be increased to obtain sufficient color within the dynamic range of the colormetric procedure. Also, once the calibration factor has been established under simulated conditions, the conditions can be modified so that the concentration of CH$_3$SH is a simple multiple of the absorbance of the colored solution.

8.2.2 The permeation sources must be stored in a wide-mouth glass bottle containing silica gel and solid sodium hydroxide to remove moisture and methyl mercaptan. The storage bottle is immersed to two-thirds its depth in a constant temperature water bath in which the water is controlled at the temperature of intended use, typically 25.0° or 30.0° ± 0.1°C. Every two weeks, the permeation sources are removed and rapidly weighed on a semi-micro (sensitivity ± 10 μg) or a micro (sensitivity ± 1 μg) balance and then returned to the storage bottle. The weight loss is recorded. The tubes are ready for use when the rate of weight loss becomes constant (within ± 2%).

9. Calculations

9.1 Determine the sample volume in m^3 from the gas meter or flow meter readings and time of sampling. Adjust volume to 101.3 kPa and 25°C (V$_s$). Compute the concentration of methyl mercaptan in the sample by one of the following formulae:

$$ppb = \frac{(A-A_0)\, 0.510 B}{V_s}$$

$$\mu g/m^3 = \frac{(A - A_0)B}{V_s}$$

where A = the sample absorbance.
  A$_0$ = the reagent blank.
  0.510 = the volume (μL) of 1 μg CH$_3$SH at 25° C, 101.3 kPa.
  B = the calibration factor, μg/ absorbance unit.
  V$_s$ = the sample volume in cubic meters corrected to 25°C, 101.3 kPa.

10. References

1. MARBACH, E.P. AND D.M. DOTY. 1956. Sulfides Released from Gamma-irradiated Meat as Estimated by Condensation with N,N-Dimethyl-*p*-phenylenediamine, J. Agr. Food Chem. 4:881.
2. MOORE, H., H.L. HELWIG AND R.J. GRAUL, 1960. A Spectrophotometric Method for the Determination of Mercaptans in Air, Am. Ind. Hyg. Assn. J. 21:466.
3. SILWINSKI, R.A. AND D.M. DOTY. 1958. Determination of Micro Quantities of Methyl Mercaptan in Gamma-irradiated Meat, J. Agr. Food Chem. 6:41.
4. ACGIH Recommended Method. 1964. Determination of Total Mercaptans in Air, July.

5. ADAMS, D.F. 1969. Analysis of Malodorous Sulfur-containing Gases, TAPPI 52:53.

Subcommittee 1

D.F. ADAMS
P.K. DASGUPTA, *Chairman*
B.R. APPEL
S.O. FARWELL

K.T. KNAPP
G.L. KOK
W.R. PIERSON
K.D. REISZNER
R.L. TANNER

121.

Determination of Phenols in the Atmosphere (Gas Chromatographic Method)

1. Principle of the Method

1.1 Phenolic compounds react with sodium hydroxide to form "phenates." Gaseous phenols or total phenols are collected by scrubbing with an alkaline solution in a standard impinger, respectively with or without a prefilter. The phenates are hydrolyzed by acid. The released phenols are separated from the acid aqueous system by steam distillation. The aqueous solution of the phenols is analyzed by gas chromatography, using a flame ionization detector and a short column. Total phenolic content is expressed as phenol **(1)**.

2. Range and Sensitivity

2.1 Using phenol as the standard test substance, and under the operating conditions given below where bubbler recovery is in the range of 88 to 95%, the sensitivity of the method is 1 ppm in solution (1 µg/mL of phenol in the final distillate) or approximately 40 µg/m^3, based upon 1.5-h sampling at 30 L/min, as described in Section 7.1.

3. Interferences

3.1 This method is not subject to interferences by other organic compounds. It involves the actual physical separation of total phenolic compounds on a highly polar substrate on an inert packing. If a longer chromatographic column is used, it is possible to separate the phenolic compounds and to determine them individually.

4. Precision and Accuracy

4.1 Generally, the precision and accuracy of gas chromatographic systems are governed by the care and technical compe-

tence of the individuals operating them. With phenol as the standard, and applying the operating conditions given above, the sensitivity of this system is 1 ppm in solution.

For a single operator, a 10 ppm standard measured over a period of several days gave a coefficient of variation of 12%. For any one day the coefficient of variation was 1% **(2)**.

5. Apparatus

5.1 A gas chromatographic system consisting of the following components:

5.1.1 A controlled-temperature injection port that will operate up to at least 300°C.

5.1.2 A column-heating chamber capable of precise temperature control (to within 0.5°C).

5.1.3 A flame ionization detector (FID) having a temperature control system.

5.1.4 A stainless steel gas chromatographic column consisting of 5% Carbowax 20M, terephthalic acid liquid phase on 60 to 80 mesh acid washed Chromosorb W. For determination of total phenols a 2-ft (61 cm) stainless steel column is used.

5.1.5 A recorder capable of 5 mV full scale response in 1 s, and of chart drive speed of 1.8 cm/min, or equivalent data system.

5.1.6 A gas chromatographic sample injection syringe capable of delivering from 1 to 10 µL with a reproducibility of 95% or better.

5.2 An air sampling system consisting of:

5.2.1 An air scrubber such as the Greenburg-Smith Impinger or its equivalent, capable of sampling at a rate of 30 L/ min.

5.2.2 An air pump capable of providing the above sampling rate.

5.2.3 A wet test meter or a rotameter.

5.3 All glass distillation apparatus, 150 mL, for steam distillation.

5.4 A pH meter or pH indicator paper.

6. Reagents

6.1 PURITY. All chemicals should be ACS analytical reagent grade. Water is ASTM reagent water, Type II, free of organic matter.

6.2 COPPER SULFATE SOLUTION. 10% $CuSO_4 \cdot 5 H_2O$ in water.

6.3 SODIUM HYDROXIDE SOLUTION. 1 N in water.

6.4 SODIUM HYDROXIDE SOLUTION. 0.1 N in water. Dilute 100 mL of 1 N sodium hydroxide to 1 L.

6.5 PHOSPHORIC ACID. 10% in water. Dilute 120 mL of phosphoric acid (85% H_3PO_4) to 1 L.

6.6 BROMATE-BROMIDE SOLUTION. Dissolve 2.784 g of potassium bromate in water. Add 10 g of potassium bromide, and dilute to 1 L.

6.7 SODIUM THIOSULFATE SOLUTION. 0.25 N in water. Prepare and standardize according to directions in Part I, Section 8.6.

6.8 PHENOL SOLUTION. 0.1% in water. Blot crystalline phenol from the reagent bottle between layers of filter paper and weigh accurately 1.00 g of the dried crystals. Transfer to a 1-L volumetric flask and fill to the mark with water.

6.9 HYDROCHLORIC ACID, concentrated.

6.10 STARCH SOLUTION. Make a paste of 1 g starch in the minimum amount of water. Add to 100 mL of boiling water.

7. Procedure

7.1 AIR SAMPLING. Draw air at 30 L/ min into 100 mL of 0.1 N sodium hydroxide in a standard impinger for a sufficient period to obtain a minimum of 1 μg/mL of phenol in the sampling solution. If only vapor phenolics are required, use a membrane filter in the sampling train to remove particulate matter. Normal air levels require sampling times of 10 min to 15 h.

7.2 ANALYSIS. (Caution — do not use stopcock grease in any apparatus.)

7.2.1 Adjust the sampling solution to a volume of 100 mL either by aliquoting or diluting to volume with water. Add 1 mL of 10% copper sulfate solution. Acidify with 10% phosphoric acid to approximately pH 3. Distill from an all-glass distillation apparatus until 90 mL have been collected. Add 10 mL of distilled water to the cooled distillation flask and continue the distillation until a total volume of 100 mL of distillate has been collected.

7.2.2 The usual precautions regarding gas chromatographic column installation and conditioning should be observed. Set the temperature of the injection port at 300°C, the column at 200°C and the detector at 260°C. Flow rate is not critical with such a short column, but should be set at a convenient value. Once the gas chromatographic apparatus has stabilized, as demonstrated by a stable baseline, inject the samples, suitably interspersed with standards. Typical sample volumes are usually in the range of 1 to 5 μL. Determine aqueous concentrations of phenols as phenol by comparison with the standard curve for the day.

8. Calibration and Standards.

8.1 STANDARDIZATION OF PHENOL SOLUTION. While neither phenol nor the stock 0.1% solution degrade rapidly, lots of Reagent phenol differ in water content; this also changes with time. Blotting the crystals of phenol largely removes this problem. However, for highest accuracy, it may be desirable to assay the material periodically, both in the reagent bottle and in the stock solution (which should be stored in the refrigerator for no more than one month). Pipet 50.0 mL of 0.1% solution into a 500-mL iodine flask. Add 100 mL of water. Pipet in 10.0 mL of bromate-bromide solution, then add 0.5 mL of concentrated hydrochloric acid. Mix. If the brown color of bromine does not persist in the solution (as it should not), add successive 10.0-mL portions of bromate-bromide solution by volumetric pipet until the bromine color persists. (This usually requires a total of four such additions.) Stopper the flask after each addition. After the final addition, allow the flask to stand, stoppered, for 10 min, then quickly add 1 g of potassium iodide, stopper, and swirl to dissolve. Titrate the liberated iodine with 0.025 N sodium thiosulfate solution in the usual way, adding starch indicator when the iodine color is nearly discharged. Carry 50.0 mL of a blank (water) through the same procedure; it will require only a single 10.0-mL portion of bromide-bromate solution. Calculate the concentration of phenol in the stock solution as follows:

$$C = [n - 1 - 0.025 \, (V_s - V_b)] \, 313.7$$

where C = phenol concentration, $\mu g/mL$;

n = number of 10 mL portions of bromide-bromate solution added to the sample;

V_s = volume of thiosulfate consumed in titrating the sample, mL;

V_b = volume of thiosulfate consumed in titrating the blank, mL; and

0.025 = the normality of the thiosulfate solution; the true normality should be substituted, if different.

8.2 PREPARATION OF WORKING STANDARD. The stock solution is approximately 1000 $\mu g/mL$. It is convenient to calculate the dilution necessary to produce a working standard that is precisely 10 $\mu g/mL$; alternatively, the stock solution may be diluted 1:100, and the measured phenol concentration carried through the balance of the calculations.

8.3 CALIBRATION OF GAS CHROMATOGRAPH. Prepare, by appropriate dilution, calibration standards containing 2, 4, 6, 8 and 10 $\mu g/mL$ of phenol. Inject a volume determined by experiment to yield good peak heights. Determine peak areas and plot against phenol concentration in the standards. Phenol concentrations as phenol in the samples are read from the plot or preferably obtained from a regression equation obtained by the method of least squares.

9. Calculations

9.1 Correct the volume of air sampled to 25°C and 101.3 kPa, if necessary. Calculate the concentration of phenols in air by the following equation:

$$C_a = \frac{C_p \times V_d \times V_r}{V_a \times V_i}$$

where C_a = concentration of phenol in air, $\mu g/m^3$;

C_p = concentration of phenol in the sample, $\mu g/mL$;

V_d = volume of distillate obtained in 7.2.1, mL, presumably 100 mL;

V_r = volume of working standard solution injected in preparation of the calibration plot, μL;

V_a = volume of air sampled, either ambient or corrected, m^3; and

V_i = volume of sample injected, μL.

10. Effects of Storage

10.1 If alkaline samples must be stored prior to analysis, their stability will be enhanced by the addition of 5 mL of copper sulfate solution. If this is done, do not add copper sulfate in 7.2.1.

10.2 Stock standard phenol solution is stable for 1 month if stored in a refrigerator. Dilute working standards should be prepared fresh daily.

11. References

1. ASTM, Phenols in Water by Gas Liquid Chromatography, ASTM D2580, American Society for Testing and Materials, Philadelphia, PA
2. KEENAN, R. G., personal communication.

Subcommittee 4/5

M. FELDSTEIN, *Chairman*
R.J. BRYAN
D.L. HYDE
D.A. LEVAGGI
D.C. LOCKE
R.A. RASMUSSEN
P.O. WARNER

122.

Determination of C_1 – C_5 Aldehydes in Ambient Air and Source Emissions as 2,4-Dinitrophenylhydrazones by HPLC

1. Principle of the Method

1.1 Formaldehyde, acrolein, and low molecular weight aldehydes are collected in 2,4-dinitrophenylhydrazine solution in two fritted impingers in series. The 2,4-dinitrophenylhydrazone derivatives (DNPH) are extracted, concentrated, and separated and quantitated by reversed phase HPLC. The DNPH derivatives of the C_1-C_5 aliphatic aldehydes are separated from the DNPH derivatives of the corresponding methyl ketones, and from the C_7 and C_8 aromatic aldehyde DNPH's (1,2). The method is applicable to ambient air, industrial emissions, and auto exhaust (1).

2. Range and Sensitivity

2.1 At a sampling rate of 0.5-1.5 L/min, the detection limits for a 30 L air sample are 1.2 ppb for HCHO, acetaldehyde, acrolein, and propionaldehyde, and 2.1 ppb for the C_4-C_6 aliphatic aldehydes and the aromatic aldehydes. These detection limits are quoted for a 5 μL injection volume and a 2 mL final sample volume. The detection limits are lower by a factor of twenty if small volume preconcentration (final sample volume 0.1 mL, section 7.4) is carried out.

3. Interferences

3.1 Depending on conditions (primarily eluent polarity), some carbonyl derivatives may not be resolved. In most cases these are isomers. Few if any of these are major atmospheric contaminants. In some cases, even these can be resolved by altering the water content of the eluent. The effects of ambient organic vapors or oxidants, SO_2, or NO_x, are not known.

4. Precision and Accuracy

4.1 The relative standard deviations for 5 replicate analyses of synthetic mixture containing 140–1000 ppb of thirteen aldehydes in air range from 1.3% to 7.0%, averaging 3.5% (1).

4.2 Although recoveries of the DNPH range from 81.5% to 103% (Table 122:I), the absolute quantitative accuracy of the method for aldehydes in ambient air is not known.

5. Apparatus

5.1 ABSORBERS. A train of two 30 mL all-glass fritted bubblers in series is used.

5.2 AIR PUMP. A pump capable of drawing 1.5 L/min of air through the sampling train is required.

5.3 AIR METERING DEVICE. A dry test meter or other suitable device is used.

5.4 PIPET. Volumetric pipet, of 2.00 mL capacity.

5.5 LIQUID CHROMATOGRAPH. Liquid chromatograph capable of operating pressures at least up to 2500 psi (17 MPa) and with an injection valve (furnished with a 5 or 10 μL loop) and an optical absorption detector capable of operation at 360 nm with a cell volume under 10 μL are required. It is possible to use a 254 nm detector, but achievable limits of detection are higher and potential of interference is increased.

5.6 CONCENTRATOR. A standard concentrator allowing evaporation of solvent to dryness under gentle (50°C) heating at

Table 122:I. Molecular Weights of Aldehydes, and Relative Retention Times and Recoveries of Their Dinitrophenylhydrazones

DNPH Derivative of	M_{ald}	Relative Retention in 70/30 ACN/H_2O	% Recovery of DNPH
formaldehyde	30.0	1.00	96.7
acetaldehyde	44.1	1.19	103.3
acrolein	56.1	1.42	81.5
propionaldehyde	58.1	1.55	98.0
crotonaldehyde	70.1	1.83	88.0
isobutyraldehyde	72.1	2.07	–
n-butyraldehyde	72.1	2.07	94.5
isovaleraldehyde	86.1	2.66	95.0
n-valeraldehyde	86.1	2.81	97.0
n-caproaldehyde	100.2	4.01	100.0
benzaldehyde	106.1	2.55	98.0
o-tolualdehyde	120.2	2.99	95.0
m-tolualdehyde	120.2	3.08	96.0
p-tolualdehyde	120.2	3.18	99.0

reduced pressure, or in a stream of purified N_2, is required.

5.7 LC COLUMN. A 200 × 4.6 mm i.d. stainless steel column slurry-packed with LiChrosorb RP-18 (5 μm particle size, E. Merck) or equivalent is required.

5.8 MICROSYRINGE. A microsyringe of 25 μL capacity is useful for sample injection.

5.9 EVAPORATION MICROVIAL. For achieving the lowest limits of detection, an optical microvial of 5 mL capacity (e.g., K-7490000, size 0005, Kontes Inc., Vineland, NJ 08360) is suggested.

5.10 MICROPIPET OR MICROSYRINGE. This device should be capable of accurately delivering 100 μL. This is intended for optional high sensitivity work.

6. Reagents

6.1 SOLVENTS. All solvents, including water, should be of HPLC grade. All reagents are ACS reagent grade, or equivalent.

6.2 ABSORBER SOLUTION. Dissolve 0.5 g of 2,4-dinitrophenylhydrazine in 500 mL of 2 N HCl. Purify by extracting twice with 5 mL CHCl₃. Discard the organic extract. Store at 5°C no longer than one week.

6.3 HPLC MOBILE PHASE. 70% acetonitrile (ACN)/30% H_2O (v/v).

7. Procedure

7.1 COLLECTION OF SAMPLE. Two impingers, each containing 10 mL of the absorber solution, are connected in series. These in turn are connected to an empty impinger (to protect the pump), a dry test meter, and an air pump. Five to 30 L of air sample should be collected at 0.5–1.5 L/min.

7.2 EXTRACTION AND EVAPORATION. The two sampling solutions are combined and extracted twice with 5 mL portions of CHCl₃. The combined extracts are washed with 20 mL of 2 N HCl and then with 20 mL H_2O, and finally evaporated to dryness at 50°C under reduced pressure, or in a stream of N_2. Proceed to Section 7.3 for routine analysis and to Section 7.4 if highest sensitivity is required.

7.3 DISSOLUTION. Dissolve the residue from step 7.2 in exactly 2.00 mL of acetonitrile and carry out chromatography as described in Section 7.5.

7.4 EVAPORATION AND DISSOLUTION. Dissolve the residue from step 7.2 in 1–2 mL of acetonitrile. Quantitatively transfer this solution to a 5.0 mL microvial; the use of an all-glass hypodermic syringe for this purpose is recommended. Evaporate again to dryness in a stream of N_2 or under reduced pressure at 50°C as in section 7.2. Dissolve the residue in 100 μL acetonitrile delivered from a suitable micropipet or microsyringe. Carry out chromatography as described in the next section.

7.5 CHROMATOGRAPHY. Carry out chromatography with the eluent described in section 6.3 at ambient temperature and a flow rate of 1.5 mL/min. Nominally, the operating pressure should be ~ 1900 psi (13 MPa). Sample volume injected into the loop valve should be large enough to completely fill the loop. While 5 and 10 µL loops are commonly used, a loop volume of up to 20 µL may be used for this application without loss of analyte resolution.

7.6 PEAK IDENTIFICATION. Peaks are identified by retention times. Table 122:I lists the retention times of the DNPH's of various carbonyls relative to formaldehyde DNPH.

8. Calibration

8.1 PREPARATION OF DNPH DERIVATIVES (3). To 4 g of 2,4-dinitrophenylhydrazine in a 250 mL Erleynmeyer flask 20 mL conc. HCl is added. Next, 30 mL of H_2O is added slowly with stirring until solution is complete. To this solution is added 10 mL 95% ethanol. To a solution containing 0.25 g of aldehyde in 10 mL 95% ethanol is added 5 mL of the 2,4-dinitrophenylhydrazine solution, and the mixture is allowed to stand until crystallization of the DNPH occurs. The DNPH is filtered and recrystallized from 95% ethanol.

8.2 PREPARATION OF STANDARD SOLUTIONS. Prepare stock solutions in 500 mL of ACN containing 0.1 g of each DNPH derivative of interest. Dilute these by adding 10, 20, 50, and 100 µL of the stock solution respectively to 10 mL of ACN. For a 5 µL sample volume, these solutions correspond respectively to 1, 2, 5 and 10 ng of DNPH derivative. Inject aliquots into the LC, and carry out chromatography as in Section 7.5

8.3 PREPARATION OF STANDARD CURVE. Plot the peak height against ng of DNPH derivative on linear graph paper. Use this plot, or preferably the equation obtained by linear least squares, to calculate ng of derivatives in the field samples.

9. Calculations

9.1 CALCULATION OF SAMPLE VOLUME. The measured air volume sampled is con-

verted to reference conditions by the usual gas law relationship:

$$V_r = V_m \times \frac{P}{T} \times 2.94$$

where V_r = volume at reference conditions, m^3; V_m = measured sample volume, m^3; P = measured mean barometric pressure, kPa; T = measured mean temperature, K; and the constant is the ratio of reference temperature, 298K, to reference pressure, 101.3 kPa.

9.2 CALCULATION OF ALDEHYDE CONCENTRATION IN AIR. The weight of DNPH derivative, W_a, in ng, is converted into aldehyde concentration in air by the following:

$$C \ (ng/m^3) = \frac{W_a}{V_r} \times \frac{M_{ald}}{M_{DNPH}}$$

$$C \ (ppbv) = \frac{W_a}{M_{DNPH}} \times \frac{0.0245}{V_r}$$

where the concentrations, C, are in the indicated units; M_{ald} = molecular weight of the specific aldehyde; and M_{DNPH} = the molecular weight of the DNPH = $M_{ald} + 180$.

10. Effect of Storage

10.1 The sampling solution should be stored at 5°C no longer than one week.

11. References

1. KUWATA, K., M. UEBORI and Y. YAMASAKI. 1979. Determination of Alphatic and Aromatic Aldehydes in Polluted Airs as Their 2,4-Dinitrophenylhydrazones by High Pressure Liquid Chromatography. J. Chromatogr. Sci., *17*:264.
2. FUNG, K. and D. GROSJEAN. 1981. Determination of Nanogram Amounts of Carbonyls as 2,4-Dinitrophenylhydrazones by High Pressure Liquid Chromatography. Anal. Chem. *53*:168.
3. SHRINER, R. L., R. C. FUSON, D. Y. CURTIN and T. C. MORRILL. 1980. "Systematic Identification of Organic Compounds." 6th Edition, John Wiley & Sons, p. 179.

Subcommittee 4/5

D.C. LOCKE
M. FELDSTEIN, *Chairman*
R.J. BRYAN
D.L. HYDE
D.A. LEVAGGI
R.A. RASMUSSEN
P.O. WARNER

128.

Determination of Continuous Carbon Monoxide Content of the Atmosphere (Nondispersive Infrared Method)

1. Principle of Method

1.1 Nondispersive infrared (NDIR) photometry provides a method of utilizing the integrated absorption of infrared energy over most of the spectrum for a given compound, to provide a quantitative determination of the concentration of that compound in a gas mixture. Specifically, the technique involves determining the difference in infrared energy absorption over all wavelengths passed by the optical system between a gas sample containing the compound of interest and a reference path. The assumption is made that the difference in infrared energy absorbed is proportional to the concentration of the subject compound in the sample gas. There are several different arrangements to detect this difference in commercially available instruments. The sample and reference paths can be separated in space or time. In the first case, the reference path is physically separate from the sample path. It may be a sealed cell containing a non-absorbing gas or a flowing cell from which the compound of interest has been selectively removed. In the time separation case, there is only one optical path: infrared energy passes sequentially through a gas filter cell containing the compound of interest and then an infrared-transparent cell prior to passing through the sample cell.

1.2 Detection of the energy difference depends on the specific instrumental principle. In the case of analyzers utilizing physically separated reference cells, gas cell detectors are used. Here detection of the energy difference is accomplished through absorption of the residual infrared energy by a mixture of the subject compound with an inert gas in a sealed detector cell. Only

that energy is absorbed that is defined by the absorption spectrum of the compound. The energy absorbed is transformed to heat causing an alternate expansion and contraction of the gas in the detector cell. Means are provided for obtaining an electrical signal from this action (1,2,3,4,5).

In the case of the gas filter correlation type analyzer incorporating cyclic sample and reference beams, a solid state detector is utilized.

2. Range and Sensitivity

2.1 The range and sensitivity of the nondispersive infrared method for carbon monoxide analysis is dependent principally upon analyzer design. Even so, considerable flexibility in range can be obtained with commercially available analyzers. The range is principally determined by the sample cell length and the operating pressure used. This derives from the application of Beer's law governing the absorption of radiation by a solute in a nonradiation-absorbing solvent (for gases, the compound of interest is the solute, and the carrier gas is the solvent). In the integrated form, Beer's law is expressed as:

$$I = I_o \exp(-KCl)$$

where I = transmitted radiation
I_o = incident radiation
K = absorption coefficient
C = molar concentration (for gases, volume concentration is equivalent to molar concentration)
l = path length

2.2 Since the available signal from the detector is directly related to the transmitted radiation, I, it can be seen that change in range can be obtained either by a change in sample pressure or in sample cell length. Most analyzers for atmospheric range service are designed to provide full scale deflection at 50 or 100 ppm (vol).

2.3 Sensitivity of the nondispersive infrared technique is determined principally by electrical and optical noise and, in addition, by characteristics and performance of the signal processing components.

For most commercial ambient analyzers the minimum detectable concentration ranges from 0.1 to 0.4 ppm.

3. Interferences

3.1 Interferences may arise from gases that absorb infrared radiation in bands that overlap that of carbon monoxide. Fortunately, this potential chance for error is relatively small for carbon monoxide. Being a heteronuclear diatomic gas, it absorbs infrared radiation but the pattern is simple. The single band involved has its absorption peak at 2165 cm^{-1} (4.6 μm). At this point the principal interference would be from water and carbon dioxide. Limited interference from methane and ethane might also be expected.

3.2 The major interference possibilities in a dual beam instrument can be eliminated by use of an interference cell in line with the sample cell, containing the principal interferents in a concentration sufficient to block the radiation from the overlapping portion of this absorption spectrum. Commercially available instruments incorporate this feature routinely. Also, it has become common practice to incorporate narrow band-pass filters to reduce still further the effect of interfering gases. The most likely interferences in ambient air are CO_2, H_2O, CH_4, and C_2H_6. However, even at the high range of expected ambient concentrations, the discrimination ratios against CO_2, CH_4, and C_2H_6 are high enough with conventional analyzers so that none of these gases should contribute more than 0.1 ppm equivalent CO interference. On the other hand, the concentration of water in the atmosphere (in volumetric or molar terms) can range from about 1500 ppm (0°C, 10% RH) to 50,000 ppm (45°C, 90% RH). Since the U.S. EPA specifies that the maximum interference from any one substance not exceed 1.0

ppm, a water discrimination ratio of 50,000:1 is required (discrimination ratio equals concentration of interferents required to give response equivalent to one ppm of carbon monoxide).

3.3 Discrimination against water vapor and carbon dioxide of the degree stated in paragraph 3.2 can be obtained with a dual beam analyzer using a flowing reference cell or with a gas filter correlation type analyzer.

In the case of a flowing reference cell, the reference gas is sample air with the CO removed by a CO converter. Thus, the CO_2 and water vapor concentration in the reference cell are essentially the same as in the sample cell and cancel out. In the gas filter correlation technique, the inherent discrimination is enhanced by use of a narrow band-pass filter between the infrared source and the remaining optical path. The discrimination against water for a dual beam gas detector cell analyzer with a flowing reference cell is approximately 50,000:1. For a gas filter correlation type analyzer, the water discrimination ratio is about 200,000:1. Enhanced discrimination against water vapor in double beam analyzers can also be achieved by use of dual detectors.

3.4 A statement from the manufacturer should be obtained as to the specifications for discrimination against expected interferents. This can change should leaks occur in either the reference or detector cells. If necessary, water can be removed by drying agents.

4. Precision and Accuracy

4.1 A great many factors can affect the expected precision and accuracy of measurements with this method. Certain of these are related to analysis uncertainties and others to instrument characteristics.

4.2 The principal analysis uncertainties affecting accuracy of measurement are the presence of unknown amounts of interfering gases in the sample and the accuracy of carbon monoxide assay in the span and zero gases. With respect to interferences it has been shown that the extent of water vapor interference may range from a negligible (< 1.0 ppm) amount to as much as 10 ppm. This latter would be an extreme case.

For most of the United States, the long-term mean interference would more likely be + 3 ppm for a double beam analyzer not incorporating water removal. Interference from water vapor for dual cell analyzers incorporating water removal (or saturation), for dual cell analyzers with flowing ambient air reference cells (CO removed), and for gas filter correlation type analyzers should be <1 ppm.

4.3 Analysis errors due to inaccurate zero and span gas assay can be minimized by using reference standards that are traceable to NBS Standard Reference Materials (SRMs). Error in zero gas assay leads to greater relative inaccuracies in low concentration samples than does error in span gas assay. If possible, a true carbon monoxide-free zero gas should be used.

4.4 Leaks in either the reference or detector cells will result in an unknown amount of absorption by carbon monoxide or interferents.

4.5 The assumption of linearity in scale preparation introduces an error because of the logarithmic nature of the absorption equation. For greatest accuracy a calibration curve should be prepared from experimental data using varying span gas concentrations. The nonlinearity effect will be accentuated if nonoptimum cell lengths are used considering the range of concentrations to be encountered. If a single concentration span gas is used for calibration, the value should be approximately two-thirds the maximum expected concentration. This will minimize the mean error for the concentration frequency distribution expected in ambient air. Some analyzers incorporate linearization circuitry in the electronics portion of the instrument.

4.6 Several factors affect both precision and accuracy. Of most concern is span and zero drift. These are instrument-dependent. The combined effect may be ± 1% over an 8 hr period. Dirty cell windows or loose dirt in the sample cell may affect precision and accuracy. Out-of-phase signals due to optical system misalignment as well as those due to scattered radiation will contribute to random noise and loss in accuracy. Sampling at higher than design rate may increase sample cell pressure to the extent that high values result.

4.7 Signal noise from the optical or electrical system may result from vibration, poor voltage regulation, or operation at too high amplifier gain. These factors contribute to loss in precision.

4.8 A formal analysis of errors involves too many assumptions for the general case in which all the above factors might operate. Under optimum conditions, however, for analyzers spanned at 50 ppm full scale, the reproducibility would not be expected to be better than ± 1 ppm.

4.9 Short-term precision, as defined by the U.S. Environmental Protection Agency, should be about ± 0.2 ppm.

5. Apparatus

5.1 Nondispersive infrared analyzers for carbon monoxide meeting the method criteria are available in several different general configurations. Only a general description of the components of each is given.

5.2 The general arrangements of NDIR carbon monoxide analyzers commercially available include:

5.2.1 A double-beam type with optical chopper and Luft type detector. This type incorporates dual infrared sources, an optical chopper to alternate the source, a sample cell, a reference cell, and a gas cell detector using a microphonic principle (Luft type). Interference filters may also be incorporated.

5.2.2 A dual cell type with optical chopper and a gas cell detector using a microflow measurement principle. This differs from the type in 5.2.1 in the detector principle.

5.2.3 A dual cell type with cross-flow modulation, and a Luft type detector. This differs from the dual cell type described in 5.2.1 in that sample and reference flow is sequentially switched from cell to cell to modulate the signal instead of using an optical chopper.

5.2.4 The gas filter correlation technique uses a single infrared source. Infrared radiation is passed through a spinning gas filter wheel containing a CO gas cell and an N_2 gas cell. The radiation passing through the CO cell has the CO absorption wavelengths essentially totally removed. The radiation passing through the N_2 cell is unaltered. The alternating beams of radiation are passed through a single multiple-pass cell to a solid state detector.

5.3 Common to all analyzers are sample handling components, zero and span controls, signal display devices, and signal output terminals for external data acquisition systems. In the double-beam type analyzers, most use flowing reference cells. The reference gas is the ambient sample air from which the CO is removed by a catalytic converter which oxidizes the CO to CO_2. Figure 128:1 shows the principal components of a double beam analyzer. Specific details should be obtained from individual manufacturers.

6. Reagents

6.1 CALIBRATION GASES. These gases are supplied in pressure cylinders, at various concentrations covering the range of interest. For routine use, working standards should be used that are traceable to an NBS SRM for CO. Certified gases that have NBS traceability documentation are available from suppliers.

6.2 ZERO GAS. Zero grade nitrogen containing less than 0.1 ppm CO should be used. Certified zero gases are available in pressure cylinders from specialty gas suppliers.

7. Procedure

7.1 Specific operating procedures are supplied by manufacturers of the commercially available analyzers. Certain general operating steps only, therefore, are given. In particular, a distinction is given between zeroing and standardizing operations as contrasted to actual calibration.

7.2 The first step common to all analyzers is to turn on the power supply and allow the recommended (or necessary) warm-up time before actual operation is begun using standard or sample gases. This time varies

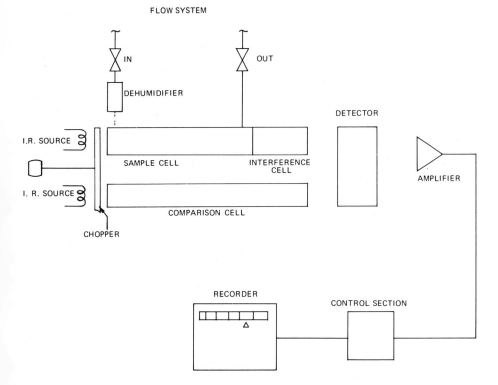

Figure 128:1 – NDIR schematic diagram.

among analyzers and with the length of time the analyzer has spent without power applied. It can range between 30 min and 3 h. Under any circumstances warm-up time should be long enough to stabilize drift.

7.3 Several classifications of procedures are actually involved. First is preliminary or initial setup; second, routine operation; and third, trouble shooting. The first and third classifications should be carried out by a competent technician or engineer. They involve the use of internal adjustments, and often require special electronic instrumentation. Routine operations involve the use of external controls only and require relatively little training.

7.4 Typical initial or start-up procedures are the setting of electrical zero, making coarse adjustments of infrared sources, oscillator tuning (if used), signal phase adjusting, and initial span and zero adjusting.

7.5 Routine operations that might ordinarily be involved are described in the sequence below.

7.5.1 Turn power on.

7.5.2 Allow analyzer to warm up for time period earlier found to be necessary.

7.5.3 Introduce "zero" gas and adjust zero control to proper reading.

7.5.4 Introduce "span" gas and adjust span control to proper reading.

7.5.5 Check sample flow rate. This must not be in excess of the rate found to cause pressure effects. Minimum rate is adjusted on the requirement for the resolution of fine sample structure.

7.5.6 Log all operations, control settings, adjustments, and the time of these operations.

7.6 Trouble shooting and maintenance are carried out both routinely and on demand. The principal routine servicing includes replacement of particle filter, water removal check and recorder servicing. Other servicing that can be expected includes sensitivity checks, analyzer signal output determination, infrared source checks, amplifier servicing, sample cell cleaning, phase adjustments, and servicing or changing detector. Recorder service requirements are the same as for any similar null-balancing instrument used for signal recording. When a flowing reference cell is used, the CO converter efficiency can be checked by introduction of a standard zero gas to the sample cell only.

8. Calibration and Standards

8.1 As contrasted with standardization, calibration is defined as the verification by independent means of the carbon monoxide concentrations present in the "zero" and "span" gases. For convenience, interference checks are also included under this heading. Absolute reliance should not be placed in "certified analysis" of commercially available gases, unless NBS traceability has been established.

8.2 CALIBRATION PROCEDURE

8.2.1 Using established operating procedure, introduce "zero" gas and adjust zero control to proper setting so recorder gives proper value. A pressure check is recommended to verify the concentration of carbon monoxide in the "zero gas." With the sample cell discharge valve closed, introduce the "zero gas" at atmospheric pressure. Set zero control at indicated carbon monoxide concentration on cylinder. Increase the pressure of the "zero gas" in the sample cell to twice atmospheric. If no carbon monoxide is present, there will be no indicated change in the recorder reading (assuming no significant interferences). If carbon monoxide is present the net increase in scale reading should be approximately equal to the carbon monoxide concentration in the "zero gas."

8.2.2 Introduce "span gas" having a concentration of about two-thirds the maximum expected concentration. Adjust span control to obtain correct recorder value.

8.2.3 Introduce several (at least three) other known concentrations of carbon monoxide in nitrogen covering the full-scale range and note recorder values. This is known as a multi-point calibration.

8.2.4 Prepare calibration chart, plotting indicated value versus true value.

8.3 INTERFERENCE CHECK. A test should be made for determining the response factor of the analyzer to water vapor. This is accomplished by sampling an atmosphere containing a known amount of water vapor with and without the water removal system. The ambient atmosphere may be used by determining water content with a psychrometer or calibrated hygrothermograph.

When so doing, it is suggested that the sample atmosphere be drawn through a mixing container sufficient to smooth out high frequency concentration variations (a 50-L glass flask is acceptable).

8.4 Another approach involves the injection of liquid water into an inert plastic sample bag, filling with air and allowing the mixture to equilibrate at room temperature for one-half hour. Sufficient water is added to insure complete saturation. As is the case of the atmospheric air sample procedure, the samples are drawn into the analyzer with and without water removal system connected. In either case, analyzer response to water is determined by calculating the volumetric water content necessary to produce one ppm response as carbon monoxide.

8.5 A record should be kept of this factor (earlier defined as discrimination ratio). Any abrupt increase in water interference should be investigated to check on effectiveness of the water removal system and to detect possible instrument malfunction.

8.6 QUALITY ASSURANCE

8.6.1 Quality assurance measures including precision checks, audits, preventive maintenance, data validation, and corrective actions should be planned and carried out.

8.6.2 Precision checks are conducted by introducing a working standard at scheduled intervals and calculating the deviation between the analyzer's indicated concentration and the known concentration using the following equation:

$$d_i = \frac{Y_i - X_i}{X_i} \times 100$$

where Y_i = analyzer's indicated concentration from ith precision check

X_i = known concentration of the test gas used in the ith precision check.

The average and standard deviation of the precision values should be calculated quarterly to develop an ongoing record and to develop control limits.

8.6.3 Audit checks are made using calibration gases and personnel not regularly used in the monitoring program. Accuracy is determined from the deviation of observed instrument concentrations from the known audit concentrations.

9. Calculation

9.1 For greatest accuracy the calibration chart should be used to reduce recorder values. Oftentimes, however, the assumption is made that over the range used in atmospheric monitoring, a linear calibration can be used. The procedure selected affects the accuracy of reported measurements. To minimize error when assuming linearity together with a single point standardization, the standardization gas concentration should be about two-thirds the expected maximum value.

9.2 Data reduction procedures used are dependent upon the parameters to be reported. For determination of mean values, graphical, mechanical or electronic integration techniques have been used. Mean values may also be obtained from a sufficient number of discrete values to give the confidence interval desired. Instantaneous peak values should be defined as having a minimum baseline time of 5 min. The value at the baseline is reported.

9.3 Concentrations in the atmosphere should be reported to the nearest ppm with the confidence level stated.

10. Effects of Storage

10.1 Not applicable.

11. References

1. Beckman Instructions 1307-C. Models IR 215, IR 315, and IR 415 Infrared Analyzers. Beckman Inst., Inc., Sci. and Proc. Inst. Div., Fullerton, California.
2. BRYAN, R.J. 1966. In Standard Methods of Chemical Analysis. Sixth edition. F.J. Welcher, Ed., Van Nostrand, Princeton, N.J. Chapter 42:850.
3. HARTZ, N.W. AND J.L. WATERS. 1952. An Improved Luft Infrared Analyzer. Instruments 25:622.
4. MSA LIRA Infrared Gas and Liquid Analyzer Instruction Book, Mine Safety Appliances Co., Pittsburgh, PA.
5. CHANEY, L. AND MCCLENNY, W. 1977. Unique Ambient Carbon Monoxide Monitor Based on Gas Filter Correlation: Performance and Application. Environ. Sci. Technol., 11, 1186.

133.

Determination of O_2, N_2, CO, CO_2, and CH_4 (Gas Chromatographic Method)

1. Principle of the Method

1.1 A dual column/dual thermal conductivity detector gas chromatograph is used to separate and determine O_2, N_2, CO, CO_2, and CH_4 in gas samples (1). The sample is introduced as a plug into the carrier gas, and after drying in a desiccant tube it passes successively through two carefully matched gas chromatography columns. The first column contains a very polar stationary liquid phase while the second is packed with molecular sieve 13-X. Detectors are placed at each end of the column. The first column retains only CO_2, which is eluted after passage of the rest of the mixture (the composite peak). The first detector thus records two peaks, one corresponding to the unresolved O_2, N_2, CH_4, and/or CO, and the second to CO_2. The gases are swept into the molecular sieve column which separates all the components. The second detector records the elution of O_2, N_2, CO, and/or CH_4. The CO_2 is irreversibly adsorbed on molecular sieve 13-X and does not elute.

The retention time of O_2 is sufficiently long to allow CO_2 to elute from the polar column before the O_2 elutes from the second column.

1.2 Peak heights are used in conjunction with calibration plots for quantitative measurements. Alternatively, electronic integration of peak areas may be used.

1.3 The separation is complete in 8.5 minutes.

2. Range and Sensitivity

2.1 The limits of detection with hot wire thermal conductivity detectors and helium carrier gas, expressed as ppm of a gas in a 1-mL sample that produces a 0.01 mV signal on a 1-mV recorder, are given below:

Gas	Limits of Detection
CO_2	250 ppm
O_2	300
N_2	300
CO	500
CH_4	300

2.2 Full scale sensitivity can be varied with an attenuator from 1 mV to 256 mV to accomodate diverse levels of different components in a sample.

3. Interferences

3.1 Argon is not separated from O_2, but is present in natural air at 0.9 volume percent. For samples with low oxygen concentrations, a correction may be necessary depending upon the preparation of the calibration standards; see Section 8.2.

3.2 Any compound present in a sample at a detectable level that elutes from either column at a time close to that of a component of interest is a potential interference. Polar compounds including acid gases are strongly retained on both columns at ambient temperatures and will not interfere. Hydrocarbons heavier than methane are retained somewhat by the polar column and elute in order of increasing molecular weight.

3.3 Hydrogen is not detected using helium carrier gas, but can be measured using argon carrier gas.

4. Precision and Accuracy

4.1 Accuracy depends upon the availability of accurate calibration standards. These may be obtained with a certificate of analysis from commercial suppliers.

4.2 Precision is controlled by the mode of sample introduction. Gas sampling valves provide precision of $\pm$ 0.3%. Reproducibility with a 1 mL gas-tight syringe is about ± 1.5%.

4.3 Precision is also affected by detector drift, which in turn depends upon the control of carrier gas flow rate and system temperature. Standard commercial gas chromatographic equipment capable of detector temperature control to ± 0.5°C and flow rate control to ± 1% is adequate.

5. Apparatus

5.1 GAS CHROMATOGRAPH. Any commercial gas chromatograph equipped with dual column fittings, a four-channel thermal conductivity detector, and both a six-port gas sampling valve and a syringe injection port, can be adapted to this analysis. A schematic diagram of one such apparatus (2) is given in Figure 133:1.

5.1.1 *Detector.* Either tungsten filament or thermistor thermal conductivity detector elements are acceptable. Gold-plated filaments are resistant to oxidation by O_2. Greatest detector stability results when the detector is thermostatted and controlled at a temperature slightly above ambient.

5.1.2 *Carrier gas.* A cylinder of purified helium with a two-stage regulator is required. Flow rate is measured at the exit of the second detector with a soap film flow meter.

5.1.3 *Sample introduction.* A six-port gas sampling valve with a 1-mL sample loop provides the better precision. Alternatively, a 1-mL precision gas-tight syringe with needle may be used.

5.1.4 *Drying tube.* A tube of 200-mL capacity with gas-tight fittings at either end is filled with 10/20 mesh Indicating Drierite. The tube is installed between the sample introduction system and the first gas chromatography column. The desiccant must be replaced when the indicator color changes from blue to pink.

5.1.5 *Gas chromatography columns.* Column 1 is a 6 ft. × 1/4 in. (1.8 m × 6 mm) column packed with 30% by weight hexamethylphosphoramide (HMPA) on 60/80 mesh Chromosorb P. Alternatively the column may be packed with 30% by weight di-2-ethyl-hexyl sebacate (DEHS) on 60/80 mesh Chromosorb P. The DEHS column has a longer lifetime than the HMPA column, but DEHS does not separate ethane and ethylene from CO_2. Column 2 is a 6.5 ft. × 3/16 in (2 m × 5 mm) column packed with 40/60 mesh molecular sieve 13-X. The columns must be carefully matched to ensure that the retention times of the components allow separation of the CO_2 from the O_2.

5.1.6 *Temperature.* The columns are operated at room temperature. As noted above, best precision results when the de-

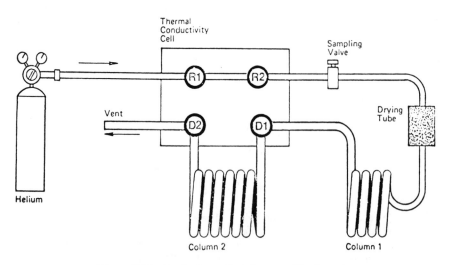

Figure 133:1 — Dual column/dual detector GC schematic.

tector is operated slightly above ambient temperature.

5.1.7 *Recorder.* Any 1 mV potentiometric strip-chart recorder with a chart speed of 1 in./min. (2.5 cm/min) is suitable.

5.1.8 *Electronic integrator.* Any suitable electronic integrator compatible with the chromatograph may be used to measure peak areas for quantitation.

6. Reagents

6.1 HELIUM. High purity grade helium (99.995%) is required.

6.2 CALIBRATION STANDARDS. Standard blends encompassing the concentration range of components in the samples can be obtained from commercial suppliers.

7. Procedure

7.1 GAS CHROMATOGRAPH. The carrier gas is turned on and the flow rate adjusted to 50 mL/min. The flow should be checked periodically. After the gas has been flowing for at least 3 min, the thermal conductivity detectors may be turned on and the currents adjusted to the values specified for the instrument by its manufacturer. About thirty minutes are required for instrument stabilization. The recorder is turned on and zeroed before samples are introduced.

7.2 INJECTION OF SAMPLE.

7.2.1 *Sampling valve.* The sample loop is flushed with several volumes of calibration standard or sample gas. The handle is then turned to divert the sample to the chromatograph.

7.2.2 *Syringe.* Sample is withdrawn from the sample vessel and quickly injected, guarding against blow-back of the plunger.

7.3 REPETITIVE ANALYSES. A new sample may be analyzed immediately after the last peak in the sample has emerged. Samples should be analyzed in duplicate.

8. Calibration and Standards

8.1 A standard curve of peak height or peak area *vs.* volume per cent is prepared for each constituent of interest, by analyzing the calibration standards. The calibration plots should bracket the sample concentrations. Linear plots should result. However, in the presence of 5 to 7% CO_2, the calibration for O_2 is not linear up to 20% (3) but the calibration plot may still be used.

8.2 If the sample source is natural air, the result for O_2 may need correction for argon present in the sample but not separated from the O_2.

8.2.1 If the O_2 calibration standard mixtures contain pure O_2 diluted with pure N_2, the apparent volume percent of O_2 in the samples must be reduced by 0.83%. The thermal conductivity detector molar response factors for Ar and O_2 are sufficiently close (4) that no appreciable error will result from assuming identical relative responses.

8.2.2 If the O_2 calibration standard mixtures contain natural air diluted with pure N_2, both samples and standards contain argon and no correction is necessary.

8.3 The standard curves should be checked periodically.

8.4 A severe loss in resolution of the CO_2/composite peaks and/or of the N_2/O_2 peaks indicates the need for replacement of columns. The polar gas chromatography column continuously loses stationary liquid phase through volatilization. These vapors are adsorbed on the molecular sieve column along with the CO_2, which leads to slow deterioration of the performance of that column. Normally this will happen slowly over a long period of time.

9. Calculations

9.1 Concentrations are determined directly from the calibration plots. The following conversion factors apply at 101.3 kPa and 25°C.

Gas	(mg/m^3)/ppm
CO_2	1.80
O_2	1.31
N_2	1.14
CH_4	0.654
CO	1.14

10. Effects of Storage

10.1 Not applicable.

11. References

1. JEFFERY, P. G., and P. J. KIPPING. 1972. *Gas Analysis by Gas Chromatography*, 2nd ed. Pergamon Press, Oxford.
2. FISHER SCIENTIFIC COMPANY, Springfield, New Jersey, Fisher-Hamilton Gas Partitioner. Models 25A and 29.
3. M. FELDSTEIN, Bay Area Air Pollution Control District, San Francisco, California, private communication.
4. DIETZ, W. A. 1967. *J. Gas Chromatog.*, 5:68.

Subcommittee 4

R. G. SMITH, *Chairman*
R. J. BRYAN
M. FELDSTEIN
D. C. LOCKE
P. O. WARNER

Subcommittee 4/5
Approved with modifications from 2nd edition

M. FELDSTEIN, *Chairman*
R. J. BRYAN
D. L. HYDE
D. A. LEVAGGI
D. C. LOCKE
R. A. RASMUSSEN
P. O. WARNER

134.

Constant Pressure Volumetric Gas Analysis for Oxygen, Carbon Dioxide, Carbon Monoxide, and Nitrogen (Orsat)

The method formerly listed under this number also gave directions for determination of hydrocarbons; today these are exclusively determined by flame ionization, with or without gas chromatography (Method 101). The basic Orsat method is still used to determine whether combustion sources are being supplied sufficient oxygen for complete combustion, and to generate correction factors routinely used in source analysis.

A number of vendors such as laboratory supply companies supply the apparatus complete. The Subcommittee is unaware of any commercially available apparatus that does not yield adequate accuracy when used carefully by reasonably experienced technicians. Typical brand names are Durrell and Hayes. Each comes with full and adequate instructions.

Since any method written here could be nothing but a copy of these instructions, and since the Subcommittee has no knowledge of any laboratory that assembled its own apparatus rather than buying the commercial version, it seems unnecessary to provide a full method here.

135.

Determination of Volatile Organic Compounds in Architectural and Industrial Surface Coatings

1. Principle of the Method

1.1 The total nonvolatile portion of the surface coating is determined by heating a representative portion for a specified time and temperature.

1.2 If the surface coating is solvent-reducible and water free, 1.1 above is the volatile organic compounds (VOC) of the sample. For water-reducible coatings the water content is determined by gas chromatography, and subtracted from 1.1 above to obtain the VOC.

2. Range and Sensitivity

2.1 The total volatiles determination covers the entire possible range of any sample, and has a sensitivity of 0.1% ^{W}/w.

2.2 The water determination easily accommodates the normally encountered 20–65% ^{W}/w in coatings, and is extensible to a range of 10–90% ^{W}/w. The sensitivity is 0.3% ^{W}/w.

3. Interferences

3.1 There are no known interferences for the water determination. For the total non-volatiles it has been found that some specialty coatings such as ultraviolet curing systems and reactive diluent types may yield erratic results. The procedure as written is applicable to approximately 90% of all applied coatings.

4. Precision and Accuracy

4.1 For the total volatiles (1) an interlaboratory study indicated the coefficient of variation was 1.7% relative at 198 degrees of freedom.

4.1.1 Based on the interlaboratory study and samples compared with vacuum distillations (2), an accuracy of ±5% should be expected.

4.2 The determination of water in surface coatings (3) has been evaluated by an interlaboratory study and found to have a coefficient of variation of 2.6% relative at 30 degrees of freedom.

4.2.1 Absolute standards have not been evaluated. However, based on the interlaboratory comparison an accuracy of ±5% should be expected.

5. Apparatus

5.1 OVEN. A forced-draft oven type 11A or 11B (1).

5.2 SYRINGE (5 ML) or DISPOSABLE BEROL PIPETTES.

5.3 BURRELL SHAKER. Or equivalent shaking device.

5.4 ALUMINUM FOIL DISH. Use 58 mm by 18 mm high with a smooth bottom. Precondition dishes and an extended paper clip for 30 min in an oven at 110°C, and store in a desiccator prior to use.

5.5 GAS CHROMATOGRAPH. Any instrument equipped with a thermal conductivity detector may be used. Temperature programming capability is preferred, but isothermal operation is acceptable.

5.5.1 COLUMN. The column should be 1.22 m long by 3.2 mm outside diameter stainless steel, or other suitable material, packed with 60/80 mesh Porapak Q. A replaceable glass sleeve packed with glass wool should be placed in the injection port to retain any nonvolatile materials, and minimize sludge build-up in the column.

5.5.2 *Column Conditioning Procedure.* The packed column is installed in the gas chromatographic unit leaving the exit end disconnected from the detector. This will prevent any contamination of the detector with the column bleed. Set the helium flow rate at 20 to 30 mL/min. Purge the column 5 or 10 min before heating. Heat the column from room temperature to 200°C at 5°C/min and hold this temperature for at least 12 h (overnight). At the end of this period of time, heat the column to 250°C (the maximum temperature for this packing) at a 5°C/min rate and hold at this temperature for several h. Cool the column to 100°C and reheat to 250°C at 5°C/min to observe the column bleed. Optimum conditioning of this column may take several cycles of the heating program before a good recorder baseline is achieved.

5.6 RECORDER. A 1–10 mV recorder with a full scale response time of 2 s or less, and a maximum noise of ± 0.03% of full scale.

5.7 SEPTUM SAMPLE VIALS. 10 mL capacity with fluorocarbon-faced septa are preferred.

5.8 LIQUID CHARGING DEVICE. Microsyringe of 5 or 10 μL capacity.

5.9 ELECTRONIC INTEGRATOR. Or its equivalent (see 8.2.3).

6. Reagents

6.1 DISTILLED WATER.

6.2 TOLUENE. Reagent grade.

6.3 2-ETHOXYETHYL ACETATE. Reagent grade.

6.4 2-PROPANOL (ANHYDROUS). (See Note 1.)

6.5 DIMETHYLFORMAMIDE (DMF) (ANHYDROUS). (See Note 1.)

6.6 HELIUM GAS. Zero grade or higher purity.

Note 1. The water content of 6.4 and 6.5 should not exceed 0.05%. The water content may be determined by a Karl Fisher Titration (4), or more simply, and acceptable, by use of the gas chromatographic procedure described herein. When using the gas chromatographic procedure use attenuations normally encountered in samples. To be acceptable there should be no response.

Note 2. Safety Precaution. Some of the above reagents as well as many of the samples may be harmful if inhaled or absorbed through the skin. Care should be taken to avoid contact with skin, and use only with adequate ventilation.

7. Procedure

7.1 TOTAL VOLATILES DETERMINATION. For solvent-reducible coatings this constitutes the total VOC.

7.1.1 Mix the sample, preferably on a mechanical shaker or roller, until homogeneous. If air bubbles become entrapped, stir by hand until the air is removed.

7.1.2 Weight accurately 0.4 to 0.6 g to the nearest mg of the coating in a preweighed aluminum dish. Disperse the coating by use of an extended paper clip using 2 mL of distilled water for a water reducible coating, and either reagent 6.2 or 6.3 for a solvent reducible coating. If the material forms a lump or cannot be dispersed, discard the sample and prepare a new one. Similarly prepare a duplicate sample.

7.1.3 Heat the aluminum foil dishes containing the dispersed samples plus the paper clip in a forced-draft oven for 1 h at 110°C±5°C.

7.1.4 Remove the dishes from the oven, place in a desiccator, cool to ambient temperature and weigh to the nearest mg.

7.2 DETERMINATION OF WATER CONTENT.

7.2.1 Prepare the sample as in 7.1.1.

7.2.2 Weigh to the nearest 0.1 mg, 0.6 g of the water-reducible coating and 0.2 g of 2-propanol into a septum vial. Add 2 mL of dimethylformamide into the vial. Seal the vial. Prepare a blank containing only 2-propanol and dimethylformamide.

7.2.3 Shake the vials on a wrist action shaker or other suitable device for 15 min. To facilitate settling of solids allow the samples to stand for 5 min just prior to injection into the chromatograph. Low speed centrifugation may also be used.

7.2.4 Inject a 2 μL sample of the supernatant from the prepared solutions onto the chromatographic column. Record the chromatograms using the conditions described in Table 135:I. The blank should not have discernible water at the attenuations used for the samples (see Note 1).

7.2.5 Measure the areas of the water and 2-propanol peaks (see 8.2.3).

Table 135:I Instrument Parameters (Typical Conditions)[a]

Detector	Thermal conductivity
Column	1.22-m by 3.2-mm outside diameter packed with 60 to 80 mesh porous polymer packing (Porapak Q).
Temperatures:	
Sample inlet	200°C
Detector	240°C
Column	
Initial	80°C
Final	170°C
Program rate	30°C/min
Carrier gas	Helium
Flow rate	50 mL/min
Detector current	150 mA
Sample size	2 μL

[a]For isothermal operation set the column temperature at 140°C. After the 2–propanol has cleared the column adjust the temperature to 170°C until DMF clears the column. Reset the temperature to 140°C for subsequent runs.

7.3 DENSITY DETERMINATION

7.3.1 Prepare sample as in 7.1.1.

7.3.2 Weigh a 25 mL volumetric flask to the nearest mg.

7.3.3 By means of a disposable Beral pipette or syringe transfer the mixed sample of coating to the volumetric flask. Fill precisely to the 25 mL graduation mark.

7.3.4 Reweigh the volumetric flask.

7.3.5 Density Calculation

$$D_c = \frac{\text{Wt of Flask + Coating} - \text{Wt of Flask}}{25}$$

$$= \text{g/mL}$$

8. Calibration

8.1 Before each calibration and series of determinations condition the chromatographic column at 200°C for 1 h with carrier gas at a flow rate of 50 mL/min.

8.2 DETERMINATION OF RELATIVE RESPONSE FACTORS. Anhydrous 2-propanol is used as an internal standard. It is essential to determine the response factor with each series of determinations. The response factor to water relative to the standard is determined by means of the following procedure.

8.2.1 Weigh about 0.2 g of water and 0.2 g of 2-propanol to the nearest 0.1 mg into a septum sample vial. Add 2 mL of dimethylformamide.

8.2.2 Inject a 2 μL aliquot of the above solution onto the column and record the chromatogram. The retention order and approximate retention times after the air peak are (1) water, about 0.7 min; (2) 2-propanol, about 2.8 min and (3) DMF, about 7 min.

8.2.3 Measure the areas of the water and 2-propanol peaks. Use of an electronic integrator is recommended. However, triangulation, planimeter, paper cut out, or ball and disk integrator may be used.

8.2.4 Calculate the response factor for water using the following equation.

$$R = \frac{(W_i)(A_{H_2O})}{(W_{H_2O})(A_i)}$$

where W_i = weight of 2-propanol
W_{H_2O} = weight of water
A_{H_2O} = area of the water peak
A_i = area of the 2-propanol peak

9. Calculations

9.1 SOLVENT REDUCIBLE COATING (g/VOC/L)

$$V_t = \frac{\text{Wt of Coating} - \text{Wt of Solids}}{\text{Wt of Coating}} \times 100$$

where V_t = weight % of total volatiles in the coating
Wt = weight of coating and solids as determined in 7.1

9.1.1

g VOC/L = (% Total Volatiles) (D$_c$) 10

where D$_c$ = Density of the coating (g/mL) as determined in 7.3

9.2 WATER REDUCIBLE COATING (G/VOC/L)

$$W = \frac{A_{H_2O} \times W_i \times 100}{A_i \times W_c \times R}$$

where W = weight % of water in the coating

A$_{H_2O}$ = area of the water peak (7.2.5)

A$_i$ = area of the 2-propanol peak (7.2.5)

W$_i$ = weight of 2-propanol added (7.2.2)

W$_c$ = weight of coating (7.2.2)

R = response factor for water (8.2.4)

9.2.1 g VOC/L = (V$_t$ – W) (D$_c$) 10

where V$_t$ = % total volatiles (9.1)

W = % water in coating (9.2)

D$_c$ = density of the coating (7.3)

9.3 The VOC in water-based coatings is sometimes calculated on the basis of the solids and organic volatile contents alone, that is, the liquid coating without its water content. The intent of such a calculation is to place the VOC of water reducible coating on a basis similar to that of a solvent reducible coating. There are many Federal, State, and local agencies whose regulations require that the VOC of a coating be expressed in this manner for regulatory purposes.

g VOC/L (minus water) =

$$\frac{(V_t - W) (D_c) (1000)}{100 - (D_c) (W)}$$

10. References

1. Proposed Test Method for Volatile Content of Coatings; ASTM Designation D2369-81.
2. Private communication, S. Balestrieri, Bay Area Air Quality Management District, San Francisco, CA.
3. Standard Test Method for Water Content of Water-Reducible Paints by Direct Injection into a Gas Chromatograph; ASTM Designation D3792-79.
4. Test for Water in Volatile Solvents; ASTM Designation D1364.

Subcommittee 4/5

D.A. LEVAGGI
M. FELDSTEIN, *Chairman*
R.J. BRYAN
D.L. HYDE
D.C. LOCKE
R.A. RASMUSSEN
P.O. WARNER

201.

Chloride Content of the Atmosphere

No separate method for chloride is presented, since chloride is generally determined by ion chromatography. For details, see Part 1, Technique 34, and Methods 720A and 720B. Chloride can also be determined by selective ion electrodes. See Part 1, Technique 24.

202.

Determination of Free Chlorine Content of the Atmosphere (Methyl Orange Method)

1. Principle of the Method

1.1 Near a pH of 3.0 the color of a methyl orange solution ceases to vary with acidity. The dye is quantitatively bleached by free chlorine, and the extent of bleaching can be determined colorimetrically. The optimum concentration range is 0.05 to 1.0 ppm Cl_2 in ambient air (145 μg to 2900 μg/m³ at 25°C and 101.3 kPa (**1,2,3**).

1.2 Such Cl_2 concentrations rarely occur in typical urban pollution, but can be experienced in emergency situations. This method is provided for this sort of emergency use (accidental Cl_2 release).

2. Sensitivity and Range

2.1 The procedure given is designed to cover the range of 5 to 100 μg of free chlorine/100 mL of sampling solution. For a 30-L air sample, this corresponds to approximately 0.05 to 1.0 ppm in air, which is the optimum range.

Increasing the volume of air sampled will extend the range at the lower end, but only within limits, since 50 L of chlorine-free air produce the same effect as about 0.01 ppm of chlorine.

By using a sampling solution more dilute in methyl orange, a concentration of 1 μg/100 mL of solution may be measured, but beyond this, problems are encountered because of the absorption of ammonia and other gases from the air and by the presence of minute amounts of chlorine-consuming materials even in distilled water.

3. Interferences

3.1 Free bromine, which gives the same reaction, interferes in a positive

direction (**4**). Manganese (III, IV) in concentrations of 0.1 ppm or above also interferes positively. In the gaseous state, interference from SO_2 is minimal, but in solution, negative interference from SO_2 is significant. Nitrites impart an off-color orange to the methyl orange reagent. NO_2 interferes positively, reacting as 20% chlorine. SO_2 interferes negatively, decreasing the chlorine by an amount equal to one-third the SO_2 concentration.

4. Precision and Accuracy

4.1 The data available (**5**) indicate that 26 chlorine concentrations produced by two different methods (flowmeter calibrated by KI absorption, and gas-tight syringe) were measured by this procedure with an average error of less than ± 5% of the amount present.

5. Apparatus

5.1 SPECTROPHOTOMETER. Suitable for measurement at 505 nm, preferably accommodating 5-cm cells.

5.2 FRITTED BUBBLER. Coarse porosity, of 250 to 350 ml capacity (Figure 202:1).

6. Reagents

6.1 Reagents must be ACS analytical grade quality. Water should conform to ASTM Standard for reagent water, Type II.

6.2 METHYL ORANGE STOCK SOLUTION, 0.05%. Dissolve 0.500 g reagent grade methyl orange in water and dilute to 1 L. This solution is stable indefinitely if freshly boiled and cooled water is used.

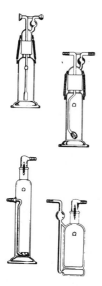

Figure 202:1:—Illustrations of fritted bubblers, coarse porosity, of 250 to 350 ml capacity.

6.3 METHYL ORANGE REAGENT, 0.005%. Dilute 100 mL of stock solution to 1 L with water. Prepare fresh for use.

6.4 SAMPLING SOLUTION. 6 mL of 0.005% methyl orange reagent is diluted to 100 mL with water and 3 drops (0.15 to 0.20 ml) of 5.0 N HCl added. One drop of butanol may be added to induce foaming and increase collection efficiency.

6.5 ACIDIFIED WATER. To 100 mL of water, add 3 drops (0.15 to 0.20 mL) of 5 N HCl.

6.6 POTASSIUM DICHROMATE SOLUTION, 0.1000 N. Dissolve 4.904 g anhydrous $K_2Cr_2O_7$, primary standard grade, in water and dilute to 1 L.

6.7 STARCH INDICATOR SOLUTION. Prepare a thin paste of 1 g of soluble starch in a few mL of water. Bring 200 mL of water to a boil, remove from heat, and stir in the starch paste. Prepare fresh before use.

6.8 POTASSIUM IODIDE, REAGENT GRADE.

6.9 SODIUM THIOSULFATE SOLUTION, 0.1 N. Dissolve 25 g of $Na_2S_2O_3 \cdot 5H_2O$ in freshly boiled and cooled water and dilute to 1 L. Add 5 mL chloroform as preservative and allow to age for 2 weeks before standardizing as follows: to 80 mL of water add with constant stirring, 1 mL concen-

trated H_2SO_4, 10.00 mL 0.1000 N $K_2Cr_2O_7$, and approximately 1 g of KI. Allow to stand in the dark for 6 min. Titrate with 0.1 N thiosulfate solution. Upon approaching the end-point (brown color changing to yellowish green), add 1 mL of starch indicator solution and continue titrating to the endpoint (blue to light green).

Normality $Na_2S_2O_3 =$

$$\frac{1.000}{\text{mL of } Na_2S_2O_3 \text{ used}}$$

6.10 SODIUM THIOSULFATE SOLUTION, 0.01 N. Dilute 100 mL of the aged and standardized 0.1 N $Na_2S_2O_3$ solution to 1 L with freshly boiled and cooled water. Add 5 mL chloroform as preservative and store in a glass-stoppered bottle. Standardize frequently with 0.0100 N $K_2Cr_2O_7$.

6.11 CHLORINE SOLUTION, APPROXIMATELY 10 PPM. Prepare by serial dilution of household bleach (approx. 50,000 ppm), or by dilution of strong chlorine water made by bubbling chlorine gas through cold water. The diluted solution should contain approximately 10 ppm of free (available) chlorine. Prepare 1 L.

7. Procedure

7.1 100 mL of sampling solution is placed in the fritted bubbler, and a measured volume of air drawn through at a rate of 1 to 2 L/min for a period of time appropriate to the estimated chlorine concentration. Transfer the solution to a 100-mL volumetric flask and make to volume, if necessary, with acidified water. Measure absorbance at 505 nm in 5-cm cells against distilled water as reference.

7.2 The volume of sampling solution, the concentration of methyl orange in the sampling solution, the amount of air sampled, the size of the absorbing vessel, and the length of the photometer cell can be varied to suit the needs of the situation as long as proper attention is paid to the corresponding changes necessary in the calibration procedure.

8. Calibration

8.1 Prepare a series of six 100-mL volumetric flasks containing 6 mL of 0.005% methyl orange reagent, 75 mL water, and 3 drops (0.15 to 0.20 mL) of 5.0 N HCl. Carefully and accurately pipet 0, 0.5, 1.0, 5.0 and 9.0 mL of chlorine solution (approximately 10 ppm) into the respective flasks, holding the pipet tip beneath the surface. Quickly mix and make to volume with water.

8.2 **Immediately** standardize the 10 ppm chlorine solution as follows: To a flask containing 1 g KI and 5 mL glacial acetic acid, add 400 mL of chlorine solution, swirling to mix. Titrate with 0.01 N $Na_2S_2O_3$ until the iodine color becomes a faint yellow. Add 1 mL of starch indicator solution and continue the titration to the endpoint (blue to colorless). One mL of 0.0100 N $Na_2S_2O_3$ = 0.3546 mg of free chlorine. Compute the amounts of free chlorine added to each flask in **8.1**.

8.3 Transfer the standards prepared in **8.1** to absorption cells and measure absorbance *vs.* micrograms of chlorine to draw the standard curve.

9. Calculation

9.1

$$\text{ppm Cl}_2 = \frac{\text{mg Cl}_2 \text{ found}}{\text{liters of air sampled}} \times \frac{24{,}450}{71}$$
$$(25° \text{ C}, 101.3 \text{ kPa})$$

For different temperatures and atmospheric pressures, proper correction for air volume should be made.

10. Effect of Storage

10.1 The color of sampled solutions is stable for at least 24 h if protected from direct sunlight, although the presence of certain interferences (Fe III) may cause slow color change.

11. References

1. TARAS, M. 1947. Colorimetric Determination of Free Chlorine with Methyl Orange. *Anal. Chem.* 19:342.
2. BOLTZ, D.F. 1958. Colorimetric Determination of Nonmetals, p. 163. Interscience Publishers, New York.
3. AMERICAN PUBLIC HEALTH ASSOCIATION. 1976. Standard Methods for the Examination of Water and Waste Water. 14th ed. Washington, D.C. 20036.
4. TRAYLOR, P.A. AND S.A. SHRADER. Determination of Small Amounts of Free Bromine in Air. Dow Chemical Co., Main Laboratory Reference MR4N, Midland, Michigan.
5. THOMAS, M.D. AND R. AMTOWER. Unpublished work.

Subcommittee 2

C.R. THOMPSON, *Chairman*
L.V. CRALLEY
L.V. HAFF
A.W. HOOK
E.J. SCHNEIDER
J.D. STRAUTHER
L.H. WEINSTEIN

Approved with modifications
from 2nd edition

Subcommittee 2
R.H. MANDL, *Chairman*
L.H. WEINSTEIN
J.S. JACOBSON

203.

Determination of Fluoride Content of the Atmosphere and Plant Tissues (Manual Methods)

PART A: *Introduction (General Precautions and Sample Preparation)*

PART B: *Isolation of Fluoride (Willard-Winter Distillation)*

PART C: *Isolation of Fluoride (Ion Exchange)*

PART D: *Isolation of Fluoride (Diffusion)*

PART E: *Determination of Fluoride (Titrimetric Method)*

PART F: *Determination of Fluoride (Spectrophotometric Method)*

PART A: *Introduction (General Precautions and Sample Preparation)*

1. General Precautions

1.1 Fluorine is one of the more common elements and occurs in at least trace amounts in virtually all natural and manufactured materials. Contamination by extraneous fluoride may, therefore, come from such sources as sampling and laboratory apparatus, reagents, and from exposure to laboratory dust and fume. Care must be exercised in the selection, purification and testing of reagents and apparatus, and only minimal exposures of samples should be permitted.

1.2 Vessels used for evaporation, ashing, or caustic fusion of samples are first rinsed with warm, dilute acid (hydrochloric or nitric) solution, then with distilled water and air dried under clean toweling. Inconel crucibles used for fusion of ash may require additional cleaning by boiling in 10% (w/v) NaOH for 1 h. Glassware is washed with hot detergent solution followed by a rinse in warm, dilute acid; it is finally rinsed with distilled water and dried (see footnote on the cleaning of distilling flasks, p. 320). All sampling devices, containers, volumetric glassware, reagent solutions, etc., are stored under suitable conditions of protection from airborne dusts and fumes, and are reserved for exclusive use in low-fluoride analysis.

1.3 Before proceeding with analysis of samples, blank determinations are repeated until satisfactorily low values (5 μg, or less, total fluoride per determination) are consistently obtained. Calibration standards are analyzed whenever new batches of reagent solutions are prepared. In addition, one blank and one standard determination are carried through the entire analytic procedure with each set of 10 or fewer samples. If samples are handled in larger sets, the ratio of one blank and one standard per 10 samples should be maintained.

2. Sample Preparation

2.1 The techniques of sample recovery and preparation will vary, as described below, with the sampling method and equipment, and will also depend upon the procedures selected for isolation and measurement of fluoride. Until proof to the contrary is established, samples are assumed to contain fluoride in refractory forms in addition to the commonly encountered interfering materials. Many details involved in the determination of gaseous and particulate fluorides in the atmosphere and in vegetation are discussed by Pack, *et al.*, **(13)** and automatic apparatus for the determination of ambient atmospheric hydrogen fluoride concentration down to 0.1 ppb has also been described **(20)**.

2.2 PARTICULATE FLUORIDES.

2.2.1 Particulate matter collected in air sampling generally requires fusion with sodium hydroxide for conversion into soluble form prior to separation of fluoride by Willard-Winter distillation **(21)**. This treatment is also necessary for materials containing fluoride associated with aluminum, for materials high in silica, and for many minerals.

2.2.2 Transfer the sample-bearing paper filter to a resistant crucible, *e.g.*, platinum, nickel, or Inconel, moisten with water and make alkaline to phenolphthalein with special low-fluoride calcium oxide.* After evaporation to dryness, ignite the paper in a muffle furnace at a temperature of 550 to 600°C until all carbonaceous material has been oxidized. Control combustion of filters of the membrane (cellulose ester) type by drenching with ethanolic sodium hydroxide and igniting with a small gas flame.

2.2.3 Remove particulate matter, collected by electrostatic precipitation, from the surfaces of both electrodes with the aid of a rubber policeman and distilled water. Make the resulting suspension alkaline and evaporate to dryness.

2.2.4 Integrated samples, *i.e.*, those containing both gaseous and particulate fluorides, which have been collected on glass fiber filters, are not amenable to fusion; filters are transferred directly to the distillation flask. Integrated samples collected in impingers are transferred to beakers made of nickel, platinum, or other resistant materials, evaporated to dryness in the alkaline condition, and the residue ashed if organic matter is present.

2.2.5 Fuse the impinger sample, residue from ashing of a filter, or electrostatic precipitator catch, with 2 g of sodium hydroxide. Dissolve the cold melt in a few mL

*Available on special order from G. Frederick Smith Chemical Co., P.O. Box 23344, Columbus, Ohio 43223.

of water, add a few drops of 30% hydrogen peroxide to oxidize sulfites to sulfates, and boil the solution to destroy excess peroxide. The sample solution is then ready for isolation of fluoride.

2.3 GASEOUS FLUORIDES.

2.3.1 *Dry Collectors.*

a. Treat filter papers impregnated with calcium-based fixative agents as described above for particulate fluorides, except that caustic fusion of the ashed residue is not required.

b. Filters impregnated with soluble alkalies are leached with water, as are fixative-coated beads or tubes. Evaporate the washings in a suitable vessel and maintain in an alkaline condition during reduction to a volume convenient for the subsequent fluoride separation procedure. Add a few drops of 30% hydrogen peroxide and boil the solution to destroy excess peroxide.

c. Gaseous fluorides collected on glass fiber filters cannot be quantitatively removed by leaching with water. Transfer such filters directly to the flasks in which a Willard-Winter distillation is to be conducted.

2.3.2 *Wet Collectors.*

a. Transfer a sample collected in water or alkaline solution to a suitably sized vessel, make alkaline to phenolphthalein with sodium hydroxide, and evaporate to the desired volume. Treat the solution with 30% hydrogen peroxide and destroy the excess peroxide by boiling before proceeding with the isolation and determination of fluoride.

2.4 VEGETATION.

2.4.1 Reduce the gross specimen to manageable size for mixing by use of hand shears or, in the case of dried materials, a Wiley cutting mill. Take a small portion (10 to 25 g) of the mixed specimen for determination of moisture by oven-drying at 80°C for 24 to 48 h.

2.4.2 Adjust the weight of material taken for fluoride determination according to condition of the specimen: 100 to 105 g of fresh or frozen vegetation is satisfactory, while for dried materials, such as cured hay, dried leaves, straw, etc., a 50-g portion is adequate.

2.4.3 Weigh the sample into a resistant vessel,* make alkaline to phenolphthalein by addition of low-fluoride calcium oxide slurry, and maintain alkalinity during evaporation to dryness on a hot plate. Raise the temperature of the hot plate until charring and partial ashing have occurred. Complete the ashing at 500 to 600°C in an electric furnace reserved for the ignition of low-fluoride materials. The ash should be white or gray, indicating removal of organic matter.

2.4.4 When the ash has cooled, pulverize it and scrape all material from the dish; mix and determine the net weight. Store in a tightly stoppered bottle.

2.4.5 In order to effect quantitative release of fluoride combined with silica in many varieties of vegetation, fusion of the limed ash with sodium hydroxide is required and is routinely performed on all vegetation specimens (**14,16,17**). Transfer approximately one g of ash into a tared nickel, Inconel, or platinum crucible and weigh accurately. Add about 5 g of sodium hydroxide pellets, cover the vessel and fuse the contents for a few min over a gas burner. After cooling the melt note its color; a blue-green color indicates the presence of manganese and treatment with hydrogen peroxide is required as described under PROCEDURE FOR SINGLE DISTILLATION, VEGETATION ASH. Disintegrate the melt with hot water, washing down the lid and walls of the crucible. Reserve the resulting material for isolation of fluoride.

PART B: *Isolation of Fluoride (Willard-Winter Distillation)*

1. Principle of the Method

1.1 The prepared sample is distilled from a strong acid such as sulfuric or perchloric, in the presence of a source of silica. Fluoride is steam-distilled as fluosilicic acid under conditions permitting a minimum of

---

*Inconel dishes 14 cm in diam and 8 cm high, obtainable from B-J Scientific Products, Inc., 1240 S.W. Alandale Ave., Albany, OR 97321, are satisfactory.

volatilization and entrainment of the liberating acid (**1**).

2. Range and Sensitivity

2.1 The Willard-Winter distillation method, on the macroscale, can accommodate quantities of fluoride ranging from 100 mg down to a few μg.

3. Interferences

3.1 Samples relatively free of interfering materials, and containing fluoride in forms from which it is easily liberated, may be subjected to a single distillation from perchloric acid at 135°C. Samples containing appreciable amounts of aluminum, boron, or silica require a higher temperature and larger volume of distillate for quantitative recovery. In this case a preliminary distillation from sulfuric acid at 165°C is commonly used. Large amounts of chloride are separated by precipitation with silver perchlorate following the first distillation. Small amounts are held back in the second distillation from perchloric acid by addition of silver perchlorate solution to the distilling flask.

4. Precision and Accuracy

4.1 Recovery data for the Willard-Winter distillation, as given in the literature, are difficult to dissociate from inaccuracies inherent in various methods of sample preparation and final evaluation of fluoride. Recovery data from field samples are further complicated by variability of interfering substances and ranges of fluoride contained. In general, recoveries should be within ± 10% of the amount of fluoride present. Under favorable circumstances of sample composition and fluoride range, mean recoveries of approximately 99% with standard deviation of about 2.5% have been reported (**19**).

5. Apparatus

5.1 STEAM GENERATOR. (Figure 203:1) 2000-mL Florence flask made of heat-resistant glass. Each flask is fitted with a stopper having at least 3 holes for inserting 6-mm OD heat-resistant glass tubing.

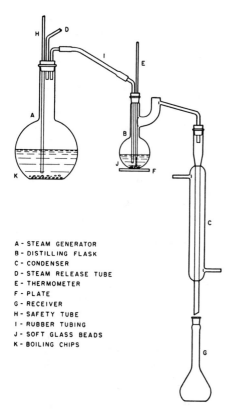

A - STEAM GENERATOR
B - DISTILLING FLASK
C - CONDENSER
D - STEAM RELEASE TUBE
E - THERMOMETER
F - PLATE
G - RECEIVER
H - SAFETY TUBE
I - RUBBER TUBING
J - SOFT GLASS BEADS
K - BOILING CHIPS

Figure 203:1 – Distillation apparatus.

Through one of the glass tubes, bent at right angles, steam is introduced into the distilling flask. The second tube is a steam release tube (Figure 203:1,D) which controls the steam pressure. The small piece of rubber tubing which is slipped over the end of the steam release tube is clamped shut during sample distillation. The third tube is a safety tube. If desired, other tubes may be added to permit the steam generator to supply a maximum of 3 distilling flasks. Any suitable heating device may be used.

5.2 DISTILLING FLASK. (Figure 203:1,B) A 250-mL modified Claisen flask made of heat-resistant glass. The auxiliary neck of this flask is sealed and the outer end of the side tube is bent downward so that it may be attached to an upright condenser. The side tube is fitted with a one-hole rubber stopper to fit the condenser and main neck with a two-hole stopper through which passes a thermometer and a 6-mm OD heat-resistant glass inlet tube for admitting the steam. Any suitable heating device may be used.

5.3 LIEBIG CONDENSER. (Figure 203:1,C) Heat-resistant glass, 300-mm jacket.

5.4 STEAM RELEASE TUBE. (Figure 203:1,D) (See STEAM GENERATOR, **Section 5.1.**)

5.5 THERMOMETER. (Figure 203:1,E) Partial immersion thermometer having a range of 0 to 200°C.

5.6 SUPPORT PLATE. (Figure 203:1,F) Metal, ceramic, or hard asbestos board. The plate shall have a perfectly round 5-cm hole in which the distilling flask is placed as shown in Figure 203:1. The Claisen flask must fit well in the 5-cm hole so that the flask wall, above the liquid level, is not subjected to direct heat. Excessive heat on the wall of the flask causes the liberating acid to be distilled.

5.7 RECEIVER. (Figure 203:1,G) 250-or 500-mL volumetric flask, or 400-mL beaker.

5.8 SAFETY TUBE. (Figure 203:1,H) A 6-mm OD heat-resistant glass tubing, 60-cm long, one end of which is 1 cm from the bottom of the steam generator flask.

5.9 RUBBER TUBING. (Figure 203:1,I) For flask connections, made from natural rubber. Lengths of rubber tubing shall be kept as short as possible.

5.10 SOFT GLASS BEADS. (Figure 203:1,J) 3-mm diam, for use in the distilling flask to prevent superheating and to supply silica for the formation of fluosilicic acid during distillation.

5.11 POROUS PUMICE STONES OR BOILING CHIPS. (Figure 203:1,K).

5.12 PINCHCOCK. To control steam supply from the generator.

6. Reagents

6.1 PERCHLORIC ACID. (70 to 72% by wt) Concentrated perchloric acid ($HClO_4$).*

6.2 SILVER PERCHLORATE SOLUTION. 50% w/v—Dissolve 100 g of silver perchlorate ($AgClO_4$) in 100 mL of water.

6.3 SULFURIC ACID. (96% by wt)—Concentrated sulfuric acid (H_2SO_4).*

6.4 WATER. All references to water shall be understood to mean ASTM reagent water, Type II.

7. Procedure

7.1 PROCEDURE FOR SINGLE DISTILLATION, MISCELLANEOUS MATERIALS.

7.1.1 Fill a steam generator about two-thirds full of water. Add to it a pellet of sodium hydroxide and a few drops of phenolphthalein indicator solution to insure that the water remains alkaline at all times. Add a piece of pumice to permit free boiling, and heat the water to boiling. Keep the steam release tube open at this time and place a pinchcock on the steam supply tubing.

7.1.2 Introduce the sample into a Claisen distilling flask containing five or six glass beads. Wash down the sides of the flask with water and bring the volume to 50 to 75 mL, the lesser volume being more desirable. Insert in the main neck of the flask the rubber stopper that contains the thermometer and steam inlet tube. Set the flask in the 5-cm diameter hole in the plate and connect the outlet to a condenser.

7.1.3 Rinse the sides of the beaker or crucible which contained the sample with 50 mL of perchloric acid (70 to 72%)† and add 1 mL of silver perchlorate. Transfer the rinsings to the distilling flask by means of a small funnel attached to the steam inlet tube. Rinse the beaker or crucible with water and add the rinsings to the flask. Mix the contents of the flask by gentle shaking and attach the flask to the steam generator. Place a 250-mL volumetric flask under the

---

*Acid giving excessively high fluoride blanks requires preboiling at 135°C, with admission of steam, prior to addition of samples.

†*Caution.* When using perchloric acid, the usual precautions should be taken. Hot concentrated perchloric acid may react explosively with reducing substances, such as organic matter. Therefore, it is wise to see that any organic matter in the sample is destroyed in the ashing process prior to distillation. Precautions for the use of perchloric acid are available in material safety data sheets, supplied by the manufacturer.

condenser to receive the distillate and begin heating the solution in the flask. Keep the pinchcock in place on the steam inlet tube until the contents of the distilling flask reach 135°C.

7.1.4 Remove the pinchcock on the steam inlet tube and place it on the steam release tube of the steam generator. Maintain the distillation temp at 135° ± 2°C. Swirl the contents of the distilling flask frequently to minimize deposition on the flask wall of any siliceous residues that might retain fluoride. After collecting 250 mL of distillate during a period of about 1 h, remove the pinchcock from the steam release tube and place it on the steam inlet tube. Disconnect the rubber tubing from the steam inlet tube, and discontinue heating.**

7.2 PROCEDURE FOR SINGLE DISTILLATION, VEGETATION ASH.

7.2.1 Transfer the disintegrated melt to a Claisen distilling flask, as described in section **7.1** for Miscellaneous Materials.

7.2.2 Rinse the sides of the crucible in which the fusion was made with 50 mL of perchloric acid (70 to 72%) and add 1 mL of silver perchlorate solution.

7.2.3 If the sample contains manganese, add sufficient (2 to 10 drops) 3% hydrogen peroxide solution to the contents of the distilling flask to reduce manganese dioxide and permanganates **(6)**.

7.2.4 Carry out the distillation as previously described, except that a 500-mL volumetric flask is used as receiver and filled with distillate during a period of about 2 h.

7.3 PROCEDURE FOR DOUBLE DISTILLATION.

7.3.1 Fill a steam generator, as directed under PROCEDURE FOR SINGLE DISTILLATION. Transfer the sample solution to a Claisen flask and rinse the sides of the beaker or crucible which contained the sample with 50 mL of concentrated sulfuric acid. Transfer the rinsings to the distilling flask through a small funnel attached to the steam inlet tube. Mix the contents of the

---

**Caution.* The distilling flasks should be cleaned using only a brush and distilled water. Repeated use of alkaline cleaning solution produces an etched surface that is difficult to clean and tends to retain fluoride.

flask by swirling, rinse and remove the funnel, and connect the distilling flask to the steam generator. Place a 400-mL beaker under the condenser and begin heating the distilling flask, and steam generator. Keep the pinchcock in place on the steam inlet tube until the contents of the distilling flask reach $165° \pm 5°C$. Swirl contents of the flask as required to prevent accumulation of insoluble material on the walls of the flask above the liquid level. Collect about 375 mL of distillate during a period of about $1\frac{1}{2}$ to 2 h.

7.3.2 Add sodium hydroxide solution (10 g/L) to the distillate until alkaline by phenolphthalein indicator. Evaporate the distillate to 10 to 15 mL by heating below the boiling point.

7.3.3 The concentrated distillate is redistilled from perchloric acid as directed under PROCEDURE FOR SINGLE DISTILLATION. Small quantities of chloride are fixed in the distilling flask by the addition of 1 mL of silver perchlorate solution. A 250-mL quantity of distillate is collected in a volumetric flask.

PART C: *Isolation of Fluoride (Ion Exchange)*

1. Principle of the Method

1.1 The sample is freed of interferences by preferential sorption on an ion exchange resin, followed by desorption of fluoride in a small volume of eluting solution. Thus, concentration of fluoride from impinger- or bubbler-collection media may be achieved without the attendant danger of contamination on prolonged exposure of solutions during evaporation (12).

2. Range of Sensitivity

2.2 The ion exchange procedure can be adapted to quantities of fluoride in the low-mg to μg range.

3. Interferences

3.1 Interfering cations may be eliminated by sorption of fluoride on an anion exchange resin. Fluoride is then eluted with sodium hydroxide solution.*

4. Precision and Accuracy

4.1 Net recoveries for quantities of fluoride of 20 μg or more should be within $\pm 5\%$ of the quantity present. Low recovery indicates incomplete preconditioning of a new column, while high recovery may be due to contamination or failure to elute completely the previous sample.

5. Apparatus

5.1 CHROMATOGRAPHIC COLUMN. Dimensions of the column are not critical and many types, available from suppliers' stocks, are usable. A column made of borosilicate glass tubing 10-mm ID and 16 cm long, having a fritted glass disc fused into the constricted base, and a reservoir of about 100 mL capacity at the top, is satisfactory. A short piece of polyvinyl chloride tubing attached to the bottom and closable with a screw hose clamp permits adjustment of flow rates and prevents complete drainage of liquid from the column.

5.2 QUARTZ SAND, WHITE. Sand of −60 + 120 mesh, purified by hot extraction with 20% sodium hydroxide solution, followed by hot 10% hydrochloric acid solution, is used as a protective layer at the top of the resin bed.

6. Reagents

6.1 ANION EXCHANGE RESIN. Intermediate-Base of the granular aliphatic polyamine type:
Duolite A-41, A-43 (Diamond Alkali Co.)
Ionac A 302 (Ionac Chem. Co., Div. of Ritter Pfaudler Corp.)

*When interferences are present in cationic as well as anionic forms, both may be removed by use of a strongly basic anion exchange resin. This is accomplished by conversion of cations into strongly-held complex anions, while the weakly-held fluoride ions are quantitatively eluted from the column (10).

Permutit A (Permutit Co., Ltd., Div. of Ritter Pfaudler Corp.)

Rexyn 205 (OH) (Fisher Scientific Co.) Mesh size is not critical but, along with column diameter and height, is a factor in controlling flow rate of solutions. Mesh sizes of –60 + 100 or –100 + 200 are usable.**

6.2 HYDROCHLORIC ACID, 2.0 N AND 1.0 N SOLUTIONS.

6.3 SODIUM HYDROXIDE, 2.0 N, 0.1 N, AND 0.01 N SOLUTIONS.

6.4 Water is ASTM reagent water, Type II.

7. Procedure

7.1 Prepare the resin column by adding a few mL of water to the chromatographic tube, then a slurry of resin (1:1) in water. Add sufficient slurry so that when the resin has settled, a layer 10 to 12 cm in height will result. Level the resin bed by twirling the tube, and before the water level has dropped below the surface of the resin, add a 2-cm layer of quartz sand. Wash the resin with 200 mL of 2.0 N hydrochloric acid solution, rinse with water; wash with 200 mL of 2.0 N sodium hydroxide solution, and, finally, rinse with 200 mL water.

7.2 Precondition the resin by passing 400 mL of a solution containing about 1 ppm hydrofluoric acid and an equal volume containing 1 ppm sodium fluoride through the column at a rate of about 5 mL/min. Follow this with 50 mL of 0.1 N sodium hydroxide solution, then 25 mL of 0.01 N sodium hydroxide solution. Discard the eluate. The resin is now ready for use.

7.3 Acidify the sample solution by addition of 0.5 mL of 1 N hydrochloric acid per 100 mL, but add no more than 3 mL of 1 N hydrochloric acid per sample. Remove, by filtration, any solids remaining in the sample after acidification. Pass the sample solution through the resin column at a rate of about 10 mL/min, followed by a water rinse of a few mL. Elute fluoride with a 25-mL portion of 0.1 N sodium hydroxide so-

lution, followed by a 25-mL portion of 0.01 N sodium hydroxide solution.

8. Effect of Storage

8.1 An ion exchange column may be preserved indefinitely if the resin is covered with water. Before the column is reused, a recovery test should be made by passage of a measured quantity of standard fluoride solution through the column; 200 mL of neutral sodium fluoride solution, at a rate of 10 mL/min, is suggested. The quantity of fluoride added should approximate that expected in the sample. Fluoride is eluted, as described under **Procedure**, and determined by the method selected for evaluation of samples.

PART D: *Isolation of Fluoride (Diffusion)*

1. Principle of the Method

1.1 An aliquot of the prepared sample is mixed with a strong acid, gently heated in a sealed container, and the liberated hydrogen fluoride is absorbed by an alkali **(7,16,18)**.

2. Range and Sensitivity

2.1 Quantities of fluoride from about 30 μg to a few tenths of a μg may be used. In routine work, blanks range from 0.5 μg to 0.0 μg.

3. Interferences

3.1 Interfering materials that volatilize from acid medium must be eliminated. Sulfites are oxidized to sulfate by preliminary treatment with 30% hydrogen peroxide solution. Relatively large amounts of chloride may be fixed in the diffusion vessel as silver chloride, by addition of 0.1 to 0.2 g silver perchlorate to the sample aliquot prior to diffusion. Samples high in carbonates require caution upon acidification, to control effervescence.

Separation of fluoride may be achieved with other resins and appropriate eluting solutions; for example, Dowex 1-X8 in the acetate form may be used with elution by sodium acetate solution **(11).

4. Precision and Accuracy

4.1 Recoveries from five sodium fluoride standards, covering the range 4 μg to 20 μg F$^-$, varied from 97.5% to 102.5%; average 99.4% (16). By a slightly modified technique, standards containing 0.2 μg, 0.5 μg, and 1.0 μg F$^-$ (five replicates of each) yielded recoveries of 94% to 101%; average 98.1% (7).

5. Apparatus

5.1 MICRODIFFUSION DISH. Disposable plastic petri dish, 48 mm ID by 8 mm deep. (Obtainable from Millipore Filter Corp., Bedford, Mass. 01730).*

5.2 OVEN. A thermostatically controlled oven capable of maintaining temperatures within ±1°C in the 50 to 60°C range.

5.3 PIPET, MOHR. Capacity 0.1 mL, 0.01 mL subdivisions.

6. Reagents

6.1 PERCHLORIC ACID, 70 to 72%.**

6.2 SILVER PERCHLORATE, ANHYDROUS, C.P. (AgClO$_4$).

6.3 SODIUM HYDROXIDE. 1 N alcoholic solution. Dissolve 4 g sodium hydroxide (NaOH) in 5 mL of water and dilute to 100 mL with ethyl, methyl, or Formula 30 denatured alcohol.

7. Procedure

7.1 Place 0.05 mL of 1 N alcoholic sodium hydroxide solution on the center of the inside top of the plastic petri dish. Use the tip of the 0.1-mL Mohr pipet to spread the droplet into a circular spot of about 3 to 4 cm diam. Dry the top for about 1 h, under slightly reduced pressure, in a desiccator containing activated alumina.

7.2 Transfer a 1.0 mL aliquot of prepared sample solution to the diffusion unit. Add 2.0 mL of perchloric acid and immediately close the dish with a prepared lid. Place the unit in an oven maintained at the

*The Conway Microdiffusion Dish, with Obrink modification, made of methyl methacrylate resin or similar plastic capable of withstanding temps up to 60° C, may also be used.

**See footnote, page 319.

selected temperature (50° to 60°C) and allow to remain for 16 to 20 h.

7.3 Carefully remove the diffusion vessel from the oven and take off the lid. Wash the alkaline absorbent into a 10- or 25-mL volumetric flask (a small funnel is helpful), the size of the flask depending upon the amount of fluoride expected and the method of measurement chosen.

PART E: *Determination of Fluoride (Titrimetric Methods)*

1. Principle of the Method

1.1 In the direct titration of fluoride with standard thorium nitrate solution, the sample solution or distillate containing sodium alizarinsulfonate is buffered at pH 3.0. Upon addition of thorium nitrate, insoluble thorium fluoride is formed. When the endpoint is reached, and all fluoride has reacted, the addition of another increment of thorium nitrate causes formation of a pink "lake" (5).

1.2 In the back titration procedure, the pink lake is first formed by addition of sodium alizarinsulfonate and a slight excess of thorium nitrate to the sample. Equal amounts of dye and thorium solution are added to a fluoride-free reference. The reference solution is then titrated with standard sodium fluoride solution until a color match is achieved with the unknown sample (15).

2. Range and Sensitivity

2.1 The direct titration procedure can accommodate 10 to 0.05 mg fluoride in the total sample. The back titration modifications can measure 50 to about 5 μg fluoride in the total sample. With photometric endpoint detection, direct titration can also be used for the lower ranges.

3. Interferences

3.1 Ions capable of forming insoluble or undissociated compounds with fluorine or with thorium interfere with these titrimetric methods and must be separated by an appropriate technique (*e.g.*, distillation, diffusion or ion exchange). Among the more common of the interfering cations are Al^{+3}, Ba^{+2}, Ca^{+2}, Fe^{+3}, Th^{+4}, TiO^{+2},

VO^{+2}, and Zr^{+4}. The principal interfering anions are PO_4^{-3} and SO_4^{-2}. However, any material which constitutes an appreciable change in total ionic strength of the sample solution will affect the endpoint color as well as stoichiometry of the reaction. Thus, excessive acidity in the distillate from a Willard-Winter distillation, as from the liberating acid or chloride content of the sample, will interfere. This effect may be reduced by careful control of temperature and rate of admission of steam, and by separation of chloride. Similarly, acidity or alkalinity of eluates from ion exchange separations must be matched with that of standards used in calibration, and with the requirements of the method of evaluation.

3.2 Sulfide and sulfite interferences are prevented by preliminary oxidation with 30% hydrogen peroxide in boiling solution, as described under **Sample Preparation**, Part A, 2. Interference by free chlorine is eliminated by addition of hydroxylamine hydrochloride solution.

4. Precision and Accuracy

4.1 These are essentially functions of the fluoride isolation techniques, $q.v.$

5. Apparatus

5.1 FLUORESCENT LAMP. To provide illumination for titrating.

5.2 MICROBURET. 5-mL capacity, 0.01-mL divisions, and a reservoir holding about 50 mL.

5.3 NESSLER TUBES. Matched set of 50-mL, tall-form tubes with shadowless bottoms. Tubes may be fitted with either ground glass or rubber stoppers. The set should be checked for optical similarity as follows: Add to the tubes 40 mL of water, 1 mL of sodium alizarinsulfonate solution, and 2 mL of 0.05 N hydrochloric acid. Add thorium nitrate solution from a buret until the color of the solution just changes to pink. Close the top of the tube and invert several times. Add the same quantity of thorium nitrate solution to the remaining tubes. Fill all the tubes to the 50-mL mark with water and mix. Compare the colors and reject any tubes showing differences in shade or intensity.

5.4 NESSLER TUBES. Matched set of 100-mL, tall-form tubes with shadowless bottoms. The set should be checked for optical similarity, using the same technique as with the 50-mL tubes, except that the quantities of reagents shall be doubled.

5.5 NESSLER TUBE RACK OR COMPARATOR.

5.6 PHOTOMETRIC TITRATOR. A Beckman Model B Spectrophotometer, equipped with an Alcoa Research Laboratories titration attachment or equivalent **(8)**. Light from the monochromator passes through a 20.3-cm (8″) sample cell to the blue-sensitive phototube mounted at the outboard end of the cell housing. A magnetic stirrer is attached under the cell compartment. The tip of a semimicroburet passes through the cell housing and is immersed in the solution to be titrated. A ball-and-socket joint connects the tip to the buret, facilitating removal of the sample cell. The titration cell is 5.1 cm (2″) wide, 7.6 cm (3″) deep, and 20.3 cm (8″) long.

6. Reagents

6.1 BUFFER-INDICATOR SOLUTION. Dissolve 0.40 g of sodium alizarin sulfonate in about 200 mL of water. Weigh 47.25 g of monochloracetic acid into a 600-mL beaker and dissolve in 200 mL of water. Add indicator solution with stirring. Dissolve 10 g of sodium hydroxide pellets in 50 mL of water, cool to approximately 15 to 20°C, and add to the above solution slowly with stirring. Filter and make to 500 mL. Prepare fresh weekly.

6.2 CHLOROACETATE BUFFER SOLUTION. Dissolve 9.45 g of monochloroacetic acid and 2.0 g of sodium hydroxide (NaOH) in 100 mL of water. This solution is stable for more than 2 weeks if stored under refrigeration.

6.3 HYDROCHLORIC ACID, STANDARD SOLUTION (0.05 N). Dilute 4.28 mL of hydrochloric acid (HCl, sp gr 1.19) to 1 L. The normality of this solution should be exactly equal to that of the sodium hydroxide (NaOH) solution (0.05 N).

6.4 HYDROXYLAMINE HYDROCHLORIDE SOLUTION. 1 g of $NH_2OH.HCl$/100 mL of water.

6.5 PHENOLPHTHALEIN INDICATOR SOLUTION (0.5 G/L). Dissolve 0.5 g of phenolphthalein in 60 mL of ethyl alcohol and dilute to 1 L with water.

6.6 SODIUM ALIZARINSULFONATE SOLUTION (0.80 G/L). Dissolve 0.40 g of sodium alizarinsulfonate in 500 mL of water.*

6.7 SODIUM ALIZARINSULFONATE SOLUTION (0.01 G/L). Dissolve 0.01 g of sodium alizarinsulfonate in 1000 mL of water.

6.8 SODIUM FLUORIDE, 100%.

6.9 SODIUM FLUORIDE, STANDARD SOLUTION (1 ML = 1.00 MG F). Dissolve 2.2105 g of sodium fluoride (NaF, 100%) in water and dilute to 1 L in a volumetric flask, mix, and transfer to a polyethylene bottle for storage.

6.10 SODIUM FLUORIDE, STANDARD SOLUTION (1 ML = 0.01 MG F). Dilute 10 ml of NaF solution (1 mL = 1.00 mg F) to 1 L with water in a volumetric flask, mix, and transfer to a polyethylene bottle for storage.

6.11 SODIUM HYDROXIDE SOLUTION (10 G/L). Dissolve 10 g of NaOH in water, dilute to 1 L and mix. Store in a polyethylene bottle.

6.12 SODIUM HYDROXIDE, STANDARD SOLUTION (0.05 N). Dissolve 2.00 g of NaOH in water and dilute to 1 L. The normality of this solution should be exactly equal to that of the standard HCl (0.05 N). Store in a polyethylene bottle.

6.13 THORIUM NITRATE, STANDARD STOCK SOLUTION (1 ML = 1.9 MG F). Dissolve 13.80 g of thorium nitrate tetrahydrate (Th(NO$_3$)$_4$ · 4H$_2$O) in water and dilute to 1 L.

6.14 THORIUM NITRATE SOLUTION (0.25 G/L). Dissolve 0.25 g of thorium nitrate tetrahydrate (Th(NO$_3$)$_4$ · 4H$_2$O) in water, dilute to 1 L and mix. Store in a polyethylene bottle.

6.15 THORIUM NITRATE SOLUTION, 0.01 N (0.19 G F/L). Dilute a 100-mL aliquot of the stock solution (**6.13**) to 1 L and store in polyethylene bottle.

---

*In the literature, this reagent is also known as alizarin Red S, alizarin Red, alizarin-S, alizarin carmine, alizarin, sodium alizarin sulfonate, sodium alizarin monosulfonate, monosodium alizarin sulfonate, and 3-alizarinsulfonic acid sodium salt. The dye is identified by Color Index No. 58005.

7. Procedure

7.1 PROCEDURE FOR DIRECT TITRATION, HIGH CONCENTRATIONS. (10 TO 0.05 MG F IN TOTAL SAMPLE).

7.1.1 Pipet an aliquot of the distillate into a 400-mL beaker and dilute to 100 mL. Add 1 mL of sodium alizarinsulfonate solution (0.80 g/L), and then sodium hydroxide solution (10 g/L) dropwise until a pink color is obtained. Discharge the pink color by adding 0.05 N hydrochloric acid dropwise. Add 1 mL of chloroacetate buffer solution dropwise, and titrate with thorium nitrate solution (1 mL = 1.9 mg F) to a faint, persistent, pink endpoint. Determine a blank obtained by carrying the same amount of all reagents through the entire procedure.

7.2 PROCEDURE FOR BACK TITRATION, MEDIUM CONCENTRATION. (0.05 TO 0.01 MG F IN TOTAL SAMPLE).

7.2.1 Transfer 50 mL of the distillate into a 50-mL Nessler tube, add 1 mL of sodium alizarinsulfonate solution (0.01 g/L) and sufficient 0.05 N sodium hydroxide solution to produce a pink color. Note precisely the volume of 0.05 N sodium hydroxide solution required for neutralization; then discard the titrated solution. If more than 4 mL of 0.05 N sodium hydroxide solution is required, make the remaining distillate alkaline to phenolphthalein, evaporate to 10 to 15 mL, and transfer it to a distilling flask. Repeat the distillation, precautions being taken to reduce the amount of perchloric acid distilled over.

7.2.2 Transfer another 50-mL portion of distillate into a 50-mL Nessler tube (sample tube) and add 1 mL of sodium alizarinsulfonate solution (0.01 g/L). Adjust the acidity with 0.05 N hydrochloric acid until the equivalent of exactly 2 mL of acid is present; that is, 2 mL minus the number of ml of 0.05 N sodium hydroxide solution required for neutralization as described. If between 2 mL and 4 mL of 0.05 N sodium hydroxide solution were required for neutralization, omit the addition of hydrochloric acid to the distillate. Add thorium nitrate solution (0.25 g/L) from a microburet until a faint pink color appears. Note the volume of thorium nitrate solution required, and save the Nessler tube for comparison with the standard.

7.2.3 Pour 50 mL of water into a 50-mL Nessler tube (standard tube) and add 1 mL of sodium alizarinsulfonate solution (0.01 g/L). If neutralization of the sample requires 2 mL or less of 0.05 N sodium hydroxide solution, pipet exactly 2 mL of 0.05 N hydrochloric acid into the standard tube. If the 50-mL aliquot of the distillate requires more than 2 mL of 0.05 N sodium hydroxide solution for neutralization, no further acidification of the distillate is necessary, but add to the standard tube a quantity of acid equivalent to that found in the sample distillate.

7.2.4 From a microburet add sodium fluoride solution (1 mL = 0.01 mg F) equivalent to about 80% of the fluoride present in the sample aliquot, as indicated by the thorium nitrate solution required. Mix thoroughly, add the same volume of thorium nitrate solution as that required for titration of the sample aliquot as described, and again mix thoroughly. The color in the standard tube will be deeper than that in the sample tube.

7.2.5 From the microburet continue to add sodium fluoride solution (1 mL = 0.01 mg F) to the standard tube until its color matches that of the sample tube. (If the colors cannot be matched, repeat the distillation.) Equalize the volumes in the sample and standard tubes by adding water. After the addition of water, mix thoroughly, then allow all bubbles to escape before making the final color comparison. Check the end point by adding 1 or 2 drops of sodium fluoride solution (1 mL = 0.01 mg F) to the standard tube. If the colors were originally matched, the color in the standard tube will be distinctly lighter in shade than that in the sample tube.

7.2.6 Determine a blank by carrying the same amount of all reagents through the procedure described. With proper attention to details, blanks of 5 μg of fluoride, or less, can be obtained.

7.3 PROCEDURE FOR BACK TITRATION, LOW CONCENTRATIONS (LESS THAN 0.01 MG F IN TOTAL SAMPLE).

7.3.1 Distill successive 85 to 90-mL portions of distillate directly into three or four 100-mL Nessler tubes. Take care to keep the amount of perchloric acid distilling over as small as possible, because the entire distillate is titrated and there is no aliquot available for a separate acidity determination. Analyze each of the distillate portions in the 100-mL Nessler tubes separately as follows:

a. Add 2 mL of sodium alizarinsulfonate solution (0.01 g/L) and neutralize the acid by adding 0.05 N sodium hydroxide solution until a pink color is produced. Add 4 mL of 0.05 N hydrochloric acid and sufficient thorium nitrate solution (0.25 g/L) to provide a faint pink color. Compare the treated distillate portion with a standard of equal total volume containing 2 mL of sodium alizarinsulfonate solution (0.01 g/L), 4 mL of hydrochloric acid and the same volume of thorium nitrate solution (0.25 g/L) as is required to produce the pink color in the sample tube. Add sodium fluoride solution to the standard tube until the color matches that of the sample tube. The sum of all significant amounts of fluoride found in each successive portion of distillate is the total amount of fluoride in the sample.

7.4 PROCEDURE FOR PHOTOMETRIC TITRATION.

7.4.1 Transfer the distillate to a 20.3 cm titration cell and add 5 mL of hydroxylamine hydrochloride solution. Adjust, if necessary, to pH 3.6 with 0.05 N perchloric acid, and then add 5 mL of buffer-indicator solution.*

7.4.2 Place the cell in the titrating attachment, immerse the buret tip, and start the stirring motor. Close the titrator lid, set the wavelength to 525 nm, and set the sensitivity knob to the proper position (usually 1). Close the shutter and adjust the slit width to give a transmittance reading of 100.

---

*The addition of buffer-indicator solution should adjust the pH to 3.0. For amounts of fluoride ordinarily encountered, the pH of the distillate should be 3.5 to 3.7 if the distillation is properly controlled. The addition of buffer-indicator solution will maintain a pH of 3.0 under these conditions. For extreme cases, where acidity of the distillate is less than pH 3.5, 0.05 N sodium hydroxide may be used to raise the pH to the proper level. However, it has been found to be the rule that distillations properly conducted will have a pH greater than 3.5. The use of sodium hydroxide for neutralization produces a slight change in the factor due to the sodium perchlorate formed.

7.4.3 Titrate with standard thorium nitrate (0.01 N solution) to a transmittance of 75%. Record the volume to the nearest 0.005 mL.

7.4.4 Deduct a blank obtained by carrying the same amount of all reagents through the entire procedure, including the preliminary distillation and titration steps. Determine the amount of fluoride present from the calibration chart.

8. Calibration and Standards

8.1 THORIUM NITRATE, STANDARD STOCK SOLUTION (1 ML = 1.9 MG F). Standardize this solution as follows:

8.1.1 Weigh 0.100 g of sodium fluoride into a distilling flask and collect 250 mL of distillate as previously described. Titrate 50 mL of the distillate (20 mg of NaF) with the solution being standardized. Carry a blank through the same procedure. Calculate the strength of the thorium nitrate solution in terms of mg of fluoride ion/mL of solution as follows:

$$\text{Fluoride ion, mg/mL} = \frac{C \times 20}{A - B}$$

where A = mL of $Th(NO_3)_4 \cdot 4H_2O$ solution required for titration of the fluoride,

B = mL of $Th(NO_3)_4 \cdot 4H_2O$ solution required for titration of the blank, and

C = 0.4524 when titrating sodium fluoride (NaF).

8.2 THORIUM NITRATE SOLUTION, 0.01 N (0.19 G F⁻/L). Prepare a calibration curve for this solution from data obtained in the following way:

8.2.1 Pipet aliquots of standard sodium fluoride solution covering the range 10 to 1000 μg of fluoride into 500-mL volumetric flasks and dilute to volume. Transfer to a 20.3-cm titration cell, add 5 mL of hydroxylamine hydrochloride solution, and adjust to pH 3.6 with 0.05 N perchloric acid. Add 5 mL of buffer-indicator solution and titrate as described in the PROCEDURE FOR PHOTOMETRIC TITRATION (7.4).

9. Calculation

9.1 Calculate the fluoride ion content of the total distillate* in mg as follows:

$$F = \frac{(A - B)\,CD}{E}$$

where F = mg of fluoride ion in total distillate,

A = mL of titrating solution** used in titration of sample aliquot,

B = mL of titrating solution** used in titrating the blank,

C = fluoride equivalent of titrating solution** in mg of fluoride ion mL of solution.

D = mL of total distillate collected, and,*

E = mL of distillate titrated.*

9.2 Calculate the fluoride concentration in the atmosphere at 25°C and 101.3 kPa, in terms of ppm of hydrogen fluoride (HF) or fluorine (F_2), or mg of particulate fluoride/m³, as follows:

Hydrogen fluoride, ppm =
$$\frac{438 \times F \times (273 + t)}{PV}$$

Fluorine, ppm =
$$\frac{219 \times F \times (273 + t)}{PV}$$

Particulate fluoride, mg/m³ =
$$\frac{340 \times F \times (273 + t)}{PV}$$

where F = mg of fluoride ion in total sample

P = sampling pressure in kPa

*The volume of total distillate collected normally is 250 mL. However, if any other volume of total distillate is collected, this volume shall be substituted for 250. The volume of the distillate titrated normally is 50 ml but may vary as described in PROCEDURE FOR BACK TITRATION, LOW CONCENTRATIONS. If this procedure applies, for each portion of distillate titrated, the value of E is equal to the value of D.

**The term "titrating solution" refers to either the $Th(NO_3)_4$ solution used in accordance with the PROCEDURE FOR DIRECT TITRATION, or the NaF solution (1 mL = 0.01 mg F) used in PROCEDURE FOR BACK TITRATION, MEDIUM CONCENTRATION and in PROCEDURE FOR BACK TITRATION, LOW CONCENTRATIONS.

t = sampling temperature in degrees Celsius, and
V = sample volume in L.

9.3 Calculate the fluoride concentration in vegetation on the oven-dry basis, as follows:

$$\text{Fluoride, ppm (dry basis)} = \frac{F \times A \times 1000}{W \times S \times [1-(M/100)]}$$

where F = mg of fluoride ion in total distillate,
A = g of total ash
W = g of ash distilled
S = g of fresh sample
M = percentage of moisture in fresh sample

10. Effects of Storage

10.1 All reagent solutions listed are stable at room temperature except as individually noted.

PART F: *Determination of Fluoride (Spectrophotometric Method)*

1. Principle of the Method

1.1 Reaction of fluoride with the metal ion moiety of a metal-dye complex results in fading (Zirconium-Eriochrome Cyanine R (9) and Zirconium-SPADNS (4) reagents) or increase (Lanthanum-Alizarin Complexone (3) reagent) in the absorbance of the solution.

2. Sensitivity and Range

2.1 Both Zirconium-Eriochrome Cyanine R and Zirconium-SPADNS reagents obey Beer's Law over the range of 0.00 μg to 1.40 μg fluoride/mL with a detection limit of the order of 0.02 μg/mL. The procedure given for the lower range, Lanthanum-Alizarin Complexone, covers the range 0.00 μg to 0.5 μg fluoride/mL with a detection limit of approximately 0.015 μg/mL.

2.2 In common with other spectrophotometric methods, these are temperature sensitive and absorbances must be read within ± 2°C of the temperature at which the respective calibration curve was established.

3. Interferences

3.1 Moderate variations in acidity of sample solutions will not interfere with the Zirconium-Eriochrome Cyanine R or Zirconium-SPADNS reagents. The Lanthanum-Alizarin Complexone reagent has greater pH sensitivity and solutions must not exceed the capacity of the buffer system to maintain an apparent pH of 4.50 ± 0.02.

3.2 Many ions interfere with these fluoride reagents, but those most likely to be encountered in analysis of ambient air are aluminum, iron, phosphate and sulfate. If these are present above the trace level their effects must be eliminated. Distillation, diffusion, or ion exchange may be employed but, in certain cases, complexation-extraction (2) may be advantageous.

3.3 In vegetation analysis, ashing and distillation by the Willard-Winter technique generally assure a sample solution sufficiently free of interfering ions for direct colorimetric evaluation. Traces of free chlorine in the distillate, if present, must be reduced with hydroxylamine hydrochloride.

4. Precision and Accuracy

4.1 Because of the wide variability in composition of samples, and in methods and conditions of sampling, no general statements of precision and accuracy for field samples can be given. Precision studies of pure sodium fluoride standards indicate that, within the concentration ranges for which the reagents follow Beer's Law, standard deviation of ± 0.015 to 0.020 μg of fluoride/mL should be expected.

5. Apparatus

5.1 SPECTROPHOTOMETER. An instrument capable of accepting sample cells of 1-cm to 2.5-cm optical path, and which is adjustable throughout the visible wavelength region, is required. Each spectrophotometer sample cell is given an identification mark and calibrated by reading a portion of the same reagent blank solution at the designated wavelength. The determined cell correction is subsequently applied to all absorbance readings made in that cell.

6. Reagents

6.1 ACETIC ACID, GLACIAL.

6.2 ACETONE, REAGENT GRADE.

6.3 ALIZARIN COMPLEXONE. (1,2-dihydroxy-3-anthraquinonylamine-N, N-diacetic acid, available from Hopkins & Williams, Ltd., Chadwell Heath, Essex, England, cat. no. 1369.4)*

6.4 AMMONIUM ACETATE SOLUTION (20% w/v). Dissolve 20.0 g ammonium acetate in water and dilute to 100 mL.

6.5 AMMONIUM HYDROXIDE, SP GR 0.880.

6.6 ERIOCHROME CYANINE R SOLUTION. Dissolve 1.800 g of Eriochrome Cyanine R (Mordant Blue 3, Color Index No. 43820) in water to make 1 L. Solution is stable for more than a year when protected from light.

6.7 LANTHANUM CHLORIDE, 99.9% ASSAY. (Available from Kleber Laboratories, Burbank, Calif.)

6.8 LANTHANUM-ALIZARIN COMPLEXONE REAGENT. Dissolve 8.2 g sodium acetate in 6 mL of glacial acetic acid, add sufficient water to permit solution, and transfer to a 200-mL volumetric flask. Dissolve 0.0479 g alizarin complexone in 1.0 mL of 20% ammonium acetate solution, 0.1 mL ammonium hydroxide and 5 mL water. Filter this solution through a Whatman #1 paper into the 200-mL volumetric flask. Wash the filter with a few drops of water and discard the residue. Add 100 mL of acetone, slowly and with mixing, to the flask. Dissolve separately 0.612 g of lanthanum chloride in 2.5 mL 2 N hydrochloric acid solution, warming slightly to promote solution, and combine this with the

flask contents.** Dilute, mix well, cool the solution to room temperature and adjust the volume to the mark with water. The reagent solution is stable for about 1 week if kept under refrigeration.

6.9 SODIUM FLUORIDE STOCK SOLUTION. (1 mL = 1.0 MG F⁻). Dissolve 2.2105 g of 100% sodium fluoride, or the equivalent weight of reagent grade sodium fluoride, in water and dilute to 1 L. Store in a polyethylene bottle.

6.10 SODIUM FLUORIDE WORKING STANDARD SOLUTION (1 mL = 10 μG F⁻). Dilute 5.0 mL of the stock solution to 500 mL. Store in a polyethylene bottle.

6.11 SPADNS SOLUTION. (4,5-dihydroxy-3-(p-sulfophenylazo)-2,7-napthalene disulfonic acid trisodium salt, available from Eastman Organic Chemicals, Rochester, N.Y., cat. no. 7309) Dissolve 0.985 g SPADNS dye in water and dilute to 500 mL.

6.12 SPADNS REFERENCE SOLUTION. Add 10 mL SPADNS Solution to 100 mL of water and acidify with a solution prepared by diluting 7 mL concentrated hydrochloric acid to 10 mL. This solution may be stored and reused repeatedly.

6.13 ZIRCONIUM SOLUTION. Dissolve 0.265 g of zirconyl chloride octahydrate (ZrOCl₂ · 8H₂O) in 50 mL of water, add 700 mL of concentrated hydrochloric acid, and dilute to 1 L with water.

6.14 ZIRCONIUM-ERIOCHROME CYANINE R REAGENT. Mix equal volumes of the Eriochrome Cyanine R and the zirconium solutions. Cool to room temperature before use. Prepare fresh daily.

6.15 ZIRCONIUM-SPADNS REAGENT. Mix equal volumes of the SPADNS and the zirconium solutions. Cool to room temperature before use. This reagent may be stored for several months, at room temperature, in a polyethylene bottle.

7. Procedure

7.1 PROCEDURE FOR INTERMEDIATE RANGE.

7.1.1 *Zirconium-Eriochrome Cyanine R Reagent.* Transfer an aliquot of the prepared sample, standard, or blank solution

---

*This reagent has been variously named in the literature, *e.g.*, alizarin complexan, alizarin complexon, alizarin complexone, and alizarin fluorine blue.

**An equimolar concentration of lanthanum nitrate may be substituted for the chloride.

to a 25-mL volumetric flask containing 4 mL of the Zirconium-Eriochrome Cyanine R Reagent. Dilute the solution to the mark, mix well, and allow to stand for 30 min for temperature equilibration. Transfer the solution to a calibrated spectrophotometer cell of about 2.5-cm light path. The spectrophotometer is set on a wavelength of 536 nm and the light control adjusted to an absorbance value of 0.500 on a Reagent Blank similarly prepared.

a. — Spectrophotometer cells of 1-cm light path may be used by making the following adjustments of volumes: transfer the aliquot of sample, standard or blank solution to a 10-mL volumetric flask, and add 3 mL of Zirconium-Eriochrome Cyanine R Reagent. Dilute the mixture to the mark, mix well and allow to stand for 30 min; then read absorbance as previously described.

7.1.2 *Zirconium-SPADNS Reagent.* Dilute a suitable aliquot of the sample solution to 25 mL and add 5.0 mL of Zirconium-SPADNS Reagent. Mix and allow to stand for 30 min to establish temperature equilibrium before transferring the solution to a spectrophotometer cell (cells of 1 cm to 2.5 cm optical path may be used). Measure the absorbance value at 570 nm with the spectrophotometer adjusted to read zero absorbance on the SPADNS Reference Solution.

7.2 PROCEDURE, LOWER RANGE. Lanthanum-Alizarin Complexone Reagent: Transfer a suitable aliquot of sample solution, containing no more than 4 μg of fluoride, to a 10-mL volumetric flask. Add 3 mL of Lanthanum-Alizarin Complexone Reagent, dilute to the mark and mix well. Allow to stand for 30 min. Measure the absorbance at 622 nm, in a calibrated 1-cm cell, using a reagent blank as reference.

8. Calibration and Standards

8.1 ZIRCONIUM-ERIOCHROME CYANINE R REAGENT. Prepare a standard series, spanning the range of zero to 20 μg of fluoride, by pipetting aliquots of the standard sodium fluoride solution (10 μg F$^-$/mL) into 25-mL volumetric flasks. Add 4 mL of Zirconium-Eriochrome Cyanine R Reagent to each flask, dilute to the mark and mix thoroughly. Allow the standards to stand

until solution temperature has equilibrated at the desired value. Measure absorbances at 536 nm, in 2.5-cm cells, against a reagent blank for which the spectrophotometer is adjusted to read 0.500 absorbance unit. Plot a calibration curve relating fluoride concentration, in μg to absorbance values at the selected working temperature.

8.1.1 If 1-cm cells are to be used, a similar standard series is prepared in 10-mL volumetric flasks, adding 3 mL of Zirconium-Eriochrome Cyanine R Reagent to each flask, and reading as described above.

8.2 ZIRCONIUM-SPADNS REAGENT. Prepare a standard series containing from zero to 35 μg of fluoride by pipetting aliquots of the standard sodium fluoride solution (10 μg F$^-$ per mL) into 25-mL volumetric flasks. Add 5 mL of Zirconium-SPADNS Reagent to each flask, dilute to the mark and mix well. Allow the standards to stand 30 min to reach temperature equilibrium at the desired value. Measure absorbances at 570 nm after zeroing the spectrophotometer on the SPADNS Reference Solution. Prepare a calibration curve relating fluoride concentration in μg to absorbance values at the selected working temperature.

8.3 LANTHANUM-ALIZARIN COMPLEXONE REAGENT. Prepare a standard series containing zero to 4 μg of fluoride by measuring portions of the standard sodium fluoride solution (10 μg F$^-$/mL) into 10-mL volumetric flasks. Add 3 mL of Lanthanum-Alizarin Complexone Reagent to each flask, dilute to the mark, mix and let stand for 30 min at the selected temperature. Measure the absorbances at 622 nm in 1-cm cells, using a reagent blank as reference.

9. Calculation

9.1 Concentrations of fluoride in air or vegetation samples are calculated by use of the formulas given under *Part E: Titrimetric Methods*, Calculations 9.2 and 9.3

10. Effects of Storage

10.1 Reagents used in these procedures are stable, except as individually noted.

11. References

1. AOAC. 1984. "Official Methods of Analysis," 14th edition. Method 25.077. Association of Official Analytical Chemists, Arlington, VA.
2. BELCHER, R. AND T.S. WEST. 1961. A study of the ceriumIII-alizarin complexan-fluoride reaction. Talanta 8:853.
3. BELCHER, R., AND T.S. WEST. 1961. A comparative study of some lanthanum chelates of alizarin complexan as reagents for fluoride. Talanta 8:863.
4. BELLACK, E. AND P.J. SCHOUBOE. 1958. Rapid photometric determination of fluoride in water. Anal. Chem. 30:2032.
5. DAHLE, D., R.U. BONNAR, AND H.J. WICHMANN. 1938. Titration of small quantities of fluorides with thorium nitrate. I. Effect of changes in the amount of indicator and acidity. J. Assoc. Official Agr. Chem. 21:459. II. Effects of chlorides and perchlorates. Ibid. 21:468.
6. DEUTSCH, S. 1955. Overcoming the effect of manganese dioxide in fluoride determinations. Anal Chem. 27:1154.
7. HALL, R.J. 1963. The spectrophotometric determination of sub-microgram amounts of fluorine in biological specimens. Analyst. 88:76.
8. MAVRODINEANU, R., AND J. GWIRTSMAN. 1955. Photoelectric end-point determination in the titration of fluorides with throium nitrate. Contrib. Boyce Thompson Inst. 18, 3:181.
9. MEGREGIAN, S. 1954. Rapid spectrophotometric determination of fluoride with Zirconium-Eriochrome Cyanine R lake. Anal. Chem. 26:1161.
10. NEWMAN, A.C.D. 1958. The separation of fluoride ions from interfering anions and cations by anion exchange chromatography. Anal. Chem. Acta. 19:471.
11. NIELSEN, H.M. 1958. Determination of microgram quantities of fluoride. Anal. Chem. 30:1009.
12. NIELSEN, J.P., AND A.D. DANGERFIELD. 1955. Use of ion exchange resins for determination of atmospheric fluorides. A.M.A. Arch. Ind. Health 11:61.
13. PACK, M.R., A.C. HILL, M.D. THOMAS, AND L.G. TRANSTRUM. 1959. Determination of gaseous and particulate inorganic fluoride in the atmosphere. A.S.T.M. Special Publication #281.
14. REMMERT, L.F., T.D. PARKS, A.M. LAWRENCE, AND E.H. MCBURNEY. 1953. Determination of fluorine in plant materials. Anal. Chem. 25:450.
15. ROWLEY, R.J., AND H.V. CHURCHILL. 1937. Titration of fluorine in aqueous solutions. Ind. Eng. Chem., Anal. Ed. 9:551.
16. ROWLEY, R.J., AND G.H. FARRAH. 1962. Diffusion method for determination of urinary fluoride. Amer. Ind. Hyg. Assoc. J. 23:314.
17. ROWLEY, R.J., J.G. GRIER, AND R.L. PARSONS. 1953. Determination of fluoride in vegetation. Anal. Chem. 25:1061.
18. SINGER, LEON AND W.D. ARMSTRONG. 1954. Determination of fluoride. Procedure based upon diffusion of hydrogen fluoride. Anal. Chem. 26:904.
19. SMITH, F.A. AND D.E. GARDNER. 1955. The determination of fluoride in urine. Amer. Ind. Hyg. Assoc. Quarterly 16:215.
20. THOMAS, M.D., G.A. ST. JOHN, AND S.W. CHAIKEN. 1958. An atmosphere fluoride recorder. A.S.T.M. Special Publication #250.
21. WILLARD, H.H. AND O.B. WINTER. 1933. Volumetric method for determination of fluorine. Ind. Eng. Chem., Anal. Ed. 5:7.

Subcommittee 2

L.V. CRALLEY, Chairman
L.V. HAFF
A.W. HOOK
E.J. SCHNEIDER
J.D. STRAUTHER
C.R. THOMPSON
L.H. WEINSTEIN

Approved with modifications
from 2nd edition
Subcommittee 2
R.H. MANDL, Chairman
L.H. WEINSTEIN
J.S. JACOBSON

204.

Determination of Fluoride Content of the Atmosphere and Plant Tissues (Semiautomated Method)

1. Principle of the Method

1.1 GENERAL. The plant material including leaf samples, washed or unwashed, is dried and ground, then ashed, alkali fused, dissolved with perchloric acid and diluted with water to 50 mL. In the case of leaf samples, an appreciable amount of fluoride may be deposited on the external leaf surfaces. This fluoride behaves differently physiologically from fluoride absorbed into the leaf and it is often desirable to wash it from the surface as a preliminary step in the analysis. Details of a leaf-washing process are given in Section **7.1**. The dissolved digest and sulfuric acid are pumped into the Teflon coil of a microdistillation device maintained at 170°C. A stream of air carries the acidified sample through a coil of Teflon tubing to a fractionation column. The fluoride and water vapor distilled from the sample are swept up the fractionation column into a condenser, and the condensate passes into a small collector. Acid and solid materials pass through the bottom of the fractionation column and are collected for disposal. In the colorimetric method, the distillate is mixed continuously with alizarin fluorine blue-lanthanum reagent, the colored stream passes through a 15-mm tubular flow cell of a colorimeter, and the absorbance is measured at 624 nm. In the potentiometric method, the distillate is mixed continuously with a buffer, the mixed stream passes through a flow-through fluoride ion electrode, and the differential millivoltage measured by an electrometer. The impulse is transmitted to a recorder. All major pieces of the apparatus are components of an automated analyzer such as the Technicon AutoAnalyzer **(15)**

(Technicon Industrial Systems, Tarrytown, NY 10591) or the CFA-200 Analyzer (Orion Scientific Instruments Corporation, Pleasantville, NY 10570) or may be constructed from various laboratory equipment available from scientific supply companies. Details of construction of the microdistillation device are given in Section **5.9**. Earlier versions of this method have been published. **(10,19,20)**.

1.2 PRINCIPLE OF OPERATION.

1.2.1 *Colorimetric System.* The absorbance of an alizarin fluorine blue-lanthanum solution is changed by very small amounts of inorganic fluoride. In addition, a number of other materials also cause changes in absorbance at 624 nm. Potential interfering substances commonly found in plant tissues are metal cations such as iron and aluminum, inorganic anions such as phosphate, chloride, nitrate, and sulfate, and organic anions such as formate and oxalate. Fortunately, metal cations and inorganic phosphate are not distilled in this system, and organic substances are destroyed by preliminary ashing. The remaining volatile inorganic anions may interfere if present in a sufficiently high concentration because they are distilled as acids. Their hydrogen ions bleach the reagent which, in addition to being an excellent complexing agent, is also an acid-base indicator. To reduce the danger of acidic interferences, a relatively high concentration of acetate buffer is used in the reagent solution despite some reduction in sensitivity. All plant tissues growing with normal nutrient supplies thus far tested have not contained concentrations of interfering substances sufficient to cause problems (Table 204:I.).

Table 204:1 Maximum Concentration of Several Anions Present in Samples at Which There Is No Analytical Interference (Colorimetric Method).

Compound Tested	Interfering Anion	Molarity Tolerated
Na_2SO_4	SO_4^{--}	2×10^{-2}
Na_2SiO_3	SiO_3^{--}	5×10^{-3}
NaCl	Cl^-	1×10^{-3}
NaH_2PO_4	PO_4^{---}	3.8
$NaNO_3$	NO_3^-	5×10^{-3}

1.2.2 *Potentiometric System.* Since the sample stream is the same for this procedure as for the colorimetric procedure, the distillation step removes all of the interfering cations. The volatile acids that remain can be buffered by mixing with the total ionic strength adjustment buffer (TISAB).

1.2.3 *Distillation System.* Since HF has a high vapor pressure, it is more efficiently distilled than the other acids previously mentioned (**1.2.1**). The factors controlling efficiency of distillation are temperature, concentration of acid in the distillation coil, and vacuum in the system. Very large amounts of dissolved solids, particularly silicates, will retard distillation. Accordingly, the smallest sample of vegetation consistent with obtaining a suitable amount of F^- should be analyzed. The aforementioned conditions must be carefully controlled, since accurate results depend on obtaining the same degree of efficiency of distillation from samples as from the standard F^- solutions used for calibration.

Temperature is maintained within $\pm$ 2° C by the thermoregulator and by efficient stirring of the silicone oil. Acid concentration during distillation is regulated by taking plant samples in the range of 0.1 to 2.0 g and by using 100 $\pm$ 10 mg of CaO and 3.0 $\pm$ 0.1 g of NaOH for ashing and fusion of each sample. Vacuum in the system is controlled with a flowmeter and a vacuum gauge. Any marked change in vacuum (greater than 0.7 kPa) over a short time period indicates either a leak or a block in the system. Distillation should take place at the same vacuum reading each day unless some other change in the system has been made. (See Section **7.3.4** for description of air flow system.)

2. Range and Sensitivity

2.1 In normal use, the procedure can detect F^- at 0.1 μg /mL. The normal range of analysis is from 0.1 to 1.6 μg /mL. Higher concentrations can be analyzed by careful dilution of samples with deionized water. If digested samples to be analyzed routinely exceed 1.6 μg /mL, the analytical portion of the pump manifold can be modified to reduce sensitivity. However, the best procedure is to analyze a smaller aliquot of the sample. The most accurate results are obtained when the F^- concentration falls in the middle part of the calibration curve.

3. Interferences

3.1 Since the air that is swept through the microdistillation unit is taken from the ambient atmosphere, airborne contaminants in the laboratory may contaminate samples. If this is a problem, a small drying bulb filled with calcium carbonate granules can be attached to the air inlet tube of the microdistillation unit.

3.2 If the Teflon distillation coil in the microdistillation unit is not cleaned periodically, particulate matter will accumulate and will reduce sensitivity.

3.3 Silicate, chloride, nitrate, and sulfate ions in high concentration can be distilled with fluoride ion and will interfere with the analysis by bleaching the alizarin fluorine blue-lanthanum reagent. Phosphate ion is not distilled and therefore does not interfere. Metals such as iron and aluminum are not distilled and will also not interfere with the analysis (most materials distilled over do not interfere with the potentiometric method). Maximum concentrations of several common anions at which there was no interference are given in Table 204:I. The sulfate concentration shown is the amount tolerated above the normal amount of sulfuric acid used in microdistillation.

4. Precision and Accuracy

4.1 GENERAL. It is essential that each laboratory occasionally perform the tests outlined below (7.3.8.1, 7.3.8.2, 7.3.8.3) to insure good results. The degree of accuracy and precision obtained in one laboratory may not be a fair representation for another laboratory, since much depends on the thoroughness with which any multi-step analytical procedure is carried out. The wide variation in results obtained in a cooperative study among 31 laboratories (6) emphasizes the importance of proper performance methods. The following information on accuracy and precision is meant, therefore, as a guide and is taken from a published report (7) in which results of analyses of tissues by the semi-automated and Willard-Winter (11,21) methods were compared. The paper should be consulted for details.

4.2 PRECISION. The standard error of a single determination is between 2 and 8 μg, depending on the kind of plant tissue used and the level of F⁻ present. With higher amounts of F, the standard deviation increases, although the coefficient of variation (standard deviation expressed as percent of F⁻ content) decreases. The coefficients of variation between 20 and 100 ppm F⁻ are generally 10% or less. Analysis of high silicate tissues (grasses) for F⁻ has always been a problem and standard deviations of results of replicate determinations on orchard grass tissues have been larger than when other tissues are analyzed. Recent modifications in procedures for transferring and distilling samples have improved analysis of high silicate samples.

4.3 ACCURACY. Since no direct means of determining accuracy are yet available, indirect means have been used. A total of 180 determinations of 4 tissues (Milo maize, gladiolus, alfalfa, and orchard grass) were performed. Within the limits of reproducibility of the results, multiple linear regression analysis indicated that no significant deviations from linearity were obtained when different amounts of F⁻ were added to a tissue. In addition, there were no significant second-order effects. Systematic

errors were not significant since the intercept values found (–3.97 to + 2.17 μg) were not significantly different from zero.

5. Apparatus

5.1 MULTICHANNEL PROPORTIONING PUMP with assorted pump tubes, nipple connectors, glass connectors, and manifold platter.

5.2 PULSE SUPPRESSORS (FOR COLORIMETRIC ANALYSIS). Pulse suppressors for the alizarin fluorine blue-lanthanum reagent stream are manufactured by Orion Scientific Instruments Corp., Part #116–6622–01. These polyethylene filter assemblies are both an effective reagent filter and a pulse suppressor. They should be discarded after one month of continuous use. The outlet ends of the suppressor tubes are forced into lengths of 0.081" ID silicone rubber tubing, which is then connected to the reagent pump tube and the other end then slipped over the "h" fitting which joins the sample and reagent streams.

5.3 AUTOMATIC SAMPLER. Sampler with plastic sample cups.

5.4 VOLTAGE STABILIZER.

5.5 COLORIMETER (FOR COLORIMETRIC ANALYSIS) with 15-mm tubular flow cell and 620 nm interference filters.

5.6 ION SELECTIVE ELECTRODE DETECTOR (FOR POTENTIOMETRIC ANALYSIS) with flow-through electrodes.

5.7 ROTARY VACUUM AND PRESSURE PUMP with continuous oiler.[1]

5.8 RECORDER.

5.9 RANGE EXPANDER (Optional).

5.10 MICRODISTILLATION APPARATUS. A schematic drawing is shown in Figure 204:1. Major components of the microdistillation apparatus include the following:

5.10.1 A 1,000-mL resin reaction flask with a conical flange and cover (Figure 204:1:A).[2]

5.10.2 Resin reaction flask clamp (Figure 204:1:B).[3]

5.10.3 Variable speed magnetic stirrer (Figure 204:1:D).[4]

---

[1] Gast #0211–V36–G 10 pump with AA 930 oiler.
[2] Fisher Scientific, Catalog #11–847B
[3] Fisher Scientific, Catalog #11–847–30A
[4] Fisher Scientific, Catalog #14–493–220T

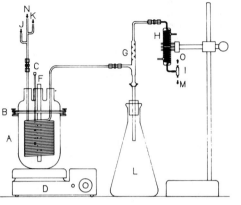

Figure 204:1 – Schematic drawing of semiautomatic microdistillation apparatus. See text.

5.10.4 Thermometer-thermoregulator, range 0–200°C (Figure 204:1:C).[5]

5.10.5 Electronic relay control box.[6]

5.10.6 Glo-Quartz immersion heater, 500 watts (Figure 204:1:F).[7]

5.10.7 Flexible Teflon TFE tubing, 3.2 mm ID, 4.8 mm OD, 0.8 mm wall. A 9.1 m length is coiled on a rigid support of such a diameter that the completed coil will fit into the resin reaction flask (**5.10.1**). Care must be taken to prevent kinking of the tubing (Figure 204:1:E).[8]

5.10.8 A flowmeter, with range of 0 to 5 L/min, with needle valve controls.

5.10.9 Vacuum gauge, with a range of 0 to 34 kPa.

5.10.10 Fractionation column, of borosilicate glass (Figure 204:1:G; also see Figure 204:5.).

5.10.11 Distillate Collector (Figure 204:1:I)[9]

5.10.12 Water-jacketed condenser (Figure 204:1:H).[10]

_____

[5] Thomas Scientific, Philadelphia, PA, Catalog #9366-K35

[6] Thomas Scientific, Philadelphia, PA, Catalog #9368-D80

[7] Thomas Scientific, Philadelphia, PA, Catalog #5932-B30

[8] Zeus Industrial Products, Inc., Raritan, NJ, Industrial specification tubing, Bulletin 7.

[9] Technicon Instrument Co., Tarrytown, N.Y., BO Electrolyte Trap.

[10] Technicon Instrument Co., Tarrytown, N.Y., Catalog #114-209.

5.10.13 Dow Corning Fluid (100 cs at 25°C).

5.11 Mechanical convection oven (Fisher Scientific 13-258-11A).

5.12 Thomas Wiley Cutting Mill (Fisher Scientific 08-338-1).

6. Reagents.

All reagents are certified ACS reagent grade unless otherwise specified. Water is ASTM reagent water type II.

6.1 REAGENTS FOR AUTOMATED FLUORIDE ANALYSIS.

6.1.1 _Sulfuric acid, 50%, v/v._ Mix 500 mL of concentrated sulfuric acid with 500 mL of water (_CAUTION – THIS MIXTURE CAN REACT VIOLENTLY. USE EYE PROTECTION AND DO IN FUME HOOD!_) Cool before use.

6.1.2 _Acetate buffer, pH 4.0._ Dissolve 60 g of sodium acetate trihydrate ($CH_3COONa.3H_2O$) in 500 mL of water. Add 100 mL of glacial acetic acid (CH_3COOH) and dilute to approximately 900 mL with water. Check the pH using the electrometer and adjust to pH 4.0 with NaOH or acetic acid. Dilute to final volume of 1 L with water.

6.1.3 _Acetone_

6.1.4 _tert-Butanol_

6.1.5 _Alizarin fluorine blue (alizarin complexone, 3-amino-ethylalizarin-N,N-diacetic acid) stock solution, 0.01 M._ Suspend 0.9634 g of alizarin fluorine blue[11] in about 100 mL of water in a clean 250-mL volumetric flask. Add 2 mL of concentrated ammonium hydroxide (NH_4OH) and shake until the dye has completely dissolved. Add 2 mL of glacial acetic acid. Dilute the solution to 250 mL volume with water and store at 4°C.

6.1.6 _Lanthanum nitrate stock solution, 0.01 M._ Dissolve 2.1652 g of lanthanum nitrate hexahydrate ($La(NO_3)_3$ · $6H_2O$ purified grade) in water in a 250-mL volumetric flask and dilute to volume with water.

_____

[11] Hopkins and Williams, Ltd., Chadwell Heath, Essex, England. Reagent also known as alizarin complexone, alizarine complexone, alizarin complexon, alizarin complexan.

6.1.7 *Brij-35[12] wetting agent, 30% w/ v.* Dissolve 30 g of Brij-35 in 100 mL of deionized water by heating over a hot plate.

6.1.8 *Working Reagent-Colorimetric.* The reagent is a modification of that reported by Yamamura, et al. **(23)**. Mix the following quantities of solutions in the order listed to make 1 L of working reagent: 300 mL of acetate buffer, 244 mL of water, 300 mL of acetone, 100 mL t-butanol, 36 mL of alizarin fluorine blue, 20 mL of lanthanum nitrate, and 40 drops of Brij-35. Unused working reagent is stable for at least 7 days. Just prior to using reagent, place under vacuum for 10 min. to remove air from solution.

6.1.9 *Diluent.* Mix 3.0 mL of standard fluoride solution, 100 ppm **(6.1.13)**, 0.5 mL Brij-35 **(6.1.7)** and water to make 1.0 L.

6.1.10 *TISAB buffer.* Add 57 mL glacial acetic acid, 58 g sodium chloride (NaCl) and 4.0 g CDTA ((1,2-cyclohexylenedinitrilo)-tetraacetic acid) to 500 mL water. Stir to dissolve and add 5 M sodium hydroxide slowly until pH is between 5.0 and 5.5. Cool and dilute to 1 L. This is also commercially available from Orion Research Inc. as cat. # 94909.

6.1.11 *Reference Electrode Reagent.* Add 1.0 mL of standard fluoride solution, 100 ppm **(6.1.13)** and 0.5 mL Brij-35 **(6.1.7)** to 1000 mL volumetric flask. Dilute to mark with TISAB **(6.1.10)**.

6.1.12 *ISA Reagent.* Add 2.0 mL Brij-35 solution **(6.1.7)** to 4.0 L of TISAB **(6.1.10)** and mix.

6.1.13 *Standard Fluoride Solutions.* Sodium fluoride (NaF) should be stored in a desiccator prior to use. Dissolve 0.2207 g of dry sodium fluoride in water and dilute to 1 L. The stock solution will contain 100 μg F^- per mL (100 ppm). Prepare working standards by taking suitable aliquots to give seven final concentrations of 0.2, 0.4, 0.6, 0.8, 1.0, 1.2, and 1.6 μg F^- per mL. Diluted working standards for plant analysis should contain 6 g of sodium hydroxide (NaOH) and 20 mL perchloric acid (70–72%, $HClO_4$) for each 100 mL of solution in order to compensate for the amounts of these substances used in alkali fusion of the ashed plant samples. Standard solutions for

fluoride analysis of water samples or of air samples absorbed in water are made up in water. Store stock solutions in clean polyethylene bottles in the cold. Since, as will be seen later, plant tissue samples are diluted to a 50 mL volume before analysis, the standard containing 0.20 μg fluoride/mL is equivalent to a sample of plant material containing 10.0 μg fluoride (0.2 μg/mL × 50 mL).

6.1.14 *Tetrasodium Ethylenediaminetetraacetate (Na₄EDTA Technical Grade) 1% (w/v).* Dissolve 1 g of Na_4EDTA in 99 mL of water.

6.1.15 *Plant Tissue Wash Solution.* Dissolve 0.5 g of Alconox (Detergent, Fisher Scientific 04–322-4) and 0.5 g sodium (tetra) ethylene diamine tetraacetate (Na_4EDTA, technical grade) in water to make 1000 mL.

6.2 REAGENTS FOR ASHING AND ALKALI FUSION OF PLANT SAMPLES.

6.2.1 *Calcium oxide (CaO), low in fluorine.*

6.2.2 *Sodium hydroxide pellets (NaOH).*

6.2.3 *Phenolphthalein solution, 1%.* Dissolve 1 g phenolphthalein in 50 mL of absolute ethanol. Add 50 mL of water and mix.

6.3 REAGENTS FOR TRANSFER OF ASHED AND FUSED SAMPLES.

6.3.1 *1:1 (v/v) perchloric acid-water solution.* Use concentrated perchloric acid (70–72% $HClO_4$).

6.4 PRECAUTIONS.

6.4.1 All containers used in the preparation of reagents are to be clean and rinsed with water prior to use.

6.4.2 When preparing solutions from concentrated reagents, such as acids and alkalies, all appropriate safety precautions should be taken.

6.4.3 All reagents except the concentrated acids and alkalies, and organic solvents may be stored in a refrigerator. The solutions should then be brought to room temperature before use.

7. Method

7.1 PREPARATION OF PLANT TISSUES FOR ANALYSIS.

7.1.1 *Purpose.* This procedure is to remove fluoride from surface tissues without

---

[12] Altas Powder, Co., Chemicals Division, Wilmington, Delaware.

altering the internal concentration of fluoride. Whether vegetation samples are to be washed will depend on the intended use of the population of plants the samples represent. For example, in forage crops or other vegetation intended for consumption by herbivores, fluoride on foliar surfaces as well as that within leaves is important and should be included in the analysis. For other kinds of vegetation, fluoride deposited on the surfaces of leaves may be unimportant with respect to the plant and it may be desirable to wash it from the surface prior to analysis.

7.1.2 *Criteria.* A standard washing procedure should meet several criteria: (1) it should be simple and gentle; (2) it should remove surface fluoride quantitatively with a minimum of leaching of internal fluoride; (3) it should not leave residues which might interfere with subsequent analysis; and (4) in the event that the tissue is to be analyzed also for nutrient status, it should not leach other internal elements. These criteria and methods are thoroughly discussed in the literature (1,2,3,4,5,7,9,12,13,14,16,17, 18,22).

7.1.3 *Procedure.* Fresh tissues are placed in a cheesecloth square, the ends folded up, and the whole washed in a polyethylene container filled with the plant tissue wash solution (6.1.15) for 30 sec with gentle agitation. The cheesecloth containing the tissue is removed and allowed to drain for a few seconds and is then rinsed for 10 sec in each of three containers of deionized water. The tissue is then placed in a plastic salad spinner and spun or blotted with dry paper towels to remove excess water.

Washed and partially dried tissue is then placed in a labelled Kraft paper bag and dried in a mechanical convection oven (5.11) at 80°C for no less than 24 h.

Grind dried tissues in a semi-micro Wiley mill (5.12) to pass a 40-mesh sieve. The sieved material is collected and placed in a labelled polyethylene container with a moisture-proof seal (Zip-lock or whirl-pak bag).

7.2 PROCEDURE FOR ASHING AND ALKALI FUSION OF PLANT TISSUES.

7.2.1 Mix the dried sample thoroughly and carefully weigh from 0.1 to 1.0 g of plant tissue, depending on the fluoride content, into a clean crucible.

7.2.2 Add 100 ± 10 mg of calcium oxide, sufficient deionized water to make a loose slurry, and 2 drops phenolphthalein. Mix thoroughly with a polyethylene policeman. The final mixture will be uniformly red in color and will remain red during evaporation to dryness.

7.2.3 Place crucibles on a hot plate and under infrared lamps. Turn on infrared lamps (do not turn on hot plate) until all liquid is evaporated. Turn on hot plate and char samples for 1 h.

7.2.4 Transfer crucibles to a muffle furnace at 600° C and ash for 2 h. **Caution:** To avoid flaming, place crucibles at front of muffle furnace with door open for about 5 min to further char samples. Crucibles may then be positioned in the furnace.

7.2.5 After ashing, remove crucibles (not more than eight at one time), add 3 ± 0.1 g of sodium hydroxide pellets, and replace in the furnace with door closed for 3 min. **Caution:** Watch for "creeping" of the molten NaOH. Remove crucibles one at a time and swirl to suspend all particulate matter until the melt is partially solidified. Allow crucible to cool until addition of small amount of water does not cause spattering. Wash down inner walls of crucible with 10 to 15 mL of water.

7.2.6 After crucibles have cooled to room temperature, suspend the melt with a polyethylene policeman and transfer to a 50 mL graduated plastic tube. Rinse crucible with 20.0 mL of a 1:1 (v/v) 70% perchloric acid-deionized water solution and add to the tube. Make sample to 50.0 mL/volume with deionized water.

7.2.7 Run several blank crucibles (about one blank for each 10 samples) containing all reagents through the entire procedure.

7.2.8 *Cleaning of Crucibles.* Crucibles should be cleaned as soon as possible after use. Inconel crucibles are soaked in 10% (w/v) sodium hydroxide solution overnight. Follow by washing with hot water and scouring with a soap-free steel wool pad and rinse three (3) times with water followed by three (3) times with deionized water. Crucibles are then immersed in 4 *N* HCl for 1.5 hours before rinsing three (3) times in tap water followed by three (3) times in deionized water.

7.3 PROCEDURE FOR AUTOMATED ANALYSIS OF ASHED AND FUSED SAMPLES.

7.3.1 *Distillation.* For the following description of the analytical system refer to Figure 204:1, 204:2, and 204:3. All flow rates given are nominal values. Standard fluoride solutions or ashed and alkali-fused samples are placed in 8.5-mL plastic cups in the sampler module. The sampler is actuated, and the sample is pumped from the cup at a net rate of 3.48 mL/min with air segmentation of 0.42 mL/min after the sampler crook (3.90 mL – 0.42 = 3.48 mL) and is pumped into the microdistillation device through the sample inlet J (Figure 204:1). Teflon or polyethylene tubing of 1.35 mm ID is employed for sample transmission. Sulfuric acid **(6.1.1)** is pumped at 2.50 mL/min through the acid inlet (K). Acid and ashed solids drop into the waste flask (L) and are discarded after the run. Distillate is pumped from the sample trap (I) at 2.50 mL/min through 1.35 mm ID

Teflon or polyethylene tubing (M), and air segmented with 0.42 mL/min air. Air enters the system at the inlet (N) and leaves the system at the top of the sample trap (O).

7.3.2 *Colorimetric Analysis.* A flow diagram of this automated procedure is shown in Figure 204:2. The sample stream is then resampled at 0.32 mL/min and the remainder of the sample stream goes to waste. The resampled stream is mixed with diluent reagent **(6.1.9)** at 0.80 mL/min and air segmented at 0.42 mL/min. The mixed stream passes through a Kel-F mixing coil and alizarin fluorine blue-lanthanum reagent **(6.1.8)** is added at 0.97 mL/min by pumping the reagent with a 1.20 mL/min silicone tubing and resampling through a 0.23 mL/min silicone tubing. The alizarin reagent should be passed through an inline filter prior to mixing with the sample stream. The two liquid streams are combined and passed through a second Kel-F coil for time delay, proper mixing, and color development. The reagent stream then passes through a debubbler fitting

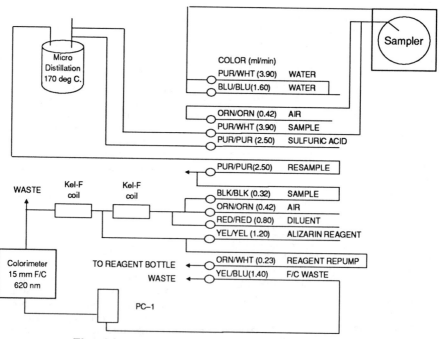

Fluoride in Plant Tissue – Colorimetric Analysis

Figure 204:2 – Block diagram of apparatus for semiautomatic colorimetric analysis.

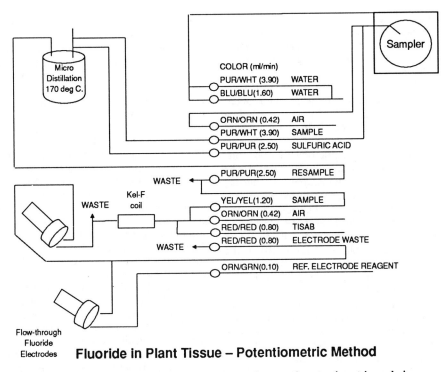

COLOR (ml/min)
PUR/WHT (3.90)   WATER
BLU/BLU(1.60)   WATER

ORN/ORN (0.42)   AIR
PUR/WHT (3.90)   SAMPLE
PUR/PUR (2.50)   SULFURIC ACID

PUR/PUR(2.50)   RESAMPLE

YEL/YEL(1.20)   SAMPLE
ORN/ORN (0.42)   AIR
RED/RED (0.80)   TISAB
RED/RED (0.80)   ELECTRODE WASTE

ORN/GRN(0.10)   REF. ELECTRODE REAGENT

Sampler

Micro Distillation 170 deg C.

WASTE

WASTE

Kel-F coil

WASTE

WASTE

Flow-through Fluoride Electrodes

Fluoride in Plant Tissue – Potentiometric Method

Figure 204:3 – Block diagram of apparatus for semiautomatic potentiometric analysis.

where a small portion of the sample stream is removed (along with any bubbles) and passes to waste. The remainder of the sample stream passes through a 15-mm tubular flow cell of the colorimeter, and the absorbance is measured at 620 nm. The sample is drawn through the flow cell and a glass pulse suppressor (PC-1, Technicon Corp.) with a silicone tubing at 1.40 mL/min. Results are plotted on a chart recorder. The lag time from sampling to the appearance of a peak on the chart recorder is about 5 min.

7.3.3 *Potentiometric Analysis.* A flow diagram of this automated procedure is shown in Figure 204:3. The sample stream (from **7.3.1**) is then resampled at 1.20 mL/min and the remainder of the sample stream goes to waste. The resampled stream is mixed with ISA reagent (**6.1.12**) (0.80 mL/min) and air segmented (0.42 mL/min). The mixed stream passes through a Kel-F mixing coil to insure proper mixing. The mixed sample stream then passes through a debubbler fitting and the air and

excess sample pass to waste. The debubbled sample stream then passes through the flow-through fluoride electrode assembly. The reference electrode reagent (**6.1.11**) is pumped at 0.10 mL/min and passed through the flow-through reference fluoride electrode. The waste from both the reference and sample electrode are resampled at 0.80 mL/min and pumped to waste.

7.3.4 *Air Flow System.* For a description of the air flow system refer to Figure 204:4. Air is drawn through the air inlet tube (a) before the Teflon microdistillation coil (b). Air sweeps through (b) to the fractionation column (c) where the air stream is diverted through the water-jacketed condenser (d) and sample trap (c) to waste bottle (f). The air then passes through a 3.2 mm ID glass tube directed against the surface of concentrated sulfuric acid contained in waste bottle (g). The partially dehydrated air passes through a gas-drying tower (h) containing 450 g of indicating silica gel. Air leaving the outlet of the drying tower passes through a T-tube (i) to which a vacuum gauge (j) (0 to 34 kPa) is connected, through a flowmeter (k) (0 to 5 L/min), and then to the vacuum pump (1). Flowmeter and vacuum gauge settings are described in **7.3.5.**

7.3.5 *Start-up Procedures.* Turn on water to condenser. Turn on colorimeter. Engage the manifold on the proportioning pump and start pump. Turn on the vacuum pump and set flow rates. Turn on the stirring motor on the microdistillation unit. Connect the lines to the sulfuric acid, the alizarin fluorine blue-lanthanum reagent, the diluent, and to the deionized water bottles. In the case of the potentiometric method, connect the lines to the TISAB and reference electrode reagent. The sampling tube of the sampler unit should be in the water reservoir. Allow the apparatus to equilibrate until the silicone oil in the microdistillation unit has reached 170°C. Be sure that all tubing connections are secure. Adjust flowmeter (Figure 204:4:k) to 4 L/min. Distillate should now fill the sample trap. Readjust flowmeter (Figure 204:4:k) to give a reading on the vacuum gauge of 17 to 20 kPa. Satisfactory setting for each instrument should be determined by trial and error. Once satisfactory value is determined, it is important that this setting be maintained each day. Turn on chart recorder, adjust the baseline to the desired level, and run a baseline for several minutes to assure that all components are operating properly. Transfer standard fluoride solutions to 8.5-mL plastic cups and place in sampler. Separate the last standard sample from unknown samples with one cup containing deionized water. Program the sampler for 20 samples/hr with 1:3 sample to wash ratio. Turn sampler on.

7.3.6 *Shut-down Procedures.* Turn off chart recorder. Disconnect the sulfuric acid line and place in deionized water. In the case of the colorimetric method, disconnect alizarin fluorine blue-lanthanum reagent line and place in 1% EDTA solution **(6.1.14)** for about 1 min. Transfer the line to deionized water and allow water to pass through the analytic system for about 5 min. or until no color remains in lines. Remove line from the diluent reagent and place in deionized water. In the case of the potentiometric method, disconnect the re-sample line and pump 10 mg/L fluoride into the sample stream for 5 minutes. Then clamp off waste lines before lifting the pump platen. Clean out Teflon distillation coil by allowing 10 to 20 mL of 4 N HCl to be drawn through the air inlet tube and then sucked through the coil. This is followed by a thorough rinse of deionized water. Turn off the heater and stirring motor in the microdistillation unit. Turn off the vacuum pump. Release pump tube manifold. Turn off water to the condensers.

7.3.7 *Maintenance.*

a. Pump tubes should be replaced after 200 working hours or prior to that if they become hard and inflexible or flattened.

b. Proportioning pumps should be oiled once a month and gain on the recorder checked and adjusted monthly.

c. All tubing containing reagent should be cleaned after each daily run with Na$_4$EDTa solution **(6.1.14)** followed by deionized water.

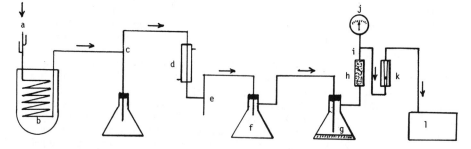

Figure 204:4 — Schematic drawing of air flow system used in semiautomatic anaylsis of fluoride in vegetation.

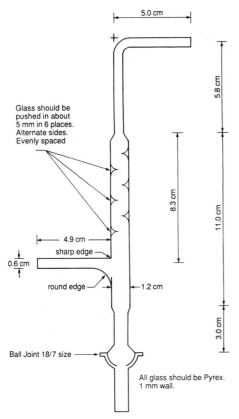

5.0 cm

5.8 cm

Glass should be
pushed in about
5 mm in 6 places.
Alternate sides.
Evenly spaced

8.3 cm

11.0 cm

4.9 cm

sharp edge

0.6 cm

round edge

1.2 cm

3.0 cm

Ball Joint 18/7 size

All glass should be Pyrex.
1 mm wall.

Figure 204:5 – Microdistillation column.

d. Pump tubes are always left in the relaxed position when not in use.

7.3.8 *Check Procedures.*

a. An estimate of the degree of fluoride contamination from reagents and equipment should always be made. Due to the ubiquity of fluoride, crucibles without sample but with all reagents should be carried through the entire procedure. Contamination from previously analyzed samples, from a contaminated muffle furnace, and from reagents can then be detected and corrective efforts made. Blank values over 5 μg F^- are considered evidence of contamination. Two blank determinations should be made with every batch of samples analyzed (approximately 2 per 40 samples). Blank values are usually found to range from 1 to 3 μg.

b. Occasionally a special calibration curve should be made by adding known amounts of sodium fluoride solution from a microburet to aliquots of a low F tissue. Recovery of added F^- should be 100 ± 10%. Low values indicate loss of F, possibly during pretreatment, and high values indicate contamination.

c. Different amounts (0.1 to 1.0 g) of a plant sample containing 50 to 65 ppm F^- should be analyzed occasionally. A linear relationship should exist between F found and amount of tissue taken. A nonlinear relationship may indicate that some component of the tissue is retarding distillation or interfering with color development.

d. Calibration curves should be repeated at least twice daily to correct for any small changes in distillation efficiency that might occur.

e. If the sample to sample absorbance ratio exceeds 5:1, then the lower sample should be repeated with a water sample preceding.

7.3.9 *Trouble Shooting.*

a. Since no method for determining μg quantities of an element in an overwhelming excess of other materials is free of occasional problems, suggestions on how to recognize the difficulties and systematically locate the problems may be of value. Fortunately, most of the potential problems in distillation and F analysis are manifested by obvious irregularities on the chart recorder.

b. Irregular fluctuations in the baseline may result from the following: (1) excessive surge pressures in the liquid streams; (2) air bubbles passing through the flow cell in the photometer; or (3) bleaching of the reagent by excess sulfuric acid carryover during distillation or insufficient buffer in the reagent. Excessive pulse pressures may be due to faulty pump tubes, the absence of surge suppressors, or the presence of pulse suppressors that are improperly placed. Air bubbles in the photometer flow cell may be due to the absence of a debubbler bypass, a blockage in the reagent pump tube, or a periodic emptying of the sample trap. The last condition will result if the air flow to the distillation trap becomes too great. Excessive sulfuric acid carryover can be caused by too high a temperature in the oil bath, improper sulfuric acid concentration, or too high a vacuum on the system. Large fluctuations or imbalances in vacuum or air flow rates in the distillation or waste systems will also produce baseline irregularities. An improper flowmeter setting, trapped air in the tubing, or a leak or block in the system should be sought as the probable cause of this type of difficulty.

c. Asymmetrical peaks, double peaks or peaks with shoulders may result from: (1) baseline irregularities; (2) interfering substances from the sample or impure reagents; (3) inadequate buffer concentration; or (4) excessive amounts of solid material in the Teflon distillation coil. The presence or accumulation of excessive solids may be due to insufficient flow of sulfuric acid, too large a sample, excessive amounts of calcium oxide or sodium hydroxide in the sample, inadequate suspension of particles in the samples, or lack of proper air segmentation in the sample tubing.

d. Poor reproducibility can be caused by improper sampling; by faulty pump tubes; by inadequate washout of Teflon distillation coils between samples; by large deviations in acid concentration, temperature or air flow in the distillation coil; or by changes in vacuum on the waste systems.

7.4 CALIBRATION.

7.4.1 Transfer two aliquots of each calibration fluoride solution (0.2, 0.4, 0.6, 0.8, 1.0, 1.2, 1.4, and 1.6 μg F/mL) to 8.5-mL sample cups and randomly distribute throughout the sample tray.

7.4.2 Proceed with analysis as described in 7.3.

7.4.3 After the calibration solutions have been analyzed and the peaks plotted by the chart recorder, draw a straight line connecting the baseline before and after the analysis. Record the absorbance of each peak and subtract the absorbance of the baseline. Compute the regression of net absorbance vs. μg/ml. by the method of least squares.

7.5 CALCULATION.

7.5.1 The fluoride content of the sample in ppm is calculated as follows:

$$F_T = \frac{F_S V_S D}{W_S}$$

where F_T = ppm of fluoride in sample
F_S = concentration of fluoride, μg/mL, of unknown sample

as calculated from the regression
V_S = volume, mL of the unknown sample (usually 50 mL)
D = dilution factor used when F in unknown sample exceeds the standard curve. For example, if the original sample was diluted from 50 mL to 100 mL, D will equal 2. (Note: All dilutions of plant samples should be made with 3 N sodium hydroxide.) If the unknown sample is not diluted, D should be dropped from calculations.
W_S = weight of sample taken for analysis, g.

7.6 EFFECT OF STORAGE.

7.6.1 The acetate buffer (6.1.2), Brij 35 (6.1.7) solution, and TISAB buffer (6.1.10) are stable at room temperature. Stock solutions of alizarin fluorine blue (6.1.5) and lanthanum nitrate (6.1.6) are stable indefinitely at 4°C. The alizarin fluorine blue-lanthanum working reagent (6.1.8) is stable at 4°C for at least 7 days. Dilute NaF solutions (6.1.13) should be stored in the cold in polyethylene bottles and are stable in the presence of sodium hydroxide. Tightly covered ashed and fused plant samples appear to be stable indefinitely.

8. References

1. ARKLEY, T.H., D.H. MUNNS AND C.M. JOHNSON. 1960. Preparation of plant tissues for micronutrient analysis. Removal of dust and spray contaminants. *J. Agr. Food Chem.* 8:318–321.
2. BOYNTON, D., J.C. CAIN AND O.C. COMPTON. 1944. Soil and season influences on the chemical composition of McIntosh apple leaves in New York. *Proc. Am. Soc. Hort. Sci.* 44:5–24.
3. CAMERON, S.H., R.T. MUELLER, A. WALLACE AND E. SARTORI. 1952. Influence of age of leaf, season of growth, and fruit production on the size and inorganic composition of Valencia orange leaves. *Proc. Am. Soc. Hort. Sci.* 60:42–50.
4. CHAPMAN, H.D. AND S.M. BROWN. 1950. Analysis of orange leaves for diagnosing nutrient status with reference to potassium. *Hilgardia* 19:501–539.
5. GRANT, CLARENCE L. 1961. Personal communication.
6. JACOBSON, J.S. AND D.C. MCCUNE. 1969. An interlaboratory study of analytical techniques for fluoride in vegetation. *J. Assoc. Offic. Anal. Chemists.* 52:894–899.
7. JACOBSON, J.S., D.C. MCCUNE, L.H. WEIN-

STEIN, R.H. MANDL AND A.E. HITCHCOCK. 1966. Studies on the measurement of F in air and plant tissues by the Willard-Winter and semi-automated methods. *J. Air Pollut. Contr. Assoc* 16(7):367-371.

8. JACOBSON, L. 1945. Iron in the leaves and chloroplasts of some plants in relation to their chlorophyll content. *Plant Physiol.* 20:233-245.

9. JACOBSON, L. AND J.J. OERTLI. 1956. The relation between iron and chlorophyll contents in chlorotic sunflower leaves. *Plant Physiol.* 31:199-204.

10. MANDL, R.H., L.H. WEINSTEIN, J.S. JACOBSON, D.C. MCCUNE AND A.E. HITCHCOCK. 1966. Simplified semi-automated analysis of fluoride. Proc. Technicon symposium "Automation in Analytical Chemistry." New York, Sept. 8, 1965, pp. 270-273.

11. MAVRODINEANU, R., J. GWIRTSMAN, D.C. MCCUNE AND C.A. PORTER. 1962. Summary of procedures used in the controlled fumigation of plants with volatile fluorides and in the determination of fluorides in air, water, and plant tissues. *Contrib. Boyce Thompson Inst.* 21:453-464.

12. NICOLAS, D.J.D., C.P. LLOYD-JONES AND D.J. FISHER. 1956. Some problems associated with determining iron in plants. *Nature* 177:336-337.

13. _____, 1957. Some problems associated with determining iron in plants. *Plant Soil* 8:367-377.

14. PARBERRY, N.H. 1935. Mineral constituents in relation to chlorosis of orange leaves. *Soil Science* 39:35-45.

15. SKEGGS, L.T., JR. 1957. An automatic method for colorimetric analyses. *Am. J. Clin. Pathol.* 28:311-322.

16. STEYN, W.J.A. 1959. Leaf analysis. Errors involved in the preparative phase. *J. Agr. Food Chem.* 7:344-348.

17. TAYLOR, G.A. 1956. The effectiveness of five cleaning procedures in the preparation of apple leaf samples for analysis. *Proc. Am. Soc. Hort. Sci.* 67:5-9.

18. THORNE, D.W. AND A. WALLACE. 1944. Some factors affecting chlorosis on high-lime soils. I. Ferrous and ferric iron. *Soil Science* 57:299-312.

19. WEINSTEIN, L.H., R.H. MANDL, D.C. MCCUNE, J.S. JACOBSON AND A.E. HITCHCOCK. 1963. A semiautomated method for the determination of fluorine in air and plant tissues. *Contrib. Boyce Thompson Inst.* 22:207-220.

20. WEINSTEIN, L.H., R.H. MANDL, D.C. MCCUNE, J.S. JACOBSON AND A.E. HITCHCOCK. 1965. Semi-automated analysis of fluoride in biological materials. *J. Air Pollut. Contr. Assoc.* 15:222-225.

21. WILLARD, H.H. AND O.B. WINTER. 1933. Volumetric method for determination of fluorine. *Ind. Eng. Chem. Anal. Ed.* 5:7-10.

22. VANSELOW, A.P. AND G.R. BRADFORD. 1961. Spectrographic Techniques. p. 212 *In*: Chapman, H.E. and P.F. Pratt. Methods of Analysis for Soils, Plants and Water. Univ. Calif. Div. of Agr. Sciences.

23. YAMAMURA, S.S., M.A. WADE AND J.H. SIKES. 1962. Direct spectrophotometric fluoride determination. *Anal. Chem.* 34:1308-1312.

Subcommittee 2

R.H. MANDL, *Chairman*
L.H. WEINSTEIN
J.S. JACOBSON

205.

Determination of Fluoride Content of Plant Tissues (Potentiometric Method)

1. Principle of the Method

1.1 The current reference method for the determination of fluorine in plant tissue involves high temperature ashing and alkaline fusion with the subsequent release of fluorine by microdistillation and either colorimetric or potentiometric analysis (1,2). These methods are complex and time-consuming. This method employs a simple extraction followed by direct analysis with the fluoride specific-ion electrode and electrometer (3). The results are comparable to the reference method although there has been no collaborative study done between referee laboratories. Plant tissue is prepared in a manner similar to the reference method, drying in a forced-draft oven, and grinding to a uniform size. The dried, ground tissue is then weighed into a plastic tube, a solution of perchloric acid is added, and the mixture is heated and shaken. The fluoride content of the solution is then measured directly in the tube by insertion of the specific-ion electrode into the stirred solution and reading the output of the electrometer.

2. Range and Sensitivity

2.1 The range of fluoride measurement of the electrode is 0.019 to 19,000 μg/mL of solution (10^{-6} to 1 M), (4); however, the recommended range of analysis for plant samples is between 0.03 and 10 μg/mL in solution. Slow response time and non-linearity of the calibration curve make fluoride measurements below 0.1 μg/mL less desirable. The minimum concentration that can be measured is 2.30 μg F/g plant tissue.

3. Interferences

3.1 The electrode measures the free fluoride ion (fluoride activity); therefore, any substances that complex the fluoride or conditions that reduce the dissociation of fluoride will produce errors. Any substance that coats or reacts with the electrode crystal or otherwise slows the response time to fluoride ion will produce analytical errors. Hydroxyl ions are measured by the electrode when the OH$^-$ to F$^-$ concentration ratio is in excess of ten to one. A high concentration of dissolved salts depresses fluoride ion activity; therefore, differences in total ionic strength between samples and standard fluoride solutions can be a source of error. There was no significant change in electrode response when standard 1 ppm solutions of fluoride were amended with 4 ppm of Cu^{+2}, Ni^{+2}, Zn^{+2}, Cd^{+2}, Pb^{+2}, Fe^{+2}, Mn^{+2}, or Al^{+3}. However, solutions containing Al^{+3} changed after standing overnight. Differences in temperature greater than 2°C between samples and standards can cause errors, particularly at low fluoride concentrations.

3.2 Analysis should be made in acid-washed polyethylene, Teflon, or polypropylene vessels. Glass surfaces should be avoided, especially with fluoride concentrations below 20 ppm.

4. Precision and Accuracy

4.1 The electrode method of fluoride analysis of plant samples was found to compare favorably with the reference method (3). The results from 75 replicates of a standard sample produced a relative standard deviation of 4.13%. Corresponding values obtained by the method of standard addition to the same sample solutions produced a standard deviation of 5.29%.

The mean recovery from 28 different samples when compared to the reference method produced a mean of 99.6% with a standard deviation of 5.57%.

5. Apparatus

5.1 A combination fluoride ion electrode (Orion 960900) or separate fluoride ion and reference electrodes (Orion 940900 and 900100 respectively).

5.2 An electrometer or an expanded scale pH meter with a mV scale for measurement of potential equivalent to the Orion model 901.

5.3 A compatible recorder that can be attached to the electrometer for obtaining a permanent record of analytical results.

5.4 Teflon-coated magnetic stirring bars and air-driven magnetic stirrers. The latter device avoids the heating effect of motor-driven stirrers.

5.5 Repeating dispenser, 25 mL capacity (Fisher Scientific, 13–688–71).

5.6 Polypropylene tubes, 50 mL capacity, graduated (Corning 25335).

5.7 Water bath with shaker (Fisher Scientific 15–450–212).

5.8 Epoxy-coated tube rack (Fisher Scientific 14–793–3).

5.9 Mechanical convection oven (Fisher Scientific 13–258–11A).

5.10 Thomas Wiley Cutting Mill (Fisher Scientific 08–338–1).

6. Reagents

6.1 All reagents should be ACS analytical grade. Water must meet the standards of ASTM Reagent Water Type II.

6.2 Perchloric acid ($HClO_4$, ACS reagent grade) 0.1 N. Dilute 8.6 mL of 70% perchloric acid, specific gravity 1.66, with water to 1000 mL.

6.3 Stock solution, 100 μg/mL F^-. Dissolve 0.222 g of dry sodium fluoride (NaF, ACS reagent grade) in water to make 1000 mL. The stock solution is stored in a plastic bottle in the cold and allowed to come to room temperature before use.

6.4 Standard fluoride, 0.2 μg F mL^{-1}. Take 2.0 mL of stock fluoride (**6.3**) and 8.6 mL of 70% perchloric acid and dilute to 1000 mL with water.

6.5 Standard fluoride, 1.0 μg F mL^{-1}. Take 10.0 mL of stock fluoride (**6.3**) and 8.6 mL of 70% perchloric acid and dilute to 1000 mL with water.

6.6 Standard fluoride, 5.0 μg F mL^{-1}. Take 50 mL of stock fluoride (**6.3**) and 8.6 mL of 70% perchloric acid and dilute to 1000 mL with water.

6.7 Plant tissue wash solution. Dissolve 0.5 g of Alconox (Detergent, Fisher Scientific 04–322–4) and 0.5 g sodium (tetra) ethylenediamine tetraacetate (Na$_4$EDTA, technical grade) in water to make 1000 mL.

7. Procedure

7.1 If plant tissue is to be washed, place fresh tissue in a cheesecloth square, fold the ends up, and wash in a polyethylene container with the plant tissue wash solution for 30 s with gentle agitation. Remove the cheesecloth containing the tissue and allow to drain for a few seconds, then rinse for 10 s in each of three containers of water. Spin the tissue in a plastic salad spinner or blot dry in paper towels to remove excess water.

Place the tissue in a labelled Kraft paper bag and dry in a mechanical convection oven at 80°C for no less than 24 h.

Grind dried tissues in a Thomas Wiley mill to pass a 40-mesh sieve. Collect the sieved material in a labelled polyethylene container with a moisture-proof seal (Ziplock or whirl-pak bag).

7.2 Weigh 0.5 g of well-mixed sample into a polypropylene tube and add 25 mL of 0.1 N perchloric acid (**6.2**). Close the tube securely with a screw cap and place in test tube rack. Prepare up to three racks of tubes in this manner containing samples, two controls, and two blanks. Place the racks in the water bath and shake at 80°C for 4 hours. Add an additional 25 mL of 0.1 N perchloric acid (**6.2**) to each of the tubes. Add a tapered tip magnetic stirring bar to each tube and place on a magnetic stirrer. Lower the specific-ion electrode (or pair of electrodes) into the solution. Allow the contents to stir thoroughly and note the reading (or record) after it has stabilized to within $\pm$ 1 digit.

8. Precautions

8.1 Proper electrode response is essential to accurate measurements. Nonlinearity of the calibration curve (as determined by linear regression using the method of least squares), poor reproducibility of replicate analyses (more than two standard deviations) or slow response time (in excess of 2 minutes) are indications that the electrode may not be operating properly. To check the electrometer, substitute a pH electrode for the fluoride electrode. If pH measurements (on expanded scale) of standard buffer solutions are satisfactory, faulty operation of the fluoride electrode is indicated.

8.2 If response of the electrode is slow, the reference solution in the reference electrode or the combination electrode should be drained and replaced. If operation is not improved, the end of the fluoride electrode can be dipped briefly in absolute methanol followed by 0.1 N HCl. After rinsing with deionized water the electrode is ready for use. This procedure should be used infrequently to avoid an effect of solvents on the plastic housing of the electrode.

9. Calibration and Standards

9.1 For calibration of the electrode(s) and electrometer, follow the instruction given in the electrometer manual using 0.2, 1.0, and 5.0 μg F mL^{-1} solutions. The instrument used reads fluoride concentration directly in μg mL^{-1} in the concentration mode. The program assumes Nernstian response of the electrode and a temperature of 20°C.

10. Calculation of Results

10.1 The following equation is used to calculate the fluoride content of the dried plant tissue:

$$F(\text{as } \mu\text{g g}^{-1}) = \mu\text{g mL}^{-1} \times 100$$

11. Effects of Storage

11.1 The 0.1 N perchloric acid and stock solutions of fluoride are stable under refrigeration.

11.2 For overnight storage, place electrode(s) in a solution containing 1 μg/mL fluoride and no salts or buffer. For long periods of storage, the fluoride electrode may be kept dry after rinsing exterior with distilled water. The reference electrode should be capped or drained.

12. References

1. MANDL, R.H., L.H. WEINSTEIN, J.S. JACOBSON, D.C. MCCUNE AND A.E. HITCHCOCK. 1966. Simplified semi-automated analysis of fluoride. Proc. Technicon Symposium "Automation in Analytical Chemistry." New York, Sept. 8, 1965, 270–273.
2. Association of Official Analytical Chemists. 1984. "Official Methods Of Analysis," fourteenth edition, 49–53.
3. VIJAN, P.N. AND B. ALDER. 1984. Determination of fluoride in vegetation by ion-selective electrode. *American Laboratory* 16(12):16–24.
4. FRANT, M.S. AND J.W. ROSS, JR. 1966. Electrode for sensing fluoride ion activity in solution. *Science* 514:1533–1555.

Subcommittee 2

RICHARD H. MANDL, *Chairman*
L. H. WEINSTEIN
J. S. JACOBSON

206.

Determination of Gaseous and Particulate Fluorides in the Atmosphere (Separation and Collection with Sodium Bicarbonate Coated Glass Tube and Particulate Filter)

1. Principle of the Method

1.1 Gaseous fluorides are removed from the air stream by reaction with sodium bicarbonate coated on the inside wall of a pyrex glass tube. Particulate fluorides are collected on a filter following the tube. The fluoride collected by the coating on the inside of the tube is removed with water or buffer and analyzed for fluoride. The particles collected by the filter are eluted with acid and analyzed for fluoride. The results are reported as μg of gaseous fluoride and μg of particulate fluoride/m³ of air at 25°C and 101.3 kPa (1,2).

1.2 Since the samples are collected on the dry tube and filter, the fluoride may be eluted with a small volume of eluent, thus concentrating the collected fluoride. This allows the determination of the collected fluoride down to fractional parts of a μg/m³.

2. Range and Sensitivity

2.1 The bicarbonate-coated tube method can be used to collect from 1 to 500 μg gaseous fluoride at a sampling rate of 0.5 CFM (14.3 L/min). The length of the sampling period can therefore be adjusted so that the amount of fluoride collected will fall within this range. For a 12-h sampling period, the detectable atmospheric concentration would be from about 0.1 to 50 μg F/m³. The actual lower limit of the method will depend upon the sensitivity of the analytical method employed and the quality of reagents used in tube preparation and analysis. It is recommended that the lower limit of detection should be considered as 2 times

the standard deviation of the monthly mean blank value. Any values greater than the blank by less than this amount should be reported as "blank value."

3. Interferences

3.1 Significant amounts of acid aerosols or gases might neutralize the sodium bicarbonate coating and prevent quantitative uptake of gaseous fluoride from the atmosphere.

3.2 The presence of large amounts of aluminum or certain other metals, or of phosphates can interfere with subsequent analyses by colorimetric or electrometric methods. This is a problem inherent with any collection method for fluoride.

4. Precision and Accuracy

4.1 PRECISION. The root mean square difference of duplicate bicarbonate-coated tubes within the range of 0.5 to 3.3 μg F/m³ is 0.051 μg F/m³.

4.2 ACCURACY. Recovery of known amounts of gaseous HF was better than 95% with amounts of F up to about 40 μg and at sampling periods of 15 to 120 min. Data on particulate F are not sufficient to establish recovery under field conditions.

5. Apparatus

5.1 GLASS TUBING. 122 cm lengths of 7 mm ID pyrex tubing. The tubing will be coated with sodium bicarbonate according to the requirements outlined in **5.8**.

5.2 FILTER AND HOLDER. The tubing will be followed by a filter holder and filter for the collection of particles for particulate fluoride analysis (Figure 206:1). Nuclepore (1 μm pore size), Acropor AN-800, and Whatman Nos. 42 and 52 (47 mm filter) have been tested and found satisfactory for this application.

5.3 The tube and filter are followed by an air sampling system which is capable of sampling at a rate of 14.3 L/min and measuring the total air sampled either on a time rate basis or with a totalizing meter.

5.4 The system should be equipped so that pressure and temperature of the gas at the point of metering also are known for correcting sample volumes to standard conditions of 101.3 kPa and 25°C. Relative humidity of the gases sampled has not been considered as a required part of the data.

5.5 The sampling system should be assembled so that the inlet of the tube is 4–6 m above ground level and protected from rain in such a manner as not to interfere with the free passage of aerosol fluorides.

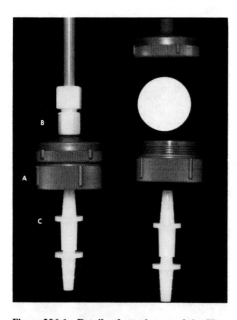

Figure 206:1 – Details of attachment of the filter assembly and limiting orifice to a bicarbonate-coated tube (7 mm. I.D.). (A) Polypropyene filter holder, (B) plastic female connector, (C) limiting orifice.

5.6 A 30-watt heating element should be installed to condition the gases to a point where condensation will not occur.

5.7 CONFIGURATION OF SAMPLING EQUIPMENT.

5.7.1 Figure 206:2 is a sketch of a suggested sampling system.

5.7.2 Other equivalent systems which meet the requirements outlined also would be satisfactory.

5.8 CRITERIA FOR COATING OF THE PYREX TUBES. Apparatus for cleaning, coating and drying the tubes may vary but the following requirements should be met.

5.8.1 The coating should be a visible uniform coating on the full length of the tube.

5.8.2 The coating should not contain any large crystals or heavy local deposits that could flake off and be collected with the aerosol fluorides.

5.8.3 The total coating should contain less than 1 μg of fluoride when analyzed without exposure, including all the reagents used in the procedure.

5.8.4 Prepared tubes should be sealed until time of use. Parafilm, serum tube caps, or other materials may be used for this purpose.

5.8.5 Any materials that may come into contact with the tubes should be checked for freedom from contamination with fluoride.

6. Reagents for Cleaning and Coating the Pyrex Tubes

6.1 All reagents should be ACS analytical grade. Water should be ASTM reagent water, Type II.

6.2 A detergent solution low in fluoride and phosphate is used for initial cleaning of the tubes.

6.3 ALCOHOLIC KOH, 10% W/V. Prepare a solution of 10% by weight of KOH in methanol by dissolving 100 g of KOH in methanol and making the volume to 1 L. Mix thoroughly.

6.4 SODIUM BICARBONATE, 5% W/V. Prepare a solution of 5% $NaHCO_3$ by dissolving 50 g of $NaHCO_3$ in water and making the volume to 1 L. Mix thoroughly.

Bicarbonate-coated Tube for Separation and Collection of Gaseous and Particulate Fluorides

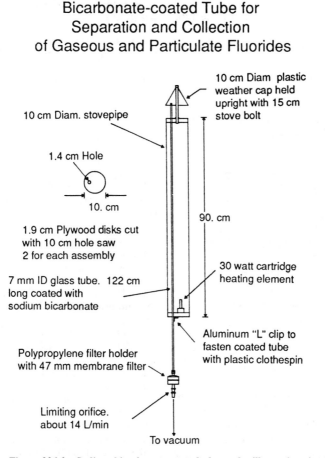

Figure 206:2 – Sodium bicarbonate-coated glass tube illustrating simple heating device.

6.5 A wetting agent should be used in the sodium bicarbonate solution to promote even wetting of the tube.*

6.6 TOTAL IONIC STRENGTH ADJUSTMENT BUFFER (TISAB). Add 57 mL glacial acetic acid, 58 g sodium chloride and 4.0 g CDTA ((1,2-cyclohexylenedinitrilo)–tetraacetic acid), to 500 mL distilled water. Stir and add 5 M sodium hydroxide slowly until pH is between 5.0 and 5.5. Cool and dilute to 1 L.

6.7 1.0 N sulfuric acid.

6.8 1.0 N sodium hydroxide.

7. Procedure

7.1 COATING THE PYREX TUBES.

7.1.1 Clean the tubes successively with detergent, alcoholic KOH solution and water.

---

*1 mL of Brij-35 (1 : 1 dilution, Atlas Powder Co., Chemicals Division, Wilmington, Delaware) may be used for each 100 mL of coating solution, or equivalent wetting agent from any other source.

7.1.2 While still wet from the cleaning, wet the internal surface of the tube with the 5% $NaHCO_3$ solution.

7.1.3 Allow the tube to drain for about 10 s and dry the coating rapidly by passing hot, dry fluoride-free air downward through the tube, hanging in a vertical position.

7.1.4 The hot fluoride-free air stream can be provided by blowing air through a 4 to 8 mesh soda lime trap and then through four feet of coiled copper tubing heated by a small gas burner or heating tape. To simplify the drying, the hot air stream can be run through a manifold terminating in several outlet ports. The flow rate through the system should be in the order of about 3 L/min/tube, and the drying should be complete in about 1 min.

7.1.5 After the tubes are dry, seal the ends and store in a clean area until used.

7.2 PREPARATION OF FILTERS.

7.2.1 The filters and filter holders should be assembled in the laboratory and sealed if not used immediately.

7.2.2 Prior to taking the tubes and filter holders to the sampling site, the tubes and filter holders should be assembled into a sampling unit. The ends should be kept sealed until installed at the sampling site.

7.3 AIR SAMPLING PROCEDURE.

7.3.1 Using this method, samples should be taken for 12-h periods to provide sufficient fluoride for accurate measurement.

7.3.2 It is recommended that sampling cover day and night conditions.

7.3.3 The sample should be taken at about 14.3 L/min (0.5 CFM) using a calibrated limiting orifice or other suitable device to control the flow at this rate (Figure 206:3). A totalizing gas meter is recommended to measure total sample volume. Pressure drop and temperature at the meter should be recorded at the beginning and end of each sampling period.

7.3.4 At the end of the sampling period, the tube and filter assemblies are capped and returned to the laboratory for analysis.

8. Preparation of Samples for F Analysis

8.1 In a laboratory free of contamination by process-generated fluorides, the tubes and filter assemblies are separated and each analyzed for total fluoride content.

8.2 PREPARATION OF TUBES FOR F ANALYSIS.

8.2.1 *For Potentiometric Analysis.*

a. With the tube in a vertical position and the lower end capped, pipet in 5.0 mL of $1/2$-strength TISAB buffer.

b. Gently agitate the tube to wet all surfaces and empty the tube into a clean plastic beaker.

c. For details of potentiometric analyses refer to Method 205.

8.3 PREPARATION OF PARTICULATE FILTERS FOR F ANALYSIS.

8.3.1 *For Potentiometric Analysis.*

a. Place particulate filters in clean 15 × 150 mm test tubes. Add 5.0 mL 1 N H_2SO_4 and mix for several seconds with a vortex mixer and allow to stand for 5 min.

b. Add 5.0 mL 1 N NaOH and 10.0 mL TISAB.

c. Analyze using potentiometric method, Method 205.

8.3.2 *For Semiautomated Analysis.*

a. Place the particulate filter in a clean test tube, pipet in 5.0 mL of 1 N H_2SO_4, mix for several seconds with a vortex mixer and allow to stand for 5 min. Filter samples to remove cellulose fibers into 8.5 mL sample cups.

b. Analyze the sample using the semiautomated method as described in Method 204, Section **7.3**. Standards for semiautomated analysis should be made up in 1 N H_2SO_4.

9. Calculation of Results

9.1 The fluoride concentration of the atmosphere is calculated by the following relationship:

$$\text{Fluoride, } \mu g/m^3 = \frac{(A - B)}{C}$$

where A = number of micrograms of F in the sample.
B = number of micrograms of F in the blank.

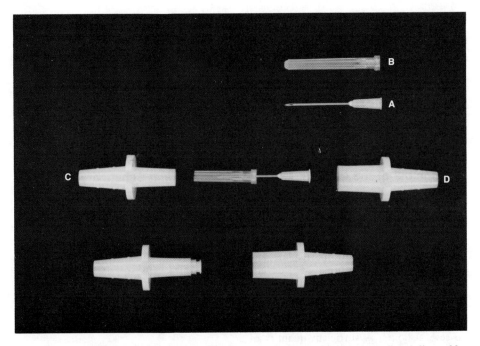

Figure 206:3 — Details of construction of a limiting orifice with "quick disconnects" and disposable hypodermic needles. (A) Disposable needle, (B) needle sheath, (C) male end of "quick-disconnect," (D) female end of "quick-disconnect."

C = volume of air sample in cubic meters at 101.3 kPa and 25°C.

10. Effect of Storage

10.1 Coated tubes and filter holders can be sealed and stored indefinitely in a clean, fluoride-free area, prior to analysis.

11. References

1. MANDL, R.H., L.H. WEINSTEIN, G.J. WEISKOPF, AND J.L. MAJOR. 1970. The separation and collection of gaseous and particulate fluorides. Paper CP-25A. 2nd Internat. Clean Air Congress, Washington, D.C., December 6–11.

2. WEINSTEIN, L.H. AND R.H. MANDL. 1971. The separation and collection of gaseous and particulate fluorides. VDI Berichte Nr. 164:53–63.

Subcommittee 2

C.R. THOMPSON, *Chairman*
G.H. FARRAH
L.V. HAFF
W.S. HILLMAN
A.W. HOOK
E.J. SCHNEIDER
J.D. STRAUTHER
L.H. WEINSTEIN

Approved with modifications
from 2nd edition
Subcommittee 2
R.H. MANDL, *Chairman*
L.H. WEINSTEIN
J.S. JACOBSON

207.

Determination of Gaseous and Particulate Fluorides in the Atmosphere (Separation and Collection with a Double Paper Tape Sampler)

1. Principle of the Method

1.1 The double paper tape sampler provides a method for the automatic separation and collection on chemically-treated paper tapes of inorganic particulate and gaseous atmospheric forms of fluoride. The device may be programmed to collect and store particulate and gaseous fluoride from individual air samples obtained over time periods ranging from several minutes to several hours. A sufficient quantity of paper tapes will allow unattended operation for the automatic collection of up to 600 samples.

1.2 Air is drawn through an air inlet tube and is first passed through an acid-treated prefilter paper tape to remove particulate fluoride and then through an alkali-treated paper tape to remove gaseous fluorides.

1.3 The exhaust air is filtered through soda lime and glass wool and the fluoride-free air is used to pressurize the front compartment, preventing fluoride contamination of the paper tapes from the ambient air.

1.4 At the end of the preset sampling period, the vacuum pump is turned off and the tapes are indexed. After indexing, the vacuum pump is again activated. Indexing results in a "dead time" of several seconds.

1.5 The individual sample spots are cut out, treated with an ionic strength adjustment buffer solution to dissolve fluoride, and the sample solution is analyzed by potentiometric or colorimetric methods **(1,2)**.

2. Range and Sensitivity

2.1 The lower limit of detection of either gaseous or particulate fluoride will depend upon the length of the sampling period selected. For a 1-h period, however, an atmospheric concentration of as low as about 0.1 μg F/m^3 (gaseous or particulate) can be detected. The upper limit is about 100 μg F/m^3

3. Interferences

3.1 Significant amounts of particulate metallic salts such as those of aluminum, iron, calcium, magnesium or rare-earth elements may react with gaseous fluoride and remove some or all of it on the prefilter.

3.2 Significant amounts of acid aerosols or gases may neutralize or acidify the alkali-treated tape and prevent quantitative uptake of gaseous fluoride from the atmosphere.

3.3 The presence of large amounts of aluminum or certain other metals, or of phosphates, can interfere with subsequent analysis of the tapes by colorimetric or electrometric methods. This is a problem inherent with any collection method for fluoride.

3.4 There are several other possible limitations of the method:

3.4.1 Although the acid-treated Whatman No. 52 prefilter has been shown to allow passage of HF, it will restrict passage of particulate F only as small as about 1 μm. Thus, smaller particles may pass through the prefilter and impinge on or pass through the alkali-treated second tape. If prefilters of higher retentivity are required for removal of submicron-size particles, citric acid-impregnated tapes of Whatman No. 42 filter paper or Acropor AN-800 may be used.

3.4.2 Sampling of the atmosphere for long periods of time may result in the collection of a sufficient amount of particulate matter to result in sorption of gaseous fluoride on the particulate matter or a change in the air sampling rate.

4. Precision and Accuracy

4.1 PRECISION.

4.1.1 *Gaseous Fluoride.* A relative standard deviation of about 16% was found in the range of 1 to 3 μg F/m^3; in the range of 12 to 45 μg F/m^3, it was about 5%. Relative humidities between 40 to 85% had no effect on precision.

4.1.2 *Particulate Fluoride.* No information is available.

4.2 ACCURACY.

4.2.1 *Gaseous Fluoride.* Recovery of known amounts of gaseous HF was better than 95% with amounts of fluoride of up to 40 μg and sampling periods of 15 to 120 min. Removal of gaseous fluoride on the prefilter was negligible at relative humidities of 85% or less. No information is available above 85% relative humidity or in very dusty atmospheres.

4.2.2 *Particulate Fluoride.* No information is available.

5. Apparatus

5.1 The double paper tape sampler is a modification of and utilizes the basic principles of the sequential paper tape sampler used for dust collection.

5.2 DESCRIPTION OF SAMPLER. (see Figure 207:1)

5.2.1 The sampler has two supply reels and two take-up reels with appropriate capstans to guide the tapes through the sampling block.

5.2.2 The sampling block and sample inlet tube are constructed of Teflon or stainless steel to minimize reactivity with gaseous F. The upper part of the sampling block has a cylindrical cavity 2.54 cm in diameter and the inlet tube is perpendicular to the paper tapes. The lower part of the sampling block is constructed of stainless steel with a 2.54 cm diameter cylindrical

cavity and with an outlet tube which passes at a right angle into the pump compartment. The lower block is spring-loaded with a total pressure of 20 kPa against the lower surface of the upper block. The surfaces of the two blocks are machined flat to insure a tight seal. The lower block is lowered by means of an electric solenoid which counteracts the spring pressure.

5.2.3 The paper tapes are 3.8 cm wide.

5.2.4 Capstans are positioned to guide the paper tapes through the sampling block and to the take-up reel.

5.2.5 The paper tapes are drawn through the sample block and wound on the take-up reels by 2 rpm synchronous motors. Indexing is accomplished either by mechanical or photoelectric means to provide even spacing between samples. Provision is made by the use of tape perforated at regular intervals, or by some other means, to locate the collected sample spots for subsequent analysis. A relay is wired in series with the indexing mechanism to turn off the vacuum pump during tape transport.

5.2.6 An interval timer is used to provide desired sampling times.

5.2.7 Air is sampled with a carbon-vane vacuum pump of nominal 28.3 L/min free-air capacity. This should provide a sampling rate through the two tapes of about 14.1 L/min.

5.2.8 Exhaust air from the pump passes through a soda lime-glass wool filter and the filtered air is used to pressurize the front compartment and prevent contamination by F from the ambient air.

5.2.9 Air flow through the sample inlet tube is measured by means of a calibrated draft tube connected to a static pressure meter. The distance from the top of the inlet tube to the point of attachment of the draft gauge should be constant for all measurements.

5.2.10 Provision is made for manual override of the tape transport mechanism.

5.2.11 *Precautions.* Strain on the sample probe (air inlet tube, Figure 207:1) should be avoided since this may displace the head (sampling block, Figure 207:1) and allow air to enter the seal (between the sampling block and solenoid, Figure 207:1). The probe should be dismantled

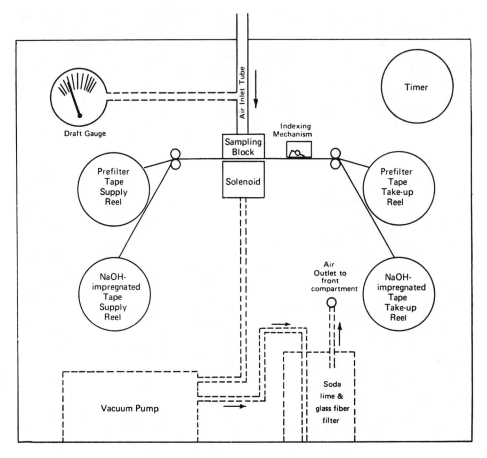

Figure 207:1 — Schematic drawing of double paper tape sampler.

and both the probe and the head (air inlet tube and sampling block, Figure 207:1) cleaned periodically in case dust has accumulated.

6. Reagents

6.1 PURITY. All reagents should be ACS analytical grade. All water should be ASTM reagent water, Type II.

6.2 Whatman No. 52 filter paper tape, 3.8 cm wide, is to be used as the prefilter. If prefilters of higher particle retention are required, Whatman No. 42 or Acropor AN-800 may be used.

6.3 Whatman No. 4 filter paper tape, 3.8 cm wide, is to be used to remove gaseous F.

6.4 0.1 M citric acid in 95% ethyl alcohol.

6.5 0.5 N NaOH in 95% ethyl alcohol containing 5% glycerol.

6.6 TOTAL IONIC STRENGTH ADJUSTMENT BUFFER (TISAB). Add 57 mL glacial acetic acid, 58 g sodium chloride and 4.0 g CDTA ((1,2-cyclohexylenedinitrilo)-tetraacetic acid) to 500 mL water. Stir and add 5 M sodium hydroxide slowly until pH is between 5.0 and 5.5. Cool and dilute to 1 L.

6.7 1.0 N sulfuric acid in water.

6.8 1.0 N sodium hydroxide in water.

7. Procedure

7.1 Treat the Whatman No. 52 paper tape with the citric acid solution in ethyl alcohol and dry.

7.2 Treat the Whatman No. 4 paper tape with the NaOH-glycerol solution in ethyl alcohol and dry.

7.3 Place the citric acid-treated tape on the upper supply reel and the NaOH-treated tape on the lower supply reel. Thread the tapes through the sampling block and to their respective take-up reels.

7.4 Set the interval timer to the desired sampling period.

7.5 Turn the tape sampler to the "on" position.

7.6 Record the rate of air flow through the inlet tube on the static pressure meter.

7.7 At convenient intervals remove the paper tapes and place each in separate clean containers.

7.8 Analysis of the individual samples is carried out in a laboratory by colorimetric or electrometric methods.

7.9 Portions of the unused tapes are also analyzed periodically to provide a blank correction.

8. Preparation of Samples for Fluoride Analysis

8.1 Analysis is carried out in a laboratory free of contamination by process-generated fluorides.

8.2 PREPARATION OF ALKALI-TREATED TAPES FOR FLUORIDE ANALYSIS.

8.2.1 *For Potentiometric Analysis.*

a. Cut out individual sample spots and place in clean 15 × 150 mm test tubes. Add 5.0 mL of 1/2-strength TISAB and mix for several seconds with a vortex mixer. Decant into a clean plastic beaker.

b. Analyze using potentiometric method (Method 205).

8.2.2 *For Semiautomated Analysis.*

a. Cut out individual sample spots and place in clean 15 × 150 mm test tubes. Add 5.0 mL of deionized water and mix for several seconds with a vortex mixer. Filter samples (to remove cellulose fibers) into 8.5 mL sample cups by a semi-micro filtration method.

b. Analyze the sample using the semi-automated method (Method 204, Section 7.3).

8.3 PREPARATION OF CITRIC ACID-TREATED (PREFILTER) TAPES FOR FLUORIDE ANALYSIS.

8.3.1 *For Potentiometric Analysis.*

a. Cut out individual sample spots and place in clean 15 × 150 mm test tubes. Add 5.0 mL 1 N H_2SO_4, mix for several seconds with a vortex mixer, and allow to stand for 5 min.

b. Add an equal volume of 1 N NaOH. Decant into a clean plastic beaker and add 10.0 mL TISAB.

c. Analyze using potentiometric method (Method 205).

8.3.2 *For Semiautomated Analysis.*

a. Cut out individual sample spots and place in clean 15 × 150 mm test tubes. Add 5.0 mL 1 N H_2SO_4, mix for several seconds with a vortex mixer, and allow to stand for 5 min. Filter samples (to remove cellulose fibers) into 8.5-mL sample cups by a semi-micro filtration method.

b. Analyze the sample using the semi-automated method (Method 204, Section 7.3).

8.4 CALIBRATION AND STANDARDS.

8.4.1 None required.

9. Calculation

9.1 The fluoride concentration of the atmosphere is calculated by the following relationship.

$$\text{Fluoride, } \mu g/m^3 = \frac{(A - B)}{C}$$

where A = number of micrograms of F in sample
B = number of micrograms of F in blank
C = volume of air sample in cubic meters at 101.3 kPa and 25°C.

10. Effect of Storage

10.1 Paper tape samples prior to analysis can be stored indefinitely under dry, fluoride-free conditions.

11. References

1. MANDL, R.H., L.H. WEINSTEIN, G.J. WEISKOPF
    AND J.L. MAJOR. 1970. The Separation and Col-
    lection of Gaseous and Particulate Fluorides. Paper
    CP-25A. 2D Internat. Clean Air Congress, Wash-
    ington, D.C., December 6–11.
2. WEINSTEIN, L.H. AND R.H. MANDL. 1971. The
    Separation and Collection of Gaseous and Particu-
    late Fluorides. VDI Berichte Nr. 164:53–63.

Subcommittee 2

C. R. THOMPSON, *Chairman*
G. H. FARRAH
L. V. HAFF
W. S. HILLMAN
A. W. HOOK
D. J. SCHNEIDER
J. D. STRAUTHER
L. H. WEINSTEIN

Approved with modifications
from 2nd edition
Subcommittee 2
R. H. MANDL, *Chairman*
L. H. WEINSTEIN
J. S. JACOBSON

301.

Determination of Particulate Antimony Content of the Atmosphere

1. Principle of the Method

1.1 Pentavalent antimony in the presence of a large excess of chloride ion **(2)** reacts with Rhodamine B to form a colored complex, which may be extracted with organic solvents such as benzene, toluene, xylene, or isopropyl ether **(2, 3, 4, 5, 6, 7, 8, 9)**. Trivalent antimony will not react with Rhodamine B and must be oxidized to the pentavalent state, (i.e., Sb_2O_5) by digestion in sulfuric, nitric, and perchloric acids. Phosphoric acid is used to minimize interference by iron. The absorbance of the pink-colored extract is measured in a spectrophotometer at 565 nm, or in a colorimeter using a green filter, and the amount of antimony present determined by reference to a calibration curve or regression equation prepared from known amounts of antimony.

2. Range and Sensitivity

2.1 With a 1-cm light path cell, the method covers a range from 0 to $10\mu g$ of antimony in 10 mL of solvent.

2.2 The sensitivity being 1.0 μg antimony (Sb)/determination (0.05 absorbance units), a concentration of 0.05 μg of antimony/m^3 can be measured if a 20-m^3 air sample is taken for analysis.

3. Interferences

3.1 According to Maren **(6)**, of the commonly encountered ions, only iron is likely to interfere. Maren suggests iron interference can be eliminated by extracting with isopropyl ether instead of benzene. For 750 μg of iron/mL, results obtained by the ACGIH show interference from iron to be insignificant **(1)**.

4. Precision and Accuracy

4.1 Three samples containing antimony

Sample No.	Antimony present $\mu g/mL$	Iron present $\mu g/mL$
1	3.0	0.0
2	5.0	750.0
3	8.0	0.0

were analyzed in triplicate by 10 collaborating laboratories in addition to the referee, and the standard deviation calculated.

	Sample 1	Sample 2	Sample 3
Standard deviation	0.113	0.237	0.245

4.2 Seven tests were made to determine the recovery of antimony added in solution by pipet to 9-cm Whatman #1 filters that were then oven dried. The table shows that recovery of the antimony ranged from 95 to 102.5% of the 2 to 10 μg added.

Sb Added	Sb Found	Deviation Percent
2.0 μg	1.90 μg	−5.0
4.0 μg	4.08 μg	+ 2.0
6.0 μg	6.15 μg	+ 2.5
6.0 μg	6.06 μg	+ 1.0
8.0 μg	8.0 μg	0
10.0 μg	10.03 μg	+ 0.03
10.0 μg	9.94 μg	−0.06

5. Apparatus

5.1 A SPECTROPHOTOMETER OR A COLORIMETER, with a green filter having maximum transmittance in the 565 nm range.

5.2 125-ML ERLENMEYER FLASKS, with ground glass stoppers.

5.3 125-ML SQUIBB SEPARATORY FUN-NELS, with Teflon stopcocks (prior to use, these should be cooled to 5°C in a refrigerator).

5.4 BOILING AIDS. Perforated pyrex beads, platinum tetrahedra or other materials may be used if necessary, to promote smooth boiling after ascertaining that no measurable contribution to the blank results from their use.

6. Reagents

6.1 REAGENTS. All reagents are ACS Reagent grade. Water is ASTM Reagent water, Type II.

6.2 SULFURIC ACID, SP GR 1.84

6.3 NITRIC ACID, SP GR. 1.42

6.4 PERCHLORIC ACID, REDISTILLED 72%.

6.5 BENZENE, TOLUENE, XYLENE, OR ISOPROPYL ETHER.

6.6 HYDROCHLORIC ACID 6N. Prepare from hydrochloric acid (S.G. 1.19) by mixing with an equal volume of water.

6.7 ORTHOPHOSPHORIC ACID 3N. Dilute 70 mL phosphoric acid to 1 L with water.

6.8 RHODAMINE B, 0.02%. Prepare solution 0.02% w/v in water.

6.9 ANTIMONY STANDARD SOLUTION. Place 0.1000 g antimony in 25 mL concentrated sulfuric acid and heat till dissolved; dilute to 1 L with distilled water. This solution contains 100 μg trivalent antimony/ mL, and from this more dilute working standards may be prepared to cover the range 1 to 10 μg/mL.

6.10 CAPRYL ALCOHOL. This may be used if excessive frothing occurs during digestion.

6.11 All the foregoing reagents and solutions should be cooled to 5°C in a refrigerator before use.

7. Procedure

7.1 Antimony collected from the atmosphere on membrane, cellulose or glass fiber filters must be oxidized to the pentavalent state before analysis with Rhodamine B is possible. The filter or an aliquot is placed in a 125-mL Erlenmeyer flask; 5 mL concentrated sulfuric acid is added, followed by 5 mL concentrated nitric acid. The flask is then heated on a hot plate or over a gas flame till brownish-red fumes of NO_2 are driven off, the solution blackens (for organic materials) and dense white fumes of sulfur trioxide are driven off. The solution is allowed to cool slightly (2 to 5 min) and 10 drops of 72% perchloric acid are added. The solution is then heated once more till white fumes of SO_3 are driven off, the solution clears and becomes colorless, and boiling stops. The resulting digest is placed in an ice bath to cool.

7.2 After temperature equilibrium is established (at least 30 min), 5 mL precooled 6 N hydrochloric acid is added slowly (taking 1 to 2 min) by pipet to minimize temperature increase in the digest. The solution is cooled in the ice bath for at least 15 min, after which 8 mL of precooled 3 N phosphoric acid is added, followed immediately by the addition of 5 mL precooled 0.02% Rhodamine B solution. Without delay, the flask is stoppered and shaken vigorously, and the contents transferred to a precooled separatory funnel.

7.3 Ten mL of precooled benzene or other solvent is added to the separatory funnel, the contents are shaken vigorously for 1 min, and after allowing the contents to separate, the aqueous layer is discarded. The benzene etc., phase (colored pinkish-red if antimony is present) is collected in a centrifuge tube and allowed to stand a few min for water to settle out or it may be centrifuged. A 1-cm light path cuvette is rinsed with several mL of the extract, then filled and read at 565 nm against a benzene etc., blank which has been taken through the entire procedure.

7.4 It is convenient to cool separately funnels, hydrochloric acid, phosphoric acid, benzene or other solvent used, and Rhodamine B solution in a refrigerator before use. The analyst should aim at maintaining a temperature of 5 to 10°C during the analysis. Since phosphoric acid will lower the results if allowed to stand, the extraction should be completed within 5 min after the phosphoric acid is added. Color development and extraction should be carried out in subdued light.

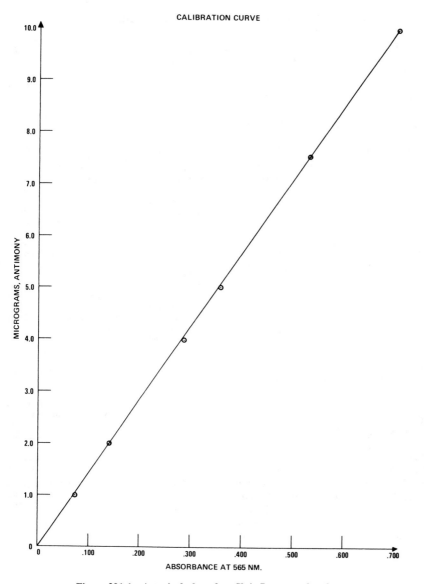

CALIBRATION CURVE

MICROGRAMS, ANTIMONY

ABSORBANCE AT 565 NM.

Figure 301:1 – A typical plot of μg Sb/mL versus absorbance.

7.5 If the color developed is too intense to measure, the extract may be diluted with the solvent used for extraction, and the antimony present found by multiplying by the aliquot factor.

8. Calibration

8.1 Dilute the antimony stock solution to prepare standards containing 1 to 10 μg Sb/mL. One to 10 μg Sb are determined by carrying not more than 1 mL of the appropriate reference solutions through the above digestion and analysis. Since NO_2 fumes may not be driven off visibly when no organic material is present, the solutions should be heated till white fumes of SO_3 are driven off strongly and the solution has stopped bubbling. The analysis will fail if any nitric acid remains in the digest. Plot μg Sb versus absorbance at 565 nm. The plot should be linear, as in Figure 301:1. Preferably, determine the calibration equation from the data by linear least squares.

Calculations

9.1 If an air sample of V cubic meters (at standard conditions) is found to contain Y μg of antimony, then the concentration of antimony in the air is

$$\mu g \text{ Sb/m}^3 \text{ air} = \frac{Y}{V}$$

If an aliquot x is taken from the total sample X, then the result must be multiplied by X/x.

10. Effect of Storage

10.1 Presumably particulate samples collected on filters may be stored indefinitely. Question: How long may the digest be stored before completing analysis? It would be useful to be able to digest the samples en masse and store them for analysis at one time.

11. References

1. AMERICAN CONFERENCE OF GOVERNMENTAL INDUSTRIAL HYGIENISTS. 1963. Manual of Recommended Analytical Methods.
2. EDWARDS, F. C., AND A. F. VOIGT. 1945. Separation of Antimonic Chloride from Antimonous Chloride by Extraction into Isopropyl Ether. *Anal. Chem.* 21:1204–1205.
3. FREEDMAN, L. D. 1947. Rhodamine Method for Microdetermination of Antimony. *Anal. Chem.* 19:502.
4. LUKE, C. L. 1953. Photometric Determination of Antimony in Lead Using the Rhodamine B Method. *Anal. Chem.* 25:674–675.
5. MAREN, T. H. 1945. The Microdetermination of Antimony. *Bull. Johns Hopkins University,* 77:338–344.
6. MAREN, T. H. 1947. Colorimetric Microdetermination of Antimony with Rhodamine B. *Anal. Chem.* 19:487–491.
7. NIELSON, W., AND G. BOLTS. 1954. Extraction and Photometric Determination of Antimony by Means of Rhodamine. *Z. Anal. Chem.* 143:(4)264–272.
8. RAMETTE, R. W., AND E. B. SANDELL. 1955. Rationalization of the Rhodamine-B Method for Antimony, *Anal. Chem. Acta,* 13:(5) 455–458.
9. WARD, F. N., AND H. W. LAKIN. 1954. Determination of Traces of Antimony in Soils and Rocks. *Anal. Chem.* 26:(7) 1168–1173.

Subcommittee 7

E. C. TABOR, *Chairman*
M. M. BRAVERMAN
H. E. BUMSTEAD
A. CAROTTI
H. M. DONALDSON
L. DUBOIS
R. E. KUPEL

Reapproved with modifications
Subcommittee 6

M. T. KLEINMAN, *Chairman*
J. R. RHODES
R. J. THOMPSON
T. C. RAINS
W. J. COURTNEY
V. P. GUINN

302.

Determination of Arsenic Content
of Atmospheric Particulate Matter

1. Principle of the Method (5)

1.1 Arsenic exists in atmospheric particulate matter primarily in the form of inorganic oxides and arsenates. The arsenic compounds are dissolved from the sample of particulate matter with hydrochloric acid followed by reduction of arsenic to the trivalent state with KI and $SnCl_2$. The trivalent arsenic is further reduced to arsine, AsH_3, by zinc in acid solution in a Gutzeit generator. The evolved arsine is then passed through an H_2S scrubber, which consists of glass wool impregnated with lead acetate, and then into an absorber containing silver diethyldithiocarbamate dissolved in pyridine. In this solution, the arsine reacts with silver diethyldithiocarbamate forming a soluble red complex which is suitable for photometric measurement with a maximum absorbance at 535 nm (2). This method replaces a Gutzeit treatment and a titrimetric method involving the absorption of arsine in mercuric chloride and subsequent iodometric determination (3).

1.2 This method is suitable for the analysis of particulate samples collected either on membrane or glass fiber filters. Other techniques such as impingers or electrostatic precipitation will not provide adequate samples within a reasonable sampling period because of the low concentrations of arsenic in the ambient air.

1.3 Since the level of arsenic in ambient air usually lies between 0.1 and 1.0 $\mu g/m^3$, an aliquot representing at least 10 m^3 of air is required if the sample absorbance is to fall on the calibration curve. The amount of filter used must be limited, or else the filter itself may physically and/or chemically interfere in the analysis. No difficulties will be encountered if the filter area used does not exceed 15 cm^2. Glass fiber filters gave better recovery of arsenic than Whatman #40 or #541 paper. In one series of tests, 7 samples of the stock solution in the range from 0.5 to 8.0 μg As_2O_3 were added to 5-cm Whatman #40 filter paper and similar additions were made to glass fiber filters. The latter were digested with hot dilute hydrochloric acid as in **7.1** and **7.2**. Average recovery of the arsenic from the paper was 86 $\pm$ 4% and from the glass fiber 103 $\pm$ 5%.

2. Range and Sensitivity

2.1 If suitable samples (10 m^3 of air) are available, concentrations as low as 0.1 μg As/m^3 can be measured by this procedure. The maximum measurable concentration with a comparable sample is 1.5 μg As/m^3. Higher concentrations can be measured if smaller samples are used.

3. Interferences

3.1 Antimony present in the sample will form stibine, SbH_3, which will interfere slightly by forming a colored complex with silver diethyldithiocarbamate which has a maximum absorbance at 510 nm. The amount of antimony found in air is so small that any interference would be insignificant.* High concentrations of nickel, copper, chromium, and cobalt interfere in the generation of arsine. In addition, certain combinations may produce a similar situation. The presence of interferences can be determined by the internal standard technique, where known amounts of arsenic are added to a sample and the recovery determined. Hydrogen sulfide would interfere,

---

*By actual test, it has been demonstrated that antimony does not interfere to any appreciable extent (4).

but it is removed by the lead acetate scrubber.

3.2 The filters used to collect atmospheric particles may contain measureable quantities of arsenic. Portions of unused filters should be analyzed and corrections made if necessary. This can be done separately or as part of the reagent blank procedure.

4. Precision and Accuracy

4.1 Spiked samples containing 0.10, 1, 5, and 10 μg arsenic were analyzed with an accuracy of ± 0.04 μg based upon 7 replicate determinations at each concentration. **(4)**.

4.2 Four samples containing known amounts of arsenic were analyzed by 8 laboratories with the following results **(1)**.

Table 302:I — Analysis of Samples Containing Known Amounts of Arsenic.

Sample	μg As/mL	Percent Average Deviation
1	0.05	9.1
2	0.50	3.5
3*	1.00	6.1
4	1.50	4.0

*Contained 0.5 μg Sb/mL.

5. Apparatus

5.1 ARSINE GENERATOR, SCRUBBER, AND ABSORBER. Fisher arsine generator, #1-405 or equivalent. See Figure 302:1.

5.2 SPECTROPHOTOMETER, for use at 535 nm with 1-cm cells, *or*

5.3 FILTER PHOTOMETER, with blue filter having a maximum transmittance at or in the range of 530 to 540 nm with 1-cm cells.

6. Reagents

6.1 PURITY. All reagents must be ACS Reagent grade. Water is ASTM Reagent water, Type II.

6.2 SILVER DIETHYLDITHIOCARBAMATE, AgSCSN$(C_2H_5)_2$ REAGENT. Dissolve 4.0 g of silver diethyldithiocarbamate (Fisher #S-666 or equivalent) in 800 mL of pyridine. The useful life of this reagent can be ex-

tended to at least 2 months by storing in a dark brown bottle or in the dark.

6.3 STANNOUS CHLORIDE REAGENT. Dissolve 10.0 g of stannous chloride dihydrate in 25 mL of 12 N (sp gr 1.19) hydrochloric acid. Place in a separatory funnel with a layer of pure mineral oil 5 mm thick on top to minimize oxidation. Drain a small quantity of the solution out of the stopcock before use. This solution is stable for 2 weeks.

6.4 LEAD ACETATE SOLUTION. Dissolve 10 g of PbC$_2$H$_3$O$_2 \cdot$3H$_2$O crystals in 100 mL of water. The solution will be slightly turbid as a small amount of the basic salt is formed, but this will not affect its usefulness.

6.5 POTASSIUM IODIDE SOLUTION. Dissolve 15 g of KI in 100 mL of water. The solution should be stored in a brown glass bottle.

6.6 STOCK ARSENIC STANDARD SOLUTION. Dissolve 0.132 g of As$_2$O$_3$ in 10 mL of 10 N NaOH and dilute with water to 1000 mL. This solution contains 100 μg As/mL.

6.7 WORKING ARSENIC STANDARD. Dilute 10 mL of stock arsenic standard solution to 1000 mL with water. This solution contains 1 μg As/mL. Prepare fresh weekly.

6.8 6N SODIUM HYDROXIDE. Dissolve 240 g of NaOH pellets in 200 mL of freshly boiled and cooled water and dilute to 1 L.

6.9 6N HYDROCHLORIC ACID. Dilute 500 mL of 12 N HCl to 1 L with distilled water.

6.10 ZINC. Granular, 20 mesh.

6.11 PYRIDINE. Some pyridine may contain colored materials that can be removed by passing the pyridine through a 2.5-cm diam by 15-cm depth alumina column at 150 to 200 mL/h. Other pyridine is sufficiently pure so that repurification is not necessary.

7. Procedure

7.1 Place a sample or aliquot of a sample representing from 0.1 to 15 μg of arsenic in a 125-mL flask (Figure 302:1) and add 30

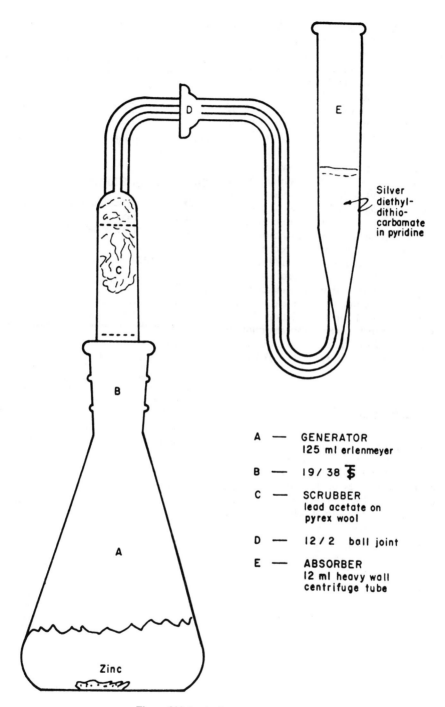

Silver
diethyl-
dithio-
carbamate
in pyridine

A — GENERATOR
    125 ml erlenmeyer

B — 19/38 $\overline{\underline{\mathbb{S}}}$

C — SCRUBBER
    lead acetate on
    pyrex wool

D — 12/2 ball joint

E — ABSORBER
    12 ml heavy wall
    centrifuge tube

Zinc

Figure 302:1 — Arsine generator.

mL of 6 N HCl. Digest at 90 to 95°C (boiling will result in loss of As) at least 1 h. Cool and add 20 mL of 6 N NaOH. After cooling the solution to room temp, add 2 mL of potassium iodide solution, mix thoroughly, and then add 8 drops of the stannous chloride reagent and mix. Allow 15 min for reduction of arsenic to the trivalent state and then cool the solution to 4°C in an ice bath.

7.2 Impregnate the glass wool in the scrubber with lead acetate solution. About 1.5 mL of lead acetate is sufficient to saturate a 4-cm length of glass wool packing. Drain off any excess of solution. Pipet exactly 3.0 mL of the silver diethyldithiocarbmate (4°C) solution into the absorber tube. Add 3 g of zinc to the flask and immediately connect the scrubber-absorber assembly, making certain that all connections are tight. Allow 90 min for complete evolution of the arsine. Pour the solution directly from the absorber into the spectrophotometer cell and determine the absorbance of the solution using the reagent blank flask as the reference.

8. Calibration

8.1 Prepare a set of standard solutions by pipetting 1, 3, 5, 7, 10, and 15 mL of the working standard arsenic solution (1 μg/mL) into a series of 125-mL Erlenmeyer generator flasks. Add sufficient 6 N HCl to each flask to give a final volume of 30 mL. Include a reagent blank.

8.2 Follow the procedure as outlined under **7.1** and **7.2**, omitting the initial digestion at 90 to 85°C.

8.3 Construct a calibration curve by plotting absorbance value *vs.* μg arsenic in the standard, or preferably obtain a calibration equation by the method of linear least squares.

9. Calculations

9.1 Use the absorbance of the sample to obtain the arsenic content from the calibration curve or equation.

9.2 Concentration as As in

$$\mu g/m^3 = \frac{\mu g \text{ As in sample analyzed}}{m^3 \text{ of air represented by sample}}$$

making appropriate corrections for any aliquot used.

10. Effect of Storage

10.1 Air particulate matter samples to be analyzed may be stored indefinitely at room temperature without loss of arsenic.

11. References

1. AMERICAN CONFERENCE OF GOVERNMENTAL INDUSTRIAL HYGIENISTS. 1955. Manual. Determination of Arsenic in Air.
2. AMERICAN PUBLIC HEALTH ASSOCIATION. 1976. *Standard Methods for the Examination of Water and Wastewater.* 14th ed. Washington, D.C. 20036.
3. CASSIL, C. C., AND H. J. WICHMANN. 1939. A rapid micromethod for determining arsenic. *J. Assoc. of Agric. Chemists,* 22:436.
4. MORGAN, GEORGE B. Personal Communication. Air Quality and Emission Data Program. National Center for Air Pollution Control, Public Health Service.
5. VASAK, V., AND V. SEDIVEC. 1952. Colorimetric determination of arsenic. *Chem. Listy.* 46:341–344.

Subcommittee 7

E. C. TABOR, *Chairman*
M. M. BRAVERMAN
H. E. BUMSTEAD
A. CAROTTI
H. M. DONALDSON
L. DUBOIS
R. E. KUPEL

Reapproved with modifications
Subcommittee 6

M. T. KLEINMAN
J. R. RHODES
R. J. THOMPSON
T. C. RAINS
W. J. COURTNEY
V. P. GUINN

303A.

General Method for the Preparation of Non-Tissue Environmental Samples for Trace Metal Analysis

1. Principle of the Method

1.1 Trace elements in environmental samples, including collected particulate matter, sludge and soils, are solubilized by ultrasonication in a heated nitric/hydrochloric acid matrix. The resulting material is centrifuged, decanted and analyzed by the method of choice (1).

2. Range, Sensitivity, and Detection Limit

2.1 The range, sensitivity, and detection limits are characteristic of the element sought and the analytical method used.

3. Interferences

3.1 Analytical interferences are those of the analytical method used.

3.2 This method is not applicable to mineral samples or other non-tissue samples in which the target elements are incorporated in a silicaceous matrix unless the dried samples are reduced mechanically to −600 mesh prior to ultrasonication/extraction.

3.3 Scrupulous attention to housekeeping is a must. Glassware must be segregated for use with blanks, low level samples and spiked samples.

4. Precision and Accuracy

4.1 The precision and accuracy are to be found in the procedures describing an instrumental approach or in the procedure for the element of interest.

5. Apparatus

5.1 *Sampling equipment.* The samples are assumed to be obtained by standard or special techniques appropriate to the medium sampled.

5.2 *Analysis.* The equipment necessary is listed under the individual analytical procedures.

5.3 *Preparative.*

5.3.1 *Equipment.*

a. Centrifuge (IEC Model K or equivalent), equipped with 12-place rotor and 30.2 × 100 mm shields.

b. Desiccator for cooling oven-dried chemicals.

c. Gravity convection type oven for drying glassware and chemicals.

d. Ultrasonic bath (Blackstone Ultrasonics, Inc. Model HT 11.2 with Model SU-8W/pc Controller or equivalent); cleaning power of at least 450 W and an operating temperature of 100°C are necessary.

e. Balance accurate to 0.1 mg or less.

f. Template to aid in sectioning glass fiber filters. See figures 303A:1 and 303A:2.

g. Pizza cutter, thin wheel, thickness less than 1 mm.

h. Rack, wire (Fisher 14-793-1, or equivalent), or polypropylene (Fisher 14-8090 or equivalent) for holding centrifuge tubes during ultrasonication.

5.3.2 *Labware.*

a. Centrifuge tubes. 50-mL linear polypropylene tubes with polypropylene screw tops. (Nalgene 3119-0050 or equivalent).

b. Bottles, linear polyethylene or polypropylene with leakproof caps.

c. Pipettes (Class A borosilicate glass).

d. Cleaning. Wash all labware with (or ultrasonicate for 30 min) in laboratory detergent, rinse, soak for a minimum of 4 h in 20 percent HNO_3, rinse 3 times with distilled-deionized water, and dry in a dust-free manner.

6. Reagents

6.1 ANALYSIS.

6.1.1 Water—all references to water are to ASTM Reagent water, Type II.

6.1.2 Concentrated (12.3M) hydrochloric acid—ACS Reagent grade.

6.1.3 Concentrated (16.0M) nitric acid, redistilled spectrographic grade for preparing samples.

6.1.4 Concentrated (15.0M) nitric acid, ACS Reagent grade, for the 20% labware cleanser only. The ACS Reagent grade nitric acid has metallic contaminants at too high a level for environmental analyses.

6.1.5 Extracting acid (1.03M HNO_3 and 2.23M HCl) is prepared with a 1000 mL volumetric flask with 500 mL of water to which 64.6 mL of concentrated distilled nitric acid (**6.1.3**) and 182 mL of concentrated HCl (**6.1.2**) are added. Shake well, cool to room temperature and dilute to volume with water. *Caution*: Acid fumes are toxic. Prepare in a well-ventilated hood.

6.1.6 Filter blanks. For batches of filters over 500, randomly select 25–30 filters per batch. For smaller batches, use a lesser number but not less than 5%. Section each filter to obtain the same fraction obtained from a sample composed of a single strip.

7. Procedure.

7.1 SAMPLING. Airborne particulate matter collection and storage is described in Method 501.

7.1.2 Other types of samples are placed in non-linear polyethylene screw cap vials or jars and stored under refrigeration until they are analyzed.

7.2 SAMPLE PREPARATION.

7.2.1 Hi-vol samples. Cut a 2 × 20 cm or 2.5 × 20 cm strip of the exposed filter using a template and a pizza cutter as shown in Figures 303A:1 and 303A:2 or use

an alternate procedure after checking reproducibility of sectioning.

7.2.2 Filters other than glass-fiber may be used (providing they contain no interferences and afford quantitative capture of the desired element). Using historical data for the locale, season and conditions (or estimating from historical data for the site) take a large enough section to afford enough of the element(s) sought to give an estimated instrumental reading 3 times the standard deviation of the background. Polymeric filters may be sectioned readily using a scalpel and paper backing. The section should be weighed with uniform particulate matter distribution assumed.

7.2.3 Sludge samples are prepared by drying prior to reweighing (to establish loss on drying) and analysis.

7.2.4 Soil and mineralogical samples containing material in a non-silicaceous matrix are dried, weighed, and prepared as described. Materials with a silicaceous matrix require reduction to –600 mesh size prior to processing.

7.3 EXTRACTION.

7.3.1 For glass fiber filters, accordion fold or tightly roll the filter strip and place on its edge in a 50-mL polypropylene centrifuge tube, using vinyl gloves or plastic forceps. Sections of other filters should be similarly rolled or folded to fit into the centrifuge tubes.

7.3.2 For other samples, introduce a weighed quantity (1.0 g or enough material to yield a sample containing a sufficient quantity of the material sought).

7.3.3 Pipette 12.0 mL of the extraction acid (**6.1.5**), which should completely cover the sample, into the centrifuge tube. Cap the tube loosely (finger tight) with the polypropylene screw top. *Caution*: Centrifuge tubes must be loosely capped to prevent elevated pressures during ultrasonication at elevated temperatures. They *will not* withstand repeated cycling to elevated pressures.

7.3.4 Label the centrifuge tube, place in a sample rack, and place upright in the preheated (100°C) ultrasonic water bath (in fume hood) so that the water level is slightly above the acid level in the centrifuge tubes but well below the centrifuge tube caps.

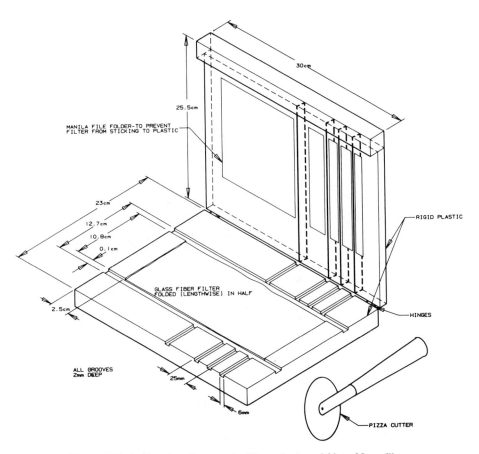

Figure 303A:1 — Template for reproducible sectioning of 20 × 25 cm filters.

This will prevent contamination of the samples during ultrasonication. Ultrasonicate the sample at 100°C for 50 min at a power setting ≥ 450 watts.

7.3.5 Remove the centrifuge tube from the ultrasonic bath and allow to cool.

7.3.6 Uncap the centrifuge tube in the fume hood and add 28.0 mL of water with pipettes or the automatic dispensing pipette. Recap the tube tightly, shake well, and centrifuge for 20 min at 2500 RPM.

7.3.7 Decant the extract into a clean polyethylene storage bottle bearing the sample I.D. Be careful not to disturb any solids in the bottom of the tube. Cap the bottle tightly and store until analysis. The final extract is now in 0.31M HNO_3 + 0.67 M HCl.

8. Analysis.

8.1 Proceed with the analysis of the sample by the appropriate method.

9. Calibration.

9.1 See the appropriate method chosen in **8.1**. Obtain the concentration in terms of micrograms of element sought per milliliter of final solution.

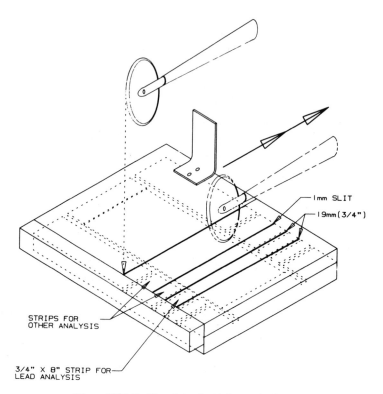

Figure 303A:2 — Template closed for sectioning.

10. Calculation

10.1 CONCENTRATION OF CHOSEN ELEMENTS.

10.1.1 *Air Samples.*

$$C = \frac{C_s \times 1/n - \overline{F}_b}{V}$$

where C = concentration of element E in the air sample, $\mu g/m^3$.

E = element determined.

C_s = concentration of E in solution, $\mu g/mL$.

n = fraction of sample per section/portion.

V = volume of air sampled, at STP or otherwise as required, m^3.

$\overline{F}_b$ = mean concentration of E in blank filter section, figure from **6.1.6**, $\mu g/filter$.

10.2.2 Soil and sludge samples

$$C = \frac{C_s \times 40}{M}$$

where C = concentration of element E in the sample, $\mu g/g$ or ppm.

m = mass of sample taken, in grams.

10.2.3 Calculate the mean, $\overline{F}_b$, the values for C and the relative standard deviation. If the value of $\overline{F}_b$ is high enough to result in a significant error in C, the filter batch should be rejected (for $\overline{F}_b \geq 10\%$ C). Below lower limit of detection (3 σ response above background) a correction is not warranted.

11. Quality Control

11.1 For particulate matter samples, controls of knowns (filter strips spiked with salts of E approximating $1/4$, $1/2$, $3/4$ and 100% of the maximum expected value, blanks, and spiked sample replicates) should be used as every n^{th} sample. (For robust quality control, 20% of the samples should be q.c. samples; usually 5-10% is used.) The extent of q.c. depends on the objectives of data use.

12. References

1. CUMMINGS, S. L., S. L. HARPER, W. A. LOSEKE AND L. J. PRANGER. 1984. "Standard Operating Procedure for the ICP-OES Determination of Trace Elements in Suspended Particulate Matter Collected on Glass-Fiber Filter." Environmental Monitoring Systems Laboratory, USEPA, Research Triangle Park, NC 27711 and references given therein. March.

311–322

Determination of a Number of Metals in Atmospheric Particulate Matter

See **Methods 303A, 822**

317.

Determination of Elemental Mercury in Ambient and Workroom Air by Collection on Silver Wool and Atomic Absorption Spectroscopy

1. Principle of the Method

1.1 Elemental mercury vapor is collected on silver wool and released by heating the wool to 400°C.

1.2 The mercury released is swept by a carrier gas through an absorption cell of an atomic absorption instrument, and the response at 253.7 nm is measured either as a peak height or peak area (1).

2. Range and Sensitivity

2.1 The procedure is applicable to the determination of elemental mercury in concentrations of 0.02 to 500 $\mu g/m^3$ at a collection flow rate of 50 to 200 mL/min over a period of 10–60 min in workroom air and 24 h in ambient air.

3. Interferences

3.1 High levels (μg quantities) of chlorine (2) and SO_2 (3) will poison the silver wool collectors. A tube packed with Ascarite preceding the collector will remove the chlorine (2) and other acid gases.

3.2 Ambient levels of dimethyl mercury, sulfur dioxide, hydrogen sulfide and nitrogen dioxide do not seriously interfere. However, hydrogen sulfide in concentrations above 13 $\mu g/m^3$ and nitrogen dioxide above 25 $\mu g/m^3$ may show an interference (1). Acetone, benzene, and ethanol do not interfere (4).

4. Precision and Accuracy

4.1 Recovery of mercury from the silver wool collectors by this method is at least 98% over the range of 0.006 to 0.6 μg with a relative standard deviation of 4% (1).

4.2 The precision of the method has been reported to be about 10% relative standard deviation for ambient mercury vapor concentrations of about 0.03 $\mu g/m^3$ (1, 5, 6).

4.3 The precision of the total method (Sampling and Analysis) at sampling flow rates of 100 mL/min, using 2 collectors in series (1 g and 2 g) at levels of 0.3 $\mu g/m^3$ to 1.96 $\mu g/m^3$, was better than 5% RSD. At these levels the second collector never contained more than 4% of the mercury. At mercury levels of 99 $\mu g/m^3$, 9.5 μg of mercury was collected on three 1-g collectors in series with 83% collected on the first tube and 17% on the second tube. Insignificant amounts (~2 ng) were detected on the third tube (7).

5. Apparatus

5.1 SAMPLE COLLECTION (see Figure 317:1).

5.1.1 *Silver wool collector.* Pyrex tube 10 cm long × 5 mm ID equipped with ball joints on the ends and packed with 1 to 2 g of clean silver wool (Fisher microanalysis grade) and permanently wound with 100 cm of 22-gauge Nichrome heating wire. Clean the silver wool initially by placing it in a furnace at 800°C for 2 h prior to packing it in the Pyrex tube.

5.1.2 30 mg Silvered Chromosorb P (commercially available), which has been recognized by NIOSH (8) as a collection substrate. The glass collection tube (2 cm long, 6 mm OD, 4 mm ID) is plugged at either end with quartz wool.

5.1.3 *Pump.* Capacity to pull from 50 to 500 mL/min of air.

5.1.4 *Flowmeter.* (50 to 500 mL/min).

5.1.5 *Flow controller.*

5.1.6 *Kel-F coated 3-way valve.*

5.1.7 *¼" glass tubing.*

5.1.8 *Caps for sealing ends of Pyrex tube.*

5.1.9 *Clamps for ball joints.*

5.2 EQUIPMENT TRAIN. (See Figure 317:2).

5.2.1 *Atomic absorption spectrophotometer.*

5.2.2 *Absorption cell.* 20 cm long by about 3 cm ID, with fused silica windows. Lag with heating tape to maintain temperature of the surface near 90°C.

5.2.3 *Mercury hollow cathode lamp.*

5.2.4 *Recorder or meter read out.*

5.2.5 *Integrator.* Digital (optional).

5.2.6 *Variable transformer.*

5.2.7 *Two charcoal filters* in drying tubes and one silver wool filter (as in **5.1.1**) without Nichrome heating wire.

5.2.8 *Teflon tubing.*

5.2.9 *Carrier gas.* (Oil free air or N_2).

5.2.10 *Oven.* Capable of maintaining 800°C.

5.2.11 *Kel-F coated 3-way valves.*

5.2.12 *Flowmeter.* (50 to 500 mL/min).

5.2.13 *Injection port.* Glass T with septum.

5.3 CALIBRATION EQUIPMENT.

5.3.1 *Pyrex bottles.* One L.

5.3.2 *Serum caps.* For Pyrex bottles.

5.3.3 *Constant temperature bath* (20.0° ± 0.1°C).

5.3.4 *Certified NBS thermometer.*

5.3.5 *Syringes.* Gas tight: 0.02 to 100 mL.

6. Reagents

6.1 MERCURY. ACS Reagent grade.

6.2 SILVER WOOL. (Fisher microanalysis grade).

6.3 CHARCOAL. 6 to 20 mesh, activated.

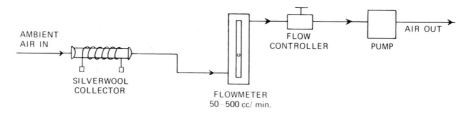

Figure 317:1 – Sample collection train.

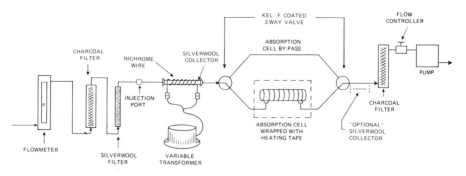

Figure 317:2 – Schematic diagram of analytical gas train.

7. Procedure

7.1 SAMPLING.

7.1.1 Connect the silver wool collector prepared in **5.1.1** into the Analytical Gas Train of Figure 317:2. Sweep with gas at about 200 mL/min while using bypass mode. Release any residual mercury from the silver wool by applying 24 V from the variable transformer across the Nichrome wire for 30 s, which yields a temperature of 400°C. After one 30-s heating period, the collector is mercury free and ready for future use. Cap both ends until ready for sampling, making sure that the caps do not touch the silver wool.

7.1.2 Set up the collection train as shown in figure 317:1 which is manifolded with 6-mm ID glass, Teflon tubing and Kel-F coated valves.* Flow rates through the manifold are produced by a pump following the absorption cell and monitored by the flowmeter. A suitable personal portable battery pack pump and sampler having all the essential parts of the sampling train can also be used.

7.1.3 Collect the mercury vapor in ambient air by pulling air through a cleaned collector containing one or two grams of silver wool at a known rate for a fixed length of time. For field sampling purposes, a flow rate of between 100 and 200 mL/min for 24 h is recommended. For workspace sampling, flow rates of 50 to 200 mL/min for up to 1 h are usually appropriate.† After use, disengage the collector from the train and cap it to prevent contamination.

7.2 ANALYSIS.

7.2.1 Set the atomic absorption spectrophotometer operating conditions as recommended by the manufacturer. The instrument should be set to the wavelength of maximum intensity for the 253.7 nm mercury line.

---

*Teflon tubing should be as short as possible and no rubber tubing or grease should be used since mercury will be absorbed by these materials.

†The flow rate, collection time, and number of collectors can be varied to suit the expected mercury level.

7.2.2 Insert a silver wool collector in the Analytical Gas Train and sweep the system for a few minutes with carrier gas until no mercury is detected. This should be done prior to running calibrations and samples.‡

7.2.3 Prepare and run standards at the beginning of each analysis period. Standards should be run to check the procedure after every tenth sample is run. Record the absorbance value or, if an integrator is used, record the peak area and prepare a calibration curve as described in Section **8.1**.

7.2.4 Clamp the collector to be analyzed into the analytical system shown in Figure 317:2 by ball-joint clamps, adjust the carrier gas flow (which is critical to the repeatability) to exactly 200 mL/min, and heat the collector for 30 s at 24 V. Silver wool and charcoal filters are inserted between the flowmeter and collector to prevent sample contamination from the laboratory air. Record the absorbance peak value or peak area obtained from the spectrophotometer and determine the amount of mercury from the calibration curve. Allow the collector to cool, remove it from the analytical system, and cap it for future use.**//

8. Standards and Calibration

8.1 STANDARD MERCURY SAMPLES. Add enough Reagent Grade mercury to cover the bottoms of 3 one-liter pyrex flasks. The flasks should be equipped with serum caps.

---

‡Residual mercury can be attributed to contamination, possibly from slow seepage of air through the joints of the apparatus during nonuse.

**If there is some uncertainty about the concentration range, place a clean (mercury free) silver wool collector between the spectrophotometer and the charcoal filter to recapture the sample in order to rerun the sample at the proper calibration range.

//The entire analysis process takes less than five min per sample at 200 mL/min carrier gas flow rate. The absorption cell bypass, shown in Figure 317:2, is convenient when collectors are being cleaned and the amount of mercury released does not need to be monitored.

Table 317:I Elemental Mercury Vapor Density at 10 to 30°C and One Atmosphere Pressure

t °C	Vapor Density ng/mL	t °C	Vapor Density ng/mL	t °C	Vapor Density ng/mL	t °C	Vapor Density ng/mL
10.0	5.57	15.5	9.01	21.0	14.33	26.5	22.39
10.5	5.82	16.0	9.41	21.5	14.93	27.0	23.30
11.0	6.09	16.5	9.82	22.0	15.56	27.5	24.24
11.5	6.36	17.0	10.25	22.5	16.21	28.0	25.22
12.0	6.65	17.5	10.69	23.0	16.89	28.5	26.23
12.5	6.95	18.0	11.15	23.5	17.59	29.0	27.28
13.0	7.26	18.5	11.63	24.0	18.32	29.5	28.36
13.5	7.58	19.0	12.13	24.5	19.07	30.0	29.49
14.0	7.92	19.5	12.65	25.0	19.86		
14.5	8.27	20.0	13.19	25.5	20.67		
15.0	8.63	20.5	13.75	26.0	21.52		

8.1.1 Place the flasks in a constant temperature bath equipped with a certified NBS thermometer and maintain the entire flask at 20.0 ± 0.1 C. One mL of air at 20°C and one atmosphere pressure at equilibrium over liquid mercury contains 13.19 ng of mercury. Concentrations of mercury at temperatures other than 20°C are given in Table 317:I.

8.1.2 *Calibrating Procedure.* Use several gas-tight syringes with volumes of from 0.02 to 100 mL and withdraw from the flasks at least five different volumes of mercury-saturated air corresponding to the expected field values. The sample should be withdrawn slowly, expelled back, and withdrawn slowly again. The temperature of the syringe must be equal to or slightly higher than 20°C to prevent possible condensation of mercury in the syringe.

Inject a given volume slowly into the analytical gas train previously cleaned as in 7.2.2 (see Figure 317:2) at the injection port while sweeping the train at exactly 200 mL/min with carrier gas. The known amount of mercury is collected on the silver wool and then released by applying 24 V to the Nichrome heater for 30 s. The system is now ready for the next injection of standard sample when the tube is slightly hot to the touch of the finger. Record the peak values or the area-time counts from all five standards and plot the calibration curve.¶

___

¶It has been found that better precision can be obtained by using the area measurement and by calibrating with one collector. It is advisable to calibrate and field test with the same collector if better than 4% relative standard deviation is necessary.

Relieve the partial vacuum in the flask reservoir by injecting the volume withdrawn with air back into the reservoir. Withdrawals should be alternated among the three reservoirs. If more than 10% of the total volume is withdrawn, equilibrium will be established between the liquid and gaseous mercury in about 15 min after replenishing the withdrawn volume with air, provided that a clean liquid mercury surface is present.

8.2 BLANK. If proper precautions to eliminate mercury contamination have been taken before the analysis step in **7.2.2** and the calibration step in **8.1.2**, the blank value is zero. In the event that the field blanks are significant, the average blank should be subtracted from the measured results on the samples.

9. Calculations

9.1 Calculate the concentration of mercury vapor in air by dividing the micrograms found by the total number of cubic meters of air taken for analysis.

9.2

$$\mu g \ Hg/m^3 = \frac{\mu g \ Hg \ found \times 10^6}{mL/min \times min \ sampled}$$

10. Storage

10.1 Collected field samples, properly capped (air tight) may be stored 6 months without gain or loss of mercury.

11. References

1. LONG, S. J., D. R. SCOTT AND R. J. THOMPSON. 1973. Atomic Absorption Determination of Elemental Mercury Collected from Ambient Air on Silver Wool. *Anal. Chem.*, 45:2227.
2. SPITTLER, T. M., R. J. THOMPSON, AND D. R. SCOTT. 1973. Division of Water, Air and Waste, 165th National Meeting, ACS, Dallas, Texas.
3. CHASE, D. L., D. L. SGONTZ, E. R. BLOSSNER, AND W. M. HENRY. 1972. Development and Evaluation of an Analytical Method for the Determination of Total Atmospheric Mercury. Battelle Columbus Laboratories, EPA Contract No. EHSD 71-32, Environmental Protection Agency.
4. SCARINGELLI, F. P., J. C. PUZAK, B. I. BENNETT, AND R. L. DENNEY. 1974. Determination of Total Mercury in Air by Charcoal Adsorption and Ultraviolet Spectrophotometry. *Anal. Chem.*, 46:278.
5. WILLISTON, S. H. 1968. *J. Geophys. Res.*, 73:7051.
6. FOOTE, R. S. 1972. *Science*, 177:513.
7. LONG, S. M. and D. R. SCOTT. Personal communication.
8. NIOSH MANUAL OF ANALYTICAL METHODS, 3rd Ed. P. Eller, Editor, US Dept. of Health and Human Services, Method 6000-Mercury. (1984).

Subcommittee 6

T. J. KNEIP, *Chairman*
R. S. AJEMIAN
J. R. CARLBERG
J. DRISCOLL
L. KORNREICH
J. W. LOVELAND
J. L. MOYERS
R. J. THOMPSON

Reapproved with modifications
Subcommittee 6
M. T. Kleinman
J. R. Rhodes
R. J. Thompson
T. C. Rains
W. J. Courtney
V. P. Guinn

319.

Determination of Molybdenum Content of Atmospheric Particulate Matter by Atomic Absorption Spectrophotometry

1. Principle of the Method

1.1 Molybdenum may exist in the atmosphere as the element, the oxide or as molybdates. Concentrations of molybdenum in ambient air range from below detectable amounts to maximum values in the order of 0.34 $\mu g/m^3$ of air (**1**). The satisfactory determination of molybdenum depends largely on the formation of an extractable green complex by the reaction of molybdenum with 4-methyl-1, 2-dimercaptobenzene, also called toluene-3, 4-dithiol or dithiol (**2**). Iron interference can be prevented by reducing ferric iron to the ferrous state with potassium iodide and decolorizing the liberated iodine with sodium thiosulfate. The dithiol complex is extracted with methyl isobutyl ketone (MIBK). The MIBK extract is analyzed with an atomic absorption spectrophotometer using the operating conditions as recommended by the manufacturer (**3**).

2. Range and Sensitivity

2.1 The detection limit for molybdenum in aqueous solutions is about 0.03 $\mu g/mL$.

2.2 The detection limit for molybdenum in organic phase is lower than that in aqueous solutions.

2.3 The sensitivity is 0.6 μg of molybdenum/mL for 1% absorption. (**4**) A concentration of 0.15 μg of molybdenum/m^3 of air can be measured if a 20 m^3 sample is taken for analysis and is extracted into 5 mL of MIBK.

2.4 To increase the sensitivity a larger air sample can be taken or a smaller amount of MIBK can be used.

2.5 Samples wet-ashed with nitric acid can be analyzed directly by electrothermal AAS with a sensitivity of 0.5 ng/mL. (**4**)

3. Interferences

3.1 No optical interferences will be encountered with this method.

3.2 Interferences may be encountered when using an air-acetylene flame to determine molybdenum.

4. Precision and Accuracy

4.1 No data are available in the literature for the range required.

5. Apparatus

5.1 An atomic absorption spectrophotometer for use at 313.3 nm.

5.2 Hollow cathode lamp for molybdenum.

5.3 Acetylene gas as fuel, nitrous oxide gas as oxidant.

5.4 125-mL separatory funnels, volumetric and general laboratory glassware.

5.5 Automatic shaker, optional.

5.6 Electrothermal AAS furnace and pyrolytic graphite tubes to fit, optional.

5.7 Argon compressed gas, pure, with regulator.

5.8 Micropipettor.

6. Reagents

6.1 PURITY. All reagents must conform to ACS Reagent specifications. Water should conform to ASTM standards for Reagent water, Type II.

6.2 STANDARD MOLYBDENUM SOLUTION (1.0 MG MO/ML). Dissolve 3.000 g of molybdenum trioxide in 200 mL of 10% sodium hydroxide and dilute to 2 L with distilled water. Store in a polyethylene container. (This solution is stable indefinitely.) Standardize this solution if necessary by titrating with standard ceric sulfate using 1, 10-phenanthroline indicator (2 drops of 0.025 M indicator are used) (5,6). Dilute an aliquot with water to obtain a solution containing approximately 25 μg of molybdenum/mL. Molybdenum trioxide of over 99.5% purity is available. Consequently sufficiently accurate standards may be prepared without standardization.

6.3 HYDROCHLORIC ACID, sp gr 1.19.

6.4 NITRIC ACID, sp gr 1.42.

6.5 POTASSIUM IODIDE. 50% (w/v) in water.

6.6 SODIUM THIOSULFATE. 10% (w/v) in distilled water.

6.7 DITHIOL REAGENT. Dissolve 1 g of dithiol (toluene-3,4-dithiol) in 500 mL of 1% sodium hydroxide solution. Stir occasionally over 1 h. Add about 8 mL of thioglycolic acid dropwise until a faint permanent opalescent turbidity forms. Store at 5°C. (**Caution:** Dithiol should not be allowed to contact the skin. Gloves should be worn when the reagent is used (2).) Discard this reagent after one week.

6.8 METHYL ISOBUTYL KETONE, (MIBK).

6.9 HYDROGEN PEROXIDE, 30%.

7. Procedure

7.1 Wet ash samples collected on membrane or cellulose filters or extract glass fiber filter samples in a Phillips beaker using enough concentrated nitric acid to cover the filter (take whole sample or suitable aliquot to give about 1.0 μg of molybdenum/mL). In the case of glass fiber filters, heat the beaker below the boiling point of the acid to extract the molybdenum, decant the liquid into a second beaker and wash the Phillips beaker and filter with nitric acid and distilled water. Repeat washing 3 times. Add all washings to the second beaker.

7.2 Place the second beaker on a low temperature hot plate (125°C) and heat until the brown fumes of oxides of nitrogen appear, then transfer the beaker to a high temperature hot plate (400°C) and heat until the fumes of nitrogen oxides disappear. Repeat the ashing treatment the second time with further additions of 2 mL of concentrated nitric acid. Addition of 3 mL of hydrogen peroxide will help complete ashing. Note: If analysis is to be performed by electrothermal AAS, go to section **7.6**.

7.3 Take up the residue with 3 mL of reagent grade HCl and dilute to 50 mL with distilled water. Add 1 mL of 50% KI solution to reduce the ferric iron, swirl, and let stand for 10 min. The appearance of a reddish brown color indicates the liberation of iodine. Decolorize the iodine by dropwise addition of 10% sodium thiosulfate solution and add 2 drops in excess.

7.4 Transfer to a 125-mL separatory funnel and add 2 mL of the dithiol reagent, shake for 30 s and let stand for 10 min; 5.0 mL of MIBK are pipetted into each funnel. The funnels are shaken vigorously and continuously for 3 min. Allow 10 min for settling after which the aqueous layer is drained off and discarded. Water is removed from the drain cock bore by inserting filter paper into the bore. The MIBK layer is drawn off and analyzed with an atomic absorption spectrophotometer.

7.5 A blank should be carried through entire procedure.

7.6 Set up the graphite furnace conditions (if that option is selected) as shown in Table 319:I.

7.7 Dilute sample as appropriate to provide a Mo concentration in the optimum range of the instrumental/furnace combination being used.

7.8 Inject 20-μL sample into furnace and process as appropriate. If matrix effects are suspected, sample should be analyzed using the method of standard additions (**Method 822; Part 1, Section 26**).

8. Calibration

8.1 Prepare a calibration curve using standards containing 0 to 10 μg of molybdenum and carry them through the complete procedure, Section 7.

Table 319:I Conditions for Graphite Furnace

Parameter	Function*				
	Dry	Char-1	Char-2	Atomize	Cool
Temperature, °C	115	1500	1500	2700	80
Ramp, s	20	20	–	0.6	12
Hold, s	10	4.6	1.0	4	–
Air flow, L/min	3	3	0	0	3

*Conditions will vary with instruments. Manufacturer's suggested settings should be tried initially, then optimized for this determination.

8.2 Plot absorbance *vs* concentration or use calibration procedure as recommended by the manufacturer. The preferred method is to obtain a calibration equation by the method of least squares.

9. Calculations

9.1 Use the absorbance of the samples to obtain the molybdenum concentration from the calibration curve or equation.

9.2 Apply the aliquot factor; correct the sampled volume to reference conditions if necessary.

9.3 Concentration of molybdenum in

$$\mu g/m^3 = \frac{\mu g \; Mo}{V}$$

where V = air sample volume, m³, corrected to 25°C, 101.3 kPa if necessary.

10. Effect of Storage

10.1 Molybdenum collected on solid filter media can be stored indefinitely.

11. References

1. AIR QUALITY DATA FROM THE NATIONAL AIR SAMPLING NETWORK AND CONTRIBUTING STATE AND LOCAL NETWORKS. 1966. U.S. Department of Health, Education and Welfare. Public Health Service, National Air Pollution Control Administration.
2. DELAUGHTER, BUFORD. 1965. The Determination of Sub-PPM Concentrations of Chromium and Molybdenum in Brines. *Atomic Absorption Newsletter*, Vol. 4, No. 5. May.
3. *Analytical Methods for Atomic Absorption Spectrophotometry*. 1971. Perkin-Elmer Corp., Norwalk, Conn.
4. HOENIG, M., Y. VAN ELSEN and R. VAN CANTER. 1986. Factors Influencing the Determination of Molybdenum Plant Samples by Electrothermal Atomic Absorption Spectrometry. *Anal. Chem.*, 58: 777–780.
5. FURMAN, N. H. and W. M. MURRY, JR. 1936. *J. Am. Chem. Soc.* 58, 1969.
6. *Scotts Standard Methods of Chemical Analyses*. 1962. Sixth Edition, I, 1205, D. Van Nostrand, N.Y.

Subcommittee 7

R. E. KUPEL, *Chairman*
M. M. BRAVERMAN
J. M. BRYANT
H. E. BUMSTEAD
A. CAROTTI
H. M. DONALDSON
L. DUBOIS
W. W. WELBON

Reapproved with modifications
Subcommittee 6

M. T. KLEINMAN, *Chairman*
J. R. RHODES
R. J. THOMPSON
T. C. RAINS
W. J. COURTNEY
V. P. GUINN

401.

Determination of Ammonia in the Atmosphere (Indophenol Method)

1. Principle of the Method

1.1 Ammonia in the atmosphere is collected by bubbling a measured volume of air through a dilute solution of sulfuric acid to form ammonium sulfate.

1.2 The ammonium sulfate formed in the sample is analyzed colorimetrically by reaction with phenol and alkaline sodium hypochlorite to produce indophenol (1), a blue dye. The reaction is accelerated by the addition of sodium nitroprusside as catalyst. The reaction has been postulated to be as follows (2).

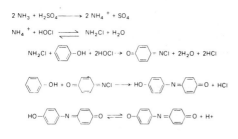

2. Range and Sensitivity

2.1 With a sampling rate of 1 to 2 L/min a concentration range of 20 to 700 $\mu g/m^3$ (0.025 to 1 ppm) in air may be determined with a sampling time of 1 h.

2.2 The limit of detection of the analysis is 0.02 μg NH_3/mL.

3. Interferences

3.1 Ammonium compounds in suspended particulate matter will be determined if they are not removed by prefiltration.

3.2 Prefilters may remove some gaseous ammonia. See section **7.1.1**.

3.3 Ferrous, chromous, and manganous ions if present in mg amounts cause positive interference in the analytical procedure because of precipitation. Copper ions inhibit color development strongly and therefore cause negative interference. Addition of ethylene diamine tetraacetic acid (EDTA) prevents these effects (1). Nitrite and sulfite interfere if present in 100-fold excess. Based on tests with solutions, formaldehyde causes a negative interference of 10 to 15%. Interfering particulate matter in the air can be removed by filtration of the air sample.

4. Precision and Accuracy

4.1 Replicate samples, collected with glass prefilters and analyzed manually showed a relative coefficient of variation of 30% in the 0.7 to 21 $\mu g/m^3$ (1 to 30 ppb) range (3). This coefficient varies with concentration of atmospheric ammonia and decreases to 5% in the 700 $\mu g/m^3$ (1 ppm) range.

4.2 No accuracy data are available.

5. Apparatus

5.1 VACUUM PUMP. Any vacuum pump that will maintain a vacuum of 60 kPa with an air flow of 5 L/min.

5.2 FLOW MEASURING DEVICE. A calibrated flow meter or a critical orifice to measure or control the air flow from 1 to 2 L/min. The flow meter should be calibrated under conditions of use.

5.3 ABSORBER. A fritted bubbler (3), midget impinger, or other gas scrubber designed for a flow rate of 1 to 2 L/min with 10 mL or more absorbing solution.

5.4 PREFILTER HOLDER. An open-face Teflon, Lexan, or similar filter holder which can be connected in line before the absorber. Metal or other plastic holder may be used if it is known that it does not absorb ammonia.

5.5 PREFILTERS. Organic-free glass fiber filters used for air sampling for suspended particulate matter are satisfactory. The filters are washed with distilled water and dried prior to use.

5.6 SPECTROPHOTOMETER. Capable of measuring the absorbance at 630 nm.

5.7 GLASSWARE. Low-actinic glassware or vessels must be used for analysis. The glassware must be rinsed with 1.2 N HCl, and washed at least 6 times with reagent water before immediate use.

6. Reagents

6.1 PURITY. All chemicals should meet specifications of ACS Analytical Reagent grade.

6.2 WATER. Water must meet specifications of ASTM Reagent water Type II, and be ammonia free.

6.3 ABSORBING SOLUTION. Dilute 3.0 mL of concentrated H_2SO_4 (18 M) to 1 L with water to obtain 0.1 N H_2SO_4.

6.4 ANALYTICAL REAGENTS.

6.4.1 *Sodium nitroprusside* [sodium nitrosylpentacyanoferrate (III)] **(4)**. Dissolve 2 g sodium nitroprusside in 100 mL of water. The solution keeps well in the refrigerator for 2 months.

6.4.2 *6.75 M sodium hydroxide.* Dissolve 270 g sodium hydroxide in about 1 L of distilled water. Boil down to 600 mL in order to volatilize the ammonia contained in the reagent. Cool and fill to 1 L. Store in polyethylene bottle. **Caution:** *This solution is extremely caustic. Prevent contact with skin or eyes.*

6.4.3 *Sodium hypochlorite solution.* Dilute 5 to 6% analytical reagent sodium hypochlorite with distilled water to give a 0.1 N solution (3.7%). Strength is determined before dilution by iodimetric titration or colorimetry after appropriate dilution. The solution keeps well for 2 months in a refrigerator.

6.4.4 *Phenol solution 45% v/v.* Melt phenol by immersing a bottle containing the material in a water bath at 60°C. Pour 45 mL (50 g) into a 100 mL warmed cylinder and fill to mark with methanol. This solution keeps well for 2 months in a refrigerator.

6.4.5 *Buffer.* Dissolve 50 g of Na_3PO_4 · $12H_2O$ and 74 mL of 6.75 M NaOH in 1 L of distilled water.

6.4.6 *Working hypochlorite solution.* Mix 30 mL of 0.1 N sodium hypochlorite and 30 mL of 6.75 M sodium hydroxide and dilute to 100 mL with distilled water. Prepare fresh daily.

6.4.7 *Working phenol solution.* Mix 20 mL of the 45% phenol solution with 1 mL of 2% sodium nitroprusside and dilute to 100 mL with distilled water. Prepare fresh every 4 hr.

6.5 AMMONIA STANDARD SOLUTION.

6.5.1 *Ammonia stock solution.* Dissolve 3.18 g of NH_4Cl or 3.88 g of $(NH_4)_2SO_4$ in 1 L of distilled water (1 mL equal to 1 mg NH_3). Add a drop of $CHCl_3$ for better preservation. The solution is stable for 2 months.

6.5.2 *Ammonia working solution.* Dilute 10 mL of the stock solution to 1 L with absorbing solution in a volumetric flask (1 mL equal to 10 μg NH_3). Prepare daily.

6.6 GLASS CLEANING SOLUTION. Dilute 10 mL of concentrated HCl (12 M) to 100 mL with distilled water (molarity approximately 1.2 M).

7. Procedure

7.1 SAMPLE COLLECTION. Place 10 mL of absorbing solution in each bubbler for samples and field blanks. Cap bubblers for transport. Assemble (in order) prefilter and holder, flowmeter, bubbler, and pump. Sample at the rate of 1 to 2 L/min for a sufficient time to obtain an adequate sample, usually 1 h. Record sampling time and flow rate. After sample collection, recap the bubblers.

7.1.1 *Prefilters.* If prefilters are not used the method will determine both gaseous ammonia and ammonium contained in particulates. At high humidity, acid gas will promote reaction on the filter causing loss of ammonia gas from the sample. In the absence of acid gases, ammonia collected momentarily on the filter during high humidity will be stripped off during sampling with little loss. The filter must be prevented from being wetted by rain.

7.2 ANALYSIS. If bubbler is marked at the 25.0 mL level, the color may be developed in the flasks. If not, transfer contents to a 25.0 mL glass stoppered graduated cylinder, being sure to blow out residual sample from the frits if they are used. Maintain all solutions and sample at 25°C. Add 2 mL buffer. Add 5 mL of the working phenol solution, mix, fill to about 22 mL, then add 2.5 mL of the working hypochlorite solution and rapidly mix. Dilute to 25 mL, mix, and store in the dark at 25°C for 30 min to develop color. Measure the absorbance of the solution against a reagent blank at 630 nm, using 1-cm cells.

7.3 FIELD BLANKS. At least one bubbler of collecting solution is carried into the field and treated in the same fashion as the actual samples except that no air is drawn into the bubbler. It is treated in analysis as if it was a sample. The value of the field blank(s) is compared with the reagent blank to determine whether sampling glassware is introducing appreciable contamination.

8. Calibration and Standardization

8.1 PREPARATION OF STANDARDS. Pipet 0.5, 1.0 and 1.5 mL of the working standard solution into 25-mL glass stoppered graduated cylinders. These correspond to 5, 10 and 15 μg of ammonia/25 mL of solution. Fill to the 10-mL mark with absorbing solution. A reagent blank with 10 mL of absorbing solution is also prepared. Add reagents to each cylinder as in the procedure for analysis. Read the absorbance of each standard against the reagent blank.

8.2 STANDARD CURVE. Plot the absorbance as the ordinate versus the concentration as the abscissa on linear graph paper. Alternatively, determine the slope by the method of least squares.

9. Calculations

$$\mu g/m^3 \ NH_3 \ = \ \frac{W}{V_O}$$

where W = μg NH$_3$ in 25 mL from standard curve.

V$_O$ = volume of air sampled in m^3 at

25° C and 101.3 kPa

$$V_O \ = \frac{F}{1000} \times t \times \frac{P_s}{101.3} \ \times \ \frac{298}{273 + T_s}$$

where F = flow rate (L/min).

t = elapsed sampling time in min.

P$_s$ = atmospheric pressure in kPa at sampling point.

T$_s$ = temperature, °C at sampling point.

10. Effects of Storage

10.1 Samples of particulate matter may be stored indefinitely if protected from contamination.

10.2 Changes in the precision of the method occur when storing the collected liquid samples during 2 days. Significantly lower values have been found on replicates stored for several days.

11. References

1. TETLOW, J.A. AND A.L. WILSON. 1964. An Absorptionmetric Method for Determining Ammonia in Boiler Feed-Water. Analyst, 89:453–465.
2. ROMMERS, P.J. AND J VISSER. 1969. Spectrophotometric Determination of Micro Amounts of Nitrogen as Indophenol. Analyst, 94:653–658.
3. AXELROD, H.D., A.F. WARTBURG, R.J. TECK AND J.P. LODGE, JR. 1971. A New Bubbler Design for Atmospheric Sampling. Anal Chem, 43:1916–1917.
4. HARWOOD, J.E. AND A.L. KUHN. 1970. A Colorimetric Method for Ammonia in Natural Waters. Water Res, 4:305.

Subcommittee 3

E. L. KOTHNY, *Chairman*
W. A. COOK
J. E. CUDDEBACK
B. DIMITRIADES
E. F. FERRAND
P. W. MCDANIEL
G. D. NIFONG
B. E. SALTZMAN
F. T. WEISS

Approved with modifications from 2nd edition.

D. A. LEVAGGI, *Chairman*
B. R. APPEL
D. W. HORSTMAN
E. L. KOTHNY
J. G. WENDT

404.

Determination of Nitrate in Atmospheric Particulate Matter (Brucine Method)

1. Principle of the Method

1.1 Particulate matter is collected from the air on a 20 × 25 cm (8″ × 10″) glass fiber filter. Nitrates are extracted with water. The nitrate in the acidified extract is reacted with brucine to form a color which is determined spectrophotometrically at 410 nm **(1,2,3,4)**.

2. Range and Sensitivity

2.1 Samples of outside atmospheres usually contain 0.1 to 10 μg nitrate/m³. Usually a 24-h sample is collected with a high-volume sampler, although much smaller samples are adequate. Results are calculated to represent 24-h averages, if a number of samples are collected within a 24-h period.

2.2 This method has a working range of 20 to 100 μg of nitrate ion and a sensitivity of 4 μg in the sample solution aliquot used for the test.

2.3 For greater sensitivity, the aliquot of sample to be extracted may be increased proportionately up to the entire sample, or the amount of diluting acid may be reduced.

3. Interferences

3.1 When the sample extract is colored, a sample blank consisting of an equal volume of the sample extract without brucine is carried through the procedure and the absorbance of the blank is subtracted from the sample.

3.2 Nitrites may cause results to be high **(2)**. However, significant amounts of nitrites (such as 4 μg nitrite ion in the solution aliquot) are not usually found in the atmospheric particulate matter.

3.2.1 The interference of nitrites is eliminated **(5)** using a modified reagent (see Section **6.4**).

3.3 Colored substances may form by interaction of organic substances with the concentrated sulfuric acid used for the reaction.

3.4 Chlorides cause erratic low results in high concentrations (1 to 60 g/L). However, significant amounts of chlorides (above 1 g/L) are not usually found in atmospheric particulate matter.

3.4.1 With suitable modifications, (see Section **6.4**) the interference of chlorides is eliminated **(5,6)**.

4. Precision and Accuracy

4.1 The method has a precision of 2 μg NO_3^- over the working range of 20 to 100 μg NO_3^-. The blank is usually negligible.

4.2 The results for nitrate by this method show excellent agreement with the phenol disulfonic acid method **(3)**.

5. Apparatus

5.1 Any device capable of collecting particulate matter from a volume of at least 20 m³ of air within the desired period of time may be employed.

5.1.1 *High-volume sampler.* A motor-operated blower with a sampling head for a 20 × 25 cm (8″ × 10″) glass fiber filter and capable of an initial air flow of 1.1 to 1.7 m³/min (40 to 60 cfm) is suitable for particulate matter sampling. A flow-measuring device, usually a rotameter, is required to indicate flow rate.

5.1.2 Membrane filters at a sampling rate of 10 L/min and electrostatic precipitators at a sampling rate of about 100 L/min may also be used. However, the high-volume sampler is recommended.

5.2 REFLUXING APPARATUS. A 125 mL flask fitted with reflux condenser and heating mantle or hot plate.

5.3 BOILING WATER BATH.

5.4 GRADUATED CYLINDERS, 50 ML.

5.5 SPECTROPHOTOMETER SUITABLE FOR MEASUREMENT AT 410 NM.

5.6 CUVETTES. 2.54 cm light path — A 1-cm light path cuvette may be used with decreased sensitivity.

6. Reagents

6.1 PURITY. All chemicals should be ACS Analytical Reagent grade. Water means ASTM Reagent water, Type II.

6.2 CONCENTRATED SULFURIC ACID, 98%, AND DILUTED SULFURIC ACID, 88%. Mix cautiously 1000 mL 98% H_2SO_4 into 250 mL of distilled water.

6.3 BRUCINE (FREE BASE), 2.5% SOLUTION IN CHLOROFORM. Filter before use. If the free base is not available, brucine hydrochloride, 1% in 0.05 N HCl may be substituted. This latter reagent requires the alternate procedure, Section **7.3.2**. Both reagents are stable for months when kept in a brown bottle **(1)** or in a refrigerator.

6.4 If nitrites and chlorides are present in more than traces, the following reagent is recommended:

6.4.1 *Stock solution.* 1 g brucine sulfate and 0.1 g sulfanilic acid dissolved in 70 mL of hot water and 3 mL concentrated HCl. Cool and dilute to 100 mL. This reagent is stable for 3 to 6 months if kept in a refrigerator.

6.4.2 *Mixed reagent.* Just prior to use mix 400 mL of a 30% NaCl solution with 100 ml of stock solution, Section **6.4.1**. Prepare fresh daily.

6.5 NITRATE STOCK SOLUTION, 100 Mg NO_3^-/ML. Dissolve 0.163 g potassium nitrate in distilled water and dilute to 1 L. Sterile solutions should stay stab'?.

6.6 DILUTE STANDARD, 10 μg NO_3^-/ML. Pipet 10 mL of stock solution into a 100-mL volumetric flask and dilute to mark. Prepare fresh.

7. Procedure

7.1 SAMPLE COLLECTION. A common procedure is to employ the high volume sampler with a 20 × 25 cm (8″ × 10″) glass fiber filter that is operated over a 24-h period. The average of the flow rates at the beginning and end of the sampling period is taken as the rate for the entire sampling period. This permits collection of nitrate from some 2,000 m^3 of air. Usually only 1% of the sample is taken for analysis. From 20 to 100 μg NO_3^- are required for the analysis.

7.2 SAMPLE PREPARATION. The sample filter may be folded along the 10″ axis for transportation and storage. When this is done, aliquots of the filter should be taken across the fold. A convenient aliquot is a 1.90 cm (3/4″) × 20.3 cm (8″) strip. This is refluxed with 50 mL of distilled water for 30 min. Insert a Whatman #42 filter paper in a funnel and wash free from nitrate with hot distilled water from a glass wash bottle. Discard the washings. Unwashed filter paper may add up to 60 μg NO_3^- to the filtrate. An adequate washing procedure is indicated by low blank values determined on previous trials. Filter the extract through this prepared paper and rinse with distilled water to obtain 100 mL of filtrate. An unexposed sample filter from the same filter batch is similarly refluxed and treated to provide a blank. This larger volume allows other analyses to be made on the same extract.

7.3 ANALYTICAL PROCEDURE **(3,5)**. Three methods are given depending on which brucine reagent has been prepared.

7.3.1 Take a 10 mL aliquot of sample filtrate into a 50 mL graduated cylinder. Add 0.2 mL of the 2.5% solution of brucine in chloroform and with caution add 20 mL of concentrated sulfuric acid. Heat in a boiling water bath. After 20 min, cool quickly in an ice bath and dilute to 50 mL with sulfuric acid.

7.3.2 Alternatively 0.5 mL of 1% solution of brucine hydrochloride is added to a cooled mixture of a 10 mL aliquot of sample filtrate with 20 mL of sulfuric acid. Place in a boiling water bath for 20 min, then cool quickly.

7.3.3 For higher-than-trace amounts of nitrites and chlorides proceed as follows: Take a 10 mL aliquot of sample filtrate in a 50-mL cylinder. Add 2.5 mL of mixed reagent, Section **6.4.2**, mix thoroughly, add 10 mL of 88% H_2SO_4 and place in a boiling water bath for 20 min. Then cool quickly.

7.3.4 Read on a suitable spectropho-tometer at 410 nm any time within 24 h against the blank.

8. Calibration and Standardization

8.1 Prepare a standard curve by diluting 2 to 10 mL aliquots of dilute standard solution to 10 mL and proceeding as in **7.3** A straight line is generally obtained by plotting the absorbance at 410 nm against the number of micrograms of nitrate ion.

9. Calculations

9.1 Micrograms of nitrate per m^3 = C/FV.

where F = sample aliquot fraction (fraction taken of exposed filter area times fraction of aqueous extract). The exposed filter area is calculated from the actual dimensions of the frame opening.

C = number of μg NO$_3^-$ found after subtraction of blank, and

V = sample air volume in m^3 at 101.3 kPa and 25°C.

10. Effect of Storage

10.1 Sample filters can be stored indefinitely.

10.2 The extractions should be performed within 4 h before the analysis.

11. References

1. HAASE, L.W. 1926. Chem Ztg, 50:372.
2. LUNGE, G. AND L. WOLFF. 1894. Z. Angew. Chem, 12:345.
3. FISHER, F.L., E.R. IBERT AND H.F. BECKMAN. 1958. Anal. Chem, 30:1972.
4. JENKINS, D. AND L. MEDSKER. 1964. Anal Chem, 36:610.
5. KAHN, L. AND F.T. BREZENSKI. 1967. Environ Sci Technol, 1:488.
6. MCFARREN, E.F. AND R.J. LISHKA. 1968. "Trace Inorganics in Water." American Chemical Society, p. 253.

Subcommittee 3

E. L. KOTHNY, *Chairman*
W. A. COOK
B. DIMITRIADES
E. F. FERRAND
G. D. NIFONG
P. W. MCDANIEL
B. E. SALTZMAN
F. T. WEISS

Approved with modification from 2nd Ed.

D. A. LEVAGGI, *Chairman*
B. R. APPEL
D. W. HORSTMAN
E. L. KOTHNY
J. G. WENDT

405.

Determination of Nitric Oxide Content of the Atmosphere

1. Principle of the Method

1.1 After chemically removing nitrogen dioxide normally present in the atmosphere, the nitric oxide is converted to an equimolar amount of nitrogen dioxide by oxidation.

1.2 A total conversion in the gas phase can be obtained with chromic oxide supported on inert inorganic materials (1,2).

1.3 The nitrogen dioxide so generated is determined by absorption in Griess-Saltzman reagent to produce a pink color (see Method 406).

2. Range and Sensitivity

2.1 This method is intended for the manual determination of nitric oxide in the atmosphere in the range of 0.005 to about 5 parts per million (ppm) by volume or 6 to 6000 $\mu g/m^3$.

2.2 The sensitivity is 0.01 μg NO/10 mL of absorbing solution.

3. Interferences

3.1 Interfering concentrations of ozone ordinarily cannot coexist with nitric oxide because of their rapid gas phase reaction (3).

3.2 Nitrogen dioxide is removed in the first absorber of the train before the oxidizer and produces a slight interference at higher concentrations (4). (3 to 4% of the incoming nitrogen dioxide is converted to nitric oxide.)

3.2.1 The use of a fritted bubbler with absorbing reagent for removal of nitrogen dioxide is not recommended (5).

3.2.2 Soda-lime not only absorbs nitrogen dioxide but also partially scrubs nitric oxide out of the gas stream and is not recommended (6).

3.3 Sulfur dioxide is removed by the oxidizer (7) and ordinarily produces no interference. If it is present in very high amounts (*e.g.,* stack gas levels), the oxidizer will deplete rapidly and must be changed more frequently. The oxidizer will indicate depletion by a change from orange to a brownish color.

3.4 Strongly oxidizing gases (*e.g.,* PAN, halogens, etc.) may affect the stability of the color of the Griess-Saltzman reagent. In such cases the absorbance of the solutions should be determined within one hour to minimize any loss (8). Interferences from oxidizing gases generally found in polluted atmosphere are negligible.

4. Precision and Accuracy

4.1 The precision of the method depends on the conversion efficiency of the oxidizer and other variables such as volume measurement of the sample, sampling efficiency of the fritted bubbler and absorbance measurement of the color. It may vary within ± 0.01 μg or ± 3%, whichever is greater.

4.2 Under controlled conditions, the conversion efficiency of the oxidizer varies within 98 to 100% (1,7).

4.3 At present, accuracy data are not available.

5. Apparatus

5.1 NITROGEN DIOXIDE ABSORBER. A 20 mm ID × 50 mm long polyethylene tube with connecting caps at both ends is filled with pellets of nitrogen dioxide absorbent. The pellets are held in place with glass wool plugs.

5.2 HUMIDITY REGULATOR. In order for the chomic oxide to maintain its 98 to 100% oxidation of nitric oxide to nitrogen dioxide it is necessary to maintain the sample air stream at 40–80% relative humidity. This is accomplished by use of a midget impinger containing 10–15 mL of water, with its inlet glass stem cut off about 2.5 cm above the liquid level (2).

5.3 OXIDIZER (**1,2,6**). A 15 mm ID glass tube with connecting ends is filled with 5–6 cm of prepared oxidizer material (Section **6.4**). The oxidizer should be changed when the material has been spent (evidenced by a change of color from yellow to the reduced green form of the chromium), or become visibly wet from prolonged sampling of extremely damp air.

5.3.1 The combined train of **5.1, 5.2, 5.3** can be used at least

$$\frac{100}{\text{ppm NO}_2}\text{ h}$$

and should be changed whenever it becomes visibly wet from sampling damp air or discolored from sampling strong reducing gases.

5.4 PLASTIC BAGS. Mylar or Tedlar plastic film bags may be used for collection and storage of gaseous samples. However, losses will occur in presence of oxygen due to the oxidation of NO to NO_2.

5.5 SYRINGES. 5-mL glass syringes with Teflon plungers are convenient for sampling NO concentrations above 100 ppm.

5.6 Evacuated bottles, syringes for sampling NO_2 and other equipment as stated in Method 406.

6. Reagents

6.1 PURITY. All chemicals should be ACS Analytical Reagent grade. Water is ASTM Reagent water Type II.

6.2 NITROGEN DIOXIDE ABSORBENT (**1,4,6**). 1.5 mm (1/16″) pellets or 10 to 20 mesh porous inert material, such as firebrick, alumina, zeolites, etc., is soaked in 20% aqueous triethanolamine, drained, spread on a wide petri dish, and dried for 30 to 60 min at 95°C in an oven. The pellets should be free flowing.

6.3 CONSTANT HUMIDITY GRAINS (CONSISTING OF A 50 + 50% ANHYDROUS AND HYDRATED SODIUM ACETATE MIXTURE). Stir slowly and add drop by drop 13 mL of water into a beaker containing 40 g of anhy-drous sodium acetate in order to obtain coarse-grained crystal pellets.

6.4 OXIDIZER (**1,2,6**). Glass, firebrick, or alumina, mesh size 15 to 40, is soaked in a solution containing 17 g of chromium trioxide in 100 mL of water. Then it is drained, dried in an oven at 105° to 115°C and exposed to 70% relative humidity. This can be done best by exposing a thin layer of pellets contained in a petri dish to a saturated solution of sodium acetate contained in a desiccator. The reddish color changes to a golden orange when equilibrated.

7. Procedure

7.1 Assemble a sampling train comprising, in order, rotameter, nitrogen dioxide absorber, humidity regulator, oxidizer, fritted absorber and pump. Pipet 10 mL of Griess-Saltzman reagent into the fritted absorber. Draw an air sample through at a rate of 0.4 L/min or less to develop sufficient final color (about 10 min). Record the total air volume sampled. If the sample temperature and pressure deviate greatly from 25°C and 101.3 kPa measure and record these values. Read the absorbance of the solution at 550 nm after 15 min color developing time.

8. Calibration and Standardization

8.1 Either of two methods of calibration may be employed. The most convenient method is standardizing with nitrite solution (**9**); the other method is by standardizing with known gas mixtures. With this latter method, stoichiometric and efficiency factors are eliminated from the calculations. Concentration of the standards should cover the expected range of sample concentrations.

8.1 STANDARDIZATION BY NITRITE SOLUTION. After conversion of NO to NO_2, standardization can be accomplished as indicated in Method 406. Construct the calibration curve by adding graduated amounts of dilute nitrite solution, equivalent to concentrations expected to be sampled, to a series of 25 mL volumetric flasks. Plot the absorbances against microliters of nitrogen dioxide per 10 mL of absorbing solution. If preferred, transmittance in per cent may be plotted on a logarithmic scale versus nitric oxide concentration on the linear coordinate on semilog graph paper. The plot follows Beer's Law up to absorbance 1 (or 10% transmittance).

8.2 GASEOUS STANDARDIZATION METHOD. Because of the reactivity of nitric oxide with oxygen and more so at high pressure when stored in a cylinder, only dynamic mixing with nitrogen in a diluting stream or static mixtures in bags prior to calibration is recommended.

8.2.1 *Dynamic dilution.* An asbestos plug dilutor **(10)** is used to feed a slow stream of 1% NO in N_2 into the air by means of a three-way stopcock placed upstream from the rotameter. One bore of the stopcock is plugged with asbestos and the gas pressure is maintained by a constant-head overflow. The 1% NO is obtained by pressurizing with N_2 a precisely measured volume of NO admitted into an evacuated stainless steel tank by means of a precision gauge. Trace amounts of oxygen present in N_2 will not cause errors when using this concentration level.

8.2.2 *Bag dilution.* A sample of pure NO is drawn with a syringe of a convenient size and then transferred to a Mylar or Tedlar bag while filling with nitrogen measured through a wet test meter. After a convenient mixing period, the procedure is repeated with a second bag to cover a lower range of concentrations. Then add to the bag contents an amount of pure oxygen corresponding to 1/4 of the amount of nitrogen and analyze within 20 min.

9. Calculations

9.1 For convenience, standard conditions are taken as 101.3 kPa and 25°C, where the molar volume is 24.47. Ordinarily the correction of the sample volume to these standard conditions is slight and may be omitted. However, if conditions deviate significantly, corrections might be made by means of the perfect gas equation.

9.2 The interpolated value of the microliters nitric oxide corresponding to an absorbance of one from the calibration graph can be used as factor K for calculating the concentration in parts per million, rather than reading all concentrations from graphs:

nitric oxide, ppm =

$$\frac{\text{absorbance in 10 mL} \times K}{\text{volume of air sample in liters}}$$

9.3 The absorbance must be corrected for fading of color when there is a prolonged interval between sampling and measurement of absorbance (Section **10.2**).

10. Effects of Storage

10.1 Atmospheric air samples can be stored in Mylar and Tedlar bags for 1 to 2 h without significant loss of their nitric oxide content.

10.2 The colors obtained by the sampling procedure must be protected from light. The well-stoppered tubes lose 3 to 4% of the absorbance daily, provided oxidizing gases have not been sampled **(7)**. In such cases the colors should be determined within one hour after sampling to minimize any loss.

11. References

1. LEVAGGI, D.A., E.L. KOTHNY, T. BELSKY, E.R. DE VERA AND P.K. MUELLER. 1971. A precise method for analyzing accurately the content of nitrogen oxides in the atmosphere. Presented at 17th Annual Meeting, Institute of Environmental Sciences, Los Angeles, California.
2. LEVAGGI, D.A., E.L. KOTHNY, T. BELSKY, E.R. DE VERA AND P.K. MUELLER. 1974. Quantitative Analysis of Nitric Oxide in Presence of Nitrogen Dioxide at Atmospheric Concentrations. Environ Sci & Tech, 6:348–350.
3. JOHNSTON, H.S., AND H.J. CROSBY. 1954. Kinetics of the Fast Gas Phase Reaction Between Ozone and Nitric Oxide. J. Chem. Phys. 22:689.
4. LEVAGGI, D.A., W. SIU, E. KOTHNY, AND M. FELDSTEIN. 1972. The Quantitative Separation of Nitric from Nitrogen Dioxide at Atmospheric Concentration Ranges. Environ Sci & Tech, 6:250–252.
5. HUYGEN, C. 1970. Reaction of Nitrogen Dioxide with Griess Type Reagents. Anal. Chem. 42:407.
6. BELSKY, T. 1970. Experimental evaluation of triethanolamine and chromium trioxide in the continuous analysis of NO in the air. AIHL Report No. 85.
7. SALTZMAN, B.E., AND A.F. WARTBURG. 1965. Absorption Tube for Removal of Interfering Sulfur Dioxide in Analysis of Atmospheric Oxidant. Anal. Chem. 37:779.
8. SALTZMAN, B.E. 1954. Colorimetric Microdetermination of Nitrogen Dioxide in the Atmosphere. Anal. Chem. 12:1919.
9. SCARINGELLI, F.P., E. ROSENBURG, AND K.A. REHME. 1970. Comparison of Permeation Tubes and Nitrite Ion as Standards for the Colorimetric Determination of Nitrogen Dioxide. Env. Sci. and Tech. 4:924.

10. SALTZMAN, B.E. 1961. Preparation and Analysis of Calibrated Low Concentrations of 16 Toxic Gases. Anal. Chem. 33:1100.

Subcommittee 3

E. L. KOTHNY, *Chairman*
W. A. COOK
B. DIMITRIADES
E. F. FERRAND
C. A. JOHNSON
L. LEVIN
P. W. MCDANIEL
B. E. SALTZMAN

Approved with modification
from 2nd edition
Subcommittee 3
D. A. LEVAGGI, *Chairman*
B. R. APPEL
D. W. HORSTMAN
E. L. KOTHNY
J. G. WENDT

406.

Determination of Nitrogen Dioxide Content of the Atmosphere (Griess-Saltzman Reaction)*

1. Principle of the Method

1.1 The nitrogen dioxide is absorbed in an azo dye forming reagent (1). A stable red-violet color is produced within 15 min which may be read visually or in an appropriate instrument at 550 nm.

2. Range and Sensitivity

2.1 This method is intended for the manual determination of nitrogen dioxide in the atmosphere in the range of 0.005 to about 5 parts per million (ppm) by volume or 0.01 to 10 μg/L, when sampling is conducted in fritted bubblers. The method is preferred when high sensitivity is needed.

2.2 Concentrations of 5 to 100 ppm in industrial atmospheres and in gas burner stacks also may be sampled by employing evacuated bottles or glass syringes. For higher concentrations, for automotive exhaust, and/or for samples relatively high in sulfur dioxide content, other methods should be applied.

3. Interferences

3.1 A tenfold ratio of sulfur dioxide to nitrogen dioxide produces no effect. A 30-fold ratio slowly bleaches the color to a slight extent. The addition of 1% acetone to the reagent before use retards the fading by forming another temporary product with sulfur dioxide. This permits reading

within 4 to 5 h (instead of the 45 min required when acetone is not added) without appreciable interferences. Interference from sulfur dioxide may be a problem in some stack gas samples (see **2.2**).

3.2 A fivefold ratio of ozone to nitrogen dioxide will cause a small interference, the maximal effect occurring in 3 h. The reagent assumes a slightly orange tint.

3.3 Peroxyacylnitrate (PAN) can give a response of approximately 15 to 35% of an equivalent molar concentration of nitrogen dioxide (2). In ordinary ambient air the concentrations of PAN are too low to cause any significant error.

3.4 The interferences from other nitrogen oxides and other gases that might be found in polluted air are negligible. However, if the evacuated bottle or syringe method is used to sample concentrations above 5 ppm, interference from NO (due to oxidation to NO_2) is possible (see **7.1.3**).

3.5 If strong oxidizing or reducing agents are present, the colors should be determined within 1 h, if possible, to minimize any loss.

4. Precision and Accuracy

4.1 A precision of 1% of the mean can be achieved with careful work (3); the limiting factors are the measurements of the volume of the air sample and of the absorbance of the color.

4.2 At present, accuracy data are not available.

5. Apparatus

5.1 ABSORBER. The sample is absorbed in an all-glass bubbler with a 60 μm maxi-

---

*This is a version of ASTM Method D 1607, adopted 1960 and revised 1976. Adapted from *Selected Methods for the Measurement of Air Pollutants*, PHS Publication No. 999-AP-11, May 1965.

mum pore diameter frit similar to that illustrated in Figure 406:1.†

5.1.1 The porosity of the fritted bubbler, as well as the sampling flow rate, affect absorption efficiency. An efficiency of over 95% may be expected with a flow rate of 0.4 L/min or less and a maximum pore diameter of 60 μm. Frits having a maximum pore diameter less than 60 μm will have a higher efficiency but will require an inconvenient pressure drop for sampling (see formula in **5.1.2**). Considerably lower efficien-

cies are obtained with coarser frits, but these may be utilized if the flow rate is reduced.

5.1.2 Since the quality control by some manufacturers is rather poor, it is desirable to measure periodically the porosity of an absorber as follows: Carefully clean the apparatus with dichromate-concentrated sulfuric acid solution and then rinse it thoroughly with distilled water. Assemble the bubbler, add sufficient distilled water to barely cover the fritted portion, and measure the vacuum required to draw the first perceptible stream of air bubbles through the frit. Then calculate the maximum pore diameter as follows:

---

†Corning Glass Works Drawing XA–8370 specifies this item with 12/5 **T** ball and socket joints. Ace Glass, Inc. specifies this item as No. 7530.

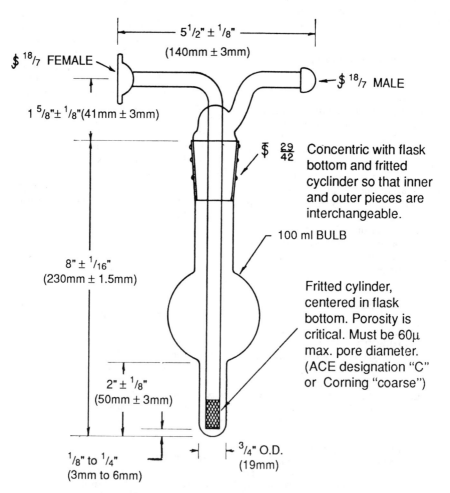

Figure 406:1 – Fritted bubbler for sampling nitrogen dioxide.

$$\text{Max pore diam } (\mu m) = \frac{4000s}{P}$$

where s = surface tension of water at the test temperature in N/m (0.073 at 18°C, 0.072 at 25°C, and 0.071 at 31°C), and

P = measured vacuum, kPa

5.1.3 Rinse the bubbler thoroughly with water and allow to dry before using. A rinsed and reproducibly drained bubbler may be used if the volume (r) of retained water is added to that of the absorbing reagent for the calculation of results. This correction may be determined as follows: Pipet into a drained bubbler exactly 10 mL of a colored solution (such as previously exposed absorbing reagent) of absorbance (A_1). Assemble the bubbler and rotate to rinse the inside with the solution. Rinse the fritted portion by pumping gently with a rubber bulb. Read the new absorbance (A_2) of the solution. Then:

$$10A_1 = (10 + r)A_2$$

$$\text{or: } r = 10\left(\frac{A_1}{A_2} - 1\right)$$

5.2 AIR METERING DEVICE. A glass rotameter capable of accurately measuring a flow of 0.4 L/min is suitable. A wet test meter is convenient to check the calibration.

5.3 SAMPLING PROBE. A glass or stainless steel tube 6 to 10 mm in diameter provided with a downward-facing intake (funnel or tip) is suitable. A small loosely fitting plug of glass wool may be inserted, when desirable, in the probe to exclude water droplets and particulate matter. The dead volume of the system should be kept minimal to permit rapid flushing during sampling to avoid losses of nitrogen dioxide on the surfaces.

5.4 GRAB-SAMPLE BOTTLES. Ordinary glass-stoppered borosilicate glass bottles of 30- to 250-mL sizes are suitable if provided with a mating ground joint attached to a stopcock for evacuation. Calibrate the volume by weighing with connecting piece, first empty, then filled to the stopcock with distilled water.

5.5 GLASS SYRINGES. Fifty or 100 mL syringes are convenient (although less accurate than bottles) for sampling.

5.6 AIR PUMP. A vacuum pump capable of drawing the required sample flow for intervals of up to 30 min is suitable. A tee connection at the intake is desirable. The inlet connected to the sampling train should have an appropriate trap and needle valve, preferably of stainless steel. The second inlet should have a valve for bleeding in a large excess flow of clean air to prevent condensation of acetic acid vapors from the absorbing reagent, with consequent corrosion of the pump. Alternatively, soda lime may be used in the trap. A filter and critical orifice may be substituted for the needle valve **(4)**.

5.7 SPECTROPHOTOMETER OR COLORIMETER–a laboratory instrument suitable for measuring the pink color at 550 nm, with stoppered tubes or cuvettes. The wavelength band width is not critical for this determination.

6. Reagents

6.1 PURITY OF CHEMICALS. All chemicals should meet ACS specifications for analytical reagents.

6.2 NITRITE-FREE WATER. All solutions are made in nitrite-free water. If available distilled or deionized water contains nitrite impurities (produces a pink color when added to absorbing reagent), redistill it in an all-glass still after adding a crystal each of potassium permanganate and of barium hydroxide.

6.3 N-(1-NAPHTHYL)-ETHYLENEDIAMINE DIHYDROCHLORIDE, STOCK SOLUTION (0.1%). Dissolve 0.1 g of the reagent in 100 mL of water. Solution will be stable for several months if kept well-stoppered in a brown bottle in the refrigerator. (Alternatively, weighed small amounts of the solid reagent may be stored.)

6.4 ABSORBING REAGENT. Dissolve 5 g of anhydrous sulfanilic acid (or 5.5 g of p-$NH_2 \cdot C_6H_4SO_3H \cdot H_2O$) in almost a liter of water containing 140 mL of glacial acetic acid. Gentle heating is permissible to speed up the process. To the cooled mixture, add 20 mL of the 0.1% stock solution of N-(1-naphthyl)-ethylenediamine dihydrochloride, and dilute to 1 L. Avoid lengthy contact with air during both preparation and use, since discoloration of reagent will result because of absorption of nitrogen dioxide. The solution will be stable for several months if kept well-stoppered in a brown bottle in the refrigerator. The absorbing reagent should be allowed to warm to room temperature before use.

6.5 STOCK SODIUM NITRITE SOLUTION. Weigh 2.03 g of sodium nitrite, and transfer to a 1 L volumetric flask. Add distilled water, dissolve, and bring to volume. This solution is stable for a year if stored in a brown bottle and kept refrigerated.

6.5.1 Take 10.0 mL of the stock solution and dilute to 1 L with distilled water. One mL of this working dilute solution produces a color equivalent to 10 μL of nitrogen dioxide. This solution must be prepared fresh when used as a calibration standard.

7. Procedure

7.1 SAMPLING. Three methods are described below. Concentrations below 5 ppm are sampled by the bubbler method. Higher concentrations may be sampled by the evacuated bottle method, or more conveniently (but less accurately) by the glass syringe method. The latter method is more useful when appreciable concentrations (*e.g.*, 20 ppm) of nitric oxide are expected.

7.1.1 *Bubbler Method.* Assemble, in order, a sampling probe (optional), a glass rotameter, fritted absorber, trap, and pump. Use ground-glass connections upstream from the absorber. Butt-to-butt glass connections with slightly greased vinyl or pure gum rubber tubing also may be used for connections without losses if lengths are kept minimal. The sampling rotameter may be used upstream from the bubbler provided occasional checks are made to show that no nitrogen dioxide is lost. The rotameter must be kept free from spray or dust. Pipet 10.0 mL of absorbing reagent into a dry fritted bubbler (see **5.1.3**). Draw an air sample through it at the rate of 0.4 L/min (or less) long enough to develop sufficient final color (about 10 to 30 min). Note the total air volume sampled. Measure and record the sample air temperature and pressure.

7.1.2 *Evacuated Bottle Method.* Sample in bottles of appropriate size containing 10.0 mL (or other convenient volume) of absorbing reagent. For 1 cm spectrophotometer cells, a 5:1 ratio of air sample volume to reagent volume will cover a concentration range up to 100 ppm; a 25:1 ratio suffices to measure down to 2 ppm. Wrap a wire screen or glass-fiber-reinforced tape around the bottle for safety purposes. Grease the joint lightly with silicone or fluorocarbon grease. If a source of vacuum is available at the place of sampling, it is best to evacuate just before sampling to eliminate any uncertainty about loss of vacuum. A three-way Y stopcock connection is convenient. Connect one leg to the sample source, one to the vacuum pump, and the third to a tee attached to the bottle and to a mercury manometer or accurate gauge. In the first position of the Y stopcock, the bottle is evacuated to the vapor pressure of the absorbing reagent. In the second position of the Y stopcock the vacuum pump draws air through the sampling line to flush it thoroughly. The actual vacuum in the sample bottle is read on the manometer. In the third position of the Y stopcock the sampling line is connected to the evacuated bottle and the sample is collected. The stopcock on the bottle is then closed. Allow 15 min with occasional shaking for complete absorption and color development. For calculation of the standard volume of the sample, record the temperature and the pressure. The latter is the difference between the filled and evacuated conditions, and the uncorrected volume is that of the bottle plus that of the connection up to the stopcock minus the volume of absorbing reagent.

7.1.3 *Glass Syringe Method.* Ten mL of absorbing reagent is kept in a capped 50 (or 100) mL glass syringe, and 40 (or 90) mL of air is drawn in at the time of sampling. The absorption of nitrogen dioxide is completed by capping and shaking vigorously for 1 min, after which the air is expelled. (When appreciable concentrations, *e.g.*, 20 ppm of nitric oxide, are suspected, interference caused by the oxidation of nitric oxide to nitrogen dioxide is minimized by expelling the air sample immediately after the absorption period.) Additional air may be drawn in and the process repeated several times, if necessary, to develop sufficient final color.

7.2 MEASUREMENT OF COLOR. After collection or absorption of the sample, a red-violet color appears. Color development is complete within 15 min at room temps. Compare with standards visually or transfer to stoppered cuvettes and read in a spectrophotometer at 550 nm, using unexposed reagent as a reference. Alternatively, distilled water may be used as a reference, and the absorbance of the reagent blank deducted from that of the sample.

7.2.1 Colors too dark to read may be quantitatively diluted with unexposed absorbing reagent. The measured absorbance is then multiplied by the dilution factor.

8. Calibration and Standardization

8.1 Standardization is performed using the diluted sodium nitrite solution prepared in Section **6.5.1**, which contains 10 μL of NO_2 per mL.

8.1.1 Standardization by nitrite solution is based upon the empirical observation **(1,5)** that 0.72 mole of sodium nitrite produces the same color as 1 mole of nitrogen dioxide.‡

8.2 Add to a series of 25 mL volumetric flasks 0, 0.2, 0.4, 0.6, and 1.0 mL of the diluted sodium nitrite solution. These standards will contain respectively 1, 2, 4, 6 and 10 μL equivalents of nitrogen dioxide.

8.2.1 Dilute the volumetric flasks to the mark with absorbing reagent **(6.4)**. Mix, allow 15 min for complete color development and read the formed colors (see **7.2**). The standards contain progressively, 0, 0.08, 0.16, 0.24 and 0.40 μL NO_2 per mL of absorbing solution.

8.2.2 Plot the respective absorbances versus the μL NO_2 per mL of absorbing reagent.

---

‡Stratmann and Buck **(6)** reported a stoichiometric relationship of 1.0. Subsequently they found **(7)** decreasing values at concentrations above 0.3 ppm, approaching approximately the 0.7 figure at a few ppm. Shaw **(8)** confirmed the 0.72 value and suggested that higher values could be obtained erroneously if inadequate corrections for blanks were made. It is recommended that no change be made in the widely used 0.72 value at present if no change is made in the construction and operation of the fritted absorber (Figure 406:1). Other types of absorbers may yield different empirical factors.

9. Calculations

9.1 Determine the μL of nitrogen dioxide per mL of absorbing reagent from the measurement made in **7.2** and the calibration curve prepared in **8.2.2**.

9.2 Compute the concentration of nitrogen dioxide in the sample as follows:

$$\text{ppm } NO_2 = \frac{\mu L \; NO_2}{mL} \times \frac{mL \text{ of absorbing reagent}}{\text{sample volume (L)}}$$

where sample volume is corrected to 25°C and 101.3 kPa.

10. Effects of Storage

10.1 Colors may be preserved, if well stoppered, with only 3 to 4% loss in absorbance per day; however, if strong oxidizing or reducing gases are present in the sample in concentrations considerably exceeding that of the nitrogen dioxide, the colors should be determined as soon as possible to minimize any loss. (See Section 3 for effects of interfering gases.)

11. References

1. SALTZMAN, B.E. 1954. Colorimetric microdetermination of nitrogen dioxide in the atmosphere. *Analytical Chemistry.* 26:1949–1955.
2. MUELLER, P.K., F.P. TERRAGLIO AND Y. TOKIWA. 1965. Chemical Interferences in Continuous Air Analysers. Presented 7th Conference on Methods in Air Pollution Studies, Los Angeles, Calif., Jan.
3. THOMAS, M.D. AND R.E. AMTOWER. 1966. Gas dilution apparatus for preparing reproducible dynamic gas mixtures in any desired concentration and complexity. *J. Air Pollut. Contr. Assoc.* 16:618–623.
4. LODGE, J.P., JR., J.B. PATE, B.E. AMMONS AND G.A. SWANSON. 1966. The use of hypodermic needles as critical orifices in air sampling. *J. Air Pollut. Contr. Assoc.* 16:197–200.
5. SALTZMAN, B.E. AND A.F. WARTBURG, JR. 1965. Precision flow dilution system for standard low concentrations of nitrogen dioxide. Analytical Chemistry. 37:1261–1264.
6. STRATMANN, H. AND M. BUCK. 1966. Messung von Stickstoffdioxid in der Atmosphäre. *Air & Water Pollut. Int. J.* 10:313–326.
7. STRATMANN, H. 1966. Personal Communication, September.
8. SHAW, J.T. 1967. The measurement of nitrogen dioxide in the air. *Atmospheric Environment.* 1:81–85.

Subcommittee 3

B. E. SALTZMAN, *Chairman*
W. A. COOK

B. DIMITRIADES
E. L. KOTHNY
L. LEVIN
P. W. MCDANIEL
J. H. SMITH

Approved with modifications from 2nd edition
Subcomittee 3

D. A. LEVAGGI, *Chairman*
B. R. APPEL
D. W. HORSTMAN
E. L. KOTHNY
J. G. WENDT

407.

Determination of
Total Nitrogen Oxides as Nitrate
(Phenoldisulphonic Acid Method)*

1. Principle of the Method

1.1 Nitric oxide (NO), nitrous anhydride (N_2O_3), nitrogen dioxide (NO_2), nitrogen tetroxide (N_2O_4), also vapor or mist of nitric acid (HNO_3) and nitrous acid (HNO_2), but not nitrous oxide (N_2O), may be collected and oxidized to the nitrate ion in an evacuated flask containing sulfuric acid and hydrogen peroxide. The yellow compound resulting from reaction of the nitrate ion with phenoldisulfonic acid is measured colorimetrically at 400 nm.

1.2 This is a long established method. The first extensive investigation of the method for nitrate determination in water was conducted by Chamot et al. (2). The method was first adapted to determination of oxides of nitrogen in air by Cook (3), although the sampling technique then suggested has since been superseded by the method here described.

2. Range and Sensitivity

2.1 The phenoldisulfonic acid method is sensitive to 1 μg of nitrate in a water sample (4).

2.2 A 1000-mL air sample as collected in an evacuated flask of this volume will permit accurate determination of 50 or more ppm (by volume) of oxides of nitrogen as nitrogen dioxide (NO_2). Accordingly, the application of this method to ambient air is appropriate only to industrial locations where the concentration may exceed 50 ppm.

---

*This procedure is comparable to ASTM Method D1608–77 (1) with inclusion of certain details from the APHA Standard Methods for the Examination of Water and Waste Water (4).

3. Interferences

3.1 Inorganic nitrates and other compounds easily oxidized to nitrates, such as nitrites and organic nitrogen compounds, give high results.

3.2 Certain reducing compounds, such as sulfur dioxide (SO_2), may consume a sufficient amount of the hydrogen peroxide in the absorbing solution that an inadequate amount is available to oxidize all of the oxides of nitrogen to the nitrate.

3.3 Chlorides and other halides lower the results, as does lead, but ordinarily insufficient amounts are present in air samples to require special treatment. If the chloride content exceeds 3 mg, silver nitrate should be used to reduce the content to about 0.1 mg (5).

3.4 Any substance that may increase the absorbance at 400 nm should be absent.

4. Precision and Accuracy

4.1 The precision of the method for ambient air can be estimated from the repeatability of 5% of the mean found on its application.

4.2 Below 50 ppm, accuracy is reduced due to incomplete absorption and relatively high blanks.

5. Apparatus

5.1 GAS SAMPLING FLASK OR BOTTLE. Standard 1000-mL round-bottom flask or bottle of borosilicate glass with a male standard taper 24/40 neck and a female cap with a sealed-on tube. Glass-fiber-reinforced tape should be wrapped around container for safety purposes. Determine the volume of the container by measuring the volume of water required to fill it.

5.2 SAMPLING ASSEMBLY. A glass or stainless steel tube 6 to 10 mm in diameter provided with a downward-facing intake (funnel or tip) is suitable. A small loosely fitting plug of glass wool may be inserted, when desirable, in the probe to exclude water droplets and particulate matter. A T-tube and three-way Y stopcock are required. One leg of the stopcock may be of small bore capillary tubing for prolonging the sampling period.

5.3 SUCTION PUMP, capable of producing vacuum to the vapor pressure of water.

5.4 PHOTOMETER. A commercial photoelectric filter photometer or, preferably, a spectrophotometer suitable for measurement at 400 nm.

5.5 LABORATORY GLASSWARE. Microburet, 10-mL capacity; buret, 50-mL capacity; pipets, 25-mL capacity; volumetric flasks, 50-mL capacity; evaporating dishes or casseroles, heat resistant glass, 200-mL capacity.

5.6 STEAM BATH.

5.7 MERCURY MANOMETER, OPEN END, or precision pressure gauge.

6. Reagents

6.1 PURITY. All chemicals should be ACS Analytical Reagent grade. Select hydrogen peroxide with a low nitrate content.

6.2 NITRATE AND NITRITE-FREE WATER. All water is to be free of nitrate and nitrite ions. If available distilled or deionized water contains nitrate or nitrite impurities, redistill it in an all-glass still after adding a crystal each of potassium permanganate and of barium hydroxide.

6.3 HYDROGEN PEROXIDE (3%). Dilute 10 mL of 30% hydrogen peroxide (H_2O_2) to 100 mL.

6.4 SULFURIC ACID (SP GR 1.84). Concentrated sulfuric acid (H_2SO_4).

6.5 SULFURIC ACID (0.3%). Mix 3 mL of H_2SO_4 (sp gr 1.84) with water and dilute to 1 L.

6.6 ABSORBENT SOLUTION. 1.0 mL of H_2O_2 (3%) in 100 mL of H_2SO_4 (0.3%). For high concentrations of oxides of nitrogen, the amount of H_2O_2 should be increased to 3 mL. Since dilute H_2O_2 may be

unstable, these solutions should not be kept for a prolonged period.

6.7 AMMONIUM HYDROXIDE (SP GR 0.90). Concentrated ammonium hydroxide (NH_4OH). Keep stoppered to avoid loss of strength.

6.8 PHENOLDISULFONIC ACID REAGENT. Dissolve 25 g of phenol in 150 mL of concentrated sulfuric acid (H_2SO_4, sp gr 1.84) by heating on a steam bath (100°C). Cool, add 75 mL of fuming sulfuric acid (15% SO_3) and heat on the water bath for 2 h. Cool and store in a brown glass bottle. The solution should be colorless; it deteriorates on long standing.

6.9 POTASSIUM NITRATE, STANDARD SOLUTION (1 ML = 0.1880 MG NO_2). Dry potassium nitrate (KNO_3) in an oven at 105 ± 1°C for 2 h. Dissolve 0.4131 g of the salt in water and dilute to 1 L in a volumetric flask.

6.10 POTASSIUM NITRATE, STANDARD SOLUTION (1 ML = 0.0188 MG NO_2). Dilute 10 mL of KNO_3 solution (1 mL = 0.1880 mg NO_2) to 100 mL with water in a volumetric flask and mix well. (0.0188 mg NO_2 in an air sample of 1000 mL = 10 ppm.)

6.11 SODIUM HYDROXIDE SOLUTION (42 G/L). Dissolve 42 g of sodium hydroxide ($NaOH$) in water and dilute to 1 L. Store in a plastic bottle and keep well stoppered.

6.12 EDTA REAGENT. Rub 50 g disodium ethylenediamine tetraacetate dihydrate with 20 mL distilled water to form a thoroughly wetted paste. Add 60 mL concentrated NH_4OH and mix well to dissolve the paste.

7. Procedure

7.1 SAMPLING. Sampling may be conducted by use of an evacuated flask or bottle. Pipet 25.0 mL of absorbent solution into the sampling container. If a source of vacuum is available at the place of sampling, it is best to evacuate just before sampling to eliminate any uncertainty about loss of vacuum. The T-tube is connected to the tube of the container cap. The mercury manometer or accurate vacuum gauge is attached to one branch of the T-tube and the three-way Y stopcock to the other. The vacuum pump is attached to a second branch of the stopcock and the sampling probe to the third. By successive positionings of the stopcock plug, the container is first evacuated to the incipient boiling point of the absorbing solution and the manometer reading recorded, then air is drawn through the sampling line or probe to flush it, and finally the sample is collected in the evacuated container. The cap of the container is turned to seal the sample and the T-tube disconnected. The temperature is recorded.

7.2 LABORATORY ANALYSIS.

7.2.1 The glass flask or bottle in which the sample has been collected should remain in contact with the absorbent overnight to complete oxidation to the nitrate. Retaining the sample in the closed sample container longer than overnight and even up to a week is desirable to increase extent of absorption (**6**). Transfer the absorbent solution quantitatively from the container into a 200-mL evaporating dish.

7.2.2 A blank should be treated in the same manner as the sample. If a 1000-mL container was used for sampling, 25 mL of the unused absorbent solution should be pipetted into a 200-mL evaporating dish and the same amount of water added as was used in transferring the sample. The blank and samples should then be treated as described in Section **7.2.3**.

7.2.3 Add NaOH solution to the sample solution and to the blank in the evaporating dish until just basic to litmus paper. Avoid adding excess NaOH as this may dissolve some silicate from the dish and later cause turbidity. Evaporate the contents of the evaporating dish to dryness on a hot water or steam bath and allow to cool. Using a glass rod, rub the residue thoroughly with 2.0 mL phenoldisulfonic acid reagent to insure solution of all solids. If necessary, heat mildly on the hot water bath a short time to dissolve the entire residue. Cool and add 20 mL distilled water, stir, then add sufficient fresh, cool NH_4OH dropwise (about 6 or 7 mL) with constant stirring to give a basic reaction with litmus. If turbidity should occur, filter the solution through 7-cm, rapid, medium-texture filter paper into a 50-mL volumetric flask. Wash the evaporating dish three times with 4 to 5 mL of water and pass the washings through the filter. Since some yellow color may be left on the filter paper, this step should be done in a reproducible manner both for the samples and the calibration curves. Instead of

filtering to remove the turbidity, the EDTA reagent may be added dropwise with stirring until the turbidity redissolves. Make up the volume to 50 mL in a volumetric flask with water and mix thoroughly.

7.2.4 Read the absorbance of the sample solution against the blank in the photometer at 400 nm. If a greater dilution is required, dilute the blank to the same volume.

7.2.5 Convert the absorbance found by means of the calibration curve to mg of NO_2.

8. Calibration and Standardization

8.1 Prepare a calibration curve of mg of NO_2 plotted against absorbance for the range of 50 to 500 ppm NO_2 based on 1000 mL samples of dry gas under standard conditions of 101.3 kPa and 25°C.

8.1.1 Using the microburet for the first 2 volumes and the 50-mL buret for the last 3, transfer 0.0, 5.0, 10.0, 35.0 and 50.0 mL of KNO_3 solution (1 mL = 0.0188 mg NO_2) into 200-mL evaporating dishes. Pipet 25 mL of the acid absorbent solution into each evaporating dish. Add NaOH solution until just basic to litmus paper. Then proceed as directed in Sections **7.2.3** and **7.2.4**.

9. Calculations

9.1 For convenience, standard conditions are taken as 101.3 kPa and 25°C, at which the molar gas volume is 24.47 L. (This is identical with standard conditions for Threshold Limit Values of the American Conference of Governmental Industrial Hygienists; it is very close to the standard conditions used for air-handling equipment, of 29.92 in Hg, 70°F, and 50% relative humidity, at which the molar gas volume is 24.76 L or 1.2% greater.)

9.1.1 The volume of the gas sample may be corrected to standard conditions by the following calculations:

$$V_s = \frac{(V_f - V_r)\, P \cdot 298.2}{101.3\,(t + 273.2)}$$

where V_s = volume of gas sample corrected to standard conditions of 101.3 kPa and 25° C, in mL,

V_f = volume of sampling flask up to stopcock, in mL,

V_r = volume of absorbent reagent,

P = vacuum in sampling container as measured by the manometer, kPa, and

t = sampling temperature, °C

9.2 CALCULATIONS OF CONCENTRATION OF NO_2 IN PARTS PER MILLION BY VOLUME

Oxides of nitrogen as NO_2

$$= \frac{24.47W \times 10^6}{46.0\ V_s}\ ppm$$

$$= \frac{532W \times 10^3}{V_s}\ ppm$$

where V_s = volume of gas sample corrected to standard conditions in mL,

W = mg of oxides of nitrogen found (as NO_2),

24.47×10^3 = standard molar volume (101.3 kPa at 25°C), in mL, and

46.0 = formula weight of NO_2.

10. Effect of Storage

10.1 The sample as collected either in the acid or alkaline absorbing solution with H_2O_2 can be held for analysis at a later time.

10.2 The yellow color formed with the phenoldisulfonic acid is stable.

11. References

1. ASTM COMMITTEE D-22. 1977. Method of Test for Oxides of Nitrogen in Gaseous Combustion Products (Phenoldisulfonic Acid Procedure) ASTM Designation D 1608-77 Book of ASTM Standards, vol. 11.03.

2. CHAMOT, E.M., D.S. PRATT AND H.W. REDFIELD. 1909, 1910, 1911. A Study on the Phenoldisulfonic Acid Method for the Determination of Nitrates in Water. *J. Amer. Chem. Soc.* *31*:922, *32*:630, *33*:336.

3. COOK, W.A. 1936. Chemical Procedures in Air Analysis. Methods for Determination of Poisonous Atmospheric Contaminants. Inorganic Substances. 1935-36 Year Book, Supplement to Amer. J. Public Health, *26*:80.

4. AMERICAN PUBLIC ASSOCIATION. 1976. Standard Methods for the Examination of Water and Waste Water. 14th ed. Washington, D.C.

5. JACOBS, M.B. 1967. The Analytical Toxicology of Industrial Inorganic Poisons, Interscience Publishers. John Wiley & Sons, New York, N.Y.

6. SALTZMAN, B.E. 1954. Colorimetric Microdetermination of Nitrogen Dioxide in the Atmosphere. (Note on phenoldisulfonic acid procedure on p. 1953) *Anal. Chem. 26*:1949.

Subcommittee 3

B.E. SALTZMAN, *Chairman*
W.A. COOK
B. DIMITRIADES
E. F. FERRAND
E. L. KOTHNY
L. LEVIN
P. W. MCDANIEL
C. A. JOHNSON, Air Conditioning & Refrigeration Institute Liaison

Approved with modifications from 2nd edition.
Subcommittee 3

D. A. LEVAGGI, *Chairman*
B. R. APPEL
D. W. HORSTMAN
E. L. KOTHNY
J. G. WENDT

408.

Analysis for Atmospheric Nitrogen Dioxide (24-H Average)

1. Principle of the Method

1.1 Nitrogen dioxide is collected by bubbling air through a solution containing triethanolamine, o-methoxyphenol (guaiacol) and sodium metabisulfite (1,2). The nitrite ion produced during sampling is determined colorimetrically by reacting the exposed absorbing reagent with sulfanilamide and 8-anilino-1-naphthalenesulfonic acid (ANSA).

1.2 The method is applicable to collection and subsequent laboratory analysis of 24-h samples.

2. Range and Sensitivity

2.1 Beer's law is obeyed from 0.025 to 4.0 μg NO_2/mL. The range of the method, assuming a sampling rate of 200 cm^3/min for 24 h, is 20 to 700 μg NO_2/m^3 (0.01–0.37 ppm).

2.2 A concentration of 0.025 μg NO^-_2/mL will produce an absorbance of about 0.025 at 550 nm using 1-cm cells.

2.3 Samples exceeding the absorbance of the highest calibration standard must be diluted with absorbing reagent until the absorbances are within the range of the highest standard.

3. Interferences

3.1 Nitrous acid (HONO) and nitrous anhydride (N_2O_3) are positive interferents.

3.2 At a nitrogen dioxide concentration of 100 μg/m^3 (0.05 ppm) the following pollutants, at the levels indicated, do not interfere: ammonia, 205 μg/m^3 (0.30 ppm); carbon monoxide, 154000 μg/m^3 (135 ppm); formaldehyde, 750 μg/m^3 (0.6 ppm); nitric oxide, 734 μg/m^3 (0.60 ppm); phenol, 150 μg/m^3 (0.04 ppm); ozone, 400 μg/m^3 (0.20

ppm); and sulfur dioxide, 439 μg/m^3 (0.17 ppm) (1).

3.3 A temperature of 40°C during sample collection has no effect on recovery (1).

3.4 Normal evaporation loss of absorbing solution during sampling has no effect on recovery (1).

4. Precision and Accuracy

4.1 Analysis of test atmospheres of 20 to 700 μg/m^3 NO_2 obtained with permeation devices result in a relative standard deviation of 2% and an overall recovery of 93% (1).

5. Apparatus

5.1 SAMPLING ASSEMBLY. A diagram of a suggested sampling apparatus is shown in Figure 408:1. It consists of a probe made of Teflon, polypropylene, or glass with a polypropylene or glass funnel attached to one end while the other end is connected to the bubbler. The bubbler is made of polypropylene tube 164 × 32 mm equipped with polypropylene two-port closures. Since rubber stoppers cause high and varying blank values, they should not be used. A glass tube restricted orifice is used to disperse the gas. The tube, approximately 8 mm O.D. and 6 mm I.D., should be 152 mm long with the end drawn out to 0.3–0.6 mm I.D. The tube should be positioned so as to allow a clearance of 6 mm from the bottom of the bubbler. The moisture trap is a polypropylene tube equipped with a two-port closure. The entrance port of the closure is fitted with tubing that extends to the bottom of the trap. The unit is loosely packed with glass wool to prevent entrainment of droplets from the bubbler. Between the trap and the flow control device, a filter holder with a 0.8–2.0 μm pore size mem-

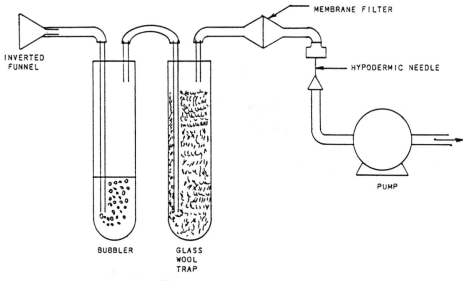

Figure 408:1 — Sampling train.

brane filter is inserted to protect the flow control device from particulate matter. The membrane filter should be changed after collecting 10 samples.

5.2 Any flow control device capable of maintaining a constant flow through the sampling solution between 180 and 220 cm³/min can be used. A recommended flow control device is a 27 gauge hypodermic needle (3) inserted through a rubber septum connected between the filter holder and a vacuum pump.

5.3 The air pump should be capable of maintaining a pressure differential of at least 0.6 to 0.7 of an atmosphere (61–71 kPa) across the flow control device. This range surpasses the downstream critical pressure for restricted orifices, i.e., 53.7 kPa (0.53 atmosphere) (3), thus assuring a safety factor to allow for variations in atmospheric pressure.

5.4 The calibration equipment consists of a flowmeter for measuring airflows up to 275 cm³/min within ± 2%, a stopwatch, and a precision wet test meter (e.g. 1 L/revolution) or a calibrated soap bubble meter.

5.5 LABORATORY GLASSWARE. One hundred, 200, 250, 500 and 1000 mL volumetric flasks; 1, 5, 10, 20 and 50 mL volumet-

ric pipets; 50 mL graduated cylinder, and 20 × 150 mm test tubes.

5.6 A spectrophotometer suitable for measuring absorbance at 550 nm.

6. Reagents

6.1 PURITY: All chemicals should be ACS Analytical Reagent grade. Water should be ASTM Reagent water, Type II.

6.2 ABSORBING SOLUTION: Dissolve 20 g triethanolamine, 0.5 g o-methoxyphenol (guaiacol), and 0.25 g sodium metabisulfite consecutively in 500 mL of water. Dilute to 1 L with water and mix. Discard any solution that is colored. The solution is stable for three weeks if kept refrigerated.

6.3 HYDROGEN PEROXIDE: Dilute 0.2 mL of 30% hydrogen peroxide to 250 mL with water. This solution can be used for a month if protected from light and refrigerated.

6.4 SULFANILAMIDE SOLUTION (2% in 4 N HCl). Dissolve 2.0 g sulfanilamide in 33 mL of concentrated HCl, dilute to 100 mL with water and mix. This solution can be used for two weeks, if refrigerated.

6.5 ANSA SOLUTION (0.1% w/v). Dissolve 0.1 g ANSA in 50 mL absolute methanol. Dilute to 100 mL with absolute methanol in a volumetric flask. Keep stoppered to minimize evaporative losses and prepare daily. Older solutions deteriorate rapidly and may result in lower absorbances.

6.6 Nitrite Standard Solutions.

6.6.1 Nitrite stock solution. Dissolve sufficient sodium nitrite, predried at 105°C for 1 h, in water and dilute to 1000 mL to obtain a solution containing 1000 μg NO_2^-/ mL. The amount of $NaNO_2$ to use is as follows:

$$G = \frac{1.5}{A} \times 100$$

where G = amount of $NaNO_2$
   1.5 = gravimetric factor $NaNO_2/NO_2$
        = 69/46.
   A = $NaNO_2$ purity in %.

This solution is stable for three months at room temperature.

6.6.2 Nitrite working solutions. Dilute 5 mL of the 1000 μg NO_2^-/mL solution to 250 mL with absorbing solution, to obtain an intermediate solution containing 20 μg NO_2/mL. Dilute 5.0 mL of the intermediate working solution to 200 mL with absorbing solution. This solution contains 0.5 μg NO_2^-/mL. Prepare both solutions daily.

7. Procedure

7.1 Sample Collection. Assemble the sampling train, as shown in Figure 408:1. Components upstream from the absorption tube may be connected either with Teflon or polypropylene tubing; glass tubing with dry ball joints; or glass tubing with butt-to-butt joints with Tygon, Teflon or polypropylene. Add 50 mL of absorbing solution to the bubbler.

7.1.1 Disconnect the funnel, connect the calibrated flowmeter, measure the flow rate before sampling and record as F_1. If the flow rate before sampling is not between 180 and 220 cm³/min, replace the flow control device and/or check the system for leaks. Start sampling only after obtaining an initial flow rate in this range.

7.1.2 Sample for 24 h. Record the sampling time in min as T.

7.1.3 Measure the flow rate after the sampling period and record as F_2.

7.1.4 Seal the collected samples and transport to the laboratory for analysis.

7.2 Laboratory Analysis.

7.2.1 Replace any water lost by evaporation during sampling by adding distilled water up to the calibrated mark on the absorption tube and mix well.

7.2.2 Pipet 5 mL of the collected sample into a test tube, add 0.5 mL of the peroxide solution and mix. After 15 s add 2.7 mL of sulfanilamide solution. After about 30 s and within 6 min of mixing the sufanilamide solution, add 3 mL of the ANSA solution and mix. Other time intervals will result in lower absorbance values.

7.2.3 A blank should be treated in the same manner as the sample using 5 mL of unexposed absorbing solution. The absorbance of the blank should be approximately the same as the y-intercept in the calibration curve described in Section 8.5.

7.2.4 Read the absorbance at 550 nm against the blank in the reference cell using 1 cm cells. The color can be read anytime from 10 to 40 min after addition of the ANSA solution.

7.2.5 The reagents used in this procedure may result in film buildup on the spectrophotometer cells. Therefore, the cells must be cleaned thoroughly after each series of analyses.

8. Calibration and Standardization

8.1 Flowmeter Calibration. Air flow rates (cm³/min) through the flowmeter at four different ball positions are determined using a wet test meter or other flowrate measuring device.

8.2 Absorption Tube Calibration. Calibrate the polypropylene absorption tube (Bubbler, Fig. 1) by first pipeting in 50 mL of water or absorbing solution. Scribe the level of the meniscus with a sharp object, and darken the area with a felt-tip waterproof marking pen, removing the excess.

8.3 Preparation of Standard Solutions. Pipet the indicated (Table 408:I) volumes of working solution into separate volumetric flasks and dilute to mark with absorbing solution.

8.4 Take 5.0 mL of each prepared standard and develop color as specified as in Section **7.2.2–7.2.4.**

Table 408:1 Preparation of Standards

Volume of working solution and concentration	Fill to mL	Concentration $\mu g\ NO_2/mL$
10 mL of 0.5 μg NO^-_2/mL	100	0.05
20 mL of 0.5 μg NO^-_2/mL	100	0.10
40 mL of 0.5 μg NO^-_2/mL	100	0.20
100 mL of 0.5 μg NO^-_2/mL	100	0.50
5 mL of 20 μg NO^-_2/mL	100	1.00
10 mL of 20 μg NO^-_2/mL	100	2.00

8.5 STANDARD CURVE. Plot the absorbances on the ordinate versus the corresponding concentrations on the abscissa using linear graph paper. Alternatively determine the linear regression equation by the method of least squares. Linearity is obtained up to 4 μg NO^-_2/mL with an approximate slope of 0.5.

9. Calculation

9.1 Volume of air sampled, V (m³). Use the following formula:

$$V = \frac{F_1 + F_2}{2} \times T \times 10^{-6}$$

where F_1 = Initial flow rate, cm³/min.
F_2 = Final flow rate, cm³/min.
T = Sampling time in min.
10^{-6} = Conversion factor of cm³ to m³

The volume of air sampled is not corrected to a reference temperature and pressure because of the uncertainity of these conditions during the 24-h sampling period.

9.2 Concentration of nitrogen dioxide (μg NO_2/m^3). Use the following formula:

$$\mu g\ NO_2/m^3 = \frac{W \times 50}{V \times 0.93}$$

where W = μg NO^-_2/mL (obtained from calibration curve)
50 = Total volume of sampling solution
V = Volume of air sampled, m³ (from **9.1**)
0.93 = Overall efficiency of the method

9.3 If desired, the concentration of nitrogen dioxide may be calculated as ppm NO_2 by using the following formula:

$$pm\ NO_2 = NO^-_2/m^3 \times 5.32 \times 10^{-4}$$

where 5.32×10^{-4} = conversion factor from μg NO_2/m^3 to ppm NO_2 (vol).

10. Effect of Storage

10.1 The collected samples are stable for three weeks if kept refrigerated.

11. References

1. MULIK, J.D., R.G. FUERST, J.R. MEEKER, M. GUYER, AND E. SAWICKI. 1973. A twenty four hour method for the collection and manual colorimetric analysis of nitrogen dioxide. Presented at the 165th ACS National Meeting in Dallas, Texas April 8–13.
2. CONSTANT, P.C., M.C. SHARP, AND G.W. SCHNEIL. 1974. Collaborative Test of The TGS-ANSA Method For Measurement of Nitrogen Dioxide in Ambient Air. EPA-650/4-74-046.
3. LODGE, J.P., J.B. PATE, B.E. AMMONS, AND G.A. SWANSON. 1966. The use of hypodermic needles as critical orifices in air sampling. J.A.P.C.A. 16:197–200.

Subcommittee 3

E.L. KOTHNY
D.A. LEVAGGI, *Chairman*
B.R. APPEL
D.W. HORSTMAN
J.G. WENDT

411.

Determination of Oxidizing Substances in the Atmosphere

1. Principle of the Method

1.1 Micro-amounts of ozone and other oxidants liberate iodine when absorbed in a 1% solution of potassium iodide buffered at pH 6.8 ± 0.2. The iodine is determined spectrophotometrically by measuring the absorption of triiodide ion at 352 nm.

1.2 The stoichiometry is approximated by the following reaction:

$$O_3 + 3 KI + H_2O \rightarrow KI_3 + 2 KOH + O_2$$

2. Range and Sensitivity

2.1 This method covers the manual determination of oxidant concentrations between 0.01 to 10 ppm (19.6 to 19620 $\mu g/m^3$) as ozone **(1)**.

2.2 When 10 mL of absorbing solution is used, between 1 and 10 μL of ozone, corresponding to absorbances between 0.1 and 1 in a 1 cm cell, are collected.

3. Interferences

3.1 Sulfur dioxide produces a negative interference equivalent to 100% of that of an equimolar concentration of oxidant.

3.1.1 Up to 100-fold ratio of sulfur dioxide to oxidant may be eliminated without loss of oxidant by incorporating a chromic acid paper absorber in the sampling train upstream from the impinger **(2)**.

3.1.2 The absorber removes sulfur dioxide without loss of oxidant but will also oxidize nitric oxide to nitrogen dioxide.

3.1.3 When sulfur dioxide is less than 10% of the nitric oxide concentration, the use of chromic acid paper is not recommended. In this case the effect of sulfur dioxide on the oxidant reading can be corrected for by concurrently analyzing for

sulfur dioxide and adding this concentration to the total oxidant value.

3.2 Nitrogen dioxide is known to give a response in 1% KI **(1)**, equivalent to 10% of that of an equimolar concentration of ozone. The contribution of the nitrogen dioxide to the oxidant reading can be eliminated by concurrently analyzing for nitrogen dioxide by an appropriate method from this volume and subtracting one-tenth of the nitrogen dioxide concentration from the total oxidant value.

3.3 Peroxyacetyl nitrate gives approximately a response equivalent to 50% of that of an equimolar concentration of ozone **(3)**. Concentrations in the atmosphere may range up to 0.1 ppm.

3.4 Other oxidizing substances besides ozone will liberate iodine with this method: *e.g.*, halogens, peroxy compounds, hydroperoxides, organic nitrites and hydrogen peroxide **(4,5)**.

3.5 Hydrogen sulfide, reducing dusts or droplets can act as negative interferences.

3.6 It has been shown that the amount of iodine formed increases with relative humidity during sampling **(6,7)**. This effect is nearly linear. Increase of iodine formation is 0–10% with RH values ranging from 0–60%. Insignificant effects were observed by increases of RH from 60 to 75%.

4. Precision and Accuracy

4.1 The precision of the method within the recommended range (Section **2.2**) is about ± 5% deviation from the mean. The major error is from loss of iodine during sampling periods; this can be reduced by using a second impinger.

4.2 The accuracy of this method has not been established for atmospheric sampling.

4.3 The method was compared against an absolute ultraviolet photometer **(8)**. In the range of 40–60% relative humidity (see Section 3.6) the ozone-iodine stoichiometry was 1.25 and not 1.00 as suggested by the equation in Section **1.2**.

5. Apparatus

5.1 SAMPLING PROBE. Sampling probes should be of Teflon, glass or stainless steel. Ozone is destroyed by contact with polyvinyl chloride tubing and rubber even after a conditioning period. Short sections of polyvinyl chloride tubing can be used to secure butt-to-butt connections of more inert tubing.

5.2 AIR METERING DEVICE. A glass rotameter capable of measuring gas flows of 0.5 to 3 L/min calibrated with a wet test meter to assure an accuracy of ± 2%.

5.3 ABSORBER. All-glass midget impingers graduated with 5 mL graduations should be used (see Figure 411:1). Impingers should be kept clean and dust free. Cleaning should be done with laboratory detergent followed by rinses with tap and distilled water.

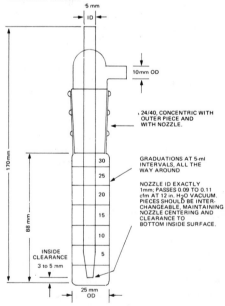

-5 mm
ID
10mm OD
24/40, CONCENTRIC WITH OUTER PIECE AND WITH NOZZLE.
GRADUATIONS AT 5-ml INTERVALS, ALL THE WAY AROUND
NOZZLE ID EXACTLY 1mm; PASSES 0.09 TO 0.11 cfm AT 12 in. H₂O VACUUM. PIECES SHOULD BE INTERCHANGEABLE, MAINTAINING NOZZLE CENTERING AND CLEARANCE TO BOTTOM INSIDE SURFACE.
170mm
88 mm
30
25
20
15
10
5
INSIDE CLEARANCE 3 to 5 mm
25 mm OD

Figure 411:1 – All-glass midget impinger (this is a commercially stocked item). Nozzle passes 2.55–3.12 L/min at 3 kPa.

5.3.1 Do not use fritted glass bubblers as these produce less iodine **(9)**.

5.4 AIR PUMP. Any suction pump capable of drawing the required sample flow for intervals up to 30 min. It is desirable to have a needle valve or critical orifice **(10)** for flow control. A trap should be installed upstream of the pump to protect against accidental flooding with absorbing solution and consequent corrosion.

5.5 SPECTROPHOTOMETER. Any laboratory instrument capable of accurately measuring the absorbance of the triiodide ion at 352 nm. Stoppered cuvettes or tubes transparent in the near ultraviolet region should be used to hold the solutions.

6. Reagents

6.1 PURITY. Reagent grade chemicals shall be used in all tests. Unless otherwise indicated, all reagents shall conform to the specifications of the Committee on Analytical Reagents of the American Chemical Society, when such specifications are available. Other grades may be used provided it is first ascertained that the reagent is of sufficiently high purity to permit its use without lessening the accuracy of the determination.

6.2 WATER. Water means ASTM Reagent water, Type II.

6.3 ABSORBING SOLUTION (1% KI IN 0.1 M PHOSPHATE BUFFER). Dissolve 13.6 g of potassium dihydrogen phosphate (KH_2PO_4), 14.2 g of disodium hydrogen phosphate (Na_2HPO_4) or 35.8 g of the dodecahydrate salt ($Na_2HPO_2 \cdot 12\ H_2O$), and 10.0 g of potassium iodide in sequence and dilute the mixture to 1 L with water. Keep at room temperature for at least 1 day before use. Measure pH and adjust to 6.8 ± 0.2 with NaOH or KH_2PO_4. This solution can be stored for several months in a glass stoppered brown bottle at room temperature without deterioration. It should not be exposed to direct sunlight.

6.4 STOCK SOLUTION 0.025m I_2 (0.05N). Dissolve 16 g of potassium iodide and 3.173 g of resublimed iodine successively and dilute the mixture to exactly 500 mL with water. Keep at room temperature at least 1 day before use. Standardize shortly before use against 0.025 M $Na_2S_2O_3$. The sodium thiosulfate is standardized against primary standard biiodate $(KH(IO_3)_2)$ or potassium dichromate $(K_2Cr_2O_7)$.

6.4.1 *0.001M I_2 solution.* Pipet exactly 4.00 mL of the 0.025 M stock solution into a 100-mL low actinic volumetric flask and dilute to the mark with absorbing solution. Protect from strong light. Discard after use.

6.4.2 *Calibrating iodine solution.* For calibration purposes exactly 5.11 ml of the 0.001 M I_2 solution (or equivalent volume for other molarity) is diluted with absorbing solution just before use to 100 mL (final volume) to make the final concentration equivalent to 1 μL of O_3/mL. This solution preparation accounts for the stoichiometry described in Section **4.3** at standard conditions of 101.3 kPa and 25°C. Discard this solution after use.

6.5 SULFUR DIOXIDE ABSORBER. Flashfired glass fiber paper is impregnated with chromium trioxide, as follows (2): Drop 15 mL of aqueous solution containing 2.5 g chromium trioxide and 0.7 mL concentrated sulfuric acid uniformly over 400 cm² of paper, and dry in an oven at 80 to 90°C for 1 h; store in a tightly capped jar. Half of this paper suffices to pack one absorber. Cut the paper in 6 × 12 mm strips, each folded once into a V-shape, pack into an 85 mL U-tube or drying tube, and condition by drawing air that has been dried over silica gel through the tube overnight. The absorber is active for at least one month. When it becomes visibly wet from sampling humid air, it must be dried with dry air before further use.

7. Procedure

7.1 Assemble a train consisting of a rotameter, U-tube with chromium trioxide paper (optional), midget impinger, needle valve or critical orifice (10) and pump. Connections upstream from the impinger should be ground glass or inert tubing butt joined with polyvinyl tubing. Fluorosilicon or fluorocarbon grease should be used sparingly. Pipet exactly 10 mL of the absorbing solution into the midget impinger. Sample at a rate of 0.5 to 3 L/min for up to 30 min. The flow rate and the time of sampling should be adjusted to obtain a sufficiently large concentration of oxidant in the absorbing solution. Approximately 1 μL of

ozone can be obtained in the absorbing solution at an atmospheric concentration of 0.01 ppm by sampling for 30 min at 3 L/min. Calculate the total volume of the air sample. Also measure the air temperature and pressure. Do not expose the absorbing reagent to direct sunlight.

7.2 MEASUREMENT OF COLOR. If appreciable evaporation of the absorbing solution occurs during sampling, add water to bring the liquid volume to 10 mL.

7.3 Within 30 to 60 min after sample collection, read the absorbance in a cuvette or tube at 352 nm against a reference cuvette or tube containing water.

7.4 BLANK CORRECTION. Measure the absorbance of the unexposed reagent and subtract the value from the absorbance of the sample.

8. Calibration and Standardization

8.1 Calibrating solutions are made up to 10 mL to facilitate the calculations.

8.1.1 Obtain a range of calibration points containing from 1 μL to 10 μL of ozone equivalent per 10.0 mL of solution. Prepare by individually adding 1.0, 2.0, 4.0, 6.0, 8.0 and 10.0 mL of the calibrating iodine solution (Section **6.4.2**) to 10.0 mL volumetric flasks. Bring each to the calibration mark with absorbing reagent.

8.2 Read the absorbance of each of the prepared calibration solutions as described in Sections **7.3** and **7.4**.

8.3 Plot the absorbances of the obtained colors against the concentration of O_3 in μL/10 mL absorbing reagent. The plot follows Beer's law. Draw the straight line through the origina giving the best fit, or fit by least squares. Do not extrapolate beyond the highest concentration.

9. Calculations

9.1 Standard conditions are taken as 101.3 kPa and 25°C, at which the molar gas volume is 24.47 liters.

9.2 Record the volume of sample collected in liters. Generally the correction of the sample volume to standard conditions is slight and may be omitted. However, for greater accuracy corrections may be calculated by means of the perfect gas laws.

9.3 The total μL of $O_3/10$ mL of reagent are read from the calibration curve.

9.4 The concentration of O_3 in the gas phase in μL/L or ppm is given by:

$$O_3 \text{ ppm} = \frac{\text{total } \mu\text{L ozone per 10 mL}}{\text{volume of air sample, L}}$$

9.5 The concentration of O_3 in terms of μg/m^3 at 101.3 kPa and 25°C is obtained when desired from the value of μL/L (Section **9.4**) by:

$$\mu\text{g } O_3/\text{m}^3 = \frac{\text{ppm} \times 48.00}{24.47} \times 10^3$$
$$= 1962 \times \text{ppm}$$

10. Effects of Storage

10.1 Ozone liberates iodine through both a fast and a slow set of reactions. Some of the organic oxidants also have been shown to cause slow formation of iodine **(4,5)**. Some indication of the presence of such oxidants and of gradual fading due to reductants can be obtained by making several readings during an extended period of time, e.g., every 20 min.

10.2 Occasionally mold may grow in the absorbing reagent. When this occurs discard the reagent because reducing substances and a change in pH make it useless.

11. References

1. BYERS, D.H. AND B.E. SALTZMAN. 1958. Determination of Ozone in Air by Neutral and Alkaline Iodide Procedures. *J. Am. Indust. Hyg. Assoc. 19*:251–7.
2. SALTZMAN, B.E. AND A.F. WARTBURG. 1965. Absorption Tube for Removal of Interfering Sulfur Dioxide in Analysis of Atmospheric Oxidant. *Anal. Chem. 37*:779.
3. MUELLER, P.K., F.P. TERRAGLIO AND Y. TOKIWA. 1965. Chemical Interferences in Continuous Air Analysis. Proc., 7th Conference on Methods in Air Pollution Studies. State of California Department of Public Health, Berkeley, California.
4. SALTZMAN, B.E. AND N. GILBERT. 1959. Iodometric Microdetermination of Organic Oxidants and Ozone. *Anal. Chem. 31*:1914–20.
5. ALTSCHULLER, A.P., C.M. SCHWAB AND M. BARE. 1959. Reactivity of Oxidizing Agents with Potassium Iodide Reagent. *Anal. Chem. 31*:1987–90.
6. HIGUCHI, J.E., F.R. LEH AND R.D. MACPHEE. 1976. Comparison of Oxidant Measurement Methods, Ultraviolet Photometry, and Moist Effects. APCA Technical Speciality Conference on Ozone/Oxidants-Interaction With The Total Environment. Dallas, TX.
7. California Air Resources Board Report (Joint Study of the Los Angeles County APCD, CARB and the U.S. Environmental Protection Agency). 1975. A Study of the Effect of Atmospheric Humidity on Analytical Oxidant Measurement Methods. Sacramento, CA.
8. DE MORE, W.B., J.C. ROMANOVSKY, W.J. HAMMING, M. FELDSTEIN and P.K. MUELLER. 1976. Interagency Comparison of Iodometric Methods for Ozone Determination. Calibration in Air Monitoring, ASTM-STP 598. American Society of Testing and Materials.
9. COHEN, I.C., A.F. SMITH and R. WOOD. 1968. A Field Method for the Determination of Ozone in the Presence of Nitrogen Dioxide. *Analyst, 93*:509.
10. LODGE, J.P., JR., J.B. PATE, B.F. AMMONS and G.A. SWANSON. 1966. The Use of Hypodermic Needles as Critical Orifices in Air Sampling. *Air Poll Control Assoc J, 16*:197–200.

Subcommittee 3

B.E. SALTZMAN, *Chairman*
W.A. COOK
B. DIMITRIADES
E. F. FERRAND
E. L. KOTHNY
L. LEVIN
P. W. MCDANIEL
C. A. JOHNSON, Air Conditioning
& Refrigeration Institute Liaison

Approved with modifications from
2nd edition.
Subcommitte 3

D. A. LEVAGGI, *Chairman*
B. R. APPEL
D. W. HORSTMAN
E. L. KOTHNY
J. G. WENDT

413.

Determination of Ozone in the Atmosphere by Gas-Phase Chemiluminescence Instruments

1. Principle of Method

1.1 The method is based on the gas-phase chemiluminescent reaction of ozone with ethylene to produce an excited species that emits light in the visible range **(1,2,3,4)**. The reaction is rapid and specific for ozone. The chemiluminescent reaction takes place in a chamber that has a light-transparent end-face coupled to a photomultiplier tube. The resulting signal produced by the photomultiplier tube is proportional to ozone concentration. The signal is further amplified and either read directly on a recorder or converted into a digital display, depending upon the construction of the specific instrument.

1.2 The chemiluminescent instrument is calibrated using an ozone generator that is, in turn, standardized against the buffered neutral 1% potassium iodide method (Method 411).

2. Range and Sensitivity

2.1 Chemiluminescent instruments are available with several measurement ranges in the same instrument. Generally these are 0 to 40, 0 to 400, and 0 to 4000 μg ozone/m³ (0 to 0.02, 0 to 0.2 and 0 to 2 ppm, respectively, by volume).

2.2 The lower limit of detection has been reported at 6 or 8 μg ozone/m³ (0.003 or 0.004 ppm, respectively) in air **(3,4)**.

3. Interferences

3.1 Exhaust gases from the instrument should be vented at a safe distance from the intake to prevent interference from the excess ethylene used when operating the instrument.

3.2 A slight humidity dependence is observed when operating the instrument at the lowest measurement range.

3.3 Other components normally found in ambient air do not interfere **(3,4)**.

4. Precision and Accuracy

4.1 Precision is defined in terms of stability and repeatability of response to ozone. Response instability due to baseline drift should be equivalent to no more than ± 20 μg ozone per m³ (± 0.01 ppm by volume) **(4)**. Noise level should be equivalent to no more than ± 16 μg ozone per m³ (± 0.008 ppm by volume) **(4)**. Repeatability of response to a standardized ozone-air mixture should have a coefficient of variation not greater than 1% **(4)**. This latter precision figure represents the composite error that is associated with both the ozone measurement and the standardization operations.

4.2 Accuracy data are not available.

5. Apparatus

5.1 DETECTOR CELL, Figure 413:1 **(5)**, is a drawing of a typical detector cell showing flow paths of gases, the mixing zone, and placement of the photomultiplier tube. Other flow paths in which the air and ethylene streams meet at a point near the photomultiplier tube are also usable.

5.2 VACUUM PUMP. A vacuum pump that will maintain a vacuum of 67 kPa with an air flow of 5 L/min is suitable.

5.3 BAROMETER. Any device capable of measuring atmospheric pressures with an error of ± 0.1%.

5.4 AIR FLOWMETER. A calibrated device for measuring air flows between 0 and 1.5 L/min.

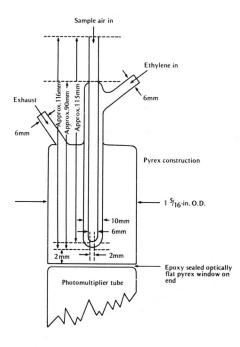

Figure 413:1 — Ozone detector cell. Cell diameter is 3.33 cm.

5.5 ETHYLENE FLOWMETER AND CONTROL. A device capable of metering and controlling ethylene flows between 0 and 80 mL/min. The actual flow should follow the manufacturer's recommendation. At any flow in this range, the device should be capable of maintaining constant flow rate within ± 3 mL/min.

5.6 AIR INLET FILTER. A Teflon filter capable of removing all particles greater than 5 μm in diameter.*

5.7 PHOTOMULTIPLIER TUBE. A high-gain, low dark current (not more than 10^{-9} A) photomultiplier tube having its maximum gain at about 430 nm. Follow manufacturer's instructions for tube replacement.

5.8 DIRECT CURRENT AMPLIFIER, capable of full scale amplification of currents from 10^{-10} to 10^{-7} A; an electrometer is commonly used with an output of 0 to 10 mV or greater.

*Mace Corporation, 1810 Floradale Ave., South El Monte, California 91733 or equivalent.

5.9 RECORDER, capable of full scale display of voltages from the DC amplifier. These voltages commonly are in the 1 mV to 1 V range.

5.10 OZONE SOURCE AND DILUTION SYSTEM. The ozone source (5,6) consists of a quartz tube into which purified air is introduced (Figure 413:2, 3) and then irradiated with a very stable low pressure mercury lamp.† The level of irradiation is controlled by an adjustable metal sleeve that fits around the lamp (6). Ozone concentrations are varied by adjustment of the sleeve. At a fixed level of irradiation and at constant temperature and humidity, ozone is produced at a constant rate. By careful control of the flow of air through the quartz tube, atmospheres can be generated that contain constant concentrations of ozone. The concentration of ozone in the test atmosphere is determined by the neutral buffered potassium iodine method.

5.11 APPARATUS FOR CALIBRATION. The calibration equipment, described in Method 411, consists of the following:

5.12.1 *Air metering device.*

5.12.2 *Absorber.*

5.12.3 *Air pump.*

5.12.4 *Spectrophotometer.*

6. Reagents

6.1 PURITY. All reagents shall conform to the specifications for ACS analytical reagent chemicals.

6.2 ETHYLENE.

†Pen-Ray model KCQ9G-1, Ultraviolet Products Inc., San Gabriel, California 91771, manufactured in lengths from 62 to 200 mm (2½" to 8"), or equivalent.

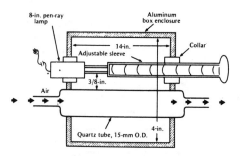

Figure 413:2 — Ozone generator.

6.3 FILTERED AIR. Purified by passing through an activated charcoal filter.

6.4 ACTIVATED CHARCOAL FILTER. A 13 cm tube × 1 cm ID or a cartridge filled with granular activated charcoal can be used to remove traces of oxidants and other interfering gases from the air.‡

6.5 CALIBRATING REAGENTS. Prepare as described in method 411. These include: Reagent water; 1% neutral potassium iodide solution; and calibrating iodine solution.

7. Procedure

7.1 Operate any commercial instrument following procedures given in the manufacturer's manual. Perform calibration as directed in Section **8**. By proper adjustments of zero and span controls, direct reading of ozone concentration may be obtained.

8. Calibration

The calibration procedure is applicable to all types of chemiluminescent instruments.

8.1 GENERATION OF TEST ATMOSPHERES. Assemble the apparatus as shown in Figure 413:3. The concentration of ozone produced by the generator (Figure 413:2), can be varied by changing the position of the adjustable sleeve, which covers more or less of the full length of the quartz-mercury lamp. The ultraviolet radiation of the lamp traverses the quartz wall and ozonizes the air or oxygen flowing through the tube. The quartz-mercury lamp must be fed through a power stabilizer for maintaining a constant output. For calibration of ambient air analyzers, the ozone source should be capable of producing ozone concentrations in the range of 100 to 2000 μg ozone per m^3 (0.05 to 1 ppm, by volume) at an air flow rate of at least 5 L/min. At all times the air flow through the generator must be greater than the total flow required by the sampling systems. Zero air can be obtained by charcoal filtration of ambient air or of zero cylinder air.

‡Filter type DZ 78006, Mine Safety Appliances, Allison Park, Pennsylvania 15101 or equivalent.

8.2 SAMPLING AND ANALYSIS OF TEST ATMOSPHERES. The chemiluminescent instrument should be conditioned before use (when using commercial instruments, follow manufacturer's instructions; conditioning time may vary from 5 min at the maximum output from the ozone generator to 30 min at atmospheric levels). The air flow rate of the analyzer should be the same as used for sampling. Sampling lines should be Teflon or glass only (7). Ozone is destroyed by contact with polyvinyl chloride and rubber tubing even after a conditioning period, and some metals show a constant small demand (7) which affects the measurement of atmospheric levels of ozone. The manifold that distributes the test atmospheres must be sampled simultaneously by the KI sampling train and the instrument to be calibrated. Check the assembled system for leaks. Record the chemiluminescent instrument response at five or six evenly-spaced ozone concentrations starting with zero. Establish the ozone concentrations by analysis, using the 1% neutral buffered potassium iodide method (Method 411). Record temperature and pressure.

8.3 BLANK. With the ozone lamp off, flush the system for several min to remove residual ozone. Pipet 10 mL of absorbing reagent into each absorber. Sample the manifold with the sampling train at 1 L/min for 10 min. Immediately transfer the exposed iodide solution to a clean (previously flushed with absorbing solution and drained for a min) 1-cm cell. Determine the absorbance at 352 nm against unexposed absorbing reagent as the reference. When consistent low blanks are obtained then proceed to calibrate the chemiluminescent instrument by the following section.

8.4 TEST ATMOSPHERES. With the ozone lamp well equilibrated, adjust the ozone concentration to the range desired for calibration. Wait 10 min for equilibration at this setting and measure according to the preceding section **8.3**.

8.5 INSTRUMENT CALIBRATION CURVE. Instrument response from the photomultiplier tube is ordinarily in current or voltage. Plot the response as the y-axis against the corresponding ozone concentrations as determined by the 1% neutral buffered potassium iodide method, in $\mu g/m^3$ (or ppm) as the x-axis.

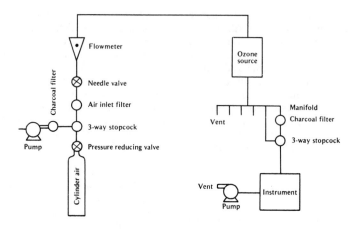

Figure 413:3 — Ozone calibration system.

9. Calculations

9.1 Calculate the total volume, in L, of air sampled into the absorbers corrected to standard conditions as follows:

$$V_R = F \times T \times \frac{P}{101.3} \times \frac{298}{t + 273}$$

where V_R = volume of air at standard condition, L,
F = flowmeter air rate, L/min,
T = time, min.
P = barometric pressure, kPa.
t = sampling temperature, °C.

9.2 The total μL of ozone are read from the spectrophotometric calibration curve (Section **8.2**).

9.3 The concentration of ozone in the gas phase in $\mu L/L$ or ppm is given by:

$$O_3 \text{ ppm} = \frac{\text{Total } \mu L \text{ ozone per 10 mL*}}{V_R}$$

9.4 The concentration of O_3 in units of $\mu g/m^3$ can be made as follows:

$$\mu g\ O_3/m^3 = \frac{\text{ppm} \times 48.00}{24.47} \times 10^3$$

$$= 1962 \times \text{ppm}$$

where 24.47 = molar volume of gas at 25°C and 101.3 kPa
48.00 = gram molecular weight of ozone

10. References

1. NEDERBRAGT, G.W., A. VAN DER HORST AND J. VAN DUNN, 1965. Rapid Ozone Determination near an Accelerator. *Nature* 206:87.

2. WARREN, G.J. AND G. BABCOCK. 1970. *Rev. Sc. Instru.* 41:280.

3. HODGESON, J.A., B.E. MARTIN AND R.E. BAUMGARDNER. 1970. Comparison of Chemiluminescence Methods for Measurement of Atmospheric Ozone. Eastern Anal. Symposium. New York City, Preprint No. 77. (National Environmental Research Center, EPA, Research Triangle Park, NC 27711).

4. CARROLL, H., B. DIMITRIADES AND M. RAY. 1972. A Chemiluminescence Detector for Ozone Measurement. Bur. of Mines, Dept. of Interior, R.I. 7650.

5. FEDERAL REGISTER. Vol. 36, No. 228, Thursday, Nov. 25, 1971, pp. 22391–22395.

6. HODGESON, J.A., R.K. STEVENS AND B.E. MARTIN. 1972. A Stable Ozone Source Applicable as a Secondary Standard for Calibration of Atmospheric Monitors. pp. 149–158. In *Air Quality Instrumentation*, Vol. I, John W. Scales, Ed., Pittsburgh, Instrument Society of America.

7. ALTSHULLER, A.P. AND A.F. WARTBURG. 1961. The Interaction of Ozone with Plastic and Metallic Materials in a Dynamic Flow System. *Int. J. Air and Water Poll.* 4:70.

---

*Calculations based on samples taken at approximately 50% RH. Additional accuracy is attainable by making slight corrections for lower RH values (see Method 411).

Subcommittee 3

E. L. KOTHNY, *Chairman*
W. A. COOK
J. E. CUDDEBACK
B. DIMITRIADES
E. F. FERRAND
R. G. KLING
P. W. MCDANIEL
G. D. NIFONG
F. T. WEISS

Approved with modifications from
2nd edition.
Subcommittee 3

D. A. LEVAGGI, *Chairman*
B. R. APPEL
D. W. HORSTMAN
E. L. KOTHNY
J. G. WENDT

415.

Determination of Atmospheric Nitric Acid

1. Principle of the Method

1.1 Atmospheric nitric acid is separated from particulate nitrates by an inert prefilter and collected on a sodium chloride-impregnated cellulose filter (1). Collection presumably involves conversion of the acid to non-volatile sodium nitrate.

1.2 Samples may be analyzed for nitrate by the automated, copper-cadmium reduction, diazotization method (2) or alternative methods with a comparable working range.

1.3 The sampling method is simple and especially applicable to routine sampling.

2. Range and Sensitivity

2.1 The lower limit of measurement is dependent on the air sampling rate, sampling duration and filter size. Sampling at 20 L/min for eight hours with 47 mm filters, the lower limit for reliable quantitation is 0.2 ppb HNO_3 (0.5 $\mu g/m^3$).

2.2 An upper limit of measurement is not given since aqueous extracts can be diluted into the working range of the analytical method.

3. Interferences

3.1 Positive interferences result from the dissociation of nitrate salts on the prefilter (e.g. NH_4NO_3) liberating nitric acid that is retained on the after-filter (3,4). The magnitude of this error depends on the level of particulate nitrate and increases with increasing ambient temperature and decreased relative humidity.

3.2 Negative interferences are produced by sorption of nitric acid on previously collected particulate matter on the prefilter (3). This error can be minimized by relatively frequent prefilter changes.

3.3 Retention of nitric acid on clean Teflon filters is not significant (9,10).

3.4 Nitrogen dioxide shows no interference at up to 90% R.H. (3,5).

4. Precision and Accuracy

4.1 In 13 laboratory trials with 20 to 115 ppb HNO_3, the median coefficient of variation (C.V.) was 3.5% for parallel filter sampling (6). The C.V. of the automated nitrate method was < 2% (3).

4.2 The accuracy of the method depends upon the positive and negative interferences described above, the efficiency of retention of nitric acid on the NaCl-impregnated filters and the accuracy of nitrate extraction and analysis. The single filter efficiency of retention of HNO_3 at R.H. $\geq$ 50% for NaCl-impregnated filters at $\leq$ 20 L/min (face velocity $\leq$ 24 cm/s) was > 97% (3). At 25% R.H., this decreased to 50% (7). At 2% R.H., the overall efficiency of two NaCl-impregnated filters in series was about 40% (6). The accuracy of nitrate extraction and analyses averaged 98%, employing glass fiber filter strips impregnated with known amounts of nitrate (3).

4.3 In short term summer-time atmospheric studies with nitric acid levels ranging from 2 to 19 ppb, results were high by a minimum of 2 ppb (4).

5. Apparatus

5.1 FILTER HOLDER. Nuclepore Corp. multiple filter holders for 47 mm filters equipped with rubber gaskets to eliminate air leaks between stages. Initial stage is equipped for open-face sampling and may employ a polycarbonate funnel (3.2 cm ID, 15 cm length) as an inlet.

5.2 AIR PUMP. Minimum capacity of 30 L/min through a particle-free sampler.

5.3 AIR VOLUME MEASUREMENT. Rotameter able to measure the air flow of 20 L/min to ± 2%.

5.4 PREFILTER. Teflon, 47 mm diameter, with high efficiency for particle collection (e.g., 2 μm pore size Zefluor, Membrana Corp., Pleasanton, CA).

5.5 AFTER-FILTERS. Immerse Whatman 41 (or equivalent) cellulose, 47 mm filter discs in 5% NaCl solution, dry in an oven at 105°C and store in a desiccator. With unwashed Whatman 41 filters, nitrate blanks were below detection (< 3 μg/filter).

5.6 PLATFORM SHAKER. Eberbach platform shaker Model 2000 (or equivalent) for aqueous extraction. An ultrasonic bath may be substituted (8).

5.7 TEST TUBES. 16 × 150 mm with Teflon lined screw caps.

5.8 Technicon II Autoanalyzer equipped for nitrate analysis (2). Equivalent technique may be used providing a working range down to 1 μg/mL.

6. Reagents

6.1 PURITY. All chemicals should meet ACS Analytical Reagent specifications. Water is ASTM Reagent water, Type II.

6.2 Reagents required for the automated nitrate determinations are detailed elsewhere (2).

7. Procedure

7.1 Assemble the sampler comprising, in order, the Teflon prefilter, NaCl-impregnated Whatman 41 after-filter and pump. Briefly attach rotameter to sampler inlet and adjust the flow to 20 L/min. After sampling, reattach rotameter and record final flow rate. Calculate total air sampled using the average flow rate. Nitric acid readily adsorbs on surfaces. Therefore, do not sample air from a manifold unless the efficiency of nitric acid penetration through the sampling manifold and tubing is established.

7.2 Blank after-filters should be carried out into the field, returned to the laboratory, and analyzed along with the atmospheric samples.

7.3 Cut up after-filter and filter blanks and insert in separate 16 × 150 mm test tubes. Add 10 mL water; cap tightly.

7.4 Mount tubes horizontally in batches on platform shaker. Shake at 90 oscillations/min for 60 minutes. Alterna-

tively, immerse in 10 mL water and ultrasonicate for 30 min.

7.5 Decant or centrifuge sample extracts and analyze for nitrate.

7.6 If desired, extract and analyze the Teflon prefilter to provide a measure of atmospheric particulate nitrate. To insure complete extraction with these hydrophobic filters, hold the filters submerged for 30 min in an ultrasonic bath (8). The resulting particulate nitrate values are subject to positive and negative errors discussed in Section 3; factors producing positive HNO_3 errors cause negative particulate nitrate errors.

8. Calibration

8.1 Prepare working curve for the Technicon Autoanalyzer using potassium nitrate standards according to the manufacturer's instructions.

9. Calculations

9.1 For convenience, assume standard conditions to be 101.3 kPa and 25°C. Ordinarily the correction for sample volume to these standard conditions is small relative to other sources of error and may be omitted. However, if conditions deviate significantly, correct the calculations using the perfect gas equation. Report the volume sampled in m^3.

9.2 Calculate the nitric acid concentration as follows:

$$\mu g/m^3 \ HNO_3 =$$

$$\frac{\text{sample } (\mu g \ NO_3^-) - \text{Field Blank } (\mu g \ NO_3^-)}{m^3} \times 1.02$$

where 1.02 is the coversion factor for NO_3^- to HNO_3.

$$\text{ppb } HNO_3 \ (25°C) = \frac{\mu g/m^3 \ HNO_3}{2.57}$$

where 2.57 is the factor for converting $\mu g/m^3 \ HNO_3$ to ppb HNO_3 (vol).

10. Storage

10.1 Loss of nitrate from NaCl-impregnated filters in room temperature storage has not been evaluated but is unlikely.

10.2 Loss of atmospheric nitrate from Teflon filters in room temperature storage within Millipore disposable plastic petri dishes (PD 10047000) sealed, in turn, inside polyethylene bags was found to be negligible over a two month period (6).

11. References

1. OKITA, T., S. MORIMOTO, S. IZAWA AND W. KONNO, 1976. The Measurement of Gaseous and Particulate Nitrates in the Atmosphere. Atmos. Environ **10**:1085-1089.
2. TECHNICON INDUSTRIAL SYSTEMS. 1973. Nitrate and Nitrite in Water and Wastewater. Method No. 100-70W. Technicon Instruments Corp., Tarrytown, NY.
3. APPEL, B.R., S.M. WALL, Y. TOKIWA AND M. HAIK. 1980. Simultaneous Nitric Acid, Particulate Nitrate and Acidity Measurements in Ambient Air. Atmos. Environ. **14**:549-554.
4. APPEL, B.R., Y. TOKIWA AND M. HAIK. 1981. Sampling of Nitrates in Ambient Air. Atmos. Environ. **15**:283-289.
5. GRENNFELT, P. 1980. Investigation of Gaseous Nitrates in an Urban and a Rural Area. Atmos. Environ. **14**:311-316.
6. APPEL, B.R., Y. TOKIWA, E.M. HOFFER, E.L. KOTHNY, M. HAIK AND J.J. WESOLOWSKI. 1980. Evaluation and Development of Procedures for Determination of Sulfuric Acid, Total Particulate Phase Acidity and Nitric Acid in Ambient Air-Phase II. Final Report to the California Air Resources Board, Contract A8-111-31.
7. FORREST, J., R.L. TANNER, D. SPANDAU, T. D'OTTAVIO AND L. NEWMAN. 1980. Determination of Total Inorganic Nitrate Utilizing Collection of Nitric Acid on NaCl-impregnated Filters. Atmos. Environ. **14**:137-144.
8. APPEL, B.R., E.M. HOFFER, Y. TOKIWA, M. HAIK AND J.J. WESOLOWSKI. 1981. Sampling and Analytical Problems in Air Pollution Monitoring. Final Report to the EPA for Grant No. R806734-01-0.
9. SPICER, C.W. AND P. SCHUMACHER. 1977. Interference in Sampling Atmospheric Particulate Nitrate. Atmos. Environ. **11**:873-876.
10. APPEL, B.R., S.M. WALL, Y. TOKIWA AND M. HAIK. 1979. Interference Effects in Sampling Particulate Nitrate in Ambient Air. Atmos. Environ. **13**:319-325.

Subcommittee 3

B.R. APPEL
D. A. LEVAGGI, *Chairman*
D. W. HORSTMAN
E. L. KOTHNY
J. G. WENDT

416.

Continuous Monitoring of Atmospheric Nitric Oxide and Nitrogen Dioxide by Chemiluminescence

1. Principle

1.1 The measurement method is based upon the rapid chemiluminescent reaction of nitric oxide (NO) with excess ozone (O_3) (1,2). The reaction is made to take place in a light-free chamber.

$$NO + O_3 \rightarrow NO^*_2 + O_2$$

$$K = 1.0 \times 10^7 \text{ L mol}^{-1}\text{sec}^{-1}$$

A portion of the resultant nitrogen dioxide is produced in a highly excited energy state (NO^*_2) and subsequently decays to the ground level state emitting light in a broad frequency band with a peak at about 1200 nm:

$$NO^*_2 \rightarrow NO_2 + \text{photons (h}\nu)$$

The intensity of the light emitted is linearly proportional to the nitric oxide concentration and is measured by a photomultiplier tube.

1.2 Atmospheric nitric oxide is determined directly in a sample stream as described in Section **1.1**.

1.3 Atmospheric nitrogen dioxide in a sample stream is measured indirectly after conversion to nitric oxide. (Converters capable of this reduction are described in Section **5.1**.) The detection and determination of the total oxides of nitrogen (NO + NO from NO_2) then proceeds as shown in Section **1.1**. The NO_2 concentration is calculated by the subtraction of the measured NO concentration from the measured total oxides of nitrogen.

2. Range and Sensitivity

2.1 Instruments usually have multiple measurement ranges; typically 0 to 0.2, 0.5, 1.0 or 2.0 ppm v/v. For ambient monitor-

ing, the most commonly used ranges are the 0 to 0.5 and 0 to 1.0 ppm v/v.

2.2 The lower detectable limit is determined by the instrument range being used. For the 0 to 0.5 ppm range a lower detectable limit of 0.01 ppm can be achieved (3).

3. Interferences

3.1 The chemiluminescent reaction of nitric oxide and ozone is not generally subject to interferences from commonly found pollutant species such as ozone, carbon monoxide and sulfur dioxide (4). However, any compound capable of being converted to nitric oxide in the instrument converter (Section **5.1.3**) could be a possible interferent in the measurement of NO_x. In this regard gaseous ammonia could present a problem at elevated temperatures (600°C) in thermal converters; also certain organic nitrogen-containing compounds such as peroxyacetyl nitrate, amines, organic nitrates and nitrites have been shown to decompose stoichiometrically to nitric oxide (5,6,7). Nitric acid reaching the converter is also decomposed to NO. A continuous method for measuring nitric acid at ambient levels based on a modified chemiluminescence analyzer has been reported (8).

Low temperature chemical converters eliminate possible ammonia interference, and the organic nitrogen compounds do not usually exist at atmospheric concentrations sufficient to merit attention.

4. Precision and Accuracy

4.1 Precision is defined by the stability and repeatability of response to NO. Response instability due to baseline drift should be equivalent to no more than ± 0.020 ppm v/v over a 24-h period, and the noise level to no more than ± 0.005 ppm v/ v in the 0 to 0.5 ppm instrument range. Repeatability of response to a standard NO-air mixture should have a coefficient of variation not greater than 2%. This latter precision figure represents the composite error that is associated with both the NO measurement and the standardization operations **(9)**.

4.2 Accuracy is dependent on the calibration equipment flow systems, and the absolute concentration of the NO standard gas cylinder. The calibration flow system specifications contained in Section **7.2**, together with a NO standard gas cylinder of Standard Reference Material (SRM) or Certified Reference Material (CRM) quality (Section **6.2**) are capable of performing instrument calibrations to an accuracy of ± 5%.

5. Apparatus

5.1 CHEMILUMINESCENCE INSTRUMENT. Two basic instrument designs have been developed for the measurement of total oxides of nitrogen, nitric oxide, and the indirect determination of nitrogen dioxide. In both cases the determination of NO_x (NO_x = NO + NO_2) and NO must be accomplished and the nitrogen dioxide calculated by subtraction of NO from the NO_x. The two instrument configurations utilize the cyclic or dual mode of operation.

5.1.1 The cyclic mode instrument is shown schematically in Figure 416:1, and has a single reaction chamber and detector. The incoming sample air is alternately cycled directly to the reaction chamber to determine NO, or through the instrument converter to determine NO_x. A normal cycle, which is approximately thirty (30) seconds, is accomplished by means of a timer controlled solenoid valve. Separate NO_x and NO values are determined every thirty seconds. The photomultiplier tube outputs are amplified and stored in memory circuits; the difference output, nitrogen dioxide, is updated electronically after each cycle and similary stored. Recorder outputs are available for all three measurement channels, NO, NO_2 and NO_x.

5.1.2 The dual mode instrument is shown schematically in Figure 416:2, and has two reaction chambers and a single detector. The incoming air sample is split into two separate streams. The NO sample stream is routed directly to the reaction chamber, while the NO_x sample stream first passes through the instrument converter before entering the second reaction chamber. An optical chopper alternately exposes the detector to the respective chemiluminescent outputs, generating continuous outputs for both NO and NO_x. The outputs are amplified and stored in memory circuits; realtime NO_2 data obtained by difference are continuously generated. Recorder outputs are available for the NO, NO_2, and NO_x concentrations.

5.1.3 *Converters.* For the accurate determination of nitrogen dioxide it is essential that the instrument converters have a high degree of efficiency (95% +) for the conversion of NO_2 to NO. The converters employed in commercially available instruments are of two basic types.

5.1.3.1 Thermal Converters are made of a high grade stainless steel and operate at elevated temperatures, 600–800°C. At these temperatures the breakdown of NO_2 into NO and O_2 occurs readily. These converters, though adequate for the breakdown of NO_2 to NO, have the obvious disadvantage of converting ammonia into NO (See Section **3**).

5.1.3.2 Chemical converters are to be found in the majority of chemiluminescence instruments used for ambient monitoring. These converters have the advantage of a much lower operating temperature, 200–400°C, with efficient NO_2 conversion. Molybdenum and carbon converters have been in general use and are available in commercial instruments.

5.2 RECORDER. Capable of full-scale display of instrument output voltages.

5.3 AIR INLET FILTER. A Teflon filter capable of removing all particulate matter greater than 5 μm in diameter.

5.4 SAMPLE LINES. The sample lines and all parts of the instrument that come in contact with the sample stream should be made of glass, Teflon or stainless steel.

5.5 VACUUM PUMP. A pump capable of a minimum vacuum of 78 kPa.

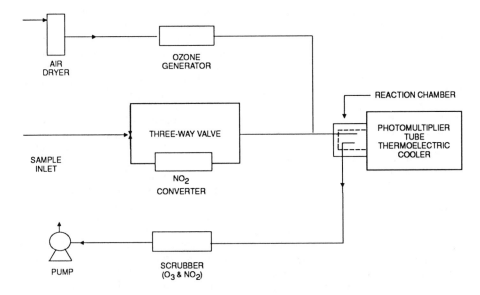

Figure 416:1 – Cyclic mode chemiluminescent analyzer.

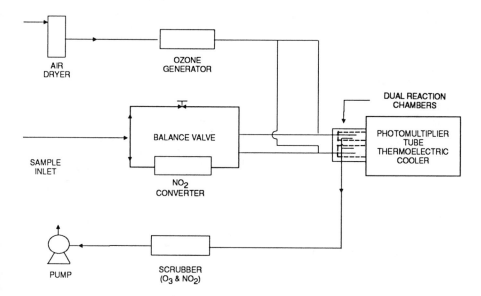

Figure 416:2 – Dual mode chemiluminescent analyzer.

6. Reagents

6.1 NO CYLINDER, CALIBRATION STANDARD. The NO standard should be trace-

able to a National Bureau of Standards NO Standard Reference Material (SRM) or a commercially available Certified Reference Material (CRM). Selection of the NO standard concentration is dependent on the operating range of the analyzer to be calibrated and on the dilution capability of the calibration system. NO cylinders normally used are in the 25–50 ± 2% ppm v/v range in N_2. The NO calibration cylinder *must* be free of any nitrogen dioxide, and should be reanalyzed on a regular basis, preferably every six months.

6.2 ZERO AIR. The air supply must be free of contaminants that would cause a detectable analyzer response, or react independently with NO.

7. Calibration

7.1 In the procedure that follows, NO and NO_2 calibrations are performed using a calibration system such as the one shown in Fig. 416:3. Nitric oxide calibrations are performed by dynamic flow dilution of a NO standard with a clean air stream. Nitrogen dioxide calibrations are performed by the rapid gas phase reaction between NO and O_3 to provide a stoichiometric quantity of NO_2, equal to the decrease in the NO concentration. The reaction is the same as shown as Section **1.1**, except that the NO remains in excess rather than the ozone as described in **1.1**. This reaction is commonly referred to as Gas Phase Titration (GPT) **(9)**. An alternative NO_2 procedure, not described herein, is the generation of known test atmospheres by means of a NO_2 permeation device **(10)**.

7.2 CALIBRATION SYSTEM. All components in the calibration system such as the one in Fig. 416:3 should be made of glass, Teflon or stainless steel. The system is designed to provide dynamic dilution for NO and GPT for NO_2. The dilution section comprises two independent flow controls that can be varied individually to provide a dilution ratio of up to 1,000 to 1. The GPT section comprises a current-regulated ozone generator through which a portion of the dilution air flows even when the ozonator is not in operation. For dynamic dilution, the metered NO combines with this portion of the dilution air and passes through the reaction chamber. It then combines with the balance of the dilution air and passes through the sampling manifold. For GPT the flow path is the same except that a por-

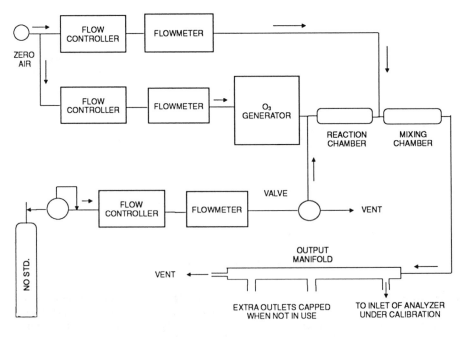

Figure 416:3 — Calibration system components.

tion of the oxygen in the air passing through the ozone generator is converted to ozone.

7.2.1 *Air Flow Controller.* A device capable of maintaining constant clean-air flows up to 5 L/min within ±2% of the required flowrate.

7.2.2 *Air Flowmeter.* A calibrated flowmeter capable of measuring air flow-rates within ± 2%.

7.2.3 *Nitric Oxide Flow Controller.* A device capable of maintaining constant NO flows within ± 2% of the required flow-rate.

7.2.4 *Nitric Oxide Flowmeter.* A calibrated flowmeter capable of measuring NO flowrates within ± 2%.

7.2.5 *Two-Stage Regulator.* The two-stage pressure regulator for the standard NO cylinder must be of stainless steel to prevent any reaction of the NO gas.

7.2.6 *Ozone Generator.* The generator must be capable of generating stable levels of O_3 for the GPT of NO to provide NO_2 concentrations throughout the calibration range.

7.2.7 *Reaction Chamber.* The chamber used for the reaction of O_3 with excess NO should be of sufficient volume that the residence time is not less than 2 minutes **(11)**.

7.2.8 *Mixing Chamber.* A chamber used to allow thorough mixing of reaction products and dilution air.

7.2.9 *Sampling Manifold.* The sampling manifold should be of adequate design to insure against any pressure buildup. It must have a vent to insure atmospheric pressure at the manifold, and for safety considerations to direct excess calibration air streams to an exhaust system.

7.3 PROCEDURE. Prior to start of calibration, for safety purposes, insure proper venting of the analyzer exhaust and the calibration system excess flow. Insure that the analyzer and the calibration system have been on for a time sufficient to provide stable operation.

7.3.1 *Flow Conditions.* Insure that the air and gas flow systems are calibrated under the conditions of use against an authoritative standard. Different output calibration gas concentrations are obtained simply by changing the ratios of flowrates between the NO and dilution air channels. It is pref-erable to maintain a constant dilution air flow and to vary the NO flow. The total flow required at the sampling manifold should equal the analyzer demand plus at least 50% excess. The following equations can be used to precalculate the specific gas dilution air-flow rates required for the desired calibration points, usually 20, 40, 60 and 80% of the instrument range.

$$S = \frac{STD \times FS}{FS + FD} \qquad (1)$$

where S = desired output concentration of NO in ppm

STD = NO standard cylinder concentration in ppm

FS = NO standard cylinder flowrate in cm^3/min

FD = dilution air flowrate in cm^3/min

Solving equation **(1)** for the NO standard flowrate (FS) that will produce the desired concentration for a given dilution flowrate (FD) gives:

$$FS = \frac{S \times FD}{STD - S} \qquad (2)$$

EXAMPLE: For a corrected dilution flow FD = 5.00 L/min, a NO cylinder concentration STD = 100 ppm v/v, and a desired output concentration S = 0.90 ppm, we substitute into equation (2) and obtain for the NO standard flow rate.

$$FS = \frac{(0.90) \times (5,000)}{(100) - (0.90)} = 45.4 \ cm^3/min$$

7.3.2 *Zero Calibration.* Activate the zero air source and allow the analyzer to sample the zero air until a stable zero response is obtained. Adjust the analyzer NO_x, NO and NO_2 zero controls as described in the instrument manual. It is good practice to recheck the zero at the end of the multipoint calibration, especially if large span adjustments were made.

7.3.3 *Preparation for the NO and NO_x Calibration.* Set the zero air and NO standard flow rates as determined in 7.3.1 for generating a NO concentration at 80% of the instrument range setting. Sample this NO concentration for a minimum of 15 minutes or until the NO and NO_x responses are stable.

7.3.3.1 *NO and NO$_x$ Span Adjustment.* Adjust as necessary the analyzer NO and NO$_x$ span controls to obtain recorder responses equal to the NO (NO$_x$ in this case as well) concentration generated. NOTE: Some analyzers may have separate span controls for NO, NO$_x$ and NO$_2$. Other analyzers may only have separate controls for NO and NO$_x$, while others may have one span control common to all channels. Always refer to the instrument manual for any necessary adjustments.

7.3.3.2 Generate additional concentrations evenly spaced across the remainder of the instrument operating scale to establish linearity by decreasing the flow of the NO standard. For each concentration generated, calculate the NO and NO$_x$ concentrations and insure that the respective recorder outputs of the NO and NO$_x$ channels are correct.

7.3.4 *Preparation for the NO$_2$ Calibration.* Set the dilution air and NO standard flow rates as determined in **7.3.1** for generating a NO concentration of about 80% of the instrument range setting. Sample this NO concentration for a minimum of 15 minutes or until the NO, NO$_x$ and NO$_2$ recorder responses are stable. Record the readings. NOTE: The NO$_2$ calibration is conveniently performed by reestablishing the 80% of scale NO-NO$_x$ calibration point, using the same dilution air and NO standard flow rates used in **7.3.3**.

7.3.4.1 *Gas Phase Titration.* Activate the ozone generator and adjust the ozone output so as to decrease the NO concentration by approximately 80%. The decrease must not exceed 90% of the NO concentration being sampled prior to the GPT. Sample this NO-NO$_2$ mixture for a minimum of 15 minutes or until the NO, NO$_x$ and NO$_2$ recorder responses are stable. Record the readings. Calculate the indicated NO$_2$ concentration as per Section **8.1**.

7.3.4.2 *Nitrogen Dioxide Span Adjustment.* Adjust as necessary the analyzer NO$_2$ span control to obtain a recorder response equal to the calculated NO$_2$ concentration.

7.3.4.3 Generate at least two additional calibration points evenly spaced across the remainder of the instrument operating scale by decreasing the O$_3$ output while maintaining the dilution air and NO

standard flow rates constant. For each calibration point generated, calculate the NO$_2$ concentration, and insure that the NO$_2$ recorder responses are correct.

7.4 DETERMINATION OF CONVERTER EFFICIENCY. Calculate the analyzer converter efficiency as per Section **8.2** for the NO$_2$ concentration generated in Section **7.3.4.1**. The converter efficiency must be 95% or greater to be acceptable.

8. Calculations

8.1 Calculation of NO$_2$ Concentration.

$$NO_2 = NO_x - NO$$

where NO$_x$ = instrument reading of the NO$_x$ channel **(7.3.4)**.

NO = instrument reading of the NO channel **(7.3.4.1)**.

8.21 Calculation of Converter Efficiency (CE).

$$\text{Converter Efficiency} = \frac{[NO_x]}{[NO_x]_{GPT}} \times 100$$

where [NO$_x$] = instrument response for original NO$_x$ conc. prior to GPT **(7.3.4)**.

[NO$_x$]$_{GPT}$ = instrument response for NO$_x$ during the GPT runs **(7.3.4.1)**.

9. References

1. A. FONTIJN, A.J. SABADELL AND R.J. RONCO. 1970. "Homogeneous Chemiluminescent Measurement of Nitric Oxide With Ozone," *Anal. Chem.,* 42:575.
2. D.H. STEDMAN, E.E. DABY, F. STUHL AND H. NIKI. 1972. "Analysis of Ozone and Nitric Oxide by a Chemiluminescent Method in Laboratory and Atmospheric Studies of Photochemical Smog," *J. Air Poll. Control Assoc., 22*:260.
3. B.E. MARTIN, J.A. HODGESON AND R.K. STEVENS. 1972. "Detection of Nitric Oxide Chemiluminescence at Atmospheric Pressure," Presented at 164th National ACS Meeting, New York City.
4. R.K. STEVENS AND J.A. HODGESON. 1973. "Application of Chemiluminescent Reactions to The Measurement of Air Pollutants," *Anal. Chem.* 45:443A.
5. J.A. HODGESON, K.A. REHME, B.E. MARTIN AND R.K. STEVENS. 1972. "Measurement for Atmospheric Oxides of Nitrogen and Ammonia by Chemiluminescence," Reprint, Presented at 1972 APCA Meeting, Miami, Florida.
6. L.P. BREITENBACH AND M. SHELEF. 1973. "Development of a Method for the Analysis of NO$_2$

and NH$_3$ by NO-Measuring Instruments," *J. Air Poll. Control Assoc., 23*:128.

7. A.M. WINER, J.W. PETERS, J.P. SMITH AND J.N. PITTS, JR., 1974. "Response of Commercial Chemiluminescent NO-NO$_2$ Analyzers to Other Nitrogen-Containing Compounds," *Envir. Sci. Tech., 8*:1118.

8. D.W. JOSEPH AND C.W. SPICER. 1978. "Chemiluminescence Method for Atmospheric Monitoring of Nitric Acid and Nitrogen Oxides. *Anal. Chem. 50*:1400–1403.

9. K.A. REHME, B.E. MARTIN AND J.A. HODGESON. 1974. "Tentative Method for the Calibration of Nitric Oxide, Nitrogen Dioxide, and Ozone Analyzers by Gas Phase Titration," EPA-R2-73-246.

10. F.P. SCARINGELLI, A.E. O'KEEFFE, E. ROSENBERG AND J.P. BELL. 1970. Preparation of Known Concentrations of Gases and Vapors with Permeation Devices Calibrated Gravimetrically, *Anal. Chem., 42*:871.

11. E.C. ELLIS. 1975. "Technical Assistance Document for the Chemiluminescence Measurement of Nitrogen Dioxide," EPA-E600/4-75-003.

Subcommittee 3

J.G. WENDT
D. A. LEVAGGI, *Chairman*
B. R. APPEL
D. W. HORSTMAN
E. L. KOTHNY

417.

Continuous Monitoring of Ozone in the Atmosphere by Ultraviolet Photometric Instruments

1. Principle of the Method

1.1 The method is based on the photometric assay of ozone (O_3) concentrations in a dynamic flow system. The concentration of O_3 is determined in an absorption cell from the measurement of the amount of light absorbed at a wavelength of 254 nm. The method is based on the absorption coefficient of O_3 at 254 nm, the optical path length through the sample, and the transmittance, temperature and pressure of the sample (1,2,3,4,5,6,7). The quantities above are related by the Beer-Lambert absorption law,

$$\text{Transmittance} = \frac{I}{I_o} = e^{-\alpha cl} \quad (1)$$

where α = absorption coefficient of O_3 at 254 nm = 310 atm^{-1} cm^{-1} at 0°C and 101.3 kPa

    c = O_3 concentration in units of atmospheres

    l = optical path of absorption cell length in cm

    I = intensity of light passing through cell with an ozone sample

    I_o = intensity of light passing through cell with zero air

1.2 Typically, an air sample is first directed through a scrubber that removes any O_3 present, but otherwise does not affect the sample. The ozone-free sample then flows through the absorption cell, and its transmittance is measured. This constitutes the zero cycle. At a preset time, a solenoid switches and another air sample flows directly into the absorption cell, bypassing the scrubber, and its transmittance is measured. This constitutes the ozone measurement cycle. The difference in transmittance between the two cycles is a measure of the O_3 concentration. The complete measurement cycle takes about 20 to 30 s. Microprocessor-controlled electronics perform timing functions, condition the signal and perform arithmetic operations in commercially available analyzers. Figures 417:1 and 417:2 show typical flow systems for both single and dual cell O_3 analyzers.

2. Range and Sensitivity

2.1 O_3 analyzers are commercially available for measurement in the 0.00 to 1.00 ppm (1962 μg/m^3) range.

2.2 The lower limit of detection has been reported at 1 ppb (1.96 μg/m^3) (8).

3. Interferences

3.1 Any gaseous component or fine particle that absorbs or scatters light at 254 nm is a potential interferent. Gaseous components normally found in ambient air do not interfere, and particles are largely removed by the Teflon filter described in Section 5.2. Specific interference from nitrogen dioxide and sulfur dioxide has been evaluated and found to be negligible (9).

4. Precision and Accuracy

4.1 Precision audits of thirty-three analyzers operated by the State of California revealed an average standard deviation of 4.3% from a reference sample. The average percent difference from the known value was –3.0% with 95% probability limits of +8% and –13% (10).

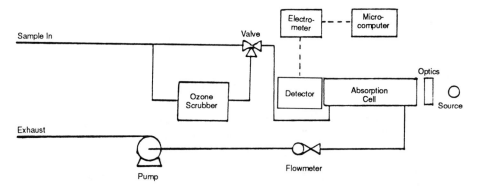

Figure 417:1 — Schematic diagram of a typical single cell ozone analyzer.

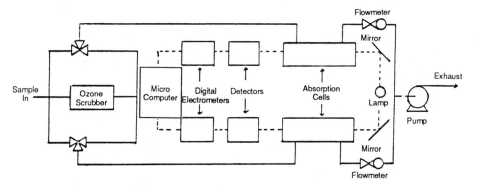

Figure 417:2 — Schematic diagram of a typical dual cell ozone analyzer.

4.2 In a controlled experiment, four photometers operated by the National Bureau of Standards, California Air Resources Board, the Jet Propulsion Laboratory, and the Environmental Protection Agency indicated a total variation of 2.8% when measuring O_3 concentrations from 0.05 to 0.70 ppm (98 to 1373 $\mu g/m^3$). Compared to the NBS photometer, correlation coefficients were 1.0000, 1.0000 and 0.9999 respectively **(11)**. For ozone analyzers, the accuracy is determined by the calibration. When calibrations are performed using photometers, ± .01 ppm is obtainable. If the neutral buffered potassium iodide spectrophotometric procedure (Method 411) is used, an overall average of ± 5% can be used.

5. Apparatus

5.1 OZONE PHOTOMETRIC ANALYZER, commercially available, complete with sample pump and sample flowmeter.

5.1.1 All connections to the ozone analyzer must be constructed of glass, Teflon, or other inert materials **(12)**.

5.2 AIR INLET FILTER. A Teflon filter capable of removing all particulate matter greater than 5μm in diameter.*

5.3 RECORDER, capable of full scale display of voltages from the instrument DC amplifier. These are commonly found in full scale ranges of 10 mV to 1V.

---

*Mace Corporation, 1810 Floradale Ave., South El Monte, California 91733 or equivalent.

5.4 CALIBRATION APPARATUS.

5.4.1 *Ultraviolet Photometer,* (UV Photometer), commercially available. The UV Photometers are primary standards for determinations of ozone in air. The units differ from the ozone photometric analyzer in Section **5.1** in that the UV Photometers do not contain an ozone scrubber, and are designed to make pressure and temperature corrections for the measured ozone to standard conditions (25°C and 101.3 kPa).

5.4.2 *Ozone Transfer Standard.* An ozone analyzer that has been precalibrated against a UV Photometer **(13)**.

5.4.3 *Ozone Source and Dilution System.* The ozone source consists of a quartz tube into which purified air is introduced and then irradiated with a stable low pressure mercury lamp. The level of irradiation is controlled by an adjustable metal sleeve that fits around the lamp **(14)**. At a fixed level of irradiation and at a constant temperature and humidity, ozone is produced at a uniform rate. By careful control of the flow of air through the quartz tube, and/or adjustment of the irradiation level, test atmospheres can be generated that contain stable but variable concentrations of ozone. An output manifold with a vent is attached to the ozonator (Fig 417:3). Ozone outputs must be available to cover the complete analyzer operating range, typically 0.00 to 1.00 ppm. The dilution system should have a total flow capability of a least 5 L/min. Any alternative system capable of these outputs is acceptable **(15)**.

5.4.4 Spectrophotometric determination of the calibration ozone streams may be performed. This manual method is based on the formation of free iodine in a neutral buffered potassium iodide solution upon exposure to ozone (Method 411), and requires the following equipment.

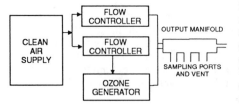

Figure 417:3 — Ozone source and dilution system.

5.4.4.1 *Air metering device.*
5.4.4.2 *Midget impingers.*
5.4.4.3 *Air pump.*
5.4.4.4 *Spectrophotometer.*

6. Reagents

6.1 PURITY. All reagents shall conform to specifications for Reagent grade chemicals of the American Chemical Society.

6.2 ZERO AIR. Ambient air purified by passing through an activated charcoal filter, or by another appropriate manner **(13)**.

7. Procedure

7.1 Operate any commerical ozone analyzer following procedures given in the manufacturer's manual. Perform the calibration as directed in Section **8**. Upon completion of a satisfactory calibration the analyzer is acceptable for ozone monitoring.

8. Calibration

8.1 Connect the ozone analyzer to the output manifold of the calibration system (Fig 417:3). Check to insure proper operating parameters according to the instrument manual.

8.2 Connect either a UV Photometer (Section **5.4.1**), or an ozone transfer standard (Section **5.4.2**) to the output manifold. Check to insure proper operating parameters according to the instrument manuals. Either of these instruments can determine the true ozone concentration of the calibration air streams.

8.2.1 As an alternative, the ozone concentration of the calibration air streams may be determined by using the spectrophotometric procedure of Method 411.

8.3 ZERO AIR. With the O_3 lamp off, flush the system for about ten minutes to remove residual O_3. While the analyzers sample the zero air, record ten consecutive digital display values. Calculate, for the ozone analyzer being calibrated and either the UV Photometer or ozone transfer standard, the sum and average of the ten values and record. If an ozone transfer standard is used, note and record temperature and pressure.

8.3.1 If the spectrophotometric procedure is used, a manual sample of the zero air should be taken. Record the ppm of ozone found, performing the calculations described in the procedure (Method 411).

8.4 TEST ATMOSPHERES. With the O_3 lamp well equilibrated, adjust the O_3 concentration to the range desired for a calibration point. Wait 10 minutes for equilibration at each setting. Measure and record the analyzer digital outputs as in the preceeding Section **8.3**. Test atmospheres of approximately 80, 60, 40 and 20% of the range of the analyzer should be run for a calibration.

8.4.1 If the spectrophotometric procedure is used, a manual sample is taken of each calibration point. Record the ppm of ozone found for each calibration level by performing the calculations described in the procedure.

9. Calculations

9.1 If a UV Photometer was used in the calibration, the ozone readings are the true ozone concentrations already corrected to standard conditions (25°C and 101.3 kPa).

$$\text{True Ozone (ppm)} = \text{Ozone Reading} - \text{Zero Reading}$$

where
Ozone Reading = The UV Photometer ozone readout for each calibration point test atmosphere (**8.4**).
Zero Reading = The UV Photometer ozone readout for the zero air stream (**8.3**).

9.2 If a transfer standard was used in the calibration, its ozone readings must be corrected to standard conditions (25°C and 101.3 kPa).

True Ozone (ppm) =

$$\left(\begin{matrix}\text{Ozone} \\ \text{Reading}\end{matrix} - \begin{matrix}\text{Zero} \\ \text{Reading}\end{matrix}\right) \times \frac{P}{101.3} \times \frac{298}{t + 273}$$

where
Ozone Reading = the transfer standard readout for each calibration point test atmosphere (**8.4**).
Zero Reading = the transfer standard ozone readout for the zero air stream (**8.3**).
P = the barometric pressure during the calibration in kPa (**8.2**).
t = the temperature during the calibration in °C (**8.2**).

9.3 If the spectrophotometric procedure was used in the calibration the true ozone is calculated as follows:

$$\text{True Ozone (ppm)} = \text{Ozone Conc.}_{CTA} - \text{Ozone Conc.}_{ZAS}$$

where
Ozone Conc._{CTA} = the ozone concentration calculated for each calibration test atmosphere (**8.4.1**).
Ozone Conc._{ZAS} = the ozone concentration calculated for the zero air stream (**8.3.1.**).

9.4 INSTRUMENT CALIBRATION CURVE. Plot the net ozone analyzer outputs, i.e., each calibration point (**8.4**) less the zero air (**8.3**), versus the respective net true ozone readings from either **9.1**, **9.2** or **9.3**. If the average percent difference between the ozone analyzer observations and the true ozone concentration is within ± 10%, the analyzer is considered to be in calibration.

10. References

1. INN, E.C.Y. AND Y. TANAKA. 1953. Absorption Coefficient of Ozone as the Ultraviolet and Visible Regions. J. Opt. Soc. Am., 43:870.
2. HEARN, K.G. 1961. Absorption of Ozone in the Ultraviolet and Visible Region of the Spectrum. Proc. Phys. Soc., London, 78:932.
3. DEMORE, W.B. AND O. REAPER. 1964. Hartley Band Extinction Coefficients of Ozone in the Gas Phase and in Liquid Nitrogen Carbon Monoxide and Argon. J. Phys. Chem. 68:412.
4. GRIGGS, M. 1968. Absorption Coefficients of Ozone in the Ultraviolet and Visible Regions. J. Chem. Phys. 49:857.
5. BECKER, K.H., U. SCHURATH AND H. SEITZ. 1974. Ozone Olefin Reactions in the Gas Phase 1.

Rate Constants and Activation Energies. Int'l Jour of Chem. Kinetics, VI:725.

6. CLYRE, M.A.A. AND J.A. COXON. 1968. Kinetic Studies of Oxyhalogen Radial Systems. Proc. Roy. Soc. A303:207.

7. SIMONS, J.A., R.J. PAUR, H.A WEBSTER AND E.J. BLAIR. 1973. Ozone Ultraviolet Photolysis VI. The Ultraviolet Spectrum. J. Chem. Phys. 59:1203.

8. Operating and Instruction Manual, Dasibi Ozone Monitor, Model 1003 AH, Dasibi Environmental Corp., Glendale, CA.

9. CHERNIACK, I. AND R.J. BRYAN. 1965. A Comparison Study of Various Types of Ozone and Oxidant Detectors Which Are Used for Atmospheric Air Sampling. JAPCA 15 (8):351.

10. California Air Quality Data, California Air Resources Board, Vol. XIII No. 1, January-March, 1981. p. 3.

11. WENDT, J., J. KOWALSKI, A.M. BASS, C. ELLIS AND M. PATAPOFF. 1978. Interagency Comparison of Ultraviolet Photometric Standards for Measuring Ozone Concentrations. NBS Special Publication 529.

12. ALTSHULLER, A.P. AND A.F. WARTBURG. 1961. The Interaction of Ozone with Plastic and Metallic Materials in a Dynamic Flow System. Int'l. Jour. Air and Water Pollution. 4:70.

13. "Transfer Standards for Calibration of Ambient Air Monitoring Analyzers for Ozone," EPA 600/4-79-056 June 1979, Department E (MD-77) Research Triangle Park, N.C. 27711.

14. HODGESON, J.A., R.K. STEVENS AND B.E. MARTIN. 1972. A Stable Ozone Source Applicable as a Secondary Standard for Calibration of Atmospheric Monitors, pp. 149-158. In *Air Quality Instrumentation*, Vol. 1, John W. Scales, Ed. Pittsburg, Instrument Society of America.

15. "Technical Assistance Document for the Calibration of Ambient Ozone Monitors," EPA 600/4-79-057 July 1979, Department E (MD 77) Research Triangle Park, N.C. 27711.

Subcommittee 3

D.W. HORSTMAN
D.A. LEVAGGI, *Chairman*
B.R. APPEL
E.L. KOTHNY
J.G. WENDT

501.

High-Volume Measurement of Size Classified Particulate Matter

1. Principle of Method

1.1 PRINCIPLE. Air is drawn through a size-selective inlet and through a 20.3 × 25.4 cm (8 × 10 in) filter at a flow rate which is typically 1132 L/min (40 ft^3/min). Particles with aerodynamic diameters less than the cut-point of the inlet are collected by the filter. The mass of these particles is determined by the difference in filter weights prior to and after sampling. The concentration of suspended particulate matter in the designated size range is calculated by dividing the weight gain of the filter by the volume of air sampled **(1,2)**.

1.2 APPLICABILITY. The high volume air sampling method is applicable to all measurements of ambient suspended particulate matter. It is accepted as a measurement of total suspended particulate matter (TSP) **(3)** and particles in the 0 to 10 μm size range (PM$_{10}$) **(4)** for determining compliance with national ambient air quality standards.

1.3 OTHER ANALYSES. Depending on the type of filter media used, filter samples can be analyzed for lead **(5)**, ions **(6)**, organic and elemental carbon **(7)**, extractable organic material **(8)**, elements **(9,10,11)**, radioactive materials **(12)**, inorganic compounds **(13)**, and single particles **(14)**. These additional analyses are often used in source apportionment studies **(15)**.

1.4 DETAILED PROCEDURES. Equipment-specific operating manuals are available from the equipment manufacturers. The United States Environmental Protection Agency (EPA) and various state agencies have issued several complete standard operating procedures **(16,17)** with suggestions for data forms, maintenance schedules, and audit schedules.

1.5 ALTERNATIVE METHODS. Medium volume samplers operating at 113 L/min (4 ft^3/min) **(18,19,20)** and low volume samplers operating at 16.7 L/min (0.59 ft^3/min) **(21,22)** have been developed to measure suspended particulate mass concentrations in size ranges comparable to those of the high volume methods. Intercomparisons of many sampling devices demonstrate that the mass measurements made by these alternatives are not always equivalent **(23,24,25)**.

2. Range and Sensitivity

2.1 LOWER QUANTIFIABLE LIMIT. For a 24-h sample duration at 1132 L/min, the detection limit is determined by the reproducibility of the filter weight difference which shows a standard deviation (sigma) of approximately ± 2 mg. The three-sigma detection limit is then approximately 3.5 μg/m^3. The three-sigma lower quantifiable limit depends on the filter used and may be as high as 5 μg/m^3 **(26)**.

2.2 UPPER QUANTIFIABLE LIMIT. For a 24-h sample duration at 1132 L/min, this limit is in the range of 400 to 1000 μg/m^3. The exact value depends on the nature of the aerosol being sampled: very small particles will clog the filter at a relatively low mass loading while larger particles will fall off during sample transport at high concentrations.

3. Interferences

3.1 PASSIVE DEPOSITION. Passive deposition occurs when windblown dust deposits on a filter both prior to and after sampling. Positive biases of approximately 10 to 15% have been found with peaked roof inlets **(27,28,29,30)**. Studies for other inlets are insufficient for quantification. This interference can be minimized by shortening the passive period or by using commercially available covers that protect the filter before and after the active sampling period.

3.2 INLET LOADING AND RE-ENTRAINMENT. Material collected in size-selective inlets can become re-entrained in the sample flow. Controlled studies are insufficient to quantify this interference. It can be minimized by greasing or oiling inlet impaction surfaces, though this may change the size-selective properties **(31)**.

3.3 RECIRCULATION. Recirculation occurs when the blower exhaust, which contains carbon and copper particles from the armature and brushes, is entrained in the sampled air. Positive biases of 0.15 $\mu g/m^3$ have been measured **(32)**, which are insignificant mass interferences but which may affect carbon and copper measurements. Recirculation can be minimized by assuring a tight seal between the blower and the sampler housing **(33)** or by ducting blower exhaust away from the sampler.

3.4 FILTER ARTIFACT FORMATION. Sulfur dioxide, nitrogen oxides, nitric acid, and organic vapors can be absorbed on the filter medium along with the suspended particles thereby causing positive biases. Samples taken in the presence of high SO_2 concentrations have been shown to yield up to 10 $\mu g/m^3$ of excess sulfate on glass fiber filters **(34,35)**. Quartz, Teflon membrane, and Teflon coated glass fiber filters have been shown to minimize NO_x and SO_2 absorption **(36,37)**, and denuder inlets **(38)** have been used to minimize nitric acid absorption. Very little information is currently available regarding organic artifacts.

3.5 FILTER CONDITIONING. Filter conditioning environments can result in different mass measurements as a function of relative humidity (RH). Soluble particles take on substantial quantities of water as RH increases, especially above the deliquescence point of approximately 70% RH **(39)**. Increased mass deposits of 50% or more have been observed as RH increases to 100% **(40)**. The hysteresis effect as RH decreases precludes a constant mass measurement unless samples are equilibrated at a RH of less than 30% prior to and after exposure. EPA procedures call for less than 50% **(16)** or 20 to 45 $\pm 5\%$ **(4)** RH during equilibration. Twenty-four hours at a con-

stant temperature and RH is considered adequate for sample equilibration.

3.6 FILTER INTEGRITY. Filter integrity is compromised by handling, which causes pieces of the filter to be lost after the pre-exposure weighing. The result is a negative bias to the mass deposit. Quartz filters are the most prone to loss of material while Teflon membrane filters retain their integrity with normal handling **(26)**. Filters can be loaded into filter cassettes in the laboratory for transport to the field to minimize losses of the filter material.

3.7 SHIPPING LOSSES. Particle loss during transport occurs when filters are heavily loaded with large dry aerosols. It is more prevalent on membrane than on fiber filters. Particle loss is minimized by shorter sample duration in heavily polluted environments, use of fiber as opposed to membrane filters, folding the filter prior to transport, and careful shipping procedures.

3.8 ELECTROSTATIC CHARGE. Both blank and exposed filters may acquire an electrostatic charge which will bias the mass measurement owing to electrical forces in the weighing chamber. This bias depends on filter type, atmospheric conditions, and the electrical environment of the sample **(41)**. Electrically charged filters can be neutralized by exposure to a low-level radioactive source prior to and during weighing.

4. Precision and Accuracy

4.1 Mass of the filter deposit, flow rate through the filter, and sampling time have typical precisions of ± 2 mg, $\pm 5\%$, and ± 1 min, respectively, as determined from performance tests **(37)**. The accuracy of these measurements can be well within these tolerances when determined with independent standards. These uncertainties combine to yield a propagated precision of approximately $\pm 13\%$ at 10 $\mu g/m^3$ and approximately $\pm 5\%$ at 100 $\mu g/m^3$. The filter deposit mass measurement precision dominates at low concentrations while the flow rate precision dominates at high concentrations.

4.2 Precisions determined from collocated measurements using identical high volume samplers vary depending on the inlet and the test conditions. Average and mean squared collocated precisions are typically in the range of 4 to 10% (23,24,25,42,43). Collocated precisions are observed to be roughly twice propagated precisions (25).

5. Apparatus

5.1 SAMPLER. The sampler consists of an inlet, a filter holder, an air mover, a flow controller, and a timer. Figure 501:1 illustrates a generic configuration with a peaked roof inlet and an automatic mass flow controller. A measurement of the exhaust pressure in the plenum by a flow recorder chart or a pressure gauge is related to flow rate through the filter when manual flow control is used. Several options exist for inlets, flow control, and filters.

5.2 SIZE SELECTIVE INLETS.

5.2.1 *Peaked Roof Inlet (Figure 501:2a).* The peaked roof inlet is the oldest inlet and consists of a right triangular structure with an open hypotenuse placed over the filter. Over 50% of the particles smaller than 30 μm to 50 μm diameter penetrate this inlet (at 566 to 1698 L/min flow rates) and deposit on the filter (44,45). The peaked roof inlet does not have a sharp sampling effectiveness curve and is intended primarily to protect the filter from dustfall. The sampling effectiveness of this inlet varies depending on its orientation with respect to wind direction and on the wind speed (44).

5.2.2 *Opposed-Jet Inertial Separation (Figure 501:2b).* Sampled air passes through successive arrays of opposed jets. Depending on the specific dimensions of the inlet and its jets, impaction removes particles larger than 10 μm diameter. These particles are trapped in the inlet while smaller particles are transmitted to the filter (46,47).

5.2.3 *Cyclonic Flow Inlet (Figure 501:2c).* Air enters a vertical cylinder with a swirling (vortex) motion and particles larger than the design cutoff (e.g., 10 μm diameter) are deposited on the inner surface of the cylinder (48).

5.2.4 *Cascade Impactor Inlet (Figure 501:2d).* A series of circular or slotted impactor stages is located between the filter and any one of the inlets specified in Sections 5.2.1 to 5.2.3. Perforated greased foils or filters are placed on each stage to collect particles of intermediate size (49).

5.3 FLOW CONTROLLERS.

5.3.1 *Manual Volume Flow Control.* A variable voltage transformer placed in series with the blower controls the blower motor power. The motor speed varies with the voltage supplied, and the flow rate through a filter can be adjusted by increasing or decreasing the voltage to obtain the desired value for the resistance of the filter being used. The flow rate decreases as filter deposit increases, but this change is normally less than 10% and is quantifiable via pre- and post-exposure flow measurements.

5.3.2 *Automatic Volume Flow Control by Critical Throat.* Sonic (critical) flow is established in a throat section between the filter and the blower. The presence of a stable shock wave in the throat limits the air flow to a design value (50).

5.3.3 *Automatic Mass Flow Control.* A heat sensing element provides a signal proportional to the number of air molecules passing the sensing probe. A feedback signal is sent to the blower to accelerate or decelerate the flow, keeping the heat transfer constant at the sensing element (51,52).

5.4 LABORATORY EQUIPMENT.

5.4.1 *Controlled Environment.* A clean laboratory environment is required for filter inspection, equilibration, and weighing. EPA recommends a temperature in the range of 15 to 30°C with ±3°C variability (4,16) and a relative humidity of 20 to 45% with ± 5% variability (4).

5.4.2 *Light Table.* A photographic slide viewing table is used for filter inspection.

5.4.3 *Analytical Balance.* The balance must be equipped with an expanded weighing chamber to accommodate 20.3 × 25.4 cm (8 × 10 in) filters and must have a sensitivity of 0.1 mg.

5.4.4 *Equilibration Rack.* This rack separates filters from one another so that the equilibration air can reach all parts of the filter surface. A phonograph record rack serves this purpose well.

5.4.5 *Numbering Machine.* Though filter ID numbers can be written on the edge of filters with a pen, an incrementing numbering machine that prints 4 to 8 digit ID numbers is more efficient and is less likely to damage the filter.

5.4.6 *Polonium Charge Neutralizer.* Electrostatic charges on certain filter mate-

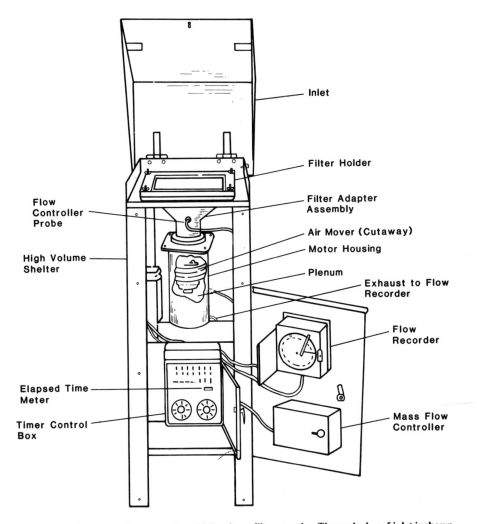

Figure 501:1 – Typical configuration for a high volume filter sampler. The peaked roof inlet is shown.

rials may interfere with accurate weighing. Two or three charge neutralizers placed in the balance weighing chamber can minimize these effects. Radioactive charge neutralizers should be replaced yearly and disposed of according to instructions.

5.4.7 *Wet Bulb/Dry Bulb Psychrometer.* The temperature and relative humidity of the controlled filter processing environment is measured and recorded before and after each filter processing session. Adjustments are made to the environmental con-

trol system when equilibration conditions exceed pre-set tolerances.

5.5 CALIBRATION AND AUDITING EQUIPMENT.

5.5.1 *Primary Flow Rate Standard.* A positive volume displacement device serves as a primary standard. A spirometer, a "frictionless" piston meter, or a Roots meter can serve as such a standard **(53)**.

5.5.2 *Orifice Transfer Standard.* The high volume sampler calibration orifice consists of a 3.175 cm (1.25 in) diameter

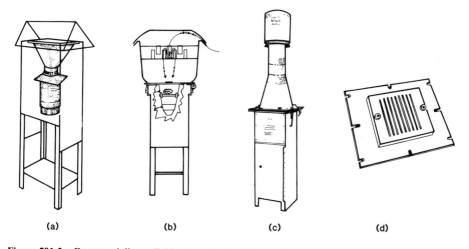

Figure 501:2 — Commercially available size selective inlets for high volume sampling: a) peaked roof inlet, b) opposed jet, c) cyclonic flow, and d) cascade impactor.

hole in the end cap of 7.62 cm (3 in) diameter by 20.3 cm (8 in) long hollow metal cylinder. This orifice is mounted tightly to the filter support in place of the inlet during calibration. A small tap on the side of the cylinder is provided to measure the pressure drop across the orifice. A flow rate of 1132 L/min through the orifice typically results in a pressure difference of several inches of water. The relationship between pressure difference and flow rate is established via a calibration curve derived from measurements against a primary standard at standard temperature and pressure (54). Flow resistances that simulate filter resistances are introduced at the end of the calibrator opposite the orifice by a variable aperture or by a set of perforated circular disks.

5.5.3 *Manometer.* A calibrated pressure gauge or water manometer spanning 0 to 15 inches of water (0–4 kPa) is used to determine the pressure drop across the orifice.

5.5.4 *Barometer.* The atmospheric pressure at the time of calibration and at the time of measurement is determined with a barometer. Flow rate corrections are made if these two pressures differ by more than 4 kPa (4% of standard 101.3 kPa).

5.5.5 *Thermometer.* The atmospheric temperature at the time of calibration and at the time of measurement is determined with a thermometer. Flow rate corrections

are made if these two temperatures differ by more than 15°C (5% of standard 298 K).

5.5.6 *Class S Weights.* A 3 g standard mass of Class S or Class M quality is used to verify the span of the analytical balance.

6. Reagents and Supplies

6.1 FILTER MEDIA. A 20.3 × 25.4 cm (8 × 10 in) fiber or membrane filter is used to collect particles. The choice of filter type results from a compromise among the following filter attributes: 1) mechanical stability, 2) chemical stability, 3) particle sampling efficiency, 4) flow resistance, 5) clogging level, 6) blank values, 7) artifact formation, and 8) cost and availability. EPA filter requirements specify 0.3 μm DOP sampling efficiency in excess of 99%, weight losses or gains due to mechanical or chemical instability of less than a 5 μg/m^3 equivalent, and alkalinity of less than 25 microequivalents/g to minimize sulfur dioxide (SO_2) and nitrogen oxides (NO_x) absorption(4). Summaries and experimental evaluations of these attributes for various filter media have been prepared (33,34,35,55–70). The most appropriate filter media for high volume sampling are cellulose fiber, glass fiber, quartz fiber, Teflon coated glass fiber, and Teflon membrane. None of these materials is perfect for all purposes.

6.1.1 Cellulose fiber filters meet requirements in most categories with the exception of sampling efficiency and water vapor artifacts. Sampling efficiencies below 50% in the submicron region have been observed, but these are highly dependent on the filter weave **(56,71,72)**. Cellulose fiber is hygroscopic and requires precise relative humidity control in the filter processing environment to obtain accurate mass measurements **(63,73,74)**. This substrate has low elemental blanks and is commonly used for chemical speciation of the deposit **(65,75)**. Because this substrate is carbon-based, it is not normally used for carbon analysis.

6.1.2 Glass fiber filters meet requirements in most categories with the exception of artifact formation and blank levels. Sampling efficiency is very high for all particle sizes **(56,76)**. The high alkalinity of these substrates causes sulfur dioxide, nitrogen oxides, and gaseous nitric acid to be absorbed **(62–70,77,78)**. Blank levels for several elements of interest are high and variable **(79,80)**. Particulate nitrate and ammonium losses have been observed when these samples are stored at room temperature for long periods **(81)**, but this is probably true of deposits on all types of filter media. Glass fiber filters may exhibit organic carbon artifacts **(82)**.

6.1.3 Quartz fiber filters meet requirements in most categories **(83)** with artifact properties which are significantly lower than those of glass fiber **(62–70)**. Quartz substrates may exhibit organic carbon artifacts **(82)**. Trace element blank levels are too high and variable for several elemental analyses **(64)**, though these filters are widely used for carbon analyses **(84)**. The greatest drawback of quartz fiber filters is their fragility **(26)**, which requires extremely careful handling for accurate mass measurements. New formulations are under development to minimize this disadvantage **(85)**.

6.1.4 Teflon coated glass fiber filters meet requirements in all categories except blank element and carbon levels **(36)**. Though a small nitric acid artifact has been observed **(36)**, it is tolerable in most situations. These filters are excellent for ion

analysis but not for carbon analyses owing to their Teflon coating.

6.1.5 Teflon membrane filters meet requirements in all categories except flow resistance and cost. Because of their low porosity, it is not usually possible to attain the flow rates needed by the size selective inlets in high volume sampling. These filters are not usually analyzed for carbon because of its presence in the filter material, though they have very low blank levels for ions and elements. The normal cost of 20.3 $\times$ 25.4 cm (8 $\times$ 10 in) filters is in the tens of dollars per sheet.

6.2 FILTER JACKET. A smooth, heavy paper folder or envelope is used to protect the filter between the lab and field and during storage. Manila envelopes are in common use. Filter and sampling data are often recorded on the outside of the jacket, but this should not be done while the filter is in the jacket to prevent damage.

7. Procedure

7.1 Figure 501:3 presents a flow diagram of the routine operating procedure described in the following sub-sections.

7.2 FILTER INSPECTION. Clean the light table surface with a methanol soaked wiper and allow to dry. Filters should be handled with gloved hands to prevent contamination. Place each filter on the light table and examine it for pinholes, loose particles, tears, creases, lumps, or other defects. Loose particles may be removed with a soft brush. Filters not meeting visual criteria should not be used. If chemical analyses are to be performed, one or two filters from each lot should be analyzed for blank levels and the lot should be rejected if pre-set specifications are not met.

7.3 FILTER IDENTIFICATION. Apply an ID number to the upper right hand corner on the smoothest side of each filter with the incrementing numbering machine. Gentle pressure must be used to avoid damaging the filter. Record this number in a chain-of-custody log book and on a filter jacket. The chain-of-custody log book contains columns opposite every filter ID to record dates and technician initials for filter inspection, equilibration, pre-weighing, shipment to field, receipt from field, re-equilibration, post-weighing, and storage.

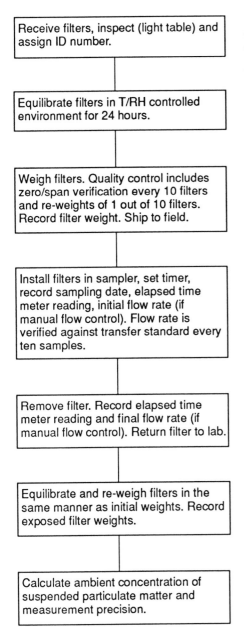

Figure 501:3 — Flow diagram for routine hivol operations.

These records identify the disposition of each sample and prevent the creation of two samples with the same ID.

7.4 FILTER EQUILIBRATION. Place blank or exposed filters in a storage rack in the controlled temperature and relative humidity environment for 24 h prior to weighing. The rack should separate filters such that all surfaces are exposed to the equilibration environment. Measure the temperature and relative humidity of the controlled environment and record the values in the equilibration column of the chain-of-custody log book.

7.5 FILTER WEIGHING. It is best to weigh filters in groups of ten to fifty. Wear gloves for all filter handling. Stack filter jackets with data forms printed on them in the same order (in ascending order of filter ID numbers, if possible) as the order of filters in the equilibration rack. Adjust the balance tare to read zero with nothing in the weighing chamber and adjust the span to read (or verify that it reads) 3.00000 g with the 3-g standard weight on the weighing pan. Place a filter in the weighing chamber and adjust the balance to its equilibrium position. If a stable reading cannot be obtained, it may be necessary to neutralize electrostatic charges with a radioactive source prior to and during weighing. Record the weight on the data form in the blank or exposed filter column. Verify the zero and span every ten filters. If these differ from their normal values by more than ± 1.0 mg, readjust them and re-weigh the previous ten filters. Place each filter in its filter jacket when weighing is complete, but do not seal the jacket opening. A separate technician randomly selects four filters or ten percent of all filters in the batch (whichever is larger), re-weighs them, and subtracts this check-weight value from the corresponding routine weight. If any check-weight differs by more than ± 5.0 mg from the routine weight, re-weigh the entire batch of filters. Seal filter jackets and ship blank filters to the field or place exposed filters into storage.

7.6 FIELD SAMPLING. Tilt back the inlet and secure it according to manufacturer's instructions. Loosen the faceplate wingnuts and remove the face plate. Remove the filter from its jacket and center it on the support screen with the rough side of the filter facing upwards. Replace the faceplate and tighten the wingnuts to secure the rubber gasket against the filter edge. Gently lower the inlet. Inertial jet and cyclonic inlets must have their seals in contact with the top of the faceplate. Look underneath the inlet just as it is coming into contact with the faceplate to assure that this contact is being made. It may be necessary to readjust the position of the filter/motor assembly in the sampler housing to obtain such a seal. Excessively windy and wet conditions should be avoided when changing samples. Pre-loading in a filter cartridge assembly, temporary removal of the sampler to a protected area, or a wind or rain shield may be used if the sample must be changed in inclement weather. Set the timer for the desired start and stop time. Replace the chart paper in the flow recorder, if there is one, set the proper time, and mark the time and date on the chart. For a manually flow controlled sampler, turn on the motor for five minutes and measure the exhaust pressure with a pressure gauge or rotameter. Read the flow rate corresponding to this exhaust pressure from the calibration curve and record it on the data sheet. Turn off the motor and assure that the timer is in its automatic mode. For automatically flow-controlled units, record the designated flow rate on the data sheet. Record the reading of the elapsed time meter.

After sampling is complete, record the final flow rate and the elapsed time in the same manner. Subtract the initial elapsed time from the final elapsed time to determine the sample duration. Remove the faceplate by removing the wingnuts. Fold the filter in half lengthwise by handling it along its edge with the exposed side inward. Insert the filter in its jacket. Note the presence of insects on the deposit, loose particles, non-centered deposits, evidence of leaks, and unusual meteorological conditions on the data sheet. Mark the flow recorder chart, if any, and return it with the data sheet.

7.7 Maintenance.

7.7.1 Wipe the filter support screen and the sampler housing with a damp cloth weekly to remove dirt. Allow the cleaned area to dry before loading a sample.

7.7.2 Disassemble and clean inlets monthly according to manufacturer's instructions.

7.7.3 Motor brushes are inspected and replaced every 500 h of operation according to manufacturer's instructions. Motors are replaced when they burn out.

7.7.4 The filter faceplate gasket is replaced when it shows signs of wear or when dark streaks on the filter border indicate that leaks are present. The gasket is fixed to the faceplate with rubber cement or double adhesive tape after the previous adhesive has been removed.

7.7.5 Rubber seals between blower components need to be replaced when leaks are encountered. Drastic changes in calibration curves indicate that such leaks might be present.

8. Calibration

8.1 Field Calibration of High Volume Sampler Flow Rate. Calibration and performance testing of the high volume sampler is accomplished with the orifice transfer standard. The sampler motor should be warmed up for at least five minutes prior to calibration. The orifice will not return valid results under very windy conditions owing to rapid variations in atmospheric pressure. The orifice is attached to the filter support instead of the faceplate and is fastened with the four wingnuts. One arm of the manometer is attached to the orifice pressure tap while the other arm is exposed to atmospheric pressure. The manometer scale is adjusted to zero with the sampler motor turned off. Calibration and performance test procedures differ for samplers with no flow control, manual flow control, automatic mass flow control, and automatic volume flow control by critical throat. Single point performance tests should be performed monthly or after every ten samples, whichever is more frequent. Samplers should be recalibrated whenever the performance test flow differs from the measured flow by more than 7% or whenever motor service has been performed.

8.1.1 *No Flow Control.* In this configuration, the sampler motor is plugged directly into the power outlet and the flow rate, typically between 1132 and 1698 L/min, is dictated by the resistance of the filter medium. Flow rate is related to the ram pressure in the plenum as measured by a pressure gauge, the height of the float in a small rotameter, or a flow recorder chart. These are attached to a tap in the blower plenum per manufacturer's instructions. Place one of the perforated disks (there are typically five metal disks with the number of perforations ranging from 5 to 19; some orifices have an iris that can be externally controlled) in the bottom of the calibration orifice and turn on the sampler. Record the reading of the orifice manometer and the blower pressure sensor. Determine the flow rate through the orifice from its calibration curve and convert this flow rate to the temperature and pressure prevailing during the calibration using the formula in Section 9. This correction need not be made if the ambient pressure is within ± 15% of the calibration pressure (in kPa) or of the calibration temperature (in K). Plot the flow rate as a function of plenum pressure and draw a curve through the points. If this curve is not smoothly varying, then the sampler should be checked for leaks and the calibration procedure should be repeated. Place a filter of the type to be used during routine monitoring on the sampler and assure that the exhaust pressure reading falls within the range of the calibration points. Select the perforated plate which offers a flow rate closest to that which is drawn through the filter and use this flow resistance to verify flow rates against the calibration curve monthly or every ten samples, whichever is more frequent.

8.1.2 *Manual Flow Control.* The flow rate must be adjusted to within ± 10% of the specified flow rate for an inlet at ambient temperatures and pressures in order to maintain the size fractionating properties of the inlet. A variable transformer with a voltage scale can be attached to the blower power cord and a relationship between the voltage supplied and the flow rate through the sample can be determined. Place a blank filter or one of the perforated plates which approximates a typical filter resistance underneath the calibration orifice assembly. Adjust the variable transformer until the desired flow rate is obtained through the inlet and record the plenum pressure reading. Increase and decrease the voltage in 10 to 20 V increments and record the plenum pressure and the flow rate through the orifice. Plot the corrected flow rate vs. plenum pressure. Whenever a new filter is installed, the variable transformer is adjusted such that the plenum pressure

(as measured by the rotameter, pressure gauge, or flow recorder chart) corresponds to the required inlet flow rate. The relationship between the transformer voltage reading and the flow rate is NOT constant because the added resistance of the orifice is in the system when the calibration points are derived. It is only the pressure in the blower exhaust plenum that is related to flow rate through the filter.

8.1.3 *Automatic Mass Flow Control.* A filter must be installed underneath the calibration orifice in order to present the same flow pattern to the probe that is experienced in routine sampling. Without such a filter, the orifice directs air to the probe at a different velocity, which results in a controlled flow rate that differs from the specified rate. With the orifice and filter in place, the flow controller is adjusted per manufacturer's instructions until the flow through the orifice equals the required inlet flow rate. The efficiency of flow control as filter resistance increases is determined by placing two and three filters in series between the calibration orifice and the blower. The flow rate through the orifice should not differ from that measured with a single filter by more than ± 5%. The mass flow controller should be adjusted at the median temperature of a typical diurnal cycle (typically late morning) and at least once per season. This will minimize the differences between the mass and the volume of the air sampled **(51)**. The flow should be reset whenever the controller probe is moved, when the sampler motor is serviced, and when performance test results differ by more than ± 7% from the designated flow rate.

8.1.4 *Critical Throat Volumetric Flow Controller.* This device will maintain a constant flow rate that is dictated by the dimensions of the throat. The flow rate is verified by placing the calibration orifice over a filter and determining the corrected flow rate through it from the orifice calibration curve. Differences of more than ± 7% from the designated flow rate may be caused by leaks, a faulty motor, or a filter resistance that is too high to obtain the necessary pressure drops across the throat.

8.2 LABORATORY CALIBRATION.

8.2.1 *Analytical Balance.* Some analytical balances can be calibrated by the operator while others require specialized skills to recalibrate. In general, analytical balances should be calibrated when first purchased, any time the balance is moved, at least every twelve months, or whenever an NBS traceable 3.0000 g weight registers outside ± 0.5 mg of its designated weight. At each weighing session a balance calibration check is performed using a Class S or Class M weight.

9. Calculations

9.1 NOMENCLATURE.

C = Mass concentration of suspended particulate matter, $\mu g/m^3$.
M_i = Weight of blank filter, g.
M_f = Weight of exposed filter, g.
N = Number of replicate weights or flow performance checks.
P_1 = Atmospheric pressure of orifice calibration, kPa.
P_2 = Atmospheric pressure during sampler calibration, kPa.
T_1 = Absolute temperature applicable to calibration orifice calibration, K.
T_2 = Absolute temperature during sampler calibration, K.
Q_1 = Sampler flow rate during sampler calibration, m^3/min.
Q_2 = Corrected sampler flow rate, m^3/min.
R_{Mj} = Difference between the jth routine and replicate weights, g.
R_{Qj} = Difference between the jth routine and performance test flow rates, m^3/min.
t = Sampling time, min.
V = Volume of air sampled, m^3
S_C = Precision of mass concentration measurements, $\mu g/m^3$.
S_M = Precision of mass measurements, g.
S_Q = Precision of air flow rate measurements, m^3/min.
S_V = Precision of air volume measurement, m^3.

9.2 CONVERSION OF ORIFICE FLOW RATE TO SAMPLING FLOW RATES

$$Q_2 = Q_1 \left[\frac{T_2 \times P_1}{T_1 \times P_2} \right]^{1/2}$$

9.3 CALCULATION OF SAMPLING VOLUME.

$$V = \left[\frac{Q_i + Q_f}{2} \right] \times t$$

If the sampler is equipped with automatic flow control, then $Q_i = Q_f$ = setpoint flow rate.

9.4 CALCULATION OF AMBIENT PARTICLE CONCENTRATIONS.

$$C = \left[\frac{M_f - M_i}{V} \right] \times 10^6$$

The 10^6 multiplier converts g to μg.

9.5 CALCULATION OF MASS AND FLOW RATE PRECISIONS.

The following formulae estimate the average precisions of mass and flow rate measurements from N sets of replicate measurements **(36,86)**.

$$S_M = \frac{1}{N} \sum_{j=1}^{N} \frac{R_{Mj}}{1.128} \times 10^6$$

$$S_Q = \frac{1}{N} \sum_{j=1}^{N} \frac{R_{Qj}}{1.128}$$

9.6 CALCULATION OF AMBIENT CONCENTRATION PROPAGATED PRECISIONS **(87)**.

$$S_C = C \left[\frac{2 S_M^2}{(M_f - M_i)^2} + \frac{S_Q^2 t^2}{2 V^2} \right]^{1/2}$$

which assumes flow rates are measured both before and after sampling, that the sampling time measurement is much more precise than the flow rate and mass measurements, and that precisions are equal for exposed and unexposed filters.

9.7 SAMPLE VALIDATION. A calculated concentration should be invalidated or flagged as suspect if any of the following criteria is not met.

9.7.1 Initial or final flow rates differ by more than $\pm$ 10% from the flow rate designated for the inlet.

9.7.2 Sampling time differs by more than $\pm$ 60 min from 1440 min for a 24-h sample or by more than $\pm$ 5% of the designated duration for shorter sampling times.

9.7.3 The filter shows evidence of damage, sampling leaks, or overloading.

10. Effects of Sample Storage

Volatilizable material may be released from the sample during prolonged storage at room temperatures **(81)**. These changes depend on the chemical character of the sampled aerosol and storage environment. Losses may be minimized by refrigeration.

11. References

1. ROBSON, C. D. AND K. E. FOSTER, 1962. Evaluation of air particulate sampling equipment. *Amer. Indust. Hyg. Assoc. Journal*, 23:404.
2. OLIN, J. G. AND J. L. KURZ, 1975. High volume air samplers. *Poll. Eng.*, *7 (1)*:30.
3. EPA, 1971. "National primary and secondary ambient air standards Appendix B: Reference method for the determination of suspended particulates in the atmosphere." *Federal Register*, 36 (84): Part II, April 30, 1971.
4. EPA, 1987. "40 CFR Part 53: Ambient Air Monitoring Reference and Equivalent Methods." *Federal Register*, 52 FR 24724; "40 CFR Part 50: Revisions to the National Ambient Air Quality Standards for Particulate Matter, Appendix J — Reference Method for the Determination of Particulate Matter as PM_{10} in the Atmosphere." *Federal Register*, 52 FR 24664; "Ambient Air Monitoring Reference and Equivalent Methods; Reference Method Designation." *Federal Register*, 52 FR 37366 and 54 FR 45684.
5. EPA. 1981. "Determination of lead concentration in ambient particulate matter by wavelength dispersive x-ray fluorescence spectrometry." *Federal Register*, April, 1981.
6. MUELLER, P. K., B. V. MENDOZA, J. C. COLLINS AND E. S. WILGUS. 1978. Application of ion chromatography to the analysis of anions extracted from airborne particulate matter. *Ion Chromatographic Analysis of Environmental Pollutants*. (Ann Arbor Science Publishers, Inc., Ann Arbor, MI.)
7. CADLE, S. H., P. J. GROBLICKI AND P. A. MULAWA, 1983. Problems in the sampling and analysis of carbon particulate. *Atmos. Environ.* 17:593.
8. GROSJEAN, D. 1975. Solvent extraction and organic carbon determination in atmospheric particulate matter: The OE-OCA technique. *Anal. Chem.*, 47:797.
9. ZOLLER, W. H., AND G. E. GORDON. 1970. Instrumental neutron activation analysis of atmospheric pollutants utilizing Ge (Li) x-ray detectors. *Anal. Chem.*, 42:257.
10. RANWEILER, L. E., AND J. L. MOYERS. 1974. Atomic absorption procedure for analysis of metals in atmospheric particulate matter. *Environ. Sci. & Tech.*, 2:152.
11. COOPER, J. A. 1973. Comparison of particle and photon excited x-ray fluorescence applied to trace element measurements of environmental samples. *Nuc. Instr. & Methods*, 106:525.
12. COOPER, J. A., L. A. CURRIE AND G. A. KLOUDA, 1981. Assessment of contemporary carbon combustion source contributions to urban air particulate levels using carbon-14 measurements. *Environ. Sci. & Tech.*, 9:1045.
13. DAVIS, B. L. 1980. Standardless x-ray diffraction quantitative analysis. *Atmos. Environ.*, 14:217.
14. CASUCCIO, G. S., P. B. JANOCKO, R. J. LEE, J. F. KELLY, S. L. DATTNER AND J. S. MGEBROFF, 1983. The use of computer controlled scanning electron microscopy in environmental studies. *J. Air Poll. Cont. Assoc.*, 30:1116.
15. COOPER, J. A., AND J. G. WATSON. 1980. Receptor oriented methods of air particulate source apportionment. *J. Air Poll. Cont. Assoc.*, 30:1116.
16. U.S. ENVIRONMENTAL PROTECTION AGENCY, 1988: "Quality assurance handbook for air pollution measurement systems: Volume II: Ambient air specific methods." EPA-600/4-77-027a, Research Triangle Park, NC.
17. CALIFORNIA AIR RESOURCES BOARD, 1979. "Standard operating procedures for air quality monitoring: Appendix E-1, station operators procedures for the high volume sampler. In "Air Monitoring Quality Assurance," Volume II, California Air Resources Board, Sacramento, CA.
18. OLIN, J. G. AND R. R. BOHN. 1983. A new PM-10 medium flow sampler. Presented at the 76th annual meeting of the Air Pollution Control Association, Atlanta, GA.
19. WEDDING, J. B., M. A. WEIGAND, M. W. LIGOTKE AND R. BAUMGARDNER. 1983. Wedding ambient aerosol sampling inlet for an intermediate flow rate (4 cfm) sampler. *Environ. Sci. & Tech.*, 17:379.
20. MCFARLAND, A. R. AND C. A. ORTIZ. 1982. A 10 μm cutpoint ambient aerosol sampling inlet. *Atmos. Environ.*, 16:2959.
21. OLIN, J. G. 1978. A new virtual impactor (dichotomous sampler) for fine particle air monitoring. Presented at the 71th annual meeting of the Air Pollution Control Association.
22. WEDDING, J. B., M. A. WEIGAND AND T. C. CARVEY. 1982. A 10 micron cutpoint inlet for the dichotomous sampler. *Environ. Sci. & Tech.*, 16:602.
23. CAMP, D. C., A. L. VAN LEHN AND B. W. LOO. 1978. "Intercomparison of samplers used in the determination of aerosol composition." EPA 600/7-78-118, Research Triangle Park, NC.
24. RODES, C. E., D. M. HOLLAND, L. J. PURDUE AND K. A. REHME. 1985. A field comparison of PM_{10} inlets at four locations. *J. Air Poll. Contr. Assoc.*, 35:345.
25. MATHAI, C. V., I. K. TOMBACH, J. G. WATSON AND C. F. ROGERS. 1985. Intercomparison of ambient aerosol samplers used in western visibility and air quality studies. Presented at the 78th annual meeting of the Air Pollution Control Association, Detroit, MI.
26. REHME, K. A., C. F. SMITH, M. E. BEARD AND T. FITZSIMMONS. 1984. "Investigation of filter media for use in the determination of mass concentrations of ambient particulate matter." EPA-600/S4-84-048, Research Triangle Park, NC.
27. BRUCKMAN, L. AND R. A. RUBINO. 1976. High

volume sampling: Errors incurred during passive deposition exposure periods. *J. Air Poll. Contr. Assoc.*, 26:881.

28. CHAHAL, H. S., AND D. J. ROMANO. 1978. High volume sampling: Effect of windborne particulate matter deposited during idle periods. *J. Air Poll. Contr. Assoc.*, 26:895.

29. BLANCHARD, G. E. AND D. J. ROMANO. 1978. High volume sampling: Evaluation of an inverted sampler for ambient TSP measurements. *J. Air Poll. Contr. Assoc.*, 28:1142.

30. SWINFORD, R. 1980. The assessment of passive loading effects on TSP measurements in attainment areas. *J. Air Poll. Contr. Assoc.*, 30:1322.

31. MCFARLAND, A. R. AND C. A. ORTIZ. 1985. Response to comments on 'A field comparison of PM-10 inlets at four locations.' *J. Air Poll. Cont. Assoc.*, 35:950.

32. COUNTESS, R. J. 1974. Production of aerosol by high volume samplers. *J. Air Poll. Contr. Assoc.*, 24:605.

33. KING, R. B. AND J. TOMA, 1975. "Copper emissions from a high-volume air sampler." NASA Technical Memorandum, NASA-TM X-71693.

34. COUTANT, R. W. 1977. Effect of environmental variables on collection of atmospheric sulfate. *Environ. Sci. & Tech.*, 11:875.

35. APPEL, B. R., Y. TOKIWA, M. HAIK AND E. L. KOTHNY. 1984. Artifact particulate sulfate and nitrate formation on filter media. *Atmos. Environ.*, 18:409.

36. SPICER, C. W., AND P. M. SCHUMACHER. 1977. Interference in sampling atmospheric particulate nitrate. *Atmos. Environ.*, 11:873.

37. MUELLER, P. K., G. M. HIDY, R. L. BASKETT, K. K. FUNG, R. C. HENRY, T. F. LAVERY, K. K. WARREN AND J. G. WATSON. 1983: "The sulfate regional experiment: Report of findings." Report EA-1901, Electric Power Research Institute, Palo Alto, CA.

38. MULAWA, P. A. AND S. H. CADLE. 1985. A comparison of nitric acid and particulate nitrate measurements by the penetration and denuder difference methods. *Atmos. Environ.*, 19:1317.

39. TANG, I. N. 1980. Deliquescence properties and particle size change of hygroscopic aerosols. In *Generation of Aerosols and Facilities for Exposure Experiments*, edited by K. Willeke, Ann Arbor Science Publishers, Ann Arbor, MI.

40. TIERNEY, G. P., AND W. CONNOR. 1967. Hygroscopic effects on weight determinations of particulates collected on glass fiber filters. *Amer. Ind. Hyg. Journal*, 28:363.

41. ENGELBRECHT, D. R., T. A. CAHILL AND P. J. FEENEY. 1980. Electrostatic effects on gravimetric analysis of membrane filters. *J. Air Poll. Cont. Assoc.*, 30:391.

42. MCKEE, H. C., R. E. CHILDERS, O. I. SAENZ, T. W. STANLEY AND J. MARGESON. 1972. Collaborative testing of methods to measure air pollutants: The high volume method for suspended particulate matter. *J. Air Poll. Contr. Assoc.*, 22:342.

43. WATSON, J. G., J. C. CHOW AND J. J. SHAH. 1981. "Analysis of inhalable particulate matter measurements." EPA-450-4-81-035, Research Triangle Park, NC.

44. MCFARLAND, A. R., C. A. ORTIZ AND C. E. RODES. 1979. Characteristics of aerosol samplers used in ambient air monitoring. 86th National Meeting of the American Institute of Chemical Engineers, Houston, TX.

45. WEDDING, J. B., A. R. MCFARLAND AND J. E. CERMAK. 1977. Large particle collection characteristics of ambient aerosol samplers. *Environ. Sci. & Tech.*, 4:387.

46. MCFARLAND, A. R., C. A. ORTIZ AND R. W. BERTCH, JR. 1984. A 10 μm cutpoint size selective inlet for hivol samplers. *J. Air Poll. Contr. Assoc.*, 34:544.

47. WEDDING, J. B., M. A. WEINGAND AND Y. J. KIM. 1985. Evaluation of the Sierra-Andersen 10-μm inlet for the high-volume sampler. *Atmos. Environ.*, 19:539.

48. WEDDING, J. B. AND M. A. WEINGAND. 1985. The Wedding ambient aerosol sampling inlet (D_{50} = 10 μm) for the high volume sampler. *Atmos. Environ.*, 19:535.

49. WILLEKE, K. 1975. Performance of the slotted impactor. *Amer. Ind. Hyg. Assoc. Journal*, 36:683.

50. WEDDING, J. B., M. A. WEIGAND, Y. J. KIM, D. L. SWIFT AND J. P. LODGE. 1987. A critical flow device for accurate PM_{10} sampling and correct indication of PM_{10} dosage to the thoracic region of the respiratory tract. *JAPCA*, 27:254.

51. WEDDING, J. B. 1985. Errors in sampling ambient concentrations employing setpoint temperature compensated mass flow transducers. *Atmos. Environ.*, 19:1219.

52. ANDERSEN SAMPLERS, INC. 1985. "Comparison of mass flow control versus critical flow volumetric flow control."

53. BERNSTEIN, D. M., R. T. DREW, AND M. LIPPMAN, 1983. "Calibration of air sampling instruments." In *Air Sampling Instruments for Evaluation of Atmospheric Contaminants, 6th Edition*, edited by P. J. Lioy and M. J. Y. Lioy, American Conference of Governmental Industrial Hygienists, Cincinnati, OH.

54. BAKER, W. C., AND J. F. POUCHOT. 1983. The measurement of gas flow: Part I. *J. Air Poll. Contr. Assoc.*, 33:66.

55. LIU, B. Y. H., D. Y. H. PUI AND K. L. RUBOW. 1981. Characteristics of air sampling filter media. *Proceedings: International Symposium of Aerosols in the Mining and Industrial Work Environment*, Minneapolis, MN.

56. LIPPMAN, M. 1983. Sampling aerosols by filtration. In *Air Sampling Instruments for Evaluation of Atmospheric Contaminants, 6th Edition*, edited by P. J. Lioy and M. J. Y. Lioy, American Conference of Governmental Industrial Hygienists, Cincinnati, OH.

57. DAVIES, C. N. 1970. The clogging of fibrous aerosol filters. *J. Aerosol Sci.*, 1:35.

58. KRAMER, D. N., AND P. W. MITCHEL. 1967. Evaluation of filters for high volume sampling of atmospheric particulates. *Amer. Ind. Hyg. Assoc. Journal*, 28:224.

59. RIMBERG, D. 1969. Penetration of IPC 1478, Whatman 41, Type 5G filter paper as a function of particle size and velocity. *Amer. Ind. Hyg. Assoc. Journal*, 30:394.

60. LEE, K. W. AND B. Y. H. LIU. 1980. On the minimum efficiency and the most penetrating particle size for fibrous filters. *J. Air Poll. Contr. Assoc.*, 30:377.

61. JOHN, W. AND G. REISCHL. 1978. Measurement of the filtration efficiencies of selected filter types. *Atmos. Environ.* 12:2015.

62. APPEL, B. R., S. M. WALL, T. TOKIWA AND M. HAIK. 1979. Interference effects in sampling particulate nitrate in ambient air. *Atmos. Environ.,* *13*:319.

63. CHARELL, P. R., AND R. E. HAWLEY. 1981. Characteristics of water adsorption on air sampling filters. *Amer. Ind. Hyg. Assoc. Journal,* 42:353.

64. COUTANT, R. W. 1977. Factors affecting the collection efficiency of atmospheric sulfate. EPA-600/2-77-076, Research Triangle Park, NC.

65. DAMS, R., K. A. RAHN AND J. W. WINCHESTER. 1972. Evaluation of filter materials and impaction surfaces for nondestructive neutron activation analysis of aerosols. *Environ. Sci. & Tech.,* 6:441.

66. MESEROLE, F. B., B. F. JONES, L. A. ROHLACK, W. C. HAWN, K. R. WILLIAMS AND T. P. PARSONS. 1979. Nitrogen oxide interferences in the measurement of atmospheric particulate nitrates. EPRI Report EA-1031, Palo Alto, CA.

67. SPICER, C. W., P. M. SCHUMACHER, J. A. KOUYOUMIJIAN AND D. W. JOSEPH. 1978. Sampling and analytical methodology for atmospheric particulate nitrates. EPA-600/2-78-067, Research Triangle Park, NC.

68. SPICER, C. W., AND P. M. SCHUMACHER. 1979. Particulate nitrate: Laboratory and field studies of major sampling interferences. *Atmos. Environ.,* *13*:543.

69. WITZ, S. AND J. G. WENDT. 1981. Artifact sulfate and nitrate at two sites in the South Coast Air Basin. *Environ. Sci. & Tech.,* *15*:74.

70. WITZ, S. 1985. Effect of environmental factors on filter alkalinity and artifact formation. *Environ. Sci. & Tech.,* *19*:831.

71. BILES, B. AND J. ELLISON. 1975. The efficiency of cellulose fiber filters with respect to lead and black smoke in urban aerosol. *Atmos. Environ.,* *9*:1030.

72. STAFFORD, R. G. AND H. J. ETTINGER. 1972. Filter efficiency as a function of particle size and velocity. *Atmos. Environ.* 6:353.

73. DEMUYNCK, M. 1975. Determination of irreversible absorption of water by cellulose filters. *Atmos. Environ.,* *9*:523.

74. LINDEKEN, C. L., R. L. MORGIN AND K. F. PETROCK. 1963. Collection efficiency of Whatman 41 filter paper for submicron aerosols. *Health Phys.,* *9*:305.

75. GELMAN, C., D. V. MEHTA, AND T. H. MELTZER. 1979. New filter compositions for the analysis of airborne particulate and trace metals. *Amer. Ind. Hyg. Assoc. Journal,* 40:926.

76. SCHLEIEN, B., J. A. COCHRAN AND A. G. FRIEND. 1966. Calibration of glass fiber filters for particle size studies. *Amer. Ind. Hyg. Assoc. Journal,* 27:253.

77. WITZ, S., AND R. D. MACPHEE. 1977. Effect of different types of glass filters on total suspended particulates and their chemical composition. *J. Air Poll. Contr. Assoc.,* *27*:239.

78. MESEROLE, F. B., K. SCHWITZGEBEL, B. F. JONES, C. M. THOMPSON AND F. G. MESICH. 1976. Sulfur dioxide interferences in the measurement of ambient particulate sulfates. EPRI Research Project 262, Palo Alto, CA.

79. CHOW, J. C., A. FLAHERTY, E. MOORE AND J. G. WATSON. 1980. "Filter analysis for TSP SIP Development." EPA-901/9-78-003, EPA Region I, Boston, MA.

80. PATE, J. B., AND E. C. TABOR. 1962. Analytical aspects of the use of glass fiber filters for the collection and analysis of atmospheric particulate matter. *Amer. Ind. Hyg. Assoc. Journal,* 23:145.

81. SMITH, J. P., D. GROSJEAN AND J. N. PITTS. 1978. Observation of significant losses of particulate nitrate and ammonia from high volume glass fiber filter samples stored at room temperature. *J. Air Poll. Contr. Assoc.,* *28*:930.

82. CADLE, S. H., P. J. GROBLICKI AND P. A. MULAWA. 1983. Problems in the sampling and analysis of carbon particulate. *Atmos. Environ.,* *17*:593.

83. LUNDGREN, D. A. AND T. C. GUNDERSON. 1975. Efficiency and loading characteristics of EPA's high temperature quartz fiber filter media. *Amer. Ind. Hyg. Assoc. Journal,* 36:806.

84. STEVENS, R. K., W. A. MCCLENNY, T. G. DZUBAY, M. A. MASON AND W. J. COURTNEY. 1982. Analytical methods to measure the carbonaceous content of aerosols. In *Particulate Carbon Atmospheric Life Cycle,* edited by G. T. Wolff and R. L. Klimisch, Plenum Press, NY.

85. WEST, L. G. 1985. A new air monitoring filter for PM-10 collection and measurement techniques. In *Quality Assurance in Air Pollution Measurements,* Air Pollution Control Association, Pittsburgh, PA.

86. UNITED STATES ENVIRONMENTAL PROTECTION AGENCY, 1976. "Quality assurance handbook for air pollution measurement systems: Volume I principles." EPA 600/9-76-005, Research Triangle Park, NC.

87. WATSON, J. G., P. J. LIOY AND P. K. MUELLER. 1983. The measurement process: Precision, accuracy, and validity. In *Air Sampling Instruments for Evaluation of Atmospheric Contaminants, 6th Edition,* edited by P. J. Lioy and M. J. Y. Lioy, American Conference of Governmental Industrial Hygienists, Cincinnati, OH.

Subcommittee 10

J. G. WATSON, *Chairman*
J. L BOWEN
J. C. CHOW
C. F. ROGERS
M. G. RUBY
M. J. ROOD
R. T. EGAMI

502.

Particle Fallout Container Measurement of Dustfall from the Atmosphere

1. Principle of Method

1.1 PRINCIPLE. Large solid and liquid particles (typically greater than 10 μm in aerodynamic diameter) are collected via gravitational settling in an open-mouth container for a designated period of time. The container is washed with a known amount of distilled water, which is filtered and then evaporated. The mass of insoluble particles is determined by the weight-gain of the filter after filtration. The mass of soluble particles is determined by the weight gain of a crucible after evaporation. These weight gains are translated into insoluble, soluble, and total particle deposition fluxes by normalizing the mass measurements by the collection area of the container and the sampling time (1,2,3).

1.2 APPLICABILITY. This method is applicable to area surveys for determining particle fallout nuisances (4) and has been adapted to the measurement of wet and dry deposition (5).

1.3 OTHER ANALYSES. Soluble and insoluble deposits have been subjected to analyses for ions (6), elements (7,8,9), inorganic and organic compounds (10,11). These additional analyses have been used in source apportionment studies (12,13).

1.4 DETAILED PROCEDURES. The American Society for Testing and Materials (3) and the Air Pollution Control Association (2,14) have issued operating procedures for particle fallout measurements. Other procedures have been developed for wet and dry sampling (15–19).

1.5 ALTERNATIVE METHODS. Wet and dry buckets (5), man-made surfaces of various compositions (20–23), natural surfaces, (24), optical attenuation (25,26), and micrometeorological (27,28) methods have also been applied to the measurement of particle fallout.

2. Range and Sensitivity

2.1 LOWER QUANTIFIABLE LIMIT. For a month-long sample duration, the lower quantifiable limit is determined by the reproducibility of the filter and drying crucible weight differences. These exhibit standard deviations (sigma) of approximately $\pm$ 2 mg. The three-sigma lower quantifiable limit for a container with a 125 cm^2 collection area is then approximately 0.35 g/m^2 mo.

2.2 UPPER QUANTIFIABLE LIMIT. For a month-long sample duration there is practically no upper quantifiable limit unless the capacity of the particle fallout container is exceeded.

3. Interferences

3.1 Bird droppings, insects, algae, fungi, and large plant material are not considered to be a portion of particle fallout. Bird droppings are minimized by a bird ring. Fungicides and algicides can be added to the container collection solution. Particles larger than 1 mm diameter are removed by sieving.

3.2 SOLUBLE GASES. Gases can be dissolved by the container solution and may manifest themselves as soluble particulate matter. This interference will be negligible for mass concentrations but may bias measurements of particulate chemical species.

3.3 WIND SPEED. The sampling efficiency of cylindrical containers has been shown to be nearly zero for wind speeds of greater than 7 m/s for particles on the order of 100 μm diameter. The efficiency is approximately 60% for 100 μm particles in 1 m/s winds (29). Alternative container designs have been proposed to minimize this interference (22).

3.4 PARTICLE BLOWOUT. Dry particles collected in a container can be resuspended by wind with subsequent removal. A quantity of water in the container minimizes this interference (30,31).

3.5 LEACHING TO CONTAINER SURFACES. Certain metal ions have been found to leach into various materials when in solution. This interference is negligible for soluble mass measurements, but it may affect the concentrations of various chemical species (32,34).

3.6 LEACHING FROM CONTAINER SURFACES. Contaminants on the surface of the collection container will be measured as particle fallout. Rinsing of containers prior to use will minimize this interference (33,34).

4. Precision and Accuracy

4.1 PRECISION. Intercomparison studies in different urban areas have found average collocated precisions (one sigma) of 1.46 g/ m^2 mo among different laboratories and average collocated precisions of 1.03 g/ m^2 mo within the same laboratory for total dustfall. This precision appears to be independent of the range of dustfall rates measured between 3 and 11 g/ m^2 mo. For the water-insoluble fraction, the between-laboratory average standard deviation was 1.8 g/ m^2 mo and the within laboratory average standard deviation was 0.78 g/ m^2 mo. For the water-soluble fraction, the between-laboratory average standard deviation was 1.64 g/ m^2 mo and the within-laboratory average standard deviation was 0.59 g/ m^2 mo (35).

4.2 ACCURACY. Interlaboratory analyses of samples spiked with known quantities of salt, sand, and granular polystyrene yielded average recoveries of $96 \pm 16\%$ for total dustfall and $91 \pm 18\%$ for insoluble dustfall (35).

5. Apparatus

5.1 PARTICLE FALLOUT COLLECTOR. A cylindrical container is located on a stand at least four feet (1.2 m) from the support surface. A bird ring extends above the rim of the container to prevent birds from perching on the container. Containers can be made of glass, plastic, or stainless steel,

should have a circular opening of greater than 15 cm diameter, and should have a length of two to three times this diameter. Fig. 502:1 shows the collector configuration.

5.2 SIEVE. A No. 18 (1 mm mesh) stainless steel sieve is used to remove large particles from the sample.

5.3 CRUCIBLES. Non-porous 50-to100-mL ceramic or glass crucibles are used for evaporation of soluble species in determining soluble particle fallout.

5.4 FILTER UNIT OR FILTER FUNNEL. A 200 mL or greater filter funnel capable of accepting a 47 mm or larger filter is used to separate soluble from insoluble particles. Either gravity or a vacuum assist can be used to draw the solute through the filter.

5.5 ANGLE-EDGED RUBBER SPATULA. The spatula (sometimes called a policeman) is used to remove deposits that adhere to the sides and bottoms of the container.

5.6 ANALYTICAL BALANCE. The balance should have a sensitivity of 0.1 mg and a range adequate for weighing of both crucibles and filters. Class S weights are used for calibration.

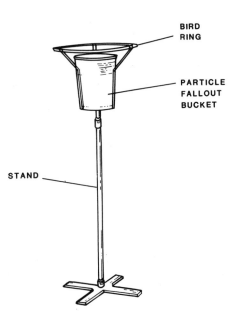

Figure 502:1 – Particle fallout measurement apparatus.

5.7 LABORATORY OVEN. The oven is used to evaporate water from a portion of the filtered liquid placed in a crucible.

5.8 GRADUATED CYLINDER, 1 L. The graduated cylinder is used to determine the volume of liquid in the container. An alternative to this method involves weighing the container before and after sampling and equating the change in mass to the liquid volume.

6. Reagents and Supplies

6.1 WATER. Water should conform to ASTM Specifications D1193 for Reagent Water, Type II.

6.2 FILTER. A quartz or glass fiber filter is most appropriate for the removal of solid particulate matter. Chemical analyses of particles deposited on the filter may require other types of media.

6.3 ALGICIDE AND FUNGICIDE. Copper sulfate will inhibit the growth of algae. Mercuric chloride will inhibit the growth of both algae and fungi, but it is toxic. One tenth of a gram (sufficient to provide a concentration of 15 mg/L when the container is full) of these reagents should be sufficient, but they should only be added if algae and fungi growth have been observed when no chemicals are added. These reagents should be fully dissolved in the collection fluid and should never attain concentration levels which would cause them to precipitate and be mistaken for insoluble particles.

6.4 ANTIFREEZE. Isopropyl alcohol can be added to reduce the freezing temperature of the water. This is necessary only when temperatures are expected to fall below 0°C. A 50% solution is usually adequate, though higher proportions of antifreeze may be necessary in colder climates or ones with higher precipitation levels.

6.5 GUMMED IDENTIFICATION LABELS. These labels are affixed to the collection container to identify it and associate it with other pertinent data.

7. Procedure

7.1 Fig. 502:2 presents a flow diagram of the routine operating procedures described in the following sub-section.

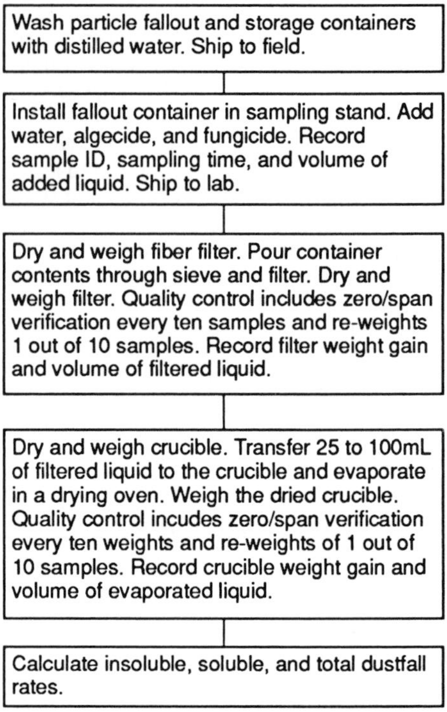

Wash particle fallout and storage containers with distilled water. Ship to field.

Install fallout container in sampling stand. Add water, algecide, and fungicide. Record sample ID, sampling time, and volume of added liquid. Ship to lab.

Dry and weigh fiber filter. Pour container contents through sieve and filter. Dry and weigh filter. Quality control includes zero/span verification every ten samples and re-weights 1 out of 10 samples. Record filter weight gain and volume of filtered liquid.

Dry and weigh crucible. Transfer 25 to 100mL of filtered liquid to the crucible and evaporate in a drying oven. Weigh the dried crucible. Quality control incudes zero/span verification every ten weights and re-weights of 1 out of 10 samples. Record crucible weight gain and volume of evaporated liquid.

Calculate insoluble, soluble, and total dustfall rates.

Figure 502:2 — Flow diagram for routine particle fallout measurement.

7.2 SAMPLER SITING. A sufficient number of sampling sites should be chosen to obtain samples representative of an area under study. An initial survey with many sampling sites may be conducted to determine whether or not a smaller number of sites is adequate. A minimum of four stations is considered necessary to be representative (36,37). Each sampling site should be located at least 100 m from fugitive dust area sources and major point sources. The rim of the container should be between 2.4 m and 15 m above ground level and at least 1.3 m above the supporting surface. The top of any nearby obstruction should not exceed a line which is 30° from horizontal at the height of the container opening.

7.3 CONTAINER PREPARATION. Thoroughly wash the container with soap and water with final rinses in Reagent water prior to sample collection. Cover the container before transporting it to the field to protect it from contamination.

7.4 FIELD SAMPLING. Place the container in its stand and remove the cover. Place the cover in a protective bag and store it until the container is retrieved from the field. Pour 0.2 to 1.0 L of water (which may contain algicide or antifreeze as specified in Sections **6.3** and **6.4**) into the container. The amount of this liquid should be sufficient to prevent evaporation over the sampling period, but not so great that the container will overflow after receiving a normal amount of precipitation. Measure this volume of liquid with a graduated cylinder and record it on the ID label. It is often convenient to pre-measure this liquid in the laboratory and to transport it to the field in a reagent bottle. Place the gummed data label on the side of the container and assign an identification label to the sample. Note the sampling site, date, and start time of dustfall sampling on the label. The standard sampling interval is 30 $\pm$ 2 days. After completion of the sampling interval, note the removal date and time on the ID label. Record any observations, such as nearby emissions activities, the presence of insects, bird droppings, or other unusual material in the container, algae formation, and freezing, on the label. Securely place the lid on the container and return it to the laboratory.

7.5 LABORATORY ANALYSIS.

7.5.1 *Filtration and Volumetric Measurement.* Dry a quartz or glass fiber filter in the oven at approximately 105°C overnight and weigh it on the analytical balance. Place it in the filtration funnel or filter unit. Channel the liquid exiting the filtration unit into a graduated cylinder. Carefully pour the liquid in the container through the No. 18 sieve and into the filtration unit. If the container contains less than 0.2 L of liquid, add this amount of deionized distilled water and mix it thoroughly with the contents of the container. When the container has been emptied of all liquid, record the volume collected in the graduated cylinder on the data label. Scrape the walls of the container with a clean rubber spatula while rinsing them with water. Filter this liquid until no visible signs of dust remain in the container. If the volume of liquid collected is to be determined gravimetrically, weigh the full container with its

lid removed both before and after the filtration process.

7.5.2 *Insoluble Dustfall Measurement.* Dry the filter in the oven for at least 2 h at 105°C. Re-weigh the filter on the analytical balance and calculate the increase in mass resulting from the filtration process.

7.5.3 *Soluble Dustfall Measurement.* Weigh a clean crucible on the analytical balance. Transfer 25 to 100 mL of the filtered liquid from the graduated cylinder to the crucible. Heat this liquid in the oven at 105°C for a sufficient period of time to evaporate all of the liquid. Re-weigh the crucible and calculate the increase in mass attributable to soluble species. Record the volume of liquid evaporated on the identification label.

7.6 MAINTENANCE. No additional maintenance is required other than that already specified in the procedures.

7.7 REPLICATES. Re-weigh one out of ten filters or crucibles. These replicate analyses are used to calculate propagated precisions of the measurements.

8. Calibration

8.1 GRAVIMETRIC ANALYSIS. Place a 3-g weight on the analytical balance prior to and after each weighing session and verify the balance calibration to within $\pm$ 0.1 mg. The balance should be serviced and recalibrated if this tolerance is not met.

8.2 VOLUMETRIC MEASUREMENTS. Scales on graduated cylinders should be verified upon receipt by weighing them before and after filling to a desired level with deionized distilled water. The increase in weight in grams should equal the number of milliliters to within $\pm$ 2%.

9. Calculations

9.1 NOMENCLATURE.

C_t = Total particle fallout rate, g/m^2 mo.

C_u = Insoluble particle fallout rate, g/m^2 mo.

C_s = Soluble particle fallout rate, g/m^2 mo.

A = Area of particle fallout container opening, m^2

T = Sampling duration, 30-d mo.

M_{iu} = Weight of blank filter, g.
M_{is} = Weight of blank crucible, g.
M_{fu} = Weight of loaded filter, g.
M_{fs} = Weight of loaded crucible, g.
R_{Mj} = Difference between the jth routine and replicate weights, g.
V_1 = Volume of all liquid collected, L.
V_2 = Volume of liquid removed for evaporation, L.
S_M = Precision of weight measurements, g.
S_{Ct} = Precision of total particle fallout rate, g/m² mo.
S_{C_u} = Precision of insoluble particle fallout rate, g/m₂ mo.
S_{C_s} = Precision of soluble particle fallout rate, g/m² mo.

9.2 CALCULATION OF INSOLUBLE PARTICLE FALLOUT RATES.

$$C_u = \frac{(M_{fu} - M_{iu})}{AT}$$

9.3 CALCULATION OF SOLUBLE PARTICLE FALLOUT RATE.

$$C_s = \frac{(M_{fs} - M_{is})}{AT} \frac{V_1}{V_2}$$

9.4 CALCULATION OF TOTAL PARTICLE FALLOUT RATE.

$$C_t = C_u + C_s$$

9.5 CALCULATION OF MASS MEASUREMENT PRECISIONS. The following formula estimates the average precision of mass measurements from N sets of replicates **(38)**.

$$S_M = \frac{1}{N} \sum_{j=1}^{N} \frac{R_{Mj}}{1.128}$$

9.6 CALCULATION OF PARTICLE FALLOUT RATE PRECISIONS.

$$S_{C_u} = \frac{\sqrt{2} \, C_u S_M}{(M_{fu} - M_{iu})}$$

$$S_{C_s} = \frac{\sqrt{2} \, C_s S_M}{(M_{fs} - M_{is})}$$

$$S_{Ct} = \left[S_{C_u}^2 + S_{C_s}^2 \right]^{1/2}$$

These calculations assume that the precisions of sample volumes and sample durations are much less than the weighing precisions.

9.7 SAMPLE VALIDATION. Samples are invalidated when large quantities of wind-blown garbage, insects, and bird droppings are observed in the container liquid.

10. Effects of Sample Storage

10.1 Algae and other biological material may grow in containers which are stored for long periods of time. Soluble species may leach from or adhere to the container surface. Samples should be analyzed as soon as possible following collection.

11. References

1. CHAPMAN, H. M. 1955. Recommended Standard Method for Continuing Dustfall Survey (APM1-A). *J. Air Poll. Contr. Assoc.*, 5:176.
2. HERRICK, R. A. 1966: Recommended Standard Method for Continuing Dust Fall Survey (APM-1, Revision 1). *J. Air Poll. Contr. Assoc.*, 16:372.
3. ASTM, 1982. "Standard Method for Collection and Analysis for Dustfall (Settleable Particulates)." ASTM D-1739-82, *Annual Book of ASTM Standards* Vol. 11.03. American Society for Testing and Materials, Philadelphia, PA.
4. ENGDAHL, R. B. 1953. The Role of the Area Survey in the Control of Dust. *Air Repair*, 3(1):11.
5. VOLCHOK, C. F. AND R. T. GRAVESON. 1976. Wet/Dry Fallout Collection. Proceedings of the Second Federal Conference on the Great Lakes, Argonne National Labs, Argonne, IL.
6. FEELY, H. W., D. C. BOGEN, S. J. NAGOURNEY, AND C. C. TORQUATO. 1985. Rates of Dry Deposition Using Wet/Dry Collectors. *J. Geophys. Res.*, 90:2161.
7. AYLING, G. M. AND H. BLOOM. 1976. Heavy Metals Analysis to Characterize and Estimate Distribution of Heavy Metals in Dust Fallout. *Atmos. Environ.* 10:61.
8. ELIAS, R. W. AND C. I. DAVIDSON. 1980. Mechanisms of Trace Element Deposition from the Free Atmosphere to Surfaces in a Remote High Sierra Canyon. *Atmos. Environ.* 14:1427.
9. DAVIDSON, C. I. AND R. W. ELIAS. 1982. Dry Deposition and Resuspension of Trace Elements in the Remote High Sierra. *Geophys. Res. Lett.*, 9:91.
10. ABDEL SALAM, M. S. AND M. A. SOWELIM. 1967. Dustfall Caused by the Spring Khamasin Storms in Cairo: A Preliminary Report. *Atmos. Environ.*, 1:221.
11. ABDEL SALAM, M. S. AND M. A. SOWELIM. 1967. Dust Deposits in the City of Cairo. *Atmos. Environ.*, 1:211.
12. FEELEY, J. A. AND H. M. LILJESTRAND. 1983. Source Contributions to Acid Precipitation in Texas. *Atmos. Environ.*, 17:807.
13. GATZ, D. F. 1984. Source Apportionment of Rain Water Impurities in Central Illinois. *Atmos. Environ.*, 18:1895.
14. CHAPMAN, H. M. 1955. Recommended Standard

Method for Continuing Dustfall Survey (APM1-A). *J. Air Poll. Contr. Assoc.*, 5:176.

15. EPA 1981. "Quality Assurance Handbook for Air Pollution Measurement Systems, Volume V Manual for Precipitation Measurement Systems, Part I, Quality Assurance Manual." EPA-600/4-82-042a, Research Triangle Park, NC.

16. TOPOL, L. E. 1983. "Utility Acid Precipitation Study Program: Field Operator Instruction Manual." Report UAPSP 104, Electric Power Research Institute, Palo Alto, CA.

17. PEDEN, M. E. 1983: "Sampling, Analytical, and Quality Assurance Protocols for the National Atmospheric Deposition Program." ASTM STP 823, S. A. Campbell, Ed., p.72.

18. SCHRODER, C. J. AND B. A. MALO. 1984. "Quality Assurance Program for Wet Deposition Sampling and Chemical Analyses for the National Trends Network." *Quality Assurance in Air Pollution Measurements*, T. R. Johnson, S. J. Penkala, Ed., Air Pollution Control Association, Pittsburgh, PA.

19. PEDEN, M. E., S. R. BACHMAN, C. J. BRENNAN, B. DEMIR, K. O. JAMES, B. W. KAISER, J. M. LOCKARD, J. E. ROTHERT, J. SAVER, L. M. SKOWRON AND M. J. SLATER. 1986. "Development of Standard Methods for the Collection and Analysis of Precipitation." Illinois State Water Survey, Champaign, IL, EPA Contract No. CR810780-01.

20. DASCH, J. M. 1985. Direct Measurement of Dry Deposition to a Polyethylene Bucket and Various Surrogate Surfaces. *Environ. Sci. & Tech.*, 19:271.

21. VANDENBERG, J. J. AND K. R. KNOERR. 1985. Comparison and Surrogate Surface Techniques for Estimation of Sulfate Dry Deposition. *Atmos. Environ.*, 19:627.

22. HALL, D. J. AND R. A. WATERS. 1986. An Improved Readily Available Dustfall Gauge. *Atmos. Environ.*, 20:219.

23. DAVIDSON, C. I., S. E. LINDBERG, J. A. SCHMIDT, L. G. CARTWRIGHT AND L. R. LANDIS. 1985. Dry Deposition of Sulfate onto Surrogate Surfaces. *J. of Geophys. Res.*, 90:2123.

24. JOHN, W., S. M. WALL AND J. L. ONDO. 1985. "Dry Acid Deposition on Materials and Vegetation Concentrations in Ambient Air." Air and Industrial Hygiene Laboratory Report No. CA/DOH/AIHL/SP-34, Berkeley, CA.

25. CORN, M., R. QUINLAN AND J. KATZ. 1967. The Optical Evaluation of Atmospheric Dustfall. *Atmos. Environ.*, 1:227.

26. ESMEN, N. A. 1973. A Direct Measurement Method for Dustfall. *J. Air Poll. Contr. Assoc.*, 23:34.

27. SIEVERING, H. 1982. Profile Measurements of Particle Dry Deposition Velocity at a Air/Land Interface. *Atmos. Environ.*, 16:301.

28. SIEVERING, H. 1986. Gradient Measurement of Sulfur and Soil Mass Dry Deposition Rates Under Clean Air and High-Wind-Speed Conditions. *Atmos. Environ.*, 20:341.

29. RALPH, M. O. AND C. F. BARRETT. 1984. "A Wind Tunnel Study of the Efficiency of Three Deposit Gauges." Report No. LR 499, Warren Springs Laboratory, Stevanage, U.K.

30. NADER, J. S. 1958. Dust Retention Efficiencies of Dustfall Collectors. *J. Air Poll. Contr. Assoc.*, 8:35.

31. STOCKHAM, J., S. RADNER AND E. GROVE. 1966. The Variability of Dustfall Analysis Due to the Container and the Collecting Fluid. *J. Air Poll. Contr. Assoc.*, 16:263.

32. CHAN, W. H., F. TOMASSINI AND B. LOESCHER. 1983. An Evaluation of Sorption Properties of Precipitation Constituents on Polyethylene Surfaces. *Atmos. Environ.*, 17:1779.

33. GOOD, A. B. AND L. J. SCHRODER, 1984. Evaluation of Metal Ion Absorptive Characteristics of Three Types of Plastic Sample Bags Used for Precipitation Sampling. *J. of Environ. Sci. & Health*, 19(5)631.

34. MAHENDRAPPA, M. K. 1985. Precipitation Chemistry Affected by Difference in Location of Collection Sites Storage Methods. *Atmos. Environ.*, 19:1681.

35. FOSTER, J. F., G. H. BEATTY AND J. E. HOWES. 1974: "Final Report on Interlaboratory Cooperative Study of the Precision and Accuracy of the Measurement of Dustfall Using ASTM Method D1739." ASTM Publication DS 55–S4, American Society for Testing and Materials, Philadelphia, PA.

36. SANDERSON, H. P., P. BRADT AND M. KATZ. 1963. A Study of Dustfall on the Basis of Replicated Latin Square Arrangements of Various Types of Collectors. *J. Air Poll. Contr. Assoc.*, 13:463.

37. KEAGY, D. M., W. W. STALKER, C. E. ZIMMER AND R. C. DICKERSON. 1961. Sampling Station and Time Requirements for Urban Air Pollution Survey: Part I: Lead Peroxide Candles and Dustfall Collectors. *J. Air Poll. Contr. Assoc.*, 11:270.

38. EPA, 1976. "Quality Assurance Handbook for Air Pollution Measurement Systems: Volume I—Principles." EPA-600/9-76-005, Research Triangle Park, NC.

Subcommittee 10

R. T. EGAMI
J. G. WATSON, *Chairman*
C. F. ROGERS
M. G. RUBY
M. J. ROOD
J. C. CHOW

503.

Continuous Tape Sampling of Coefficient of Haze

1. Principle of Method

1.1 PRINCIPLE. Light is transmitted through a section of filter tape before and after ambient air is drawn through it. The optical density of the particle deposit on the tape is determined from the logarithm of the ratio of intensities measured through the filter tape with and without the deposit. A clean portion of the filter tape is periodically (usually hourly) moved into the sampling position to obtain a short-term measurement of optical density. This short-term sampling frequency allows diurnal changes in particulate optical density to be monitored. If the optical properties of the aerosol are constant, the optical density is related to the mass concentration in the atmosphere. This optical density can also be related to atmospheric light absorption and Coefficient of Haze (COH). COH is defined as one hundred times the optical density of the deposit. It is usually normalized to COH/1000 linear meters or feet.

1.2 APPLICABILITY. The continuous tape sampling method is appropriate for determining short-term variations in particulate matter optical characteristics. It is only applicable to determining variations in mass concentrations when the particle size, shape, refractive index, and color can be shown to be constant over the sampling period.

1.3 OTHER ANALYSES. Chemical measurements of continuous tape aerosol are not generally appropriate since the deposits may be contaminated by contact with other deposits on the takeup reel.

1.4 DETAILED PROCEDURES. Equipment-specific operating manuals are available from the equipment manufacturers. A standard operating procedure has been issued by ASTM (1).

1.5 ALTERNATIVE METHODS. Several laboratory measurements of high, medium, and low volume particulate samples have been developed to measure the optical density and optical absorption of aerosol deposits. These include: integrating plate (2,3,4), laser transmittance (5,6) and the multipass transmissometer (7). A continuous monitoring device for making filter transmittance measurement, called an aethalometer, has been developed at Lawrence Berkeley Laboratories. Continuous tape measurements of mass concentrations in different particle size fractions have been demonstrated using beta attenuation (8,9,10,11).

2. Range and Sensitivity

2.1 LOWER QUANTIFIABLE LIMIT. The lower quantifiable limit is an optical density of about 0.05.

2.2 UPPER QUANTIFIABLE LIMIT. The upper quantifiable limit is of the order of 0.30, which is a 50% attenuation of the incident light. Sampling times and flow rates are adjusted to obtain deposits with these upper and lower limits.

3. Interferences

3.1 COLOR OF THE FILTER TAPE. The filter tape should be as nearly transparent to light as possible. White glass or quartz fiber or Teflon tapes are the most appropriate media, though cellulose paper is widely used, and polycarbonate membrane filters give good results.

3.2 STRAY LIGHT. The light measured by the detector should be only that which is transmitted through the filter tape. Light entering the detector by other routes will not be attenuated by the aerosol deposit and will bias the measurement.

3.3 CHANGES IN LIGHT INTENSITY OR DETECTOR RESPONSE. Changes in the measured intensity over the filter exposure period will bias the optical density readings.

4. Precision and Accuracy

4.1 The precision and accuracy of this method are unquantified. They are functions of the precision and accuracy of the air flow rate, filtration efficiency and the optical system.

5. Apparatus

5.1 SAMPLER. The continuous tape sampler consists of an inlet, a continuous tape holder to accommodate a roll of filter material, an interval timer and clock, a light source with a wavelength of approximately 400 nm, a photomultiplier detector, a calibrated flow meter, a pump, and a filter tape advance and positioning mechanism. The sampler is schematically illustrated in Figure 503:1. Requirements for the individual components are listed below.

5.1.1 *Flow Meters.* A suitable meter capable of measuring air flow in the range of 3 to 15 L/min is located downstream of the filter tape.

5.1.2 *Pump.* A suitable positive displacement pump capable of maintaining flow rates in the 3 to 15 L/min range with changes of less than $\pm$ 10% over the sampling period is located downstream of the flow meter.

5.1.3 *Clamping Device.* A clamping device securely fastens a portion of the filter tape over the flow opening so that air from the inlet passes only through the filter, not around it. The clamping devices should be made of materials which can be easily cleaned such as stainless steel, plastic, or other corrosion-resistant, noncontaminating materials.

5.1.4 *Size Selective Inlet.* An inlet that passes particles in the desired size range is located upstream of the filter and clamping device. The sampling duct between the inlet

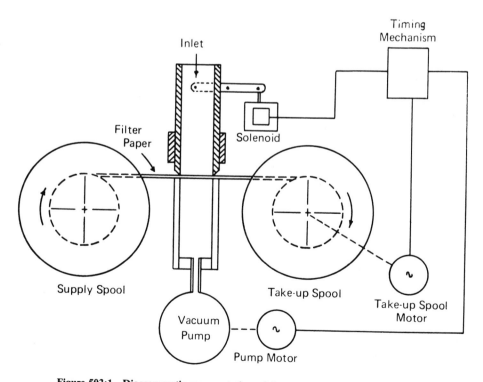

Figure 503:1 – Diagrammatic representation of the automatic filter tape sampler.

and the filter should not exceed 3 m in length and should not contain any sharp bends in which particles might be deposited.

5.1.5 *Interval Timer.* The timer releases the filter tape after each sampling cycle and advances it to an unexposed portion. Timing cycles of 1, 2, 3, 4, and 6 h are the usual options provided.

5.1.6 *Densitometer.* The densitometer consists of a stable light source with a wavelength peaking at about 400 nm and a photoelectric detector. The light source may consist of a broadband incandescent source and a filter with a bandpass in the 375 to 450 nm region.

6. Reagents and Supplies

6.1 FILTER MEDIA. Whatman No. 4 cellulose fiber filter tapes are available for this type of measurement. This medium is not generally suitable for chemical analysis of deposits, has particle sampling efficiencies of 33 to 99% (depending on filter batch, face velocity, and particle size), and is hygroscopic (12).

7. Procedure

7.1 Figure 503:2 presents a flow diagram of the routine operating procedure described in the following sub-sections.

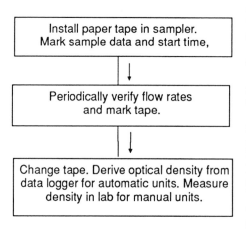

Figure 503:2.—**Flow diagram for continuous tape sampling.**

7.1.1 After assembling the paper tape sampler per manufacturer's instructions, place a roll of clean filter tape in the holder. Set the flow rate and sample duration to acquire deposits which do not exceed 0.30 in optical density. Mark the sample start date and time on a portion of the filter tape that is not to be exposed to an airstream.

7.1.2 Verify the air flow rate on a periodic basis and readjust it as necessary. Mark the time and date on the filter tape.

7.1.3 Some samplers have built-in densitometers. The output from these can be connected to a data logging device, and the tape need only be changed when it is near its end.

7.1.4 For units which do not have built-in densitometers, remove the tape at the end of the sampling period and take it to a laboratory environment. Measure the light transmission through a portion of the filter tape near each particle deposit to determine the transmission of the unexposed filter. Measure the light transmission through the spot that contains the aerosol deposit. Record the transmission of the clean filter and the filter deposit with the time and date of the sample on a data sheet. The time and date is determined by counting deposit spots from the time and date markings on the tape and adding an appropriate number of sample durations.

8. Calibration

8.1 Flow rate measurement devices are calibrated against field transfer standards which are traceable to primary flow standards.

8.2 Interval timers are calibrated against a quartz crystal watch.

8.3 Densitometers are calibrated against neutral density filters of known optical density.

8.4 A continuous filter tape sampler is not calibrated in the absolute sense because an optical density or COH reading cannot be explicitly related to atmospheric particulate or gas measurements (except light absorption, if certain assumptions are met). There is no direct relation between optical density and mass concentration of suspended particles, though empirical relationships may be established for certain situations. If particle size, shape, refractive index, and other factors are reasonably constant, these empirical relationships may be useful in comparing deposits from different types of sources and different locations. These relationships are not adequate for regulatory purposes.

9. Calculations

9.1 NOMENCLATURE.

OD	= Optical Density, no units.
I_o	= Intensity of transmitted light through clean filter tape, W/m^2.
I	= Intensity of transmitted light through the filter tape and the aerosol deposit, W/m^2.
F	= Average flow rate, m^3/s.
t	= sampling interval, s.
A	= area of the aerosol deposit on the filter tape, m^2.
COH	= Coefficient of haze, no units.
COHs/1000 m	= Coefficient of haze per 1000 meters of path length, m^{-1}.

9.2 FORMULAE.

9.2.1 The Optical Density is:

$$OD = \log(I_o/I)$$

9.2.2 The Coefficient of Haze is

$$COH = 100 \times OD$$

9.2.3 The number of COH units per 1000 meters is

$$COHs/1000 \text{ m} = 10^3 (COH \times A)/(F \times t)$$

10. Effects of Sample Storage

Filter tapes may undergo color changes with time when exposed to light.

11. References

1. ASTM, 1985. "Standard Method of Test for Particulate Matter in the Atmosphere: Optical Density of Filtered Deposit." ASTM Designation D 1704-78.
2. LIN, C. I., M. BAKER AND R. J. CHARLSON, 1973. Absorption Coefficient of Atmospheric Aerosol: A Method for Measurement. *Applied Optics, 12*:1356.
3. WEISS, R. E., A. P. WAGGONER, R. CHARLSON, D. L. THORSELL, J. S. HALL AND L. A. RILEY, 1979. Studies of the Optical, Physical, and Chemical Properties of Light Absorbing Aerosols. In *Proceedings of Carbonaceous Particles in the Atmosphere Conference*, Report 9037, Lawrence Berkeley Laboratory. Edited by T. Novakov, p. 257.
4. JAPAR, S. M., W. W. BRACHACZEK, R. A. GORSE, J. H. NORBECK AND W. R. PIERSON, 1986. The Contribution of Elemental Carbon to the Optical Properties of Rural Atmospheric Aerosols. *Atmos. Environ., 20*:1281.
5. ROSEN, H., A. D. A. HANSEN, L. GUNDEL AND T. NOVAKOV, 1978. Identification of the Optically Absorbing Component in Urban Aerosols. *Appl. Optics, 17*:3859.
6. MUELLER, P. K. AND J. G. WATSON, 1982. "Eastern Regional Air Quality Measurements: Volume 1. Final Report." EA-1914. Electric Power Research Institute, Palo Alto, CA.
7. GERBER, H. E. 1982. Optical Techniques for the Measurement of Light Absorption by Particulates. In *Particulate Carbon: Atmospheric Life Cycle*. Edited by G. T. Wolff and B. L. Klimisch. Plenum Publishing Corporation, New York, NY.
8. HUSAR, R. B. 1974. Atmospheric Particulate Mass Monitoring with a Beta Radiation Detector. *Atmos. Environ., 8*:183.
9. LILIENFIELD, P. 1970. Beta-Absorption-Impactor Aerosol Mass Monitor. *J. Am. Ind. Hygiene Assn., 31*:722.
10. SHIMP, D. 1988. Field Comparison of Beta Attenuation PM_{10} Sampler and High-Volume PM_{10} Sampler. In PM_{10}: Implementation of Standards, APCA, Pittsburgh, PA.
11. RODDY, W. J. 1988. Beta Attenuation Monitoring for PM_{10}. In PM_{10}: Implementation of Standards, APCA, Pittsburgh, PA.
12. LIPPMAN, M. 1983. Sampling Aerosols by Filtration. In *Air Sampling Instruments for Evaluation of Atmospheric Contaminants, 6th Edition*, edited by P. J. Lioy and M. J. Y. Lioy, American Conference of Governmental Industrial Hygienists, Cincinnati, OH.

Subcommittee 10

R. A. HERRICK
S. KINSMAN
J. P. LODGE
D. LUNDGREN
C. PHILLIPS
R. S. SHOLTES
E. STEIN
J. WAGMAN
J. G. WATSON, *Chairman*

507.

Integrating Nephelometer Measurement of Scattering Coefficient and Fine Particle Concentrations

1. Principle of the Method

1.1 PRINCIPLE. The integrating nephelometer (**1,2,3**) measures the optical scattering coefficient (σ_s) from the light scattered by particles (σ_{sp}) and gases (σ_{sg}) in a sample volume integrated over essentially all scattering angles (hence the name) over a weighted range of visible wavelengths (**4**). For a given particle shape (**5**), size distribution (**6**), and chemical composition (**7**), this scattering extinction coefficient is proportional to the mass concentration of suspended particles in the sample volume and the scattering due to atmospheric gases. The gaseous or "Rayleigh" scattering is a function of atmospheric temperature and pressure and can be accurately calculated to determine the fraction of the integrated scattering due to suspended particulate matter (**8**).

The sample volume is illuminated by a lamp mounted behind a lambertian diffusing window (milk glass). The light scattered by the gas and suspended particulate matter in the sample volume is measured by a photomultiplier tube. The instrument may be calibrated by filling the enclosed sample volume with a higher density, particle-free gas (e.g., Refrigerant-12) having a previously measured scattering coefficient (**9**). Since the light scattered by the air and the particles is a function of both concentration and wavelength, the value of the scattering coefficient measured by the integrating nephelometer will depend on the temperature and pressure of the gas, the concentration of the particulate matter, and the spectral sensitivity of the instrument. A narrow-band (20–50 nm) spectral sensitivity centered between 520 and 580 nm is recommended for measuring visual air quality

and estimating fine particle concentrations (**10**).

The measured scattering coefficient can be related to the visual range. This requires a number of simplifying assumptions (which may be difficult to justify) that should be clearly stated if an estimated meteorological range is calculated and reported. However, several studies have shown an approximate correlation between the scattering coefficient and observed visual range, particularly at relative humidities below 60% (**11,12**).

The measured scattering coefficient can be related to the mass concentration of fine (less than 2.5 μm in diameter) particulate matter. Several studies have shown a correlation between the scattering coefficient and measured fine particle mass concentration. The actual value of the ratio between the variables will depend on the local character of the aerosol. The particle scattering coefficient to fine particle mass ratios generally vary from about 2.8 to 4.0 m^{2}g^{-1} (**13,14**). Measurements of particle size ranges that include large particles (e.g., 1 to 10 μm and 0 to 30 μm) are not as strongly correlated with the scattering coefficient (**15**).

1.2 APPLICABILITY. The integrating nephelometer measures the scattering coefficient for nearly all normal ambient atmospheric aerosols. The major exception is aerosols composed primarily of large particles (e.g., windblown dust or wet fogs) that tend to deposit, impact, or evaporate in the air supply system. An open instrument designed to be operated in the ambient environment without an air supply system or an enclosed sample volume is more appropriate for suspended particles that should be measured in situ.

1.3 OTHER ANALYSES. Temperature and humidity conditioning of the sampled air can be applied to estimate aerosol composition, volatility, and size **(16,17,18,19)**. Multiwavelength scattering coefficients can also be measured and will provide information about the particle size distribution **(20,21)**.

1.4 DETAILED PROCEDURES. Instrument-specific standard operating procedures have been developed for long-term visibility studies **(13,22,23)**. Operational recommendations have been derived from research studies **(24)**. This procedure follows that adopted by the TE-5 Visibility Committee of the Air Pollution Control Association **(25)**.

1.5 ALTERNATIVE METHODS. Total extinction coefficients (scattering plus absorption coefficient) are measured in a horizontal path by transmissometers **(26)** and teleradiometers, **(27)** and in a vertical path by photometers **(28)**. These long-path measurements require numerous assumptions that are not always valid in practice. Intercomparisons of integrating nephelometer measurements with extinction measurements derived from these long-path measurements have shown significant differences **(29–33)**.

2. Range and Sensitivity

2.1 The integrating nephelometer can be used to measure the particle scattering coefficient, σ_{sp}, over the range of approximately 1×10^{-6} to 1×10^{-2} m^{-1}. Measurements at the lower limit require frequent checks against filtered, particle-free air. The sensitivity of commercial instruments is about 0.5×10^{-6} m^{-1}. The specific range and sensitivity is defined by the individual instrument's optics, electronic design, and mode of operation.

3. Interferences

3.1 Atmospheric moisture may interfere with the measurement under some circumstances, especially when the ambient relative humidity exceeds 40% **(34–39)**. Because many particles found in the ambient atmosphere are hygroscopic, a change in the relative humidity will result in a change in the particle diameter and, thus, the particle scattering extinction coefficient. If an uninsulated instrument is operated in an air-conditioned room in a hot, humid climate, water vapor can condense in the sample volume. A sample air preheater is often used in the sample line to reduce the relative humidity of the sample. An increase in the air temperature by 15°C is sufficient to reduce the relative humidity of the sample air below 40% under almost all conditions. In addition to drying the sample this temperature rise may partially evaporate volatile components of the aerosol such as organic compounds and ammonium nitrate. The scattering extinction coefficient measured in an unheated instrument may be as much as 1.6 times the value obtained after heating **(40)**. The decision to heat the sample will involve a trade-off between reducing the effects of humidity and the potential removal of important volatile components of the aerosol. A temperature increase of 8–9°C is sufficient in most circumstances. For optical extinction or visual purposes, a measurement of the atmospheric light scattering coefficient should be made at ambient conditions. If the measured particle scattering coefficient is to be related to the fine particle mass concentration a sample air preheater should be used, as this will generally improve the correlation with mass measurements.

3.2 The continuously-lit incandescent lamp used in some instruments can heat the air in the sample volume by 5 to 10°C, although this can be reduced by installing baffles/insulation to reduce heating of the inlet air tubing. Heating will also be reduced by minimizing the sample's residence time within the nephelometer and by bathing the entire unit with ambient air. The sample air preheater supplied by the manufacturer with some instruments has been observed to heat the air sample by as much as 37°C. The actual temperature increase will vary with flow rate, which will depend on the arrangement of the air supply system. If an air preheater is used, it should be controlled so that the increase in air temperature is no more than 15°C **(41)**. An instrument installed in a closed shelter and drawing its sample from a long inlet line may result in a measurement that differs from that at ambient conditions.

4. Precision and Accuracy

4.1 PRECISION. In actual use the 1-σ precision is on the order of 3–4% of the scattering coefficient for the measurement of normal ambient atmospheric aerosols **(42)**. Some slight degradation in precision may occur for readings in extremely clean air, with the exact degree depending on the zero drift of the instrument.

4.2 ACCURACY. Most designs of the integrating nephelometer have an intrinsic potential bias in σ_{sp} of up to –5% for particles less than 2 μ in diameter. This bias is caused by an optical design that only collects light scattered between approximately 9° and 170°**(43)**. The measurement of the scattering coefficient for large particles will be reduced by about one-half due to the light lost in the omitted forward angle **(44,45)**. In addition, a significant fraction of the large particles can be lost by impaction and deposition in the air supply system.

The dominant cause of short term inaccuracy is instrument zero drift. In clean-air locations with σ_{sp} below 5×10^{-5} m^{-1}, it is necessary to reestablish the zero calibration every 1 to 6 h. Ambient temperatures above 20°C may increase the zero drift due to changes in the photomultiplier tube dark current. A once-a-day zero check will fail to observe a significant intra-day drift, giving a false confidence in the instrument stability. An automated system can be used to accomplish the zero calibration routinely.

5. Apparatus

5.1 NEPHELOMETER. An enclosed integrating nephelometer generally consists of an optical assembly, an electronics and control unit, and an air blower and filter, which may be packaged as separate units or in a single container. A typical system is shown in Fig. 507:1. A sample air preheater may also be installed in the sampling line. The air preheater should not be operated without sufficient air flow, as some units are not temperature-protected and may ignite. Flexible tubing is commonly used to connect the instrument to the sampling manifold. The inlet tubing should be of

large enough diameter that particles are not lost to the walls of the tubing. In a typical instrument the air sample is drawn continuously through an enclosed sample volume at a rate of about 140 L/min. The measurement is independent of the flow rate. The typical sample air blower supplied by the manufacturer cannot operate against a sampling manifold pressure greater than –2 cm of water. Interferences from stray light are minimized by using a chopped light source and synchronized detector.

5.2 AUXILIARY EQUIPMENT. Calibration equipment consists of a container of Refrigerant-12 (CCl$_2$F$_2$) or Refrigerant-22 (CHClF$_2$) with a control valve, a particle filter, and sufficient tubing to connect these to the optical assembly. The value of the scattering coefficient for Refrigerant-12 is about twice that of Refrigerant-22, but the latter is believed to be potentially less harmful to the atmosphere and its use is recommended whenever possible. A data recording device such as a stripchart or digital data logger is needed for unattended operation.

6. Reagents

6.1 No laboratory reagents are required for the operation of the integrating nephelometer.

6.2 A high density gas, such as Refrigerant-12 or Refrigerant-22, is required for the primary calibration of the instrument. Ordinary refrigerant quality cans of pressurized gas are adequate if the gas is filtered to remove any particulate matter that may be present in the gas. Refrigerant gases that contain a tracer dye should not be used for calibration.

7. Procedure

7.1 The integrating nephelometer will operate without operator intervention once it has been installed and turned on. A routine field visit consists of systems tests, performance tests, and replacement of the data recording medium, as illustrated in the flow diagram of Fig. 507:2.

7.2 Check the air supply system and the optical assembly (enclosed sample volume) to ensure that it is airtight and that the sample is drawn from the sampling manifold

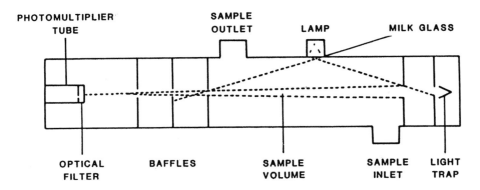

Figure 507:1 Schematic diagram of integrating nephelometer.

and not from the instrument room. Puffs of tracer smoke (from a cigarette or cigar) blown along the air supply line and along the optical housing should not produce any significant changes in the nephelometer's output. Many commercial instruments cannot be made sufficiently airtight. In that case, the instrument should be housed in a room without any other activity. Smoking is not permitted in the instrument room. An initial primary calibration is required after installation.

7.3 After an adequate period to allow the instrument to warm up and the electronics to stabilize (at least 1 h), perform, as necessary, either a primary or secondary calibration of the instrument as described in Section **8**.

7.4 Turn on the sample air blower. If the sample air preheater is used, turn it on and allow the sample temperature to stabilize. Set the appropriate scale for the output, depending on the estimated particulate loading of the air.

7.5 The instrument may now be left on for recording with no further adjustments. Depending on the average particle concentrations, a zero check should be performed daily, if it is provided with the instrument. These recommended performance test frequencies may be extended or changed as indicated by experience in a specific situation. Recalibration with refrigerant gases

may be restricted to a schedule of once every 3 mo.

7.6 The expected life of lamps is 3 mo to 1 y. Spare lamps should be kept on hand. Lamps should be replaced per manufacturer's instructions.

8. Calibration and Standards

8.1 Primary calibration of the integrating nephelometer is performed by using the scattering coefficient of filtered, particle-free air for the zero calibration and the previously measured scattering coefficient of a high density refrigerant gas for the span calibration. Because the scattering coefficient is a function of wavelength and the different models of commercially available integrating nephelometer have differing spectral sensitivity, it is necessary to use reference values specific to the instrument being used. These values should be further adjusted to the actual temperature and pressure in the sample volume. Reference values for the two most common models of integrating nephelometer have been published (**24,41**) or may be obtained from the manufacturer. After the primary calibration is established, the measured scattering coefficient of an internal optical path (if provided) may be used as a secondary mechanical span calibration.

8.2 PRIMARY CALIBRATION. Turn off the air preheater and the sample blower. Dis-

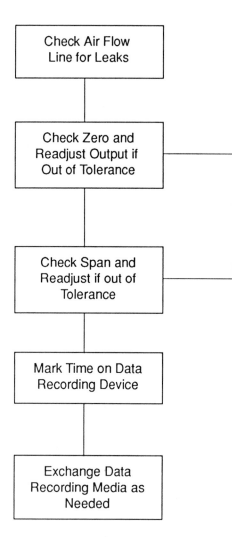

Figure 507:2 Flow diagram of routine nephelo-meter operations.

connect the air supply and exhaust tubing from the optical assembly, and prepare to close the openings (with a rubber stopper or similar plug). Partially seal the upper opening of the nephelometer with tissue or a soft cloth. The seal of the upper opening should allow air to pass through the opening, but not light.

8.2.1 Draw air from the sampling manifold and supply it to the lower opening of the optical assembly through a high-efficiency ("absolute") filter, applying a slight positive pressure of filtered, particle free air in the scattering volume. Allow the instrument to warm up, set the switches in their desired positions, and wait for output to stabilize. If a photomultiplier tube offset adjustment is provided, turn the lamp off, adjust the offset of the dark current from the photomultiplier tube so that the output reads zero, and turn the lamp back on. Adjust the instrument output zero offset so the output value is zero.

8.2.2 Disconnect the clean air supply. Plug the optical assembly's inlet openings. Connect the refrigerant gas container through a filter to one of the optical assembly inlet nipples with suitable tubing. Open the refrigerant container valve and slowly purge the sample value for 10 min or until the output has stabilized. The flow rate may be regulated at approximately 10 L/min. When the output is stable, reduce the flow of the refrigerant gas to a minimum. Make the span calibration by setting the output equal to the difference between the scattering coefficient of the refrigerant gas and the scattering coefficient of clean air, both at the temperature and pressure in the sample volume. Close the refrigerant container valve. Be certain to completely purge the viewing volume with clean air after making the span calibration.

8.2.3 Recheck the zero calibration after making a primary (or secondary) span calibration adjustment. It may be necessary to make a second check of the span calibration if the zero calibration is changed. After the zero and span calibration points are established, the zero offset may be adjusted to an arbitrary positive value before recording data. A positive offset is recommended to ensure the output will remain positive despite changes in atmospheric temperature and pressure or normal instrument zero drift. This offset is particularly important for instruments used in airborne sampling. The positive zero offset should be measured by again filling the sample volume with filtered, particle-free air. If a positive zero offset is not used, the zero will be at the scattering coefficient of clean air, so the output will be the particle scattering coefficient (σ_{sp}). If the output of the arbitrary zero offset is set to display the value of the scattering coefficient of clean air, the reported values will be the scattering coefficient. The particle scattering coefficient will then be obtained by subtracting this offset from the measured values.

8.2.4 Remove the plugs at the optical assembly's inlet and outlet, reseal the optical assembly's nipples, and reconnect the air supply, exhaust tubing and sample air pre-heater, if used.

8.3 SECONDARY MECHANICAL CALIBRATION. Most integrating nephelometers include an internal optical path that provides a fixed amount of light to the photomultiplier tube. After the primary calibration is completed, the output of the instrument in this mode may be measured and recorded for future comparison. This may be used as a secondary span calibration point. With some instruments it is necessary to make the secondary mechanical calibration with filtered, particle-free air in the sample volume. In this case it is also necessary to note if the measured value includes a positive zero offset and if so, the value of the offset.

8.3.1 Check and adjust the zero calibration as described in **8.2.1**. With some instruments the positive zero offset is not automatically cancelled during calibration and it is sufficient to check the value of the offset unless an adjustment must be made to the span value. When the span is adjusted, the zero calibration point must be first set to zero.

8.3.2 Check and adjust the span calibration by activating the internal mechanical calibration. With some instruments it will be necessary to supply filtered, particle-free air to the sample volume while making this check. With other instruments the optical path does not include the sample volume and the contents of the sample volume are not relevant. If the span calibration is to be adjusted, the positive zero offset must be cancelled (if not done automatically) and the value of the mechanical calibration without the zero offset must be used.

8.3.3 Recheck the zero calibration after making a span calibration adjustment. It may be necessary to make a second check of the span calibration if the zero calibration is changed.

9. Calculations

9.1 No calculations are required to obtain the particle scattering coefficient (other than a correction for a positive zero offset). When reporting data it is important to state clearly if an artificial value representing the Rayleigh scattering is included in the reported value. The model of nephelometer used should also be reported.

9.2 Measured scattering coefficients are often converted to and reported as meteorological range. This requires significant assumptions that are frequently not met in practice. For this reason this practice is discouraged.

9.2.1 If it is assumed that the point measurement is the same as the spatially averaged extinction coefficient (σ_e in meters^{-1}) between the point and the limit of visibility, the meteorological visual range (L_v in meters) can be estimated from the average extinction coefficient with the Koschmeider equation: **(4)**

$$L_v = \frac{3.9}{\sigma_e}$$

The derivation of this equation assumes, among other things, that the critical contrast for limiting visibility is 2% (values as high as 7% have been used). Further, the equation requires a value of the extinction coefficient, which is the sum of the scattering coefficient (measured by the integrating nephelometer) and the absorption coefficient. Absorption by gases is usually not important except directly in NO_2-rich plumes. Absorption by particles may account for approximately 5% of the total extinction in rural areas and has been measured as high as 40% in a polluted urban atmosphere **(40)**. Even if it is assumed that the reported scattering coefficient is approximately equal to the extinction coefficient, significant inaccuracies remain in translating the measured scattering coefficient to a meteorological range value. Since the spectral sensitivity of the nephelometer is not necessarily the same as the spectral sensitivity of the human eye the difference in the amount of light scattered at the different wavelengths must be taken into account. The correction factor depends on the size distribution of the aerosol, the magnitude of the scattering coefficient, and the spectral sensitivity of the integrating nephelometer. Correction factors that have been supplied with some instruments do not take all these elements into account and will give erroneous results (41). In addition, the Koschmeider equation assumes the illumination is uniform between the viewer and the distant object and beyond. A time-averaged meteorological range estimate must be calculated from the average of the inverse of the measured scattering coefficient rather than the average of the scattering coefficient (46).

10. Effect of Storage

10.1 NOT APPLICABLE.

11. References

1. AHLQUIST, N. C., AND R. J. CHARLSON, 1967. A new instrument for evaluating the visual quality of air. *J. Air Poll. Cont. Assoc., 17*:467.
2. CHARLSON, R. J., H. HORVATH AND R. F. PUES-CHEL, 1967. The direct measurement of atmospheric light scattering coefficient for studies of visibility and pollution. *Atmos. Environ., 1*:469.
3. BEUTTEL, R. G. AND A. W. BREWER, 1949. Instruments for the measurement of the visual range. *J. Sci. Inst., 26*:357.
4. MIDDLETON, W. E. K. 1952. *Vision through the atmosphere.* University of Toronto Press, Toronto.
5. JAGGARD, D. L., C. HILL, R. W. SHORTHILL, D. STUART, M. GLANTZ, F. ROSSWOG, B. TAGGART AND S. HAMMOND. 1981. Light scattering from particles of regular and irregular shape. *Atmos. Environ., 15*:2511.
6. WILLEKE, K. AND J. E. BROCKMAN, 1977. Extinction coefficient for multimodal atmospheric particle size distributions. *Atmos. Environ., 11*:995.
7. SLOANE, C. S. 1984. Optical properties of aerosols of mixed composition. *Atmos. Environ., 18*:871.
8. BODHAINE, B. A. 1979. Measurement of the Rayleigh scattering properties of some gases with a nephelometer. *Appl. Optics, 18*:121.
9. CHARLSON, R. J., N. C. AHLQUIST, H. SELVIDGE AND P. B. MACCREADY, 1969. Monitoring of atmospheric aerosol parameters with the integrating nephelometer. *J. Air Poll. Cont. Assoc., 19*:937.
10. PRESLE, G. AND H. HORVATH, 1978. Visibility in turbid media with colored illumination. *Atmos. Environ., 12*:2455.
11. HORVATH, H. AND K. E. NOLL, 1969. The relationship between atmospheric light scattering coefficient and visibility. *Atmos. Environ., 3*:543.
12. HARRISON, A. W. 1979. Nephelometer estimates of visual range. *Atmos. Environ., 13*:645.
13. DZUBAY, T. G., R. K. STEVENS, C. W. LEWIS, D. H. HERN, W. J. COURTNEY, J. W. TESCH AND M.

A. MASON, 1982. Visibility and aerosol composition in Houston, Texas. *Environ. Sci. & Tech., 16*:514.
14. WAGGONER, A. P. and R. E. WEISS, 1980. Comparison of fine particle mass concentration and light scattering in ambient aerosol. *Atmos. Environ., 14*:623.
15. CHARLSON, R. J., N. C. AHLQUIST AND H. HORVATH, 1968. On the generality of correlation of atmospheric aerosol mass concentration and light scatter. *Atmos. Environ., 2*:455.
16. CHARLSON, R. J., A. H. VANDERPOL, D. S. COVERT, A. P. WAGGONER AND N. C. AHLQUIST, 1974. $H_2SO_4/(NH_4)_2SO_4$ background aerosol: Optical detection in St. Louis region. *Atmos. Environ., 8*:1257.
17. WEISS, R. E. AND A. P. WAGGONER, 1982. In-situ rapid response measurement of $H_2SO_4/(NH_4)_2SO_4$ aerosols in rural Virginia. *Environ. Sci. & Tech., 16*:525.
18. WAGGONER, A. P., R. E. WEISS AND T. V. LARSON. 1983. In-situ rapid response measurements of $H_2SO_4(NH_4)_2SO_4$ aerosols in Houston: A comparison with rural Virginia. *Atmos. Environ., 17*:1723.
19. ROOD, M. J., D. S. COVERT AND T. V. LARSON. 1987. Temperature and Humidity Controlled Nephelometry: Improvements and Calibration. *Aerosol Sci. Tech., 7*:57.
20. AHLQUIST, N. C. AND R. J. CHARLSON. 1969. Measurement of the wavelength dependence of atmospheric extinction due to scatter. *Atmos. Environ., 3*:551.
21. ENSOR, D. S. AND K. T. WHITBY. 1972. Multi-wavelength nephelometer measurements in Los Angeles smog aerosol I: Comparison of calculated and measured light scattering. *J. Colloid and Interface Sci., 39*:240.
22. BABSON, B. L., R. W. BERGSTROM, M. A. SAMUELSON, C. SEIGNEUR, A. WAGGONER AND W. C. MALM. 1982. Statistical analysis of nephelometer regional field data. *Atmos. Environ., 16*:2335.
23. WATSON, J. G., T. E. HOFFER AND J. C. CHOW. 1988. "Guidance on methods to investigate existing visibility impairment and attribute it to sources." DRI document No. 8090.F. prepared by the Desert Research Institute for the National Parks Service, Contract No. CX-0001-4-0055, Reno, NV.
24. WAGGONER, A. P. 1985. Quality assurance for integrating nephelometers. In *Quality Assurance in Air Pollution Measurements*, edited by T. R. Johnson and S. J. Penkala, Air Pollution Control Association, Pittsburgh, PA.
25. RUBY, M. G. 1985. Visibility measurement methods I: Integrating nephelometer. *J. Air Poll. Contr. Assoc., 35*:244.
26. LEE, P. H., T. E. HOFFER, D. E. SCHORRAN, E. C. ELLIS AND J. W. MOYER. 1981. Laser transmissometer: A description. *Sci. of the Total Environ., 23*:321.
27. HORVATH, H., O. PREINING AND R. PIRICH. 1982. Measurements of the spatial distribution of particulate pollution with a telephotometer. *Atmos. Environ., 16*:1457.
28. VOLZ, F. E. 1973. Economical multispectral sun photometer for measurements of aerosol extinction from 0.44 μm to 1.6 μm and precipitable water. *Appl. Optics, 121*732.
29. HOFFMAN, H. E. AND W. KUEHNEMANN. 1979.

Comparison of the results of two measuring methods determining the horizontal standard visibility with the visual visibility range. *Atmos. Environ.*, *13*:1629.

30. TOMBACH, I. AND D. ALLARD. 1980. Intercomparison of visibility measurement methods. *J. Air Poll. Cont. Assoc.*, *30*:134.

31. RICHARDS, L. W., G. R. MARKOWSKI AND N. WATERS. 1981. Comparison of nephelometer, telephotometer, and aerosol data in the southwest. Presented at the 74th Annual Meeting of the Air Pollution Control Association, Philadelphia, PA.

32. MALM, W., A. PITCHFORD, R. TREE, E. WALTHER, M. PEARSON AND S. ARCHER. 1980. The visual air quality predicted by conventional and scanning teleradiometers and an integrating nephelometer. *Atmos. Environ.*, *15*:2547.

33. HITZENBERGER, R. M., R. B. HUSAR AND H. HORVATH. 1984. An intercomparison of measurements by a telephotometer and an integrating nephelometer. *Atmos. Environ.*, *18*:1239.

34. COVERT, D. S., R. J. CHARLSON AND N. C. AHLQUIST. 1972. A study of the relationship of chemical composition and humidity to light scattering by aerosols. *J. Appl. Meteor.*, *11*:968.

35. HANEL, G. 1972. Computation of the extinction of visible radiation by atmospheric aerosol particles as a function of the relative humidity, based upon measured properties. *J. Aerosol Sci.*, *3*:377.

36. MESZAROS, A. 197. An attempt to explain the relation between visual range and relative humidity on the basis of aerosol measurements. *J. Aerosol Sci.*, *8*:31.

37. TUOMI, J. T. 1980. Light scattering by aerosols with layered humidity-dependent structure. *J. Aerosol Sci.*, *11*:367.

38. SLOANE, C. S. AND G. T. WOLFF. 1985. Prediction of ambient light scattering using a physical model response to relative humidity: Validation with measurements from Detroit. *Atmos. Environ.*, *19*:669.

39. ROOD, M. J., D. S. COVERT AND T. V. LARSON. 1987. Hydroscopic properties of atmospheric aerosol in Riverside, California. *Tellus*, *39B*:383.

40. WAGGONER, A. P., R. E. WEISS, N. C. AHLQUIST, D. S. COVERT, S. WILL AND R. J. CHARLSON. 1981. Optical characteristics of atmospheric aerosols. *Atmos. Environ.*, *15*:1891.

41. RUBY, M. G. AND A. P. WAGGONER. 1981. Intercomparison of integrating nephelometer measurement. *Environ. Sci. & Tech.*, *15*:109.

42. TOMBACH, I. H. 1971. "Procedure for particulate measurement with the integrating nephelometer." Meteorology Research Inc., Pasadena, CA.

43. RABINOFF, R. A. AND B. M. HERMAN, 1973. Effect of aerosol size distribution on the accuracy of the integrating nephelometer. *J. Appl. Meteorol.*, *12*:184.

44. ENSOR, D. S. AND A. P. WAGGONER. 1970. Angular truncation error in the integrating nephelometer. *Atmos. Environ.*, *4*:481.

45. QUENZEL, H., G. H. RUPPERSBERG AND R. SCHELLHASE. 1975. Calculations about the systematic error of visibility-meters measuring scattered light. *Atmos. Environ.*, *9*:587.

46. RUBY, M. G. AND I. TOMBACH. 1981. Protection of visibility by an ambient air quality standard. Presented at the 74th Annual Meeting of the Air Pollution Control Association, Philadelphia, PA.

Subcommittee 10

M. G. RUBY
M. J. ROOD
A. P. WAGGONER
E. ROBINSON
D. L. BLUMENTHAL
J. G. WATSON, *Chairman*

601.

Determination of Gross Alpha Radioactivity
Content of the Atmosphere

1. Principle of the Method

1.1 Air particulate matter collected on a filter of high surface retention is counted with an alpha-sensitive detector to establish the gross concentration of alpha emitters present in the sampled ambient air. Since a given alpha emitter cannot be present in a greater concentration than the gross alpha concentration of a mixture of unidentified radionuclides, a gross analysis may eliminate the need for a more time-consuming and expensive analysis for specific radionuclides. While a gross alpha analysis does not yield specific information on the radionuclide composition of a sample, it is simple, rapid, and inexpensive. The method is therefore useful as a screening technique.

2. Range and Sensitivity

2.1 The method has the advantage of being suitable over the entire range of concentrations of airborne alpha emitters ordinarily encountered in the environment. Since most alpha counters have inherently low backgrounds, sensitivity is primarily a function of the volume of air sampled. For example, a sample from 100 m^3 of air with an alpha particulate concentration of 0.006 pCi/m^3 (0.22 mBq/m^3) will yield 1.32 dpm, sufficient to produce acceptable precision in less than an hour's counting time in most alpha counters.

3. Interferences

3.1 The principal interference is from the progeny of naturally occurring radon and thoron that are usually present in the outdoor atmosphere in concentrations of about 100 pCi/m^3 and 1 pCi/m^3, respec-
tively. They may build up to as much as ten-fold greater concentrations in closed indoor environments. Since the effective radioactive half-lives of the progeny of these naturally occurring radionuclides are controlled by relatively short half-life radionuclides, namely 26.8 min/^{214}Pb and 19.7 min ^{214}Bi for radon, and 10.6 h ^{212}Pb for thoron, their interferences may be nullified by waiting until these daughter activities are negligible before counting. This would require about 3 days for the long-lived thoron progeny. Less accurate, but more rapid, evaluations of long-lived gross alpha concentrations can be secured by a dual counting procedure (**1, 2**).

3.2 Another way to differentiate between these natural progeny and other alpha emitters is to employ a solid-state semiconductor detector, the signal from which is proportional to the energy of the detected alpha particles. Most long-lived alpha emitters have energies below 6 MeV, whereas the short-lived daughters of radon and thoron have greater energies. It is therefore possible to gate the associated electronics to count only the pulses from low-energy alpha emissions (**3**).

3.3 Inert dust loadings and the use of improper filter media are also potential sources of error. Alpha absorption and energy degradation (in the case of solid-state counting) become pronounced in samples with total dust loading larger than 1 mg/cm^2. The total volume of air that can be sampled in a dust-laden atmosphere is thereby limited. Absorption of alpha activity takes place if particles have penetrated into the filter matrix. The use of membrane-type filters is recommended to avoid this problem.

4. Precision and Accuracy

4.1 The precision of the method is essentially a function of the volume of air sampled, the inert dust loading of the air, the length of time for which the sample is counted, and the background of the counter. For example, the precision of count of the very low-activity sample described above (with an activity of only 1.32 dpm), counted for an hour at 50% geometry in a counter with a background of 0.20 cpm, would be 1.32 ± 0.52 dpm (at the 95% confidence level).

4.2 The accuracy of a gross alpha analysis is dependent on the accuracy of measurement of the volume of air sampled, and the appropriateness of the self-absorption correction.

5. Apparatus

5.1 The basic requirement for gross alpha analysis is an air mover, a calibrated and properly situated device for measuring air flow rate or sample volume, a particle filter, an alpha detector, and a counter. The size of the sample should not be any larger than the sensitive area of the alpha detector. Due to the limited range of alpha radiation, a high surface-retention filter (*i.e.*, low penetration of particles into the filter matrix) such as a membrane type filter, on which collection is essentially on the surface, is recommended.

5.2 A windowless or very thin window alpha-sensitive detector should be used to count the filter sample. Spurious counts may be produced in windowless counters by the buildup of static charge on filters. This may be eliminated by the use of anti-static fluid. ZnS-coated photomultiplier tubes are widely used for alpha counting. Although presently limited to small diameter samples, solid-state detectors are extremely attractive for gross alpha counting because of their stability and low backgrounds (in the order of 0.01 cpm).

6. Reagents

6.1 None.

7. Procedure

7.1 Operate the sampler in a location representative of the atmosphere for which the concentration is to be established and for a period long enough to collect measurable alpha activity, considering the minimum detectable concentration.

7.2 Place the filter in a protective cover or container upon removal from its holder to minimize the possibility of dislodging the collected activity.

7.3 Prior to counting, place the filter sample in a planchet or holder for reproducible positioning under the detector.

7.4 Count the sample for an interval (or total count) sufficient for a statistically significant result.

8. Calibration and Standards

8.1 The efficiency of an alpha counter should be determined with an alpha standard as close as possible to the sample in size and composition. If the latter is unknown, a standard may be prepared from uranium solution.

8.2 The counter's background should be determined over an interval similar to or longer than the sample counting time. Alpha counters are usually quite stable and it is satisfactory to make only daily determinations of background to ascertain that major shifts due to malfunction, drift, or contamination have not occurred, unless anomalous counting results are encountered.

9. Calculation of Concentration

9.1 CALCULATION.

Counter Efficiency =

$$\frac{\text{net counting rate of standard (cpm)}}{\text{disintegration rate of standard (dpm)}}$$

(1)

Net Sample Disintegration Rate (dpm) =

$$\frac{\dfrac{\text{gross counts}}{\text{sample count}}{\text{time (min.)}} - \dfrac{\text{background counts}}{\text{background count}}{\text{time (min.)}}}{\text{counter efficiency}}$$

(2)

Activity (pCi/m³) =

$$\frac{\text{sample disintegration rate}}{\text{volume sample (m}^3) \times 2.2 \text{ (dpm/pCi)}}$$

(3)

9.2 · The statistical significance of this result should also be indicated. It can be determined from the standard deviation (σ) of the count as follows:

σ (cpm) =   (4)

$$\pm \sqrt{\frac{\text{cpm (gross)}}{\text{sct (min.)}} + \frac{\text{cpm (background)}}{\text{bct (min.)}}}$$

where sct = sample count time
      bct = background count time

$$\sigma \text{ (dpm)} = \frac{\pm \, \sigma \text{ (cpm)}}{\text{counter efficiency}}$$

(5)

9.3 The statistical significance of a count should be expressed as the uncertainty (or error) of the account at a specified confidence level, such as 95% (1.96 σ).

10. Effects of Storage

10.1 Most properly protected samples should experience no effect other than ra-dioactive decay during storage. Some alpha emitters, such as ^{210}Po, are found to "creep" by recoil action and thereby lose activity from the sample to the immediate surroundings.

11. References

1. SETTER, L.R. AND G.I. COATS. 1961. The Determination of Airborne Radioactivity. *AIHA J.* 22:64.
2. SCHULTE, H.F. 1968. Chapter on Monitoring Airborne Radioactivity in *Air Pollution*, 2nd Edition. Vol. 11, Academic Press, New York.
3. LINDEKEN, C.L. AND K.F. PETROEK. 1966. Solid-State Pulse Spectroscopy of Airborne Alpha Radioactivity Samples. *Health Physics* 12:683.

Subcommittee 8

B. SHLEIEN, *Chairman*
R. G. HEATHERTON
P. F. HILDEBRANDT
L. B. LOCKHART, JR.
R. S. MORSE
W. STEIN
D. F. VAN FAROWE
A. HULL, *Health Physics Society Liaison*

Approved with modifications from 2nd Edition
Subcommittee 8

A. P. HULL, *Chairman pro tem*
J. A. BROADWAY
M. L. MAIELLO
R. P. MILTENBERGER

602.

Determination of Gross Beta Radioactivity Content of the Atmosphere

1. Principle of Method

1.1 Air particulate matter collected on a filter of high retention is counted with a beta-sensitive detector to establish the gross concentration of the beta emitters present in the sampled ambient air. Since a given beta emitter cannot be present in a greater concentration than the gross beta concentration of a mixture of unidentified radionuclides, a gross analysis may eliminate the need for a more time-consuming and expensive analysis for specific radionuclides. While the gross beta analysis does not yield specific information on the radionuclide composition of the sample, it is simple, rapid, and inexpensive. Since it can be performed in the field using relatively simple equipment, gross beta measurement of air samples is widely employed as a screening technique.

2. Range and Sensitivity

2.1 The method has the advantage of being suitable over the entire range of concentrations of airborne beta emitters ordinarily encountered in the environment. With proper control of counter background, sensitivity is primarily a function of the volume of air sampled. A sample from 10 m^3 of air with a beta particulate concentration of 3 pCi/m^3 (111 mBq/m^3) will yield 66.6 dpm, sufficient to produce acceptable precision in 10 min counting time.

3. Interferences

3.1 The principal interference is from the progeny of naturally occurring radon and thoron usually present in the outdoor atmosphere in concentrations of about 100 pCi/m^3 and 1 pCi/m^3 respectively. They may build up to as much as ten-fold greater concentrations in closed indoor environments. Since the effective radioactive half-lives of the progeny of these naturally occurring radionuclides are controlled by relatively short half-life radionuclides, namely 26.8 min ^{214}Pb and 19.7 min ^{214}Bi for radon, and 10.6 h ^{212}Pb for thoron, their interferences may be nullified by waiting until these daughter activities are negligible before counting. This would require about 3 days for the long-lived thoron progeny. Less accurate, but more rapid, evaluations of long-lived gross beta concentrations can be secured by a dual counting procedure (1, 2).

3.2 Inert dust loadings may cause serious interference, particularly for measuring low energy beta emitters. Reasonably accurate self-absorption corrections may be made providing these loadings do not exceed a few mg/cm^2.

4. Precision and Accuracy

4.1 The precision of the method is essentially a function of the volume of air sampled, the inert dust loading of the air, the length of time for which the sample is counted, and the background of the counter. For example, the precision of count of the sample described above (with an activity of 66.6 dpm), counted for 10 min at 30% overall efficiency in a counter with a background of 10 cpm, would be 66.6 ± 13.2 dpm (at the 95% confidence level).

4.2 The accuracy of a gross beta analysis is dependent on the accuracy of measurement of the volume of air sampled, the appropriateness of the self-absorption corrections, and the similarity of energy of the calibration standard to the energy of the unknown sample radionuclides. Since beta energies vary over an order of magnitude, the gross beta activities (and concentrations) must be reported in terms of the calibration standard employed.

5. Apparatus

5.1 The basic requirements for gross beta analysis are an air mover, a calibrated and properly situated device for measuring flow rate or air volume sampled, a particulate filter, a beta detector and a counter. A suitable high-efficiency filter should be employed.

5.2 If filters of appreciably less than 100% collection efficiency are employed, it is necessary that their collection efficiency as a function of filter face velocity and the physical properties of the material being sampled be known (3).

5.3 A beta-sensitive detector such as an end-window Geiger-Muller tube, thin window proportional flow counter, plastic scintillator, or solid state detector can be used for counting the sample.

6. Reagents

6.1 None.

7. Procedures

7.1 Operate the air sampler in a location representative of the atmosphere for which the gross beta concentration is to be determined, for a period long enough to collect measurable beta activity considering the minimum expected concentration.

7.2 To minimize the possibility of dislodging collected activity, store the sample in a protective cover or container upon removal from its holder.

7.3 For counting, place the filter sample in a planchet or holder at a reproducible distance under the window or sensitive surface of a beta detector. Care should be taken to avoid contaminating the detector. This can be done by covering the sample with thin mylar sheet.

7.4 Count the filter sample for an interval (or total count) sufficient for a statistically significant result.

8. Calibration and Standards

8.1 The efficiency of a beta counter should be determined with a beta standard resembling the sample in size and matrix, and containing the beta radionuclide of principal concern or one of similar energy. If the latter is unknown, ^{137}Cs is a desirable standard, since its average energy corresponds to that of aged mixed fission products in fallout. It is desirable that a number of other beta radionuclides covering a range of energies be counted in order to determine the energy dependence of the counter. If counting is done through a thin mylar cover, **7.3**, calibration must be done through an equivalent absorber.

8.2 Determine the background of the counter over an interval similar to or longer than the sample counting time. Most beta counters are relatively stable, and it is sufficient to make one daily determination of background to ascertain that major shifts due to malfunction, drift or contamination have not occurred.

9. Calculation of Concentration

9.1 CALCULATION.

Counter Efficiency =

$$\frac{\text{net counting rate of standard (cpm)}}{\text{disintegration rate of standard (dpm)}}$$

Net Sample Disintegration Rate (dpm) =

$$\frac{\dfrac{\text{gross counts}}{\substack{\text{sample count} \\ \text{time (min.)}}} - \dfrac{\text{background counts}}{\substack{\text{background count} \\ \text{time (min.)}}}}{\text{counter efficiency}}$$

Activity (pCi/m³) =

$$\frac{\text{sample disintegration rate}}{\text{volume sample (m}^3) \times 2.2 \text{ (dpm/pCi)}}$$

9.2 The statistical significance of this result should also be indicated. It can be determined from the standard deviation (σ) of the count as follows:

$$\sigma \text{ (cpm)} = \qquad\qquad (4)$$

$$\pm \sqrt{\frac{\text{cpm (gross)}}{\text{sct (min.)}} + \frac{\text{cpm (background)}}{\text{bct (min.)}}}$$

where sct = sample count time
      bct = background count time

$$\sigma \text{ (dpm)} = \frac{\pm \ \sigma \text{ (cpm)}}{\text{counter efficiency}}$$

9.3 The statistical significance of a count should be expressed as the uncertainty (or error) of the count at a specified confidence level, such as 95% (1.96σ).

10. Effects of Storage

10.1 Samples should be stored so they are protected from the loss of loosely embedded particles. A properly protected sample should experience no effects other than radioactive decay during storage.

11. References

1. SETTER, L.R. AND G.I. COATS. 1961. The Determination of Airborne Radioactivity, *AIHA J.* 22:64.

2. SCHULTE, H.F. 1976. "Radionuclide Surveillance," in *Air Pollution*, A. C. Stern, ed., 3rd ed., pp 414–451, Vol III. Academic Press, New York.

Subcommittee 8

B. SHLEIEN, *Chairman*
R.G. HEATHERTON
P.F. HILDEBRANDT
L.B. LOCKHART, JR.
R.S. MORSE
W. STEIN
D.F. VAN FAROWE
A. HULL, *Health Physics Society Liaison*

Approved with modifications from 2nd edition
Subcommittee 8

A.P. HULL, *Chairman Pro Tem*
J.A. BROADWAY
M.L. MAIELLO
R.P. MILTENBERGER

603.

Determination of the Iodine-131 Content of the Atmosphere (Particulate Filter Charcoal Gamma)

1. Principle of the Method

1.1 Radioactive iodine may be released to the atmosphere in the elemental state as I_2, in an inorganic combined state as HOI or CsI or in an organically combined state as CH_3I. All except CsI may be present in the vapor form or may be in part attached to particles. CsI would be expected to be present in a dissolved aqueous state following core damage at a water-cooled reactor and could be discharged to the atmosphere entrained in released steam, which then would condense into a particulate form. Thus, sampling for radioiodines in the atmosphere presents a unique problem.

The proposed method involves sampling for radioiodine in its solid and gaseous states with a high efficiency particulate prefilter in series with a chemically activated treated charcoal cartridge. If radiogases are also expected to be present (as would be the case following a reactor accident), the use of a silver-zeolite cartridge is recommended.

The collected radioiodine activity is then quantitively determined in the laboratory by gamma spectroscopy. Alternatively, it may be determined in the field using a GM survey meter. Iodine attached to solid particles and possibly some vaporous iodine will be deposited on the particulate filter, while the remaining vaporous iodine will be adsorbed on the charcoal or silver-zeolite cartridge.

Although other fission product radioiodines (^{132}I-^{135}I) may be also present in an atmospheric release from an operating reactor or during the first few days following a reactor accident, ^{131}I is dosimetrically the most significant form. So this method is principally directed toward its evaluation.

1.2 For routine environmental surveillance, samples may be collected continuously at low flow rates for periods of up to one or two weeks. In order to make a rapid evaluation of suspected or known incidents or accidents in which radioiodines may have been released to the atmosphere, samples may be collected at higher flow rates for periods as short as five minutes.

2. Sensitivity and Range

2.1 At a short-term sampling rate of 140 L/min (5 cfm), the minimum detectable concentration of nonparticulate ^{131}I by laboratory analysis of a five minute sample is about 2×10^{-10} Ci/m^3. By gross-gamma count with a GM survey meter, the minimum detectable concentration is about 1×10^{-8} Ci/m^3. For a one week sample at 28 L/min (1 cfm), the minimum detectable concentration is about 5×10^{-15} Ci/m^3.

2.2 While in principle the upper range is dependent on the state of the iodine and the absorption capacity of the collected medium, in practice it is limited by the dead time of the detector. This would become appreciable for samples containing in excess of 1 mCi of activity. However, this exceeds the amount likely to be collected in any environmental sample, even under the circumstances of a very large atmospheric release following a major reactor accident.

3. Interferences

3.1 The major interference may be the presence of other collected gamma emitting radionuclides, particularly on the particulate filters and when analyzed using a NaI detector. This effect can be diminished or eliminated by permitting a sufficiently long decay time (3–4 days) prior to counting.

Most gamma spectroscopy is currently performed with high-resolution GeLi detectors and computer assisted interpretation. When provided with an adequate library of potentially interfering nuclides, such a system is essentially interference-free.

3.2 Another interference may be due to adsorption of radioactive gases or vapors on the charcoal cartridge. This may be minimized by the use of silver-zeolite collecting medium (1). If radioactive gases are present in much larger concentrations than radioactive iodines, they may *still* interfere with the GM survey meter gross-gamma activity method.

4. Precision and Accuracy

4.1 A flow measurement accuracy of about 3% is obtainable with most commonly employed devices when properly calibrated.

4.2 A minimum detectable quantity of approximately 1.2 pCi of ^{131}I can be determined at the 95% confidence level by the employment of a well-shielded counting system and a sufficiently long counting interval (approximately 50,000 s).

4.3 With the employment of charcoal treated with KI or triethylene diamine (TEDA), the collection of all forms of non-particulate iodine will approach 100%.

4.4 A calibration accuracy of ± 5 to 25% is achievable. In the calibration of the cartridges, this depends on how well the distribution of ^{131}I in the calibration standard conforms to the actual distribution of ^{131}I on the sample cartridge.

4.5 For short sampling times, it may be assumed that collected radioiodines are on the upstream face of the cartridges. For longer sampling times, some migration of the iodine may occur, so that the assumption of a uniform distribution would be more conservative.

5. Apparatus

5.1 The sampling apparatus consists of a particulate prefilter, charcoal (or silver-zeolite) cartridge and an air sampling pump in series, associated holders and connections, and a flow measuring device. Suit-able air samplers are available from several vendors.

5.2 Prefabricated plastic or metal encased cartridges containing charcoal treated with either KI or TEDA (to assure the collection of organic iodine vapor) or containing silver zeolite are available from several commercial vendors. They correspond in size to the charcoal cartridges employed in face breathing masks for respiratory protection, which were initially adapted for use in air sampling for radioiodines.

Most of the commercial vendors of these cartridges can also supply suitably sized high-efficiency particulate prefilters and filter holders. Since there are several "standard" sizes, all about 2″ diameter × 1″ thick (5 × 2.5 cm), the user should be careful to purchase a match of particulate filter, cartridge and filter holder.

5.3 As indicated above, NaI detectors may be employed. However, in view of their much superior resolution of photopeaks, GeLi detectors are preferable. Although single-channel analyzers set to the 0.364 MeV photopeak of ^{131}I may be employed, current state-of-the-art procedures employ a multi-channel analyzer that can obtain information on other radioiodines or other radionuclides that may be present. Most laboratories also employ associated computer data reduction systems which incorporate libraries of nuclides to facilitate the interpretation of complex spectra.

Alternatively, for quick "field" evaluation, a survey meter with an enhanced-response cylindrical or pancake GM type detector may be employed (2).

7. Procedure

7.1 SAMPLING. Air is drawn by the sampling system into the filter holder air inlet. Sampling should be performed in a location free of major restrictions such as trees and structures. Permanent samplers should be located in protective structures. Air flow may be determined by use of a rotameter at the initiation and termination of the sampling period with a simple average being taken as the characteristic flow rate. An alternate method employs a dry gas meter inserted in the line. If the air flow measurement is made between the air sampler and the filtering media, correction must be made for the reduced and variable density of the gas passing through the meter (3). Vacuum readings may be taken with a gauge at the initiation and termination of the sampling period with a simple average being taken as the characteristic pressure drop.

For long-term sampling (one day up to one week), an air sampler that will provide a flow rate of about 28 L/min (1 cfm) should be employed. Due to the possibility of the appreciable buildup of atmospheric dust on the particulate filter, a positive displacement air sampler should be employed. Constant-flow samplers are available from several vendors.

For short-term sampling (5–10 minutes up to 1–2 hours), a greater flow rate of at least 140 L/min (5 cfm) is desirable.

7.2 COUNTING.

7.2.1 For equipment see paragraph **5.3**. If only ^{131}I ($t_{1/2}$ = 8.05 d) is known to be present, the use of a NaI detector and a single-channel analyzer will yield accurate results. However, the relatively short-lived radioiodines ^{132}I – ^{135}I should also be anticipated to be present in a sample of the effluents from an operating reactor or from one that has recently been shut down. If present in appreciable quantities or if the radioactive noble gases (Kr and Xe) have been adsorbed in the cartridge they will produce a background in the 0.364 MeV region of the principal ^{131}I photopeak.

7.2.2. Counting of the cartridge must be performed in a counting geometry approximating that used for calibration.

7.2.3 While the predominant ^{131}I photopeak at 0.364 MeV may be employed for ^{131}I quantification, most computer-assisted systems include comprehensive libraries of nuclides which enable them to identify other radioiodines (^{132}I-^{135}I) as well as other nuclides that may be present on the particulate sample or the cartridge.

8. Calibration

8.1 Calibration for ^{131}I is achieved by counting a ^{131}I standard (periodically available from the NBS or from commercial vendors) in the same geometry in which the particulate sample or the cartridge is to be counted. For the computer assisted interpretation of complex spectra, a number of standards of nuclides of various energies are employed and counting efficiencies for other nuclides determined by interpolation. Standardized samples of ^{133}Ba, which has a prominent photopeak at 0.356 MeV and a half-life of 1.05 years, are sometimes employed as a substitute for ^{131}I.

8.2 For the particulate prefilter calibration, liquid standards of ^{131}I (and other nuclides) should be distributed as uniformly as possible on the particulate filter medium.

8.3 For the calibration of charcoal or silver-zeolite cartridges, it may be assumed that only radioiodines have penetrated the particulate filter to any extent. Their distribution within the cartridge will affect the calibration and may vary, depending on the length of the sampling period, their amount and chemical state, the flow rate, and the relative humidity. For short sampling periods, it may be assumed that most of the radioiodines have been adsorbed close to the upstream surface of the cartridge. For longer sampling periods, it is more conservative to assume a uniform distribution.

8.4 To prepare a calibration standard for the charcoal or silver zeolite cartridge, a particulate filter standard prepared as in 8.2 above should be placed within a container of the same dimensions as the cartridge at a depth of $^{1}/_{8}''$ (3 mm) in the collecting medium to simulate surface collection, or midway to simulate uniform collection in the medium.

9. Calculation

9.1 If in addition to ^{131}I the collected sample contains a mixture of radioiodines or other radionuclides and if their immediate quantification is desired, then it should be analyzed in a laboratory equipped with a multi-channel analyzer and associated computer interpretation capability that incorporates a library of the major nuclides expected to be present in the sample.

The associated analysis programs usually provide for the user input of variable parameters such as flow rate or total flow, sampling time, detector efficiency, and time of analysis. They also provide outputs that include the necessary corrections for radioactive decay between the end of the sampling period and time of analysis and statistical error in the calculated activities or concentrations of the identified nuclides in the sample.

9.2 If the user's interest is solely in ^{131}I and there is no time pressure for results, single-channel analysis and decay measurements of the ^{131}I photopeak can be used to determine ^{131}I in the presence of long-lived gamma emitters without resorting to more complex techniques. The filter and/or cartridge is counted at appropriate time intervals (every 2 to 4 days is sufficient). The observed counting rate (counts/min) at each counting time under the ^{131}I photopeak (approximately from 0.32 to 0.40 MeV) is plotted on the logarithmic axis of semi-log paper, while the time elapsed since the end of collection is plotted on the arithmetic axis. Connect the points with a straight line. The line should have a slope indicating the half life of ^{131}I (8.05 days). Extend the line to the counting rate axis. The point at which the line intercepts the logarithmic axis indicates the counting rate due to ^{131}I at the end of collection.

9.3 In order to correct for interfering radionuclides having energies greater than 0.364 MeV, a method of "background" (Comptom continuum) subtraction may be used **(4)**. A second single channel analyzer is set for an equivalent width region just above 0.40 MeV to establish the base line counting rate. Sum up the observed counting rate under the photopeak (approximately from 0.32 to 0.40 MeV) and subtract the sum of the base line counting rate over the same width energy interval to obtain the net counting rate due to ^{131}I. The activity is then determined by dividing this net counting rate by the counting efficiency factor (net counting rate of the standard divided by the disintegration rate of the standard).

9.4 For manual calculations, the activity per unit volume of air at the time of counting (A_c) is determined as follows:

$$A_c = \frac{s \times A_{STD}}{STD \times T \times r} \qquad (1)$$

where s(cpm) = net gamma counting rate of sample (total sample counting rate minus background counting rate) for 0.364 MeV photopeak
STD(cpm) = net gamma counting rate

of standard (total standard counting rate minus background counting rate) for 0.364 MeV photopeak.
A_{STD} = activity of standard (pCi)
T = total sampling time (min)
r = average volume sampling rate (m^3/min)
A_c = activity of sample at time of count (pCi/m^3)

9.5 Correct for activity at the end of sampling (A_1) due to decay of ^{131}I as follows:

$$A_1 = \frac{A_c}{e^{-\lambda t_1}} \qquad (2)$$

where e = 2.72
t_1 = time between end of sampling and counting (days)
λ = 0.0858 day^{-1} (^{131}I decay factor)

If ^{131}I is determined by the decay method, indicated in 9.2, the activity at end of sampling may be directly discerned from the resultant graph.

9.6 If the sampling period is a significant portion of the half-life of ^{131}I and its concentration was uniform throughout the sampling period, a correction can be made for the fraction of the sample activity which has decayed during the sampling period as follows:

$$A = A_1 \frac{t\lambda}{e^{-\lambda t}} \qquad (3)$$

where t = sampling time (days)

This correction may be incorporated into the computer analysis program.

9.7 The statistical significance of the result should be indicated. It can be determined from the standard deviation (σ) of the count rate as follows:

$$\sigma(cpm) = \qquad (4)$$

$$\pm \sqrt{\frac{gcr}{sct} + \frac{bcr}{bct}}$$

where σ = Standard deviation of the net counting rate.
gcr = gross counting rate (cpm)
bcr = background counting rate (cpm)
sct = sample count time (min)
bct = background count time (min)

The statistical significance of a count should be expressed as the uncertainty (or error) of the count at a specified confidence level, such as 95% (1.96σ). This may be established by the computer program.

9.8 Depending on their sophistication, computer-assisted data interpretation programs are able to make most, if not all of these calculations on the basis of input parameters such as sample times, sample flow rate, and counting efficiency.

10. Effect of Storage

10.1 A properly protected sample should experience no effects other than radioactive decay during storage.

11. References

1. CLINE, J. E. 1981. Retention of Noble Gases by Silver-Zeolite Cartridges. *Health Physics* 40:61-83.
2. FEDERAL EMERGENCY MANAGEMENT AGENCY. 1980. "Guidance on Off-site Emergency Radiation Measurement Systems: Phase 1, Airborne Systems," FEMA-REP-2, September.
3. CRAIG, D. K. 1971. The Interpretation of Rotameter Air Flow Reading, *Health Physics* 21:328-332.
4. CLINE, J. E. 1980. Methods for the Analysis of Radioactive Sample Cartridges for ^{131}I (Content), Lecture Notes, Science Applications, Inc., Rockville, MD, 20850.

Subcommittee 8

A. P. HULL, *Chair (pro tem)*
J. A. BROADWAY
M. L. MAIELLO
R. P. MILTENBERGER

606.

Determination of Radon-222 Content of the Atmosphere

Part A: Estimation of Airborne Radon-222 by Filter Paper Collection and Alpha Activity Measurement of Its Daughters

Part B: Determination of Airborne Radon-222 by Its Adsorption on Charcoal and Gamma Measurements

1. Introduction

1.1 A summary of the techniques presently used for measurement of airborne radon is presented in this Section. These methods are useful in determining the potential hazard of radon in certain occupational environments, in evaluating the radiation hazard of ambient radon to general population groups, and in atmospheric tracer studies involving radon.

1.2 Interest in naturally occurring radioactivity arises from the fact that it has always been part of man's radiation environment and thus may serve as a standard against which radiation hazards from other sources can be compared. Studies of gaseous radon*, together with its particulate daughter products, provide a source of natural radiation data with which to evaluate this inhalation exposure to general populations from technically-enhanced, artificially created radioactivity.

1.3 Following the identification in 1984–85 of unusually high levels of radon in homes located in the Reading Prong geological region of Eastern Pennsylvania, indoor radon has been recognized as a major contributor to the background radiation

dose from naturally occurring radioactivity. In its 1987 report on the exposure of the U.S. population to ionizing radiation, the National Council of Radiation Protection estimated that the average concentration of radon in U.S. homes is about 1 pCi/L, (37 Bq/m³) which results in an equivalent whole body dose of 200 mrem/yr. (2 mSv/a) (1). This makes it the largest single contributor from natural background to the total radiation dose to the U.S. population.

1.4 The Environmental Protection Agency (EPA) has established a guide for indoor radon of 4 pCi/L, above which remedial action is recommended (2). Surveys have been conducted in several regions of the U.S., from which it now appears that about 20% of the homes in the U.S. may exceed EPA's remedial action standards (3). The identification of housing with elevated radon concentrations has become a matter of public health concern. As a consequence of this concern, studies of decontamination problems involving radon control and hazard evaluation have been initiated. In addition to studies related to health effects, radon may also be used as an atmospheric tracer in meteorological, geophysical and air pollution studies.

1.5 Radon is the immediate daughter product of radium-226, which is intermediate in the fourteen steps in the natural uranium-238 decay chain. It is ubiquitous in the terrestrial environment. Gaseous radon atoms, released to the atmosphere by emanation from soil, decay to lead, bismuth and polonium daughter atoms. The decay series for radon is presented in Figure 606:1.

1.6 Outdoor concentrations for radon and radon daughters in a given locale vary from season to season, day to day and even

---

*Radon in these methods refers to the isotope ^{222}Rn and is not to be confused with other radon isotopes; ^{220}Rn (thoron) or ^{219}Rn (actinon).

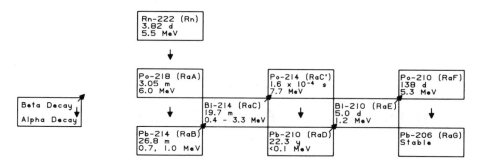

Figure 606:1 — Decay series of radon 222.

hour to hour, depending principally on meteorological conditions. Beyond this, indoor concentrations also depend on building materials and ventilation rates, and may vary from room to room within a structure.

1.7 For the estimation of the effective dose equivalent due to the inhalation of ^{222}Rn and its daughters, it is necessary to measure their mean concentrations over an extended period of time. While in principle this can be estimated from a series of grab samples or short-term measurements, methods that integrate over the same period of time are preferable. This is especially so for the screening residences with regard to the EPA remedial action levels.

2. Measurement of Indoor Radon

2.1 Over the past few years, practical methods have evolved for monitoring radon levels in occupied residences. Simple, unobtrusive devices placed in the home can collect samples for analysis at a central laboratory. Two types of passive monitoring devices have come into widespread use: the alpha-track detector and the charcoal canister. Both of these methods are called "passive" because they require no pumps or other powered systems to collect radon. Instead, the radon is collected through natural diffusion.

2.2 The passive alpha-track detector uses a small section of special film that is sensitive to alpha radiation. This film resides in a container that physically defines the volume for measurement and screens out interferences. As the radon and its progeny decay, alpha particles are emitted

that create damage tracks on the film. Calculating the track density (number of damage tracks per unit area) gives the average radon concentration over the sampling period.

2.3 Because exposure is a function of both concentration and time, it is usually defined in terms of picocuries of radon per liter of air averaged over a day or a month: (pCi/L) · d or (pCi/L) · mo. Brief samples taken at high concentrations and extended samples taken at low concentrations can yield the same results. For example, an average concentration of 1 pCi/L for one month would give the same track density as 0.5 pCi/L for two months, or any other combination totaling one (pCi/L) · mo.

2.4 As indicated in Table 606:I, the statistical uncertainty of these measurements is inversely proportional to the concentration. For radon concentrations at or above 4 (pCi/L) · mo that are monitored for at least three months, or 12 (pCi/L) · mo, the uncertainty is never more than about 10%, and it becomes less as the concentration or the sampling period increases. This uncertainty can also be reduced by using a larger area for counting tracks, but that increases counting costs.

2.5 Passive alpha-track detectors have a detection range from 0.4 to 50,000 (pCi/L) · mo. The detectors, which are available only from commercial laboratories, are generally used for sampling periods of one month to one year. Screening a house for radon usually requires one to three samples, depending on the number of floors and the presence of a basement. The cost of

Table 606:I Statistical uncertainty (one standard deviation) of radon measurements[+]

Radon Concentration (pCi/L)	Passive Charcoal Canister %[*]		Passive Alpha-Track Detector %[***]	
	3 days	7 days[**]	90 days	1 year
0.5	32	46	24 to 32	10 to 14
1	17	24	15 to 21	7 to 10
4	5	7	7 to 10	3.5 to 5
10	2.3	3.2	4.5 to 6	2.2 to 3.1
30	1.2	1.5	2.6 to 3.6	1.3 to 1.8

[*] Charcoal canisters tested at GEOMET Technologies, Inc. under RP2034–1.
[**]Diffusion barrier charcoal adsorption collector developed at the University of Pittsburgh (B. L. Cohen, "Comparison of Nuclear Track and Diffusion Barrier Adsorption Methods for Measurement of Rn–222 Levels in Indoor Air," *Health Physics* Vol. 50, No. 6, pp. 828–829, June, 1986).
[***]Calculations based on performance specifications of commercially available devices.
[+] Data from EPRI Technical Briefs, EMU-101, February, 1977. Electric Power Research Institute, Palo Alto, CA, 94303.

using the alpha-track method for radon detection is from $18 to $25 per sample.

2.6 The charcoal canister is the newest development in passive radon monitoring. Filled with activated charcoal, the container has a lid that can be removed to expose the charcoal to the atmosphere. Radon diffuses into the charcoal and is adsorbed. At the end of the sampling period, the lid is replaced and the sealed canister is returned to the laboratory where the sample is analyzed for gamma radiation emitted by radon progeny in the canister.

2.7 Measured gamma activity is related to the amount of radon trapped in the canister. The amount of trapped radon is related to the length of the sampling period, as well as to the time elapsed since its end. Working backward, the analysis uses known decay rates to determine the average radon concentration during the sampling period.

2.8 Optimum sampling periods range from one to seven days. Practical measurement uncertainty at 4 pCi/L is generally 10% or less, provided that the sampling period does not exceed one week. In contrast with the alpha-track detector, the charcoal canister's performance declines over extended sampling periods.

2.9 Lower limits of detection for a four-day charcoal canister sample fall between 0.1 and 0.2 pCi/L. A number of commercial and research laboratories offer sampling services.

3. Purpose

3.1 The purpose of this Section is to provide an investigator with a broad outline of methods employed in the determination of airborne radon, as well as their applications and limitations, so there may be a basis for choosing one method for a particular application. At the same time, detailed descriptions for two of the more commonly employed procedures are presented that can serve as a guide for those charged with routine determination of radon air concentrations.

3.2 It is beyond the scope of this method to describe all analytical techniques for airborne radon determination in detail. For example, microcomputer-based portable grab samplers are commercially available. Some may be operated in a quasi-continuous mode. However, they are principally intended for the evaluation of short-term concentrations in the industrial setting or the investigation of unusually high indoor levels in homes. Their expense and labor requirements makes them unsuitable for the routine measurement of long-term average concentrations.

3.3 Choice of the method to use in determining radon air concentrations from the various techniques available depends on: (a) the level of radon concentration to be measured, (b) the accuracy required, (c) the equipment available, and (d) complexity of the technique. Table 606:II summa-

Table 606:II. Methods of ^{222}Rn measurement

Collection Mode	Theoretical Basis	Counting Method	Principle	Sensitivity	Application
I. Filter paper	Particulate daughter products collected on filter paper and their activity related to radon concentration	Alpha (2,18,20, 24,26,31-32,36,38)	Filter collection for 5-10 min.	0.1 pCI/L	Grab sample
		Beta (25)	Filter collection for 5-10 min.	0.1 pCi/L	Grab sample
		TLD (3,16,33)	Filter collection for 1-2 weeks	0.01 pCi/L	Integrating
		Surface Barrier (21)	Filter collection measurement with silicon surface barrier alpha detector	0.001 pCi/L	Integrating
II. Passive diffusion into sensitive volume		Plastic (10,27-28)	Direct radon and thoron daughter tracks in alpha sensitive plastic (which register tracks of radon and thoron daughters)	0.1pCi/L	Integrating
		Electret collection TLD (21)	TLD measurement of radiation from collected daughters	0.03 pCi/L	Integrating
		Plastic (17)	Alpha track measurement of radiation from collected daughters	100 (pCi/L) · day	Integrating
III. Entrapment of air containing radon	Radon and daughters not isolated from air	A. Scintillation cell (18,23)	1. Scintillation Alpha Counting	<0.1-1.0 pCi/L	Grab sample
		(4,27,35,37)	2. Flow-through scintillation alpha cell	<0.1-1.O pCi/L	Continuous monitoring
		B. Ionization chamber (11)	Sample transferred into ion chamber. Pulse or current counting	<0.05 pCi/L	Grab sample
IV. Adsorption on activated carbon	Gaseous radon collected by adsorption from atmosphere by passage over activated charcoal	Gamma (5,7,13-14)	Adsorption on activated charcoal Gamma count for ^{214}Pb and ^{214}Bi gamma rays	0.2 pCi/L for 100-h exposure	Integrating

rizes suggested methods together with comments on each method. Those methods which are completely described below offer the advantages of simplicity, rapidity and easy field application and/or the advantages of high sensitivity and lack of dependence on radon and radon-daughter equilibrium.

Part A: Estimation of Airborne Radon-222 by Filter Paper Collection and Alpha Activity Measurements of Its Daughters (Thomas Method or Modified Kusnetz Method)

1. Principle of Method

1.1 This method is based on the collection of the short-lived radon daughters whose activity is then related to the parent gaseous radon concentration. The alpha activity of ^{214}Po (Ra C′) and to a lesser extent that of ^{218}Po (Ra A) is measured. Assuming either equilibrium conditions or a specific disequilibrium factor between radon and its short-lived daughters, the concentration of the parent radionuclide, ^{222}Rn, can be calculated. A factor of 0.5 is widely utilized for indoor radon. However, 0.4 is being used increasingly as more realistic. Depending upon sensitivity requirements, collection times can be as short as 6 min. This method is simple, rapid and inexpensive.

2. Range and Sensitivity

2.1 LOWER RANGE. Assuming a 5-min sampling period, a sampling rate of 5 L/min, a counter efficiency of 0.30, a counter background of 1 cpm, a counting time of 30 min and a delay of 2 min from sampling to start of counting, the lower detection limit for this method is approximately 1.0 pCi/L.

2.2 UPPER RANGE. Not Limited.

3. Interferences

3.1 The method assumes that only ^{218}Po (Ra, A), ^{214}Pb (Ra B) and ^{214}Bi (Ra C) are present, that their concentrations do not change during the sampling period and that the counter has the same efficiency for the ^{218}Po and ^{214}Po alphas.

3.2 Other long-lived alpha-emitting radionuclides, such as the daughters of ^{220}Rn, thoron, have to be corrected for by recounting 4 h later. ^{210}Po, $^{239, 240}$Pu, $^{235, 238}$U and ^{226}Ra are usually present in low enough concentrations that they offer minimal interference with the determination of ^{222}Rn.

4. Precision and Accuracy

4.1 A precision of 10% at the 95% confidence level is obtainable due to counting errors. Flow measurements should be correct to ± 2%.

5. Apparatus

5.1 COLLECTION. A vacuum pump with a sampling rate of 28–56 L/min (1 to 2 cfm), a 25 or 47 mm diameter membrane filter with a 0.8 μm pore size, a flow rate meter and filter holder.

5.2 ALPHA COUNTING EQUIPMENT. The scintillation counter consists of a silver-activated zinc sulfide screen mounted on a 2.54 cm (1″) or 5.08 cm (2″) photomultiplier tube, preamplifier, high voltage power supply, scaler, timer and a light-tight counting enclosure. This type of counter is commercially available. Typical counter backgrounds range from 0.05–0.5 counts/min depending on detector size and contamination-free construction material. Counter efficiency should be determined with a suitable standard spread over a membrane filter or a plated source of the same geometry as the collected sample. Efficiencies of 40–50% are achievable.

5.3 Alternate counting systems to the one described above employ either an internal gas flow proportional counter or a solid state surface barrier detector. These counting systems are not generally applicable for field use but are employed in a laboratory environment.

6. Reagents

6.1 None.

7. Procedure

7.1 Sample for collection period of 5 min. Although it is in principle practicable to sample for 4 h or longer so that radon daughter decay on the filter and radon daughter buildup on the filter will be in equilibrium, the convenience of this method is that it allows for a short sampling period.

7.2 Remove sample from filter holder. Record sampling rate (L/min), duration of sampling, and the time of sample removal.

7.3 Count the sample for the time intervals specified in Section **9**.

7.4 Recount the sample after the decay of the radon daughters, about 4 h after end of sampling. The second count represents activity of longer-lived alpha emitters (see Interference, Section **3**).

8. Calibration

8.1 The alpha counter should be calibrated with a standard such as ^{242}Cm of known disintegration rate distributed over a geometric area similar to the area of the air filter sample. The "counting efficiency" is the ratio of the count rate of the standard to the known disintegration rate. Suitable standards are commercially available and should be traceable to the National Bureau of Standards.

9. Calculations

9.1 The total count method for determining the air concentrations of the ^{218}Po, ^{214}Pb, and ^{214}Po daughters of ^{222}Rn requires counting in three different time intervals after the end of counting. The number of counts in the time intervals, 2–5 min, 6–20 min, and 21–30 min, counter efficiency, flow rate, and sampling time, are used for calculating the concentrations of the three daughters. For a sampling time of 5 min, the following equations apply:

$$C\,^{218}Po = \frac{1}{VE}\,[0.1689\,G(2,5) - 0.0820\,G(6,20) + 0.0775\,G(21,30) - 0.0566R]$$

$$C\,^{214}Pb = \frac{1}{VE}\,[0.0012\,G(2,5) - 0.0206\,G(6,20) + 0.0491\,G(21,30) - 0.1575R]$$

$$C\,^{214}Bi = \frac{1}{VE}\,[0.0225\,G(2,5) - 0.0332\,G(6,20) - 0.0377\,G(21,30) - 0.0576R]$$

E = Counter efficiency; G = Number of alpha counts during time periods specified (uncorrected); V = Sampling flow rate (L/min); R = background count; in the case of samples with long-lived alpha emitters, the count 4 h after the end of sampling (Section **7.4**). Then, to a good approximation, for indoor air, $C^{222}Rn = 0.103\,C^{218}Po + 0.507\,C^{214}Pb + 0.373\,C^{214}Bi$ where all concentrations, C, are in pCi/L **(28,37)**

10. Effect of Storage

10.1 Not applicable.

Part B: Determination of Airborne Radon-222 by Its Adsorption from the Atmospheric and Gamma Measurement

1. Principle of the Method

1.1 Gaseous radon, quantitatively adsorbed on activated carbon (usually coconut charcoal) is counted on a gamma spectrometer to establish the activity of the ^{214}Bi and ^{214}Pb daughters. The count rate of the 0.61 MeV ^{214}Bi and 0.295 and 0.35 MeV ^{214}Pb peaks are determined. The efficiencies for counting these peaks are sufficient to detect ^{214}Bi and ^{214}Pb from their quantities that have reached equilibrium with environmentally collected radon.

There are two ways of employing charcoal-based collection devices. In the first and most popular, radon is allowed to diffuse into the activated carbon canister passively by molecular diffusion. Sampling periods of from 1 d to 1 wk may be employed. However, radon may diffuse out from the canister with a reduction of the ambient concentration. The EPA recommends a period of not more than 48 h. In the second way, an activated carbon cartridge is placed in an air sampling stream and up to 1 L of air is sampled for each gram of activated carbon in the cartridge.

2. Range and Sensitivity

2.1 This method is useful in field analysis of radon. Measurement is normally performed in a laboratory after collection. It is not dependent upon radon-radon daughter equilibrium in the atmosphere. Concentrations lower than 0.1 pCi/L of radon can be determined by the active method and from 0.2–0.3 pCi/L by the passive method.

3. Interferences

3.1 Charcoal is an excellent adsorbent of many gases and of water vapor. Its adsorption of Rn at high humidities is about half of that at low humidity. To minimize this effect, charcoal used for radon sampling should be dried before use and should be sealed from the atmosphere except during sampling.

While charcoal used in the passive mode has obvious advantages over a pump-driven system, it is subject to back- as well as forward-diffusion of radon. This may lead to large errors when used over extended sampling periods during which there has been a large change in radon concentrations.

The back-diffusion rate increases and, therefore, the effective sampling rate decreases as a function of temperature. Therefore, the effective sampling rates of a passive sampler should be established at the anticipated temperature at which they will be exposed. Back-diffusion may be minimized at the cost of sensitivity by the employment of a diffusion barrier between the charcoal and the atmosphere being sampled.

4. Precision and Accuracy

4.1 When standard calibration factors are employed with corrections for humidity, a concentration of 0.2–0.3 pCi/L can be determined with an accuracy of ± 20%.

4.2 With the active method, 0.1 pCi/L radon can be determined with an error of 10% at the 95% confidence level.

5. Apparatus

5.1 Investigators have fabricated passive activated-carbon monitors in different sizes and configurations. The passive radon charcoal canister used by the EPA consists of the following materials: (1) 8-ounce (236 mL) metal can with lid (4 in diameter by 1 1/8 in deep, 10.2 × 2.86 cm) (2) 70 ± 1 g of 6 × 16 mesh PCB activated charcoal, (3) metal screen with an openness of at least 30 to 50%, (4) removable, internally expanding retaining ring, (5) pad material attached to the inner surface of the lid, and (6) a 13-in (33-cm) strip of pliant vinyl tape. The materials are assembled as seen in Figure 606:2. Prefabricated charcoal canisters for passive radon sampling are available from several vendors.

5.2 The active method requires a vacuum pump with a sampling rate of 1-2 L/min, a flow rate meter, and an activated carbon cartridge, 100-g capacity. Total volume collected should not exceed 100 L.

5.3 A gamma-ray spectrometer may be coupled to either a scintillation NaI (Tl) detector or a solid-state Ge(Li) detector. While better detection capability is normally obtained with a solid-state detector, this may be offset by its limited size, relative to the geometry of practicable charcoal canisters.

The following components make up the counting system used by the EPA's Eastern Environmental Radiation Facility (4): (1) 3-in by 3-in (7.6-cm by 7.6-cm) sodium iodide detector and photomultiplier tube inside counting shield, (2) high voltage power supply, (3) pre-amplifier and amplifier, (4) single-channel analyzer and scaler, (5) timer, (6) time-of-day clock, and (7) printer.

Four sodium iodide detectors are served by a single high voltage power supply, timer, printer, and time-of-day clock. Each detector is used in conjunction with its own pre-amplifier, amplifier, single-channel analyzer, and scaler. Four detectors are arranged together inside a shield which has steel walls approximately 8 in (20 cm) thick. Within the shield the detectors are separated from each other by 5 in (13 cm) of lead. A wooden jig is used on each detector to assure consistent counting geometry for the canisters on the detector.

6. Reagents

6.1 None.

7. Procedure

7.1 The activated carbon collector devices are counted with a NaI or GeLi detector and a gamma spectrometer. Following sampling, at least 4 h should elapse prior to counting to allow for the daughter to reach equilibrium with collected radon. A correction for the decay of radon from the midpoint of the sampling period to the time of counting should be applied.

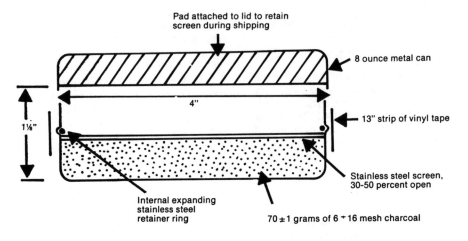

Figure 606:2 — Charcoal canister assembly from Reference 15.

8. Calibration and Standards

8.1 EFFECTIVE SAMPLING RATE. For the passive method, it is necessary to establish an "effective sampling rate" (in L/min). It is based on the exposures of samples of charcoal to a constant concentration of radon over different periods of time under controlled conditions of temperature and humidity in laboratory exposure chambers. The effects of humidity are established by making runs for different sampling times at various relative humidities. This can be related to the amount of water adsorbed by weighing the canister before and after exposure.

The basic equation for the effective sampling rate is

$$S_{eff} = \frac{net\ cpm}{t_s \cdot E \cdot Rn \cdot \exp{(-0.693\ t/5501)}}$$

where S_{eff} = effective radon absorption rate (L/min)

t_s = exposure time of canister (min)

E = detector efficiency (cpm/pCi)

Rn = radon concentration (pCi/L)

t = time from the mid-point of sampling to the time of counting (min).

8.2 DETECTOR CALIBRATION. To calibrate the detector, a known quantity of NBS certified radium solution may be added to the charcoal in a sampling canis-

ter, which should then be sealed. Following a sufficient time for radon (and its immediate daughters) to achieve equilibrium, the canister may be placed on the detector and the peaks from the ^{214}Bi or the ^{214}Pb integrated. The count-rate is determined, counter background subtracted, and the net count rate is divided by the amount of ^{222}Rn (equal to the ^{226}Ra) in the sampling device to establish the efficiency for the specific counting geometry.

9. Calculations

9.1 The activity of radon in the sample, counted 4 h after the end of sampling so that correction for disequilibrium need not be made, is calculated as follows:

Rn in pCi/L =

$$= \frac{net\ cpm\ under\ 0.61\ MeV\ peak}{counting\ efficiency\ (0.61\ MeV)\ (2.22)\ volume\ (L)\ \exp{(-0.693t/5501)}}$$

$$= \frac{net\ cpm\ under\ 0.295\ MeV\ peak}{counting\ efficiency\ (0.295\ MeV)\ (2.22)\ volume\ (L)\ \exp{(-0.693t/5501)}}$$

$$= \frac{net\ cpm\ under\ 0.35\ MeV\ peak}{counting\ efficiency\ (0.35\ MeV)\ (2.22)\ volume\ (L)\ \exp{(-0.693t/5501)}}$$

where t = elapsed time from midpoint of sampling period (min).

Note that the sampled volume is the product of S_{eff} and t_s for the conditions of the sample.

Calibration data, as obtained by EPA, for varying exposure times and relative humidities are shown in Figure 606.3. For more accurate results, calibration factors may be established as a function of the weight of water absorbed by a cartridge for a standardized sampling period. That for the 48-h sampling period utilized by EPA is shown in Figure 606.4.

Governmental agencies have access to facilities in which charcoal canisters can be exposed to known concentrations of radon over a range of relative humidities and/or temperatures in exposure rooms that have been constructed at DOE's Environmental Measurement Laboratory in New York

City and at EPA's Eastern Environmental Radiation Facility of Montgomery, AL. Nongovernmental entities may purchase similar exposure services from EG&G, Mound Applied Technologies, P.O. Box 32, Miamisburg, OH, 45342, or Radon QC, 1528 Don Gaspar, Santa Fe, NM, 87505. For the passive method, the volume is the product of the effective sampling rate and the elapsed sampling time. The effective sampling rate should be that for the exposure time and relative humidity (as established from the amount of collected moisture, see Figure 606.4).

References

1. ALTER, H. W. AND R. L. FLEISCHER, 1981. Passive integrating radon monitor for environmental monitoring. *Health Phys.* 40:693.
2. BIGU, J., R. RAZ, K. GOLDEN AND P. DOMINQUEZ, 1983. A computer-based continuous monitor

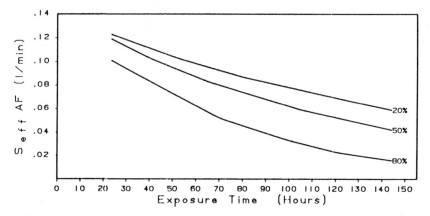

Figure 606:3 — Exposure time versus effective sampling rate (S_{eff}) for low, medium and high humidity from Reference 15.

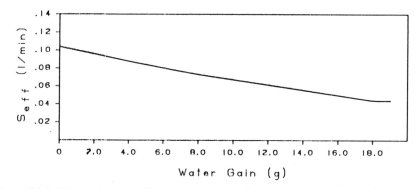

Figure 606:4 — Water gain versus effective sampling rate (S_{eff}) for a two day exposure from Reference 15.

for the determination of the short-lived decay prod-ucts of radon and thoron. Canada Centre for Mineral and Energy Technologies, Division Report MRP/MRL 83 (OP)J.

3. BRESLIN, A. J., S. F. GUGGENHEIM, A. C. GEORGE AND R.T. GRAVESON, 1977. A working level dosimeter for uranium miners. U.S. DOE Report EML-333.

4. CHITTAPORN, P., M. EISENBUD AND N. HARLEY, 1981. A continuous monitor for the measurement of environmental radon. Health Phys. 41:405.

5. COHEN, B. L. AND E. S. COHEN, 1983. Theory and practice of radon monitoring with charcoal adsorption. Health Phys. 45:501.

6. COHEN, B. L. AND NASO, R., 1986. A diffusion barrier charcoal adsorption collector for measuring Rn concentration in indoor air, Health Phys. 50:47.

7. COUNTESS, R. J., 1976. Radon flux measurement with a charcoal canister. Health Phys. 32:455.

8. EPA. 1986. A Citizens Guide to Radon: What It Is and What to Do About It. USEPA Public Information Center, Mail Code PM-211B, 820 Quincy St., NW, Washington, DC 20011.

9. EPA. 1987. Press briefing, August, 4, 1987. (See Environmental Reporter, August 6, 1987.)

10. FLEISCHER, R. L., 1980. Dosimetry of environmental radon, methods and theory for low-dose integrated measurements. G.E. Research and Development Center, unpublished report.

11. FISENNE, I. M. AND H. KELLER, 1985. The EML pulse ionization chamber systems for ^{222}Rn measurements. U.S. DOE Report EML-437.

12. GEORGE, A. C., 1976. Scintillation flasks for the determination of low-level concentrations of radon. Proceedings of Ninth Midyear Health Physics Symposium.

13. GEORGE, A. C., 1984. Passive integrated measurement of indoor radon using activated carbon. Health Phys. 46:867.

14. GEORGE, A. C. AND A. J. BRESLIN, 1977. Measurement of environmental radon with integrating instruments. Workshop on Methods for Measuring Radiation In and Around Uranium Mills. Atomic Industrial Forum, Inc., Vol. 3, No. 9.

15. GRAY, D. J. AND S. T. WINDHAM, 1987. EERF Standard Operating Procedures for Radon-22 Measurement USEPA Eastern Environmental Radiation Facility, EPA 520/5-87-005.

16. GUGGENHEIM, S. F., A. C. GEORGE, R. T. GRAVESON AND A. J. BRESLIN, 1979. A time-integrating environmental radon daughter monitor. Health Phys. 36:452.

17. IIDA, T., Y. IKEBE, T. HATTORI, H. YAMANISHI, S. ABE, K. OCHIFUJI AND S. YOKOYAMA. 1988. An electrostatic integrating ^{222}Rn monitor with cellulose nitrate film for environmental monitoring. Health Phys. 54:139-148.

18. JONASSEN, N. AND E. I. HAYES, 1974. The measurement of low-concentrations of the short lived radon-222 in the air by alpha spectrometry. Health Phys. 26:104.

19. KOTRAPPA, P., J. C. DEMPSEY, J. R. HICKEY AND L.R. STIEFF. 1988. An electret passive environmental ^{222}Rn monitor based on ionization measurement. Health Phys. 54:47.

20. KUSNETZ, H. L. 1956. Radon daughters in mine atmospheres. A field method for determining concentrations. Am. Ind. Hyg. Assoc. J. 17:85.

21. LATNER, N., S. WATNICK AND R. T. GRAVESON, 1981. Integrating working level monitor EML type TF-11. U.S. DOE Report EML-389.

22. LUCAS, H. F., 1957. Improved low-level alpha scintillation counter for radon. Rev. of Sci. Instrum. 28:680.

23. MAIELLO, M. L. AND N. H. HARLEY, 1987. Egard: An environmental gamma ray and ^{222}Rn detector, Health Physics 53:3, 301-305.

24. MARTZ, D. E., D. E. HOLLEMAN, D. E. MCCURDY AND K.J. SCHIAGER, 1969. Analysis of atmospheric concentrations of RaA, RaB and RaC by alpha spectroscopy. Health Phys. 17:131.

25. MCDOWELL, W. B., D. J. KEEFE, P. G. GROOR AND R.T. WITELE, 1977. A microprocessor assisted calibration for a remote working level monitor. IEE Trans. Nucl. Sci. NS-24:1.

26. NAZAROFF, W. W., 1984. Optimizing the total-alpha three count technique for measuring concentrations of radon progeny in residences. Health Phys. 46:395.

27. NAZAROFF, W. W., F.J. OFFERMAN AND A. W. ROBB, 1983. Automated system for measuring air exchange rate and radon concentration in houses. Health Phys. 45:525.

28. NCRP 1984. Evaluation of Occupational and Environmental Exposures to Radon and Radon Daughters in the United States. NCRP Report NO. 78.

29. NCRP 1987. Exposure of the Population of the United States and Canada from Natural Background Radiation. NCRP Report No. 94.

30. PRICHARD, H. M. AND K. MARIEN, 1985. A passive diffusion ^{222}Rn sampler based on activated charcoal adsorption, Health Phys. 48:797.

31. RAABE, O. G. AND M. E. WRENN, 1969. Analysis of the activity of radon daughter samples by weighted least squares. Health Phys. 17:593.

32. ROLLE, R., 1972. Rapid working level monitoring. Health Phys. 22:233.

33. SCHIAGER, K. J., 1974. Integrating radon progeny air sampler. Am. Ind. Hyg. Assoc. J. 35:165.

34. SCOTT, A. G., 1981. A field method for measurement of radon daughters in air. Health Phys. 41:403.

35. SPITZ, H. AND M. W. WRENN, 1977. Design and application of a continuous digital-output environmental radon measuring instrument. U.S. ERDA Report HASL-325.

36. THOMAS, J. W., 1972. Measurement of radon daughters in air. Health Phys. 23:783.

37. THOMAS, J. W. AND R. J. COUNTESS, 1979. Continuous radon monitor. Health Phys. 36:734.

38. TSIVOGLOU, E. C., H. E. AYER AND D. A. HOLLADAY, 1953. Occurrence of nonequilibrium atmospheric mixtures of radon and its daughters. Nucleonics 1:40.

39. URBAN, M. AND E. PIESCH, 1982. Low level environmental radon dosimetry with a passive track etch device. Radiation Protection Dosimetry 1:97.

Subcommittee 8

A.P. HULL, Chairman Pro Tem
A. BROADWAY
L. MAIELLO
P. MILTENBERGER

609A.

Determination of Tritium Content of Water Vapor in the Atmosphere

1. Principle of the Method

1.1 Air is pumped at a sampling rate of approximately 100–200 mL/min through a trap containing 200 g of desiccant that collects moisture from the air. This method describes the use of silica gel as the desiccant. However, other common desiccants such as molecular sieves (alumino-silicates), anhydrous calcium sulfate, and activated alumina can be used. The stated flow rate is sufficient to allow continuous air sampling for 1–2 wk at 50% relative humidity. Other methods of moisture collection such as cold traps, bubblers and dehumidifiers may also be used for short-term sampling with various effects on measurement accuracy, depending on completeness of collection (**6**).

The collected moisture is then removed from the trap and analyzed for tritium by mixing a suitable volume of it with a scintillation solution and counting the mixture in a liquid scintillation beta counter. The air flow rate and total volume of moisture collected during the sampling period, or the absolute humidity, must be determined to convert the concentration of tritium in the collected moisture sample to its concentration in air. For long sampling periods, the determination of sampling rate may be more convenient than humidity measurements.

2. Range and Sensitivity

2.1 The method is suitable for sampling periods of up to 1–2 wk over the entire significant range of tritium concentrations in the air. The sampling period can be extended if low relative humidity conditions exist or if the column size is increased. Sensitivity is a direct function of the volume of air sampled and an inverse function of the prevailing temperature and relative humidity of the air. The amount of water that silica gel can retain is also a function of relative humidity. Typically, the adsorptive capability of silica gel at 10% relative humidity is 0.05 g of water per g of adsorbent while at 50% relative humidity the adsorptive capability is 0.26 g of water per g of adsorbent (**5,6**). A sample of moisture from air at an average temperature of 30°C (86°F) and at 100% relative humidity, and containing tritium in a concentration of 2×10^{-2} μCi/m^3 (10% of the Maximum Permissible Concentration) would have a concentration of about 700 pCi/mL, sufficient to produce acceptable precision in less than 2 min of counting in most liquid scintillation counting systems.

3. Interferences

3.1 None.

4. Precision and Accuracy

4.1 The precision of the method is, over the range of concentrations of interest, a function of counter efficiency and counting statistics. For example, the precision of the sample described above would be 2%, counted for 2 min at an efficiency of 25% using a moisture aliquot of 7 mL and a liquid scintillation counter that had a 7 cpm background in the tritium channel.

5. Apparatus

5.1 Liquid scintillation beta counter.
5.2 Traps, 12 in $\times$ 1.25 in (30.5 $\times$ 3.2 cm) diameter. The trap construction can be aluminum, plastic or glass. If the sampling device is to be used to sample air streams with very high tritium concentrations, tritium can exchange with plastic, thus causing cross contamination of samples.

5.3 Low volume air pump, flow rate indicator, critical or limiting orifice, and flow integration device (or temperature and relative humidity device).

5.4 Distilling Flask and Condenser.

5.5 Pipets.

5.6 Vials, Polyethylene, 20 mL for liquid scintillation counters.

5.7 Rubber Stoppers, Tubing.

6. Reagents

6.1 Insta-gel or equivalent scintillator (United Technology). Other brands of commercial scintillators can be used provided that the moisture aliquot is changed to reflect the scintillator's ability to hold water. A reduction in moisture aliquot size will decrease the sensitivity of the method. A scintillation "cocktail" may also be produced in-house **(1,2)**.

6.2 Silica gel, 6–16 mesh, non-indicating.

6.3 Silica gel, 6–16 mesh, indicating.

7. Procedure

7.1 Dry 6–16 mesh, non-indicating silica gel for 24 h at 150°C prior to use.

This step may be omitted if the user is willing to accept a 5% uncertainty in the total moisture content. Omission of this step may be highly desirable if the drying process cannot be performed in a low-background environment. The silica gel can be contaminated by ambient air in the bake-out/cool-down step of the procedure.

7.2 Fill one trap (see Figure 609A:1 for typical design) with nonindicating silica gel. Fill the second cylinder with a mixture of 10% indicating and 90% non-indicating silica gel. Since silica gel can adsorb moisture up to 40% of its own weight at 100% relative humidity **(3, 5, 6)**, 200 g should not saturate during a sampling period as long as 1 wk in extremely warm and humid conditions. Seal inlet and outlet.

If a cold finger or de-humidifer is used to collect sample, start procedure at 7.8 (substituting collected moisture for distillate).

7.3 Unseal traps. Connect the outlet port of the sample trap to the inlet port of the back-up trap (containing the indicating silica gel).

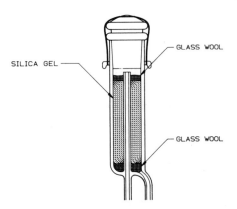

Figure 609A:1 – Adsoprtion cell for determination of elemental tritium.

7.4 Connect the outlet port of the back-up trap to the low volume air pump. If necessary, connect the inlet part of the sample trap to the sample line (this is used if the air stream is being sampled from a remote location). Start pump, set flowrate, and sample for appropriate time, usually 1 wk.

7.5 Following sample collection, remove non-indicating silica gel trap and seal ends for transport to laboratory. Check back-up trap for change in the color of the indicating silical gel. If change of color has occurred, remove back-up trap, seal ends and transport to the laboratory for analysis.

7.6 Pour silica gel into 250–500 mL distilling flask.

7.7 Heat until all moisture is removed (500°C). Weigh or measure total moisture collected. This maximum temperature seems needed for some types of silica gel, but it makes reuse of the adsorbent impossible. If back-up trap was sent for analysis, heat its contents similarly, and add distillate from it to that from the first trap.

7.8 Add 9 mL of scintillation solution to a polyethylene vial containing 7 mL of distillate.

7.9 Count for not less than 2 min in a liquid scintillation beta counter. (Deviation from environmental concentration can easily be detected with sample count times of 100 min.)

8. Calibration and Standards

8.1 Using distilled tritium-free water, dilute an NBS* tritiated water primary standard to obtain known calibration standards for the liquid scintillation counter. A concentration between 10^3 to 10^4 dpm/mL is suggested to give a practicable calibration standard.

8.2 To determine counting efficiency mix 7 mL of the tritiated water standard with 9 mL of scintillation solution, and count under the same conditions as the sample. Counting efficiency should be verified with each batch of samples.

8.3 To determine background, add 7 mL of distilled tritium-free water to 9 mL of scintillation mix and count. Background samples should be included at the start of every counting run and between groups of not more than ten samples.

9. Calculations

9.1 Efficiency:

$$E = \frac{S - B}{D} \qquad (1)$$

where E = counting efficiency.
    S = counts per minute (cpm) of standard (in scintillation solution).
    B = background cpm of distilled water (in scintillation solution).
    D = disintegrations per minute (dpm) of standard.

9.2 To determine tritium concentration of moisture:

$$C_m \ (\mu Ci/mL) = \frac{C_s - B}{E \times V \times K} \qquad (2)$$

where C_m = concentration of tritium in collected moisture in $\mu Ci/mL$.
    C_s = gross sample cpm; i.e., sample count plus background.
    V = volume (mL) of sample counted.
    K = 2.2×10^6 to convert dpm to μCi).

---

*National Bureau of Standards, SRM No. 4927, or standard of equivalent calibration accuracy.

9.3 Error due to counting (at the 95% confidence level) is calculated as follows:

$$E_{95} = $$

$$1.96 \ \sqrt{\frac{C_s}{\text{Time counted}} + \frac{B}{\text{Time counted}}}$$

$$(3)$$

9.4 The concentration of tritium oxide in air is calculated as follows if silica gel (having 100% collection efficiency) is used and if the flow rate is known.

$$C_a \ (\mu Ci/mL) = \frac{C_m \times W}{F \times t} \qquad (4)$$

where C_a = concentration of tritium oxide in air in $\mu Ci/mL$.
    F = sampling rate (mL/min).
    t = sampling time (min).
    W = total volume of water collected (mL).

9.5 If a less than 100% efficient collector, such as a cold trap or dehumidifier is used, or the flow rate is unknown, the following equation is used to determine tritium oxide concentration in air:

$$C_a(\mu Ci/cm^3) = C_m A \times 10^{-6} \qquad (5)$$

where A = absolute humidity during sample collection (g/m^3)**

10. Effects of Storage

10.1 No effects are encountered by storage if silica gel collection cylinders are sealed and stored in a non-tritiated atmosphere.

**This is the mass of water vapor present in unit volume of the atmosphere, usually expressed as grams per cubic meter. Representative values applicable at sea level are indicated in Table 609:I (4).

Table 609:I. Weight of a Cubic Meter of Aqueous Vapor at Different Temperatures and Percentages of Saturation*

Temperature (°F)	Percentage of Saturation									
	10	20	30	40	50	60	70	80	90	100
					g/m^3					
-20	0.039	0.076	0.114	0.151	0.190	0.229	0.266	0.304	0.341	0.380
-15	0.050	0.101	1.149	0.199	0.250	0.300	0.350	0.398	0.449	0.499
-10	0.064	0.130	0.197	0.261	0.325	0.392	0.458	0.522	0.586	0.653
-5	0.085	0.169	0.254	0.339	0.424	0.508	0.593	0.678	0.762	0.847
0	0.110	0.220	0.330	0.440	0.550	0.662	0.772	0.882	0.992	1.101
5	0.140	0.279	0.419	0.559	0.698	0.838	0.978	1.118	1.257	1.397
10	0.179	0.355	0.534	0.710	0.888	1.067	1.243	1.422	1.598	1.777
15	0.227	0.451	0.678	0.90	1.129	1.356	1.580	1.807	2.031	2.258
20	0.284	0.566	0.847	1.131	1.415	1.697	1.978	2.262	2.546	2.828
25	0.355	0.710	1.065	1.420	1.777	2.132	2.487	2.842	3.197	3.552
30	0.444	0.886	1.328	1.772	2.217	2.659	3.101	3.545	3.989	4.431
35	0.543	1.083	1.626	2.166	2.709	3.252	3.792	4.335	4.875	5.418
40	0.653	1.305	1.958	2.611	3.261	3.914	4.566	5.219	5.872	6.524
45	0.781	1.564	2.345	3.128	3.909	4.690	5.473	6.254	7.037	7.818
50	0.934	1.866	2.801	3.733	4.667	5.601	6.533	7.468	8.400	9.334
55	1.111	2.221	3.332	4.443	5.551	6.662	7.772	8.883	9.994	11.104
60	1.314	2.631	3.948	5.262	6.577	7.984	9.210	10.525	11.839	13.156
65	1.553	3.105	4.660	6.213	7.765	9.318	10.871	12.426	13.978	15.531
70	1.827	3.655	5.482	7.310	9.137	10.964	12.792	14.619	16.447	18.274
75	2.143	4.284	6.428	8.569	10.713	12.856	14.997	17.141	19.282	21.425
80	2.503	5.008	7.511	10.016	12.519	15.022	17.528	20.031	22.536	25.039
85	2.917	5.833	8.750	11.665	14.583	17.500	20.415	23.333	26.248	29.165
90	3.387	6.774	10.161	13.548	16.934	20.321	23.708	27.095	30.482	33.869
95	3.920	7.843	11.764	15.686	19.607	23.527	27.450	31.371	35.293	39.214
100	4.527	9.052	13.580	18.105	22.632	27.159	31.684	36.212	40.737	45.264

*Adapted from Reference (4).

11. References

1. BUTLER, F.F 1961. Determination of Tritium in Water and Urine. *Anal. Chem.* 33:3, 409–414.
2. MOGHISSI, A.A., H.L. KELLEY, J.I. REGNIER, AND M.W. CARTER. 1969. Low Level Counting by Liquid Scintillation – I. Tritium Measurement in Homogeneous Systems. *Int'l. Jrl. Appl. Rad. and Isotopes* 20:145–56.
3. PERRY, R.H., C.H. CHILTON, AND S.D. KIRKPATRICK. 1963. Chemical Engineers Handbook, Section 16.
4. MARVIN, C.F. 1941. Psychrometric Tables. Weather Bureau Publication #735.
5. NATIONAL COUNCIL ON RADIATION PROTECTION AND MEASUREMENTS, 1976. Tritium Measurement Techniques. NCRP Report No. 47.
6. BUDNITZ, R.J., A.V. NERO, D.S. MURPHY AND R. GRAVEN. 1983. Instrumentation for Environmental Monitoring. Volume 1, Radiation. Wiley Interscience, New York.

Subcommittee 8

R. P. MILTENBERGER
A. P. HULL, *Chairman*
J. A. BROADWAY
M. L. MAIELLO

609B.

Determination of Elemental Tritium Content of the Atmosphere

1. Principle of the Method

1.1 Air is pumped at a sampling rate of approximately 100 cm^3/min through 2 traps each containing about 200 g of desiccant that collects moisture from the air. The air is then mixed with enough additional hydrogen to sustain catalytic oxidation, and proceeds into a catalytic converter that oxidizes free hydrogen and tritium. The resulting water (H$_2$O + HTO) is then collected in a desiccant trap (**6**). This method describes the use of a palladium sponge catalyst, and silica gel as the desiccant; however, other catalysts and desiccants are available. (See Method 609A for discussion of alternative desiccant and reference for alternate catalysts.) The stated flow rate is sufficient to allow continuous air sampling for one week at 50% relative humidity. The collected moisture is then removed from the traps and analyzed for tritium by mixing a suitable volume of the moisture with a scintillation solution and counting the mixture in a liquid scintillation beta counter. The air sampling rate, and total volume of moisture collected during the sampling period, must be determined to convert the concentration of tritium in the collected moisture samples to its concentration of air.

2. Range and Sensitivity

2.1 The method is suitable for sampling periods of 1–2 wk over the entire significant range of elemental and tritiated water vapor concentrations in the air. Sampling can be extended if low relative humidity conditions exist or if the column size is increased. Sensitivity is a direct function of the volume of air sampled and amount of carrier gas used. Typically, one would expect to collect 7–8 g of water produced from the catalytic reaction between palladium and elemental hydrogen and tritium. An air sample containing elemental tritium in a concentration of 2×10^{-2} μCi/m^3 (0.001 % of the Maximum Permissible Concentration) would have a concentration of about 4000 pCi/mL; sufficient to produce acceptable precision in less than 2 min of counting in most liquid scintillation counting systems.

3. Interferences

3.1 Water vapor not collected by the first two desiccant traps from the air stream of interest may cause the elemental tritium concentration to be overestimated.

4. Precision and Accuracy

4.1 The precision of the method is, over the range of concentrations of interest, a function of counter efficiency and counting statistics. For example, the precision of the sample described above would be less than ± 2%, counted for 2 min at an efficiency of 25% using a moisture aliquot of 7 mL and a liquid scintillation counter that had a 7 cpm background in the tritium channel.

5. Apparatus

5.1 Liquid scintillation beta counter.
5.2 Traps 12″ × 1.25″ (30.5 × 3.2 cm) diameter. The trap construction can be aluminum, plastic or glass. If the sampling device is to be used to sample air streams with very high tritium concentrations, tritium can exchange with plastic, thus causing cross contamination of samples.
5.3 Low volume air pump with flow rate indicator, critical or limiting orifice, or flow integration device.
5.4 Distilling Flask and Condenser.
5.5 Pipets.

5.6 Vials, Polyethylene, (20 mL for liquid scintillation counters).

5.7 Rubber Stoppers, Tubing.

5.8 3% hydrogen and 97% nitrogen carrier gas (size H cylinder lasts about 4 months).

5.9 Palladium sponge catalyst (2 g).

5.10. Glass wool plugs.

6. Reagents

6.1 Insta-gel (or equivalent) scintillator (United Technology). Other brands of commercial scintillators can be used provided that the moisture aliquot is changed to reflect the scintillator's ability to hold water. A reduction in moisture aliquot size will decrease the sensitivity of the method. A scintillator "cocktail" may also be produced in-house **(1,2)**.

6.2 Silica gel, 6–16 mesh, non-indicating.

6.3 Silica gel, 6–16 mesh, indicating.

7. Procedure

7.1 Dry 6–16 mesh, non-indicating silica gel for 24 h at 150°C prior to moisture collection use. This step may be omitted if the user is willing to accept a 5% uncertainty in the total moisture content. Omission of this step may be highly desirable if the drying process cannot be performed in a low-background environment. The silica gel can be contaminated by ambient air in the bake-out/cool-down step of the procedure.

7.2 Fill two 12″ traps (see Figure 609B:1 for typical design) with non-indicating silica gel. Fill the third trap with a mixture of 10% indicating and 90% non-indicating silica gel. Since silica gel can absorb moisture up to 40% of its own weight, at 100% relative humidity **(3,4,5)**, 200 g should not saturate during a sampling period as long as 1 wk except in extremely warm and humid conditions. Seal inlet and outlet.

7.3 Introduce the palladium sponge into a length of 6-mm Pyrex tubing, holding it in place with small portions of glass wool.

7.4 Unseal traps. Assemble in series the sampling line, if any; a T-connector, one arm of which is attached to the tank of H_2-in-N_2; a trap containing non-indicating silica gel; a back-up trap containing 10% indicating silica gel; the tube containing palladium catalyst; the second trap containing non-indicating silica gel; and the sampling pump. Appropriate arrangements must be added to regulate and measure sample and H_2-in-N_2 flows.

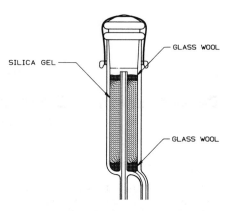

Figure 609B:1—Absorption cell for determination of elemental tritium.

7.5 Set H_2-in-N_2 flow at 33 mL/min, then turn on sampling pump and adjust sample flow to 100 mL/min (total pump flow 133 mL/min).

7.6 Following sample collection, remove non-indicating silica gel traps, label appropriately, and seal ends. Check back-up trap for change in the color of the indicating silica gel. If a change of color has occurred, remove back-up trap, seal ends and transport to the laboratory for analysis.

7.7 It if is desired to determine tritiated water content of the atmosphere, process first trap (and back-up if necessary) according to Method 609A, sections 7.6–7.9.

7.8 To determine elemental tritium, process final trap according to Method 609A, sections **7.6–7.9.**

8. Calibration and Standards

8.1 Using distilled tritium-free water, dilute an NBS* tritiated water primary standard to obtain known calibration standards for the liquid scintillation counter. A concentration between 10^3 and 10^4 dpm/mL is

---

*National Bureau of Standards, SRM No. 4927, or standard of equivalent calibration accuracy.

suggested to give a practicable calibration standard.

8.2 To determine counting efficiency, mix 7 mL of the tritiated water standard with 9 mL of scintillation solution, and count under the same conditions as the sample. Counting efficiency should be verified with each batch of samples.

8.3 To determine background, add 7 mL of distilled tritium-free water to 9 mL of scintillation mix and count. Background samples should be included at the start of every counting run and between groups of not more than ten samples.

9. Calculations

9.1 Efficiency:

$$E = \frac{S - B}{D} \qquad (1)$$

where E = counting efficiency.
S = counts per minute (cpm) of standard (in scintillation solution).
B = background cpm of distilled water (in scintillation solution).
D = disintegrations per minute (dpm) of standard.

9.2 To determine tritium concentration of moisture:

$$C_m \ (\mu Ci/mL) = \frac{C_s - B}{E \times V \times K} \qquad (2)$$

where C_m = concentration of tritium in collected moisture in $\mu Ci/$ mL.
C_s = gross sample cpm; i.e., sample count plus background.
V = volume (mL) of water sample counted.
K = 2.22×10^6 to convert disintegrations per minute to μCi.

9.3 Error due to counting at the 95% confidence level) is calculated as follows:

$$E_{95} =$$

$$1.96 \ \sqrt{\frac{C_s}{\text{Time counted}} + \frac{B}{\text{Time counted}}}$$

$$(3)$$

9.4 The concentration of tritium oxide in air is calculated as follows:

$$C_a \ (\mu Ci/mL) = \frac{C_m \times W}{F \times t} \qquad (4)$$

where C_a = concentration of tritium oxide in air in $\mu Ci/mL$.
F = sampling rate (mL/min).
t = sampling time (min).
W = total volume of water collected (mL).

9.5 The same calculation, applied to the water content of the final silica gel trap, yields the total elemental tritium content of the atmosphere. Since the water content of this trap derives principally from the tank of H_2-in-N_2, the intermediate value calculated by eq. (2) has no physical meaning other than as an intermediate in the final calculation.

10. Effects of Storage

10.1 No effects are encountered by storage if silica gel collected cylinders are sealed and stored in a non-tritiated atmosphere.

11. References

1. BUTLER, F.F. 1961. Determination of Tritium in Water and Urine. *Anal. Chem. 33*:409–414.
2. MOGHISSI, A.A., H.L. KELLEY, J.I. REGNIER AND M.W. CARTER. 1969. Low Level Counting by Liquid Scintillation—I. Tritium Measurement in Homogeneous Systems. *Int'l. J. Appl. Rad. and Isotopes 20*:145–56.
3. PERRY, R.H., C.H. CHILTON AND S.D. KIRKPATRICK. 1963. Chemical Engineers Handbook Section 16.
4. NATIONAL COUNCIL ON RADIATION PROTECTION AND MEASUREMENT. 1976. Tritium Measurement Techniques. NCRP Report No. 47.
5. BUDNITZ, R.J., A.V. NERO, D.S. MURPHY AND R. GRAVEN. 1983. Instrumentation for Environmental Monitoring. Volume 1, Radiation. Wiley Interscience, New York.
6. GRIESBACH, O. AND J.R. STENCEL. 1988. The PPPL Differential Atmospheric Tritium Sampler (DATS). *Health Physics*, in press.

Subcommittee 8

R. P. MILTENBERGER
A.P. HULL, *Chairman*
J. A. BROADWAY
M. L. MAIELLO

701.

Determination of Hydrogen Sulfide Content of the Atmosphere

1. Principle of the Method

1.1 Hydrogen sulfide may be present in the ambient atmosphere at concentrations of a few ppb or less. The reported odor detection threshold is in the 0.7 to 8.4 $\mu g/m^3$ (0.5 to 6.0 ppb) range **(1,2)**. Concentrations in excess of 139 $\mu g/m^3$ (100 ppb) are seldom encountered except as a result of an accident. Very high concentrations (of the order of several percent) of H_2S may be present in geothermal gases **(3)**. For the workplace environment, the ACGIH suggested exposure limit for H_2S is 14 mg/m³ (10 ppm) at the time of preparation of this volume.

1.2 Hydrogen sulfide is collected by aspirating a measured volume of air through an alkaline suspension of cadmium hydroxide **(4,5)**. The sulfide is precipitated as cadmium sulfide to prevent air oxidation of the sulfide which occurs rapidly in an aqueous alkaline solution. Arabinogalactan is added to the cadmium hydroxide slurry prior to sampling to minimize photo-decomposition of the precipitated cadmium sulfide **(5)**. The collected sulfide is subsequently determined by spectrophotometric measurement of the methylene blue produced by the reaction of the sulfide with a strongly acid solution of N,N-dimethyl-*p*-phenylenediamine and ferric chloride **(4,7)**:

The analysis should be completed within about 24 h following collection of the sample.

2. Range and Sensitivity

2.1 This method is intended to provide a measure of hydrogen sulfide in ambient air in the range of 2.2 to 200 $\mu g/m^3$ (1.6–144 ppb). For concentrations above 70 $\mu g/m^3$ (50 ppb), the sampling period can be reduced or the liquid volume increased either before or after aspirating. When sampling air at the maximum recommended rate of 1.5 L/min for 2 h, the minimum detectable sulfide concentration is 1.1 $\mu g/m^3$ (0.8 ppb) at 101.3 kPa and 25°C. Although a limit of detection of 0.20 μg H_2S/m^3 (0.14 ppb) may be computed for a 1 m³ air sample based on the aqueous limit of detection of 8 ng/mL and a solution volume of 25 mL, losses averaging 46 percent have been reported for 24 h sampling of calibration atmospheres containing 25 to 56 μg H_2S/m^3 (18–40 ppb). **(8)**.

2.2 This method is also suitable for the measurement of H_2S in the mg/m³ range. One hundred milliliters of the cadmium hydroxide-arabinogalactan medium in 250 mL fritted glass bubblers may be used with excellent results for 5 min sampling of at-

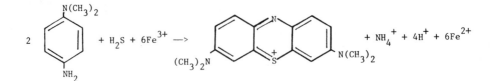

mospheres containing 7 to 70 mg H_2S/m^3 (5–50 ppm). Upper limits are set by kinetics rather than stoichiometry; the amount of cadmium ion in the medium is sufficient to react with 6 mg/mL of H_2S.

3. Interferences

3.1 The methylene blue reaction is highly specific for sulfide at the low concentrations usually encountered in ambient air. Reducing agents such as sulfite retard color development if present in significant concentrations ($\geq$ 10 µg sulfite/mL) even in the case of solutions containing several µg $S^=$/mL. Up to 40 µg/mL of sulfite interference can however be overcome by adding 0.1–0.3 mL of ferric chloride instead of 0.05 mL for color development and extending the reaction time to 50 min.

3.2 At levels of $\geq$ 0.5 µg/mL, nitrite produces a pale yellow color with the reagents. No interference is encountered when 0.3 ppm NO_2 is aspirated through a midget impinger containing a slurry of cadmium hydroxide-cadmium sulfide-arabinogalactan. If H_2S and NO_2 are simultaneously aspirated through the cadmium hydroxide-arabinogalactan slurry, lower H_2S results are obtained, probably because of gas-phase oxidation of H_2S by NO_2.

3.3 Ozone at 57 ppb reduces the recovery of sulfide previously precipitated as CdS by 15% **(9)**.

3.4 Cadmium sulfide decomposes significantly when exposed to light. Addition of 1% arabinogalactan greatly reduces such loss but does not completely eliminate it **(6)**. Consequently, the absorber vessel needs to be protected from light.

3.5 When sampling H_2S concentrations $\geq$ 7 mg/m³, the sampling time must be limited to 5 min. Prolonged sampling of high concentrations of H_2S may result in deposition of sulfur in the frits of the absorber due to oxidation of sulfide. Such deposition causes a gradual decrease in sampling rate in the absence of a device to maintain the flow rate constant. If sulfur deposit does occur, the frit may be cleaned in boiling concentrated HNO_3.

4. Precision and Accuracy

4.1 The typical coefficient of variation at an H_2S concentration of 60 µg/m³ generated with permeation tubes is 3.5% **(6)**. Careful studies with gravimetric calibration of a permeation tube or with parallel flame photometric measurement (see method 709 in this volume) have shown that even if calibration is based on aqueous sulfide, the recovery of collected H_2S (0.125–1 mg/m³) in 30 min samples is quantitative within experimental error **(8)**. However, with aqueous sulfide calibrant, glassware identical to that used for sample analysis must be used **(8)**.

4.2 The overall coefficient of variation of the method, including both sampling and analysis, for occupational health applications has been reported to be 12.1% **(10)**.

4.3 At low concentrations (0.6–1.4 µg/m³), the collection efficiency at sampling rates up to 1.7 L/min is typically 95.1 ± 1.5% **(11)**. Essentially quantitative collection has been reported at a concentration level of 1 mg/m³ with a sampling rate of 1L/min **(8)**.

4.4 Results that average 54% of the true values have been observed for 24-h sampling of standard atmospheres containing 25–56 µg H_2S/m^3 **(8)**.

5. Advantages and Disadvantages

5.1 EFFECT OF LIGHT AND STORAGE.

5.1.1 Hydrogen sulfide is readily volatilized from aqueous solution when the pH is below 7.0. Alkaline aqueous sulfide solutions are very unstable because sulfide ion is rapidly oxidized by exposure to the air.

5.1.2 Cadmium sulfide is not appreciably oxidized even when aspirated with pure oxygen in the dark. However, exposure of a bubbler containing cadmium sulfide to laboratory or to more intense light sources produces an immediate and variable photo-decomposition. Losses of 50 to 90% of added sulfide have been routinely reported by a number of laboratories. Even though the addition of arabinogalactan to the absorbing solution controls the photodecomposition, it is necessary to protect the bubbler from light at all times. This is achieved by the use of low actinic glass bubblers, paint on the exterior of the impingers, or an aluminum foil wrapping.

6. Apparatus

6.1 ABSORBER. Midget bubbler fitted with coarse-porosity frit.

6.2 AIR PUMP with a flow meter and/or gas meter having a minimum capacity of drawing 2 L/min of air through a midget bubbler.

6.3 SPECTROPHOTOMETER. Capable of operation at 670 nm. A red light emitting diode-based colorimeter with the center wavelength of emission at 660 nm is acceptable. A long path length cell (2–5 cm) may be used to improve sensitivity.

6.4 AIR VOLUME MEASUREMENT. The air meter must be capable of measuring the air flow within ± 2%. Wet or dry gas meters, specially calibrated rotameters or critical orifices may be used.

7. Reagents

7.1 PURITY. Reagents must meet specifications of the American Chemical Society except as mentioned. Water should conform to the standards of ASTM reference reagent water, Type II.

7.2 Solutions should be refrigerated when not in use.

7.3 AMINE-SULFURIC ACID STOCK SOLUTION. Add 50 mL conc. H_2SO_4 to 30 mL water and cool. Dissolve 12 g of N,N-dimethyl-p-phenylenediamine dihydrochloride (p-aminodimethylaniline dihydrochloride; if the compound is very dark in appearance, purification by distillation under reduced pressure is necessary) in the sulfuric acid solution.

7.4 SULFURIC ACID, 50%. Add 500 mL conc. H_2SO_4 slowly to 400 mL water and allow to cool. Quantitatively transfer to a 1-L volumetric flask and make up to the mark.

7.5 AMINE TEST SOLUTION. Dilute 25 mL of the stock solution in **7.3** to 1 L with 50% H_2SO_4.

7.6 FERRIC CHLORIDE SOLUTION, 3.7 M. Dissolve 100 g of $FeCl_3 \cdot 6H_2O$ in 30 mL water and 9 mL of conc. HCl. The volume of the solution will be approximately 100 mL.

7.7 AMMONIUM PHOSPHATE, 40% w/v. Dissolve 400 g of diammonium phosphate in water and make up to 1 L in a volumetric flask.

7.8 ARABINOGALACTAN. The Grade 1 material from Sigma Chemical Company (POB 14509, St. Louis MO) or the surfactant STRactan 10 (Stein-Hall and Co., 385 Madison Ave., New York, NY) is suitable.

7.9 ABSORBING SOLUTION. Dissolve 4.3 g of 3 $CdSO_4 \cdot 8H_2O$ and 1.8 g NaOH to separate 250-mL portions of water. Add the two solutions to each other, add 10 g of arabinogalactan and dilute to 1 L. Shake the resultant suspension vigorously before removing an aliquot. Prepare fresh every 3 days. Previous versions of this method prescribe a smaller amount of NaOH in the absorber solution. The pH of the absorber solution recommended here is ~10 and yields superior results **(12,13)**.

7.10 CALIBRATION STANDARDS.

7.10.1 *Gas phase calibration.* Sources emitting ≤ 100 ng/min are needed for calibration at realistic levels. With an air flow rate of 2 L/min, emission rates of 4–200 ng/min are necessary to directly generate H_2S concentrations in the range 2–50 μg/m^3. Permeation tubes which emit H_2S at the rate of ≥ 40 ng/min (at 30°C) are available (with certified emission rates if desired) from VICI/Metronics (2991 Corvin Dr., Santa Clara, CA 95051). From the same vendor, wafer-type permeation devices are available. These exhibit emission rates even lower than the tubular devices. However, the magnitude of the permeation rate makes gravimetric calibration commercially impractical; the wafer devices are not available calibrated and must be calibrated by the user. The permeation sources must be stored in a wide-mouth glass bottle containing silica gel and sodium hydroxide to remove moisture and the emitted H_2S. The storage bottle is immersed to two-thirds its height in a constant-temperature water bath in which the water is controlled at the temperature of intended use, typically 25.0° or 30.0° ± 0.1°C.

7.10.2 *Aqueous calibration.*

7.10.2.1 Concentrated standard sulfide solution. Transfer ~ 1 L of freshly boiled and cooled 0.1 M NaOH to a 1-L volumetric flask. Flush with nitrogen to remove oxygen and adjust to volume [Commercially available, compressed nitrogen contains trace quantities of oxygen in sufficient concentration to oxidize the small concentrations of sulfide contained in the standard and dilute standard sulfide solutions. Trace quantities of oxygen should be removed by passing the stream of tank nitrogen through a pyrex or quartz tube containing copper turnings heated to 500° to 450°C or through a bubbler containing alkaline pyrogallate (6 g pyrogallol and 40 g KOH in 200 mL of solution).] Immediately stopper the flask with a serum cap. Inject 300 mL of H_2S gas through the septum. Shake the flask. Withdraw measured volumes of standard solution with a 20-mL hypodermic syringe and fill the resulting void with an equal volume of nitrogen. Standardize with standard iodine and thiosulfate solution in an iodine flask under a nitrogen atmosphere to minimize air oxidation. The approximate concentration of the sulfide solution will be 420 μg $S^=$/mL of solution. The exact concentration must be determined by iodine-thiosulfate standardization immediately prior to dilution (Refer to any standard text, e.g., ref. 14, for details of iodimetric standardization.)

For the most accurate results in the iodometric determination of sulfide in aqueous solution, the following general procedure is recommended:

a. Replacement of oxygen from the flask with an inert gas such as carbon dioxide or nitrogen.

b. Addition of an excess of standard iodine, acidification and back titration with standard thiosulfate and starch indicator (14).

7.10.2.2 Dilute standard sulfide solution. Dilute 10 mL of the concentrated sulfide solution in section 7.10.2.1 to 1 L with freshly boiled, distilled water. Protect the boiled water under an oxygen-free nitrogen atmosphere while cooling. Transfer the deoxygenated water to a flask previously purged with oxygen-free nitrogen and immediately stopper the flask. This sulfide solution is unstable. Therefore, prepare this solution immediately prior to use. This test solution will contain approximately 4 μg $S^=$/mL.

8. Procedure

8.1 CLEANING OF EQUIPMENT. All glassware should be thoroughly cleaned. The following procedure is recommended:

8.1.1 Wash with a detergent and tap water solution followed by tap water and distilled water rinses.

8.1.2 Soak in 1:1 or concentrated nitric acid for 30 minutes and then follow with tap, distilled, and reagent water rinses.

8.2 COLLECTION OF SAMPLES.

8.2.1 Pipet 10 mL of the absorbing solution (Section 7.9) into the midget bubbler. The addition of 5 mL of 95 percent ethanol to the absorbing solution just prior to aspiration controls foaming for 2 h (induced by the presence of arabinogalactan) (8). In addition, 1 or 2 Teflon discs, which serve as demisters, may be slipped up over the bubbler air inlet tube to a height approximately 2.5 cm from the top of the tube (8).

8.2.2 The sampling set up is the same as that for method 704A. Connect the bubbler (via the absorption tube) to the vacuum pump with a short piece of flexible tubing. The minimum amount of tubing necessary should be used. The air being sampled should not be passed through any other tubing or other equipment before entering the bubbler.

8.2.3 Wrap up the bubbler entirely in foil or otherwise make arrangements to prevent exposure to light.

8.2.4 Turn on the pump to begin sample collection. Care should be taken to measure the flow rate, time and/or volume as accurately as possible. The sampling rate should not exceed 1.5 L/min.

8.2.5 After sampling, the bubbler stem can be removed and cleaned. Tap the stem gently against the inside wall of the bubbler bottle to recover as much of the sampling solution as possible. Wash the stem with a small amount (1–2 mL) of unused absorbing solution and add the wash to the bubbler. Then the bubbler is sealed with a hard, non-reactive stopper (preferably Teflon). Do not seal with rubber. The stoppers on the bubblers should be tightly sealed to prevent leakage during shipping. If it is preferred to ship the bubblers with the stems in, the outlets of the stem should be sealed with Parafilm or other non-rubber covers, and the ground glass joints should be sealed (i.e., taped) to secure the top tightly.

8.2.6 Care should be taken to minimize spillage or loss by evaporation at all times. Refrigerate samples if analysis cannot be done within a day.

8.2.7 Whenever possible, hand delivery of the samples is recommended. Otherwise, special bubbler shipping cases should be used to ship the samples.

8.2.8 A "blank" bubbler should be handled as the other samples (fill, seal and transport) except that no air is sampled through this bubbler.

8.3 ANALYSIS.

8.3.1 Add 1.5 mL of the amine-test solution to the midget bubbler through the air inlet tube and mix.

8.3.2 Add 1 drop (50 μL) of ferric chloride solution and mix. (See Section 3.1 if SO_2 is likely to exceed 10 μg/mL in the absorbing medium.)

8.3.3 Transfer the solution to a 25-mL volumetric flask. Discharge the color due to the ferric ion by adding 1 drop ammonium phosphate solution. If the yellow color is not destroyed by 1 drop ammonium phosphate solution, continue dropwise addition until solution is decolorized. Make up to volume with distilled water and allow to stand for 30 min.

8.3.4 Prepare a zero reference solution in the same manner using a 10 mL volume of absorbing solution, through which no air has been aspirated.

8.3.5 Measure the absorbance of the color at 670 nm in a spectrophotometer set at 100 percent transmission against the zero reference.

9. Calibration

9.1 CALIBRATION WITH AQUEOUS SULFIDE. It is essential for aqueous calibration that the same type of glassware and reagent volumes be used as recommended in section 8.3 for sample analysis **(8)**.

9.1.1 Place 10 mL of the absorbing solution in each of a series of midget bubblers and add 0.25, 0.50, 0.75, 1.0 and 2.0 mL of the diluted sulfide standard to individual bubblers. Mix the contents of each bubbler by shaking. Proceed through steps 8.3.1, 8.3.3–8.3.5.

9.2 CALIBRATION WITH GASEOUS H_2S. Calibration with a gaseous standard is preferred because this tests the proper operation of the sampling procedure as well.

9.2.1 *Generation of standard gaseous H_2S.* Refer to Section 3, Part I, for the generation of standard gaseous atmospheres using permeation sources.

9.2.2 *Procedure for preparing simulated curves.* Obviously one can prepare a multitude of curves by selecting different combinations of sampling rate and sampling time. The following description represents a typical procedure for ambient air sampling of short duration.

The system is designed to provide an accurate measure of hydrogen sulfide in the 1.4–84 μg/m^3 (1–60 ppb) range. It can be easily modified to meet special needs.

The dynamic range of the colorimetric procedure fixes the total volume of the sample at 186 L; then, to obtain linearity between the absorbance of the solution and the concentration of hydrogen sulfide in ppb, select a constant sampling time. This fixing of the sampling time is desirable also from a practical standpoint: In this case, select a sampling time of 120 minutes. Then to obtain a 186-L sample of air requires a flow rate of 1.55 L/minute. The concentration of standard H_2S in air is computed as follows:

$$C = \frac{P_r x\ M}{R}$$

where:

   C = Concentration of H_2S in ppb
   P_r = Permeation rate in ng/minute
   M = Reciprocal of vapor density, 0.719 nL/ng
   R = Total rate of diluent air/nitrogen through generation system, L/min.

The data for a typical calibration are listed in Table 701:I.

A plot of the concentration of hydrogen sulfide in ppb (x – axis) against absorbance of the final solution (y – axis) will yield a straight line, the reciprocal of the slope of which is the factor for conversion of absorbance to ppm. This factor includes the correction for collection efficiency. Any deviation from linearity at the lower concentration range indicates a change in collec-

Table 701:I Typical Calibration Data

Concentration H$_2$S, ppb	Amount of H$_2$S in μL/186 L	Absorbance of sample
1	0.186	.010
5	0.930	.056
10	1.86	.102
20	3.72	.205
30	5.58	.307
40	7.44	.410
50	9.30	.512
60	11.16	.615

tion efficiency of the sampling system. Such deviations are best detected by actually plotting the data; however, the slope is more accurately determined by linear regression. If the range of interest is below the dynamic range of the method the total volume of air collected should be increased to obtain sufficient color within the dynamic range of the colorimetric procedure. Also, once the calibration factor has been established under simulated conditions, the conditions can be modified so that the concentration of H$_2$S is a simple multiple of the absorbance of the colored solution.

The remainder of the analytical procedure is the same as that described for aqueous sulfide.

9.3 LONGER TERM SAMPLING. The applicability of the method to long-term sampling (e.g., 8 h or 24 h) is subject to calibration at the intended sampling rate and concentration range for the intended period of time.

9.4 STORAGE OF PERMEATION SOURCES. The permeation sources must be stored in a wide-mouth glass bottle containing silica gel and solid sodium hydroxide to remove moisture and hydrogen sulfide. The storage bottle is immersed to two-thirds its depth in a constant temperature water bath in which the water is controlled at the temperature of intended use, typically 25.0° or 30.0° ± 0.1°C. Every two weeks, the permeation sources are removed and rapidly weighed on a semimicro (sensitivity ± 10 μg) or a micro (sensitivity ± 1 μg) balance and then returned to the storage bottle. The weight loss is recorded. The sources are ready to use when the rate of weight loss becomes constant (within ±2 percent).

10. Calculations

Determine the sample volume in m³ from the gas meter or flow meter readings and time of sampling. Adjust volume to 101.3 kPa and 25°C (V$_s$). Compute the concentration of H$_2$S in the sample by one of the following equations:

$$ppb = \frac{(A-A_o)0.719B}{V_s}$$

$$\mu g/m^3 = \frac{(A-A_o)B}{V_s}$$

where:

A is the sample absorbance.

A$_o$ is the reagent blank.

0.719 is the volume (μL) of 1 μg H$_2$S at 25°C, 101.3 kPa

B is the calibration factor, μg/absorbance unit.

V$_s$ is the sample volume in m³ corrected to 25°C, 101.3 kPa (by PV = nRT).

11. Effects of Light and Storage

11.1 After color development, the methylene blue color is stable at least for 48 hours at room temperature and in the dark **(15)**. If ethanol is added to inhibit foaming, accelerated decay of color may occur **(8)**.

11.2 See Section 5.2 for photooxidation of cadmium sulfide.

12. References

1. ADAMS, D.F., F.A. YOUNG and R.A. LUHR. 1968. Evaluation of an Odor Perception Threshold Test Facility, Tappi 51:62A–67A.
2. LEONARDOS, G., D. KENDALL and N. BARNARD. 1969. Odor Threshold Determination of 53 Odorant Chemicals, *J. Air Pollut. Contr. Assoc.* 19:91–95.
3. TSAI, F., S. JUPRASERT and S. SANYAL. 1978. A Review of the Chemical Composition of Geothermal Effluents, Proceedings of the Second Workshop on Sampling Geothermal Effluents, EPA-600/7-78-121.
4. JACOBS, M.B., M.M. BRAVERMAN and S. HOCHHEISER. 1957. Ultramicro Determination of Sulfides in Air, *Anal. Chem.* 29:1349–1351.
5. JACOBS, M.B. 1965. Recommended Standard Method for Continuing Air Monitoring for Hydrogen Sulfide. Ultramicrodetermination of Sulfides in the Air, *J. Air Pollut. Contr. Assoc.* 15:314–315.
6. BAMESBERGER, W.L. and D.F. ADAMS. 1969. Improvements in the Collection of Hydrogen Sulfide

in Cadmium Hydroxide Suspension, *Environ. Sci. Technol.*, 3:258-261.

7. MECKLENBURG, W. and R. ROZENKRANZER. 1914. Colorimetric Determination of Hydrogen Sulfide, *Z. Anorg. Chem.* 86:143-153.

8. MCCURDY, R.A. and S.L. ALTSHULER. 1978. A Review of the Determination of Hydrogen Sulfide in Air by the Cadmium Hydroxide-STRactan Colorimetric Method: Current Practices and Modifications, Proceedings of the Second Workshop on Sampling Geothermal Effluents, EPA-600/7-78-121.

9. THOMAS, B.L. and D.F. ADAMS. 1972. Unpublished data.

10. Documentation of NIOSH Validation Tests. Contract No. CDC 99-74-45.

11. BOSTRÖM, C.-E. 1966. The Absorption of Low Concentrations (pphm) of Hydrogen Sulfide in a $Cd(OH)_2$-Suspension as studied by an Isotopic Tracer Method, *Air and Water Pollut. Int. J.*, 10:435-441.

12. MOEST, R.R. 1975. Hydrogen Sulfide Determination by the Methylene Blue Method, *Anal. Chem.* 47:1204-1205.

13. MARBACH, E.P. and D.M. DOTY. 1956. Sulfides Released from Gamma-irradiated Meats as Estimated by Condensation with N,N-dimethyl-*p*-phenylenediamine, *J. Agr. Food. Chem.* 4:881-884.

14. KOLTHOFF, I.M. and P.J. ELVING. 1961. Treatise on Analytical Chemistry, Part II. Analytical Chemistry of the Elements. Vol. 7. Interscience, New York.

15. ALTSHULER, S.L. and S.G. SARP. 1973. Geysers Air Monitoring Program; July 1970-November 1972; Progress Report. Report No. 7485.4-75, Pacific Gas and Electric Company, Department of Engineering Research, San Ramon, CA.

Subcommittee 1

P.K. DASGUPTA, *Chairman*
D.F. ADAMS
B.R. APPEL
S.O. FARWELL
K.T. KNAPP
G.L. KOK
W.R. PIERSON
K.D. REISZNER
R.L. TANNER

704A.

Determination of Sulfur Dioxide Content of the Atmosphere (Tetrachloromercurate Absorber/Pararosaniline Method)

1. Principle of the Method

1.1 Sulfur dioxide, in an air sample, is absorbed into a solution of potassium or sodium tetrachloromercurate, TCM, by aspirating a measured air sample through an absorber vessel. This procedure results in the formation of a monochlorosulfonatomercurate (II) complex (3), which resists oxidation by the oxygen in the air (1,2,3). Ethylenediamine-tetraacetic acid disodium salt (EDTA) is added to this solution to complex heavy metals that catalyze the oxidation of the collected sulfur dioxide (4,5). Once this monochlorosulfonatomercurate complex is formed, it is stable to strong oxidants (e.g. ozone, oxides of nitrogen and hydrogen peroxide) (2). After sampling is completed, any ozone in the solution is allowed to decay (5). The liquid is treated first with a solution of sulfamic acid to destroy the nitrite anion formed from the absorption of oxides of nitrogen present in the atmosphere (6). It is then treated with solutions of formaldehyde and specially purified, acid-bleached pararosaniline containing phosphoric acid to control pH. Pararosaniline, formaldehyde, and the bisulfite anion react to form the intensely colored pararosanilinemethylsulfonic acid, which behaves as a two-color pH indicator ($\lambda_{max.}$ = 548 nm at pH 1.6 ± 0.1, ϵ = 47.7 × 10^3). The ϵ value assumes quantitative production of the absorbing species. The pH of the final solution is adjusted to 1.6 ± 0.1 by the addition of a prescribed amount of 3M phosphoric acid to the pararosaniline reagent (5). This technique is the basis of the reference method for the

determination of sulfur dioxide by the U.S. Environmental Protection Agency (7).

1.1.1 Two variations are given, differing only in the pH of the final solution. The variation described above is designated Variation A and is the method of choice. It gives the highest sensitivity. In Variation B, a larger quantity of phosphoric acid is added to yield a pH of 1.2 ± 0.1 in the final solution. The wavelength of maximum absorbance under these conditions is 575 nm, and the compound has a molar extinction of 37.0 × 10^3, assuming quantitative production of the absorbing species. Variation B is less sensitive, but has the advantage of a lower blank. It is pH-dependent, and may be more suitable with less expensive spectrophotometers.

2. Range and Sensitivity

2.1 Atmospheric sulfur dioxide concentrations are measurable by this technique in the range from 10 ppbv to a few ppmv. Use smaller gas samples when analyzing higher concentrations (5 to 500 ppmv), found in special studies. A rapid redox reaction occurs between Hg (II) and the sulfite ion, if concentrations of the latter exceed 500 μg/mL (8).

2.2 Collection efficiency falls off rapidly below 10 ppbv, and varies with the geometry of the absorber, the size of the gas bubbles, and the contact time with the solution (9,10,11).

2.3 The lower limit of detection of sulfur dioxide in 10 mL of TCM is 0.3 μL (based on twice the standard deviation representing a concentration of 10 ppbv (26 μg/m³) SO_2 in an air sample of 30 L. One cannot extrapolate to lower values by taking larger volumes of air (e.g., 100 L at 3 ppbv). The method is only applicable to concentrations below 10 ppbv if the absorption efficiency of the particular system is determined.

2.4 Beer's Law is followed through the working range from 0.1 to 1.0 absorbance units (0 to 35 μg in 25 mL final solution).

3. Interferences

3.1 The effects of the principal potential interferences [oxides of nitrogen, ozone and transition metals (e.g. iron, manganese, and chromium)] have been minimized or eliminated. The interferences by oxides of nitrogen are eliminated by sulfamic acid **(5,6)**, the ozone by time delay **(4)**, and the transition metals by EDTA and phosphoric acid **(4,5)**. At least 60 μg of Fe(III), 10 μg of Mn (II), 10 μg of Cr(III), 10 μg of Cu(II) and 10 μg of V(V) in 10 mL of absorbing reagent cause no interference.

4. Precision and Accuracy

4.1 The precision at the 95% confidence level is 4.6% **(5)**.

5. Apparatus

5.1 ABSORBER. Satisfactory absorbers are (a) the midget or standard fritted bubbler: (b) the midget impinger; (c) the Greenburg-Smith impinger; and (d) the multiple jet bubbler **(12)**.

5.2 AIR VOLUME MEASUREMENT. The sample air volume must be measured within ± 2%. See Section 5 of Part I for details of air volume measurement. A critical orifice is recommended **(13)**.

5.3 SPECTROPHOTOMETER OR COLORIMETER. The instrument must be suitable for measurement of color at 548 nm or 575 nm. With Variation A, reagent blank problems may result with spectrophotometers or colorimeters having greater spectral band width than 16 nm. The wavelength calibration of the spectrophotometer should be verified.

6. Reagents

6.1 PURITY OF CHEMICALS. All chemicals must be ACS analytical reagent grade. The pararosaniline dye should meet the specifications described in 6.9.

6.2 WATER. Water must conform to the ASTM Standard for Reference Reagent Water, Type II. This can be produced either by distillation or from a cartridge deionization system.

6.3 ABSORBING REAGENT, 0.04M POTASSIUM TETRACHLOROMERCURATE (TCM), K_2HgCl_4. Dissolve 10.86 g mercuric chloride (CAUTION: Highly poisonous and corrosive. If spilled on skin, flush off with water immediately), 5.96 g of potassium chloride, and 0.066 g of EDTA in water, and bring to mark in a 1-L volumetric flask. Sodium chloride (4.68 g) may be substituted for the potassium chloride, but potassium chloride is usually available in purer form. The pH of this reagent should be 4 ± 1 **(14)**. The absorbing reagent is normally stable for six months, but if a precipitate forms, discard the solution.

6.4 SULFAMIC ACID, 0.6%. Dissolve 0.6 g of sulfamic acid in 100 mL of water. This reagent can be kept for 10 days if it is stored in a stoppered bottle.

6.5 BUFFER STOCK SOLUTION (pH 4.7). In a 100-mL volumetric flask, dissolve 13.61 g of sodium acetate trihydrate in water. Add 5.7 mL of glacial acetic acid and dilute to volume with water.

6.6 1.0N HYDROCHLORIC ACID. Dilute 83 mL of 12.1M acid (36% HCl) to 1 L.

6.7 3M PHOSPHORIC ACID (H_3PO_4). Dilute 205 mL of 14.6M acid (85% H_3PO_4) to 1 L.

6.8 BUTANOL. 1-Butanol is required for the purification of the pararosaniline dye. The butanol should be checked for oxidants that can consume SO_2. Check it by shaking 20 mL of 1-butanol with 5 mL of 20% aqueous KI. A yellow color in the alcohol phase indicates the presence of oxidants, and the butanol must be redistilled from silver oxide.

6.9 PURIFIED PARAROSANILINE, 0.2% (NOMINAL) STOCK SOLUTION. Pararosaniline dye needed to prepare this reagent must meet the following performance specifications: The dye must have a wavelength of maximum absorbance of 540 nm when assayed in a 0.1M sodium acetate-acetic acid buffer; the absorbance of the reagent blank (Section 7.3), which is temperature sensitive (0.015 A. U./°C), should not exceed 0.170 absorbance unit at 22°C; and the reagents should give a calibration curve with a slope of 0.746 ± 0.04 absorbance units/(μg SO_2/mL) for a 1-cm cell. Prepare a solution of the dye: Weigh 0.200 g and completely dissolve by shaking with 100 mL of lN HCl in a 100-mL glass-stoppered graduated cylinder. If the pararosaniline dye is obtained in purified form as 0.2% solution, proceed to 6.9.5. Specially purified dye, 0.2% in lN HCl solution, is available from Eastman Kodak, J.T. Baker or Harleco.

6.9.1 *Dye Purification Procedure.* The pararosaniline dye (PRA) is purified by extraction of impurities into 1-butanol. In a 250-mL separating funnel, shake 100 mL each of 1-butanol and 1 N HCl and allow the layers to separate. Collect each layer separately. Add 0.1 g of pararosaniline to 50 mL of the butanol-saturated acid and allow it to stand for several minutes. Add this acid pararosaniline solution to 50 mL of acid-saturated 1-butanol in a 125-mL separatory funnel. The impurities, violet colored, will be transferred to the organic phase. Save the lower (aqueous) phase and extract again with 20 mL of 1-butanol. Repeat this procedure three times using 10-mL portions of 1-butanol. If a violet color still appears in the 1-butanol after 5 extractions, discard this lot of dye. After the final extraction, filter the aqueous phase through a glass wool plug into a 50-mL volumetric flask and bring to volume with 1N HCl. The final solution is nominally 0.2% pararosaniline in 1N HCl saturated with 1-butanol.

6.9.2 *1.0M Sodium Acetate−Acetic Acid Buffer.* Dissolve 57.3 mL glacial acetic acid and 136 g sodium acetate dihydrate in water and make up to the mark in a 1-L volumetric flask.

6.9.3 *Assay Procedure.* The concentration of PRA need be assayed only once for each lot of dye in the following manner: Dilute 1 mL of the stock reagent (0.2%) to the mark in a 100-mL volumetric flask with distilled water. Transfer a 5-mL aliquot to a 50-mL volumetric flask. Add 5 mL of 1M sodium acetate-acetic acid buffer, and dilute the mixture to 50 mL with water. After

1 h, determine the absorbance at 540 nm with a spectrophotometer. Determine the percent of the nominal concentration of PRA by the formula:

$$\% \text{ PRA } = \frac{\text{Absorbance} \times \text{K}}{\text{grams taken}}$$

For 1-cm cells and a narrow-bandwidth spectrophotometer, K should equal 21.3 (Mean value after extensive purification of dye).

6.9.4 *Pararosaniline Reagent.* To a 250-mL volumetric flask add 20 mL of stock pararosaniline reagent. Add an additional 0.2 mL of stock for each percent the stock assays below 100%. For Variation A, add 25 mL of 3M H_3PO_4 and dilute to volume with water. For Variation B, add 200 mL of 3M H_3PO_4 and dilute to volume. These reagents are stable for at least 9 months.

6.10 FORMALDEHYDE, 0.2%. Dilute 5 mL of 37% formaldehyde to 1 L with water. Prepare daily.

6.11 REAGENTS FOR STANDARDIZATION.

6.11.1 *Stock Iodine Solution, 0.1N.* Place 12.7 g of iodine in a 250-mL beaker, add 40 g of potassium iodide and 25 ml of water. Stir until all the solid is dissolved, then dilute to 1 L with water. If desired, prestandardized solutions of iodine can be purchased from laboratory supply dealers.

a) Working Iodine Solution, 0.01N: Prepare approximately 0.01N iodine solution by diluting 50 mL of the stock solution to 500 mL with distilled water.

6.11.2 *Starch Indicator Solution.* Triturate 0.4 g of soluble starch and 0.002 g of mercuric iodide (preservative) with a little water, and add the paste slowly to 200 mL of boiling water. Continue boiling until clear; cool, and transfer to a glass-stoppered bottle. If the indicator solution is only going to be kept for a few days, omit the mercuric iodide.

6.11.3 *Standard 0.1N Sodium Thiosulfate Solution.* Dissolve 25 g of sodium thiosulfate ($Na_2S_2O_3 \cdot 5 H_2O$) in 1 L of freshly boiled, cooled distilled water and add 0.1 g of sodium carbonate to the solution. Allow the solution to stand for one day before standardizing. To standardize, accurately weigh 1.5 g of potassium iodate, primary standard grade, that was dried at 180°C, and dilute to volume in a 500-mL volumetric flask. Into a 500-mL glass-stoppered Erlenmeyer flask, pipet 50 mL of the iodate solution. Add 2 g of potassium iodide and 10 mL of a 1:10 dilution of concentrated hydrochloric acid. Stopper the flask. After 5 minutes, titrate with thiosulfate to a pale yellow color. Add 5 mL of starch indicator solution and complete the titration.

$$\text{Normality of} \atop \text{Thiosulfate} = \frac{\text{Wt. (grams KIO}_3)}{\text{mL of thiosulfate}} \times 2.804$$

Prestandardized solutions of sodium thiosulfate can be purchased from laboratory supply dealers.

6.11.4 *Standard Sulfite Solution.* Dissolve 0.400 g of sodium sulfite (Na_2SO_3) or 0.300 g of sodium metabisulfite ($Na_2S_2O_5$) in 500 mL reagent quality deaerated water. For a reagent purity of 100% and no loss of sulfite in making up the solution, the respective SO_2 content of this solution will be 406 μg/mL (for Na_2SO_3), and 404 μg/mL (for $Na_2S_2O_5$) as SO_2. In practice, SO_2 concentration will be 0 to 10% below the theoretical value. If this level of uncertainty is acceptable it is not necessary to standardize the sulfite solution. To minimize loss of sulfite, the dilute sulfite solution should be prepared immediately as indicated in 6.10.5. The actual concentration of sulfite in the standard solution can be determined by adding excess iodine and back-titrating with sodium thiosulfate that has been standardized by iodometric titration against potassium iodate or dichromate (primary standard). Aqueous solutions of sulfite are unstable due to air oxidation of the sulfite or loss of volatile SO_2.

a) Back-titration is performed in the following manner: into each of two 500-mL Erlenmeyer flasks with ground glass stoppers, pipet accurately 50 mL of the 0.01N iodine. Into flask A (blank) add 25 mL of distilled water, and into flask B (sample) pipet 25 mL of the standard sulfite solution. Stopper the flasks and allow to react for 5 minutes. By means of a buret containing standard 0.01N thiosulfate, titrate each flask, in turn, to a pale yellow color. Then add 5 mL of starch solution and continue the titration to the disappearance of the blue color. Calculate the concentration of sulfur dioxide in the standard solution as follows:

$$SO_2(\mu g/ml) = \frac{(X - Y)\,N\,Z}{S}$$

Where:
X = number of mL for blank,
Y = number of mL for sample,
N = normality of thiosulfate solution,
Z = micro-equivalent weight for SO_2, 3.203 × 10⁴
S = sample volume taken in mL.

6.11.5 *Dilute Sulfite Solution.* Immediately after standardization of sulfite solution or preparation of an acceptable unstandardized solution, pipet accurately 2 mL of the freshly standardized solution into a 100-mL volumetric flask and bring to mark with 0.04M TCM. This solution is stable for 30 days if stored at 5°C.

7. Procedure

7.1 Collection of Sample. Place 10 mL of 0.04M TCM (20 mL for sampling of long duration) absorbing solution in a midget impinger, or 75 to 100 mL in a larger absorber.

The sampling probe-to-absorber linkage will depend on the installation. This should be kept as short and direct as possible using glass or Teflon tubing. Care must be taken to avoid condensation in the sample inlet, which can occur when warm humid air is brought into an air conditioned location. After the absorber, a trap, such as a flask tightly packed with glass wool, should be placed to prevent corrosion damage to downstream components from aerosolized TCM droplets, which are highly corrosive. Provision must be made for the flow volume measurement farther downstream of the absorber. The duration of sampling will depend on the concentration of SO_2. With midget impingers, rates of 0.5 to 2.5 L/min are satisfactory; with large absorbers the rate can be 5 to 15 L/min. Rates of sampling within the above ranges generally have an absorption efficiency of 98% or greater. For best results, rates and sampling time should be chosen to absorb 0.5 to 30 μg/mL of SO_2. Shield the absorber from direct sunlight by covering with a suitable wrapping. If the sample must be stored for more than a day before analysis, keep it at 5°C, if possible.

7.2 CENTRIFUGATION. If a precipitate is observed, remove it by centrifugation.

7.3 ANALYSIS. After collection, transfer the sample quantitatively to a 25-mL volumetric flask; use about 5 mL of water for rinsing. Aliquots may be taken, at this point, if the concentration or volume of reagent is larger. If the sample has been freshly collected, delay analyses for 20 min to allow any ozone present to decompose. For each set of determinations, prepare a reagent blank by adding 10 mL of the unexposed absorbing reagent to a 25-mL volumetric flask. To each flask add 1 mL of 0.6% sulfamic acid and allow to react for 10 min to destroy the nitrite from oxides of nitrogen. Accurately pipet in 2 mL of the 0.2% formaldehyde, then 5 mL of pararosaniline reagent prescribed for Variation A or Variation B. Start a laboratory timer that has been set for 30 min. Bring all flasks to volume with freshly boiled water. After the 30 min, determine the absorbances of the sample and of the blank at the wavelength of maximum absorbance, 548 nm for Variation A or 575 nm for Variation B. Use water (*not* the reagent blank) in the reference cell. Do not allow the colored solution to stand in sample absorbance cell; a film of dye will be deposited on the cell windows.

7.3.1 If the absorbance of the sample solution ranges between 1.0 and 2.0, the sample can be diluted 1:1 with a portion of the reagent blank and read within a few minutes. Solutions with higher absorbance can be diluted up to six-fold with the reagent blank in order to obtain on-scale readings within 10% of the true absorbance value.

7.3.2 An automated procedure can be used as an alternative to manual addition of reagents and colorimetric analysis (15).

8. Calibration and Standards

8.1 Accurately pipet graduated amounts of the diluted sulfite solution (such as: 0, 1, 2, 3, 4, and 5 mL) into a series of 25-mL volumetric flasks. Add sufficient 0.04M TCM to each flask to bring the volume to approximately 10 mL. Then add the remaining reagents as described in the procedure. For greatest precision a constant temperature bath is preferred. The temperature

of calibration should not differ from the temperature of analysis by more than a few degrees.

8.2 The total absorbances of the solutions are plotted (as ordinates) against the total micrograms of SO_2. A linear relationship is obtained. The intercept with the vertical axis of the line best fitting the points usually is within 0.02 absorbance units of the blank (zero standard) reading. Under these conditions the plot need be determined only once to evaluate the calibration factor (reciprocal of the slope of the line). More accurate values of slope and intercept can be obtained by linear regression. This calibration factor can be used for calculating results provided there are no radical changes in temperature or pH. At least one control sample is recommended per series of determinations to insure the reliability of this factor.

8.3 ALTERNATIVE CALIBRATION PROCEDURE. Permeation tubes containing liquefied sulfur dioxide are calibrated gravimetrically and used to prepare standard concentrations of sulfur dioxide in air (16–18). Sampling of these known concentrations and subsequent analysis of collected samples give calibration curves that simulate all of the operational conditions during the sampling and analytical procedure. Such calibration curves include the important correction for collection efficiency at various concentrations of sulfur dioxide. For details on the use of permeation tubes as calibration sources see Section 3, Part I.

9. Calculations

9.1 Compute the concentration of sulfur dioxide in the sample by the following formula:

$$ppm = \frac{(A-A_o)0.382\ B}{V}$$

Where:

A = the sample absorbance
A_o = the reagent blank absorbance
0.382 = the volume (μL) of 1 μg SO_2 at 25°C, 101.3 kPa
B = the calibration factor, μg/absorbance unit
V = the sample volume in liters cor-

rected at 25°C, 101.3 kPa (by PV = nRT)

10. Effects of Storage

Solutions of monochlorosulfonatomercurate (II) are stable towards the loss of SO_2 when stored at 5°C for 30 days. At 25°C the loss of SO_2 ranges from 1–5% per day (3,19,20). These losses of SO_2 follow a first order reaction; the rate constant is independent of concentration. For storage of samples under conditions where SO_2 loss may be important, e.g., without refrigeration, correction factors must be applied. Alternately, an increase in the Cl^- concentration will improve the stability of the monochlorosulfonatomercurate (II) species (3,19). This modification leads to some difficulties in sampling and a loss of sampling efficiency. Thermoelectrically cooled enclosures for housing the absorber vessels during or after sampling are commercially available from RAC, Division of Andersen Samplers, Inc. Atlanta, Georgia.

11. References

1. WEST, P.W. and G.C. GAEKE. 1956. Fixation of Sulfur Dioxide as Sulfitomercurate (II) and Subsequent Colorimetric Determination, *Anal. Chem.* 28:1816–1819.
2. EPHRAIMS, F. 1948. *Inorganic Chemistry.* Thorne, P.C.L. and E.R. Roberts, Eds., 5th ed. p. 562, Interscience, N.Y.
3. DASGUPTA, P.K. and K.B. DECESARE, 1982. Stability of Sulfur Dioxide in Formaldehyde Absorber and its Anomalous Behavior in Tetrachloromercurate (II), *Atmos. Environ.* 16:2927–2934.
4. ZURLO, N. and A.M. GRIFFINI. 1962. Measurement of the SO_2 Content of Air in the Presence of Oxides of Nitrogen and Heavy Metals, *Med. Lavoro*, 53:325–329.
5. SCARINGELLI, F.P., B.E. SALTZMAN and S.A. FREY, 1967. Spectrophotometric Determination of Atmospheric Sulfur Dioxide, *Anal. Chem.* 39:1709–1719.
6. PATE, J.B., B.E. AMMONS, G.A. SWANSON and J.P. LODGE, JR. 1965. Nitrite Interference in Spectrophotometric Determination of Atmospheric Sulfur Dioxide. *Anal. Chem.* 37:942–945.
7. Federal Register. 1971. 36(84):8187–91, April 30.
8. LYLES, G.R., F.B. DOWLING and V.J. BLANCHARD. 1965. Quantitative Determination of Formaldehyde in Parts Per Hundred Million Concentration Level, *J. Air. Pollut. Contr. Assoc.* 15:106–108.
9. URONE, P., J.B. EVANS and C.M. NOYES. 1965. Tracer Techniques in Sulfur Dioxide–Air Pollution Studies. Apparatus and Studies of Sulfur Dioxide Colorimetric and Conductometric Methods. *Anal Chem,* 9:1104–1107.
10. BOSTRÖM, C.E. 1965. The Absorption of Sulfur Dioxide at Low Concentrations (pphm) Studied by an Isotopic Tracer Method, *Air & Water Pollut. Int. J.* 9:333–341.
11. BOSTRÖM, C.E. 1966. The Absorption of Low Concentration (pphm) of Hydrogen Sulfide in a $Cd(OH)_2$ Suspension as Studied by an Isotopic Tracer Method, *Air & Water Pollut. Int. J.* 10:435–441.
12. STERN, A.C. 1968. *Air Pollution.* Vol. II. 2nd ed. Academic Press, N.Y.
13. LODGE, J.P., JR., J.B. PATE, B.E. AMMONS and G.A. SWANSON. 1966. The Use of Hypodermic Needles as Critical Orifices in Air Sampling, *J. Air Pollut. Contr. Assoc.* 16:197–200.
14. SCARINGELLI, F.P., L.A. ELFERS, D. NORRIS and S. HOCHHEISER. 1970. Enhanced Stability of Sulfur Dioxide in Solution, *Anal. Chem.* 42:1818–1820.
15. LOGSON, O.J. II and M.J. CARTER. 1975. Comparison of Manual and Automated Analysis Methods for Sulfur Dioxide in Manually Impinged Ambient Air Samples, *Environ. Sci. Technol.,* 9:1172–1174.
16. O'KEEFFE, A.E. and G.C. ORTMAN. 1966. Primary Standards for Trace Gas Analysis, *Anal. Chem.,* 38:760–763.
17. SCARINGELLI, F.P., S.A. FREY and B.E. SALTZMAN. 1967. Evaluation of Teflon Permeation Tubes for Use with Sulfur Dioxide, *Amer. Ind. Hyg. Assoc. J.* 28:260–266.
18. THOMAS, M.D. and R.E. AMTOWER. 1966. Gas Dilution Apparatus for Preparing Reproducible Dynamic Gas Mixtures in Any Desired Concentration and Complexity, *J. Air Pollut. Contr. Assoc.* 16:618–623.
19. FUERST, R.G. 1976. Improved Temperature Stability of Sulfur Dioxide Samples Collected by the Federal Reference Method, EPA-600/4-76-024, U.S. Environmental Protection Agency, Research Triangle Park, N.C., May.
20. FUERST, R.G., F.P. SCARINGELLI and J.H. MARGESON. 1976. Effects of Temperature on the Stability of Sulfur Dioxide Samples Collected by the Federal Reference Method, EPA-600/4-76-024, U.S. Environmental Protection Agency, Research Triangle Park, N.C., May.

Subcommittee 1

G.L. KOK
P.K. DASGUPTA, *Chairman*
D.F. ADAMS
B.R. APPEL
S.O. FARWELL
K.T. KNAPP
W.R. PIERSON
K.D. REISZNER
R.L. TANNER

704B.

Determination of Sulfur Dioxide Content of the Atmosphere (Formaldehyde Absorber/Pararosaniline Method)

1. Principle of the Method

1.1 Sulfur dioxide, in an ambient air sample, is collected in a buffered formaldehyde absorber solution, which enhances the collection efficiency and stabilizes the resulting sulfite from oxidative loss. Potassium hydrogen phthalate is used as the buffer in the absorber and formaldehyde as the stabilizer for the sulfite species, which forms the oxidation resistant species, hydroxymethanesulfonate (1,2). The procedure is similar to method 704A in absorption and recovery efficiency, sensitivity and precision, and avoids the use of the toxic and costly tetrachloromercurate (II). In addition, stability of the collected SO_2 is greatly enhanced over method 704A. The analytical measurement is based on the formation of the intensely colored pararosanilinemethylsulfonic acid which is produced from sulfite, pararosaniline and formaldehyde.

2. Range and Sensitivity

2.1 Atmospheric sulfur dioxide concentrations measurable by this technique range from 10 ppbv to a few ppmv. Higher concentrations (5 to 500 ppmv), employed in special studies, must be analyzed by using smaller gas samples.

2.2 Collection efficiency falls off rapidly below 10 ppbv, and varies with the geometry of the absorber, the size of the gas bubbles, and the contact time with the solution (3,4,5).

2.3 The lower limit of detection of sulfur dioxide is 0.3 μL gaseous SO_2 in 15 mL of formaldehyde absorber (based on twice the standard deviation) representing a concentration of 10 ppbv (26 μg/m³) SO_2 in an air sample of 30 L. One cannot extrapolate to lower values by taking larger volumes of air, e.g., 100 L at 3 ppbv. The method is applicable to concentrations below 10 ppbv if the absorption efficiency of the particular system is determined.

2.4 Beer's Law is followed through the working range from 0.1 to 0.7 absorbance units (up to 30 μg in 25 mL final solution).

3. Interferences

3.1 The effects of the principal known interferences, oxides of nitrogen, ozone and heavy metals (e.g., iron, manganese, and copper) have been minimized or eliminated. The interferences by oxides of nitrogen are eliminated by sulfamic acid, the ozone by time-delay, and the transition metals by CDTA. Compared to method 704A this procedure is less susceptible to interferences by NO_2 even without added sulfamic acid; however, the latter must be used when large concentrations of NO_2 are present. At least 100 μg of Fe(III), 20 μg of Mn(II), 100 μg of Cu(II), and 200 μg of (V) in 15 mL of absorbing reagent can be tolerated in the procedure.

4. Precision and Accuracy

4.1 The precision at the 95% confidence level is 2.7% for SO_2 concentrations of 1 μg/mL in the absorber (1).

5. Apparatus

5.1 ABSORBER. Satisfactory absorbers are (a) the midget or standard fritted bubbler: (b) the midget impinger; (c) the Greenburg-Smith impinger; and (d) the multiple jet bubbler (6).

5.2 AIR VOLUME MEASUREMENT. The sample air volume must be measured within ± 2%. See Section 5, Part I for details of air volume measurement. The use of a critical orifice is recommended (7).

5.3 SPECTROPHOTOMETER OR COLORIMETER. The instrument must be suitable for measurement of color at 580 nm. The wavelength calibration of the spectrophotometer should be verified.

5.4 CULTURE VIALS. 25-mL PTFE-lined screw cap glass culture vials.

6. Reagents

6.1 PURITY OF CHEMICALS. All chemicals must be ACS analytical reagent grade. The paraosaniline dye should meet the specifications described in Method 704A-6.9.

6.2 WATER. Water must conform to the ASTM Standard for Reference Reagent Water, Type II. This can be produced either by distillation or from a cartridge deionization system.

6.3 SODIUM HYDROXIDE, 4.5 M. In a 100-mL volumetric flask place 18 g solid NaOH and about 80 mL water. After the NaOH is dissolved and the solution has cooled, dilute to the mark. This solution should be stored in a plastic bottle.

6.4 STOCK CDTA SOLUTION. Dissolve 18.2 g (trans-1,2 cyclohexylenedinitrilo)tetraacetic acid (CDTA) and ~0.4 g NaOH in water and dilute to 100 mL. This provides a 5×10^{-2} M stock solution of CDTA.

6.5 BUFFERED FORMALDEHYDE ABSORBING REAGENT, 10X. Dilute 5.30 mL of 37% formaldehyde solution, 2.04 g of potassium hydrogen phthalate and 20 mL of the 0.05 M Na_2 CDTA solution to 1 L. This solution will keep for one year.

6.6 BUFFERED FORMALDEHYDE ABSORBER, WORKING REAGENT. Dilute the 10X absorbing reagent tenfold with water to produce the buffered formaldehyde absorbing reagent.

6.7 SULFAMIC ACID, 0.6%. Dissolve 0.6 g of sulfamic acid in 100 mL of water. Add dilute NaOH dropwise to adjust to pH 4. This reagent can be kept for 10 days if protected from air.

6.8 HYDROCHLORIC ACID, 1.0N. Dilute 83 mL of 12.1M acid (37% HCl) to 1 L.

6.9 PURIFIED PARAROSANILINE. There are three sources for purified pararosaniline: (a) Several chemical companies sell specially purified dye in 0.2% concentration in 1N HCl solution. This dye is ready for use without any additional treatment, (b) Commercially available, solid unpurified pararosaniline can be purified by one of two methods. Method 704A-6.9.1 gives the 1-butanol extraction procedure, which produces the pararosaniline as a 0.2% solution in 1N HCl, (c) As an alternative the dye can be recrystallized to form the trichloroacetate salt as detailed below (8). The purity of the dye must be checked by the procedure given in Method 704A-Section 6.9.4.

6.9.1 *Dye Purification Procedure.* Dissolve 1 g of dye in 100 mL of approximately 1M trichloroacetic acid solution. Bring the pH up to 2.0 by the addition of 4.5M NaOH. Allow the solution to stand one hour and filter to remove any precipitate. Add additional 4.5M NaOH to the filtrate and bring the pH up to 8-9. The trichloroacetate salt of pararosaniline will precipitate and can be filtered from the liquid under vacuum. The collected precipitate should be washed with 10 mL of 0.1N $NaHCO_3$. The precipitate can be dried by storage in a desiccator. To prepare a 0.2% nominal solution of pararosaniline in acid, weigh 0.28 g of the pararosaniline trichloracetate and dissolve in 100 mL of 1N HCl. This will give a concentration equivalent of 0.2% as purchased or prepared by the butanol extraction procedure.

6.9.2 *Pararosaniline Reagent.* Dilute 133 mL of the 0.2% purified pararosaniline solution and 114 mL of concentrated HCl to 1 L with distilled water.

6.10 STANDARD SULFITE SOLUTION. Dissolve 0.400g of sodium sulfite (Na_2SO_3) or 0.300 g of sodium metabisulfite ($Na_2S_2O_5$) in 500 mL reagent quality deaerated water. For a reagent purity of 100% and no loss of sulfite in making up the solution, these solutions will contain 406 μg/mL (for Na_2SO_3), and 404 μg/mL (for $Na_2S_2O_5$) as SO_2. In practice SO_2 concentration will be 0 to 10% below the theoretical value. If this level of uncertainty is acceptable it is not necessary to standardize the sulfite solution. To minimize loss of sulfite, the dilute sulfite solution should be prepared immediately as indicated in 6.10.1. The actual concentration of sulfite in the standard solution can be determined by adding excess iodine and back-titrating with sodium thiosulfate that has been standardized by iodometric titration against potassium iodate or dichromate (primary standard). This procedure is detailed in Method 704A, Section 6.11.4. Procedures for the titrimetric analysis of sulfite can be found in standard analytical texts (9). Solutions containing sulfite are unstable due to air oxidation of the sulfite or loss of volatile SO_2.

6.10.1 *Dilute Sulfite Solution.* Immediately after standardization of sulfite solution or preparation of an acceptable unstandardized solution, pipet 2 mL of the standard sulfite solution and 10 mL of the 10X absorbing reagent in a 100-mL volumetric flask and fill to the mark with distilled water. This sulfite solution is stable for one month and is used to prepare working sulfite standards as indicated in section 8.1.

7. Procedure

7.1 COLLECTION OF SAMPLE. Place 15 mL (20 mL for sampling of long duration) of the buffered formaldehyde absorbing solution in a midget impinger, or 75 to 100 mL in one of the larger absorbers.

The sampling probe-to-absorber linkage will be variable depending on the installation. This should be kept as short and direct as possible using glass or Teflon tubing. Care must be taken to avoid condensation of moisture in the sample inlet, which can occur when warm, humid air is brought into an air conditioned location. Farther downstream of the absorber, provision must be made for the flow volume measurement. The duration of sampling will depend on the concentration of SO_2. With midget impingers, a sampling rate up to 0.5 L/min is satisfactory; with large absorbers the sampling rate can be up to 12 L/min. Rates of sampling within the above ranges generally have an absorption efficiency of 98% or greater. For best results, rates and sampling time should be chosen to obtain an absorber concentration of 0.5 to 1.0 μg/mL of SO_2. Shield the absorber from direct sunlight by covering it with a suitable wrapping.

7.2 ANALYSIS. After collection, 1 mL of 0.6% sulfamic acid solution should be added to the sample if NO_2 interference is expected to be present. Subsequently, the sample can be stored at room temperature for up to one month with no significant degradation of the collected sulfite (1,2). For analysis, 1 mL of 4.5M NaOH is added to the collected sample. The NaOH is best added from a dispensing pipet. After NaOH addition and careful mixing, the sample is rapidly added to 5 mL of the pararosaniline working reagent in a 25-mL capped culture tube. The sample-pararosaniline solution should be immediately capped and mixed by inverting several times. It is important that the sample be added to the acidic pararosaniline solution and not in the reverse order, to prevent lower absorbance values and poor precision. After mixing, water is added to the contents of the culture tube to bring the level up to the neck, giving 25 mL of total volume. The absorbance of the resulting color should be measured between 10–15 minutes after preparation at a wavelength of 580 nm. Use distilled water (*not* the reagent blank) in the reference cell. Do not allow the colored solution to stand in the sample cell; a film of dye will be deposited on the cell windows.

7.3 As an alternative to manual addition of reagents and colorimetric analysis, the method can be automated (10).

8. Calibration and Standards

8.1 WORKING SULFITE STANDARDS. The dilute sulfite solution prepared in 6.10.1 contains approximately 8 μg/mL as SO_2. To prepare a standard calibration series dilute x mL of the sulfite standard with (*15-x*) mL of the absorbing solution. This series of standards is then analyzed as described in Section 7.2.

8.2 The total absorbance of each solution is plotted (as ordinates) against the total micrograms of SO_2. A linear relationship is obtained. The intercept with the vertical axis of the line best fitting the points usually is within 0.03 absorbance units of the blank (zero standard) reading. Under these conditions the plot need be determined only once to evaluate the calibration factor (reciprocal of the slope of the line). This factor can be used for calculating results provided there are no radical changes in temperature or pH. At least one control sample is recommended per series of determinations to insure the reliability of this factor.

8.3 ALTERNATIVE CALIBRATION PROCEDURE. As an alternative to calibration with standard solutions, permeation tubes can be used to prepare calibration curves which simulate all of the operational conditions performed during sampling and chemical analysis. For details on the use of permeation tubes as calibration sources see Section 3, Part I.

9. Calculations

9.1 Compute the concentration of sulfur dioxide in the sample by the following formula:

$$\text{ppm} = \frac{(A-A_o)0.382\ B}{V}$$

Where

A = the sample absorbance

A_o = the reagent blank absorbance

0.382 = the volume (μl) of 1 μg SO_2 at 25°C, 101.3 kPa

B = the calibration factor, $\mu g/$absorbance unit

V = the sample volume in liters corrected to 25°C, 101.3 kPa (by PV = nRT)

10. Effects of Storage

Solutions of collected sulfur dioxide, as the hydroxymethanesulfonate, are very stable from loss of sulfite when protected from direct sunlight. Samples of hydroxymethanesulfonate stored at 22°C for 40 days had a loss of 2.6% sulfite. This loss increased to 12% in 19 days for samples stored in direct sunlight (2).

11. References

1. DASGUPTA, P.K., K. DECESARE and J.C. ULLREY. 1980. Determination of Atmospheric Sulfur Dioxide Without Tetrachloromercurate (II) and the Mechanism of the Schiff Reaction, *Anal. Chem.* 52:1912–1922.

2. DASGUPTA, P.K. and K.B. DECESARE. 1982. Stability of Sulfur Dioxide in Formaldehyde Absorber and its Anomolous Behavior in Tetrachloromercurate (II), *Atmos. Environ.* 16:2927–2934.

3. URONE, P., J.B. EVANS and C.M. NOYES. 1965. Tracer Techniques in Sulfur Dioxide – Air Pollution Studies. Apparatus and Studies of Sulfur Dioxide Colorimetric and Conductometric Methods. *Anal Chem.* 9:1104–1107.

4. BOSTRÖM, C.E. 1965. The Absorption of Sulfur Dioxide at Low Concentrations (pphm) Studied by an Isotopic Tracer Method, *Air & Water Pollut. Int. J.* 9:333–341.

5. BOSTRÖM, C.E. 1966. The Absorption of Low Concentration (pphm) of Hydrogen Sulfide in a $Cd(OH)_2$ Suspension as Studied by an Isotopic Tracer Method, *Air & Water Pollut. Int. J.* 10:435–441.

6. HENDRICKSON E.R. 1968. Air Sampling and Quantity Measurement in "Air Pollution," Vol. II, A.C. Stern, Ed., 2nd Ed. Academic Press, N.Y., pp 3–51.

7. LODGE, J.P., JR., J.B. PATE, B.E. AMMONS and G.A. SWANSON. 1966. The Use of Hypodermic Needles as Critical Orifices in Air Sampling, *J. Air Pollut. Contr. Assoc.* 16:197–200.

8. DASGUPTA, P.K. 1981. Determination of Atmospheric Sulfur Dioxide Without Tetrachloromercurate (II): Further Refinements of Pararosaniline Method and Field Application, *J. Air Pollut. Control Assoc.* 31:779–782.

9. MEITES, L. 1963. "Handbook of Analytical Chemistry," pp. 3–70, McGraw Hill, New York.

10. KOK, G.L., S.N. GITLIN, B.W. GANDRUD and A.L. LAZRUS. 1984. Automated Determination of Sulfur (IV) Using the Schiff Reaction, *Anal. Chem.* 56:1993–1994.

Subcommittee 1

G.L. KOK
P.K. DASGUPTA, *Chairman*
D.F. ADAMS
B.R. APPEL
S.O. FARWELL
K.T. KNAPP
W.R. PIERSON
K.D. REISZNER
R.L. TANNER

704C.

Determination of Sulfur Dioxide Content of the Atmosphere (Hydrogen Peroxide Method)

1. Principle of the Method

Sulfur dioxide is absorbed by aspirating a measured volume of air through a solution of dilute hydrogen peroxide (1). Sulfur dioxide is oxidized by hydrogen peroxide to sulfuric acid. The oxidation of S(IV) (dissolved SO_2) to S(VI) (sulfate) by H_2O_2 is very fast (2). The sulfate collected in the absorber solution may be determined by any method suitable for aqueous sulfate (see 720. DETERMINATION OF AQUEOUS SULFATE).

2. Range and Sensitivity

2.1 Atmospheric SO_2 concentrations typically range from a maximum of a few mg/m^3 to less than 1 $\mu g/m^3$ (1 $\mu g/m^3$ = 0.382 ppbv SO_2 at 25°C). Using 50 mL of 0.6 % H_2O_2 as absorber with small orifice-size impingers, at a sampling rate of 0.2 L/min, collection is known to be essentially quantitative at 24 $\mu g/m^3$ SO_2 and above (1). Ambient sampling results suggest that the collection under these conditions is probably quantitative at 10 $\mu g/m^3$ SO_2 and above (1). The collection efficiency is dependent on the geometry of the bubbler/impinger, the size of the gas bubbles and the contact time with the solution. Although the collection efficiency with this absorber has been measured only at a sampling rate of 0.2 L/min and an absorber volume of 50 mL (1), based on comparison with other absorbers, the collection should be essentially quantitative with fritted midget bubblers containing 10 mL or more absorber up to a sampling rate of 0.5 L/min. The collection is quantitative with 40 mL 1% H_2O_2 absorber in a protruding end bubbler (e.g., type 7530, Ace Glass, Vineland, NJ) at a flow rate of 2 L/min (3).

2.2 The minimum detectable concentra-

tion of SO_2 will depend on both the air sample volume and the limit of detection of the specific analytical method used. With a typical air sample volume of 0.288 m^3 (0.2 L/min for 24 h) drawn into 10 mL absorbing solution, 26 $\mu g/m^3$ (10 ppbv) SO_2 can be detected by most of the methods given for analysis of aqueous sulfate (see 720. DETERMINATION OF AQUEOUS SULFATE).

3. Interferences

3.1 Unlike the other methods for the determination of atmospheric SO_2 (704A and 704B), this analysis relies on the determination of sulfate produced from SO_2, rather than SO_2 itself. Consequently, it is imperative that particulate sulfate be removed from the air stream, before it reaches the absorbing solution, with a filter that removes particulate matter without removing SO_2 from the air stream. For this purpose, the only suitable filters at present are polytetrafluoroethylene (PTFE) or polycarbonate membrane filters and acid-treated quartz-fiber or PTFE-coated glass fiber filters (see Section 3, method 824).

3.2 The collection of particulate material by the absorber is neither quantitative nor particularly reproducible. Therefore it is not possible to determine SO_2 + total particulate sulfate by omitting the filter.

3.3 Some of the analytical methods given in the section on DETERMINATION OF AQUEOUS SULFATE are susceptible to interference by heavy metal ions. These are removed by cation exchange of the sample prior to sulfate determination. In the present collection procedure, the presence of the filter prevents any particulate metal compounds entering the absorber; the cation exchange step may therefore be omitted.

3.4 Other sulfur-containing gaseous species, e.g. H_2S, CH_3SH or COS do not significantly interfere. Stated recoveries were 97.5 ± 2.5% for 132–208 $\mu g/m^3$ SO_2 in the presence of 71–200 $\mu g/m^3$ of the above gases (1). Ozone does not interfere significantly. In the presence of 980 $\mu g/m^3$ O_3, stated recoveries were 103 ± 1% for 54–139 $\mu g/m^3$ SO_2 (1).

3.5 The prefilter, if properly chosen (see 3.1), does not affect the efficiency of SO_2 collection. The collection efficiency of the impinger is reported to be 95% with 50–174 $\mu g/m^3$ SO_2 when a 5-μm pore PTFE membrane filter is used (1).

3.6 Relative humidity does not have any significant effect on the efficiency of SO_2 collection. The collection efficiency is reported to be 92–95% at SO_2 concentrations of 50–174 $\mu g/m^3$ and 100% relative humidity when a 5-μm pore PTFE membrane filter is used (1).

3.7 Effect of temperature on SO_2 collection efficiency is negligible. The collection efficiency remains at 99.5 ± 4.5% in a collection temperature range of 5 to 40°C (1).

4. Precision and Accuracy

The precision and accuracy of this method are excellent when combined with ion chromatographic analysis. The method yielded a mean concentration of 404 $\mu g/m^3$ at a level of 400 $\mu g/m^3$ SO_2 (1). In comparison, the U.S. Environmental Protection Agency reference method (3), which is essentially the method described in 704A, yielded a mean concentration of 392 $\mu g/m^3$. The mean coefficient of variation for SO_2 levels ranging from 24.3 to 1291 $\mu g/m^3$ was 5.7% for 24-h sampling at 0.2 L/min through 50 mL of the absorber. The worst coefficient of variation of 23% occurred at the lowest SO_2 concentration. The relative standard deviation is typically below 5%, often below 1% with SO_2 concentrations above 100 $\mu g/m^3$ (1).

5. Apparatus

This description is based on the only validated experiments conducted (1). These experiments used a 47 mm, 5-μm pore, PTFE membrane filter, a 100-mL capacity impinger containing 50 mL of absorbing solution and a flow rate of 0.2 L/min. It is likely that other judicious combinations of filter material, impinger/bubbler size and type, absorber volume and air flow rate will lead to satisfactory results. However, the user must validate the setup for the particular conditions employed.

5.1 FILTER HOLDER AND FILTER. The filter holder and support screen should be made of inert material and should accept a 47 mm diameter filter. Polycarbonate filter holders (e.g., Gelman type 1119, Gelman Sciences, Ann Arbor, MI) are suitable. The filter should be a 47 mm diameter PTFE membrane filter. A 5-μm pore size (e.g., Millipore type LSWP 047-00, Millipore Corp, Bedford, MA) filter has been prescribed (1); however, filters of smaller pore size should be satisfactory. Indeed, the use of smaller pore size filters will increase the efficiency of removal of particulate sulfate.

5.2 IMPINGER. The capacity of the impinger is 100 mL. The size of the orifice at the tip is 0.38 mm (0.015 in.; No. 79 jeweller's drill passes through it and No. 78 does not). Such impingers are available from a number of vendors (e.g., RAC Div., Andersen Samplers Inc., Atlanta, GA).

5.3 TRAP. A flask or other suitable vessel is tightly packed with glass wool with the air inlet end extended to the bottom and the outlet end at the top. This is placed after the impinger to trap any aerosolized absorber droplets.

5.4 FLOW MEASURING DEVICE. A calibrated rotameter with a needle valve for flow control. The use of a calibrated critical orifice is recommended in lieu of a rotameter. A 27-gauge hypodermic needle permits a flow of approximately 0.2 L/min.

5.5 PUMP. Leak-free, vacuum type. Must be capable of maintaining a vacuum of at least 53 kPa (40 cm Hg) if a critical orifice is used.

6. Reagents

6.1 WATER. Water must be sulfate-free and should conform to the ASTM standard for Reference Reagent Water, Type II. Such water may be produced either by distillation or preferably a cartridge deionization system.

6.2 HYDROGEN PEROXIDE, 30%. Reagent grade. Store refrigerated.

6.3 HYDROCHLORIC ACID, 0.6N. Pipet 5 mL of concentrated HCl (approximately 12 N) into a 100-mL volumetric flask and dilute to the mark with water.

6.4 ABSORBING SOLUTION. Pipet 20 mL of 30% H_2O_2 solution into a 1-L volumetric flask. Pipet 0.1 mL of 0.6 N HCl solution into it and dilute to the mark with water. Storage of the absorbing solution in capped bottles up to 30 days and up to a temperature of 40°C has no effect on the collection efficiency of the absorber.

7. Procedure

7.1 SAMPLING ARRANGEMENT. Use PTFE tubes for all connections up to the impinger. Connect the air inlet tube to the filter holder containing the membrane filter. Connect the exit tube from the filter holder to the impinger inlet with a glass-to-PTFE connection. Connect the impinger outlet to the trap inlet. The trap outlet is connected to the critical orifice or the bottom end of the rotameter. The other side of the critical orifice or the rotameter top is connected to the pump. Avoid condensation problems at sample inlet (see 7.1, method 704A).

7.2 COLLECTION OF SAMPLE. Fill the impinger with 50 mL of the absorbing solution (see 6.4) and begin drawing air through the assembly. Adjust the rotameter valve to obtain the desired flow rate. Record the time at which sampling is commenced. After a typical sampling time of 24 h, stop the air flow and quantitatively transfer the contents of impinger and washings into a 50-or 100-mL volumetric flask, depending on the extent of evaporative absorber loss. Dilute to the mark with water.

7.3 ANALYSIS. Refer to 720. DETERMINATION OF AQUEOUS SULFATE.

8. Calibration and Standards

8.1 For the calibration of sulfate analysis in the collected sample, consult the method being used in DETERMINATION OF AQUEOUS SULFATE.

8.2 For calibration of the complete experimental arrangement with known concentrations of gaseous SO_2, see Section 8.3 in method 704A.

9. Calculations

Compute the concentration of SO_2 in the air sample as:

$$SO_2(\mu g/m^3) =$$

$$\frac{\text{Total no. of } \mu\text{moles of sulfate in impinger sample}}{\text{air flow rate (L/min)} \times \text{sampling time (min)}} \times 64{,}062$$

$$SO_2(\text{ppbv}) = SO_2(\mu g/m^3) \times 0.382$$

The air flow rate should be corrected to reference conditions of 25°C and 101.3 kPa.

10. Effects of Storage

After SO_2 collection the absorber solution may be stored in capped bottles up to 30 days without any significant effect on the amount of sulfate found (1).

11. References

1. MULIK, J.D., A. TODD, E. ESTES, R. PUCKETT, E. SAWICKI and D. WILLIAMS. 1978. Ion Chromatographic Determination of Atmospheric Sulfur Dioxide. *In* "Ion Chromatographic Analysis of Environmental Pollutants", Vol. 1., E. Sawicki, J.D. Mulik, and E. Wittgenstein eds. Ann Arbor Science, Ann Arbor, MI, p 23.
2. MARTIN, L.R. and D.E. DAMSCHEN. 1981. Aqueous Oxidation of Sulfur Dioxide at Low pH. *Atmos. Environ. 15*:1615-1621.
3. PIERSON, W.R., Research Staff, Ford Motor Company 1985. Unpublished data.

Subcommittee 1

P.K. DASGUPTA, *Chairman*
D.F. ADAMS
B.R. APPEL
S.O. FARWELL
K.T. KNAPP
G.L. KOK
W.R. PIERSON
K.D. REISZNER
R.L. TANNER

707.

Continuous Monitoring of Atmospheric Sulfur Dioxide with Amperometric* Instruments

1. Principle of the Method

1.1 This procedure describes an automated analyzer intended for the continuous detection and determination of sulfur dioxide in the atmosphere. It is applicable to the determination of sulfur dioxide when other sulfur compounds or other interferents do not exceed 5% of the sulfur dioxide concentrations. If this limit is exceeded, then appropriate sample pre-treatment procedures must be used to remove the interferences.

1.2 Sulfur dioxide and other reducing gases present in the air sample react with electrically generated free iodine or bromine (titrant) in the detector cell. The detection cell contains two pairs of electrodes that function as anode/cathode and sensor/reference. The cell and its associated electronics operate to maintain a constant titrant concentration in the electrolyte. Any compound introduced into the cell that reacts with the titrant will change the titrant concentration and produce a potential change in the solution that is detected by the sensor/reference electrode pair. Current is supplied to the anode/cathode generator pair until the original potential and titrant concentration are restored. The quantity of electricity required is directly proportional to the reactant equivalents introduced into the cell. Details on the construction of a thin layer cell for the amperometric analysis of sulfur dioxide are available in the literature (1).

One Faraday (96,492 C) passing through an electrolyte produces one gram-equivalent of chemical change of the material in the solution. The number of coulombs (Q) expended is equal to the current

(i) in A multiplied by the titration time (t) in s or $Q/t = i$.

In air monitoring, the electrolyte composition is selected to promote the most favorable reaction of sulfur dioxide with regard to reaction rate and specificity. Iodine or bromine is used for the titration of sulfur-containing compounds. During a titration, the current passing through the cell is measured as the potential drop (iR) across a precision resistor. A potentiometric strip-chart recorder may be used to record the potential-time curve. The area under this curve [(i) × (t)] is equivalent to the number of coulombs used.

2. Range and Sensitivity

2.1 The optimum operating range for amperometric sulfur dioxide detectors is 26 to 5200 $\mu g/m^3$ (0.01 to 2 ppmv). The range provided by any given analyzer is affected by cell configuration, electrolyte formulation, and associated electronics.

3. Interferences

3.1 The most commonly encountered interfering pollutants include oxidizing, reducing, and olefinic compounds.

Any compound in the air sample capable of oxidizing bromide to bromine or iodide to iodine under the conditions in the cell will be sensed as excess titrant. Such compounds as ozone or oxides of nitrogen act in this manner. Similarly, compounds that reduce bromine to bromide or iodine to iodide will create a positive interference. Other compounds including unsaturated hydrocarbons, α-hydroxy acids, some amines and phenols may consume titrant by addition or substitution reactions.

---

*The term amperometric more accurately describes this process which is commonly called coulometric.

3.1.1 Common oxidizing compounds such as ozone, hydrogen peroxide, chlorine, and nitrogen dioxide, when absorbed in the detecting cell electrolyte, oxidize the sulfur dioxide. The presence of such oxidizing pollutants causes a negative titration error, and less titrant will be generated than corresponds to the amount of sulfur dioxide present.

3.1.2 Amperometric sulfur dioxide detectors also respond to any compounds that will reduce free halogen. Common interfering reductants include hydrogen sulfide, alkyl mercaptans, and alkyl sulfides and disulfides. These will cause positive interference.

Differences in the stoichiometry of sulfur-containing compounds may also create a problem in the calculation of concentration should the analyzer be used as a total sulfur gas analyzer. Hydrogen sulfide requires four times as much titrant per sulfur atom as does sulfur dioxide. Thus sulfur dioxide-hydrogen sulfide mixtures of varying ratios, all of equal sulfur atom content, will yield titration values varying over a range of 4.

3.1.3 Olefins will react, but with less than quantitative stoichiometry, primarily because of a slow and non-stoichiometric addition reaction with the titrant (bromine or iodine) coupled with a short gas residence time in the detection cell.

Interferences in the analytical technique may also occur from α-hydroxy acids, some amines, and phenols, by addition or substitution reactions with the halogen titrant.

3.2 Interferences may be removed from atmospheric mixtures by use of appropriate scrubbers. Several systems having varying degrees of removal efficiencies for selected compound types have either been reported in the literature (2) or remain proprietary information of certain instrument manufacturers (3). Chemically impregnated filters or heated silver gauze (4) are the types most frequently used and may be supplied as a built-in component or accessory in commercially available analyzers. Selection of the interference-removing filter should be based on detailed knowledge of the range of potential interfering compounds present within the intended survey area. Any filters that are used should be individually tested for removal of known interferences and the ability to pass sulfur dioxide quantitatively.

3.3 Amperometric sulfur dioxide analytical instrumentation is currently not known to be manufactured in the United States, although widely used in Europe. With the small number of instruments that have been produced and consequently the limited data set of experience with this technique, it is best applied under conditions where the potential interferences are well characterized. A careful cross-comparison of the amperometric technique with the barium perchlorate-thorin indicator titration technique (see Method 720) using spiked ambient air shows a 20–25% negative error for the amperometric technique (5). The cause of this low reading is not known; however, it suggests that care must be taken to avoid artifacts in measurements with the amperometric technique.

4. Precision and Accuracy

4.1 Although the coulomb is a primary standard, the precision and accuracy of the method will be determined by the sample delivery train, sample pretreatments, detector/cell response, interfering pollutants, and the recording system.

4.1.1 *Accuracy.* An analyzer calibrated within the range of 1 to 2 ppm full scale should have an overall accuracy of 0.02 ppm (1% relative). Short term accuracy of 0.5% of full scale should be attainable with standardization against a known dilution flow from a certified permeation tube.

4.1.2 *Precision.* Repeatability shall be within 1% relative standard deviation for successive identical samples and within 3% over several 24-h periods.

5. Apparatus

5.1 A general description of instrument components is given. Not all instruments may meet the specifications described below for the measurement of sulfur dioxide. The basic components are shown schematically in Figure 707:1. Generally, an automated amperometric analyzer system will consist of a probe containing a heated filter for removal of particulate matter, a sampling valve for introducing span and cali-

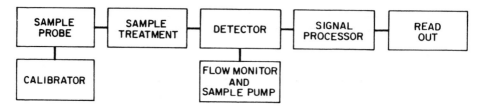

Figure 707:1—Basic components of the automated amperometric analyzer system.

bration gases automatically or on demand, zero and calibration gases, an interference scrubber, the detector cell-electrolysis power supply, signal processor, a means for recording the output signal, and an air sampling pump.

5.1.1 The sample probe and delivery train should be fabricated from Teflon or borosilicate glass. The inlet should be protected by a Teflon filter to remove small insects from the sample. The entire system should be maintained at a temperature to prevent condensation and should include a heated SO_2-inert filter (see Method 824) for particulate matter removal.

5.1.2 The calibrator (Figure 707:1) includes a constant temperature gas dilution system for permeation tubes. Details on the use of permeation tubes for calibration are given in Section 3, Part I.

5.1.3 The flow monitor and sample pump (Figure 707:1) must provide a reliable and accurate flow of sample air to the detector. A visual display of the air sampling rate is desirable. The air pump must be designed for continuous-duty operation.

5.1.4 Sample treatment (Figure 707:1) includes procedures for removing potential interfering substances while allowing SO_2 to pass through without loss. These devices should be evaluated in line by using a dynamic source of an SO_2 reference concentration to establish per cent SO_2 loss on a real-time basis throughout the anticipated range of concentration. Similarly, the effectiveness of the pre-treatment devices for removal of potential interfering gases should be tested at ten times the highest range of concentrations anticipated to determine retention efficiency and breakthrough times. Pretreatment devices must pass 95% or more of the SO_2 in the concentration range of interest and must pass no more than 2% (or to a lower value of 0.02 ppm interfer-

ence equivalent) of the interfering compounds for which the device was designed.

5.1.5 The detection cell (Figure 707:1) contains the electrolyte, sensor and generator electrodes, a mechanism to maintain constant electrolyte level and concentration, a sample dispersion device, and a method for stirring the solution. The total system must be carefully evaluated to establish its lower limit of detection; linearity of response; zero and span drift; signal-to-noise ratio; and lag, rise, and fall times as defined in the equivalency document published in the Federal Register (6); see Section 5.3.

5.1.6 The signal processor and electrolysis power supply (Figure 707:1) should be of solid-state design, temperature-compensated to give less than 0.5% drift in 24 h and should respond by less than 1% of full scale to a 10% fluctuation in line voltage. The circuit should incorporate a standard millivolt source so that an "electric function" check may be performed by actuating a "Function Test" switch. Signal output modes should be sufficiently flexible to permit analog recording within the range of 1 to 1000 mV as well as analog-to-digital data conversion.

5.1.7 Signal display (Figure 707:1) is most commonly provided by a potentiometric recorder. For some systems having a wide concentration range measuring capability, an automatic signal attenuator may be used in conjunction with the recorder to permit, for example, 5.2 mg/m³ (2 ppm) SO_2 full scale recording during periods of peak SO_2 concentrations.

5.2 Variable losses of trace concentrations of sulfur gases occur on metals, glass and plastic surfaces (7,8); one must use care in selecting materials that come into contact with the air sample or reference gas. FEP Teflon is preferable to all other plastic because it quickly conditions to give repeatable results (9). In addition, all components of the gas flow system must be closely coupled and dead volume kept to a minimum.

5.3 Summary of Minimum Specifications for Amperometric Instruments. (6)

		Units	SO₂
5.3.1	Range (full scale)	(ppm)	10 to 1
5.3.2	Minimum detectable limit	(ppm)	0.01
5.3.3	Rise time, 95%	(sec)	15
5.3.4	Fall time, 95%	(sec)	15
5.3.5	Lag time, 95%	(sec)	30
5.3.6	Zero drift, 12h†	(ppm)	±0.02
	Zero drift, 24h†	(ppm)	±0.02
5.3.7	Span drift, 24h†	(ppm)	±0.02
5.3.8	Precision at 20% full scale for each attenuation range	(ppm)	±0.02
5.3.9	Precision at 80% of full scale for each attenuation range	(ppm)	±0.04
5.3.10	Total interference equivalent	(ppm)	±0.02
5.3.11	Noise	(ppm)	0.005
5.3.12	Linearity§	(ppm)	±3%

†Determined under conditions of continuous operation while measuring zero gas and span gas over the stated period.

§Each value obtained from calibration gases must be within ±3% of the actual concentration predicted by the line of best fit determined by the method of least squares.

6. Reagents

6.1 CLEAN AIR. Commercial grade, dry, in a cylinder. Clean air may alternatively be supplied from ambient air through a low pressure-drop, multi-component filter consisting of activated charcoal and soda lime. Proprietary filters are also available (2).

6.2 ELECTROLYTE. The electrolyte should be either prepared from ACS Reagent Grade chemicals and distilled water according to directions furnished by the instrument manufacturer or purchased in prepackaged containers from the manufacturer. The electrolyte should be maintained or exchanged according to the manufacturer's instructions. Distilled water should conform to the ASTM specifications for Reference Reagent Water, Type II.

7. Procedure

7.1 For specific operating instructions, refer to the manufacturer's manual. Turn the power on and allow the instrument to warm up for the manufacturer's recommended time. The instrument should meet minimum specifications for drift at the end of this warmup period.

7.2 Ambient air flows through the Teflon probe and the multiport valve into the detector. Adjust the air flow to the recommended rate. Energize the multiport valve on a programmed or demand time schedule to sample zero air and span gas. Adjust the instrument and recording system with the zero air and span gas.

7.3 Install and evaluate the instrument on-site to establish compliance with the minimum specifications described herein.

8. Calibration and Standards

8.1 Details on the use of permeation tubes for calibration are given in Section 3, Part I.

8.2 Introduce at least four different concentrations, greater than zero, of the standard SO₂ reference gas into the probe system to give responses from 10 to 95% of full scale. Compare the response of the instrument to the same step concentration standards introduced directly into the detection cell to establish probe line loss. Adjust the recorder span to give the known calibration value. If a span control is not available on the instrument, prepare a calibration curve by plotting the several added SO₂ dilutions against the recorded instrument output.

8.3 Prepare a calibration curve, plotting standard concentrations against recorded instrument output.

8.4 To minimize errors when assuming linearity and a single point quality control sample, the standard gas concentration should be selected so as to optimize the most important segment of the data. For example, if peak excursions are important, the instrument should be calibrated at full scale or at the anticipated peak value. If no particular range of values is specially important, calibration should be carried out at mid-range. In some amperometric instruments, reagent deterioration is initially reflected in a falloff at near full-scale response. Such instruments should not be respanned to accommodate this behavior. The reagent should be replaced and the total instrument system completely recalibrated.

9. Calibration

9.1 Correct the flow rate of the sample to an ambient air condition of 101.3 kPa and 25°C (298.1K).

9.2 Calculate the concentration of standard gas mixtures in ppm by volume as follows:

$$ppm \; SO_2 = (0.382) \; B/A$$

A = flow rate of air + dilution air, L/min.

B = output of permeation tube, μg/min.

0.382 = the volume (μL) of 1 μg SO_2 at 25°C, 101.3 kPa.

9.3 Calculate the concentration of sulfur dioxide in air by multiplying the net recorder response (sample less baseline reading on recorder chart) by the slope of the linear calibration curve.

9.4 Data reduction procedures used are dependent upon the parameters to be reported.

9.4.1 For determination of mean values, for any given time period, graphical, mechanical or electronic integration of analog strip charts is used. Hourly mean values may also be obtained from a sufficient number of discrete values to give the confidence level desired.

9.4.2 For determination of fumigation values, valid peak values must have a duration of two times the sum of the rise and fall times of the instrument used.

9.5 Concentrations in the atmosphere should be reported to the nearest 0.02 ppm, or 1% relative to the measured value, whichever is larger. A corresponding confidence level should accompany the result. The latter value is best taken from field installation evaluations of overall precision and accuracy. These values must meet minimum specifications (Section 5.3).

10. References

1. BRUCKENSTEIN, S., K.A. TUCKER and P.R GIFFORD. 1980. Determination of Sulfur Dioxide by Reaction with Electrogenerated Bromine in a Thin-Layer Cell Having a Gas-Porous Wall. *Anal. Chem.* 52:2396–2400.
2. ADAMS, D.F., W.L. BAMESBERGER, and T.J. ROBERTSON. 1968. Analysis of Sulfur-Containing Gases in the Ambient Air Using Selective Prefilters and a Micro Coulometric Detector. *J. Air Poll. Control Assoc.* 18:145–148.
3. Beckman Instruments, Inc., Instruction Manual 82043-A, Fullerton, California, 92634.
4. Phillips Electronic Instruments, Mount Vernon, New York, 10550.
5. Ford Motor Company, 1980. Research report "Development of Continuous SO_2 and Sulfate Sampling and Analysis Methodology for Vehicle Emissions," Report to U.S.E.P.A. under Contract 68-02-2787, January.
6. *Federal Register.* 1973. Ambient Air Monitoring Equivalent and Reference Methods. 38:28438, Part II, Friday, Oct. 12, Washington, D.C.
7. KOPPE, R.K. and D.F. ADAMS. 1967. Evaluation of Gas Chromatographic Columns for Analysis of Sub-parts Per Million Concentration of Gaseous Sulfur Compounds. *Environ. Sci. Tech.* 1:479–481.
8. PESCAR, R.E. and C.H. HARTMANN. 1971. Automated Gas Chromatographic Analyses of Sulfur Compounds, 17th Annual Analysis Instrumentation Symposium, Instrument Society of America, Houston, Texas.
9. STEVENS, R.K., J.D. MULIK, A.E. O'KEEFFE, and K.J. KROST. 1971. Gas Chromatography of Reactive Sulfur Gases in Air at the Parts Per Billion Level, *Anal. Chem.* 43:827–831.

Subcommittee 1

G.L. KOK
W.R. PIERSON
D.F. ADAMS
B.R. APPEL
P.K. DASGUPTA, *Chairman*
S.O. FARWELL
K.T. KNAPP
K.D. REISZNER
R.L. TANNER

708.

Determination of Mercaptan Content of the Atmosphere

See **Method 118**

709.

Determination of Sulfur-Containing Gases in the Atmosphere (Continuous Method with Flame Photometric Detector)

709A.

Determination of Sulfur-Containing Gases in the Atmosphere (Following Chromatographic Separation, with the FPD)

709B.

Determination of Sulfur-Containing Gases in the Atmosphere (Total Gaseous Sulfur with the FPD)

1. Principle of Operation of a Flame Photometric Detector for Sulfur Compounds

1.1 The flame photometric detector (FPD) detects compounds in a flowing gas stream by burning those compounds in a flame and sensing the increase in light emission from the flame (flame chemiluminescence) during combustion (1,2). The FPD is a flame optical emission detector consisting of a hydrogen-air flame in an enclosed chamber, an optical window for observing flame emissions, an optical filter for spectroscopically selecting the wavelength range of emissions to be observed, a photomultiplier tube (PMT) for measuring the intensity of emitted light, and an electrometer for measuring the current output of the PMT.

1.2 The intensity and wavelength of light emitted from the FPD flame depend on the geometry of the flame burner as well as the absolute and relative flows of hydrogen and air supplied to the burner (3,4). Adjustment of these flows can enhance emission from the analyte species relative to other interferent species. Typical FPD flames produce luminescence due to molecular band emissions or continuum emissions from recombination reactions, but are not hot enough to generate substantial emissions from atomic species.

1.3 Sulfur compounds are detected by the FPD in atmospheric samples, either directly in continuously flowing ambient air streams (5) or following elution of a discrete sample from a sample loop (6,7). Hydrogen-air or hydrogen-oxygen diffusion flames are employed in this scheme. Sulfur compounds are detected by flame chemiluminescence produced by recombination of sulfur atoms to form "excited-state" S_2 molecules (S_2^*) in the flame region. The S_2* emission has a band structure extending from about 300 to 425 nm, with the most intense band centered at 394 nm.

Formation of S_2* is favored in flame regions in which hydrogen atoms are abundant, thus optical emission from sulfur compounds is maximized in hydrogen-rich diffusion flames. In contrast, interferent emissions from carbon-containing compounds, due principally to CH and C_2 band emissions, are maximized in 0-atom-rich regions of the flame, thus providing an additional means of discriminating sulfur compounds from atmospheric hydrocarbons. This is particularly significant in total gaseous sulfur determinations in ambient air, or in gas chromatographic applications of the FPD when certain hydrocarbon species are not resolved from sulfur-containing compounds being analyzed. A dual-flame FPD burner has been introduced for GC applications; this greatly reduces the hydrocarbon quenching effect on sulfur emissions which occurs even in fuel-rich hydrogen-air diffusion flames (8).

Phosphorus-containing compounds also produce flame chemiluminescence involving the HPO molecular species in fuel-rich hydrogen-air diffusion flames. However, the band center for HPO emission is about 530 nm, and, by selection of an optical filter centered at about 394 nm, interference from phosphorus-containing compounds is rendered negligible because volatile phosphorus compounds in the atmosphere are vanishingly low in most areas except in the vicinity of certain pesticide manufacturing plants.

1.4 The FPD response from phosphorus-containing compounds is derived from HPO molecular emission and hence is linearly proportional to the P-atom flow into the flame. Sulfur-compound emission is, however the result of S_2* emissions in the flame and is non-linearly related to S-atom flow. In general S_2* emissions are approximately proportional to the square of sulfur compound concentrations. The FPD response in either case depends on the P-or S-atom flow per unit time and the FPD is thus a mass flow-rate type of detector. The upper limit of concentrations that may be determined is limited by the onset of self-absorption effects in the flame; that is, the concentration of ground-state S_2 molecules

becomes large enough to reabsorb light emitted by excited-state S_2 molecules.

1.5 The FPD may be used for determination of the total concentration of sulfur-containing gases in the atmosphere. For this application sulfur-containing particulate matter is removed by placing an inert filter (Teflon) in the inlet stream to the detector. It is applicable for determination of sulfur dioxide in atmospheres containing other sulfur gases in amounts not more than 5% of the SO_2 concentration.

1.6 In the presence of other sulfur compounds at ppb levels, the FPD may be used for determination of individual sulfur-containing compounds following their separation on a gas chromatographic column (6). Care must be taken in the selection of column and column-packing materials to prevent significant loss of SO_2 and other sulfur compounds prior to their detection in the FPD; several column packings have been used successfully in this application.

1.7 The FPD analyzer or GC detector is usually calibrated using H_2S, SO_2, CH_3SH and dimethyl sulfide permeation tubes together with a dual-flow gas dilution device capable of producing reference standard atmospheres down to the limits of detection of the method (9). The FPD analyzer (Method 709B) is calibrated by continuous admission of standard atmospheres of sulfur gases in zero air to the detector. The FPD used as a GC detector (Method 709A) is calibrated by repetitive admissions of samples of standard atmospheres of sulfur gases in the sample air, using a continuously flushed Teflon or other inert sampling valve.

2. Sensitivity, Limit of Detection and Range

2.1 The sensitivity, range and limit of detection of the FPD, used as a continuous analyzer or as a GC detector, are dependent on several factors: type and materials of the sample handling system, GC column materials, hydrogen, air and carrier flow rates and their variability, PM tube type and operating voltage, and the electronic amplification system. Several general statements can be made.

Using the FPD as a GC detector, with sulfur response during peak elution varying as the nth power of S-atom mass flow rate, the sulfur sensitivity is determined as follows:

$$S_s = (A_i/m_s)(1/\sqrt{m_s^*})^{n-1} \qquad (1)$$

where:

S_s = sulfur sensitivity or response, A/ (g S/s)^n

A_i = integrated peak area, A s

m_s = mass of S in the injected sulfur gas, g S,

and

m_s^* = mass-flow rate of S in the eluting sulfur gas.

If the sulfur gas response is a pure quadratic function ($n = 2$), the sensitivity function reduces to:

$$S_s = A_i/m_s \cdot m_s^* \qquad (2)$$

More typically, $n = 1.7$ to 1.8 for sulfur gases found in the atmosphere.

Using the FPD as a continuous monitor of total gaseous sulfur compounds in the atmosphere, a sulfur response varying as the nth power of S-atom mass flow rate leads to an alternative form of the sulfur sensitivity, S_s, defined as the differential change in response (A) with respect to concentration of the sulfur-containing test gas:

$$\frac{dI_s}{dC_s} = \frac{d(KC_s^n)}{dC_s} = nKC_s^{n-1} \qquad (3)$$

where:

I_s = sulfur response, A

K = instrumental constant, A/ppb S

C_s = sulfur gas concentration, ppb S

and

n = power function of sulfur response.

This is parallel to the above formulation for FPD use as a GC detector if we recall that:

$$C_s = \frac{m_s^* (R/MW)}{F} \qquad (4)$$

where:

R = gas constant,

MW = molecular weight of the sulfur gas

and

F = total gas flow rate to the FPD.

Under conditions of constant F, Eq. (3) becomes

$$\frac{dI}{dm_s^*} = \frac{nkR}{(MW)F} (m_s^*)^{n-1} = nK' (m_s^*)^{n-1} \qquad (5)$$

Sensitivity of an individual FPD used for sulfur gas determinations should be measured experimentally using continuous or batch introduction of sulfur hexafluoride or other suitable standard materials. Specification of the sensitivity of an FPD should be accompanied by the following additional information:

(a) test substance
(b) concentration range tested (20 to 200x the noise level), within the range of concentration over which the power law function (n) is constant
(c) flame background
(d) detector geometry (single or dual flame)
(e) wavelength and bandpass
(f) hydrogen flow rate
(g) air or oxygen flow rate
(h) carrier gas and flow rate as applicable
(i) detector temperature
(j) electrometer
(k) method of measurement

The sensitivity of an FPD is most frequently measured using permeation devices (see Section 8 below on Calibration and Standards) although exponential dilution, dynamic dilution, and diffusion dilution methods may be used.

2.2 The limit of detection of the FPD is defined as the mass flow rate of S-atoms or concentration of sulfur-containing compound generating a signal that is two times the noise. The noise is defined as the $\pm \sigma$ envelope of short-term fluctuations in background current measured for a period of the order of 1 min. For chromatographic applications the limit of detection of the FPD for the sulfur compound is defined as that sample size yielding a peak height at least four times the noise. The minimum quantifiable amount may be larger depending on the required precision and accuracy.

The limit of detection for continuous analysis of sulfur gases in the atmosphere by the FPD method is usually expressed in terms of SO_2 and is about 2 ppbv (1.0 ng S/ min). The addition of a constant amount of a stable sulfur compound to the fuel gas of a continuous FPD analyzer will reduce the limit of detection to about 0.25 ppb (11), because the sensitivity of the FPD increases with increasing concentration, as shown by the sensitivity functions in Equation (3). Concentrations of 30–50 ppb of sulfur hexafluoride (SF_6) in the analyzed gas stream increase the signal-to-noise ratio for 1 ppb SO_2 in air by about a factor of ten. The principle is applicable to use of the FPD as a GC detector although no applications have been reported to date using SF_6 in the fuel gas.

2.3 The range of the FPD extends from a lower limit of the detection limit to an upper limit defined by the departure of the detector from a constant power-function relationship between response and mass flow rate or concentration (constant n). This range, for single-burner instruments using conventional burner design, extends from about 2 ppb to 1 ppm (10–5000 ng S/ min at 0.4 L/min total gas flow). The upper limit derives from the onset of self-absorption of emitted S_2^* radiation by ground-state S_2 in the flame.

3. Interferences

3.1 For a substance to exhibit a positive interference in the determination of gaseous sulfur compounds it must: (a) emit light of frequencies within the bandpass of the filter; (b) be present in the sample at detectable concentrations; and (c) in the case of GC-FPD operation, have an elution time close to those of other sulfur compounds, e.g., SO_2, H_2S, CH_3SH, $(CH_3)_2S$, etc.

3.2 The interference from phosphorus compounds via HPO* emission at about 530 nm is effectively reduced by the optical filter; the discrimination ratio is about $10^3:1$ or greater for most FPDs (12).

3.3 The interference from hydrocarbons originates principally from the CH band emission near 400 nm, within the bandpass of the optical filter. This leads to a discrimination ratio of about $10^4:1$ for hydrocarbons (12). Shielding of the emissions from the 0-atom-rich region near the burner tip from observation by the PMT reduces the positive interference from hydrocarbons still further. Hydrocarbon compounds can also, in a single-burner FPD, suppress the

signal of the sulfur compounds leading to negative interference.

3.4 The response of the FPD to sulfur compounds is also affected by the concentration of CO_2 in the air stream, possibly due to continuum emission from the CO + 0-atom $\rightarrow CO_2$ reaction in the flame.

3.5 Sampling of ambient atmospheres for sulfur compounds is hampered by a negative interference due to water vapor in the sampled air (11). The FPD background current is inversely proportional to the water vapor concentration in the air. This should not affect the determination of trace sulfur compounds by the GC-FPD method.

3.6 The GC-FPD or continuous FPD detection of trace levels of sulfur compounds is complicated by the fact that such compounds are often reactive and are irreversibly sorbed on many surfaces. Care must be taken to ensure that the entire sampling and analysis system is free of active sorption sites for these compounds. The use of Teflon sampling lines and silanized glass GC columns and inlets is good practice (6).

4. Precision and Accuracy

4.1 Precision of the flame photometric detector depends principally on control of the sample and hydrogen flows to the detector. An airflow change of 1% changes the response by approximately 2%. The use of mass flow controllers to control the flow of H_2 and of sample air is recommended; it improves the reproducibility to ± 5% or better, and also reduces the long-term drift in FPD response. The use of FPD analyzers for continuous analysis of sulfur compounds from aircraft is possible only with instruments modified to control mass-flow of fuel and sample air (13).

4.2 Reproducible peak heights and/or areas from FPDs used as GC detectors are principally dependent on control of the materials of construction, operating parameters and column preparation techniques. It is usually necessary to precondition the analysis system with a series of standard injections to achieve elution equilibrium. Once equilibrium is established, repetitive sampling of standard reference gases containing 30 ppbv H_2S and 40 ppb SO_2 have given a relative standard deviation of <3% over several 24-h periods (14).

4.3   Accuracy of continuous FPD or GC-FPD methods will depend on the ability to control the flow and temperature of dilution gases over certified or calibrated permeation tubes maintained in a gas dilution device **(9)**. Certified permeation sources for H_2S, SO_2 and other sulfur-containing gases are available from the National Bureau of Standards.

5. Apparatus

5.1   A description of instrument components required for an automated GC-FPD system is given in this section. Generally (Figure 709:1) the automated GC-FPD system will consist of a regulated system for delivery of hydrogen to the FPD burner block (A); an automatic multiport valve (B) for reproducible injection of the air sample and reference gases into the GC; a GC column (C) in an isothermal controlled or temperature-programmable oven; a flow measurement and control system (D) to admit known flows of carrier gas (E) and ambient air (F) to the multiport valve (B), or clean air (G) to the calibration system (H) and thence to the multiport valve; the flame photometric detector (I) consisting of the burner block light-collection optics and photomultiplier tube (PMT); the power supply for the PMT (J); the amplifier (K) and readout (L)—recorder and/or data logger—for the PMT signal.

5.1.1   The sample delivery train for ambient air and calibration standards [Figure 709:1, (D) to (H)] should be fabricated from FEP Teflon. Ambient air should be filtered to remove atmospheric particulate matter and all lines should be heated to about 50°C to prevent condensation of water and significant sorption of gaseous sulfur compounds.

5.1.2   The calibration system [Figure 709:1, (H)] should include a constant-temperature oven or bath and a gas dilution system for permeation tubes. Clean, temperature-controlled dilution air is passed over permeation tubes containing the sulfur-containing compounds of interest, and further diluted with secondary clean air flow to provide sample concentrations of up to 80% of full scale during calibrations. Calibration standards are provided alternately with sample air to the multiport valve through appropriate solenoid switching.

5.1.3   Hydrogen gas is provided by differential pressure directly to the FPD from a pressurized source (cylinder or H_2 generator); likewise, the carrier gas is pro-

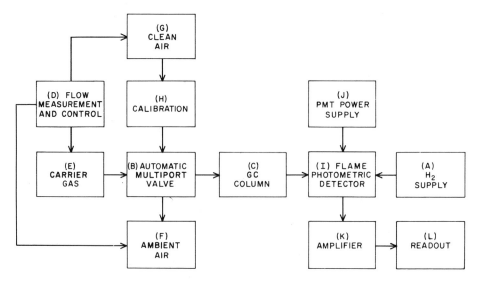

Figure 709:1—Automated Gas Chromatograph-Flame Photometric Detector System

vided to the multiport valve and thence to the column from a pressurized source. Calibration standards and sample air may be provided either by positive pressure through an inert sampling pump, or by pulling the appropriate sample through the multiport valve using a downstream pump.

5.1.4 Sample pretreatment may be required depending on the application. For example, high concentrations of alkyl sulfides and disulfides can be removed by passage of the sample air through a small precolumn containing Deactigel in applications where only SO_2 and H_2S are to be determined. Use of this column in a backflush arrangement will prevent eventual breakthrough of the unwanted sulfur compounds.

5.1.5 The FPD detects the luminescence produced near 394 nm by S_2^* formed in the hydrogen-rich flame by sulfur-containing compounds. The carrier gas from the GC (often N_2) acts as a diluent in the flame, so the air/fuel ratio and the total flow should be optimized for each individual burner to maximize the signal/noise ratio. For GC-FPD systems in which N_2 is the carrier, oxygen is added to the gas stream to approximate air composition at the point of mixing with fuel gas at the burner tip.

A narrow bandpass (5 nm) optical filter with transmission maximum at 394 nm is placed between the burner and the PM tube. The sensitivity of the PM tube depends on its type, operating temperature and voltage; temperature isolation from the burner is thus required together with a well-regulated high voltage supply.

5.1.6 An electrometer (amplifier, Figure 709:1) is used to amplify the PMT signal to levels required by the signal readout devices. Either a logarithmic or a linear amplifier may be used depending on the range of concentrations expected in the intended application. Temperature compensation of the amplifier is required to reduce the 24-h drift to $<0.5\%$. Automatic re-zeroing of background current between chromatograms should be considered for automated operation.

5.1.7 Signal readout by both potentiometric recorder and by automated data recording systems (microprocessors) is now standard in most automated operations. A-D conversion with tape, floppy disc or hard-disc storage and hard-copy or paper tape printout of preprocessed data using integration programs and stored calibration factors is presently a commonly used procedure. Telemetering to a control data processing center is another option. Microprocessor control of hardware operations (such as sample introduction) through TTL switches can be considered to augment automated data reduction and fully automate the GC-FPD analytical process.

5.1.8 Care must be devoted to the materials of construction used throughout the sample and reference gas streams, with special attention to proper preconditioning procedures. Heated metal surfaces must be avoided since they catalyze the conversion of mercaptans to disulfides. Teflon is the preferred construction material since it conditions more rapidly and with lower irreversible sorption losses than other plastics such as PVC, polystyrene, and even polypropylene and linear polyethylene (6,15).

In addition, all components in the gas flow system must be closely coupled as part of concerted efforts to keep dead volume to a minimum.

5.1.9 Sample injection of air samples and standard calibration gas mixtures is accomplished sequentially with an automated multiport valve constructed of special stainless steel with the core often Teflon coated. The valve incorporates two or more Teflon sample loops — matched or with calibrated volumes — along with an appropriate timing mechanism for valve actuation at precise intervals.

5.1.10 A gas chromatographic column often used for separation of H_2S and SO_2 is a 5 ft $\times$ 1/8 in (1.3 m $\times$ 3 mm) OD Teflon column filled with 100/120 mesh Deactigel (16). A column used for five commonly analyzed atmospheric sulfur compounds — H_2S, SO_2, CH_3SH, $(CH_3)_2S$, and $(CH_3)_2S_2$ — is 10 ft $\times$ 1/8 in. (3 m $\times$ 3 mm) OD Teflon filled with 5% Triton X-305 on Chromosorb G/HP, 100/120 mesh (17). An alternate column material used is Carbopack B HT 100; and the separation of H_2S, COS, SO_2 and CH_3SH on this column is shown in Figure 709:2 (7). Certified columns may be purchased or packed in the laboratory by well-established techniques.

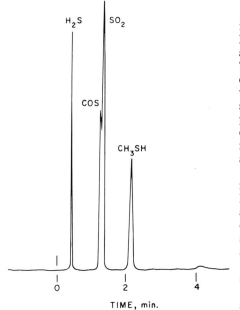

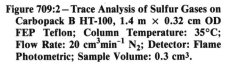

Figure 709:2 — Trace Analysis of Sulfur Gases on Carbopack B HT-100, 1.4 m × 0.32 cm OD FEP Teflon; Column Temperature: 35°C; Flow Rate: 20 cm^3min^{-1} N_2; Detector: Flame Photometric; Sample Volume: 0.3 cm^3.

5.1.11   Summary of Minimum Specifications

Category	Units	Specification for SO_2
Range	ppb	2–1000
Limit of Detection	ppb	2
Rise Time, 95%	s	10
Fall Time, 95%	s	10
Lag Time*	min	5–10
Zero Drift, 24-h**	ppb	±10
Span Drift, 24-h**	%	±10
Precision at 10 × LOD	%	±10
Precision at 80% of Full Range	%	±5
Total Interference Equivalent	ppb	10
Noise	ppb	<2
Linearity	%	3

*Time between sampling and readout.
**Determined during continuous operation while measuring zero air or span gas over stated time period.

5.2   The apparatus described here is for Method 709B (Total Gaseous Sulfur with the FPD). A general description is given since more than one commercial analyzer is available; in practice most analyzers used for Method 709B will contain FPDs of the Meloy design. The basic components of an automated system are shown in Figure 709:3. Generally, they include a regulated delivery system for hydrogen; a sampling valve for automatically delivering atmospheric air, zero air, or calibration gas samples; an FPD; an electrometer for amplification of the FPD current; a means of recording the electrometer output; and an air sampling pump.

The descriptions of instrumental components below parallel those in Section 5.1, Method 709A, except for differences such as the absence of a chromatographic column in Method 709B. Hence only brief summaries are given; for details consult the appropriate paragraph in Section 5.1.

5.2.1   The sample delivery train should be constructed of FEP Teflon.

5.2.2   The calibration gas source consists of permeation sources at constant temperature to be diluted with clean air.

5.2.3   The flow producing, measuring and regulating system (Figure 709:3) measures and controls the flows of gases to the detector, sampling valves and calibration source. Mass flow controllers are recommended for hydrogen and air flows (the latter indirectly by controlling exhaust gas flows) for stable operation under fluctuating pressure conditions.

5.2.4   Sample pretreatment includes facilities for scrubbing or filtering unwanted sulfur compounds from the gas stream, e.g., H_2S by a heated Ag screen or particulate sulfur compounds by a quartz filter.

5.2.5   The FPD measures the signal from excited S_2 molecules near 394 nm in an H_2-rich flame produced by mixing tank

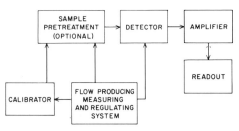

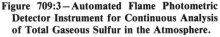

Figure 709:3 — Automated Flame Photometric Detector Instrument for Continuous Analysis of Total Gaseous Sulfur in the Atmosphere.

H_2 directly with sample air under sensitivity-optimized conditions.

5.2.6 A temperature-stabilized amplifier (logarithmic or linear) is used to amplify the PMT signal. In some FPD applications for sulfur compounds, a so-called square-root amplifier is used (usually a logarithmic amp/signal divider/exponential amp combination) to provide a signal which is directly linear with respect to sulfur mass flow through the detector **(3)**. This provides a direct reading of concentration in the gas stream and simplifies integration of GC-FPD peaks. It does require careful compensation for background current since linear calibrations will not be obtained if uncompensated background current is "linearized" during amplification **(11)**.

5.2.7 The electrometer signal is recorded on a potentiometric recorder and/or automated data processing equipment.

5.2.8 Sample losses and equilibration times are minimized by use of Teflon construction materials.

5.2.9 Summary of Minimum Specification for SO_2

Category	Units	Specification for SO_2
Range	ppb	2–1000
Limit of Detection	ppb	2
Rise Time	s	10
Fall Time	s	20
Lag Time	s	30
Zero Drift, 24–h	ppb	± 10
Span Drift, 24–h	%	± 10
Precision at 10× LOD	%	± 10
Precision at 80% of Full Range	%	± 5
Total Interference Equivalent	ppb	10
Noise	ppb	< 2
Linearity	%	± 3

6. Reagents

6.1 HYDROGEN. Dry, Matheson UHP grade H_2 or equivalent, is required in a pressurized cylinder equipped with a two-stage regulator and a temperature compensated flow restrictor. A hydrogen generator capable of producing at least 125 cm³/min of STP H_2 is an alternate source.

6.2 AIR. Commercial zero air, dried with Drierite and filtered to remove any particles, is required. Zero air produced by scrubbing ambient air of sulfur compounds without removing moisture (i.e. with a diffusion denuder tube) is also recommended since it avoids the problem, noted above, of negative interference of H_2O vapor on zero air response in the FPD.

6.3 PERMEATION TUBES. Prepare or obtain Teflon permeation tubes that emit SO_2 or other sulfur gases at rates of 3–100 ng/min to prepare calibration standards in the range of 2–500 ppb, the normal operating range of the FPD for ambient samples determined by Method 709A or 709B.

6.4 DEACTIGEL, 100/120 MESH. The material must be further purified by washing with concentrated HCl, then rinsing with H_2O and acetone.

6.5 TRITON X-305. Alkyl aryl polyether alcohol.

6.6 CHROMOSORB G/MP, 100/120 MESH. Obtainable from many chromatography supply houses.

6.7 CARBOPACK B, 40–80 MESH. Obtainable from Supelco, Inc.

7. Procedure

7.1 This section describes procedures for use in Method 709A (Sulfur Gases by GC-FPD). Details concerning the specific operating instructions should be obtained by referring to the manufacturer's manual. The basic principles of gas chromatography and their application to the separation and analysis of air pollutants are described in Part I, Section 17.

7.1.1 Ambient air and standard reference gas mixtures are continuously flushed through their respective sample loops in the multiport sampling valve, the latter mixtures during calibration periods only. The multiport valve is activated on an assigned time schedule to flush the ambient air sample from the sample loop onto the GC column; likewise, during calibration periods or span checks, the contents of the calibration gas loop is flushed onto the column.

7.1.2 The compliance of an instrument to the specifications listed in this method must be affirmed on-site on an individual basis.

7.1.3 Temperature dependencies of several components of the GC-FPD system require that the instrument be maintained in a controlled-temperature environment at $20° \pm 5°C$.

7.1.4 Column conditions should be maintained as follows:

7.1.4.1 *DEACTIGEL COLUMN:* isothermal operation at $85°C$ for H_2S + SO_2 analysis. Elution is complete in 4 min under these conditions and the multiport valve should be set for one analysis every 5 min.

7.1.4.2 *TRITON X-305 COLUMN:* temperature-programmed operation from 50 to $100°C$ at $12°C/min$ for separation and determination of H_2S, SO_2, CH_3SH, alkyl sulfides and disulfides. Loss of SO_2 on this column at low ppb levels has been reported. Elution of these compounds (through dimethyl disulfide) will be completed in approximately 6 min. The multiport valve should be set to cycle once every 12 min to permit cooling of the oven and equilibration of column temperature between injections.

7.1.4.3 *CARBOPACK B HT 100 COLUMN:* isothermal operation at $35°C$ for separation and determination of H_2S, SO_2, COS and CH_3SH in less than 2 min. Coating this column with 1.5% XE-60 + 1.0% H_3PO_4 and operating at $50°C$ allows separation of ethyl mercaptan from dimethyl sulfide in less than 8 min.

7.2 This section describes procedures for use of Method 709B (Total Sulfur Gases by FPD). The manufacturer's manual should be consulted for specific operating instructions.

7.2.1 Filtered, ambient air flows continuously through the multiport sampling valve and Teflon connections to the detector.

7.2.2 The sampling valve is activated at prescribed intervals to admit zero air or calibration gas mixtures to the detector.

7.2.3 Due to the observed dependencies of zero air (background) response on water vapor and CO_2 concentrations in the air sample, zero air and calibration gases should be prepared with the same water vapor and CO_2 concentrations as in the ambient air being sampled. This is most conveniently done by using filtered ambient air scrubbed of sulfur compounds as the zero air source.

8. Calibration and Standards

8.1 The sample handling system must be tested for conformance to the required performance standards at least once a week during periods of continuous operation. All flow measurement devices should be calibrated against a wet test meter. Use of mass flow meters for precise flow measurement and control is recommended; however, calibration of mass flow meters against a wet test meter or other absolutely calibrated device is still required for accurate flow measurement.

8.2 Calibrate the instrument under the same conditions as those used in the automatic determination of sulfur gases in the environment using a permeation-tube dilution system or equivalent.

8.3 Introduce at least four different concentrations, not including zero, of the standard reference gases to give responses from 10 to 80% of full scale on the sensitivity range(s) to be used in ambient concentration measurements. For Method 709A (Sulfur Gases by GC-FPD) the standard reference gases should be added as a mixture to assure adequate column performance for the determination.

For a linear amplifier, the difference in response between calibration standards and zero air is plotted versus concentration of the standards on logarithmic coordinate paper. For Method 709A this response has the units of peak area (A-s) measured for eluted peaks from injection of fixed samples. For Method 709B, in which calibration standards are continuously and directly admitted to the FPD, the response is the difference in current (A) observed for standards and zero air. A typical calibration curve for sulfur dioxide standards, admitted continuously according to Method 709B, is shown in Figure 709:4. With a log amplifier response, or a "linearized" response, plot response vs log concentration on rectangular coordinate paper.

8.4 The linearity of the electrometer from one attenuation scale to another should be checked with standards for each range used.

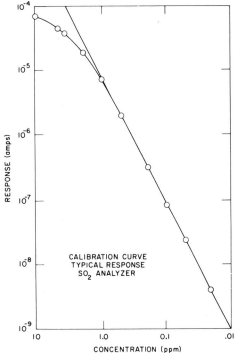

Figure 709:4 – FPD Calibration Curve for SO$_2$; Flame-on Current: 4.2 × 10^{-9} A; Flame-On Noise: 5.0 × 10^{-11}A; Exhaust Temperature: 125°C; Burner Block Temperature: 95°C; Air Flow Rate: 200 cm^3 min^{-1}; H$_2$ Flow Rate: 200 cm^3 min^{-1}.

9. Calculations

9.1 For Method 709A the response from the FPD is recorded by means of a recorder, integrator or other readout device. The elution times are used to identify peaks for each sulfur gas.

9.1.1 A peak area function is calculated for each sulfur gas. This function is proportional to the amount of sulfur gas present, and it is more accurate for all but very rapidly eluting components than an analogous function derived from peak heights.

Peak Area Function, $R = W(HA)^{1/n}$

where

W = peak width at half weight, mm
H = peak height, mm
A = attenuation factor from the FPD electrometer setting
n = power function of sulfur re-

sponse, derived from the slope of the calibration curve (See Section 8.3)

9.1.2 The concentration of sulfur gas is then calculated from the peak area function, R (example given for SO$_2$):

$$[SO_2] = f \times R$$

where f = calibration response factor for SO$_2$ derived from calibration data for SO$_2$ converted to peak area function

This response factor is inversely proportional to the sulfur sensitivity, S_s, appearing in Equation (2)-(4), Section 2.1. Alternately, the concentrations of sulfur gases can be read directly from the log-log calibration curves for each sulfur gas.

9.2 For Method 709B the average response from the FPD for the time period of interest is recorded. The total concentration of sulfur gases is derived from the calibration curve of response vs concentration constructed as described in Section 8.3. In the absence of further knowledge concerning the composition of sulfur gases in the sample atmosphere, it is recommended that the calibration curve for SO$_2$ be used, since SO$_2$ is the most abundant sulfur gas in the ambient atmosphere in most locations.

10. Effect of Storage

10.1 Permeation tubes should be discarded when approximately 10% of the original volume of permeant liquid remains. Replacement permeation tubes may be stored prior to use in a refrigerator in a sealed bottle containing an absorbant to remove the permeants. Re-equilibrate the tubes to operating temperature for 48 h prior to use.

10.2 Gas chromatographic columns must be conditioned with the sulfur-containing gases of interest just prior to use as directed in Section 4.

11. References

1. BRODY, S.S. and J.E. CHANEY. 1966. Flame Photometric Detector. *J. Gas Chromatog.* 2:42–46.
2. CRIDER, W.L. 1965. Hydrogen Flame Emission Spectrophotometry in Monitoring for Sulfur Dioxide and Sulfuric Acid. *Anal. Chem.* 37:1770–1773.
3. ECKHARDT, J.G., M.B. DENTON and J.L. MOY-

ERS. 1975. Sulfur FPD Flow Optimization and Response Normalization with a Variable Exponential Function Device. *J. Chromatog. Sci. 13*:133–138.

4. SUGIYAMA, T., Y. SUZUKI and T. TAKEUCHI. 1973. Characterization for the Selective Detection of Sulphur Compounds by Flame Photometry. *J. Chromatog. 85*:45–51.

5. STEVENS, R.K., A.E. O'KEEFFE and G.C. ORTMAN. 1969. Absolute Calibration of a Flame Photometric Detector to Volatile Sulfur Compounds at Sub-Part-Per-Million Levels. *Environ. Sci. Technol. 3*:652–655.

6. STEVENS, R.K., J.D. MULIK, A.E. O'KEEFFE and K.J. KROST. 1971. Gas Chromatography of Reactive Sulfur Gases in Air at the Parts-Per-Billion Level. *Anal. Chem. 43*:827–831.

7. BRUNER, F., A. LIBERTI, M. POSSANZINI and I. ALLEGRINI. 1972. Improved Gas Chromatographic Method for the Determination of Sulfur Compounds at the PPB Level in Air. *Anal. Chem. 44*:2070–2074.

8. PATTERSON, P.L., R.L. HOWE and A. ABU-SHUMAYS. 1978. Dual-Flame Photometric Detector for Sulfur and Phosphorus Compounds in Gas Chromatograph Effluents. *Anal. Chem. 50*:339–344.

9. SCARINGELLI, F.P., A.E. O'KEEFFE, E. ROSENBERG and J.P. BELL. 1970. Preparation of Known Concentrations of Gases and Vapors with Permeation Devices Calibrated Gravimetrically. *Anal. Chem. 42*:871–876.

10. American Society for Testing and Materials. 1981. Standard Practice for Using Flame Photometric Detectors in Gas Chromatography, E840–81. Annual Book of ASTM Standards, Vol. 14.01.

11. D'OTTAVIO, T., R. GARBER, R.L. TANNER and L. NEWMAN. 1981. Determination of Ambient Aerosol Sulfur Using a Continuous Flame Photometric Detection System. II. The Measurement of Low-Level Sulfur Concentrations Under Varying Atmospheric Conditions. *Atmos. Environ. 15*:197–203.

12. PESCAR, R.E. and C.H. HARTMAN. 1973. Automated Gas Chromatographic Analysis of Sulfur Pollutants. *J. Chromatog. Sci. 11*:492–502.

13. GARBER, R.W., P.H. DAUM, R.F. DOERING, T.

D'OTTAVIO and R.L. TANNER. 1983. Determination of Ambient Aerosol and Gaseous Sulfur Using a Continuous FPD – III. Design and Characterization of a Monitor for Airborne Applications. *Atmos. Environ. 17*:1381–1385.

14. STEVENS, R.K. and A.E. O'KEEFFE. 1970. Modern Aspects of Air Pollution Monitoring. *Anal. Chem. 42*:143A-148A.

15. KOPPE, R.K. and D.F. ADAMS. 1967. Evaluation of Gas Chromatographic Columns for Analysis of Sub-Parts-Per-Million Concentrations of Gaseous Sulfur Compounds. *Environ. Sci. Technol. 1*:479–481.

16. THORNSBERRY, W.L., JR. 1971. Isothermal Gas Chromatographic Separation of Carbon Dioxide, Carbon Oxysulfide, Hydrogen Sulfide, Carbon Disulfide and Sulfur Dioxide. *Anal. Chem. 43*:452–453.

17. ADAMS, D.F. and R.K. KOPPE. 1959. Gas Chromatographic Analysis of Hydrogen Sulfide, Sulfur Dioxide, Mercaptans, and Alkyl Sulfides and Disulfides. *Tappi 42*:601–605.

ACKNOWLEDGMENT

One of the committee members (RLT) acknowledges support of the Office of Health and Environmental Research, U.S. Department of Energy through the Processing of Emissions by Clouds and Precipitation (PRECP) program during the preparation of this document.

Subcommittee 1

R.L. TANNER
P.K. DASGUPTA, *Chairman*
D.F. ADAMS
B.R. APPEL
S.O. FARWELL
K.T. KNAPP
G.L. KOK
W.R. PIERSON
K.D. REISZNER

711.

Determination of Gaseous Sulfuric Acid and Sulfur Dioxide in Stack Gases

1. Principle of the Method

1.1 Sulfur gases are emitted from a typical combustion stack as H$_2$SO$_4$ (not SO$_3$) and SO$_2$. The gas is withdrawn through a heated glass or glass-lined probe. Particulate matter in the sampled gas is removed with a silica-fiber filter maintained 10–20°C above the stack gas temperature and placed behind the probe. Sulfuric acid and sulfur dioxide are then collected from the gaseous sample.

1.2 Gaseous sulfuric acid is collected as condensed droplets in a temperature-controlled condenser behind the heated silica-fiber filter. With the condenser temperature maintained above the dew point of water but below the dew point of the acid, sulfuric acid is collected on the coils and the frit of the condenser. This procedure reduces the artifact production of H$_2$SO$_4$ from the much larger quantities of SO$_2$ usually present in the gas stream, and this provides improved precision and accuracy in the collection of H$_2$SO$_4$. Downstream of the condenser are two impingers in series containing dilute H$_2$O$_2$ for the collection of SO$_2$. In these impingers the collected SO$_2$ is oxidized to H$_2$SO$_4$. The sulfate concentration in the condensed H$_2$SO$_4$ sample and in the impinger samples can be determined by any method suitable for the determination of aqueous sulfate (see 720. DETERMINATION OF AQUEOUS SULFATE). A less direct and less sensitive (but generally adequate) analysis involves the alkalimetric titration of the collected H$_2$SO$_4$ with NaOH with bromophenol blue as the indicator.

The concentrations of H$_2$SO$_4$ and SO$_2$ in the stack gas are calculated from the measured concentrations of H$_2$SO$_4$ in the corresponding samples, the sampling time and the volumetric flow rate. The total emission is then computed from the stack gas concentration and total stack gas flow rate, the latter being best determined by a Pitot tube traverse (1).

1.3 COLLECTION EFFICIENCY

1.3.1 *Sulfuric Acid.* The controlled condensation method for H$_2$SO$_4$ was originally proposed by Knol (2). The method was further developed and studied by Goksøyr and Ross (3) and others (4–5). These studies indicated a 98% collection efficiency for H$_2$SO$_4$ in the condenser and frit for H$_2$SO$_4$ passing the filter. More recent studies show that 95 ± 6% of the H$_2$SO$_4$ entering the probe is eventually recovered from the condenser (6). In experiments with a filter preloaded with an amount of fly ash equivalent to a mass concentration of 1300 mg/m³ in the stack gas, the recovery of H$_2$SO$_4$ from the condenser at the 40 mg/m³ (10 ppm) level was 80–85%. The user should be aware that the collection efficiency may be very significantly affected if parameters such as temperature, sampling rate, etc. are altered from those recommended (7).

1.3.2 *Sulfur Dioxide.* The collection efficiency of SO$_2$ in 3% H$_2$O$_2$ solution as absorber in two impingers connected in series has been determined (8). With 15 mL peroxide solution contained in each midget impinger, collection efficiency was found to be 96%. Over the temperature and concentration range examined, no change in the collection efficiency was observed. Collection efficiency dropped to 90% at a sampling rate of 3 L/min. When using much higher sampling rates, e.g. 12 L/min, in conjunction with a 47 mm diameter silica-fiber filter, and two 250 mL capacity impingers in series, each containing 150 mL of 3% H$_2$O$_2$, a collection efficiency 98% of the SO$_2$ was obtained (9).

1.4 SAMPLING RATES. The sampling rate must correlate with the size of the filter and the impingers used. It is critical that the combination of the filter size and sampling rate provide a sample face velocity at the filter of 11–12 cm/s **(9)**. At lower face velocities, H_2SO_4 may condense on the filter. The most convenient and the recommended combination is a 47-mm filter, a sampling rate of 12 L/min and two 250 mL impingers in series, each containing 150 mL of 3% H_2O_2. An alternate smaller-scale arrangement involves a 25-mm filter, a sampling rate of 1 L/min and two 25 mL midget impingers in series, each containing 15 mL of 3% H_2O_2.

2. Range and Sensitivity

This method is applicable to the determination of SO_2 in the range 26 to 15,600 mg/m³ (10 to 6000 ppm) and H_2SO_4 in the range 40 to 24,250 mg/m³ (or 10 to 6000 ppm) using the alkalimetric titration described here, for a total air sample volume of 360 L (30 min @ 12 L/min). Below 10 ppm SO_2 or H_2SO_4, the color change at the end point cannot be visually detected. The limits of detection of sulfate in the various methods given for the determination of aqueous sulfate (see 720. DETERMINATION OF AQUEOUS SULFATE) vary substantially; most of these methods are at least as sensitive as the alkalimetric titration method given here. Ion chromatography is particularly sensitive and allows much lower detection limits for equivalent sample volumes, or better time resolution (shorter sampling times) at the same detection limit.

3. Interferences (8,10)

3.1 Ammonia, ammonium compounds and fluorides are known to interfere with the alkalimetric titration. High concentrations of HCl will also produce positive errors in the alkalimetric determination. This can be corrected for by a previous determination of Cl⁻ in the sample. However, such a correction is meaningful only if the Cl⁻ is known to be HCl.
3.2 Metallic compounds are generally effectively removed by the silica-fiber filter. If any such compound is not trapped by the filter, some of the prescribed analytical methods for aqueous sulfate (see 720. DETERMINATION OF AQUEOUS SULFATE) will lead to erroneous results unless cation exchange of the sample for H⁺ is carried out before analysis. Labile organosulfur compounds, if present, may also lead to erroneous results.
3.3 Particulate matter itself can be an interferent in the H_2SO_4 measurement by collecting the acid on the filter **(9)**. This can be avoided by maintaining the filter temperature above 260°C. However, this high filter temperature poses the potential risk of evaporating some of the acid that was originally adsorbed on the particles at quasi-equilibrium in the stack. Normally, the filter and probe should be heated to a temperature 10 to 20°C above that of the stack gas. According to prevailing convention, any sulfuric acid associated with the particulate material in the stack is not counted towards total gaseous H_2SO_4 emission from the stack.

4. Precision and Accuracy

4.1 At a sample concentration of 0.001 M H_2SO_4, the analytical precision of the alkalimetric titration is ± 0.5%.
4.2 The overall precision of the method, based on actual stack experiments, is ± 11%. However, in most stacks, the variation of the source concentration itself is much higher; ± 65% in a coal-fired facility is typical **(6)**. Consequently, to obtain valid average measurements from this type of source, a larger number of samples spread out over a period of days is needed.

5. Apparatus

5.1 The following sections describe an integrated, modular flue-gas sampling apparatus for collection of H_2SO_4 by the condenser method and collection of SO_2 in impingers. The description given here assumes a 47 mm filter, sampling rate of 12 L/min, and 250 mL capacity impingers. For a 25 mm filter, sampling rate of 1 L/min and 25 mL impingers, the 18/9 standard ball/socket joints described below should be replaced with 12/5 standard ball/socket joints.

5.1.1 *Probe and Probe Heating.* The probe is usually a 1-to 2-m length of 5 mm bore fused silica or borosilicate tubing with an 18/9 standard socket joint on the downstream (filter) end. The inlet end of the probe is left flat and open. The glass probe is wrapped with heating tape and inserted into a metal sheath (stainless steel is recommended) along with a thermocouple. The thermocouple and heating tape are connected to a temperature controller and heater power supply. The temperature controller is set to provide a probe temperature above the acid dew point.

5.1.2 *Filter Holder and Filter.* The filter holder is constructed of silica or borosilicate glass. The back half of the filter holder contains a coarse frit that supports the silica-fiber filter, and an 18/9 standard socket joint leading to the condenser. The front half of the filter holder has a standard 18/9 ball joint to accept the probe exit end. A heating device such as a heated box or insulated heating tape is placed around the filter holder. The heating device and a thermocouple attached to the filter holder are connected to a temperature controller and power supply. See Fig. 711:1 for a diagram of the filter holder.

5.1.3 *Sulfuric Acid Condenser.* The condenser is constructed of a borosilicate glass or quartz coil with a medium porosity sintered glass frit at the downstream end. The upstream end of the condenser has an 18/9 standard ball joint to mate with the filter holder exit end. The downstream end of the condenser has an 18/9 standard socket joint to mate with the SO₂ impinger system. The condenser is water jacketed;

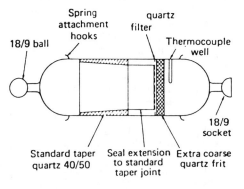

Figure 711:1 – Quartz filter holder.

temperature control is provided by a thermostated, circulating water bath.

5.1.4 *Impingers.* Two impingers of capacity between 250 and 500 mL are modified by addition of 18/9 standard ball/socket joint connectors. Such modified impingers are commercially available from scientific glassware suppliers. Plastic or rubber tubing cannot be used for connecting the impingers.

5.1.5 *Calibrated Dry or Wet Test Meter.*

5.1.6 *Pump.* Leak-free vacuum type, capacity of at least 15 L/min.

5.1.7 *Flowmeter.* Rotameter or equivalent to measure 0 to 15 L/min air flow.

5.1.8 *Stopwatch.* Accurate to 0.1 min or better for measurement of sampling duration.

5.1.9 *Thermometer.* A dial thermometer or thermocouple capable of accurately reading in the range of 100° to 260°C for measuring stack gas temperature.

5.1.10 *Plastic Bottles.* Polypropylene or polyethylene, screw capped, 250–500 mL for storage of impinger and condenser samples. Reference to "plastic bottles" in the following description means such bottles.

5.1.11 *Rubber Squeeze Bulb.* 75 mL capacity, or large enough to attach to 18/9 ball joint.

6. Reagents

6.1 PURITY. ACS Reagent grade chemicals shall be used in all tests.

6.2 WATER. Water must conform to the ASTM standard for Reference Reagent Water, Type II. This can be produced by distillation or from a cartridge deionization system. It is essential that the water be sulfate-free.

6.3 HYDROGEN PEROXIDE (3%) FOR SO₂ COLLECTION. Prepare by tenfold dilution of 30% hydrogen peroxide. This reagent should be prepared fresh daily and stored in plastic bottles.

6.4 PHENOLPHTHALEIN. Dissolve 50 mg phenolphthalein in 50 mL ethanol and dilute to 100 mL with water.

6.5 SODIUM HYDROXIDE, APPROX. 1 N. Slowly add 40 g of NaOH pellets to 800 to 900 mL of water in a 2-L beaker while stirring until all pellets are dissolved. Dilute to 1000 mL with water and mix well. Store in a tightly capped plastic bottle. Standardize against potassium acid phthalate using 2–4 drops of phenolphthalein indicator. Dilute 10 mL of this standardized solution to 1000 mL immediately before use to prepare the 0.01 N titrant.

6.6 POTASSIUM ACID PHTHALATE. Primary standard grade dried to constant weight and stored in a desiccator. Weigh approximately 2 g to an accuracy of 0.0002 g and dissolve in 1 L of water.

6.7 BROMOPHENOL BLUE. Dissolve 50 mg of the dye in 50 mL ethanol and dilute to 100 mL with water.

6.8 ISOPROPYL ALCOHOL.

7. Procedure

7.1 SAMPLING RATES.

7.1.1 *Sulfuric Acid.* The flow rate is critical for the collection of H_2SO_4 by the controlled condensation method. The flow rate affects the amount of vapor phase H_2SO_4 collected by the filter and particles during sampling. A filter face velocity of 11–12 cm/s must be maintained. See 1.4.

7.1.2 *Sulfur Dioxide.* Any convenient sampling rate between 1 and 12 L/min may be used with the appropriate size impingers. The filter size is critical if the method is used for the simultaneous determination of H_2SO_4. See 1.4.

7.2 SAMPLING TIME. The expected SO_2 concentration should be calculated. For oil- and coal-fired units, SO_2 concentration may be estimated (650 to 700 ppm SO_2/1% S) from the fuel analysis (C, H, S), the fuel feed rate and the amount of excess air. The H_2SO_4 content is usually 1 to 3% of the SO_2 concentration. Sampling for extended periods of time may result in H_2SO_4 loss on the filter. When the alkalimetric titration procedure described here is chosen as the final analytical method, the H_2SO_4 concentration in the final sample should be at least 0.001 M.

7.3 SAMPLE COLLECTION. Fit the probe module to the stack flue. Connect power cords between the probe heater through the variable transformer to the power control panel. Load a filter into the filter holder. Connect the filter holder to the probe exit and heat to the probe temperature. Heat the probe to the operating temperature of 160°C. Connect the H_2SO_4 condenser to the water bath. After checking the water level in the H_2SO_4 condenser jacket (adding water, if necessary), switch on the condenser heater and allow to come to operating temperature of 60° to 70°C. Fill each SO_2 impinger with 150 ml of 3% peroxide solution. After the probe and H_2SO_4 condenser have reached operating temperatures, assemble the collector module as shown in Figure 711:2.

Connect the pump to the last impinger, close the filter inlet and leak-check the filter, condenser and impingers. Admit air slowly at the filter holder inlet and reattach the filter holder to the probe. Turn on the control module. Record the initial reading of the dry gas meter, and record the atmospheric pressure and temperature values during sampling. Determine the stack gas temperature and moisture content. At the end of the 20-to 30-min sampling period, switch off the pump and record the gas meter reading and gas meter temperature. Perform final leak check after determining the sample volume by difference from the dry gas meter final reading. The sampling setup is shown in Figure 711:3.

Disassemble the sample module. Rinse the H_2SO_4 collector (the water can be forced through the frit by applying slight pressure from a squeeze bulb attached to a 18/9 ball joint) with several portions of sulfate-free water. Transfer the sample to a plastic bottle for transport to the laboratory. Rinse the H_2SO_4 collector with isopropanol. Then draw clean SO_2-free air through the collector for a short period of time to dry the system to ready it for reuse. Transfer the contents of the two impingers (which contain the SO_2 sample) into a plastic bottle. Rinse the impingers several times with sulfate-free water and add these washings to the contents of the plastic bottle. For immediate collection of additional samples, refill the impingers.

7.4 ANALYSIS

7.4.1 *Analysis for Sulfuric Acid.* Quantitatively, transfer the contents of the plastic bottle containing the condenser washings into a 100-mL graduated beaker. Evaporate the solution to approximately 15 mL. Transfer to a 50-mL volumetric flask and dilute to the mark with sulfate-free water. Refer to 720. DETERMINATION OF AQUEOUS SULFATE for analysis, or use

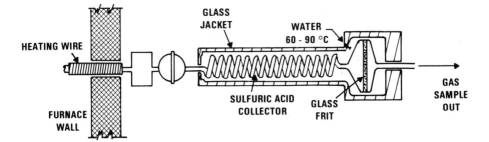

Figure 711:2 – Probe and H$_s$SO$_4$ condenser collection assembly.

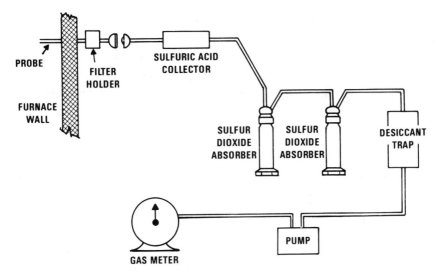

Figure 711:3 – Sulfur oxides gas analysis apparatus.

the alkalimetric titration procedure given below.

Pipet 10 mL of the above sample into a 125-mL Erlenmeyer flask. (Use a larger sample volume if the concentration is low.) Add 2–4 drops of bromophenol blue indicator and titrate to a blue-gray end point with 0.01 N NaOH. Run a blank with each series of samples (7,8).

7.4.2 *Analysis for Sulfur Dioxide.* The procedure given below assumes that a sampling rate of 12 L/min with impingers of corresponding size was used. For smaller sampling rates the volumes given below should be appropriately scaled down.

Quantitatively transfer the solution from the bottle containing the liquid from the SO$_2$ impingers into a 500-mL volumetric flask and dilute to the mark with sulfate-free water. Refer to 720. DETERMINATION OF AQUEOUS SULFATE for analysis or proceed as follows.

Pipet a 50-mL aliquot into a 250 mL Erlenmeyer flask. Add 5 drops of bromophenol blue indicator and titrate to a blue-gray end point with 0.01 N NaOH. It may be necessary to use a smaller sample aliquot when the SO$_2$ concentrations are high. Run a blank with each series of samples.

8. Calibration and Standards

8.1 For calibration of the sulfate analysis method see the appropriate method in 720. DETERMINATION OF AQUEOUS SULFATE being used.

8.2 The limit of detection for H_2SO_4 can be increased by collecting a larger sample volume or by concentrating the condenser washing.

8.3 Large amounts of particulate matter on the filter may act as a H_2SO_4 collector. To assess the extent of loss of H_2SO_4 on the filter due to too low a filter temperature or too large a particle loading or both, extract the filter (see method 824 for recommended extraction procedure) and determine the sulfate in the extract. Since some particulate sulfate is normally present in the stack gas, some sulfate should be expected. However, this sulfate concentration should be significantly smaller than the H_2SO_4 concentration measured from the condensate in the coil. If this is not the case, the validity of the gaseous H_2SO_4 measurement is suspect.

9. Calculations

9.1 DRY GAS VOLUME. Correct the sample volume measured by the dry gas meter to standard conditions (25°C, 101.3 kPa) as follows:

$$V_s = \frac{PV}{T} \frac{298}{101.3}$$

where V_s is the corrected volume, V is the volume registered at the meter, T is the temperature at the meter in K and P is the barometric pressure in kPa at the meter inlet.

9.2 SO_x CONCENTRATION CALCULATIONS. Use the following equations:

$$H_2SO_4(mg/m^3) =$$

$$\frac{\text{total no. of mmoles of } H_2SO_4 \text{ in sample}}{\text{total sample volume } (V_s) \text{ in liters}} \times$$

$$9.81 \times 10^4$$

$$H_2SO_4(ppm) = H_2SO_4(mg/m^3) \times 0.249$$

$$SO_2(mg/m^3) =$$

$$\frac{\text{total no. of mmoles of } SO_2 \text{ in sample}}{\text{total sample volume } (V_s) \text{ in liters}} \times$$

$$6.41 \times 10^4$$

$$SO_2(ppm) = SO_2(mg/m^3) \times 0.382$$

10. Effects of Storage

Field samples are stable for two weeks when stored in capped plastic bottles.

11. References

1. American Society for Testing and Materials. 1972. Standard Test Method for Average Velocity in a Duct (Pitot Tube Method), D-3154.
2. KNOL, B.P. 1960. Improvements in Determination of SO_3 and SO_2 in Combustion Gases, *Riv. Combustibile* 4:542.
3. GOKSØYR, H. and K. ROSS. 1962. Determination of Sulfuric Trioxide in Flue Gases, J. Inst. Fuel (London) *35*:177.
4. HISSINK, M. 1963. Determination of SO_2 and SO_3 in (Flue) Gases, *J. Inst. Fuel* (London) *36*:372.
5. LISLE, E.S. and J.D. SENSENBAUGH. 1965. The Determination of Sulfur Trioxide and Acid Dew Point in Flue Gases, *Combustion 36*:12.
6. MADDALONE, R.F., S.F. NEWTON, R.G. RHUDY and R.M. STATNICK. 1979. Laboratory and Field Evaluation of the Controlled Condensation System for SO_3 Measurements in Flue Gas Streams, *J. Air Pollut. Contr. Assoc.* 29:626.
7. MATTY, R.E. and E.K. DIEHL. 1957. Measuring Flue-Gas SO_2 and SO_3, *Power 101*:94–97.
8. DRISCOLL, J.N. and A.W. BERGER. 1971. "Improved Chemical Methods for Sampling and Analysis of Gaseous Pollutants from the Combustion of Fossil Fuels", Final Report, Vol. I-Sulfur Oxides, Walden Research Corporation.
9. CHENEY, J.L. and J.B. HOMOLYA. 1979. Sampling Parameters for Sulfate Measurement and Characterization, *Environ. Sci. Technol.* 13:584.
10. CORBETT, P.F. 1951. The determination of SO_2 and SO_3 in Flue Gases, *J. Inst. Fuel* (London) 24:247–251.

Subcommittee 1

K.T. KNAPP
W.R. PIERSON
P.K. DASGUPTA, *Chairman*
D.F. ADAMS
B.R. APPEL
S.O. FARWELL
G.L. KOK
K.D. REISZNER
R.L. TANNER

713.

Semi-Continuous Determination of Atmospheric Particulate Sulfur, Sulfuric Acid and Ammonium Sulfates

1. Principle of the Method

1.1 Air containing gaseous and particulate sulfur compounds is passed through a heater and then through a diffusion denuder coated with lead dioxide followed by a flame photometric detector (FPD).

1.2 The response of the FPD to sulfur-containing compounds is based on the light emitted near 400 nm from the excited S_2 band in a hydrogen-rich flame. Net current from the photomultiplier tube is approximately proportional to the square of the sulfur concentration.

1.3 With the heater at close to ambient temperature, only normally gaseous sulfur species are removed by the denuder, providing an FPD response due to particulate sulfur species.

1.4 With the heater at about 120°C, sulfuric acid aerosol volatilizes and is removed by the denuder together with normally gaseous sulfur species. In the absence of interferents, the decrease in FPD response, relative to that near ambient temperature, is relatable to the concentration of sulfuric acid present.

1.5 With the heater at about 300°C, sulfuric acid (H_2SO_4), ammonium sulfate (($NH_4)_2SO_4$) and ammonium acid sulfate (NH_4HSO_4) volatilize and are removed together with the normally gaseous sulfur species. In the absence of interferents, the change in FPD response relative to that at about 120°C is related to the sum of the concentrations of $(NH_4)_2SO_4$, NH_4HSO_4 and intermediate compounds (e.g. $(NH_4)_3H(SO_4)_2$).

1.6 The residual sulfur signal at 300°C, relative to that sampling clean air, is relatable to the concentration of non-volatile sulfur particulate species (e.g. metal sulfates).

1.7 The sensitivity of the FPD for sulfur is enhanced by continuous addition of a constant concentration of sulfur.

1.8 The method provides semi-continuous measurements with a 10–15 minute temperature programming cycle.

2. Range and Sensitivity

2.1 A limit of detection of 1 to 2 $\mu g/m^3$ was reported for H_2SO_4, implying that 5 to 10 $\mu g/m^3$ H_2SO_4 is required for a precision of 10% (1).

3. Interferences

3.1 Any sulfur compound in the particle phase that volatilizes at ≤ 120 or $\leq 300°C$ is an interferent with respect to measuring the corresponding inorganic sulfate species. Potential interferents include S-containing organic compounds (e.g. bishydroxymethane sulfone (2), hydroxymethanesulfonic acid (3), and alkyl sulfonic acids (4).

3.2 The ammonium salts $(NH_4)_2SO_4$ and NH_4HSO_4 volatilize at about the same temperature and are, therefore, not resolved.

3.3 Ammonia addition to the sample inlet can aid in detecting interferents. A decrease in apparent H_2SO_4 upon NH_3 addition is consistent with the presence of H_2SO_4 and may exclude some potential interferents.

3.4 The response of the FPD is altered by change in levels of carbon dioxide (5,6), relative humidity (7), and barometric pressure (7).

3.5 Doping the H_2 supply with 66 ppb SF_6 increases the signal/noise ratio by a factor of 13 at 1 ppb S (11).

4. Precision and Accuracy

An atmospheric intermethod comparison, for total particulate sulfur (TPS), including three systems similar to that described, showed agreement within ± 5% for four of the six systems at concentrations in the range 1–10 $\mu g/m^3$. This implies relatively good precision for TPS. Precision and accuracy have not been reported for measurement of H_2SO_4 or ammonium sulfates by this technique (8).

5. Apparatus

5.1 Several investigators have constructed systems of this type (1,7,9–11). The unit described by Allen et al. (1) includes the best features of each of the prior investigators (Figure 713:1). The present description is based largely on Reference 1.

5.2 The unit includes a source of NH_3-free air and an NH_3 permeation tube. Ammonia is used to neutralize all sulfate species prior to their entry to the FPD. This minimizes errors resulting from differing responses for various inorganic sulfates (7). Ammonia can also be directed into the sample inlet.

5.3 Particulate sulfur-free air is supplied by filtration of ambient air.

5.4 The heater consists of a 15-cm section of 5-mm I.D. stainless steel tubing. Heating is provided by passing an 80-A current (at 1.3 VAC) through the tubing. The various temperatures in the thermal cycle may be controlled by a proportioning-type electronic temperature controller. A single controller can be made to control at two temperatures (e.g. 120°C and 300°C) by wiring a second potentiometer across the original temperature-setting potentiometer and switching between them with a relay.

5.5 The lead oxide denuder is prepared (13) by coating the inner surface of a 15-cm section of 5-mm I.D. stainless steel tubing with a thin slurry of litharge (6 g PbO in 6 mL glycerol-water). Complete removal of the glycerol under conditions that oxidize PbO to PbO_2 and minimize charring is essential. The glycerol-water is removed by passing air through the tube heated initially to 110°C. Temperature is increased to 200°C gradually over a period of 4 h. The orange-colored coating of PbO is replaced by a white to tan coating.

5.6 A microprocessor or controller-timer (e.g., Chrontrol Model CT, Lindberg Enterprises, Inc., San Diego, CA), not shown, can be used to operate the solenoid valves and heater.

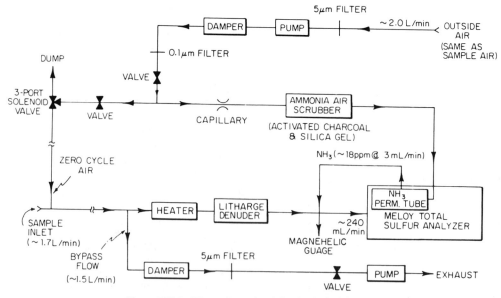

Figure 713:1 — Flow schematic of the Analytical System.

6. Reagents

6.1 Activated charcoal.
6.2 Silica gel.
6.3 PbO powdered.
6.4 5:1 v/v glycerol-water solution.
6.5 Ammonium sulfate.
6.6 H_2 doped with 50 to 75 ppb SF_6 (available from Scott Specialty Gases, Plumsteadville, PA)

7. Procedure

7.1 For a 3–5 min period, the sample flows through the heater at <50°C. The FPD responds to all particle-phase S. (Point #1, Figure 713:2).

7.2 For a 4-min period the heater is maintained at 120–130°C to volatilize H_2SO_4. The FPD responds to all particle-phase S species not volatilized at that temperature (Point #2, Figure 713:2).

7.3 For a 4-min period the heater is maintained at 300°C to volatilize ammonium sulfates. The FPD responds only to non-volatile S species (Point #3, Figure 713:2).

7.4 For a 3–7 min period the system is cooled down while sampling particulate sulfur-free air (Point #4, Figure 713:2).

7.5 The difference in FPD response, Pt 1-Pt 2, Pt 2-Pt 3, and Pt 3-Pt 4, is used to measure H_2SO_4, ammonium sulfates and metal sulfates, respectively.

8. Calibration and Standards

8.1 Ammonium sulfate and H_2SO_4 aerosols are generated from dilute aqueous solutions employing a nebulizer or other appropriate system. The response of the analyzer to each aerosol is optimized by adjusting volatilization temperatures.

8.2 Primary calibration of the FPD can be done with known concentrations of SO_2 in air containing 315 to 385 ppm CO_2. Comparison of FPD total sulfate values, calculated from calibration with SO_2, and filter-collected sulfate concentrations showed relatively good agreement. Linear regression yielded the equation:

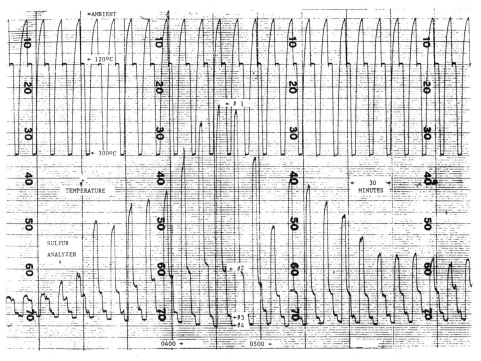

Figure 713:2 — Typical Data from the Analytical System.

FPD SO_4 = 0.982 (Dichot. SO_4) −0.45

r^2 = 0.93

9. Calculations

Sulfate concentrations are calculated using zero air response (e.g. Point #4) as a baseline. Data are reduced as follows:

$$\mu g/m^3 SO_4^= = 3.93 \times A \text{ (net chart div.} \times B \text{ Volts/chart div.)}^C$$

Parameters A and C are obtained from the calibration with SO_2. B is the volts per chart division of the recorder. The factor 3.93 converts ppb SO_2 (at 25°C) to $\mu g/m^3 SO_4^=$.

It may be necessary to correct the observed H_2SO_4 concentration to allow for the fraction not volatilized at the midpoint temperature in the system. For the system described by Allen et al. **(1)**, this correction was 8%.

For the example given in Figure 2, with A = 3.06, B = 0.05 and C = 0.929:

Point #1 = 52.0 $\mu g/m^3$ Total Sulfate
Point #3 = 2.7 $\mu g/m^3$ Non-Volatile Sulfate
Point #2 = 14.2 $\mu g/m^3$ Sulfate. Thus:
Sulfate as sulfuric acid = (52.0 − 14.2) × 1.08 = 40.8 $\mu g/m^3$
Sulfate as ammonium sulfate plus ammonium bisulfate = 52.0 − 40.8 − 2.7 = 8.5 $\mu g/m^3$

10. References

1. ALLEN, G.A., W.A. TURNER, J.M. WOLFESON and J.D. SPENGLER. 1984. Description of a Continuous Sulfuric Acid Sulfate Monitor, Presented at the Fourth Annual National Symposium, "Recent Advances in Pollutant Monitoring of Ambient Air and Stationary Sources". Raleigh, N.C.
2. EATOUGH, D.J. and L.D. HANSEN. 1984. Bishydroxymethyl Sulfone: A Major Aerosol Product of Atmospheric Reactions of $SO_2(g)$, *Sci. Tot. Environ.* 36:319-328.
3. RICHARDS, L.W., J.A. ANDERSON, D.L. BLUMENTHAL, J.A. MCDONALD, G.L. KOK and A.L. LAZRUS. 1983. Hydrogen Peroxide and Sulfur (IV) in Los Angeles Cloud Water, *Atmos. Environ.* 17:911-914.
4. PANTER, R. and R.D. PENZHORN. 1980. Alkyl Sulfonic Acids in the Atmosphere, *Atmos. Environ.* 14:149-151.
5. VON LEHMDEN, D.J. 1978. Suppression Effect of CO_2 on FPD Total Sulfur Air Analyzers and Recommended Corrective Action. EPA Report 600/J-78-096.
6. WEBER, D., K.B. OLSEN and J.D. LUDWICK. 1980. Field Experience with Ambient-Level Flame Photometric Sulfur Detectors, *Talanta* 27:665-668.
7. TANNER, R.L., T. D'OTTAVIO, R. GARBER and L. NEWMAN. 1980. Determination of Ambient Aerosol Sulfur Using a Continuous Flame Photometric Detection System I. Sampling System for Aerosol Sulfate and Sulfuric Acid. *Atmos. Environ.* 14:121-127.
8. CAMP, D.C., R.K. STEVENS, W.G. COBOURN, and R.B. HUSAR. 1982. Intercomparison of Concentration Results from Fine Particle Sulfur Monitors, Atmos. Environ. 16:911-916.
9. HUNTZICKER, J.J., R.S. HOFFMAN, and C.S. LING. 1978. Continuous Measurement and Speciation of Sulfur-Containing Aerosols by Flame Photometry, *Atmos. Environ.* 12:83-88.
10. COBOURN, W.G., R.B. HUSAR and J.D. HUSAR. 1978. Continuous *In Situ* Monitoring of Ambient Particulate Sulfur using Flame Photometry and Thermal Analysis, *Atmos. Environ.* 12:89-98.
11. D'OTTAVIO, T., R. GARBER, R.L. TANNER and L. NEWMAN. 1981. Determination of Ambient Aerosol Sulfur Using a Continuous Flame Photometric Detection System II. The Measurement of Low Level Sulfur Concentrations Under Varying Atmospheric Conditions, *Atmos. Environ.* 15:197-203.
12. HUNTZICKER, J.J., R.S. HOFFMAN, and R.A. CARY. 1984. Aerosol Sulfur Episodes in St. Louis, MO., *Environ. Sci. Technol.* 18:962-967.
13. WOLFESON, J.M. 1985. Private Communication.
14. GARBER, R.W., P.H. DAUM, R.F. DOERING, T. D'OTTAVIO, and R.L. TANNER. 1983. Determination of Ambient Aerosol and Gaseous Sulfur Using a Continuous FPD−III. Design and Characterizations of a Monitor for Airborne Applications, *Atmos. Environ.* 17:1381-1385.

Subcommittee 1

B.R. APPEL
R.L. TANNER
D.F. ADAMS
P.K. DASGUPTA, *Chairman*
K.T. KNAPP
G.L. KOK
W.R. PIERSON
K.D. REISZNER

714.

Determination of Sulfur Dioxide Emissions in Stack Gases by Pulsed Fluorescence

1. Principle of the Method

1.1 This procedure describes an automatic analyzer for the continuous detection and determination of sulfur dioxide in stack gases.

1.2 Flue gas is withdrawn from the stack through a heated probe, filter and sample pump, a heat-traced sample transfer line to a refrigerated cooler (sample conditioner), and thence to the analyzer. Approximately 90% of the heated sample is vented just prior to the sample cooler through a back pressure regulator to assure delivery of a constant-pressure sample to the analyzer. Conversely, the sample may be diluted with clean air to bring it below its dew-point at ambient temperature, thereby eliminating the sample cooler and pressure regulator. The setting of the back pressure regulator will be determined by the sampling requirements of the SO_2 analyzer used **(1)**.

1.3 The continuously-sampled gas stream is irradiated by pulsed ultraviolet illumination that has passed through an interference filter peaking at 216 nm with a half width of 23 nm. The 90° emitted fluorescent light is passed through a broadband optical filter (240 to 420 nm) and is detected by a photomultiplier (PM) tube. The emitted light is linearly proportional to the concentration of the SO_2 in the sample **(2)**.

1.4 The current from the PM tube is amplified, conditioned to eliminate signal noise, and the magnitude of the electronic response is provided to an instrument meter for visual information and to an analog recorder for hard copy printout.

1.5 The automated sampling and analysis system includes a multi-port valve to enable selection of either calibration or stack gas, and a pressurized air source to back-flush the stack gas sample transfer line on a routine basis as required (at least twice daily) to purge the filter and probe of particulate matter.

2. Sensitivity and Range

2.1 This method is applicable to the determination of SO_2 in the range of 2.6 to 13,000 mg/m³ (1 to 5000 ppm). Depending on the instrument configuration, the linear response range may cut off at about 650 mg/m³ (250 ppm) **(3)**.

2.2 The sensitivity, repeatability and accuracy of the method are dependent upon many variables, including the materials of construction of the stack-gas handling system from the probe to the analyzer detection chamber.

2.3 The sensitivity of the method will vary from instrument to instrument within the same manufacturing production run. The observed differences in sensitivity are no doubt related to a normal probability distribution curve for the sensitivities of the PM tubes used. Maximum sensitivity may be achieved through selection of production PM tubes from the sensitive end of the manufacturer's frequency distribution curve.

3. Interferences

3.1 The build-up of particulate matter and condensed water on all surfaces in contact with the sampled gases will produce a gradual decrease in response to the sample and calibration gases.

3.2 The pulsed fluorescence analyzer is based upon a spectroscopic principle. An interference filter isolates the excitation wavelength at 216 nm with a half-band of 23 nm. Scattering is rejected and the SO_2 fluorescence is isolated by a broadband filter that transmits light between 240 and

420 nm. There are no other compounds reportedly producing u.v. fluorescence at 216 nm within the same concentration range. However, carbon disulfide, nitric oxide and ethylene fluoresce weakly (2). Other organic compounds may absorb the fluorescent light. The rejection ratio generally exceeds 5000:1; thus, interference from these compounds appears unlikely.

4. Precision and Accuracy

4.1 Precision and repeatability of the pulsed ultraviolet fluorescence method for SO_2 depends primarily upon proper maintenance of the stack gas transfer line and the instrument optics. Stack gas monitors are subjected to hostile environments and require more frequent maintenance than is required for ambient air monitors.

4.2 Rapid fluctuations in the SO_2 content of the stack gas may be damped by the recognized and varying "memory" properties of all plastic sampling lines (including Teflon) such as absorption, adsorption and permeation.

4.3 Repetitive analysis with one instrument for three SO_2 standard reference gases containing 208 mg/m³ (80 ppm), 936 mg/m³ (360 ppm) and 2127 mg/m³ (818 ppm), respectively, showed a mid-scale error of approximately 4.4% (of full-scale) and a full-scale error of nearly 1.4% (1).

Similar analyses with a second instrument from the same manufacturer for two SO_2 standard reference gases of 83 mg/m³ (32 ppm) and 208 mg/m³ (80 ppm) showed a mid-scale error of approximately 10.7% (of full-scale) and a full-scale error of 6.4% (1). The instrument used for the determination of the lower range of SO_2 in scrubbed stack gas did not satisfy the EPA Performance Specification 2 (4). There was no apparent explanation for the observed differences in performance of these two supposedly comparable instruments.

5. Apparatus

The following sections describe an integrated sampling and monitoring system for SO_2 (which also provides for the sampling of other gaseous compounds in conditioned stack gas).

5.1 PROBE AND PROBE HEATING. The probe should be constructed of Teflon, borosilicate (Pyrex) glass, Vycor, or quartz, with a minimum of stainless steel. The tube should have an I.D. in the range of 0.5 to 1.9 cm (0.2 to 0.75 in.), depending upon the total volume of stack gas required. If a 90° nozzle is used, it should be pointed downstream from the stack gas flow to minimize the intake of the larger particulate matter. The probe is heat-traced to maintain the temperature of the sampled stack gas between 130°C and 200°C. The probe temperature is controlled by a variable transformer preset to provide a temperature of 160°C.

5.2 HEATED FILTER. An in-line Alundum thimble is placed in the stack, or externally in an insulated box heated to between 130°C and 200°C.

5.3 SAMPLE PUMP AND SAMPLE TRANSFER TUBE. The hot, filtered stack gas is pumped from the heated probe-filter to the flue gas cooler and SO_2 analyzer through a heat-traced sample line. For convenience, flexible Teflon tubing is generally used. This transfer line should be as short as possible to minimize sample loss. Sudden increases and decreases in stack gas SO_2 concentrations will be masked by the "memory" characteristics of Teflon (or other) sample transfer lines.

The sample pump must be capable of handling hot stack gases containing corrosive compounds including SO_2, SO_3, and NO_2 at a pressure of 10 to 15 psi (70–100 kPa). Pressurized systems are preferred to vacuum gas transfer lines because of their freedom from sample dilution by influx air around system leaks.

5.4 The sample transfer line is equipped with a 3-way, electrically controlled valve between the heated filter and sample pump. Purge air at 15 to 25 psi (100–170 kPa) is automatically admitted between the sample pump and filter into the filter and probe section for approximately 30 s, twice daily, to purge the filter and sample probe of collected particulate matter. The purge air may be provided from a cylinder of compressed air equipped with a two-stage pressure regulator, by a separate air pressure pump activated by the automatic purge system timer, or by plumbing into a convenient source of pressure-regulated process air (15–25 psi, 100–170 kPa).

5.5 SAMPLE GAS BACK PRESSURE RELIEF. A 0 to 15 psi (0–100 kPa) back pressure regulator set at 5 psi (35 kPa) is installed in the heated stack gas sample line just ahead of the sample cooler inlet. The pressure regulator vents approximately 80 to 90% of the heated stack gas sample, assures that a constant pressure is maintained in the sample distribution line coming from the cooler (important when operating multiple stack gas analyzers from a single sample probe) and to minimize the sample transfer line wall effects (hysteresis) due to changing flue gas SO_2 concentrations.

5.6 PULSED FLUORESCENCE SO_2 ANALYZER. The configuration of the analyzer may vary with the manufacturer's specifications. The instrument should be equipped with its own sampling pump and sample flow regulator. If the analyzer has a range selector switch, the range selected should be that just greater than the maximum anticipated SO_2 concentration. Most instrument specifications indicate the ambient operating temperature range. A weatherproof shelter must be provided to protect the instrument from temperature extremes, water and from fugitive dust. In some applications, the instrument shelter may require continuous purging with "instrument" air (dust-and corrosive gas-free air).

5.7 CALIBRATED WET TEST METER. Accurate to ± 1% for calibration of instrument sample flow meter.

6. Reagents

6.1 CALIBRATION GASES. Cylinders equipped with two-stage regulators, containing dilute mixtures of SO_2 in air are required. The diluted gases should be provided by the vendor in clean, interior-passivated aluminum cylinders to assure maximum purity and long-term stability. The vendor's stated concentrations must be verified by triplicate analysis by the purchaser using the reference analytical method (4).

6.2 ZERO GAS. A pressure cylinder containing SO_2-free nitrogen and equipped with a two-stage regulator.

7. Procedure

7.1 For specific operating instructions, refer to the manufacturer's manual.

7.2 The sampling probe and particulate filter must be back-flushed twice daily for 30 to 60 seconds. The probe and filter should be inspected occasionally just prior to a back-flushing cycle to determine whether the frequency and duration of the cleaning cycle is adequate to permit free passage of the sampled stack gas. The pressure drop across the probe and filter could be monitored with a manometer installed between the stack and the downstream side of the filter.

7.3 Commercially-available instruments must be evaluated on-site in the anticipated monitoring mode to determine compliance with the EPA performance criteria (4,5).

7.4 Perform all routing and major maintenance of the instrument as described in the manufacturer's manual. If the calibration drifts slowly, or sudden changes develop in response to calibration gases, determine the cause within the instrument or the sample transfer train and take immediate corrective action. Consult the instrument manufacturer, if necessary, to correct the problem.

The construction materials should generally be Teflon, Pyrex glass, and a minimum of stainless steel.

7.5 SAMPLE CONDITIONING. Sampling for gaseous pollutants in stacks is often complicated by high temperatures and high moisture levels. The dewpoint temperature is usually greater than the operating temperature of most instruments and consequently must be lowered. Most methods for cooling samples require use of condensing system.

When a non-heated gas handling system is used, a condenser must be used; however, a significant portion of the SO_2 may be lost. The potential may be reduced by adding sufficient acid to the condensate dropout bottle to maintain the condensate at a low pH (6).

Another system described by O'Keeffe uses the principle of permeation of gases through a membrane material and into a clean gas sample stream. This procedure (7) and a similar one in which the water vapor is drawn through a silicone or Nafion tube into the dry air surrounding the tube bundle (**8,9**) may be used.

Dilution of the stack gas sample with a known flow of clean, dry air provides a third method of reducing the water vapor of a heated gas to below the dewpoint at room temperature.

8. Calibration and Standards

8.1 Introduce the "zero" gas at the same flow rate used to monitor SO_2 in stack gas, and adjust the zero control to the proper setting to give the proper recorder or meter value. Frequently, the "zero" setting is intentionally offset to a 5% or 10% scale reading to permit adequate evaluation in the event that the analytical instrument should exhibit a downscale drift.

8.2 Introduce the 80% to 90% of full scale calibration gas at the same flow rate used to monitor SO_2 in stack gas. Adjust the span control to obtain the correct meter or recorder response.

8.3 Repeat "zero" gas step 8.1 to determine any possible interaction between the instrument zero and span controls. If there is interaction, repeat steps 8.1 and 8.2 until the correct instrument readings are obtained without further adjustment of these two controls.

8.4 Introduce the one or two low and/or medium scale calibration gases and record the associated instrument meter and recorder readings.

8.5 Plot the calibration gas concentrations against the instrument responses. This plot should give a straight line. Calculate the instrument response factor (K) from the slope of the line of the calibration plot K = C/R where K is the response factor, C is the concentration of SO_2 and R is the instrument response.

8.6 Perform the total calibration of the entire sample handling system several times per year. First, remove the probe from the stack, turning off the sample pump at the filter, and close the back pressure relief valve.

Introduce the usual flow of the 80% to 90% full-scale calibration gas into the sample probe at the pressure required to produce the usual sample flow rate through the SO_2 analyzer. Compare the instrument response obtained in this manner to the response obtained when the same concentration calibration gas is introduced directly into the analyzer. The percentage loss of SO_2 to the gas handling system is determined from

$$\% \text{ Loss} = \frac{IR-SR}{IR} \times 100$$

where IR is the instrument response to the calibration gas introduced directly into the analyzer sample inlet and SR is the instrument response for the same gas introduced into the sample probe.

8.7 Correct all continuous stack gas monitoring data for the observed SO_2 loss to the gas handling system.

9. Calculations

9.1 Instrument response and the recorder chart design or instrument data processor should be designed to provide a minimum amount of manual work for data compilation and evaluation.

9.2 The instrument shall display at least one measurement per 5 minutes (**4**).

9.3 The instrument response (R) can be converted to the SO_2 concentration (C) by the relationship $C = R \cdot K$ where K is the instrument response factor determined from the calibration procedure in Section 8.5.

10. Effect of Storage

10.1 The rate of possible loss of SO_2 calibration gas through reactions with the inner walls of the pressure cylinders is established by weekly determinations over a period of several weeks of these SO_2 concentrations by the recommended procedure outlined in 704A. Under most circumstances, the SO_2 calibration mixtures will remain relatively constant, provided that the calibration mixtures are contained in clean, passivated aluminum cylinders. A slow reduction in SO_2 concentration can be treated in a linear manner, once several reference analyses have been accomplished. If a rapid loss of SO_2 concentration is encountered, return the cylinder to the vendor and specify delivery of a different, clean cylinder of the SO_2 standard.

10.2 There are no other storage problems.

11. References

1. PEDUTO, E.F., R.B. STRONG and W.B. KUYKENDAL. 1981. Continuous Monitoring for Sulfur Dioxide at a Utility Boiler Equipped with a Limestone Scrubber. *J. Air Pollut. Contr. Assoc. 31*:192–196.
2. OKABE, H., P.L. SPLITSTONE and J.J. BALL. 1973. Ambient and Source SO_2 Detector Based on a Fluorescence Method. *J. Air Pollut. Contr. Assoc. 23*:514–516.
3. CHATHA, J.P.S., P.K. ARORA and P.B. KULKARNI. 1977. Fluorescent Technique for Monitoring Sulphur Dioxide. *Chem. Age (India) 28*:814–816.
4. U.S. Environmental Protection Agency. *Federal Register 40*:(182) 42178–42184 (Sept. 19, 1978).
5. Ibid. *44*:(113) 33580–33624 (June 11, 1979).
6. *The State of the Art of Sulfur Dioxide Monitoring in Pulp and Paper Industry Stack Gases.* Atmospheric Quality Improvement Technical Bulletin No. 78, May 1975. National Council of the Paper Industry for Air and Stream Improvement, Inc., New York, NY 10016.
7. O'KEEFFE, A.E. 1972. Sampling Gases from a Hostile Environment. *Anal. Chem. 44*:1102.
8. CASSIDY, F.J. 1974. Evaluation of the Perma Pure Dryer for Drying Stack Gases. AID.8BA.74. Exxon Research and Engineering Company, Linden, NJ 07036.
9. Perma Pure Products, Inc. Box 2105, Toms River, NJ 08754.

Subcommittee 1

D.F. ADAMS
B.R. APPEL
P.K. DASGUPTA, *Chairman*
S.O. FARWELL
K.T. KNAPP
G.L. KOK
W.R. PIERSON
K.D. REISZNER
R.L. TANNER

720A.

Suppressed Anion Chromatography

1. Principle of the Method

1.1 Since the introduction of the technique in 1975 by Small, Stevens and Bauman (1), (suppressed) ion chromatography (IC) has become a very powerful tool for the general analysis of ionic species, especially common anions. A number of available monographs (2–5) address suppressed and non-suppressed IC in detail.

1.2 Figure 720A:1 shows the block diagram of the basic elements of an Ion Chromatograph (suppressed Ion Chromatography is patented technology of Dionex Corporation and "Ion Chromatograph" is a trademark). Pump P pumps the eluent E, a solution of an alkali metal salt of a weak acid, typically a mixture of sodium carbonate and bicarbonate, through a loop-type injection valve V and a separator column SE containing a latex-agglomerated pellicular anion exchange resin. The effluent from the separator column is passed through a suppressor device SU which exchanges all cations for H$^+$. The suppressor device may be a column packed with high capacity cation exchange resin in H$^+$-form; it is necessary to regenerate the resin periodically (typically after every 6–8 h) to convert it back to the H$^+$-form. However, the preferred suppressor device is membrane-based.

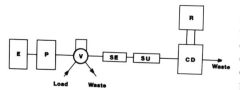

Figure 720A:1 — Flow schematic of an ion chromatography. E: eluent reservoir, P: high pressure pump, V: loop-type injection valve, SE: separator column, SU: suppressor device, CD: conductivity detector, R: recorder/data acquisition system.

The effluent from the separator flows on one side of a thin cation exchanger membrane while a dilute acid regenerant flows on the other side. The cations in the effluent are transported across the membrane via ion-exchange sites in the membrane to the regenerant side and thence to waste while the ion exchange sites in the membrane are continuously regenerated to the H$^+$-form by the acid regenerant. Currently available commercial membrane suppressors are not only continuously rejuvenated; they exhibit much less band dispersion than a packed column, and thus permit better chromatographic resolution and greater sensitivity.

Following the suppressor, the effluent flows through a conductivity detector CD and the output is processed by a strip-chart recorder R and/or a data acquisition system.

1.3 If a sodium carbonate eluent is being used, SE is in the carbonate form. As a sample containing various anions, e.g., Cl$^-$, NO$_3^-$, SO$_4^{2-}$ is injected on the column, the anions are retained by SE in varying degrees and appear at the column outlet at different times. The eluent Na$_2$CO$_3$ is quantitatively converted to very weakly ionized H$_2$CO$_3$ by the suppressor device SU and thus the detector registers a low conductivity background. The residual conductivity is offset by the detector zero control to establish a zero baseline on the recorder. As Cl$^-$, NO$_3^-$, SO$_4^{2-}$ etc. emerge from SU, they enter the detector as the completely ionized corresponding strong acids, and the increased conductivity due to the eluting peaks is registered by the detector. A typical chromatogram of a mixture of common anions at low concentrations is shown in Fig. 720A:2.

1.4 Identification of an eluted anion is based solely on the retention time (the interval between injection and the peak maxi-

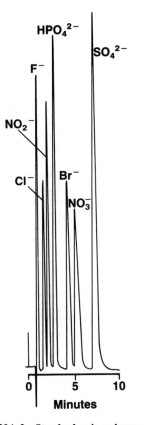

Figure 720A:2 — Standard anion chromatogram of seven common anions. Column: Dionex HPIC AS-4, Eluent 2.8 mM NaHCO$_3$ + 2.2 mM Na$_2$CO$_3$, 2 mL/min. Sample concentration, in μg/mL: F$^-$, 2; CL$^-$, 2; NO$_2^-$, 5; PO$_4^{3-}$, 15; Br$^-$, 15; NO$_3^-$, 15; and SO$_4^{2-}$, 20. 50 μL sample, 10 μS full scale.

2. Range and Sensitivity (6)

For the common anions shown in Figure 720A:2, the limits of detection in μg/L, based on a 50 μL sample size and a signal/noise ratio of 3, are: F$^-$,5; Cl$^-$,5; NO$_2^-$,10; PO$_4^{3-}$,25; Br$^-$,30; NO$_3^-$,50; and SO$_4^{2-}$,30. The slope of the calibration plot for peak height, in microsiemens vs. concentration, varies from anion to anion; for sulfate the slope is ≈ 0.25 μS/ppm sulfate at 22°C (for a 50 μL sample). A better linear correlation is observed for injected concentration vs. peak area rather than vs. peak height, especially for asymmetric peaks, such as that of nitrate. The linear relationship between peak area and injected concentration holds for all of the above-listed anions from the respective limits of detection to a concentration of ≈ 25 mg/L. At higher concentrations, a linear relationship with a different slope is generally observed. When the total anionic content in the injected sample volume significantly exceeds 0.2 microequivalents, the system begins to show the effect of column overloading; peak shapes and retention times are perceptibly affected. If high-concentration samples are to be analyzed routinely, a smaller sample (loop) volume is an attractive alternative to sample dilution.

3. Interferences

3.1 With the recommended column and elution conditions and a membrane suppressor (as described in sections 5-7), no interferences due to co-elution are observed in real samples for the common anions shown in Figure 720A:2, with the possible exception of fluoride.

3.2 Fluoride elutes virtually without any retention; consequently other unretained species or very weakly retained species (notably acetate) may co-elute. The fluoride peak is also partially merged with the water dip (from the water associated with the sample matrix, see Fig. 720A:2) making quantitation difficult at very low concentrations.

3.3 The water dip does not exist when the sample matrix is the same as the IC eluent. Filters used for collecting particulate matter, for example, can be extracted with the IC eluent rather than with water.

mum sensed by the detector). For example, with a new column, a fiber suppressor, and conditions as described in sections 5-7, typical retention times ($\pm$ 15%) are 1.4 min for Cl$^-$, 1.7 min for NO$_2^-$, 5.1 min for NO$_3^-$ and 6.6 min for SO$_4^{2-}$ with the water dip appearing at 0.65 min. Retention times change with the aging of a column (generally decrease). The order of elution however remains the same as long as the eluent composition is not changed. If there is any ambiguity as to the identity of an eluting peak, the retention time for the ion of interest is determined by injecting the corresponding standard solution.

Thus, when the sample matrix is the IC eluent itself, quantitation of very low levels of F⁻ and Cl⁻ is facilitated since the water dip is eliminated.

3.4 The water dip problem is considerably worse with the packed column suppressor because the location of the dip depends on the state of exhaustion of the suppressor, making low level quantitation of both F⁻ and Cl⁻ difficult. Additionally, problems due to undesirable reactions in the suppressor column have been reported for the determination of nitrite (7).

3.5 There are no cationic interferences. A number of heavy metal cations are, however, very strongly retained by the membrane suppressor and thus slowly poison the membrane. In some cases, it may be possible to rejuvenate a poisoned membrane; the manufacturer's recommendation should be followed for this purpose.

4. Precision and Accuracy (6)

At a concentration level of 1 mg/L, the coefficient of variation of peak area for each of the anions shown in Figure 720A:2 is $\leq \pm$ 3%, typically $\pm$ 1.5% for standards. At concentrations varying from approximately three times the limit of detection to 25 mg/L, the relative error in determining the concentration from a linear equation based on peak area (see section 2) is $\pm$ 3%.

5. Apparatus

5.1 ION CHROMATOGRAPH. Any of the various models of Ion Chromatographs (available from Dionex Corp., Sunnyvale, CA) equipped with a conductivity detector is suitable.

5.2 COLUMN. Dionex type HPIC AS-4. It is likely that more efficient columns will be developed; the recommended eluent and flow rates for such a column may also be different. The manufacturer's recommendations for this specific analysis should be followed.

5.3 GUARD COLUMN. The use of a guard column, Dionex type HPIC AG-4, is recommended for prolonging the useful life of the more expensive main column (5.2). The guard column is connected in series ahead

of the main column. The use of the guard column will increase the respective retention times cited in Section 1.4 by ~20%.

5.4 MEMBRANE SUPPRESSOR. Dionex Anion Fiber Suppressor AFS 1-2 or Anion Micro Membrane Suppressor AMMS are both suitable. The latter is capable of a higher dynamic ion exchange capacity and exhibits lower band dispersion.

5.5 DATA ACQUISITION SYSTEM. A strip-chart recorder, or preferably a digital integrator.

5.6 SYRINGE FILTERS. Millex-HV 0.2 μm pore size (Millipore Corporation, Bedford, MA) or equivalent. These are useful for filtration of sample solutions containing suspended matter.

6. Reagents

All reagents are ACS reagent grade. Water must meet the standards of ASTM reference reagent water, Type II.

6.1 SODIUM CARBONATE, ANHYDROUS. Heat at 110°C overnight and allow to cool in a desiccator.

6.2 SODIUM BICARBONATE. Do *not* attempt to oven dry.

6.3 ELUENT, 2.8 mM NaHCO₃ + 2.2 mM Na₂CO₃. Weigh 0.941g NaHCO₃ and 0.933g Na₂CO₃ to the nearest mg and dissolve in 4 L of water.

6.4 REGENERANT, 12.5 mM H₂SO₄. Add 2.8 mL conc. H₂SO₄ (J.T. Baker "Instra-Analyzed" grade or equivalent) to 4 L of water. Mix thoroughly.

6.5 ANION STANDARD STOCK SOLUTIONS. Dry the following solid reagents at 105°C for 30 min. Each of the following stock solutions contains 1000 μg/mL of the designated anion.

Fluoride (F⁻): Dissolve 2.2100g NaF in water and dilute to the mark in a 1-L volumetric flask.

Chloride (Cl⁻): Dissolve 1.6485 g NaCl in water and dilute to the mark in a 1-L volumetric flask.

Nitrite (NO₂⁻): Dissolve 1.4998g NaNO₂ in water and dilute to the mark in a 1-L volumetric flask.

Phosphate (PO₄³⁻): Dissolve 2.4329g KH₂PO₄ in water and dilute to the mark in a 1-L volumetric flask.

Bromide (Br⁻): Dissolve 1.4893g KBr in water and dilute to the mark in a 1-L volumetric flask.

Nitrate (NO_3^-): Dissolve 1.3706g $NaNO_3$ in water and dilute to the mark in a 1-L volumetric flask.

Sulfate (SO_4^{2-}): Dissolve 1.8141g K_2SO_4 in water and dilute to the mark in a 1-L volumetric flask.

6.6 MIXED ANION STANDARD SOLUTION. Pipet the following amounts of the 1000 μg/mL stock standard solutions (6.5) into a single 1-L volumetric flask and dilute to the mark: F^-(2 mL), Cl^-(2 mL), NO_2^-(5 mL), PO_4^{3-}(15 mL), Br^-(15 mL), NO_3^-(15 mL) and SO_4^{2-}(20 mL).

7. Procedure

7.1 The manufacturer's instructions for general operation should be followed. The specific operating parameters are listed below.

7.2 Eluent flow rate: 2 mL/min. With the HPIC AS-4 column, either of the membrane suppressors described in 5.4 and the recommended eluent (6.3), the pressure gauge reading should not exceed 1200 psi (8300 kPa). If the pressure exceeds 1500 psi (10,000 kPa), take the column and the suppressor out of the flow path, one at a time, to determine the source of the problem and replace the defective component. A common cause of increased column pressure is fouled column bed supports.

7.3 Regenerant flow rate: 3.5 mL/min. The eluent flow rate through the fiber suppressor must never exceed 3 mL/min to avoid permanent damage.

7.4 Detector sensitivity: 10 μS full scale or below depending on the concentration of the samples to be analyzed.

7.5 Recorder/Integrator chart speed: 0.5 cm/min. Follow manufacturer's recommendations to set the slope sensitivity of the integrator.

7.6 Injection loop volume: 50 μL.

7.7 Within 15–30 minutes of beginning eluent and regenerant flow, the base line should be stable at a sensitivity of 10 μS full scale. At more sensitive detector settings, attainment of a stable baseline will take correspondingly longer. After a stable baseline reading is obtained, use detector zero control to offset residual conductance and start recorder/integrator. To check column

performance, inject the mixed anion standard (6.6). For injection of standards or samples, introduce at least 0.5 mL, or three times the loop volume (whichever is greater), through the injection valve in the load position and then switch to inject position. Instrument response to injected standards should be checked daily. Adequate resolution of the mixed anion standard should be checked at least twice weekly. If the retention time of a given anion has shifted by more than 20%, the system will need to be recalibrated. If calibration is conducted in terms of peak heights, a retention time shift of only 10% can be tolerated before the response changes by more than 5% and recalibration is necessary.

7.8 Analyze samples by duplicate injections; wait for the last anion (typically sulfate, in samples pertinent to the present volume) to elute before repeating the injection. Use the mean peak area (or height) for calculating the sample concentration.

7.9 If there is any particulate matter in the sample solution, introduce the sample into the injection valve through a small-volume, disposable membrane filter (5.6). For trace determination (at concentrations below 1 μg/mL), any contamination originating from such a filter must be determined by injecting the eluent blank through the filter, and the resulting signal must be subtracted from the results obtained for the sample.

7.10 The following is a brief outline for troubleshooting instrument malfunction and refers to the results of injecting the mixed anion standard; a normal chromatogram under these conditions will resemble Figure 720A:2. If no peaks or only small peaks are observed, check if the injection valve is properly operating and for continuity of flow. If flow is present at the column outlet but not at the suppressor outlet, the membrane has burst (AFS), or leaking due to high backpressure (AMMS). If continuity of flow is maintained through the column, suppressor and detector, and background conductivity is high, the suppressor is malfunctioning. The reason may be inadequate regenerant flow or a poisoned membrane.

If the column provides inadequate resolution for the mixed anion standard, it can still be used for some time, in many cases.

The columns are expensive and have a limited lifetime. In some cases, the efficiency of a column can be restored, at least partially. Follow the manufacturer's recommendations for such rejuvenation attempts.

8. Calibration and Standards

8.1 Prepare, by serial dilution of the stock standards (6.5), solutions containing 25, 10, 5, 2, 1, 0.5, 0.2, 0.1 and 0.05 $\mu g/mL$ of the anion of interest.

8.2 Obtain the mean peak area (or peak height if an integrator is not available) for each injected concentration (triplicate injections) and plot the peak area (or peak height) against the injected concentration. Calculate the best-fit equation using the method of linear least squares. Examine the plot for linearity. The deviation from linearity may be too large to be acceptable for a peak height vs. concentration plot, especially for asymmetric peaks, over a large range of concentration. However, acceptable accuracy is obtained if the plot is divided into two or more segments, each described by a linear equation.

9. Calculations

Calculate the concentration of the sample from the observed mean peak area by means of the best linear equation obtained in 8.2. If the calibration was conducted in terms of peak height data, calculate the concentration by means of the best linear equation describing the appropriate segment.

10. Effects of Storage

Stock standards (6.5) are stable for at least one month when stored at 4°C. Dilute working standards should be prepared weekly, except nitrite and phosphate which must be prepared fresh daily. The eluent and regenerant solutions are stable indefinitely when stored in capped plastic containers at room temperature.

11. References

1. SMALL, H., T.S. STEVENS and W.C. BAUMAN. 1975. Novel Ion Exchange Chromatographic Method Using Conductimetric Detection, *Anal. Chem.* 47:1801–1809.
2. MULIK, J.D., E. SAWICKI and E. WITTGENSTEIN, eds. 1978. "Ion Chromatographic Analysis of Environmental Pollutants", Vol. I, Ann Arbor Science, Ann Arbor.
3. FRITZ, J.S., D.T. GJERDE and C. POHLANDT. 1982. "Ion Chromatography", Alfred Huthig, Heidelberg.
4. SMITH, F.C. JR. and R.C. CHANG. 1983. "The Practice of Ion Chromatography", Wiley, New York.
5. TARTER, J.G., ed. 1986. "Ion Chromatography", Marcel Dekker, New York.
6. STILLIAN, J.R. 1985. Dionex Corp., private communication.
7. KOCH, W.F. 1979. Complication in the Determination of Nitrite by Ion Chromatography, *Anal. Chem. 51*:1571–1573.

Subcommittee 1

P.K. DASGUPTA, *Chairman*
D.F. ADAMS
B.R. APPEL
S.O. FARWELL
K.T. KNAPP
G.L. KOK
W.R. PIERSON
K.D. REISZNER
R.L. TANNER

720B.

Nonsuppressed Anion Chromatography

1. Principle of the Method

1.1 Nonsuppressed anion chromatography (also called single column ion chromatography, SCIC) with conductimetric detection was introduced by Fritz *et al.* A detailed description of principle and scope is given in the monograph by Fritz, Gjerde and Pohlandt **(1)**.

1.2 In nonsuppressed ion chromatography, the system setup is identical to that shown in Fig. 720A:1, except that there is no suppressor device, SU.

1.3 Detection of eluting ions in this method relies on the fact that the total ionic concentration, in equivalents, eluting from the column always remains constant. If, for example, 5 mM NaE is being used as eluent, where E^- is the eluent anion, when a sample anion S^- elutes with a peak concentration of 1 mM, the effluent composition at the peak is 4 mM NaE + 1 mM NaS. The detector registers this change by virtue of the difference in conductance between 5 mM NaE and 4 mM NaE + 1 mM NaS; i.e., this change is proportional to the difference in equivalent conductance between E^- and S^- as well as the concentration of eluting S^-.

1.4 To perform trace analysis by this principle, it is essential that the background conductance be kept low, which necessitates a low eluent concentration. For successful chromatography in a reasonable time period, this, in turn, requires the eluent anion to have high eluting ability and the column to have relatively low ion exchange capacity. A large organic anion is generally chosen as the eluent ion — the sensitivity is also maximized by such a choice since the equivalent conductance of a large organic anion is substantially lower than those of common inorganic anions of interest.

1.5 Two variations are given here. One uses a polymeric anion exchange column with a *p*-hydroxybenzoate eluent to achieve complete separation of F^-, Cl^-, NO_2^-, Br^-, NO_3^-, PO_4^{3-} and SO_4^{2-} in about 8 min (variation 1). The second method (variation 2) uses a silica-based anion exchange column with a phthalate eluent for fast analysis of sulfate (~ 70 s) for samples which contain sulfate as essentially the only anion (e.g., samples from methods 704C and 711).

1.6 Identification of eluted anions is based solely on retention times under a given chromatographic condition. Retention times vary from column to column ($\pm 20\%$) for a given column type and are acutely dependent on the eluent pH. A general decrease in retention time is also observed as the column ages. For these reasons, the retention times cited below should be considered approximate. In variation 1, the respective retention times of F^-, HCO_3^-, Cl^-, NO_2^-, Br^-, NO_3^-, PO_4^{3-} and SO_4^{2-} are 1.5, 1.8, 2.1, 2.5, 3.0, 3.6, 4.3 and 5.5 min, with two system peaks appearing at 1.1 and 8 min. (Figure 720B:1). The system peaks arise from eluent ion redistribution between the mobile and the stationary phases as a result of sample injection and may be either negative or positive depending on the total ionic equivalents present in the sample as well as its pH. Peaks due to HCO_3^- are often observed in standard mixtures which did not contain any intentionally added HCO_3^-. The presence of HCO_3^- stems from the intrusion of atmospheric CO_2. Negative peaks may also be observed at the retention time of HCO_3^- indicating that the concentration of HCO_3^- in the eluent is greater than that in the sample. In variation 2, SO_4^{2-} and NO_3^- elute at 60 and 40 s respectively while the system peak occurs at 10 s.

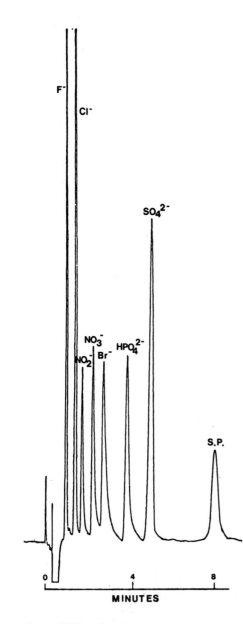

Figure 720B:1 — Standard ion chromatogram of seven common anions by variation 1. Eluent 7mM PHBA, pH 8.40, flow rate 2.5 mL/min. Column: Wescan Anion/R (4.1 × 250 mm). with guard cartridge Wescan Ion Guard RP/R (4.6 × 30 mm). Sample volume 500 μL. Sample concentration, in μg/mL: F⁻, 4; Cl⁻, 4; NO_2^-, 4; NO_3^-, 4; BR⁻, 4; PO_4^{3-}, 6; SO_4^{2-}, 6. Attenuation × 10.

2. Range and Sensitivity (2)

The limit of detection (S/N = 3) of sulfate for a 100 μL sample by variation 2 is 200 μg/L. The respective limits of detection for F⁻, Cl⁻, NO_2^- Br⁻, NO_3^-, PO_4^{3-} and SO_4^{2-} for a 500 μL sample by variation 1 are 8, 15, 75, 60, 60, 130 and 60 μg/L. The above-listed detection limits are attainable with an insulated column and detector compartment when the instrument is located in an air-conditioned laboratory without drafts. Thermal fluctuations are the primary source of detector noise and drift in SCIC; if the instrument is operated under worse conditions than that indicated above, it must be equipped with an active thermostatic accessory (available from the manufacturer) to attain the stated detection limits. The sensitivity (i.e., slope of the calibration plot) varies from one anion to another and increases with increasing equivalent conductance of the anion.

A better linear relationship is observed with injected sample concentration vs. peak area rather than vs. peak height for both variations of the method. The linear relationship with peak area holds for all anions for at least three orders of magnitude range in concentration beginning from the limit of detection. The linear range may be moved to higher concentrations by decreasing the sample loop volume, which is recommended if high-concentration samples are to be routinely analyzed.

For concentration vs. peak height plots, although the fit of a single straight line over the entire range of interest may not be acceptable, multiple linear segments may be satisfactorily used as in method 720A.

3. Interferences

3.1 In variation 1, with the column and elution conditions as described in section 6–8, no interferences due to co-elution are observed in real samples for the common anions shown in Fig. 720B:1, with the possible exception of fluoride.

3.2 Certain weakly retained organic anions, notably acetate, may co-elute with fluoride.

3.3 At very low fluoride concentrations, the fluoride peak is not completely separated from the water dip (from the water associated with the sample matrix) which makes accurate quantitation difficult.

3.4 The water dip does not exist when the sample matrix is the same as the eluent. Filters used for collecting particulate matter, for example, can be extracted with the p-hydroxybenzoate eluent rather than with water.

3.5 Variation 2 is intended for the determination of sulfate only (Figure 720B:2). However, it is possible to determine nitrate under these conditions. Bromide, if present at significant levels, interferes with the nitrate determination.

3.6 There are no cationic interferences.

4. Precision and Accuracy (2)

4.1 In variation 1, for all anions except phosphate, the coefficient of variation of peak area is typically $\leq \pm 1\%$ at sample concentrations in the range of 0.1-1 mM (500 μL injection). The coefficient of varia-tion for phosphate is substantially worse, typically 7%. In variation 2, the peak area precision for sulfate is typically 3% at sulfate concentrations in the range 0.01-1 mM (100 μL injection).

4.2 For the respective conditions for each variation as stated in 4.1, the relative error in determining the concentration from a linear equation based on peak area (see section 2) is $\pm 3\%$.

4.3 At low sample concentrations ($\leq$ 0.01 mM) peak height precision is superior to peak area precision. However, the linearity of response in terms of peak area extends over a greater dynamic range.

4.4 Because of the use of very low-capacity columns in SCIC, column overloading effects are observed at lower sample concentrations than in suppressed IC. Retention times and peak shapes change, which can cause errors in peak height-based quantitation. Peak area-based quantitation, however, is relatively unaffected.

5. Advantages and Disadvantages

5.1 ADVANTAGES. SCIC involves nonproprietary technology; the necessary equipment is available from a number of instrument manufacturers. The system setup is simpler and equipment cost is lower compared to suppressed IC. SCIC displays better sensitivities for very weak acid anions, e.g. CN^-, CO_3^{2-}, SiO_3^{2-}, etc., compared to suppressed IC and permits a wider latitude in selecting eluent pH.

5.2 DISADVANTAGES. For the majority of common anions, detection limits attainable by SCIC are not as low as those attainable by suppressed IC for identical sample volumes. SCIC is more susceptible to thermally-induced detector noise and column overloading effects compared to suppressed IC (see sections 2 and 4.3).

6. Apparatus

6.1 HIGH PRESSURE PUMP. Capable of pumping up to 5 mL/min at pressures up to 5000 psi (34.5 MPa). This is available from a number of manufacturers. The use of a pulse damper is recommended for variation 1 and is essential for variation 2.

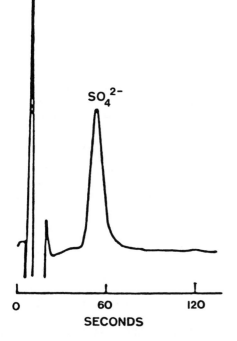

Figure 720B:2 – Chromatogram of sulfate by variation 2. Eluent 15 mM phthalic acid, flow rate 4 mL/min. Column: Wescan 269-003 (4.6 × 30 mm). Sample volume: 100 μL.

6.2 INJECTOR, DETECTOR, COLUMN COMPARTMENT. Ion Chromatography Module, ICM-100 (Wescan Instruments, Santa Clara, CA) or equivalent. The injector should be equipped with a 500 μL loop for variation 1 and with a 100 μL loop for variation 2.

6.3 COLUMN AND GUARD COLUMN. Variation 1: Wescan Anion/R (4.1 × 250 mm) with guard cartridge Wescan Ion Guard RP/R (4.6 × 30 mm). Variation 2: Wescan 269-003 (4.6 × 30 mm). (CAUTION: The silica-based column used in variation 2 is incompatible with alkaline eluents such as that used in variation 1. Use of an eluent with a pH greater than about 7.5 for any significant period of time will irreversibly damage a silica column.)

6.4 DATA ACQUISITION SYSTEM. A strip-chart recorder, or preferably a digital integrator.

6.5 PH METER. Capable of reading accurately to 0.01 pH unit.

6.6 SYRINGE FILTERS. Millex-HV (Millipore Corp, Bedford, MA) or equivalent, for filtration of sample solutions containing suspended matter.

7. Reagents

All reagents are ACS reagent grade. Water must meet the standards of ASTM reference reagent water, Part II.

7.1 ELUENT, VARIATION 1. Dissolve 1.93 g of p-hydroxybenzoic acid (PHBA) in 25 mL HPLC grade methanol and swirl or ultrasonicate until dissolved. Transfer the contents to a 2-L volumetric flask containing deionized water and rinse the beaker several times with water to ensure complete transfer and dilute to the mark. This solution contains 7 mM PHBA.

7.1.1 *pH Adjustment*. Prepare a sodium hydroxide solution by dissolving 1 g (approx 5–6 pellets) NaOH in 25 mL of deionized water. To avoid excessive CO_2 contamination, this solution must be prepared immediately before use. Transfer approximately 1800 mL of the PHBA solution into a large beaker and add the NaOH solution dropwise with stirring until the pH as read by the (properly calibrated) meter reads 8.40 ± 0.02. If too much NaOH has been added (pH > 8.42), add some of the PHBA

solution remaining in the volumetric flask to adjust the pH to 8.40 ± 0.02.

7.2 ELUENT, VARIATION 2. Dissolve 4.98 g phthalic acid in 2 L deionized water. This solution contains 15 mM phthalic acid. The retention time of sulfate for the conditions given for variation 2 is 60 s. However, because column capacity varies, the retention time may vary. If the retention time is greater than 80 s, a small amount of NaOH may be added to the eluent to increase eluting power.

7.3 ANION STANDARD STOCK SOLUTIONS. See section 6.5, method 720A.

7.4 MIXED ANION STANDARD SOLUTION. See section 6.6, method 720A.

7.5 MIXED SULFATE/NITRATE STANDARD. Pipet 25 mL each of the 1000 μg/mL nitrate and sulfate solution (7.3) into a single 1-L volumetric flask and dilute to the mark. This solution contains 25 μg/mL each of sulfate and nitrate.

8. Procedure

8.1 The manufacturer's instructions should be followed for general operation. The specific operating parameters are listed below.

8.2 ELUENT FLOW RATE. Variation 1: 2.5 mL/min. Variation 2: 4 mL/min. With the respective columns for each variation, the system operating pressure should be about 1500–3000 psi (10–20 MPa) for variation 1 depending on the age of the column and 200–400 psi (1.4–2.8 MPa) for variation 2.

8.3 INJECTION VOLUME. 500 μL for variation 1 and 100 μL for variation 2.

8.4 DETECTOR SENSITIVITY. Typically the attenuation is set at 10; the setting may need to be altered depending on sample concentration.

8.5 RECORDER/INTEGRATOR CHART SPEED. 0.5 cm/min. The manufacturer's recommendations should be followed to set the slope sensitivity of the integrator.

8.6 Within 15–30 min of beginning eluent flow, the baseline should be stable at an attenuation setting of 10. At more sensitive detector settings, attainment of a stable baseline will require correspondingly longer. After a stable baseline reading is obtained, use detector autozero to offset the residual conductance and start the recorder/integrator.

8.7 To check column performance, inject the mixed anion standard (7.4) for variation 1 and the mixed sulfate/nitrate standard (7.5) for variation 2. In each case, all components should be baseline-resolved. For injection of standards and samples, use at least 1.5 mL for variation 1 and 0.5 mL for variation 2 to flush the loop completely of its previous contents. If loop volumes other than those specified in section 8.3 are used, use at least 0.5 mL, or three times the loop volume, whichever is greater. Instrument response and adequate resolution of standards should be checked daily. If the retention time of a given anion has shifted by more than 20% since last calibration, the system will need to be recalibrated. If calibration is conducted in terms of peak heights, a retention time shift of only 10% can be tolerated before the response changes by more than 5%, and recalibration is necessary.

8.8 Analyze sample by duplicate injections, wait for the last peak (in variation 1 the system peak; in variation 2 sulfate) to elute before repeating the injection. Use the mean peak area (or height) for calculating the sample concentration.

8.9 If there is any particulate material in the sample solution, prefilter the sample through a syringe-adaptable disposable membrane filter (6.6). Any ionic contribution due to the use of the filter must be checked and accounted for; see Section 7.9, method 720A.

9. Calibration and Standards

Proceed in the same fashion as described in Section 8, method 720A.

10. Calculations

Proceed in the same fashion as described in Section 9, method 720A.

11. Effects of Storage

See Section 10, method 720A for stability of standards. The phthalic acid eluent (variation 2) is stable indefinitely when stored in capped plastic containers at room temperature. The PHBA eluent (variation 1) should be prepared fresh every week.

12. References

1. FRITZ, J.S., D.T. GJERDE and C. POHLANDT. 1982. "Ion Chromatography", Alfred Huthig, Heidelberg.
2. GJERDE, D.T. 1985. Wescan Instruments, Santa Clara, CA., Personal Communication.

Subcommittee 1

P.K. DASGUPTA, *Chairman*
D.F. ADAMS
B.R. APPEL
S.O. FARWELL
K.T. KNAPP
G.L. KOK
W.R. PIERSON
K.D. REISZNER
R.L. TANNER

720C.

Flow Injection Determination of Aqueous Sulfate (Methylthymol Blue Method)

1. Principle of the Method

Methylthymol blue (MTB) is a metallochromic indicator which forms metal complexes different in spectral absorption characteristics from the free indicator. The barium complex is formed at relatively high pH with maximum absorption at 608 nm; free MTB at the same pH has a maximum absorption at 460 nm. When sulfate is added to a solution of the Ba-MTB complex, the following displacement reaction occurs:

$$Ba-MTB + Sulfate \rightarrow BaSO_4 + MTB$$

Sulfate can be determined by monitoring either the free MTB or the Ba-MTB, the latter approach being more sensitive.

1.1 IMPURITIES IN COMMERCIAL MTB. Commercial MTB(IV, See Figure 720C:1) is synthesized from thymolsulfonphthalein(I) and iminodiacetic acid(II). The commercial product contains, in addition to significant amounts of the original reactants, the monosubstituted product semimethylthymol blue (SMTB, III) and the desired disubstituted compound, methylthylmol blue (IV). A "dye content" of 75–80% is frequently specified by the supplier; this specification merely implies that the remaining 20–25% are nonabsorbing impurities, primarily iminodiacetic acid or its sodium salt. Commercial MTB preparations rarely exceed 50% in MTB content (1).

Pure MTB, as well as SMTB, can be obtained from the commercial product by column chromatography on cellulose and ion-exchange columns (2) or more conveniently on silica gel (1).

Purification of commercial MTB is necessary only if a linear working range larger than that described in Section 2 and optimal sensitivity are required.

The procedure given here is based on commercial MTB. Since the exact composition of commercial MTB varies from batch to batch, optimal sensitivity may not be obtained by exact replication of the reagent composition given here. However, the procedure for determining the optimum reagent composition is included.

1.2 INSTRUMENTAL APPROACHES. The automated MTB procedure for measuring sulfate, based on a segmented flow continuous analysis system (Autoanalyzer, Technicon Instruments, Tarrytown, N.J.), was introduced by Lazrus, Hill, and Lodge (3). Further refinements have been made (4-9). For a MTB-based Autoanalyzer procedure, these references should be consulted. The procedure described here is for unsegmented continuous flow analysis (also referred to as Flow Injection Analysis, FIA). FIA adaptation of the MTB procedure has been described by Madsen and Murphy (10) and is the primary basis of the procedure described. Commercial flow injection analyzers are available from a number of manufacturers; it is also possible to construct them from available components as described in Section 5.

If the decrease in absorption due to the Ba-MTB complex is monitored, it is possible to use a very simple and inexpensive photodetector, based on a photodiode detector and a light emitting diode (LED) as source. The signal/noise characteristics of such single wavelength detectors are excellent. A detector based on a red LED typically emits at a center wavelength of 600–630 nm with a bandwidth of 30 nm. Circuit description and construction details are available (11).

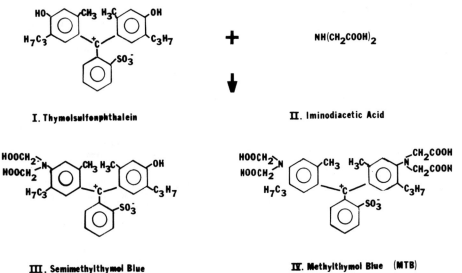

I. Thymolsulfonphthalein

II. Iminodiacetic Acid

III. Semimethylthymol Blue (SMTB)

IV. Methylthymol Blue (MTB)

Figure 720C:1—Synthesis of Methylthymol Blue.

2. Range and Sensitivity

2.1 The limit of detection is 0.1 μg/mL sulfate for the system described in Sections 5-7. The linear working range extends to about 6 μg sulfate/mL, depending on reagent purity. The linear working range can be extended at the expense of the detection limit by using a proportionately smaller sample volume. The sample volume is most easily changed by changing the length of the sample loop tubing. When detection at 608 nm is used with a bandwidth of $\leq$ 4 nm, the absorbance decrease is −0.4 absorbance units/(μg sulfate/mL) **(10)**. For larger spectral bandwidths, the absorbance decrease is lower.

2.2 LINEARITY OF RESPONSE. Even with pure MTB, the response of the system is nonlinear at low sulfate concentration. This is a characteristic of all double-decomposition reactions, and arises from the fact that the extent of completion of the reaction varies with the amount of sulfate added. This also occurs, for example, to the barium chloranilate procedure (method 720 F, reference 4). A second source of nonlinearity is present at higher sulfate concentrations with impure MTB. SMTB displays lower absorptivity and also forms weaker complexes than MTB. Consequently the Ba-SMTB complex first reacts with sulfate, liberating SMTB. When sufficient sulfate is added to react with most of the Ba-SMTB, the Ba-MTB reacts with additional sulfate and liberates the more intensely absorbing MTB, thus further increasing the response. The situation is the same when the Ba-ligand complex is monitored rather than the free ligand because the Ba-MTB complex absorbs more intensely than the Ba-SMTB complex.

Another source of nonlinearity is present if the Ba-MTB reagent contains a greater than 1:1 molar ratio of barium to MTB + SMTB. Because MTB is capable of forming 2:1 metal-ligand complexes, the low end response is nonlinear and is greatly decreased.

3. Interferences

Cations, especially heavy metal cations, produce severe negative interference because they complex with MTB. To prevent cationic interference, the sample is introduced into the loop through a H^+-form cation exchanger minicolumn. This procedure is used for all samples, so all samples (whether from method 704C, 711 or 824) are treated in the same fashion.

Common anions, present in real samples from the above methods, do not interfere. For samples from other sources, note that some glass-fiber filters contain phosphate, which interferes **(9)**.

4. Precision and Accuracy (10)

Average coefficient of variation within the working range for a sample throughput rate of 20/h was 4.1%. Results for real samples (rainwater) showed a mean agreement of 97% with results from suppressed ion chromatography.

5. Apparatus

5.1 MULTICHANNEL PERISTALTIC PUMP. The pump must have a minimum of 3 pumping channels. The choice of the pump is critical because the extent of pulsations in the output flow of these pumps varies greatly. The amplitude of pulsations produced by the pump is in turn a principal factor in achieving low baseline noise and a low detection limit. The pump must deliver the stated flow rate (2.0 mL/min) at its highest rotation rate against a minimum back pressure of 20 psi (140 kPa). Model Minipuls 2 (Gilson Medical Electronics, Middleton, WI) equipped with a 4-channel head is recommended because it produces relatively low pulsations.

Minimum pulsation is generally achieved by the combination of a relatively large pump rotation rate and a relatively small bore pump tubing, rather than the reverse. For the Gilson Minipuls 2, a pump setting of 780 with 1.0 mm i.d. pump tubing delivers 2.0 mL/min. The manufacturer's instructions should be followed to adjust the pressure plate position to optimize "occlusion" and thereby minimize pulsations.

5.2 PNEUMATIC FLUID DELIVERY SYSTEMS. An inert gas pressurized fluid delivery system in which each reagent reservoir is individually pressurized by cylinder nitrogen, argon, or helium, via individual pressure regulators, may also be used. This fluid delivery system is pulseless and superior to peristaltic pump-based systems. Some commercial flow injection analyzers (e.g., from Control Equipment Corp., Lowell MA) use such pneumatic solvent delivery systems. They can also be constructed in-house fairly simply (CAUTION: Glass bottles must not be used due to the danger of explosion). Gas bubbles may form in the spectrophotometer flow cell when gas pressurization is used; and significant back pressure must be applied (by connecting a length of tubing) at the flow cell outlet to eliminate bubble formation. Because helium is relatively insoluble in aqueous media, its use as the pressurizing gas minimizes bubble formation in the cell. A convenient pneumatic delivery bottle, based on a 1-L screw capped polypropylene plastic bottle is shown in Figure 720C:2; this will withstand 30 psi (205 kPa) **(12)**. (CAUTION: Pressurized bottles must be completely enclosed in a heavy cardboard or wooden box, filled with shock-absorbing packing, e.g., polystyrene or polyurethane foam.)

5.3 FLOW-THROUGH PHOTOMETRIC DETECTOR. The detector should be capable of operation at a wavelength of 608 nm. Filter photometers which have relatively large bandwidths are usable, provided that the noise level is proportionately lower. At constant instrumental noise level, the attainable limit of detection (LOD) decreases with decreasing bandwidth, to about a spectral bandwidth of 4 nm. The photometric noise level, at background absorbance values between 0 and 0.5 A.U., for a bandwidth of 4 nm and integration or damping time no greater than 5 s, must be equal to or lower than 0.001 A.U. to attain the stated limit of detection. For a LED- or filter-based photodetector, with a typical spectral band-

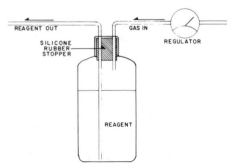

Figure 720C:2 – Pneumatic Fluid Delivery System.

width of 10–30 nm, the photometric noise level should be $\leq$ 0.0005 A.U. to attain the stated LOD. As long as these detectors include a current-controlled source or a reference detector to monitor the source output, acceptably low noise levels are generally attained.

Flow-through filter photometers or spectrophotometric detectors are available from a large number of manufacturers of liquid chromatographic instrumentation. Such detectors, equipped with an 8–20 μL volume, 10 mm pathlength flow cell, preferably equipped with detachable PTFE connecting tubing, are particularly suitable (for example, type 2900–0072 from Kratos Instruments, Ramsey, N.J.). Stainless steel flow cells with very narrow-bore welded stainless steel inlet/outlet tubing should be avoided. Precipitated barium sulfate may clog the narrow-bore tubing, rendering them almost impossible to clean.

If a conventional spectrophotometer is to be equipped with a flow-through cell, choose a 10 mm path length cell of 8–20 μL cell volume. Such cells are available from a number of precision optical cell manufacturers (e.g., Hellma Cells, Jamaica, NY; Spectrocell Inc., Oreland, PA). Because of the small optical aperture of such cells (typically 1 mm for an 8-μL cell), it is essential that the spectrophotometer beam be exactly aligned with the cell aperture. These flow-through cells are externally of the same dimension as a standard 10-mm path length cuvette (12.5 mm × 12.5 mm) and will be correctly positioned in regard to x- and y-dimensions in cell holders designed to accept standard cells. It is the remaining dimension, the z-dimension (defined to be the distance from the bottom of the cell compartment to the vertical center of the beam) that is critical and typically varies from 8.5 to 15 mm for most commercial spectrophotometers. Obtain the z-dimension specification from the manufacturer of the spectrophotometer (it is rarely noted in the instrument manual) and specify the same while ordering the cell. It is possible to use a low-z cell (e.g. z = 8.5 mm) in a spectrophotometer designed for a higher z-value (e.g., 15 mm) by putting in a spacer of appropriate thickness at the bottom of the cell compartment. The reverse is not possi-

ble, however, without permanently altering the spectrophotometer cell compartment.

Most spectrophotometers are designed to work with a higher beam throughput than a 1-mm cell aperture allows. Consequently, in spite of exact z-matching, even with only water in the flow-through cell, the background absorbance may register as high as 1.6 A.U., which the spectrophotometer/recorder electronics may not be able to offset completely. A simple means to offset this background absorbance is to machine an aluminum block (preferably anodized black, or a wood block painted black) to the exact dimensions of a standard cell (12.5 × 12.5 × 40–50 mm) and drill a hole (of the same size as the optical aperture of the flow-through cell) at the appropriate z-location, and place this in the reference cell compartment. A less attractive solution is to fill a cuvette with a solution of a stable dye that absorbs in this region (methylene blue is recommended), stopper the cuvette tightly and place this cuvette in the reference compartment. The concentration of the dye in the cell may be adjusted to obtain the desired background absorbance reading.

5.4 STRIP-CHART RECORDER. Recorder sensitivity and spectrophotometer voltage output should provide a full-scale recorder deflection span of 0.25 A.U. The pen should be set at 100% recorder deflection for the background absorbance (approximately 0.50–0.55 A.U. with respect to water) since injection of samples will produce negative (lower than background) peaks. If these span/offset settings cannot be achieved by the use of the spectrophotometer output offset and/or recorder zero adjust, the light throughput through the reference beam side may be adjusted by controlling the size of the aperture in the reference beam or, more conveniently, the dye concentration in the reference cell (see 5.3). The principal objective is to attain a span of 0.25 A.U. with the background absorbance set at 95–100% recorder deflection.

5.5 INJECTION VALVE. A loop-type injection valve with easily interchangeable loops is necessary. Six-port rotary valves made of inert material (e.g., PTFE) are usable, e.g., type 50, with 0.8 mm connecting passages, with optional pneumatic actuator, Rheodyne Inc., Cotati, CA; or type 86916 with optional electrical actuator, Hamilton Co., Reno, NV; however, injectors made from two 4-way valves generally produce less valve actuation-related baseline disturbance. Such valves with 0.8 mm connecting passages (e.g., Altex type 201–06, Rainin Instrument Co, Woburn, MA; type S-500, MER chromatographic, Mountainview, CA) are available from suppliers of chromatographic equipment and accessories. Connections to all valves above are made with 1/4–28 threaded male nuts and tubing equipped with gripper fittings.

5.6 TUBING. PTFE tubing (1.5 mm o.d., 0.8 mm i.d. or 20 AWG, standard wall, Zeus Industrial Products, Raritan, NJ) is suitable.

5.7 MALE NUTS, 1/4–28 THREADED. These are used for connecting tubing and are available from chromatographic suppliers. Nuts made from polypropylene, and fluorocarbon (Kel-F) are suitable.

5.8 GRIPPER FITTINGS. These are used for tube-to-tube connections and connections to the valve and to tees. They are available from Rainin Instruments (Woburn, MA) and other suppliers. Make a slanted cut at the end of the piece of tubing to be connected. First slip a male nut, head first, over the tube and then slip the end of the tube through the gripper, PTFE face outward. Pull the end of the tubing through the gripper with a pair of pliers and then cut the end of the tubing flush with the PTFE face of the gripper fitting. The tubing can now be connected to any 1/4–28 threaded port.

5.9 TEES. Two low dead-volume tees connect the flow channels. Threaded (1/4-28) 0.8 mm bore tees are available from chromatographic suppliers (e.g., type 200–22, Rainin Instrument Co., Woburn, MA).

5.10 MINI-PRECOLUMN FOR CATION EXCHANGE. 50 × 3 mm glass columns with 1/4–28 threaded ends and complete with frits (e.g. Omni type 45–6310, available from Rainin Instrument Co., Woburn, MA) are suitable. One is necessary; a second one for rapid replacement is recommended.

5.11 SYRINGES. Disposable, 5 mL capacity, Luer-tip.

5.12 COLLAPSIBLE PLASTIC CONTAINERS, ONE-LITER. If a peristaltic pumping system is used, collapsible containers (e.g., from Cole-Parmer, Northfield, IL) make ideal reagent reservoirs because the reagent is not exposed to air. Three are required.

5.13 ULTRASONIC BATH. It is recommended for degassing reagent solutions.

5.14 ASSEMBLED SYSTEM. Refer to Figure 720C:3.

5.14.1 *Cation Exchange Precolumn.* Fill the empty column with an aqueous slurry of 400-mesh size Dowex 50W-X8 (or equivalent) H^+-form cation exchange resin. After connecting the top end fittings, pump some water through the column at the highest flow rate possible. Inspect the column for any settling of the resin bed at the top. If there is any void space, add more resin-slurry and repeat the process until it is completely filled. Pack a spare second column if available.

5.14.2 *Connections to the Injection Valve.* Connect nuts and grippers to each end of a 2.2 m tube (see Sec. 5.6) and connect between the loop ports of the valve. This represents a loop volume of ca. 1.1 mL. Always coil long lengths of tubing into a convenient size (ca. 5 cm dia.) coil, using wire or plastic ties to hold the coil together. Connect the precolumn, using the minimum length of tubing possible, to the injection port of the valve. Connect a 1/4–28 threaded female Luer adapter (usually supplied with the injection valve) to the top of the column. Connect a drain line from the drain port of the valve to a waste bottle. Connect a 1.2 m tube between the "pump" port of the valve and the pump channel designated for pumping water. Connect a 0.4 m tube between the "system" port of the valve and one of the long arms of a threaded tee-Connector. Connect the opposite port of the tee, by a minimum length of tubing, to the pump channel designated for pumping the MTB reagent. Connect the short arm port of the tee to one of the long arm ports of a second tee with a 4.8 m tube. Connect the opposing port of the second tee to the pump channel designated for pumping NaOH with a 2.4 m tube. Connect the remaining (short arm) port of the second tee to the detector using a 2.4 m tube. Connect at least a 1 m length of tubing to the detector cell outlet to provide adequate

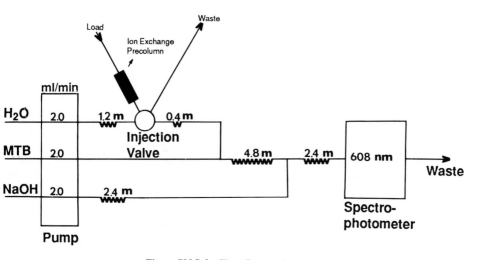

Figure 720C:3 – Flow System Schematic.

backpressure. If pneumatic fluid delivery is used, use at least a 5 m length of tubing to decrease bubble formation in the detector cell. Smaller lengths of narrower-bore tubing may be substituted for this purpose. Calculate the necessary length based on the fact that the pressure drop for a given length of tube is inversely dependent on the fourth power of the tube i.d., all other conditions remaining the same.

6. Reagents

Water must meet the standards of ASTM Reference Reagent Water Type II, and must be sulfate-free.

6.1 SODIUM HYDROXIDE, 0.035 N IN 50% ETHANOL. Dissolve 1.4 g solid NaOH (approximately 7 pellets) in 100 mL water. Add 525 mL 95% reagent grade ethanol and make up with water to 1 L in a volumetric flask. Degas thoroughly before use by ultrasonication.

6.2 METHYLTHYMOL BLUE, TETRASODIUM SALT (MTB). The product from Eastman Kodak Co. is recommended.

6.3 BARIUM-MTB REAGENT. Dissolve 116 mg MTB in 80 mL water. Add 0.5 mL of concentrated (37%) HCl. Add 21.4 mg solid barium chloride dihydrate and swirl until dissolved. Add 95% reagent grade ethanol to make up to 1 L in a volumetric flask. Degas thoroughly before use, preferably by ultrasonication.

6.4 DEAERATED WATER. Boil sulfate-free water vigorously for one hour and transfer into a collapsible container while hot and cap it while excluding most of the air in the container. (CAUTION: Use care when handling hot water!)

6.5 STANDARD SULFATE SOLUTION. Heat anhydrous sodium sulfate at 105–110° C for 4 hours and cool in a desiccator. Dissolve 0.148 g of this salt in sulfate-free water and make up to the mark in a 1-L volumetric flask. This solution contains 100 μg sulfate/mL.

6.6 ION EXCHANGE RESIN. Strong acid form cation exchange resin, Dowex 50W-X8 (or equivalent), 400-mesh.

7. Procedure

7.1 When the system is set up as described in section 5, start pumping Ba-MTB reagent, NaOH and water through designated pumping channels and set spectrophotometer at 608 nm. For a filter photometer, use the filter with a center wavelength of 605 or 610 nm (bandpass $\leq$ 30 nm).

7.2 After 10–15 min, the baseline should stabilize with a noise level $\leq$ 0.002 A.U. Adjust spectrophotometer output, recorder sensitivity and offset/zero adjusts to set the baseline at approximately 100% recorder deflection and a full scale recorder span of 0.25 A.U. (Section 5).

7.3 Carry out calibration procedure (Section 8). Additionally, determine the extent of baseline disturbance caused by actuation of the valve from load to inject position and vice-versa. The valve actuation pulse, if present, should appear on the recorder trace between 1 and 1.5 min after actuating the valve. For a good fast-actuating injection valve with a flow bypass, the pulse may be essentially undetectable and this is desired. If the actuation pulse produces a baseline deflection equal to or greater than that observed with a 0.1 μg/mL sulfate standard, provide additional pulse damping in the water flow channel. (The primary purpose of the 1.2 m and 0.4 m tubes in this flow channel is to accomplish this damping.) Additional damping is obtained by connecting a segment (ca. 5 cm) of thin walled silicone rubber tubing before the valve, in addition to the existing tubing. A similar segment of silicone rubber tubing may also be connected after the valve, provided that the bore of the tubing is not greater than 1 mm.

7.4 Inject samples in the same manner as standards. Injections can be repeated at minimum intervals of 2.5 min. At least 3 times the loop volume or 3 mL, whichever is greater, must be put through the system to flush the precolumn and the loop completely of their previous contents.

7.5 Replace the precolumn, or the resin in the injector precolumn, after every 100 standards and/or samples. More frequent replacement may be necessary if the cation content (other than H$^+$) of the samples is relatively high.

7.6 OPTIMIZATION OF THE BA-MTB REAGENT COMPOSITION. If optimum sensitivity and linearity are required, make six different compositions of the Ba-MTB reagent following the directions given in 6.2 except using 100, 120, 140, 160, 180 and 200 mg MTB in each batch. Determine the peak height obtained for a 1-μg/mL sulfate standard for each reagent composition using at least triplicate injections. Use the reagent composition that produces the largest peak.

8. Calibration

8.1 Prepare a 10 μg/mL sulfate standard from the 100 μg/mL sulfate stock by diluting 100 mL of this standard to 1 L in a volumetric flask. Prepare 0.1, 0.2, 0.4, 0.6, 0.8, 1.0, 2.0, 4.0 and 6.0 μg/mL standards by pipetting 1.0, 2.0, 4.0, 6.0, 8.0, 10.0, 20.0, 40.0 and 60.0 mL of the 10 μg/mL standard into individual 100 mL volumetric flasks and making up to the mark with water.

8.2 Switch the injection valve to "load" position. Begin with the lowest concentration standard. Load the loop with the standard solution by injecting at least 3 times the loop volume (or 3 mL, whichever is greater) through the precolumn. Switch the valve to "inject" position; the recorder trace should show the peak due to the standard approximately 1.5 min from valve actuation. Repeat injection of the same standard at least once more; injections can be repeated at minimum intervals of 2.5 min.

8.3 Wash the syringe with the next higher concentration standard and repeat the procedure of 8.2. Continue with higher concentration standards. Typical response obtained with the system is shown in Figure 720C:4.

8.4 Change precolumn or precolumn resin after a maximum of 100 injections.

8.5 Measure mean peak heights obtained for each concentration and plot against the concentration of the sample injected. A linear plot should result. Compute the best fit of the data to a straight line using standard linear least-squares regression. A typical equation obtained is peak height (A.U.) = (0.0383 ± 0.0003) × μg/mL (SO$_4^{2-}$) + (–0.0009 ± 0.0006)

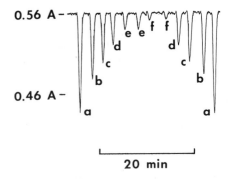

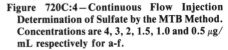

Figure 720C:4 — Continuous Flow Injection Determination of Sulfate by the MTB Method. Concentrations are 4, 3, 2, 1.5, 1.0 and 0.5 μg/mL respectively for a-f.

8.6 If the system is to be used for the measurement of sulfate concentrations higher than 6 $\mu g/mL$, reduce the sample loop length proportionally. For example, to work in the 1–60 $\mu g/mL$ range, use a 22 cm loop. Prepare standards of suitable concentration and calibrate the system in the same fashion as before.

9. Calculations

Compute the sulfate content in $\mu g/mL$ from the measured peak height obtained for the sample either from the calibration plot or from the best fit equation.

10. Effects of Storage

10.1 Sulfate standards of concentrations 10 $\mu g/mL$ and below should be prepared weekly as needed. The 100 $\mu g/mL$ stock sulfate standard is indefinitely stable if stored refrigerated in a capped bottle.

10.2 The ethanolic sodium hydroxide should be prepared fresh every week.

10.3 The Ba-MTB reagent must be prepared fresh daily.

11. References

1. VITHANAGE, R.S. 1985. Texas Tech University, Private Communication.
2. YOSHINO, T., H. IMADA, T. KUWANO and K. IWASA. 1969. Studies on Methylthymol Blue–I. Separation and Purification of Methylthymol Blue and Semimethylthymol Blue, *Talanta 16*:151–156.
3. LAZRUS, A.L., E. LORANGE and J.P. LODGE, JR. 1968. New Automated Microanalyser for Total Inorganic Fixed Nitrogen and For Sulfate Ion in Water. *In* "Trace Inorganics in Water", *ACS Adv. Chem. Ser. 73*:164–171.
4. MCSWAIN, M.R., R.J. WATROUS and J.E. DOUGLAS. 1974. Improved Methylthymol Blue Procedure for Automated Sulfate Determinations, *Anal. Chem. 46*:1329–1331.
5. COLOVOS, G., M.R. PANESAR and E.P. PARRY. 1976. Linearizing the Calibration Curve in Determination of Sulfate by the Methylthymol Blue Method, *Anal. Chem. 48*:1693–1696.
6. Environmental Protection Agency, Quality Assurance Branch, 1977. Tentative Method for the Determination of Sulfates in the Atmosphere (Automated Technicon II Methylthymol Blue Procedure).
7. Environmental Protection Agency, Environmental Monitoring and Support Laboratory. 1978. Tentative Method for the Determination of Sulfates in the Atmosphere (Automated Technicon II Methylthymol Blue Procedure).
8. APPEL, B.R., E.L. KOTHNY, E.M. HOFFER and J.J. WESOLOWSKI. 1977. Comparison of Wet Chemical and Instrumental Methods of Measuring Airborne Sulfate. Final Report, EPA-600/7-77-128.
9. APPEL, B.R., E.M. HOFFER, M. HAIK, W. WEHRMEISTER, E.L. KOTHNY and J.J. WESOLOWSKI. 1979. Improvement and Evaluation of Methods for Sulfate Analysis. Final Report, EPA-600/4-79-028.
10. MADSEN, B.C. and R.J. MURPHY. 1981. Flow Injection and Photometric Determination of Sulfate in Rainwater with Methylthymol Blue, *Anal. Chem. 53*:1924–1926.
11. SLY, T.J., D. BETTERIDGE, T.J. WIBBERLEY and D.G. PORTER. 1982. An Improved Flow-Through Phototransducer, *J. Autom. Chem. 4*:186–189.
12. MCDOWELL, W.L. 1985. Texas Tech University, Private Communication.

Subcommittee 1

P.K. DASGUPTA, *Chairman*
D.F. ADAMS
B.R. APPEL
K.T. KNAPP
G.L. KOK
W.R. PIERSON
K.D. REISZNER
R.L. TANNER

720D.

Barium Sulfate Turbidimetry

1. Principle of the Method

1.1 Sulfate ion reacts with barium ion to form insoluble barium sulfate. A colloid stabilizing agent is added to stabilize the suspension of $BaSO_4$. The absorbance of the sample at 500 nm is read before and after the addition of $BaCl_2$ to compensate for any coloration or turbidity present in the sample. This differential absorbance is related to the amount of sulfate present in the sample using a previously constructed calibration plot **(1–3)**. This is a manual procedure requiring no specialized equipment.

1.2 Glycerol is added to stabilize the colloid and ethanol is added to promote precipitation by reducing solubility; it also increases sensitivity of the method.

2. Range and Sensitivity

The lower detectable concentration is 40 μg sulfate in a 20 mL sample (2 μg/mL). The calibration plot is approximately linear in the range 30–75 μg/mL. There is considerable nonlinearity at sulfate concentrations below about 30 μg/mL.

3. Interferences

3.1 Sulfur-containing anions other than sulfate generally interfere via oxidation to sulfate.

3.2 Any color or turbidity present in the sample represents a potential interference. This is accounted for by measuring the absorbance before and after $BaCl_2$ addition and constructing the calibration plot and determining the concentration from the absorbance difference.

4. Precision and Accuracy

4.1 Measurement is dependent upon the stability of the suspension of $BaSO_4$ particles, size of particles, barium ion concentration, pH, and temperature. These variables must be fully controlled to obtain precise and accurate results. Use of solid $BaCl_2$ addition leads to a high barium ion concentration without further dilution. The use of hydrochloric acid keeps the solution at a desirably low pH and prevents the precipitation of any other barium salt. Variations in temperature between 20° to 30°C do not have a significant effect. Vigorous and reproducible shaking of the solution after addition of $BaCl_2$ promotes the formation of finer $BaSO_4$ particles that stay in suspension longer.

4.2 For sulfate concentrations between 2 and 12 μg/mL, the coefficient of variation ranges from 2.1 to 12.2%, generally increasing with decreasing sulfate concentration. Accuracy in this range is within ± 13.5%. With 15–100 μg/mL sulfate, the coefficient of variation ranges from 0.7 to 8.4%, typically about 3.5%. Accuracy within this range is within ± 6.3% **(2)**.

4.3 This method has the poorest sensitivity, precision, and accuracy at low levels among the methods described under 720. A sample volume of 2 mL is required. The slope of the calibration plot is dependent on the specific batch of reagents, especially on the barium chloride particle size. The stated precision and accuracy are obtainable only with exacting care and under closely controlled conditions. Despite these deficiencies, the method has been widely used and requires only simple equipment.

5. Apparatus

5.1 SPECTROPHOTOMETER. The instrument should be capable of measuring at 500 nm with a band pass of less than 16 nm. The limit of detection as stated in Section 2 was obtained with a cell of path length 20 mm **(2)**. The limit of detection is correspondingly higher with shorter path length cells.

5.2 GRADUATED CYLINDERS. Stoppered, 25 mL capacity.

5.3 SCOOP. Capacity ca. 250 mg $BaCl_2 \cdot 2H_2O$. "Quaver," a Teflon-coated scoop of appropriate size (Bel-Art F36715, American Scientific Products S1590-2) is suitable.

6. Reagents

All reagents are analytical reagent grade chemicals. Water is ASTM reagent water, Type II.

6.1 HYDROCHLORIC ACID, 10N. Dilute 80 mL conc. HCl (approx. 12 N) to 100 mL with water.

6.2 GLYCEROL-ALCOHOL-ACID SOLUTION. Mix 20 mL of glycerol with 40 mL of 95% reagent grade ethanol and 15 mL 10N HCl.

6.3 BARIUM CHLORIDE. $BaCl_2 \cdot 2H_2O$. 20-30 mesh material if available; this yields the highest sensitivity.

6.4 STANDARD SULFATE SOLUTION. Dry anhydrous sodium sulfate at 105-110°C for 4 h and cool in a desiccator. Dissolve 0.148 g of this salt in sulfate-free water and make up to the mark in a 1-L volumetric flask. This solution contains 100 μg sulfate/mL.

7. Procedure

7.1 Pipet an aliquot of the sample, normally 20 mL, into a clean dry 25-mL stoppered graduated cylinder. Use a smaller aliquot for sulfate concentrations greater than 100 μg/mL, adding sulfate-free water to make up to 20 mL. Add 5 mL of the glycerol-alcohol-acid solution, stopper, and mix by inverting three times.

7.2 Pour a portion of the solution into a clean dry cuvette. Use 20-50 mm pathlength cuvettes if the cuvettes can be accurately and reproducibly positioned in the spectrophotometer. Determine the absorbance at 500 nm against air as blank (do not insert a cuvette or anything at all in the reference compartment) and record as A_1.

7.3 Pour as much of the sample solution as possible back into the graduated cylinder from the cuvette. Do not wash the cuvette into the cylinder. Add one scoop (approximately 250 mg) of $BaCl_2$ crystals, stopper the flask and shake vigorously for approximately 45 s to dissolve the crystals.

7.4 Let the sample stand for exactly 40 min at room temperature (20-30°C). Then gently mix the suspension by inverting the graduated cylinder once, transfer into a cuvette and measure the absorbance (A_2) of the solution against air as blank as in 7.2. When a large number of samples are to be analyzed it is convenient to add the $BaCl_2$ crystals (Sec. 7.3) at timed intervals (e.g., 1 min apart) with similar intervals between spectrophotometric measurement.

7.5 Analyze at least one standard sulfate sample as control in each batch of samples to detect gross variations in the analysis. Deviations of up to 5% from the original calibration plot are within the limits of the precision of the method.

7.6 A preparation of barium chloride containing a proprietary colloid stabilizing agent is available commercially as SulfaVer IV® (Hach Co., Loveland, CO), in bulk or as individual "pillows" ready for addition to a 20 mL sample. Substitution of this material for reagents in Sections 6.2 and 6.3 produces precision and accuracy comparable to the procedure described.

7.7 Screw-capped (PTFE-lined cap) 25 mL Pyrex tubes may be used in lieu of graduated cylinders since they can be used as cuvettes in spectrophotometers such as the Spectronic 20 (Bausch and Lomb). A_1 and A_2 are then read in the same tube before and after SulfaVer IV® addition; no liquid transfer is involved. Each tube should be marked so that it can be reproducibly positioned in the spectrophotometer (**4**).

8. Standards and Calibration

8.1 Pipet 0.0, 0.4, 1.0, 2.0, 3.0, 5.0, 10.0, 15.0 and 20.0 mL of the standard sulfate solution containing 100 μg sulfate/mL into individual stopper-equipped graduated cylinders and pipet the necessary amount of sulfate-free water to obtain a total volume of 20 mL. Proceed as in Sections 7.2-7.4.

8.2 Plot the differential absorbance A_2-A_1 against μg sulfate. The plot should be linear between 300 and 1500 μg sulfate. Below 300 μg, the relationship is nonlinear.

9. Calculations

An amount of sulfate between 300 and 1500 μg in the sample aliquot may be calculated from a linear regression equation representing this portion of the calibration plot. Below 300 μg, the relationship is nonlinear and the sulfate concentration must be determined graphically from the original calibration plot.

10. Effect of Storage

The glycerol-alcohol-acid solution is stable for at least 15 months when stored refrigerated in a capped bottle (2).

11. References

1. American Society for Testing and Materials. 1982. Standard Test Methods for Sulfate Ion in Water, D516–82.

2. APPEL, B.R., E.M. HOFFER, M. HAIK, W. WEHRMEISTER, E.L. KOTHNY and J.J. WESOLOWSKI. 1979, 1980. Improvement and Evaluation of Methods for Sulfate Analysis, EPA Report, Part I: EPA-600/4–79–028, April 1979; Part II: EPA-600/4–80–024, April, 1980.

3. U.S. Public Health Service. 1965. Publication No. 997-AP-11. Determination of Sulfate in Atmospheric Suspended Particulates: Turbidimetric Barium Sulfate Method, in "Selected Methods for the Measurement of Air Pollutants".

4. Air and Industrial Hygiene Laboratory, California Department of Health Services. 1978. Method 75: Evaluation and Improvement of a Turbidimetric Method for Sulfate Using SulfaVer IV®.

Subcommittee 1

R.K. DASGUPTA, *Chairman*
D.F. ADAMS
B.R. APPEL
S.O. FARWELL
K.T. KNAPP
G.L. KOK
W.R. PIERSON
K.D. REISZNER
R.L. TANNER

720E.

Barium Perchlorate Microtitration

1. Principle of the Method

1.1 Sulfate ion is titrated with barium perchlorate to produce insoluble barium sulfate. A metallochromic indicator, thorin, is added to the sample. The titration of sulfate is conducted in 80% 2-propanol to reduce the solubility of $BaSO_4$. As the titrant is added to the sample, the barium preferentially reacts with sulfate. When all of the sulfate is reacted, barium reacts with the free yellow indicator to form the pink barium-indicator complex, signifying the end-point of the titration **(1,2)**.

1.2 The change of color at the end point for thorin indicator at sulfate concentrations ≤ 10 $\mu g/mL$ is not especially sharp. A colorimeter dip probe (e.g., model PC/600, Brinkmann Instruments Co., Div. of Sybron Corp., Cantiague Rd., Westbury, NY) is useful in determining the end point at low sulfate levels. A number of alternative indicators, which provide sharper end points, have been suggested **(3–9)**. However, with the exception of Beryllon II **(9)**, none of the studies were carried out at sufficiently low sulfate levels to judge their applicability. Although thorin, alizarin red S, beryllon II, sulfonazo III, and fast sulfon black F all lead to reasonably sharp color changes at high sulfate levels, within a specific optimum pH range for each indicator, only thorin or beryllon II are useful at trace ($\leq$ 10 $\mu g/mL$) sulfate levels **(9)**. Although beryllon II as indicator reportedly permits determination of sulfate down to 0.25 $\mu g/mL$, many commercial batches of the dye were found to be unsatisfactory and the pure dye must be synthesized in the laboratory **(9)**.

2. Range and Sensitivity

The lower limit of the determination is 2.5 $\mu g/mL$ sulfate. The titrant composition described is adequate for sulfate levels up to 500 $\mu g/mL$. Above this concentration, a tenfold more concentrated titrant should be used, or the sample appropriately diluted.

3. Interferences

3.1 The method is susceptible to both cationic and anionic interference. Cations, especially heavy metal ions, produce severe negative interference. These cations, if present, must be removed by cation exchange on a strong acid H^+-form cation exchange resin. Unfortunately, these resins contain sulfonic acid functionalities and produce a small amount of sulfate even after repeated washings with water. Of the three methods (704C, 711, 824) that require determination of aqueous sulfate, only samples obtained by method 824 can potentially contain metal ions. The samples resulting from methods 704C and 711 are essentially dilute solutions of H_2SO_4. For these samples, the cation exchange step should therefore be omitted unless metal ion contamination is suspected (sulfate results, as obtained by the present method, will be lower than true values). Metal ion contamination in samples from methods 704C and 711 means failure of the prefilter used in these methods and invalidates the results.

3.2 Several anions, namely phosphate, fluoride, chloride and nitrate, produce positive errors in the determination. Phosphate produces the most severe interference, but is unlikely to be present in the samples from methods 704C and 711. Presence of phosphate in samples from method 824 is possible if airborne soil concentration is high and in the vicinity of phosphorus-related industries. In this case, method 720E is not recommended for samples obtained by method 824.

The error in sulfate determination is below +3% up to a fluoride/sulfate ratio of 2, but rises to over +6% at a fluoride/sulfate ratio of 3. A higher fluoride/sulfate ratio in real samples from methods 704C, 711 and 824 is unlikely. Chloride produces less than 1% interference at chloride/sulfate ratios up to 6. The samples from method 704C will contain approximately 2 μg/mL chloride as added HCl. Even after allowing for incorporation of chloride from the sampled air, samples obtained by methods 704C and 711 are unlikely to exceed a chloride/sulfate ratio of 6. Chloride/sulfate for samples obtained over maritime regions or coastal areas by method 824 may, however, exceed this ratio. Even then, chloride interference is unlikely to be large. Nitrate produces a positive interference of ca. 2.5% at a nitrate/sulfate ratio of 1. The interference increases to ca. 4% at a nitrate/sulfate ratio of 6. The extent of interference in real samples is unlikely to be significant.

4. Precision and Accuracy (10)

At the level of 2.7 μg/mL sulfate, the coefficient of variation is ± 5.6%, which decreases to ± 2.8% at the level of 5.4 μg/mL, and to ± 1.1% at higher sulfate levels. The method yields results that are slightly but consistently lower than true values. The error amounts to –8.1% at 2.7 μg/mL sulfate, –4.2% at 5.4 μg/mL sulfate, –2.9% at 13.6 μg/mL sulfate and below –1.1% at sulfate levels above 25 μg/mL.

5. Apparatus

5.1 MICROBURET. 5 mL capacity, graduated in 0.01 mL divisions.
5.2 MAGNETIC STIRRER. Preferably with illuminated base. A micro stirring bar (10 × 2 mm) is recommended.
5.3 PH METER. Capable of accurately reading at least to 0.1 pH unit. The use of a semimicro or micro combination electrode is recommended.
5.4 COLORIMETER DIP PROBE. (Optional) Such as Brinkmann model PC/600, Brinkmann Instrument Co., Div. of Sybron Corp., Cantiague Rd., Westbury, New York. This should be equipped with a narrow band-pass filter of center wavelength between 500 and 540 nm.

5.5 DISPOSABLE PASTEUR PIPETS. 23 cm (9 in) long.
5.6 PYREX GLASS WOOL.

6. Reagents

All reagents are of ACS reagent grade, except thorin and modified methyl orange which are indicator grade. Water must be sulfate-free and meet the standards of ASTM Reference Reagent Water, Type II.

6.1 THORIN 0.1%. 2-(3,6-disulfo-2-hydroxy-1-naphthylazo)benzenearsonic acid, disodium salt. Dissolve 100 mg thorin in 20 mL sulfate-free water and add 80 mL of 2-propanol.
6.2 MODIFIED METHYL ORANGE. "IndicatAR", pH 3–4 (Mallinckrodt Inc.).
6.3 SODIUM CARBONATE, ANHYDROUS. Dry at 110°C overnight and allow to cool in a desiccator.
6.4 SULFURIC ACID, 0.1 M. Dilute 5.5 mL of conc. H_2SO_4 (18 M) to 1 L with water. Standardize against sodium carbonate using modified methyl orange indicator.
6.5 SULFURIC ACID, 0.005M. Dilute 50 mL of standardized 0.1 M sulfuric acid to 1 L with sulfate-free water.
6.6 2-PROPANOL. Reagent grade.
6.7 BARIUM PERCHLORATE, 0.005 M, in 80% 2-propanol. Dissolve 1.68 g of the anhydrous salt or 1.95 g of the trihydrate in 200 mL water. After complete dissolution, add 800 mL 2-propanol. Mix thoroughly. Standardize against 0.005 M sulfuric acid using thorin as indicator (see 8.1).
6.8 PERCHLORIC ACID, 0.2 M, in 80% 2-propanol. Add 17 mL of 70% $HClO_4$ to 185 mL sulfate-free water in a 250 mL graduated cylinder. Mix thoroughly with the aid of a glass rod and transfer quantitatively into a 1-L bottle. Do not wash. Add 800 mL 2-propanol and mix thoroughly.
6.9 SODIUM HYDROXIDE, 0.2 M, in 80% 2-propanol. Dissolve 8 g (approximately 40 pellets) NaOH in 200 mL water. Dissolve, mix thoroughly and add 800 mL 2-propanol. Mix again.
6.10 2-PROPANOL, 80% V/V. Add 100 mL water to 400 mL 2-propanol in a 1-L wash bottle. The wash bottle tip should be capped when not in use.

6.11 STRONG ACID FORM CATION EX-CHANGE RESIN. 100–200 mesh, e.g., Dowex 50W-X8.

7. Procedure

7.1 CATION EXCHANGE OF SAMPLE. This is only necessary for samples from method 824 which did *not* employ any cation exchange resin during the extraction of the filter.

7.1.1 *Preparation of Cation-exchange Microcolumn.* Insert a wad of (prewashed and dried) Pyrex glass wool into the top end of a long Pasteur pipet. Push it down with a suitable tool, e.g., a second pipet, to form a plug (*ca.* 1 cm) at the bottom. Clamp the pipet on a suitable stand. Make a slurry of the cation exchange resin in water and pour through the top of the pipet to form a ca. 5 cm tall resin bed. Wash several times with sulfate-free water immediately before use with the sample.

7.1.2 Place a waste flask at the bottom of the microcolumn and transfer the sample to the top of the column by using another clean dry Pasteur pipet. Discard the first 2–3 mL of effluent and then substitute a clean dry Erlenmeyer flask as receiver. Ion exchange requisite amount of sample without exceeding a total sample volume of 50 mL. Use a fresh microcolumn for another sample.

7.2 PH ADJUSTMENT. Pipet 4 mL of the cation exchanged sample (if from method 824) or the direct sample (if from method 704C or 711) into a 25 mL beaker. The sample should contain 10–500 μg/mL sulfate. Determination can be carried out down to 2.5 μg/mL sulfate but this requires considerable skill and experience on the part of the analyst to determine the end point visually. Pipet 16 mL 2-propanol with a graduated pipet or from a dispensing pipet into the beaker containing the sample. Calibrate the pH meter with pH 4 standard aqueous buffer. Wipe the electrode dry and measure the apparent pH of the sample (in 80% 2-propanol). To adjust the apparent pH to 3.5, add 0.2 N $HClO_4$ or NaOH (also in 80% 2-propanol) as appropriate (most commonly the former), dropwise, with swirling. Wash the electrode with a minimum amount of 80% 2-propanol from a

wash bottle and collect the washings in the sample beaker.

7.3 TITRATION. Add 3–4 drops of the thorin indicator and a micro stirring bar, and place the sample on the (illuminated) stirrer under the microburet and stir at low speed. If a colorimeter probe is available, dip the clean and dry probe into the sample. Titrate slowly with 0.005 M $Ba(ClO_4)_2$ from the microburet. At the end point, the color will change from yellow to pink. Record the titrant volume (V_1). If a colorimeter probe is used, the end point is indicated by a significant and abrupt change in the meter reading. Carry a blank sulfate-free water sample through identical steps and record the amount of the titrant required to reach the end point (V_2). The titrant consumed is then V_1–V_2.

8. Calibration and Standards

8.1 STANDARDIZATION OF THE BARIUM PERCHLORATE TITRANT. Pipet a 4 mL sample of the 0.005 M standardized H_2SO_4 solution into a 25-ml beaker, add 2-propanol and adjust pH as in 7.3. Titrate this solution and a blank with the barium perchlorate titrant as in 7.4, recording the volumes as V_3 and V_4. Repeat twice. Molarity of the barium perchlorate titrant (M_{Ba}) is given by:

$$M_{Ba} = \frac{4 \times \text{molarity of } H_2SO_4}{\text{mean } V_3(\text{mL}) - \text{mean } V_4(\text{mL})}$$

9. Calculations

Calculate the sulfate molarity (M_{SO_4}) in the sample as:

$$M_{SO_4} = \frac{[V_1 \text{ (mL)} - V_2 \text{ (mL)}] \times M_{Ba}}{4}$$

Calculate the sulfate concentration in μg/mL from SO_4^{2-} (μg/mL) = $M_{SO_4} \times 9.61 \times 10^4$.

10. Effects of Storage

All reagents are stable for at least 30 days when stored in capped bottles at room temperature.

11. References

1. FRITZ, J.S. and S.S. YAMAMURA. 1955. Rapid Microtitration of Sulfate, *Anal. Chem. 27*:1461.
2. HAARTZ, J.C., P.M. ELLER and R.W. HORNUNG. 1979. Critical Parameters in the Barium Perchlorate/Thorin Titration of Sulfate, *Anal. Chem. 51*:2293.
3. FRITZ, J.S. and M.Q. FREELAND. 1954. Direct Titrimetric Determination of Sulfate, *Anal. Chem. 26*:1593.
4. SAVVIN, S.B., YU. M. DEDKOV and V.P. MAKAROVA. 1961. New Metalloindicators for Barium Ions. Determination of Sulfates, *Zh. Anal. Khim. 17*:43.
5. KARALOVA, Z.K. and N.P. SHIBAEVA. 1964. Determination of Microamounts of Sulfate Ions in Highly Pure Water, *Zh. Anal. Khim. 19*:258.
6. BUDESINSKY, V. 1965. Modification of Flask Method for Sulfur Determination. Determination of Sulfates with Sulfonazo III, *Anal. Chem. 37*:1159.
7. BUDESINSKY, B. and L. KRUMLOVA. 1967. Determination of Sulphur and Sulphate by Titration with Barium Perchlorate. Comparison of Various Color Indicators, *Anal. Chim. Acta 39*:375.
8. BUDESINSKY, B. 1968. Barium Titration of Sulfate with Chlorophosphonazo III as Indicator, *Microchem. J. 14*:242.
9. HWANG, H. and P.K. DASGUPTA. 1984. Microtitration of Sulfate with Beryllon II as Indicator: Determination of Sulfate in Environmental Samples, *Microchim. Acta 159*.
10. HWANG, H. 1983. "Microdetermination of Sulfate in Environmental Samples", M.S. Thesis, Texas Tech University.

Subcommittee 1

P.K. DASGUPTA, *Chairman*
D.F. ADAMS
B.R. APPEL
S.O. FARWELL
K.T. KNAPP
G.L. KOK
W.R. PIERSON
K.D. REISZNER
R.L. TANNER

720F.

Barium Chloranilate Spectrophotometry

1. Principle of the Method

1.1 When a sparingly soluble barium salt of a strongly absorbing anion, D^{2-}, is reacted with a sample containing sulfate, insoluble $BaSO_4$ is formed and the free anion D^{2-} is liberated. In order to make this conversion of sulfate to $BaSO_4$ via

$$BaD + SO_4^{2-} \rightarrow BaSO_4 + D^{2-}$$

virtually stoichiometric, the reaction is conducted in 80% 2-propanol, to reduce the solubility of $BaSO_4$. The sulfate is then determined indirectly, by measuring the absorbance due to the liberated D^{2-} anion in the supernate after centrifugation to remove the suspended $BaSO_4$ and excess BaD.

1.2 Barney et al. (1–3) introduced the use of barium chloranilate (barium salt of chloranilic acid, 2,5-dichloro-3,6-dihydroxy-*p*-benzoquinone) as the sparingly soluble salt BaD. A generalized theoretical treatment of the procedure is available (4–6). Schafer (7) improved the method by using an 80% 2-propanol reaction medium and by measuring of absorbance at 312 nm, the isosbestic wavelength for chloranilate and hydrogen chloranilate. The 2-propanol medium increases sensitivity, and measurement at the isosbestic wavelength minimizes any effects due to changes in pH. Schafer's account (7) forms the basis of the method described.

1.3 An automated (Autoanalyzer) procedure for the barium chloranilate method has been described (8). Versions of the barium chloranilate procedure have appeared previously as suggested methods (9). Although specific advantages, compared to barium chloranilate, have been listed for the use of other barium salts such as rhodizonate (10), violurate (11) or beryllonate (12), extensive use of the barium chloranilate method in the past and the benefit of the available experience therefrom makes the barium chloranilate method the one of choice.

1.4 Most barium chloranilate procedures use a chloroacetate buffer. If measurement is made at the isosbestic wavelength of 312 nm, there is no improvement in accuracy and precision for buffer-containing systems over those that do not contain a buffer (13). This description therefore does not use a buffer.

2. Range and Sensitivity

The limit of detection of the method is approximately 0.25 μg/mL aqueous sulfate. The working linear range of the method extends from 1 to 50 μg/mL aqueous sulfate. The absorbance (over that of a blank) for a 1 cm cell is 0.1 A.U. for an aqueous sulfate sample containing 3.7 μg sulfate/mL.

3. Interferences

Cationic interference, especially from heavy metal ions, can be severe. Such cations should not be present in samples obtained by methods 704C and 711. Samples from method 824 may contain such cations (unless cation exchange resin was used during the extraction of the filter) and cations must be removed by cation exchange.

Chloride and nitrate (the major anions likely to be present in the samples obtained by methods 704C, 711 and 824) do not interfere. Fluoride does not interfere. Phosphate and oxalate interfere to a small degree at high levels which are unlikely to be present in real samples.

4. Precision and Accuracy

Detailed data on precision and accuracy of the method in 80% 2-propanol and isosbestic wavelength measurement are not available. Single laboratory precision is about 2.5% (**7**). However, such data are available for 50% ethanol medium with measurement at 530 nm (**1**). At levels of 20 μg/mL and 200 μg/mL sulfate, the coefficient of variation and the relative error were within about 2%. The accuracy was (98 ± 2%) in comparison with a $BaSO_4$ turbidimetric method for potable water samples containing 20–50 μg/mL sulfate.

5. Apparatus

5.1 ULTRAVIOLET SPECTROPHOTOMETER. Capable of measurement at 312 nm, with a bandwidth $\leq$ 4 nm. The noise level at this wavelength, after an appropriate warmup period and no cuvettes inserted in either reference or sample compartment, must be less than 0.001 absorbance unit.

5.2 LABORATORY CENTRIFUGE. Capable of accepting at least four 10–12 mL capacity centrifuge tubes and operation at $\geq$ 2000 rpm.

5.3 SHAKER, WRIST-ACTION. (e.g., Model S-1061-1A American Scientific Products, McGraw Park, IL).

5.4 CENTRIFUGE TUBES. Screw capped, 10–12 mL capacity (e.g., Corning 8142 or Kimble 3985).

5.5 DISPOSABLE PASTEUR PIPETS. 23 cm (9 in) long.

5.6 PYREX GLASS WOOL.

5.7 GRADUATED PIPET. 1 mL capacity, with 0.01 mL divisions.

5.8 SMALL SCOOP. Practice reproducibly delivering 25–30 mg of barium chloranilate with this.

6. Reagents

All reagents are of ACS reagent grade except modified methyl orange, which is indicator grade. Water must be sulfate-free and meet the standards of ASTM Reference Reagent Water Type II.

6.1 2-PROPANOL. Reagent grade.

6.2 2-PROPANOL. 80% V/V. Add 100 mL water to 400 mL 2-propanol. Mix thoroughly.

6.3 BARIUM CHLORANILATE, TRIHYDRATE. Must be crystalline and not amorphous. The product from J.T. Baker

Chemical Co. has been found satisfactory. Wash at least 5 times with 80% 2-propanol. Oven-dry overnight at 60°C.

6.4 MODIFIED METHYL ORANGE. "IndicatAR" pH 3–4, (Mallinckrodt Inc.).

6.5 SODIUM CARBONATE, ANHYDROUS. Dry at 110°C overnight and allow to cool in a desiccator.

6.6 SULFURIC ACID, 0.1 M. Dilute 5.5 mL of H_2SO_4 (18 M) to 1 L with water. Standardize against sodium carbonate using modified methyl orange indicator.

6.7 SULFURIC ACID, 0.0005 M. Dilute 5 mL of the standardized 0.1 M sulfuric acid to 1 L with water. This solution contains approximately 50 μg/mL sulfate.

6.8 STRONG ACID FORM CATION EXCHANGE RESIN. 100–200 mesh, e.g., Dowex 50W-X8.

7. Procedure

7.1 CATION EXCHANGE OF SAMPLE. This is only necessary for samples from method 824 that did *not* use cation exchange resin during the extraction of the filter.

7.1.1 *Preparation of Cation Exchange Microcolumn.* See Section 7.1.1, method 720E.

7.1.2 ION EXCHANGE PROCEDURE. See Section 7.1.2, method 720E.

7.2 ANALYSIS. Pipet 2 mL of sample (cation exchanged sample, if from method 824) into a clean dry centrifuge tube. Pipet 8 mL of 2-propanol. Add a scoop (25–30 mg) of barium chloranilate to the contents of the tube. Cap the tube and shake for 20 min with the wrist-action shaker. Centrifuge the mixture for 5 min. Transfer, using a clean dry Pasteur pipet, the supernatant solution into a clean dry cuvette and measure the absorbance at 312 nm against air as reference (no cuvettes in reference compartment). Record the absorbance as A_1. Carry out a blank (water or absorber blank) through an identical protocol and record this absorbance as A_2.

8. Calibration and Standards

Two separate calibration plots are to be constructed for the ranges 5–100 μg sulfate and 0.5–10 μg sulfate (each in a 2-mL sample, i.e., in terms of concentration the ranges are 2.5–50 μg/mL and 0.25–5.0 μg/mL).

For the first calibration plot, pipet 0.10, 0.20, 0.50, 1.00 and 2.00 mL of 0.0005 M H_2SO_4 (6.7) into individual centrifuge tubes. Next pipet 1.90, 1.80, 1.50, 1.00 and 0.0 mL of water into the respective centrifuge tubes, to attain a total volume of 2 mL. Then add 8 mL of 2-propanol and proceed as in Section 7.2. Record the absorbances as A_3. Treat a blank in identical fashion; record the absorbance as A_4. Repeat three times and calculate means of individual A_3 and A_4 values. Plot the net absorbance (mean A_3–A_4) against μg sulfate from the exact standardized value of the H_2SO_4 standard as obtained in Section 6.6. The plot should be linear with zero intercept and with a slope of 1 absorbance unit per 75 (±20) μg sulfate. Calculate the best-fit equation by standard linear regression.

For the second calibration plot, proceed exactly as the first part except using a ten-fold diluted H_2SO_4 standard (0.00005 M H_2SO_4) prepared from the 0.0005 M standard. The plot will be non-linear.

9. Calculations

If the sulfate content in the 2 mL sample is above 5 μg (concentration $\geq$ 2.5 μg/ml), calculate the sulfate content of the sample from the net absorbance (A_2–A_1) and the linear regression equation obtained in Section 8. If the sulfate content in the 2 mL sample is below 5 μg, obtain the sulfate content by direct comparison with the low-range calibration plot.

10. Effects of Storage

All solutions except the most dilute sulfate standard are stable for at least 30 days when stored in capped bottles.

11. References

1. BERTOLACINI, R.J. and J.E. BARNEY, II. 1957. Colorimetric Determination of Sulfate with Barium Chloranilate, *Anal. Chem.* 29:281–283.
2. BERTOLACINI, R.J. and J.E. BARNEY, II. 1958. Ultraviolet Spectrophotometric Determination of Sulfate, Chloride, and Fluoride with Chloranilic Acid, *Anal. Chem.* 30:202–205.
3. KLIPP, R.W. and J.E. BARNEY, II. 1959. Determination of Sulfur Traces in Naphthas by Lamp Combustion and Spectrophotometry, *Anal. Chem.* 31:596–597.
4. AGTERDENBOS, J. and N. MARTINIUS. 1964. Theoretical Considerations on the Indirect Determination of Anions. Determination of Sulphate with Barium Chloranilate, *Talanta* 11:875–885.
5. BARNEY, J.E., II. 1965. Determination of Sulphate with Barium Chloranilate. Letter to the Editor, *Talanta* 12:425–426.
6. AGTERDENBOS, J. and N. MARTINIUS. Determination of Sulphate with Barium Chloranilate. Author's Reply, Talanta 12:426–429.
7. SCHAFER, H.N.S. 1967. An Improve Spectrophotometric Method for the Determination of Sulfate with Barium Chloranilate as Applied to Coal Ash and Related Materials, *Anal. Chem.* 39:1719–1726.
8. GALES, M.E., W.H. KAYLOR and J.E. LONGBOTTOM. 1968. Determination of Sulphate by Automatic Colorimetric Analysis. Analyst 93:97–100.
9. Intersociety Committee. 1977. 712. Tentative Method of Analysis of Sulfur Trioxide and Sulfur Dioxide Emissions from Stack Gases (Colorimetric Procedure), in "Methods of Air Sampling and Analysis", 2nd ed. pp. 744–751.
10. BABKO, A.K. and V.A. LITVINENKO. 1963. Barium Rhodizonate Complexes and Their Use for The Photometric Determination of Sulfate, Zh. *Anal. Khim.* 18:237–244.
11. LAMBERT, J.L. and J. RAMASAMY. 1975. Colorimetric Determination of Sulfate in Water with Barium Violurate, *Anal. Chim. Acta* 75:460–463.
12. HWANG, H. and P.K. DASGUPTA. 1985. Spectrophotometric Determination of Trace Aqueous Sulfate with Barium-beryllon II. *Mikrochim. Acta I*, 313–324.
13. DUVALL, T.R. 1980. California Primate Research Center, Davis, CA. Private Communication.

Subcommittee 1

P.K. DASGUPTA, *Chairman*
D.F. ADAMS
B.R. APPEL
S.O. FARWELL
K.T. KNAPP
G.L. KOK
W.R. PIERSON
K.D. REISZNER
R.L. TANNER

730.

Determination of Sulfur in Particulate Material by X-Ray Fluorescence

1. Principle of the Method

1.1 X-ray fluorescence (XRF) spectroscopy is a powerful technique for elemental analysis. All elements with atomic numbers 12 (Mg) and higher can be measured by most commercially available XRF units; many instruments can measure from atomic number 9 (F) and up. In general, the technique is fast, non-destructive, permits multi-element measurement, and is very sensitive. While XRF is most commonly used for multi-elemental, non-destructive analyses of solid and liquid samples, single element determinations, for example sulfur, can be run with speed, sensitivity, and minimum sample preparation. Sulfur determination by XRF in particulate matter collected on filters, a routine procedure in many laboratories, requires only 1 to 3 min per sample and little or no sample preparation.

1.2 Two general types of XRF spectrometers are commercially available, wavelength-dispersive and energy-dispersive instruments. The general approach to XRF and the two techniques are described in detail in many publications (1–5).

2. Sensitivity, Detectability, and Range

2.1 Sensitivity in XRF spectrometric analysis may be defined as the rate of change of the analyte-line intensity with change in amount of the analyte. For thin film samples, sensitivity is the intensity per unit area density [counts/($\mu g/cm^2$)]. Sensitivity is dependent on the excitation source strength, fluorescence yield, the geometry of the spectrometer and the detector, and the detector efficiency.

Limit of Detection (LOD) is defined as $3\sqrt{B}/S$ where B is the number of counts for background and S is the sensitivity, the same counting time being used for B and S measurement.

3. Interferences

3.1 The aerosol deposit should be limited to a thin layer. This eliminates the need for matrix corrections. The usual criterion for acceptable "thinness" of the sample deposit is:

$$\text{mass deposit density (g/cm}^2\text{)} \times$$

$$\text{mean mass attenuation coefficient} \leq 0.1.$$

For sulfur determination at the K_α emission line in a typical sample of stack fly ash composition, the deposit density for an acceptably thin sample would be less than 80 $\mu g/cm^2$.

3.2 For sulfur (K_α) determinations, the only significant overlap inferences are from the lines of the following elements.

Element	Percent Interference (by mass)
Mo	3.1
Bi	0.27
Pb	0.24

The interference coefficients shown above are typical for a wavelength-dispersive instrument. Corrections need to be applied as follows: For example, for each microgram of lead present, 0.0024 micrograms of sulfur should be subtracted from the raw value.

4. Precision and Accuracy

4.1 Precision, or repeatability, is limited by counting statistics for concentrations less than about 20 times the 3σ-detection limit. At higher concentrations, factors such as instrument drift and equivalence of sample positions in individual placements are more important.

Accuracy depends on the accuracy of the standards, homogeneity of the deposit and weighing accuracy of both standard and analyte, particle size of standard and analyte, and attainable instrumental precision.

In a typical analysis, precision and accuracy levels better than 3% are attainable by a careful analyst.

5. Apparatus

Commercially available XRF instruments are either wavelength dispersive or energy dispersive. Both types produce x-ray fluorescence spectra for qualitative and quantitative analysis.

Wavelength-dispersive spectrometers are capable of higher resolution than energy-dispersive instruments and also have higher count rate capability. The basic components of a wavelength-dispersive instrument are: an x-ray tube to irradiate the sample and induce fluorescence, a crystal monochromator to disperse the emitted x-ray, and a detector to determine the intensity of the characteristic monochromatic x-ray. Both sequential and simultaneous wavelength-dispersive instruments are available.

Energy-dispersive spectrometers usually induce sample fluorescence with gamma rays or filtered x-rays from a secondary target. Advantages of such an instrument are that there are no moving parts, and simultaneous multielement measurement is feasible. Further, the geometry allows high detector efficiency which makes low-power excitation possible.

A lithium-drifted silicon [Si(Li)] solid-state detector with associated amplifiers and a multichannel analyzer is used to sort the emitted x-rays according to their energies.

6. Reagents

6.1 AMMONIUM HYDROXIDE, CONC.

7. Procedure

7.1 SAMPLING. Membrane filters (nylon, polycarbonate or polytetrafluoroethylene) of the appropriate size (typically 47 mm) are preweighed, using a balance of sensitivity $\pm$ 10 μg or better, after equilibration at constant temperature and humidity. Particle samples are collected by drawing known volumes of the air sample of interest through the membrane filters housed in appropriate holders. See sections 4 and 29, Part 1 for particle sampling.

7.2 ANALYSIS. Filters with the collected particle deposits are equilibrated in a constant temperature and humidity environment, identical to that used for preweighing the filter, and then weighed.

7.2.1 *Ammonia Exposure.* Most XRF spectrometers utilize a vacuum chamber to minimize absorption of the emitted fluorescent radiation. If the particle deposit on the filter contains free sulfuric acid, the acid will slowly volatilize during the counting period. This loss is eliminated by exposing the filters to ammonia. Exposing the filter for 10 s to an open container containing conc. NH_4OH is ample to convert the acid to the less volatile ammonium salt.

7.2.2 Measurement of the sulfur K_α fluorescence from the filter sample is made with a wavelength- or energy-dispersive spectrometer according to the instructions provided by the instrument manufacturer.

8. Calibration

8.1 Thin film standards ($\leq$ 80 μg/cm^2 deposit density) are used for calibration. If the deposit density in the actual samples is greater than 80 μg/cm^2, it is necessary to make corrections for attenuation in the sample matrix. Both the incident and transmitted x-rays can be attenuated by the filter medium, individual particles, and layers of particles. The effects of attenuation are most pronounced for low energy x-rays.

The sensitivity or calibration factor for the element sulfur is determined with thin standard films. Several types of thin samples, available commercially, are used to calibrate x-ray spectrometers. These include standards deposited by thermal evaporation on Mylar films, and dried solution deposits on membrane filters prepared by a multi-drop technique.

$$\text{Air concentration of sulfate in } \mu g/m^3 = \frac{C_f \times \text{exposed filter area in cm}^2 \times 3}{\text{Volume of sampled air, m}^3}$$

9. Calculations

9.1 Generally, calculations are carried out as prescribed in the operational procedures for the individual x-ray systems. This is typically done by the systems operating computer. A typical calculation procedure for particulate sulfate collected on a filter is as follows:

$$C_f(\mu g/cm^2) = \frac{\text{peak counts--background counts}}{\text{sensitivity factor} \times \text{correction factor}}$$

where C_f is the weight of sulfur in μg per cm^2 of exposed filter, the sensitivity factor is counts/($\mu g/cm^2$) of sulfur as determined from a standard, and the correction factor includes absorption corrections for particle size and for non-homogeneous layering for samples that are not acceptably thin.

$$\text{Percent S} = \frac{C_f \times \text{exposed filter area in cm}^2 \times 100}{\text{Sample weight in } \mu g}$$

For cases where it can be assumed that sulfates are the only sulfur compounds present, the sulfate concentration may be computed as:

10. References

1. BENNETT, R.L., J. WAGMAN and K.T. KNAPP. 1975. The Application of a Multichannel Fixed and Sequential Spectrometer System to the Analysis of Air Pollution Particulate Samples from Source Emissions and Ambient Air, *Adv. X-Ray Anal.* 19:393–402.
2. BIRKS, L.S. 1969. "X-Ray Spectrochemical Analysis," Second Edition. Interscience Publishers, New York, NY.
3. GIAUQUE, R.D., F.S. GOULDING, J.M. JAKLEVIC and R.H. PEHL. 1975. Trace-Element Determination with Semiconductor-Detector X-Ray Spectrometers, *Anal. Chem. 45*:671–681.
4. HUDSON, G.M., H.C. KAUFMANN, J.W. NELSON and M.A. BONACCI. 1980. Advances in the Use of PIXE and PESA for Air Pollution Sampling, *Nucl. Instrum. Methods 168*:259–263.
5. JENKINS, R. and J.L. DEVRIES. 1970. "Practical X-Ray Spectrometry," Second Edition. Springer-Verlag, New York, NY.

Subcommittee 1

K.T. KNAPP
P.K. DASGUPTA, *Chairman*
D.F. ADAMS
B.R. APPEL
S.O. FARWELL
G.L. KOK
W.R. PIERSON
K.D. REISZNER
R.L. TANNER

PART III
METHODS FOR CHEMICALS
IN AIR OF THE WORKPLACE
AND IN BIOLOGICAL SAMPLES

801.

Determination of Ammonia in Air

See **Method 401.**

804.

As, Se, and Sb in Urine and Air by Hydride Generation and Atomic Absorption Spectrometry

1. Principle of Method

1.1 Urine and air particulate samples are wet ashed with a mixture of nitric, perchloric and sulfuric acids to destroy the organic matrix. Volatile metal hydrides are generated from the acidic solution by the addition of sodium borohydride. The trace metal content of the evolved gas is determined by background-corrected atomic absorption spectrometry.

2. Range and Sensitivity

2.1 For a 25-mL urine sample, the optimum range extends from 1 $\mu g/L$ to 50 $\mu g/L$ for arsenic and selenium, and from 10 $\mu g/L$ to 500 $\mu g/L$ for antimony.

2.2 For a 60-L air sample, the optimum range extends from 0.4 $\mu g/m^3$ to 50 $\mu g/m^3$ for arsenic and selenium and from 4 $\mu g/m^3$ to 500 $\mu g/m^3$ for antimony.

2.3 The sensitivity of the hydride generation and measurement systems is such that 5 ng of As and Se and 50 ng of Sb may be determined with a relative standard deviation of 20% or better if interferences are not present.

3. Interferences

3.1 The presence of high concentrations of several metals may significantly reduce the efficiency of the hydride generation for As, Sb and Se (1,2). Such interferences may be checked and corrected for by using the method of standard additions described in Section **8.1.6.** The sensitivity of the method will be reduced if metal interferences are present.

3.2 Interference from moderate amounts of copper can be reduced or eliminated by using thiourea as a masking agent **(3)**.

3.3 Background correction techniques **(4,5)** must be used with this method since the changes in flame composition resulting from the H_2 generated by the addition of sodium borohydride to the sample will change the amount of radiation absorbed by the flame.

4. Precision and Accuracy

4.1 The precision of the method for As, Se and Sb has not been reported using urine and air samples. The precision of the hydride generation procedure has been reported to be as good as 1–2% RSD for 0.1 μg of Sb, Se and As **(6,7)**.

4.2 No accuracy data are available at this time for As, Se and Sb in air and urine samples.

5. Apparatus

5.1 AIR SAMPLING EQUIPMENT.

5.1.1 *Cellulose acetate membrane filters*, 37 mm, e.g., Millipore Type AA (0.8 μm pore diameter) or equivalent.

5.1.2 *Filter unit*, consisting of filter medium **(5.1.1)** and appropriate cassette filter holder, either a 2- or 3-piece filter cassette.

5.1.3 *A vacuum pump* such as a personal sampling pump. The pump must be sufficient to maintain an air face velocity of approximately 7 cm/s (flow rate of 2–5 L/min through a 37 mm diameter filter).

5.1.4 *A flow measurement device* such as a calibrated rotameter is required to monitor and control the sampling rate. A critical orifice can be used if the sampling pump creates sufficient vacuum.

5.2 GLASSWARE, BOROSILICATE. Before use, all glassware must be cleaned in hot 1:1 nitric acid and rinsed several times with water.

5.2.1 *125-mL beakers* with watch glass covers (either Phillips or Griffin-style beakers can be used).

5.2.2 *25-mL volumetric flasks.*

5.2.3 *500-mL volumetric flasks.*

5.2.4 *1-L volumetric flasks.*

5.2.5 *125-mL polyethylene bottles.*

5.2.6 *Hydride generator* (see Figure 804:1, or equivalent).

5.2.7 *Additional glassware* such as pipettes and different sized volumetric glassware may be required.

5.3 EQUIPMENT.

5.3.1 *Atomic absorption spectrophotometer (AAS)* with burner head for air-entrained argon-hydrogen flame, and equipped for simultaneous background correction. Electrodeless Discharge Lamps (EDL) are recommended (hollow cathode lamps may be used with reduced sensitivity). Lamps are required for each element to be tested.

5.3.2 *A continuum lamp* or source is needed for background correction with most AAS units. Those units equipped with Zeeman-type background correction do not require background correction lamps.

5.3.3 *Hotplate* (suitable for operation at 140°C to 250°C).

5.3.4 *Two-stage regulators* for argon and hydrogen.

5.3.5 *Rotameter* for hydride generator (0–5 L/min).

5.4 SUPPLIES.

5.4.1 *Hydrogen gas* (cyclinder), electrolytic grade.

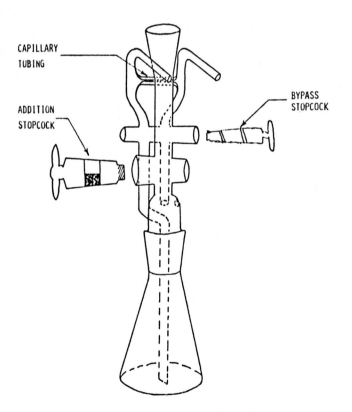

CAPILLARY TUBING

ADDITION STOPCOCK

BYPASS STOPCOCK

Figure 804:1 — Hydride generator.

5.4.2 *Argon gas* (cylinder), high purity.

6. Reagents

6.1 Purity—All chemicals used in these tests shall be ACS reagent grade or equivalent. References to water shall be understood to mean ASTM reagent water, Type II. Care in selection of reagents and in following the listed precautions is essential if low blank values are to be obtained.

6.2 Concentrated ammonium hydroxide (28%) specific gravity 1.090.

6.3 Concentrated nitric acid (68-71%), redistilled, specific gravity 1.42.

6.4 Concentrated sulfuric acid (90%), specific gravity 1.84.

6.5 Perchloric acid (72%), specific gravity 1.67.

6.6 Concentrated hydrochloric acid (36-38%), specific gravity 1.18.

6.7 Ammonium oxalate—saturated aqueous solution.

6.8 Sodium borohydride, 8-9 mm pellets (approximately 200 mg).

6.9 Standard stock solutions (1000 μg/mL) for each metal element. Commercially prepared or prepared from the following after desiccator drying as described in Section **8.1** below:

6.9.1 Sodium arsenate ($Na_2HAsO_4 \cdot 7H_2O$).

6.9.2 Antimony metal (Sb).

6.9.3 Selenium dioxide (SeO_2).

7. Procedure

7.1 COLLECTION OF SAMPLES.

7.1.1 Urine samples are collected in polyethylene bottles which were precleaned in nitric acid. About 0.1 g EDTA is added as a preservative. At least 75 mL should be collected.

7.1.2 Air samples are collected on 37 mm membrane filters using a plastic filter holder in a closed-faced configuration. The flow rate, times, and/or volumes must be measured as accurately as possible. Record atmospheric pressure and temperature at beginning and end of the sampling period. The sample should be taken at a flow rate of 2 to 3 L/min. Sampling rate can be monitored using a calibrated rotameter or controlled using a critical orifice. A minimum

of 60 L should be collected. Larger sample volumes are strongly encouraged provided the filters do not become loaded with dust to the point that either loose material might fall off or the filter becomes plugged. The sample cassette should be sealed after collection, to protect the sample during shipment to the laboratory.

7.2 STORAGE.

7.2.1 Urine samples are stored or shipped in the polyethylene containers. Care should be taken to maintain the containers in an upright position to avoid leakage.

7.2.2 Air filters are stored and shipped in the sample cassettes. The sample cassettes should be shipped in a suitable container designed to minimize contamination and to prevent damage in transit. Care must be taken during storage and shipping that no part of the sample is dislodged from the filter, nor that the sample surface be disturbed in any way. Loss of sample from extremely heavy deposits on the filter may be prevented by mounting a clean filter in the cassette on top of the sample filter (care must be taken to avoid contamination when this is done). When samples are analyzed the protective filter is included as part of the sample. Blank filters (minimum of 1 filter blank for every 10 samples) should be mounted as the samples. (Note: No air is drawn through the blank filter and blanks are treated exactly as samples in the storage, shipping, and analysis procedures.)

7.3 PREPARATION OF SAMPLES.

7.3.1 Determine the specific gravity of the urine sample at room temperature. This may be done with the use of a calibrated specific gravity meter or urinometer.

7.3.2 For urine samples, transfer 25 mL into a 125 mL beaker. For air samples, place the sample filter, and cover filter if any, in a 125 mL beaker. If Se is to be determined in urine samples, 0.1 g of solid HgO should be added to the sample to prevent volatilization of Se during the ashing of sample.

7.3.3 Wet-ash sample by adding first 4 mL of a mixture of 3 parts HNO_3 and 1 part H_2SO_4. Carefully add 1 mL $HClO_4$, swirl, and heat on a hot plate at 130°-150°C. Add dropwise small amounts of redistilled HNO_3 (Caution: Safety precautions for perchloric acid digestions must be followed at all times) until the digest is colorless. Do not allow the sample to dry. If the sample volume is less than 1 mL and ashing is still incomplete, add additional ashing acid mixture (3:1:1 = HNO_3, H_2SO_4, $HClO_4$), dropwise.

7.3.4 If ashing appears complete, heat to appearance of fumes of SO_3. After cooling the colorless or nearly colorless liquid, add 10 mL of distilled water and 5 mL of a saturated solution of ammonium oxalate. Heat the mixture again, until dense white fumes of SO_3 appear, to free the ashed solution of all traces of nitric acid. Do not heat the resulting solution for more than a few minutes after SO_3 evolves, to ensure that none of the elements being analyzed are lost.

7.3.5 Add 10 mL of concentrated HCl to the sample and heat at 90°C for 15 minutes (this step reduces Se(VI) to Se(IV) and improves sensitivity) **(8)**.

7.3.6 Allow the mixture to cool, quantitatively transfer to a 25 mL volumetric flask, and make up to volume with water.

7.4 ANALYSIS.

7.4.1 Set the instrument operating conditions as recommended by the manufacturer. The instrument should be set at the radiation intensity maximum for the wavelength listed below for the element being determined.

As – 193.7 nm
Sb – 217.6 nm
Se – 196.0 nm

7.4.2 The position of maximum atom population (As, Se, or Sb) in the flame (thus position of maximum radiation absorption) when using the introduction system described in **7.4.3** will not necessarily be identical to that obtained when sample solutions are aspirated in the conventional manner. For this reason, it is necessary to adjust the EDL (or hollow cathode) beam position in the flame. The horizontal burner position can be optimized by aspirating a high-concentration solution by the conventional pneumatic nebulizer and by adjusting the horizontal placement until a maximum absorbance is achieved. This step assures centering the output of the radiation source with respect to the flame. Vertical adjustment of the beam passing through

the flame is made by introducing a series of identical standards (using procedures in **7.4.3** or **7.4.4**) at increasingly lower burner positions (these positions are noted). The vertical burner position is then adjusted to that position giving maximum absorbance for the analysis of samples and standards as described in the following section.

7.4.3 *Use of solid Na BH$_4$.*

a. Be sure the stopcocks of the hydride generation equipment are turned so that the argon (Ar) flow bypasses the reaction vessel. (Note: The pressure drop across the capillary tubing in Figure 804:1 must be greater than that resulting from the liquid head present in the generation flask. If this is not the case, the argon used to purge the system will not be swept through the reservoir in a reproducible manner.)

b. Add 50 mL of 6 N HCl* and 5 mL of a standard solution (from Section **8.1.3**) to the reaction flask. Add a magnetic stirring bar. Connect the sample flask to the generating system and mix well with the magnetic stirrer. Turn the bypass stopcock so that the Ar passes through the reaction vessel. Allow the argon flow to purge the air from the vessel.

c. Add a single sodium borohydride pellet (8 mm diameter, 200 mg) to the sample solution via the addition stopcock. Record the absorption signal on a rapid-response strip chart recorder (0.5 s full scale response) or use the electronic peak integration facility built into most modern AAS units.

d. Standard solutions should match the sample matrix as nearly as possible and should be run in duplicate. Prepare a calibration graph as described in Section **8.1.4**. (Note: All combustion products from the AA flame must be directly exhausted through a flame ventilation hood system.)

7.4.4 Recently, automated reagent introduction systems based upon the use of aqueous $NaBH_4$ have become available (e.g., see Reference 7). Such systems can be used in place of that described in Section **7.4.3**. Follow manufacturer's procedures for the use of such automated systems. Make sure, however, that in the procedure, the sample is acidified with HCl prior to the

---

*Note: Sensitivity may be greatly improved for Sb if 1N HCl is used instead of 6N HCl **(8)**.

addition of the reducing agent. Interferences are more pronounced if the reducing agent is added to the sample prior to the addition of HCl (2).

7.4.5 Analyze 5 mL of the sample from Section **7.3.5** following the instructions in **7.4.3** or **7.4.4** and record the absorbance of peak integrals for comparison with results obtained with calibration standards. Should the absorbance be outside the calibration range, analyze an appropriately smaller or larger aliquot. A mid-range standard must be analyzed with sufficient frequency (i.e., once every 10 samples) to assure the precision of the sample determinations. To the extent possible, all determinations are to be based on replicate analysis.

8. Standards and Calibrations

8.1 STANDARD SOLUTIONS.

8.1.1 Standard stock solutions of each element. Standard stock solutions are made to a concentration of 1.000 mg of element/mL (1000 ppm). These standard stock solutions are stable for at least 3 months when stored in polyethylene bottles.

a. *Arsenic Standard Stock Solution, 1000 ppm.* Dissolve 0.4165 of $Na_2HAsO_4 \cdot 7H_2O$ in a mixture of water, 5 mL of concentrated H_2SO_4 and 20 mL of concentrated HCl and bring volume to 100 mL with water.

b. *Selenium Standard Stock Solution, 1000 ppm.* Dissolve 1.405 g of dried SeO_2 (heated at 105°C for 1 h and cooled in a dessicator) in water containing 5 mL of 5 N NaOH. Dilute to volume in a 1-L flask with water.

c. *Antimony Standard Stock Solution, 1000 ppm.* Dissolve 1.000 g of Sb in a minimum volume of concentrated H_2SO_4. Dilute to volume in a 1-L flask with water.

8.1.2 Dilute standard solutions of each element. Dilute standards are made to a concentration of 1 $\mu g/mL$ for As and Se and 10 $\mu g/mL$ for Sb. Dilute standards should be prepared weekly.

a. *Dilute Arsenic Standard.* Pipette 1 mL of the As master standard into a 1000-mL volumetric flask and dilute to volume with water. This solution contains 1 $\mu g/mL$ As.

b. *Dilute Selenium Standard.* Pipette 1 mL of the Se master standard into a 1000-mL volumetric flask and dilute to volume with water. This solution contains 10 $\mu g/mL$ Se.

c. *Dilute Antimony Standard.* Pipette 10 mL of the Sb master standard into a 1000-mL flask and dilute to volume with water. This solution contains 10 $\mu g/mL$ Sb.

8.1.3 *Working Standards.*

a. Working standards of 0.025, 0.05, 0.1, 0.25, and 0.50 μg of As and Se per 25 mL of solution are prepared. Working standards of 0.25, 0.5, 1.0, 2.5 and 5.0 μg of Sb per 25 mL of solution are prepared. The working standards are prepared daily.

b. Add 0 mL, 0.5 mL, 1 mL, 2 mL, 5 mL, and 10 mL of the dilute standard for As, Se and Sb to separate 500-mL volumetric flasks. Add to these flasks 300 mL distilled water (approximately), 100 mL of concentrated HCl and 25 mL concentrated H_2SO_4. Dilute to volume with water. These solutions will contain As, Se, and Sb in the concentrations specified in **8.1.3.a.** For urine samples in which Se is being determined, the working standards must have HgO added such that the concentration of Hg^{++} matches that added to the samples. The flask to which no dilute standards are added is used as the reagent blank.

8.1.4 The standard solutions are analyzed as described in **7.4.3** or **7.4.4** and the absorbance area (or peak height) recorded. If the instrument used displays transmittance, these values must be converted to absorbance. After the subtraction of the reagent blank signal from all standard signals, a calibration curve is prepared by plotting absorbance versus metal concentration. The best-fit line (calculated by linear least square regression analysis) is fitted to the data points. This line or the equation describing the line is used to obtain the metal concentration in the samples being analyzed.

8.1.5 To ensure that the preparation procedure is being properly followed, clean blank membrane filters are spiked with known amounts of the elements being determined by adding appropriate amounts of the previously described standards and carried through the entire procedure. The amount of metal is determined and the percent recovery calculated. These tests will provide recovery and precision data for the procedure as it is carried out in the laboratory for the soluble compounds of the elements being determined.

8.1.6 Analysis by the method of standard additions. In order to check for interferences (which may not have been fully investigated for use in this method), samples are initially and periodically analyzed by the method of standard additions and the results compared to those obtained by the conventional analytical determination. For this method the sample is divided into three 5-mL aliquots. One 5-mL aliquot is added to the reaction flask and analyzed as described in Section **7.4.5**. The second 5-mL aliquot is added to the reaction flask along with an amount of element (As, Se or Sb) approximately equal to that in the sample (typically 0.5 to 5 mL of the highest-concentration working standard described in **8.1.3.a**) and analyzed. The third 5-mL aliquot, along with an amount of element approximately equal to twice the sample concentration, is added to the reaction vessel and analyzed. A 5-mL portion of the sample blank is also analyzed. The blank reading is subtracted from the three aliquot readings and these corrected values are plotted against metal added to the original sample. The line obtained from such a plot is extrapolated to 0 absorbance and the intercept on the concentration axis is taken as the amount of metal in the original sample (4). If the result of this determination does not agree to within 10% of the values obtained with the procedure described in Section **8.1.4**, an interference is indicated and standard addition techniques should be utilized for sample analysis.

8.2 BLANK. Blank filters must be carried through the entire procedure each time samples are analyzed. Reagent blanks are used for making blank corrections in the case of urine samples.

9. Calculations

9.1 The concentration of As, Se, and Sb in the urine sample should be expressed as mg element per liter of urine.

$$\text{mg element/L} = \frac{\mu\text{g element}}{\text{mL of urine}}$$

where μg element = (μg of element in total sample used in **7.3.2**: normally 25 mL) − (reagent blank value)

mL of urine = volume of urine sample used in **7.3.2**: normally 25 mL

The use of specific gravity correction factor to normalize values of 1.024, the average specific gravity of urine, is recommended.

corrected mg element/L =

$$\text{mg element/L} \times \frac{(1.024 - 1.000)}{(\text{Sp G} - 1.000)}$$

9.2 The concentration of arsenic in air can be expressed as mg As per cubic meter of air, which is numerically equal to μg As per liter of air.

$$\text{mg element/m}^3 = \mu\text{g element}/V_s$$

where μg element = (μg element in total sample) − (blank filter value)

V_s = volume of air sampled in L at 25°C and 101.3 kPa

10. Storage

10.1 Untreated filter samples may be stored indefinitely if sealed in an air tight container.

10.2 Urine samples treated with Na_2H_2EDTA and refrigerated are biologically stable for 1 wk. The specific gravity of the sample should be measured within 1 wk of sample collection.

11. References

1. SMITH, E. E. 1975. *Analyst, 100*, 300
2. PIERCE, F. D., AND H. R. BROWN. 1976. Inorganic Interferences Study of Automated Arsenic Selenium Determination with Atomic Absorption Spectrometry. *Anal. Chem., 48*, 693 Thiourea as a Complexing Agent for Reduction of Copper Interference in the Determination of Selenium by Hydride Generation/Atomic Absorption Spectrometry.
3. BYE, R., L. ENGVIK, AND W. LUND. 1983 *Anal. Chem.*, 55, 2457–2460.
4. SLAVIN, W., *Atomic Absorption Spectroscopy*, Interscience Publisher, N.Y., N.Y. (1968).

5. RAMIREZ-MUNOZ, J. *Atomic Absorption Spectroscopy*, Elsevier Publishing Co., N.Y., N.Y. (1968).
6. VIJAN, P. N., AND G. R. WOOD, *Atomic Absorption Newsletter, 13*, 33 (1974).
7. FIORINO, J. A., J. W. JONES, AND S. G. CAPAR. 1976. Sequential Determination of Arsenic Selenium Antimony, and Tellurium in Foods Via Rapid Hydride Evolution and Atomic Absorption Spectrometry. *Anal. Chem., 48*, 120.
8. OLIVEIRA, E ĎE, J. W. MCLAREN, AND S. S. BERMAN. 1983. *Anal. Chem., 55*, 2047–2050. Simultaneous Determination of Arsenic, Antimony, and Selenium in Marine Samples of Inductively Coupled Plasmic Atomic Emission Spectrometry.

Subcommitte 6

M. T. KLEINMAN, *Chairman*
W. J. COURTNEY
V. P. GUINN
T. C. RAINS
J. R. RHODES
R. J. THOMPSON

805.

Determination of Chloride in Air

1. Principle of the Method

1.1 Atmospheric samples are taken using midget impingers containing 10 mL of 0.5 M sodium acetate.

1.2 Samples are analyzed using the chloride ion selective electrode (**1,2**).

2. Range and Sensitivity

2.1 The range and sensitivity of this method have not been established. The electrode recommended range is 0.35 to 3,500 μg/mL.

3. Interferences

3.1 Sulfide ion must be absent because it poisons the chloride ion electrode. Touch a drop of the sample to a piece of lead acetate paper to check for sulfide ion.

3.2 If sulfide is present, it is removed by addition of a small amount of powdered cadmium carbonate to the sample. Swirl to disperse the solid and recheck a drop of the sample with lead acetate paper. Avoid a large excess of cadmium carbonate and long contact time with the solution. Filter the sample through a small plug of glass wool and proceed with the analysis.

3.3 For interference-free operation, the chloride ion level must be at least 300 times the bromide ion level, 100 times the thiosulfate level and 8 times the ammonia level.

4. Precision and Accuracy

4.1 The precision and accuracy of this method have not been established. No collaborative tests have been performed on this method.

5. Advantages and Disadvantages of the Method

5.1 Advantages are the simplicity, specificity, speed and accuracy of the method. Disadvantages are the interferences from sulfide ion and possibly ammonia.

6. Apparatus

6.1 ORION 94–17 CHLORIDE ION ELECTRODE, OR EQUIVALENT.

6.2 ORION 90-02 DOUBLE JUNCTION REFERENCE ELECTRODE, OR EQUIVALENT.

6.3 EXPANDED SCALE MILLIVOLT PH METER.

6.4 MIDGET IMPINGERS.

6.5 SOURCE OF VACUUM, electric or battery operated pumps.

6.6 STOPWATCH.

6.7 ASSOCIATED LABORATORY GLASSWARE.

6.8 MAGNETIC STIRRER AND STIRRING BARS.

7. Reagents

7.1 PURITY. The reagents described must be ACS Reagent grade. Water is ASTM Reagent water Type II.

7.2 SODIUM ACETATE 0.5 M. Dissolve 41 g sodium acetate in water and dilute to 1 L. Adjust pH to 5 with acetic acid.

7.3 SODIUM CHLORIDE STANDARDS.

7.3.1 Dissolve 5.84 g NaCl in distilled water and dilute to 1 L for 10^{-1} M (Cl⁻) (3500 μg/mL). This solution is stable for about 2 months. The following more dilute standards should be prepared fresh weekly.

7.3.2 Dilute 10 mL 10^{-1} M (Cl⁻) to 100 mL with 0.5 M sodium acetate for 10^{-2} M (Cl⁻) (350 μg/mL).

7.3.3 Dilute 10 mL 10^{-2} M (Cl⁻) to 100 mL with 0.5 M sodium acetate for 10^{-3} M (Cl⁻) (35 μg/mL).

7.3.4 Dilute 10 mL 10^{-3} M (Cl⁻) to 100 mL with 0.5 M sodium acetate for 10^{-4} M (Cl⁻) (3.5 $\mu g/mL$).

7.3.5 Dilute 10 mL 10^{-4} (Cl⁻) to 100 mL with 0.5 M sodium acetate for 10^{-5} M (Cl⁻) (0.35 $\mu g/mL$).

8. Procedure

8.1 CLEANING EQUIPMENT. All glassware is washed in detergent solution, rinsed in tap water followed by a rinse with reagent water.

8.2 COLLECTION OF SAMPLES.

8.2.1 Atmospheric samples are collected in midget impingers containing 10 mL of 0.5 M sodium acetate solution.

8.2.2 A sampling rate of about 2.5 L/ min is used.

8.2.3 A total air volume of about 200 L drawn through the impingers.

8.2.4 After sampling, the openings of the midget impingers are sealed with masking tape and the impingers returned to the laboratory for analysis.

8.2.5 A "blank" impinger should be treated in the same way as other samples, except that no air is passed through this impinger.

8.3 ANALYSIS OF SAMPLES.

8.3.1 The volume of the sampling solution in the impinger is measured and recorded.

8.3.2 The solution is tested for sulfide and treated if necessary as described in **3.1–3.2.**

8.3.3 The solution is transferred to a 50-mL beaker and the pH adjusted to 5.0 to 5.2, if necessary, with 0.5 M sodium acetate or acetic acid. Dilute to a volume of 25 mL with water.

8.3.4 The chloride ion electrode and the double junction reference electrode are placed in the stirred solution and the resulting millivolt reading recorded. The reading should be taken to the nearest 0.5 mV after the meter has stabilized.

9. Calibration and Standards

9.1 Prepare a series of chloride standards solutions in 50-mL beakers by dilut-

ing 10 mL of each of the chloride standards, prepared as in **7.3.2** to **7.3.5**, to a volume of 25 ml with double distilled water, starting with the most dilute standard. Place the chloride ion electrode and the double reference electrode in the stirred solution. Record the resultant millivolt readings to the nearest 0.5 mV.

9.2 Plot the mV readings *vs* the chloride ion concentration of the standards on semi-log paper. The chloride ion concentration in $\mu g/mL$ is plotted on the logarithmic axis. Preferably, determine the calibration equation by the method of least squares.

10. Calculations

10.1 The mV reading obtained from the analysis of the sample is translated to μg Cl per mL of solution using the calibration curve or equation.

10.2 The Cl content of the sample is multiplied by the sample volume to obtain the total μg Cl in the sample. The blank analysis result, if any, is deducted from total μg Cl in the sample.

10.3 The total μg Cl is divided by the volume of air sampled in liters to obtain μg Cl per L or mg Cl per m³. (The volume of air sampled is converted to standard conditions of 25°C and 101.3 kPa.)

11. References

1. ORION RESEARCH INC., Chloride Ion Specifications, Cambridge, Mass.
2. FRANT, M.S. 1974. Detecting Pollutants with Chemical Sensing Electrodes, *Environmental Science and Technology*, 8:224.

Revised by Subcommittee 2

C. R. THOMPSON, *Chairman*
G. H. FARRAH
L. V. HAFF
A. W. HOOK
J. S. JACOBSON
E. J. SCHNEIDER
L. H. WEINSTEIN

Approved with modifications
Subcommittee 2

R. A. MANDL, *Chairman*
L. H. WEINSTEIN
J. S. JACOBSON

806.

Determination of Free Chlorine in Air

See **Method 202.**

807.

Determination of Chromic Acid Mist in Air

1. Principle of Method

1.1 A known amount of air is drawn through a polyvinyl chloride (2) filter. The filter is washed with 0.005 M hydrochloric acid and the solution brought to a pH of about 2.8 after addition of ammonium pyrrolidine dithiocarbamate (APDC).

1.2 The Cr(VI) in the solution is separated from Cr(III) by extraction with a methylisobutyl ketone (MIBK)–hexane solvent which selectively extracts the Cr(VI) ion (3).

1.3 The MIBK-hexane phase is aspirated into an atomic absorption spectrophotometer and the absorbance at 357.9 nm is read and compared to known standards of Cr(VI).

2. Range and Sensitivity

2.1 This method is applicable to Cr(VI) over the range of 0.01 mg/m^3 to 0.25 mg/m^3 (calculated as CrO$_3$); higher concentrations can be determined by dilution (4).

2.2 The limit of detection for Cr(VI) (as CrO$_3$) is about 0.02 μg/mL of extract which corresponds to 0.01 mg/m^3 on the basis of a 20-L sample of air (1).

3. Interferences

3.1 Extraction with MIBK and ammonium pyrrolidine dithiocarbamate (APDC) complexing reagent avoids Cr(III) interference by transferring Cr(VI) into the organic layer. Under the conditions specified, the same reagent will extract zinc, cadmium, iron, manganese, copper, nickel, silver and lead. The interference from manganese can be minimized by extraction at a pH between 2.5 and 3.0 where manganese is not extracted to any great extent. The other metals do not affect the chromium determination unless concentrations are a few orders of magnitude greater than chromium, in which case iron and nickel may suppress the chromium absorption (5). Where interferences are suspected, use the method of standard additions to correct for the inferences (5,6).

3.2 Reducing agents will lower the result for Cr(VI). The most notable interferences are hydrogen, olefins and acetylenes and hydrocarbons of C$_3$ or higher (7). Reduced valence states of most metals will also reduce the Cr(VI) to Cr(III) on the filter.

4. Precision and Accuracy

4.1 Precision of the method has not been reported. However, in a method similar to this, a relative standard deviation of 11% at a concentration of 5 μg/L in water was reported (8).

4.2 Accuracy of the method has not been established.

5. Advantages and Disadvantages

5.1 Polyvinyl chloride filters are an advantage in collection of samples.

5.2 Cr(VI) is separated from Cr(III) by an extraction procedure.

5.3 No ashing procedure is required.

5.4 Reducing agents, such as unsaturated hydrocarbons, and reduced valence states of heavy metals will yield lower results from Cr(VI).

6. Apparatus

6.1 SAMPLING APPARATUS.

6.1.1 *Polyvinyl chloride filter.* Diameter to fit the filter cassette (37 mm suggested). (Gelman type VM-1, 5 μm pore size, or equivalent).

6.1.2 *Battery operated personal sampling pump.* To pull up to 2 L/min of air (Monitaire Sampler, MSA, Pittsburgh, PA or equivalent).

6.1.3 *2-Piece filter cassette.* (Polystrene, open face preferred). (Millipore M-000-037-00 or equivalent with cellulose backing pad) (AP-10-037-00 or equivalent).

6.2 ATOMIC ABSORPTION SPECTROPHOTOMETER. With an acetylene burner head.

6.3 VOLUMETRIC GLASSWARE.

6.4 pH METER.

6.5 FORCEPS. Teflon tipped.

7. Reagents

7.1 PURITY. ACS Reagent grade chemicals shall be used in all tests. References to water shall be understood to mean ASTM Reagent water, Type II. Care in selection of reagents is essential if low blank values are to be obtained.

7.2 CONC NITRIC ACID (69.0–71.0%). sp gr 1.42.

7.3 CONC HYDROCHLORIC ACID (36.5%–38.0%). sp gr 1.18.

7.4 STOCK DILUTE HYDROCHLORIC ACID 0.5 N. Dilute 43 mL of concentrated acid to 1 L with water.

7.5 WORKING DILUTE ACID 0.005 N. Dilute 10 mL of 0.5 N HCl to 1 L with water.

7.6 AMMONIUM PYRROLIDINE DITHIOCARBAMATE (APDC) SOLUTION. Dissolve 1.0 g APDC in water and dilute to 100 mL. Prepare fresh solution daily. Filter if necessary.

7.7 3:1 METHYLISOBUTYL KETONE (MIBK)–HEXANE SOLUTION. Mix 1 part hexane with 3 parts of MIBK and shake to ensure homogeneity of solvent. This solution should be saturated with 0.005 N HCl prior to use.*

7.8 ETHANOL.

7.9 BROMPHENOL BLUE INDICATOR SOLUTION. Dissolve 0.10 g bromphenol blue in 100 mL of 50-50 ethanol-water.

7.10 SODIUM HYDROXIDE SOLUTION, 0.1 N. Dissolve 4.0 g NaOH in 1 L of water.

---

*Pure MIBK is slightly soluble in acidic aqueous solutions. Use of the 3:1 mixture decreases the solubility in the aqueous phase without any significant change in the extraction efficiency (3).

7.11 $K_2Cr_2O_7$.

7.12 CLEANING SOLUTION. Add 150 mL of conc hydrochloric acid to 200 mL of water, then add 50 mL of conc nitric acid with stirring. Use to clean all glassware and store in a borosilicate, glass stoppered bottle. (Never use chromic acid cleaning solution).

8. Procedure

8.1 Collect the chromic acid mist on the filter at a flow rate of 2 L/min for a minimum period of one h. For longer periods of collection, flow rates less than 2 L/m can be used as dictated by the expected Cr(VI) concentrations. Some minimum sampling volumes for various conc levels based on the limit of detection are as follows:

Concentration to be Measured (mg CrO_3/m^3)	Minimum Required Sample Size (L)
0.01	20
0.05	4
0.1 (TLV)	2
0.2	1

8.2 Prepare two filter blanks in the same manner as for sampling and transport to sampling area, but keep tightly capped at all times.

8.3 Remove the filter with forceps and place on the fritted glass of a funnel held in a suction flask, apply suction and wash sides of holder and the filter with 3 separate 20-mL washings of 0.005 N hydrochloric acid solution. Be careful not to combine the backing pad with the filter.

8.4 Pour washings into a 200-mL volumetric flask. Rinse the funnel and suction flask with 2 separate 20-mL washings of 0.005 N hydrochloric acid and add washings to volumetric flask.

8.5 Add 2 drops of bromphenol blue indicator solution and add sufficient 0.1 N sodium hydroxide or 0.5 N HCl solution with stirring to bring the pH to between 2.2 and 2.4. Check the pH with a pH meter on the first few samples to ensure proper indicator color change with pH value.

8.6 Add 5.0 mL of APDC solution and mix. This should raise the pH to about 2.8 ± 0.1. Check with a pH meter occasionally to ensure the pH is within these limits.

8.7 Add exactly 10.0 mL of the MIBK-hexane solvent saturated with 0.005 N HCl and shake vigorously for 3 min.

8.8 Allow the two layers to separate and add 0.005 N HCl until the MIBK-hexane layer is completely in the neck of the flask.

8.9 Aspirate the MIBK-hexane layer directly into the atomic absorption apparatus. Measure the absorbance or other suitable scale reading at 357.9 nm after adjusting the instrument to the operating conditions specified by the manufacturer.

8.10 Adjust the fuel-to-air ratio prior to running samples and standards with MIBK-hexane solution until a steady blue flame is obtained. Do not reduce the fuel flow to the extent that the flame will blow out when no solvent is being aspirated.

8.11 Compare the reading to that of a calibration curve previously obtained as described in **9.2** after subtracting the average of the blank samples taken to the sampling site.

8.12 Should the reading be off scale, dilute the MIBK-hexane with an appropriate amount of solvent that has been saturated with 0.005 N HCl and run again.

9. Calibration and Standards

9.1 CHROMIC ACID STANDARDS.

9.1.1 *Master Solution A*: 1 mL = 1.000 mg chromium = 1.92 mg CrO_3. Dissolve exactly 2.828 g of dry potassium dichromate in 0.005 N HCl and dilute to 1 L with 0.005 N HCl. Keeps for 1 year.

9.1.2 *Dilute Standard Solution B*: 1 mL = 0.010 mg chromium = 0.0192 mg CrO_3. Dilute 10 mL of Solution A to 1 L with 0.005 N HCl. Prepare monthly.

9.1.3 *Working Standard Solution C:* 1 mL = 0.1 μg chromium = 0.192 μg CrO_3. Dilute 10 mL of Solution B to 1 L with 0.005 N HCl. Prepare weekly.

9.2 CALIBRATION PROCEDURE.

9.2.1 Pipet onto separate polyvinyl chloride filters in holders inserted into separate suction flasks the following amounts of working standards from Solution C: 0, 2, 5, 10 and 25 mL.

9.2.2 Wash each filter with 3 successive 20-mL portions of 0.005 N HCl as done in the procedure in step **8.3**.

9.2.3 Perform the analysis from step **8.3** through **8.10** of the procedure. Plot the calibration curve of μg of Cr(VI) versus absorbance (or appropriate scale readings linear with absorbance) using the best curve as determined by the least squares method. Subtract the calibration blank value so that 0 mL equals 0 μg $CrCO_3$†

10. Calculations

10.1 Convert all readings to μg of chromium trioxide (CrO_3) per 10.0 mL of MIBK-hexane solution by the following factor: μg CrO_3 = 1.92 × μg Cr(VI).

10.2 Concentration of CrO_3 in air is calculated as follows:

$$mg\ CrO_3/m^3 =$$

$$\frac{(\mu g\ CrO_3\ found\ per\ 10\ ml\ of\ solvent - mg\ CrO_3\ in\ sample\ blank)}{L\ of\ air\ sampled}$$

---

†The hydrochloric acid solutions used should be the same for samples and calibration and blanks. Use of a different bottle of concentrated hydrochloric acid or a new batch of filters requires recalibration of the procedure to ensure that any variation in metal impurities does not affect the results.

11. References

1. U.S. DEPARTMENT OF HEALTH, EDUCATION AND WELFARE, Public Health Service. 1973. Occupational Exposure to Chronic Acid. NIOSH.
2. ABELL, M. T., AND J. R. CARLBERG, 1974. *Am. Ind. Hyg. Assoc. J.* 35:229–233.
3. ENVIRONMENTAL PROTECTION AGENCY, NATIONAL ENVIRONMENTAL RESEARCH CENTER. 1971. Methods for Chemical Analysis of Water and Wastes, Analytical Quality Control Lab, Cincinnati, Ohio 45268.
4. MIDGETT, M. R., AND M. J. FISHMAN, 1967. *At. Abs. Newsletter*, 6:128.
5. SLAVIN, W., 1968. Atomic Absorption Spectroscopy, Interscience Publishers, N.Y., N.Y.
6. DEAN, J. A. AND T. C. RAINS, 1969. Flame Emission and Atomic Absorption Spectrometry, Vol. 1, p. 377, Marcel Dekker, New York.
7. PRIVATE COMMUNICATION, John Kralgic, Allied Chemical, Syracuse, N.Y.
8. BROWN, E., M. W. SKOUGSTAD, AND M. J. FISHMAN, 1970. Techniques of Water Resources Investigations of the U.S. Geological Survey, Chapter A1, Methods for Collection & Analysis of Water Samples for Dissolved Minerals and Gases. Dept. of Interior, Laboratory Analysis, Book 5, pp. 76–80.

Subcommittee 6

T. J. KNEIP, *Chairman*
R. S. AJEMIAN
J. N. DRISCOLL
F. I. GRUNDER
L. KORNREICH
J. W. LOVELAND
J. L. MOYERS
R. J. THOMPSON

Approved with modifications
Subcommittee 6

M. T. KLEINMAN, *Chairman*
J. R. RHODES
R. J. THOMPSON
T. C. RAINS
W. J. COURTNEY
V. P. GUINN

808.

Determination of Cyanide in Air

1. Principle of the Method

1.1 Atmospheric samples are taken using midget impingers that contain 10 mL of 0.1 M NaOH.

1.2 Samples are analyzed using the cyanide ion selective electrode (**1,2**).

2. Range and Sensitivity

2.1 The range and sensitivity of the method have not been established. The recommended range of the method is 0.013 to 13 mg/m^3 in air.

3. Interferences

3.1 Sulfide ion irreversibly poisons the cyanide ion selective electrode and must be removed if found to be present in the sample. Check for the presence of sulfide ion by touching a drop of sample to a piece of lead acetate paper. The presence of sulfide is indicated by discoloration of the paper.

3.2 Sulfide is removed by the addition of a small amount (spatula tip) of powdered cadmium carbonate to the pH 11 to 13 sample. Swirl to disperse the solid, and recheck the liquid by again touching a drop to a piece of lead acetate paper. If sulfide ion has not been removed completely, add more cadmium carbonate; avoid a large excess of cadmium carbonate and long contact time with the solution.

3.3 When a drop of liquid no longer discolors a strip of lead acetate paper, remove the solid by filtering the sample through a small plug of glass wool contained in an eye dropper and proceed with the analysis.

4. Precision and Accuracy

4.1 The precision and accuracy of this method have not been determined. No collaborative tests have been performed on this method.

5. Advantages and Disadvantages of the Method

5.1 Advantages are the simplicity and speed of the method. Specificity depends upon the removal of sulfide prior to analysis.

6. Apparatus

6.1 SAMPLING EQUIPMENT. The sampling unit for the impinger collection method consists of the following components:

6.1.1 *A prefilter unit (if needed).* Consists of the filter medium and cassette filter holder.

6.1.2 *A midget impinger.* Containing the absorbing solution or reagent.

6.1.3 *A pump suitable for delivering desired flow rates.* The sampling pump is protected from splashover or water condensation by an absorption tube loosely packed with a plug of glass wool, inserted between the exit arm of the impinger and the pump.

6.1.4 *An integrating volume meter such as a dry gas or wet test meter.*

6.1.5 *Thermometer.*

6.1.6 *Manometer.*

6.1.7 *Stopwatch.*

6.2 ORION 94-06 CYANIDE ION SELECTIVE ELECTRODE, OR EQUIVALENT.

6.3 ORION 90-01 SINGLE JUNCTION REFERENCE ELECTRODE, OR EQUIVALENT.

6.4 EXPANDED SCALE MILLIVOLT-pH METER.

6.5 ASSOCIATED LABORATORY GLASSWARE.

6.6 PLASTIC BOTTLES.

6.7 MAGNETIC STIRRER AND STIRRING BARS.

7. Reagents

7.1 PURITY. The reagents described must be ACS Reagent grade. Water is ASTM Reagent water, Type II.

7.2 SODIUM HYDROXIDE 0.1 M. Dissolve 2.0 g NaOH in water and dilute to 500 mL.

7.3 POTASSIUM CYANIDE STANDARDS.

7.3.1 Dissolve 0.65 g KCN in 0.1 M NaOH and dilute to 100 mL with additional 0.1 M NaOH for 10^{-1} M [CN$^-$] (2600 μg/mL). This solution should be checked on a weekly basis. All other dilutions should be made up fresh daily.

7.3.2 Dilute 10 mL of 10^{-1} M [CN$^-$] to 100 mL with 0.1 M NaOH for 10^{-2} M [CN$^-$] (260 μg/mL).

7.3.3 Dilute 10 mL of 10^{-2} M [CN$^-$] to 100 mL with 0.1 M NaOH for 10^{-3} M NaOH for 10^{-3} M [CN$^-$] (26 μg/mL).

7.3.4 Dilute 10 mL of 10^{-3} M [CN$^-$] to 100 mL with 0.1 M NaOH for 10^{-4} M [CN] (2.6 μg/mL).

7.3.5 Dilute 10 mL of 10^{-4} M [CN$^-$] to 100 mL with 0.1 M NaOH for 10^{-5} M [CN^{-1}] (0.26 μg/mL).

7.4 LEAD ACETATE PAPER.

7.5 CADMIUM CARBONATE.

8. Procedure

8.1 CLEANING OF EQUIPMENT. All glassware is washed in detergent solution, rinsed in tap water and then rinsed with Reagent water.

8.2 COLLECTION AND SHIPPING OF SAMPLES.

8.2.1 Pour 10 mL of the absorbing solution (0.1 M NaOH) into the midget impinger, using a graduated cylinder to measure the volume.

8.2.2 Connect the impinger (via the absorption tube) to the vacuum pump and the prefilter assembly (if needed) with a short piece of flexible tubing. The minimum amount of tubing necessary to make the joint between the prefilter and impinger should be used. The air being sampled should not be passed through any other tubing or other equipment before entering the impinger.

8.2.3 Turn on pump to begin sample collection. Care should be taken to measure the flow rate, time and/or volume as accurately as possible. The sample should be taken at a flow rate of 2.5 L/min. A sample size of not more than 200 L and no less than 10 L should be collected. The minimum volume of air sampled will allow the measurement at least 1/10 times the TLV, 0.5 mg/m^3 (101.3 kPa, 25°C).

8.2.4 After sampling, the impinger stem can be removed and cleaned. Tap the stem gently against the inside wall of the impinger bottle to recover as much of the sampling solution as possible. Wash the stem with a small amount (1 to 2 mL) of unused absorbing solution and add the wash to the impinger. Seal the impinger with a hard, non-reactive stopper (preferably Teflon). Do not seal with rubber. The stoppers on the impingers should be tightly sealed to prevent leakage during shipping. If it is preferred to ship the impingers with the stems in, the outlets of the stem should be sealed with Parafilm or other non-rubber covers, and the ground glass joints should be sealed (*i.e.*, taped) to secure the top tightly.

8.2.5 Care should be taken to minimize spillage or loss by evaporation at all times. Refrigerate samples if analysis cannot be done within a day.

8.2.6 Whenever possible, hand delivery of the samples is recommended. Otherwise, special impinger shipping cases designed by NIOSH should be used to ship the samples.

8.2.7 A "blank" impinger should be handled as the other samples (fill, seal and transport) except that no air is sampled through the impinger.

8.2.8 Where a prefilter has been used, the filter cassettes are capped and placed in an appropriate cassette shipping container. One filter disc should be handled like the other samples (seal and transport) except that no air is sampled through, and this is labeled as a blank.

8.3 ANALYSIS OF SAMPLES.

8.3.1 The solution is quantitatively transferred from the impinger to a graduated cylinder and the volume recorded. Check for the presence of sulfide, and treat if necessary, in accordance with 3.1–3.3.

The solution is made up to 25 mL with water and transferred to a 50-mL beaker.

8.3.2 The cyanide ion electrode and the single junction reference electrode are placed in the solution and the resulting mV reading recorded. The reading should be taken after the meter has stabilized. Both the samples and standards should be stirred while the readings are being taken.

9. Calibration and Standards

9.1 Pour 10 mL of each potassium cyanide standard, Sections **7.3.2** to **7.3.5**, into a graduated cylinder and make up to 25 mL with water. Transfer the standard solution in each case to 50-mL beaker and place the cyanide ion electrode and the single junction reference electrode in the solution. Obtain the mV readings from each of the cyanide standards, commencing with the weakest standard.

9.2 Plot the mV readings vs. the cyanide ion concentration of the standards on semilog paper. The cyanide ion concentration in $\mu g/mL$ is plotted on the log axis. Preferably obtain the calibration equation by the method of least squares.

10. Calculations

10.1 The mV readings from the analysis of the sample are converted to μg CN/mL of solution using the calibration curve or equation.

10.2 The cyanide concentration of the sample is multiplied by the sample volume to obtain the total μg CN in the sample.

10.3 Convert the volume of air sampled to standard conditions of 25°C and 101.3 kPa:

$$V_s = V \times \frac{P}{101.3} \times \frac{298}{T + 273}$$

where V_s = volume of air in L at 25°C and 101.3 kPa

V = volume in air in L as measured

P = barometric pressure in kPa

T = temperature of air in °C.

10.4 The concentration of CN in the air sampled can be expressed in μg CN/L or mg CN/m^3.

$$mg/m^3 = \mu g/L$$

$$mg/m^3 = \frac{\text{total } \mu g \text{ CN}}{V_s} \quad \begin{array}{l}\text{(Section } \textbf{10.2}\text{)}\\ \text{(Section } \textbf{10.3}\text{)}\end{array}$$

10.5 The concentration of CN can also be expressed in ppm, defined as μL of component per liter of air.

$$ppm = \mu L \text{ CN}/V_s = R/MW \cdot \mu g \text{ CN}/V_s$$
$$= 0.94 \times \mu g \text{ CN}/V_s$$

where R = 24.45 at 25°C, 101.3 kPa
MW = 26

11. References

1. CYANIDE ION SPECIFICATIONS. Orion Research Inc., Cambridge, Mass.
2. FRANT, M. S. 1974. Detecting Pollutants with Chemical Sensing Electrodes. *Environmental Science & Technology* 8:224.

Revised by Subcommittee 3

B. DIMITRIADES, *Chairman*
J. CUDDENBACK
E. L. KOTHNY
P. W. MCDANIEL
L. A. RIPPERTON
A. SABADELL

Approved with modifications
Subcommittee 3

D. LEVAGGI, *Chairman*
B. R. APPEL
D. W. HORSTMAN
E. L. KOTHNY
J. G. WENDT

809.

Determination of Fluorides and Hydrogen Fluoride in Air

A sample of 10–200 L of air is passed through 10 mL of 0.1 M NaOH at 2.5 L/min. The resulting solution is suitable for analysis by Method 205.

810.

Determination of Gaseous and Particulate Fluorides in Air

1. Principle of the Method

1.1 Particulate material from a measured volume of air is collected by means of a membrane filter.

1.2 Gaseous fluoride, from the same sample of air, is absorbed by an alkali-impregnated cellulose pad placed immediately behind the membrane filter.

1.3 The membrane filter and collected solids are made alkaline, ashed, the residue fused with additional alkali, and fluoride determined in a solution of the melt by use of a fluoride ion selective electrode.

1.4 Gaseous fluoride is determined in an aqueous extract of the cellulose pad, also by means of the fluoride ion selective electrode.

2. Range and Sensitivity

2.1 The method is capable of collection and analysis of both particulate and gaseous fluorides over a wide range (**1, 2, 3**). The recommended range of the method is 0.005 to 5 mg F$^-$/m^3 air. The sensitivity has not been established.

3. Interferences

3.1 Because a selective ion electrode responds to ionic activity, insoluble and complexed forms of fluoride must be released by appropriate combinations of fusion, adjustment of pH, and addition of complexing agent.

3.2 Acidity (pH) and ionic strengths of fluoride standard solutions must be matched to those of samples.

3.3 Temperature of sample and standard solutions must be controlled within ± 2°C.

4. Precision and Accuracy

4.1 Precision and accuracy of this method have not been determined.

5. Advantages and Disadvantages of the Method

5.1 Advantages include simplicity, elimination of distillation or diffusion, speed and specificity.

5.2 No significant disadvantages are known at present.

6. Apparatus

6.1 PERSONAL SAMPLING PUMP.

6.2 FILTER HOLDER. Plastic holders of the pre-loaded personal monitor type, which accept filters of 37 mm diameter, are preferred. The holder is to be numbered for identification. (Millipore Corporation Field Monitor, Cat. No. MAWP 037 AO, or equivalent. This is supplied with a membrane filter and cellulose back-up pad).

6.3 MEMBRANE FILTER OF MIXED CELLULOSE ESTERS. 0.8 μm pore size, and of diameter to fit the filter holder (**6.2**).

6.4 CELLULOSE PAD. Of size to fit the filter holder (**6.2**).

6.5 CRUCIBLES. 20 mL, of nickel or Inconel.

6.6 FLUORIDE ION SELECTIVE ELECTRODE.

6.7 REFERENCE ELECTRODE, CALOMEL TYPE, Preferably combined with the fluoride ion selective electrode. (Orion Research Inc. combination fluoride electrode, No. 96-09, or equilvalent).

6.8 AN ELECTROMETER OR EXPANDED SCALE pH METER. With a mV scale for measurement of potentials.

6.9 MAGNETIC STIRRER AND TEFLON COATED STIRRING BARS. For 50-mL plastic beakers.

6.10 PLASTIC BEAKERS, 50-mL.

6.11 REFERENCE FLOWMETER, bubble flowmeter or wet test meter.

7. Reagents

7.1 PURITY. Reagents must be ACS Reagent grade. Water is ASTM Reagent water, Type II. Polyethylene beakers and bottles should be used for holding and storing all fluoride-containing solutions.

7.2 ALKALINE FIXATIVE SOLUTION. Dissolve 25 g sodium carbonate (Na_2CO_3) in water, add 20 mL glycerol, and dilute to 1 L.

7.3 TOTAL IONIC STRENGTH ACTIVITY BUFFER (TISAB). Dissolve 37 g potassium chloride (KCl), 68 g sodium acetate ($NaC_2H_3O_2 \cdot 3 H_2O$), and 294 g sodium citrate ($Na_3C_6H_5O_7 \cdot 2 H_2O$) in water, adjust to pH 8 $\pm$ 0.1 with addition of a small amount of 1:1 hydrochloric acid, and dilute to 1 L.

7.4 STANDARD FLUORIDE SOLUTION, 100 mg/L. Dissolve 0.2211 g sodium fluoride (NaF), dried at 105°C for 2 h, in water and dilute to volume in a 1 L volumetric flask.

7.5 ETHANOL, DENATURED. Formula 30 denatured alcohol is satisfactory.

7.6 BORATE-CARBONATE FUSION MIXTURE. Intimately mix a 1:2 (w/w) combination of sodium tetraborate ($Na_2B_4O_7$) and sodium carbonate (Na_2CO_3).

8. Procedure

8.1 PREPARATION OF MONITORS. Disassemble the personal monitor, **6.2**, removing the membrane filter and cellulose pads. Moisten the pad with a measured volume of alkaline fixative solution, **7.3**; 0.8 mL is required for a pad of 37 mm diameter. Dry the pad at 105°C for 30 to 45 min. Preparation of alkali-impregnated pads must be carried out in fluoride-free environment with minimum exposure.

8.2 Reassemble the filter monitor, inserting an impregnated pad and membrane filter, and closing with the filter retaining ring and front cover. The assembly is sealed against air leakage by a wrap of masking tape covering the crevice between retaining ring and back cover. Inlet and outlet openings of the monitor are closed with plastic plugs.

8.3 CALIBRATION OF PERSONAL MONITORING PUMP.

8.3.1 Select several of the prepared monitors at random, for calibration of air flow rate with the personal sampling pump **6.1**. Connect the monitor exit to the sampling pump by means of a 75 cm (30″) length of hose. Connect the flowmeter (preferably a bubble flowmeter or wet test meter) to the inlet port of the monitor. Start the pump and adjust its rate, noting the position of the sampling pump rotameter ball when a sampling rate of 2.5 L/min is indicated by use of the calibration flowmeter and a timer.

8.4 SAMPLING.

8.4.1 A worker whose exposure is to be evaluated is equipped with a personal monitor connected by a 75 cm length of hose to a belt-supported sampling pump. The monitor is attached to the worker's collar and the cover removed. Air is drawn through the filter at the calibrated rate of 2.5 L/min and maintained at that rate by occasional checking and adjustment. On termination of sampling, the duration of sampling is noted, the monitor is resealed and returned to the laboratory. A minimum air sample of 250 L should be filtered. This will allow the measurement of 1/10 of the TLV, 0.2 mg/m^3 (101.3 kPa, 25°C).

8.4.2 Total particulate loading may be determined, if required, by pre- and post weighing of the membrane filter.

8.5 ANALYSIS OF SAMPLES.

8.5.1 *Particulate Fluoride.* Carefully remove the membrane filter from the filter holder and place it in a nickel or Inconel crucible containing about 0.5 g borate-carbonate fusion mixture. Transfer any dust from inside the filter cover and retaining ring to the crucible. Drench the filter with ethanol and ignite with a small gas flame. Heat the residue to fusion temperature for 1 to 2 min, then cool and dissolve in a few mL of water. Transfer the sample solution into a plastic beaker by means of 25 mL TISAB followed by a rinse of the crucible with a few drops of 1:1 hydrochloric acid. Adjust to pH 8 $\pm$ 0.1 by further addition of 1:1 hydrochloric acid. Dilute to

50 mL in a volumetric flask, mix, and bring to standard temperature (25°C). Pour about 20 mL of the solution into a plastic beaker and read the potential while the electrodes are immersed in the gently stirred solution. Convert potential (mV) to fluoride concentration (μg/ml) by means of the calibration chart or equation determined from the standard fluoride series containing borate-carbonate flux **9.3**.

8.5.2 *Gaseous Fluoride.* Transfer the impregnated cellulose pad to a 100-mL plastic beaker containing 25 mL water and 25 ml TISAB. Allow the pad to soak for about 30 min with sufficient stirring to reduce it to a pulp. Bring the solution to standard temperature (25°C), insert the electrodes, and read potential of the gently stirred mixture after two minutes. Convert potential (mV) to fluoride concentration (μg/ml) by means of the calibration chart or equation, determined from the standard fluoride series **9.5**.

9. Calibration and Standards

9.1 FLUORIDE STANDARDS, PARTICULATE. Add 1.0 g borate-carbonate fusion mixture to each of four 250-mL beakers containing 10 mL water and 50 mL TISAB. Add various size aliquots (0.1, 1, 5, 10, and 25 mL) of 100 mg/L standard fluoride solution to produce a series of working standards (0.1, 1, 15, 10 and 25 μg F$^-$/mL). Adjust each solution to pH 8 $\pm$ 0.1 with 1:1 hydrochloric acid, transfer to a 100-mL volumetric flask, and dilute to volume with water. These standards may be stored for several months in tightly capped polyethylene bottles, under refrigeration.

9.2 For calibration of the electrode, pour about 20 mL of the working standard solution into a plastic beaker containing a Teflon-coated stirring bar. Adjust solution to within $\pm$ 2°C of the 25°C standard temperature. Insert the fluoride and reference electrodes into the constantly stirred solution and record the potential after 2 min. Repeat for each of the working standards.

9.3 Prepare a calibration curve, on three-cycle semi-log paper, relating potential, in millivolts (linear scale), to concentration of fluoride, in μg/mL (log scale). Reproducibility of each point should be $\pm$ 1 mV. A linear calibration curve is obtained

in the 0.1 to 25 μg/mL range, with a slope of between 57 and 59 mV per tenfold change in fluoride concentration. Preferably determine the calibration equation by the method of least squares. If solutions containing less than 0.1 μg/mL are measured, additional standards must be prepared since the calibration curve is not linear at low fluoride concentrations.

9.4 FLUORIDE STANDARDS, GASEOUS. In each of four 250-mL beakers, place 10 mL water and 50 mL TISAB. Add various size aliquots (0.1, 1, 5, 10 and 25 mL) of 100 mg/L standard fluoride solution to produce a series of working standards (0.1, 1, 5, 10 and 25 μg F$^-$/mL). Adjust each solution to pH 8.0 $\pm$ 0.1 with 1:1 hydrochloric acid, transfer to a 100-mL volumetric flask, and dilute to volume with water.

9.5 A separate calibration curve or equation for gaseous fluoride is obtained from potential measurements of these standards **9.4** by the same procedure as used for particulate fluorides, **9.2** and **9.3**.

10. Calculations

10.1 The concentration of particulate fluorides in air, integrated over the sampling period, is calculated as follows:

$$C_p = 0.05 \times \frac{C_1}{V}$$

where C_p = concentration of particulate fluoride, in mg/m^3
C_1 = concentration of fluoride in particulate sample solution, in μg/mL
V = volume of air sample, in m^3, corrected to 25°C and 101.3 kPa.

10.2 The concentration of gaseous fluoride in air, integrated over the sampling period, is calculated as follows:

$$C_G = 0.05 \times \frac{C_2}{V}$$

where C_G = concentration of gaseous fluoride, in mg/m^3

C_2 = concentration of fluoride in gaseous sample solution, in $\mu g/mL$

V = volume of air sample, in m^3, corrected to 25°C and 101.3 kPa.

10.3 If desired, gaseous fluoride concentration in air, in mg/m^3, may be converted to equivalent concentration expressed as parts per million: C_G mg/m^3 $\times$ 1.29 = C_G, ppm

11. References

1. ELFERS, L. A. AND C. E. DECKER. 1968. Determination of Fluoride in Air and Stack Gas Samples by Use of an Ion Specific Electrode. *Anal. Chem.* 40:1658.
2. ASTM METHOD D3269 Standard Methods for Analysis for Fluoride Content of the Atmosphere and Plant Tissues.
3. WEINSTEIN, L. H. and R. H. MANDL. 1971. The Separation and Collection of Gaseous and Particulate Fluorides. *VDI Berichte Nr.* 164, 53–63.

Revised by Subcommittee 2

C. R. THOMPSON, *Chairman*
G. H. FARRAH
L. V. HAFF
A. W. HOOK
J. S. JACOBSON
E. J. SCHNEIDER
L. H. WEINSTEIN

Approved with modification
Subcommittee 2

R. H. MANDL
L. H. WEINSTEIN
J. S. JACOBSON

811.

Determination of Fluoride in Urine

Urine samples stabilized by the addition of about 0.2 g/100 mL of EDTA disodium salt may be analyzed directly by the procedure of Method 205.

812.

Determination of Hydrogen Sulfide in Air

See **Method 701.**

815.

Determination of Mercury in Urine

1. Principle of the Method

1.1 After initial decomposition of the urine sample with nitric acid, the mercury is reduced to its elemental state with stannous chloride.

1.2 Air is bubbled through the resulting solution, carrying the mercury vapor through an absorption cell of an atomic absorption spectrometer and the response at 253.7 nm is measured either as peak height or peak area.

1.3 The mercury content of the sample is calculated from calibration data for the system, obtained by carrying known amounts of ionic mercury standards through the analysis procedure.

2. Range and Sensitivity

2.1 For a 1.0 mL urine sample the sensitivity is 0.003 mg/L or below. This corresponds to an absolute sensitivity of 3 ng of mercury.

2.2 The range extends up to 0.3 mg/L for a 1.0 mL aliquot. The range can be extended beyond 0.3 mg/L by taking an aliquot of urine that is less than 1.0 mL.

3. Interferences

3.1 Metals such as gold, platinum and copper will interfere since they form an alloy with the reduced Hg.

3.2 Certain organic solvents, such as benzene, that absorb 253.7 nm radiation would interfere if present in significant amounts as would any substance with broad absorption such as dust, water droplets, etc.

4. Precision and Accuracy

4.1 Comparisons of bubbler recoveries from water standards and urine samples spiked with identical amounts of radioactive or stable mercury indicate that the urine samples give a 10% lower signal (Figure 815:1). Although the use of water standards may reduce the accuracy of the method, results may be acceptable for routine analyses. Where more accurate values are needed, a multiple addition technique using the same urine spiked with Hg standard is recommended.

4.2 Absolute deviations from the means between sets of identical samples varied from 0.000 to 0.005 mg/L. The average relative standard deviation (1σ) was ± 3%.

4.3 The temperature of the water used to dilute the sample should be controlled to within ± 1°C.

4.4 At least 5 mL of nitric acid per mL sample should be used to ensure the release of metabolized mercury.

4.5 Only aliquots of urine less than 1 mL should be used if water standards are used. *Larger* aliquots of urine would increase the difference in transfer efficiency between water standards and urine (Figure 815:1).

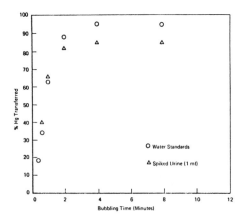

Figure 815:1–Transfer efficiency in the bubbler system.

5. Advantages and Disadvantages of the Method

5.1 A trained technician can analyze 20 to 40 samples a day.

6. Apparatus

6.1 MERCURY EVOLUTION TRAIN.
    6.1.1 *Bubbler Flask.* 300-mL BOD sample bottle with coarse fritted bubbler extending to within 0.3 cm of bottom of bottle.
    6.1.2 *Connecting Tubing.* Minimum lengths of glass.
    6.1.3 *Rotameter*, 0.5 to 3 L/min range.
    6.1.4 *Drying Tube.* Approximately 8-cm length (optional).
    6.1.5 *Midget Impinger.* To act as safety trap.
6.2 ANALYTICAL EQUIPMENT.
    6.2.1 Atomic absorption spectrometer for mercury measurement, as described in Method 317, Sections **5.2.1–5.2.3**
    6.2.2 *Pipets.*

7. Reagents

7.1 All reagents must meet ACS specifications for Analytical Reagents. Water is ASTM Reagent water, Type II.
7.2 NITRIC ACID, CONC.
7.3 STANNOUS CHLORIDE SOLUTION, 20% (w/v) in 6 N HCl. Freshly prepared.
7.4 STANDARD STOCK MERCURY SOLUTION. Add 0.1000 g metallic mercury or 0.1713 g mercuric nitrate. $Hg(NO_3)_2 \cdot H_2O$, into a clean, dry 100-mL volumetric flask. Add 10 mL concentrated HNO_3, dissolve the mercury and then dilute to the mark with water. Pipet exactly 10.0 mL of this solution into a 1-L volumetric flask. Add 50 mL concentrated HNO_3 and dilute to the mark with water. The final concentration is 10 μg/mL. This solution is stable for 6 months.
7.5 WORKING STANDARD MERCURY SOLUTION. Transfer 10.0 mL of the standard stock solution to a 1-L volumetric flask. Add 50 mL concentrated HNO_3 and dilute to the mark with water. The final concentration is 0.1 μg/mL. This solution should be prepared fresh daily.

7.6 POTASSIUM PERSULFATE. Low nitrogen.
7.7 MAGNESIUM PERCHLORATE, ANHYDROUS (FOR DRYING TUBE). The drying tube must be repacked with fresh magnesium perchlorate after 20 analyses.
7.8 ANTIFOAM SOLUTION (**1**). Suspend 5 g of General Electric "Antifoam 60" in 95 mL of water. Antifoam 60 is available from General Electric Company, Silicone Products Division, Waterford, New York.

8. Procedure

8.1 CLEANING OF EQUIPMENT. Acid-clean all glassware before use. This can be done by using the following procedure.
    8.1.1 Wash in 1% Na_3PO_4 solution and follow with tap water.
    8.1.2 Soak in conc HNO_3 for 30 min and follow with tap, distilled and Reagent water rinses.
8.2 COLLECTION AND SHIPPING OF SAMPLES.
    8.2.1 Urine samples must be collected in acid-cleaned borosilicate bottles. At least 25 mL should be collected.
    8.2.2 The samples must be preserved by the addition of 0.1 g potassium persulfate per 100 mL of urine. Urine treated with potassium persulfate is stable at room temperature for 2 weeks (**2**).
8.3 ANALYSIS OF SAMPLES.
    8.3.1 Transfer a 1.0 mL urine sample to the BOD bottle. If a sediment is present in the collection bottle, shake well immediately before removing the aliquot. Add 5 mL of conc HNO_3 to the BOD bottle and allow the sample to stand at room temperature for 3 min.
    8.3.2 Add 95 mL of distilled water (at the same temperature) blowing out any mist that may form above the liquid with gentle puffs of air or nitrogen. Add 2 drops antifoam solution. Add 5.0 mL of $SnCl_2$ solution and then immediately connect the flask to the bubbler tube in the generating train, shown in Figure 815:2. The samples and standards should be at the same temperature ($\pm$ 1°C) since variations in water temperatures produce variations in peak height of nearly 1%/°C (**1**).
    8.3.3 Turn on pump and blow air through the flask at a constant rate. Recommended air flow is between 0.5 to 3 L/

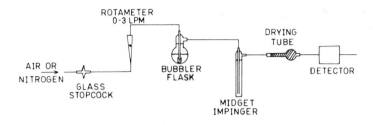

Figure 815:2 — Schematic diagram of test apparatus.

min to obtain maximum peaks. Optimum rate of flow must be determined experimentally and thereafter held constant.

8.3.4 Record the peak heights or areas. If the reading goes beyond calibration range, repeat the analysis using a smaller sample volume or if too low take larger quality up to 5 mL of urine.

8.3.5 Remove the BOD bottle and allow the signal to return to zero. Analyze each sample in duplicate using two different volumes of urine.

9. Calibration and Standards

9.1 Standards for instrument calibration are analyzed exactly as samples except that known amounts of mercury are added to the sample.

9.2 The calibration curve for the multiple addition method is constructed by subtracting the reading of the unspiked urine from that of the spiked aliquots.

9.3 Prepare a calibration curve by plotting maximum peak height or area versus μg mercury added. Standards of 0.0, 0.05, 0.1, 0.2, 0.3 and 0.4 μg should be plotted, or preferably a calibration equation obtained by the method of least squares.

9.4 Run a calibration before analysis of samples, then insert a standard after every 5th or 6th sample to maintain a check on method integrity.

10. Calculations

10.1 Subtract the background signal, if any, from the sample signal.

10.2 Micrograms of mercury are determined directly from the calibration curve or regression equation.

10.3 The mercury concentration is calculated by dividing the μg of mercury by the sample volume analyzed to give μg Hg per liter.

10.4 The mercury level may be corrected to a standard urine specific gravity (S.G.) of 1.024 by use of the factor

$$\frac{(1.024 - 1.000)}{(S.G. - 1.000)}$$

11. References

1. RATHJE, ARNOLD O. 1969. A Rapid Ultraviolet Absorption Method for the Determination of Mercury in Urine. *Am. Ind. Hyg. Assoc. J.* 30:126–132.
2. TRUJILLO, P., STEIN, P. AND CAMPBELL, E. 1974. The Preservation and Storage of Urine Samples for the Determination of Mercury. *Am. Ind. Hyg. Assoc. J.* 35:257–261.

Revised by Subcommittee 6

T. J. KNEIP, *Chairman*
R.S. AJEMIAN
J. N. DRISCOLL
F. I. GRUNDER
L. KORNREICH
J. W. LOVELAND
J. L. MOYERS
R. J. THOMPSON

Approved with modifications
Subcommittee 6

M. T. KLEINMAN, *Chairman*
J. T. RHODES
R. J. THOMPSON
W. J. COURTNEY
T. C. RAINS
V. P. GUINN

816.

Determination of Nitric Oxide in Air

See **Method 405.**

817.
Determination of Nitrogen Dioxide in Air

See **Method 406.**

818.

Determination of Nitrogen Dioxide in Air

1. Principle of the Method

1.1 Nitrogen dioxide is absorbed from the air by a solid absorber containing triethanolamine (TEA). Subsequent analysis is performed using an azo-dye forming reagent (1, 2, 6) The color produced by the reagent is measured in a spectrophotometer at 540 nm.

2. Range and Sensitivity

2.1 The atmospheric concentration for which this method may be used with confidence is 10 to 1000 $\mu g/m^3$ (0.005 to 0.50 ppm). Method performance at NO_2 levels between 1000 $\mu g/m^3$ (0.5 ppm) and 6200 $\mu g/m^3$ (3.1 ppm) has not yet been established (2, 6).

2.2 The sensitivity of the method is dependent on that of the Griess-Saltzman reagent. For a 1 cm sample cell path and 0.1 absorbance units this is equivalent to 0.14 $\mu g/mL$ of NO_2 in the absorbing solution.

3. Interferences

3.1 Sulfur dioxide in concentrations up to 250 $\mu g/m^3$ (0.13 ppm) does not interfere and negative interference caused by concentrations of up to 3000 $\mu g/m^3$ (1.6 ppm) is about 20% of the collected NO_2, if hydrogen peroxide is added after sampling and before color development (as described in Section 8.2). Ozone causes no interference at atmospheric concentration ranges (up to 1000 $\mu g/m^3$). Nitric oxide at concentrations up to 800 $\mu g/m^3$ (0.6 ppm) as an 8-h average causes no interference (2). The interference of TLV levels of SO_2 and nitric oxide have not been established.

3.2 Organic nitrites and peroxyacyl nitrate (PAN) which might be present in the air would produce a positive interference. There are no data on the magnitude of the interference; however, in view of the average concentrations of organic nitrites and PAN that have been reported in the literature, it would appear that the interference would be negligible (2).

4. Precision and Accuracy

4.1 The precision of the method in the field, expressed as standard deviation, was ± 4 $\mu g/m^3$ (± 0.002 ppm) on a 4-day sampling for a mean of 95 $\mu g/m^3$ (0.050 ppm) as compared with a continuous colorimetric analyzer (2). The precision at higher levels has not been established. The TLV is 5.0 ppm.

4.2 At present, accuracy data are not available. However, the method is comparable to that of continuous air monitoring instrumentation using the modified Griess-Saltzman reagent (2, 3). The NO_2 (gas) to NO_2^- (ion) factor was found to average 0.85 in 25 tests, evaluated against the Saltzman factor of 0.72 (2), for ambient air sampling.

4.3 Absorption efficiency is over 95%, using either the liquid or solid absorbers (2, 6), in the concentration range up to 0.5 ppm.

5. Advantages and Disadvantages of the Method

5.1 The method is simple and samples collected in the field can be stored without loss for up to 4 weeks.

5.2 The samples will not spill.

5.3 The disadvantage of the method is the extracting step in the central laboratory.

6. Apparatus

6.1 SPECTROPHOTOMETER, UTILIZING 10-MM CELLS.

6.2 OBSERVATION TUBE. Becton-Dickinson No. 420 LST, 55-cm × 5-mm glass, with a standard Luer lock fitting permanently attached to one end.

6.3 AIR METERING DEVICE. A rotameter capable of measuring flow rates up to 0.4 L/min, calibrated against a wet test meter, is suitable. Calibrated orifices, such as hypodermic needles can also be used for controlling the flow. In such case, a minimum pressure differential of 60 kPa across the needle is required. A membrane filter should be placed in front of the hypodermic needle to prevent water droplets from adhering to the needle, causing alteration of the gas flow.

6.4 SAMPLING LINES. Sampling lines should be constructed of Teflon, glass or stainless steel.

6.5 AIR PUMP. A vacuum pump capable of sampling air through the absorber at the required rate of 0.2 L/min for 8 h is suitable. The pump should be equipped with a needle valve for accurate control of sampling rate, unless an orifice is used.

6.5.1 If a critical orifice is used as a flow control device, the pump should be also capable of maintaining a minimum vacuum of 60 kPa across the orifice at the required rate of 0.2 L/min.

7. Reagents

7.1 PURITY. Reagents should be ACS analytical reagent grade. Water should be free of nitrite and conform to ASTM standards for Reagent water, Type II.

7.2 LIQUID ABSORBER. Add 15 g of triethanolamine to approximately 500 mL of water; than add 3 mL of 1-butanol. Mix well and dilute to 1 L with water. The 1-butanol acts as a surfactant. The reagent is stable for 2 months if kept in a brown bottle, preferably in the refrigerator.

7.3 SOLID ABSORBER. Place 25 g of triethanolamine in a 250-mL beaker, add 4.0 g of glycerol, 50 mL of acetone, and sufficient water to dissolve. Dilute to 100-mL total volume with water. To the mixture add about 50 mL of 12 to 30 mesh molecular sieve 13 X. Stir and let stand in the covered beaker for about 30 min. Decant the excess liquid and transfer the molecular sieve to a porcelain pan. Place under a heat-

ing lamp until the bulk of moisture has evaporated, and then in an oven at 110°C for 1 h until dry. Store in a closed glass container.

7.4 HYDROGEN PEROXIDE. Dilute 0.2 mL of 30% hydrogen peroxide to 250 mL with water. **Caution:** Hydrogen peroxide is a strong oxidant and will damage skin and clothing.

7.5 SULFANILAMIDE SOLUTION. Dissolve 10 g of sulfanilamide in 400 mL of water, then add 25 mL of conc (85%) phosphoric acid. Make up to 500 mL. This solution is stable for several months if stored in a brown bottle and up to 1 year in the refrigerator.

7.6 N-(1-NAPHTHYL)-ETHYLENEDIAMINE DIHYDROCHLORIDE (NEDA). Dissolve 0.1 g of NEDA in 100 mL of water. The solution is stable for a month, if kept in a brown bottle in the refrigerator.

7.7 NITRITE STOCK SOLUTION. Dissolve 0.135 g of sodium nitrite in distilled water and make up to 1 L. The sodium nitrite should be of known purity or analyzed before use. The solution contains 90 μg of nitrite per mL which corresponds to 100 μg NO_2 (gas) per mL (sec. **9.2**). The solution should be made up once a month.

8. Procedure

8.1 SAMPLING. Fill an observation tube with the solid absorber, using small glasswool plugs for retainment of the solid. The sampling train consists of (in tandem) a sampling probe, observation tube, a critical orifice (27 gauge hypodermic needle) and a pump having a capacity of at least 60 kPa vacuum. Turn pump on and rapidly check flow rate entering system and record. This flow rate should be between 150 and 200 mL/min. Sample for 8 h. At the end of the sampling period recheck flow rate and record.

8.2 COLOR DEVELOPMENT. Transfer the molecular sieve and glass wool plugs to a 50-mL test tube. Wash the observation tube with approximately 10 mL of water and add washings to the molecular sieve. The volume occupied by the molecular sieve and glass wool is so small (1 %) it may be ignored. Dilute to 50 mL with liquid absorber. Cap and shake vigorously for about 1 min. Let stand and shake again after 10 min. Allow the solids to settle and transfer 10 mL to a 25-mL graduated cylinder.

Similarly, treat an unexposed portion of solid absorber as a blank as directed above. To each cylinder add 1.0 mL of dilute hydrogen peroxide solution and mix well. Then add 10 mL of the NEDA solution to sample and blank, mix well and allow color to develop for 10 min. Measure the absorbance of the sample solution against the blank at 540 nm. The NO_2 is then determined from the calibration curve.

9. Calibration and Standardization

9.1 A calibration curve may be obtained by one of two different methods. The simplest and most convenient method is standardization with nitrite solutions. However, accurately known gas mixtures would provide more realistic standards for calibration of the entire analytical system. The known gas mixtures may be prepared by use of a flow dilution system (4) or permeation tubes (5).

9.2 STANDARDIZATION WITH NITRITE SOLUTION. Transfer 1.0 mL of the nitrite stock solution into a 50-mL volumetric flask and dilute to the mark with absorbing solution. The working solution so obtained contains 1.8 μg NO^-_2/mL (corresponding to 2 μg NO_2 (gas) per ml). To a series of 25-mL graduated cylinders introduce 0, 1, 3, 5, and 7 mL of the nitrite working solution. Dilute to 10 mL with absorbing solution. From this point treat the standards as described in **8.2**. A graph of absorbance versus NO_2 concentration is constructed or, preferably a calibration equation is obtained by the method of least squares. Weekly checks should be made to determine if the response of the reagents has changed.

10. Calculations

10.1 The air sampling volume is corrected to 25°C and 101.3 kPa. The normal deviations from these conditions add only small corrections.

10.2 Standardization by the nitrite solution requires use of an empirical factor that relates μg of NO_2 (gas) to μg of nitrite ion. The conversion factor in this method is 0.90, *i.e.* 90% of the NO_2 in the air sample is eventually converted to nitrite ion which reacts with the color producing azo-dye reagent (2, 3). This factor has been incorporated in the nitrite stock solution (Section **7.7**).

10.3 Computation of the average concentration of NO_2 in air for the 8-h sample period may be made as follows:

Read the μg NO_2 directly from the graph, or calculate from the calibration equation.

$$c = \frac{\mu g\ NO_2 \times 10^3}{r \times t \times k}\ \mu g\ NO_2/m^3$$

$$p = \frac{\mu g\ NO_2 \times 0.532}{r \times t \times k} = \frac{c \times 0.532}{10^3}\ ppm$$

where c = concentration in μg NO_2/m^3
p = concentration in ppm NO_2 $(\mu L/L)$
r = sampling rate in L/min
t = sampling time in min
k = dilution factor (i.e. 0.2 if one fifth of the sample is analyzed)
0.532 = μl $NO_2/\mu g$ NO_2 at 25°C and 101.3 kPa
10^3 = conversion factor, L/m^3

11. References

1. SALTZMAN, B.E. 1954. Colorimetric microdetermination of nitrogen dioxide in the atmosphere. *Anal. Chem.* 26:1949.
2. LEVAGGI, D.A., W. SIU, AND M. FELDSTEIN. 1973. A new method for measuring average 24-hour nitrogen dioxide concentrations in the atmosphere. *J. Air Poll. Cont. Assoc.* 23:30.
3. SCARINGELLI, F.P., E. ROSENBURG, AND K.A. REHME. 1970. Comparison of permeation tubes and nitrite ions as standards for the colorimetric determination of nitrogen dioxide. *Env. Sci. Tech.* 4:924.
4. SALTZMAN, B.E. 1961. Preparation and analysis of calibrated low concentrations of 16 toxic gases. *Anal. Chem.* 33:1100.
5. SALTZMAN, B.E., W.R. BURG, AND G. RAMASWAMY. 1971. Performance of permeation tubes as standard gas sources. *Env. Sci. Tech.* 5:1121.

6. BLACKER, J.H. 1973. Triethanolamine for collecting NO_2 in the TLV range. *Amer. Ind. Hyg. Assoc. J.* 34:390.

Revised by Subcommittee 3

B. DIMITRIADES, *Chairman*
J. CUDDEBACK
E.L. KOTHNY
P.W. MCDANIEL
L.A. RIPPERTON
A. SABADELL

Approved with modification
Subcommittee 3

D. LEVAGGI, *Chairman*
B.R. APPEL
D.W. HORSTMAN
E.L. KOTHNY
J.G. WENDT

819.

Determination of Ozone in Air

See **Method 411.**

821.

Determination of Phosgene in Air

1. Principle of the Method

1.1 4-(4'-nitrobenzyl) pyridine in diethyl phthalate reacts with traces of phosgene to produce a brilliant red color. The addition of an acid acceptor such as N-phenylbenzylamine stabilizes the color and increases the sensitivity. The absorbance is determined at 475 nm (1–3).

2. Range and Sensitivity

2.1 Sampling efficiency is 99% or better. Five μg of phosgene can be detected; the minimum sample size is 25 L. Therefore, 0.02 mg/m^3 phosgene should be detectable with a 250-L sample. A 20 mg/m^3 upper limit should be within the range of most laboratory photometers for one 25-L air sample.

3. Interferences

3.1 Other acid chlorides, alkyl and aryl derivatives which are substituted by active halogen atoms, and sulfate esters, are known to produce color with this reagent. However, most of these interferences can be removed in a prescrubber containing an inert solvent such as "Freon-113" cooled by an ice bath.

3.2 This method is not subject to interference from likely concentrations of chloride, hydrogen chloride, chlorine dioxide or simple chlorinated hydrocarbons such as carbon tetrachloride, chloroform, and tetrachlorethylene. A slight depression of color density has been noted under high humidity conditions.

4. Precision and Accuracy

4.1 The accuracy and precision of this method has not been determined. Calibration is a major problem because of the reactivity of phosgene. Flow dilution systems can be used but require anhydrous air for dilution. Permeation-type calibration standards may be employed, if available.

5. Advantages and Disadvantages of the Method

5.1 Advantages include high sensitivity, standard laboratory equipment, and relative simplicity.

5.2 Disadvantages include potential interferences, relative changes in color formation with various lots of reagents and the need for frequent calibration checks.

6. Apparatus

6.1 MIDGET IMPINGERS.

6.2 VACUUM SOURCE, Aspirator or pump

6.3 FLOW METER (Wet test meter)

6.4 POLYETHYLENE BOTTLES AND BEAKERS

6.5 BAUSCH AND LOMB SPEC. 20, OR EQUIVALENT PHOTOMETER.

6.6 ASSOCIATED LABORATORY GLASSWARE AND TUBING. No plasticized tubing (such as Tygon) should be used.

7. Reagents

7.1 4-(4'-NITROBENZYL) PYRIDINE. (Aromil Chemical Co., Baltimore, Md., or Aldrich Chemical Co., Milwaukee, Wis.)

7.2 N-PHENYLBENZYLAMINE. (ACS reagent grade or better).

7.3 DIETHYL PHTHALATE. (Special selection may be required since some lots of the phthalate produce unstable color).

7.4 COLOR REAGENT. Reagent is made up of a solution of 2.5 g 4-(4'-nitrobenzyl)-pyridine, 5 g N-phenylbenzylamine and 992.5g diethylphthalate. The concentration of color forming reagents in diethylphthalate is critical as less color is developed with either a decreased or increased concentration.

8. Procedure.

8.1 CLEANING OF EQUIPMENT. All glassware and plastic ware are to be doubly rinsed with 2-propanol (IPA) and dried. If IPA rinse(s) does not thoroughly clean apparatus, use standard glassware cleaning procedures before IPA rinses.

8.2 COLLECTION OF SAMPLES.

8.2.1 Atmospheric samples are collected in midget impingers loaded with 10 mL of reagent prepared as described in Section **7.4** above.

8.2.2 A calibrated sampling rate of 1 L/min is used.

8.2.3 A known volume of about 50 L is drawn through the impingers for a range of 0.04 to 1 ppm phosgene. (For ranges above 1 ppm, 25 L should be drawn.)

8.2.4 The final solution volume is recorded and the sample transferred to a small polyethylene or glass bottle or directly to the cuvette.

8.2.5 The red color formed is stable for at least 4 h, but should be measured within 9 h as a 10% to 15% loss in color density after 8 h has been reported.

8.2.6 If the presence of interfering substances is suspected, add a "pre-scrubber" as described in Section **3.1**.

8.3 ANALYSIS OF SAMPLE.

8.3.1 The sample is transferred to a cuvette for the photometric measurement at 475 nm.

8.3.2 The photometer should be "zeroed" (at 100% T or 0 absorbance) with a matched cuvette containing unreacted reagent at the 475 nm wavelength.

8.3.3 The absorbance (or transmittance) of the sample is measured with the photometer at the 475 nm wavelength.

9. Calibration and Standards

9.1 Phosgene is highly reactive, and standards are difficult to prepare. However, dynamic standards may be prepared from a flow dilution system using a standard cylinder of 1000 ppm phosgene in nitrogen. (Lower concentrations of phosgene in cylinders are generally unstable. In any case, the "standard" cylinder concentration should be checked periodically by mass spectrometric or other methods.) A double dilution system with $COCl_2$-free air to dilute the phosgene about 1000:1 should be used. Permeation-type calibration techniques can be employed if phosgene standard permeation tubes are available.

10. Calculations

10.1 A calibration curve or regression equation should be prepared of absorbance at 475 nm versus the phosgene concentration in air with 50 L of gas sample.

10.2 For a gas sample of volume X, the $COCl_2$ concentration as determined by the calibration curve should be multiplied by

$$\frac{50}{X} \text{ for the corrected value}$$

11. References

1. AMERICAN INDUSTRIAL HYGIENE ASSOC. 1969. Akron, Ohio. Analytical Guides.
2. LINCH, A.L., S.S. FORD, K.A. KUBITZ, AND M.R. DEBRUNNER. 1965. Phosgene in Air—Development of Improved Detector Procedures, *Amer. Ind. Hyg. Assoc. J.* 26:465.
3. NOWEIR, M.H., AND E.A. PFITZER. 1964. The Determination of Phosgene in Air. Unpublished paper presented at the Amer. Ind. Hyg. Assoc. Conference in Philadelphia, Pennsylvania April 28.

Subcommittee 2

C.R. THOMPSON, *Chairman*
G.H. FARRAH
L.V. HAFF
A.W. HOOK
J.S. JACOBSON
E.J. SCHNEIDER
L.H. WEINSTEIN

Approved with modifications
Subcommittee 2

R. A. MANDL, *Chairman*
L.H. WEINSTEIN
J.S. JACOBSON

822.

General Atomic Absorption Procedure for Trace Metals in Airborne Material Collected on Filters

1. Principle of Method

1.1 This procedure describes a general method for the collection, dissolution and determination of trace metals in industrial and ambient airborne material. The samples are collected on filters and treated with nitric acid to oxidize the organic matrix and to dissolve the metals present in the sample. The analysis is subsequently made by flame atomic absorption spectrophotometry (AAS).

1.2 Samples and standards are aspirated into the appropriate AAS flame. A hollow cathode lamp for the metal being measured provides a source of characteristic radiation energy for that particular metal. The absorption of this characteristic energy by the atoms of interest in the flame is related to the concentration of the metal in the aspirated sample. The flames and operating conditions for each element are listed in Table 822:I.

1.3 This method can also be applied to non-flame electrothermal atomization atomic absorption spectrophotometry (ETA). ETA can be extremely valuable for measuring low solution concentration levels, in the range of 0.001 to 0.5 $\mu g/mL$ if matrix interferences and high blank levels are not a problem. To overcome these problems, several different strategies have been developed by the various instrument manufacturers. Detailed discussion of these strategies is beyond the scope of this method.

2. Range and Sensitivity

2.1 The sensitivity, detection limit and optimum working range for each metal are given in Table 822:II. The sensitivity is defined as that concentration of a given element which will absorb 1% of the incident radiation (0.0044 absorbance units) when aspirated into the flame. The detection limit is defined as that concentration of a given element which produces a signal equivalent to three times the standard deviation of the blank signal (Note: The blank signal is defined as that signal that results from all added reagents and a clean filter which has been ashed exactly as the samples.) The working range for an analytical precision better than 3% is generally defined as those sample concentrations that will absorb 10% to 70% of the incident radiation (0.05 to 0.52 absorbance units). The values for the sensitivity and detection limits are instrument-dependent and may vary from instrument to instrument. Higher blank levels and, hence, worsened sensitivities may be found for other filter matrices. The values in Table 822:II should be taken only as a guide.

3. Interferences

3.1 In atomic absorption spectrophotometry the occurrence of interferences is less common than in many other analytical determination methods. Interferences can occur, however, and when encountered are corrected for as indicated in the following sections. The known interferences and correction methods for each metal are indicated in Table 822:I. The methods of standard additions and background monitoring and correction (1–4) are used to identify the presence of an interference problem. Insofar as possible the matrix of the samples and standards are matched to minimize the possible interference problems.

3.1.1 Background or non-specific absorption can be caused by particles produced in the flame that can scatter the incident radiation, causing an apparent

Table 822:I Flame and Operating Condition for Metals

Element	Type of Flame	Analytical Wavelength nm	Interferences†	Remedy†	References
Ag	Air-C_2H_2 (oxidizing)	328.1	IO_3^-, WO_4^{-2}, MnO_4^{-2}	‡	(5)
Al*	N_2O-C_2H_2 (reducing)	309.3	Ionization, SO_4^{-2}, V	‡,§,‖	(4)
Ba	N_2O-C_2H_2 (reducing)	553.6	Ionization, large conc. of Ca	§, #	(1, 4)
Be*	N_2O-C_2H_2 (reducing)	234.9	Al, Si, Mn	‡	(4)
Bi	Air-C_2H_2 (oxidizing)	223.1	None known		
Ca	Air-C_2H_2 (reducing) N_2O-C_2H_2	422.7	Ionization & chemical	§, ‖	(1, 4)
Cd	Air-C_2H_2 (oxidizing)	228.8	None known		
Co*	Air-C_2H_2 (oxidizing)	240.7	None known		
Cr*	Air-C_2H_2 (oxidizing)	357.9	Fe, Ni	‡	(4)
Cu	Air-C_2H_2 (oxidizing)	324.8	None known		
Fe	Air-C_2H_2 (oxidizing)	248.3	High Ni conc., Si	‡	(1, 4)
In	Air-C_2H_2 (oxidizing)	303.9	Al, Mg, Cu, Zn, $H_xPO_4^{x-3}$	‡	(10)
K	Air-C_2H_2 (oxidizing)	766.5	Ionization	§	(1, 4)
Li	Air-C_2H_2 (oxidizing)	670.8	Ionization	§	(10)
Mg	Air-C_2H_2 (oxidizing) N_2O-C_2H_2 (oxidizing)	285.2	Ionization & chemical	§, **	(1, 4)
Mn	Air-C_2H_2 (oxidizing)	279.5	None known		
Na	Air-C_2H_2 (oxidizing)	589.6	Ionization	**	(1, 4)
Ni	Air-C_2H_2 (oxidizing)	232.0	None known		
Pb	Air-C_2H_2 (oxidizing)	217.0 283.3	Ca, high conc. SO_4^{-2}	‡	(7)
Rb	Air-C_2H_2 (oxidizing)	780.0	Ionization	§	(1, 8)
Sr	Air-C_2H_2 (reducing) N_2O-C_2H_2 (reducing)	460.7	Ionization & chemical	§, **	(1, 8)
Tl	Air-C_2H_2 (oxidizing)	276.8	None known		
V*	N_2O-C_2H_2 (reducing)	318.4	None known in N_2O-C_2H_2 flame		
Zn	Air-C_2H_2 (oxidizing)	213.9	None known		

*Some compounds of these elements will not be dissolved by the procedure described here. When determining these elements one should verify that the types of compounds suspected in the sample will dissolve using this procedure. (See Section 3.2)

†High concentrations of Si in the sample can cause an interference for many of the elements in this table and may cause aspiration problems. No matter what elements are being measured, if large amounts of silica are extracted from the samples, the samples should be allowed to stand for several hours and centrifuged or filtered to remove the silica.

‡Samples are periodically analyzed by the method of additions to check for chemical interferences. If interferences are encountered, determinations must be made by the standard additions method or, if the interferent is identified, it may be added to the standards.

§Ionization interferences are controlled by bringing all solutions to 1000 ppm Cs (samples and standards).

**1000 ppm solution of La as a releasing agent is added to all samples and standards.

#In the presence of very large Ca concentrations (greater than 0.1%) a molecular absorption from $Ca(OH)_2$ may be observed. This interference may be overcome by using background correction when analyzing for Ba.

absorption signal. Light scattering problems may be encountered when solutions of high salt content are being analyzed. Light scattering problems are most severe when measurements are made at the lower wavelengths (i.e., below about 250 nm). Background absorption may also occur as the result of the formation of various molecu-

lar species that can absorb light. The background absorption can be accounted for by the use of background correction techniques (1).

3.1.2 Spectral interferences are those interferences that occur as the result of an atom different from that being measured absorbing a portion of the incident radia-

Table 822:II Sensitivity Detection Limit, and Optimum Working Range of Metals

Element	Sensitivity* $\mu g/mL$	Range $\mu g/mL$	Detection Limits* $\mu g/mL$	Detection Limits* $\mu g/m^3$	Minimum‡ TLV $\mu g/m^3$
Ag	0.04	0.5–5.0	.003	0.1	10 (metal and soluble compounds)
Al	0.40	5–50	.04	2	1,000 (metal and oxides)
Ba	0.20	1–10	.01	0.4	500 (soluble compounds)
Be	0.01	0.1–1.0	.002	0.08	2
Bi	0.20	1–10	.07	3	10,000 (as bismuth telluride)
Ca	0.02	0.1–1.0	.002	0.03	2,000 (Ca0)
Cd	0.01	0.1–1.0	.002	0.03	50 (metal dust & soluble salts)
Co	0.96	0.5–5.0	.005	0.3	50 (metal fume and dust)
Cr	0.05	0.5–5.0	.003	0.2	50 (water soluble Cr (VI) compounds)
Cu	0.03	0.5–5.0	.004	0.1	200 (fume)
					1,000 (dusts and mists)
Fe	0.05	0.5–5.0	.003	0.2	10,000 (iron oxide fume, as iron oxide)
					1,000 (soluble compounds)
K	0.02	0.1–1.0	.003	0.1	2,000 (as potassium hydroxide)
Li	0.02	0.1–1.0	.002	0.08	24 (as lithium hydride)
Mg	0.004	.05–.50	.0002	0.01	10,000 (as magnesium oxide fume)
Mn	0.026	0.5–5.0	.003	0.1	5,000 (metal and compounds)
Na	0.005	.05–.05	.0002	0.01	2,000 (as sodium hydroxide)
Ni	0.05	0.5–5.0	.007	0.3	100 (soluble compounds)
Pb	0.10	1–10	.01	0.8	159 (inorganic compounds, fumes & dusts)
Rb	0.02	0.5–5.0	.003	0.1	NL§
Sr	0.06	0.5–5.0	.00	0.2	NL§
Tl	0.23	5–50	.03	0.8	100 (soluble compounds)
V	0.96	10–100	0.5	4	50 (V_2O_5 dust and fume)
Zn	0.007	0.1–1.0	.002	0.08	1,000 ($ZnCl_2$ fume)
					5,000 (ZnO fume)

*Sensitivity and solution detection limits are taken from Reference 10. The atmospheric concentrations were calculated assuming a collection volume of 0.24 m^3 (2 L/min for 2 h) and an analyte volume of 10 mL for the entire sample. (NOTE: These detection limits represent the ultimate values since the blanks resulting from the reagents and the filter material have not been taken into account.)

‡Threshold limit values of airborne contaminants and physical agents with intended changes adopted by ACGIH for 1987–1988. All values listed are expressed as elemental concentrations except as noted.

§NL signifies No Limit expressed for this element or its compounds.

tion. Such interferences are extremely rare in atomic absorption. In some cases multi-element hollow cathode lamps may cause a spectral interference by having closely adjacent radiation lines from two different elements. In such instances multi-element hollow cathode lamps should not be used.

3.1.3 Ionization interferences can occur when easily ionized atoms are being measured. The degree to which such atoms are ionized is dependent upon the atomic concentration and presence of other easily ionized atoms in the sample. Ionization interferences can be controlled by the addition to the sample of a high concentration of another easily ionized element that will buffer the electron concentration in the flame.

3.1.4 Chemical interferences occur in atomic absorption spectrophotometry when species present in the sample cause variations in the degree to which atoms are formed in the flame. Such interferences may be corrected for by controlling the sample matrix or by using the method of standard additions (2).

3.1.5 Physical interferences may result if the physical properties of the samples vary significantly. Changes in viscosity and surface tension can affect the sample aspiration rate and thus cause erroneous results. Sample dilution and/or the method of standard additions are used to correct such interferences. High concentrations of Si in the sample can cause an interference for many of the elements and may cause aspiration problems. No matter what elements are being measured, if large amounts of Si are extracted from the samples, the samples should be allowed to stand for several hours and centrifuged or filtered to remove the Si.

3.2 This procedure describes a generalized method for sample preparation which is applicable to the majority of samples of interest. There are, however, some relatively rare chemical forms of a few of the elements listed in Table 822:I that will not be dissolved by this procedure. If such chemical forms are suspected, results obtained using this procedure should be compared with those obtained using an appropriately altered dissolution procedure. Alternatively, the results may be compared with values obtained utilizing a nondestructive technique that does not require sample dissolution (e.g., x-ray fluorescence, activation analysis).

4. Precision and Accuracy

4.1 The relative standard deviation of the analytical measurement is approximately 3% when measurements are made in the ranges listed in Table 822:II. The overall relative standard deviation will be somewhat larger than this value due to errors associated with the sample collection and preparation steps.

4.2 No data are presently available on the accuracy of this method for actual air samples.

5. Advantages and Disadvantages of the Method

5.1 The sensitivity is adequate for all metals in air samples but only for certain metals in biological matrices. The sensitivity of this direct aspiration method is not adequate for Be, Cd, Ca, Cr, Mn, Mo, Ni, and Sn in biological samples.

5.2 A disadvantage of the method is that at least 1 to 2 mL of solution is necessary for each metal determination. For small samples, the necessary dilution would decrease sensitivity.

5.3 To a great extent, these difficulties can be overcome using non-flame ETA methods. Adequate sensitivity for all of the above samples can be obtained and typical analyses require from 15 μL to 75 μL per element. (Assuming 3 atomizations per analysis; 5 μL to 25 μL per atomization.)

5.4 Disadvantages to ETA include the relatively high cost of instrumentation, reliance on sophisticated background correction schemes, and decreased atomization-to-atomization precision, unless an automated sample injector is used (at additional cost).

6. Apparatus

6.1 SAMPLING EQUIPMENT.

6.1.1 Membrane filters with a pore size of 0.8 μm appropriately sized for the sampling holder are satisfactory for air sampling. Quartz and glass fiber filters, fluorocarbon filters and cellulose filters can also be used. The selection will depend on the specific application.

6.1.2 Sampling pump must be of sufficient capacity to maintain a face velocity of 7 cm/s (5). A personal sampling pump must be calibrated with a representative filter unit in the line.

6.1.3 Flow measurement device such as a calibrated rotameter is required to monitor, or a critical flow orifice is used to control the sampling rate.

6.1.4 Typical personal sampling configurations might include a two-piece open-face cassette holder or a filter holder preceded by an impaction or cyclone stage to sample size-segregated aerosols. Other devices such as multi-stage cascade impactors may be used, and area samples may be taken with devices ranging in dimension from 47 mm diameter to 20 × 25 cm rectangular high volume units. Air volumes sampled might range from a few hundred cubic centimeters to thousands of cubic meters (over 7 orders of magnitude).

6.2 GLASSWARE, BOROSILICATE. Before use all glassware must be cleaned in 1:1 diluted nitric acid and rinsed several times with distilled water.

6.2.1 125-mL Phillips or Griffin beakers with watch glass covers.

6.2.2 15-mL graduated centrifuge tubes.

6.2.3 10-mL volumetric flasks.

6.2.4 100-mL volumetric flasks.

6.2.5 1-L volumetric flasks.

6.2.6 125-mL polyethylene bottles.

6.2.7 Additional auxiliary glassware such as pipettes and different size volumetric glassware will be required depending on the elements being determined and dilutions required to have sample concentrations above the detection limit and in the linear response range (i.e., see Table 822:II). All pipettes and volumetric flasks required in this procedure should be calibrated class A volumetric glassware.

6.3 EQUIPMENT.

6.3.1 Atomic absorption spectrophotometer, with burner heads for air-acetylene and nitrous oxide-acetylene flames.

6.3.2 Hollow cathode lamps for each metal, and a continuum lamp.

6.3.3 Hotplate, suitable for operation at 140°C and 250°C.

6.3.4 Two-stage regulators, for air, acetylene and nitrous oxide.

6.3.5 Heating tape and rheostat, for nitrous oxide regulator (second regulator stage may need to be heated to approximately 60°C to prevent freeze-up).

6.4 SUPPLIES.

6.4.1 Acetylene gas (cylinder). A grade specified by the manufacturer of the instrument. (Replace cylinder when pressure decreases below 100 psi or 700 kPa.)

6.4.2 Nitrous oxide gas (cylinder).

6.4.3 Air supply, with a minimum pressure of 40 psi (275 pKa), filtered to remove oil and water.

7. Reagents

7.1 PURITY. ACS reagent grade chemicals or equivalent shall be used in all tests. References to water shall be understood to mean ASTM reagent water Type II. Care in selection of reagents and in following the listed precautions is essential if low blank values are to be obtained.

7.2 CONC NITRIC ACID (68 to 71%). Redistilled, sp.gr. 1.42

7.3 STANDARD STOCK SOLUTIONS (1000 μg/mL) for each metal in Table 822:I. Commercially prepared or prepared from the following:

7.3.1 *Silver nitrate (AgNO$_3$).*

7.3.2 *Aluminum wire.*

7.3.3 *Barium chloride (BaCl$_2$ · 2H$_2$O).* *

7.3.4 *Beryllium metal.*

7.3.5 *Bismuth metal.*

7.3.6 *Calcium carbonate (CaCO$_3$).* *

7.3.7 *Cadmium metal.*

7.3.8 *Cobalt metal.*

7.3.9 *Copper metal.*

7.3.10 *Potassium chromate (K$_2$CrO$_4$).*

7.3.11 *Iron wire.*

7.3.12 *Indium metal.*

7.3.13 *Potassium chloride (KCl).* *

7.3.14 *Lithium carbonate (Li$_2$CO$_3$).* *

7.3.15 *Magnesium ribbon.*

7.3.16 *Manganese metal.*

7.3.17 *Sodium chloride (NaCl).* *

7.3.18 *Nickel metal.*

7.3.19 *Lead nitrate [Pb(NO$_3$)$_2$].* *

7.3.20 *Rubidium chloride (RbCl).* *

7.3.21 *Strontium nitrate [Sr(NO$_3$)$_2$].* *

7.3.22 *Thallium nitrate (TlNO$_3$).*

7.3.23 *Vanadium metal.*

7.3.24 *Zinc metal.*

7.4 LANTHANUM NITRATE [LA(NO$_3$)$_3$ · 6H$_2$O].

7.5 CESIUM NITRATE (CsNO$_3$) .

8. Procedure

8.1 CLEANING OF EQUIPMENT.

8.1.1 Before initial use, glassware is cleaned with a saturated solution of sodium dichromate in concentrated sulfuric acid (Note: Do not use for chromium analysis) and then rinsed thoroughly with warm tap water, conc nitric acid, tap water and reagent water, in that order, and then dried.

8.1.2 All glassware is soaked in a mild detergent solution immediately after use to remove any residual grease or chemicals.

8.2 COLLECTION AND SHIPPING OF SAMPLES.

8.2.1 Ambient atmospheric particulate matter and industrial dusts and fumes are sampled with cellulose ester membrane filters of 0.8 μm average pore size (Millipore Type AA or equivalent). The pump used with any membrane filter must be sufficient to maintain a face velocity of at least 7 cm/s throughout the sampling period.

---

*These salts must be stored in a desiccator to avoid pick-up of water from the atmosphere.

Sample flow rate is monitored with a calibrated rotameter (5) or the equivalent. The flow rate, ambient temperature and barometric pressure are recorded at the beginning and the end of the sample collection period.

8.2.2 For personal sampling, 37-mm diam filters in holders (Millipore Filter Type AA or equivalent) are used. The personal sampler pumps for this application are operated at 1.5 L/min. In general, a 2-h sample at 1.5 L/min will provide enough sample to detect the elements sought at air concentrations of 0.2 × TLV.

8.2.3 After sample collection is complete the exposed filter surface should be covered with a clean filter. Losses of sample due to overloading of the filter must be avoided.

8.3 SHIPMENT OF SAMPLES.

8.3.1 Filter samples (with clean filter covers) should be sealed in individual plastic filter holders for storage and shipment.

8.4 PREPARATION OF SAMPLES. Most workplace or environmental air samples collected on membrane filters or cellulose-based filters can be prepared by the total dissolution method described in sections **8.4.1** through **8.4.3**. Filters of quartz or glass fiber or fluorocarbon-based media are treated via the extraction method which begins at Section **8.4.4**.

8.4.1 Samples suitable for total dissolution including the clean filter covers and blanks (minimum of 1 filter and cover blank for every 10 filter samples) are transferred to clean 125-mL Phillips or Griffin beakers and sufficient concentrated HNO_3 is added to cover the sample. Each beaker is covered with a watch glass and heated on a hot plate (140°C) in a fume hood until the sample dissolves and a slightly yellow solution is produced. Approximately 30 min of heating will be sufficient for most air samples. However, subsequent additions of HNO_3 may be needed to completely ash and destroy high concentrations of organic material, and under these conditions longer ashing times will be needed. Once the ashing is complete as indicated by a clear solution in the beaker, the watch glass is removed; and the samples are allowed to evaporate to near dryness.

8.4.2 Place the sample beaker on a hot plate controlled at 250°C for several minutes. If evidence of charring is observed, remove from hot plate, cool and repeat the procedure in **8.4.1**. If the residue in the beaker is a light whitish material the beaker is cooled and 1 mL of HNO_3 and 2 to 3 mL of distilled H_2O are added. The beaker is replaced on the hot plate and swirled occasionally until the residue is dissolved (as indicated by a light clear solution). The beaker is then removed and the solution is quantitatively transferred with distilled water to a 10-mL volumetric flask. If any elements are being determined that require the ionization buffer, 0.2 mL of 50 mg/mL Cs (**9.1.2**) is added to the volumetric flask (see Table 822:I, footnote §). If any elements requiring the releasing agent are being determined, 0.2 mL of 50 mg/mL La (**9.1.1**) is added to each volumetric flask (see Table 822.I, footnote **). The samples are then diluted to volume with water.

8.4.3 The 10 mL solution may be analyzed directly for any element of very low concentration in the sample. Aliquots of this solution may then be diluted to an appropriate volume for the other elements of interest present at high concentrations (Note: Approximately 2 mL of solution are required for each element being analyzed.) The dilution factor will depend upon the concentration of elements in the sample and the number of elements being determined by this procedure.

8.4.4 Samples on non-ashable filter media (e.g., glass or quartz fiber, or fluorocarbon filters), including blanks and appropriate quality assurance samples, are placed into clean, 125 mL Phillips or Griffin beakers. Samples should lie face-up on the bottom of the beaker; they may be cut if necessary. Add 15 mL of hot (85°C), mixed acid extractant (65 mL HNO_3 and 182 m HCl diluted to 1 L with water and cover the beakers with parafilm. Agitate the samples in a heated (85°C) ultrasonic bath at 100 W power for 60 min. Remove the beakers from the bath and allow them to cool to room temperature.

8.4.5 Assemble the vacuum filtration unit (Figure 822:I) with a 25 mm glass filter and prerinse the apparatus with 10 mL of clean, hot mixed acid extractant, carefully rinsing the walls of the funnel, and discard the filtrate. Place a clean 25-mL volumetric flask in the bell jar and express the sample

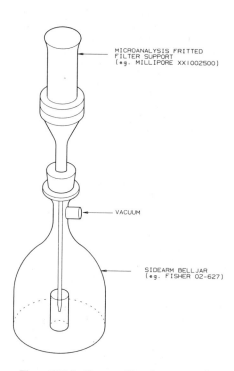

Figure 822:1 – Vacuum filtration apparatus.

through the filter using vacuum, retaining the insoluble sampling medium in the beaker. Rinse the beaker and insoluble materials 3 times with 3 mL portions of water and express the rinse liquid through the filter. Use a clean glass stirring rod to agitate the insoluble material, to ensure complete recovery of trapped liquid. If any of the metals being detected require the use of ionic buffer (see Table 822:I), add 0.5 mL of 50 mg/mL Cs (**9.1.2**) to the volumetric flask. If releasing agent is required add 0.5 mL of 50 mg/mL La (**9.1.1**). Dilute to volume with water and proceed with analysis in Section **8.5**. Note: Samples small enough to be adequately treated in 25 mL beakers can be extracted with 5 mL of hot extractant, washed with 1.5 mL portions and collected in 10 mL volumetric flasks. Reduce additions of releasing agent and ionization buffer to 0.2 mL.

8.5 ANALYSIS.

8.5.1 Set the instrument operating conditions as recommended by the manufacturer. The instrument should be set at the radiation intensity maximum for the wavelength listed in Table 822:I for the element being determined.

8.5.2 Standard solutions should match the sample matrix as closely as possible and should be run in duplicate. Working standard solutions, prepared fresh daily, are aspirated into the flame and the absorbance recorded. Prepare a calibration graph as described in Section **9.2.4** (Note: All combustion products from the AA flame must be removed by direct exhaustion through use of a good separate flame ventilation system.)

8.5.3 Aspirate the appropriately diluted samples directly into the instrument and record the absorbance for comparison with standards. Should the absorbance be above the calibration range, dilute an appropriate aliquot to 10 mL. Aspirate water between samples. A mid-range standard must be aspirated with sufficient frequency (i.e., once every 10 samples) to assure the accuracy of the sample determinations. To the extent possible, all determinations are to be based on replicate analysis.

9. Calibration and Standards

9.1 IONIZATION AND CHEMICAL INTERFERENCE SUPPRESSANTS.

9.1.1 *Lanthanum solution (50 mg/mL)*. Dissolve 156.32 g of lanthanum nitrate [La(NO$_3$)$_3$ • 6H$_2$O] in 2% (v/v)HNO$_3$. Dilute to volume in a 1-L volumetric flask with 2% (v/v)HNO$_3$. When stored in a polyethelene bottle this solution is stable for at least one year.

9.1.2 *Cesium solution (50 mg/mL)*. Dissolve 73.40 g of cesium nitrate (CsNO$_3$) in distilled water. Dilute to volume in a 1-L volumetric flask with distilled water. When stored in a polyethylene bottle this solution is stable for at least one year.

9.2 STANDARD METAL SOLUTIONS.

9.2.1 *Standard stock solutions.* All standard stock solutions are made to a concentration of 1.0 mg of metal per mL. Except as noted these standard stock solutions are stable for at least one year when stored in polyethylene bottles.

a. Master silver standard. Dissolve 1.574 g silver nitrate ($AgNO_3$) in 100 mL distilled water. Dilute to volume in a 1-L volumetric flask with 2% (v/v) HNO_3. The silver nitrate solution will decompose in light and must be stored in an amber bottle away from direct light. New master silver standards should be prepared every few months.

b. Master aluminum standard. Dissolve 1.000 g of Al wire in a minimum volume of 1:1 HCl. Dilute to volume in a 1-L flask with 10% v/v HNO_3.

c. Master barium standard. Dissolve 1.779 g of barium chloride ($BaCl_2 \cdot 2H_2O$) in water. Dilute to volume in a 1-L volumetric flask with distilled water.

d. Master beryllium standard. Dissolve 1.000 g of Be metal in a minimum volume of 1:1 HCl. Dilute to volume in a 1-L flask with 2% (v/v) HNO_3.

e. Master bismuth standard. Dissolve 1.000 g of bismuth metal in a minimum volume of 6 N HNO_3. Dilute to volume in a 1-L volumetric flask with 2% (v/v) HNO_3.

f. Master calcium standard. To 2.498 g of primary standard calcium carbonate ($CaCO_3$) add 50 mL of deionized water. Add dropwise a minimum volume of HCl (approximately 10 mL) to dissolve the $CaCO_3$. Dilute to volume in a 1-L volumetric flask with distilled water.

g. Master cadmium standard. Dissolve 1.000 g of cadmium metal in a minimum volume of 6 N HCl. Dilute to volume in a 1-L volumetric flask with 2% (v/v) HNO_3.

h. Master cobalt standard. Dissolve 1.000 g of Co metal in a minimum volume of 1:1 HCl. Dilute to volume in a 1-L flask with 2% (v/v) HNO_3.

i. Master copper standard. Dissolve 1.000 g of copper metal in a minimum volume of 6 N HNO_3. Dilute to volume in a 1-L volumetric flask with 2% (v/v) HNO_3.

j. Master chromium standard. Dissolve 3.735 g of potassium chromate (K_2CrO_4) in distilled water. Dilute to volume in a 1-L flask with distilled water.

k. Master iron standard. Dissolve 1.000 g of iron wire in 50 mL of 6 N HNO_3. Dilute to volume in a 1-L volumetric flask with distilled water.

l. Master indium standard. Dissolve 1.000 g of indium metal in a minimum volume of 6 N HCl. Addition of a few drops of HNO_3 and mild heating will aid in dissolving the metal. Dilute to volume in a 1-L volumetric flask with 10% (v/v) HNO_3.

m. Master potassium standard. Dissolve 1.907 g of potassium chloride (KCl) in distilled water. Dilute to volume in a 1-L volumetric flask with distilled water.

n. Master lithium standard. Dissolve 5.324 g of Li_2CO_3 in a minimum volume of 6 N HCl. Dilute to volume in a 1-L volumetric flask with distilled water.

o. Master magnesium standard. Dissolve 1.000 g of magnesium ribbon in a minimum volume of 6 N HCl. Dilute to volume in a 1-L volumetric flask with 2% (v/v) HNO_3.

p. Master manganese standard. Dissolve 1.000 g of manganese metal in a minimum volume of 6 N HNO_3. Dilute to volume in a 1-L volumetric flask with 2% (v/v) HNO_3.

q. Master sodium standard. Dissolve 2.542 g of sodium chloride (NaCl) in distilled water. Dilute to volume in a 1-L volumetric flask with distilled water.

r. Master nickel standard. Dissolve 1.000 g of nickel metal in a minimum volume of 6 N HNO_3. Dilute to volume in a 1-L volumetric flask with 2% (v/v) HNO_3.

s. Master lead standard. Dissolve 1.598 g of lead nitrate [$Pb(NO_3)_2$] in 2% (v/v) HNO_3. Dilute to volume in a 1-L volumetric flask with 2% (v/v) HNO_3.

t. Master rubidium standard. Dissolve 1.415 g of rubidium chloride (RbCl) in distilled water. Dilute to volume in a 1-L volumetric flask with distilled water.

u. Master strontium standard. Dissolve 2.415 g of strontium nitrate [$Sr(NO_3)_2$] in distilled water. Dilute to volume in a 1-L volumetric flask with distilled water.

v. Master thallium standard. Dissolve 1.303 g of thallium nitrate ($TlNO_3$) in 10% (v/v) HNO_3. Dilute to volume in a 1-L volumetric flask with 10% HNO_3.

w. Master vanadium standard. Dissolve 1.000 g of vanadium metal in minimum volume of 6 N HNO_3. Dilute to volume in a 1-L volumetric flask with 10% (v/v) HNO_3.

x. Master zinc standard. Dissolve 1.000 g of zinc metal in a minimum volume of 6 N HNO_3. Dilute to volume in a 1-L volumetric flask with 2% (v/v) HNO_3.

9.2.2 *Dilute standards.* Diluted standard mixtures of the elements listed in 9.2.1 are prepared according to the directions in the following sections a to c. The mixed dilute standards are prepared such that the accuracy of the working standard preparation (section 9.2.3) is maximized. Only those elements being determined in the samples need to be prepared as dilute and working standards.

a. Mixed calcium, cadmium, potassium, lithium, magnesium, sodium and zinc standard (0.010 mg/mL for each metal). Pipet 10 mL of the master standards for calcium, cadmium, potassium, lithium, magnesium, sodium, and zinc into a 1-L flask, add 100 mL of conc HNO_3, and dilute to volume with distilled water. Prepare fresh monthly.

b. Mixed barium, bismuth, cobalt, chromium, copper, iron, manganese, nickel, lead, rubidium and strontium standard (0.100 mg/mL for each metal). Pipet 10 mL of the master standards for barium, bismuth, cobalt, chromium, copper, iron, manganese, nickel, lead, rubidium and strontium into a 100-mL volumetric flask, add 10 mL of conc HNO_3 and dilute to volume with distilled water. Prepare fresh monthly (Note: Due to volume considerations, if more than 8 elements are to be prepared one must prepare 2 dilute standards.)

c. Dilute silver standard (1.100 mg of silver per mL). Pipet 10 mL of the master silver standard and 10 mL conc HNO_3 into a 100-mL volumetric flask, and dilute to volume with distilled water. Store in an amber bottle away from direct light. Prepare fresh weekly.

9.2.3 *Working standards.*

a. Mixed working standards. Working standards are prepared by pipetting appropriate amounts of the dilute standards from 9.2.2.a, 9.2.2.b and the master standards for Al, In, Tl and V. Pipet 1 mL of the dilute standard from section 9.2.2.a and 1 mL from the dilute standard from 9.2.2.b into a 100-mL volumetric flask. Pipet into this same volumetric flask, 1 mL from each of the master standards for Al, In, Tl and V. Add 2 mL of 50 mg/mL Cs solution, 2 mL of 50 mg/mL La solution and 10 mL of HNO_3 to the volumetric flask and dilute to

volume with distilled water.* This solution contains the following metals at the indicated concentrations: Ca, Cd, K, Li, Mg, Na and Zn—0.1 ppm; Rb, Ba, Bi, Co, Cr, Cu, Fe, Mn, Ni, Pb, & Sr—1.0 ppm; Al, In, Tl, and V—10.0 ppm. (Note: dilute and working standards need to be prepared only for the elements being determined in the sample. If Cr and K are being determined in the sample, separate working solutions must be prepared since the chromium standard contains K. Using the above described preparation procedures a standard containing 1.0 ppm Cr will also contain 1.5 ppm K). This procedure is repeated using 2, 3, 4, and 5 mL of the same standard metal solutions indicated above. These standards must be prepared fresh daily.

b. Working silver standard. Pipet 1 mL of the dilute silver standard (from section 9.2.2.c) into a 100-mL volumetric flask and dilute to volume with distilled water. This solution contains 1.0 ppm silver ion. Repeat this procedure using 2, 3, 4, and 5 mL of the dilute silver standard. The working silver standards must be prepared fresh daily.

9.2.4 The standard solutions are aspirated into the flame and the absorbance recorded. If the instrument used displays transmittance, these values must be converted to absorbance. A calibration curve is prepared by plotting absorbance versus metal concentration. The best fit curve (calculated by linear least squares regression analysis) is fitted to the data points. This line or the equation describing the line is used to obtain the metal concentration in the samples being analyzed.

9.2.5 To ensure that the preparation is being properly followed, clean membrane filters are spiked with known amounts of the elements being determined by adding appropriate amounts of the previously described standards and carried through the entire procedure. The amount of metal is

---

*The procedure as described has been designed to match the nitric acid concentration of samples and standards (i.e., 10% v/v HNO_3). If the sample solution from section 8.4.3 is to be diluted prior to analysis the amount of acid added to the standards must be reduced by an amount equal to the sample dilution factor.

determined and the percent recovery calculated. These tests will provide recovery and precision data for the procedure as it is carried out in the laboratory for the soluble compounds of the elements being determined.

9.2.6 Analysis by the method of standard additions. In order to check for interferences, samples are initially and periodically analyzed by the method of standard additions and the results compared to those obtained by the conventional analytical determination. For this method the sample is divided into three 2-mL aliquots. To one of the aliquots an amount of metal approximately equal to that in the sample is added. To another aliquot twice this amount of metal is added. (Note: Additions should be made by micropipetting techniques such that the added volume does not exceed 1% of the original aliquot volume—i.e., 10 μL and 20μL additions to a 2-mL aliquot.) The solutions are then analyzed and the absorbance readings (ordinate) are plotted against metal added to the original sample (abscissa). The line obtained from such a plot is extrapolated to 0 absorbance and the intercept on the concentration axis is taken as the amount of metal in the original sample (2). If the result of this determination does not agree to within 20% of the values obtained with the procedure described in section 9.2.4, an interference is indicated and standard addition techniques should be utilized for sample analysis.

9.3 BLANK. Blank filters must be carried through the entire procedure each time samples are analyzed.

10. Calculations

10.1 The uncorrected air volume sampled by the filter is calculated by averaging the beginning and ending sample flow rates, converting to cubic meters and multiplying by the sample collection time.

$$V = \frac{F_B + F_E}{2 \times 1000} t$$

where V = uncorrected sample volume (m^3)

F_B = sample flow rate at beginning of sample collection (L/min)

F_E = sample flow rate at end of sample collection (L/min)

t = sample collection time (minutes)

10.2 If required, the volume is corrected to 25°C and 101.3 kPa by using the formula

$$V_{corr} = \frac{(298)\,(P)\,(V)}{(101.3)\,(T)}$$

where V_{corr} = corrected sample volume (m^3)

P = average barometric pressure during sample collection period (kPa)

T = average temperature during the sample collection period (K)
(Note: K = °C + 273.)

V = uncorrected volume calculated from 10.1 (m^3).

10.3 After any necessary correction for the blank has been made, metal concentrations are calculated by multiplying the μg of metal per mL in the sample aliquot by the aliquot volume and dividing by the fraction which the aliquot represents of the total sample and the volume of air collected by the filter.

$$\mu\text{g metal/m}^3 = \frac{(C \times V) - B}{V_{corr} \times F}$$

where C = concentration (μg metal/mL) in the aliquot

V = volume of aliquot (mL)

B = total μg of metal in the blank

F = fraction of total sample in the aliquot used for measurement (dimensionless)

V_{corr} = corrected volume of air sample and calculated from 10.2.

10.4 Untreated filter samples may be stored indefinitely.

11. References

1. SLAVIN, W. 1968. Atomic Absorption Spectroscopy. Interscience Publishers, N.Y., N.Y.
2. RAMIREZ-MUNOZ, J. 1968. Atomic Absorption Spectroscopy, Elsevier Publishing Co., N.Y., N.Y.
3. DEAN, J. A. AND T. C. RAINS, ed. 1969. Flame Emission and Atomic Absorption Spectrometry— Volume I. Theory. Marcel Dekker, N.Y., N.Y.
4. WINEFORDNER, J. E. ed. 1971. Spectrochemical Methods of Analysis, John Wiley & Sons, Inc.

5. AMERICAN CONFERENCE OF GOVERNMENTAL INDUSTRIAL HYGIENISTS. 1983. Air Sampling Instruments for Evaluation of Atmospheric Contaminants. 6th ed. P. J. Lioy and M. J. Y. Lioy, eds.

6. KNEIP, T. J. AND M. T. KLEINMAN 1982, Analysis of airborne particles in the workplace and ambient atmospheres, in Atomic Absorption Spectrometry, J. E. Cantle, ed, Elsevier Scientific Publishing Co, Oxford, pp. 123–138.

7. BELCHER, C. B., R. M. DAGNALL AND T. S. WEST. 1964. Examination of the Atomic Absorption Spectroscopy of Silver. *Talanta* 11:1257.

8. MULFORD, C. E. 1966. Gallium and Indium Determinations by Atomic Absorption. *Atomic Absorption Newsletter*. 5:28.

9. ROBINSON, J. W. 1966. Atomic Spectroscopy. Marcel Dekker, Inc., N.Y., N.Y.

10. CANTLE, J. E., 1982. Instrumental requirements and optimisation, in Atomic Absorption Spectrometry, J. E. Cantle ed, Elsevier Scientific Publishing Co., Oxford. pp. 15–66.

Subcommittee 6

M. T. KLEINMAN, *Chairman*
J. R. RHODES
V. P. GUINN
R. J. THOMPSON

822A.

General Method for Preparation of Tissue Samples for Analysis for Trace Metals

1. Principle of the Method

1.1 Tissue samples including bone, soft tissue, blood, serum, feces or urine, are completely ashed in hot concentrated nitric acid **(1)**. The resulting solution is analyzed for trace metals by atomic absorption spectrophotometry (AAS). The entire sample may be analyzed to avoid problems inherent in attempting to obtain representative tissue aliquots.

2. Range, Sensitivity and Detection Limit

2.1 The range, sensitivity and detection limits for selected elements are listed in Table 822A:I, based on the concentration in the final solution.

2.2 Sensitivity or detection limits may be expressed on a $\mu g/g$ tissue basis by calculations based on sample weights. (See Table 822A:II)

3. Interferences

3.1 Incomplete wet digestion may result in severe matrix effects during the final AAS determination. The method of standard additions may be used to define the presence or absence of matrix interferences.

3.2 Interferences for the individual metals are the same as those covered for air filter samples in Method 822.

3.3 Random contamination is a potentially serious problem. Glassware must be segregated for use with blanks, low level samples, and spiked samples. Smoking should be prohibited in the laboratory or laboratories where the sample preparations and analyses are performed.

3.4 Calcareous samples such as bones or teeth may contain sufficient calcium to precipitate calcium nitrate during ashing. This

is easily redissolved but may cause "bumping" during the ashing stage and also require the final solution to be somewhat dilute. If concentration is necessary to achieve adequate sensitivity an extraction, ion exchange or other enrichment scheme may be employed.

4. Precision and Accuracy

4.1 The precision of this method using AAS for the determination of Pb, Cd, Cu, Mn, Zn, Fe, and V is typically from 5–10% Relative Standard Deviation at the 95% confidence level **(2)**.

4.2 The results obtained by this method for National Bureau of Standards (NBS) Bovine Liver samples agreed ($\alpha \leq 0.01$) with the reported NBS concentrations for Fe, Cu, Mn and Zn **(2,3)**.

5. Apparatus

5.1 Glassware—Borosilicate glassware (or Teflon ware) should be used throughout the analysis. The glassware should be cleaned by soaking overnight in 1:1 HNO_3, rinsed with water and soaked overnight in distilled water.

5.2 EQUIPMENT.

5.2.1 Centrifuge: Capable of operation in the range 0–2000 rpm.

5.2.2 Hot Plates—Stirring.

5.2.3 Teflon coated magnetic stirring bars.

5.2.4 Fume hood—During this procedure copious amounts of nitrogen dioxide and nitric acid vapor are emitted. Hood exhausts should be adequately vented and construction materials must be corrosion resistant.

Table 822A:I Measurement Limits for Selected Elements by AAS

Elements	Sensitivity* $\mu g/mL$	Range of Optimum Instrumental Precision $\mu g/mL$	Detection Limits** $\mu g/mL$
Ag	0.036	0.5–5.0	0.003
Al	0.76	5–50.0	0.04
Ba	0.20	1–10	0.01
Be	0.017	0.1–1.0	0.002
Bi	0.22	1–10	0.06
Ca	0.021	0.1–1.0	0.0005
Cd	0.011	0.1–1.0	0.0006
Co	0.066	0.5–5.0	0.007
Cr	0.055	0.5–5.0	0.005
Cu	0.040	0.5–5.0	0.005
Fe	0.062	0.5–5.0	0.005
In	0.38	5–50	0.05
K	0.010	0.1–1.0	0.003
Li	0.017	0.1–1.0	0.002
Mg	0.003	0.05–0.50	0.0003
Mn	0.026	0.5–5.0	0.003
Na	0.003	0.05–0.50	0.0003
Ni	0.066	0.5–5.0	0.008
Pb	0.11	1–10	0.02
Rb	0.042	0.5–5.0	0.003
Sr	0.044	0.5–5.0	0.004
Tl	0.28	5–50	0.02
V	0.88	10–100	0.1
Zn	0.009	0.1–1.0	0.002

  * Sensitivity data are taken from "Analytical Data for Elements Determined by Atomic Absorption Spectrometry," Varion Techtron, Walnut Creek, California, 1971.
 ** Solution detection limits are taken from "Detection Limits for Model AA–5 Atomic Absorbtion Spectrophotometer," Varion Techtron, Walnut Creek, California. (NOTE: These detection limits represent the ultimate values since the blanks resulting from the reagents and the filter material have not been taken into account.)

Table 822A:II Typical Tissue Sample Weights

Tissue	Wet Weight* (g)	Dry Weight** (g)	Ash Weight** (g)	Initial Volume of HNO_3 Used for Ashing (mL)	Final Sample Volume for Analysis (mL)
Lung (right)	450	103	5	1000	100
Liver	400	140	5	1000	100
Kidney (right)	150	38	2	1000	100
Bone (two vertebrae)	150	—	—	100	250
Blood	20	—	—	100	25
Lymph Nodes	10	—	—	100	25

  * Average size of samples obtained for trace metal analyses by Bernstein, et al. (2).
 ** Tipton and Cook, *Health Physics* 9:103–145 (1963).

6. Reagents

6.1 Water — all references to water refer to ASTM Reagent water, Type II.

6.2 Purity — all reagents should be ACS reagent grade or the equivalent.

6.3 Nitric acid, concentrated doubled distilled in glass or redistilled (G. Frederick Smith Chemical Co., Redistilled Grade or equivalent).

6.4 For reagents needed for the preparation of standard metal solutions see Method 822.

7. Procedure

7.1 SAMPLE COLLECTION AND STORAGE.

Tissue samples are weighed and frozen as soon as possible after collection. Samples may be weighed in the frozen state for convenience since weight changes have been found negligible over the first few days after freezing. The samples are placed in clean polyethylene bags and stored until they are analyzed.

7.2 SAMPLE PREPARATION.

7.2.1 A blank must be run with every set of samples.

7.2.2 The sample is removed from the freezer, placed into a tared glass tray and allowed to thaw.

7.2.3 Large specimens, such as a whole liver or lung are cut into smaller fragments (15–20 cm^3) using stainless steel surgical implements. Knives made of glass or tantalum should be used if Fe, Ni, Cr or other components of steel are found at levels indicating contamination.

7.2.4 Blood, serum or urine samples may be diluted with water to a volume convenient for sample handling in step **7.2.5**. The following steps are designed to be performed in a fume hood. See alternate step **8**.

7.2.5 The sample is added slowly with constant stirring to a beaker containing hot concentrated HNO$_3$ at about 100°C. The initial amount of acid depends on sample size. Small samples can be ashed in about 100 mL, but 1000 mL is required for whole lungs, livers or kidneys (Table 822A:II). A

small amount of foaming is expected after the addition of each fraction of tissue. After the foaming has subsided, additional tissue can be added to the beaker. (Note: excessive foaming can be controlled with the addition of a few drops of 1-octanol.) Minimal volumes of all reagents are used to reduce the blanks.

7.2.6 Reweigh the glass tray to determine the weight of sample ashed.

7.2.7 After the entire sample has been added, the beaker is covered with a watch glass, and the tissue is allowed to digest for 1–2 h at about 100°C.

7.2.8 The volume of the solution is slowly reduced by evaporation to near dryness. (Caution: The sample should not be allowed to go to complete dryness, as the partially nitrated organic material is highly flammable and may ignite.)

7.2.9 Small volumes of concentrated HNO$_3$ are added until a clear, straw-colored solution is obtained. (Note: Small amounts of H$_2$O$_2$ may be added to accelerate oxidation; however, some H$_2$O$_2$ has been found to be highly contaminated, and the trace metals content of the reagent batch should be checked before use. Iida, et al. (4) have reported that addition of 1 mL of perchloric acid was essential for quantitative Ca recovery in the analysis of 500 mg of NBS Bovine Liver standard.)

7.2.10 The solution is reduced to a volume of 2–3 mL for small samples and 10 mL for large samples (Table 822A:II), quantitatively transferred to a volumetric flask, and diluted to volume with water. The final volume of solution will depend on the weight of the original tissue sample and the expected concentration of trace elements in the sample. The final volume is selected to avoid interference in the AAS determination due to the viscosity of the solution.

7.3 ANALYSIS.

The analytical procedure is described in Method 822. Solutions from step **7.2.10** are introduced at step **8.5** of that method. It is good practice to check the results of tissue analyses using the method of standard additions as described in that method.

8. Alternate Procedures for Small Samples

If small samples are to be analyzed, or if it is important to minimize the amounts of acid required to ash a sample to lower the blanks, a refluxing method can be employed. The advantages are that less reagent will be used and that ashing can be performed without using a fume hood, if necessary. The disadvantages are that the method takes more time and that the reflux glassware, heating mantles and controllers are expensive.

8.1 The tissue sample is placed into an appropriately sized round bottom flask with 2 or 3 glass boiling beads, and sufficient concentrated nitric acid is added to completely cover the sample.

8.2 The condenser is attached and the system connected to a suitable fume removal unit.

8.3 The sample is permitted to stand for several hours to allow time for the oxidation of labile components (overnight, if possible).

8.4 Water flow to the condenser is started and heat is applied gradually to avoid excessive foaming. Continue increasing the temperature until rapid boiling is achieved.

8.5 Reflux until a clear solution is obtained.

8.6 Discontinue refluxing, remove the condenser and begin to gradually evaporate the acid solution to near dryness. The evaporated acid can be recondensed, or it can be trapped in the fume removal unit.

8.7 Treat the sample as described from step **7.2.8** to step **7.3**.

References

1. GORSUCH, T. T. 1970. *The Destruction of Organic Matter*, Pergamon Press, New York.
2. BERSTEIN, D. B., T. J. KNEIP, M. T. KLEINMAN, R. RIDDICK AND M. EISENBUD. 1974. "Uptake and Distribution of Airborne Trace Metals in Man," Proceedings of Symposium Trace Substances in Environmental Health VIII, University of Missouri.
3. CLEGG, M. S., C. L. KEEN, B. LONNERDAL AND L. S. HURLEY. 1981. "Influence of Ashing Techniques on the Analysis of Trace Elements in Animal Tissue: I. Wet Ashing," *Biological Trace Element Research* 3:107–115.
4. IIDA, C., T. UCHIDA AND I. KOJIMA. 1980. "Decomposition of Bovine Liver in a Sealed Teflon Vessel for Determination of Metals by Atomic Absorption Spectrometry," *Anal. Chim. Acta.* 113:365–368.

Subcommitte 6

M. T. KLEINMAN, *Chairman*
W. J. COURTNEY
V. P. GUINN
T. C. RAINS
J. R. RHODES
R. J. THOMPSON

822B.

X-Ray Fluorescence Spectrometry for Multielement Analysis of Airborne Particulate and Biological Material

1. Principle of Method

1.1 X-ray fluorescence (XRF) spectrometry consists of irradiating several cm^2 of a preferably flat solid or liquid specimen with X-rays from a suitable source so as to excite the characteristic X-ray lines of the elements in the specimen. Each line energy (or wavelength*) is characteristic of the element atomic number, and the line intensity is proportional to element concentration.

1.2 The X-ray energies (or wavelengths) from the specimen are resolved using either an energy dispersive (EDX) or wavelength dispersive (WDX) spectrometer. Excitation sources normally employed in EDX systems are low power (1-100w) X-ray tubes or radioisotope sources. WDX systems invariably employ high power (1-5 kw) X-ray tubes.

1.3 In an EDX system the spectrometer is usually a LN_2-cooled solid state detector plus electronic pulse processor. The detector absorbs the incident X-rays and the processor produces output pulses of charge proportional to the energy of each X-ray detected. X-rays of all energies are processed one by one at a rate of 10^3–10^4/s until a statistically significant sample of the complete spectrum is collected, which typically takes 100–1000 seconds.

1.4 In a WDX system a crystal analyzer diffracts each incident X-ray through an angle corresponding to the X-ray wavelength. The incident and diffracted X-rays must be highly collimated. In a "simultaneous" WDX system up to 20 crystal-detector combinations are each preset to an element

line, so allowing simultaneous multielement analysis.

1.5 The wide range of elements measured in pollution analysis cannot be excited with optimum signal to background ratio using a single X-ray source. It is common practice in WDXRF to employ a dual-target X-ray tube, one target (e.g. Cr) for low Z elements and a second (e.g. W) for medium and high Z elements. In EDXRF up to four excitation energies are used, either as one X-ray tube with multiple targets or as separate radioisotope sources. Up to about fifteen elements having adjacent characteristic X-ray energies can be excited efficiently by a single source energy.

1.6 A typical analysis procedure is to irradiate each specimen sequentially with the sources while measuring simultaneously the elements excited by each source.

1.7 In the main procedure to be described, the specimen is in the form of air particulate material uniformly deposited on a filter substrate. The method involves taking the whole or a part of the filter, mounting it in a suitable XRF spectrometer and directly measuring the mass per unit area ($\mu g/cm^2$) of each element. If the air volume sampled per unit area of filter (i.e. m^3 air/ cm^2 filter) is known, the mass of each element per unit volume of air (μg element/ m^3) can be calculated directly.

1.8 Also, plant and animal tissues can be analyzed for trace elements. The tissue must first be homogenized and presented in solid or liquid form. The preferred method is to freeze-dry, pulverize and press into a thin homogeneous pellet.

1.9 X-ray fluorescence analysis measures total element content in the specimen. In the method described here, it does not

---

*Wavelength $\propto \dfrac{1}{\text{Energy}}$; 1 nm = 1.24 kev

differentiate with respect to chemical binding nor, in the case of particulate material on filters, does it differentiate between the particulate deposit and the filter substrate. The latter should, therefore, be free from significant quantities of the elements to be measured.

1.10 Quantitative data are obtained after calibration of the equipment with known standards and correction for interferences, if any (usually spectral background, line overlap, matrix absorption and particle size).

1.11 Air filter specimens are nearly transparent to the X-rays. This results in an absence of matrix absorption and particle size effects and makes for linear calibrations for all except low atomic number elements. However, if the particles are embedded in the filter substrate, corrections may be necessary for X-ray absorption due to the substrate.

1.12 Quantitative analysis of thin pellets generally requires corrections to be made for X-ray absorption within the specimen.

2. Range and Sensitivity

2.1 Table 822B:I lists most of the elements measured or sought on air particulate filters, together with their preferred analytical lines, examples of excitation energies, typical values for Minimum Detectable Limit (MDL), and concentration ranges found in ambient air samples. The elements are listed in order of increasing line energy.

2.2 The nominal elemental range is Al-U. Determination of low Z elements becomes rapidly more difficult as the atomic number decreases below about 20 (Ca). Different XRF instruments have different low Z cutoffs, some below Al and others above. The K lines (the most energetic) are used from the low Z limit to Mo K (in WDX systems) or of Ba K (in EDX systems). The L lines are used for higher Z elements. In practice not more than about 20 of the most abundant elements listed in Table 822B:I are seen in ambient air samples. Workplace and stack samples usually contain fewer elements but their concentrations could be much higher, and the elements are not restricted to those listed in the table.

2.3 The X-ray line intensity of each element listed in Table 822B:I is, to a first order, a linear function of element mass per unit area (ng/cm^2) of the total deposit weight and the filter weight. Also, interelement interferences (i.e. absorption, enhancement and particle size) are small or negligible. As a result, the interference-free values of MDL quoted in Table 822B:I are close to the values obtained in practice, except for rare cases of severe line overlap.

The reason for the linearity and absence of interferences is that the total deposit mass per unit area* for ambient air particulate collected on filters is normally below the "thin specimen limit" for all except low Z elements or below the "filled monolayer limit" for most low Z elements (3). This is not necessarily true for heavier deposits found in stack and workplace sampling, in which case precautions must be taken to check that the abovementioned criteria for linearity have not been exceeded.

2.4 The values of MDL quoted in Table 822 B:I are for an EDXRF system using radioisotope sources for excitation. The measurement time was 10 min per source. Note that the MDL varies smoothly with element Z for a given excitation energy but has discontinuities between excitation energies. Most EDX systems have similar sensitivities (differing mainly through choice of excitation energies) because the instrumental sensitivity is limited by the maximum spectral intensity that the detector and associated pulse processor can handle. Most EDX systems operated in this "count rate limited" mode.

Increasing the source strength will decrease the MDL, in proportion to the square root of the source strength, for all WDX and EDX systems not operating in the "count rate limited" mode. Increasing the spectrum accumulation time will decrease the MDL for all systems, in proportion to the square root of the time.

WDX systems have much superior resolution for low Z elements and as a result are capable of significantly better sensitivity for these elements, as long as the spectrometer has been properly modified to analyze thin specimens. Jaklevic and Walter (1)

---

*A typical upper limit is 400 $\mu g/cm^2$ for a 24-h Hi-vol sample.

Table 822B:I Typical Ranges and Sensitivities for Particle Deposits on Filters

Element	Analytical Line	Excitation Line	MDL [a] (ng/cm^2)	Concentration Range for Ambient Air Samples [b] (ng/cm^2)	
Al	Kα	Mn K	190	8×10^2	$- 4 \times 10^4$
Si	Kα	Mn K	50	2×10^4	$- 3 \times 10^5$
P	Kα	Mn K	24	2×10^2	$- 4 \times 10^3$
S	Kα	Mn K	15	8×10^3	$- 6 \times 10^4$
Cl	Kα	Mn K	9	4×10^2	$- 4 \times 10^4$
K	Kα	Mn K	3	1×10^3	$- 2 \times 10^4$
Ca	Kα	Mn K	3	3×10^3	$- 1 \times 10^5$
Ti	Kα	Mn K	2	40	$- 3 \times 10^5$
V	Kα	Pu	2	4	$- 1 \times 10^4$
Cr	Kα	Pu	100	8	$- 1 \times 10^3$
Mn	Kα	Pu	80	40	$- 8 \times 10^3$
Fe	Kα	Pu	60	2×10^3	$- 8 \times 10^4$
Co	Kα	Pu	50	0.8	$- 80$
Ni	Kα	Pu	40	4	$- 2 \times 10^3$
Cu	Kα	Pu	30	80	$- 2 \times 10^4$
Ta	Lα	Pu	70	< 1	
	Lβ	Pu	70		
W	Lα	Pu	70	0.4	$- 4$
	Lβ	Pu	70		
Zn	Kα	Pu	30	80	$- 4 \times 10^4$
Ga	Kα	Pu	30	2	$- 20$
Au	Lα	Pu	110	< 0.1	
	Lβ	Pu	110		
Ge	Kα	Pu	20	2	$- 20$
Hg	Lα	Pu	100	< 1	$- 20$
	Lβ	Pu	100		
Tl	Lα	Pu	90	< 0.1	$- 0.4$
	Lβ	Pu	90		
As	Kα	Pu	40	8	$- 3 \times 10^2$
Pb	Lα	Pu	40	4×10^2	$- 4 \times 10^4$
	Lβ	AgKα	80		
Bi	Lα	AgKα	70	2	$- 30$
	Lβ	AgKα	70		
Se	Kα	AgKα	40	0.4	$- 40$
Br	Kα	AgKα	40	20	$- 4 \times 10^3$
Th	Lα	AgKα	70	0.2	$- 4$
	Lβ	AgKα	70		
Rb	Kα	AgKα	30	20	$- 3 \times 10^2$
U	Lα	AgKα	60	0.2	$- 8$
	Lβ	AgKα	60		
Sr	Kα	AgKα	25	20	$- 4 \times 10^2$
Y	Kα	AgKα	25	4	$- 20$
Zr	Kα	AgKα	20	20	$- 3 \times 10^2$
Nb	Kα	AgKα	20	1	$- 8$
Mo	Kα	AgKα	20	8	$- 3 \times 10^2$
Ru	Kα	Eu K	80	< 0.1	
Rh	Kα	Eu K	75	< 0.1	
Pd	Kα	Eu K	70	< 1	
Ag	Kα	Eu K	70	1	$- 40$
Cd	Kα	Eu K	90	1	$- 2 \times 10^2$
In	Kα	Eu K	70	4×10^{-2}	$- 4$

Table 822B:I (Cont)

Element	Analytical Line	Excitation Line	MDL [a] (ng/cm^2)	Concentration Range for Ambient Air Samples [b] (ng/cm^2)	
Sn	Kα	Eu K	70	8	$- 2 \times 10^2$
Sb	Kα	Eu K	70	2	$- 4 \times 10^2$
Te	Kα	Eu K	80	<0.4	$- 4$
I	Kα	Eu K	100	<0.2	$- 20$
Cs	Kα	Eu K	100	0.8	$- 10$
Ba	Kα	Eu K	160	1×10^2	$- 4 \times 10^2$
La	Kα	Eu K	180	4	$- 80$

[a]Interference-free Minimum Detection Limit defined as 3 standard deviations of the blank signal (1,2), for Teflon membrane substrate, using an EDXRF system (3).
[b]Typical Urban Aerosols. Concentrations in ng/m^3 of air converted to ng/cm^2 of filter using the factor 4 m^3/cm^2 (typical for 24-h Hi-vol sample).

compare values of MDL for typical WDX and EDX systems.

2.5 The total specimen mass per unit area and its average atomic number affect the MDL by changing the spectral background. In ambient air samples the filter substrate represents at least 90% of the specimen mass per unit area. Jaklevic and Walter (1) compare MDL's of the EDX and WDX methods for different substrates.

2.6 Table 822B:II lists results for the determination of 13 elements in NBS SRM 1571 Orchard Leaves, using an EDX system (4). Each specimen was prepared as a pressed pellet 2.54 cm in diameter with a mass per unit area of 30 mg/cm^2. The extra amount of sample over that available on an

air particulate filter makes for improved sensitivity, while the total material used (150 mg) is still small enough to be compatible with the demands of biological sample analysis.

Similar results are obtained with pressed pellets of freeze-dried tissue and body fluids, the freeze-drying affording a preconcentration factor.

3. Interferences and Background

3.1 Two types of interferences (5) exist, spectral and matrix. Some spectral interferences such as line overlap are mainly instrumental in nature. Other spectral interferences are due to background from either

Table 822B:II Typical Sensitivities for Biological Specimens Analyzed as Thin Pellets

Element	Concentration (μg/g)	Experimental Error (μg/g)	Interference-Free MDL (μg/g)	NBS Certified Value (μg/g)
Ti	–	–	3	–
Cr	2.5	1.6	1	2.3
Mn	88.6	2.2	–	91 ± 4
Fe	276	8	0.6	300 ± 20
Ni	1.3	0.4	–	1.3 ± 0.2
Cu	12.6	0.6	0.3	12 ± 1
Zn	23.7	0.8	–	25 ± 3
As	10.6	0.8	–	14 ± 2
Br	9.3	0.6	0.3	10
Rb	11.0	0.8	0.4	12 ± 1
Sr	36.6	1.2	–	37
Hg	–	–	0.4	–
Pb	45.4	2.0	0.4	45 ± 3

the instrument or the specimen. Matrix interferences occur solely in the specimen. They consist of changes in the line intensity of the analyte that are due to causes other than a change in the concentration of the analyte.

3.2 LINE OVERLAP.

3.2.1 EDX spectrometers have more line overlaps than do WDX systems because their resolution (below 15 kev) is not as good. The most common line overlaps are the $K\alpha$ of element $Z + 1$, $Z + 2$ or $Z + 3$ with the $K\beta$ of element Z. This occurs to a greater or lesser extent throughout the element range K to Ba. Other possible overlaps are between the L or M X-rays of a high Z element and the K X-rays of a lower Z element. Important examples are Pb $L\alpha$/As $K\alpha$, Pb $M\alpha$/S $K\alpha$, Ba $L\alpha$/Ti $K\alpha$, Ba $L\beta$/V $K\alpha$ and Ba $L\gamma$/Cr $K\alpha$. In the case of lead and barium the alternate Pb $L\beta$ and Ba $K\alpha$ lines are relatively free of interference. However the determination of As and S in the presence of excess Pb, and of Ti, V and Cr in the presence of excess Ba can be difficult because these elements do not have an alternate strong spectral line.

Correction for line overlaps can be made accurately for "thin specimens" because the $K\alpha/K\beta$, $L\alpha/L\beta$ and K/L/M intensity ratios of the interfering elements do not vary appreciably with sample composition. The ratios can be measured during calibration and used in subsequent corrections.

Since overlap corrections rely at least implicitly on subtraction of two intensities each having a statistical error, at some point the interference correction will cause a significant decrease in sensitivity and precision of the analyte measurement. The overlapping line intensity must be at least an order of magnitude greater than the analyte line intensity before this happens. Since $K\alpha/K\beta$ intensity ratios are about 6 for all elements, the interfering element concentration must be nearly two orders of magnitude greater than the analyte concentration before the sensitivity is degraded significantly. However the $L\alpha$ and M lines of interfering high Z elements are excited with 50 to 100% of the efficiency of the $K\alpha$ lines of analytes, so K/L and K/M interference can be somewhat more of a problem.

3.2.2 Escape peaks are significant in EDX spectrometers at X-ray energies just above the absorption edge energy of the detector element. In Si(Li) detectors this is from 1.84 to about 5 kev. After the detector absorbs the incident photon the detector characteristic X-ray (in this case Si K) is emitted, with significant probability of escape from the sensitive volume of the detector. The net energy deposited for an incident energy E is thus E – Si $K\alpha$ or E – 1.74 kev. The result is that each X-ray line of energy E is detected as a doublet, E and E – 1.74 kev, with the escape peak having an intensity of up to a few percent of the main peak.

The most important interferences caused by this in Si(Li) EDX spectrometers are K $K\alpha$ – Si $K\alpha$ overlapping Al $K\alpha$, and Ca $K\alpha$ – Si $K\alpha$ overlapping P $K\alpha$. The escape fraction is an instrumental constant and so corrections can be programmed at calibration time. Since the intensities of K and Ca K X-rays in environmental samples are usually one to two orders of magnitude higher than those of Al and P K X-rays, the corrections can be quite large even though the escape fraction is small.

3.2.3 WDX spectrometers can suffer from spectral overlap caused by X-rays from another element being diffracted at a higher order from the crystal and so falling at the same wavelength as the analyte. For example, the first order reflection for Ni $K\alpha$ X-rays (0.166 nm) is at the same Bragg angle as the second order reflection for Y $K\alpha$ X-rays (0.083 nm). This form of interference is eliminated in well-designed spectrometers by using an energy selective detector, such as a proportional counter.

3.3 BACKGROUND.

3.3.1 Several causes of instrumental background exist, most of which are reduced by good design. Scattered and characteristic X-rays from the inside of the spectrometer chamber have a much greater chance of being detected when analyzing thin specimens than they do when the more usual thick samples are measured. As a result "standard" WDX and EDX spectrometers may have to be modified for air particulate analysis.

3.3.2 Characteristic lines from the X-ray tube anode are an important source of background. Good instrument design will avoid problems due to this. Similar source-related background can arise from insufficiently pure radioisotope sources, or sources encapsulated in the wrong materials. Experienced XRF source manufacturers avoid this type of background.

3.3.3 A well-designed, "clean" EDXRF system will have background from two origins only. The first is source radiation scattered by the specimen. With monochromatic excitation this radiation will be concentrated at and just below the source energy, that is, resolved from the analytical lines. The remaining background consists of a fairly flat continuous spectrum extending from the source energy down to zero energy. This is produced in the detector as a result of incomplete charge collection from a small fraction of the incident photons. It can be minimized by the use of electronic guard rings and/or mechanical apertures. Note that this background is proportional to the intensity of scattered source radiation because the latter normally represents at least 90% of the radiation from the specimen.

3.3.4 The scattered background from the specimen is proportional to its mass per unit area and varies inversely as its average atomic number.

The filter substrate is the main origin of specimen-related background in analysis of ambient particle deposits. Glass fiber, quartz fiber and cellulose fiber, with values of mass per unit area of 8 to 10 mg/cm^2, produce the highest background, while Teflon membrane and polycarbonate (1 mg/cm^2) produce the lowest. Cellulose membrane (4 mg/cm^2) is intermediate.

3.3.5 Elements present as impurities or major constituents of filter substrates can produce unacceptably high background spectra. Certain glass fiber filters contain significant concentrations of some or all of the following elements: K, Ca, Fe, Zn, Sr and Ba **(3)**. In many cases these impurity concentrations vary from filter to filter, making accurate corrections impossible. Quartz fiber (spectrograde) is clean but unsuitable for measuring elements below S in the Periodic Table, because of its Si content. PVC membranes are likewise unsuitable for measuring elements below K, because of their Cl content. The Cl signal is so strong that they may be unsuitable for use at all in some spectrometers.

Teflon membrane, cellulose membrane, cellulose fiber and polycarbonate (e.g., Nuclepore) are essentially free from impurities causing X-ray line interference.

3.4 MATRIX EFFECTS.

Matrix absorption is the attenuation of the exciting and fluorescent radiation within the specimen. Since X-ray attenuation coefficients are strongly dependent on atomic number, any change in specimen composition will change the matrix absorption. Matrix absorption effects are negligible in airborne particle analysis because (i) the overall X-ray attenuation is small in these "thin specimens" and (ii) the concentrations of the higher-Z elements are low.

In thin pellets matrix absorption effects are small and can be corrected for by a preliminary measurement. Methods of correction have been reviewed **(5)**.

Matrix enhancement is secondary excitation of the analyte line by another spectral line in the sample. This is a smaller effect than matrix absorption and is negligible in air filter and biological material analysis.

3.5 PARTICLE SIZE.

Particle size effects cause changes in analyte line intensity due to X-ray attenuation within individual grains of particulate specimens.

3.5.1 Most airborne particle deposits contain particles less than about 10 μm diameter, with mean diameters of the order of 1 μm. Particle size corrections are only necessary below about 6 kev (Fe K X-rays) for this material. The magnitude of the correction increases rapidly with decreasing Z below Cl.

3.5.2 It has been shown that particle deposits on filters are monolayers of particles either in one plane on the surface of membrane filters, or effective monolayers (not in one plane) distributed in filters that allow particle penetration **(3,6,7)**. That is, the deposit mass per unit area does not exceed the limit where particle stacking occurs (except for extremely fine particles where particle attenuation is negligible). In this case the particles behave independently and the correction is simply a factor to be applied to the calibration for each element. No interelement effects occur unless the particles are combinations of fluorescing elements. An additional feature of the monolayer model is that the response (X-ray intensity vs element mass per unit area) remains linear even though the average mass per unit area exceeds the thin specimen limit (the thin specimen criterion assumes homogeneous, not particulate, material). This means that even for low-Z elements the calibration for particulate deposits is linear and independent of other element concentrations.

3.5.3 The particle size must be known in order to make the correction and find the slope of the calibration line. The particle size of specimens from dichotomous samplers (fine fraction 0–2.5 μm, coarse fraction 2.5–10 μm) is usually well-defined enough to make the correction to within $\pm$ 10% or better. The correction may be less accurate for samples taken with less particle size control.

3.5.4 When individual particles are combinations of elements, correction for particle size effects is more difficult. One important example is sulfur, which may occur in several combined forms such as $(NH_4)_2SO_4$, FeS_2 and $Pb\,SO_4$. If the combined form of sulfur is completely unknown, the error in sulfur determination may be as large as $\pm$ 34% **(8)**. However, the maximum error is reduced to a few percent for the fine fraction of a dichotomous sampler collected on a membrane filter **(7,8)**.

3.5.5 The particle size of biological material, even if ground to –400 mesh, is considerably greater than that of air filter samples. Also, specimens in the form of thin pellets are not monolayers. Particle size effects are expected, therefore, to be both greater and more complex, involving interelement interferences **(9)**. One mitigating factor is the low Z matrix which considerably reduces the intraparticle attenuation. In practice, errors due to particle size effects in thin pellets have been found to be acceptable **(5)**. Furthermore, the experimental procedure for measuring pellet absorption (Section **8.3.4**) to a large extent corrects for particle size effects.

3.6 SUBSTRATE ABSORPTION.

Errors due to substrate absorption can occur if some of the particle deposit penetrates below the surface of the collecting filter. Studies of this effect have been reviewed **(5)**. A correction can be made if the depth distribution is known **(8)**. However, not only is the precise depth distribution

unknown but it can vary with the compound and particle size collected. Fortunately the magnitude of the correction is not very sensitive to the depth distribution, especially for thin substrates. A preliminary measurement of "front-to-back" X-ray intensity ratio of all the elements to be determined, for the proposed substrate, will usually yield enough information to decide on the acceptability of the substrate and to estimate any substrate absorption corrections necessary (see Section **8.2.4**).

Substrate absorption effects are significant only for low-Z elements (line energies below about 5 kev) and for fibrous (nonmembrane) substrates. The front-to-back ratio method fails when insufficient signal can be obtained from the back of the filter. This occurs for elements below about phosphorus (line energies below 2 kev).

4. Precision and Accuracy

4.1 Precision is the repeatability of a given measurement, made under specified conditions. In XRF analysis the two main sources of imprecision are counting statistics and instrumental changes such as electronic drifts, mechanical drifts and sample placement errors. A measure of precision can be obtained by repeatedly analyzing one sample at least ten times, removing it and replacing it in different positions of the sample changer between analyses, while also exercising other instrument functions, as appropriate (e.g. source changer; helium/vacuum path on/off; amplifier gain changer; crystal spectrometer scanner). Assuming the errors follow a normal distribution, the precision can be defined as the relative standard deviation (RSD) of the mean concentration č, for each element monitored. (More than one element will be monitored to test the system properly). The RSD is the sum of the two sources of error:

$$\text{RSD} = \frac{100}{\bar{c}} \sqrt{\sigma_{STAT}^2 + \sigma_{INST}^2} \qquad (1)$$

where σ_{STAT} is the standard deviation due to counting statistics and σ_{INST} is the standard deviation due to instrumental changes.

The RSD due to instrument changes, $(100\ \sigma_{INST})/\bar{c}$, is a constant percentage of the signal and therefore of the element concentration. A well-designed and properly adjusted XRF spectrometer can be expected to have an instrumental RSD of 1% or less.

The RSD due to counting statistics, $(100\ \sigma_{STAT})/\bar{c}$, is inversely proportional to the square root of the signal (concentration). At high concentrations $100\ \sigma_{STAT}/\bar{c} < 100\ \sigma_{INST}/\bar{c}$ and the overall precision tends to 1%. At the Minimum Detectable Limit $(100\ \sigma_{STAT})/\bar{c} = 33\ 1/3\%$, by definition (see Section 2.1) and the precision is therefore 33 1/3%. The precision varies between these limits at intermediate element concentrations.

4.2   Accuracy is the degree of agreement between the analytical result and the true value or an accepted reference value.

The accuracy of XRF for analysis of airborne particles on filters has been assessed by means of interlaboratory comparisons using replicated synthetic samples (10). The concentrations of the samples were known independently of any analytical technique. The results indicated that an accuracy of ± 10% RSD or better could be expected for elements K and above in the Periodic Table. Larger errors were experienced for elements Al-K, but at the time the intercomparison was done (1974) many of the low-Z element corrections were developmental.

More recently, attempts to analyze NBS SRM 1648 (Air Particulate) have been unsuccessful owing to the fact that the SRM contains particles over 100 μm in diameter and is too heterogeneous to be accurately sampled in the necessary sub-milligram quantities onto membrane filter substrates. The accuracy of the thin pellet technique for analyzing biological samples is exemplified by Giauque's results (4) for NBS SRM 1571, Orchard Leaves, which are shown in Table 822B:II, Section 2.6.

5. Advantages and Disadvantages

5.1   XRF is the only analytical technique that combines all the following features: It is multielemental, non-destructive and rapid. A suitably calibrated system can analyze upwards of 100 samples per day for 20 elements per sample. The results are available immediately. The sample can be reanalyzed in case of questions. The method is not susceptible to errors associated with dissolution or extraction common to many other methods.

5.2   The sensitivity is adequate for direct analysis of all the most important pollution elements on air filters with the possible exception of Se and Cd in ambient air samples.

5.3   The capital cost of XRF equipment possessing all the necessary features is high (at least $100,000). This limits the use of the technique to central laboratories where the cost savings associated with high sample throughput can be realized. In this case costs are of the order of U.S. $1.00 per element, for direct air filter analysis.

5.4   Labor and other operating costs are quite low since the instrumentation is usually automated and sample preparation requirements are minimal. However it is quite important that an XRF specialist be available to handle maintenance and service problems and non-routine analyses.

6. Apparatus

6.1   An XRF spectrometer equipped for air filter analysis is required, having appropriate sensitivity and multielement capability (Section 2), and ability to correct for interferences with acceptable precision. Results of tests using filter specimens should be available. EDXRF systems should have a live time correction method of proven accuracy. All systems should have a suitable automatic sample changer, if large scale analysis projects are contemplated, with appropriate facilities for mounting air filter specimens or thin (fragile) pressed pellets. Automatic selection of excitation energy and system gain, and initiation of helium or vacuum paths (if used) is also desirable.

Vacuum X-ray spectrometers must be capable of being pumped down without damaging the most fragile membrane filter likely to be analyzed.

An X-ray specialist must be available who is qualified to advise on setting up, calibration and maintenance of the XRF analyzer.

The XRF analyzer should be installed in

a clean, dust-free area to avoid sample contamination.

6.2 Filter storage and handling equipment should include stainless steel, flat-tipped tweezers; a non-shear filter cutter or stainless steel scissors (if it is required to cut filters to fit the spectrometer); a supply of plastic petri dishes with labels (to hold specimens); and storage space. The filter handling and storage area should be clean, dust-free and draft-free. If air filters are pre-mounted for sampling, the mounts must be tested for compatibility with the X-ray analyzer, from the points of view of mechanical fit and X-ray background.

6.3 Supplies needed include cooling water for high-power X-ray tubes, liquid nitrogen for Si(Li) EDX detectors and compressed reagent grade helium for helium-filled spectrometer paths.

6.4 To prepare thin pellets from fine powders the following accessories are required: a 40,000 psi (275 MPa) pellet press with suitable die, and microbalance (10 μg sensitivity) with weighing papers and spatulas. Targets for pellet absorption correction measurements consist of metal foils or 1:1 mixtures of suitable element compounds (e.g. oxides, carbonates) with polyethylene powder, hot pressed into thin discs. The targets can contain several elements. Lucite spacer rings are also needed of thickness equal to the average pellet thickness, ID slightly greater than the pellet OD and OD small enough to fit in the spectrometer specimen charger.

6.5 Appropriate calibration standards are required (see Section **7**).

7. Calibration and Standards

7.1 Calibration involves (i) determination of the instrument response for all the analytes; (ii) identification and measurement of correction factors for spectral interferences; (iii) establishment of correction procedures for any matrix interferences, particle size effects and substrate absorption effects that might occur.

Different workers have employed different combinations of calibration standards and calculation methods to achieve these objectives. The approach described here is one self-consistent strategy.

Note that only calibration standards and methods are described in this section. Quality assurance standards and methods are described in Section **9**. Standard reference materials and interlaboratory comparisons, used for assessing accuracy, are described in Section **4**.

7.2 AIR FILTER STANDARDS.

Suitable calibration standards for air filter analysis must have the following properties: They should comprise known quantities of elements deposited uniformly on or in thin substrates so as to have a similar physical form to the specimens.

7.2.1 A substantial backing for mechanical support is not allowed on EDX standards, because of count rate limitations, but is allowed on WDX standards. The standards must have long-term stability of response (less than 1% of variation over the life of the standard) and should be accurately reproducible. They should be mechanically strong enough to withstand handling. WDX standards must also withstand repeated irradiation with a high power X-ray tube. They must also be able to withstand pressure cycling if used in vacuum spectrometers. The deposit uniformity should be such that different XRF analyzers (having different response profiles) will yield the same overall response (to approximately 1%) when the standard is repeatedly measured.

7.2.2 The element concentrations should be known independently of any analytical method. The standards should be prepared from pure stoichiometric materials using gravimetric or volumetric methods, with accuracy preferably better than the precision of the instrument. The concentration of each element should be far enough above the MDL to enable precise measurement (see Section **4.1**) while still being within the linear concentration range. If the standard contains more than one element, there should be no spectral or matrix interferences between the elements. Spectral analysis procedures that employ library spectra will require single-element standards.

7.2.3 Standards need not have a defined particle size as long as the particle size is small enough to avoid particle attenuation. However particulate standards with known particle size should be used to calibrate low Z element responses. The particulate deposit must not penetrate the substrate.

7.2.4 The standards need not be made on the same substrate as the specimens to be analyzed. Precise corrections for different substrates can be made as long as the substrates are available, thin and free from impurity spectral lines.

7.2.5 Thin pellets can be analyzed using the same standards and calibration procedures as described for particulate deposits on filters, with additional steps for measuring pellet absorption.

7.2.6 Calibration standards can be made using published methods **(4,11–14)** or obtained commercially **(15,16)**. The commercially-available standards have been tested in interlaboratory comparisons **(10)**.

7.3 CALIBRATION PROCEDURE.

7.3.1 Acquire at least one standard for each analyte and for each interfering element that may be present. The standards should be in the concentration range 10–100 $\mu g/cm^2$ (lower limit at least 10 × MDL, upper limit below the spectrometer count rate limit or the linear response limit, whichever is less). Substrates should be organic, free from elements that might give interfering X-rays and of total mass per unit area less than 10 mg/cm^2. A blank substrate of each type used by the standards must be available.

If some elements are not available, their response factors and spectral interference factors may be found by interpolation (see Section **7.3.5**) as long as the spectrum analysis algorithm employed by the instrument will allow this.

Standards for Al-Ca should be particulate deposits having a mean particle size close to the mean particle size of the samples to be analyzed. The substrates should be membrane filters that do not allow significant particle penetration.

Standards for S-U need not be particulate deposits with defined particle size as long as any particles present are less than 2 μm diameter. In the S-Ca range both particulate and non-particulate standards should be used to establish correspondence.

7.3.2 Measure each of the standards to determine the net analyte response (I_i) and also the net response for all possible spectrally interfering elements (I_j). I_j is the

signal due to element j in element i's window. Subtract background using a background sample of the same substrate and the method indicated by the instrument manufacturer. Calculate all spectral overlap factors (R_{ij}) using the method indicated by the instrument manufacturer. Note that all standards must be counted long enough to assure a preselected precision goal for each element.

7.3.3 If the non-particulate standards are homogeneously impregnated into their substrates, the substrate absorption factors must be found **(17)**. This can be done for each analyte by making two measurements with the standard and a blank substrate closely sandwiched together. In the first measurement the blank substrate is nearer the source-detector and the net intensity, I_i, for element i is given by

$$I_i = k \, I_p \exp(-\bar{\mu}m)_{i,s} \qquad (2)$$

where k is an instrumental constant; I_p is the source intensity; $\bar{\mu}$ is the substrate mass absorption coefficient for element; and m is the substrate mass per unit area.

In the second measurement the sandwich is turned over so that the standard is nearer the source-detector. The net intensity is now

$$I_{i,o} = k \, I_p \qquad (3)$$

The substrate absorption is

$$F_{i,s} = \frac{1 - \exp(-\bar{\mu}m)_{i,s}}{(\bar{\mu}m)_{i,s}} = \frac{1 - I_i/I_{i,o}}{\ln(I_{i,o}/I_i)} \qquad (4)$$

Note that this procedure assumes that the analyte is homogeneously impregnated throughout the thickness of the substrate. This can be checked by measuring the standard, then re-measuring it after turning it over. The two results should be equal to within the precisions of the standard and the instrument.

7.3.4 The particle size correction factor (P_i) is calculated for particulate deposits as follows

$$P_i = \frac{1 - \exp(-\bar{\mu}\rho d)}{\bar{\mu}\rho d} \qquad (5)$$

where $\bar{\mu}$ = is the average mass absorption coefficient of the fluorescent particles (cm^2/g); ρ is their density (g/cm^3); and d is the physical (not the aerodynamic) particle size (cm)

$$\bar{\mu} = \Sigma_k \, (\bar{\mu}_e \cosec \, \phi_e + \mu_f \cosec \, \phi_f)_k r_k \quad (6)$$

where ϕ_e and ϕ_f are the angles between the excitation and fluorescent beams, respectively, and the sample surface; $\mu_{e,k}$ is the mass absorption coefficient of element k at the exciting energy, $\mu_{f,k}$ is that for the fluorescent energy and r_k is the weight fraction of element k in the fluorescent particles. The summation is over all elements, k, in the fluorescent particles. Mass absorption coefficients are tabulated in Reference **18**.

This calculation is not necessary if it is assumed that the particle size of the standards and unknowns are equal and if it is not desired to compare the responses of the particulate and nonparticulate standards.

7.3.5 The absorption free, thin specimen sensitivities (i.e., slope or response factors) (S_i) for each element, i, are now calculated. They are

$$S_i = \frac{I_i}{m_i \, F_{i,s} \, P_i} \quad (7)$$

Note that $P_i = 1.00$ for non-particulate standards and $F_{i,s} = 1.00$ for particulate deposits on the surface of their substrates. A plot of S_i vs analyte Z is now made. It should be a series of smooth curves. Discontinuities will exist between L and K analyte lines and for different excitation energies.

If more than one standard is used for some analytes the S values can be averaged before plotting on the S vs Z curves, or plotted directly.

If no standard is available for a given analyte the S-factor can be obtained from the S vs Z curve by linear or non-linear interpolation. Any irregularities in the S vs Z curve are cause for questioning the accuracy of the particular standard, or the measurements or the calculations.

7.3.6 Correction procedures for non-thin specimens (i.e. pellets), finite particle size, and substrate absorption in analyzed specimens are described in Section **8**.

8. Analytical Procedure

8.1 SPECIMEN PREPARATION.

8.1.1 Incoming air filter samples may be of any size and shape from 20 cm × 25 cm Hi-vol filters to 37 mm diameter membrane filters. The XRF analyzer measures an area usually about 2 cm in diameter with, in most cases, a non-uniform sensitivity. It is, therefore, important that the particulate deposit on both large and small filters be uniform, otherwise the measurement will not be representative. Deposit uniformity on large filters can be checked by cutting out and analyzing several areas.

8.1.2 If possible, the substrate type should be selected beforehand, bearing in mind the interferences described in Section **3**.

8.1.3 Preparation of large air filters for analysis involves cutting out suitable areas without disturbing the particles on the area to be measured. Most XRF analyzers accept samples 2.5 to 5 cm in diameter. A shearless filter punch or a pair of scissors can be used to cut the specimens. Filters should be handled with flat-tipped tweezers.

8.1.4 It is important to keep track of individual specimens and their identification throughout handling and analysis. They should be stored in labelled plastic petri dishes when not actually in the sample changer.

8.1.5 37 and 47 mm diameter air filters can usually be analyzed directly. Some filters are pre-mounted in 5 cm square photographic slide holders for sampling and subsequent analysis. The XRF analyzer accepts a slide tray holding up to 80 slideholders.

8.1.6 Certain particle samples are highly non-uniform and small in area, such as those from impactor-type samplers. It is usually necessary to remove the particles from such samples and re-deposit them uniformly on a 37 mm diameter filter. Standard procedures have not been established for this.

8.1.7 If the final results are required in units of μg element per m^3 of air originally sampled, the mean air flow through the sample in m^3/cm^2 must be obtained from the sampling records.

8.1.8 Thin pellets of biological material are prepared as follows, starting from a fine (preferably – 400 mesh), homogenized, dry powder. 150 ± 1 mg of powder is weighed out and poured into a steel die approximately 2.5 cm in diameter. The diameter should be known to ±0.1 mm. The powder is stirred and the die tapped on a level surface to ensure an even distribution of the powder. The plunger is inserted slowly so as to avoid blowing powder out. A pellet is pressed using a pressure of at least 20000 psi (140 MPa). The pellet is carefully extracted and its mass per unit area calculated (m mg/cm^2).

8.2 ANALYSIS OF AIR FILTER SPECIMENS.

8.2.1 The elements to be analyzed are decided upon and the necessary excitation conditions and spectrometer settings are selected. S factors and R factors (see Section 7) are stored in memory for all elements to be analyzed. Count times are chosen for each excitation energy so as to achieve the desired element sensitivities in the minimum overall measurement time. Possible spectral overlap, particle size and substrate absorption problems are recognized and the analysis strategy planned.

8.2.2 The spectrometer is checked for performance to its specification (background, resolution, source strength, X-ray tube stability etc. as per manufacturer's instructions).

8.2.3 A blank substrate is analyzed to check for spectral interferences from the substrate.

8.2.4 The following procedure can be used to determine whether or not unacceptable substrate absorption effects will occur for any low Z analytes. First the maximum attenuation factor, F_{max} is found. A sample of blank substrate is sandwiched with a standard filter containing the analyte, i. Two measurements are made (as in Section 7.3.3), one with the substrate, and the second with the blank, facing the source/detector. If the standard is a particulate deposit, it should be turned over so that the deposited side faces the source detector for both measurements. The two X-ray intensities I_i and $I_{i,o}$ are given, as before, by equations 2 and 3, Section 7.3.3. $(F_{i,s})_{max}$, the maximum attenuation factor for element i in substrate s is defined as:

$$(F_{i,s})_{max} = \exp{(-\mu m)_{i,s}}$$
$$= I_i/I_{i,o} \tag{8}$$

This is the attenuation factor from front to back of the filter if the whole deposit is concentrated on the surface.

Next, several specimens typical of the batch to be analyzed are measured once

Table 822B:III Values of Maximum Front to Back Ratio for a Cellulose Membrane Filter

Element	$(F_{i,s})_{max}$
Al	0.0010
Si	0.0108
P	0.0564
S	0.1332
Cl	0.2129
K	0.4281
Ca	0.5073
Ti	0.7006
V	0.7321

Filter used was Millipore type SMWP, 5.0 mg/cm^2.

with their deposits facing the source/detector ("front") and a second time with the deposits facing away from the source/detector ("back"). The "front" to "back" ratio for element i, $(I_f/I_b)_i$, will approach the value $1/(F_{i,s})_{max}$ if the deposit is superficial and will approach unity if the deposit is homogeneous throughout the substrate thickness. Intermediate values result from indeterminate particulate distributions.

Analytical results for element i on or in substrate s will be acceptable if the "front" to "back" ratio is close to that expected for either a surface deposit or a homogeneous volume deposit. Table 822B:III lists experimental values of $(F_{i,s})_{max}$ for elements A1-V in a cellulose membrane substrate using an EDX system. Values for other substrates, elements, and instruments can be measured or calculated.

8.2.5 The following procedure can be employed to determine what particle size corrections should be applied, if any, to any low-Z analytes. If sampling is not size selective (e.g., Hi-vol), assume that the mean particle size is 3 μm and calculate the expected particle size correction P_i (see equation 5, Section 7.3.4) for each low Z analyte, using mass attenuation data from Reference **18** and assumed compositions for each particle (Al_2O_3, SiO_2, $NH_4H_2PO_4$, $(NH_4)_2SO_4$, NaCl, KNO_3 and $CaCO_3$ are reasonable for the seven lowest-Z elements in ambient air). A mean particle size must be assumed, according to the type of samples used. Calculate P_i as before. Table 822B:IV lists values of P_i for the above-mentioned 7 components and 3 particle sizes, assuming an excitation energy of 5.90

Table 822B:IV Calculated Values for the Particle Size Correction Factor

Analyte	Compound	Particle Size (μm)		
		1	3	5
Al	Al_2O_3	0.825	0.584	0.435
Si	SiO_2	0.907	0.753	0.634
P	$NH_4H_2PO_4$	0.951	0.861	0.783
S	$(NH_4)_2SO_4$	0.964	0.897	0.837
Cl	NaCl	0.938	0.830	0.738
K	KNO_3	0.970	0.912	0.860
Ca	$CaCO_3$	0.963	0.893	0.830

kev and normal incidence and take-off angles (cosec ϕ_e = cosec ϕ_f = 1).

Note that particulate low-Z standards are available (15) whose particle size matches that of the fine fraction from a dichotomous sample.

8.2.6 Load the XRF sample changer with specimens, taking care not to disturb any of the particulate deposits. The deposits should face the source/detector unless it has previously been determined that substrate absorption is negligible. A blank substrate must be included. Also quality control standards must be included to provide a check of precision at the upper and lower end of the elemental range analyzed by each exciting energy.

8.2.7 The background subtraction procedure must provide for subtraction of the blank for all elements, i, in addition to any spectral background stripping included in the instrument spectrum analysis algorithm. Spectral overlap (R factors) will be corrected according to the instrument spectrum analysis algorithm. The element mass per unit area, m_i μg/cm^2, is given by

$$(I_i)_{corr} = (S_i)_{corr}\, m_i \qquad (9)$$

where $(I_i)_{corr}$ counts/s is the count rate for element i corrected for spectrometer deadtime, background, blank substrate spectrum and spectral overlap; and $(S_i)_{corr}$ counts/s per μg/cm^2 is the response factor obtained at calibration time (Section 7.3.5), updated for source decay if radioactive source excitation is used.

The concentration, C_i, of element i in μg/m^3 is given by

$$C_i = m_i/Q \ \mu g/m^3 \qquad (10)$$

where Q is the air flow through the samples in m^3/cm^2 of deposit.

The computer coupled to the XRF analyzer should be programmed to type out this result along with an estimate of the instrumental error (derived from equation 1, as modified by the background and overlap correction procedures used by the particular instrument).

Correction factors for substrate absorption and particle size should be applied according to the procedures described in Sections **8.2.4** and **8.2.5**.

8.3 ANALYSIS OF THIN PELLETS.

8.3.1 Decide upon the elements to be analyzed and select the necessary excitation conditions and spectrometer settings according to the manufacturer's instructions. Read into memory the necessary R and S factors (see Sections **7.3.2** and **7.3.5**). Choose count times for each excitation energy so as to achieve the desired element sensitivities in the minimum overall measurement time.

8.3.2 Check the XRF analyzer for performance according to its specification (background, resolution, source strength, accuracy of live time correction, X-ray tube stability etc.).

8.3.3 Load the XRF sample changer with thin pellets, taking care not to allow any broken pellets or pellet fragments to compromise the measurement. A "blank" pellet pressed from clean cellulose powder should be included. Also enough quality control standards should be included to provide precision checks at the upper and lower end of the elemental range analyzed by each exciting energy. Run the analysis.

8.3.4 In order to produce the data required for the pellet absorption correction, repeat the procedure described in Section **8.3.3** for a selected number of pellets, with the following modifications. Place a target foil (Section **6.4**) on top of each pellet, selecting enough foils and pellets to enable pellet absorption corrections to be made for all analytes and each different pellet matrix. Rerun the analysis. Then remove the pellets, using the spacers (Section **6.4**) so that the target foils are left exactly in place. Rerun the analysis.

The pellet mass absorption factor, A_i, is given by the following equation, for each element, i:

$$A_i = \frac{(I_{PT} - I_P)_i}{(I_T)_i} = \exp(-\bar{\mu}m_P)_i \quad (11)$$

where $(I_{PT})_i$, $(I_P)_i$ and $I_T)_i$ are the count rates (after background, blank and spectral overlap corrections) of element i for the pellet plus target foil, pellet alone and target foil alone, respectively; $\bar{\mu}$ is the average mass absorption coefficient of the pellet for both excitation and fluorescent radiation (inclusive of excitation and "take-off" angles); and m_P is the pellet mass per unit area (mg/cm^2).

8.3.5 Calculate the concentration, C_i, $\mu g/g$, of each element as follows:

$$C_i = \frac{10^3 m_i \; \ln(1/A_i)}{m_P \; (1 - A_i)} \; \mu g/g \quad (12)$$

where m_i is the apparent element mass per unit area, before absorption correction, as given by equation 9.

9. Quality Control

9.1 Reference **(19)** describes quality control policies and practices that are required to assure reliable, error-free results, especially for large-scale analysis projects in which X-ray fluorescence is often employed.

The main aspects of quality control are (i) prevention of sample mixups; (ii) maintenance of instrument precision, both short term and long term; and (iii) periodic testing of instrument accuracy using externally-supplied audit samples; (iv) validation of analytical results and (v) record keeping.

9.2 Samples should be logged in and individually inspected immediately on receipt in the XRF laboratory. Acknowledgement of receipt and description of any discrepancies and damage should be made. Each analytical specimen should be kept in an individually-labelled container at all times except when actually being analyzed. Sam-

ples should be stored under conditions that prevent deterioration or damage while under control of the X-ray laboratory. Packing and shipping samples from the X-ray laboratory should be done in a fashion that prevents damage. Chain-of-custody documentation should be kept (logging out, acknowledgements of receipt, etc.).

9.3 The instrument must be maintained uncontaminated, accurately calibrated and in satisfactory operating condition.

9.3.1 To test for spectrometer contamination, a "master blank" should be kept for re-analysis with each sample load. In addition, at least one specimen of blank substrate per batch of air filters should be analyzed to check for specimen blank drift.

9.3.2 Instrument precision is tested by analyzing standards (which can be synthetic and must be of proven long term stability) with each sample load as per Sections **8.2.6** and **8.3.3**

9.3.3 If the "precision" standards are the same as the calibration standards, instrument calibration will be checked at the same time. Otherwise the calibration should be checked before each sample load.

9.3.4 Instrument gain, resolution and source performance should also be checked periodically. Many instruments perform these checks automatically during or after each sample analysis. Otherwise the checks should be made before each sample load.

9.3.5 All the abovementioned checks should also be performed after any routine maintenance operation, such as refilling the EDX detector with LN_2.

9.3.6 The criterion for satisfactory instrument precision is based on equation 1 (Section **4.1**). The actual relative standard deviation of a series of repeat measurements should not be significantly greater than the specifications established during initial commissioning of the analyzer. Individual outlying results (greater than 2σ or 3σ from the running mean, depending on the criteria adopted) will be cause for investigation. Running means for each standard and blank should be kept for the duration of every analysis project.

9.3.7 The criterion for accuracy of instrument calibration is based on comparison of the calibration factors obtained in the above QC operations with corresponding ones obtained during the most recent establishment (or revision) of the master calibration (see Section **7.3.5**). Acceptance/rejection criteria should be established as part of the planning of the analysis project.

9.3.8 The master calibration (Section 7.3.5) should be revised or completely repeated after every major maintenance operation (e.g. change of X-ray tube, radioactive source, detector, amplifier). It should also be done after every change in the inventory of calibration standards. Otherwise the master calibration should be updated at regular intervals, say 2 or 3 times per year.

9.3.9 Instrument accuracy should be periodically tested using externally supplied audit samples if available.

Another method is to take part in interlaboratory comparisons with other laboratories employing the same and different analytical techniques. To be of value, interlaboratory comparisons must be carefully planned and executed (**10**).

9.4 Procedures for validation of analytical results include analyzing blind replicate samples and having a third party check the data at various stages. Audit samples are also used. At least one specimen per sample load should be selected at random and re-analyzed in the next sample load. Out-of-tolerance results will be cause for examination of sample handling procedures and instrument performance.

Diagnostic data should be included in the analyzer print out to test the functioning of the data analysis and output.

9.5 Records should be kept of every maintenance operation and of the results of every gain, resolution, blank, precision and calibration check, preferably using a standard format. Records should also be kept of action taken to remedy out-of-tolerance results. The most recent master calibration should be kept on file together with earlier versions of it, as appropriate. Results of all interlaboratory comparisons and analyses of audit samples should be kept on file.

10. References

1. JAKLEVIC, J. M. AND R. L. WALTER, (1977) "Comparison of Minimum Detectable Limits Among X-Ray Spectrometers" in "X-Ray Fluores-

cence Analysis of Environmental Samples", T. G. Dzubay, Ed., Ann Arbor Science, pp. 63–75.
2. CURRIE, L. A. (1977) "Detection and Quantitation in X-ray Fluorescence Spectrometry", ibid., pp. 289–306.
3. RHODES, J. R. (1973) "Energy Dispersive X-ray Spectrometry for Multielement Pollution Analysis", American Lab., July, pp. 57–73.
4. GIAUQUE, R. D. (1973) "Trace Element Determination with Semiconductor Detector X-Ray Spectrometers", Anal. Chem. *45*, 671–681.
5. RHODES, J. R. (1979) "Techniques for Minimising Interference in Energy Dispersive X-Ray Emission Analysis" (a review), in "Practical Aspects of Energy-Dispersive X-Ray Fluorescence Analysis", IAEA Report 216, International Atomic Energy Agency, Vienna.
6. RHODES, J. R. AND C. B. HUNTER, (1972) "Particle Size Effects in X-Ray Emission Analysis", X-Ray Spectrometry *1*, 113–117 and 107–111.
7. LOO, B. W., R. C. GATTI, B. Y. H. LIUS, C. S. KIM, AND T. G. DZUBAY, (1977) "Absorption Corrections for Submicron Sulfur Collected on Filters", in "X-Ray Fluorescence Analysis of Environmental Samples", T. G. Dzubay, Ed., Ann Arbor Science, pp 181–202.
8. DZUBAY, T.G. AND R. O. NELSON, (1975) "Self Absorption Corrections for X-Ray Fluorescence Analysis of Aerosols" in "Advances in X-Ray Analysis", W.L. Pickles, Ed., Vol 18 pp 619–631, Plenum Press, New York.
9. BERRY, P. F., T. FURUTA AND J. R. RHODES, (1969), Particle Size Effects in X-Ray Spectrometry", in Advances in X-Ray Analysis, Vol 12, pp 612–632; C. S. Barrett et al, Eds., Plenum Press, New York,
10. CAMP, D. C., A. L. VAN LEHN, J. R. RHODES, AND A. H. PRADZYNSKI, (1975) "Intercomparisons of Trace Element Determinations in Simulated and Real Air Particulate Samples", X-Ray Spectrometry *4* 123–137.
11. PRADZYNSKI, A. H. AND J. R. RHODES (1976) "Development of Synthetic Standard Samples for Trace Analysis of Air Particulates", A.S.T.M. Special Technical Publication 598, pp 320–336, A.S.T.M., 1916 Race St, Philadelphia, PA 19103.
12. BAUM, R. M., R. O. WILLIS, R. L. WALTER, W. F. GUTKNECHT, AND A. R. STILES, (1977) "Solution Deposited Standards Using a Capillary Matrix and Lyophilization", in "X-ray Fluorescence Analysis of Environmental Samples" T. G. Dzubay, Ed., Ann Arbor Science, pp 165–173.
13. SEMMLER, R. A., R. D. DRAFTZ AND J. PURETZ, (1977) "Thin Layer Standards for The Calibration of X-Ray Spectrometers", ibid, pp. 181–184.
14. DZUBAY, T. G., N. MOROSOFF, G. L. WHITAKER, AND H. YASUDA, (1980). "Evaluation of Polymer Film Standards for X-Ray Spectrometers" in Proceedings of the 4th Symposium on Electron Microscopy and X-Ray Applications to Environmental Health and Occupational Health, The Pennsylvania State University, State College, PA.
15. COLUMBIA SCIENTIFIC INDUSTRIES, P.O. Box 203190, Austin, TX 78720, U.S.A. Also X-Ray Spectrometry *6* 171–173, 1977.
16. MICROMATTER CORP., Rtel, Box 72-B, Eastsound, WA, 98245, U.S.A.
17. RHODES, J. R. "Recommended Procedures for the

use of C.S.I. Thin Standards for X-Ray Fluorescence Spectrometry," ARD Internal Report 206, available from Columbia Scientific Industries, P.O. Box 203190, Austin, TX 78720, U.S.A.

18. LEROUX, J. AND T. P. THINH, (1977) "Revised Tables of X-Ray Mass Attenuation Coefficients", Corporation Scientifique Claisse Inc. 7-1301 Place de Mérici, Québec Canada Q1S 4N8.

19. DUX, J. P. (1983) "Quality Assurance in the Analytical Laboratory", American Laboratory, July, pp 54–63.

Subcommittee 6

J. R. RHODES
M. T. KLEINMAN, *Chairman*
R. J. THOMPSON
T. C. RAINS
W. J. COURTNEY
V. P. GUINN

824.

Determination of Airborne Sulfates

1. Principle of the Method

1.1 SCOPE. The method is intended for the measurement of atmospheric sulfate concentrations. In certain environments, e.g., in a sulfuric acid manufacturing facility, H_2SO_4 is known to be the dominant airborne aerosol sulfur compound. In such cases, this method provides a measurement of ambient aerosol sulfuric acid.

1.2 SAMPLING. Sulfuric acid/sulfate aerosol is collected by filtration.

1.3 SAMPLE PREPARATION. The collected sulfate is leached from the filter with aqueous extractant, typically pure water.

1.4 ANALYSIS. Any method listed under 720. DETERMINATION OF AQUEOUS SULFATE is suitable, provided the sample size is adequate for the method.

2. Range and Sensitivity

2.1 The range of interest for ambient-atmosphere sulfuric acid/sulfate is from < $1 \mu g/m^3$ to $100 \mu g/m^3$. The range of interest of sulfuric acid mist for occupational safety is $100 \mu g/m^3$ to $3 mg/m^3$ with maximum precision and accuracy at $H_2SO_4 = 1 mg/m^3$ **(1)**.

2.2 At the lowest ambient concentrations a 1-m^3 to 1000-m^3 air sample may be required, depending on the analytical method chosen and especially on the filter blank and reagent blank. At the concentrations of interest for occupational safety, 1 m^3 should be ample for any of the listed analytical methods.

3. Interferences

3.1 Any filter that causes SO_2 to be retained and oxidized to sulfate presents a potential positive interference. These include glass-fiber filters **(1–5)**, quartz-fiber filters that have been strengthened by alkali treatment **(2–4)**, and nylon filters **(6)**. Suitable filter media include polytetrafluoro-ethylene (PTFE Teflon) membrane, Teflon-impregnated glass fiber, or pH-neutral quartz fiber. The last-named does generate a small amount of artifact sulfate **(2–5)**. Acid-treated quartz-fiber filters have, however, been reported to be artifact-free **(7)**. Polycarbonate can also be used **(8)**.

3.2 Alkaline particulate matter collected on the filter can cause a positive interference, by causing SO_2 to be retained and oxidized. Ordinarily this effect is insignificant but can be suppressed if necessary by using denuder tubes coated with MgO or Na_2CO_3 to remove the SO_2 upstream of the filter **(6)**.

3.3 Aerosol sulfur in lower oxidation states is a potential positive interference. This is particularly true of S(IV) compounds, which in aqueous solution are easily oxidized to sulfate. Aerosol S(IV) has been identified in ambient air **(9–12)**, though at levels significantly less than concurrently present sulfate concentrations.

3.4 Any cationic species that forms an insoluble sulfate, notably barium or lead, is a potential negative interference. Cation exchange often can be employed to remove the interfering species during extraction **(13)** but this may generate a positive error from sulfate formed by decomposition of the resin and, in some cases **(14)**, may not even be effective.

3.5 Hydrophobic organic particulate material can cause a negative interference by preventing complete extraction of the sulfate by water. In that case, extraction with aqueous (1:1 v/v) 2-propanol is recommended **(11)**.

3.6 Although no filter can quantitatively collect particles of all sizes, quartz-filters intended for atmospheric sampling, or membrane filters of mean pore diameter of $\leq 2 \mu m$, are highly efficient for the collection of H_2SO_4/sulfate aerosols.

3.7 For interferences specific to the analytic method employed, see 720. DETERMINATION OF AQUEOUS SULFATE.

4. Precision and Accuracy

The sampling, extraction, and analysis steps all contribute to the precision and accuracy of the calculated airborne concentration.

4.1 The standard high-volume (Hi-Vol) sampler provides the greatest sample size for a given sampling period (flow rate 1–2 m^3/min) at the cost of flow accuracy. Calibration of the Hi-Vol, by one of the orifice plate or Venturi-type calibration kits commercially available for this use, can give an accuracy of 5% or better if the calibration includes the flow rates at which the sampling is conducted. Flow chart recorders are useful on Hi-Vols but cannot give accurate flow readings unless the whole assembly is checked against the calibration unit, and the manufacturer's calibration of the Hi-Vol/recorder loses accuracy with continued use of the Hi-Vol. The Hi-Vol is neither precise nor stable unless equipped with a flow controller to compensate for line voltage fluctuations and for filter-loading effects. Commercially available flow controllers can hold the flow constant to within a few percent in the flow rate region of 1–2 m^3/min. The ill-defined inlet particle size cutoff and the dependence of inlet particle size cutoff upon wind speed and direction are major limitations of the standard Hi-Vol samplers. However, commercially available inlet cyclone preseparators and Hi-Vol cascade impactor stages may be used as countermeasures.

4.2 The best sampling accuracy is obtained with a bellows-type positive displacement pump; this design is inherently free of air leakage paths to the outside. These pumps are commercially available for ambient sampling at unrestricted flow capacities up to 0.16 m^3/min. However, because of the pressure drop across the sample filter, the unrestricted capacity will not be realized, especially with membrane filters. With such a pump, all air drawn through the filter can be measured continuously through a calibrated wet or dry test meter attached to the pump exhaust port. Commercially available dry test meters can be readily calibrated against a spirometer or meter prover (bell prover, see Ref. **16**) with an accuracy better than 1%. Temperature-compensated dry test meters are available for use in variable-temperature environments, but normalization to STP conditions still requires consideration of the barometric pressure and also the meter's specified reference temperature. To obtain a true time-averaged sample, the air flow should be held constant throughout the sampling operation. Flow control can be accomplished within a few percent by means of a vacuum regulator on the pump inlet to maintain the inlet at a constant pressure and hence, at a constant flow rate, as the filter loads up. A flow regulator on the pump outlet serves the same purpose; however, the sacrifice in maximum attainable flow rate is greater with an outlet flow regulator than an inlet vacuum regulator.

4.3 The dichotomous sampler or virtual impactor, as presently designed, with a positive displacement pump and rotameters to indicate the flows, provides a flow accuracy intermediate between options **4.1** and **4.2**. The flows are neither recorded nor totalized, so the accuracy and precision depend on the operator's ability to set and read the flow, usually about ±5%. The stability depends on frequent checking and adjusting of the rotameter readings; unattended, the instrument has a flow stability no better than ±10% in 24 hours. Inlets with well-defined particles size cutoffs have been designed for these instruments.

4.4 Extraction uncertainties generally stem from incomplete extraction. Good recovery of an internal standard does not guarantee that the extraction procedure is adequate. If cation exchange resin is employed during extraction to remove cationic interferences **(13)**, sulfate from partial decomposition of the resin may add a significant and variable blank; therefore it is important to wash the resin thoroughly with water shortly before use. Even without resin, certain filter types indigenously contain extractable sulfate which must be evaluated as a blank for each filter lot used. Finally, some insoluble sulfates are not extracted even if resin is used **(13)**. In that event, total aerosol sulfur as determined by x-ray fluorescence or combustion methods, which do not require extraction, is a good approximation for aerosol sulfate (14); ambient aerosol sulfur is principally sulfate (9–12).

5. Advantages and Disadvantages

5.1 ADVANTAGES. The method is straightforward and employs standard apparatus.

5.2 DISADVANTAGES.

5.2.1 The method involves collection and subsequent analysis. Therefore, temporal resolution is poor and the results are not available until later.

5.2.2 The method gives no information as to what the sulfate species are.

5.2.3 If the analytical method chosen requires extraction (as in 720. DETERMINATION OF AQUEOUS SULFATE), incomplete extraction is a problem. Analytical methods that do not require extraction give total aerosol sulfur, not just sulfate.

6. Apparatus

6.1 SAMPLING. The apparatus is that generally used for airborne particulate sampling.

6.1.1 *Hi-Vol.* Calibrated standard Hi-Vol sampler, preferably with internal motor cooling, flow controller, and motor voltage meter (0–120 V A.C.). Resettable A.C. elapsed timer, wired across the on/off switch. Circular-chart flow-rate recorder (wired directly to the A.C. line rather than across the Hi-Vol on/off switch). Exchangeable filter-holder sampling heads (a number of designs are suitable). Filters 8″ × 10″ (20 × 25 cm., choice of type will depend on intended analyses other than sulfate). Cassettes or other means of transporting filter to and from Hi-Vol. Spare flow-recorder charts, preferably to be changed with each sample or at least once per 24-h chart revolution. Recorder ink. Spare motor brushes. Depending on application intended, a cyclone preseparator or a Hi-Vol impactor may be mounted on the inlet to define the inlet particle size cutoff. Hi-Vol calibration kit.

6.1.2 *Filter with Positive-displacement Pump.* In order, from inlet to exhaust: Size-selective inlet (if used), filter in holder, vacuum regulator (if used) with a vacuum gauge or absolute-pressure gauge on the upstream side, bellows-type positive displacement pump (e.g., from Metal Bellows Corp., Sharon, MA) equipped with on/off switch and resettable elapsed timer (wired across the switch), and calibrated dry gas meter (preferably temperature-compensated). Tubing from filter to pump must be strong enough to withstand the vacuum generated (often over 0.5 atmosphere, 51 kPa). Also needed: Spare filters. To change filters it is convenient to have spare holders already loaded and in the field change holders only. It is also desirable to install easily disengageable connectors (e.g., Quick Connect, Crawford Fitting Co., Cleveland, OH) at filter inlet and outlet.

6.1.3 *Virtual Impactor (Dichotomous Sampler).* In order, from inlet to exhaust: Size-selective inlet (10 or 15 μm, supplied with the instrument), cassette containing prelabelled filters, virtual impactor of either the automatic or manual type with on/off switch, elapsed timer, rotameters to read coarse and total flows, and valves to adjust coarse and total flows. Also needed: a second cassette for uninterrupted sampling, a carrier to transport cassettes to and from the sampler, and spare filters.

6.2 EXTRACTION. For the analytical methods requiring extraction, apparatus used depends on the procedure adopted but will generally include an ultrasonic bath and/or a vortex shaker, a hot plate or microwave oven, pipettes, volumetric flasks, beakers, and a 0–100°C thermometer. Wide-mouth screw-capped polypropylene or polymethylpentene jars (100 mL capacity) are suitable as extraction vessels. Teflon digestion vessels (e.g., from Savillex Corp, Minnetonka, MN), or PTFE-lined stainless steel digestion bombs (e.g., from Parr Instrument Co., Moline, IL) can also be used. CAUTION: Do not use chromic acid ($H_2SO_4 \cdot XCrO_3$) cleaning solution on any of the apparatus as the sulfate contamination is very difficult to remove.

6.3 ANALYSIS. Depends on the analytical method. See 720. DETERMINATION OF AQUEOUS SULFATE.

7. Reageants

7.1 SAMPLING. None.
7.2 EXTRACTION (for analytical methods requiring extraction).
7.2.1 *Water.* Deionized, meeting ASTM specifications for Reagent Water, Type II.
7.2.2 *2-Propanol.* 50% by volume in water, if required.
7.2.3 *Ion Exchange Resin.* (If used). Any strong-acid cation exchange resin, hydrogen form, washed thoroughly with deionized water shortly before use.
7.3 ANALYSIS. See 720. DETERMINATION OF AQUEOUS SULFATE.

8. Procedure

8.1 SAMPLING. Procedures are those generally used for airborne particulate filter sampling.
8.1.1 *Hi-Vol.* The flow should be set low enough so that the flow controller will be able to control the flow throughout the sample collection period. (Once the flow controller has increased the motor voltage to the line voltage, it can control no further.) The motor voltage, timer, and flow chart should be checked after the run commences, and preferably from time to time thereafter. Care should be taken in mounting the flow chart to make sure that the correct time is lined up on the stylus. A clean place without drafts should be chosen for loading filters into holders and removing them. The filters should be handled with forceps or gloved hands, not with bare hands.
8.1.2 *Filter and Positive Displacement Pump.* The vacuum regulator, if used, should be set low enough to allow it to control throughout the run as the filter loads up. Before the sampling commences, a test should be conducted with the filter holder inlet stoppered to make sure there are no leaks upstream of the pump outlet (zero flow should be observed). Also before sampling begins, the unit should be run to make sure that the flow rate is as expected. The unit should not be operated in such a confined space that the heat buildup from pump operation unduly raises the temperature of the gas displacement meter.

8.1.3 *Virtual Impactor (Dichotomous Sampler).* It is imperative to maintain the total flow and coarse flow at their nominal values (for example, on some units 16.7 and 1.67 L/min, respectively) in order to have the unit operate as designed, with a cut point between coarse and fine at 2.5 μm. The coarse-particle filter is meant to collect 10% of the total flow and 10% of the fine particles along with 100% of the coarse particles; it is imperative that the ratio of flows be accurately known so that the calculation of coarse-particle sulfate makes proper correction for fine sulfate in the coarse-particle filter. Ordinarily, most of the sulfate will be on the "fine" filter; nevertheless, the "coarse" filters should be handled gently as the large particles are easily dislodged.
8.2 EXTRACTION.
8.2.1 For most purposes the following is adequate (17): Place the filter sample (or a known portion thereof; e.g., a quarter) in 10–100 mL deionized water (enough to cover it). Agitate the sample on the vortex mixer or in the ultrasonic bath for 30 min. Digest at 65°C for 3 h. Repeat the ultrasonic or vortex agitation for another 30 min. Remove the filter and wash it with deionized water, adding the washings to the sample. If the filter is quartz fiber it will have disintegrated by this point; the fibers can be removed by filtration through a filter apparatus that has been thoroughly washed beforehand; after the sample has been filtered, more deionized water is passed through the filter apparatus and the wash is added to the sample (filtrate). It is important that the blank be carried through this procedure. Dilute the extract to a desired known volume (typically 100 mL for $^1/_4$ of a 8″ × 10″ filter) in a volumetric flask.
8.2.2 Teflon membrane filters are difficult to extract because of their hydrophobic nature and tendency to float on the water surface. The most convenient method to quantitatively extract sulfate from such filters involves the following method (18,19). Place the uncut filter (up to 47 mm dia.) unfolded, sample side up, in a 100 mL capacity plastic jar. Weigh down the filter with short sections of 3-mm dia. glass rod bent into a "V", touching the filter at two points. Add 10–20 mL deionized water, suf-

ficient to cover the filter completely. The apex of the glass rod should extend above the liquid level. Cap the jar. Extract the filters for 30 min at room temperature, up to batches of eight, distributed uniformly within an ultrasonic bath. Adjust the liquid level of the bath to be equal to that in the samples. After extraction, filter through a prewashed filter apparatus as in **8.2.1.**

8.2.3 When cationic interference is expected, add 3 to 5 mL of a deionized water-slurry of cation-exchange resin to the sample at the beginning of extraction. The resin is removed during the filtration step in **8.2.1.**

8.2.4 When organic material is expected to interfere with complete extraction, conduct the extraction in aqueous 2-propanol (1:1 v/v) instead of water. In this case, ion-exchange resin should not be used without extensive prior washing of the resin with aqueous 2-propanol.

8.2.5 When the analysis is to be conducted by suppressed ion chromatography, a better chromatogram may be obtained if the extraction is conducted in the anion-chromatographic eluent solution, i.e., $Na_2CO_3/NaHCO_3$ (See Method 720A). If the determination is by nonsuppressed ion chromatography (see Method 720B), the p-hydroxybenzoate or phthalate used as eluent is conveniently used as extractant. These extractants cannot be used if certain other species in the extract, particularly H^+, are to be determined.

8.2.6 Regardless of the extraction procedure used, blank filters of the same type (and preferably of the same lot) must be carried through the extraction and analysis procedures along with the samples.

8.3 ANALYSIS. Depends on the analytical method. See 720. DETERMINATION OF AQUEOUS SULFATE.

9. Calibration and Standards

9.1 SAMPLING. Procedures are those generally used for filter sampling of airborne particulate matter. A few details are as follows.

9.1.1 *Hi-Vol.* The flow chart or other flow-reading device on the unit must be calibrated against a commercially available Hi-Vol calibration kit at the flow rates used

for sampling. This should be performed, at the very least, whenever the Hi-Vol is moved.

9.1.2 *Positive Displacement Pump*: The dry gas meter must be calibrated against a spirometer or meter prover at the flow rates used in sampling. If the discrepancy exceeds a few percent, the meter should be adjusted to read nearer the correct displacement. In carrying out the calibration, care should be taken that the meter is not hooked up in such a way as to be operating significantly above or below atmospheric pressure. Once calibrated, the meter should be handled with care.

9.1.3 *Virtual Impactor (Dichotomous Sampler).* The flowmeters are generally considered calibrated. The total-flow reading can be checked by connecting the virtual impactor inlet to a spirometer. Individual calibration for the coarse or fine fraction is more involved and should be conducted according to the manufacturer's instructions. Corrections for barometric pressure, if applied, should be carried out in accordance with data supplied for the flowmeters.

9.2 ANALYSIS. Depends on the analytical method. See 720. DETERMINATION OF AQUEOUS SULFATE.

10. Calculations

10.1 HI-VOL. The whole filter is seldom utilized for sulfate analysis. The deposit is assumed to be uniform and the sulfate on the whole filter (millimoles, μg, etc.) is calculated from that on the section analyzed. The whole-filter sulfate blank is subtracted. The net sulfate is divided by the calibration-corrected air sample volume, the latter having been determined as average calibration-corrected flow rate multiplied by sampling time as read on the elapsed timer.

10.2 POSITIVE-DISPLACEMENT PUMP. The sulfate on the filter is corrected for blank and divided by the calibration-corrected total gas displacement determined from difference between initial and final dry gas meter readings.

10.3 *Virtual Impactor (Dichotomous Sampler).* Subtract the sulfate blank from the coarse-filter sulfate measurements. Subtract 1/9 of the blank-corrected fine-filter sulfate value from the blank-corrected coarse-filter sulfate value; the difference is the coarse-particle sulfate mass corrected for the 10% of the fine-particle sulfate that collects on the coarse-particle filter. Multiply the blank-corrected fine-filter sulfate value by [10/9]; the product is the fine-particle sulfate mass including that which was collected on the coarse-particle filter. Divide the coarse- and fine-particle sulfate masses by total air volume to obtain coarse and fine sulfate atmospheric concentrations (μ^3). Total air volume is the calibrated flowmeter reading (e.g., m^3/h) corrected for the barometric-pressure effect applicable at the sampling site, multiplied by elapsed collection time. These calculations of coarse and fine sulfate concentrations presuppose that $1/10$ of the total flow passed through the coarse filter as designed.

11. References

1. LEE, R. E., JR. AND J. WAGMAN. 1966. A Sampling Anomaly in the Determination of Atmospheric Sulfate Concentration, J. Amer. Ind. Hyg. Assoc. 27:266–271.
2. PIERSON, W. R., R. H. HAMMERLE AND W. W. BRACHACZEK. 1976. Sulfate Formed by Interaction of Sulfur Dioxide with Filters and Aerosol Deposits, Anal. Chem. 48:1808–1811.
3. PIERSON, W. R. 1977. Spurious Sulfate in Aerosol Sampling: A Review, Paper #59, Division of Environmental Chemistry, American Chemical Society 173rd National Meeting, New Orleans, LA, March 20–25, 1977.
4. PIERSON, W. R., W. W. BRACHACZEK, T. J. KORNISKI, T. J. TRUEX AND J. W. BUTLER. 1980. Artifact Formation of Sulfate, Nitrate, and Hydrogen Ion on Backup Filters: Allegheny Mountain Experiment, J. Air Pollut. Control Assoc. 30:30–34.
5. APPEL, B. R., Y. TOKIWA, M. HAIK AND E. L. KOTHNY. 1984. Artifact Particulate Sulfate and Nitrate Formation on Filter Media, Atmos. Environ. 18:409–416.
6. JAPAR, S. M. AND W. W. BRACHACZEK. 1985. Artifact Sulfate Formation from SO_2 on Nylon Filters, Atmos. Environ. 18:2479–2482.
7. TANNER, R. L., R. CEDERWALL, R. GARBER, D. LEAHY, W. MARLOW, R. MEYERS, M. PHILLIPS AND L. NEWMAN. 1977. Separation and Analysis of Aerosol Sulfate Species at Ambient Concentrations. Atmos. Environ. 11:955–966.
8. FROHLIGER, J. F. Private communication, Graduate School of Public Health, University of Pittsburgh, Pittsburgh, PA 15213.
9. EATOUGH, D. J., T. MAJOR, J. RYDER, M. HILL, N. F. MANGELSON, N. L. EATOUGH, L. D. HANSEN, R. G. MEISENHEIMER, AND J. W. FISCHER. 1978. The Formation and Stability of Sulfite Species in Aerosols, Atmos. Environ. 12:263–271.
10. EATOUGH, D. J., B. E. RICHTER, N. L. EATOUGH AND L. D. HANSEN. 1981. Sulfur Chemistry in Smelter and Power Plant Plumes in the Western U.S., Atmos. Environ. 15:2241–2253.
11. EATOUGH, D. J., J. J. CHRISTENSEN, N. L. EATOUGH, M. W. HILL, T. D. MAJOR, N. R. MANGELSON, M. E. POST, J. R. RYDER, L. D. HANSEN, R. G. MEISENHEIMER AND J. W. FISCHER. 1982. Sulfur Chemistry in a Copper Smelter Plume, Atmos. Environ. 16:1001–1015.
12. EATOUGH, D. J. AND L. D. HANSEN. 1983. Organic and Inorganic S(IV) Compounds in Airborne Particulate Matter, in "Advances in Environmental Science and Technology", Vol. 12, S. E. Schwartz, ed., p. 268. John Wiley & Sons, New York.
13. SAMUELSON, O. 1963. Ion Exchange Separations in Analytical Chemistry, pp. 247–249. John Wiley & Sons, New York.
14. TRUEX, T. J., W. R. PIERSON, D. E. MCKEE, M. SHELEF AND R. E. BAKER, 1980. Effects of Barium Fuel Additive and Fuel Sulfur Level on Diesel Particulate Emissions, Environ. Sci. Technol. 14:1121–1124.
15. SCHUETZEL, D., L. M. SKEWES, G. E. FISHER, S. P. LEVINE AND R. A. GORSE, J. 1981. Determination of Sulfates in Diesel Particulates, Anal. Chem. 53:837–840.
16. PORTER, R. W. AND D. M. CONSIDINE. 1950. in "Process Control, Chemical Engineer's Handbook;", Perry, J. H., ed.; 3rd ed., p. 1283. McGraw-Hill, New York.
17. PIERSON, W. R. 1985. Research staff, Ford Motor Company, Dearborn, MI 48121. Unpublished Data.
18. APPEL, B. R., E. M. HOFFER, W. WEHRMEISTER, M. HAIK AND J. J. WESOLOWSKI. 1980. Improvement and Evaluation of Methods for Sulfate Analysis. Part II, EPA Report 600/4-80-024.
19. STEVENS, R. K., T. G. DZUBAY, G. RUSSWURM AND D. RICKEL. 1978. Sampling and Analysis of Atmospheric Sulfates and Related Species, Atmos. Environ. 12:55–68.

Subcommitte 1

W. R. PIERSON
P. K. DASGUPTA, *Chairman*
D. F. ADAMS
B. R. APPEL
S. O. FARWELL
K. T. KNAPP
G. L. KOK
K. D. REISZNER
R. L. TANNER

825.

Determination of Acrolein in Air

See **Method 114.**

826.
Determination of Acrolein in Air

1. Principle of the Method

1.1 Air is drawn through two midget impingers containing 1% $NaHSO_3$.

1.2 The reaction of acrolein with 4-hexylresorcinol in an alcoholic trichloroacetic acid solvent medium in the presence of mercuric chloride results in a blue colored product with strong absorption maximum at 605 nm **(1–3)**.

2. Range and Sensitivity

2.1 The absorbances at 605 nm are linear for at least 1 to 30 μg of acrolein in the 10 mL portions of mixed reagent.

2.2 A concentration of 10 ppb of acrolein can be determined in a 50-L air sample based on a difference of 0.05 absorbance unit from the blank using a 1-cm cell. Greater sensitivity could be obtained by use of a longer path-length cell.

3. Interfaces

3.1 There is no interference from ordinary quantities of sulfur dioxide, nitrogen dioxide, ozone and most organic air pollutants. A slight interference occurs from dienes: 1.5% for 1,3-butadiene and 2% for 1,3-pentadiene. The red color produced by some other aldehydes and undetermined materials does not interfere in spectrophotometric measurement.

4. Precision and Accuracy

4.1 Known standards can be determined to within ± 5% of the true value **(4)**. No data are available on precision and accuracy of air samples.

5. Advantages and Disadvantages of the Method

5.1 The method is sensitive to the extent that 0.01 ppm of acrolein can be determined in a 50-L air sample.

5.2 Both the solid trichloroacetic acid and the solution are corrosive to the skin. Mercuric chloride is highly toxic. This reagent and the former should be handled carefully.

6. Apparatus

6.1 ABSORBERS. All glass standard midget impingers with fritted glass inlets are acceptable. The fritted end should have a porosity approximately equal to that of Corning EC (170 to 220 μm maximum pore diameter). A train of 2 bubblers is needed.

6.2 AIR PUMP. A pump capable of drawing at least 2 L of air/min for 20 min through the sampling train is required. A trap at the inlet to protect the pump from the corrosive reagent is recommended.

6.3 AIR METERING DEVICE. Either a limiting orifice of approximately 1 or 2 L/min capacity or a glass meter can be used. If a limiting orifice is used, regular and frequent calibration is required.

6.4 SPECTROPHOTOMETER. This instrument should be capable of measuring the developed color at 605 nm. The absorption band is rather narrow and thus a lower absorptivity may be expected in a broad-band instrument.

7. Reagents

7.1 PURITY. All reagents must be analytical reagent grade. Water is ASTM reagent water, Type II.

7.2 $HGCL_2$-4-HEXYLRESORCINOL. Dissolve 0.30 g $HgCl_2$ and 25 g 4-hexylresorcinol in 50 mL of 95% ethanol (stable at least 3 weeks in refrigerator).

7.3 TCAA. To a 1 lb. (454 g) bottle of trichloracetic acid add 23 mL of distilled water and 25 mL of 95% ethanol. Mix until all the TCAA has dissolved.

7.4 COLLECTION MEDIUM. 1% sodium bisulfite in water.

7.5 ACROLEIN, PURIFIED. Freshly prepare a small quantity (less than 1 mL sufficient) by distilling 10 mL of the purest grade of acrolein commercially available. Reject the first 2 mL of distillate. (The acrolein should be stored in a refrigerator to retard polymerization.) The distillation should be done in a hood because the vapors are irritating to the eyes.

7.6 ACROLEIN, STANDARD SOLUTION "A" (1 MG/ML). Weigh 0.1 g (approximately 0.12 mL) of freshly prepared, purified acrolein into a 100-mL volumetric flask and dilute to volume with 1% aqueous sodium bisulfite. This solution may be kept for as long as a month if properly refrigerated.

7.7 ACROLEIN, STANDARD SOLUTION "B" (10 µg/ML). Dilute 1 mL of standard solution "A" to 100 mL with 1% sodium bisulfite. This solution may be kept for as long as a month if properly refrigerated.

8. Procedure

8.1 CLEANING OF EQUIPMENT. No specialized cleaning of glassware is required. However, since known interferences occur with dienes, cleaning techniques should ensure the absence of all organic materials.

8.2 COLLECTION OF SAMPLES. Two midget impingers, each containing 10 mL of 1% $NaHSO_3$ are connected in series with Tygon tubing. These are followed by and connected to an empty impinger (for meter protection) and a dry test meter and a source of suction. During sampling the impingers are immersed in an ice bath. Sampling rate of 2 L/min should be maintained. Sampling duration will depend on the concentration of aldehydes in the air. One hour sampling time at 2 L/min is adequate for ambient concentrations.

8.3 STORAGE OF SAMPLES. After sampling is complete, the impingers are disconnected from the train, the inlet and outlet tubes are capped, and the impingers stored in an ice bath or at 0°C in a refrigerator until analyses are performed. Cold storage

is necessary only if the acrolein determination cannot be performed within 4 h of sampling.

8.4 SHIPPING OF SAMPLES. Because of the 2 h limit for initiation of color development, it is best to analyze the samples soon after completion of sampling. However, where a refrigerated sample can be shipped and analyzed within 48 h after collection, then samples can be shipped or carried to a central lab at some distance from the sampling area.

8.5 ANALYSIS OF SAMPLES. (Each impinger is analyzed separately). To a 25-mL graduated tube add an aliquot of the collected sample in bisulfite containing no more than 30 µg acrolein. Add 1% sodium bisulfite (if necessary) to a volume of 4.0 mL. Add 1.0 mL of the $HgCl_2$-4-hexylresorcinol reagent and mix. Add 5.0 mL of TCAA reagent and mix again. Insert in a boiling water bath for 5 to 6 min, remove, and set aside until tubes reach room temperature. Centrifuge samples at 1500 rpm for 5 min to clear slight turbidity. One h after heating, read in a spectrophotometer at 605 nm against a bisulfite blank prepared in the same fashion as the samples. Determine the acrolein content of the sampling solution from a curve previously prepared from the standard acrolein solutions.

9. Calibration and Standards

9.1 PREPARATION OF STANDARD CURVE.

9.1.1 Pipet 0, 0.5, 1.0, 2.0, and 3.0 mL of standard solution "B" into glass stoppered test tubes.

9.1.2 Dilute each standard to exactly 4 mL with 1% aqueous sodium bisulfite.

9.1.3 Develop color as described above, in Section **8.5**.

9.1.4 Plot data on semi-log paper, of absorbance values against µg of acrolein in the color developed solution. The color after forming is stable for about 2 h.

10. Calculations

10.1 The concentration of acrolein in the sampled atmosphere may be calculated by using the following equations.

$\mu g/m^3 =$

$$\frac{\text{total } \mu g \text{ of acrolein in test sample} \times 1000}{\text{air sample volume in liters.}}$$

$$\text{ppm} = \frac{\text{total } \mu g \text{ of acrolein in sample}}{2.3 \times \text{sample volume in liters}}$$

Correct the air sample volume to 25°C and 101.3 kPa.

11. References

1. ROSENTHALER, L. AND G. VEGEZZI, 1954. Determination of Acrolein. *Z. Lebensm. Untersuch. Forsch.* 99:352.
2. COHEN, I. R. AND A. P. ALTSHULLER. 1961. A New Spectrophotometric Method for the Determination of Acrolein in Combustion Gases and the Atmosphere. *Anal. Chem.* 33:726.
3. ALTSHULLER, A. P. AND S. P. MCPHERSON. 1963. Spectrophotometric Analysis of Aldehydes in the Los Angeles Atmosphere. *J. Air Poll. Control Assoc.* 13:109.
4. LEVAGGI, D. A. AND M. FELDSTEIN. 1970. The Determination of Formaldehyde, Acrolein and Low Molecular Weight Aldehydes in Industrial Emissions on a Single Collected Sample. *JAPCA* 20:312.

Revised by Subcommittee 5

E. SAWICKI, *Chairman*
T. BELSKY
R. A. FRIEDEL
D. L. HYDE
J. L. MONKMAN
R. A. RASMUSSEN
L. A. RIPPERTON
L. D. WHITE

Approved with Modifications from 2nd edition
Subcommittee 4/5

M. FELDSTEIN, *Chairman*
R. J. BRYAN
D. L. HYDE
D. A. LEVAGGI
D. C. LOCKE
R. A. RASMUSSEN
P. O. WARNER

827.

Determination of Aromatic Amines in Air

1. Principle of the Method

1.1 A known volume of air is drawn through a tube containing silica gel to trap the aromatic amines present.

1.2 The silica gel in the tube is transferred to a glass-stoppered tube and desorbed with ethanol.

1.3 An aliquot of the resulting solution of desorbed aromatic amines in ethanol is injected into a gas chromatograph.

1.4 Peak areas are determined and compared with calibration curves obtained from the injection of standards.

2. Range and Sensitivity

2.1 The lower limit of this method using a flame ionization detector is 0.01 mg/sample of any one compound when the analyte is desorbed with 5 mL ethanol and 10 μL of the resulting solution is injected into the gas chromatograph. Sensitivity for p-nitroaniline is 0.002 mg/sample.

2.2 For air samples of 100% relative humidity, containing 140 mg aniline/m^3, at least 13 mg aniline/sample is retained without breakthrough by the primary absorbing section of sampling Tube A after 8 h of sampling at 200 mL/min. For Tube B, the primary absorbing section will retain 0.9 mg aniline/sample without breakthrough for air samples of 100% relative humidity, containing 90 mg aniline/m^3 after 10 min of sampling at 1 L/min.

3. Interferences

3.1 The most common sampling interference is water vapor. The sampling Tube A has been designed so that 96 L of 100% humidity air can be sampled over an 8-hour period at 200 mL/min without displacement of collected aromatic amines from the initial sorbent section by water vapor at room temperature.

3.2 Any compound that has nearly the same retention time as one of these aromatic amines at the gas chromatograph analytical conditions described in this method is an interference. This type of interference can often by overcome by changing the operating conditions of the gas chromatograph or selecting another column. Retention time data on a single column, or even on a number of columns, cannot be considered as conclusive proof of chemical identity in all cases. For this reason, it is important that whenever practical a sample of the bulk compound or mixture be submitted at the same time as the sample tube (but shipped separately) so that chemical identification(s) can be made by other means.

4. Precision and Accuracy

4.1 The accuracy of the method depends appreciably upon collection efficiencies and desorption efficiencies. If a negligible amount of aromatic amine is detected on the backup section, the collection efficiency of the tube must be essentially 100%. Desorption efficiencies for the range of 1 to 8 mg have been found to be 100% within experimental error of $\pm$ 5%.

4.2 The precision of the chromatographic analysis is quite dependent upon the precision and sensitivity of the technique used to quantitate gas chromatographic peaks of samples and standards. More accurate peak integration techniques (normal or electronic) can maximize precision, particularly at lower concentrations.

4.3 Precision of standard preparations can be $\pm$ 1% (calibration standards).

4.4 The precision of the overall method, $2\sigma = \pm$ 9%, has been determined from eight consecutive identical samples taken with a personal sampling pump.

4.5 Analytical precision can be improved by reducing or eliminating the error associated with syringe injection into the gas chromatograph. This is best accomplished by the addition of an internal standard to the ethanol used to both prepare standards and elute samples from the silica gel. Therefore, it is recommended that a 0.1% solution of n-heptanol in ethanol be used. For isothermal analyses at temperatures above 120°C, n-octanol may be preferable.

5. Advantages and Disadvantages

5.1 The sampling method uses a small, portable device involving no liquids. It is not affected by the humidity of the air. A sample of up to 8 h can be taken for an average work day concentration or a 15-min sample can be taken to test for excursion concentrations. Desorption of the collected sample is simple and is accomplished with a solvent with low toxicity. The analysis is accomplished by a quick instrumental method. Most analytical interferences which occur can be eliminated by altering gas chromatographic conditions. Since several aromatic amines can be collected and analyzed simultaneously, this method is useful where the composition of the aromatic amine vapors may not be known.

5.2 A major disadvantage of the method is the limitation on its precision due to the use of personal sampling pumps currently available. After initial adjustment of flow any change in the pumping rate will affect the volume of air actually sampled. Furthermore, if the pump used is calibrated for one tube only, as is often the case, the precision of the volume of air samples will be limited by the reproducibility of the pressure drop across the tubes.

6. Apparatus

6.1 SAMPLING EQUIPMENT. The sampling unit for the silica adsorption method has the following components:

6.1.1 *The glass—silica gel sampling tubes.* (Refer to Section **6.2**).

6.1.2 *One or more personal sampling pumps*, whose flow can be set at and maintained at 1 L/min for 15 min or 200 mL/min for 8 h. A bypass attachment or a syringe needle may be needed, depending on

the pressure drop of the pumps, to allow the lower flow rate.

6.1.3 *Thermometer.*

6.1.4 *Manometer.*

6.1.5 *Stopwatch.*

6.2 Two types of *pyrex glass sampling tubes* of the dimensions shown in Figure 827:1, packed with two sections of 45/60 mesh silica gel. The silica gel can be purchased from commercial sources presieved or can be ground from larger particles and sieved before use. No other preparation appears necessary. Glass wool or woven glass cloth should be used for the plug in front of the primary sorbent section. More porous urethane foam plugs should be used between the sorbent sections and after the backup section. For reproducible pressure drops, the consistency of plug size and packing is critical. A 12-mm long piece of 5-mm OD pyrex tubing between the primary and backup sections greatly reduces migration of the sample to the backup section prior to analysis. After packing, the

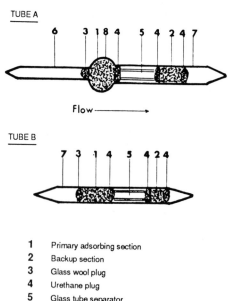

1	Primary adsorbing section
2	Backup section
3	Glass wool plug
4	Urethane plug
5	Glass tube separator
6	6 mm O.D. pyrex tube
7	8 mm O.D. pyrex tube
8	16 mm O.D. "bubble," 12 mm wide

Figure 827:1 — Silica gel sampling tubes for aromatic amines.

ends of each tube should be flame sealed to prevent contamination before sampling.

6.2.1 *Tube A.* This sampling tube contains a primary adsorbing section of approximately 1 g of 45/60 mesh silica gel packed firmly into a nearly spherical volume of 1.4 cm³. The backup section contains 100 mg of silica gel in 8-mm OD tubing. The pressure drop of such tubes should not exceed 85 mm of water (0.83 kPa) at air flows of 200 mL/min. (Figure 827:1)

6.2.2 *Tube B.* This sampling tube contains a primary adsorbing section of 150 mg and a backup section of 100 mg of 45/60 mesh silica gel in a straight 8-mm OD pyrex tube. The pressure drop of such tubes should not exceed 85 mm of water (0.83 kPa) at air flows of 200 mL/min or 25 mm Hg (3.4 kPa) at 1 L air/min. (Figure 827:1).

6.3 GAS CHROMATOGRAPH, equipped with a flame ionization detector. Linear temperature programming capability is desirable but not essential.

6.4 COLUMN (1.22 m × 3.2 mm OD S.S.). Packed with Silicone OV-25 liquid phase, 10% on 80/100 mesh Supelcoport (or equivalent support). A column (61 cm × 3.2 mm OD) packed with Chromosorb 103 80/100 mesh) will also work except for p-nitroaniline.

6.5 RECORDER, and some method for determining peak height or area.

6.6 GLASS-STOPPERED TUBES OR FLASKS, 1- AND 10-ML.

6.7 SYRINGES, 10-μL.

6.8 PIPETTES AND VOLUMETRIC FLASKS. For preparation of standard solutions.

7. Reagents

7.1 PURITY. All reagents used must be ACS reagent grade or better.

7.2 ETHANOL, 95%.

7.3 N-HEPTANOL OR N-OCTANOL.

7.4 DESORBING SOLUTION. Prepare a 0.1% solution of n-heptanol or n-octanol in ethanol.

7.5 AROMATIC AMINE STANDARDS.

7.6 GRADE A HELIUM.

7.7 PREPURIFIED HYDROGEN, "ELECTROLYTIC GRADE."

7.8 FILTERED COMPRESSED AIR.

8. Procedure

8.1 CLEANING OF EQUIPMENT. All glassware used for the laboratory analysis should be detergent-washed followed by tap and distilled water rinses.

8.2 CALIBRATION OF PERSONAL PUMPS. Each pump must be calibrated with a representative tube in the line to minimize errors associated with uncertainties in the sample volume collected.

8.3 COLLECTION AND SHIPPING OF SAMPLES.

8.3.1 Immediately before sampling break the ends of the tube so as to provide an opening at least one half of the internal diameter of the tube.

8.3.2 Attach the sampling tube (backup section nearer the pump) to the sampling pump using a short piece of flexible tubing. Air being sampled should not be passed through any hose or tubing before entering the sampling tube.

8.3.3 Sample the atmosphere at the desired flow rate for the desired period of time. Suggested sampling rates and times are given in Table 827:I. The flow rate and sampling time or the volume of air sampled must be measured as accurately as possible.

8.3.4 Measure and record the temperature and pressure of the atmosphere being sampled.

8.3.5 Seal the sampling tubes with plastic caps immediately after sampling. Under no circumstances should rubber caps be used.

8.3.6 One tube should be handled in the same manner as the sample tube (break, seal, and transport), except that no air is sampled through this tube. This tube should be labeled as a blank.

8.3.7 Capped tubes should be packed tightly before shipping to minimize tube breakage during shipping.

8.3.8 Samples of the bulk liquids or solids from which the aromatic amine vapors arise should be submitted to the laboratory also, but *not* in the same container as the air samples or blank tubes.

8.3.9 *Storage.* Tubes after sampling should be tightly capped and not subjected to extremes of high temperature or low pressure, if avoidable. If the analysis is to be delayed beyond one week after sam-

Table 827:I. Aromatic Amines For Which the Method Has Been Tested

Compound	OSHA Standard*		15-min Tube	Sampling Rate (L/min)†	8–h Tube	Sampling Rate (L/min)†
	(ppm)	(mg/m³)				
Aniline	5	19	B	0.2	A	0.2
N,N-Dimethylaniline	5	25	B	0.2	A	0.2
o-Toluidine	5	22	B	0.2	A	0.2
o-Chloroaniline	Not established		B	0.2	A	0.2
Xylidines (mixed)	5	25	B	0.2	A	0.2
o-Anisidine	0.1	0.5	B	1	B	0.2
p-Anisidine	0.1	0.5	B	1	B	0.2
p-Nitroaniline	1	6	B	0.2	A	0.2

*Federal Register, 37, #202, 22139 (18 October 1972).
†See Section **8.3.3**.

pling, each tube should be filled with an inert gas (helium, nitrogen, etc.) to retard the loss of sample by oxidation. Refrigeration is also recommended.

8.4 ANALYSIS OF SAMPLES.

8.4.1 Carefully transfer the silica gel of the primary adsorbing section and the glass wool that precedes it to a glass-stoppered tube or flask (10 mL for Tube A or 2 mL for Tube B).

8.4.2 Remove and discard the separating sections of foam and glass tubing. Exercise due care to avoid any loss of silica gel particles.

8.4.3 Then transfer the silica-gel in the back-up section to a 2-mL glass-stoppered tube or flask.

8.4.4 These two silica gel sections are desorbed and analyzed separately according to the succeeding sections.

8.4.5 Pipet an aliquot of the desorbing solution – 5 mL for the primary adsorbing section of Tube A and 1 mL for all other sections – into each tube or flask. Allow the samples to stand for 30 min. Tests indicate that desorption is complete in 30 min if the sample is stirred or shaken occasionally.

8.4.6 Inject an aliquot of the sample into the gas chromatograph.* With an internal standard in the eluent, direct injection of up to 10 μL with a microliter syringe

*Gas chromatograph conditions: Typical operating conditions for the gas chromatograph are: carrier flow (25 mL He/min); injection port (150°C); flame ionization detector (250°C, 50 mL/min H₂, 470 mL/min air); oven temperature program (100°C for 4 min, then increase at 8°C/min to 225°C).

is acceptably precise. At least duplicate injections of the same sample or standard are recommended.

8.4.7 Measure the areas of the sample peak and the internal standard peak by an electronic integrator or some other suitable method of area measurement. The ratio of these areas is calculated and used to determine sample concentration in the eluent by using a standard curve prepared as discussed in Section **9**.

9. Calibration and Standards

9.1 For good accuracy in the preparation of standards, it is recommended that one standard be prepared in a relatively large volume and at a high concentration. Aliquots of this standard can then be diluted to prepare other standards. The solvent and diluent used must be the same desorbing solution used for the elution of samples. For example, to prepare 100 mL of standard corresponding to a 96-L sample of air containing 95 mg aniline/m³ desorbed with 5 mL of desorbing solution, 178 μL [eq. (1)] of aniline (density = 1.022 g/mL) is added to a 100-mL volumetric flask and diluted to the mark. The resulting concentration is 1.82 mg/mL [eq. (2)]. If 2 mL of this solution is diluted to the mark with the desorbing solution in a 10-mL volumetric flask, the resulting concentration is 0.365 mg/mL [eq. (3)]

$$\frac{95 \text{ mg/m}^3 \times 0.096 \text{ m}^3 \times 100 \text{ mL}}{5 \text{ mL} \times 1.022 \text{ mg/}\mu\text{L}}$$

$$= 178 \ \mu\text{L} \ (1)$$

$$\frac{178 \ \mu L \ \times \ 1.022 \ mg/\mu L}{100 \ mL} = 1.82 \ mg/mL \quad (2)$$

$$1.82 \ mg/mL \ \times \frac{2 \ mL}{10 \ mL} = 0.365 \ mg/mL \quad (3)$$

When microliter pipettes are used instead of microliter syringes, it is better to prepare standards using a round number of microliters (*e.g.*, 200 μL aniline instead of 178 μL). For solids the amount of compound used for the first standard should be weighed on an analytical balance.

9.2 Prepare a series of standards of varying concentration over the range of interest.

9.3 The standards prepared as above should be analyzed under the same GC conditions and during the same day as the unknown samples.

9.4 Prepare a standard curve for each compound, plotting ratios of peak areas of the compound to the internal standard against concentration of the compound. From the resulting curve the concentration of an eluted sample is determined. This concentration (in mg/mL) is then converted to total sample weight by multiplying with the volume of desorbing solution used for that silica gel section (1 or 5 mL).

10. Calculations

10.1 Corrections for the blank must be made for each sample.

$$\text{Corrected mg} = mg_s - mg_b$$

where mg_s = mg found in primary adsorbing section of sample tube

mg_b = mg found in primary adsorbing section of blank tube

A similar procedure is followed for the backup sections.

10.2 Add the corrected amounts present in the primary and backup sections of the same sample tube to determine the total measured amount (w) in the sample.

10.3 Convert the volume of air sampled to standard conditions of 25°C and 101.3 kPa.

10.4 The concentration of the organic solvent in the air sampled can be expressed in mg per m³, which is numerically equal to μg per liter of air

$$mg/m^3 = \mu g/L = \frac{w(mg) \times 1000 \ (\mu g/mg)}{V_s}$$

where V_s is sample volume at standard conditions.

10.5 Another method of expressing concentration is ppm, defined as μL of compound vapor per L of air

$$ppm = \mu L \text{ of vapor}/V_s$$

$$ppm = \frac{\mu g \text{ of vapor}}{V_s} \times \frac{24.45}{MW}$$

where 24.45 = molar volume at 25°C and 101.3 kPa in μL per μmole.

MW = molecular weight of the compound, in μg per μmole.

11. References

1. E. E. CAMPBELL, G. O. WOOD AND R.G. ANDERSON, 1972, 1973, 1974. Los Alamos Scientific Laboratory Progress Reports LA-5104-PR, LA-5164-PR, LA-5308-PR, LA-5389-PR, LA-5484-PR, and LA-5634-PR, Los Alamos, N. M., Nov., Jan., June, Aug., Dec., and June.

Revised by Subcommittee 5

E. SAWICKI, *Chairman*
T. BELSKY
R. A. FRIEDEL
D. L. HYDE
J. L. MONKMAN
R. A. RASMUSSEN
L. A. RIPPERTON
L. D. WHITE

Approved with Modifications from 2nd edition
Subcommittee 4/5

M. FELDSTEIN, *Chairman*
R. J. BRYAN
D. L. HYDE
D. A. LEVAGGI
D. C. LOCKE
R. A. RASMUSSEN
P. O. WARNER

828.

Determination of Bis-(Chloromethyl) Ether (Bis-CME) in Air

1. Principle of the Method

1.1 Bis-CME is concentrated from air by adsorption on Chromosorb 101 in a short tube. It is desorbed by heating and purging with helium through a gas chromatography column where it is separated from most interferences. The concentration of bis-CME is measured from the mass spectrometric signal at m/e 79 and 81.

2. Range and Sensitivity

2.1 The range of the mass spectrometric signal for the conditions listed corresponds to 0.5 and 10 ppb.

2.2 A concentration of 0.5 ppb of bis-CME can be determined in a 5-L air sample. Greater sensitivity can be obtained by using a larger sample.

3. Interferences

3.1 Interferences resulting from materials having retention times similar to bisCME or simply background such as ions C_6H_7, $C_5{}^{13}CH_6$, C_4H_3Si and Br giving rise to m/e 79 can be encountered. However, the DuPont 21–491 mass spectrometer **(5)** has sufficient resolution to resolve at least three of these ions at m/e 79, particularly the $C_5{}^{13}CH_6$ and C_6H_7 ions, that are normally encountered in the plant samples. These are completely resolved from the bis-CME $C_2H_4{}^{35}ClO$ ion.

4. Precision and Accuracy

4.1 The precision of this method has been determined to be $\pm$ 10% relative standard deviation when different sampling tubes were spiked with 133 ng (correspond-ing to 1.4 ppb in 20 L of air) at intervals of approximately 1 h. These data were obtained using 10 cm $\times$ 6 mm stainless steel sampling tubes packed with 60/80 mesh Chromosorb 101.

4.2 The accuracy of the analysis is approximately $\pm$ 10% of the amount reported as determined from repeated analysis of several standards **(6,7)**.

5. Advantages and Disadvantages of the Method

5.1 The gas chromatography-mass spectrometry technique interfaced with a Watson-Biemann helium separator **(4)** is extremely sensitive and specific for the analysis of bis-CME. The gas chromatographic separation yields a retention time that is characteristic for bis-CME, but it is not highly specific for positive assignment of the signal as bis-CME. The mass spectrometer in combination with the gas chromatograph provides the high degree of specificity. The most intense ion in the mass spectrum of bis-CME is formed at m/e 79 with its chlorine-37 isotope at m/e 81. The chlorine isotopic abundances require that the intensity ratio of peaks at m/e 79:81 be approximately 3:1. The mass spectrometer is set to monitor these ions continuously as the sample elutes through the gas chromatograph. It is absolutely necessary that the retention time, observed masses at m/e 79 and 81 and the intensity ratio be correct in order to assign the signal to bis-CME.

5.2 Bis-CME, an impurity in chloromethyl methyl ether, has been reported to be a pulmonary carcinogen **(1–3)**. Extreme safety precautions should be exercised in the preparation and disposal of liquid and gas standards and the analysis of air samples. Bis-CME may be destroyed in a methanol-caustic solution.

5.3 Sampling must be carried out away from the mass spectrometer.

5.4 GC-MS is not a good approach for continuous monitoring.

6. Apparatus

6.1 SAMPLING TUBES.

6.1.1 The sampling tubes are prepared by packing a 6.4 cm × 6 mm stainless steel tubing with 4.4 cm of 60/80 mesh Chromosorb 101 with glass wool in the ends. These are conditioned overnight at 200°C with helium or nitrogen flow set at 10 to 30 mL/min. The conditioned sampling tubes are then cooled under flow and capped immediately upon removal. An identification number is scribed on one of the nuts. The stainless steel tubing should be rinsed internally with water, acetone and methylene chloride before packing with Chromosorb 101.

6.1.2 Longer sampling tubes may be prepared with a proportional amount of Chromosorb 101.

6.2 GAS CHROMATOGRAPHY COLUMN.

6.2.1 A 1.22 m × 3.2 mm stainless steel separator column is rinsed internally with water, acetone and methylene chloride and air dried. The cleaned tubing is then packed with Chromosorb 101. The column is conditioned overnight at 200°C and 10 to 30 mL/min helium or nitrogen flow.

6.3 A WATSON-BIEMANN HELIUM SEPARATOR.

6.4 GAS CHROMATOGRAPH.

6.4.1 A Hewlett-Packard 5750 gas chromatograph or equivalent. A gas chromatograph employing a single column oven and a temperature programmer is adequate.

6.5 MASS SPECTROMETER.

6.5.1 A mass spectrometer with a resolution of 1000 to 2000, equipped with a repetitive scan attachment, can be used in conjunction with a gas chromatograph. A CEC (now DuPont Analytical Instruments Div.) Model 21–491 modified by adding a follower amplifier with a long time constant (0.03 s) to the output of the electron multiplier to increase the gain and signal to

noise ratio has been found satisfactory for this purpose.

6.6 SYRINGES.

6.6.1 *Syringe*, 1-mL. Gas tight (The Hamilton Co., Inc.)

6.6.2 *Syringe*, 10-μL. (The Hamilton Co., Inc.)

7. Reagents and Materials

7.1 PURITY. All reagents must be analytical reagent grade.

7.2 BIS-CHLOROMETHYL ETHER (EASTMAN KODAK).

7.3 ACETONE.

7.4 DICHLOROMETHANE.

7.5 N-PENTANE.

7.6 METHANOL.

7.7 SODIUM HYDROXIDE.

7.8 PLASTIC FILM GAS BAG, 13 L. Saran brand (Trademark of The Dow Chemical Company abroad) plastic film gas bag has proven to be satisfactory for this purpose (The Anspec Co., Inc.). Polyethylene film bags are not suitable.

7.9 CHROMOSORB 101 (JOHNS-MANVILLE).

8. Procedure

8.1 CLEANING OF EQUIPMENT. NONE REQUIRED.

8.2 COLLECTION OF SAMPLES.

8.2.1 Spot samples may be obtained by attaching the sampling tube to a manual syringe pump. The sample is drawn into the end of the sampling tube with the marked nut.

8.2.2 Continuous sampling of air may be accomplished by attaching a vacuum pump to a throttling valve and manometer. Each sample tube is calibrated so that the pressure drop is determined. Then the flow is established by adjusting the pressure drop with the throttling valve.

8.2.3 Continuous or cumulative sampling at fixed points may be accomplished by attaching a vacuum pump to a throttling valve and calibrated rotameter. The flow is then adjusted with the throttling valve through the calibrated rotameter. This sampling is much preferred to **8.2.2**

8.2.4 For larger sample size it is important to realize that larger flows of air over a long period of time may cause elution of bis-CME through the sampling tube. It has been shown that a total of 25 L of air can be pumped through 10 cm × 6 mm packing of Chromosorb 101 over 24 h without any elution of the bis-CME. This corresponds to a flow of 17 cm³/min maximum. A maximum air volume of 5 L at a rate of 80 cm³/min has been found to be safe for the 6.4 cm × 6 mm sampling tube.

8.2.5 Bis-CME has been found to be stable in the sampling tubes for at least 10 days when kept tightly closed and at room temperature.

8.3 ANALYSIS OF SAMPLE.

8.3.1 *Instrument Conditions and Setup.* The mass spectrometer should be set to sweep peaks at m/e 79 to 81 with the recorder speed at 2 cm/s for recording the resulting signal.

8.3.2 Hold the temperature of the Watson-Biemann helium separator at 160°C. Set the helium flow at 30 cm³/min. Balance the helium flow so that there is no vacuum applied to the outlet of the gas chromatographic column.

8.3.3 Attach the sampling tube to the gas chromatograph via an adapter to the septum nut. The inscribed sample tube nut should be closest to the inlet of the analytical column.

8.3.4 A 1/4″ (6 mm) Swagelok Tee fitting is attached on the other end of the sampling tube to accommodate the carrier gas and a silicone rubber septum so that standards may be injected.

8.3.5 After thoroughly checking for leaks, wrap the sampling tube with heat tape and heat to 150°C for 5 min to assure complete desorption of the organic components onto the analytical column which is at ambient temperature.

8.3.6 Temperature program the analytical column to 130°C at 30°C/min and hold to yield a retention time for bis-CME of 8 to 9 min. Turn on recorder 30 s before elution of bis-CME.

8.3.7 The analytical column should be cooled to ambient temperature before changing sampling tubes.

9. Calibration and Standards

9.1 PREPARATION OF GAS STANDARD.

9.1.1 Rinse out a 13-L Saran bag with prepurified nitrogen.

9.1.2 Introduce 5 L of prepurified nitrogen through a dry test meter into the Saran bag and cap it with a cork.

9.1.3 Place plastic tape on the Saran bag and inject 1 μL of bis-CME through the tape into the bag. The tape should prevent any cracking of the bag at the injection point.

9.1.4 Tape the injection point at once upon removing the syringe needle from the bag.

9.1.5 Clean the syringe immediately with methanol-caustic solution and then water and acetone.

9.1.6 Knead the bag for approximately 2 min.

9.1.7 Carefully connect the bag with the meter to introduce an additional 5 L of prepurified nitrogen (the end of the Saran tubing inside the bag should be kept tight with a finger from the outside in order to avoid any loss of bis-CME while connecting the bag with the meter).

9.1.8 Knead the bag for additional 2 min to assure complete mixing.

9.1.9 With the 10-L bag this procedure can prepare a ppm level standard (*e.g.*, 1 μL of liquid bis-CME yields a 28.3 ppm by volume standard). This standard should not be kept for more than 3 days.

9.2 PREPARATION OF LIQUID STANDARD.

9.2.1 Bis-CME Standard (0.132 μg bis-CME/μL). Dissolve 1 μL of bis-CME liquid in 10 mL n-pentane. This standard solution may be kept for as long as desired if evaporation is not allowed.

9.3 CALIBRATION.

9.3.1 Inject a 1-mL portion of the 28.3 ppm standard bis-CME (0.132 μg) through a standard septum onto a sampling tube attached to the inlet of the analytical column. This may be done routinely after every two or three samples to ensure accuracy. Also one μL of the liquid standard (0.132 μg/μL) could be used for the calibration.

9.3.2 Repeat steps as described in paragraphs **8.3.3–8.3.7**.

10. Calculations

10.1 The total ng of bis-CME in the air sampled is determined from the ratio of the peak height of m/e 79 of the sample and the standard multiplied by 132.

10.2 Concentration in ppb is calculated as follows:

$$ppb = \frac{ng\ bis\text{-}CME}{V} \times \frac{24.45}{MW}$$

where ng bis-CME = total ng concentration as determined in **10.1**

V = volume of air in liters sampled at 25°C and 101.3 kPa

24.45 = molar volume of an ideal gas at 25°C and 101.3 kPa

MW = molecular weight of bis-CME, 115

11. References

1. VAN DUUREN, B. L., A. SINAK, B. M. GOLDSCH-MIDT, C. KATZ AND S. MELCHIONNE. 1969. *J. Nat. Cancer Inst.* 43:481.
2. LASKIN, S., M. KUSCHNER, R. T. DREW, V. P. CAUPPIELLO AND N. NELSON. 1971. Technical Report, New York University Medical Center, Department of Environmental Medicine, August.
3. LEONG, K. J., H. N. MACFARLAND AND W. H. REESE. 1971. *Arch. Environmental Health* 22:663.
4. WATSON, J. T. AND K. BIEMANN. 1964. *Anal. Chem.* 36:1135.
5. DUPONT DE NEMOURS, E. I. Analytical Instrument Division, 1500 S. Shamrock Ave., Monrovia, Calif.
6. SHADOFF, L. A., G. J. KALLOS AND J. S. WOODS. 1973. Analysis for bis-Chloromethyl Ether in Air. *Anal. Chem.* 45:2341.
7. FEDERAL REGISTER. 1974. Vol. 39, No. 20, Tuesday, January 29, pp. 3773-3776.

Subcommittee 5
E. SAWICKI, *Chairman*
T. BELSKY
R. A. FRIEDEL
D. L. HYDE
J. L. MONKMAN
R. A. RASMUSSEN
L. A. RIPPERTON
L. D. WHITE

Approved with Modifications from 2nd edition
Subcommittee 4/5

M. FELDSTEIN, *Chairman*
R. J. BRYAN
D. L. HYDE
D. A. LEVAGGI
D. C. LOCKE
R. A. RASMUSSEN
P. O. WARNER

829.

Determination of Chloromethyl Methyl Ether (CMME) and Bis-Chloromethyl Ether (Bis-CME) in Air

1. Principle of the Method

1.1 A known volume of air is drawn through glass impingers containing a methanolic solution of the sodium salt of 2,4,6-trichlorophenol.

1.2 CMME and bis-CME react with the derivatizing reagent to produce stable derivatives.

1.3 Sample is heated on a steam bath for 5 min, cooled, diluted with an equal volume of distilled water and 2 mL of hexane for extraction.

1.4 An aliquot of the hexane is analyzed by electron capture gas chromatography.

1.5 The peak heights of the CMME and bis-CME derivatives are measured and the concentration determined from standard curves **(1)**.

2. Range and Sensitivity

2.1 A relative standard deviation of 10% can be expected within the concentration range of 0.02 to 0.3 ng/μL of both CMME and bis-CME.

2.2 The sensitivity of the method is 0.5 ppb (v/v) when a 10-L air sample is used.

3. Interferences

3.1 The known components used in chloromethylation processes do not interfere with the determination of CMME or bis-CME.

3.2 Interferences can be expected from highly halogenated organic compounds or compounds that may produce the same derivatives.

3.3 The quality of 2,4,6-trichlorophenol is important since impurities can be extracted with hexane and seriously interfere with the chromatographic analysis.

4. Precision and Accuracy

4.1 The precision of the sampling technique with an accurate, calibrated air sampling pump and analysis of the derivative samples is 10% relative standard deviation.

4.2 The accuracy of the method is typically affected by the efficiency in sampling, extraction, calibration and data handling.

5. Advantages and Disadvantages

5.1 The major advantage of this method is the sensitivity and the simultaneous analysis of both CMME and bis-CME.

5.2 The derivative stabilizes both CMME and bis-CME while significantly increasing the sensitivity.

5.3 Preparation of the derivatives eliminates further hazardous handling of both CMME and bis-CME during the analysis.

5.4 Disadvantages of the method are the handling of liquids, extractions and the dilution of the samples by the hexane extraction.

6. Apparatus

6.1 SAMPLING EQUIPMENT. The sampling unit for the impinger collection method consists of the following components:

6.1.1 *Two standard air impinger assemblies.* With fritted glass inlets.

6.1.2 *Calibrated battery-powered pump.* Capable of drawing an accurate and reproducible volume of air through the impingers at a flow rate of 0.5 L/min is required.

6.1.3 *Rotameter*, calibrated for air. Parts should include only the sapphire ball and small Teflon tubing, inside at both ends.

6.1.4 *Thermometer*.

6.1.5 *Stopwatch*.

6.2 STEAM BATH. Any bath capable of maintaining a temperature of 65° to 90°C is adequate.

6.3 GAS CHROMATOGRAPH. Equipped with a ^{63}Ni electron capture detector.

6.4 GAS CHROMATOGRAPH COLUMN. A 6-foot long (1.83 m) × ¼" (6.35 mm) glass column is packed with 100/120 mesh textured glass beads (GLC-100) coated with a two-component stationary phase consisting of 0.1% by weight of QF-1 and 0.1% by weight OV-17. The column is equipped for on-column injection. The packed column is preconditioned at 160°C overnight with nitrogen carrier at a flow rate of 30 to 50 cm^3/min.

6.5 STRIP CHART RECORDER, 1.0 MV FULL SCALE RANGE.

6.6 HAMILTON MICROSYRINGES.

6.7 ASSORTED LABORATORY GLASSWARE, pipettes, graduated cylinders, etc.

7. Reagents

PURITY. All reagents are analytical reagent grade. Water means ASTM reagent water, Type II.

7.1 DERIVATIZING REAGENT, prepared in the laboratory as described in the procedure **7.8**.

7.2 SODIUM METHOXIDE, AR GRADE.

7.3 2,4,6-TRICHLOROPHENOL, M.P. 67° TO 68°C.

7.4 METHANOL AND HEXANE, DISTILLED IN GLASS.

7.5 CHLOROMETHYL METHYL ETHER, B.P. 55° TO 58°C.

7.6 BIS-CHLOROMETHYL ETHER, B.P. 100°C TO 102°C.

7.7 SODIUM HYDROXIDE.

7.8 PREPARATION OF DERIVATIZING REAGENT. Twenty-five g of sodium methoxide are weighed into a beaker and dissolved in 1 L of methanol. Five g of 2,4,6-trichlorophenol are weighed into the beaker and allowed to dissolve in the methanol-

sodium methoxide solution. The reagent is stable for 3 or 4 weeks when stored in a dark brown bottle. The sodium methoxide should be added slowly to the methanol since the reaction is very exothermic.

7.9 The sensitivity of bis-CME can be increased 6- to 8-fold by using stoichiometric quantities of sodium methoxide and 2,4,6-trichlorophenol (16.0 g of 2,4,6-trichlorophenol and 4.4 g of sodium methoxide in 1 L of methanol). When using this formulation, 2.0 N sodium hydroxide is used in place of distilled water prior to the hexane extraction of the derivative (see Section **8.3**).

8. Procedure.

8.1 CLEANING OF EQUIPMENT. All glassware used for the analysis must be thoroughly washed, rinsed with distilled water and dried. The impinger assemblies can be rinsed with reagent grade methanol for repeated use.

8.2 COLLECTION OF SAMPLES.

8.2.1 CMME and bis-CME in air are sampled at a rate of 0.5 L/min, (up to 2 h if necessary) through the two impinger assemblies each containing 10 mL of the derivatizing solution.

8.2.2 Teflon connections should be used for the attachment of the two impinger assemblies in series and the rotameter to the first impinger. Rubber tubing may be used for the connection of the second impinger to the intake of the pump.

8.2.3 A bypass flow of air can be established in case the pump does not operate at low flow rates. An appropriate syringe needle injected into the rubber tubing between the impinger and the intake of the pump should provide satisfactory flow rates. Also a needle valve connected through a T should produce adjustable flow rates.

8.3 ANALYSIS OF SAMPLE.

8.3.1 After sampling, the solutions are transferred to a vial that is capped loosely and placed on a steam bath for five minutes (any bath capable of maintaining a temperature of 65 to 90°C is suitable). The samples are allowed to cool and an equal amount of distilled water and 2 mL of hexane is pipetted into the vial. Then the sample is shaken for 5 min. The mixture is allowed to stand for a few minutes to allow the phases to separate.

8.3.2 *GC conditions.* The following are the recommended starting instrumental conditions. A gas chromatograph with a ^{63}Ni electron capture detector is equipped with a 6-foot long (1.83 m) by ¼″ (6.35 mm) glass column packed with 100 to 120 mesh textured glass beads (GLC-100) coated with a two component stationary phase consisting of 0.1% by weight QF-1 and 0.1% by weight OV-17. The column is equipped for on-column injection. The flow rate of the prepurified nitrogen carrier gas is set at 30 cm^3/min. The temperature of the sample injection zone is adjusted at 175°C and that of the detector at 250°C. The column oven is operated isothermally at 140°C. An oxygen filter is required on the carrier gas.

8.3.3 *Injection.* A 2-μL aliquot of the hexane extract is injected into the GC. A complete chromatogram should be obtained in about 6 min. Duplicate injections of each sample and a standard should be made.

9. Calibration and Standards

9.1 PREPARATION OF STANDARD CURVE.

9.1.1 Two μL each of CMME and bis-CME are added to 50 mL of hexane. The weights of the components are obtained by using their respective specific gravities, 1.03 g/mL for CMME and 1.32 g/mL for bis-CME. This concentrated standard is then used for preparing a standard curve. Both of these compounds should be handled in a good hood only!!

9.1.2 Ten mL of the derivatizing reagent (Section **8.3**) is pipetted into five 75 mL screw cap vials. Ten, 5, 2, 1 and 0 mL of the concentrated standard are added. These volumes are equivalent to 0.50, 0.25, 0.10, and 0.05 μg of bis-CME and 0.40, 0.20, 0.08 and 0.04 μg of CMME.

9.1.3 The vials are capped loosely and placed on a steam bath for 5 min. The standard is cooled and 10 mL of distilled water and 2 mL of hexane are pipetted into the vials. Then the standards are shaken for 5 min.

9.2 Standard curves are established by plotting the peak height by recorder response versus concentration in nanograms.

10. Calculations

10.1 Determine from the calibration curve the concentration of the desired component in nanograms.

$$\text{Component (ng)} = \frac{A \times B}{C}$$

where A = response for the component of interest in the sample

B = weight of CMME or BCME in the standard expressed in nanograms

C = response for the component of interest in the standard

10.2 The concentration of CMME and BCME in the sampled atmosphere can be calculated in ppb (v/v).

$$\text{ppb (v/v)} = \frac{D \times 24.45}{V_s \times MW}$$

where D = total ng concentration as determined in Section **10.1**

24.45 = molar volume of an ideal gas at 25°C and 101.3 kPa

V_s = volume of air sampled in liters at 25°C and 101.3 kPa

MW = molecular weight of CMME and bis-CME, 80.5 and 115, respectively.

11. Reference

1. SOLOMON R. A. AND G. J. KALLOS. 1975. *Anal. Chem.* 47:955.

Subcommittee 5
E. SAWICKI, *Chairman*
T. BELSKY
R. A. FRIEDEL
D. L. HYDE
J. L. MONKMAN
R. A. RASMUSSEN
L. A. RIPPERTON
L. D. WHITE

Approved with Modifications from 2nd edition
Subcommittee 4/5

M. FELDSTEIN, *Chairman*
R. J. BRYAN

D. L. HYDE
D. A. LEVAGGI
D. C. LOCKE
R. A. RASMUSSEN
P. O. WARNER

830.

Determination of 3,3'-Dichloro-4,4'-Diaminodiphenylmethane (MOCA)* in Air

1. Principle of Method

1.1 A known volume of air is drawn through a Gas-chrom S tube with a personal air sampling pump.

1.2 Sections of the Gas-chrom S tube are transferred individually to stoppered, 1-mL volumetric flasks and extracted with 0.5 mL acetone.

1.3 Sample is permitted to stand for 25 min with occasional agitation, and an aliquot is analyzed by flame ionization gas chromatography.

1.4 The peak area of MOCA is determined and its concentration found from a calibration curve (1,2).

2. Range and Sensitivity

2.1 Within the concentration range 0.002 to 0.15 $\mu g/\mu L$ of MOCA, a relative standard deviation of 5% can be expected.

2.2 A detection limit with 1-μL aliquot of sample is 0.002 μg or about 2 $\mu g/m^3$ for a 500-L air sample.

3. Interferences

3.1 Fortuitously, no impurity yet encountered in field trials has the same retention time as MOCA to cause interference in this concentration range.

3.2 Interference can be expected from organic compounds with similar functional groups.

3.3 Isomers of chloraniline commonly associated with MOCA do not interfere as

*MOCA—Trade name E. I. DuPont de Nemours.

they are completely resolved by the gas chromatograph.

3.4 The solvent effect is pronounced since MOCA appears on the tailing edge of the acetone peak. It is therefore important to maintain an injection volume of less than 2 μL.

4. Precision and Accuracy

4.1 The precision of the sampling method with a personal air sampling pump and analysis of synthetic MOCA samples varies from 1.3 to 5% relative standard deviation.

5. Advantages and Disadvantages of the Method

5.1 The sampling device is small, portable, involves no liquids, and is relatively free of interference from commonly associated molecular species.

5.2 Both short and long term (8-h) samples may be taken.

5.3 The collection efficiency of Gas-chrom S in the tubular configuration approaches 100% for MOCA in air, while the desorption efficiency is near 92%.

5.4 Disadvantages of the method are the exacting experimental conditions of hydrogen flow rate and the large solvent effect when the injection volume is greater than 2 μL.

5.5 MOCA is a carcinogenic substance and precautions must be taken to avoid skin or other contact.

6. Apparatus

6.1 Gas-chrom S (40/60 mesh) is sieved to 45 mesh and packed into 63 mm × 6 mm OD tapered pyrex tubes as two separate sections of 10 and 50 mg., Figure 830:1. These sections are held in place with a thin layer of quartz wool.

6.2 AIR-SAMPLING PUMP. An air pump capable of drawing a constant flow of air through a limiting orifice (1 L/min) is suitable for area samples. For personal samplers, a calibrated battery-operated portable pump capable of aspirating an accurate and reproducible volume of air through the tube is required.

6.3 GC EQUIPPED WITH FLAME IONIZATION DETECTOR.

6.4 GAS CHROMATOGRAPH COLUMN. A 30-cm long by 3-mm OD (2.34-mm ID) stainless steel tube is packed with 10% by weight of Dexsil 300 GC coated on 80/90 mesh ABS Anakrom. The packed column is pre-conditioned at 300°C for 24 h with helium at a flow rate of 40 to 50 cm^3/min.

6.5 Silylation of the packed column is accomplished by installing the column and a 3-mm OD pyrex injector liner on the GC. A column temperature of 150°C and a helium flow rate of 30 cm^3/min is maintained while introducing three 10-μL samples of Silyl-8 with a microsyringe. The detector is removed from the GC to avoid contamination.

6.6 A STRIP CHART RECORDER. 1 mV sensitivity is acceptable.

6.7 MICROSYRINGES (10- and 25-μL) and 1- and 10-mL stoppered volumetric flasks are required to handle samples.

7. Reagents

7.1 AR GRADE ACETONE AND SILYL-8.

7.2 HELIUM CARRIER GAS. Highest purity.

7.3 HYDROGEN. Ultrapure hydrogen.

7.4 AIR. Prepared by mixing oxygen and nitrogen to 21% O_2.

7.5 MOCA, purified chromatographically to 99.5%.

8. Procedure

8.1 CLEANING OF GLASSWARE. All glassware used for the laboratory analysis should be thoroughly cleaned, washed and rinsed with distilled water. Volumetric ware should be dried in vacuo at 50°C, and allowed to cool.

8.2 CALIBRATION OF PERSONAL PUMP. Each personal pump must be calibrated with a representative Gas-chrom S tube.

8.3 COLLECTION.

8.3.1 MOCA in air is sampled at a rate of 1 L/min (up to 8 h if necessary) through a Gas-chrom S tube with a personal air sampling pump.

8.3.2 The tapered end of the tube is attached to the pump so that the initial air sampling occurs in the 50-mg section.

8.4 ANALYSIS OF SAMPLE.

8.4.1 After sampling, each section of the Gas-chrom S tube with its plug is transferred into a 1-mL volumetric flask and 0.5 mL of acetone is added to initiate extraction of the collected MOCA.

The sample is allowed to stand 25 min with occasional agitation prior to analysis by gas chromatography. Acetone is a very volatile solvent, hence volume changes can occur during the desorption process depending on the ambient temperature and the seal of the volumetric flask. In all cases proper safety precautions should be taken while working with hazardous chemicals.

8.4.2 *GC conditions.* Operating conditions for the GC may vary from instrument to instrument and must be optimized. The following are the recommended starting instrumental conditions. A gas chromatograph with a flame ionization detector (FID) is equipped with a silylated glass liner and a 30-cm 10% Dexil 300 GC column operated at column and inlet temperatures of 200° and 250°C, respectively. The flow

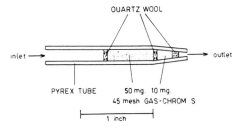

QUARTZ WOOL

inlet → outlet

PYREX TUBE   50 mg. 10 mg.
          45 mesh GAS-CHROM S

1 inch

Figure 830:1 — MOCA sampling tube.

rate of the helium carrier gas is adjusted to 35 cm^3/min, and the flow ratio of helium: hydrogen: air is maintained at 1:0.71:10. After the recorder is turned on, the electrometer is balanced and the FID is ignited.

8.4.3 *Injection*. After desorption, a 2-μL aliquot of the sample solution is injected into the GC. The 2 μL of the acetone extract is injected into the GC with the microsyringe held in the sample inlet until the acetone peak, as displayed on the recorder, goes off-scale and then returns on-scale. At this point, the syringe is withdrawn. After 2 to 3 min the range switch is reset to 0.1 and a complete chromatogram is obtained in about 13 min. At low MOCA concentrations an increase in the sensitivity is achieved with a 1-μL sample, provided the calibration curve is prepared with the same volume, 1 μL Triplicate injection of each sample and a standard should be made.

9. Calibration and Standards

9.1 MEASUREMENTS OF AREA. The peak area of MOCA is measured by the peak height multiplied by peak width at half height, or by use of a suitable electronic integrator. The unknown concentration is determined from a calibration curve prepared with identical injection volumes.

9.2 Since no internal standard is used in this method, standard solutions must be analyzed at the same time and with the same injection volume as the test sample. This will minimize the effect of known day-to-day variations of the FID response. Standard curves are established by plotting concentrations in μg versus peak area or integrator output.

10. Calculations

10.1 It is convenient to express concentration of MOCA in terms of milligrams per cubic meter of air, mg MOCA per m^3. By use of the following equation the concentrations of MOCA in the air sampled can be expressed in mg per m^3, which is numerically equal to μg/L of air.

$$\frac{(\mu g \ MOCA) \ (A) \ (1.087)}{[Sampling \ rate \ (L/min)] \ [Sampling \ time \ (min)]}$$

$$= mg \ MOCA/m^3$$

where A = 250 for 2-μL or 500 for 1-μL aliquot sample

10.2 Another method of expressing concentration is ppm, defined as μL of compound per liter of air.

$$ppm = \mu L \ of \ compound/V_s$$

$$ppm = \frac{\mu g \ of \ compound}{V_s} \times \frac{24.45}{MW}$$

where 24.45 = molar volume at 25°C and 101.3 kPa

MW = molecular weight of the compound

V_s = volume of air in liters at 25°C and 101.3 kPa

11. References

1. FEDERAL REGISTER 38, 1973. No. 85, 10929–10930 May 3.
2. A. L. LINCH, G. B. O'CONNOR, J. B. BARNES, A. S. KILLIAN, JR. AND W. E. NEELD, JR. 1971. *Amer. Ind. Hyg. Assoc. J.* 802–819. Dec.

Subcommittee 5

E. SAWICKI, *Chairman*
T. BELSKY
R. A. FRIEDEL
D. L. HYDE
J. L. MONKMAN
R. A. RASMUSSEN
L. A. RIPPERTON
L. D. WHITE

Approved with Modifications from 2nd edition
Subcommittee 4/5

M. FELDSTEIN, *Chairman*
R. J. BRYAN
D. L. HYDE
D. A. LEVAGGI
D. C. LOCKE
R. A. RASMUSSEN
P. O. WARNER

831.

Determination of p,p-Diphenylmethane Diisocyanate (MDI) in Air

1. Principle of the Method

1.1 MDI is hydrolyzed by a solution of hydrochloric and acetic acids to methylene dianiline. The method is a modification of the Marcali method **(2)**.

1.2 The methylene dianiline is diazotized by sodium nitrite-sodium bromide solution.

1.3 The diazo compound is coupled with N-(1-Naphthyl)-ethylenediamine to form the colored complex.

1.4 The amount of colored complex formed is in direct proportion to the amount of MDI present. The amount of colored complex is determined by reading the absorbance of the solution of 555 nm **(1)**.

2. Range and Sensitivity

2.1 The range of the standards used is 1.5 to 15 μg MDI. In a 20-L air sample, this range is equal to 0.007 ppm to 0.073 ppm.

2.2 For samples of high concentration whereby absorbance is greater than the limits of the standard curve, dilution of the sample with absorber solution and rereading the absorbance can extend the upper limit of the range.

2.3 The amount of MDI that would saturate 15 mL of the absorber solution has not been determined. Therefore, it is possible that, in extremely high concentrations, some of the MDI would not be absorbed by the absorber solution. In such cases two impingers should be used in series, and the appropriate corrections made for efficiency.

3. Interferences

3.1 Any free aromatic amine may be diazotized and coupled forming a positive interference.

3.2 Toluenediisocyanate (TDI) also interferes.

4. Precision and Accuracy

4.1 The precision and accuracy are unknown.

4.2 There has been no collaborative testing.

5. Advantages and Disadvantages of the Method

5.1 The acidity for the coupling reaction has been changed in this method so as to reduce the final reaction time to 15 min compared with 2 h for the Marcali method.

5.2 Since any free aromatic amine may interfere and TDI definitely interferes, the method is not specific for MDI.

6. Apparatus

6.1 SAMPLING EQUIPMENT. The sampling unit for personal samples by the impinger collection method consists of the following components:

6.1.1 *A midget impinger.* Containing the absorbing solution or reagent.

6.1.2 *Battery operated personal sampling pump—MSA Model G, or equivalent.* The sampling pump is protected from splashover or water condensation by an absorption tube loosely packed with a plug of glass wool and inserted between the exit arm of the impinger and the pump.

6.1.3 *An integrating volume meter such as a dry gas or wet test meter.*

6.1.4 *Thermometer.*

6.1.5 *Manometer.*

6.1.6 *Stopwatch.*

6.1.7 *Various clips, tubing, spring connectors, and belt*, for connecting sampling apparatus to worker being sampled.

6.2 BECKMAN MODEL B SPECTROPHOTOMETER, OR EQUIVALENT.

6.3 CELLS, 5-CM MATCHED QUARTZ CELLS.

6.4 VOLUMETRIC FLASKS. Several of each: 100-mL and 1-L.

6.5 BALANCE, capable of weighing to at least three places, preferably four places.

6.6 PIPETS. Delivery: 0.5-, 1-, 2-, 5-, 10-, 15-mL, graduated: 2 mL.

6.7 GRADUATED CYLINDERS; 50-, 100-ML.

7. Reagents

7.1 PURITY. All reagents must be made using ACS reagent grade chemicals or a better grade.

7.2 Water means ASTM reagent water, Type II.

7.3 SODIUM NITRITE.

7.4 SODIUM BROMIDE.

7.5 SULFAMIC ACID.

7.6 CONCENTRATED HYDROCHLORIC ACID, 11.7 N.

7.7 GLACIAL ACETIC ACID, 17.6 N.

7.8 N-(1-NAPHTHYL)-ETHYLENEDIAMINE DIHYDROCHLORIDE.

7.9 SODIUM CARBONATE.

7.10 METHYLENE DIANILINE (MDA).

7.11 SODIUM NITRITE-SODIUM BROMIDE SOLUTION. Dissolve 3.0 g sodium nitrite and 5.0 g sodium bromide in water and dilute to 100 mL. The solution may be stored in the refrigerator for one week.

7.12 SULFAMIC ACID SOLUTION. Dissolve 10.0 g sulfamic acid in 90 mL water.

7.13 ABSORBING SOLUTION. Add 35 mL conc hydrochloric acid and 22 mL glacial acetic acid to about 600 mL water and dilute to 1 L with water.

7.14 COUPLING SOLUTION. Dissolve 1.0 g N-(1-Napthyl)-ethylenediamine dihydrochloride in 50 mL water, add 2 mL concentrated hydrochloric acid and dilute to 100 mL with water. This solution is stable for about 10 days.

7.15 SODIUM CARBONATE. Dissolve 16.0 g sodium carbonate in water and dilute to 100 mL.

7.16 SOLUTION A. Dissolve 0.3000 g MDA in 700 mL glacial acetic acid. Dilute to 1 L with water.

7.17 SOLUTION B. Immediately after making solution A, pipet 10 mL solution A into a 1-L volumetric flask. Add 35 mL hydrochloric acid, 15 mL glacial acetic acid, and dilute to volume with water. This solution is equivalent to 3.8 μg MDI/mL.

8. Procedure

8.1 CLEANING OF EQUIPMENT.

8.1.1 Wash all glassware in hot detergent solution, such as Alconox, to remove any oil.

8.1.2 Rinse thoroughly with hot tap water.

8.1.3 Rinse thoroughly with double distilled water. Repeat this rinse several times.

8.2 COLLECTION AND SHIPPING OF SAMPLES.

8.2.1 Pipet 15 mL of the absorbing solution (7.13) into the midget impinger.

8.2.2 Connect the impinger (via the absorption tube) to the personal sampling pump with a short piece of flexible tubing. The minimum amount of tubing should be used between the sampling zone and impinger. The air being sampled should not be passed through any other tubing or other equipment before entering the impinger.

8.2.3 Turn on pump to begin sample collection. Care should be taken to measure the flow rate, time and/or volume as accurately as possible. Record atmospheric pressure and temperature. The sample should be taken at a flow rate of 1 L/min. Sample for 20 min making the final volume 20 L.

8.2.4 After sampling, the impinger stem can be removed and cleaned. Tap the stem gently against the inside wall of the impinger bottle to recover as much of the sampling solution as possible. Wash the stem with a small amount (1 to 2 mL) of unused absorbing solution and add the wash to the impinger. Seal the impinger with a hard, non-reactive stopper (preferably Teflon or glass). Do not seal with rubber. The stoppers on the impingers should be tightly sealed to prevent leakage during shipping. If it is preferred to ship the impingers with the stems in, the outlets of the stem should be sealed with Parafilm or other non-rubber covers, and the ground glass joints should be sealed (*i.e.*, taped) to secure the top tightly.

8.2.5 Care should be taken to minimize spillage or loss by evaporation at all times. Refrigerate samples if analysis cannot be done within a day.

8.2.6 Whenever possible, hand delivery of the samples is recommended. Otherwise, special impinger shipping cases designed by NIOSH should be used to ship the samples.

8.2.7 A "blank" impinger should be handled as the other samples (fill, seal and transport) except that no air is sampled through this impinger.

8.3 ANALYSIS OF SAMPLES.

8.3.1 Remove the bubbler tube, if it is still attached, taking care not to remove any absorbing solution.

8.3.2 Start reagent blank at this point by adding 15 mL fresh absorbing solution to a clean bubbler tube.

8.3.3 To each tube, including the blanks, add 0.5 mL sodium nitrite-bromide solution, stir well, and allow to stand for 2 min.

8.3.4 Add 1 mL 10% sulfamic acid solution to each tube, stir for 30 s, and allow to stand for 2 min.

8.3.5 Add 1.5 mL sodium carbonate solution to each tube and stir.

8.3.6 Add 1 mL coupling solution to each tube, make up to volume of 20.0 mL with water and stir. Allow color to develop for 15 to 30 min.

8.3.7 Transfer each solution to a 5-cm quartz cell.

8.3.8 Using the blank, adjust the spectrophotometer to 0 absorbance at 555 nm.

8.3.9 Determine the absorbance of each sample at 555 nm.

9. Calibration and Standards

9.1 To a series of 5 impinger tubes, add the following amounts of absorbing solution: 15.0, 14.5, 14.0, 13.0, 10.0 mL respectively.

9.2 To each tube add standard solution B in the same order as the absorbing solution was added: 0.0, 0.5, 1.0, 2.0, and 5.0 mL, so that the final volume is 15 mL (*i.e.*, 0.0 mL of standard is added to the 15 mL absorbing solution: 0.5 mL of standard is added to the 14.5 mL absorber solution, etc.). The cylinders now contain the equivalent of 0.0, 1.9, 3.8, 7.6 and 18.9 μg MDI, respectively. The standard containing 0.0 mL standard solution is a blank.

9.3 To each tube, add 0.5 mL sodium nitrite solution, stir well and allow to stand for 2 min.

9.4 Add 1 mL of 10% sulfamic acid solution, stir for 30 s, and allow to stand for 2 min.

9.5 Add 1.5 mL sodium carbonate solution and stir.

9.6 Add 1 mL coupling solution, make up to volume of 20.0 mL with water and stir. Allow color to develop for 15 to 30 min.

9.7 Transfer each solution to a 5-cm quartz cell.

9.8 Using the blank, adjust the spectrophotometer to 0 absorbance of each standard at 555 nm.

9.9 Determine the absorbance of each standard at 555 nm.

9.10 A standard curve is constructed by plotting the absorbance against micrograms MDI.

10. Calculations

10.1 Blank values (Section **8.2.7**), if any, should first be subtracted from each sample.

10.2 From the calibration curve (Section **9.10**), read the μg MDI corresponding to the absorbance of the sample.

10.3 Calculate the concentration of MDI in the air sampled in ppm, defined as μL MDI per liter of air.

$$ppm = \frac{\mu g}{V_s} \times \frac{24.45}{MW}$$

where ppm = parts per million MDI
   μg = micrograms MDI (Section **10.2**)
   V_s = volume of air (L, corrected to 25°C and 101.3 kPa)
   24.45 = molar volume of an ideal gas at 25°C and 101.3 kPa
   MW = molecular weight of MDI, 250.27.

11. References

1. GRIM, K. E. AND A. L. LINCH. 1964. Recent Isocyanate-in-Air Analysis Studies, *American Industrial Hygiene Association Journal*, 25:285.
2. MARCALI, K. 1957. Microdetermination of Toluene diisocyanates in Atmosphere, *Anal. Chem.* 29:552.

Revised by Subcommittee 3

B. DIMITRIADES, *Chairman*
J. CUDDEBACK
E. L. KOTHNY
P. W. MCDANIEL
L. A. RIPPERTON
A. SABADELL

Approved with modifications from 2nd edition.

D. A. LEVAGGI, *Chairman*
B. R. APPEL
D. W. HORSTMAN
E. L. KOTHNY
J. G. WENDT

832.

Determination of Nitroglycerin and Ethylene Glycol Dinitrate (Nitroglycol) in Air

1. Principle of the Method

1.1 A known volume of air is drawn through a glass tube packed with Tenax-CG (a porous organic polymer) to collect the vapors of the nitrate esters.

1.2 The sorbent is transferred to a small test tube and extracted with ethanol.

1.3 An aliquot of the ethanol solution is injected into a gas chromatograph with an electron-capture detector.

1.4 The areas of the resulting peaks are measured and the amount of each nitrate ester is determined from calibration curves. The concentration of the nitrate ester in the air sample is calculated from the amount found by GC, the size of the aliquot, and the volume of air sampled (1).

2. Range and Sensitivity

2.1 The minimum of EGDN detectable by the overall procedure as described is substantially smaller than 0.1 μg. This sensitivity easily permits the detection of EGDN in a 10-L air sample containing 0.01 mg/m^3. The minimum amount of NG detectable by the overall procedure is smaller than 1 μg. This sensitivity permits the detection of NG in a 100-L air sample containing 0.01 mg/m^3.

2.2 The sorbent tubes will retain at least 1 mg of either EGDN or NG.

3. Interferences

3.1 Ethylene glycol mononitrate, if present in a high concentration relative to that of EGDN, produces a GC peak that tails into the EGDN peak. This interference, although troublesome, does not prohibit a reasonably satisfactory analysis for EGDN. The combined selectivity of Tenax-GC and electron-capture detection practi- cally eliminates other interference. Water vapor does not interfere.

4. Precision and Accuracy

4.1 The precision (2σ) of the analytical method is estimated to be about $\pm$ 6%.

4.2 The precision of the overall sampling and analytical method has not been adequately determined. However, since both EGDN and NG are sorbed on Tenax-GC at ambient temperature with 100% efficiency, it is believed that the overall precision will be determined by variations in the measurement of sample volume.

4.3 In laboratory measurements not involving a personal sampling pump, the average recovery of EGDN from a vapor sample was 104% and that of NG was 102%.

5. Advantages and Disadvantages of the Method

5.1 The general technique of sampling and analysis is similar to that already established for organic solvents in air in which a charcoal tube and gas chromatography are used (Method 834). The advantages of that method also apply to the present method.

5.2 The principal difficulties are those associated with the use of electron-capture detection. They include the limited linear range of electron-capture detectors, and operating variables such as the effects of column bleed, moisture, and oxygen. The effects of these variables may be minimized by analyzing standards at the same time as samples. The electron-capture detector is about 500 times as sensitive to the nitrate esters as a hydrogen-flame ionization detector, and it provides a substantial advantage in specificity.

5.3 The principal limitation on precision and accuracy in this method is that associated with the use of small personal sampling pumps for the measurement of sample volume.

5.4 The range of sample size is very wide. However, repeated dilutions of the ethanol extract may be necessary to bring the amount of sample injected into the proper range for the electron-capture detector.

6. Apparatus

6.1 An approved and properly calibrated personal sampling pump is required.

6.2 The sorbent tubes are 70 mm long and 5 mm ID. They contain two sections of 35/60 mesh Tenax-GC separated and held in place by glass-wool plugs. The front section contains 100 mg of sorbent and the backup section contains 50 mg. Since the pressure drop must be limited to 1 in. of Hg (3.4 kPa) at 1.0 L/min, it is necessary to avoid overpacking with glass wool. Limited laboratory evaluation indicates that 40/60-mesh Porapak Q may be substituted for the Tenax-GC. (Tenax-GC is distributed by Applied Science Laboratories, Inc., State College, Pa., and Porapak Q by Waters Associates, Inc., Milford, Mass.)

6.3 A gas chromatograph equipped with an electron-capture detector, a 76 cm by 6 mm glass column packed with 10% of OV-17 on 60/80-mesh Gas Chrom Q, and an integrator are required.

6.4 A volumetric flask and pipets, a 10-μL syringe, and other normally available laboratory supplies and equipment are needed.

7. Reagents

7.1 The only reagent required is absolute ethanol for extraction of the sorbent. The gas chromatograph requires a supply of helium and an argon-methane mixture, or other carrier or purge gases as required for the particular instrument used.

8. Procedure

8.1 CLEANING OF EQUIPMENT. All glassware for the laboratory analysis should be washed with detergent and thoroughly rinsed with tap water and distilled water. Particular attention should be paid to the cleaning of the microliter syringe with ethanol.

8.2 CALIBRATION OF PERSONAL SAMPLING PUMP. Each pump should be calibrated with a Tenax-GC sorbent tube in the line. This will minimize errors associated with uncertainties in the volume of sample collected.

8.3 COLLECTION AND SHIPPING OF SAMPLES.

8.3.1 Immediately before sampling, break the ends of the tube to provide an opening of at least 2 mm.

8.3.2 The smaller section of sorbent is used as a backup and should be positioned nearest the sampling pump.

8.3.3 The sorbent tube should be placed in a vertical direction during sampling.

8.3.4 Air being sampled should not be passed through any hose or tube before entering the sorbent tube. This is particularly important with the nitrate esters since they are strongly absorbed on most surfaces, including glass.

8.3.5 The flow, time, and/or volume must be measured as accurately as possible. The sample should be taken at a flow rate of 1.0 L/min or less to attain the total sample volume required. The minimum volume that must be collected to permit detection of NG at the TLV concentration (or EGDN at a still lower concentration) is only 0.1 L. At the TLV concentration, sample volumes as large as 1000 L can be collected without causing trouble in the analysis.

8.3.6 The temperature and pressure of the atmosphere being sampled should be measured and recorded.

8.3.7 The Tenax-GC tubes should be capped with the supplied plastic caps immediately after sampling. Under no circumstances should rubber caps be used.

8.3.8 One tube should be handled in the same manner as the sample tube (break, seal, and transport), except that no air is sampled through this tube. This tube should be labeled as blank.

8.3.9 Capped tubes should be packed tightly before they are shipped to minimize tube breakage during shipping.

8.4 ANALYSIS OF SAMPLES.

8.4.1 The sorbent tube is scored with a file near the front end and is broken open. The first glass-wool plug and the front (100-mg) section of sorbent are transferred to a small vial or test tube. The remaining glass-wool plug and the backup section of sorbent are transferred to another small vial. Two mL of ethanol are added to each vial, the vials are stoppered, and the contents are shaken for about 1 min. Aliquots of the ethanol solutions are injected into the gas chromatograph. It is convenient to take a 5-μL aliquot, but it may be necessary to dilute the ethanol solution and take another aliquot if the concentrations of the nitrate esters are high.

8.4.2 Typical operating conditions for the gas chromatographic analysis are:

Column: 10% OV-17 on 60/80-mesh
    Gas Chrom Q, 76 cm by 6 mm glass
    column
Carrier gas: Helium, 100 mL/min
Purge gas: Argon/methane, 125 mL/min
Column Temperature: 130°C
Injection port temperature: 160°C
Detector temperature: 280°C

9. Calibration and standards

9.1 The use of pure NG and pure EGDN as reference standards is considered to be impractical for general application because of the safety hazard and the lack of availability of the compounds in pure form. In the course of the development of this method, several reference standards were obtained from the quality control laboratories of manufacturers of nitrate ester products. These materials included the following:

(1) A material called "Nitroglycerin Lactose Trituration" was provided by Eli Lilly and Company, Indianapolis, Indiana. This material is used for quality control of sublingual nitroglycerin tablets that are manufactured for medicinal use and has a stated nitroglycerin content. It consists of nitroglycerin dispersed in an inert, ethanol-soluble powder. To prepare a GC calibration curve, a weighed amount of the material is dissolved in a known volume of ethanol and appropriate aliquots are injected into the gas chromatograph.

(2) Solutions of known concentrations of NG and EGDN were provided by Hercules, Inc. These solutions were quantitatively prepared from analyzed mixtures of the compounds and are used for plant quality control in the manufacture of explosive products and in industrial hygiene applications. Aliquots of these solutions, appropriately diluted, may be injected directly into the gas chromatograph. Department of Transportation regulations may limit the shipment of such solutions by public transportation.

(3) A sample of analyzed dynamite containing EGDN but no NG was provided by Atlas Chemical Industries, ICI America, Inc. A weighed quantity of the dynamite is dispersed in ethanol, filtered, washed with ethanol, and diluted to a known volume. Aliquots of this solution are injected. Since dynamite is a heterogeneous material, care must be taken to obtain a representative sample.

(4) The pharmaceutical-grade sublingual NG tablets may be obtained from wholesale drug dealers. These tablets are manufactured to meet United States Pharmacopeia specifications, which require that the actual NG content be between 80 and 112% of the stated volume. The actual NG content of a specific lot of tablets may be obtained from the manufacturer. For use, the tablets are weighed, crushed, and extracted with ethanol, the solution is filtered and diluted to a fixed volume, and aliquots are injected.

An experimental comparison of the NG and EGDN contents of these standard materials gave a relative standard deviation of 3.6% for the NG contents and 1.6% for the EGDN contents.

It is necessary that standards be analyzed concurrently with samples to minimize the effects of variations in detector response and other factors.

10. Calculations

10.1 From the measured peak areas for NG and EGDN and the calibration curve, determine the amounts of each compound in each section of the sample sorbent tube (usually, no NG or EGDN will be found in the backup section). Subtract from these values the amounts found in the corresponding sections of the blank solvent tube.

10.2 Add the corrected amounts found in the two sections of the sample tube to determine the total amount of each nitrate ester.

10.3 Calculate the concentration of each compound in the air sample by the following equation:

Concentration in mg/m³ (or μg/L) =

$$\frac{\text{Total amount of compound, } \mu g}{\text{volume of air sampled, L}}$$
$$(V_s)$$

The volume of air sampled is corrected to standard conditions (defined as 25° C and 101.3 kPa) by the equation:

$$V_s = V \times \frac{P}{101.3} \times \frac{298}{T + 273}$$

where V_s = corrected volume of air sampled

V = volume of air in liters, as measured

P = pressure in kPa

T = temperature in °C

10.4 Concentrations in mg/m³ may be converted to parts per million (ppm) by volume by the following equations:

For NG,   ppm = mg/m³ × 0.108
For EGDN,   ppm = mg/m³ × 0.161

11. Reference

1. SOUTHERN RESEARCH INSTITUTE, 1974. Contract No. HSM-99-73-63, Final Report, September.

Subcommittee 3

B. DIMITRIADES, *Chairman*
J. CUDDEBACK
E. L. KOTHNY
P. W. MCDANIEL
L. A. RIPPERTON
A. SARADELL

Approved with modifications from 2nd edition

D. A. LEVAGGI, *Chairman*
B. R. APPEL
D. W. HORSTMAN
E. L. KOTHNY
J. G. WENDT

833.

Determination of N-Nitrosodimethylamine in Ambient Air

1. Principle of the Method

N-nitrosodimethylamine (DMN) is concentrated from ambient air on Tenax-GC in a short glass tube (1,2). It is desorbed by heating and purging with helium into a liquid nitrogen-cooled nickel capillary trap and then introduced onto a high resolution gas chromatographic column where it is separated from interferences. The concentration of DMN is measured from the mass spectrometric signal at m/e 74 (3).

2. Range and Sensitivity

2.1 The range of the mass spectrometric signal for the conditions listed corresponds to 0.5 ppt to 10 ppb.

2.2 A concentration of 0.5 ppt of DMN can be determined in a 150-L air sample.

3. Interferences

Inteferences may result from materials having ions of m/e 74 ($C_2H_8N_3$, $C_2H_4NO_2$, $C_2H_6N_2O$, C_3H_3Cl, C_3H_6S, $C_3H_{10}N_2$) if at the same retention time as DMN.

4. Precision and Accuracy

4.1 The precision of this method has been determined to be $\pm 10\%$ relative standard deviation when replicate sampling cartridges were spiked with 50 ng (corresponding to 10 ppb in 150 L of air). These data were obtained using 10.0-cm long glass tubes (1.5-cm ID) packed with 35/60 mesh of Tenax-GC (bed dimensions: 1.5 cm × 6 cm depth).

4.2 The accuracy of the analysis is approximately $\pm 10\%$ of the amount reported as determined from repeated analysis of several standards.

5. Advantages and Disadvantages of the Method

5.1 The gas chromatography-mass spectrometry technique interfaced with a Finnigan glass jet separator (Model 01512-42158 Finnigan Corp., Sunnyvale, CA) is extremely sensitive and specific for the analysis of DMN. The high resolution gas chromatographic separation yields a retention time that is characteristic for DMN, and relatively specific for positive assignment of the signal as DMN. The mass spectrometer in combination with high resolution gas chromatography yields a very high degree of specificity. The base peak of DMN is at m/e 74 which is also the parent ion. In order to assign the signal at m/e 74 to DMN it is absolutely necessary that the retention time matches with the signal.

5.2 Collected samples can be stored up to 1 month with less than 10% losses.

5.3 Because DMN is a suspected carcinogen in man, it is extremely important to exercise safety precautions in the preparation and disposal of liquid and gas standards, cleaning of used glassware, etc., and the analysis of air samples.

5.4 Since the mass spectrometer can not be conveniently mobilized, sampling must be carried out away from the instrument.

5.5 High resolution gas chromatography/low resolution mass spectrometry is not a convenient technique for handling a large number of samples (> 100/wk).

5.6 Efficiency of air sampling increases as the ambient air temperature decreases (i.e. sensitivity increases).

5.7 Ambient air sampling is limited to cases where the NO_x levels are less than 3 ppm when dimethylamine is also present.

6. Apparatus

6.1 SAMPLING TUBES.

6.1.1 The sampling tubes are prepared by packing a 10-cm long × 1.5-cm ID glass tube with 6.0 cm of 35/60 mesh Tenax GC with glass wool in the ends. Cartridge samplers are conditioned at 270°C with helium flow at 30 mL/min for 20 min. The conditioned cartridges are transferred to Kimax (2.5-cm × 150-cm) culture tubes, immediately sealed using Teflon-lined caps, and cooled.

6.1.2 Cartridge samplers with longer beds of sorbent may be prepared using a proportional amount of Tenax GC.

6.2 GAS CHROMATOGRAPHIC COLUMN.

6.2.1 A 0.35-mm ID × 50-m glass SCOT capillary coated with DEGS stationary phase and 0.1% benzyl triphenylphosphonium chloride is used. The capillary column is conditioned (detector end disconnected) for 48 h at 210°C @ 1.5 to 2.0 mL/min helium flow.

6.3 A Finnigan-type glass jet separator on a magnetic or quadrupole instrument is used at 200°C.

6.4 INLET MANIFOLD.

6.4.1 An inlet manifold is fabricated and employed (Figure 833:1; **1, 2, 4**).

6.5 GAS CHROMATOGRAPH.

6.5.1 A Varian 1700 gas chromatograph or equivalent. *A gas chromatograph employing a single column oven and a temperature programmer is adequate.*

6.6 MASS SPECTROMETER.

6.6.1 A mass spectrometer with a resolution of 500 to 2000 equipped with single ion monitoring capabilities must be used in conjunction with a gas chromatograph. A Varian-MAT CH-7 has been found to be satisfactory for this purpose (**2, 3**).

6.7 SYRINGES.

6.7.1 *Syringes,* 1-mL gas tight (Precision Sampling, Inc.) and 10 μL (The Hamilton Co., Inc.)

7. Reagents and Materials

All reagents must be analytical reagent grade.

7.1 N-NITROSODIMETHYLAMINE

7.2 ACETONE

7.3 ISOCLEAN

7.4 TENAX GC (35/60 mesh, Applied Science)

7.5 TWO 2-L ROUND BOTTOM FLASKS, fitted with injection ports.

7.6 SOXHLET APPARATUS

8. Procedure

8.1 CLEANING OF GLASSWARE. All glassware, glass sampling tubes, cartridge holders, etc., should be washed in Isoclean/water, rinsed with doubly distilled water and acetone and air dried. Glassware is heated to 230°C for 2 h.

8.2 PREPARATION OF TENAX GC.

8.2.1 Virgin Tenax GC is extracted in a Soxhlet apparatus overnight with acetone prior to its use.

8.3 COLLECTION OF DMN IN AMBIENT AIR.

8.3.1 Continuous sampling of ambient air may be accomplished using a Nutech Model 221-A portable sampler (Nutec Corp., Durham, NC) or its equivalent (**2**). Flow rates are adjusted with a metering valve through a calibrated rotameter. Total flow is registered by a dry gas meter.

8.3.2 For larger sample sizes it is important to realize that a larger total volume of air may cause elution of DMN through the sampling tube. It has been demonstrated that exceeding a total of 385, 332, 280, 242, 224, 204, 163, 156, 148, 127, 107, 93, or 79 L of air at ambient temperatures of 10, 13, 16, 18, 21, 24, 27, 29, 32, 35, 38, 41, or 43°C, respectively, will result in elution of DMN from the cartridge sampler. A flow of 10 mL /min to 30L/min may be used with the sampler described in **6.1**.

8.3.3 DMN has been found to be stable and quantitatively recoverable from cartridge samplers after 4 weeks when tightly closed in cartridge holders, protected from light and stored at 0°C.

8.4 ANALYSIS OF SAMPLE.

8.4.1 *Instrument Conditions and Setup.* The thermal desorption chamber and six-port valve are set to 200°C. The glass jet separator is maintained at 200°C. The mass spectrometer is set to monitor *m/e* 74 (Figure 833:2).

8.4.2 Adjust the He purge gas through the desorption chamber to 50 mL/

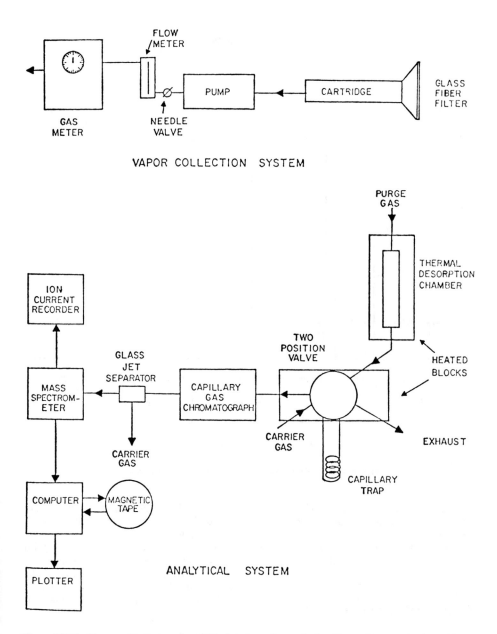

FLOW
METER

GAS
METER

NEEDLE
VALVE

PUMP

CARTRIDGE

GLASS
FIBER
FILTER

VAPOR COLLECTION SYSTEM

PURGE
GAS

THERMAL
DESORPTION
CHAMBER

HEATED
BLOCKS

ION
CURRENT
RECORDER

TWO
POSITION
VALVE

GLASS
JET
SEPARATOR

CAPILLARY
GAS
CHROMATOGRAPH

MASS
SPECTROM-
ETER

CARRIER
GAS

EXHAUST

CARRIER
GAS

CAPILLARY
TRAP

COMPUTER

MAGNETIC
TAPE

PLOTTER

ANALYTICAL SYSTEM

Figure 833:1 — Vapor collection and analytical systems for analysis of hazardous vapors in ambient air.

min. Cool the Ni capillary trap at the inlet manifold with liquid nitrogen.

8.4.3 Place the cartridge sampler in the desorption chamber and desorb for 5 min.

8.4.4 Rotate the six-port valve on the inlet-manifold to position "B", heat the Ni capillary trap to 180°C with a wax bath.

8.4.5 Temperature program the glass capillary column from 75 to 205°C at 4°C/

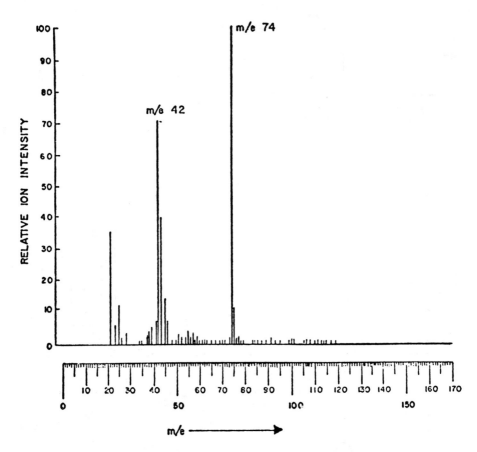

Figure 833:2 — Mass spectrum of N-nitrosodimethylamine.

min and hold at upper limit for 10 min. The retention time of DMN is approximately 26 min (Figure 833:3).

8.4.6 The analytical column is cooled to ambient temperature and the next sample is processed.

9. Calibration and Standards

9.1 PREPARATION OF GAS STANDARD.

9.1.1 Purge two 2-L round bottom flasks with helium, warm flasks to 50°C with heating mantles and use magnetic bar to stir vapors.

9.1.2 Inject 0.1 to 1 μL of DMN into flask and let stir for 30 min. Make further dilutions into second flask by transferring milliliter gas volumes as needed.

9.1.3 Purge air/vapor mixtures from second flask onto cartridge samplers.

9.2 CALIBRATION.

9.2.1 Prepare standard curve (with ten concentration points) by thermally desorbing cartridge samplers loaded with 3 ng to 30 μg of DMN. Plot *m/e* 74 response *vs* ng of DMN. A linear response is observed.

10. Calculations

10.1 The total quantity of DMN in ambient air is determined by comparing *m/e* 74 response for samples of DMN with standard curve.

$$\text{ppb} = \frac{\text{ng DMN}}{\text{V}} \times \frac{24.45}{74}$$

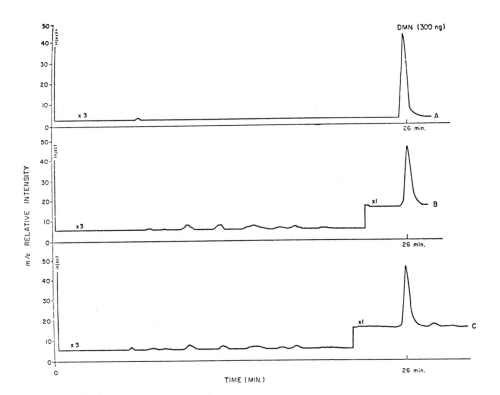

Figure 833:3 — Mass (m/e 74) chromatograms. A = standard DMN; B,C = replicate air samples.

where ng DMN = total ng concentration as determined in **9.2.1**

V = volume of air in liters sampled at 25°C and 101.3 kPa

24.45 = molar volume of an ideal gas at 25°C and 101.3 kPa

74 = molecular weight of DMN.

References

1. PELLIZZARI, E. 1974. Development of Method for Carcinogenic Vapor Analysis in Ambient Atmosphere, Pub. No. EPA-650/2-74-121, 148pp.
2. PELLIZZARI, E. 1974. Development of Analytical Techniques for Measuring Ambient Atmospheric Carcinogenic Vapors. Pub. No. EPA-600/2-75-076, 187pp.
3. PELLIZZARI, E., J. E. BUNCH, R.E. BERKLEY AND J. T. BURSEY. 1976. Identification of N-Nitrosodimethylamine in Ambient Air by Capillary Gas-Liquid Chromatography Mass Spectrometry Computer. Biomedical Mass Spec. 3, 196–200.
4. PELLIZZARI, E. D., B. H. CARPENTER, J. E.

BUNCH AND E. SAWICKI. 1975. Collection and Analysis of Trace Organic Vapor Pollutants in Ambient Atmospheres-Thermal Desorption of Organic Vapors from Sorbent Media. Environ Sci & Technol, 9, 556–560.

834.

Determination of Organic Solvent Vapors in Air

1. Principle of the Method

1.1 A known volume of air is drawn through a charcoal adsorption tube to trap the organic solvent vapors present (**4,5**).

1.2 The charcoal is transferred to a small stoppered tube and the organic compounds are desorbed with carbon disulfide (**5,8**).

1.3 An aliquot of the desorbed sample is injected into a gas chromatograph (G.C.) equipped with a flame ionization detector (FID).

1.4 The area of the resulting peak(s) is (are) measured and compared with areas obtained from the injection of standards.

2. Range and Sensitivity

2.1 The approximate breakthrough capacity in milligrams (mg) of selected compounds per 100 mg of activated charcoal is given in Table 834:I. This value is the approximate number of milligrams of the compound which the front section will hold before a significant amount of compound is found on the backup section (The charcoal tube consists of two sections of activated charcoal separated by a section of urethane foam. (See Section **6.2**). If a particular atmosphere is suspected of containing a large amount of contaminant, a smaller sampling volume should be used.

2.2 The lowest detectable limit in mg/sample for specific compounds using this method has not been determined. Values as low as 0.001 mg/sample (about 25 ppb in an integrated 10-L air sample) were reported during collaborative testing of this method (**1**). It was also reported that when the weight of the compound on the charcoal section was less than 0.1 mg, analytical errors were substantially higher (greater than ± 20%). Therefore, if a contaminant is present at relatively low concentrations in the industrial atmosphere, a large volume of air should be sampled.

3. Interferences

3.1 When the amount of water vapor or mist in the air is so great that condensation occurs in the sampling tube, organic vapors will not be trapped. Preliminary experiments indicate that high temperatures, high humidity, and high sampling flow rates cause a decrease in the adsorption capacity of activated carbon for organic solvent vapors.

3.2 When two or more solvents are known or suspected to be present in the air, such information, including their suspected identities, should be transmitted with the sample since, with differences in polarity, one may displace another from the charcoal.

3.3 It must be emphasized that any compound that has the same retention time as the specific compound under study at the operating conditions described in this method is an interference. Hence, retention time data on a single column, or even on a number of columns, cannot be considered as proof of chemical identity. For this reason it is important that a sample of the bulk solvent(s) be submitted at the same time so that identity(ties) can be established by other means (**2**).

3.4 If the possibility of interference exists, G.C. separation conditions (column packing, temperature, etc.) must be changed to circumvent the problem.

3.5 Activated charcoal cannot be stored, transported or handled in a contaminated atmosphere. The adsorbed compounds will contribute to the background and may prove to be an interference. The charcoal tubes should be flame sealed or otherwise sealed from the atmosphere.

Table 834:I Selected Parameters

Organic Solvent	Method Classification**	Approximate Breakthrough Capacity* † (mg/100 mg charcoal)	Suggested Sample † Volume (L)	Federal Standard‖ ppm
Benzene	A	6	10	1
2–Butanone	B	9	10	200
1,2–Dichloroethane	A	–	10§	50
Dichloromethane	B	4.5‡	4‡	500
1,4–Dioxane	A	–	10§	100
Styrene	D	15	10	100
Tetrachloroethane	B	21	10	5
Tetrachloromethane	A	–	10§	10
Toluene	B	28‡	10‡	200
1,1,1–Trichloroethane	B	11	5	350
1,1,2–Trichloroethane	B	10	10	10
Trichloromethene	A	–	10§	50
Trichloroethane	A	13	10§	100
Xylene	A	–	10§	100
(1,2–dimethylbenzene 1,3–dimethylbenzene 1,4–dimethylbenzene) mixture				

* See Section 2.1.
† Based on breakthrough experiments with activated petroleum charcoal, unless otherwise stated.
‡ Based on breakthrough experiments with activated coconut charcoal.
§ Estimated.
‖ Reference 7.
** Method classification: A, Recommended; B, Accepted; D, Operational.

4. Precision and Accuracy

4.1 Seven selected organic solvents (benzene, carbon tetrachloride, chloroform, p-dioxane, ethylene dichloride, trichlorethylene, and xylene) have been collaboratively tested (1). The precision of the sampling method using an approved personal sampling pump plus the analytical method varied from ± 6 to ± 17% relative standard deviation, depending upon the compound, mixture and concentration.

4.2 The accuracy of the sampling and analytical method is about ± 10% when the personal sampling pump is calibrated with a charcoal tube in the line. Accuracy is affected primarily by sampling error and the extent of desorption of each compound from the activated charcoal. Sampling precision can be improved by using a more powerful vacuum pump with associated gas-volume integrating equipment (e.g., limiting orifice).

5. Advantages and Disadvantages of the Method

5.1 The sample device is small, portable, and involves no liquids. The samples are analyzed by means of a relatively fast, instrumental method. The method is applicable for the sampling and analysis of a number of organic vapors. Compounds that have been determined by National Institute for Occupational Safety and Health (NIOSH) Laboratories using this method are listed in Tables 834:I and 834:II (8, unpublished data). A special analytical procedure has been developed for carbon disulfide (6), since it is used as the desorbent in this procedure.

5.2 The occupational exposure standards (3) established by authority of the Occupational Safety and Health Act (OSHA) of 1970 are classified as acceptable ceiling concentrations and as 8-h time weighted averages. An acceptable grab sampling technique may be used to deter-

Table 834:II Chemicals That Have Greater Than 80% Desorption But Have Not Been Thoroughly Tested by NIOSH
Class E

Acetic acid, ethyl ester	Isophorone
Allyl glycidyl ether	Isopropyl glycidyl ether
Butanoic acid, ethyl ester	
Butanol	3–Methylbutanoic acid, ethyl ester
2–Butoxyethanol	4–Methyl–2–pentanone
Butyl glycidyl ether	2–Methylpropanoic acid, ethyl ester
Chlorobenzene	2–Methyl–1–propanol
2–Chloro–1,3–epoxypropane	2–Methylpropanoic acid, methyl ester
Cyclohexane	α–Methyl stryrene
Cyclohexanone	4–Methyl styrene
1,2–Dichlorobenzene	Octane
1,4–Dichlorobenzene	3–Octanone
Diethyl ether	Pentane
N,N–Dimethylaniline	2–Pentanone
2,6–Dimethylpyridine	Pentanoic acid, ethyl ester
2–Ethoxyacetic acid, ethyl ester	α–Pinene
Ethylbenzene	Propanoic acid, ethyl ester
Formic acid, ethyl ester	2–Propanone
Furfural	Propenenitrile
Heptane	Propenoic acid, methyl ester
3–Heptanone	1,1,2,2–Tetrachloroethane
Hexane	Tetrahydrofuran
2–Hexanone	1,1,2–Trichloro–1,2,2–trifluoroethane

mine peak concentrations of chemicals for which a ceiling value has been specified. However, for those chemicals that have standards based on 8-h time weighted average concentrations, a sampling procedure capable of collecting an integrated sample over a longer time span is required. The activated charcoal sampling tube is particularly suited for taking integrated solvent vapor samples. The Sipin pump (Section **6.2**) can be used to take an integrated charcoal tube sample over a time period of more than 3 h (10-L sample).

5.3 Interferences are minimal, and most of those which do occur can be eliminated by altering G. C. operating parameters. The method can also be used for the simultaneous analysis of two or more solvents suspected to be present in the same sample by simply changing G.C. conditions from isothermal to a temperature-programmed mode of operation (**8**).

5.4 The greatest selectivity of activated charcoal is toward nonpolar organic solvent vapors. Organic compounds that are gaseous at room temperature, reactive, polar, or oxygenated (aldehydes, alcohols and some ketones) are either not adsorbed (rela-

tively early breakthrough) or desorbed efficiently. Such compounds may require individual testing to determine whether the charcoal tube may be satisfactorily employed for sampling.

5.5 A disadvantage of the method is that the amount of sample that can be collected is limited by the number of milligrams that the charcoal will adsorb before overloading. When the sample value obtained for the backup section of the charcoal trap exceeds 25% of that found on the front section, the possibility of sample loss exists.

5.6 The precision of the method is limited by the reproducibility of the pressure drop across the tubes. Variable pressure drops will affect the flow rate and cause the sampled volume to be imprecise, because the personal sampling pump is usually calibrated for one tube only.

5.7 The desorption efficiency of a particular compound can vary from one batch of charcoal to another. Whenever a new batch of charcoal is used, it is therefore necessary to determine at least once the percentage of each specific compound removed in the desorption process.

6. Apparatus

6.1 CHARCOAL TUBES. A glass tube with both ends flame sealed, 7 cm long with a 6-mm OD and a 4-mm ID, containing 2 sections of 20/40 mesh activated charcoal separated by a 2-mm portion of urethane foam **(4)**. The activated charcoal* is fired at 600°C with nitrogen purge for 1 h prior to packing. The adsorbing section contains 100 mg of charcoal, the backup section 50 mg. A 3-mm portion of urethane foam is placed between the outlet end of the tube and the backup section. The pressure drop across the tube must be less than one inch of mercury (3.4 kPa) at a flow rate of 1 L/min. Such tubes are commercially available from SKC, Inc. and from Mine Safety Appliances Co., both of Pittsburgh, PA.

6.2 AIR SAMPLING PUMP. An air or vacuum pump capable of drawing a constant air flow through a limiting orifice (1 L/min) is suitable for area samples. For personal samples, a calibrated battery-operated portable pump capable of aspirating an accurate and reproducible volume of air through a charcoal tube is required. A pocket-sized pump capable of accurately drawing from 50 to 200 cm³ of air/min through the charcoal tube is commercially available from Anatole J. Sipin Co., 385 Park Avenue South, New York City. OSHA approved coal-mine-dust personal sampling pumps are also recommended for collection of breathing zone samples.

6.3 GAS CHROMATOGRAPH (G.C.). Equipped with a flame ionization detector (FID) **(8)**.

6.4 GAS CHROMATOGRAPHIC COLUMN. Any column capable of separating the organic vapor mixture under study is acceptable. Stainless steel tubing is recommended, especially if reactive chemicals or high temperatures are involved. A stainless steel column with 10% FFAP stationary phase on 80/100 mesh AW-DMCS chromosorb W solid support has been used to separate fourteen selected organic solvents **(8)**.

6.5 A strip chart recorder with a one mV range and one second response is acceptable. A mechanical or electronic integrator is also recommended.

6.6 A 10-μL syringe is recommended for sample injection into the G.C., and convenient (10 to 100 μL) sizes for preparation of standards.

6.7 GLASS OR POLYETHYLENE STOPPERED TUBES. 2.5-mL graduated microcentrifuge tubes are recommended.

6.8 PIPETS. 0.5-mL delivery pipets or 1.0 mL type graduated in 0.1 mL increments.

6.9 VOLUMETRIC FLASKS. 10-mL or convenient sizes for making standard solutions.

6.10 A calibrated flowmeter is required to accurately measure the flow rate of the air drawn through the charcoal tube during calibration of the sampling pumps.

7. Reagents

7.1 PURITY. All organic compounds utilized to make G.C. standards should be of ACS analytical grade or spectrograde quality.

7.2 SPECTROQUALITY CARBON DISULFIDE.

7.3 CARRIER GAS. Compressed nitrogen, argon, or helium should be prepared from water-pumped sources or be of guaranteed high purity.

7.4 HYDROGEN. High quality electrolytic generators that produce ultrapure hydrogen under pressure are recommended. They provide some degree of safety when compared to the explosive hazard of bottled hydrogen. All connections should be made with thoroughly cleaned stainless steel tubing and properly tested for leaks. If hydrogen in compressed gas cylinders is used, the hydrogen should be prepared from water-pumped sources.

7.5 AIR. Filtered compressed air, low in hydrocarbon content.

8. Procedure

8.1 CLEANING OF GLASSWARE. All glassware used for the laboratory analysis should be detergent-washed and thoroughly rinsed with tap water and distilled water.

---

*Activated charcoal is prepared from coconut shells.

8.2 CALIBRATION OF PERSONAL PUMPS. Each personal pump must be calibrated with a representative charcoal tube in the line. This will minimize errors associated with uncertainties in the sample volume collected.

8.3 COLLECTION AND SHIPPING OF SAMPLES.

8.3.1 Immediately before sampling, break the ends of the tube to provide an opening at least one-half the ID of the tube (2 mm).

8.3.2 The small section of charcoal is used as a back-up and should be positioned nearest the sampling pump.

8.3.3 The charcoal tube should be placed in a vertical direction during sampling.

8.3.4 Air being sampled should not be passed through any hose or tubing before entering the charcoal tube.

8.3.5 The flow, time, and/or volume must be measured as accurately as possible. The sample should be taken at a flow rate of 1 L/min or less. The suggested sample volume that should be collected for each solvent at its OSHA standard **(3)** or threshold limit value **(8)** is shown in Table 834:I.

8.3.6 An identification number is marked on the tube. The temperature and pressure of the atmosphere being sampled should be measured and recorded along with other appropriate field data (location of worker, distance from operation, type of operation, wind direction, etc.).

8.3.7 The charcoal tubes should be capped with masking tape or plastic caps immediately after sampling. Under no circumstances should rubber caps be used. If the sample tube must be stored for more than a week, refrigeration is recommended (or the ends may be flame sealed).

8.3.8 One tube should be handled in the same manner as the sample tube (break, seal, and transport), except that no air is sampled through this tube. This tube should be labeled as a blank.

8.3.9 Capped tubes should be packed tightly before they are shipped to minimize tube breakage during shipping.

8.3.10 Samples of the suspected solvent(s) should be submitted to the laboratory (take appropriate safety precautions). These liquid bulk samples should **not** be transported in the same container as the samples or blank tube. If possible a bulk *air* sample (*i.e.*, a charcoal tube used to sample a *large* volume of air) should be shipped for qualitative identification purposes.

8.3.11 Due to the high resistance of the charcoal tube, this sampling method places a heavy load on the sampling pump. Therefore, no more than ten charcoal tube samples should be taken without fully recharging the battery.

8.4 ANALYSIS OF SAMPLES.

8.4.1 *Preparation of samples.* In preparation for analysis, each charcoal tube is scored with a file in front of the first section of charcoal and broken open. The glass wool is removed and discarded. The charcoal in the first (larger) section is transferred to a small stoppered tube. The separating section of foam is removed and discarded; the second section is transferred to another test tube. These two sections are analyzed separately.

8.4.2 *Desorption of samples.* Prior to analysis, 0.5 mL of carbon disulfide is pipetted into each tube. All work with carbon disulfide *must* be performed in a hood because of its high toxicity. Tests indicate that desorption is complete in 30 min if the sample is agitated occasionally during this period. An ultrasonic probe suitable for agitation of the sample is available from Heat Systems Ultrasonics, Inc., New York. The use of graduated, stoppered microcentrifuge tubes is recommended so that the analyst can observe any apparent change in volume during the desorption process. Carbon disulfide is a very volatile solvent, so volume changes can occur during the desorption process depending on the surrounding temperature and how the tube is stoppered. The time allowed for desorption should be as consistent as possible and should not exceed 3 h **(8)** to minimize volume and concentration changes.†

8.4.3 *G.C. Conditions.* Operating conditions for the G.C. may vary depending upon the type of instrument used and requirements of the sample being analyzed. It is recommended that the injector and detector temperatures be maintained at a level

---

†The initial volume occupied by the charcoal plus the 0.5 ml CS$_2$ should be noted and corresponding volume adjustments made whenever necessary just before GC analysis.

that will completely vaporize the sample and prevent undesirable condensation of combustion products (about 200°C). Simple mixtures and individual compounds often may be determined with a column at constant temperature (isothermal). However, complex mixtures will require the testing of different column materials and temperature programs to resolve the components (8). The analyst should use the bulk liquid sample or bulk air sample to establish optimum G.C. operating conditions and the column that will completely separate the components of the sampled mixture *prior* to the analysis of the charcoal sample tubes.‡

8.4.4 *Injection.* After desorption, an aliquot of the sample solution is injected into the G.C. To eliminate difficulties arising from blowback or distillation within the syringe needle, one should employ the solvent flush injection technique. The 10 μL syringe is first flushed with carbon disulfide several times to wet the barrel and plunger. Three μL of solvent are drawn into the syringe to increase the accuracy and reproducibility of the injected sample volume. The needle is removed from the carbon disulfide, and the plunger is pulled back about 0.2 μL to separate the solvent flush from the sample with a pocket of air to be used as a marker. The needle is then immersed in the sample, and a 5-μL aliquot is withdrawn, taking into consideration the volume of the needle, since the sample in the needle will be completely injected. After the needle is removed from the sample and prior to injection, the plunger is pulled back a short distance to minimize evaporation of the sample from the tip of the needle. Triplicate injections of each sample and standard should be made. No more than a 3% difference in area is to be expected.

8.4.5 *Measurement of area.* The area of the sample peak is measured by an electronic integrator or some other suitable

‡Typical operating conditions for GC are:
(1) helium carrier gas flow, 85 cm³/min (70 psig, 480 kPa gauge)
(2) hydrogen gas flow to detector, 65 cm³/min (24 psig, 165 kPa gauge)
(3) air flow to detector, 500 cm³/min (50 psig, 345 kPa gauge)
(4) injector temperature 200°C.

form of area measurement, and preliminary results are read from a standard curve prepared as discussed below.

8.5 DETERMINATION OF DESORPTION EFFICIENCY.

8.5.1 *Importance of determination.* The desorption efficiency of a particular compound can vary from one laboratory to another, from one batch of activated charcoal to another, and can also vary with the amount of compounds adsorbed on the charcoal. Thus, it is necessary to determine at least once the percentage of each specific compound that is removed in the desorption process each time a different batch of charcoal is used. The Physical and Chemical Analysis Branch of NIOSH has found that the average desorption efficiencies for the compounds in Tables 834:I and 834:II are between 81% and 100% and vary with each batch of charcoal.

8.5.2 *Procedure for determining desorption efficiency.* Activated charcoal equivalent to the amount in the first section of the sampling tube (100 mg) is measured into a 5-cm 4-mm ID glass tube, flame-sealed at one end. This charcoal must be from the same batch as that used in obtaining the samples and can be obtained from unused charcoal tubes. The open end is capped with Parafilm. A known amount of the compound is injected directly into the activated charcoal with a μL syringe, and the tube is capped with more Parafilm. The amount injected is usually equivalent to that present in a 10-L sample at a concentration equal to the OSHA standard (3). (For the conversion formula, see Section **10.7**).

a. At least 5 tubes are prepared in this manner and allowed to stand for at least overnight to assure complete adsorption of the specific compound onto the charcoal. These 5 tubes are referred to as the samples. A parallel blank tube should be treated in the same manner except that no sample is added to it. The sample and blank tubes are desorbed with carbon disulfide (CS_2) and analyzed in exactly the same manner as the sampling tube described in Section **8.4**.

b. Two or three standards are prepared by injecting the same volume of compound into 0.5 mL of CS_2 with the same syringe used in preparation of the sample. These are analyzed with the samples.

c. The desorption efficiency equals the difference between the average peak area of the samples and the peak area of the blank divided by the average peak area of the standards, or

Desorption efficiency =

$$\frac{\text{Area sample - Area blank}}{\text{Area standard}}$$

9. Calibration and Standards

9.1 It is convenient to express concentration of standards in terms of mg/0.5 mL CS_2 because samples are desorbed in this amount of CS_2. To minimize error due to the volatility of carbon disulfide, one can inject 20 times this weight into 10 mL of CS_2. For example, to prepare a 0.3 mg/0.5 mL standard, one would inject 6.0 mg into exactly 10 mL of CS_2 in a stoppered flask. The density of the specific compound is used to convert 6.0 mg into μL for easy measurement with a μL syringe. A series of standards, varying in concentration over the range of interest, is prepared and analyzed under the same G.C. conditions and during the same time period as the unknown samples. Standard curves are established by plotting concentration in mg/0.5 mL versus peak area or integrator output.

Note: Since no internal standard is used in the method, standard solutions must be analyzed at the same time that the sample analysis is done. This will minimize the effect of known day-to-day variations and variations during the same day of the FID response. Standard solutions should be made fresh each day they are required for G.C. analysis. However, this does not preclude the use of an internal standard in this method.

10. Calculations

10.1 Read the weight, in mg, corresponding to each peak area from the standard curve for the particular compound. No volume corrections are needed, because the standard curve is based on mg/0.5 mL CS_2 and the volume of sample injected is identical to the volume of the standards injected into the G.C.

10.2 Corrections for the blank must be made for each sample.

$$\text{Corrected mg} = \text{mg}_s - \text{mg}_b$$

where mg_s = mg found in front section of sample tube
mg_b = mg found in front section of blank tube

A similar procedure is followed for the backup sections.

10.3 Add the corrected amounts present in the front and backup sections of the same sample tube to determine the total measured amount in the sample.

10.4 Divide this total weight by the determined desorption efficiency (See Section **8.5.2**) to obtain the total mg per sample.

10.5 Convert the volume of air sampled to standard conditions of 25°C and 101.3 kPa.

10.6 The concentration of the organic solvent in the air sampled can be expressed in mg per m³, which is numerically equal to μg per liter of air.

$$\text{mg/m}^3 = \mu\text{g/L} =$$

$$\frac{\text{total mg (Section } \mathbf{10.4}) \times 1000 \ (\mu\text{g/mg})}{\text{corrected sample volume, L}}$$

10.7 Another method of expressing concentration is ppm, defined as μL of compound per liter of air

ppm = μL of compound/corrected sample volume

$$\text{ppm} = \frac{\mu\text{g of compound}}{\text{corrected sample volume}} \times \frac{24.45}{\text{MW}}$$

where 24.45 = molar volume at 25°C and 101.3 kPa
MW = molecular weight of the compound

Note: Activated charcoal adsorbs nearly all organic vapors, requiring precautions be taken to prevent it from becoming contaminated after the sampling process. Masking tape or plastic caps may be used to cap the ends for short term storage. If the tubes are to be stored for a week or longer, refrigeration or flame sealing of the tube ends is recommended to prevent loss of the more volatile adsorbed vapors.

Many organic vapors are chemically stable on activated charcoal at room temperature and can be stored for months. Highly volatile, oxygenated, and reactive compounds should be analyzed as soon as possible; preliminary experiments indicate that recovery efficiencies of such compounds decrease with time.

11. References

1. SCOTT RESEARCH LABORATORIES, INC. 1973. Contract No. HSM 99-72-98. Collaborative Testing of Activated Charcoal Sampling Tubes for Seven Organic Solvents, SRL 1316 10 0973.
2. COPPER, C. V., L. D. WHITE AND R. E. KUPEL. 1971. Qualitative Detection Limits for Specific Compounds Utilizing Gas Chromatographic Fractions, Activated Charcoal and a Mass Spectrometer. *Amer. Ind. Hyg. Assoc. J.* 32:383.
3. FEDERAL REGISTER. 1972. Vol. 37, No. 202, Part 1910, 93, Wednesday, October 18, Washington, D.C. Occupational Safety and Health Act of 1970, Public Law 91-596.
4. KUPEL, R. E. AND L. D. WHITE. 1971. Report on a Modified Charcoal Tube. *Amer. Ind. Hyg. Assoc. J.* 32:456.
5. OTTERSON, E. J. AND C. U. GUY. 1964. A Method of Atmospheric Solvent Vapor Sampling on Activated Charcoal in Connection with Gas Chromatography. Transactions of the Twenty-Sixty Annual Meeting of the American Conference of Governmental Industrial Hygienists, Philadelphia, Pa., p. 37, American Conference of Governmental Industrial Hygienists, Cincinnati, Ohio.
6. QUINN, P. M., C. S. McCAMMON AND R. E. KUPEL. 1975. A Charcoal Sampling Method and a Gas Chromatographic Analytical Procedure for CS_2. Paper presented at the American Industrial Hygiene Conference in Boston, Massachusetts, May. *Amer. Ind. Hyg. Assoc. J.* 36:618.
7. MACKISON, F. W., R. S. STRICOFF AND J. L. PARTRIDGE, JR., eds. 1978. NIOSH/OSHA Pocket Guide to Chemical Hazards. Government Printing Office, Washington, DC.
8. WHITE, L. D., D. G. TAYLOR, P. A. MAUER, R. E. KUPEL. 1970. A Convenient Optimized Method for the Analysis of Selected Solvent Vapors in the Industrial Atmosphere. *Amer. Ind. Hyg. Assoc. J.* 31.225.

Revised by Subcommittee 5

E. SAWICKI, *Chairman*
T. BELSKY
R. A. FRIEDEL
D. L. HYDE
J. L. MONKMAN
R. A. RASMUSSEN
L. A. RIPPERTON
L. D. WHITE

Approved with Modifications from 2nd edition
Subcommittee 4/5

M. FELDSTEIN, *Chairman*
R. J. BRYAN
D. L. HYDE
D. A. LEVAGGI
D. C. LOCKE
R. A. RASMUSSEN
P. O. WARNER

835.

Determination of EPN, Malathion and Parathion in Air

Principle of the Method

1.1   SAMPLING. EPN (phenylphosphonothioic acid 0-ethyl 0-*p*-nitrophenyl ester; CAS#2104-64-5), malathion {[(dimethoxyphosphinothioyl)thio] butanedioic acid diethyl ester; CAS#121-75-5}, and parathion [phosphorothioic acid, 0,0-diethyl 0-(4-nitrophenyl) ester; CAS#56-38-2] are effectively removed and retained from an air sample by passing a measured volume through a glass fiber filter.

1.2   SAMPLE PREPARATION. The analyte(s) is (are) extracted from the filter with isooctane. (The filters must be extracted immediately after collection if malathion is to be determined.)

1.3   ANALYSIS. An aliquot of the isooctane extract is gas chromatographed and the analyte determined by a flame photometric detector (FPD) designed to measure the emissivity of phosphorus at 526 nm in a hydrogen-rich flame.

1.4   APPLICABILITY. The procedure described is applicable for the measurement of EPN and parathion at 0.1 to 3 times the OSHA exposure limit of 0.5 mg/m^3 for EPN and 0.1 mg/m^3 for parathion, and for measurement of malathion at 0.08 to 2 times the OSHA exposure limit of 15 mg/m^3, when a 120-L sample is taken.

2. Range and Sensitivity

2.1   RANGE. The working range of the method is 0.05 to 1.5 mg/m^3 for EPN, 1 to 30 mg/m^3 for malathion, and 0.01 to 0.3 mg/m^3 for parathion for 120-L air samples.

2.2   SENSITIVITY. The sensitivity and linearity are dependent on the particular gas chromatograph (GC) and auxiliary equipment used; however, the estimated limit of detection is 2 ng/sample for EPN, and 4 ng/sample for malathion and parathion.

3. Interferences

3.1   Although there are no known interferences, all equipment and reagents should be scrupulously free of any traces of phosphate detergents and other contaminants that might display retention volumes close to those of the insecticide of interest.

4. Precision and Accuracy

4.1   The precision and accuracy data were obtained by generating air concentrations of the analytes at one-half, one and two times their OSHA standards (1,2,3,9). The coefficient of variation of the analytical method ranged from 6 to 8%, depending on the compound, for ranges of 0.3 to 1.2 mg/m^3 for EPN, 8 to 35 mg/m^3 for malathion and 0.07 to 0.26 mg/m^3 for parathion. Recoveries were 100% for EPN, 103% for malathion and 100% for parathion. Test atmospheres were generated using Trion-6-EPN (Wilbur Ellis Co.), Malathion-50 (Ortho) and Aqua-8 Parathion (FMC Corp.). Gelman Type AE glass-fiber filters were used for all sampling and measurement recovery studies. Collection efficiency of the glass-fiber filter was determined to be 100% for all three analytes and losses from vaporization of known amounts of the analytes deposited on a glass fiber filter were negligible (8,10). This was confirmed later for parathion by Hill and Arnold using a different sampling device (5). Filters spiked with solutions of EPN or parathion gave quantitative recoveries after storage for 7 days at 25°C. Filter samples containing malathion gave 88% recovery if stored for 7 days and 100% if the samples were extracted with isooctane soon after sampling.

5. Advantages and Disadvantages of the Method

5.1 ADVANTAGES.

5.1.1 The method is very sensitive (1 $\times$ 10^{-12} g/sec of P).

5.1.2 The detector is highly selective for phosphorus compounds and the GC column increases selectivity.

5.1.3 Separation and determination are rapid.

5.2 DISADVANTAGES.

5.2.1 Cost of equipment and supplies may tax the budget of some laboratories.

5.2.2 The sensitivity of the equipment depends on careful adjustments of operating parameters.

5.2.3 The procedure requires a skilled operator.

5.2.4 Equipment and reagents are easily contaminated.

6. Apparatus

6.1 SAMPLER. Glass-fiber filter (Gelman Type AE or equivalent), 37 mm, in a two-piece polystyrene cassette supported by a backup pad.

6.2 PUMP. A personal sampling pump, equipped with flexible connecting tubing, with a capacity of drawing 1 to 2 L/min at constant flow.

6.3 GAS CHROMATOGRAPH. Gas chromatograph equipped with a column bypass valve, phosphorus flame photometric detector, integrator and column **(6.4)**.

6.4 COLUMN. 2m $\times$ 6mm glass column packed with 3% OV-1 on 100/120 mesh Gas Chrom Q (or equivalent) for EPN and parathion. 2m $\times$ 6mm glass column packed with 1.5% OV-17 + 1.95% OV-210 on 80/100 mesh Gas Chrom Q (or equivalent) for malathion.

6.5 VIALS. 20 mL, glass with PTFE-lined caps or PTFE-lined silicone septum caps, labels.

6.6 SYRINGE. 10μL, readable to 0.1μL.

6.7 VOLUMETRIC FLASK. 15 mL.

6.8 PIPETS. 15 mL with pipet bulb.

6.9 TWEEZERS.

6.10 FLOWMETER. Calibrated rotameter or soap-bubble meter.

Reageants

7.1 EPN, MALATHION AND PARATHION (KNOWN PURITY). CAUTION: EPN, malathion, and parathion are highly toxic cholinesterase inhibitors with cumulative effects **(6,7,8)**. Special care must be taken to avoid inhalation or skin contact.

7.2 ISOOCTANE, Chromatographic grade. CAUTION: Flammable, handle in well-ventilated hood.

7.3 CALIBRATION STOCK SOLUTIONS. (prepare fresh daily).

a. EPN, 15 mg/mL, in isooctane.

b. Malathion, 450 mg/mL, in dimethylformamide.

c. Parathion, 3.3 mg/mL, in isooctane.

Note: 4 μL of each solution contains a mass of analyte equivalent to a 120-L air sample at the OSHA exposure limit (0.06 mg EPN, 1.8 mg malathion and 0.0132 mg parathion).

7.4 NITROGEN, purified. Helium may also be used as the carrier.

7.5 HYDROGEN, prepurified.

7.6 OXYGEN, purified.

7.7 AIR, filtered, compressed.

8. Procedure

8.1 COLLECTION OF SAMPLE. Calibrate each personal sampling pump with a representative sampler in line. Sample at 1 to 2 L/min for 1 to 2 h (120-L sample is recommended). Do not exceed 2 mg total dust loading on the filter. Within 1 h after sampling, transfer the glass fiber filter with tweezers to a clean 20-mL vial. If sampling for malathion, add 15 mL isooctane to the vial immediately after sampling. Mark the level of isooctane.

8.2 SAMPLE PREPARATION. Pipet 15 mL isooctane into each vial. Cap each vial and swirl the contents for 1 h. NOTE: Do not add isooctane to the malathion sample vials which have had isooctane added in the field. If any malathion sample vial is found to contain less than 15 mL on receipt at the laboratory, it should be rejected prior to analysis.

8.3 ANALYSIS. Prior to sample preparation, fit the gas chromatograph with the appropriate column (**6.4**). Start the carrier gas flow through the column several hours before analysis with the column temperature set for the intended analysis. Optimize air, hydrogen and oxygen flow rates according to the gas chromatograph manufacturer's instructions. Set the carrier flow to 60 mL/min. Other critical parameters are as follows:

EPN
Column temperature: 205°C
Injection port temperature: 215°C
Detector temperature: 200°C

Malathion
Column temperature: 185°C
Injection port temperature: 240°C
Detector temperature: 200°C

Parathion
Column temperature: 165°C
Injection port temperature: 215°C
Detector temperature: 200°C

Inject 5-μL sample aliquots using solvent flush technique. Vent the solvent peak by opening the column bypass valve at the time of injection. Close the column bypass valve after the solvent peak has eluted (ca. 30 sec) and before the analyte peak elutes. Make replicate injections of samples and standards. Measure the peak area.

9. Calibration and Standards

9.1 STANDARDS. Prepare at least five working standards covering the analytical range of the method, by diluting aliquots of the calibration stock solutions to 15 mL with isooctane.

9.2 CALIBRATION. Analyze calibration standards in triplicate with the unknowns, blanks and other quality control spiked samples. Prepare a calibration graph by plotting peak area vs. mg of analyte.

9.3 QUALITY CONTROL STANDARDS. Analyze duplicate control samples with each sample set. To prepare control samples, place a blank filter in a clean 20-mL vial. Inject analyte calibration stock solution directly onto the filter with a microliter syringe. If analyzing for malathion, extract the filters with 15 mL isooctane before allowing the filter to stand overnight. Cap

the vial and allow to stand overnight. Prepare the sample (8.2) and analyze it (8.3). Calculate the recovery (mg found on the filter divided by mg added to filter).

Calculations

10.1 SAMPLING. Compute the volume of air sampled using the following formula:

$$V = Qt$$

where V = Volume of air sampled, liters at 25°C and 101.3 kPa.
Q = Mean flow rate in L/min.
t = time, min.

10.2 ANALYSIS. Read the mass, mg per sample, of analyte found on the sample filters, W, and average the media blank filters, B, from the calibration graph. Calculate the concentration, C, of analyte in the air volume sampled, V (in liters):

$$C = \frac{(W - B) \cdot 10^3}{V}, \text{ mg/m}^3$$

References

1. NIOSH MANUAL OF ANALYTICAL METHODS, 3rd ed. 1984. No. 5012, U.S. Dept. of Health and Human Services, Publ. DHHS (NIOSH) 84-100.
2. DOCUMENTATION OF THE NIOSH VALIDATION TESTS, S285 and 295 1977. U.S. Dept. of Health, Education, and Welfare, Publ. (NIOSH) 77-185.
3. BACKUP DATA REPORT S370 1976. Prepared under NIOSH Contract 210-76-0123, available as Order No. PB 271-712 from NTIS, Springfield, VA 22161.
4. NIOSH MANUAL OF ANALYTICAL METHODS, 2nd ed. 1977. V. 3, S285, S370 and S295, U.S. Dept. of Health, Education, and Welfare, Publ. (NIOSH) 77-157-C.
5. HILL, R. H., JR. AND J. E. ARNOLD. 1979. A Personal Air Sampler for Pesticides, Arch. Environ. Contam. Toxicol. 8: 621–628.
6. U.S. DEPT. of HEALTH, EDUCATION, AND WELFARE. 1976. Criteria for a Recommended Standard . . . Occupational Exposure to Malathion. Publ. (NIOSH) 76-205.
7. U.S. DEPT. OF HEALTH, EDUCATION, AND WELFARE. 1976. Criteria for a Recommended Standard . . . Occupational Exposure to Parathion. Publ. (NIOSH) 76-190.
8. U.S. DEPT. OF HEALTH, EDUCATION, AND WELFARE. 1978. Criteria for a Recommended Standard . . . Occupational Exposure During the Manufacture and Formulation of Pesticides. Publ. (NIOSH) 78-174.
9. U.S. DEPARTMENT OF HEALTH AND HUMAN SERVICES. 1980. NIOSH Research Report — Development and Validation of Methods for Sam-

pling and Analysis of Workplace Toxic Substances. Publ. (NIOSH) 80-133.

10. AMERICAN CONFERENCE OF GOVERNMENTAL INDUSTRIAL HYGIENISTS. 1983. Threshold Limit Values for Chemical Substances and Physical Agents in the Workroom Environment with Intended Changes for 1983–84.

11. U.S. DEPT. OF HEALTH, EDUCATION, AND WELFARE. 1978. Pocket Guide to Chemical Hazards. Publ. (NIOSH) 78-210, pp 94, 122, 148.

Subcommittee 1

K. D. REISZNER
P. K. DASGUPTA, *Chairman*
D. F. ADAMS
B. R. APPEL
S. O. FARWELL
K. T. KNAPP
G. L. KOK
W. R. PIERSON
R. L. TANNER

836.

Determination of Total Particulate Aromatic Hydrocarbons (TpAH) in Air: Ultrasonic Extraction Method

1. Principle of the Method

1.1 Airborne particles collected from polluted atmospheres on glass fiber filters are extracted ultrasonically in the presence of silica powder (**1,2,3**). The TpAH in the filtered extract are separated by high performance liquid chromatography on a column of Corasil II with a non-polar solvent, and the absorbance is measured by a UV detector at 254 nm. Compounds responding to the detector are shown in Tables 836:I, II, III, IV. The extract is suitable also for the analysis of the aliphatic hydrocarbons (**4**).

2. Range and Sensitivity

2.1 Minimum reproducible level of standard benzo(a)pyrene at 254 nm is approximately 3 μg.

2.2 The minimum detectable TpAH [in terms of benzo(a)pyrene] for particles collected on one glass fiber filter of approximately 452 cm^2 is approximately 5 μg, or 3.3 ng/m^3 air if 1500 m^3 of air are sampled in the ambient atmosphere.

2.3 The upper range of TpAH concentration can be increased by dilution of the extract and/or analyzing smaller samples. Sensitivity for low concentrations can be increased by injecting larger samples into the chromatograph. Thus, very high levels of TpAH can be measured.

3. Interferences

3.1 Any compound that is not retained on the silica column and absorbs light at 254 nm is measured in this procedure.

Table 836:I Elution of PAH*

Compound	% Eluted Through Column	PA‡/μg × 10^3
Mono-, dicyclics		
Benzene	99	0.4
N-Hexylbenzene	100	0.5
N-Heptylbenzene	100	0.7
Naphthalene	101	0.7
Azulene	93	3.0
Tricyclics		
Anthracene	100	36.0
9-Methylanthracene	99	15.0
Xanthene	102	1.3
Phenoxathiin	92	0.2
Phenanthrene	100	10.0
Tetracyclics		
Naphthacene	95	4.7
Chrysene	105	4.5
Pyrene	96	3.6
4-Methylpyrene	100	1.7
1,3-Dimethylpyrene	96	0.9
Triphenylene	100	9.0
Benz(a)anthracene	96	4.3
7,12-Dimethylbenz-(a)anthracene	102	3.3
Pentacyclics		
Dibenz(a,h)anthracene	96	0.6
Benzo(a)pyrene	100	5.3
Benzo(e)pyrene	92	2.2
Picene	99	5.0
Perylene	96	5.8
Hexacyclics		
Benzo(ghi)perylene	99	1.8
Anthanthrene	93	2.6
Dibenzo(fg,op) naphthacene	93	0.6
Coronene	91	0.5
Dibenzo(g,p)chrysene	96	1.0
Naphtho(2,1,8-qra) naphthacene†	100	0.7

*Retention time is approximately 2 minutes.
†Or naphtho(2,3-a)pyrene
‡PA = Peak Area

Table 836:II Elution of Fluorene, Analogues and Derivatives

Compound	t_R Min.	% Eluded Through Column	PA/μg × 10^{-3}
Fluorene	2.0	100	2.9
Dibenzothiophene	2.0	96	1.8
Dibenzofuran	2.0	98	0.3
Fluoranthene	2.0	95	2.5
Benzo(k)fluoranthene	1.8	97	1.4
Benzo(b)fluoranthene	1.0	99	1.6
2-Ethylfluorene	1.0	95	1.6
11-H-Benzo(b)fluorene	1.0	110	5.6
2-Nitrofluorene	4.8	104	0.2
2,5-Dinitrofluorene	7.0	71	0.3
9-Fluorenal	8.5	14	0.2
3,6-Dinitrodibenzoselenophene	18.2	38	0.2
3-Aminofluorene	18.2	68	0.4
4-Fluorenecarboxylic acid		Retained on column	
2-Hydroxyfluorene		Retained on column	
2-Nitro-7-hydroxyfluorene		Retained on column	
Fluorenone		Retained on column	

3.2 Fluorene and some of its analogues and derivatives listed in Table 836:II, and polychloro derivatives of some di- and tricyclic hydrocarbons in Table 836:III are examples of such compounds.

3.3 Amino, carbonyl, hydroxy and nitro compounds elute after the PAH, so do not interfere. See Table 836:II.

3.4 Carbazoles and aldehydes are either retained or have retention times larger than the PAH, except N-alkyl substituted derivatives, which elute with the PAH. See Table 836:IV.

3.5 Oxygenated compounds, some phenols and aza- and imino-heterocyclics (except some members of the indole series) are

retained. Examples are benzoquinone, o-ethylphenol, acridine and quinoline.

3.6 Most interfering compounds have quite low peak area per μg values, which decreases their significance, as shown in Tables 836:II, III, IV.

4. Precision and Accuracy

4.1 Homogeneous glass fiber filter samples containing air particles were analyzed by Soxhlet and ultrasonic extraction. See Table 836:V. The relative standard deviation for 6 ultrasonic extracts was ± 1.33% and for 4 Soxhlet extracts ± 26.1%. The ratio of ultrasonic to Soxhlet recovery was 1.14.

Table 836:III Elution of Polychloro Derivatives of Di- and Tricyclic Hydrocarbons*

Compound	% Eluted Through Column	PA/μg × 10^{-3}
1,1-Dichloro-2,2-bis(p-chlorophenyl)ethane (p,p'-DDD)	94	0.02
1,1-Dichloro-2,2bis(p-chlorophenyl)ethylene (DDE)	97	0.50
1,1,1-Trichloro-2,2-bis(p-chlorophenyl)ethane (p,p'-DDT)	85	0.02
Aroclor 1260 (chlorinated biphenyls, 60% chlorine)	100	0.13
Aroclor 5432 (chlorinated triphenyls, 32% chlorine)	104	0.61
Halowax 1099 (mixture of tri- and tetrachloronaphthalenes, 52% chlorine)	101	0.25
1,2,3,4,5,6,7,8-Octachloronaphthalene	97	0.64
2,3,4,5,6,2,3,4,5,6-Decachlorobiphenyl	95	0.19
1,2,3,4,5,6,7,8-Octachlorodibenzofuran	93	0.33
1,2,3,4,6,7,8,9-Octochlorodibenzo-p-dioxin	98	0.85
Tetradecachloro-p-terphenyl	95	0.22

* Retention times from 1 to 2 min.

Table 836:IV Elution of Some Indoles, Carbazoles and Aromatic Aldehydes

Compound	t_R Min.	% Eluted Through Column	PA/μg $\times$ 10-3
Indole	5.3	82	1.1
Carbazole	11.8	67	0.7
4-H-Benzo(def)-carbazole	8.0	98	2.0
11-H-Benzo(a)carbazole	14.5	55	3.0
7-H-Dibenzo(c,g)-carbazole	18.0	92	2.1
N-Phenyl-carbazole	2.3	74	1.8
N-Ethyl-carbazole	2.5	98	0.5
5-Methyl-5,10-dihydroindeno-(1,2-b)indole	2.8	103	1.9
2,3-Dimethyl-indole	5.3	90	5.5
2-Methyl-carbazole	6.8	100	0.8
2-Hydrozy-carbazole		Retained on column	
N-Ethyl-3-amino-carbazole		Retained on column	
Benzaldehyde	12.8	56	1.1
2-Naphthalde-hyde	8.2	78	0.3

4.2 Recovery of PAH added to glass fiber filter blanks and extracted ultrasonically was 95% for anthracene; 97.5% for phenanthrene; and 98.2% for benzo(a)pyrene, Table 836:VI.

Table 836:V Comparison of Ultrasonic and Soxhlet Extractions

Sample No.	Ultrasonic PA/μg	Ultrasonic % Eluted*	Soxhlet PA/μg	Soxhlet % Eluted*
1.	0.575	51	0.449	28
2.	0.562	53	0.509	—
3.	0.567	50	0.500	—
4.	0.579	48	0.545	31
5.	0.560	44	—	—
6.	0.573	44	—	—
Average	0.569	49	0.509	30
Rel. Std. Dev.	± 1.33%		± 26.1%	
Ultrasonic/Soxhlet Recovery = 1.14				

*Refers to % of TpAH in the extracted material.

5. Advantages and Disadvantages of the Method

5.1 The extraction is done at room temperature. Complete extraction of the TpAH is assured by the fine shredding of the glass fibers and the breaking up of clumps of particles.

5.2 Only a relatively small sample of air particles is required. Complete analysis time is well under an hour, most of which is waiting time.

5.3 Most of the polar constituents are removed by adsorption in the homogenizing vessel. The remainder are removed by the fast simple chromatographic analysis.

5.4 The method can accommodate a wide range of hydrocarbon pollutant concentration, since sample extract volumes ranging from 0.1 to 2 mL can be chromatographed.

5.5 Time and work are saved by not weighing the samples or the soluble organics.

5.6 A disadvantage is that a blank correction must be made for the fiber glass filter. Also care must be taken to avoid evaporation of the extract to dryness.

5.7 A further disadvantage is that the ultrasonic extraction must be done in a Sonabox to reduce the unacceptably high noise level.

6. Apparatus

6.1 SONIFIER CELL DISRUPTOR. 20 KHz power ultrasonic generator capable of dialing 70 W accurately, with a 1.27 cm (1/2″) horn disruptor and Sonabox.

6.2 LIQUID CHROMATOGRAPH. With stainless steel column 2.6 $\times$ 300 mm, UV Detector with 254 mm filter and loop injector with a capacity ranging from 0.1 to 2 mL.

6.3 STRIP CHART RECORDER WITH DISC INTEGRATOR.

6.4 APPROVED AND CALIBRATED PERSONAL SAMPLING PUMP for collection of particulate matter. Any vacuum pump whose flow can be determined accurately to within 1 L/min or less.

6.5 COLUMN BYPASS.

6.6 FISHER FILTRATOR AND MEDIUM SINTERED-GLASS FILTER.

Table 836:VI Recovery of Added PAH

Compound	Sample, μg	Peak Area Sonified Filter + Std.	Standard Solution	% Recovery
Anthracene	0.035	1005	1055	95.0
Phenanthrene	0.147	1155	1185	97.5
Benzo(a)pyrene	0.355	1846	1880	98.2

6.7 U.S. STANDARD SIEVE SERIES NO. 120, WITH 125 μm OPENINGS.

7. Reagents

7.1 CYCLOHEXANE, A.C.S. Spectroanalyzed, distilled once from glass.
7.2 POLYNUCLEAR AROMATIC HYDROCARBONS.
7.3 GLASS POWDER. Spherical, non-wettable 38 to 53 μm in diam.
7.4 CORASIL II.

8. Procedure

8.1 EXTRACTION.
8.1.1 The 1.27-cm horn of the sonifier cell disruptor is supported in a sonabox to reduce noise. The sonifying vessel is a beaker 3.8 cm ID × 10 cm tall. The end of the horn is set about 0.6 cm above the bottom of the beaker to ensure adequate "stirring" of the mixture and equal exposure to areas of intense cavitation. Approximately 16 cm² of the exposed fiber glass filter and blank are cut into roughly 1.3 cm squares to facilitate shredding. The sonifying vessel is surrounded by an ice water bath up to the level of the solvent mixture.
8.1.2 Homogeneous replicate samples of approximately 16 cm² of exposed and blank glass fiber filters are prepared and adjusted to exactly 100 mg. This weight

Table 836:VII Analysis of Particular Samples

Description	Corrected Peak Area	m³ Air Sampled	PA/m³ Air*	TpAH† μg/m³ air
Urban I	1200	1500	1120	.211
Urban II	620	1500	580	.109
Urban III	545	1500	509	.096
Mt. Storm	0	1673	0	.000

* See *Calculations* — Section **10.1**.
† See *Calculations* — Section **10.2**.

necessarily includes both the particles and the glass fiber. These samples were used to maximize parameters and for comparison of ultrasonic and Soxhlet extractions, shown in Table 836:V.
8.1.3 Samples for routine analysis are not weighed. Only the areas of the sample (16 cm²) and the whole filter, the volume of air sampled and the volume of extract injected need to be determined. Sample at rate of at least 2 L/min for 1 h or more.
8.1.4 *Extraction Procedure.* The sample, 60 mL cyclohexane and 5 mL silica powder are placed in the sonifying vessel, and sonified for 8 min at 70 W. The supernatant is decanted into the sintered glass filter supported on a Fisher Filtrator. Cyclohexane is added to the sonifying vessel to the level of the original mixture (usually about 50 mL). Sonification is carried on for an additional 4 min. The contents are filtered and combined with the first fraction, and rinsed with 50 mL cyclohexane. The filtrates and rinsings are collected in an Erlenmeyer flask and evaporated to about 5 mL, transferred quantitatively to a 10-mL volumetric flask and made to the mark.
8.1.5 Sample and blank filters, **8.1.2**, are extracted by Soxhlet with 80 mL cyclohexane for 6 to 8 h for comparison with the ultrasonic extraction. See Table 836:V. After filtering, the extracts are evaporated in the same manner as the ultrasonic extracts.
8.1.6 The glass fiber filters used for air sampling should be as free as possible of soluble compounds that absorb at 254 nm. It may be necessary to flash fire or extract them, and care should be taken to avoid contaminating them.
8.2 CHROMATOGRAPHIC SEPARATION.
8.2.1 A schematic of the chromatographic system is shown in Figure 836:1. The stainless steel column is 2.6 × 300 mm; the packing is Corasil II; the eluent is cyclo-

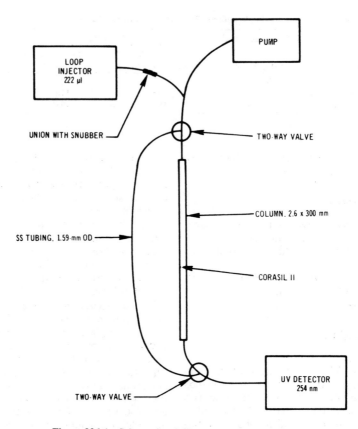

Figure 836:1 – Schematic of Chromatographic System.

hexane. Two two-way valves are installed in the chromatographic line, one before the column, the other after it. They are connected with stainless steel tubing. This enables the sample to be pumped through either the column or the tubing (column bypass) into the UV detector. A union with snubber may be placed in the line to prevent clogging.

8.2.2 To test the performance of the column, the percent of PAH that elutes is calculated from the peak areas through the column and the column bypass. Typical chromatograms from column and tubing are shown in Figure 836:2. Recovery of benzo(ghi)perylene was 99%. The percent of other hydrocarbons that eluted through the column ranged from 91 to 105, Table 836:I.

8.3 ANALYSIS PROCEDURE.

8.3.1 An appropriate volume of extract is injected through the loop injector. A flow rate of 1.6 mL/min gives a pressure drop of less than 200 PSI (1380 kPa). The peak area is measured with a disk integrator, driven by 0 to 10 mV servo strip chart recorder with a 0.5″/min (1.27 cm/min) chart speed. The PAH elute in 3 to 5 min. Benzo(a)pyrene is used as the standard. Polar compounds are retained on the column. Samples can be chromatographed every 5 to 10 min.

8.3.2 The column bypass is also used to determine the per cent of PAH in the organic material of the extract. Chromatograms of sample extracts made on the column and column bypass are shown in Figure 836:3. On the basis of absorbance measurements at 254 nm, approximately

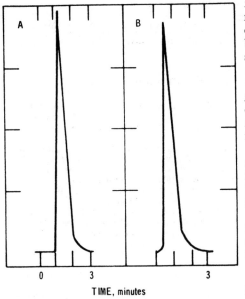

TIME, minutes

Figure 836:2 – Chromatograms of benzo(ghi)perylene, 0.629 μg in cyclohexane, through the column (A) and through the column bypass (B). Stationary phase, Corasil II; eluent, cyclohexane; flow rate, 1.6 mL/minute. Peak area on column, 1165; on column bypass, 1180; recovery on column, 99%; peak area/μg, 1850.

50% of the organic material in the unchromatographed extract is PAH. This procedure is not necessary for routine analyses, but is helpful in elucidating the analytical situation.

8.4 EFFECTS OF STORAGE.

8.4.1 Urban particulate matter on glass fiber filters stored in the dark in an envelope for one year lost 32% of its benzo(a)pyrene. Losses of some other PAH ranged from 1 to 88% **(5)**.

8.4.2 Benzene-soluble extracts evaporated to dryness and stored in closed bottles in a refrigerator were stable [in terms of benzo(a)pyrene concentrations] for 4 years. See Method 102A.

8.4.3 The ultrasonic extract is stable in the dark at room temperature for several days, longer in the refrigerator. However losses usually occur after about 2 weeks.

9. Calibration and Standards

9.1 The benzo(a)pyrene (BaP) standard is made in cyclohexane and is chromatographed when the samples are run, and repeated whenever a parameter such as solvent lot is changed. Both standard and samples are run at concentrations which do not overload the detector and give reproducible results when diluted. For example,

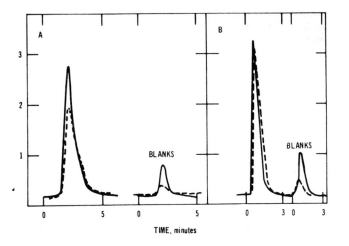

TIME, minutes

Figure 836:3 – Chromatograms of ultrasonic and Soxhlet extracts of composited sample No. 1, Table 836:VI and blanks, through the column (A) and through the column bypass (B). Stationary phase, Corasil II; eluent, cyclohexane; flow rate, 1.6 mL/min. Solid lines are ultrasonic extracts; broken lines are Soxhlet extracts. Extracts were diluted × 3.3 for column bypass.

0.4 μg BaP gave a peak area of about 2000 and fulfilled the above criteria.

The standard is expressed in terms of peak area per microgram (PA/μg). The unit of measurement for the samples is corrected peak area per cubic meter of air (PA/m^3). The BaP equivalent of the TpAH is calculated from these data (**10.2**). The standard is kept in the dark and is stable for more than 30 days when refrigerated nights and weekends.

10. Calculations

10.1 The peak area of the TpAH in a cubic meter of air is given by the equation.

$$PA/m^3 \quad \frac{PA \times A \times B}{V \times a \times b}$$

where PA = Peak area, corrected for the blank

      V = Volume of air sampled in m^3 corrected to 25°C and 101.3 kPa

      A = Area of glass fiber filter in cm^2

      B = Volume of extract in mL

      a = Area of glass fiber filter sample in cm^2

      b = Volume of extract injected in mL

10.2 The concentration of the TpAH may be expressed in terms of their equivalent in benzo(a)pyrene.

$$TpAH(\mu g/m^3 \text{ air} = \frac{PA/m^3 \text{ air (See Table 836:VII)}}{PA/\mu g \text{ benzo(a)pyrene (See Table 836:I)}}$$

11. References

1. BROWN, B. AND J. E. GOODMAN. 1965. High Intensity Ultrasonics, Industrial Applications, Chapter 2, p. 30–35. Van Nostrand Co., Inc., Princeton, N.J.
2. CHATOT, G., M. CASTEGNARO, J. L. ROCHE AND R. FONTANGES. 1971. *Anal. Chem. Acta* 53:259.
3. CHATOT, G., R. DANGY-CAYE AND R. FONTANGES. 1972. *J. Chromatog.* 72:202.
4. WITTGENSTEIN, E. and E. SAWICKI. 1972. *Intern. J. Environ. Anal. Chem.* 2:11.
5. COMMINS, B. T. 1962. In Analysis of Carcinogenic Air Pollutants, E. Sawicki and K. Cassel Jr., Eds. Nat'l. Cancer Inst. Monograph, No. 9, p. 225.

Subcommittee 5

E. SAWICKI, *Chairman*
T. BELSKY
R. A. FRIEDEL
D. L. HYDE
J. L. MONKMAN
R. A. RASMUSSEN
L. A. RIPPERTON
L. D. WHITE

Approved with Modifications from 2nd edition
Subcommittee 4/5

M. FELDSTEIN, *Chairman*
R. J. BRYAN
D. L. HYDE
D. A. LEVAGGI
D. C. LOCKE
R. A. RASMUSSEN
P. O. WARNER

837.

Determination of 2,4-Toluenediisocyanate (TDI) in Air

1. Principle of the Method

1.1 TDI is hydrolyzed by HCl-acetic acid solution to the corresponding toluenediamine derivative. This method is a modification of the Marcali method (1).

1.2 The diamine is diazotized by the sodium nitrite-sodium bromide solution.

1.3 The diazo compound is coupled with N-(naphthyl)-ethylenediamine to form a colored complex.

1.4 The amount of colored complex formed is in direct proportion to the amount of TDI present and is determined by reading the absorbance of the solution at 550 nm.

1.5 Toluenediamine is formed via hydrolysis of TDI on a mole-to-mole basis. This amine is used in place of the TDI for standards. It has the advantages of being less toxic than the TDI and can be weighed because it is solid at room temperature. Both compounds have been tested by this method and the results compare favorably (2).

2. Range and Sensitivity

2.1 The range of the standards used is equivalent to 1.0 to 20.0 μg TDI. In a 20-L air sample, this range is equivalent to 0.007 to 0.140 ppm.

2.2 For samples of high concentration with absorbance greater than the limits of the standard curve (1.0 to 20.0 μg TDI), dilution of the sample with absorbing solution and rereading of the absorbance can extend the upper limit of the range.

2.3 The amount of TDI that would saturate 15 mL of absorbing solution has not been determined. It is possible that, in extremely high concentrations, some of the TDI would not be absorbed by the absorbing solution. Therefore, if a sample is diluted and reread, it could give an erroneously low value.

2.4 A single bubbler absorbs 95% of the diisocyanate if the air concentration is below 2 ppm. Above 2 ppm, about 90% of the diisocyanate is recovered. At high levels, it is suggested that two impingers in series be used.

3. Interferences

3.1 Any free aromatic amine may give a coupling color and thus may be a positive interference.

3.2 Methylene-di-(4-phenylisocyanate) (MDI) will form a colored complex in this reaction. However, its color development time is about 1 to 2 h compared with 5 min for TDI. Therefore, MDI will generally not interfere.

4. Precision and Accuracy

4.1 The precision and accuracy of this method are unknown.

4.2 There has been no collaborative testing.

5. Advantages and Disadvantages of the Method

5.1 Few methods are available.

5.2 Any free aromatic amine may interfere. The method is not specific for TDI.

6. Apparatus

6.1 SAMPLING EQUIPMENT. The sampling unit for personal samples by the impinger collection method consists of the following components.

6.1.1 *An all glass, calibrated midget-impinger containing the absorbing solution or reagent.*

6.1.2 *Battery operated personal sampling pump—MSA Model G, or equivalent.* The sampling pump is protected from splashover or water condensation by an adsorption tube loosely packed with a plug of glass wool and inserted between the exit arm of the impinger and the pump.

6.1.3 *An integrating volume meter such as a dry gas or wet test meter.*

6.1.4 *Thermometer.*

6.1.5 *Manometer.*

6.1.6 *Stopwatch.*

6.1.7 *Various clips, tubing, spring connectors, and belt,* for connecting sampling apparatus to worker being sampled.

6.2 Spectrophotometer Capable of Reading Absorbance at 550 nm.

6.3 Cells, 1-cm and 5-cm Matched Quartz Cells.

6.4 Volumetric Flasks, (several of each), glass stoppered: 50-, 100-, 1000-mL.

6.5 Analytical Balance, capable of weighing 200 g with a sensitivity of 1 mg or less.

6.6 Pipets; 0.5-, 1-, 15-mL.

6.7 Graduated Cylinders, 25-, 50-mL.

7. Reagents

7.1 Purity. All reagents must be made using ACS reagent grade or a better grade.

7.2 Water means ASTM reagent water, type II.

7.3 2,4-Toluenediamine.

7.4 Hydrochloric Acid, Conc 11.7 N.

7.5 Glacial Acetic Acid, Conc 17.6 N.

7.6 Sodium Nitrite.

7.7 Sodium Bromide.

7.8 Sodium Nitrite Solution. Dissolve 3.0 g sodium nitrite and 5.0 g sodium bromide in about 80 mL water. Adjust volume to 100 mL with water. This solution is stable for one week if refrigerated.

7.9 Sulfamic Acid.

7.10 Sulfamic Acid Solution, 10% w/v. Dissolve 10 g sulfamic acid in 100 mL water.

7.11 N-(1-Naphthyl)-ethylene-di-amine Dihydrochloride.

7.12 N-(1-Naphthyl)-ethylene-diamine Solution. Dissolve 50 mg in about 25 mL water. Add 1 mL conc hydrochloric acid and dilute to 50 mL with water. Solution should be clear and colorless; coloring is due to contamination by free amines, and solution should not be used. The solution is stable for 4 days.

7.13 Absorbing Solution. Add 35 mL conc hydrochloric acid and 22 mL glacial acetic acid to approximately 600 mL water. Dilute the solution to 1 L with water. 15 mL is used in each impinger.

7.14 Standard Solution A. Weigh out 140 mg of 2,4-toluenediamine (equivalent to 200 mg of 2,4-toluenediisocyanate). Dissolve in 660 mL of glacial acetic acid, transfer to a 1-L glass-stoppered volumetric flask, and make up to volume with distilled water.

7.15 Standard Solution B. Transfer 10 mL of standard solution A to a glass-stoppered 1-L volumetric flask. Add 27.8 mL of glacial acetic acid so that when solution B is diluted to 1 L with distilled water, it will be 0.6 N with respect to acetic acid. This solution contains an equivalent of 2 μg TDI/mL.

8. Procedure

8.1 Cleaning of Equipment.

8.1.1 Wash all glassware in a hot detergent solution such as Alconox to remove any oil.

8.1.2 Rinse well with hot tap water.

8.1.3 Rinse well with reagent water. Repeat this rinse several times.

8.2 Collection and Shipping of Samples.

8.2.1 Pipet 15 mL of the absorbing solution (Section **7.13**) into the midget impinger.

8.2.2 Connect the impinger (via the absorption tube) to the vacuum pump and to the prefilter assembly (if needed) with short pieces of flexible tubing. The minimum amount of tubing necessary should be used between the breathing zone and impinger. The air being sampled should not be passed through any other tubing or other equipment before entering the impinger.

8.2.3 Turn on pump to begin sample collection. Care should be taken to measure the flow rate, time and/or volume as accurately as possible. Record atmospheric pressure and temperature. The sample should be taken at a flow rate of 1 L/min. Sample for 20 min, making the final air volume 20 L.

8.2.4 After sampling, the impinger stem can be removed and cleaned. Tap the stem gently against the inside wall of the impinger bottle to recover as much of the sampling solution as possible. Wash the stem with a small amount (1 to 2 mL) of unused absorbing solution and add the wash to the impinger. Seal the impinger with a hard, non-reactive stopper (preferably Teflon or glass). Do not seal with rubber. The stoppers on the impingers should be tightly sealed to prevent leakage during shipping. If it is preferred to ship the impingers with the stems in, the outlets of the stem should be sealed with Parafilm or other non-rubber covers, and the ground glass joints should be sealed (*i.e.*, taped) to secure the top tightly.

8.2.5 Care should be taken to minimize spillage or loss by evaporation at all times. Refrigerate samples if analysis cannot be done within a day.

8.2.6 Whenever possible, hand delivery of the samples is recommended. Otherwise, special impinger shipping cases designed by NIOSH should be used to ship the samples.

8.2.7 A "blank" impinger should be handled as the other samples (fill, seal and transport) except that no air is sampled through this impinger.

8.3 ANALYSIS OF SAMPLES.

8.3.1 Remove bubbler tube, if it is still attached, taking care not to remove any absorber solution.

8.3.2 Start reagent blank at this point by adding 15 mL fresh absorbing solution to a clean bubbler tube.

8.3.3 To each bubbler, including the blank, add 0.5 mL of 3% sodium nitrite solution, gently agitate, and allow solution to stand for 2 min.

8.3.4 Add 1 mL of 10% sulfamic acid solution to each tube, agitate for 30 s and allow solution to stand 2 min to destroy all the excess nitrous acid acid present.

8.3.5 Add 1 mL of 0.1% N-(1-naphthyl)-ethylenediamine solution to each tube. Agitate and allow color to develop. Color will be developed in 5 min. A reddish-blue or pink color indicates the presence of TDI.

8.3.6 Add double distilled water to adjust the final volume to 20 mL in bubbler tube. Mix.

8.3.7 Transfer each solution to 1-cm or 5-cm quartz cell.

8.3.8 Using the blank, adjust the spectrophotometer to 0 absorbance at 550 nm.

8.3.9 Determine the absorbance of each sample at 550 nm.

9. Calibration and Standards

9.1 To each of a series of eight graduated cylinders add 5 mL of 1.2 N hydrochloric acid.

9.2 To these cylinders add the following amounts of 0.6 N acetic acid: 10.0, 9.5, 9.0, 8.0, 7.0, 6.0, 5.0 and 0.0 mL respectively.

9.3 To the cylinder add standard solution B in the same order as the acetic acid was added: 0.0, 0.5, 1.0, 2.0, 3.0, 4.0, 5.0 and 10 mL, so that the final volume is 15 mL (*i.e.*, 0.0 mL of the standard is added to the 10 mL acetic acid; 0.5 mL of the standard is added to the 9.5 mL acid; etc. The cylinders now contain the equivalent of 0.0, 1.0, 2.0, 4.0, 6.0, 8.0, 10.0 and 20.0 μg TDI, respectively. The standard containing 0.0 mL standard solution is a blank).

9.4 Add 0.5 mL of the 3.0% sodium nitrite reagent to each cylinder. Mix. Allow to stand 2 min.

9.5 Add 1 mL of the 10% sulfamic acid solution. Mix. Allow to stand for 2 min.

9.6 Add 1 mL of the N-(1-Naphthyl)-ethylenediamine solution. Mix. Let stand for 5 min.

9.7 Make up to 20 mL with water.

9.8 Transfer each solution to a 1-cm or 5-cm quartz cell.

9.9 Using the blank, adjust the spectrophotometer to 0 absorbance at 550 nm.

9.10 Determine the absorbance of each standard at 550 nm.

9.11 A standard curve is constructed by plotting the absorbance against micrograms TDI.

10. Calculations

10.1 Subtract the blank absorbance, (Section **8.2.7**), if any, from the sample absorbance.

10.2 From the calibration curve (Section **9.11**), read the micrograms TDI corresponding to the absorbance of the sample.

10.3 Calculate the concentration of TDI in air in ppm, defined as μL TDI per liter of air.

$$\text{ppm} = \frac{\mu g}{V_s} \times \frac{24.45}{MW}$$

where ppm = parts per million TDI.

μg = micrograms TDI (Section **10.2**)

V_s = volume of air in L corrected to 25°C and 101.3 kPa

24.45 = molar volume of an ideal gas at 25°C and 101.3 kPa

MW = molecular weight of TDI, 174.15

11. References

1. MARCALI, K. 1957. Microdetermination of Toluene-diisocyanates in Atmosphere. *Anal. Chem.* 29:552.

2. LARKIN, R. L. AND R. E. KUPEL. 1969. Microdetermination of Toluenediisocyanate Using Toluenediamine as the Primary Standard. *Am. Indus. Hyg. Assoc. J.* 30:640.

Revised by Subcommittee 3

B. DIMITRIADES, *Chairman*
J. CUDDEBACK
E. L. KOTHNY
P. W. MCDANIEL
L. A. RIPPERTON
A. SABADELL

Approved with modifications from 2nd edition
D. A. LEVAGGI, *Chairman*
B. R. APPEL
D. W. HORSTMAN
E. L. KOTHNY
J. G. WENDT

PART IV
STATE-OF-THE-ART REVIEWS

1. STATE-OF-THE-ART REVIEW*

The Measurement of Strong Acids in Atmospheric Samples

1. Introduction

1.1 Atmospheric strong acids consist principally of nitric and sulfuric acids derived from the oxidation of sulfur dioxide and nitrogen oxides which in turn were emitted from stationary and (for NO_x) mobile combustion sources **(1)**. Some evidence suggests that hydrochloric acid may also be present in the atmosphere **(2)** derived from primary, coal-fired utility emissions or evolved due to interactions between sea salt and acidic sulfate aerosols. These species are neutralized principally by atmospheric gaseous ammonia (the latter mostly originated from surface biogenic and anthropogenic sources) and soil-derived particulate matter to produce the observed composition of sulfate and nitrate aerosols and the levels of nitric acid found in the atmosphere **(3,4)**. Despite the acknowledged general validity of the above summary statements, many questions remain to be fully addressed: What is the relative strong acid content of atmospheric aerosols, especially its spatial and temporal variability? Is aerosol strong acid content associated only with aerosol sulfate and nitrate? Are ammonia and nitric acid in equilibrium with atmospheric aerosols? Are atmospheric gaseous strong acids (HNO_3, HCl, etc.) present in amounts comparable with aerosol species? Do weak acids contribute significantly to the acidity of atmospheric samples? What are the relationships of gaseous and aerosol strong acids to the composition of clouds and precipitation?

1.2 Active research efforts are in progress in all of the above question areas. These efforts are dependent on the development and application in the last 10–20 years of techniques for rapid, accurate, microscale determination of total strong acid and of individual acidic species in atmospheric aerosol and gaseous samples. It is the purpose of this state-of-the-art review first to detail these technique developments, which consist in large measure of filter-based schemes but with some continuous and/or real-time approaches. The analytical complications derived from sampling anomalies and artifact formation problems will be reviewed, and some sampling/analysis protocols recommended that minimize these complications. Lastly, some results of aerosol and gaseous composition studies in which strong acid measurements were critically important will be briefly summarized and the implication of these studies for future atmospheric pollution studies reviewed.

2. Methodologies

Descriptions of methodologies for strong acids are arranged according to the measurement principle—filter collection and post-collection extraction, derivatization and analysis, or real-time, in situ analysis—and further segregated according to whether total strong acid content is measured or individual species (H_2SO_4, HNO_3, HCl) are determined.

2.1 FILTER COLLECTION.

    2.1.1 *Thermal Volatilization.* Thermal volatilization schemes were very popular for several years for speciation of acidic

*Prepared for the Intersociety Committee, *Methods of Air Sampling and Analysis,* 3rd Edition by Roger L. Tanner
Environmental Chemistry Division
Department of Applied Science
Brookhaven National Laboratory
Upton, NY 11973

sulfate compounds in aerosols. Filter samples were heated and H_2SO_4 collected by microdiffusion (5,6) or determined directly by flame photometry after volatilization from Teflon filters (7,8). In one method, H_2SO_4 was distinguished from other volatile sulfates [e.g., NH_4HSO_4 and $(NH_4)_2SO_4$], and non-volatile sulfates (e.g. Na_2SO_4) by heating consecutively at 2 different temperatures (9). 2-Perimidinylammonium sulfate was formed from acid sulfates and thermally decomposed to SO_2 for West-Gaeke analysis in another approach (10).

These methodologies were stimulating attempts to analyze acidic sulfate aerosols for individual species. However, due to serious recovery problems (9) and limited success in distinguishing the two major aerosol species [NH_4HSO_4 and $(NH_4)_2SO_4$] from each other (11), they have fallen into disfavor in recent years, with three exceptions. One technique uses a temperature-cycled diffusion denuder tube in connection with a real-time flame photometric detector to determine H_2SO_4 in ambient aerosols (12,13) (cf. also Method 713, this volume). A related technique uses a series of denuder tubes at varying temperatures to collect nitric acid, sulfuric acid and bisulfate constituents separately for integrated analysis with several-h time resolution (14). Recently the heated denuder system has been mated with a flame photometric detector to produce a computed-controlled system for H_2SO_4, ammonium acid sulfates and non-volatile sulfate with time resolution of 5 min to 1 h depending on requisite sensitivity (15). A photoionization detector has also been used for the determination of H_2SO_4 after preconcentration in a denuder and gas chromatographic separation (16). A related technique determines nitric acid by denuder tube collection with subsequent decomposition to NO_x and determination by ozone-chemiluminescence (17).

2.1.2 *Extraction with pH Measurement or H^+ Titration.* Filter-collected samples may be analyzed for net strong acid content by extraction into water or dilute mineral acid. The strong acid may be determined by simple measurement of pH; however, this is potentially erroneous because of the potential presence of buffering agents such as weak carboxylic acids or hy-

drated forms of heavy metal ions, e.g., Fe(III) and Al(III) (18,19).

Titration procedures for strong acid employing a logarithmic display of data points (Gran titration) were originated by Junge and Scheich (19), perfected by Brosset and coworkers (20,21), and used widely by other groups (22,23). Coulometric generation of strong base for Gran titrations has been used by several groups (23–25). Dissolution of filter samples in ~0.1mM mineral acid followed by Gran titration with correction for blank (25) allows for titration of 1 μmole levels of strong acid with precision and accuracy better than $\pm 10\%$ (22,26).

2.1.3 *Specific Extraction of Atmospheric Acids.* Most of the effort in specific extraction of atmospheric acids has related to aerosol H_2SO_4 analysis. Benzaldehyde has been shown to be specific for H_2SO_4 in dried acidic aerosol sulfate/nitrate samples with analysis for sulfate in aqueous back-extracts (9). 2-Propanol (27) quantitatively extracts H_2SO_4 from quartz filter media but also removes ammonium bisulfate phases (9). The behavior of 2-propanol as an extractant is not well characterized for mixed nitrate/sulfate aerosols. In addition, difficulties have been reported with respect to quantitative removal and selectivity of extraction using the benzaldehyde extraction technique by two groups (28,29). Since, as noted below, free H_2SO_4 is not a common constituent of ambient aerosols, use of specific extractant methodologies has decreased in recent years in favor of generic strong acid determinations.

A related technique is now in wide use — gaseous nitric acid determination as nitrate by IC after collection onto nylon or NaCl-impregnated paper filters from a particle-filtered air stream (30–32). This is specific for nitric acid in the presence of most other nitrogen oxides and oxyacids and, in particular, NO_2, peroxyacyl nitrates (PANs), and alkyl nitrates are not retained by nylon under usual ambient temperature and humidity conditions (31,33,34).

Recently nylon filters have also been used to collect "volatile chloride" — thought to be largely HCl(g) — after prefiltration of particulate chloride (35). Presumably the method of retention in both cases is by reaction of strong acid with the amide linkage in the nylon, forming a nitrate or chloride salt, which can be removed by extraction (displacement) under mild basic conditions such as the carbonate/bicarbonate solution used as an IC eluent (35). HCl methodologies employing derivatization and gas chromatography have also been reported (36).

2.1.4 *Specific Extraction with Derivatization.* A method has been proposed for derivatization of collected H_2SO_4 by dry diethylamine followed by reaction with CS_2 and cupric ion to form a colored complex for spectrophotometric determination (37). This method suffers from a large ammonium bisulfate interference. Likewise an approach in which filter-collected H_2SO_4 is converted to dimethyl sulfate by reaction with diazomethane, with subsequent analysis by gas chromatography-flame photometric detection, does not specifically determine H_2SO_4 in the presence of ammonium bisulfate and sulfate salts (38,39). A related method, by which H_2SO_4 and other aerosol strong acids are converted to ^{14}C-labelled bis(diethylammonium) sulfate and analogs, does serve as a useful technique for determination of low levels of strong acid in aerosols (40).

2.2 CONTINUOUS AND/OR REAL-TIME ANALYSIS.

2.2.1 *Sulfuric Acid.* Sulfuric acid may be determined using a continuous flame photometric detector (FPD) although the measurement is not real-time and represents an average concentration over a few-minute time period (12,13,15). The technique uses a diffusion denuder tube for SO_2 removal attached to an FPD, identical to the instrumentation for a continuous aerosol sulfur analyzer as described by several groups (41–43). However, the temperature of the denuder tube or a zone just upstream therefrom is cycled between room temperature and about 120°C. At ambient temperatures sulfuric acid remains in the aerosol phase, but at ~120°C it is volatilized and removed in the denuder tube. The difference in response at ambient temperatures and 120°C represents ambient H_2SO_4 levels. The minimum cycle time and hence time resolution for the technique is about 6–8 min. Sensitivity-enhanced FPD measurements using SF_6-doped H_2 fuel gas is required for ambient measurements (44).

Acidic sulfates including sulfuric acid may also be differentiated using the humidograph technique of Charlson (45) and thermidograph variations developed more recently (46,47). This latter technique involves heating an aerosol-containing air stream progressively from 20°C to 380°C in 5 min cycles, rapidly cooling it to the dry bulb temperature and measuring the light scattering at 65–70% RH with a nephelometer. By comparing with the thermidograms of test aerosols, the fractional acidity can be measured and approximate level of H_2SO_4 determined.

The fractional acidity can also be determined using an impactor: attenuated total reflectance (ATR), Fourier-transform infrared (FTIR) spectroscopic technique (48). Impactor samples are pressed into a KBr matrix and the IR spectrum used to determine relative acid and qualitatively identify aerosols with molar H^+SO_4 ratios greater than 1, the condition for the presence of H_2SO_4 in the aerosol samples.

2.2.2 *Nitric Acid.* The most common technique for the continuous measurement of nitrogen oxides uses the chemiluminescent reaction of ozone with NO following reduction of NO_2 and other oxides and oxyacids to NO. This method can be used to determine nitric acid by exploiting the specific sorption properties of nylon (49). A two-channel, chemiluminescent NO_x instrument is now used, but with a nylon filter in the second inlet line. The difference in signal can be measured continuously and is due to nitric acid alone. Because of the low levels of nitric acid in the atmosphere and the fact that this is a difference technique, even with a sensitivity-enhanced "super" NO_x instrument the detection limit (~0.4 ppbv) of the instrument is often above ambient concentration levels (50). There is certainly room for improvement in continuous, near-real-time monitoring techniques for nitric acid.

3. Sampling Anomalies

Sampling anomalies have plagued the measurement technologies for atmospheric strong acids present in the gas phase or as aerosols. Hence, a separate section is devoted to the nature of these artifact problems together with suggested sampling techniques which will minimize their effects. These sampling problems generally fall into two classes: (a) reversible or irreversible sorption losses onto filter materials in integrative methodologies or onto sampling lines in continuous and/or real-time techniques; (b) equilibrium-driven loss or gain of species due to non-steady-state conditions in the sampled atmosphere over the time period of the measurement. Both of these phenomena affect sampling/analysis techniques for strong acids in the atmosphere and will be discussed below.

3.1 SORPTION LOSSES. Studies of sorption losses on filters have centered on three major areas, as they relate to the measurement of strong acid content in atmospheric samples. One area is the loss of strong acid in aerosol particles by reaction with basic sites in the filter matrix used to collect the particles (48). Filter matrices used for high-volume sampling include glass fiber or cellulose. These media, in particular glass fiber filters of all types, are unsuitable for collection of acidic aerosol particles for subsequent extraction and titrimetry (52). This is true even if the glass fiber filters are pretreated with acid as well as fired to high temperature, since subsequent rinsing exposes additional free basic sites in the glass resulting in neutralization of the sample (25). High-purity quartz filters can be pretreated to remove basic sites for high-volume sampling, hence treated quartz and Teflon filter media have generally replaced glass fiber or cellulose filters for sampling of acid aerosols by "high-volume" and "low volume" techniques, respectively.

Filter treatments as described above also eliminate a positive source of error in sulfate measurements—the artifact sulfate formed by base-catalyzed oxidation of SO_2 (sorbed on the filter surface) to form sulfuric acid (52–54). The acid so formed is neutralized on the filter surface, but the residual sulfate remains and is measured using standard extraction/analysis techniques.

A related neutralization problem surfaced during experiments designed to elucidate problems in aerosol nitrate sampling made evident during an intercomparison study done in 1977 (55). Subsequent studies confirmed that many filter media, including all glass fiber media, will collect nitric

acid, neutralizing it to form a stable metal nitrate which is extracted and analyzed as nitrate by standard analytical techniques (56). Although not formally an interference with respect to aerosol strong acid analysis, since the latter cannot be performed with glass fiber-collected samples anyway, it is a surface sorption phenomenon, the result of which is to render most historical measurements of nitrate invalid. The "nitrate" measured is minimally the sum of particulate nitrate and nitric acid; for some media, there is a partial interference from NO_2 as well.

A third problem is the loss of strong acid content by topochemical reactions between co-collected basic and acidic particles on the filter surface. This most frequently occurs as the result of coarse ($>2.5\ \mu m$), alkaline, soil-derived particles interacting with fine ($<2.5\ \mu m$), acidic sulfate particles (57). This problem can be eliminated by sampling with a dichotomous sampler, by high-volume sampling with a cyclone to remove coarse particles, or by sampling for shorter time durations. In connection with the latter solution, sampling in most ambient environments for 3 h or less results in low enough surface coverage to prevent topochemical reactions, but does not prevent subsequent neutralization during extraction procedures.

3.2 EQUILIBRIUM-DRIVEN LOSSES. Equilibria involving gas- and particulate-phase species may lead to artifactual errors in sampling strong acids as well as other species in the atmosphere, both under steady-state and non-steady-state conditions. For example, equilibrium may exist between sorbed and gaseous forms for such atmospheric species as polynuclear aromatic hydrocarbons (PAHs) and aldehydes, hence, transport of condensed-phase species on particles to environs with lower gas-phase concentrations may lead to loss of particulate species to reestablish existing equilibria. The example of greatest import for atmospheric acid determinations is the equilibrium between ammonium nitrate-containing particles and gaseous nitric acid and ammonia.

Several groups have demonstrated that observed levels of ammonia and nitric acid are in order-of-magnitude agreement with calculated values under ambient temperature (T) and humidity conditions (RH) when a solid ammonium nitrate aerosol phase is present (**58–61**). Early comparisons were made using averaged aerosol conditions with sampling periods long compared with the mean time for thermodynamically significant changes in T and RH. More recent short-term-sampling results confirm that equilibrium probably exists (**61**) although the position of equilibrium may shift as a function of distance from the earth's surface: i.e., the fraction of nitrate present as gaseous nitric acid may increase with altitude because of the vertical gradient in neutralizing ammonia (**61a**).

The principal implication of these findings for measurement of strong acids in the atmosphere is that sampling must be performed in a way that does not disturb the equilibrium significantly in the process. This has led to the development of the diffusion sampler approach (**62**) to nitric acid measurement by which nitric acid is removed to the walls of a diffusion denuder in which the residence time is short enough ($\tau \leq$ ca 1 sec at ~1L/min sampling rate) that reequilibration cannot significantly proceed. Nitrate is sampled downstream on a filter that captures both particulate NO_3^- and evolved HNO_3. Nitric acid is measured by extraction of denuder-captured HNO_3, or by difference between sampling lines with and without an upstream denuder. In locations at which coarse nitrate levels are significant a preliminary cyclone has been used to prevent sedimentation or impactive loss of nitrate in the denuder tube (**63**).

It should be noted that collection on simple filter packs (front filter for removal of particulate matter followed by nylon or NaCl-implemented cellulose to collect HNO_3) does not always lead to equilibrium-driven loss or gain of nitric acid, i.e., negative or positive artifact formation. However, the potential for such artifact exists, especially for long sampling times or in environments with rapidly changing HNO_3 or nitrate levels.

Thermodynamic considerations suggest that aerosol sulfuric acid/sulfate mixtures should also be in dynamic equilibrium with atmospheric ammonia (**64**). Indeed, as noted initially, ammonia is believed to be the principal neutralizing agent for sulfuric acid formed by heterogeneous or homoge-

neous oxidation of SO_2. However, the equilibrium level of NH_3 for even slightly acid sulfate is much below the usually observed ambient NH_3 levels. This suggests that a non-steady state, mixing-limited situation normally exists (**51**) or, alternately, that mixed nitrate/sulfate salts are usually present with their concomittantly much greater equilibrium NH_3 concentrations.

4. Recommended Protocols

4.1 Strong Acid in Aerosols.

The recommended method for strong acid in aerosols is collection on inert filter media, ultrasonic extraction into weak acid (ca 0.1 mM), and titration with base using a Gran plot to determine the equivalence point. For sampling periods longer than about 6 h (depending on ambient levels), a virtual impactor-based sampler collecting particles on Teflon filter medium should be used. With shorter term sampling for which high-volume sampling apparatus is required, acid-treated quartz filters can be used, preferably with a cyclone preseparator for coarse particle removal. Automation of the titration procedure using coulometric generation of hydroxide is a practical necessity when large numbers of samples must be processed. Precision and accuracy approaching $\pm 10\%$ is possible with careful flow calibration and for sample sizes exceeding 0.5 μeq ($\geq 25\mu$g as H_2SO_4) (**26**).

4.2 Specific Determination of H_2SO_4.

No method is fully satisfactory for determining the low levels of H_2SO_4 occasionally found in ambient aerosols. Two approaches that may be used with reasonable success are the following. Filter pack samples, collected on treated quartz filters which are then thoroughly dried over desiccant, may be extracted into benzaldehyde, the sulfate therein then back-extracted into water and determined by ion chromatography or other soluble sulfate methodologies (**9,25**). Careful drying of the filter is required, as noted above, to prevent significant interference from ammonium bisulfate (usually present in excess of H_2SO_4); in addition, impurity benzoic acid in the benzaldehyde may be present in amounts sufficient to interfere with IC or other sulfate determinations. Sulfuric acid may also be determined by flame photometry using a temperature-cycled diffusion denuder tube (12,13). Time resolution is limited by the minimum temperature cycle time of the denuder (5–10 min) and is quite adequate. The limit of detection—only about 1 $\mu g/m^3$ H_2SO_4 even with sensitivity enhancement through use of SF_6-doped H_2 unless denuder concentration is used (15)—is, however, not adequate for many ambient applications. Direct denuder-tube collection of H_2SO_4 in heated denuders is a viable alternative, but the time resolution is several hours if extraction and wet chemical analysis is used, due to the necessarily slow flow rate through denuder tubes.

4.3 DETERMINATION OF GASEOUS STRONG ACIDS.

Nitric acid may be determined utilizing its unique sorption properties on nylon filters (31). Direct collection downstream of a filter for particle removal, denuder difference determination (62), or real-time chemiluminescence determination (49) are all recommended procedures that utilize nylon filters. Limitations include the following: the direct collection method is subject to artifact formation problems; the denuder difference method requires twice as many analyses; and the chemiluminescent method is limited by the sensitivity of the instrument to ca 0.4 ppb, higher than the average HNO_3 level in many locations. A new approach employing denuder tube concentration of HNO_3 followed by thermal evolution and chemiluminescent detection may prove to be the preferred approach for applications requiring <ca 3-h time resolution (17).

Hydrochloric acid methodologies for application to atmospheric samples have received less attention than those for nitric acid. Collection on nylon or base-coated cellulose filters with IC determination of chloride appears to be a reasonably sound analytical approach.

5. Aerosol and Gaseous Strong Acids—Ambient Data

5.1 AEROSOL ACTIVITY.

A substantial number of aerosol characterization studies have been conducted, using techniques sufficiently sensitive and specific for ambient concentrations of

strong acids in ambient air, that certain conclusions can be made, at least for certain locations and seasons. Data are most complete for NE USA and the Los Angeles area, adequate for SE and Midwest USA, SE Canada, N. Europe and Japan and relatively sparse elsewhere. Most of the data have been collected in the summer season, but there are extensive results for NE USA and SE Canada for other seasons. A summary of U.S. and Canadian surface data is shown in Table 1:I.

The most significant finding reported is that for NE USA data the molar ratio of NH_4^+/SO_4^{2-} varies generally from 1 to 2 and that the H^+/SO_4^{2-} ratio is generally <0.5 and averages approximately 0.25. Data taken in the summer season yields H^+/SO_4^{2-} ratios higher than those in other seasons, with occasional excursions in regional pollution episodes to values >1.0, suggesting the presence of free H_2SO_4 if the aerosol is homogenously mixed. There is a highly significant correlation with sulfate for H^+, NH_4^+, and especially for their sum ($H^+ + NH_4^+$ in molar units) for all studies. Ratios of NH_4^+/SO_4 exceeding 2.0 are believed to occur only when significant levels of aerosol nitrate are present.

Results from European studies and from measurements made in Midwest and Western USA locations (28) usually show the presence of nearly completely neutralized sulfate/nitrate aerosols and, in studies in which gaseous ammonia (NH_3) was measured, higher NH_3 levels. This is consistent with the hypothesis that observed ambient aerosol acidity is the residual from sulfuric and nitric acids, formed from homogenous oxidation of SO_2 and NO_2, after these acids have been mostly neutralized by surface sources of NH_3.

Several studies, using both integrated and continuous, real-time sampling techniques, have demonstrated the occasional presence of free sulfuric acid aerosol in the atmosphere (15,16,70,73,75–78), usually under conditions of episodically high sulfate levels in slow-moving high pressure areas. Acid-to-sulfate ratios are generally higher in rural areas during these episodes than in nearby urban areas (66). Analytical interferences from coarse particle constituents or higher levels of ambient NH_3 in urban areas may cause these differences.

Table 1:I Strong Acid- and Ammonia-to-Sulfate Molar Ratios in Ambient Aerosols

Location	Date	Sampling Period, h	N of Samples	Mean Sulfate, $\mu g/m^3$	Molar H^+/SO_4^{2-} ($\pm$sd)	Molar NH_4^+/SO_4^{2-} ($\pm$sd)	Reference No
Brookhaven HiVol							
New York, NY	8/76	12	37	12.2	0.24 ($\pm$0.20)	1.37	65
Brookhaven (LI), NY	10–11/76	3,6	19	9.8	0.20 ($\pm$0.24)	1.61	—
New York, NY	2/77	6	46	10.8	0.26	1.61	65
Brookhaven (LI), NY	8/77	6	22	16.7	0.13 ($\pm$0.15)	1.15	66
High Point, NJ	8/77	12	55	11.7	0.57 ($\pm$0.40)	1.02	66
New Haven, CT	8/77	6	56	11.7	0.08 ($\pm$0.06)	1.90	66
New York, NY	8/77	6	35	14.1	NA	1.34	66
Brookhaven (LI), NY	1/78	6	11	5.5	0.25	1.01	—
Whiteface Mtn., NY	7/82	12	43	1.46	0.48	1.77	67
Argonne Impactor							
State College, PA	1/77–2/80	4	1369	1.33	0.31	2.7	68
Charlottesville, VA	5/77–2/80	4	988	1.48	0.24	2.7	68
Rockport, IN	7/77–12/79	4	819	2.17	0.07	3.2	68
Brookhaven (LI), NY	12/77–1/80	4	843	1.26	0.21	2.7	68
Raquette Lake, NY	10/78	4	442	0.78	0.27	2.7	68
EPA HiVol, LoVol							
New York, NY	8/77	4	48	NA	NA	1.52 ($\pm$0.30)	69
RTP, NC	8/77	4	16	17	0.6	1.4	69
Cedar Is., NC	8/78	24	4	NA	0.22 ($\pm$0.06)	1.18 ($\pm$0.58)	69
Great Smoky Mtns.	9/78	12	14	11.2	0.86 ($\pm$0.18)	1.10 ($\pm$0.26)	70
Shenandoah V., VA	7–8/80	12	28	14.3	0.50 ($\pm$0.34)	1.68 ($\pm$0.48)	71
Houston, TX	9/80	12	15	13.7	0.38 ($\pm$0.12)	1.38 ($\pm$0.12)	72
Abastumani, USSR	7/79	24	17	4.7	NA	1.9	71
Other Data, Locations							
Allegheny Mtn., PA	7–8/77	12	35	14.1	1.06	0.88	73
Greater Los Angeles	2/79	4	18	~ 2.5	0.01–1.0	1–4	28
Shenandoah V., VA	7–8/80	5 min	>10³	Day 14.6 / Night 13.4	NA	1.44 / 1.71	74
Houston, TX	8/80	5 min	>10³	Day 13.9 / Night 11.1	NA	1.25 / 1.46	74

NA = data not available.

However, episodes in which free sulfuric acid has been determined in urban aerosol samples as well as rural samples have been documented **(13,79,80)**. Concentrations as high as 20 $\mu g/m^3$ H_2SO_4 have been observed **(73)**, but these are infrequent occurrences. Average H_2SO_4 levels even in the summer season in general are < 1 $\mu g/m^3$.

Measurements of aerosol acidity have largely been confined to the earth's surface which coincides with the source area of the principal atmospheric base, ammonia. Measurements of the vertical variability of strong acid aerosols and of the acid-to-sulfate ratio have been made in only a few cases. A recent report **(80)** presents evidence that within the surface boundary layer, acid-to-sulfate ratios in aerosols increase with altitude or remain unchanged — they very seldom decrease with height. Acid-to-sulfate ratios at 0.5 to 2 km are frequently approximately unity in NE North American samples, and since cloud scavenging of aerosols frequently occurs in this layer, aerosol contribution of strong acids to cloud water may be much greater than predicted from surface measurements. These data appear to suggest that reequilibration of aerosol acidity with gaseous acids and bases (principally nitric acid and ammonia) occurs rapidly compared to the typical eddy mixing time in the atmosphere.

5.2 GASEOUS STRONG ACIDS.

Nitric acid measurements have been made in a variety of locations and seasons, including southern California **(31,81)**, the Rocky Mountains **(82)**, urban, rural and remote sites in NE North America **(67,83)**, and even in global remote locations **(32)** using methodologies of reasonable validity and ability to distinguish HNO_3 from particulate nitrate (cf. Sampling Anomalies section above). The most complete set derives from an intercomparison at Claremont, CA during August, 1979 **(81)**, although an interesting additional set from Ontario, Canada has recently been reported **(84)**. Most HNO_3 data fall into the range of 0.1 to 5 ppb although urban plume episodes with concentrations ~ 50 ppb have been reported. In many data sets, nitric acid levels exceed particulate nitrate levels **(67,83)**. In areas of high surface ammonia concentrations, however, the reverse is true, due to the dynamic equilibrium thought to exist

between particulate ammonium nitrate and gaseous HNO_3 and NH_3 in the atmosphere **(56,60)**.

Consistent with this picture is the observation **(80)** that (nitric acid + nitrate) sums decrease with altitude in the boundary layer, but the fraction of nitric acid in that sum increases with altitude, and ammonia levels are generally below detection limits at 0.5 km or more above the earth's surface. Nitric acid is expected to be rapidly scavenged by clouds and, in fact, has never been observed above detection limits in cloud interstitial air (in contrast to NO_2 which is present at clear air levels within clouds).

More measurements of gaseous HCl in the ambient atmosphere, especially in North America, are needed before any clear idea of average HCl levels can emerge.

No discussion is included in this review on the levels or significance of primary sulfuric acid emissions on ambient aerosol acid sulfate levels, but it is believed that primary emissions constitute a small contribution compared to that derived from secondary aerosol formation involving oxidation of SO_2. It has also been suggested that significant emissions of both HCl and NH_3 are derived from coal-burning utility sources, but confirmatory data are presently rather sparse.

6. Implications for Atmospheric Pollution Studies

Many of the implications of measurements of strong acids (gases and aerosols) in the ambient atmosphere have already been introduced in the preceding sections. This author's perceptions of the most important of these implications are summarized below.

1. Most of the strong acid content of the atmosphere derives from partially NH_3-neutralized sulfuric acid aerosol and from nitric acid.

2. Acid sulfate aerosols and nitric acid are predominantly formed by secondary oxidation processes in the atmosphere, with only a small fraction from primary emissions.

3. Techniques now exist that are in large part sufficiently sensitive and selective for ambient measurements with adequate time

resolution for surface measurements. Improvements are still needed for certain airborne applications.

4. Summertime and daytime levels of strong acids tend to be higher than those observed at night or in other seasons.

5. Local ammonia concentrations largely control the surface levels of atmospheric strong acids to which populations are exposed. Additional measurements of this important species are recommended.

7. Acknowledgment

The support of the Office of Health and Environmental Research, U.S. Department of Energy, in the preparation of this review is gratefully acknowledged.

8. References

1. NATIONAL RESEARCH COUNCIL. 1983. Acid Deposition: Atmospheric Processes in Eastern North America, National Academy Press, Washington, DC, pp. 35–40.
2. RAHN, K. A., R. D. BORYS, E. L. BUTLER AND R. A. DUCE. 1979. Gaseous and Particulate Halogens in the New York City Atmosphere, Ann. NY Acad. Sci., 322: 143–152.
3. BROSSET, C., K. ANDREASSON AND M. FERM. 1975. The Nature and Possible Origin of Acid Particles Observed on the Swedish West Coast, Atmos. Environ., 9: 631–642.
4. NATIONAL RESEARCH COUNCIL. 1977. Ammonia, U.S. Environmental Protection Agency, Research Triangle Park, NC, Report EPA-600/1-77-054, November.
5. SCARENGELLI, F. P. AND K. A. REHME. 1969. Determination of Atmospheric Concentrations of Sulfuric Acid Aerosol by Spectrophotometry, Coulometry, and Flame Photometry, Anal. Chem., 41:707–713.
6. DUBOIS, L., C. J. BAKER, T. TEICHMAN, A. ZDROJEWSKI AND J. L. MONKMAN. 1969. The Determination of Sulphuric Acid in Air: A Specific Method, Mikrochim. Acta (Wein), 269–279.
7. RICHARDS, L. W. AND P. S. MUDGETT. 1974. Methods and Apparatus for Sulfuric Acid Aerosol Analysis, U.S. Patent No. 3,833,972, October 1.
8. RICHARDS, L. W., K. R. JOHNSON AND L. S. SHEPARD. 1978. Sulfate Aerosol Study, Rockwell International, Newbury Park, CA, Report AMC8000.13FR, August.
9. LEAHY, D., R. SIEGEL, P. KLOTZ AND L. NEWMAN. 1975. The Separation and Characterization of Sulfate Aerosol, Atmos. Environ. 9:219–229.
10. MADDALONE, R. F., A. D. SHENDRIKAR AND P. W. WEST. 1974. Radiochemical Evaluation of the Separation of H_2SO_4 Aerosol by Microdiffusion from Various Filter Media, Mikrochim. Acta (Wien), 391–401.
11. THOMAS, R. L., V. DHARMARAJAN, G. L. LUNDQUIST AND P. W. WEST. 1976. Measurement of Sulfuric Acid Aerosol, Sulfur Trioxide, and the Total Sulfate Content of the Ambient Air, Anal. Chem., 48:639–642.
12. TANNER, R. L., T. D'OTTAVIO, R. GARBER AND L. NEWMAN. 1980. Determination of Ambient Aerosol Sulfur Using a Continuous Flame Photometric Detection System. I. Sampling System for Aerosol Sulfate and Sulfuric Acid, Atmos. Environ., 14:121–127.
13. ALLEN, G. A., W. A. TURNER, J. M. WOLFSON AND J. D. SPENGLER. 1984. Description of a Continuous Sulfuric Acid/Sulfate Monitor. Presented at the 4th National Symposium on Recent Advances in Pollutant Monitoring of Ambient Air and Stationary Sources, Raleigh, NC, May.
14. SLANINA, J., L. VAN LAMOEN-DOORNENBAL, W. A. LINGERAK, W. MEILOF, D. KLOCKOW AND R. NIESSNER. 1981. Application of a Thermo-Denuder Analyzer to the Determination of H_2SO_4, HNO_3, and NH_3 in Air, Intern. J. Environ. Anal. Chem., 9:59–70.
15. SLANINA, J., C. A. M. SCHOONEBEEK, D. KLOCKOW AND R. NIESSNER. 1985. Determination of Sulfuric Acid and Ammonium Sulfates by Means of a Computer-Controlled Thermodenuder System, Anal. Chem., 57:1955-1960.
16. LINDQVIST, F. 1985. Determination of Ambient Sulfuric Acid Aerosol by Gas Chromatography/ Photoionization Detection After Preconcentration in a Denuder, Atmos. Environ., 19:1671-1680.
17. KLOCKOW, D., M. RICHTER AND R. NIESSNER. 1982. Combination of Diffusion Separation and Chemiluminescence Detection for Collection and Selective Determination of Airborne Nitric Acid. Presented at the 12th Annual Symposium on the Analytical Chemistry of Pollutants, Amsterdam, the Netherlands, April 14–16.
18. COMMINS, B. T. 1963. Determination of Particulate Acid in Town Air, Analyst, 88:364–367.
19. JUNGE, C. AND G. SCHEICH. 1971. Determination of the Acid Content of Aerosol Particles, Atmos. Environ., 5:165–175.
20. BROSSET, C. AND M. FERM. 1978. Man-Made Airborne Acidity and Its Determination, Atmos. Environ., 12:909–916.
21. ASKNE, C. AND C. BROSSET. 1972. Determination of Strong Acid in Precipitation, Lake-Water and Air-Borne Matter, Atmos. Environ., 6:695–696.
22. STEVENS, R. K., T. G. DZUBAY, G. RUSSWURM AND D. RICKEL. 1978. Sampling and Analysis of Atmospheric Sulfates and Related Species, Atmos. Environ., 12:55–68.
23. LIBERTI, A., M. POSSANZINI AND M. VICEDOMINI. 1972. The Determination of the Non-Volatile Acidity of Rain Water by a Coulometric Procedure, Analyst, 97: 352–356.
24. KRUPA, S., M. R. COSCIO, JR. AND F. A. WOOD. 1976. Evaluation of a Coulometric Procedure for the Detection of Strong and Weak Acid Components in Rainwater, J. Air. Pollut. Contr. Assoc., 26:221–223.
25. TANNER, R. L., R. CEDERWALL, R. GARBER, D. LEAHY, W. MARLOW, R. MEYERS, M. PHILLIPS AND L. NEWMAN. 1977. Separation and Analysis of Aerosol Sulfate Species at Ambient Concentrations, Atmos. Environ., 11:955–967.

26. PHILLIPS, M. F., J. S. GAFFNEY, R. W. GOOD-RICH AND R. L. TANNER. 1984. Computer-Assisted Gran Titration Procedure for Strong Acid Determination, Brookhaven National Laboratory, Upton, NY, Report BNL 35734, October.

27. BARTON, S. C. AND W. G. MCADIE. 1970. A Specific Method for the Automatic Determination of Ambient H_2SO_4 Aerosol, *Proc. 2nd International Clean Air Congress*, Air Pollution Control Association, Washington, DC, 379-382.

28. APPEL, B. R., S. M. WALL, M. HAIK, E. L. KOTHNY AND Y. TOKIWA. 1980. Evaluation of Techniques for Sulfuric Acid and Particulate Strong Acidity Measurements in Ambient Air, Atmos. Environ., 14:559-563.

29. EATOUGH, D. J., S. IZATT, J. RYDER AND L. D. HANSEN. 1978. Use of Benzaldehyde as a Selective Solvent for Sulfuric Acid: Interferences by Sulfate and Sulfite Salts, Environ. Sci. Technol., 12:1276-1279.

30. OKITA, T., S. MORIMOTO, M. IZAWA AND S. KONNO. 1976. Measurement of Gaseous and Particulate Nitrates in the Atmosphere, Atmos. Environ., 10:1085-1089.

31. SPICER, C. W. 1977. Photochemical Atmospheric Pollutants Derived from Nitrogen Oxides, Atmos. Environ., 11:1089-1095.

32. HUEBERT, B. J. AND A. L. LAZRUS. 1978. Global Tropospheric Measurements of Nitric Acid Vapor and Particulate Nitrate, Geophys. Res. Lett., 5:557-580.

33. TANNER, R. L., Y.-N. LEE, T. J. KELLY AND J. S. GAFFNEY. 1983. Ambient Nitric Acid Measurements—Interference from PAN and Organonitrogen Compounds, Presented at the 25th Rocky Mountain Conference, Denver, CO.

34. GAFFNEY, J. S. 1984. Brookhaven National Laboratory, Upton NY, unpublished data.

35. ANLAUF, K. AND H. WIEBE. 1984. Atmospheric Environment Service, Downsview, Ontario, unpublished data.

36. MATUSCA, P., B. SCHWARZ AND K. BÄCHMANN. 1984. Measurements of Diurnal Concentration Variations of Gaseous HCl in Air in the Sub-Nanogram Range, Atmos. Environ., 18:1667-1674.

37. HUYGEN, C. 1975. A Simple Photometric Determination of Sulfuric Acid Aerosol, Atmos. Environ., 9:315-319.

38. PENZHORN, R.-D. AND W. G. FILBY. 1976. Eine Methode zur Spezifischen Bestimmung von Schwefelhaltigen Säuren in Atmosphärischen Aerosol, Staub-Reinhalt Luft, 36:205-207.

39. TANNER, R. L. AND R. FAJER. 1981. Brookhaven National Laboratory, Upton, NY, unpublished results.

40. DZUBAY, T. G., G. K. SNYDER, D. J. REUTER AND R. K. STEVENS. 1979. Aerosol Acidity Determination by Reaction with ^{14}C Labelled Amine, Atmos. Environ., 13:1209-1212.

41. HUNTZICKER, J. J., R. S. HOFFMAN AND G. S. LING. 1978. Continuous Measurement and Speciation of Sulfur-Containing Aerosols by Flame Photometry, Atmos. Environ., 12:83-88.

42. COBOURN, W. G., R. B. HUSAR AND J. D. HUSAR. 1978. Continuous, in situ Monitoring of Ambient Particulate Sulfur Using Flame Photometry and Thermal Analysis, Atmos. Environ., 12:89-98.

43. CAMP, D. C., R. K. STEVENS, W. G. COBOURN AND R. B. HUSAR. 1982. Intercomparison of Concentration Results from Fine Particulate Sulfur Monitors, Atmos. Environ., 16:911-916.

44. D'OTTAVIO, T., R. GARBER, R. L. TANNER AND L. NEWMAN. 1981. Determination of Ambient Aerosol Sulfur Using a Continuous Flame Photometric Detection System. II. The Measurement of Low-Level Sulfur Concentrations Under Varying Atmospheric Conditions, Atmos. Environ., 15:197-204.

45. CHARLSON, R. J., A. H. VANDERPOHL, D. S. COVERT, A. P. WAGGONER AND N. C. AHLQUIST. 1974. $H_2SO_4/(NH_4)_2SO_4$ Aerosol: Optical Detection in the St. Louis Region, Atmos. Environ., 8:1257-1268.

46. LARSON, T. V., N. C. AHLQUIST, R. E. WEISS, D. S. COVERT AND A. P. WAGGONER. 1982. Chemical Speciation of $H_2SO_4/(NH_4)_2SO_4$ Particles Using Temperature and Humidity Controlled Nephelometry, Atmos. Environ., 16:1587-1590.

47. ROOD, M. J., T. V. LARSON, D. S. COVERT AND N. C. AHLQUIST. 1985. Measurement of Laboratory and Ambient Aerosols with Temperature and Humidity-Controlled Nephelometry, Atmos. Environ., 19:1181-1190.

48. CUNNINGHAM, P. T. AND S. A. JOHNSON. 1976. Spectroscopic Observation of Acid Sulfate in Atmospheric Particulate Samples, Science, 191:77-79.

49. JOSEPH, D. W. AND C. W. SPICER. 1978. Chemiluminescence Method for Atmospheric Monitoring of Nitric Acid and Nitrogen Oxides, Anal. Chem., 50:1400-1403.

50. TANNER, R. L., P. H. DAUM AND T. J. KELLY. 1983. New Instrumentation for Airborne Acid Raid Research, Intern. J. Environ. Anal. Chem., 13:323-335.

51. APPEL, B. R., Y. TOKIWA, S. M. WALL, M. HAIK, E. L. KOTHNY AND J. J. WESOLOWSKI. 1979. Determination of Sulfuric Acid, Total Particle-Phase Acidity and Nitric Acid in Ambient Air, Air and Industrial Hygiene Laboratory, Berkeley, CA, Report CA/DOH/AIHL/SP-18, June.

52. COUTANT, R. W. 1977. Effect of Experimental Variables on Collection of Atmospheric Sulfate, Environ. Sci. Technol., 11:873-878.

53. PIERSON, W. R., R. H. HAMMERLE AND W. W. BRACHACZEK. 1976. Sulfate Formed by Interaction of Sulfur Dioxide with Filters and Aerosol Deposits, Anal. Chem., 48:1808.

54. PIERSON, W. R., W. W. BRACHACZEK, T. J. KORNISKI, T. J. TRUEX AND J. W. BUTLER. 1980. Artifact Formation of Sulfate, Nitrate and Hydrogen Ion on Backup Filters: Allegheny Mountain Experiment, J. Air Pollut. Contr. Assoc., 30:30-34.

55. CAMP, D. C. 1980. An Intercomparison of Results from Samplers Used in the Determination of Aerosol Composition, Environ. Intern., 4: 83-100. Cf. also Report EPA-600/7-78-118, July 1978.

56. SPICER, C. W., P. M. SCHUMACHER, J. A. KOUYOUM-JIAN AND D. W. JOSEPH. 1978. Sampling and Analytical Methodology for Atmospheric Nitrates, Battelle Columbus Laboratories, Columbus, OH. Final Report for Contract No. 68-02-2213 with U.S. Environmental Protection Agency.

57. TANNER, R. L., R. GARBER, W. MARLOW, B. P. LEADERER AND M. A. LEYKO. 1979. Chemical

Composition of Sulfate as a Function of Particle Size in New York Summer Aerosol, Ann. NY Acad. Sci., 322:99–114.

58. STELSON, A. W., S. K. FRIEDLANDER AND J. H. SEINFELD. 1979. A Note on the Equilibrium Relationship Between Ammonia and Nitric Acid and Particulate Ammonium Nitrate, Atmos. Environ., 13:369–371.

59. DOYLE, G. J., E. C. TUAZON, R. A. GRAHAM, T. M. MISCHKE, A. M. WINER AND J. N. PITTS, JR. 1979. Simultaneous Concentrations of Ammonia and Nitric Acid in a Polluted Atmosphere and Their Equilibrium Relationship to Particulate Ammonium Nitrate, Environ. Sci. Technol., 13:1416–1419.

60. HARRISON, R. M. AND C. A. PIO. 1983. An Investigation of the Atmospheric HNO_3-HN_3-NH_4NO_3 Equilibrium Relationship in a Cool, Humid Climate, Tellus, 35B: 155–159.

61. TANNER, R. L. 1982. An Ambient Experimental Study of Phase Equilibrium in the Atmospheric System: Aerosol H^+, NH_4^+, SO_4^{2-}, NO_3^- – $NH_{3(g)}$, $HNO_{3(g)}$ Atmos. Environ., 16:2935–2942.

61a. TANNER, R. L., R. KUMAR AND S. JOHNSON. 1984. Vertical Distribution of Aerosol Strong Acid and Sulfate in the Atmosphere. J. Geophys. Research, 89:7149–7158.

62. SHAW, R. W., T. G. DZUBAY AND R. K. STEVENS. 1979. The Denuder Difference Experiment, *Current Methods to Measure Atmospheric Nitric Acid and Nitrate Artifacts*, R. K. Stevens, Ed., U.S. EPA, Research Triangle Park, NC, Report EPA-600/2-79/051, March, pp. 79–84.

63. APPEL, B. R., S. M. WALL, Y. TOKIWA AND M. HAIK. 1980. Simultaneous Nitric Acid, Particulate Nitrate and Acidity Measurements in Ambient Air, Atmos. Environ., 14:549–554.

64. LEE, Y.-H. AND C. BROSSET. 1979. Interaction of Gases with Sulfuric Acid in the Atmosphere, Presented at the WMO Symposium on the Long-Range Transport of Pollutants and Its Relation to General Circulation Including Stratospheric/Tropospheric Exchange Processes, Sofia, Bulgaria, October 1–5.

65. TANNER, R. L., W. H. MARLOW AND L. NEWMAN. 1979. Chemical Composition Correlations of Size-Fractionated Sulfate in New York City Aerosol, Environ. Sci. Technol., 13:75–78.

66. LEADERER, B. P., R. L. TANNER AND T. R. HOLFORD. 1982. Diurnal Variations, Chemical Composition and Relation to Meteorological Variables of the Summer Aerosol in the New York Subregion, Atmos. Environ., 16:2075–2087.

67. KELLY, T. J., R. L. TANNER, L. NEWMAN, P. J. GALVIN AND J.A. KADLECEK. 1984. Trace Gas and Aerosol Measurements at a Remote Site in the Northeast U.S., Atmos. Environ., 18:2565–2576.

68. JOHNSON, S. A., R. KUMAR, P. T. CUNNINGHAM AND T. A. LANG. 1981. The MAP3S Aerosol Acidity Network: a Progress Report and Data Summary, Argonne National Laboratory, Argonne, IL, Report ANL-81-63, November.

69. STEVENS, R. K. 1983. Atmospheric Sciences Research Laboratory, U.S. Environmental Protection Agency, Research Triangle Park, NC, unpublished data.

70. STEVENS, R. K., T. G. DZUBAY, R. W. SHAW, JR., W. A. MCCLENNY, C. W. LEWIS AND W. E. WILSON. 1980. Characterization of the Aerosol in the Great Smoky Mountains, Environ. Sci. Technol., 14:1491–1498.

71. STEVENS, R. K., T. G. DZUBAY, C. W. LEWIS AND R. W. SHAW, JR. 1984. Source Apportionment Methods applied to the Determination of the Origin of Ambient Aerosols That Affect Visibility in Forested Areas, Atmos. Environ., 18:261–272.

72. DZUBAY, T. G., R. K. STEVENS, C. W. LEWIS, D. H. HERN, W. J. COURTNEY, J. W. TESCH AND M. A. MASON. 1982. Visibility and Aerosol Composition in Houston, Texas, Environ. Sci. Technol., 16:514–525.

73. PIERSON, W. R., W. R. BRACHACZEK, T. J. TRUEX, J. W. BUTLER AND T. J. KORNISKI. 1980. Ambient Sulfate Measurements on Allegheny Mountain and the Question of Atmospheric Sulfate in the Northeastern United States, Ann. NY Acad. Sci., 388:145–173.

74. WAGGONER, A. P., R. E. WEISS AND T. V. LARSON. 1983. In Situ, Rapid Response Measurement of H_2SO_4/$(NH_4)_2SO_4$ Aerosols in Urban Houston: A Comparison with Rural Virginia, Atmos. Environ., 17:1723–1731.

75. TANNER, R. L. AND W. H. MARLOW. 1977. Size Discrimination and Chemical Composition of Ambient Airborne Sulfate Particles by Diffusion Sampling, Atmos. Environ., 11:1143–1150.

76. APPEL, B. R., E. M. HOFFER, Y. TOKIWA AND E. L. KOTHNY. 1982. Measurement of Sulfuric Acid and Particulate Strong Acidity in the Los Angeles Basin, Atmos. Environ., 16:589–593.

77. MORANDI, M. T., T. J. KNEIP, W. G. COBOURN, R. B. HUSAR AND P. J. LIOY. 1983. The Measurement of H_2SO_4 and Other Sulfate Species at Tuxedo, NY with a Thermal Analysis Flame Photometric Detector and the Analysis of Simultaneously Collected Quartz Filter Samples, Atmos. Environ., 17:843–848.

78. COBOURN, W. G. AND R. B. HUSAR. 1982. Diurnal and Seasonal Patterns of Particulate Sulfur and Sulfuric Acid in St. Louis, July 1977–June 1978. Atmos. Environ., 16:1441–1490.

79. TANNER, R. L., B. P. LEADERER AND J. D. SPENGLER. 1981. Acidity of Atmospheric Aerosols, Environ. Sci. Technol., 15:1150–1153.

80. TANNER, R. L., R. KUMAR AND S. JOHNSON. 1984. Vertical Distribution of Aerosol Strong Acid and Sulfate in the Atmosphere, J. Geophys. Res., 89:7149–7158.

81. SPICER, C. W., J. E. HOWES, JR., T. A. BISHOP, L. H. ARNOLD AND R. K. STEVENS. 1982. Nitric Acid Measurement Methods: An Intercomparison, Atmos. Environ., 16:1487–1500.

82. HUEBERT, B. J., R. B. NORTON, M. J. BOLLINGER, D. D. PARRISH, P. C. MURPHY, D. L. ALBRITTON AND F. C. FEHSENFELD. 1982. HNO_3 and NO_3^- Particulates Measured at a Remote Site, Proc. 2nd Symp. on the Composition of the Non-Urban Troposphere, American Meteorological Society, Boston, MA, pp. 163–165.

83. SHAW, R. W., R. K. STEVENS, J. BOWERMASTER, J.W. TESCH AND E. TEW. 1982. Measurements of Atmospheric Nitrate and Nitric Acid: The Denuder

Difference Experiment, Atmos. Environ., 16:845–853.

84. ANLAUF, K. G., P. FELLIN, H. A. WIEBE, H. I. SCHIFF, G. I. MacKAY, R. S. BRAMAN AND R. GILBERT. 1985. A Comparison of Three Methods for Measurement of Atmospheric Nitric Acid and Aerosol Nitrate and Ammonium, Atmos. Environ., 19:325–333.

2. STATE-OF-THE-ART-REVIEW*

Determination and Speciation of Ambient Particulate Sulfur Compounds

1. Introduction

Although evidence has been mounting documenting a correlation between sulfates and deleterious health effects, environmental damage and ecological deterioration, further studies are still required for adequate comprehension of these relationships. Of special importance is the current controversy on the contribution of sulfur species, specifically sulfuric acid, to the acidification of water bodies (Acid Rain) in the U.S. as well as abroad. As a consequence, sulfate in the atmosphere has become an area of intense interest by many investigators. A growing awareness has also arisen of the necessity to measure accurately not only total sulfate but also to differentiate quantitatively among the various sulfate species to assess their respective roles in atmospheric chemistry. In response to perceived needs, the technology for sulfate measurement has developed rapidly in recent years.

Several critical reviews of the state of the art for measuring sulfate have appeared in the literature (1–4). In this report we will emphasize those procedures that have been more widely accepted and that are currently applied by researchers in atmospheric sulfur chemistry.

*Prepared for Intersociety Committee
Methods of Air Sampling and Analysis,
3rd Edition
Joseph Forrest and Leonard Newman
Environmental Chemistry Division
Department of Applied Science
Brookhaven National Laboratory
Upton, NY 11973
August 1985
This research was performed under the auspices of the United States Department of Energy under Contract No. DE-AC02-76CH00016.

2. Collection Methods

2.1 Although some instrumental procedures for continuous monitoring have been reported, they require specialized and often expensive equipment. The most widely practiced approach is to collect samples on filters for a specific period of time and perform laboratory analyses by one or more of a variety of methods. Sampling frequency and duration will depend upon the detail desired, sample size requirements for analysis, type of equipment used, flow rate, and nature of filter collectors. In practice, atmospheric particles containing sulfate are usually collected on filters for subsequent measurement. Most filters have been evaluated for quantitative retention, but an often neglected question in the past has been an evaluation of the extent of chemical transformations of collected particles with the spurious formation of sulfate from sulfur dioxide adsorption and subsequent oxidation on the filter matrix. Not surprisingly, many past data on sulfates are unreliable due to a lack of consideration of these sampling difficulties.

Problems related with sampling for sulfate measurement and speciation generally fall within several categories:

1. Reaction of ambient SO_2 with the filter media to form sulfate.

2. Neutralization or transformation of collected acidic sulfate by reaction with filter media.

3. Neutralization of collected acidic sulfate by atmospheric NH_3.

4. Reaction of acidic sulfate with collected particles.

2.2 Monitoring of ambient particulate usually includes the measurement of several species in addition to sulfate. Recourse is often made to high volume samplers (5), with glass fiber as the filter material most commonly used. These filters are highly efficient and have small pressure drops. Although they usually contain relatively large amounts of residual sulfate, blanks can be reduced by washing with acid and distilled water, and still further by heating to 475°C followed by final washing (6). Meserole (7) reduced sulfate blanks from 6.9 to 0.87 μg cm^{-2} by pretreatment with acid and multiple rinses.

A more troublesome property of glass fibers is the tendency to adsorb SO_2 (8–10) and convert it to sulfate by reaction with, or catalysis by, one or more substances in the filter behaving as reactive sites (6).

Pierson et al. (11) found that egregious amounts of spurious sulfate were a consequence of filter alkalinity as determined by aqueous extracts. Meserole et al. (7) concluded that sulfate resulting from SO_2 conversion on a glass-fiber filter, expressed as equivalent ambient sulfate concentration, ranged from 1.6–8.9 μg SO_4^{2-} m^{-3} on a 24-hour basis, representing 25–50% of expected atmospheric sulfate levels. A model of the adsorption process of SO_2 in the presence of basic filter media components was developed by Coutant (12). The model evaluates the effect of relative humidity, temperature, SO_2 concentration, filter alkalinity and blank sulfate content. The validity of Coutant's chemical model for predicting artifact sulfate formation under a given set of conditions was investigated by Witz and Wendt (13) in a field study in California, with artifact sulfate ranging up to 10.9 μg m^{-3}. At Allegheny Mountain, PA, artifact sulfate was observed by Pierson et al. (14) to be directly related to filter alkalinity, relative humidity, and cumulative exposure to SO_2 with the more alkaline types converting $\geq 10\%$ of the SO_2 to which they were exposed.

2.3 A complicating factor in filter collection is that acid sulfates may be neutralized wholly or in part by reaction with basic sites on filter surfaces. "Neutral" glass fiber filters, Teflon-coated glass fiber filters and many plastic filter materials (e.g. Acropore, Millipore, Nuclepore, etc.) may be suitable in some cases for total sulfate measurements but are unreliable for ambient sulfate speciation.

Several filter materials are non-reactive with acidic sulfate particles, the most widely used being Mitex and Fluoropore. Both are constructed from Teflon but differ in their fabrication method. However, particles of the order of 0.1 μm diameter are not collected efficiently by Mitex (15). Difficulties have been observed in removing particles from the hydrophobic Teflon surfaces, requiring ultrasonic extraction procedures. A major difficulty with Teflon membrane filters is their high pressure drop, seriously limiting flow rate and rendering them unsuitable for high volume sampling.

A quartz material, Pallflex QAO 2500, although unsuitable in its original form because of high blanks and extreme delicacy, has been modified by Leahy et al. (16) to make it acceptable for sulfate collection and speciation. The material is washed with deionized water, immersed in hot (75–90°C) 1.2 N HCl for 15 min, rinsed with deionized water and ignited at 750°C for 10 min. The cooled filters are sonicated for 30 min in pH 4 phosphoric acid and dried. The author's studies show no artifact sulfate formation on the modified filters, and they can be used for high volume, high efficiency particle collection of atmospheric aerosols without interference in molecular composition determinations.

2.4 Another problem affecting filtration sampling is the neutralization of collected acidic particles by ammonia absorbed from the sampled air stream, especially during long-term sampling periods. Neutralization was reduced by employing a diffusion denuder for ammonia removal prior to particle sampling (17). Hara, et al. (18) used four parallel diffusion tubes coated with oxalic acid at flow rates of 20 L min^{-1} to remove >95% of gaseous ammonia for field sampling of sulfuric acid and acidic sulfate particles.

2.5 Reaction of acidic sulfate with filtered particles could also change the molecular composition of the sulfate. Harker et al. (19) presented evidence that sulfuric acid released nitrate, presumably as nitric acid, from filtered particles. This process must obviously change the molecular composition of the sulfuric acid, possibly to form ammonium sulfate. In similar studies by Forrest et al. (20), quartz filters, pre-loaded with ambient aerosol, were subsequently subjected to a sulfuric acid mist at somewhat higher than actual atmospheric concentrations. After 24 h contact, from 10 to 15% of the deposited acid had reacted with nitrate with presumable changes in molecular composition.

2.6 The most effective method for reducing particle formation and transformation reactions on filters is by minimizing particle sample size but consistent with the needs of available analytical methods. Sampling of a minimum quantity also reduces collision-induced interaction of particles on the filter surface, and thus actual chemical composition of species are less likely to be affected. However, determination of particulate sulfate *in situ* without filter collection would be the only way to completely eliminate confounding particle reactions.

3. Total Particulate Sulfur Determination

Several techniques are available for determining total aerosol sulfur, present predominantly as sulfate, in atmospheric particles. Although techniques for oxidation state differentiation of aerosol sulfur are available, most of the reported data are given as sulfate or total soluble sulfate.

3.1 In X-Ray Fluorescence (XRF), samples collected on a filter are subjected to X-rays, and the emitted fluorescence is diffracted in a wavelength dispersive system. A more convenient system, especially for multi-element analysis, is energy dispersive, whereby all secondary fluorescence is collected within a solid angle and the number of pulses is plotted versus energy (**21**). Fluorescence may also be induced by protons, a technique utilized by Ahlberg et al. (**22**) to quantitate sulfur as a function of particle size. In all cases, uniform distribution of particles on the filter is mandatory. Extensive data for total aerosol sulfur were reported by Dzubay and Stevens (**23**) using a non-dispersive spectrometer. Detection limits were *ca* 0.075 μg m^{-3} for a 2-h sample collected on a 37 mm filter. Corrections of XRF analyses for particle penetration into collection media were made by Davis et al. (**24**) who recommend membrane filters rather than fiber or thick Teflon filters. Reasonable agreement was found with soluble sulfate extracted from the filters and

measured by Thorin titration. Jaklevic et al. (**25**) automated a continuous tape filter to collect fine particulate samples from a dichotomous sampler. Analyses for sulfur were performed immediately following collection by a sensitive XRF spectrometer. Sampling times of 20 min or less were feasible.

3.2 Electron Spectroscopy for Chemical Analysis (ESCA) measures the kinetic energy of photoelectrons emitted from a sample irradiated with monoenergetic X-rays. The kinetic energy is related to the binding energy of electrons in the atomic subshells, which in turn is characteristic of a particular element. Since penetration depths of the photon sources are only in the tens of angstroms, nonhomogeneous samples may yield erroneous results if elemental analyses, rather than surface studies, are desired. This method has been pioneered by Novakov (**26**) and also reported by Craig et al. (**27**) and Barbaray et al. (**28**). A more important aspect of this technique is sulfur speciation according to oxidation state, to be discussed later.

3.3 The Flame Photometric Detector (FPD) is widely used for monitoring ambient levels of gaseous sulfur compounds, after first filtering out particulate sulfate. (See Method 709B.) The device is based upon measurement of the band emission of excited S_2 molecules formed from sulfur-containing compounds in a hydrogen-rich flame. Particulate sulfur, collected on filters or impactors, is subjected to elevated temperatures in an air stream and the vaporized sulfur compounds introduced into a FPD. Roberts and Friedlander (**29**) flash volatilized samples collected on steel strips or glass fiber filters, thus measuring total ambient aerosol sulfur, but encountered difficulty in recovering H_2SO_4 from glass fiber discs. An electrostatically pulsed FPD system alternatively detected total sulfur (gaseous and particulate) and only gaseous by removing aerosol sulfur with an electrostatic precipitator (**30**). Some of the problems encountered were SO_2 interference, low sensitivity to sulfuric acid compared to other sulfates, sodium interference and a response time of ~1.5 min. Several groups have removed gaseous sulfur compounds, principally SO_2, in incoming air by passage through diffusion denuder tubes, prior to the FPD, without significant loss of aerosol sulfur compounds. In a closed experimental system, Durham et al. (31) "denuded" SO_2 through a tube coated internally with PbO_2. A cylinder of filter paper impregnated with $Pb(CH_3CO_2)_2$ served as a SO_2 and H_2S stripper for Huntzicker et al. (32). They overcame differences in sensitivity of the flame photometer between $(NH_4)_2SO_4$ and NH_4HSO_4 compared to H_2SO_4 by addition of NH_3 to the sample air just upstream to the detector. Cobourn et al. (33) monitored the sulfate component of the St. Louis ambient aerosol using a PbO_2 denuder immediately upstream of the PFD. Lowered sensitivity to H_2SO_4 was attributed to its volatilization at the 150°C temperature of the burner block and subsequent absorption on the metal walls before reaching the flame. Although reduction of the burner block temperature improved sensitivity, differences were still troublesome. Detection levels of 0.1 μg m^{-3} were claimed. Tanner et al. (34) selectively absorbed H_2SO_4 in a heated (135°C) Na_2CO_3-coated denuder prior to FPD, and then cyclically added NH_3 before the denuder to neutralize and detect H_2SO_4 plus ammonium sulfates, thus eliminating the sensitivity problem of H_2SO_4. The effects of changes in ambient water vapor concentration and barometric pressure on the FPD sulfur response were quantitatively evaluated, and it was shown that other environmental effects on sulfur response could be eliminated by frequent and proper zeroing. Plots and data were presented demonstrating the specificity and sensitivity of the system to <1.0 ppb aerosol sulfur.

3.4 REDUCTION OF SULFATE to H_2S followed by detection of hydrogen sulfide by any one of many sensitive methods offers a basis for determination of total sulfate (35). Various combinations of hydriodic, hypophosphorous and hydrochloric acids as reducing agents have been used on different materials. Subsequent detection has involved iodimetry or photometry of PbS (36), ammonium molybdate and KCNS (37), methylene blue (38,39) and ferric ion and 1,10 phenanthroline (40). Forrest and Newman (6) applied the reduction technique to aerosol sulfate samples by collecting the evolved H_2S in cadmium acetate solution, metathesizing the CdS formed with

^{110}AgNO$_3$ and counting the ^{110}Ag$_2$S in a conventional NaI(T1) well counter. This method was originally designed for isotopic sulfur analysis, but in addition showed good precision and accuracy as a quantitative procedure in the 1–10 μg S range (41). The reduction digestion steps were performed directly on the filter sample for total aerosol analysis, or on evaporated aqueous extracts of the sample for soluble sulfate determination.

Cathodic stripping voltammetry was applied by Warriner et al. (42) to detect H_2S, liberated by chemical reduction, in samples containing as little as 35 ng sulfate. The polarographic cell-supporting electrolyte solution, which was also the H_2S absorber, consisted of 0.1 M NaCl–0.1 M NaOH. The sulfide peak appeared at –0.76 V vs. SCE. Another variation of the reduction technique involved converting the liberated H_2S to SO_2 in a combustion tube under an oxygen atmosphere (43). The sulfur dioxide was then titrated in a Dohrman microcoulometric titration system. Limit of detection was 10 ng/mL; maximum amount of sulfur per analysis was 3 μg, and reproducibility was ± 5%. Swaim and Ellebracht (44) swept the H_2S from their reduction generation apparatus with argon into a dc plasma jet and monitored the emitted sulfur radiation at 180.7 nm with a vacuum ultraviolet atomic emission spectrometer. The practical detection limit was 10 ng S; linear working range was 0.02 to 300 μg with a 5% relative error.

4. Soluble Sulfate

4.1 All wet chemical analyses are preceded by sample dissolution from a collecting matrix. Differences may arise from solubility considerations, such as type of solvent (water, acid) and temperature of solvent (cold, hot). Consideration should be given to the possible presence of sulfur(IV) compounds, which may be oxidized merely by hot water dissolution (45). One should also give thought as to whether the method of choice determines sulfur in all oxidation states, for otherwise the states must either be preserved if differentiation is desired or converted to the analytically desired state for determining total sulfur.

For most locations, water-soluble sulfate

is generally synonymous with total aerosol sulfate. A direct comparison was made using a turbidimetric determination of water-soluble sulfate with an X-ray induced X-ray fluorescence method. Results yielded a correlation coefficient R = 0.955 with the 20% average difference being considered insignificant (47). Similar studies by Appel et al. (48) comparing XRF results for total sulfur with wet chemical soluble sulfate analyses found the degree of agreement significantly affected by the choice of the wet chemical method. Their report provides a comprehensive comparison among four techniques for determining water soluble sulfate.

4.2 SPECTROPHOTOMETRY. Low concentrations of sulfate have been analyzed gravimetrically in the past by concentrating large volumes and precipitating with $BaCl_2$. These procedures have been simplified and the capabilities for lower concentration expanded by measuring the light-scattering properties of colloidal $BaSO_4$ by turbidimetry (180° source attenuation) or nephelometry (90° scattering) (46,49). Automation of the turbidimetric procedure using gelatin as a suspending agent was introduced by Technicon with their Autoanalyzer (50). Concentration ranges of aqueous extracts are 0.5–5 µg/mL. However, any extraneous source of turbidity will produce positive errors.

An increasingly popular automated method developed by Lazrus et al. (51) with further improvements (52–54) is the methyl thymol blue (MTB) procedure. Heavy metals are removed by ion exchange and sulfate is precipitated by $BaCl_2$ in the presence of equimolar amounts of MTB. Unreacted barium forms a non-absorbing chelate with the dye; remaining excess MTB is measured spectrophotometrically. Limit of detection is about 0.2 µg/mL sulfate. This method has been found by the U.S. EPA to be superior in range and extremely reliable for long-term, low-volume sampling (48) (Method 720C).

Displacement of the loosely bound chloranilate ion from barium by sulfate and subsequent spectrophotometry of the liberated chloranilate is the basis of the popular chloranilate method (55) (Method 720F). It has been modified to minimize changing pH and cation interferences (56) and was later adapted for automated analysis (57). Another variation is based upon barium iodate exchange with sulfate and oxidation of iodide by solubilized iodate to iodine. The I_3^- ion is measured either directly (58) or as the starch complex (59).

In a multiple exchange reaction (60), SO_4^{2-} was reacted with Ba^{2+} and remaining excess Ba^{2+} set free an equivalent amount of Cu^{2+} from a Cu^{2+} EDTA complex. The freed Cu^{2+} reacted with pyridyl azonaphthol to form an intense red complex. Barium-dye exchange was the basis for a procedure devised by Hoffer et al. (61) Ba-dye + SO_4^{2-} → $BaSO_4$ + dye. The barium-dye nitrosulfonazo III complex, which absorbs strongly at 640 nm, was added in excess to the sample solution; and the decrease in absorbance measured spectrophotometrically. Working range was 1–12.5 µg/mL.

4.3 TITRIMETRIC methods for sulfate use dilute barium ion with various indicators for end point determination. In one of the most popular procedures sulfate is titrated in 80% ethanol with 0.005 M Ba^{++} using Thorin as either a visual (62) or spectrophotometric (63) indicator. The method has been markedly improved (64,65), with one of the changes being replacement of Thorin by dimethylsulfonazo III for sharper visual end point indication (64,66). A further modification of the Thorin titration permitted rapid estimation of soluble sulfate; limit of detection was ≤ 0.2 µg/mL (67).

A new indicator (Beryllon II, DSNADNS) for microtitrimetry was proposed by Hwang and Dasgupta (68) with an end point color change sharper and faster than Thorin. Sensitivity was increased (LOD = 0.25 ppm) and interferences minimized. Comparison of analyses with those from ion chromatography for environmental samples was highly favorable.

4.4 ION CHROMATOGRAPHY (IC). Ion exchange resins are capable of providing excellent separation of ions, but until recently the automated determination of eluted species was not possible because of the interference from the electrolyte used as eluent. Small et al. (69) succeeded in neutralizing or suppressing this background by using a combination of resins, thus permitting use of a conductivity cell as a universal and sensitive monitor for both anionic and cationic species. Commercial Ion Chromatographs (IC) are now available, using this system for trace anion analysis, including sulfate (70). A strong base anion exchanger is used as an analytical column, followed by a strong acid exchange suppressor column that converts the eluent, typically 0.003 M Na_2CO_3 + 0.024 M $NaHCO_3$, into a nonconducting H_2CO_3 solution. Another commercial instrument (71) uses a single exchange column and has eliminated the suppressor column, balancing the eluent conductivity signal electrically.

Ion chromatography was shown to yield results comparable to those obtained by an automated methyl thymol blue method for ambient particulate filter samples (72). Appel and Wehrmeister (73) reported a precision of 4% for atmospheric samples, and excepting samples with high nitrate levels, accuracy within 2%, using EPA audit strips, was attained. IC sulfate results with atmospheric samples agreed within 10% with those by five wet chemical methods. Recent improvements in columns and techniques have reduced levels of detection to 0.01 μg/mL (74). IC has become the preferred routine method for particulate sulfate analysis in increasing numbers of installations because of its accuracy, unambiguity, sensitivity and applicability to simultaneous multiple analyses for several anions. Disadvantages of IC are high initial capital cost of the equipment and 3 to 7 times longer analysis time as compared to automated spectrophotometric procedures for sulfate.

4.5 FLAME PHOTOMETRIC DETECTION (FPD) of SO_2 emitted by flash volatilization of aqueous filter extracts was used by Roberts and Friedlander (75) for a low level sulfate measurement. Evaporated extracts were heated to >1000°C via a capacitive discharge. Husar et al. (76) flash pyrolized their samples in tungsten boats at ca 1100°C. Recoveries exceeded 90% at concentrations of 1 to 10 μg.mL sulfur. However, in a comparison of ambient aerosol samples, flash pyrolysis-FPD results were 23% lower than those by XRF. Tanner et al. (47) further refined the technique, employing platinum sample containers, reduced chamber and transfer line volumes and a modified FPD, achieving reliable precision at >0.5 ng/mL sulfur. On the other hand, application of the FPD method is limited to relatively clean matrices requiring very low extract volumes since it does not determine refractory sulfates. It is also subject to erratic results when certain constituents are present in the extract.

4.6 THERMOMETRIC methods have been applied to determine sulfate as well as other sulfur oxyanions of environmental interest (77). In this technique, titrant is continuously added and the heat output of the associated reaction is continuously monitored. Improvements were reported by Williams and Janata (78) and partial automation of the titration approach was described by Dube and Kimmerle (79). In the direct enthalpimetric injection variation, Sajo and Sipos (80) rapidly injected an excess of titrant and monitored the total temperature rise. This method was extensively modified by Hansen and co-workers (81) to determine microgram quantities of sulfate in environmental samples. Concentrated $BaCl_2$, upon direct addition, created a temperature rise of about 0.0032°C/μmol. The amount of sulfate in solution was determined by reference to a standard curve. Thermometric titrimetry has also been used to determine the sulfur oxidation state, as discussed later.

4.7 MISCELLANEOUS. In the determination of sulfate by substoichiometric isotopic dilution analysis (82), an ammoniacal EDTA solution of barium sulfate tagged with ^{35}S is the combined isotope dilution reagent and precipitation agent. Upon addition of the reagent to the sample solution, sulfate exchanges between sample and reagent. Acidification of the solution dissociates the Ba-EDTA complex and $BaSO_4$ precipitates. From the activity of the supernatant solution measured with a scintillation counter, the sulfate concentration may be calculated.

A combination gravimetric-pyrolysis procedure was reported by Maddalone et al. (83). Microgram quantities of sulfate were precipitated by perimidylammonium bromide (PDA-Br), and the precipitate thermally decomposed at 500°C to produce SO_2 which was then absorbed in West-Gaeke solution or introduced to a FPD. Dasgupta et al. (84) reacted particulate sulfate on glass fiber filters with PDA-Br reagent to form $(PDA)_2SO_4$. Excess reagent was removed by frontal elution chromatography with methanol, and remaining (PDA)SO_4 treated with HNO_3 to form a colored trinitro derivative, monitored spectrophotometrically.

A thin film technique described by Ayers (85) collected particles by impaction or by electrostatic precipitation onto 3 mm electron microscope screens. A coat of vacuum-evaporated $BaCl_2$ was applied. Rings of $BaSO_4$ around individual particles, indicating the presence of soluble sulfate, were photographed and the diameter related to the amount of sulfate in the original particles, irrespective of molecular species.

Proton induced γ-ray analysis was used by Macias et al. (86) to measure the sulfur composition, plus carbon and nitrogen, of atmospheric aerosols. Samples collected on quartz filters were nondestructively irradicated with 7.0 MeV protons. The γ-rays emitted by inelastic scattering of the irradiating beam were indicative of elemental composition.

5. Speciation of Sulfate

5.1 SOLVENT EXTRACTION. Selective removal of sulfuric acid in the presence of other particulate sulfates by solvent extraction was reported by Barton and McAdie (87). 2-Propanol (IPA) selectively removed H_2SO_4 from aerosol collected on Nuclepore filters, followed by chloranilate analysis for sulfate. The extraction procedure was subsequently automated (88) and interferences reduced by buffer control of the IPA extract. However, further investigation by Leahy et al. (89) demonstrated that ammonium bisulfate was also extracted quantitatively by IPA, which partially removed other bisulfates. Since ammonium bisulfate is a recognized constituent of ambient aerosol (47), this shortcoming compromises the use of IPA as a selective solvent for H_2SO_4.

Benzaldehyde was shown to be a selective extractant for H_2SO_4 in the presence of bisulfates and sulfates (89), without neutralization from co-collected basic materials, including e.g. ammonium sulfate. Sulfuric acid was reproducibly removed from several different filter media (Mitex, Fluoropore, H_3PO_4-treated quartz) at 75% to 95% recoveries for 10 μg and 100μg

amounts in tracer experiments using $H_2^{35}SO_4$ (47). Inconsistent selectivity for $(NH_4)_2SO_4$ and NH_4HSO_4 was attributed to the presence of water in commercial benzaldehyde by Barrett et al. (90) but was rectified by vacuum distillation of the extractant and storage under nitrogen. Subsequently, they confirmed the selectivity of the benzaldehyde extraction without interferences from bisulfate and sulfate in 5 μg laboratory aerosol H_2SO_4 samples. Eatough et al. (91) confirmed the selectivity of benzaldehyde, but their results indicated that bivalent metal bisulfate salts and some sulfite salts are also extracted with benzaldehyde. If the benzaldehyde contains benzoic acid, significant quantities of bivalent metal sulfate salts are also extracted.

Tanner et al. (92) extended the solvent extraction method whereby H_2SO_4 is first extracted from the dried filter surface by benzaldehyde, IPA extraction next removes bisulfates in the presence of sulfates, and a water leach removes the remaining soluble sulfate. Alternatively an aqueous extraction may be performed on a separate portion of the filter for total sulfate and $(NH_4)_2SO_4$ is obtained by difference. Although his recovery of H_2SO_4 in the absence of other materials compared with Tanner's (92) results, the presence of $(NH)_2SO_4$ or NH_4HSO_4 caused positive errors for Appel (93) with apparent co-extraction of NH_4HSO_4, possibly due to not pre-drying the aerosol samples before extraction. Recovery of laboratory-generated H_2SO_4 when added to atmospheric particulate-loaded samples varied inversely with particulate loading (94), emphasizing the usefulness of short-term sampling.

5.2 THERMAL VOLATILIZATION. Sulfuric acid was determined by thermal volatilization at 400°C from glass fiber filters by Scarengelli and Rehme (95), but it later became obvious that $(NH_4)_2SO_4$ and NH_4HSO_4 were also volatilized. Concurrently, Dubois et al. (96) developed a microdiffusion technique, whereby H_2SO_4 was volatilized at 200°C from a filter sample in a closed petri dish and was absorbed on the NaOH-coated dish cover. However, ammonium sulfate was a potential interferent. At reduced diffusion temperature of 125°C, Maddalone et al. (97) concluded that glass fiber filters retained H_2SO_4, and that

Teflon or graphite were more suitable filters for collecting H_2SO_4 aerosol. Perimidylammonium bromide (PDA-Br) replaced NaOH as the immobilizing agent for microdiffusing H_2SO_4 at 125°C (98). The PDA sulfate was decomposed at 400°C and the released SO_2 determined by FPD or the West-Gaeke method (99). Removal of H_2SO_4 from ambient particulate-loaded filters was still a problem.

Controlled pyrolysis for thermal speciation was reported by Leahy et al. (89). Sulfuric acid was removed from Teflon filters or quartz boats at 190°C in a pre-moistened N_2 stream, NH_4HSO_4 samples likewise at 275°C and $NaHSO_4$ at 325°C. Although not reported, $(NH_4)_2SO_4$ should be decomposed with NH_4HSO_4. Experiments with mixed aerosols in a 90 m^3 environmental chamber were less successful, due probably to reactions of H_2SO_4 with particulate matter, and indicated the difficulty in transferring H_2SO_4 at elevated temperatures in the presence of airborne particles.

Mudgett et al. (100) claim a more successful approach to thermal volatilization by passing heated (150°C) dry N_2 through ambient aerosol collected on Fluoropore filters and determining released H_2SO_4 by FPD. A prototype instrument developed by EPA for monitoring applications (101) has been tested in several field applications (102). Based on the combined experience from many investigators in removing H_2SO_4 from particulate filters at elevated temperatures, one would expect difficulties utilizing this technique and also for resolution of H_2SO_4 from NH_4HSO_4.

Continuous measurement of sulfur-containing aerosols by flame photometric detection was described by Huntzicker et al. (32,103). Sulfur-containing gases were removed by a diffusion tube stripper, followed by injection of NH_3 to increase the FPD response of H_2SO_4 to equal that of $(NH_4)_2SO_4$ and NH_4HSO_4. Speciation was achieved by heating the incoming aerosol stream at temperatures characteristic of each sulfate and stripping the product gas in the denuder, thereby decreasing the FPD output. Thermograms of mixed sulfates were not very specific, limiting the method's usefulness. Cobourn, et al. (33,104) used a similar procedure (without NH_3 addition), cycling the temperature controller

to initiate a 10-min temperature ramp from ambient to 300°C every 15 min. The decrease in sulfur concentration measured at 175°C was interpreted as the minimum amount of sulfur in the form of H_2SO_4. Their technique did not distinguish between NH_4HSO_4 and $(NH_4)_2SO_4$, and was suitable only for measuring total particulate sulfur and minimum H_2SO_4. For airborne monitoring, a fixed temperature of 100°C was selected by Cobourn, et al. (105). At this setting, H_2SO_4 and $(NH_4)SO_4$ aerosols were 50% and 5% volatilized respectively, and an appropriate correction factor was applied toward H_2SO_4 readings. A prototype instrument developed for EPA (106) used stagewise condensation and revaporization to separate H_2SO_4 from other ambient air components. In the sampling mode, incoming air was rapidly heated above the dew point temperature of H_2SO_4 and the vapor then condensed on the cold walls of a collector tube, the remaining gases being ejected. Interference from SO_2 and NH_3 was eliminated by addition of HCl gas upstream of the hot inlet tube. In the analysis mode, clean air saturated with moisture was introduced as the collector walls were rapidly heated, and the vaporized H_2SO_4 was measured by a FPD. Ammonium sulfate, if present on the collector walls, comes off at a time later than the H_2SO_4. A FPD difference procedure for H_2SO_4 measurement was described by Tanner et al. (34). Sampled air was passed through a Na_2CO_3-coated diffusion tube heated at 130°C where >95% of aerosol H_2SO_4 but <5% of $(NH_4)_2SO_4$ and NH_4HSO_4 was removed. Cyclic addition of NH_3 upstream of the denuder permitted the H_2SO_4 to reach the detector, adding to the original signal. Since free H_2SO_4 was never permitted to reach the detector, the problem of reduced sensitivity for this species was eliminated. Niessner and Klockow (107) sampled ambient air through a heated (410°K) thermodenuder coated with NaCl to preferentially absorb H_2SO_4. The aqueous extract of the denuder tube was then analyzed for sulfate by an isotope dilution technique employing ^{35}S. This procedure was further enlarged (108) into a series of coated tubes operated at different temperatures to selectively collect nitrate and sulfate-containing atmospheric trace compounds as well as free ammonia. Sulfuric acid was collected by the NaF coating of a denuder tube heated to 390–401°K; a similarly coated tube heated to 490–510°K trapped the ammonium sulfates. The tubes were water extracted and SO_4^{2-} determined by ion chromatography.

5.3 IMMOBILIZATION ON FILTERS. As a deterrent to topochemical reactions of H_2SO_4 with collected particles, (PDA-Br)-impregnated filters were used to stabilize H_2SO_4 (109). Pyrolysis then produced stoichiometric amounts of SO_2. Total sulfate was determined by extracting the filter with PDA-Br solution, followed by pyrolysis. Ammonium sulfate (and presumably ammonium bisulfate) were acknowledged to interfere with the method. To determine H_2SO_4 in the presence of other particulate sulfates, including $(NH_2)_2SO_4$, Dasgupta et al. (110) heated the sample stream to reduce relative humidity to ~40% such that $(NH_4)_2SO_4$ and other particulate sulfates were collected as solid particles and did not react with the (PDA-Br)-impregnated glass fiber filter, as did H_2SO_4. Treatment with barium acetate solution in methanol-acetone converted the remaining particulate sulfates to $BaSO_4$. Excess PDA-Br was removed by frontal elution chromatography with methanol and the organic sulfate (PDA-Br + H_2SO_4) reacted with HNO_3 to form a colored compound that was measured spectrophotometrically. No attempts were made to differentiate H_2SO_4 from NH_4HSO_4; also the method cannot be applied to high volume samplers without major engineering modifications in order to obtain the required 40% r.h.

Huygen (111) passed a gaseous reactant, diethylamine, through atmospheric aerosols collected on polypropylene filters to immobilize H_2SO_4. The amount of reactant absorbed was determined photometrically by conversion to the copper-dithio-carbamate complex. Interferences by other acids, NH_4HSO_4 and $(NH_4)_2SO_4$ were observed. Dzubay et al. (112) analyzed acidity of aerosol particles collected on Teflon filters by exposure to ^{14}C-labeled trimethylamine vapor, then counting the resultant β-ray emission. Ammonium sulfate interference was minimized by storing collected samples over silica gel to remove adsorbed moisture; NH_4HSO_4 interference was not evaluated.

5.4 GAS-PHASE AMMONIA TITRATION. A gas-phase NH_3 titration approach to obtain qualitative information on sulfate speciation has been developed and used by Covert et al. (113) and Charlson et al. (114,115). A "humidograph" measured the light-scattering coefficient (σ_{scat}) of airborne particles as a function of relative humidity (RH). Hygroscopic particles such as H_2SO_4 or NH_4HSO_4 exhibit a monotonic humidogram (a plot of σ_{scat} vs. RH); deliquescent substances yield a relatively flat curve followed by a rapid rise in σ_{scat} to an inflection point, e.g. 80% RH for $(NH)_2SO_4$. Further information may be gleaned by adding ppm quantities of NH_3 before recording the humidogram. Production of $(NH_4)_2SO_4$ from NH_3 addition to acid sulfate aerosol can be detected by appearance of the inflection point near 80% RH if at least 30 to 50% of the aerosol mass is present as sulfate. In addition to humidification, sampled air was rapidly heated to a controlled temperature, up to a maximum of 380°C, and then rapidly cooled such that the rh was between 65 and 70% (116). An integrating nephelometer measured the scattering coefficient, resulting in a "thermidogram" when plotted against temperature. Distinctions in thermidograms were observed for NH_4^+/SO_4^{2-} ratios from 0 to 2. A system of instruments was devised with time resolution of 5 minutes, using separate nephelometers, which measured both humidograms and thermidograms (117). Gaseous NH_3 was periodically mixed with sampled air to neutralize acidic aerosol. From the thermidograph system, quantitative information on the mass concentration of particulate sulfur compounds was obtained. Semi-quantitative information on the extent of NH_3 neutralization of acid sulfate compounds was gleaned from the humidograms. Chemical compositions were assigned based upon NH_4^+/SO_4^{2-} ratios in seven categories, ranging from 0.5 to 2. This system was used in the field to measure fine particle mass and composition in terms of NH_4^+/SO_4^{2-} ratio in urban Houston and rural Virginia (118).

5.5 INFRARED SPECTROSCOPY. Infrared spectroscopy as applied to identification of species in airborne particles was described by Blanco et al. (119,120). Size-segregated samples were incorporated into a KBr pellet and analyzed by infrared microspectrophotometry. Submicrometer fractions of urban samples displayed bands attributed to ammonium and sulfate. Samples scraped from a Lundgren impactor and analyzed as KBr pellets in the 400 to 3600 cm^{-1} region by Fourier transform spectroscopy were dominated by ammonium and sulfate bands from the submicrometer stage (121). Cunningham and Johnson (122) reported the appearance of a series of bands which they attributed to changes in the acid content of the samples and assigned them to bisulfate and "sulfate more acidic than bisulfate." Quantitative determination of any of the identified species has not yet been developed. In addition, neutralization reactions may occur during the multi-hour sampling period and also by pelletizing with KBr. Direct measurement of the spectra on the collection matrix has been made using internal reflection infrared spectroscopy (123). Johnson et al. (124) described an attenuated total internal reflection (ATR) device containing a head-on cascade virtual impactor to remove particles larger than 1.2 μm. The airstream was then split and impinged on opposite sides of an internal reflection element (IRE), depositing particles between 0.5 and 1.2 μm on the IRE surfaces. The samples on the IRE were then analyzed using attenuated total internal reflection FTIR spectroscopy, which, because of the multiple internal reflections, provided a very sensitive analytical technique. This procedure has been applied to airborne detection of acidic sulfate aerosol (125,126).

5.6 LASER RAMAN SPECTROSCOPY. The origin of the Raman effect lies in the ability of chemical species (molecules, ions, etc.) to scatter incident electromagnetic radiation inelastically. Some of the incident light energy is given up to the chemical species or else the species imparts some of its energy to the light. Both cases result in a change of the scattered radiation, characteristic of the chemical identity and physical environment of the species.

Stafford et al. (127) were able to measure artificially produced ammonium sulfate aerosol down to ~8 ppb SO_4^{2-} (considering the species as a dispersed gas) with a 1000-s integration time and 31W of effective laser power for excitation. The Raman spectra of H_2SO_4, NH_4HSO_4 and $(NH_4)_2SO_4$ were scanned by Rosen and Novakov (128). They detected a peak at 976 cm^{-1}, associated with sulfate from $(NH_4)_2SO_4$, which was not observed in the other two spectra. A sample of ambient aerosol was found to contain the 976 cm^{-1} peak which they ascribed to the presence of $(NH_4)_2SO_4$. Adamowicz et al. (129) found a 981 cm^{-1} Raman band common to 1 M H_2SO_4 and $(NH_4)_2SO_4$ solutions, while a 1054 cm^{-1} band was detected only in the H_2SO_4 solution and attributed to HSO_4^-. The Raman spectrum of a dry ammonium sulfate powder was found to be strikingly different, especially for small Raman shifts. A recent study of the detection limit of Raman spectroscopy for aqueous solutes (130) indicated that the SO_4^{2-} moiety can be detected at a concentration of 5×10^{-5} M, which is well below the expected sulfate concentration in wet ambient aerosol. Since the intensity of the Raman effect is notoriously weak, its ultimate utility for atmospheric applications rests on more powerful lasers, intra-cavity laser techniques, efficient signal collection, processing electronics, and fluorescence reduction.

5.7 MISCELLANEOUS. Metal (iron and nickel) films deposited on glass slides were used by Hazrati (131) to collect aerosol and measure the H_2SO_4 content of individual droplets. The reduction in thickness of film, caused by chemical reaction between acid and metal, was determined by density or intensity differences between background and spot via photometry or microdensitometry. From the weight of metal removed, the equivalent weight of acid was calculated.

The sulfur-nitrogen ratio of ambient aerosols served as a qualitative indicator for the presence of acid sulfate (132). Elemental concentrations of sulfur and nitrogen were measured by a technique involving γ-ray analysis for light elements (GRALE). Because of the possible presence of particulate nitrate, the method is not quantitative, but a sulfur/nitrogen ratio >2.28 (NH_4HSO_4) is a positive indicator of sulfuric acid even in the presence of nitrates.

6. Determination of Oxidation State of Aerosol Sulfur

6.1 ESCA. The principles and application of ESCA to the determination of total aerosol sulfur were discussed earlier. Novakov (133) has pointed out that elemental oxidation states may be differentiated by electron binding energy shifts, dependent upon the effective charge of the atom in its molecular environment. Hulett et al. (134) reported three chemical states of sulfur present on coal smoke particles, including a single reduced state assigned to H_2S or a mercaptan, and two higher oxidation species corresponding to sulfite and sulfate. Airborne particulate samples from Los Angeles and San Francisco were shown to contain sulfur in seven oxidation states, SO_3, SO_4^{2-}, SO_2, SO_3^{2-}, $S°$, and two kinds of S^{2-} (27). Sulfates were always found to be the dominant species, although at times reduced forms of sulfur were present at comparable concentrations. Novakov et al. (133) in a series of laboratory experiments demonstrated that graphite and soot particles could oxidize SO_2 in air. ESCA spectra of these particles revealed two photoelectron peaks corresponding to sulfate and sulfide. Adsorption and oxidation of SO_2 over V_2O_5 and carbon was studied by X-Ray Photoelectron Spectroscopy by Barbaray et al. (28). It was shown that V_2O_5 aerosols do not catalyze SO_2 oxidation at temperatures under 150°C, whereas over carbon particles, sulfate formation is detected from 25°C upwards. At higher temperatures, secondary pollutants such as elementary sulfur, carbon disulfide or oxysulfide were obtained over carbon. Eatough et al. (135) used ESCA, among other techniques, to verify the existence, in smelter flue dusts, of sulfur in the +6 and +4 oxidation states, and of elemental sulfur and sulfide species in some of the samples.

6.2 THERMOMETRIC METHODS. Hansen et al. (81) described a system for determining S(IV) and S(VI) in extracts of atmospheric aerosols by thermometric titration with dichromate. Collected aerosols were dissolved under air-free conditions in 0.1 N HCl plus 5 mM $FeCl_3$ to prevent oxidation of sulfite and to form an iron-sulfite complex. The S(IV) as well as other dichromate-oxidizable constituents, including organic sulfites, were distinguished by their characteristic ΔH value, obtained by measuring the overall change in heat production up to the equivalence point. After completion of oxidation, an excess of Ba^{2+} is rapidly introduced and the heat developed by $BaSO_4$ precipitation is measured to determine total sulfur. Sulfate was calculated as the difference between total and oxidizable sulfur. Hansen et al. (136) showed the existence of sulfite in samples near a copper smelter and sulfite and organic sulfites in New York City aerosol (137). Eatough et al. (135) found sulfite in lead smelter flue dust.

Several discrepancies in the above-described thermometric titrations were reported by Dasgupta et al. (138). Their experiments seemed to indicate that the species actually being determined by the dichromate titration was Fe^{2+}, or some complex thereof such as S(V) intermediates, which had a comparable heat of reaction, or else played a role in the titration. Failure by Schlesinger et al. (139) to produce sharp end points in their thermometric titrations of laboratory-produced Fe(III)-S(IV) aerosol was attributed to redox reactions of free SO_3^{2-} and Fe(II) in 0.1 M HCl + 5 mM $FeCl_3$ extracts. They supported the contention that the oxidizable species were not sulfite complexes but rather Fe(II) resulting from the redox reaction involving oxidizable sulfur and Fe(III) in the presence of Cl^-.

6.3 MISCELLANEOUS TECHNIQUES. Sulfur (IV) in fly ash and smelter dust was measured by Dasgupta et al. (140) by extraction in a formate-buffered solution containing formaldehyde and CDTA [(trans-1,2 cyclohexenedinitrilo) tetraacetic acid], adding NaOH, and pouring an aliquot into pararosaniline hydrochloride solution. The resulting color was measured spectrophotometrically at 575 nm. As an alternative procedure for determining S(IV) and also other soluble anions by absorption in formaldehyde, Dasgupta (141) described a modified ion chromatographic technique for resolving chloride, sulfite, nitrate and sulfate.

Stabilization of sulfur(IV) aerosols during collection on filters was accomplished by impregnating cellulose filters with formalin (142). Sulfite artifact production from SO_2 was minimal. Extracts in 0.1%

formalin were analyzed by ion chromatography.

7. Summary

Our knowledge of atmospheric sulfur chemistry has been vastly enlarged during the past decade, attributable partially to the significant amount of progress which has been made in the measurement of ambient particulate sulfur compounds. Previous errors associated with sample collection have been more clearly defined, and procedures to minimize filter artifact sulfate formation and particle reactions have been devised. Ion chromatography has rapidly arisen as the popular method for analysis of routine soluble sulfate simultaneous with other anions as well. Real-time aerosol sulfate instruments utilizing flame photometric detection have been constructed, with modifications even for sulfate speciation. Information on sulfate speciation has been further enhanced by application of the principle of light scattering by airborne particles as a function of relative humidity, and through temperature programming and addition of ammonia, the degree of quantification may be enhanced. Although significant inroads have been established in determining the oxidation state of aerosol sulfur, further progress is still needed to clarify some of the existing discrepancies.

References

1. FORREST, J. AND L. NEWMAN. 1973. Ambient Air Monitoring for Sulfur Compounds; A Critical Review. JAPCA, 23:761.
2. TANNER, R. L., J. FORREST AND L. NEWMAN. 1978. Determination of Atmospheric Gaseous and Particulate Sulfur Compounds; in Sulfur in the Environment, Part I, J. O. Nriagu, John Wiley and Sons.
3. TANNER, R. L. AND L. NEWMAN. 1976. The Analysis of Airborne Sulfate. JAPCA, 26:737.
4. NEWMAN, L. 1977. Techniques for Determining the Chemical Composition of Aerosol Compounds. Atmos. Environ., 12:113.
5. METHODS OF AIR SAMPLING AND ANALYSIS, SECOND EDITION, 1977. American Public Health Assoc. Method 501, p. 578.
6. FORREST, J. AND L. NEWMAN. 1973. Sampling and Analysis of Atmospheric Sulfur Compounds for Isotope Ratio Studies. Atmos. Environ., 7:561.
7. MESEROLE, F. B., K. SCHWITZGEBEL, B. F. JONES, C. M. THOMPSON AND F. G. MESICH. 1976. Sulfur Dioxide Interferences in the Measurement of Ambient Particulate Sulfates. Report of

Radian Corp., Austin, Texas. Prepared for Electric Power Research Institute, Report No. 262.
8. LEE, R. L., JR. AND J. WAGMAN. 1966. A Sampling Anomaly in the Determination of Atmospheric Sulfate Concentration. Amer. Ind. Hyg. Assoc. J., 27:266.
9. BYERS, R. L. AND J. W. DAVIS. 1970. Sulfur Dioxide Adsorption and Desorption on Various Filter Media. JAPCA, 20:236.
10. AXELROD, H. D., R. J. TECK AND J. P. LODGE, JR. 1971. Further Study on Prefilter Interference for Sulfur Dioxide Measurements. JAPCA, 21:218.
11. PIERSON, W. R., R. H. HAMMERLE AND W. W. BRACHACZEK. 1976. Sulfate Formed by Interaction of Sulfur Dioxide with Filters and Aerosol Deposits. Anal. Chem., 48:1808.
12. COUTANT, R. W.. 1977. Effect of Environmental Variables on Collection of Atmospheric Sulfate. Envir. Sci. Technol., 11:873.
13. WITZ, S. AND J. G. WENDT. 1981. Artifact Sulfate and Nitrate Formation at Two Sites in the South Coast Air Basin. Envir. Sci. Technol., 15:79.
14. PIERSON, W. R., W. W. BRACHACZEK, T. J. KORNISKI, T. J. TRUEX AND J. W. BUTLER. 1980. Artifact Formation of Sulfate, Nitrate and Hydrogen Ion on Backup Filters: Allegheny Mountain Experiment. JAPCA, 30:30.
15. LIU, B. Y. H. AND K. W. LEE. 1976. Efficiency of Membrane and Nuclepore Filters for Submicrometer Aerosols. Environ. Sci. Technol., 10:345.
16. LEAHY, D. F., M. F. PHILLIPS, R. W. GARBER AND R. L. TANNER. 1980. Filter Material for Sampling of Ambient Aerosols. Anal. Chem., 52:1779.
17. DURHAM, J. L., E. B. BAILEY AND W. WILSON. Measurement of Particulate Sulfur by Flame Photometry. A Sulfur Balance. 1975 EPA/NERC/CPL Report.
18. HARA, H., M. KURITA AND T. OKITA. 1982. Ammonia Denuder for Field Sampling of Sulfuric Acid Particles. Atmos. Environ., 16:1565.
19. HARKER, A. B., L. W. RICHARDS AND W. E. CLARK. 1977. The Effect of Atmospheric SO_2 Photochemistry Upon Observed Nitrate Concentrations in Aerosols. Atmos. Environ., 11:87.
20. FORREST, J. R. L. TANNER, D. SPANDAU, T. D'OTTAVIO AND L. NEWMAN. 1980. Determination of Total Inorganic Nitrate Utilizing Collection of Nitric Acid on NaCl-Impregnated Filters. Atmos. Environ., 14:137.
21. GIAUQUE, R. D., F. S. GOULDING, J. M. JAKLEVIC AND R. H. PEHL. 1973. Trace Element Detection with Semiconductor Detector X-ray Spectrometers. Anal. Chem., 45:671.
22. AHLBERG, M. S., A. C. D. LESLIE AND J. W. WINCHESTER. 1978. Characteristics of Sulfur Aerosol in Florida as Determined by PIXE Analysis. Atmos. Environ., 12:773.
23. DZUBAY, T. G. AND R. K. STEVENS. 1975. Ambient Air Analysis with Dichotomous Sampler and X-ray Fluorescence Specrometer. Environ. Sci. Technol., 7:663.
24. DAVIS, D. W., R. L. REYNOLDS, G. C. TSOU AND L. ZAFONTE. 1977. Filter Attenuation Corrections for the X-ray Fluorescence Analysis of Atmospheric Aerosols. Anal. Chem., 49:1990.
25. JAKLEVIC, J. J., B. W. LOO AND T. Y. FUJITA. 1981. Automatic Particulate Sulfur Measurements

with a Dichotomous Sampler and On-Line X-ray Fluorescence Analysis. Environ. Sci. Technol., 15:687.

26. NOVAKOV, T. S. G. CHANG AND A. B. HARKER. 1974. Sulfates as Pollution Particulates: Catalytic Formation on Carbon (Soot) Particles. Science, 186:259.

27. CRAIG, N. L., A. B. HARKER AND T. NOVAKOV. 1974. Determination of the Chemical States of Sulfur in Ambient Pollution Aerosols by X-ray Photoelectron Spectroscopy. Atmos. Environ., 8:15.

28. BARBARAY, B., J. P. CONTOUR AND G. MOUVIER. 1977. Sulfur Dioxide Oxidation over Atmospheric Aerosol − X-ray Photoelectron Spectra of Sulfur Dioxide Adsorbed on V_2O_5 and Carbon. Atmos. Environ., 11:351.

29. ROBERTS, P. T. AND S. K. FRIEDLANDER. 1976. Analysis of Sulfur in Deposited Aerosol Particles by Vaporization and Flame Photometric Detection. Atmos. Environ., 10:403.

30. KITTELSON, D. B., R. MCKENZIE, M. VERMEERSCH, M. LINNE, F. DORMAN, D. PUI, B. LIU AND K. WHITBY. 1978. Total Sulfur Aerosol Concentration with an Electrostatically Pulsed Flame Photometric Detector System. Atmos. Environ., 12:105.

31. DURHAM, J. L., W. E. WILSON AND E. B. BAILEY. 1978. Application of an SO_2-Denuder for Continuous Measurement of Sulfur in Submicrometric Aerosols. Atmos. Environ., 12:883.

32. HUNTZICKER, J. J., R. S. HOFFMAN AND C. S. LING. 1978. Continuous Measurement and Speciation of Sulfur-Containing Aerosols by Flame Photometry. Atmos. Environ., 12:83.

33. COBOURN, W. G., R. B. HUSAR AND J. D. HUSAR. 1978. Continuous in situ Monitoring of Ambient Particulate Sulfur Using Flame Photometry and Thermal Analysis. Atmos. Environ., 12:89.

34. TANNER, R. L., T. D'OTTAVIO, R. GARBER AND L. NEWMAN. 1980. Determination of Ambient Aerosol Sulfur Using a Continuous Flame Photometric Detection System. I. Sampling System for Aerosol Sulfate and Sulfuric Acid. Atmos. Environ., 14:121.

35. LUKE, C. L. 1945. Determination of Total Sulfur in Rubber. Anal. Chem., 17:298.

36. LUKE, C. L. 1949. Photometric Determination of Sulfur in Metals and Alloys. Anal. Chem., 21:1369.

37. PEPKOWITZ, L. P. AND E. L. SHIRLEY. 1951. Microdetection of Sulfur. Anal. Chem., 23:1709.

38. JOHNSON, C. M. AND H. NISHITA. 1952. Microestimation of Sulfur. Anal. Chem., 24:736.

39. GUSTAFSON, L. 1960. Determination of Ultramicro Amounts of Sulfate as Methylene Blue − II. Talanta, 4:236.

40. DAVIS, J. B. AND F. LINDSTROM. 1972. Spectrophotometric Microdetermination of Sulfate. Anal. Chem., 44:524.

41. FORREST, J. AND L. NEWMAN. 1977. Silver-110 Microgram Sulfate Analysis for the Short Time Resolution of Ambient Levels of Sulfur Aerosol. Anal. Chem., 49:1579.

42. WARRINER, J. P., G. S. WILSON AND J. L. MOYERS. 1976. The Determination of Ultramicro Amounts of Sulfate by Cathodic Stripping Voltammetry. Personal Communication.

43. VAN GRONDELLE, M. C., F. VAN DE CROOTS AND J. D. VAN DER LAARSE. 1977. Microcoulometric Determination of Total Inorganic Sulfur in Water by a Hydriodic Acid Reduction Method. Analyt. Chim. Acta, 92:267.

44. SWAIM, P. D. AND S. R. ELLEBRACHT. 1979. Determination of Sulfur by Vacuum Ultraviolet Atomic Emission Spectrometry with Hydrogen Sulfide Evolution. Anal. Chem., 51:1605.

45. HANSEN, L. D., L. WHITING, D. J. EATOUGH, T. E. JENSEN AND R. M. IZATT. 1976. Determination of Sulfur IV and Sulfate in Aerosols by Thermometric Methods. Anal. Chem., 48:634.

46. U.S. DEPT. OF HEALTH, EDUCATION AND WELFARE, PUBLIC HEALTH SERVICE, DIV. OF AIR POLLUTION, PUBLICATION 999-AP-11, SELECTED METHODS FOR THE MEASUREMENT OF AIR POLLUTANTS. 1965. Determination of Sulfate in Atmospheric Suspended Particulates: Turbidimetric Barium Sulfate Method.

47. TANNER, R. L., R. CEDARWALL, R. GARBER, D. LEAHY, W. MARLOW, R. MEYERS, M. PHILIPS AND L. NEWMAN. 1977. Separation and Analysis of Aerosol Sulfate Species at Ambient Concentrations. Atmos. Environ., 11:955.

48. APPEL, B. R., E. L. KOTHNY, E. M. HOFFER AND J. J. WESOLOWSKI. 1976. Comparison of Wet Chemical and Instrumental Methods for Measuring Airborne Sulfate. EPA-600/2-76-059, U.S. Environmental Protection Agency, Research Triangle Park, NC.

49. KOLTHOFF, I. M., E. B. SANDELL, E. J. MEEHAN AND S. BRUCKENSTEIN. Quantitative Chemical Analysis, 4th Ed., MacMillan, Toronto, Can., 1969, pp. 992–994.

50. Sulfate Method VIb via Turbidimetry, Technicon Corp., Tarrytown, NY. 1959.

51. LAZRUS, A., E. LORANGE AND J. P. LODGE, JR. 1968. New Automated Microanalyses for Total Inorganic Fixed Nitrogen and for Sulfate Ion in Water. Advances in Chemistry Series, No. 73, American Chemical Society.

52. McSWAIN, M. R., R. J. WATROUS AND J. E. DOUGLASS. 1974. Improved Methylthymol Blue Procedure for Automated Sulfate Determinations. Anal. Chem., 46:1329.

53. COLOVOS, G., M. R. PANESAR AND E. P. PARRY. 1976. Linearizing the Calibration Curve in Determination of Sulfate by the Methylthymol Blue Method. Anal. Chem., 48:1693.

54. ADAMSKI, J. M. AND S. P. VILLARD. 1975. Application of the Methylthymol Blue Sulfate Method to Water and Wastewater Analysis. Anal. Chem., 47:1191.

55. BERTOLACINI, R. J. AND J. E. BARNEY. 1957. Colorimetric Determination of Sulfate with Barium Chloranilate. Anal. Chem., 29:281; see also ibid. 30:202 (1958).

56. SCHAFER, H. N. S. 1967. An Improved Spectrophotometric Method for the Determination of Sulfate with Barium Chloranilate as Applied to Coal Ash and Related Materials. Anal. Chem., 39:1719.

57. GALES, M. E., JR., W. H. KAYLOR AND J. E. LONGBOTTOM. 1968. Determination of Sulfate by Automatic Colorimetric Analysis. Analyst (London), 93:97.

58. KLOCKOW, D. AND G. RÖNICKE. 1973. An Amplification Method for the Determination of Particle

Sulfate in Background Air. Atmos. Environ., 7:163.

59. HINTZE, W. L. AND R. E. HUMPHREY. 1973. Spectrophotometric Determination of Sulfate Ion with Barium Iodate and the Linear Starch Iodine System. Anal. Chem., 45:814.

60. PÖTZL, K. 1974. Staub-Reinholt. Luft, 34:55-59. New Trace Analytical Methods for Sulfate Chloride Nitrate Anions in Relation to a Long Term Investigation into the Nature and Composition of Atmospheric Aerosols.

61. HOFFER, E. M., E. L. KOTHNY AND B. R. APPEL. 1979. Simple Method for Microgram Amounts of Sulfate in Atmospheric Particulates. Atmos. Environ., 13:303.

62. FRITZ, J. S. AND S. S. YAMAMURA. 1955. Rapid Microtitration of Sulfate. Anal. Chem., 27:1461.

63. RODES, C. E. 1974. A Colorimeter System for Determination of Method 6 Thorin Titration Endpoint. 167th National Meeting, American Chemical Society, Los Angeles, CA, April 1-5. Paper ENVR 62.

64. BUDESINSKY, B. AND L. KRUMLOVA. 1967. Determination of Sulfur and Sulfate by Titration with Barium Perchlorate. Anal. Chim. Acta, 39:375.

65. PERSSON, G. A. 1966. Automatic Colorimetric Determination of Low Concentrations of Sulfate for Measuring Sulfur Dioxide in the Ambient Air. Intern. J. Air Water Poll., 10:845.

66. SCROGGINS, L. H. 1974. Collaborative Study of the Microanalytical Oxygen Flask Sulfur Determination with Dimethylsulfonazo III as Indicator. J. Assoc. Offic. Anal. Chem., 57:22.

67. BROSSET, C. AND M. FERM. 1976. An Improved Spectrophotometric Method for the Determination of Low Sulfate Concentrations in Aqueous Solutions. Personal Communication.

68. HWANG, H. AND P. K. DASGUPTA. 1984. Microtitration of Sulfate with Beryllon II as Indicator: Determination of Sulfate in Environmental Samples. Microchim. Acta, II:159.

69. SMALL, H., T. S. STEVENS AND W. C. BAUMAN. 1975. Novel Ion Exchange Chromatographic Method Using Conductimetric Detection. Anal. Chem., 47:1801.

70. "Dionex Analytical Ion Chromatorgaphs," Technical Brochure, Dionex Corp., Palo Alto, Calif., 1975.

71. "An Introduction to Single Column Ion Chromatography," Technical Brochure, Wescan Instruments, Santa Clara, Calif.

72. FUNG, K. K., S. L. HEISLER, A. PRICE, B. V. NEUSCA AND P. K. MUELLER. 1979. Comparison of Ion Chromatography and Automated Wet Chemical Methods for Analysis of Sulfate and Nitrate in Ambient Particulate Filter Samples: in Ion Chromatographic Analysis of Environmental Pollutants, Vol. 2, pp. 203-209. Ann Arbor Science, Ann Arbor, Mich.

73. APPEL, B. R. AND W. J. WEHRMEISTER. 1979. An Evaluation of Sulfate Analyses of Atmospheric Samples by Ion Chromatography; in Ion Chromatographic Analysis of Environmental Pollutants, Vol. 2, pp. 223-233. Ann Arbor Science, Ann Arbor, Mich.

74. KLOTZ, P. Private communication.

75. ROBERTS, P. T. AND S. K. FRIEDLANDER. 1976. Analysis of Sulfur in Deposited Aerosol Particles by Vaporization and Flame Photometric Detection. Atmos. Environ., 10:403.

76. HUSAR, J. D., R. B. HUSAR AND P. K. STUBITS. 1975. Determination of Submicrogram Amounts of Atmospheric Particulate Sulfur. Anal. Chem., 47:2062.

77. NAKANISHI, M. AND S. FUJIEDA. 1974. Correction for Heats of Mixing and Dilution in Injection Enthalpimetric Analysis. Anal. Chem., 46:119.

78. WILLIAMS, M. B. AND J. JANATA. 1970. Thermometric Titration of Sulfate. Talanta, 17:548.

79. DUBE, G. AND M. KIMMERLE. 1975. Semi-Automated Thermometric Titration of Sulfate. Anal. Chem., 47:285.

80. SAJO, I. AND B. SIPOS. 1967. Rapid Analysis of Plating Baths by a Direct-Reading Thermometric Method. Talanta, 14:203.

81. HANSEN, L. D., L. WHITING, D. J. EATOUGH, T. E. JENSEN AND R. M. IZATT. 1976. Determination of Sulfur IV and Sulfate in Aerosols by Thermometric Methods. Anal. Chem., 48:634.

82. KLOCKOW, D., H. DENZINGER AND G. RÖNICKE. 1976. Determination of Sulfate by Substoichiometric Isotopic Dilution Analysis, in European Monitoring Program — Manual for Sampling and Chemical Analysis Procedures. Norwegian Inst. for Air Research, Lillestrom, Norway.

83. MADDALONE, R. F., G. L. MCCLURE AND P. W. WEST. 1975. Determination of Sulfate by Thermal Reduction of Perimidylammonium Sulfate. Anal. Chem., 47:316.

84. DASGUPTA, P. K., L. G. HANLEY, JR. AND P. W. WEST. 1978. Spectrophotometric Determination of Trace Sulfate in Water. Anal. Chem., 50:1793.

85. AYERS, G. P. 1978. Quantitative Determination of Sulfate in Individual Aerosol Particles. Atmos. Environ., 12:1613.

86. MACIAS, E. S., C. D. RADCLIFFE, C. W. LEWIS AND C. R. SAWICKI. 1978. Proton Induced γ-Ray Analysis of Atmospheric Aerosols for Carbon, Nitrogen and Sulfur Composition. Anal. Chem., 50:1120.

87. BARTON, S. C. AND H. G. MCADIE. 1971. A Specific Method for the Automatic Determination of Ambient H_2SO_4 Aerosol. Proc. 2nd Intern. Clean Air Cong., H. M. Englund and W. T. Beery, Academic Press, New York, pp. 379-382.

88. BARTON, S. C. AND H. G. MCADIE. 1973. An Automated Instrument for Monitoring Ambient H_2SO_4 Aerosol. Proc. 3rd Intern. Clean Air Cong., Dusseldorf, W. Ger., p. c25.

89. LEAHY, D., R. SIEGAL, P. KLOTZ AND L. NEWMAN. 1975. The Separation and Characterization of Sulfate Aerosol. Atmos. Environ., 9:219.

90. BARRETT, W. J., H. C. MILLER, I. E. SMITH, JR. AND C. H. GWIN. 1977. Development of a Portable Device to Collect Sulfuric Acid Aerosol. U.S. Environmental Protection Agency, Rept. EPA 600/2-77-027, Feb.

91. EATOUGH, D. L., S. IZATT, I. RYDER AND L. D. HANSEN. 1978. Use of Benzaldehyde as a Selective Solvent for Sulfuric Acid: Interferences by Sulfate and Sulfite Salts. Environ. Sci. Tech., 12:1276.

92. TANNER, R. L., R. W. GARBER AND L. NEWMAN. 1977. Speciation of Sulfate in Ambient Aerosols by Solvent Extraction with Flame Photometric Detection, 173rd National Meeting, Amer. Chem. Soc., New Orleans, La., March 20-25, Paper ENVR-41.

93. APPEL, B. R., S. M. WALL, M. HAIK, E. L. KOTHNY AND Y. TOKIWA. 1980. Evaluation of Techniques for Sulfuric Acid and Particulate Strong Acidity Measurements in Ambient Air. Atmos. Environ., 14:559.

94. APPEL, B. R., E. M. HOFFER, Y. TOKIWA AND E. L. KOTHNY. 1982. Measurement of Sulfuric Acid and Particulate Strong Acidity in the Los Angeles Basin. Atmos. Environ., 16:589.

95. SCARENGELLI, F. P. AND K. A. REHME. 1969. Determination of Atmospheric Concentrations of Sulfuric Acid Aerosol by Spectrophotometry, Coulometry and Flame Photometry. Anal. Chem., 41:707.

96. DUBOIS, L, C. I. BAKER, T. TEICHMAN, A. ZDROJEWSKI AND J. L. MONKMAN. 1969. The Determination of Sulfuric Acid in Air: A Specific Method. Mikrochim. Acta (Wien):269; ibid:1268.

97. MADDALONE, R. F., A. D. SHENDRIKAR AND P. W. WEST. 1974. Radiochemical Evaluation of the Separation of H_2SO_4 Aerosol by Microdiffusion from Various Filter Media. Mikrochim. Acta (Wien):391.

98. MADDALONE, R. F., R. L. THOMAS AND P. W. WEST. 1976. Measurement of Sulfuric Acid Aerosol and Total Sulfate Content of Ambient Air. Environ. Sci. Tech., 10:162.

99. WEST, P. W. AND G. C. GAEKE. 1956. Fixation of Sulfur Dioxide as Disulfitomercurate II, Subsequent Colorimetric Estimation. Anal. Chem., 28:1916.

100. MUDGETT, P. S., L. W. RICHARDS AND J. R. ROEHRIG. 1974. A New Technique to Measure Sulfuric Acid in the Atmosphere; in Analytical Methods Applied to Air Pollution Measurement, R. Stevens and W. Herget, eds., Ann Arbor Science Publ. Inc., Ann Arbor, Mich.

101. HARRIS, B. 1975. Second Stage Development of an Automated Field Sulfuric Acid Sampler and the Development of a Sulfuric Acid Analyzer. Cabot Corp., Monthly Technical Progress Narrative 6 for EPA Contract 68-02-2238, Billerica, Mass., Dec.

102. LAMOTHE, P. J., T. G. DZUBAY AND R. K. STEVENS. 1976. Chemical Characterization of Aerosols Present During the General Motors Sulfate Dispersion Experiment, in U.S. Environmental Protection Agency, Rept. EPA-600/3-76-035, April, pp. 1–28.

103. HUNTZICKER, J. J., R. A. CARY AND C. S. LING. 1980. Neutralization of Sulfuric Acid Aerosol by Ammonia. Environ. Sci. Tech., 14:819.

104. COBOURN, W. G., J. DJUKIC-HUSAR AND R. B. HUSAR. 1980. Monitoring of Sulfuric Acid Episodes in St. Louis, Missouri, J. Geophys. Res., 85:4487.

105. COBOURN, W. G., J. DJUKIC-HUSAR, R. B. HUSAR AND S. KOHLI. 1980. Airborne In-Situ Measurement of Particulate Sulfur and Sulfuric Acid with Flame Photometry and Thermal Analysis. Presented at Symposium on Plumes and Visibility: Measurements and Model Components, Grand Canyon, Ariz., Nov. 10–14.

106. BARDEN, J. D. 1981. Analysis System for Total Sulfuric Acid in Ambient Air-Development and Preliminary Evaluation, U.S. Environmental Protection Agency, Rept. EPA-600/S2-81-013, May.

107. NIESSNER, R. AND D. KLOCKOW. 1980. A Thermoanalytical Approach to Speciation of Atmospheric Strong Acids. Intern. J. Environ. Anal. Chem., 8:163.

108. SLANINA, J., L. V. LAMOEN-DOORNENBAL, W. A. LINGERAK, W. MEILOF, D. KLOCKOW AND R. NEISSNER. 1981. Application of a Thermo-Denuder Analyser to the Determination of H_2SO_4, HNO_3 and NH_3 in Air. Intern. J. Environ. Anal. Chem., 9:59.

109. THOMAS, R. L., V. DHARMARAJAN, G. L. LUNDQUIST AND P. W. WEST. 1976. Measurement of Sulfuric Acid Aerosol, Sulfur Trioxide and the Total Sulfate Content of the Ambient Air. Anal. Chem. 48:639.

110. DASGUPTA, P. K. G. L. LUNDQUIST AND P. W. WEST. 1979. Specific Determination of Aerosol Sulfuric Acid in the Presence of Ammonium Sulfate. Atmos. Environ., 13:767.

111. HUYGEN, C. 1975. A Simple Photometric Determination of Sulfuric Acid Aerosol. Atmos. Environ., 9:315.

112. DZUBAY, T. G., G. K. SNYDER, D. J. REUTTER AND R. K. STEVENS. 1979. Aerosol Acidity Determination by Reaction with ^{14}C Labelled Amine. Atmos. Environ., 13:1209.

113. COVERT, D. S., CHARLSON, R. J. AND N. C. AHLQUIST. 1972. A Study of the Relationship of Chemical Composition and Humidity to Light Scattering by Aerosols. J. Appl. Meteorol., 11:968.

114. CHARLSON, R. J., A. H. VANDERPOL, D. S. COVERT, A. P. WAGGONER AND N. C. AHLQUIST. 1974. Sulfuric Acid-Ammonium Sulfate Aerosol: Optical Detection in the St. Louis Region. Science, 184:156.

115. CHARLSON, R. J., A. H. VANDERPOL, D. S. COVERT, A. P. WAGGONER AND N. C. AHLQUIST. 1974. $H_2SO_4/(NH_4)_2SO_4$ Background Aerosol: Optical Detection in St. Louis Region. Atmos. Environ., 8: 1257.

116. LARSON, T. V., N. C. AHLQUIST, R. E. WEISS, D. S. COVERT AND A. P. WAGGONER. 1982. Chemical Speciation of $H_2SO_4–(NH_4)_2SO_4$ Particles Using Temperature and Humidity Controlled Nephelometry. Atmos. Environ., 16:1587.

117. WEISS, R. E., T. V. LARSON AND A. P. WAGGONER. 1982. In-Situ Rapid-Response Measurement of $H_2SO_4/(NH_4)_2SO_4$ Aerosols in Rural Virginia. Environ. Sci. Tech., 16:525.

118. WAGGONER, A. P., R. E. WEISS AND T. V. LARSON. 1983. In-Situ Rapid Response Measurement of $H_2SO_4/(NH_4)_2SO_4$ Aerosols in Urban Houston: A Comparison with Rural Virginia. Atmos. Environ., 17:1723.

119. BLANCO, A. J. AND G. B. HOIDALE. 1968. Microspectrophotometric Technique for Obtaining the Infra-Red Spectrum of Microgram Quantities of Atmospheric Dust. Atmos. Environ., 2:327.

120. BLANCO, A. J. AND R. G. MCINTYRE. 1972. An Intra-Red Spectroscopic View of Atmospheric Particulates over El Paso, TX. Atmos. Environ., 6:557.

121. CUNNINGHAM, P. T., S. A. JOHNSON AND R. T. YANG. 1974. Variations in Chemistry of Airborne Particulate Material with Particle Size and Time. Environ. Sci. Tech. 8:131.

122. CUNNINGHAM, P. T. AND S. A. JOHNSON. 1976. Spectroscopic Observation of Acid Sulfate in Atmospheric Particulate Samples. Science, 191:77.

123. NOVAKOV, T., S. G. CHANG, R. L. DOD AND H.

ROSEN. 1976. Chemical Characterization of Aerosol Species Produced in Heterogeneous Gas-Particle Reactions. LBL-5215. Presented at the Air Poll. Control Assoc. Annual Meeting, Portland, OR, June 27–July 1.

124. JOHNSON, S. A., D. G. GROCZYK, R. KUMAR AND P. T. CUNNINGHAM. 1981. Analytical Techniques for Ambient Sulfate Aerosols. Argonne National Laboratory, Rept. ANL-81-12, June.

125. JOHNSON, S. A., R. KUMAR AND P. T. CUNNINGHAM. 1983. Airborne Detection of Acidic Sulfate Aerosol Using an ATR-Impactor. Aerosol. Sci. Technol., 2:401.

126. TANNER, R. L., R. KUMAR AND S. JOHNSON. 1984. Vertical Distribution of Aerosol Strong Acid and Sulfate in the Atmosphere. J. Geophys. Res., 89:7149.

127. STAFFORD, R. G., R. K. CHANG AND P. J. KINDLMANN. 1976. Laser-Raman Monitoring of Ambient Sulfate Aerosols. Presented at the 8th Materials Research Symposium of Methods and Standards for Environmental Measurement, National Bureau of Standards, Gaithersburg, MD.

128. ROSEN, H. AND T. NOVAKOV. 1978. Identification of Primary Particulate Carbon and Sulfate Species by Raman Spectroscopy. Atmos. Environ., 12:923.

129. ADAMOWICZ, R. S. SCHWARTZ AND R. E. MEYERS. 1977. Unpublished Brookhaven National Laboratory results.

130. CUNNINGHAM, P. T. 1977. Private Communication.

131. HAZRATI, A. M. 1979. The Use of Metal Films in Studying Sulfuric Acid Droplets in the Atmosphere. JAPCA, 29:372.

132. DELUMYEA, R., E. S. MACIAS AND W. G. COBOURN. 1979. Detection of the Presence of Ambient Acid Sulfate Aerosols from the Sulfur/Nitrogen Ratio. Atmos. Environ., 13:1337.

133. NOVAKOV, T. 1974. Sulfates in Pollution Particulates. Lawrence Berkeley Laboratories Report LBL-3035, presented at 67th APCA Annual Meeting, Denver, Col., June 9–13.

134. HULETT, L. D., T. A. CARLSON, B. R. FISH AND J. L. DURHAM. 1971. Studies of Sulfur Compounds Adsorbed on Smoke Particles and Other Solids by Photoelectron Spectroscopy. Proceedings of the Symposium on Air Quality, 161st National Meeting, Amer. Chem. Soc., Los Angeles. Plenum, Washington, D.C.

135. EATOUGH, D. J., N. L. EATOUGH, M. W. HILL, N. F. MANGELSON, J. RYDER, L. D. HANSEN, R. G. MEISENHEIMER AND J. W. FISCHER. 1979. The Chemical Composition of Smelter Flue Dusts. Atmos. Environ., 13:489.

136. HANSEN, L. D., D. J. EATOUGH, N. F. MANGELSON, T. E. JENSEN, D. CANNON, T. J. SMITH AND D. E. MOORE. 1975. Sulfur Species and Heavy Metals in Particulates from a Copper Smelter. Proceedings of International Conference on Environmental Sensing and Assessment, Vol. 2, Las Vegas, Nev.

137. HANSEN, L. D., T. MAJOR, J. RYDER, B. RICHTER, N. F. MANGELSON AND D. J. EATOUGH. 1977. Reduced Species and Acid-Base Components of The New York City Aerosols, presented at the Industrial Hygiene Conference, New Orleans, La., May 24–27.

138. DASGUPTA, P. K., P. A. MITCHELL AND P. W. WEST. 1979. Study of Transition Metal Ion-S(IV) Systems. Atmos. Environ., 13:775.

139. SCHLESINGER, R. B., J. L. GURMAN AND L. C. CHEN. 1980. The Production and Characterization of a Transition Metal (Fe(III))-S(IV) Aerosol. Atmos. Environ., 14:1279.

140. DASGUPTA, P. K., K. B. DE CESARE AND M. BRUMMER. 1982. Determination of S(IV) in Particulate Matter. Atmos. Environ., 16:917.

141. DASGUPTA, P. K. 1982. On the Ion Chromatographic Determination of S(IV). Atmos. Environ., 16:1265.

142. FORTUNE, C. R. AND B. DELLINGER. 1982. Stabilization and Analysis of Sulfur (IV) Aerosols in Environmental Samples. Environ. Sci. Tech., 16:62.

3. STATE-OF-THE-ART-REVIEW

Source Sampling for Regulated Air Pollutants from Stationary Sources*

1. Introduction

The purpose of this paper is to present the state-of-the-art of source sampling for regulated air pollutants from stationary sources. Based on the committee's experience and their joint agreement, the current state-of-the-art for measuring regulated air pollutants from Stationary Sources was determined to be the United States Environmental Protection Agency's (U.S. EPA) Reference Test Methods. The major reason for this agreement is the significant effort that has been expended to ensure that these are the most suitable test methods and that they are kept current with the latest sampling technology. Some key aspects of the Reference Method (RM) system include:

- RMs are developed for and validated at concentrations representative of well-controlled source emission levels.
- All RMs are subjected to rulemaking procedures which include public comment.
- The RM system allows for method modification as the technology changes and as improvements are developed.
- Many of the RMs have active performance audit programs which are sponsored by the Federal Government.
- A Quality Assurance Handbook has been developed and is updated for many of the RMs (1).
- The U.S. EPA headquarters personnel can respond to technical questions and problems regarding the RMs.
- There are numerous training courses,

workshops, and technical publications available concerning various aspects of the RMs (2).
- The RMs use commercially available equipment and/or components and provide detailed construction specifications.

This paper provides background information on the Reference Methods, the intent of the methods, precautions for method applications different than their intended use, and the necessary listings and references to allow the user to obtain copies of the detailed sampling and analytical procedures for each method.

2. Background

As a result of the Clean Air Act, the U.S. Environmental Protection Agency (U.S. EPA) was given the task of developing regulations for new facilities (for all significant source categories) and sources of hazardous air pollutants. Reference Method(s) which detail sampling and analytical procedures for compliance testing are a required part of each regulation.

The U.S. EPA has currently developed regulations for a majority of the significant source categories and, in support of the regulations, has developed about seventy Reference Methods. These regulations are published under two categories. The sources of hazardous air pollutants are regulated under the National Emission Standards for Hazardous Air Pollutants (NESHAPS); these are published in the *Federal Register* under Title 40, Part 61 and summarized in the Code of Federal Regulations (40 CFR 61). The new sources are regulated by the New Source Performance Standards (NSPS) which are published in the *Federal*

*Prepared for the Intersociety Committee,
Methods of Air Sampling and Analysis
3rd Edition by William G. DeWees
Entropy Environmentalists, Inc.
Research Triangle Park, N.C. 27709

Register under Title 40, Part 60 and summarized in the Code of Federal Regulations (40 CFR 60).

3. Intent of U.S. EPA Reference Methods

The intent of a Reference Method is to provide industry with a reasonable and reliable means of demonstrating compliance with applicable regulations. Since under Federal Regulations, industry has the burden of demonstrating its compliance with applicable regulations, the Reference Methods must satisfy several criteria under standard applications:

(1) Their precision and accuracy should be acceptable when measuring pollutant concentrations at or near the emissions standard (concentration) **(3)**;
(2) The procedures should guard against errors that produce a low bias of the measured results;
(3) The constituents that are routinely present in the emissions of the regulated source should not have a significant effect on the measured results; and
(4) The methods should be reasonable in cost, use commercially available equipment or components, and be simple enough to be performed by skilled technical personnel.

Most of the Reference Methods are partially or totally developed by the U.S. EPA. All Reference Methods are also evaluated/validated by the U.S. EPA to ensure that the sampling and analytical procedures are both reasonable and reliable for their intended use.

4. Application of U.S. EPA Reference Methods

The validation procedures followed by the U.S. EPA generally provide good assurance that the Reference Methods will give accurate and precise results with the standard application to an applicable source. There are several factors which must be considered when applying a Reference Method to both applicable and nonapplicable sources.

(1) Any time the tester uses the allowable options or alternative procedures written into the Reference Methods, the measured results will generally be unaffected or biased high. Since industry has the burden of proving compliance, the U.S. EPA has given industry flexibility in the methods through the use of a biasing technique. Any alternative procedure or method that would tend either to have no effect on the results or provide higher than true measured results can be acceptable as a means of demonstrating compliance. Both the tester and the industry should always be aware of the potential effect of the measured results of these allowable changes in the procedures, prior to using them.
(2) Any time a Reference Method is applied to a source having pollutant concentrations which differ greatly from the regulated level, the precision and accuracy of the method may be greatly affected. All procedures, equipment, and sampling materials are designed to collect and analyze pollutant concentrations within a specific range (close to the regulated level).
(3) When a Reference Method is applied to sources with constituents in the effluent stream different than the applicable source categories, the results may be affected. The method description will generally list all of the known interferents for the applicable source category. However, the tester should be aware that even a change in pH of the collected materials or the addition of some compound in the effluent can create erroneous measured values.
(4) To provide an average result, each of the Reference Methods requires a specified number of sample collections over a specified length of sampling time. If the number of samples, sample time, or method of averaging is reduced, the precision and accuracy of the results may be affected.

5. Listing of U.S. EPA Reference Methods

This section lists all Reference Methods currently promulgated under the New Source Performance Standards (NSPS) and National Emission Standards for Hazardous Air Pollutants (NESHAPS). These methods are published in the Code of Federal Regulations, 40 CFR 60 and 40 CFR 61, respectively.

NSPS

Method 1	SAMPLE AND VELOCITY TRAVERSES FOR STATIONARY SOURCES
Method 1A	SAMPLE AND VELOCITY TRAVERSES FOR STATIONARY SOURCES WITH SMALL STACKS OR DUCTS
Method 2	DETERMINATION OF STACK GAS VELOCITY AND VOLUMETRIC FLOW RATE (TYPE S PITOT TUBE)
Method 2A	DIRECT MEASUREMENT OF GAS VOLUME THROUGH PIPES AND SMALL DUCTS
Method 2B	DETERMINATION OF EXHAUST GAS VOLUME FLOW RATE FROM GASOLINE VAPOR INCINERATORS
Method 2C	DETERMINATION OF STACK GAS VELOCITY AND VOLUMETRIC FLOW RATE FROM SMALL STACKS OR DUCTS (STANDARD PITOT TUBE)
Method 2D	MEASUREMENT OF GAS VOLUME FLOW RATES IN SMALL PIPES AND DUCTS
Method 3	GAS ANALYSIS FOR CARBON DIOXIDE, EXCESS AIR, AND DRY MOLECULAR WEIGHT
Method 3A	GAS ANALYSIS FOR CARBON DIOXIDE OR OXYGEN – INSTRUMENTAL METHOD
Method 4	DETERMINATION OF MOISTURE CONTENT IN STACK GASES
Method 5	DETERMINATION OF PARTICULATE EMISSIONS FROM STATIONARY SOURCES
Method 5A	DETERMINATION OF PARTICULATE EMISSIONS FROM THE ASPHALT PROCESSING AND ASPHALT ROOFING INDUSTRY
Method 5B	DETERMINATION OF NON-SULFURIC ACID PARTICULATE MATTER FROM ELECTRIC UTILITY STEAM GENERATORS
Method 5C	DETERMINATION OF PARTICULATE EMISSIONS FROM STATIONARY SOURCES WITH SMALL STACKS OR DUCTS
Method 5D	DETERMINATION OF PARTICULATE EMISSIONS FROM POSITIVE PRESSURE FABRIC FILTERS
Method 5E	DETERMINATION OF PARTICULATE EMISSIONS FROM WOOL FIBERGLASS INSULATION MANUFACTURING INDUSTRY
Method 5F	DETERMINATION OF NON-SULFATE PARTICULATE MATTER FROM STATIONARY SOURCES
Method 6	DETERMINATION OF SULFUR DIOXIDE EMISSIONS FROM STATIONARY SOURCES
Method 6A	DETERMINATION OF SULFUR DIOXIDE AND CARBON DIOXIDE EMISSIONS FROM FOSSIL FUEL COMBUSTION SOURCES
Method 6B	DETERMINATION OF SULFUR DIOXIDE AND CARBON DIOXIDE DAILY AVERAGE EMISSIONS FROM FOSSIL FUEL COMBUSTION SOURCES
Method 6C	DETERMINATION OF SULFUR DIOXIDE EMISSIONS FROM STATIONARY SOURCES – INSTRUMENTAL METHOD
Method 7	DETERMINATION OF NITROGEN OXIDES EMISSIONS FROM STATIONARY SOURCES
Method 7A	DETERMINATION OF NITROGEN OXIDES EMISSIONS FROM STATIONARY SOURCES: ION CHROMATOGRAPHIC METHOD
Method 7B	DETERMINATION OF NITROGEN OXIDES EMISSIONS FROM STATIONARY SOURCES: ULTRAVIOLET SPECTROPHOTOMETRY
Method 7C	DETERMINATION OF NITROGEN OXIDES EMISSIONS FROM STATIONARY SOURCES: ALKALINE-

PERMANGANATE/
COLORIMETRIC METHOD

Method 7D   DETERMINATION OF NI-
TROGEN OXIDES EMIS-
SIONS FROM STATIONARY
SOURCES: ALKALINE-
PERMANGANATE/ION
CHROMATOGRAPH
METHOD

Method 7E   DETERMINATION OF NI-
TROGEN OXIDES EMIS-
SIONS FROM STATIONARY
SOURCES: INSTRUMENTAL
METHOD

Method 8   DETERMINATION OF SUL-
FURIC ACID MIST AND
SULFUR DIOXIDE EMIS-
SIONS FROM STATIONARY
SOURCES

Method 9   VISUAL DETERMINATION
OF THE OPACITY OF EMIS-
SIONS FROM STATIONARY
SOURCES

Method 9A   ALTERNATE 1 –
DETERMINATION OF THE
OPACITY OF EMISSIONS
FROM STATIONARY
SOURCES REMOTELY BY
LIDAR

Method 10   DETERMINATION OF CAR-
BON MONOXIDE EMIS-
SIONS FROM STATIONARY
SOURCES: INSTRUMENTAL
METHOD

Method 10A   DETERMINATION OF CAR-
BON MONOXIDE EMIS-
SIONS FROM STATIONARY
SOURCES: COLORIMETRIC
METHOD

Method 11   DETERMINATION OF HY-
DROGEN SULFIDE EMIS-
SIONS FROM STATIONARY
SOURCES

Method 12   DETERMINATION OF INOR-
GANIC LEAD EMISSIONS
FROM STATIONARY
SOURCES

Method 13A   DETERMINATION OF TO-
TAL FLUORIDE EMISSIONS
FROM STATIONARY
SOURCES – SPADNS ZIRCO-
NIUM LAKE METHOD

Method 13B   DETERMINATION OF TO-
TAL FLUORIDE EMISSIONS
FROM STATIONARY
SOURCES – SPECIFIC ION
ELECTRODE METHOD

Method 14   DETERMINATION OF FLUO-
RIDE EMISSIONS FROM
POTROOM ROOF MONI-
TORS FOR PRIMARY ALU-

MINUM PLANTS

Method 15   DETERMINATION OF HY-
DROGEN SULFIDE, CAR-
BONYL SULFIDE, AND
CARBON DISULFIDE EMIS-
SIONS FROM STATIONARY
SOURCES: INSTRUMENTAL
METHOD

Method 15A   DETERMINATION OF TO-
TAL REDUCED SULFUR
EMISSIONS FROM STATION-
ARY SOURCES

Method 16   SEMICONTINUOUS DETER-
MINATION OF SULFUR
EMISSIONS FROM STATION-
ARY SOURCES

Method 16A   DETERMINATION OF TO-
TAL REDUCED SULFUR
EMISSIONS FROM STATION-
ARY SOURCES

Method 17   DETERMINATION OF PAR-
TICULATE EMISSIONS
FROM STATIONARY
SOURCES (IN-STACK FIL-
TRATION METHOD)

Method 18   MEASUREMENT OF GAS-
EOUS ORGANIC COM-
POUND EMISSIONS BY GAS
CHROMATOGRAPHY

Method 19   DETERMINATION OF SUL-
FUR DIOXIDE REMOVAL
EFFICIENCY AND PARTICU-
LATE, SULFUR DIOXIDE
AND NITROGEN OXIDES
EMISSION RATES FROM
ELECTRIC UTILITY STEAM
GENERATORS

Method 19A   DETERMINATION OF SUL-
FUR OXIDE EMISSION
RATES FROM FOSSIL FUEL-
FIRED STEAM GENERA-
TORS

Method 20   DETERMINATION OF NI-
TROGEN OXIDES, SULFUR
DIOXIDE, AND OXYGEN
EMISSIONS FROM STATION-
ARY GAS TURBINES

Method 21   DETERMINATION OF VOLA-
TILE ORGANIC COM-
POUNDS LEAKS

Method 22   VISUAL DETERMINATION
OF FUGITIVE EMISSIONS
FROM MATERIAL PROCESS-
ING SOURCES

Method 23   DETERMINATION OF HA-
LOGENATED ORGANICS
FROM STATIONARY
SOURCES

Method 24   DETERMINATION OF VOLA-
TILE MATTER CONTENT,
WATER CONTENT, DEN-

6. References

1. QUALITY ASSURANCE HANDBOOK FOR AIR POLLUTION MEASUREMENT SYSTEMS, VOLUME III, STATIONARY SOURCE SPECIFIC METHODS. August 1977. U.S. EPA, Publication No. EPA 600/4-77-027b, Research Triangle Park, N.C.
2. STACK SAMPLING TECHNICAL INFORMATION, A COLLECTION OF NOMOGRAPHS AND PAPERS, Volumes I-IV. October 1978. U.S. EPA Office of Air Quality Planning and Standards, Emission Standards and Engineering Division, Publication Nos. EPA 450/2-78-042a through d, Research Triangle Park, N.C.
3. COLLABORATIVE STUDY OF PARTICULATE EMISSIONS MEASUREMENTS BY EPA METHODS 2, 3, AND 5 USING PAIRED PARTICULATE SAMPLING TRAINS. March 1976. By H. F. Hamil and R. E. Thomas, U.S. EPA, Publication No. EPA 600/4-76-014, Research Triangle Park, N.C.

Subcommittee 11

W. S. SMITH, *Chairman*

PART V
CONVERSION FACTORS

PART V
Conversion Factors

Previous editions of the Manual have included lengthy tables of conversion factors and formulas. However, any book of instruction on methods, analytical or otherwise, must begin with assumptions concerning the knowledge the reader brings with him, and the context in which it will be used.

Thus, the reader of this volume will surely have been introduced to the elements of the metric system, and will use these methods in an environment that is tuned to that system of weights and measures; he is unlikely to need to convert from, say, tablespoons to milliliters. He is unlikely to encounter, in his professional work, many of the worst problems of the English (or other) systems, such as rods, furlongs, slugs, and pecks. Conversion factors for these are, then, scarcely needed.

However, the old metric system is itself being redefined, and has emerged as the International System of Weights and Measures, usually called SI from its French name. In most cases this has caused few problems; changes in value of units have generally been beyond the limits of accuracy of the usual laboratory work. A few conventions are changed; densities are to be reported, not in g/cm^3, but in kg/m^3, with an awkward factor of 10^3. The degree Kelvin is now simply the kelvin. On the positive side, we are rid of some awkward "metric" units such as statcoulombs and abamperes.

In two particular areas we are faced with "new" units. Pressure is reported in pascals, or newtons per square meter, and a newton is one kilogram meter per second squared (force, mass times acceleration). In fact, this is (save a factor of 10^5) the old bar. Radioactivity is now measured in becquerels, that amount of radioactive material yielding 1 disintegration per second, surely more fundamental than the curie. However, it is admittedly new. The curie is 3.7×10^{10} Bq.

Finally, a uniform set of symbols for the units is specified. These have been used, as have the SI units themselves, throughout this Manual, with few exceptions. Those interested in the details will find them in the current version of ASTM E380.

Meanwhile, outside the laboratory, the U.S. remains on the English system, even though the English units have all been redefined in terms of SI units. (An inch, for example, is now precisely 0.0254 m.) It may well be necessary to order metal tubing by fractional-inch diameter and feet of length. Barometric pressures may only be available in millibars or millimeters (or inches) of mercury.

To cover these and other contingencies, the following table was extracted from the somewhat truncated version of ASTM E380 that is reproduced in Volume 11.03 of the Annual Book of ASTM Standards, with kind permission of the American Society for Testing and Materials.

Selected Conversion Factors

To convert from	to	multiply by
atmosphere (760 mm Hg)	pascal (Pa)	$1.013\ 25 \times 10^5$
board foot	cubic metre (m^3)	$2.359\ 737 \times 10^{-3}$
Btu (International Table)	joule (J)	$1.055\ 056 \times 10^3$
Btu (International Table)/h	watt (W)	$2.930\ 711 \times 10^{-1}$
Btu (International Table $\cdot$ in./s $\cdot$ ft^2 $\cdot$ °F (k, thermal conductivity)	watt per metre kelvin [W/(m $\cdot$ K)]	$5.192\ 204 \times 10^2$
calorie (International Table)	joule (J)	$4.186\ 800*$
centipoise	pascal second (Pa $\cdot$ s)	$1.000\ 000* \times 10^{-3}$
centistokes	square metre per second (m^2/s)	$1.000\ 000*\ 10^{-6}$
circular mil	square metre (m^2)	$5.067\ 075 \times 10^{-10}$
degree Fahrenheit	degree Celsius	$t_{°C} = (t_{°F} - 32)/1.8$
foot	metre (m)	$3.048\ 000* \times 10^{-1}$
ft^2	square metre (m^2)	$9.290\ 304*\ 10^{-2}$
ft^3	cubic metre (m^3)	$2.831\ 685 \times 10^{-2}$
ft $\cdot$ lbf	joule (J)	$1.355\ 818$
ft $\cdot$ lbf/min	watt (W)	$2.259\ 697 \times 10^{-2}$
ft/s^2	metre per second squared (m/s^2)	$3.048\ 000* \times 10^{-1}$
gallon (U.S. liquid)	cubic metre (m^3)	$3.785\ 412 \times 10^{-3}$
horsepower (electric)	watt (W)	$7.460\ 000* \times 10^{+2}$
inch	metre (m)	$2.540\ 000* \times 10^{-2}$
$in.^2$	square meter (m^2)	$6.451\ 600* \times 10^{-4}$
$in.^3$	cubic metre (m^3)	$1.638\ 706 \times 10^{-5}$
inch of mercury (60°F)	pascal (Pa)	$3.376\ 85 \times 10^3$
inch of water (60°F)	pascal (Pa)	$2.488\ 4 \times 10^2$
kgf/cm^2	pascal (Pa)	$9.806\ 650* \times 10^4$
kip (1000 1bf)	newton (N)	$4.448\ 222 \times 10^3$
kip/$in.^2$ (ksi)	pascal (Pa)	$6.894\ 757 \times 10^6$
ounce (U.S. fluid)	cubic metre (m^3)	$2.957\ 353 \times 10^{-5}$
ounce-force	newton (N)	$2.780\ 139 \times 1^{-4}$
ounce (avoirdupois)	kilogram (kg)	$2.834\ 952 \times 10^{-2}$
oz (avoirdupois)/ft^2	kilogram per square metre (kg/m^2)	$3.051\ 517 \times 10^{-1}$
oz (avoirdupois)/yd^2	kilogram per square metre (kg/m^2)	$3.390\ 575 \times 10^{-2}$
oz (avoirdupois)/gal (U.S. liquid)	kilogram per cubic metre (kg/m^3)	$7.489\ 152$
pint (U.S. liquid)	cubic metre (m^3)	$4.731\ 765 \times 10^{-4}$
pound-force (lbf)	newton (N)	$4.448\ 222$
pound (lb avoirdupois)	kilogram (kg)	$4.535\ 924 \times 10^{-1}$
lbf/in^2 (psi)	pascal (Pa)	$6.894\ 757 \times 10^3$
lb/$in.^3$	kilogram per cubic metre (kg/m^3)	$2.767\ 990 \times 10^4$
lb/ft^3	kilogram per cubic metre (kg/m^3)	$1.601\ 846 \times 10$
quart (U.S. liquid)	cubic metre (m^3)	$9.463\ 529 \times 10^{-4}$
ton (short, 2000 lb)	kilogram (kg)	$9.071\ 847 \times 10^2$
torr (mm Hg, 0°C)	pascal (Pa)	$1.333\ 22 \times 10^2$
W $\cdot$ h	joule (J)	$3.600\ 000* \times 10^3$
yard	metre (m)	$9.144\ 000* \times 10^{-1}$
yd^2	square metre (m^2	$8.361\ 274 \times 10^{-1}$
yd^3	cubic metre (m^3)	$7.645\ 549 \times 10^{-1}$

*Exact

INDEX

Editor's Note:

No index is ever perfect, or even fully satisfactory; it would require omniscience on the part of the compiler. What has been attempted here is (a) a reference to every chemical species mentioned in the volume, under a reasonably familiar name, (b) reference to all definitions, (c) reference to most pieces of apparatus not available in the usual laboratory, (d) reference to uses of basic techniques, (e) reference to certain concepts, and (f) reference to named phenomena and equipment, such as Compton Scattering and Kuderna-Danish Apparatus. Other items have been included that were felt to be of interest to the user of the index.

Certain items were omitted because of their probable marginal interest, and others because of their ubiquity. Obviously, for example, items such as "air" and "water" would appear on virtually every page. This is basically a key word index with some collapsing of synonyms and closely related words. Thus, for example, "gas chromatograph" was subsumed into the heading of "Gas Chromatography."

I am sure I have missed some prospective entries, and perhaps confused some others. It is quite possible that the entry for "filters" includes optical filters as well as air and water filters. For this I can only apologize.

Organic compound nomenclature is invariably a jungle, and the names selected were personal judgments as to the most widely used names. Trivial names were used where these are widely known, but trade names were generally avoided. Hence "Propenal" was entered as "Acrolein," but "Freon 12" will be found under "Difluorodichloromethane."